W9-BVB-248

Applied STATISTICS

To Ed and Penny, my beloved family.

Applied STATISTICS

From Bivariate Through Multivariate Techniques

Rebecca M. Warner
University of New Hampshire

For information:

Sage Publications, Inc.
2455 Teller Road
Thousand Oaks, California 91320
E-mail: order@sagepub.com

Sage Publications Ltd.
1 Oliver's Yard
55 City Road
London EC1Y 1SP
United Kingdom

Sage Publications India Pvt. Ltd.
B 1/I 1 Mohan Cooperative Industrial Area
Mathura Road, New Delhi 110 044
India

Sage Publications Asia-Pacific Pte. Ltd.
33 Pekin Street #02-01
Far East Square
Singapore 048763

Printed in the United States of America

Library of Congress Cataloging-in-Publication Data

Warner, Rebecca M.
Applied statistics : from bivariate through multivariate techniques/Rebecca M. Warner.
p. cm.
Includes bibliographical references and index.
ISBN-13: 978-0-7619-2772-3 (cloth)
1. Social sciences—Statistical methods. 2. Psychology—Statistical methods. 3. Multivariate analysis. I. Title.

HA31.35.W37 2007
519.5′35—dc22 2006033700

This book is printed on acid-free paper.

07 08 09 10 11 10 9 8 7 6 5 4 3 2 1

Acquisitions Editor: Vicki Knight
Associate Editor: Sean Connelly
Editorial Assistant: Lauren Habib
Production Editor: Laureen A. Shea
Copy Editors: Linda Gray and QuADS
Typesetter: C&M Digitals (P) Ltd.
Indexer: Will Ragsdale
Cover Designer: Candice Harman
Marketing Manager: Stephanie Adams

Contents

Preface

This book has been written to provide a bridge between the many excellent statistics books that already exist at introductory and advanced levels. I have been persuaded by years of teaching that most students do not have a very clear understanding of statistics after their first one or two courses. The concepts covered in an introductory course include some of the most difficult and controversial issues in statistics, such as level of measurement and null hypothesis significance testing. Until students have been introduced to the entire vocabulary and system of thought, it is difficult for them to integrate all these ideas. I believe that understanding statistics requires multiple-pass learning. This is the reason I have included a review of basic topics (such as null hypothesis significance test procedures) along with an introduction to more advanced topics in statistics. Students need varying amounts of review and clarification; this textbook is designed so that each student can review as much basic material as necessary prior to the study of more advanced topics such as multiple regression. Some students need a review of concepts involved in bivariate analyses (such as partition of variance), and most students can benefit from a thorough introduction to statistical control in simple three-variable research situations. This textbook differs from many existing textbooks for advanced undergraduate- and beginning graduate-level statistics courses because it includes a review of bivariate methods that clarifies important concepts and a thorough discussion of methods for statistically controlling for a third variable (X_2) when assessing the nature and strength of the association between an X_1 predictor and a Y outcome variable. Later chapters present verbal explanations of widely used multivariate methods applied to specific research examples.

This textbook consists of three sections. The first nine chapters review material covered in most introductory textbooks; they provide a fresh perspective on important concepts, including variance partitioning, effect size, and statistical power. Chapters 10 through 13 examine three-variable research situations in detail and discuss many of the possible interpretations for outcomes that arise when we examine how an X_1 predictor is associated with a Y outcome variable while statistically controlling for an X_2 variable. Diagrams are used to illustrate partition of variance, and path models are used to illustrate different theoretical "causal" models. Important concepts such as suppression, spuriousness, mediation, and moderation or interaction are discussed extensively in the context of three-variable research situations. The goal is to make these concepts as clear as possible before moving on to more advanced analyses. The remaining chapters introduce several widely used multivariate analyses: These include multiple linear regression, analysis of covariance, discriminant analysis, multivariate analysis of variance, factor analysis, and binary logistic regression. A unique feature of this book is a chapter (Chapter 19) about the development of multiple-item scales and measurement quality issues such as reliability and validity.

When I teach an advanced undergraduate statistics course, I typically cover Chapter 1 through Chapter 14 (omitting Chapter 12 on dummy variables). This provides students with a complete review of basic statistics and an introduction to multiple linear regression. When I teach a graduate-level course in multivariate statistics, I typically begin with a brief review of material in Chapters 7 through 9 (bivariate correlation and regression) and spend a few weeks on the three-variable research situations covered in Chapters 10 through 13. We then go on to the more advanced topics in Chapters 14 through 21. Graduate students who studied introductory statistics a long time ago or who never felt confident about their understanding of basic concepts such as null hypothesis significance tests can use the material in Chapters 1 through 6 for remediation or review.

Each chapter concludes with comprehension questions; answers to these comprehension questions are available to instructors on the companion study site for this textbook (www.sagepub.com/warnerstudy). These can be used as teaching aids or for evaluation. Other instructor resources for each chapter include PowerPoint presentations that summarize the major ideas and include SPSS screen shots to guide menu selections and to illustrate output. Annotated SAS input and output that corresponds to each SPSS example are also provided on the study site. Datasets used as examples in the textbook are available on the site in two file formats (SPSS and Excel). Additional datasets for student data analysis projects are also available on the textbook study site.

Writing a textbook requires difficult decisions about what topics to include and what to leave out and how to handle topics where there is disagreement among authorities. This textbook does not cover nonparametric statistics or complex forms of factorial analysis of variance, nor does it cover all the advanced topics found in more encyclopedic treatments (such as time series analysis, multilevel modeling, survival analysis, and log linear models). The topics that are included provide a reasonably complete set of tools for data analysis at the advanced undergraduate or beginning graduate level along with explanations of some of the fundamental concepts that are crucial for further study of statistics. For example, comprehension of structural equation modeling (SEM) requires students to understand path or "causal" models, latent variables, measurement models, and the way in which observed correlations (or variances and covariances) can be reproduced from the coefficients in a model. This textbook introduces path models and the tracing rule as a way of understanding linear regression with two predictor variables. The explanation of regression with two predictors makes it clear how estimates of regression slope coefficients are deduced from observed correlations and how observed correlations among variables can be reproduced from the coefficients in a regression model. Explaining regression in this way helps students understand why the slope coefficient for each *X* predictor variable is context dependent (i.e., the value of the slope coefficient for each *X* predictor variable changes depending on which other *X* predictor variables are included in the regression analysis). This explanation also sets the stage for an understanding of more advanced methods, such as SEM, that use model parameters to reconstruct observed variances and covariances.

I have tried to develop explanations that will serve students well whether they use them only to understand the methods of analysis covered in this textbook or as a basis for further study of more advanced statistical methods.

One additional note for instructors: All SPSS examples in this textbook were produced using SPSS 14. Starting with SPSS version 15, one additional menu selection is required for the Graphics procedures shown here. After making the menu selection <Graphics>, choose <Legacy Dialogs>; the drop-down menu that appears next provides access to the graphics procedures shown in this book, such as Scatter and Box Plot.

Acknowledgments

Writers depend on many other people for intellectual preparation and moral support. My understanding of statistics was shaped by several exceptional teachers, including the late Morris de Groot at Carnegie Mellon University, and my dissertation advisers at Harvard, Robert Rosenthal and David Kenny. Several of the teachers who have most strongly influenced my thinking are writers I know only through their books and journal articles. I want to thank all the authors whose work is cited in the reference list. Authors whose work has greatly influenced my understanding include Jacob and Patricia Cohen, Barbara Tabachnick, Linda Fidell, James Jaccard, Richard Harris, Geoffrey Keppel, and James Stevens.

I wish to thank the University of New Hampshire (UNH) for sabbatical leave time and Mil Duncan, director of the Carsey Institute at UNH, for release time from teaching. I also thank my department chair, Ken Fuld, who gave me light committee responsibilities while I was working on this book. These gifts of time made the completion of the book possible.

Thanks are due to the SPSS Corporation for providing up-to-date versions of SPSS for use in preparation of the data analysis examples in this book and to the Computing and Information Services at UNH for making a copy of SAS 9.1 available.

Special thanks are due to the reviewers who provided exemplary feedback on the first drafts of the chapters:

David J. Armor, *George Mason University*

Michael D. Biderman, *University of Tennessee at Chattanooga*

Susan Cashin, *University of Wisconsin–Milwaukee*

Ruth Childs, *University of Toronto*

Young-Hee Cho, *California State University, Long Beach*

Jennifer Dunn, *Center for Assessment*

William A. Fredrickson, *University of Missouri–Kansas City*

Robert Hanneman, *University of California, Riverside*

Andrew Hayes, *Ohio State University*

Lawrence G. Herringer, *California State University, Chico*

Jason King, *Baylor College of Medicine*

Patrick Leung, *University of Houston*

Scott E. Maxwell, *University of Notre Dame*

W. James Potter, *University of California, Santa Barbara*

Kyle L. Saunders, *Colorado State University*

Joseph Stevens, *University of Oregon*

James A. Swartz, *University of Illinois at Chicago*

Keith Thiede, *University of Illinois at Chicago*

Their comments were detailed and constructive. I hope that revisions based on their reviews have improved this book substantially. The publishing team at Sage, including Lisa Cuevas Shaw, Margo Crouppen, Laureen Shea, Karen Greene, Karen Wiley, and Stephanie Adams, provided extremely helpful advice, support, and encouragement. Special thanks are also due to the copy editors for their diligence and care, including the supervising copy editor, Linda Gray, and the copy editors from QuADS.

I wish to thank Stan Wakefield, the agent who persuaded me to write the textbook that I had been contemplating for many years and who negotiated my contract with Sage.

Many people provided moral support, particularly my husband, Ed Tedesco; my parents, David and Helen Warner; my sister, Amy Warner; and friends and colleagues at UNH, including Ellen Cohn, Ken Fuld, Jack Mayer, and Kerryellen Vroman. Ed has seen me through many difficult times in the writing process; he upgraded my computer and software and rescued my files from several system crashes; without his help, parts of this book might have been lost beyond recovery.

I hope that this book is worthy of the support that all of these people and many others have given me and that it represents the quality of the education that I have received. Of course, I am responsible for any errors and omissions that remain. I will be most grateful to readers, teachers, and students who point out mistakes; if this book goes into subsequent editions, I want to make all possible corrections. In addition, information about any citations of sources that have been unintentionally omitted will help me to improve subsequent editions.

Last but not least, I want to thank all my students, who have also been my teachers. Their questions continually make me search for better explanations—and I am still learning.

Dr. Rebecca M. Warner
Professor
Department of Psychology
University of New Hampshire

CHAPTER 1

Review of Basic Concepts

1.1 ♦ Introduction

This textbook assumes that students have taken basic courses in statistics and research methods. A typical first course in statistics includes methods to describe the distribution of scores on a single variable (such as **frequency distribution** tables, **medians, means, variances**, and standard deviations) and a few widely used **bivariate statistics**. Bivariate statistics (such as the Pearson **correlation**; the independent samples *t* test; and one-way **analysis of variance,** or **ANOVA**) assess how pairs of variables are related. This textbook is intended for use in a second course in statistics; the presentation in this textbook assumes that students recognize the names of statistics such as ***t* test** and Pearson correlation but may not yet fully understand how to apply and interpret these analyses.

The first goal in this course is for students to develop a better understanding of these basic bivariate statistics and the problems that arise when these analytic methods are applied to real-life research problems. Chapters 1 through 3 deal with basic concepts that are often a source of confusion because actual researcher behaviors often differ from the recommended methods described in basic textbooks; this includes issues such as sampling and statistical significance testing. Chapter 4 discusses methods for preliminary data screening; before doing any statistical analysis, it is important to remove errors from data, assess whether the data violate assumptions for the statistical procedures, and decide how to handle any violations of assumptions that are detected. Chapters 5 through 9 review familiar bivariate statistics that can be used to assess how scores on one *X* **predictor variable** are related to scores on one *Y* **outcome variable** (such as the Pearson correlation and the independent samples *t* test). Chapters 10 through 13 discuss the questions that arise when a third variable is added to the analysis. Later chapters discuss analyses that include multiple predictor and/or multiple outcome variables.

When students begin to read journal articles or conduct their own research, it is a challenge to understand how textbook knowledge is applied in real-life research situations. This textbook provides guidance on dealing with the problems that arise when researchers apply statistical methods to actual data.

1.2 ♦ A Simple Example of a Research Problem

Suppose that a student wants to do a simple experiment to assess the **effect** of caffeine on anxiety. (This study would not yield new information; there has already been substantial research on the effects of caffeine; however, this is a simple research question that does not require a complicated background story about the nature of the variables.) In the United States, before researchers can collect data, they must have the proposed methods for the study reviewed and approved by an **institutional review board (IRB)**. If the research poses unacceptable risks to participants, the IRB may require modification of procedures prior to approval.

To run a simple experiment to assess the effects of caffeine on anxiety, the researcher would obtain IRB approval for the procedure, recruit a sample of participants, divide the participants into two groups, give one group a beverage that contains some fixed dosage level of caffeine, give the other group a beverage that does not contain caffeine, wait for the caffeine to take effect, and measure each person's anxiety, perhaps by using a self-report measure. Next, the researcher would decide what statistical analysis to apply to the data to evaluate whether anxiety differs between the group that received caffeine and the group that did not receive caffeine. After conducting an appropriate data analysis, the researcher would write up an interpretation of the results that takes the design and the limitations of the study into account. The researcher might find that participants who consumed caffeine have higher self-reported anxiety than participants who did not consume caffeine. Researchers generally hope that they can generalize the results obtained from a sample to make inferences about outcomes that might conceivably occur in some larger population. If caffeine increases anxiety for the participants in the study, the researcher may want to argue that caffeine would have similar effects on other people who were not actually included in the study.

This simple experiment will be used to illustrate several basic problems that arise in actual research:

1. Selection of a **sample** from a **population**
2. Evaluating whether a sample is **representative** of a population
3. **Descriptive** versus **inferential** applications of statistics
4. **Levels of measurement** and **types of variables**
5. Selection of a statistical analysis that is appropriate for the type of data
6. **Experimental design** versus **nonexperimental design**

The following discussion focuses on the problems that arise when these concepts are applied in actual research situations and comments on the connections between research methods and statistical analyses.

1.3 ♦ Discrepancies Between Real and Ideal Research Situations

Terms that appear simple (such as *sample* vs. *population*) can be a source of confusion because the actual behaviors of researchers often differ from the idealized research process described in introductory textbooks. Researchers need to understand how compromises that are often made in actual research (such as the use of convenience samples) affect the interpretability of research results. Each of the following sections describes common practices in actual research in contrast to idealized textbook approaches. Unfortunately, because of limitations in time and money, researchers often cannot afford to conduct studies in the most ideal manner.

1.4 ♦ Samples and Populations

A sample is a subset of members of a population.[1] Usually, it is too costly and time-consuming to collect data for all members of an actual population of interest (such as all registered voters in the United States), and therefore researchers usually collect data for a relatively small sample and use the results from that sample to make inferences about behavior or attitudes in larger populations. In the ideal research situation described in research methods and statistics textbooks, there is an actual population of interest. All members of that population of interest should be identifiable; for example, the researcher should have a list of names for all members of the population of interest. Next, the researcher selects a sample from that population using either random sampling or other sampling methods (Cozby, 2004).

A sample is random if every member of the population has an equal chance of being included in the sample. Random sampling can be done in a variety of ways. For a small population, the researcher can put each participant's name on a slip of paper, mix up the slips of paper in a jar, and draw names from the jar. For a larger population, if the names of participants are listed in a spreadsheet such as Excel, the researcher can generate a column of random numbers next to the names and make decisions about which individuals to include in the sample based on those random numbers. For instance, if the researcher wants to select $\frac{1}{10}$ of the members of the population at random, the researcher may decide to include each participant whose name is next to a random number that ends in one arbitrarily chosen value (such as "3").

In theory, if a sample is chosen randomly from a population, that sample should be representative of the population from which it is drawn. A sample is representative if it has characteristics similar to those of the population. Suppose that the population of interest to a researcher is all the 500 students at Corinth College in the United States. Suppose the researcher randomly chooses a sample of 50 students from this population by using one of the methods just described. The researcher can evaluate whether this **random sample** is representative of the entire population of all Corinth College students by comparing the characteristics of the sample with the characteristics of the entire population. For example, if the entire population of Corinth College students has a mean age of 19.5 years and is 60% female and 40% male, the sample would be representative of the population with respect to age and gender composition if the sample had a mean age close to 19.5 years and gender composition of about 60% female and 40% male. Representativeness of a sample can be assessed for many other characteristics, of course. Some characteristics may be particularly relevant to a research question; for example, if a researcher were primarily interested in the political attitudes of the population of students at Corinth, it would be important to evaluate whether the composition of the sample was similar to that of the overall Corinth College population in terms of political party preference.

Random selection may be combined with systematic sampling methods such as stratification. A stratified random sample is obtained when the researcher divides the population into "strata," or groups (such as Buddhist/Christian/Hindu/Islamic/Jewish/Other religion or Male/Female), and then draws a random sample from each stratum or group. **Stratified sampling** can be used to ensure equal representation of groups (such as 50% women and 50% men in the sample) or to ensure that the proportional representation of groups in the sample is the same as in the population (if the entire population of students at Corinth College consists of 60% women and 40% men, the researcher might want the sample to contain the same proportion of women and men).[2] Basic sampling methods are reviewed in Cozby (2004); more complex survey sampling methods are discussed by Kalton (1983).

In some research domains (such as the public opinion polls done by the Gallup and Harris organizations), sophisticated sampling methods are used, and great care is taken to ensure that the sample is representative of the population of interest. In contrast, many behavioral and social science studies do not use such rigorous sampling procedures. Researchers in education, psychology, medicine, and many other disciplines often use **accidental** or **convenience samples** (instead of random samples). An accidental or convenience sample is not drawn randomly from a well-defined population of interest. Instead, a convenience sample consists of participants who are readily available to the researcher. For example, a teacher might use his class of students or a physician might use her current group of patients.

A systematic difference between the characteristics of a sample and a population can be termed ***bias***. For example, if 25% of Corinth College students are in each of the 4 years of the program, but 80% of the members of the convenience sample obtained through the subject pool are first-year students, this convenience sample is biased (it includes more first-year students, and fewer second-, third-, and fourth-year students, than the population).

The widespread use of convenience samples in disciplines such as psychology leads to underrepresentation of many types of people. Convenience samples that consist primarily of first-year North American college students typically underrepresent many kinds of people, such as persons younger than 17 and older than 30 years, persons with serious physical health problems, people who are not interested in or eligible for a college education, persons living in poverty, and persons from cultural backgrounds that are not numerically well represented in North America. For many kinds of research, it would be highly desirable for researchers to obtain samples from more diverse populations, particularly when the outcome variables of interest are likely to differ across age and cultural background. The main reason for the use of convenience samples is the low cost. The extensive use of college students as research participants limits the potential generalizability of results (Sears, 1986); this limitation should be explicitly acknowledged when researchers report and interpret research results.

1.5 ♦ Descriptive Versus Inferential Uses of Statistics

Statistics that are used only to summarize information about a sample are called descriptive statistics. One common situation where statistics are used only as descriptive information occurs when teachers compute summary statistics, such as a mean for exam scores for students in a class. A teacher at Corinth College would typically use a mean exam score only to describe the performance of that specific classroom of students and not to make inferences about some broader population (such as the population of all students at Corinth College or all college students in North America).

Researchers in the behavioral and social sciences almost always want to make inferences beyond their samples; they hope that the attitudes or behaviors that they find in the small groups of college students who actually participate in their studies will provide evidence about attitudes or behaviors in broader populations in the world outside the laboratory. Thus, almost all the statistics reported in journal articles are inferential statistics. Researchers may want to estimate a population mean from a **sample mean** or a population correlation from a sample correlation. When means or correlations based on samples of scores are used to make inferences about (i.e., estimates of) the means or correlations for broader populations, they are called inferential statistics. If a researcher finds a strong correlation between self-esteem and popularity in a convenience sample of Corinth College students, the researcher typically hopes that these variables are similarly related in broader populations, such as all North American college students.

In some applications of statistics, such as political opinion polling, researchers often obtain representative samples from actual, well-defined populations by using well-thought-out sampling procedures (such as a combination of stratified and random sampling). When good sampling methods are used to obtain representative samples, it increases researcher confidence that the results from a sample (such as the stated intention to vote for one specific candidate in an election) will provide a good basis for making inferences about outcomes in the broader population of interest.

However, in many types of research (such as experiments and small-scale surveys in psychology, education, and medicine), it is not practical to obtain random samples from the entire population of a country. Instead, researchers in these disciplines often use convenience samples when they conduct small-scale studies.

Consider the example introduced earlier: A researcher wants to run an experiment to assess whether caffeine increases anxiety. It would not be reasonable to try to obtain a sample of participants from the entire adult population of the United States (consider the logistics involved in travel, for example). In practice, studies similar to this are usually conducted using convenience samples. At most colleges or universities in the United States, convenience samples primarily include persons between 18 and 22 years of age.

When researchers obtain information about behavior from convenience samples, they cannot confidently use their results to make inferences about the responses of an actual, well-defined population. For example, if the researcher shows that a convenience sample of Corinth College students scores higher on anxiety after consuming a dose of caffeine, it would not be safe to assume that this result is generalizable to all adult humans or to all college students in the United States. Why not? For example, the effects of caffeine might be quite different for adults older than 70 than for 20-year-olds. The effects of caffeine might differ for people who regularly consume large amounts of caffeine than for people who never use caffeine. The effects of caffeine might depend on physical health.

Although this is rarely explicitly discussed, most researchers implicitly rely on a principle that Campbell (cited in Trochim, 2001) has called "proximal similarity" when they evaluate the potential generalizability of research results based on convenience samples. It is possible to imagine a **hypothetical population**—that is, a larger group of people that is similar in many ways to the participants who were included in the convenience sample—and to make cautious inferences about this hypothetical population based on the responses of the sample. Campbell suggested that researchers evaluate the degree of similarity between a sample and hypothetical populations of interest and limit generalizations to hypothetical populations that are similar to the sample of participants actually included in the study. If the convenience sample consists of 50 Corinth College students who are between the ages of 18 and 22 and mostly of Northern European family background, it might be reasonable to argue (cautiously, of course) that the results of this study potentially apply to a hypothetical broader population of 18- to 22-year-old U.S. college students who come from similar ethnic or cultural backgrounds. This hypothetical population—all U.S. college students between 18 and 22 years from a Northern European family background—has a composition fairly similar to the composition of the convenience sample. It would be questionable to generalize about response to caffeine for populations that have drastically different characteristics from the members of the sample (such as persons who are older than age 50 or who have health problems that members of the convenience sample do not have).

Generalization of results beyond a sample to make inferences about a broader population is always risky, so researchers should be cautious in making generalizations. An example involving research on drugs highlights the potential problems that can arise when researchers are too quick to assume that results from convenience samples provide accurate information about the effects of a treatment on a broader population. For example, suppose that a researcher conducts a series of studies to evaluate the effects of a new antidepressant drug on depression. Suppose that the participants are a convenience sample of depressed young adults between the ages of 18 and 22. If the researcher uses appropriate experimental designs and finds that the new drug significantly reduces depression in these studies, the researcher might tentatively say that this drug may be effective for other depressed young adults in this age range. It could be misleading, however, to generalize the results of the study to children or to older adults. A drug that appears to be safe and effective for a convenience sample of young adults might not be safe or effective in patients who are younger or older.

To summarize, when a study uses data from a convenience sample, the researcher should clearly state that the nature of the sample limits the potential generalizability of the results. Of course, inferences about hypothetical or real populations based on data from a single study are never conclusive, even when random selection procedures are used to obtain the sample. An individual study may yield incorrect or misleading results for many reasons. Replication across many samples and studies is required before researchers can begin to feel confident about their conclusions.

1.6 ♦ Levels of Measurement and Types of Variables

A controversial issue introduced early in statistics courses involves types of measurement for variables. Many introductory textbooks list the classic levels of measurement defined by Stevens (1946): **nominal, ordinal, interval,** and **ratio** (see Table 1.1 for a summary

Table 1.1 ♦ Levels of Measurement, Arithmetic Operations, and Types of Statistics

Stevens's Levels of Measurement	*Logical and Arithmetic Operations That Can Be Applied (According to Stevens)*	*Traditional or Conservative Recommendation*	*Simpler Distinction Between Two Types of Variables*[a]
Nominal	$=, \neq$	Only nonparametric statistics	Categorical
Ordinal	$=, \neq, <, >$	Only nonparametric statistics	Quantitative
Interval[b]	$=, \neq, <, >, +, -$	Parametric statistics	Quantitative
Ratio	$=, \neq, <, >, +, -, \times, \div$	Parametric statistics	Quantitative

a. Jaccard and Becker (2002).

b. Many variables that are widely used in the social and behavioral sciences, such as 5-point rating scales for attitude and personality measurement, probably fall short of satisfying the requirement that equal differences between scores represent exactly equal changes in the amount of the underlying characteristics being measured. However, most authors (such as Harris, 2001) argue that application of parametric statistics to scores that fall somewhat short of the requirements for interval level of measurement does not necessarily lead to problems.

and Note 3 for a more detailed review of these levels of measurement).[3] Strict adherents to the Stevens theory of measurement argue that the level of measurement of a variable limits the set of logical and arithmetic operations that can appropriately be applied to scores. That, in turn, limits the choice of statistics. For example, if scores are nominal or categorical level of measurement, then according to Stevens, the only things we can legitimately do with the scores are count how many persons belong to each group (and compute proportions or percentages of persons in each group); we can also note whether two persons have equal or unequal scores. It would be nonsense to add up scores for a nominal variable such as eye color (coded 1 = Blue, 2 = Green, 3 = Brown, 4 = Hazel, 5 = Other) and calculate a "mean eye color" based on a sum of these scores.[4]

In recent years, many statisticians have argued for a much less strict application of level of measurement requirements. In practice, there are many common types of variables (such as 5-point ratings of degree of agreement with an attitude statement) that probably fall short of meeting the strict requirements for equal interval level of measurement. A strict enforcement of the level of measurement requirements outlined in many introductory textbooks creates a problem: Can researchers legitimately compute statistics (such as mean, *t* test, and correlation) for scores such as 5-point ratings when the differences between these scores may not represent exactly equal amounts of change in the underlying variable that the researcher wants to measure (in this case, strength of agreement)? Many researchers implicitly assume that the answer to this question is "yes."

The variables that are presented as examples when ordinal, interval, and ratio levels of measurement are defined in introductory textbooks are generally classic examples that are easy to classify. In actual practice, however, it is often difficult to decide whether scores on a variable meet the requirements for interval and ratio levels of measurement. The scores on many types of variables (such as 5-point rating scales) probably fall into a fuzzy region somewhere between the ordinal and interval levels of measurement. How crucial is it that scores meet the strict requirements for interval level of measurement?

Many statisticians have commented on this problem, noting that there are strong differences of opinion among researchers. Vogt (1999) noted that there is considerable controversy about the need for a true interval level of measurement as a condition for the use of statistics such as mean, variance, and **Pearson *r***, stating that "as with constitutional law, there are in statistics strict and loose constructionists in the interpretation of adherence to assumptions" (p. 158). Although some statisticians adhere closely to Stevens's recommendations, many authors argue that it is not necessary to have data that satisfy the strict requirements for interval level of measurement in order to obtain interpretable and useful results for statistics such as mean and Pearson *r*.

Howell (1992) reviewed the arguments and concluded that the underlying level of measurement is not crucial in the choice of a statistic:

> The validity of statements about the objects or events that we think we are measuring hinges primarily on our knowledge of those objects or events, not on the measurement scale. We do our best to ensure that our measures relate as closely as possible to what we want to measure, but our results are ultimately only the numbers we obtain and our faith in the relationship between those numbers and the underlying objects or events . . . the underlying measurement scale is not crucial in our choice of statistical techniques . . . a certain amount of common sense is required in interpreting the results of these statistical manipulations. (pp. 8–9)

Harris (2001) says,

> I do not accept Stevens's position on the relationship between strength [level] of measurement and "permissible" statistical procedures . . . the most fundamental reason for [my] willingness to apply multivariate statistical techniques to such data, despite the warnings of Stevens and his associates, is the fact that the validity of statistical conclusions depends only on whether the numbers to which they are applied meet the distributional assumptions . . . used to derive them, and not on the scaling procedures used to obtain the numbers. (pp. 444–445)

Gaito (1980) reviewed these issues and concluded that "scale properties do not enter into any of the mathematical requirements" for various statistical procedures, such as ANOVA. Tabachnick and Fidell (2007) address this issue in their multivariate textbook: "The property of variables that is crucial to application of multivariate procedures is not type of measurement so much as the shape of the distribution" (p. 6). Zumbo and Zimmerman (1993) used computer simulations to demonstrate that varying the level of measurement for an underlying empirical structure (between ordinal and interval) did not lead to problems when several widely used statistics were applied.

Based on these arguments, it seems reasonable to apply statistics (such as the sample mean, Pearson *r*, and ANOVA) to scores that do not satisfy the strict requirements for interval level of measurement. (Some teachers and journal reviewers continue to prefer the more conservative statistical practices advocated by Stevens; they may advise you to avoid the computation of means, variances, and Pearson correlations for data that aren't clearly interval/ratio level of measurement.)

When making decisions about the type of statistical analysis to apply, it is useful to make a simpler distinction between two types of variables: *categorical* versus *quantitative* (Jaccard & Becker, 2002). For a categorical or nominal variable, each number is merely a label for group membership. A **categorical variable** may represent naturally occurring groups or categories (the categorical variable gender can be coded 1 = Male, 2 = Female). Alternatively, a categorical variable can identify groups that receive different treatments in an experiment. In the hypothetical study described in this chapter, the categorical variable treatment can be coded 1 for participants who did not receive caffeine and 2 for participants who received 150 mg of caffeine. It is possible that the outcome variable for this imaginary study, anxiety, could also be a categorical or nominal variable; that is, a researcher could classify each participant as either 1 = Anxious or 0 = Not anxious, based on observations of behaviors such as speech rate or fidgeting.

Quantitative variables have scores that provide information about the magnitude of differences between participants in terms of the amount of some characteristic (such as anxiety in this example). The outcome variable, anxiety, can be measured in several different ways. An observer who does not know whether each person had caffeine could observe behaviors such as speech rate and fidgeting and make a judgment about each individual's anxiety level. An observer could rank order the participants in order of anxiety: 1 = Most anxious, 2 = Second most anxious, and so forth. (Note that ranking can be quite time-consuming if the total number of persons in the study is large.)

A more typical measurement method for this type of research situation would be self-report of anxiety, perhaps using a 5-point rating scale similar to the one below. Each participant would be asked to choose a number from 1 to 5 in response to a statement such as "I am very anxious."

1	2	3	4	5
Strongly disagree	Disagree	Neutral	Agree	Strongly agree

Conventionally, 5-point rating scales (where the five response alternatives correspond to "degrees of agreement" with a statement about attitude, belief, or behavior) are called **Likert scales**. However, rating-scale questions can have any number of response alternatives, and the response alternatives may have different labels—for example, reports of the frequency of a behavior. (See Chapter 19 for further discussion of self-report questions and response alternatives.)

What level of measurement does a 5-point rating scale similar to the one above provide? The answer is that we really don't know. Scores on 5-point rating scales probably do not have true equal interval measurement properties; we cannot demonstrate that the increase in the underlying amount of anxiety represented by a difference between 4 points and 3 points corresponds exactly to the increase in the amount of anxiety represented by the difference between 5 points and 4 points. It is not clear whether the "strongly disagree" response represents a true 0 point. When we try to decide whether a 5-point scale is ordinal, interval, or ratio, we can give, at best, only an approximate answer. Five-point rating scales similar to the example above probably fall into a fuzzy category somewhere between the ordinal and interval levels of measurement. In practice, many researchers apply statistics such as means and standard deviations to rating-scale data in spite of the fact that these rating scales may fall short of the strict requirements for equal interval level of measurement; the arguments made by Harris and others above suggest that this common practice is not necessarily problematic.

This is another instance where actual researcher behaviors differ from the guidelines suggested in many statistics textbooks. If Stevens's level of measurement requirements are strictly enforced, statistics such as means and Pearson correlations could be computed only for variables that can be proved to have a true interval/ratio level of measurement. Unfortunately, many of the variables used in behavioral and social sciences (such as attitude-rating scales, personality trait measures, and so forth) probably fall somewhere between ordinal and interval in terms of level of measurement. A conservative interpretation of the guidelines about levels of measurement would lead to the conclusion that statistics such as mean and Pearson *r* should not be applied to rating-scale data. However, in practice, many researchers do compute means, standard deviations, and Pearson correlations for variables that probably do not satisfy the strict requirements for equal interval level of measurement, such as attitude-rating scales. Thus, there is a discrepancy between the implicit standards for choice of statistics based on the theory of measurement that is still presented in many introductory statistics books and the data analysis practices of many researchers. When level of measurement falls into a gray area (where the information provided by scores is possibly better than ordinal but probably falls short of the requirements for a true interval level of measurement), many data analysts go ahead and apply statistics such as Pearson correlation and the *t* test in spite of the fact that Stevens's measurement model is often interpreted as a prohibition of this practice. Stevens (1951) himself acknowledged that in some situations, violation of the measurement model can lead to reasonable results.

Tabachnick and Fidell (2007) and the other authors cited above have argued that it is more important to consider the distribution shapes for scores on quantitative variables (rather than their levels of measurement). Many of the statistical tests covered in introductory statistics books were developed based on assumptions that scores on quantitative variables are normally distributed. To evaluate whether a batch of scores in a sample has a nearly **normal distribution** shape, we need to know what an ideal normal distribution looks like. The next section reviews the characteristics of the standard normal distribution.

1.7 ♦ The Normal Distribution

Introductory statistics books typically present both empirical and theoretical distributions. An empirical distribution is based on scores from a sample, while a theoretical distribution is defined by a mathematical function or equation.

A description of an empirical distribution can be presented as a table of frequencies or in a graph such as a **histogram**. Sometimes it is helpful to group scores in order to obtain a more compact view of the distribution. **SPSS®** makes reasonable default decisions about grouping scores and the number of intervals and interval widths to use; these decisions

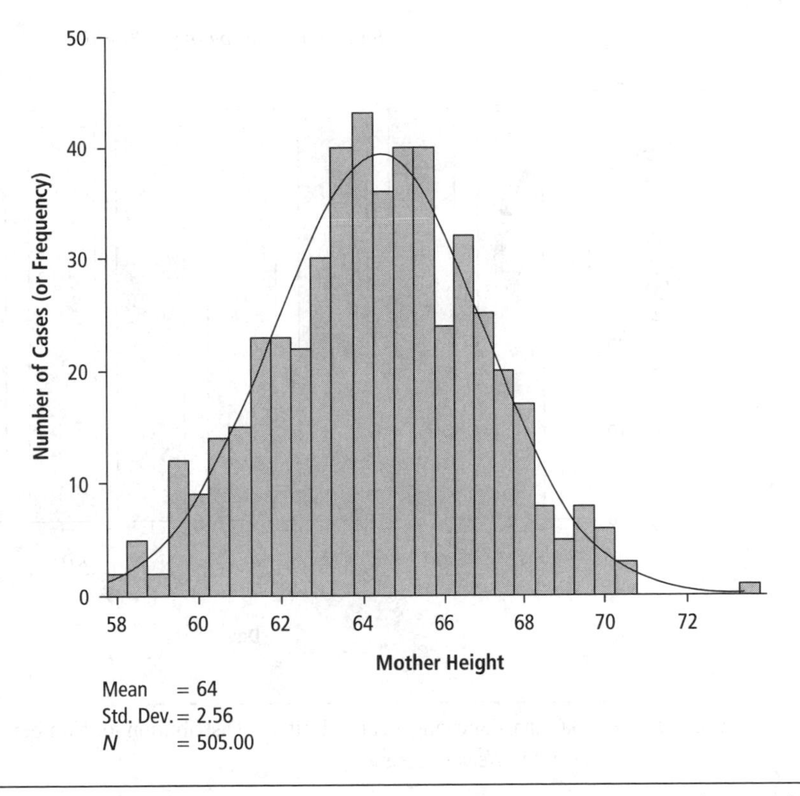

Figure 1.1 ♦ A Histogram Showing an Empirical Distribution of Scores That Is Nearly Normal in Shape

can be modified by the user. Details about the decisions involved in grouping scores are provided in most introductory statistics textbooks and will not be discussed here. Thus, each bar in an SPSS histogram may correspond to an interval that contains a group of scores (rather than a single score). An example of an empirical distribution appears in Figure 1.1. This shows a distribution of measurements of women's heights (in inches). The height of each bar is proportional to the number of cases; for example, the tallest bar in the histogram corresponds to the number of women whose height is 64 in. For this empirical sample distribution, the mean female height $M = 64$ in., and the standard deviation for female height (denoted by s or SD) is 2.56 in. Note that if these heights were transformed into centimeters ($M = 162.56$ cm, $s = 6.50$ cm), the shape of the distribution would be identical; the labels of values on the X axis are the only feature of the graph that would change.

The smooth curve superimposed on the histogram is a plot of the mathematical (i.e., theoretical) function for an ideal normal distribution with a population mean $\mu = 64$ and a population standard deviation $\sigma = 2.56$.

Empirical distributions can have many different shapes. For example, the distribution of number of births across days of the week in the **bar chart** in Figure 1.2 is approximately uniform; that is, approximately one seventh of the births take place on each of the seven days of the week (see Figure 1.2).

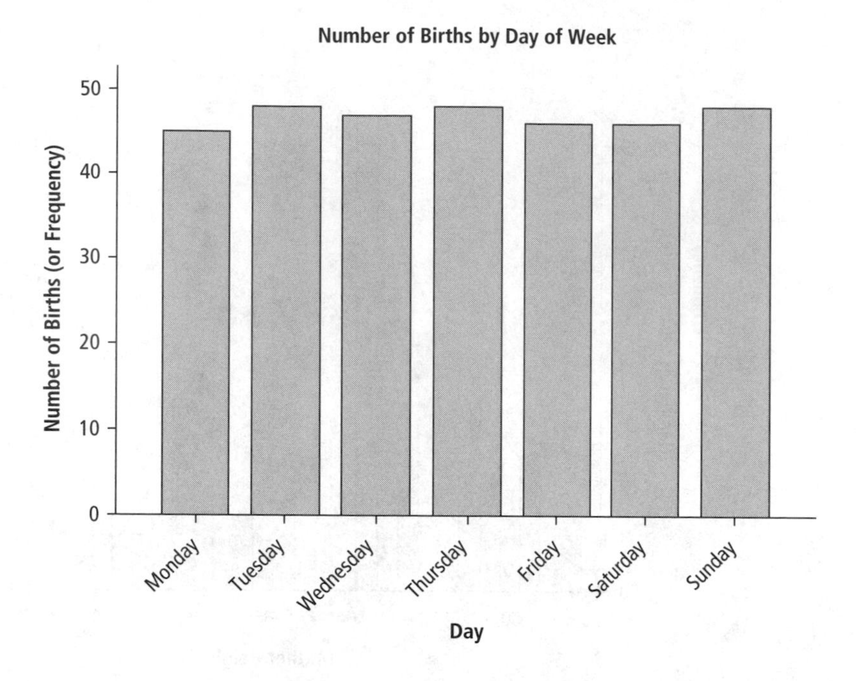

Figure 1.2 ♦ A Bar Chart Showing a Fairly Uniform Distribution for Number of Births (*Y* Axis) by Day of the Week (*X* Axis)

Some empirical distributions have shapes that can be closely approximated by mathematical functions, and it is often convenient to use that mathematical function and a few parameters (such as mean and standard deviation) as a compact and convenient way to summarize information about the distribution of scores on a variable. The proportion of area that falls within a slice of an empirical distribution (in a bar chart or histogram) can be interpreted as a probability. Thus, based on the bar chart in Figure 1.2, we can say descriptively that "about one seventh of the births occurred on Monday"; we can also say that if we draw an individual case from the distribution at random, there is approximately a one-seventh probability that it is a birth that occurred on a Monday.

When a relatively uncommon behavior (such as crying) is assessed through self-report, the distribution of frequencies is often a J-shaped, or roughly exponential, curve, as in Figure 1.3, which shows responses to the question "How many times did you cry last week?" (data from Brackett, Mayer, & Warner, 2004). Most people reported crying 0 times per week, a few reported crying 1 to 2 times a week, and very few reported crying more than 11 times per week.

A theoretical distribution shape that is of particular interest in statistics is the normal (or Gaussian) distribution illustrated in Figure 1.4. Students should be familiar with the shape of this distribution from introductory statistics. The curve is symmetrical, with a peak in the middle and tails that fall off gradually on both sides. The normal curve is often described as a bell-shaped curve. A precise mathematical definition of the theoretical normal distribution is given by the following equations:

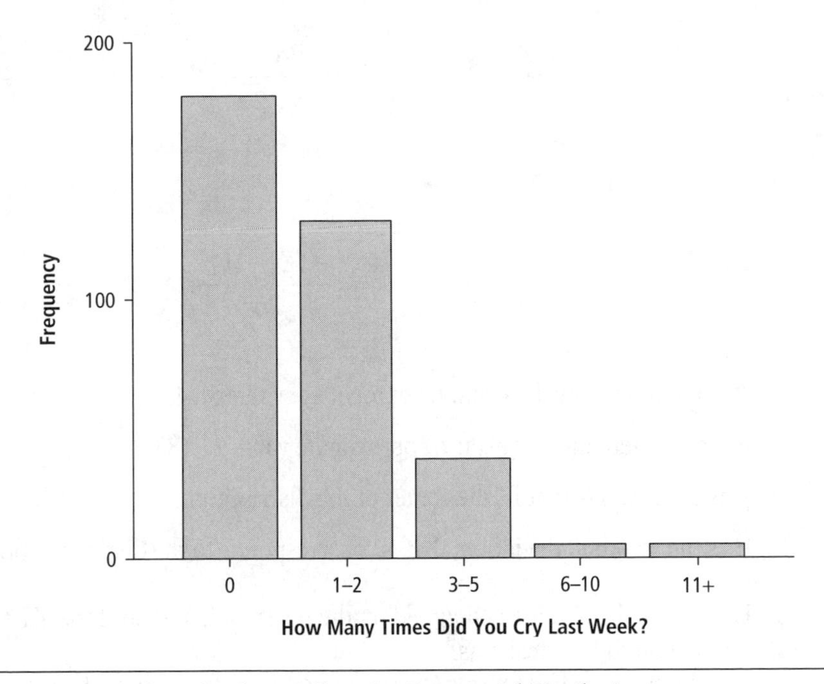

Figure 1.3 ♦ A Bar Chart Showing a J-Shaped or Exponential Distribution

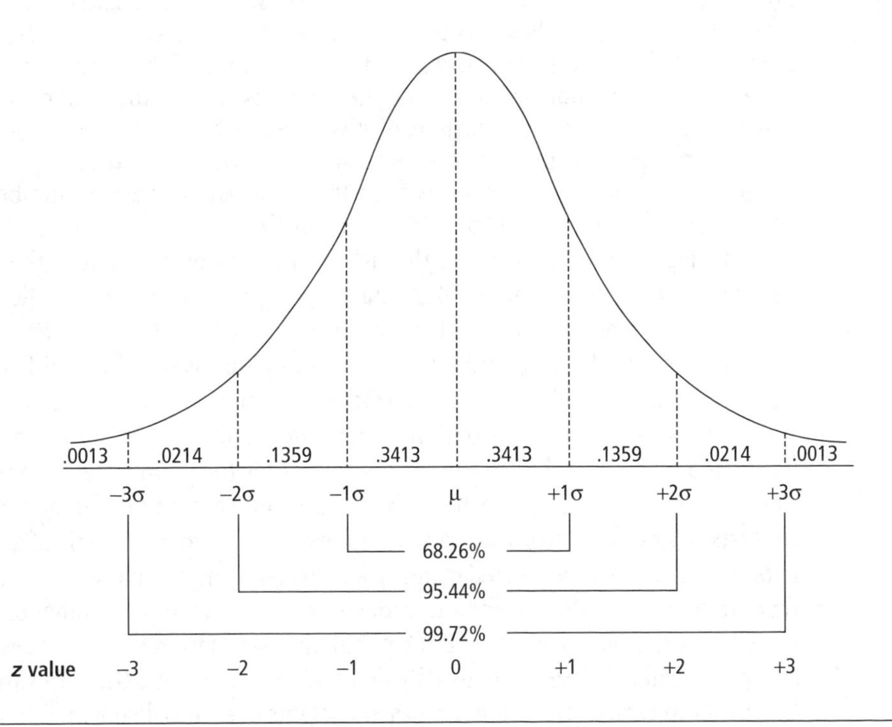

Figure 1.4 ♦ A Standard Normal Distribution, Showing the Correspondence Between Distance From the Mean (Given as Number of σ Units or *z* Scores) and Proportion of Area Under the Curve

$$Y = \frac{e^k}{\sigma\sqrt{2\pi}}, \tag{1.1}$$

where

$$k = \frac{-(X - \mu)^2}{2\sigma}. \tag{1.2}$$

π is a mathematical constant, approximate value 3.1416 . . .

e is a mathematical constant, approximate value 2.7183 . . .

μ is the mean—that is, the center of the distribution.

σ is the standard deviation; that is, it corresponds to the dispersion of the distribution.

In Figure 1.4, the *X* value is mapped on the horizontal axis, and the *Y* height of the curve is mapped on the vertical axis.

For the normal distribution curve defined by this mathematical function, there is a fixed relationship between the distance from the center of the distribution and the area under the curve, as shown in Figure 1.4. The *Y* value (the height) of the normal curve asymptotically approaches 0 as the *X* distance from the mean increases; thus, the curve theoretically has a range of *X* from $-\infty$ to $+\infty$. In spite of this infinite range of *X* values, the area under the normal curve is finite. The total area under the normal curve is set equal to 1.0 so that the proportions of this area can be interpreted as probabilities. The standard normal distribution is defined by Equation 1.1 with μ set equal to 0 and σ set equal to 1. In Figure 1.4, distances from the mean are marked in numbers of standard deviations—for example, $+1\sigma$, $+2\sigma$, and so forth.

From Figure 1.4, one can see that the proportion of area under the curve that lies between 0 and $+1\sigma$ is about .3413; the proportion of area that that lies above $+3\sigma$ is .0013. In other words, about 1 of 1,000 cases lies more than 3σ above the mean, or, to state this another way, the probability that a randomly sampled individual from this population will have a score that is more than 3σ above the mean is about .0013.

A "family" of normal distributions (with different means and standard deviations) can be created by substituting in any specific values for the population parameters μ and σ. The term *parameter* is (unfortunately) used to mean many different things in various contexts. Within statistics, the term generally refers to a characteristic of a population distribution that can be estimated by using a corresponding sample statistic. The parameters that are most often discussed in introductory statistics are the population mean μ (estimated by a sample mean *M*) and the population standard deviation σ (estimated by the **sample standard deviation**, usually denoted by either *s* or *SD*). For example, assuming that the empirical distribution of women's heights has a nearly normal shape with a mean of 64 in. and a standard deviation of 2.56 in. (as in Figure 1.1), a theoretical normal distribution with $\mu = 64$ and $\sigma = 2.56$ will approximately match the location and shape of the empirical distribution of heights. It is possible to generate a family of normal distributions by using different values for μ and σ. For example, intelligence quotient (IQ) scores are normally distributed with $\mu = 100$ and $\sigma = 15$; heart rate for a population of healthy young adults might have a mean of $\mu = 70$ beats per minute (bpm) and a standard deviation σ of 11 bpm.

When a population has a known shape (such as "normal") and known parameters (such as $\mu = 70$ and $\sigma = 11$), this is sufficient information to draw a curve that represents the shape of the distribution. If a population has an unknown distribution shape, we can still compute a mean and standard deviation, but that information is not sufficient to draw a sketch of the distribution.

The standard normal distribution is the distribution generated by Equations 1.1 and 1.2 for the specific values $\mu = 0$ and $\sigma = 1.0$ (i.e., population mean of 0 and population standard deviation of 1). When normally distributed X scores are rescaled so that they have a mean of 0 and a standard deviation of 1, they are called **standard scores** or ***z* scores.** Figure 1.4 shows the standard normal distribution for z. There is a fixed relationship between distance from the mean and area, as shown in Figure 1.4. For example, the proportion of area under the curve that lies between $z = 0$ and $z = +1$ under the standard normal curve is always .3413 (34.13% of the area).

Recall that a proportion of area under the uniform distribution can be interpreted as a probability; similarly, a proportion of area under a section of the normal curve can also be interpreted as a probability. Because the normal distribution is widely used, it is useful for students to remember some of the areas that correspond to z scores. For example, the bottom 2.5% and top 2.5% of the area of a standard normal distribution lie below $z = -1.96$ and above $z = +1.96$, respectively. That is, 5% of the scores in a normally distributed population lie more than 1.96 standard deviations above or below the mean. We will want to know when scores or test statistics have extreme or unusual values relative to some distribution of possible values. In most situations, the outcomes that correspond to the most extreme 5% of a distribution are the outcomes that are considered extreme or unlikely.

The statistics covered in introductory textbooks, such as the t test, ANOVA, and Pearson r, were developed based on the assumption that scores on quantitative variables are normally distributed in the population. Thus, distribution shape is one factor that is taken into account when deciding what type of statistical analysis to use.

Students should be aware that although normal distribution shapes are relatively common, many variables are not normally distributed. For example, income tends to have a distribution that is asymmetric; it has a lower limit of 0, but there is typically a long tail on the upper end of the distribution. Relatively uncommon behaviors, such as crying, often have a J-shaped distribution (as shown in Figure 1.4). Thus, it is important to examine distribution shapes for quantitative variables before applying statistical analyses such as ANOVA or Pearson r. Methods for assessing whether variables are normally distributed are described in Chapter 4.

1.8 ♦ Research Design

Up to this point, the discussion has touched on two important issues that should be taken into account when deciding what statistical analysis to use: the types of variables involved (the level of measurement and whether the variables are categorical or quantitative) and the distribution shapes of scores on quantitative variables. We now turn from a discussion of individual variables (e.g., categorical vs. quantitative types of variables and the shapes of distributions of scores on variables) to a brief consideration of research design.

It is extremely important for students to recognize that a researcher's ability to draw causal inferences is based on the nature of the research design (i.e., whether the study is an experiment) rather than the type of analysis (such as correlation vs. ANOVA). This section briefly reviews basic research design terminology. Readers who have not taken a course in research methods may want to consult a basic research methods textbook (such as Cozby, 2004) for a more thorough discussion of these issues.

1.8.1 ♦ Experimental Design

In behavioral and social sciences, an experiment typically includes the following elements:

1. Random assignment of participants to groups or treatment levels (or other methods of assignment of participants to treatments, such as **matched samples** or **repeated measures,** that ensure equivalence of participant characteristics across treatments).
2. Two or more researcher-administered treatments, dosage levels, or interventions.
3. **Experimental control** of other **"nuisance" or "error" variables** that might influence the outcome: The goal is to *avoid confounding* other variables with the different treatments and to minimize random variations in response due to other variables that might influence participant behavior; the researcher wants to make certain that no other variable is confounded with treatment dosage level. If the caffeine group is tested before a midterm exam and the no-caffeine group is tested on a Friday afternoon before a holiday weekend, there would be a **confound** between the effects of the exam and those of the caffeine; to avoid a confound, both groups should be tested at the same time under similar circumstances. The researcher also wants to make sure that random variation of scores within each treatment condition due to nuisance variables is not too great; for example, testing persons in the caffeine group at many different times of the day and days of the week could lead to substantial variability in anxiety scores within this group. One simple way to ensure that a nuisance variance is neither confounded with treatment nor a source of variability of scores within groups is to "hold the variable constant"; for example, to avoid any potential confound of the effects of cigarette smoking with the effects of caffeine and to minimize the variability of anxiety within groups that might be associated with smoking, the researcher could "hold the variable smoking constant" by including only those participants who are not smokers.
4. Assessment of an outcome variable after the treatment has been administered.
5. Comparison of scores on outcome variables across people who have received different treatments, interventions, or dosage levels: Statistical analyses are used to compare **group means** and assess how strongly scores on the outcome variable are associated with scores on the treatment variable.

In contrast, nonexperimental studies usually lack a researcher-administered intervention, experimenter control of nuisance or error variables, and random assignment of participants to groups; they typically involve measuring or observing several variables in naturally occurring situations.

The goal of experimental design is to create a situation in which it is possible to make a causal inference about the effect of a manipulated treatment variable (*X*) on a measured outcome variable (*Y*). A study that satisfies the conditions for causal inferences is said to have **internal validity**. The conditions required for causal inference (a claim of the form "*X* causes *Y*") (from Cozby, 2004) include the following:

1. The *X* and *Y* variables that represent the "cause" and the "effect" must be systematically associated in the study. That is, it only makes sense to theorize that *X* might cause *Y* if *X* and *Y* covary (i.e., if *X* and *Y* are statistically related). Covariation between variables can be assessed using statistical methods such as ANOVA and Pearson *r*. Covariation of *X* and *Y* is a necessary, but not sufficient, condition for causal inference. In practice, we do not require perfect covariation between *X* and *Y* before we are willing to consider causal theories. However, we look for evidence that *X* and *Y* covary "significantly" (i.e., we use statistical significance tests to try to rule out chance as an explanation for the obtained pattern of results).

2. The cause, *X*, must precede the effect, *Y*, in time. In an experiment this requirement of **temporal precedence** is met by manipulating the treatment variable *X* prior to measuring or observing the outcome variable *Y*.

3. There must not be any other variable confounded with (or systematically associated with) the *X* treatment variable. If there is a confound between *X* and some other variable, the **confounded variable** is a **rival explanation** for any observed differences between groups. Random assignment of participants to treatment groups is supposed to ensure equivalence in the kinds of participants in the treatment groups and to prevent a confound between individual difference variables and treatment condition. Holding other situational factors constant for groups that receive different treatments should avoid confounding treatment with situational variables, such as day of the week or setting.

4. There should be some reasonable theory that would predict or explain a cause-and-effect relationship between the variables.

Of course, the results of a single study are never sufficient to prove causality. However, if a relationship between variables is replicated many times across well-designed experiments, belief that there could be a potential causal connection tends to increase as the amount of supporting evidence increases. Compared with quasi-experimental or nonexperimental designs, experimental designs provide relatively stronger evidence for causality, but causality cannot be proved conclusively by a single study even if it has a well-controlled experimental design.

There is no necessary connection between the type of design (experiment vs. nonexperimental) and the type of statistic applied to the data (such as *t* test vs. Pearson *r*) (see

Table 1.2 ♦ Statistical Analysis and Research Designs

	Experimental Design	*Nonexperimental Design*
t test or analysis of variance (ANOVA)	Very common: Many experiments compare scores across groups that receive different treatments	Fairly common—for example, comparison of means across naturally occurring groups (e.g., men vs. women, smokers vs. nonsmokers)
Pearson *r*	Uncommon, but Pearson *r* might be applied if the manipulated treatment variable has many levels	Very common because surveys often include many quantitative variables
χ^2 test of association	Uncommon, but χ^2 can be applied in experiments where the measured outcome variable is categorical (such as "Does the person stop to help a victim?" coded 1 = No, 2 = Yes)	Very common because surveys often include many categorical variables

Table 1.2). Because experiments often (but not always) involve comparison of group means, ANOVA and *t* tests are often applied to experimental data. However, the choice of a statistic depends on the type of variables in the dataset rather than on experimental design. An experiment does not have to compare a small number of groups. For example, participants in a drug study might be given 20 different dosage levels of a drug (the **independent variable** *X* could be the amount of drug administered), and a response to the drug (such as self-reported pain, *Y*) could be measured. Pairs of scores (for X = Drug dosage and Y = Reported pain) could be analyzed using methods such as Pearson correlation, although this type of analysis is uncommon in experiments.

Internal validity is the degree to which the results of a study can be used to make causal inferences; internal validity increases with greater experimental control of extraneous variables. On the other hand, **external validity** is the degree to which the results of a study can be generalized to groups of people, settings, and events that occur in the real world. A well-controlled experimental situation should have good internal validity.

External validity is the degree to which the results of a study can be generalized (beyond the specific participants, setting, and materials involved in the study) to apply to real-world situations. Some well-controlled experiments involve such artificial situations that it is unclear whether results are generalizable. For example, an experiment that involves systematically presenting different schedules of reinforcement or reward to a rat in a Skinner box, holding all other variables constant, typically has high internal validity (if the rat's behavior changes, the researcher can be reasonably certain that the changes in behavior are caused by the changes in reward schedule, because all other variables are held constant). However, this type of research may have lower external validity; it is not clear whether the results of a study of rats isolated in Skinner boxes can be generalized to populations of children in school classrooms, because the situation of children in a classroom is quite different from the situation of rats in a Skinner box. External validity is better when the research situation is closely analogous to, or resembles, the real-world situations that the researcher wants to learn about. Some experimental situations achieve strong internal validity at the cost of external validity. However, it is possible to conduct experiments in field settings, or to create extremely lifelike and involving situations in laboratory settings, and this can improve the external validity or generalizability of research results.

The strength of internal and external validity depends on the nature of the research situation; it is not determined by the type of statistical analysis that happens to be applied to the data.

1.8.2 ♦ Quasi-Experimental Design

Quasi-experimental designs typically include some, but not all, of the features of a true experiment. Often they involve comparison of groups that have received different treatments and/or comparison of groups before versus after an intervention program. Often they are conducted in field rather than in laboratory settings. Usually, the groups in quasi-experimental designs are not formed by random assignment, and thus, the assumption of equivalence of participant characteristics across treatment conditions is not satisfied. Often the intervention in a quasi experiment is not completely controlled by the researcher, or the researcher is unable to hold other variables constant. To the extent that a quasi experiment lacks the controls that define a well-designed experiment, a quasi-experimental design provides much weaker evidence about possible causality (and, thus, weaker internal validity). Because quasi experiments often focus on interventions that take place in real-world settings (such as schools), they may have stronger external validity than laboratory-based studies. Shadish, Cook, and Campbell (2001) provide further information about issues in the design and analysis of quasi-experimental studies.

1.8.3 ♦ Nonexperimental Research Design

Many studies do not involve any manipulated treatment variable. Instead, the researcher measures a number of variables that are believed to be meaningfully related. Variables may be measured at one point in time or, sometimes, at multiple points in time. Then, statistical analyses are done to see whether the variables are related in ways that are consistent with the researcher's expectations. The problem with **nonexperimental research design** is that any potential independent variable is usually correlated or confounded with other possible independent variables; therefore, it is not possible to determine which, if any, of the variables have a causal impact on the **dependent variable**. In some nonexperimental studies, researchers make distinctions between independent and dependent variables (based on implicit theories about possible causal connections).

However, in some nonexperimental studies, there may be little or no basis to make such a distinction. Nonexperimental research is sometimes called "correlational" research. This use of terminology is unfortunate because it can confuse beginning students. It is helpful to refer to studies that do not involve interventions as "nonexperimental" (rather than correlational) to avoid possible confusion between the Pearson *r* correlation statistic and nonexperimental design.

As shown in Table 1.2, a Pearson correlation can be performed on data that come from experimental designs, although it is much more often encountered in reports of nonexperimental data. A *t* test or ANOVA is very often used to analyze data from experiments, but these tests are also used to compare means between naturally occurring groups in nonexperimental studies. (In other words, to judge whether a study is experimental or nonexperimental, it is not useful to ask whether the reported statistics were ANOVAs or correlations. We have to look at the way the study was conducted, that is, whether it has the features typically found in experiments that were listed earlier.)

The degree to which research results can be interpreted as evidence of possible causality depends on the nature of the design (experimental vs. nonexperimental), not on the type of statistic that happens to be applied to the data (Pearson *r* vs. *t* test or ANOVA). While experiments often involve comparison of groups, group comparisons are not necessarily experimental.

A nonexperimental study usually has weak internal validity; that is, merely observing that two variables are correlated is not a sufficient basis for causal inferences. If a researcher finds a strong correlation between an *X* and *Y* variable in a nonexperimental study, the researcher typically cannot rule out rival explanations (e.g., changes in *Y* might be caused by some other variable that is confounded with *X*, rather than by *X*). On the other hand, some nonexperimental studies (particularly those that take place in field settings) may have good external validity, that is, they may examine naturally occurring events and behaviors.

1.8.4 ♦ Between-Subjects Versus Within-Subjects or Repeated Measures

When an experiment involves comparisons of groups, there are many different ways in which participants or cases can be placed in these groups. In spite of the fact that most writers now prefer to use the term *participant* rather than *subject* to refer to a person who contributes data in a study, the letter *S* is still widely used to stand for "subjects" when certain types of research designs are described.

When a study involves a categorical or group membership variable, we need to pay attention to the composition of the groups when we decide how to analyze the data. One common type of group composition is called between-subjects (between-S) or independent groups. In a between-S or independent groups study, each participant is a member of one and only one group. A second common type of group composition is called within-subjects (within-S) or repeated measures. In a repeated measures study, each participant is a member of every group; if the study includes several different treatments, each participant is tested under every treatment condition.

For example, consider the caffeine/anxiety study. This study could be done using either a **between-S** or a **within-S design.** In a between-S version of this study, a sample of 30 participants could be divided randomly into two groups of 15 each. Each group would be given only one treatment (Group 1 would receive a beverage that contains no caffeine; Group 2 would receive a beverage that contains 150 mg caffeine). In a within-S or repeated measures version of this study, each of the 30 participants would be observed twice: once after drinking a beverage that does not contain caffeine and once after drinking a beverage that contains 150 mg of caffeine. Another possible variation of design would be to use both within-S and between-S comparisons. For example, the researcher could randomly assign 15 people to each of the two groups, caffeine versus no caffeine, and then assess each person's anxiety level at two points in time, before and after consuming a beverage. This design has both a between-S comparison (caffeine vs. no caffeine) and a within-S comparison (anxiety before vs. after drinking a beverage that may or may not contain caffeine).

Within-S or repeated measures designs raise special problems, such as the need to control for order and **carryover effects** (discussed in basic research methods textbooks such as Shaughnessy, Zechmeister, & Zechmeister, 2003). In addition, different statistical tests are used to compare group means for within-S versus between-S designs. For a between-S design, one-way ANOVA for independent samples is used; for a within-S design, **repeated measures ANOVA** is used. A thorough discussion of repeated measures analyses is provided by Keppel (1991), and a brief introduction to repeated measures ANOVA is provided in Chapter 20 of this textbook. Thus, a researcher has to know whether the composition of groups in a study is between-S or within-S in order to choose an appropriate statistical analysis.

In nonexperimental studies, the groups are almost always between-S because they are usually based on previously existing participant characteristics (e.g., whether each participant is male or female, a smoker or a nonsmoker). Generally, when we talk about groups based on naturally occurring participant characteristics, group memberships are mutually exclusive; for example, a person cannot be classified as both a smoker and a nonsmoker. (We could, of course, create a larger number of groups such as nonsmoker, occasional smoker, and heavy smoker if the simple distinction between smokers and nonsmokers does not provide a good description of smoking behavior.)

In experiments, researchers can choose to use either within-S or between-S designs. An experimenter typically assigns participants to treatment groups in ways that are intended to make the groups equivalent prior to treatment. For example, in a study that examines three different types of stress, each participant may be randomly assigned to one and only one treatment group or type of stress.

In this textbook, all group comparisons are assumed to be between-S unless otherwise specified. Chapter 20 deals specifically with repeated measures ANOVA.

1.9 ♦ Parametric Versus Nonparametric Statistics

Another issue that should be considered when choosing a statistical method is whether the data satisfy the assumptions for parametric statistical methods. Definitions of the term ***parametric statistics*** vary across textbooks. When a variable has a known distribution shape (such as normal), we can draw a sketch of the entire distribution of scores based on just two pieces of information: the shape of the distribution (such as normal) and a small number of population parameters for that distribution (for normally distributed scores, we need to know only two parameters, the population mean μ and the population standard deviation σ, in order to draw a picture of the entire distribution). Parametric statistics involve obtaining sample estimates of these population parameters (e.g., the sample mean *M* is used to estimate the population mean μ; the sample standard deviation *s* is used to estimate the population standard deviation σ).

Most authors include the following points in their discussion of parametric statistics:

1. Parametric statistics include the analysis of means, variances, and **sums of squares.** For example, *t* test, ANOVA, Pearson *r*, and regression are examples of parametric statistics.

2. Parametric statistics require quantitative dependent variables that are at least approximately interval/ratio level of measurement. In practice, as noted in Section 1.6, this requirement is often not strictly observed. For example, parametric statistics (such as mean and correlation) are often applied to scores from 5-point rating scales, and these scores may fall short of satisfying the strict requirements for interval level of measurement.

3. The parametric statistics included in this book and in most introductory texts assume that scores on quantitative variables are normally distributed. (This assumption is violated when scores have a uniform or J-shaped distribution, as shown in Figures 1.2 and 1.3, or when there are extreme **outliers**.)

4. For analyses that involve comparisons of group means, the variances of dependent variable scores are assumed to be equal across the populations that correspond to the groups in the study.

5. Parametric analyses often have additional assumptions about the distributions of scores on variables (e.g., we need to assume that *X* and *Y* are linearly related to use Pearson *r*).

It is unfortunate that some students receive little or no education on **nonparametric statistics.** Most introductory statistics textbooks include one or two chapters on nonparametric methods; however, these are often at the end of the book, and instructors rarely have enough time to cover this material in a one-semester course. Sometimes nonparametric statistics do not necessarily involve estimation of population parameters; they often rely on quite different approaches to sample data—for example, comparing the sum of ranks across groups in the **Wilcoxon rank sum test**. There is no universally agreed on definition for nonparametric statistics, but most discussions of nonparametric statistics include the following:

1. Nonparametric statistics include the median, the chi-square (χ^2) test of association between categorical variables, the Wilcoxon rank sum test, the sign test, and the Friedman one-way ANOVA by ranks. Many nonparametric methods involve counting frequencies or finding medians.

2. The dependent variables for nonparametric tests may be either nominal or ordinal level of measurement. (Scores may be obtained as ranks initially, or **raw scores** may be converted into ranks as one of the steps involved in performing a nonparametric analysis.)

3. Nonparametric statistics do not require scores on the outcome variable to be normally distributed.
4. Nonparametric statistics do not typically require an assumption that variances are equal across groups.
5. Outliers are not usually a problem in nonparametric analyses; these are unlikely to arise in ordinal (rank) or nominal (categorical) data.

Researchers should consider the use of nonparametric statistics when their data fail to meet some or all of the requirements for parametric statistics.[5] The issues outlined above are summarized in Table 1.3.

Jaccard and Becker (2002) pointed out that there is disagreement among behavioral scientists about when to use parametric versus nonparametric analyses. Some conservative statisticians argue that parametric analyses should be used only when all the assumptions listed in the discussion of parametric statistics above are met (i.e., only when scores on the dependent variable are quantitative, interval/ratio level of measurement, and normally distributed and meet all other assumptions for the use of a specific statistical test). On the other hand, Bohrnstedt and Carter (1971) have advocated a very liberal position; they argued that many parametric techniques are fairly robust[6] to violations of assumptions and concluded that even for variables measured at an ordinal level, "parametric analyses not only can be, but should be, applied."

The recommendation made here is a compromise between the conservative and liberal positions. It is useful to review all the factors summarized in Table 1.3 when making the choice between parametric and nonparametric tests. A researcher can safely use parametric statistics when all the requirements listed for parametric tests are met. That is, if scores on the dependent variable are quantitative and normally distributed, scores are interval/ratio level of measurement and have equal variances across groups, and there is a minimum N per group of at least 20 or 30, parametric statistics may be used.

When only one or two of the requirements for a parametric statistic are violated, or if the violations are not severe (e.g., the distribution shape for scores on the outcome variable is only slightly different from normal), then it may still be reasonable to use a parametric statistic. When in doubt about whether to choose parametric or nonparametric statistics, many researchers lean toward choosing parametric statistics. There are several reasons for this preference. First, the parametric tests are more familiar to most students, researchers, and journal editors. Second, it is widely thought that parametric tests have better **statistical power**; that is, they give the researcher a better chance of obtaining a **statistically significant outcome** (however, this is not necessarily always the case). An additional issue becomes relevant when researchers begin to work with more than one predictor and/or more than one outcome variable. For some combinations of predictor and outcome variables, a parametric analysis exists, but there is no analogous nonparametric test. Thus, researchers who use only nonparametric analyses may be limited to working with fewer variables. (This is not necessarily a bad thing.)

When violations of the assumptions for the use of parametric statistics are severe, it is more appropriate to use nonparametric analyses. Violations of assumptions (such as the assumption that scores are distributed normally in the population) become much more problematic when they are accompanied by small (and particularly small and unequal) group sizes.

Table 1.3 ♦ Parametric and Nonparametric Statistics

Parametric Tests (Such as M, t, F, Pearson r) Are More Appropriate When	*Nonparametric Tests (Such as Median, Wilcoxon Rank Sum Test, Friedman One-Way, ANOVA, by Ranks, Spearman r) Are More Appropriate When*
The outcome variable *Y* is interval/ratio level of measurement[a]	The outcome variable *Y* is nominal or ordinal level of measurement
Scores on *Y* are approximately normally distributed[b]	Scores on *Y* are not necessarily normally distributed
There are no extreme outlier values of *Y*[c]	There can be extreme outlier *Y* scores
Variances of *Y* scores are approximately equal across populations that correspond to groups in the study[d]	Variances of scores are not necessarily equal across groups
The *N* of cases in each group is "large"[e]	The *N* of cases in each group can be "small"

a. Many variables that are widely used in psychology (such as 5-point or 7-point attitude rating scales, personality test scores, and so forth) have scores that probably do not have true equal interval level measurement properties. For example, consider 5-point degree of agreement rating scales: The difference between a score of 4 and 5 and the difference between a score of 1 and 2 probably do not correspond to exactly the same increase of change in agreement. Thus, 5-point rating scales probably do not have true equal-interval measurement properties. Based on the arguments reported in Section 1.5, many researchers go ahead and apply parametric statistics (such as Pearson *r* and *t* test) to data from 5-point rating scales and personality tests and other measures that probably fall short of satisfying the requirements for a true interval level of measurement as defined by Stevens (1946).

b. Chapter 4 discusses how to assess this by looking at histograms of *Y* scores to see if the shape resembles the bell curve shown in Figure 1.4.

c. Chapter 4 also discusses identification and treatment of outliers or extreme scores.

d. Parametric statistics such as the *t* test and ANOVA were developed based on assumption that the *Y* scores have equal variances in the populations that correspond to the samples in the study. Data that violate the assumption of equal variances can, in theory, lead to misleading results (an increased risk of Type I error, discussed in Chapter 3). In practice, however, the *t* test and ANOVA can yield fairly accurate results even when the equal variance assumption is violated, unless the *N*s of cases within groups are small and/or unequal across groups. Also, there is a modified version of the *t* test (usually called **"separate variances *t* test"** or "equal variances not assumed *t* test") that takes violations of the equal variance assumption into account and corrects for this problem.

e. There is no agreed-on standard about an absolute minimum sample size required in the use of parametric statistics. The suggested guideline given here is as follows: Consider nonparametric tests when *N* is less than 20, and definitely use nonparametric tests when *N* is less than 10 per group; but this is arbitrary. Smaller *N*s are most problematic when there are other problems with the data, such as outliers.

In practice, it is useful to consider this entire set of criteria. If the data fail to meet just one criterion for the use of parametric tests (for example, if scores on *Y* do not quite satisfy the requirements for interval/ratio level of measurement), researchers often go ahead and use parametric tests, as long as there are no other serious problems. However, the larger the number of problems with the data, the stronger the case becomes for the use of nonparametric tests. If the data are clearly ordinal or if *Y* scores have a drastically nonnormal shape and if, in addition, the *N*s within groups are small, a nonparametric test would be strongly preferred. Group *N*s that are unequal can make other problems (such as unequal variances across groups) more serious.

Almost all the statistics reported in this textbook are parametric. (The only nonparametric statistics reported are the χ^2 test of association in Chapter 8 and the **binary logistic regression** in Chapter 21.) If a student or researcher anticipates that his or her data will usually require nonparametric analysis, that student should take a course or at least buy a good reference book on nonparametric methods.

Special statistical methods have been developed to handle ordinal data—that is, scores that are obtained as ranks or that are converted into ranks during preliminary data handling. Strict adherence to the Stevens theory would lead us to use medians instead of means for ordinal data (because finding a median involves only rank ordering and counting scores, not summing them).

The choice between parametric and nonparametric statistics is often difficult because there are no generally agreed on decision standards. In research methods and statistics, generally, it is more useful to ask, What are the advantages and disadvantages of each approach to the problem? than to ask, Which is the right and which is the wrong answer? Parametric and nonparametric statistics each have strengths and limitations. Experimental and nonexperimental designs each have advantages and problems. Self-report and behavioral observations each have advantages and disadvantages. In the discussion section of a research report, the author should point out the advantages, and also acknowledge the limitations, of the choices that he or she has made in research design and data analysis.

The limited coverage of nonparametric techniques in this book should not be interpreted as a negative judgment about their value. There are situations (particularly designs with small *N*s and severely nonnormal distributions of scores on the outcome variables) where nonparametric analyses are preferable. In particular, when data come in the form of ranks, nonparametric procedures developed specifically for the analysis of ranks may be preferable. For a thorough treatment of nonparametric statistics, see Siegel and Castellan (1988).

1.10 ♦ Additional Implicit Assumptions

Some additional assumptions are so basic that they generally are not even mentioned, but these are important considerations whether a researcher uses parametric or nonparametric statistics:

1. Scores on the outcome variable are assumed to be independent of each other (except in repeated measures data, where correlations among scores are expected and the pattern of dependence among scores has a relatively simple pattern). It is easier to explain the circumstances that lead to "nonindependent" scores than to define **independence of observations** formally. Suppose a teacher gives an examination in a crowded classroom, and students in the class talk about the questions and exchange information. The scores of students who communicate with each other will not be independent; that is, if Bob and Jan jointly decide on the same answer to several questions on the exam, they are likely to have similar (statistically related or dependent) scores. Nonindependence among scores can arise due to many kinds of interactions among participants, apart from sharing information or cheating; nonindependence can arise from persuasion, competition, or other kinds of social influence. This problem is not limited to human research subjects. For example, if a researcher measures the heights of trees in a grove, the heights of neighboring trees are not independent; a tree that is surrounded by other tall trees has to grow taller in order to get exposure to sunlight.

   For both parametric and nonparametric statistics, different types of analysis are used when the data involve repeated measures than when they involve independent outcome scores.

2. The number of cases in each group included in the analysis should be reasonably large. Parametric tests typically require larger numbers per group than nonparametric tests to yield reasonable results. However, even for nonparametric analyses, extremely small *N*s are undesirable. (A researcher does not want to be in a situation where a 1- or 2-point change in score for one participant would completely change the nature of the outcome, and very small *N*s sometimes lead to this kind of instability.) There is no agreed-on absolute minimum *N* for each group. For each analysis presented in this textbook, a discussion about sample size requirements is included.

3. The analysis will yield meaningful and interpretable information about the relations between variables only if we have a "**correctly specified model**"; that is, we have included all the variables that should be included in the analysis, and we have not included any irrelevant or inappropriate variables. In other words, we need a theory that correctly identifies which variables should be included and which variables should be excluded. Unfortunately, we can never be certain that we have a correctly specified model. It is always possible that adding or dropping a variable in the statistical analysis might change the outcome of the analysis substantially. Our inability to be certain about whether we have a correctly specified model is one of the many reasons why we can never take the results of a single study as proof that a theory is correct (or incorrect).

1.11 ♦ Selection of an Appropriate Bivariate Analysis

Bivariate statistics assess the relation between a pair of variables. Often, one variable is designated as the independent variable and the other as dependent. When one of the variables is manipulated by the researcher, that variable is designated as the independent variable. In nonexperimental research situations, the decision regarding which variable to treat as an independent variable may be arbitrary. In some research situations, it may be preferable not to make a distinction between independent and dependent variables; instead, the researcher may merely report that two variables are correlated without identifying one as the predictor of the other. When a researcher has a theory that *X* might cause or influence *Y*, the researcher generally uses scores on *X* as predictors of *Y* even when the study is nonexperimental. However, the results of a nonexperimental study cannot be used to make a causal inference, and researchers need to be careful to avoid causal language when they interpret results from nonexperimental studies.

During introductory statistics courses, the choice of an appropriate statistic for various types of data is not always explicitly addressed. Aron and Aron (2002) and Jaccard and Becker (2002) provide good guidelines for the choice of bivariate analyses. The last part of this chapter summarizes the issues that have been discussed up to this point and shows how consideration of these issues influences the choice of an appropriate bivariate statistical analysis. Similar issues continue to be important when the analyses include more than one predictor and/or more than one outcome variable.

The choice of an appropriate bivariate analysis to assess the relation between two variables is often based, in practice, on the types of variables involved: categorical versus quantitative. The following guidelines for the choice of statistic are based on a discussion in Jaccard and Becker (2002). Suppose that a researcher has a pair of variables *X* and *Y*. There are three possible combinations of types of variables (see Table 1.4):

Case I: Both *X* and *Y* are categorical.

Case II: *X* is categorical, and *Y* is quantitative (or *Y* is categorical, and *X* is quantitative).

Case III: Both *X* and *Y* are quantitative.

Consider Case I: The *X* and *Y* variables are both categorical; the data are usually summarized in a **contingency table** that summarizes the numbers of scores in each *X*, *Y* group. The **chi-square test of association** (or one of many other contingency table statistics) can be used to assess whether *X* and *Y* are significantly related. This will be discussed in Chapter 8. There are many other types of statistics for contingency tables (Everitt, 1977).

Consider Case III: The *X* and *Y* variables are both quantitative variables. If *X* and *Y* are linearly related (and if other assumptions required for the use of Pearson *r* are reasonably well satisfied), a researcher is likely to choose Pearson *r* to assess the relation between the *X* and *Y* variables. There are other types of correlation (such as **Spearman *r***) that may be preferred when the assumptions for Pearson *r* are violated; Spearman *r* is an appropriate analysis when the *X* and *Y* scores consist of ranks (or are converted to ranks to get rid of problems such as extreme outliers).

Now consider Case II: One variable (usually the *X* or independent variable) is categorical, and the other variable (usually the *Y* or dependent variable) is quantitative. In this situation, the analysis involves comparing means, medians, or sums of ranks on the *Y* variable across the groups that correspond to scores on the *X* variable. The choice of an appropriate statistic in this situation depends on several factors; the following list is adapted from Jaccard and Becker (2002):

1. Whether scores on *Y* satisfy the assumptions for parametric analyses or violate these assumptions badly enough so that nonparametric analyses should be used

2. The number of groups that are compared (i.e., the number of levels of the *X* variable)

3. Whether the groups correspond to a between-S design (i.e., there are different participants in each group) or to a within-S or repeated measures design

One cell in Table 1.4 includes a decision tree for Case II from Jaccard and Becker (2002); this decision tree maps out choices among several common bivariate statistical methods based on the answers to these questions. For example, if the scores meet the assumptions for a parametric analysis, two groups are compared, and the design is between-S, the independent samples *t* test is a likely choice. Note that although this decision tree leads to just one analysis for each situation, there are sometimes other analyses that could be used.

This textbook covers only parametric statistics (i.e., statistics in the parametric branch of the decision tree for Case II in Table 1.4). In some situations, however, nonparametric statistics may be preferable (see Siegell & Castellan, 1988, for a thorough presentation of nonparametric methods).

Table 1.4 ♦ Selecting an Appropriate Bivariate Statistic Based on Type of Independent Variable (IV) and Dependent Variable (DV)

<table>
<tr><th></th><th>Categorical DV</th><th>Quantitative DV</th></tr>
<tr><td>Categorical IV</td><td>Case I:
Set up a table to report the numbers of cases within each group; use a nonparametric test (such as χ^2) to assess whether group membership on the DV is predictable from group membership on the IV</td><td>Case II: (Use This Decision Tree)
Parametric Test
<ul>
<li>IV is between-subjects
<ul><li>IV has 2 levels → Independent groups t test</li><li>IV has 3 or more levels → One-way between-subjects analysis of variance</li></ul></li>
<li>IV is within-subjects
<ul><li>IV has 2 levels → Correlated groups t test</li><li>IV has 3 or more levels → One-way repeated measures analysis of variance</li></ul></li>
</ul>
Nonparametric Test
<ul>
<li>IV is between-subjects
<ul><li>IV has 2 levels → Wilcoxon rank sum test/ Mann-Whitney U test or chi-square test</li><li>IV has 3 or more levels → Kruskal-Wallis test or chi-square test</li></ul></li>
<li>IV is within-subjects
<ul><li>IV has 2 levels → Wilcoxon signed-rank test</li><li>IV has 3 or more levels → Friedman analysis of variance by ranks</li></ul></li>
</ul></td></tr>
<tr><td>Quantitative IV</td><td>Case II:
Use the decision tree for Case II shown in the upper right corner of this table</td><td>Case III:
If scores on the IV and DV are both ordinal, use the nonparametric correlation, Spearman r.
If scores on the IV and DV are interval/ratio, use the parametric correlation, Pearson r.</td></tr>
</table>

SOURCE: Decision tree from *Statistics for the Behavioral Sciences, 4th edition* by Jaccard/Becker, 2002. Reprinted with permission of Wadsworth, a division of Thomson Learning: www.thomsonrights.com. Fax 800-730-2215.

1.12 ♦ Summary

Reconsider the hypothetical experiment to assess the effects of caffeine on anxiety. Designing a study and choosing an appropriate analysis raises a large number of questions even for this very simple research question.

A nonexperimental study could be done; that is, instead of administering caffeine, a researcher could ask participants to self-report the amount of caffeine consumed within the past 3 hr and then self-report anxiety.

A researcher could do an experimental study (i.e., administer caffeine to one group and no caffeine to a comparison group under controlled conditions and subsequently measure anxiety). If the study is conducted as an experiment, it could be done using a between-S design (each participant is tested under only one condition, either with caffeine or without caffeine), or it could be done as a within-S or repeated measures study (each participant is tested under both conditions, with and without caffeine).

Let's assume that the study is conducted as a simple experiment with a between-S design. The outcome measure of anxiety could be a categorical variable (i.e., each participant is identified by an observer as a member of the "anxious" or "nonanxious" group). In this case, a table could be set up to report how many of the persons who consumed caffeine were classified as anxious versus nonanxious, and how many of those who did not receive caffeine were classified as anxious versus nonanxious, and a chi-square test could be performed to assess whether people who received caffeine were more likely to be classified as anxious than people who did not receive caffeine.

If the outcome variable, anxiety, is assessed by having people self-report their level of anxiety using a 5-point rating scale, an independent samples t test could be used to compare mean anxiety scores between the caffeine and no-caffeine groups. If examination of the data indicated serious violations of the assumptions for this parametric test (such as nonnormally distributed scores or unequal group variances, along with very small numbers of participants in the groups), the researcher might choose to use the Wilcoxon rank sum test to analyze the data from this study.

This chapter reviewed issues that generally are covered in early chapters of introductory statistics and research methods textbooks. Based on this material, the reader should be equipped to think about the following issues, both when reading published research articles and when planning a study:

1. Evaluate whether the sample is a convenience sample or a random sample from a well-defined population, and recognize how the composition of the sample in the study may limit the ability to generalize results to broader populations.

2. Understand that the ability to make inferences about a population from a sample requires that the sample be reasonably representative of or similar to the population.

3. For each variable in a study, understand whether it is categorical or quantitative.

4. Recognize the differences between experimental, nonexperimental, and quasi-experimental designs.

5. Understand the difference between between-S (independent groups) and within-S (repeated measures) designs.
6. Recognize that research designs differ in internal validity (the degree to which they satisfy the conditions necessary to make a causal inference) and external validity (the degree to which results are generalizable to participants, settings, and materials different from those used in the study).
7. Understand why experiments typically have stronger internal validity, and why experiments may have weaker external validity compared with nonexperimental studies.
8. Understand the issues involved in making a choice between parametric and nonparametric statistical methods.
9. Be able to identify an appropriate statistical analysis to describe whether scores on two variables are related, taking into account whether or not the data meet the assumptions for parametric tests; the type(s) of variables, categorical versus quantitative; whether the design is between-S or within-S; and the number of groups that are compared. The decision tree in Table 1.4 identifies the most widely used statistical procedure for each of these situations.
10. Most important of all, readers should remember that whatever choices researchers make, each choice typically has both advantages and disadvantages. The discussion section of a research report can point out the advantages and strengths of the approach used in the study, but it should also acknowledge potential weaknesses and limitations. If the study was not an experiment, the researcher must avoid using language that implies that the results of the study are proof of causal connections. Even if the study is a well-designed experiment, the researcher should keep in mind that no single study provides definitive proof for any claim. If the sample is not representative of any well-defined, real population of interest, limitations in the generalizability of the results should be acknowledged (e.g., if a study that assesses the safety and effectiveness of a drug is performed on a sample of persons 18 to 22 years old, the results may not be generalizable to younger and older persons). If the data violate many of the assumptions for the statistical tests that were performed, this may invalidate the results.

Table 1.5 provides an outline of the process involved in doing research. Some issues that are included (such as the IRB review) apply only to research that involves human participants as subjects, but most of the issues are applicable to research projects in many different disciplines. It is helpful to think about the entire process and anticipate later steps when making early decisions. For example, it is useful to consider what types of variables you will have and what statistical analyses you will apply to those variables at an early stage in planning. It is essential to keep in mind how the planned statistical analyses are related to the primary research questions. This can help researchers avoid collecting data that are difficult or impossible to analyze.

(Text continues on page 35)

Table 1.5 ♦ Preview of a Typical Research Process for an Honors Thesis, Master's Thesis, or Dissertation

Identify one or more research questions.	Generally, researchers examine at least two variables. In an experiment, an *X* treatment variable is manipulated, and a *Y* outcome variable is measured. In nonexperimental studies, usually none of the variables are manipulated. Later chapters in this book discuss research questions that involve more than one predictor or more than one outcome variable.
Review existing research about this research question.	For most disciplines, the easiest way to identify relevant past research is to use computer databases, usually accessible through a university library. For example, a psychologist would look for relevant studies by using the PsycArticle and PsycInfo databases to search for studies that mention both caffeine and anxiety (and related keywords such as *mood, tension,* or *arousal*).
Use the review of existing research to identify methods that you can use in your own study.	For example, past research on caffeine and anxiety might have used a mood-rating list developed by Thayer (the Activation/Deactivation Checklist). The researcher might consider whether to use the same measure (so that results are directly comparable with past studies) or a different measure (to demonstrate that the effect of caffeine on mood can be replicated using different measurement methods or to choose a measure that may have advantages over the measures used in past research, such as stronger reliability or questions that are easier for participants to understand).
Decide on "something new" to include in the study.	For thesis research and journal articles, people rarely perform exact replications of existing studies. Sometimes a study examines a pair of variables that have never been examined in past research, but that level of originality is rare. More often, a beginning researcher does a "replication with a twist." For example, the researcher might use a past study that examines caffeine and anxiety as a model and change at least one important feature of the study (the type of participant, the setting, the dosage levels of caffeine, the type of mood measurement used to assess outcome, the level of control over other variables such as participant drug use, and so forth). For research to be original, it should provide some new knowledge or information. Sometimes the part of the study that is new is only a small change (e.g., if past research has only studied the effects of caffeine on 18- to 22-year-olds, the present study might examine the effects of caffeine in younger or older persons). If past studies have some clear limitation or problem, of course, the best kind of "twist" is to change the methods to get rid of that limitation or problem.

(Continued)

Table 1.5 ♦ (Continued)

Decide on the procedures to manipulate the treatment variable (if the study is an experiment or a quasi experiment).	Later chapters (particularly Chapters 5 and 6) discuss issues to be considered, such as the number of treatment groups and the choice of dosage levels for the treatment.
Decide how to measure nonmanipulated variables (in both nonexperimental and other types of designs).	A later chapter (Chapter 19) discusses some of the options available to researchers, such as making up new self-report questions versus using existing questionnaires, and the issues to consider in evaluating whether a measure has good qualities, such as reliability and validity.
Identify a setting (laboratory or field).	For a survey, it may be convenient to administer questionnaires in classrooms. Some types of research are done in field settings such as schools, hospitals, prisons, and nursing homes. Note that obtaining access to field settings can be difficult.
Identify the type of participants or cases.	Decide whether to use human (or nonhuman animal) participants. Also, decide on age, gender, and other characteristics. In practice, when researchers use convenience samples such as students in introductory psychology, it amounts to a default decision to use primarily young adults between ages 18 and 23. For some types of research, participants with special characteristics are needed (such as people who are bilingual, clinically depressed, color blind, or left-handed, or gifted in math). From this point onward in the outline, human participants are assumed, but there are analogous issues when the cases correspond to geographic regions, nonhuman animals, trees, asteroids, or other objects.
Decide how to select, sample, and recruit human participants.	In psychology, this often involves posting sign-up sheets for students who are required to participate in research for credit in their introductory psychology course. There are many other ways of recruiting participants that would yield more diverse samples, of course.
Decide how you will protect participants from harm, how you will keep data confidential, and what compensation (if any) participants will receive for participation; also, work out ways to minimize any potential risk or harm.	For research with human participants, read the university's ethical guidelines (and possibly also the ethical guidelines published by professional associations such as the American Psychological Association). These outline ethical issues that should be considered in detail.

Before you collect any data, get your thesis supervisor (or thesis committee) to approve your proposed methodology. You will typically write a much longer thesis proposal for your committee than for the institutional review board (IRB) review.	For an undergraduate honors thesis, you may have just one faculty supervisor rather than a committee. For a master's thesis or dissertation, you will have a committee. Discuss the requirements for the composition of the committee, the length and format of the proposal, and other issues with your research supervisor. A proposal has sections similar to a journal article, except that most parts of a proposal are in the future tense/hypothetical: Introduction (review of past research, leading up to the present study); Proposed Methods; Proposed Data Analyses; Discussion (of the proposed methods); References; and Appendixes, including documentation such as the complete questionnaire you will use (if you are doing a survey), informed consent forms, instruction sheets, and any other materials.
After your committee has approved your proposed research, write a briefer version of your research proposal and submit it to the university (or department or college level) IRB for review. You must have IRB approval *before* you collect any data.	Most universities now post instructions and forms for IRB review of proposals that involve human participants on their Web sites (for example, at the University of New Hampshire [UNH], this information appears at www.unh.edu/osr/compliance/irb.html). A separate set of standards and a different committee are responsible for review of research that involves nonhuman animals (at UNH, this is called the Institutional Committee for Animal Care and Use, IACUC: www.unh.edu/osr/compliance/iacuc.html). Most research methods textbooks (such as Cozby, 2004) include a chapter that reviews the most important ethical issues.
Run a pilot study.	Try out your procedures using a small number of participants. A trial run often reveals problems (e.g., the survey takes much longer to complete than you anticipated, question numbering on a survey is incorrect, or instructions are confusing). Stop and correct any problems you identify during the pilot study. Discuss any changes in methods with your research supervisor. If you make major changes in methods after your pilot study (other than just dropping questions), you may need to write a letter to the IRB requesting approval for the changes in methods.
Collect your data and get your data into computer-readable form.	If you use measuring instruments that are directly connected to a computer or use a Web form to collect survey data, data are generally stored in computer files. If you have participants write responses on a questionnaire booklet or record observations on log sheets, you may need to enter the data into the computer by hand. Computer-scorable answer sheets can be scanned, but usually the data are returned to you in a form that isn't suitable for analysis by SPSS, so you need to do some additional work to make the data into an SPSS file. Whether you use automated systems or human data entry, it's important to check for accuracy.

(Continued)

Table 1.5 ♦ (Continued)

Conduct preliminary data screening.	Chapter 4 describes preliminary data screening. You need to identify data error entries (e.g., an age of 257 years cannot be correct), evaluate whether scores on quantitative variables are normally distributed, check to see if groups formed by categorical variables have reasonably large *N*s, and so forth.
Based on the types of variables, the nature of the data, and the research question, identify an appropriate statistical analysis.	For example, in the imaginary experiment about the effects of caffeine on anxiety, there is a categorical independent variable with two groups (caffeine/no caffeine) and a quantitative dependent variable (a rating of anxiety on a 5-point scale). You would check to see if an additional assumption is met (Are the variances in the groups approximately equal?) and do an independent samples *t* test to evaluate whether mean anxiety is significantly higher for the group that received caffeine.
Compute an effect size index.	If the computer printout does not include effect size information, you can compute an effect size index by hand (e.g., see Chapter 5 for information about an effect size index for the independent samples *t* test).
Write up a first draft of your research report, including the Results section, which reports statistical findings.	The Results section should include a statement about what analysis was done, with what variables, to answer what question; a discussion of any problems detected during preliminary data screening, including what was done to remedy any problems that were identified; the test statistic and associated degrees of freedom and ***p* value;** effect size information; group means and standard deviations; and a clear verbal interpretation about the nature of group differences.
Ask your research supervisor and other people to read and critique your research report.	Revise your research report based on constructive criticism and suggestions.
Present your findings (orally or in writing or both).	Presentation of research results takes many forms. You may be required to do a formal oral defense (for a thesis or dissertation). An honors student might present an honors thesis as a talk at an undergraduate research conference. If the study was well designed, the sample size is reasonably large, and the findings are clear and interesting, the research might also be presented as a poster or talk at a professional research conference, or it might be submitted to a professional journal for review and possible publication. Note that journal reviewers often require substantial revisions before they accept a research report for publication.

Notes

1. Examples in this textbook assume that researchers are dealing with human populations; however, similar issues arise when samples are obtained from populations of nonhuman animals, plants, geographic locations, or other entities. In fact, many of these statistics were originally developed for use in industrial quality control and agriculture, where the units of analysis were manufactured products and plots of ground that received different treatments.

2. Systematic differences in the composition of the sample, compared with the population, can be corrected for by using case-weighting procedures. If the population includes 500 men and 500 women, but the sample includes 25 men and 50 women, case weights could be used so that, in effect, each of the 25 scores from males would be counted twice in computing summary statistics.

3. The four levels of measurement are called nominal, ordinal, interval, and ratio. In *nominal* level of measurement, each number code serves only as a label for group membership. For example, the nominal variable *gender* might be coded 1 = Male, 2 = Female, and the nominal variable *religion* might be coded 1 = Buddhist, 2 = Christian, 3 = Hindu, 4 = Islamic, 5 = Jewish, 6 = Other religion. The sizes of the numbers associated with groups do not imply any rank ordering among groups. Because these numbers serve only as labels, Stevens argued that the only logical operations that could appropriately be applied to the scores are = and ≠. That is, persons with scores of 2 and 3 on religion could be labeled as "the same" or "not the same" on religion. In *ordinal* measurement, numbers represent ranks, but the differences between scores do not necessarily correspond to equal intervals with respect to any underlying characteristic. The runners in a race can be ranked in terms of speed (runners are tagged 1, 2, and 3 as they cross the finish line, with 1 representing the fastest time). These scores supply information about rank (1 is faster than 2), but the numbers do not necessarily represent equal intervals. The difference in speed between Runners 1 and 2 (i.e., 2 – 1) might be much larger or smaller than the difference in speed between Runners 2 and 3 (i.e., 3 – 2), in spite of the fact that the difference in scores in both cases was one unit. For ordinal scores, the operations > and < would be meaningful (in addition to = and ≠). However, according to Stevens, addition or subtraction would not produce meaningful results with ordinal measures (because a one-unit difference does not correspond to the same "amount of speed" for all pairs of scores). Scores that have *interval* level of measurement qualities supply ordinal information and, in addition, represent equally spaced intervals. That is, no matter which pair of scores is considered (such as 3 – 2 or 7 – 6), a one-unit difference in scores should correspond to the same amount of the thing that is being measured. Interval level of measurement does not necessarily have a true 0 point. The centigrade temperature scale is a good example of interval level of measurement: The 10-point difference between 40 and 50°C is equivalent to the 10-point difference between 50 and 60°C (in each case, 10 represents the same number of degrees of change in temperature). However, because 0°C does not correspond to a complete absence of any heat, it does not make sense to look at a ratio of two temperatures. For example, it would be incorrect to say that 40°C is "twice as hot" as 20°C. Based on this reasoning, it makes sense to apply the plus and minus operations to interval scores (as well as the equality and inequality operators). However, by this reasoning, it would be inappropriate to apply multiplication and division to numbers that do not have a true 0 point. *Ratio* level measurements are interval level scores that also have a true 0 point. A clear example of a ratio level measurement is height. It is meaningful to say that a person who is 6 ft tall is twice as tall as a person 3 ft tall because there is a true 0 point for height measurements. The narrowest interpretation of this reasoning would suggest that

ratio level is the only type of measurement for which multiplication and division would yield meaningful results.

Thus, strict adherence to Stevens's measurement theory would imply that statistics that involve addition, subtraction, multiplication, and division (such as mean, Pearson *r*, *t* test, analysis of variance, and all other multivariate techniques covered later in this textbook) can only legitimately be applied to data that are at least interval (and preferably ratio) level of measurement.

4. There is one exception. When a nominal variable has only two categories and the codes assigned to these categories are 0 and 1 (e.g., the nominal variable gender could be coded 0 = Male, 1 = Female), the mean of these scores represents the proportion of persons who are female.

5. Violations of the assumptions for parametric statistics create more serious problems when they are accompanied by small *N*s in the groups (and/or unequal *N*s in the groups). Sometimes, just having very small *N*s is taken as sufficient reason to prefer nonparametric statistics. When *N*s are very small, it becomes quite difficult to evaluate whether the assumptions for parametric statistics are satisfied (such as normally distributed scores on quantitative variables).

6. A nontechnical definition of *robust* is provided at this point. A statistic is robust if it provides "accurate" results even when one or more of its assumptions are violated. A more precise definition of this term will be provided in Chapter 2.

+ − × ÷

Comprehension Questions

1. Chapter 1 distinguished between two different kinds of samples:
   a. *Random samples* (selected randomly from a clearly defined population)
   b. *Accidental* or *convenience* samples
      1. Which type of sample (a or b) is more commonly reported in journal articles?
      2. Which type of sample (a or b) is more likely to be representative of a clearly defined population?
      3. What does it mean to say that a sample is "representative" of a population?

2. Suppose that a researcher tests the safety and effectiveness of a new drug on a convenience sample of male medical students between the ages of 24 and 30. If the drug appears to be effective and safe for this convenience sample, can the researcher safely conclude that the drug would be safe for women, children, and persons older than 70 years of age? Give reasons for your answer.

3. Given below are two applications of statistics. Identify which one of these is *descriptive* and which is *inferential* and explain why.

   Case I: An administrator at Corinth College looks at the verbal Scholastic Aptitude Test (SAT) scores for the entire class of students admitted in the fall of 2005 (mean = 660) and the verbal SAT scores for the entire class admitted in the fall of 2004 (mean = 540) and concludes that the class of students admitted to Corinth in 2005 had higher verbal scores than the class of students admitted in 2004.

   Case II: An administrator takes a random sample of 45 Corinth College students in the fall of 2005 and asks them to self-report how often they engage in binge drinking. Members of the sample report an average of 2.1 binge drinking episodes per week. The administrator writes a report that says, "The average level of binge drinking among all Corinth College students is about 2.1 episodes per week."

4. We will distinguish between two types of variables: *categorical* and *quantitative*. For your answers to this question, *do not* use any of the variables used in the textbook or in class as examples; think up your own example.
   a. Give an example of a specific variable that is clearly a categorical type of measurement (e.g., gender coded 1 = Female and 2 = Male is a categorical variable). Include the groups or levels (in the example given in the text, when gender is the categorical variable, the groups are female and male).
   b. Give an example of a specific variable that is clearly a quantitative type of measurement (e.g., IQ scores).
   c. If you have scores on a categorical variable (e.g., religion coded 1 = Catholic, 2 = Buddhist, 3 = Protestant, 4 = Islamic, 5 = Jewish, 6 = Other religion), would it make sense to use these numbers to compute a mean? Give reasons for your answer.

5. Using the guidelines given in this chapter (and Table 1.4), name the most appropriate statistical analysis for each of the following imaginary studies. Also, state which variable is the predictor or independent variable and which variable is the outcome or dependent variable in each situation. Unless otherwise stated, assume that any categorical independent variable is between-S (rather than within-S), and assume that the conditions required for the use of parametric statistics are satisfied.

   Case I: A researcher measures core body temperature for participants who are either male or female (i.e., gender is a variable in this study). What statistics could the researcher use to see if mean body temperature differs between women and men?

   Case II: A researcher who is doing a study in the year 2000 obtains yearbook photographs of women who graduated from college in 1970. For each woman, there are two variables. A rating is made of the "happiness" of her facial expression in the yearbook photograph. Each woman is contacted and asked to fill out a questionnaire that yields a score (ranging from 5 to 35) that measures her "life satisfaction" in the year 2000. The researcher wants to know whether "life satisfaction" in 2000 can be predicted from the "happiness" in the college yearbook photo back in 1970. Both variables are quantitative, and the researcher expects them to be linearly related, such that higher levels of happiness in the 1970 photograph will be associated with higher scores on life satisfaction in 2000. (Research on these variables has actually been done; see Harker & Keltner, 2001.)

   Case III: A researcher wants to know whether preferred type of tobacco use (coded 1 = No tobacco use, 2 = Cigarettes, 3 = Pipe, 4 = Chewing tobacco) is related to gender (coded 1 = Male, 2 = Female).

   Case IV: A researcher wants to know which of five drug treatments (Group 1 = Placebo, Group 2 = Prozac, Group 3 = Zoloft, Group 4 = Effexor, Group 5 = Celexa) is associated with the lowest mean score on a quantitative measure of depression (the Hamilton Depression Rating Scale).

6. Draw a sketch of the standard normal distribution. What characteristics does this function have?

7. Look at the standard normal distribution in Figure 1.4 to answer the following questions:
   a. Approximately what proportion (or percentage) of scores in a normal distribution lie within a range from -2σ below the mean to $+1\sigma$ above the mean?
   b. Approximately what percentage of scores lie above $+3\sigma$?
   c. What percentage of scores lie inside the range -2σ below the mean to $+2\sigma$ above the mean? What percentage of scores lie outside this range?

8. For what types of data would you use nonparametric versus parametric statistics?

9. What features of an experiment help us to meet the conditions for causal inference?

10. Briefly, what is the difference between internal and external validity?

11. For each of the following sampling methods, indicate whether it is random (i.e., whether it gives each member of a specific well-defined population an equal chance of being included) or accidental/convenience sampling. Does it involve stratification; that is, does it involve sampling from members of subgroups, such as males and females or members of various political parties? Is a sample obtained in this manner likely to be representative of some well-defined larger population? Or is it likely to be a **biased sample**, a sample that does not correspond to any clearly defined population?
    a. A teacher administers a survey on math anxiety to the 25 members of her introductory statistics class. The teacher would like to use the results to make inferences about the average levels of math anxiety for all first-year college students in the United States.
    b. A student gives out surveys on eating disorder symptoms to her teammates in gymnastics. She wants to be able to use her results to describe the correlates of eating disorders in all college women.
    c. A researcher sends out 100,000 surveys on problems in personal relationships to mass mailing lists of magazine subscribers. She gets back about 5,000 surveys and writes a book in which she argues that the information provided by respondents is informative about the relationship problems of all American women.
    d. The Nielson television ratings organization selects a set of telephone area codes to make sure that its sample will include people taken from every geographical region that has its own area code; it then uses a random number generator to get an additional seven-digit telephone number. It calls every telephone number generated using this combination of methods (all area codes included and random dialing within an area code). If the person who answers the telephone indicates that this is a residence (not a business), that household is recruited into the sample, and the family members are mailed a survey booklet about their television-viewing habits.

12. Give an example of a specific sampling strategy that would yield a random sample of 100 students taken from the incoming freshman class of a state university. You may assume that you have the names of all 1,000 incoming freshmen on a spreadsheet.

13. Give an example of a specific sampling strategy that would yield a random sample of 50 households from the population of 2,000 households in that particular town. You may assume that the telephone directory for the town provides a nearly complete list of the households in the town.

14. Is each of the following variables categorical or quantitative? (Note that some variables could be treated as either categorical or quantitative.)

    Number of children in a family

    Type of pet owned: 1 = None, 2 = Dog, 3 = Cat, 4 = Other animal

    IQ score

    Personality type (Type A, coronary prone; Type B, not coronary prone)

15. Do most researchers still insist on at least interval level of measurement as a condition for the use of parametric statistics?
16. How do categorical and quantitative variables differ?
17. How do between-S versus within-S designs differ? Make up a list of names for imaginary participants, and use these names to show an example of between-S groups and within-S groups.
18. When a researcher has an accidental or convenience sample, what kind of population can he or she try to make inferences about?
19. Describe the guidelines for selecting an appropriate bivariate statistic. That is, what do you need to ask about the nature of the data and the research design in order to choose an appropriate analysis?

CHAPTER 2

Introduction to SPSS

Basic Statistics, Sampling Error, and Confidence Intervals

2.1 ♦ Introduction

We will begin by examining the distribution of scores on just one individual variable. The first few chapters of a typical introductory statistics book present simple methods for summarizing information about the distribution of scores on a single variable. It is assumed that readers understand that information about the distribution of scores for a quantitative variable, such as heart rate (HR), can be summarized in the form of a frequency distribution table or a histogram and that readers are familiar with concepts such as central tendency and dispersion of scores. This chapter reviews the formulas for summary statistics that are most often used to describe central tendency and dispersion of scores in batches of data (including the mean M and standard deviation s). These formulas provide instructions that can be used for by-hand computation of statistics such as the sample mean, M. A few numerical examples are provided to remind readers of how these computations are done. The goal of this chapter is to lead students to think about the formula for each statistic (such as the sample mean, M). A thoughtful evaluation of each equation makes it clear what information each statistic is based on, the range of possible values for the statistic, and the patterns in the data that lead to large versus small values of the statistic.

Each statistic provides an answer to some question about the data. The sample mean, M, is one way to answer the question, What is a typical score value? It is instructive to try to imagine these questions from the point of view of the people who originally developed the statistical formulas and to recognize why they used the arithmetic operations that they did. For example, summing scores for all participants in a sample is a way of summarizing or combining information from all participants. Dividing a sum of scores by N corrects for the impact of sample size on the magnitude of this sum.

The notation used in this book is summarized in Table 2.1. For example, the mean of scores in a sample batch of data is denoted by M. The (usually unknown) mean of the

Table 2.1 ♦ Notation for Sample Statistics and Population Parameters

Name of Statistic	*Sample Statistic*[a]	*Corresponding Population Parameter*
Mean	M (or $\bar{X}$)	μ (Greek letter *mu*)
Standard deviation	s or SD (or $\hat{s}$)	σ (Greek letter *sigma*)
Variance	s^2 (or $\hat{s}^2$)	σ^2
Distance of individual X score from the mean	$z = (X - M)/s$ or $(X - \bar{X})/SD$	$z = (X - \mu)/\sigma$
Standard error of the sample mean	SE_M (or $SE_{\bar{X}}$)	σ_M (or $\sigma_{\bar{X}}$)

a. The first notation listed for each sample statistic is the notation most commonly used in this book.

population that the researcher wants to estimate or make inferences about, using the sample value of M, is denoted by μ (Greek letter *mu*).

One of the greatest conceptual challenges for students who are taking a first course in statistics arises when the discussion moves beyond the behavior of single X scores and begins to consider how sample statistics (such as M) vary across different batches of data that are randomly sampled from the same population. On first passing through the material, students are often so preoccupied with the mechanics of computation that they lose sight of the questions about the data that the statistics are used to answer. This chapter does not go into some of the computational details that sometimes appear in introductory statistics textbooks (e.g., calculation of a sample mean from scores in a grouped frequency distribution). Instead, it focuses on what information the researcher wants to obtain when he or she computes a sample mean or sample variance. Students are asked to look at each formula as something more than just a recipe for computation; each formula can be understood as a meaningful sentence. The formula for a sample statistic (such as the sample mean, M) tells us what information in the data is taken into account when the sample statistic is calculated. Thinking about the formula and asking what will happen if the values of X increase in size or in number makes it possible for students to answer questions such as, Under what circumstances (i.e., for what patterns in the data) will the value of this statistic be a large or a small number? What does it mean when the value of the statistic is large or when its value is small?

The basic research questions in this chapter will be illustrated by using a set of scores on HR; these are contained in the file hr130.sav. For a variable such as HR, how can we describe a typical HR? We can assess this by looking at measures of central tendency such as mean or median HR. How much does HR vary across persons? We can assess this by computing a variance and standard deviation for the HR scores in this small sample. How can we evaluate whether an individual person has an HR that is relatively high or low, compared with other people's HRs? When scores are normally distributed, we can answer questions about the location of an individual score relative to a distribution of scores by calculating a z score to provide a unit-free measure of distance of the individual HR score from the mean HR and using a table of the standard normal distribution to find areas under the normal distribution that correspond to distances from the mean. These areas can be interpreted as proportions and used to answer questions such as, Approximately what proportion of people in the sample had HR scores higher than 84?

We will consider the issues that must be taken into account when we use the sample mean, M, for a small random sample of $N = 9$ persons to estimate the population mean, μ, for a larger population. In introductory statistics courses, students are introduced to the concept of **sampling error**; that is, the value of the sample mean, M, varies across different batches of data that are randomly sampled from the same population. Because of sampling error, the sample mean, M, for a single sample is not likely to be exactly correct as an estimate of μ, the unknown population mean. When researchers report a sample mean, M, it is important to include information about the magnitude of sampling error; this can be done by setting up a **confidence interval (CI)**. This chapter reviews the concepts that are involved in setting up and interpreting CIs.

2.2 ♦ Research Example: Description of a Sample of HR Scores

In the following discussion, the population of interest consists of 130 persons; each person has a score on HR, reported in beats per minute (bpm). Scores for this hypothetical population are contained in the data file hr130.sav. Shoemaker (1996) generated these hypothetical data so that sample statistics such as the sample mean, M, would correspond to the outcomes from an empirical study reported by Mackowiak, Wasserman, and Levine (1992). For the moment, it is useful to treat this set of 130 scores as the population of interest and to draw one small random sample (consisting of $N = 9$ cases) from this population. This will provide us with a way to evaluate how accurately a mean based on a random sample of $N = 9$ cases estimates the mean of the population from which the sample was selected. (In this case, we can easily find the actual population mean, μ, because we have HR data for the entire population of 130 persons.)

SPSS has a procedure that allows the data analyst to select a random sample of cases from a data file; the data analyst can specify either the percentage of cases to be included in the sample (e.g., 10% of the cases in the file) or the number of cases (N) for the sample. In the following exercise, a random sample of $N = 9$ HR scores is selected from the population of 130 cases in the SPSS file hr130.sav.

Figure 2.1 shows the Data View for the SPSS worksheet for the hr130.sav file. Each row in this worksheet corresponds to scores for one participant. Each column in the SPSS worksheet corresponds to one variable. The first column gives each person's HR in beats per minute.

Clicking on the tab near the bottom left corner of the worksheet shown in Figure 2.1 changes to the Variable View of the SPSS dataset, displayed in Figure 2.2. In this view, the names of variables are listed in the first column. Other cells provide information about the nature of each variable—for example, variable type. In this dataset, HR is a **numerical variable**, and the variable type is "scale" (i.e., quantitative or approximately interval/ratio) level of measurement. Readers who have never used SPSS will find a brief introduction to SPSS in the appendix to this chapter; they may also want to consult an introductory user's guide for SPSS, such as Pavkov (2005). HR is conventionally reported in whole numbers; the choice of "0" in the decimal points column for this variable instructs SPSS to include no digits after the decimal point when displaying scores for this variable.

Prior to selection of a random sample, let's look at the distribution of this population of 130 scores. A histogram can be generated for this set of scores by starting in the Data

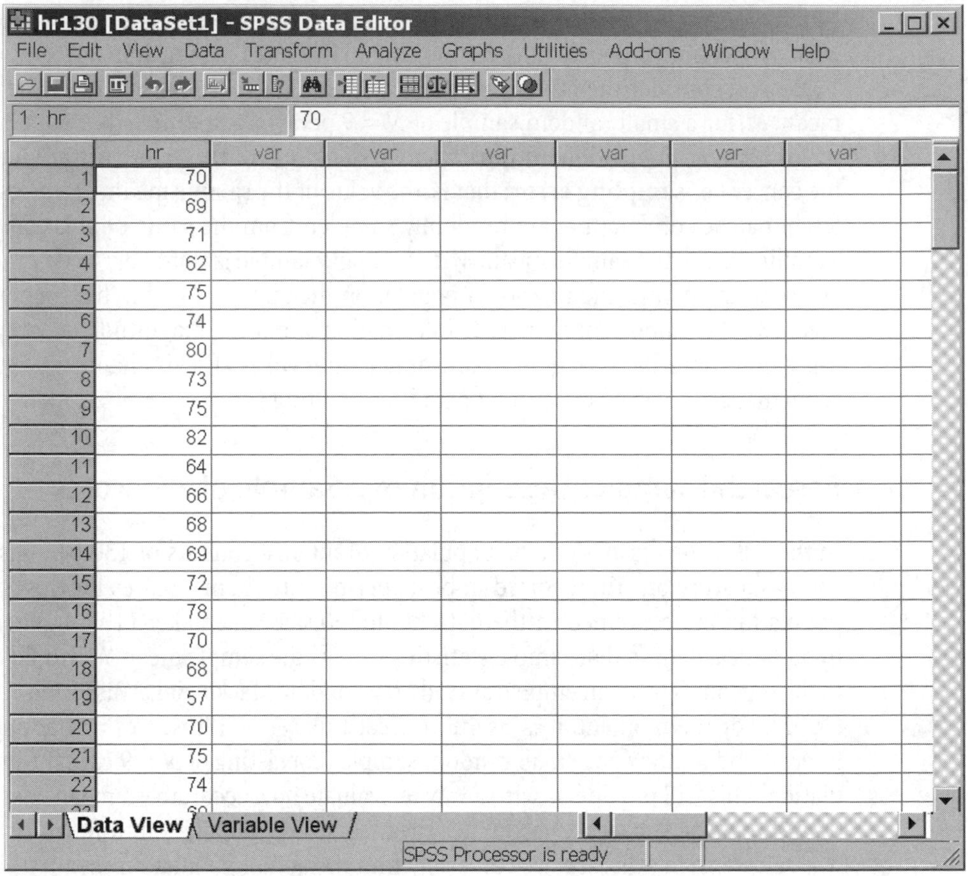

Figure 2.1 ♦ The SPSS Data View for the First 22 Lines From the SPSS Data File hr130.sav

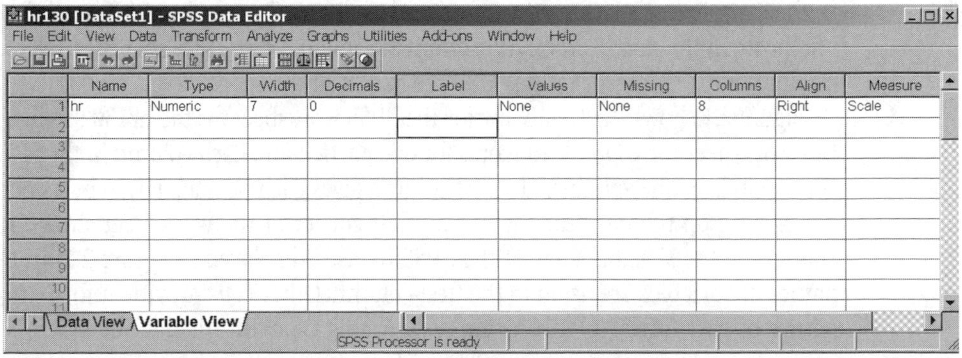

Figure 2.2 ♦ The SPSS Variable View for the SPSS Worksheet for hr130.sav

View worksheet, selecting the <Graphs> menu from the menu bar along the top of the SPSS worksheet, and then choosing <Histogram> from the pop-up menu, as shown in Figure 2.3.

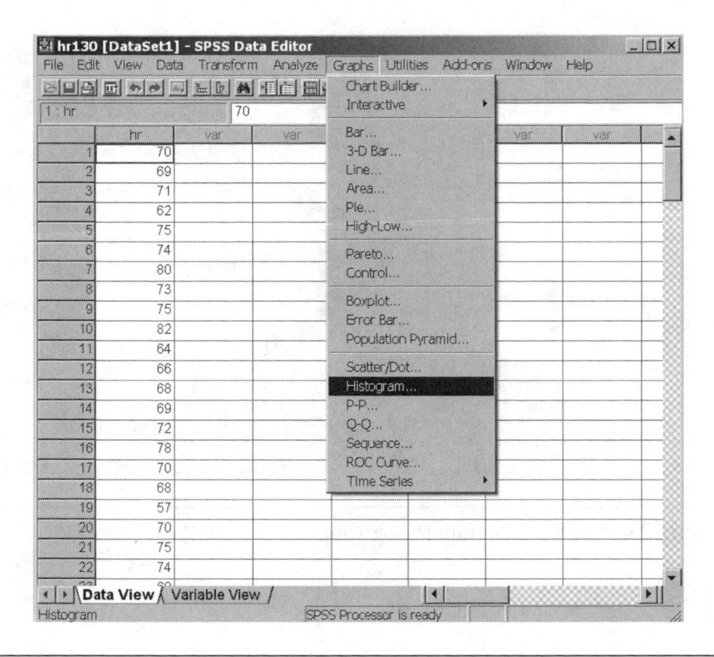

Figure 2.3 ♦ SPSS Menu Selections <Graphs> → <Histogram> to Open Histogram Dialog Window

NOTE: SPSS version 14 was used for all examples in this book. As of June 2007, the current release was SPSS 15. In version 15, one additional menu selection is required for use of the Graphics procedures described in the text. Under Graphics, the user must select the menu option <Legacy Dialogs>. From that point onward, the menu selections are the same as in these examples.

Figure 2.4 shows the SPSS dialog window for the Histogram procedure. Initially, the names of all the variables in the file (in this example there is only one variable, HR) appear in the left-hand panel, which shows the available variables. Notice that the variable HR has a "ruler" icon associated with it. This ruler icon indicates that scores on this variable are scale (i.e., quantitative or interval/ratio) level of measurement. To designate HR as the variable for the histogram, highlight it with the cursor and click on the right-pointing arrow to move the variable name HR into the small window on the right-hand side under the heading Variable. To request a superimposed normal curve, click the check box for Display normal curve. Finally, to run the procedure, click the OK button in the upper right-hand corner of the Histogram dialog window. The output from this procedure appears in Figure 2.5 along with the values for the population mean $\mu = 73.76$ and population standard deviation $\sigma = 7.06$ for the entire population of 130 scores.

To select a random sample of size $N = 9$ from the entire population of 130 scores in the SPSS dataset hr130.sav, make the following menu selections, starting from the SPSS Data View worksheet, as shown in Figure 2.6: <Data> → <Select Cases>. This opens the SPSS dialog window for Select Cases, which appears in Figure 2.7. In the Select Cases dialog window, click the radio button for Random sample of cases. Then, click the Sample button; this opens the Select Cases: Random Sample dialog window in Figure 2.8. Within this box under the heading Sample Size, click the radio button that corresponds to the word "Exactly" and enter in the desired sample size (9) and the number of cases in the entire population (130). The resulting SPSS command is, "Randomly select exactly 9 cases from the first 130 cases." Click the Continue button to return to the main Select Cases

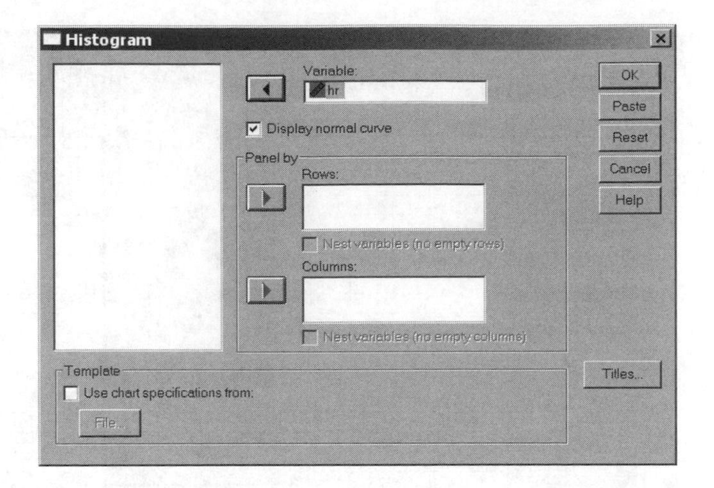

Figure 2.4 ♦ SPSS Histogram Dialog Window

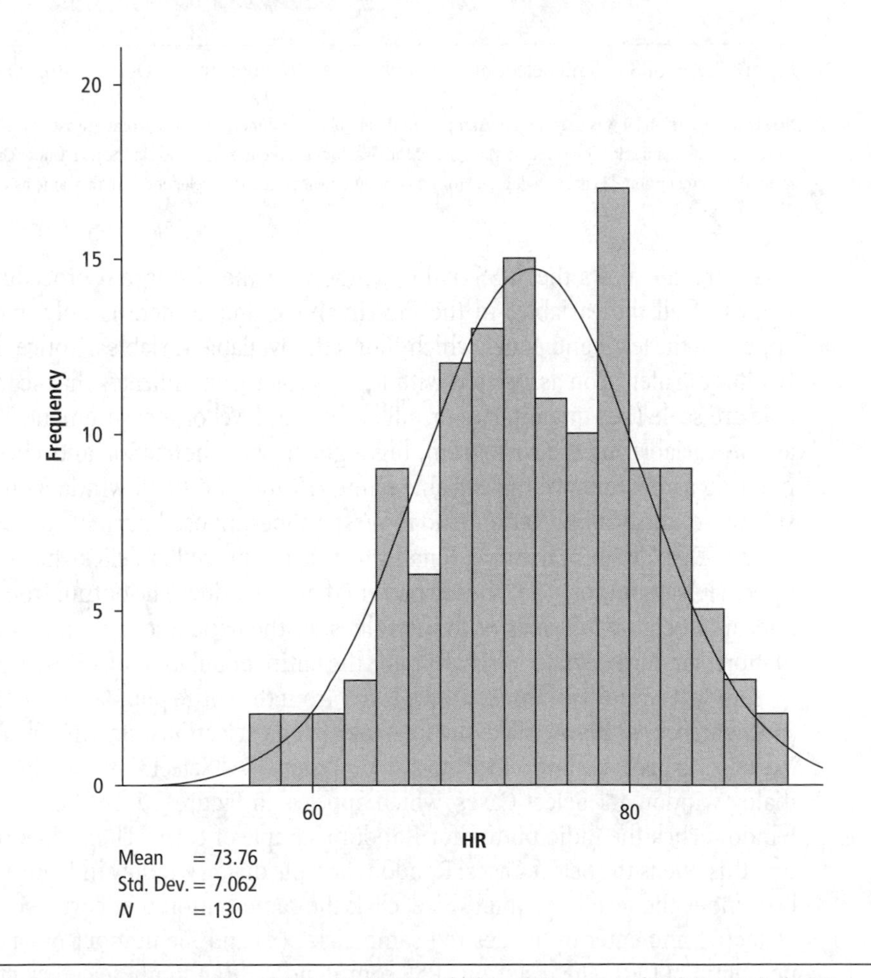

Figure 2.5 ♦ Output: Histogram for the Entire Population of Heart Rate (HR) Scores in hr130.sav

dialog window. To save this random sample of $N = 9$ HR scores into a separate, smaller file, click on the radio button for "Copy selected cases to a new dataset" and provide a name for the dataset that will contain the new sample of 9 cases—in this instance, hr9.sav. Then, click the OK button.

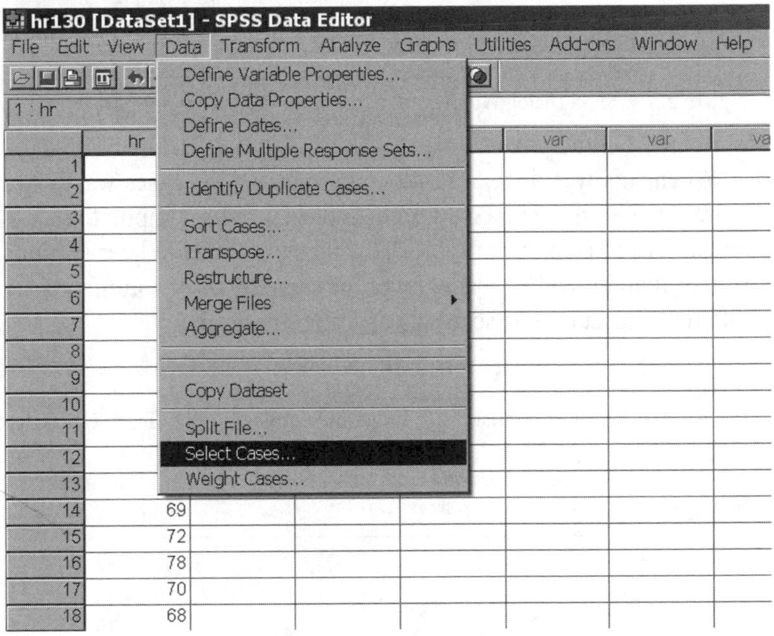

Figure 2.6 ♦ SPSS Menu Selection for <Data> → <Select Cases>

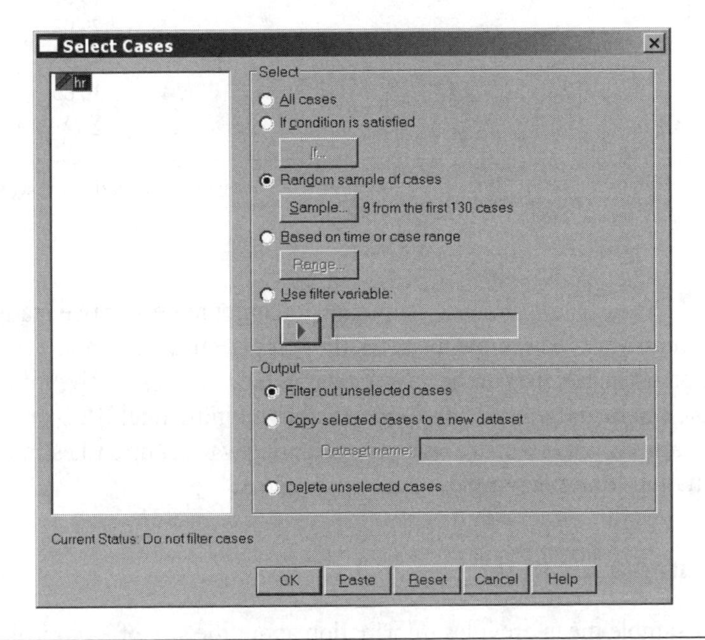

Figure 2.7 ♦ SPSS Dialog Window for Select Cases

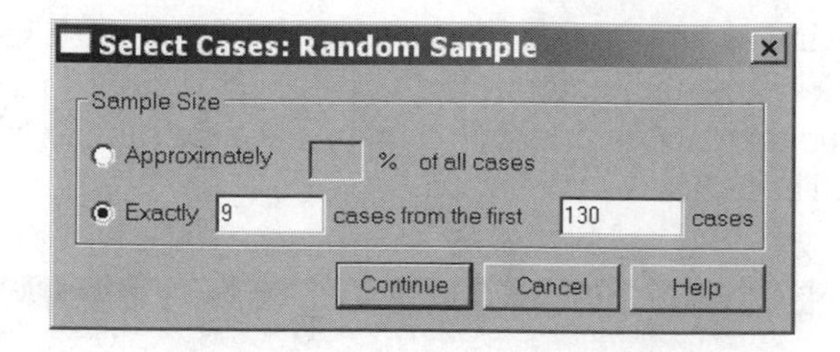

Figure 2.8 ♦ SPSS Dialog Window for Select Cases: Random Sample

When this was done, a random sample of nine cases was obtained; these nine HR scores appear in the first column of Table 2.2. (The computation of the values in the second and third columns in Table 2.2 will be explained in later sections of this chapter.) Of course, if an individual reader gives the same series of commands, he or she will obtain a different subset of nine scores as the random sample.

Table 2.2 ♦ Summary Statistics for Random Sample of $N = 9$ Heart Rate (HR) Scores

HR (X)	*Deviation From Sample Mean HR (X – M)*	*Squared Deviation From the Mean $(X - M)^2$*
70	–3.11	9.67
71	–2.11	4.45
74	.89	.79
80	6.89	47.47
73	–.11	.01
75	1.89	3.57
82	8.89	79.03
64	–9.11	82.99
69	–4.11	16.89
$\Sigma X = 658$	$\Sigma (X - M) = 0$	$\Sigma (X - M)^2 = 244.89 = SS$

NOTES: Sample mean for HR: $M = \Sigma X/N = 658/9 = 73.11$. Sample variance for HR: $s^2 = SS/(N - 1) = 244.89/8 = 30.61$. Sample standard deviation for HR: $s = \sqrt{s^2} = 5.53$.

The next few sections show how to compute descriptive statistics for this sample of nine scores: the sample mean M, the sample variance s^2, and the sample standard deviation s. The last part of the chapter shows how this descriptive information about the sample can be used to help evaluate whether an individual HR score is relatively high or low, relative to other scores in the sample, and how to set up a CI estimate for μ using the information from the sample.

2.3 ♦ Sample Mean (*M*)

A sample mean provides information about the size of a "typical" score in a sample. The interpretation of a sample mean M can be worded in several different ways. A sample mean, M, corresponds to the center of a distribution of scores in a sample. It provides us with one kind of information about the size of a typical X score (other ways of describing a typical score are provided by a median or **mode**).

Scores in a sample can be represented as $X_1, X_2, \ldots, X_n$, where N is the number of observations or participants and X_i is the score for participant number i. For example, the HR score for a person with the SPSS case record number 2 in Figure 2.1 could be given as $X_2 = 69$. Some textbooks, particularly those that offer more mathematical or advanced treatments of statistics, include subscripts on X scores; in this book, the i subscript is used only when omitting subscripts would create ambiguity about which scores are included in a computation. The sample mean, M, is obtained by summing all the X scores in a sample of N scores and dividing by N, the number of scores:

$$M = \sum X/N. \tag{2.1}$$

Adding the scores is a way of summarizing information across all participants. The size of $\sum X$ depends on two things: the magnitudes of the individual X scores and N, the number of scores. If N is held constant and all X scores are positive, $\sum X$ increases if the individual X scores are increased. Assuming all X scores are positive, $\sum X$ also increases as N gets larger. To obtain a sample mean that represents the size of a typical score, and that is independent of N, we have to correct for sample size by dividing $\sum X$ by N, to yield M, our sample mean. Equation 2.1 is more than just instructions for computation. It is also a statement or "sentence" that tells us the following:

1. What information is the sample statistic M based on? It is based on the sum of the Xs and the N of cases in the sample.

2. Under what circumstances will the statistic (M) turn out to have a large or small value? M is large when the individual X scores are large and positive. Because we divide by N when computing M to correct for sample size, the magnitude of M is independent of N.

In this chapter, we explore what happens when we use a sample mean M based on a random sample of $N = 9$ cases to estimate the population mean μ (in this case, the entire set of 130 HR scores in the file hr130.sav is the population of interest). The sample of $N = 9$ randomly selected HR scores appears in the first column of Table 2.2. For the set of the $N = 9$ HR scores shown in Table 2.2, we can calculate the mean by hand:

$$70,\ 71,\ 74,\ 80,\ 73,\ 75,\ 82,\ 64,\ 69$$

$$\sum X = [70 + 71 + 74 + 80 + 73 + 75 + 82 + 64 + 69] = 658$$

$$M = \sum X/N = 658/9 = 73.11$$

(Note that the values of sample statistics are usually reported up to two decimal places unless the original X scores provide information that is accurate up to more than two decimal places.)

The SPSS Descriptive Statistics: Frequencies procedure was used to obtain the sample mean and other simple descriptive statistics for the set of scores in the file hr9.sav. On the

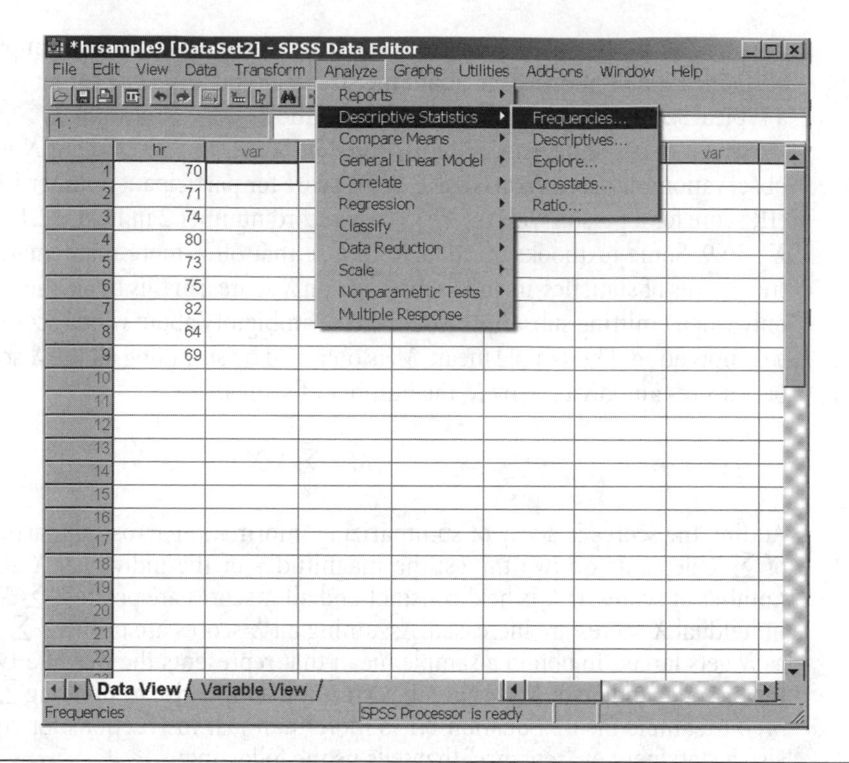

Figure 2.9 ♦ SPSS Menu Selections for the Descriptive Statistics and Frequencies Procedures Applied to the Random Sample of $N = 9$ Heart Rate Scores From the Dataset hri30.sav

Data View worksheet, find the Analyze option in the menu bar at the top of the worksheet and click on it. Select Descriptive Statistics from the pull-down menu that appears (as shown in Figure 2.9); this leads to another drop-down menu. Because we want to see a distribution of frequencies and also obtain simple descriptive statistics such as the sample mean, M, click on the Frequencies procedure from this second pull-down menu.

This series of menu selections displayed in Figure 2.9 opens the SPSS dialog window for the Descriptive Statistics: Frequencies procedure shown in Figure 2.10. Move the variable name HR from the left-hand panel into the right-hand panel under the heading Variables to indicate that the Frequencies procedure will be performed on scores for the variable HR. Clicking the Statistics button at the bottom of the SPSS Frequencies dialog window opens up the Frequencies: Statistics dialog window; this contains a menu of basic descriptive statistics for quantitative variables (see Figure 2.11). Check box selections can be used to include or omit any of the statistics on this menu. In this example, the following sample statistics were selected: Under the heading Central Tendency, Mean and Sum were selected, and under the heading Dispersion, Standard deviation and Variance were selected. Click Continue to return to the main Frequencies dialog window. When all the desired menu selections have been made, click the OK button to run the analysis for the selected variable, HR. The results from this analysis appear in Figure 2.12. The top panel of Figure 2.12 reports the requested summary statistics, and the bottom panel reports the table of frequencies for each score value included in the sample. The value for the sample mean that appears in the SPSS output in Figure 2.12, $M = 73.11$, agrees with the numerical value obtained by the earlier calculation.

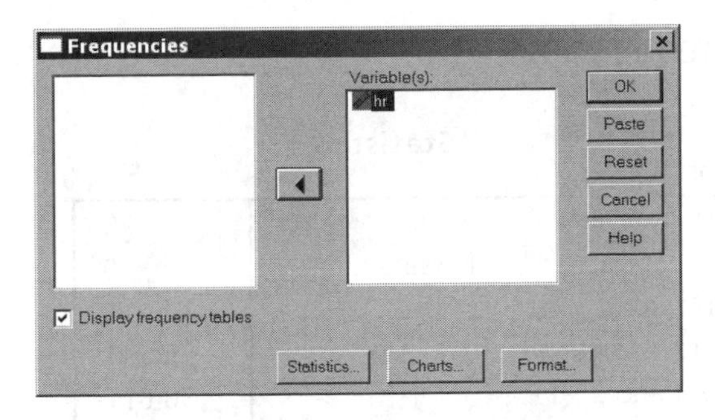

Figure 2.10 ♦ The SPSS Dialog Window for the Frequencies Procedure

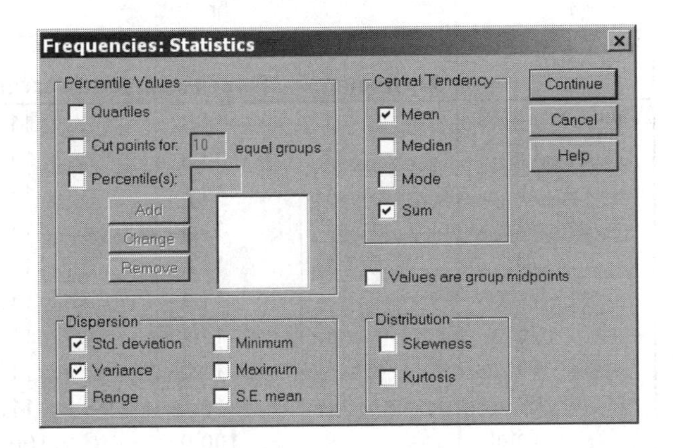

Figure 2.11 ♦ The Frequencies: Statistics Window With Check Box Menu for Requested Descriptive Statistics

How can this value of $M = 73.11$ be used? If we wanted to estimate or guess any one individual's HR, in the absence of any other information, the best guess for any randomly selected individual member of this sample of $N = 9$ persons would be $M = 73.11$ bpm. Why do we say that the mean M is the "best" prediction for any randomly selected individual score in this sample? It is best because it is the estimate that makes the sum of the prediction errors zero, and minimizes the overall sum of squared prediction errors, across all participants.

To see this, reexamine Table 2.2. The second column of Table 2.2 shows the deviation of each score from the sample mean $(X - M)$, for each of the nine scores in the sample. This deviation from the mean is the prediction error that arises if M is used to estimate that person's score; the magnitude of error is given by the difference $X - M$, the person's actual HR score minus the sample mean HR, M. For instance, if we use M to estimate Participant 1's score, the prediction error for Case 1 is $(70 - 73.11) = -3.11$; that is, Participant 1's actual HR score is 3.11 points below the estimated value of $M = 73.11$.

How can we summarize information about the magnitude of prediction error across persons in the sample? One approach that might initially seem reasonable is summing the $X - M$ deviations across all the persons in the sample. The sum of these deviations appears at the bottom of the second column of Table 2.2. By definition, the sample mean, M, is the value for which the sum of the deviations across all the scores in a sample equals 0. In that sense, using M to estimate X for each person in the sample results in the

Frequencies

Statistics

hr

N	Valid	9
	Missing	0
Mean		73.11
Std. Deviation		5.533
Variance		30.611
Sum		658

hr

		Frequency	Percent	Valid Percent	Cumulative Percent
Valid	64	1	11.1	11.1	11.1
	69	1	11.1	11.1	22.2
	70	1	11.1	11.1	33.3
	71	1	11.1	11.1	44.4
	73	1	11.1	11.1	55.6
	74	1	11.1	11.1	66.7
	75	1	11.1	11.1	77.8
	80	1	11.1	11.1	88.9
	82	1	11.1	11.1	100.0
	Total	9	100.0	100.0	

Figure 2.12 ♦ SPSS Output From Frequencies Procedure for the Sample of $N = 9$ Heart Rate (HR) Scores Randomly Selected From the File hr130.sav

smallest possible sum of prediction errors. It can be demonstrated that taking deviations of these X scores from any constant other than the sample mean, M, yields a sum of deviations that is not equal to 0. However, the fact that $\Sigma (X - M)$ always equals 0 for a sample of data makes this sum uninformative as summary information about dispersion of scores. Just calculating $\Sigma (X - M)$ will not provide information about how far individual X scores tend to be from M.

We can avoid the problem that the sum of the deviations always equals 0 in a simple manner: If we first square the prediction errors or deviations (i.e., if we square the $X - M$ value for each person, as shown in the third column of Table 2.2) and then sum these squared deviations, the resulting term $\Sigma (X - M)^2$ is a number that gets larger as the magnitudes of the deviations of individual X values from M increase.

There is a second sense in which M is the best predictor of HR for any randomly selected member of the sample. M is the value for which the sum of squared deviations (SS), $\Sigma (X - M)^2$, is minimized. The sample mean is the best predictor of any randomly selected person's score because it is the estimate for which prediction errors sum to 0, and it is also the estimate that has the smallest sum of squared prediction errors. The term ***ordinary least squares*** (**OLS**) refers to this criterion; a statistic meets the criterion for best OLS estimator when it minimizes the sum of squared prediction errors.

This empirical demonstration only shows that $\sum (X - M) = 0$ for this particular batch of data. An empirical demonstration is not equivalent to a formal proof. Formal proofs for the claim that $\sum (X - M) = 0$ and the claim that M is the value for which the SS, $\sum (X - M)^2$, is minimized are provided in mathematical statistics textbooks such as deGroot and Schervish (2001). The present textbook provides demonstrations rather than formal proofs.

Based on the preceding demonstration (and the proofs provided in mathematical statistics books), the mean is the best estimate for any individual score when we do not have any other information about the participant. Of course, if a researcher can obtain information about the participant's drug use, smoking, age, gender, anxiety level, aerobic fitness, and other variables that may be predictive of HR (or that may even influence HR), better estimates of an individual's HR may be obtainable by using statistical analyses that take one or more of these predictor variables into account. (*Note:* SPSS shows variables and value labels as they are typed in by the user; the variable "gender," for example, might appear as GENDER, Gender, or gender, depending on how the word was originally typed in.)

Two other statistics are commonly used to describe the average or typical score in a sample: the mode and the median. The mode is simply the score value that occurs most often. This is not a very useful statistic for this small batch of sample data because each score value occurs only once; there is no single score value that has a larger number of occurrences than other scores. The median is obtained by rank ordering the scores in the sample from lowest to highest and then counting the scores. Here is the set of nine scores from Figure 2.1 and Table 2.2 arranged in rank order:

$$[64,\ 69,\ 70,\ 71,\ 73,\ 74,\ 75,\ 80,\ 82].$$

The score that has half the scores above it and half the scores below it is the median; in this example, the median is 73. Because M is computed using $\sum X$, the inclusion of one or two extremely large individual X scores tends to increase the size of M. For instance, suppose that the minimum score of "64" was replaced by a much higher score of "190" in the set of nine scores above. The mean for this new set of nine scores would be given by

$$\begin{aligned} M &= \sum X/N \\ &= [70 + 71 + 74 + 80 + 73 + 75 + 82 + 69 + 190]/9 \\ &= 784/9 \\ &= 87.11. \end{aligned}$$

However, the median for this new set of nine scores with an added outlier of $X = 190$,

$$[69,\ 70,\ 71,\ 73,\ 74,\ 75,\ 80,\ 82,\ 190],$$

would change to 74, which is still quite close to the original median (without the outlier) of 73.

The preceding example demonstrates that the inclusion of one extremely high score typically has rather little effect on the size of the sample median. However, the presence of one extreme score can make a substantial difference in the size of the sample mean, M. In this sample of $N = 9$ scores, adding an extreme score of $X = 190$ raises the value of M from 73.11 to 87.11, but it changes the median by only one point. Thus, the mean is less "robust" to extreme scores or outliers than the median; that is, the value of a sample mean can be changed substantially by one or two extreme scores. It is not desirable for a sample statistic to change drastically because of the presence of one extreme score, of course.

When researchers use statistics (such as the mean) that are not very robust to outliers, they need to pay attention to extreme scores when screening the data. Sometimes extreme scores are removed or recoded to avoid situations in which the data for one individual participant has a disproportionately large impact on the value of the mean (see Chapter 4 for a more detailed discussion of identification and treatment of outliers).

When scores are perfectly normally distributed, the mean, median, and mode are equal. However, when scores have nonnormal distributions (e.g., when the distribution of scores has a longer tail on the high end), these three indexes of central tendency are generally not equal. When the distribution of scores in a sample is nonnormal (or skewed), the researcher needs to consider which of these three indexes of central tendency is the most appropriate description of the center of a distribution of scores.

In spite of the fact that the mean is not robust to the influence of outliers, the mean is more widely reported than the mode or median. The most extensively developed and widely used statistical methods, such as analysis of variance (ANOVA), use group means and deviations from group means as the basic building blocks for computations. ANOVA assumes that the scores on the quantitative outcome variable are normally distributed. When this assumption is satisfied, the use of the mean as a description of central tendency yields reasonable results.

2.4 ♦ Sum of Squared Deviations and Sample Variance (s^2)

The question we want to answer when we compute a sample variance can be worded in several different ways. How much do scores differ among the members of a sample? How widely dispersed are the scores in a batch of data? How far do individual X scores tend to be from the sample mean M? The sample variance provides summary information about the distance of individual X scores from the mean of the sample. Let's build the formula for the sample variance (denoted by s^2) step by step.

First, we need to know the distance of each individual X score from the sample mean. To answer this question, a deviation from the mean is calculated for each score as follows (the i subscript indicates that this is done for each person in the sample—that is, for scores that correspond to person number i for $i = 1, 2, 3, \ldots, N$). The deviation of person number i's score from the sample mean is given by Equation 2.2:

$$\text{Deviation of the score for person } i, X_i, \text{ from the mean} = X_i - M. \tag{2.2}$$

The value of this deviation for each person in the sample appears in the second column of Table 2.2. The sign of this deviation tells us whether an individual person's score is above M (if the deviation is positive) or below M (if the deviation is negative). The magnitude of the deviation tells us whether a score is relatively close to, or far from, the sample mean.

To obtain a numerical index of variance, we need to summarize information about distance from the mean across subjects. The most obvious approach to summarizing information across subjects would be to sum the deviations from the mean for all the scores in the sample:

$$\Sigma (X_i - M). \tag{2.3}$$

However, this sum turns out to be uninformative because, by definition, deviations from a sample mean in a batch of sample data sum to 0. We can avoid this problem by squaring the deviation for each subject and then summing the squared deviations. This *SS* is an important piece of information that appears in the formulas for many of the more advanced statistical analyses discussed later in this textbook:

$$SS = \sum (X - M)^2. \tag{2.4}$$

What range of values can *SS* have? *SS* has a minimum possible value of 0; this occurs in situations where all the *X* scores in a sample are equal to each other and therefore also equal to *M*. (Because squaring a deviation must yield a positive number, and *SS* is a sum of squared deviations, *SS* cannot be a negative number.) The value of *SS* has no upper limit. Other factors being equal, *SS* tends to increase when

1. the number of squared deviations included in the sum increases; or
2. the individual $X_i - M$ deviations get larger in absolute value.

A different version of the formula for *SS* is often given in introductory textbooks:

$$SS = \sum (X^2) - \left[\left(\sum X \right)^2 / N \right]. \tag{2.5}$$

Equation 2.5 is a more convenient procedure for by-hand computation of the *SS* than is Equation 2.4 because it involves fewer arithmetic operations and results in less rounding error. This version of the formula also makes it clear that *SS* depends on both $\sum X$, the sum of the *X*s, and $\sum X^2$, the sum of the squared *X*s. Formulas for more complex statistics often include these same terms: $\sum X$ and $\sum X^2$. When these terms ($\sum X$ and $\sum X^2$) are included in a formula, their presence implies that the computation takes both the mean and the variance of *X* scores into account. These chunks of information are the essential building blocks for the computation of most of the statistics covered later in this book.

From Table 2.2, the numerical result for $SS = \sum (X - M)^2$ is 244.89.

How can the value of *SS* be used or interpreted? The minimum possible value of *SS* occurs when all the *X* scores are equal to each other and, therefore, equal to *M*. For example, in the set of scores [73, 73, 73, 73, 73], the *SS* term would equal 0. However, there is no upper limit, in practice, for the maximum value of *SS*. *SS* values tend to be larger when they are based on large numbers of deviations and when the individual *X* scores have large deviations from the mean, *M*. To interpret *SS* as information about variability, we need to correct for the fact that *SS* tends to be larger when the number of squared deviations included in the sum is large.

2.5 ♦ Degrees of Freedom (*df*) for a Sample Variance

It might seem logical to divide *SS* by *N* to correct for the fact that the size of *SS* gets larger as *N* increases. However, the computation (*SS*/*N*) produces a sample variance that is a biased estimate of the population variance; that is, the sample statistic SS/N tends to be smaller than σ^2, the true **population variance.** (This can be empirically demonstrated by taking hundreds of small samples from a population, computing a value of s^2 for each sample by using the formula $s^2 = SS/N$, and tabulating the obtained values of s^2. When this experiment is performed, the average of the sample s^2 values turns out to be smaller than the population variance σ^2.[2] This is called bias in the size of s^2; s^2 calculated as SS/N is smaller on average than σ^2, and thus, it systematically underestimates σ^2. SS/N is a biased estimate because the SS term is actually based on fewer than N independent pieces of information. How many independent pieces of information is the SS term actually based on?

Let's reconsider the batch of HR scores for $N = 9$ people and the corresponding deviations from the mean; these deviations appear in Column 2 of Table 2.2. As mentioned earlier, for this batch of data, the sum of deviations from the sample mean equals 0; that is, $\sum (X_i - M) = -3.11 - 2.11 + .89 + 6.89 - .11 + 1.89 + 8.89 - 9.11 - 4.11 = 0$. In general, the sum of deviations of sample scores from the sample mean, $\sum (X_i - M)$, always equals 0. Because of the constraint that $\sum (X - M) = 0$, only the first $N - 1$ values (in this case, 8) of the $X - M$ deviation terms are "free to vary." Once we know any eight deviations for this batch of data, we can deduce what the remaining ninth deviation must be; it has to be whatever value is needed to make $\sum (X - M) = 0$. For example, once we know that the sum of the deviations from the mean for Persons 1 through 8 in this sample of nine HR scores is +4.11, we know that the deviation from the mean for the last remaining case must be −4.11. Therefore, we really only have $N - 1$, in this case eight, independent pieces of information about variability in our sample of 9 subjects. The last deviation does not provide new information. The number of independent pieces of information that a statistic is based on is called the **degrees of freedom,** or ***df.*** For a sample variance for a set of N scores, $df = N - 1$. The SS term is based on only $N - 1$ independent deviations from the sample mean.

It can be demonstrated empirically, and formally proved, that computing the sample variance by dividing the SS term by N results in a sample variance that systematically underestimates the true population variance. This underestimation or bias can be corrected by using the degrees of freedom as the divisor. The preferred (unbiased) formula for computation of a sample variance for a set of X scores is thus

$$s^2 = SS/df \text{ or } SS/(N-1) \text{ or } \sum (X-M)^2/(N-1). \tag{2.6}$$

Whenever a sample statistic is calculated using sums of squared deviations, it has an associated degrees of freedom that tells us how many independent deviations the statistic is based on. These df terms are used to compute statistics such as the sample variance and, later, to decide which distribution (in the family of ***t* distributions**, for example) should be used to look up **critical values** for statistical significance tests.

For this hypothetical batch of nine HR scores, the deviations from the mean appear in column 2 of Table 2.2; the squared deviations appear in column 3 of Table 2.2; the SS is 244.89; $df = N - 1 = 8$; and the sample variance, s^2, is $244.89/8 = 30.61$. This agrees with the value of the sample variance in the SPSS output from the Frequencies procedure in Figure 2.12.

It is useful to think about situations that would make the sample variance s^2 take on larger or smaller values. The smallest possible value of s^2 occurs when all the scores in the sample have the same value; for example, the set of scores [73, 73, 73, 73, 73, 73, 73, 73, 73] would have a variance $s^2 = 0$. The value of s^2 would be larger for a sample in which individual deviations from the sample mean are relatively large, for example, [44, 52, 66, 97, 101, 119, 120, 135, 151], than for the set of scores [72, 73, 72, 71, 71, 74, 70, 73], where individual deviations from the mean are relatively small.

The value of the sample variance, s^2, has a minimum of 0. There is, in practice, no fixed upper limit for values of s^2; they increase as the distances between individual scores and the sample mean increase. The sample variance $s^2 = 30.61$ is in "squared HR in beats per minute." We will want to have information about dispersion that is in terms of HR (rather than HR squared); this next step in the development of sample statistics is discussed in Section 2.7. First, however, let's consider an important question: Why is there variance? Why do researchers want to know about variance?

2.6 ♦ Why Is There Variance?

The best question ever asked by a student in my statistics class was, "Why is there variance?" This seemingly naive question is actually quite profound; it gets to the heart of research questions in behavioral, educational, medical, and social science research. The general question, Why is there variance? can be asked specifically about HR: Why do some people have higher and some people lower HR scores than average? Many factors may influence HR: for example, family history of cardiovascular disease, gender, smoking, anxiety, caffeine intake, and aerobic fitness. The initial question that we consider when we compute a variance for our sample scores is, How *much* variability of HR is there across the people in our study? In subsequent analyses, researchers try to account for at least some of this variability by noting that factors such as gender, smoking, anxiety, and caffeine use may be systematically related to and therefore predictive of HR. In other words, the question, Why is there variance in HR? can be partially answered by noting that people have varying exposure to all sorts of factors that may raise or lower HR, such as aerobic fitness, smoking, anxiety, and caffeine consumption. Because people experience different genetic and environmental influences, they have different HRs. A major goal of research is to try to identify the factors that predict (or possibly even causally influence) each individual person's score on the variable of interest, such as HR.

Similar questions can be asked about all attributes that vary across people or other subjects of study; for example, Why do people have differing levels of anxiety, satisfaction with life, body weight, or salary?

The implicit model that underlies many of the analyses discussed later in this textbook is that an observed score can be broken down into components and that each component of the score is systematically associated with a different predictor variable. Consider participant number 7 (let's call him Joe), with an HR of 82 bpm. If we have no information about Joe's background, a reasonable initial guess would be that Joe's HR is equal to the mean resting HR for the sample, $M = 73.11$. However, let's assume that we know that Joe smokes cigarettes and that we know that cigarette smoking tends to increase HR by about 5 bpm. If Joe is a smoker, we might predict that his HR would be 5 points higher than the population mean of 73.11 (73.11, the overall mean, plus 5 points, the effect of smoking on HR, would yield a new estimate of 78.11 for Joe's HR). Joe's actual HR (82) is a little higher than this predicted value (78.11), which combines information about what is average for most people with information about the effect of smoking on HR. An estimate of HR that is based on information about only one predictor variable (in this example, smoking) probably will not be exactly correct, because there are likely to be many other factors that influence Joe's HR (e.g., body weight, family history of cardiovascular disease, drug use, and so forth). These other variables that are not included in the analysis are collectively called sources of "error." The difference between Joe's actual HR of 82 and his predicted HR of 78.11 (82 – 78.11 = +3.89) is a prediction error. Perhaps Joe's HR is a little higher than we might predict based on overall average HR and Joe's smoking status because Joe has poor aerobic fitness or because Joe was anxious when his HR was measured. It might be possible to reduce this prediction error to a smaller value if we had information about additional variables (such as aerobic fitness and anxiety) that are predictive of HR.

Because we do not know all the factors that influence or predict Joe's HR, a predicted HR based on just a few variables is generally not exactly equal to Joe's actual HR, although it may be a better estimate of his HR than we would have if we just used the sample mean to estimate his score.

Statistical analyses covered in later chapters will provide us with a way to "take scores apart" into components that represent how much of the HR score is associated with each predictor variable. In other words, we can "explain" why Joe's HR of 82 is 8.89 points higher than the sample mean of 73.11 by identifying parts of Joe's HR score that are associated with, and predictable from, specific variables such as smoking, aerobic fitness, and anxiety. More generally, a goal of statistical analysis is to show that we can predict whether individuals tend to have high or low scores on an outcome variable of interest (such as HR) from scores on a relatively small number of predictor variables. We want to explain or account for the variance in HR by showing that some components of each person's HR score can be predicted from his or her scores on other variables.

2.7 ♦ Sample Standard Deviation (s)

An inconvenient property of the sample variance that was calculated in Section 2.5 ($s^2 =$ 30.61) is that it is given in squared HR rather than in the original units of measurement. The original scores were measures of HR in beats per minute, and it would be easier to talk about typical distances of individual scores from the mean if we had a measure of dispersion that was in the original units of measurement. To describe how far a typical subject's HR is from the sample mean, it is helpful to convert the information about dispersion contained in the sample variance, s^2, back into the original units of measurement (scores on HR rather than HR squared). To obtain an estimate of the sample standard deviation (s), we take the square root of the variance. The formula used to compute the sample standard deviation (which provides an **unbiased estimate** of the population standard deviation) is as follows:

$$s = \sqrt{\frac{SS}{df}} \text{ or } \sqrt{\frac{\sum (X - M)^2}{N - 1}}. \quad (2.7)$$

For the set of $N = 9$ HR scores given above, the variance was 30.61; the sample standard deviation $s = \sqrt{30.61} = 5.53$. The sample standard deviation, $s = 5.53$, tells us something about typical distances of individual *X* scores from the mean, *M*. Note that the numerical estimate for the sample standard deviation, *s*, obtained from this computation agrees with the value of *s* reported in the SPSS output from the Frequencies procedure that appears in Figure 2.12.

How can we use the information that we obtain from sample values of *M* and *s*? If we know that scores are normally distributed, and we have values for the sample mean and standard deviation, we can work out an approximate range that is likely to include most of the score values in the sample. Recall from Chapter 1 that in a normal distribution, about 95% of the scores lie within ±1.96 standard deviations from the mean. For a sample with $M = 73.11$ and $s = 5.53$, if we assume that HR scores are normally distributed, an estimated range that should include most of the values in the sample is obtained by finding $M \pm 1.96 \times s$. For this example, $73.11 \pm (1.96 \times 5.53) = 73.11 \pm 10.84$; this is a range from 62.27 to 83.95. These values are fairly close to the actual minimum (64) and maximum (82) for the sample. The approximation of range obtained by using *M* and *s* tends to work much better when the sample has a larger *N* of participants and when scores are normally distributed within the sample. What we know at this point is that the average for HR was about 73 bpm and that the range of HR in this sample was from 64 to 82 bpm. Later in the chapter, we will ask, How can we use this information from the sample (*M* and *s*) to estimate μ, the mean HR for the entire population?

However, there are several additional issues to consider before we take on the problem of making inferences about μ, the unknown population mean. These are discussed in the next few sections.

2.8 ♦ Assessment of Location of a Single *X* Score Relative to a Distribution of Scores

We can use the mean and standard deviation of a population (μ and σ, respectively) or the mean and standard deviation for a sample (*M* and *s*, respectively) to evaluate the location of a single *X* score (relative to the other scores in a population or a sample).

First, let's consider evaluating a single *X* score relative to a population for which the mean and standard deviation, μ and σ, respectively, are known. In real-life research situations, researchers rarely have this information. One clear example of a real-life situation where the values of μ and σ are known to researchers involves scores on standardized tests such as the Wechsler Adult Intelligence Scale (WAIS).

Suppose you are told that an individual person has received a score of 110 points on the WAIS. How can you decide what this score means? Does it represent a high or a low score relative to other people who have taken the test? Is it far from the mean or close to the mean of the distribution of scores? Is it far enough above the mean to be considered "exceptional" or unusual? To evaluate the location of an individual score, you need information about the distribution of the other scores. If you have a detailed frequency table that shows exactly how many people obtained each possible score, you can work out an exact percentile rank (the percentage of test takers who got scores lower than 110) using procedures that are presented in detail in introductory statistics books. When the distribution of scores has a normal shape, a standard score or z score provides a good description of the location of that single score relative to other people's scores without the requirement for complete information about the location of every other individual score.

In the general population, scores on the WAIS intelligence quotient (IQ) test have been scaled so that they are normally distributed with a mean $\mu = 100$ and a standard deviation σ of 15. The first thing you might do to assess an individual score is to calculate the distance from the mean—that is, $X - \mu$ (in this example, $110 - 100 = +10$ points). This result tells you that the score is above average (because the deviation has a positive sign). But it does not tell whether 10 points correspond to a large or a small distance from the mean when you consider the variability or dispersion of IQ scores in the population.

To obtain an index of distance from the mean that is "unit free" or standardized, we compute a z score; we divide the deviation from the mean $(X - \mu)$ by the standard deviation of population scores (σ) to find out the distance of the X score from the mean in number of standard deviations, as shown in Equation 2.8:

$$z = (X - \mu)/\sigma. \tag{2.8}$$

If the z transformation is applied to every X score in a normally distributed population, the shape of the distribution of scores does not change, but the mean of the distribution is changed to 0 (because we have subtracted μ from each score), and the standard deviation is changed to 1 (because we have divided deviations from the mean by σ). Each z score now represents how far an X score is from the mean in "standard units"—that is, in terms of the number of standard deviations. The mapping of scores from a normally shaped distribution of raw scores, with a mean of 100 and a standard deviation of 15, to a standard normal distribution, with a mean of 0 and a standard deviation of 1, is illustrated in Figure 2.13.

For a score of $X = 110$, $z = (110 - 100)/15 = +.67$. Thus, an X score of 110 IQ points corresponds to a z score of +.67, which corresponds to a distance of two thirds of a standard deviation above the population mean.

Recall from the description of the normal distribution in Chapter 1 that there is a fixed relationship between distance from the mean (given as a z score, i.e., numbers of standard deviations) and area under the normal distribution curve. We can deduce approximately what proportion or percentage of people in the population had IQ scores higher (or lower) than 110 points by (a) finding out how far a score of 110 is from the mean in standard score or z score units and (b) looking up the areas in the normal distribution that correspond to the z score distance from the mean.

The proportion of the area of the normal distribution that corresponds to outcomes greater than $z = +.67$ can be evaluated by looking up the area that corresponds to the obtained z value in the table of the standard normal distribution in Appendix A. The obtained value of z (+.67) and the corresponding areas appear in the three columns on the right-hand side of the first page of the standard normal distribution table, about eight lines from the top. Area C corresponds to the proportion of area under a normal curve that lies to the right of $z = +.67$; from the table, area $C = .2514$. Thus, about 25% of the area in the normal distribution lies above $z = +.67$. The areas for sections of the normal distribution are interpretable as proportions; if they are multiplied by 100, they can be interpreted as percentages. In this case, we can say that the proportion of the population that had z scores

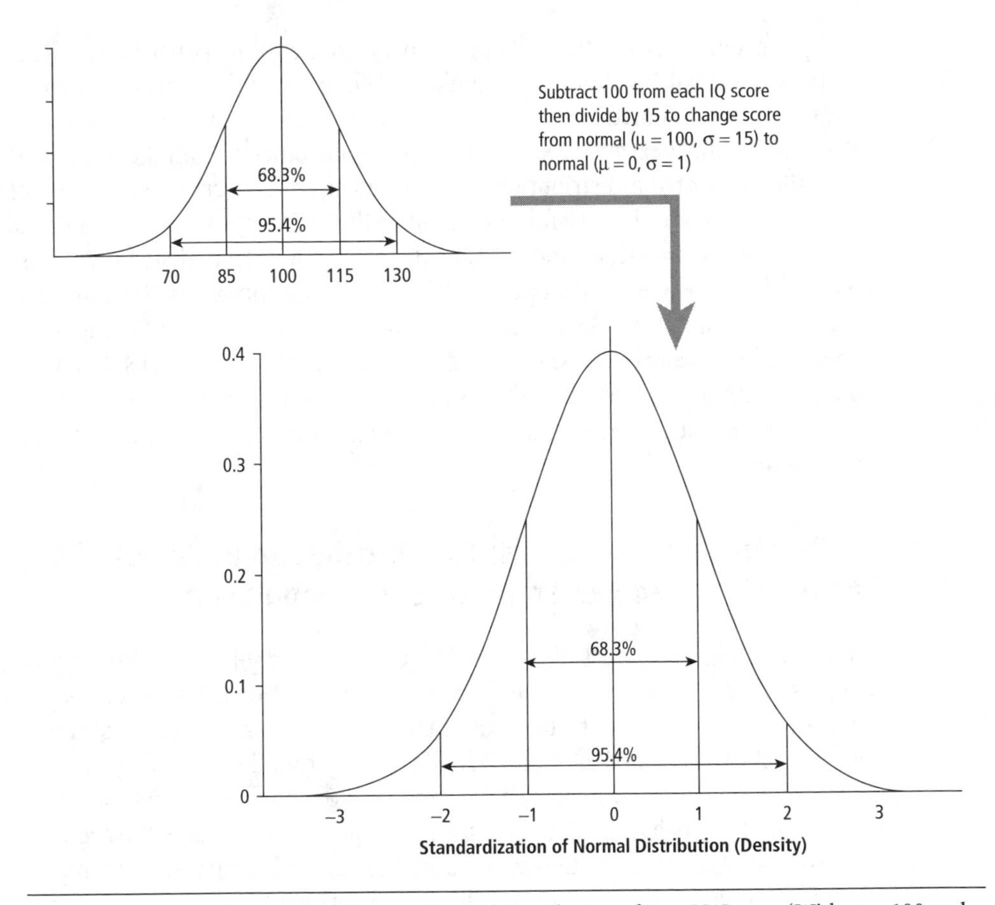

Figure 2.13 ♦ Mapping of Scores From a Normal Distribution of Raw IQ Scores (With $\mu = 100$ and $\sigma = 15$) to a Standard Normal Distribution (With $\mu = 0$ and $\sigma = 1$)

equal to or above +.67 and/or IQ scores equal to or above 110 points was .2514. Equivalently, we could say that 25.14% of the population had IQ scores equal to or above 110.

Note that the table in Appendix A can also be used to assess the proportion of cases that lie below $z = +.67$. The proportion of area in the lower half of the distribution (from $z = -\infty$ to $z = .00$) is .50. The proportion of area that lies between $z = .00$ and $z = +.67$ is shown in column B (area = .2486) of the table. To find the total area below $z = +.67$, these two areas are summed: .5000 + .2486 = .7486. If this value is rounded to two decimal places and multiplied by 100 to convert the information into a percentage, it implies that about 75% of persons in the population had IQ scores below 110. This tells us that a score of 110 is above average, although it is not an extremely high score.

Consider another possible IQ score. If a person has an IQ score of 145, that person's z score is $(145 - 100)/15 = +3.00$. This person scored 3 standard deviations above the mean. The proportion of the area of a normal distribution that lies above $z = +3.00$ is .0013. That is, only about 1 in 1,000 people have z scores greater than or equal to + 3.00 (which would correspond to IQs greater than or equal to 145).

By convention, scores that fall in the most extreme 5% of a distribution are regarded as extreme, unusual, exceptional, or unlikely. (While 5% is the most common criterion for "most extreme," sometimes researchers choose to look at the most extreme 1% or .1%.) Because the most extreme 5% (combining the outcomes at both the upper and the lower extreme ends of the distribution) is so often used as a criterion for an "unusual" or "extreme" outcome, it is useful to remember that 2.5% of the area in a normal distribution lies below $z = -1.96$, and 2.5% of the area in a normal distribution lies above $z = +1.96$. When the areas in the upper and lower tails are combined, the most extreme 5% of the scores in a normal distribution correspond to z values ≤ -1.96 and $\geq +1.96$. Thus, anyone whose score on a test yields a z score greater than 1.96 in absolute value might be judged "extreme" or unusual. For example, a person whose test score corresponds to a value of z that is greater than +1.96 is among the top 2.5% of all test scorers in the population.

2.9 ♦ A Shift in Level of Analysis: The Distribution of Values of *M* Across Many Samples From the Same Population

At this point in the discussion, we need to make a major shift in thinking. Up to this point, the discussion has examined the distributions of individual X scores in populations and in samples. We can describe the central tendency or average score by computing a mean; we describe the dispersion of individual X scores around the mean by computing a standard deviation. We now move to a different level of analysis: We will ask analogous questions about the behavior of the sample mean, M; that is, What is the average value of M across many samples, and how much does the value of M vary across samples? It may be helpful to imagine this as a sort of "thought experiment."

In actual research situations, a researcher usually has only one sample. The researcher computes a mean and a variance for the data in that one sample, and often the researcher wants to use the mean and variance from one sample to make inferences about (or estimates of) the mean and variance of the population from which the sample was drawn.

Note, however, that the single sample mean, M, reported for a random sample of $N = 9$ cases from the hr130 file ($M = 73.11$) was not exactly equal to the population mean μ of 73.76 (in Figure 2.5). The difference $M - \mu$ (in this case, $73.11 - 73.76$) represents an estimation error; if we used the sample mean value $M = 73.11$ to estimate the population mean of $\mu = 73.76$, in this instance, our estimate will be off by $73.11 - 73.76 = -.65$. It is instructive to stop and think, Why was the value of M in this one sample different from the value of μ?

It may be useful for the reader to repeat this sampling exercise. Using the <Data> → <Select Cases> → <Random> SPSS menu selections, as shown in Figures 2.6 and 2.7 earlier, each member of the class might draw a random sample of $N = 9$ cases from the file hr130.sav and compute the sample mean, M. If students report their values of M to the class, they will see that the value of M differs across their random samples. If the class sets up a histogram to summarize the values of M that are obtained by class members, this is a **"sampling distribution"** for M; that is, a set of different values for M that arise when many random samples of size $N = 9$ are selected from the same population. Why is it that no two students get exactly the same answer for the value of M?

2.10 ♦ An Index of Amount of Sampling Error: The Standard Error of the Mean (σ_M)

Different samples drawn from the same population typically yield different values of *M* because of sampling error. Just by "luck of the draw," some random samples contain one or more individuals with unusually low or high scores on HR; for those samples, the value of the sample mean, *M*, will be lower (or higher) than the population mean, μ. The question we want to answer is, How much do values of *M*, the sample mean, tend to vary across different random samples drawn from the same population, and how much do values of *M* tend to differ from the value of μ, the population mean that the researcher wants to estimate? It turns out that we can give a precise answer to this question. That is, we can quantify the magnitude of sampling error that arises when we take hundreds of different random samples (of the same size, *N*) from the same population. It is useful to have information about the magnitude of sampling error; we will need this information later in this chapter to set up CIs, and we will also use this information in later chapters to set up statistical significance tests.

The outcome for this distribution of values of *M*—that is, the sampling distribution of *M*—is predictable from the **central limit theorem**. A reasonable statement of this theorem is provided by Jaccard and Becker (2002):

> Given a population [of individual *X* scores] with a mean of μ and a standard deviation of σ, the sampling distribution of the mean [*M*] has a mean of μ and a standard deviation [generally called the "standard error," σ_M] of $\sigma / \sqrt{N}$ and approaches a normal distribution as the sample size on which it is based, *N*, approaches infinity. (p. 189)

For example, an instructor using the entire dataset hr130.sav can compute the population mean $\mu = 73.76$ and the population standard deviation $\sigma = 7.062$ for this population of 130 scores. If the instructor asks each student in the class to draw a random sample of $N = 9$ cases, the instructor can use the central limit theorem to predict the distribution of outcomes for *M* that will be obtained by class members. (This prediction will work well for large classes; e.g., in a class of 300 students, there are enough different values of the sample mean to obtain a good description of the sampling distribution; for classes smaller than 30 students, the outcomes may not match the predictions from the central limit theorem very closely.)

When hundreds of class members bring in their individual values of *M*, mean HR (each based on a different random sample of $N = 9$ cases), the instructor can confidently predict that when all these different values of *M* are evaluated as a set, they will be approximately normally distributed with a mean close to 73.76 bpm (the population mean) and with a standard deviation or standard error, σ_M, of $\sigma / \sqrt{N} = 7.062 / \sqrt{9} = 2.35$ bpm. The middle 95% of the sampling distribution of *M* should lie within the range $\mu - 1.96\sigma_M$ and $\mu + 1.96\sigma_M$; in this case, the instructor would predict that about 95% of the values of *M* obtained by individual class members should lie approximately within the range between $73.76 - 1.96 \times 2.35$ and $73.76 + 1.96 \times 2.35$, that is, mean HR between 69.15 and 78.37 bpm. On the other hand, about 2.5% of students are expected to obtain sample mean *M* values below 69.15, and about 2.5% of students are expected to obtain sample mean M values above 78.37. In other words, before the students go through all the work involved in actually drawing hundreds of samples and computing a mean M for each sample and then setting up a histogram and frequency table to summarize the values of M across the hundreds of class members, the instructor can anticipate the outcome; while the instructor cannot predict which individual students will obtain unusually high or low values of M, the instructor can make a fairly accurate prediction about the range of values of M that most students will obtain.

The fact that we can predict the outcome of this time-consuming experiment on the behavior of the sample statistic M based on the central limit theorem means that we do not, in practice, need to actually obtain hundreds of samples from the same population in order to estimate the magnitude of sampling error, σ_M. We only need to know the values of σ and N and to apply the central limit theorem to obtain fairly precise information about the typical magnitude of sampling error.

The difference between each individual student's value of M and the population mean, μ, is attributable to sampling error. When we speak of sampling error, we do not mean that the individual student has necessarily done something wrong (although students could make mistakes while computing M from a set of scores). Rather, sampling error represents the differences between the values of M and μ that arise just by chance. When individual students carry out all the instructions for the assignment correctly, most students obtain values of M that differ from μ by relatively small amounts; and a few students obtain values of M that are quite far from μ.

Prior to this section, the statistics that have been discussed (such as the sample mean, M, and the sample standard deviation, s) have described the distribution of individual X scores. Beginning in this section, we use the population standard error of the mean, σ_M, to describe the variability of a sample statistic (M) across many samples. The standard error of the mean describes the variability of the distribution of values of M that would be obtained if a researcher took thousands of samples from one population, computed M for each sample, and then examined the distribution of values of M; this distribution of many different values of M is called the sampling distribution for M.

2.11 ♦ Effect of Sample Size (*N*) on the Magnitude of the Standard Error (σ_M)

When the instructor sets up a histogram of the M values for hundreds of students, the shape of this distribution is typically close to normal; the mean of the M values is close to μ, the population mean, and the standard error (essentially, the standard deviation) of this distribution of M values is close to the theoretical value given by

$$\sigma_M = \frac{\sigma}{\sqrt{N}}. \tag{2.9}$$

Refer back to Figure 2.5 to see the histogram for the entire population of 130 HR scores. Because this population of 130 observations is small, we can calculate the population mean $\mu = 73.76$ and the population standard deviation $\sigma = 7.062$ (these statistics appeared along with the histogram in Figure 2.5). Suppose that each student in an extremely large class (500 class members) draws a sample of size $N = 9$ and computes a mean M for this sample; the values of M obtained by 500 members of the class would be normally distributed and centered at $\mu = 73.76$, with

$$\sigma_M = \frac{7.062}{\sqrt{9}} = \frac{7.062}{3} \cong 2.354,$$

as shown in Figure 2.15. When comparing the distribution of individual X scores in Figure 2.5 with the distribution of values of M based on 500 samples each with an N of 9 in Figure 2.15, the key thing to note is that they are both centered at the same value of μ (in this case, 73.76); but the variance or dispersion of the distribution of M values is less than the variance of the individual X scores. In general, as N (the size of each sample) increases, the variance of the M values across samples decreases.

Recall that σ_M is computed as $\sigma/\sqrt{N}$. It is useful to examine this formula and to ask: Under what circumstances will σ_M be larger or smaller? For any fixed value of N, this equation says that as σ increases, σ_M also increases. In other words, when there is an increase in the variance of the original individual X scores, it is intuitively obvious that random samples are more likely to include extreme scores, and these extreme scores in the samples will produce sample values of M that are farther from μ.

For any fixed value of σ, as N increases, the value of σ_M will decrease. That is, as the number of cases (N) in each sample increases, the estimate of M for any individual sample tends to be closer to μ. This should seem intuitively reasonable; larger samples tend to yield sample means that are better estimates of μ—that is, values of M that tend to be closer to μ. When $N = 1$, $\sigma_M = \sigma$; that is, for samples of size 1, the standard error is the same as the standard deviation of the individual X scores.

Figures 2.14 through 2.17 (on pages 66 through 69) illustrate that as the N per sample is increased, the dispersion of values of M in the sampling distributions continues to decrease in a predictable way. The numerical values of the standard deviations for the histograms shown in Figures 2.14 through 2.17 are approximately equal to the theoretical values of σ_M computed from σ and N:

$$\text{For } N = 4,\ \sigma_M = \frac{7.062}{\sqrt{4}} = \frac{7.062}{2} \cong 3.53.$$

$$\text{For } N = 9,\ \sigma_M = \frac{7.062}{\sqrt{9}} = \frac{7.062}{3} \cong 2.35.$$

$$\text{For } N = 25,\ \sigma_M = \frac{7.062}{\sqrt{25}} = \frac{7.062}{5} \cong 1.41.$$

$$\text{For } N = 64,\ \sigma_M = \frac{7.062}{\sqrt{64}} = \frac{7.062}{8} \cong .88.$$

The standard error, σ_M, provides information about the predicted dispersion of individual sample means around μ (just as σ provides information about the dispersion of individual X scores around the population mean and s provides information about the dispersion of individual X scores around the sample mean).

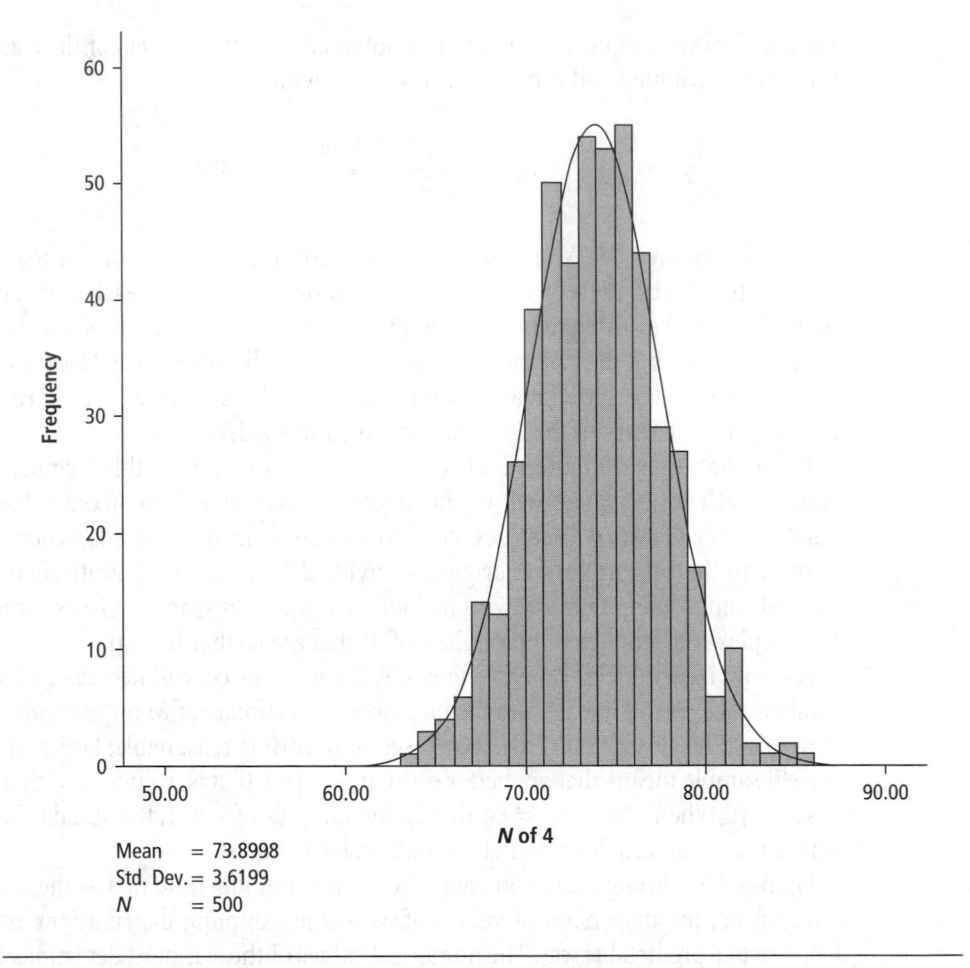

Figure 2.14 ♦ The Sampling Distribution of 500 Sample Means, Each Based on an N of 4, Drawn From the Population of 130 Heart Rate Scores in the hr130.sav Dataset

Basically, we want to know the typical magnitude of differences between M, an individual sample mean, and μ, the population mean that we want to estimate using the value of M from a single sample. When we use M to estimate μ, the difference between these two values $(M - \mu)$ is an estimation error. Recall that σ, the standard deviation for a population of X scores, provides summary information about the distances between individual X scores and μ, the population mean. In a similar way, the standard error of the mean, σ_M, provides summary information about the distances between M and μ, and these distances correspond to the estimation error that arises when we use individual sample M values to try to estimate μ. We hope to make the magnitudes of estimation errors, and therefore the magnitude of σ_M, small. Information about the magnitudes of estimation errors helps us to evaluate how accurate or inaccurate our sample statistics are likely to be as estimates of population parameters. Information about the magnitude of sampling errors is used to set up CIs and to conduct statistical significance tests.

Because the sampling distribution of M has a normal shape (and σ_M is the "standard deviation" of this distribution) and we know from Chapter 1 (Figure 1.4) that 95% of the

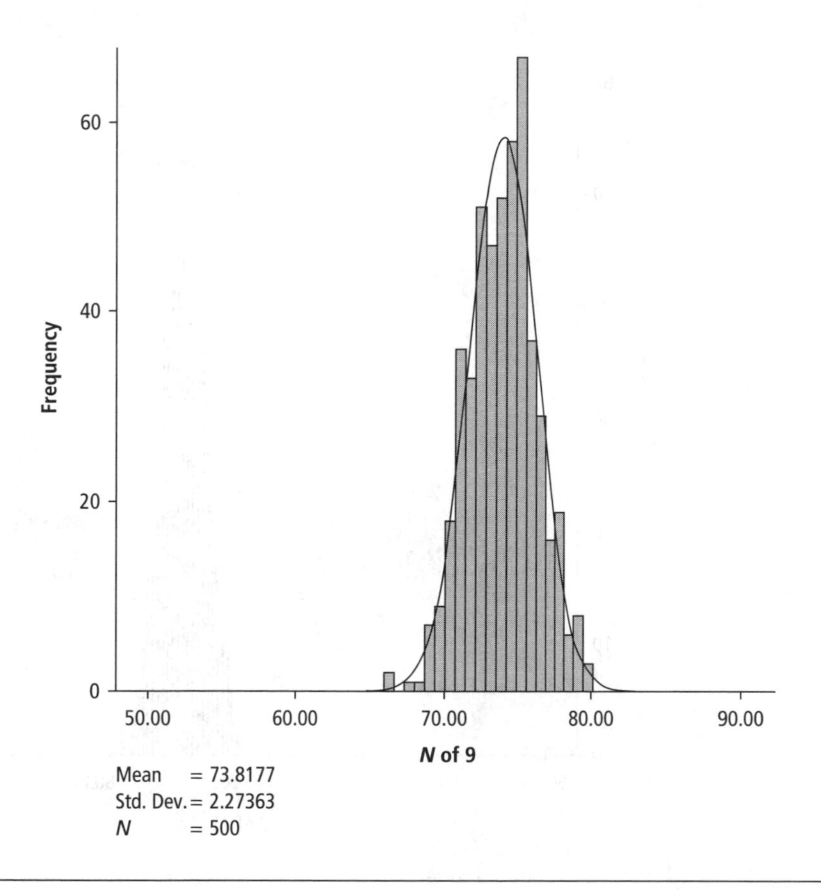

Figure 2.15 ♦ The Sampling Distribution of 500 Sample Means Each Based on an N of 9 Drawn From the Population of 130 Heart Rate Scores in the hr130.sav Dataset

area under a standard normal distribution lies between $z = -1.96$ and $z = +1.96$, we can reason that approximately 95% of the means of random samples of size N drawn from a normally distributed population of X scores, with a mean of μ and standard deviation of σ, should fall within a range given by $\mu - (1.96) \times \sigma_M$ and $\mu + 1.96 \times \sigma_M$. The critical z value of 1.96 is the distance from the mean in a standard normal distribution that corresponds to the middle 95% of the area under the normal distribution curve.

For the empirical example earlier in this section, about 95% of the students in an imaginary class of 500 students (or about 475 out of 500 students), each of whom uses a sample of size $N = 9$, should obtain values of M that lie between $\mu - 1.96 \times \sigma_M$ and $\mu + 1.96 \times \sigma_M$—that is, between $73.76 - (1.96 \times 2.35)$ and $73.76 + (1.96 \times 2.35)$ (between 69.15 and 78.37 bpm). The remaining 5% (about 25 out of 500 students) will probably obtain sample values of M that are rather far from 73.76—that is, values of M that are below 69.15 bpm or above 78.37 bpm.

2.12 ♦ Sample Estimate of the Standard Error of the Mean (SE_M)

The preceding section described the sampling distribution of M in situations where the value of the population standard deviation, σ, is known. In most research situations, the

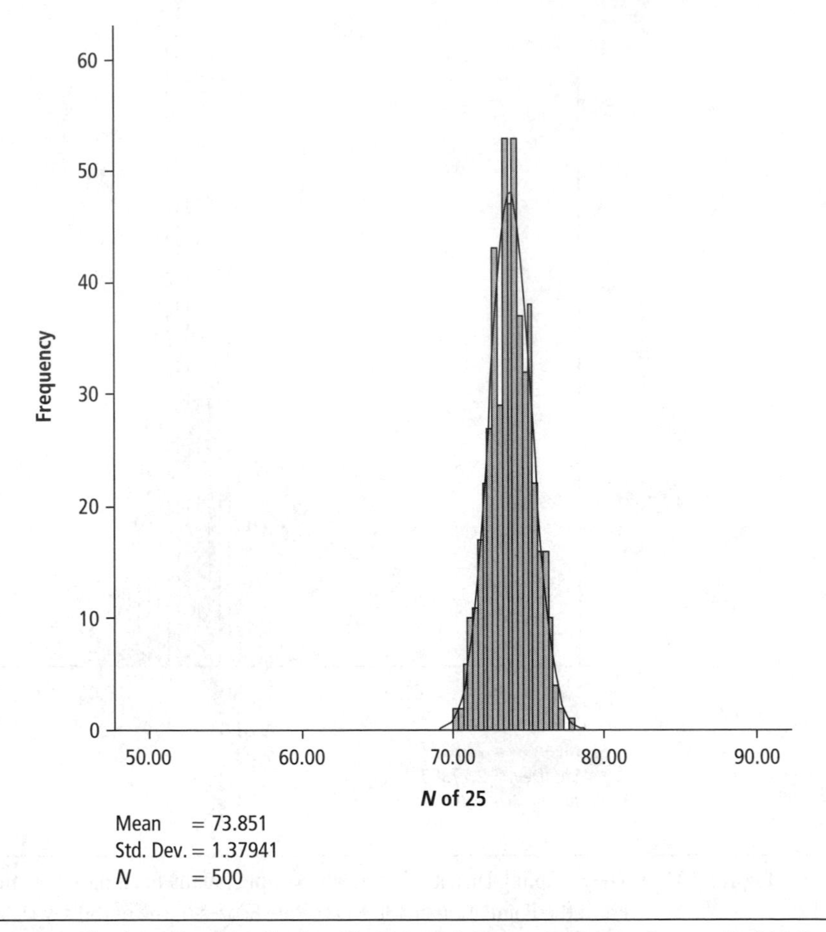

Figure 2.16 ♦ The Sampling Distribution of 500 Sample Means, Each Based on an *N* of 25, Drawn From the Population of 130 Heart Rate Scores in the hr130.sav Dataset

population mean and standard deviation are not known; instead, they are estimated by using information from the sample. We can estimate σ by using the sample value of the standard deviation; in this textbook, as in most other statistics textbooks, the sample standard deviation is denoted by *s*. Many journals, including those published by the American Psychological Association, use *SD* as the symbol for the sample standard deviations reported in journal articles.

Earlier in this chapter, we sidestepped the problem of working with populations whose characteristics are unknown by arbitrarily deciding that the set of 130 scores in the file named hr130.sav was the "population of interest." For this dataset, the population mean, μ, and standard deviation, σ, can be obtained by asking SPSS to calculate these values for the entire set of 130 scores that are defined as the population of interest. However, in many real-life research problems, researchers do not have information about all the scores in the population of interest, and they do not know the population mean, μ, and standard deviation, σ. We now turn to the problem of evaluating the magnitude of prediction error in the more typical real-life situation, where a researcher has one sample of data of size *N*

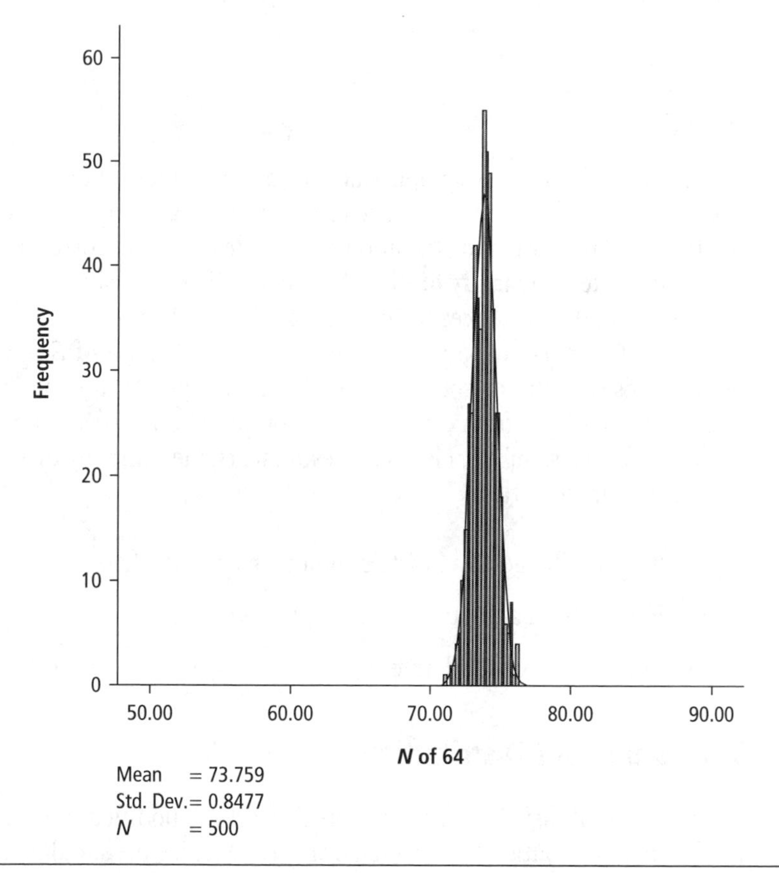

Figure 2.17 ♦ The Sampling Distribution of 500 Sample Means, Each Based on an *N* of 64, Drawn From the Population of 130 Heart Rate Scores in the hr130.sav Dataset

and the researcher can compute a sample mean, *M*, and a sample standard deviation, *s*, but the researcher does not know the values of the population parameters μ or σ. The researcher will want to estimate μ using the sample *M* from just one sample. The researcher wants to have a reasonably clear idea of the magnitude of estimation error that can be expected when the mean from one sample of size *N* is used to estimate μ, the mean of the corresponding population.

When σ, the population standard deviation, is not known, we cannot find the value of σ_M. Instead, we calculate an estimated standard error (SE_M), using the sample standard deviation *s* to replace the unknown value of σ in the formula for the standard error of the mean, as follows (when σ is known):

$$\sigma_M = \frac{\sigma}{\sqrt{N}}.$$

When σ is unknown, we use *s* to estimate σ and relabel the resulting standard error to make it clear that it is now based on information about sample variability rather than population variability of scores:

$$SE_M = \sqrt{\frac{s^2}{N}} \text{ or } \frac{s}{\sqrt{N}}. \quad (2.10)$$

The substitution of the sample statistic s as an estimate of the population σ introduces additional sampling error.[1] Because of this additional sampling error, we can no longer use the standard normal distribution to evaluate areas that correspond to distances from the mean. Instead, a **family of distributions** (called t distributions) is used to find areas that correspond to distances from the mean.

Thus, when σ is not known, we use the **sample value of SE_M** to estimate σ_M; and because this substitution introduces additional sampling error, the shape of the sampling distribution changes from a normal distribution to a t distribution. When the standard deviation from a sample (s) is used to estimate σ, the sampling distribution of M has the following characteristics:

1. It is distributed as a t distribution with $df = N - 1$.
2. It is centered at μ.
3. The estimated standard error is $SE_M = s/\sqrt{N}$.

2.13 ♦ The Family of *t* Distributions

The family of "t" distributions is essentially a set of "modified" normal distributions, with a different t distribution for each value of df (or N). Like the standard normal distribution, a t distribution is scaled so that t values are unit free. As N and df decrease, assuming that other factors remain constant, the magnitude of sampling error increases, and the required amount of adjustment in distribution shape also increases. A t distribution (like a normal distribution) is bell shaped and symmetrical; however, as the N and df decrease, t distributions become flatter in the middle compared with a normal distribution, with thicker tails (they become platykurtic). Thus, when we have a small df value, such as $df = 3$, the distance from the mean that corresponds to the middle 95% of the t distribution is larger than the corresponding distance in a normal distribution.

As the value of df increases, the shape of the t distribution becomes closer to that of a normal distribution; for $df > 100$, a t distribution is essentially identical to a normal distribution. Figure 2.18 shows t distributions for df values of 3, 6, and ∞. As df increases, the shape of the t distribution converges toward the normal distribution; a t distribution with $df > 100$ is essentially indistinguishable from a normal distribution.

For a research situation where the sample mean is based on $N = 7$ cases, $df = N - 1 = 6$. In this case, the sampling distribution of the mean would have the shape described by a t distribution with 6 df; a table for the distribution with $df = 6$ would be used to look up the values of t that cut off the top and bottom 2.5% of the area. The area that corresponds to the middle 95% of the t distribution with 6 df can be obtained either from the table of the t distribution in Appendix B or from the diagram in Figure 2.18. When $df = 6$, 2.5% of the area in the t distribution lies below $t = -2.45$, 95% of the area lies between $t = -2.45$ and $t = +2.45$, and 2.5% of the area lies above $t = +2.45$.

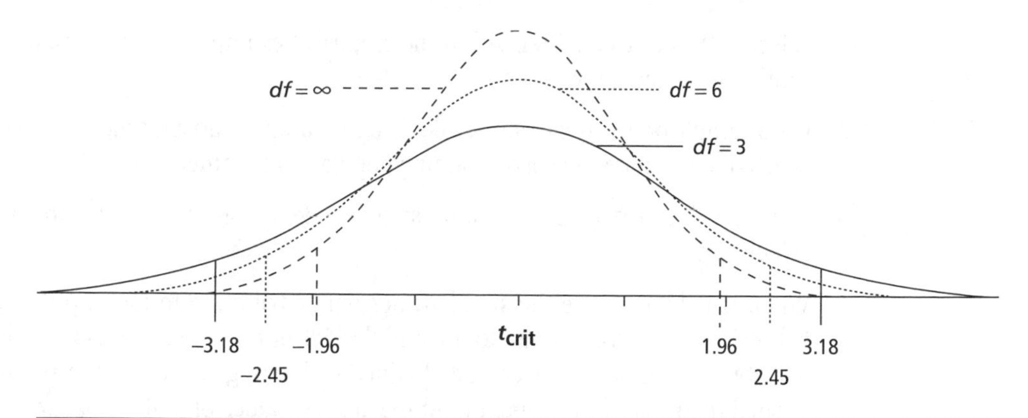

Figure 2.18 ♦ Graph of the *t* Distribution for Three Different *df* Values (*df* = 3, 6, and Infinity, or ∞)

SOURCE: www.psychstat.smsu.edu/introbook/sbk24m.htm.

2.14 ♦ Confidence Intervals

2.14.1 ♦ The General Form of a CI

If the single value of M in a sample is reported as an estimate of μ, it is called a point estimate. An interval estimate (CI) makes use of information about sampling error. The most general form of a CI is a pair of equations to calculate a lower limit and an upper limit. The level of "confidence" is an arbitrarily selected probability, usually 90%, 95%, or 99%.

The computations for a CI make use of the reasoning, discussed in earlier sections, about the sampling error associated with values of M. Based on our knowledge about the sampling distribution of M, we can figure out a range of values around μ that will probably contain most of the sample means that would be obtained if we drew hundreds or thousands of samples from the population. SE_M provides information about the typical magnitude of estimation error; that is, the typical distance between values of M and μ. Statistical theory tells us that (for values of df larger than 100) approximately 95% of obtained sample means will likely be within a range of about 1.96 SE_M units on either side of μ.

When we set up a CI around an individual sample mean, M, we are essentially using some logical sleight of hand and saying that if values of M tend to be close to μ, then the unknown value of μ should be reasonably close to (most) sample values of M. However, the language used to interpret a CI is tricky. It is incorrect to say that a CI computed using data from a single sample has a 95% chance of including μ. (It either does or doesn't.) We can say, however, that in the long run, approximately 95% of the CIs that are set up by applying these procedures to hundreds of samples from a normally distributed population with mean = μ will include the true population mean, μ, between the lower and the upper limits. (The other 5% of CIs will not contain μ.)

2.14.2 ♦ Setting Up a CI for *M* When σ Is Known

To set up a 95% **CI to estimate the mean** when σ, the population standard deviation, is known, the researcher needs to do the following:

1. Select a **"level of confidence."** In the empirical example that follows, the level of confidence is set at 95%.

2. For a sample of N observations, calculate the sample statistic (such as M) that will be used to estimate the corresponding population parameter (μ).

3. Use the value of σ (the population standard deviation) and the sample size N to calculate σ_M.

4. When σ is known, use the standard normal distribution to look up the **"critical values" of** z that correspond to the middle 95% of the area in the standard normal distribution. These values can be obtained by looking at the table of the standard normal distribution in Appendix A; the critical values of z also appear in the last line of the table of **critical values of the** t **distribution** in Appendix B. The shape of a t distribution with ∞ degrees of freedom—that is, $df > 100$—converges to the shape of the normal distribution. For a 95% level of confidence, from Appendix B, we find that the critical values of z (or t with infinite df) that correspond to the middle 95% of the area are $z = -1.96$ and $z = +1.96$.

This provides the information necessary to calculate the lower and upper limits for a CI. In the equations below, LL stands for the lower limit (or boundary) of the CI, and UL stands for the upper limit (or boundary) of the CI. Because the level of confidence was set at 95%, the critical values of z, z_{critical}, were obtained by looking up the distance from the mean that corresponds to the middle 95% of the normal distribution. (If a 90% level of confidence is chosen, the z values that correspond to the middle 90% of the area under the normal distribution would be used.)

The lower and upper limits of a **CI for a sample mean** M correspond to the following:

$$\text{Lower limit} = M - [z_{\text{critical}} \times \sigma_M]; \tag{2.11}$$

$$\text{Upper limit} = M + [z_{\text{critical}} \times \sigma_M]. \tag{2.12}$$

There are some research situations in which the population standard deviation, σ, is known. For example, during test development, standardized tests such as the WAIS are administered to tens of thousands of people; the scores from this normative population are used to compute the means and standard deviations that can be used to evaluate the location of a single IQ score relative to the population mean ($\mu = 100$). For tests, such as the WAIS, for which the population standard deviation, σ, is known, we can use σ to calculate σ_M and set up CIs using the normal distribution to obtain the distances from the mean that correspond to the middle 95% of the distribution.

As an example, suppose that a student researcher collects a sample of $N = 25$ scores on IQ for a random sample of people drawn from the population of students at Corinth College. The WAIS IQ test is known to have σ equal to 15. Suppose the student decides to set up a 95% CI. The student obtains a sample mean IQ, M, equal to 128.

The student needs to do the following:

1. Find the value of $\sigma_M = \sigma/\sqrt{N} = 15/\sqrt{25} = 15/5 = 3.0$.

2. Look up the critical values of z that correspond to the middle 95% of a standard normal distribution. From the table of the normal distribution in Appendix A, these critical values are $z = -1.96$ and $z = +1.96$ are the z_{critical} values.

3. Substitute the values for σ_M and z_{critical} into equations 2.11 and 2.12 to obtain the following results:

   Lower limit $= M - [z_{\text{critical}} \times \sigma_M] = 128 - [1.96 \times 3] = 128 - 5.88 = 122.12$;

   Upper limit $= M + [z_{\text{critical}} \times \sigma_M] = 128 + [1.96 \times 3] = 128 + 5.88 = 133.88$.

What conclusions can the student draw about the mean IQ of the population (all students at Corinth College) from which the random sample was drawn? It would *not* be correct to say that "there is a 95% chance that the true population mean IQ, μ, for all Corinth College students lies between 122.12 and 133.88." It would be correct to say that "the 95% CI around the sample mean lies between 122.12 and 133.88." (Note that the value of 100, which corresponds to the mean, μ, for the general adult population, is not included in this 95% CI for a sample of students drawn from the population of all Corinth College students. It appears, therefore, that the population mean WAIS score for Corinth College students may be higher than the population mean IQ for the general adult population.)

To summarize, the 95% **confidence level** is not the probability that the true population mean, μ, lies within the CI that is based on data from one sample (μ either does lie in this interval or does not). The confidence level is better understood as a long-range prediction about the performance of CIs when these procedures for setting up CIs are followed. When the rules for setting up 95% CIs are followed, we expect that approximately 95% of the CIs that researchers obtain in the long run will include the true value of the population mean, μ. The other 5% of the CIs that researchers obtain using these procedures will not include μ.

2.14.3 ♦ Setting Up a CI for *M* When the Value of σ Is Not Known

In a typical research situation, the researcher does not know the values of μ and σ; instead, the researcher has values of M and s from just one sample of size N and wants to use this sample mean, M, to estimate μ. In Section 2.12, I explained that when σ is not known, we can use s to calculate an estimate of SE_M. However, when we use SE_M (rather than σ_M) to set up CIs, the use of SE_M to estimate σ_M results in additional sampling error. To adjust for this additional sampling error, we use the t distribution with $N - 1$ degrees of freedom (rather than the normal distribution) to look up distances from the mean that correspond to the middle 95% of the area in the sampling distribution. When N is large (>100), the t distribution converges to the standard normal distribution; therefore, when samples are large ($N > 100$), the standard normal distribution can be used to obtain the critical values for a CI.

The formulas for the upper and lower limits of the CI when σ is not known, therefore, differ in two ways from the formulas for the CI when σ is known. First, when σ is unknown, we replace σ_M with SE_M. Second, when σ is unknown and $N < 100$, we replace z_{critical} with t_{critical}, using a t distribution with $N - 1$ *df* to look up the critical values (for $N \geq 100$, z_{critical} may be used).

For example, suppose that the researcher wants to set up a 95% CI using the sample mean data reported in an earlier section of this chapter with $N = 9$, $M = 73.11$, and $s = 5.533$ (sample statistics are from Figure 2.12). The procedure is as follows:

1. Find the value of $SE_M = s/\sqrt{N} = 5.533/\sqrt{9} = 5.533/3 = 1.844$.
2. Find the $t_{critical}$ values that correspond to the middle 95% of the area for a t distribution with $df = N - 1 = 9 - 1 = 8$. From the table in Appendix B, these are $t_{critical} = -2.31$ and $t_{critical} = +2.31$.
3. Substitute the values of M, $t_{critical}$, and SE_M into the following equations:
   Lower limit $= M - [t_{critical} \times SE_M] = 73.11 - [2.31 \times 1.844] = 73.11 - 4.26 = 68.85$;
   Upper limit $= M + [t_{critical} \times SE_M] = 73.11 + [2.31 \times 1.844] = 73.11 + 4.26 = 77.37$.

What conclusions can the student draw about the mean HR of the population (all 130 cases in the file named hr130.sav) from which the random sample of $N = 9$ cases was drawn? The student can report that "the 95% CI for mean HR ranges from 68.85 to 77.37." In this particular situation, we know what μ really is; the population mean HR for all 130 scores was 73.76 (from Figure 2.5). In this example, we know that the CI that was set up using information from the sample actually did include μ. (However, about 5% of the time, when a 95% level of confidence is used, the CI that is set up using sample data will not include μ.)

The sample mean, M, is not the only statistic that has a sampling distribution and a standard error. The sampling distributions for many other statistics are known; thus, it is possible to identify an appropriate sampling distribution and to estimate the standard error and set up CIs for many other sample statistics, such as Pearson r.

2.14.4 ♦ Reporting CIs

Based on recommendations made by Wilkinson and the Task Force on Statistical Inference (1999), the most recent revision of the *Publication Manual of the American Psychological Association* (American Psychological Association [APA], 2001) states that CI information should be provided for major outcomes wherever possible. SPSS provides CI information for many, but not all, outcome statistics of interest. In some situations, researchers may need to calculate CIs by hand (Kline, 2004).

When we report CIs, such as a CI for a sample mean, we remind ourselves (and our readers) that the actual value of the population parameter that we are trying to estimate is generally unknown and that the values of sample statistics are influenced by sampling error. Note that it may be inappropriate to use CIs to make inferences about the means for any specific real-world population if the CIs are based on samples that are not randomly selected from a specific, well-defined population of interest. As pointed out in Chapter 1, the widespread use of convenience samples (rather than random samples from clearly defined populations) may lead to situations where the sample is not representative of any real-world population. It would be misleading to use sample statistics (such as the sample mean, M) to make inferences about the population mean, μ, for real-world populations if the members of the sample are not similar to, or representative of, that real-world population. At best, when researchers work with convenience samples, they can make inferences about hypothetical populations that have characteristics similar to those of the sample.

The results obtained from the analysis of a random sample of nine HR scores could be reported as follows:

Results

Using the SPSS random sampling procedure, a random sample of $N = 9$ cases was selected from the population of 130 scores in the hr130.sav data file. The scores in this sample appear in Table 2.2. For this sample of nine cases, mean HR $M = 73.11$ bpm, and the standard deviation of HR $SD = 5.53$ bpm. The 95% CI for the mean based on this sample ranged from a lower limit of 68.85 to an upper limit of 77.37.

2.15 ♦ Summary

Many statistical analyses include relatively simple terms that summarize information across X scores, such as $\sum X$ and $\sum X^2$. It is helpful to recognize that whenever a formula includes $\sum X$, information about the mean of X is being taken into account; when terms involving $\sum X^2$ are included, information about variance is included in the computations.

This chapter reviewed several basic concepts from introductory statistics:

1. The computation and interpretation of sample statistics, including the mean, variance, and standard deviation, were discussed.

2. A z score is used as a unit-free index of the distance of a single X score from the mean of a normal distribution of individual X scores. Because values of z have a fixed relationship to areas under the normal distribution curve, a z score can be used to answer questions such as, What proportion or percentage of cases have scores higher than X?

3. Sampling error arises because the value of a sample statistic such as M varies across samples when many random samples are drawn from the same population.

4. Given some assumptions, it is possible to predict the shape, mean, and variance of the sampling distribution of M. When σ is known, the sampling distribution of M has the following known characteristics: it is normal in shape; the mean of the distribution of values of M corresponds to μ, the population mean; and the standard deviation or standard error that describes typical distances of sample mean values of M from μ is given by $\sigma / \sqrt{N}$. When σ is not known and the researcher uses a sample standard deviation s to estimate σ, a second source of sampling error arises; we now have potential errors in estimation of σ using s as well as errors of estimation of μ using M. The magnitude of this additional sampling error depends on N, the size of the samples that are used to calculate M and s.

5. Additional sampling error arises when s is used to estimate σ. This additional sampling error requires us to refer to a different type of sampling distribution when we evaluate distances of individual M values from the center of the sampling distribution—that is, the family of t distributions (instead of the standard normal distribution).
6. The family of t distributions has a different distribution shape for each degrees of freedom. As the df for the t distribution increases, the shape of the t distribution becomes closer to that of a standard normal distribution. When N (and therefore df) becomes greater than 100, the difference between the shape of the t and normal distributions becomes so small that distances from the mean can be evaluated using the normal distribution curve.
7. All these pieces of information come together in the formula for the CI. We can set up an "interval estimate" for μ based on the sample value of M and the amount of sampling error that is theoretically expected to occur.
8. Recent reporting guidelines for statistics (e.g., Wilkinson and the Task Force on Statistical Inference, 1999) recommend that CIs should be included for all important statistical outcomes in research reports wherever possible.

Appendix on SPSS

Students who have never used SPSS (or programs that have similar capabilities) may need an introduction to SPSS, such as Einspruch (2005) or Pavkov (2005). The examples in this textbook use SPSS version 14.0.

As with other statistical packages, students may either purchase a personal copy of the SPSS software and install it on a PC or use a version installed on their college or university computer network. When SPSS access has been established (either by installing a personal copy of SPSS on a PC or by doing whatever is necessary to access the college or university network version of SPSS), an SPSS® icon appears on the Windows desktop, or an SPSS for Windows folder can be opened by clicking on Start in the lower left corner of the computer screen and then on All Programs. When SPSS is started in this manner, the initial screen asks the user whether he or she wants to open an existing data file or type in new data.

When students want to work with existing SPSS data files, such as the SPSS data files on the Web site for this textbook, they can generally open these data files just by clicking on the SPSS data file; as long as the student has access to the SPSS program, SPSS data files will automatically be opened using this program. SPSS can save and read several different file formats. On the Web site that accompanies this textbook, each data file is available in two formats: as an SPSS system file (with a full file name of the form dataset.sav) and as an Excel file (with a file name of the form dataset.xls). Readers who use programs other than SPSS will need to use the drop-down menu that lists various "file types" to tell their program (such as SAS) to look for and open a file that is in Excel XLS format (rather than the default SAS format).

SPSS examples are presented in sufficient detail in this textbook so that students should be able to reproduce any of the analyses that are discussed. Some useful data-handling features of SPSS (such as procedures for handling missing data) are discussed in the context of statistical analyses, but this textbook does not by itself provide a comprehensive treatment of the features in SPSS. Students who want to know more about the basic features of SPSS will want to consult the introductory books cited at the beginning of the appendix. Students who want a more comprehensive treatment of SPSS may consult books by Norusis (2005a, 2005b).

Notes

1. Demonstrations do not constitute proofs; however, they require less lengthy explanations and less mathematical sophistication from the reader than proofs or formal mathematical derivations. Throughout this book, demonstrations are offered instead of proofs, but readers should be aware that a demonstration only shows that a result works using the specific numbers involved in the demonstration; it does not constitute a proof.

2. The population variance, σ^2, is defined as $\sigma^2 = \Sigma\ (X - \mu)^2/N$.

I have already commented that when we calculate a sample variance, s^2, using the formula $s^2 = \Sigma\ (X - M)^2/N - 1$, we need to use $N - 1$ as the divisor to take into account the fact that we only have $N - 1$ independent deviations from the sample mean. However, there is a second problem that arises when we calculate s^2; that is, we calculate s^2 using M, an estimate of μ that is also subject to sampling error.

Comprehension Questions

1. Consider the following small set of scores. Each number represents the number of siblings reported by each of the $N = 6$ persons in the sample:

   X scores are [0, 1, 1, 1, 2, 7].

   a. Compute the mean (M) for this set of six scores.
   b. Compute the six deviations from the mean ($X - M$), and list these six deviations.
   c. What is the sum of the six deviations from the mean you reported in (b)? Is this outcome a surprise?
   d. Now calculate the sum of squared deviations (SS) for this set of six scores.
   e. Compute the sample variance, s^2, for this set of six scores.
   f. When you compute s^2, why should you divide SS by $(N - 1)$ rather than by N?
   g. Finally, compute the sample standard deviation (denoted by either s or SD).

2. In your own words, what does an SS tell us about a set of data? Under what circumstances will the value of SS equal 0? Can SS ever be negative?

3. For each of the following lists of scores, indicate whether the value of SS will be negative, 0, between 0 and +15, or greater than +15. (You do not need to actually calculate SS.)

   Sample A: X = [103, 156, 200, 300, 98]

   Sample B: X = [103, 103, 103, 103, 103, 103]

   Sample C: X = [101, 102, 103, 102, 101]

4. For a variable that interests you, discuss why there is variance in scores on that variable. (In Chapter 2, e.g., there is a discussion of factors that might create variance in heart rate, HR.)

5. Assume that a population of thousands of people whose responses were used to develop the anxiety test had scores that were normally distributed with $\mu = 30$ and $\sigma = 10$. What proportion of people in this population would have anxiety scores within each of the following ranges of scores?
   a. Below 20
   b. Above 30
   c. Between 10 and 50
   d. Below 10
   e. Below 50
   f. Above 50
   g. Either below 10 or above 50

   Assuming that a score in the top 5% of the distribution would be considered extremely anxious, would a person whose anxiety score was 50 be considered extremely anxious?

6. What is a confidence interval (CI), and what information is required to set up a CI?

7. What is a sampling distribution? What do we know about the shape and characteristics of the sampling distribution for M, the sample mean?

8. What is SE_M? What does the value of SE_M tell you about the typical magnitude of sampling error?
   a. As s increases, how does the size of SE_M change (assuming that N stays the same)?
   b. As N increases, how does the size of SE_M change (assuming that s stays the same)?

9. How is a t distribution similar to a standard normal distribution score? How is it different?

10. Under what circumstances should a t distribution be used rather than the normal distribution to look up areas or probabilities associated with distances from the mean?

11. Consider the following questions about CIs.

    A researcher tests emotional intelligence (EI) for a random sample of children selected from a population of all students who are enrolled in a school for gifted children. The researcher wants to estimate the mean EI for the entire school. The population standard deviation, σ, for EI is not known.

    Let's suppose that a researcher wants to set up a 95% CI for IQ scores using the following information:

    The sample mean $M = 130$.

    The sample standard deviation $s = 15$.

    The sample size $N = 120$.

    The $df = N - 1 = 119$.

For the values given above, the limits of the 95% CI are as follows:

Lower limit $= 130 - 1.96 \times 1.37 = 127.31$;

Upper limit $= 130 + 1.96 \times 1.37 = 132.69$.

The following exercises ask you to experiment to see how changing some of the values involved in computing the CI influences the width of the CI.

Recalculate the CI above to see how the lower and upper limits (and the width of the CI) change as you vary the N in the sample (and leave all the other values the same).

a. What are the upper and lower limits of the CI and the width of the 95% CI if all the other values remain the same ($M = 130, s = 15$) but you change the value of N to 16?

For $N = 16$, lower limit = __________ and upper limit = ______________.

Width (upper limit – lower limit) = ___________________________.

Note that when you change N, you need to change two things: the computed value of σ_M and the degrees of freedom used to look up the critical values for t. *See Appendix B for the critical values of* t *that are needed for different values of df and different levels of confidence.*

b. What are the upper and lower limits of the CI and the width of the 95% CI if all the other values remain the same but you change the value of N to 25?

For $N = 25$, lower limit = __________ and upper limit = ______________.

Width (upper limit – lower limit) = ___________________________.

c. What are the upper and lower limits of the CI and the width of the 95% CI if all the other values remain the same ($M = 130, s = 15$) but you change the value of N to 49?

For $N = 49$, lower limit = __________ and upper limit = ______________.

Width (upper limit – lower limit) = ___________________________.

d. Based on the numbers you reported for sample size N of 16, 25, and 49, how does the width of the CI change as N (the number of cases in the sample) increases?

e. What are the upper and lower limits and the width of this CI if you change the confidence level to 80% (and continue to use $M = 130, s = 15$, and $N = 49$)?

For an 80% CI, lower limit = __________ and upper limit = ______________.

Width (upper limit – lower limit) = ___________________________.

f. What are the upper and lower limits and the width of the CI if you change the confidence level to 99% (continue to use $M = 130, s = 15$, and $N = 49$)?

For a 99% CI, lower limit = __________ and upper limit = ______________.

Width (upper limit – lower limit) = ___________________________.

g. How does increasing the level of confidence from 80% to 99% affect the width of the CI?

12. Data Analysis Project:

The $N = 130$ scores in the temphr.sav file are hypothetical data created by Shoemaker (1996) so that they yield results similar to those obtained in an actual study of temperature and HR (Mackowiak et al., 1992).

Use the Temperature data in the temphr.sav file to do the following:

Note that temperature in degrees Fahrenheit (tempf) can be converted into temperature in degrees centigrade (tempc) by the following: tempc = (tempf − 32)/1.8.

The following analyses can be done on tempf, tempc, or both tempf and tempc.

a. Find the sample mean M, standard deviation s, and standard error of the mean SE_M for scores on temperature.
b. Examine a histogram of scores on temperature. Is the shape of the distribution reasonably close to normal?
c. Set up a 95% CI for the sample mean, using your values of M, s, and N ($N = 130$ in this dataset).
d. The temperature that is popularly believed to be "average" or "healthy" is 98.6°F (or 37°C). Does the 95% CI based on this sample include this value, which is widely believed to represent an "average/healthy" temperature? What conclusion might you draw from this result?

CHAPTER 3

Statistical Significance Testing

3.1 ♦ The Logic of Null Hypothesis Significance Testing (NHST)

Chapter 2 reviewed the procedures used to set up a confidence interval (CI) to estimate an unknown population mean (μ) using information from a sample, including the sample mean (M), the number of cases in the sample (N), and the population standard deviation (σ). Based on what is known about the magnitude of sampling error, we expect values of the sample mean, M, to be normally distributed around a mean of μ with a standard deviation or standard error of σ_M. The value of σ_M depends on the values of σ and N:

$$\sigma_M = \sigma/\sqrt{N}. \tag{3.1}$$

Because the theoretical distribution of sample means has a normal shape when σ is known, we can use the standard *normal* distribution to evaluate distances from the mean of this sampling distribution in research situations where the value of σ is known. To set up a CI with a 95% level of confidence, we use the values of z that correspond to the middle 95% of the area in a normal distribution ($z = -1.96$ and $z = +1.96$) along with the value of σ_M to estimate the lower and upper boundaries of the CI. About 95% of the CIs set up using these procedures should contain the actual population mean, μ, across many replications of this procedure.

However, there is another way that the sample mean, M, can be used to make inferences about likely values of an unknown population mean, μ. Values of the sample mean, M, can be used to test hypotheses about a specific value of an unknown population mean through the use of **NHST** procedures.

What is the logic behind NHST? At the simplest level, when a researcher conducts a **null hypothesis significance test,** the following steps are involved.

First, the researcher makes a "guess" about the specific value of μ for a population of interest; this guess can be written as a formal **null hypothesis (H_0)**. For example, suppose that the variable of interest is the driving speed on Route 95 (in mph) and the population of interest is all the drivers on Route 95. A researcher might state the following null hypothesis:

$$H_0\text{: } \mu = \mu_{\text{hyp}}, \text{ for example, } H_0\text{: } \mu = 65 \text{ mph.}$$

In words, this null hypothesis corresponds to the assumption that the unknown population mean driving speed, μ, is 65 mph. In this example, the value of 65 mph corresponds to the posted speed limit. In a study that tests hypotheses about the value of one population mean, using the mean from a sample drawn from the population, the "effect" that the researcher is trying to detect is the difference between the unknown actual population mean μ and the hypothesized population mean μ_{hyp}.

Next, the researcher selects a random sample from the population of all passing cars on Route 95, measures the driving speed for this sample, and computes a sample mean, M (e.g., the sample mean might be $M = 82.5$ mph).

Then, the researcher compares the observed sample mean, M ($M = 82.5$ mph), with the hypothesized population mean, μ ($\mu = 65$ mph), and asks the following question: If the true population mean driving speed, μ, is really 65 mph, is the obtained sample mean ($M = 85$ mph) a likely outcome or an unlikely outcome? A precise standard for the range of outcomes that would be viewed *unlikely* is established by choosing an **alpha** level and deciding whether to use a **one-** or **two-tailed test,** as reviewed later in this chapter; in general, outcomes that would be expected to occur less than 5% of the time when H_0 is true are considered unlikely.

The theoretical unlikeliness of a specific value of M (when H_0 is true) is evaluated by calculating how far M is from the hypothesized value of μ_{hyp} in the number of standard errors. The distance of M from a hypothesized value of μ is called a z ratio, and the areas that correspond to values of z are evaluated using the table of the standard normal distribution in these situations where the population standard deviation, σ, is known:

$$z = \frac{M - \mu_{hyp}}{\sigma_M}. \tag{3.2}$$

As discussed in Chapters 1 and 2, for a normal distribution, there is a fixed relationship between a distance from the mean given by a z score and the proportion of the area in the distribution that lies beyond the z score. Recall that in Chapter 2 a similar z ratio was used to evaluate the location of an individual X score relative to a distribution of other individual X scores. In Chapter 2, a z score was calculated to describe the distance of an individual X score from a sample or population mean. In this chapter, the z ratio provides information about the location of an individual sample mean, M, relative to a theoretical distribution of many different values of M across a large number of independent samples. Once we convert an observed sample mean M into a z score, we can use our knowledge of the fixed relationship between z scores and areas under the normal distribution to evaluate whether the obtained value of M was "close to" $\mu = 65$ and, thus, a likely outcome when $\mu = 65$ or whether the obtained value of M was "far from" $\mu = 65$ and, thus, an unlikely outcome when $\mu = 65$.

When a researcher obtains a sample value of M that is very far from the hypothesized value of μ, this translates into a large value of z. A z ratio provides precise information about the distance of an individual sample mean M from μ in the number of standard errors.

The basic idea behind NHST is as follows. The researcher assumes a value for the unknown population mean, μ (in this example, H_0: $\mu = 65$). The researcher evaluates the obtained value of the sample mean, M (in this case, $M = 82.5$), relative to the distribution of values of M that would be expected if μ were really equal to 65. The researcher wants to make a decision whether to reject H_0: $\mu = 65$ as implausible, given the value that was obtained for M. If the sample mean, M, is a value that is likely to occur by chance when H_0 is true, the decision is, Do not reject H_0. If the sample mean, M, is an outcome that is very unlikely to occur by chance when H_0 is true, the researcher may decide, Reject H_0.

Subsequent sections review the details involved in the NHST decision process. How can we predict what outcomes of M are likely and unlikely to occur, given an assumed value of μ for a specific null hypothesis along with information about σ and N? How can we evaluate the likelihood of a specific observed value of M relative to the distribution of values of M that would be expected if H_0 were true? What conventional standards are used to decide when an observed value of M is so unlikely to occur when H_0 is true that it is reasonable to reject H_0?

There is a bit of logical sleight of hand involved in NHST, and statisticians disagree about the validity of this logic (for further discussion of these issues, see Cohen, 1994; Greenwald, Gonzalez, Harris, & Guthrie, 1996). The logic that is generally used in practice is as follows: If a researcher obtains a value of M that would be unlikely to occur if H_0 is true, then the researcher rejects H_0 as implausible.

The logic involved in NHST is confusing and controversial for several reasons. Here are some of the issues that make NHST logic potentially problematic.

1. In everyday thinking, people have a strong preference for stating hypotheses that they believe to be correct and then looking for evidence that is consistent with their stated hypotheses. In NHST, researchers often (but not always) hope to find evidence that is inconsistent with the stated null hypothesis.[1] This in effect creates a double negative: Researchers often state a null hypothesis that they believe to be incorrect and then seek to reject that null hypothesis. Double negatives are confusing; and the search for "disconfirmatory" evidence in NHST is inconsistent with most people's preference for "confirmatory" evidence in everyday life.

2. NHST logic assumes that the researcher has a random sample from a well-defined population of interest and that, therefore, the mean, M, and standard deviation, s, based on the sample data should be good estimates of the corresponding population parameters, μ and σ. However, in many real-life research situations, researchers use convenience samples rather than random samples selected from a well-defined actual population. Thus, in many research situations, it is unclear what population, if any, the researcher is in a position to make inferences about. NHST logic works best in research situations (such as industrial quality control studies) where the researcher is able to draw a random sample from the entire population of interest (e.g., a random sample from all the products that come off the assembly line). NHST is more problematic in situations where the study uses a convenience sample; in these situations, the researcher can at best make tentative inferences about some hypothetical broader population that has characteristics similar to those of the sample (as discussed in Chapter 1).

3. As pointed out by Cohen (1994) and many other critics, NHST does not tell researchers what they really want to know. Researchers really want to know, given the value of M in a batch of sample data, how likely it is that H_0, the null hypothesis, is correct. However, the probability estimate obtained using NHST refers to the

probability of something quite different. A p value is *not* the conditional probability that H_0 is correct, given a sample value of M. It is more accurate to interpret the p value as the theoretical probability of obtaining a value of M farther away from the hypothesized value of μ than the value of M obtained in the study, given that the null hypothesis H_0 is correct. When we use a z ratio to look up a p value, as discussed later in this chapter, the p value is the (theoretical) probability of obtaining an observed value of M as large as or larger than the one in the sample data, given that H_0 is correct and all the assumptions and procedures involved in NHST are correct. If we obtain a value of M that is very far away from the hypothesized value of the population mean, μ (and that, therefore, would be very unlikely to occur if H_0 is correct), we typically decide to reject H_0 as implausible. Theoretically, setting up the decision rules in this manner yields a known risk of committing a **Type I error** (rejecting H_0 when H_0 is correct).

Given these difficulties, and the large number of critiques of NHST that have been published in recent years, why do researchers continue to use NHST? NHST logic provides a way to make yes/no decisions about null hypotheses with a (theoretically) known risk of Type I error; that is, a theoretically known probability of rejecting H_0 when H_0 is correct. Note, however, that this estimated risk of Type I error is only correct when all the assumptions involved in NHST procedures are satisfied and all the rules for conducting significance tests are followed. In practice, because of violations of assumptions and departures from the ideal NHST procedure, the actual risk of Type I error is often higher than the expected risk.

There are real-world situations, such as medical research to assess whether a new drug significantly reduces disease symptoms in a sample compared with an untreated control population, where researchers and practitioners need to make yes/no decisions about future actions: Is the drug effective, and should it be adopted as a medical treatment? For such situations, the objective yes/no decision standards provided by NHST may be useful, provided that the limitations of NHST are understood (Greenwald et al., 1996). On the other hand, when research focuses more on theory development, it may be more appropriate to think about the results of each study as cumulative evidence, in the manner suggested by Rozeboom (1960): "The primary aim of a scientific experiment is not to precipitate decisions, but to make an appropriate adjustment in the degree to which one . . . believes the hypothesis . . . being tested" (p. 420).

3.2 ♦ Type I Versus Type II Error

Two different types of errors can occur when statistical significance tests are used to make binary decisions about the null hypothesis. In other words, a researcher who uses NHST typically reports one of two possible decisions about the status of H_0: either "Reject H_0" or "Do not reject H_0."[2] In actuality, H_0 may be either true or false. When we evaluate whether an NHST decision is correct, there are thus four possible outcomes; these are summarized in Table 3.1. For example, a researcher may decide to reject H_0 when H_0 is actually false; a researcher may decide not to reject H_0 when H_0 is true. On the other hand, there are two different possible types of error. A researcher may reject H_0 when it is actually correct; this is called a Type I error. A researcher may fail to reject H_0 when H_0 is actually false; this is called a **Type II error.**

Table 3.1 ♦ Type I Versus Type II Error

	Actual State of the World	
Researcher Decision	*H_0 Is True*	*H_0 Is False*
Reject H_0	Type I error (α)	Correct decision ($1 - \beta$)
Do not reject H_0	Correct decision ($1 - \alpha$)	Type II error (β)

When the null hypothesis is actually correct, a researcher may obtain a large difference between the sample mean, M, and the actual population mean, μ (because of sampling error), that leads to an (incorrect) decision to reject H_0; this is called a Type I error. When a researcher commits a Type I error, he or she rejects the null hypothesis when in fact the null hypothesis correctly specifies the value of the population mean.

The theoretical risk of committing a Type I error is related to the choice of alpha level; *if* a researcher sets the alpha level that is used to look up values for the reject regions at .05 and *if* all other assumptions of NHST are satisfied, then in theory, the use of NHST leads to a 5% risk of rejecting H_0 when H_0 is actually correct. The theoretical risk of Type I error can be reduced by making the alpha level smaller—for example, setting α at .01 rather than the conventional .05.

On the other hand, when the null hypothesis is incorrect (i.e., the population mean is actually different from the value of μ that is stated in the null hypothesis), the researcher may fail to obtain a statistically significant outcome. Failure to reject H_0 when H_0 is incorrect is called a Type II error. The probability of failing to reject H_0 when H_0 is incorrect—that is, the risk of committing a Type II error—is denoted by $\boldsymbol{\beta}$.

Researchers want the risk of both types of error (α and β) to be reasonably low. The theoretical risk of Type I error, α, is established when a researcher selects an alpha level (often $\alpha = .05$) and uses that alpha level to decide what range of values for the test statistic such as a z or ***t* ratio** will be used to reject H_0. If all the assumptions involved in NHST are satisfied and the rules for significance testing are followed, then in theory the risk of committing a Type I error corresponds to the alpha level chosen by the researcher. However, if the assumptions involved in NHST procedures are violated or the rules for conducting hypothesis tests are not followed, then the true risk of Type I error may be higher than the nominal alpha level, that is, the alpha level named or selected by the investigator as the desired standard for risk of Type I error.

The risk of committing a Type II error depends on several factors, including sample size. Factors that influence the theoretical risk of committing a Type II error are discussed in Section 3.9 on statistical power. Statistical power is the probability of correctly rejecting H_0 when H_0 is false, denoted as $(1 - \beta)$. Therefore, the statistical power $(1 - \beta)$ is the complement of the risk of Type II error (β); factors that increase statistical power also decrease the risk of Type II error.

3.3 ♦ Formal NHST Procedures: The *z* Test for a Null Hypothesis About One Population Mean

Let's consider an application of NHST to a research question about the value of one unknown population mean, μ, in a situation where the population standard deviation, σ, is known. The first step is to identify a variable and population of interest. Suppose that a researcher wants to test a hypothesis about the mean intelligence score for a population of all the residents in a large nursing home. The variable of interest is a score on a widely used intelligence quotient (IQ) test, the Wechsler Adult Intelligence Scale (WAIS). These scores are normally distributed and are scaled to have a mean of $\mu = 100$ points and a population standard deviation of $\sigma = 15$ points for the general adult population.

3.3.1 ♦ Obtaining a Random Sample From the Population of Interest

Suppose that the entire nursing home population consists of hundreds of residents and the director can afford to test only $N = 36$ people. Therefore, the researcher plans to obtain a random sample of $N = 36$ persons from the population of the entire nursing home and to compute the mean IQ score, M, for this sample. The mean IQ score for the sample, M, will be used to decide whether it is plausible to believe that the mean IQ for the entire nursing home population is equal to 100, the mean value of the IQ for the general adult population.

3.3.2 ♦ Formulating a Null Hypothesis (H_0) for the One-Sample *z* Test

Next the researcher sets up a null hypothesis that specifies a "guessed" value for an unknown population mean. One possible null hypothesis for the nursing home study is based on the general adult population mean IQ of 100 points. The researcher could state the following null hypothesis:

$$H_0: \mu = \mu_{\text{hyp}}. \tag{3.3}$$

In this example, H_0: $\mu = 100$, where μ is the unknown population mean for the variable of interest (IQ) and the population of interest (all the nursing home residents). In words, then, the researcher hypothesizes that the IQ for the nursing home population corresponds to an average ability level compared with the intelligence for the general population.

Note that there are several likely sources of values for μ_{hyp}, the hypothesized population mean. When the variable of interest is a standardized psychological test, the value of μ_{hyp} might be selected based on knowledge of the normative values of scores. A researcher might select the overall population mean as the point of reference for hypothesis tests (as in this example), or the researcher might use a **cutoff value** that corresponds to a clinical diagnosis as the point of reference for hypothesis tests. For example, on the Beck Depression Inventory (Beck, Ward, Mendelson, Mock, & Erbaugh, 1961), a score of 19 or more corresponds to moderate to severe depression. A researcher might set up the null hypothesis that the mean depression score for a population of nursing home patients equals or exceeds this cutoff value of 19 points. Similarly, for physiological measures, such as systolic blood pressure, a clinical cutoff value (such as 130 mmHg, which is sometimes used to diagnose borderline hypertension) might be used as the value of μ_{hyp}. A legal standard (such as the posted speed limit on a highway, as in the earlier example) is another possible source for a specific value of μ_{hyp}. In industrial quality control applications, the hypothesized population mean, μ_{hyp}, often corresponds to some production goal or standard. For example, Warner and

Rutledge (1999) tested the claim of a cookie manufacturer that the population mean number of chocolate chips in each bag of cookies was equal to or greater than 1,000.

Most applications of NHST presented subsequently in this book involve testing null hypotheses that correspond to an assumption that there is no difference between a pair of means for populations that have received different treatments, or no relationship between a pair of variables,[3] rather than hypotheses about the value of the population mean for scores on just one variable.

3.3.3 ♦ Formulating an Alternative Hypothesis (H_1)

The **alternative hypothesis** (sometimes called the **research hypothesis**), H_1, can take one of three possible forms. The first form is called a nondirectional or two-tailed alternative hypothesis:

$$H_1: \mu \neq 100. \tag{3.4}$$

Using this alternative hypothesis, the researcher will reject H_0 for values of M that are either much higher or much lower than 100. This is called a **nondirectional** or two-tailed **test;** when we figure out a decision rule (the ranges of values of M and z for which we will reject H_0), the reject region includes outcomes in both the upper and lower tails of the sampling distribution of M as shown in Figure 3.1. For $\alpha = .05$, two-tailed, we would reject H_0 for obtained values of $z < -1.96$ and $z > +1.96$.

The second and third possible forms of H_1, the alternative hypothesis, are called one-tailed or directional alternative hypotheses. They differ in the direction of the inequality that is stated—that is, whether the true population mean is predicted to be lower than or higher than the specific value of μ stated in the null hypothesis. The second form of H_1 is a one-tailed test that corresponds to a reject region in the lower tail of the distribution of outcomes for M.

$$H_1: \mu < 100. \tag{3.5}$$

If we use the alternative hypothesis stated in Equation 3.5, we will reject H_0 only for values of M that are much lower than 100. This is called a **directional** or one-tailed **test.** The decision to reject H_0 will be made only for values of M that are in the lower tail of the sampling distribution, as illustrated in Figure 3.2.

The third and last version of the alternative hypothesis is stated in Equation 3.6.

$$H_1: \mu > 100. \tag{3.6}$$

If we use the alternative hypothesis specified in Equation 3.6, we will reject H_0 only for values of M that are much higher than 100. This is also called a directional or one-tailed test; the decision to reject H_0 will be made only for values of M that are in the upper tail of the sampling distribution of values of M, as shown in Figure 3.3.

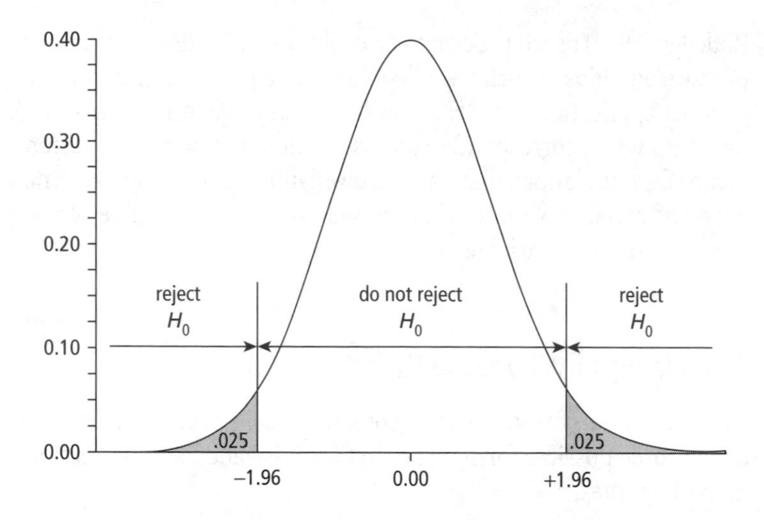

Figure 3.1 ♦ Standard Normal Distribution Curve Showing the Reject Regions for a Significance Test Using $\alpha = .05$, Two-Tailed

NOTES: Reject regions for a z ratio for $\alpha = .05$, two-tailed: Reject H_0 for $z < -1.96$ and for $z > +1.96$; do not reject H_0 for $z \geq -1.96$ and $z \leq +1.96$.

For a normal distribution, the proportion of the area that lies in the tail below $z = -1.96$ is .025, and the proportion of the area that lies in the tail above $z = +1.96$ is .025. Therefore, the part of the distribution that is "far" from the center of the distribution, using $\alpha = .05$ as the criterion for far, corresponds to the range of z values less than −1.96 and greater than +1.96.

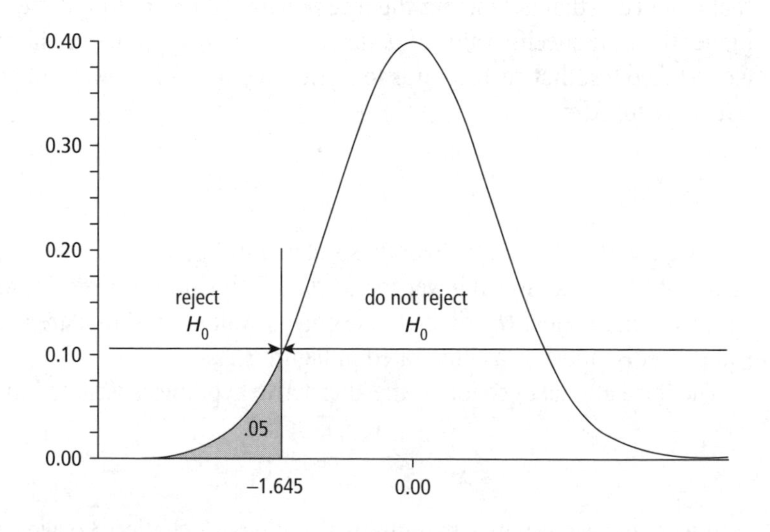

Figure 3.2 ♦ Standard Normal Distribution Curve With Reject Region for Significance Test With $\alpha = .05$, One-Tailed, H_1: $\mu < \mu_{hyp}$

NOTES: Reject region for a one-tailed test, $\alpha = .05$, shown as the shaded area in the lower tail of the normal distribution. This reject region is for the directional alternative hypothesis H_1: $\mu < 100$. Reject H_0 if obtained $z < -1.645$. Do not reject H_0 if obtained z value is ≥ -1.645.

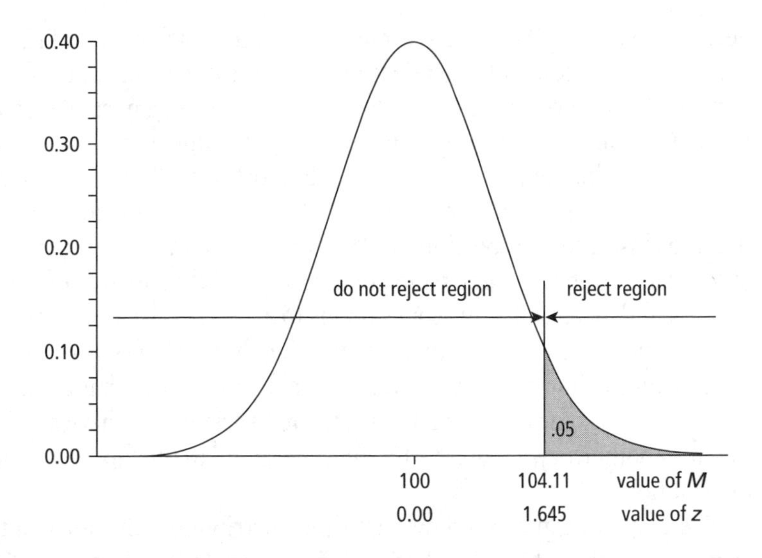

Figure 3.3 ♦ Normal Distribution Curve With Reject Region for Significance Test With $\alpha = .05$, One-Tailed, H_1: $\mu > \mu_{hyp}$

NOTE: X axis shows values of M and corresponding values of z. The reject regions for the one-tailed test, $\alpha = .05$, correspond to the shaded area in the upper tail of the normal distribution. For the directional test (with H_1: $\mu > 100$), we would reject H_0 for values of $M > 104.11$ that correspond to values of $z > +1.645$.

In the following example, we will test the null hypothesis H_0: $\mu = 100$ against the alternative hypothesis H_1: $\mu > 100$. The choice of H_1: $\mu > 100$ as the alternative hypothesis means that we will reject H_0 only if the sample mean, M, is substantially greater than 100 IQ points; that is, we will use the reject region shown in Figure 3.3.

3.3.4 ♦ Choosing a Nominal Alpha Level

Next the researcher must choose a nominal alpha level or **level of significance.** The nominal alpha level is a theoretical risk of committing a Type I error, that is, the probability of rejecting the null hypothesis H_0: $\mu = 100$ when the null hypothesis H_0: $\mu = 100$ is actually correct. We call this a *nominal* alpha level because it is named or *nominated* as a standard for making judgments about statistical significance by the researcher. A nominal alpha level is chosen arbitrarily by the researcher. Following the example set by Sir Ronald Fisher, most users of statistics assume that an α value of .05 represents an acceptably small risk of Type I error in most situations. However, in exploratory research, investigators are sometimes willing to use alpha levels (such as $\alpha = .10$) that correspond to a higher risk of Type I error. Sometimes, investigators prefer to use smaller alpha levels; $\alpha = .01$ and $\alpha = .001$ are common choices when researchers want to keep the theoretical risk of Type I error very small.

3.3.5 ♦ Determining the Range of z Scores Used to Reject H_0

Next we need to use the alpha level and the choice of a nondirectional or directional alternative hypothesis to formulate a decision rule: For what range of values of z will we decide to reject H_0? When we set $\alpha = .05$ as the acceptable level of risk of Type I error and use a one-tailed test with the reject region in the upper tail, our *reject region* is the range of *z* scores that corresponds to the top 5% of the area in a normal distribution. The *z* value that corresponds to a .05 proportion of area in the upper tail of a normal distribution can be found in the table in Appendix A, "Proportions of Area Under a Standard Normal Curve." The column that corresponds to "Area C" is examined to locate the table entries that are closest to a proportion of .05 of the area in the upper tail; from the second page of this table in the far right-hand column, we find that an area of .0505 corresponds to $z = +1.64$ and an area of .0495 corresponds to $z = +1.65$. An area of exactly .05 corresponds to a *z* value of +1.645.[4] In other words, when a variable (such as the value of *M* across many samples drawn randomly from the same population) has a normal distribution, 5% of the outcomes for the normally distributed variable will have *z* values $> +1.645$; or to say this another way, *z* scores $> +1.645$ correspond to the top 5% of outcomes in a normal distribution.

The graph in Figure 3.3 shows a normal distribution; the tail area that lies beyond the *critical value* of $z = +1.645$ is shaded. A decision rule based on $\alpha = .05$ and a one-tailed or directional version of the alternative hypothesis H_1 that should theoretically give us a 5% risk of Type I error is as follows:

Reject H_0 for obtained values of $z > +1.645$.

Do not reject H_0 for obtained values of $z \leq +1.645$.

Recall that the *z* value shown in Equation 3.2 provides information about the direction and magnitude of the difference between an obtained sample mean, *M*, and the hypothesized population mean, μ, stated in the null hypothesis. A large value of *z* corresponds to a value of *M* that is "far away" from the hypothesized value of μ.

3.3.6 ♦ Determining the Range of Values of *M* Used to Reject H_0

It is helpful to understand that the "reject" regions for the outcome of the study can also be stated in terms of values obtained for *M*, the sample mean. For a critical value of *z* (such as +1.645) that is obtained from a table of the standard normal distribution, we can figure out the corresponding value of *M* that would be used to make a decision whether to reject H_0 if we know the hypothesized value of μ and the calculated value of σ_M.

Recall that from Equation 3.2, $z = (M - \mu_{\text{hyp}})/\sigma_M$. In this example, the hypothesized population mean $\mu_{\text{hyp}} = 100$ and $\sigma_M = s/\sqrt{N} = 15/\sqrt{36} = 2.50$. We can translate the reject regions (given above in ranges of values for the *z* ratio) into reject regions given in terms of values of *M*, the sample mean. In this problem, we obtain a *z* score to evaluate the location of any specific sample mean, *M*, relative to μ_{hyp} by computing the corresponding *z* value: $(M - \mu_{\text{hyp}})/\sigma_M = (M - 100)/2.5$. We can rearrange this equation to calculate the value of *M* (M_{critical}) that corresponds to a specific critical value of *z*, z_{critical}.

If

$$z_{\text{critical}} = (M_{\text{critical}} - \mu_{\text{hyp}})/\sigma_M,$$

then

$$z_{\text{critical}} \times \sigma_M = (M_{\text{critical}} - \mu_{\text{hyp}}).$$

Rearranging the expression above to isolate the value of M on one side of the equation yields the following equation for M_{critical}, the boundary of the reject region in terms of outcome values for M, the sample mean:

$$M_{\text{critical}} = [z_{\text{critical}} \times \sigma_M] + \mu_{\text{hyp}}. \tag{3.7}$$

Equation 3.7 tells us the value of the sample mean, M (M_{critical}), that corresponds to a specific critical value of z (such as $z = +1.645$). Using the specific numerical values of μ_{hyp} and σ_M in this situation, and the critical value $z = +1.645$, we can calculate the boundary for the reject region in terms of values of the sample mean, M:

$$M_{\text{critical}} = [+1.645 \times 2.5] + 100 = 104.11.$$

Figure 3.3 shows the value of M that corresponds to the critical value of z for a directional test with $\alpha = .05$. A critical value of M was calculated given specific numerical values of μ and σ_M, and the critical value of M appears on the X axis of the normal distribution. Given $\sigma = 15$, $N = 36$, and $\sigma_M = 2.50$, and H_0: $\mu = \mu_{\text{hyp}} = 100$, we can state the decision rule in terms of obtained values of z (as at the end of the previous section):

Reject H_0 for any sample mean that corresponds to a value of $z > +1.645$.

But we can also state our decision rule directly in terms of values of the sample mean. In this case, we would reject H_0 for obtained sample values of $M > +104.11$.

It may be helpful to stop and think about the reasoning behind this decision rule. Knowledge about sampling error that was reviewed in Chapter 2 makes it possible for us to predict the amount of variation in the magnitude of M when samples of N cases are randomly drawn from a population with known values of μ and σ. In other words, given specific numerical values of N and σ and an assumed value of μ_{hyp} stated in a null hypothesis, we can predict the distribution of values of M that are expected to occur if H_0 is true. If all our assumptions are correct, if $N = 36$ and $\sigma = 15$, and if $\mu = 100$, then the values of M should be normally distributed around a mean of 100 with a standard deviation or standard error of σ_M ($\sigma_M = 15/\sqrt{36} = 2.5$). This corresponds to the distribution that appears in Figure 3.3. If μ is really 100, then most sample values of M are expected to be "fairly close to" this population mean of 100. We use an alpha level (such as $\alpha = .05$) and an alternative hypothesis (such as H_1: $\mu > 100$) to decide what set of outcomes for the sample mean, M, would be less consistent with the null hypothesis than with the alternative hypothesis. In this example, using the alternative hypothesis H_1: $\mu > 100$, we will reject H_0 and prefer H_1 only for values of M that are substantially greater than 100. Because we have set the nominal alpha level at $\alpha = .05$ and this alternative hypothesis corresponds to a one-tailed test with a reject region that corresponds to values of M that are greater than 100, we identify the top 5% of possible values of M as the set of outcomes that are least likely to occur when H_0 is correct. Because the top 5% of the area in a normal distribution corresponds to a z score location of +1.645 standard deviations or standard errors above the mean, we decide to use only those values of M that correspond to distances from the mean greater than $z = +1.645$ as evidence to reject H_0. We can convert this z score distance into a corresponding value of the sample mean, M, using Equation 3.7. If H_0 is true, then 95% of the time we would expect to observe an outcome value for M that is less than or equal to $M = 104.11$. If H_0 is true, we would expect to observe values of $M > 104.11$ only 5% of the time.

The basic idea behind NHST can be summarized as follows. The researcher chooses a nominal alpha level and formulates a null and an alternative hypothesis. The alpha level and hypotheses are used to decide whether to use a one- or two-tailed reject region for the decision about H_0 and how much area to include in one or both tails. For example, when we use a one-tailed test with H_1: $\mu > 100$ and $\alpha = .05$, we reject H_0 for the range of values of M that corresponds to the top 5% of values for M that we would expect to see across many samples if H_0 were true.

Note that the probability that we can assess when we conduct a significance test is how likely the obtained value of M is, given the assumption that H_0 is correct. If we obtain a value of M in our sample that is very unlikely to arise just by chance due to sampling error when H_0 is correct, it is reasonable to make the decision "Reject H_0." If we obtain a value of M in our sample that is very likely to arise just by chance due to sampling error when H_0 is correct, it is more reasonable to make the decision "Do not reject H_0." The specific criterion for "very unlikely" is determined by the nominal alpha level (usually, $\alpha = .05$) and the choice of an alternative hypothesis (H_1) that corresponds to a reject region that includes one tail or both tails of the distribution.

Suppose the researcher collected IQ scores for a sample of nursing home patients and obtained a sample mean $M = 114.5$ points. In the situation described in this section, with $N = 36$, $\sigma = 15$, H_0: $\mu = 100$, and H_1: $\mu > 100$, this value of M would lead to a decision to reject H_0. If the population mean were really equal to 100 points, the probability of obtaining a sample mean greater than 104.11 would be, in theory, less than 5%. Thus, based on our understanding of sampling error, a sample mean of $M = 114.5$ falls within a range of values of M that are theoretically very unlikely outcomes if H_0 is correct.

3.3.7 ♦ Reporting an "Exact" *p* Value

Most introductory textbooks present NHST as a yes/no decision rule about H_0. If the obtained value of z exceeds the critical values of z (that correspond to the chosen alpha level and directional or nondirectional alternative hypothesis) or if the obtained value of M exceeds the critical value of M that corresponds to this critical value of z, the researcher decision is, "Reject H_0." If the obtained value of z does not exceed the critical value(s) of z that correspond to the criteria (such as $\alpha = .05$, one-tailed), the researcher decision is, "Do not reject H_0." However, there is another way of reporting the outcome for a significance test; an **"exact" *p* value** can be reported, in addition to or instead of a decision whether to reject H_0.

What is an exact p value? The exact p value is the (theoretical) probability of obtaining a sample mean, M, farther away from the hypothesized value of μ specified in the null hypothesis than the value of M in the sample in the study, *if H_0 is actually correct.* For a sample mean, the exact p value for a one-tailed test corresponds to the proportion of outcomes for the sample mean that would be theoretically expected to be greater than the obtained specific value for the sample mean if H_0 were true. We can determine an exact p

value that corresponds to a sample value of M, given that we have the following information: the value of μ specified in the null hypothesis, information whether the test is one- or two-tailed, and the value of σ_M.

To determine an exact one-tailed p value for an obtained sample mean, M, of 106 (we continue to use H_0: $\mu = 100$, H_1: $\mu > 100$, and $\sigma_M = 2.5$), we first compute the corresponding value of z that tells us how far the sample mean, M, is from the hypothesized value of μ (in the number of standard errors):

$$z = (M - \mu)/\sigma_M = (106 - 100)/2.5 = 2.40.$$

The use of a directional or one-tailed alternative hypothesis (H_1: $\mu > 100$) means that we need to look at only one tail of the distribution, in this case the area in the upper tail, the area that lies above a z score of +2.40. From the table of areas that correspond to z score distances from the mean in Appendix A, we find that the tail area to the right of $z = 2.40$ is .0082, or .8% of the area. This value, .0082, is therefore the exact one-tailed p value that corresponds to a z value of +2.40; that is, exactly .0082 of the area in a normal distribution lies above $z = +2.40$. This can be interpreted in the following manner. If H_0 is correct, the (theoretical) likelihood of obtaining a sample mean larger than the one in this sample, $M = 106$, is .0082. Because this is a very unlikely outcome when H_0 is correct, it seems reasonable to doubt that H_0 is correct.

When a program such as SPSS is used to compute a significance test, an exact p value is generally reported as part of the results (often this is denoted by ***sig***, which is an abbreviation for "statistical significance level"). Where there is a possibility of using a one-tailed or two-tailed test, it is usually possible to select either of these by making a check box selection in one of the menus.

In practice, users of SPSS and other statistical programs who want to make a binary decision about a null hypothesis (i.e., either "Reject H_0" or "Do not reject H_0") make this decision by comparing the obtained p value that appears on the SPSS printout with a preselected alpha level. If the p value reported on the SPSS printout is less than the preselected nominal alpha level (usually $\alpha = .05$), the decision is "Reject H_0." If the p value reported on the SPSS printout is greater than the preselected alpha level, the decision is "Do not reject H_0."

Recent changes in publication guidelines in some academic disciplines such as psychology call for the reporting of exact p values. The fifth edition of the *Publication Manual of the American Psychological Association* (APA, 2001) summarizes recommendations as follows:

> Two types of probabilities are generally associated with the reporting of significance levels in inferential statistics. One refers to the a priori probability you have selected as an acceptable risk of falsely rejecting a given null hypothesis. This probability, called the "alpha level" (or "significance level") is the probability of a Type I error in hypothesis testing and is commonly set at .05 or .01. The other kind of probability, the [exact] p value (or significance probability) refers to the a posteriori likelihood of obtaining a result that is as extreme as or more extreme than the observed value you obtained, assuming that the null hypothesis is true. . . . Because most statistical packages now report the [exact] p value (given the null and alternative hypotheses provided) and

because this probability can be interpreted according to either mode of thinking, in general it is the exact probability (p value) that should be reported. (pp. 24–25)

When it is inconvenient to report exact p values—for example, in large tables of correlations—it is common practice to highlight the subset of statistical significance tests in the table that have p values below conventional prespecified alpha levels. It is fairly common practice to use one asterisk (*) to indicate $p < .05$, two asterisks (**) for $p < .01$, and three asterisks (***) for $p < .001$. When an exact p value is reported, the researcher should state what alpha level was selected. To report the numerical results presented earlier in this section, a researcher could write either of the following.

Results Version 1 (statement whether the obtained exact p value was less than an a priori alpha level): The null hypothesis that the mean IQ for the entire population of nursing home residents was equal to the general population mean for all adults (H_0: $\mu = 100$) was tested using a directional alternative hypothesis (H_1: $\mu > 100$) and $\alpha = .05$. The obtained sample mean IQ was $M = 106$ for a random sample of $N = 36$ residents. The z value for this sample mean was statistically significant; $z = +2.40$, $p < .05$, one-tailed. Thus, given this sample mean, we reject the null hypothesis that the mean IQ for the entire population of nursing home residents is equal to 100.

Results 2 (report of exact p value): For all significance tests in this research report, the significance level that was used was $\alpha = .05$, one-tailed. The sample mean IQ was $M = 106$ for a random sample of $N = 36$ residents. The z value for this sample mean was statistically significant, $z = +2.40$, $p = .008$, one-tailed.

Most statistical programs report p values to three decimal places. Sometimes an exact obtained p value has zeros in the first three decimal places and appears as .000 on the printout. Note that p represents a theoretical probability of incorrectly rejecting H_0 and this theoretical risk is never 0, although it becomes smaller as the value of z increases. When SPSS shows a significance value of .000 on the printout, this should be reported as $p < .001$ rather than $p = .000$.

3.4 ♦ Common Research Practices Inconsistent With Assumptions and Rules for NHST

NHST assumes that we have a random sample of N independent observations from the population of interest and that the scores on the X variable are quantitative and at least approximately interval/ratio level of measurement. It is desirable but not essential that the scores on the X variable be approximately normally distributed. If scores on X are at least approximately normally distributed, then the distribution of values of M approaches a normal distribution shape even for rather small values of N and rather small numbers of samples; however, even when scores on the original X variable are nonnormally distributed, the distribution of values of M is approximately normal, provided that the value of N in each sample is reasonably large and a large number of samples are obtained.

The rules for NHST involve the following steps. First select a nominal alpha level and state a null and an alternative hypothesis. Next, take a random sample from the population of interest and compute the sample mean, *M*; evaluate how far this sample mean, *M*, is from the hypothesized value of the population μ_{hyp} by calculating a *z* ratio. Then use a decision rule based on the alpha level and the nature of the alternative hypothesis (i.e., a one-tailed or two-tailed reject region) to evaluate the obtained *z* value to decide whether to reject H_0; or report the exact one- or two-tailed *p* value that corresponds to the obtained value of *z* and state the alpha level that should be used to decide whether this obtained *p* value is small enough to judge the outcome "statistically significant." The risk of Type I error should, theoretically, correspond to the nominal alpha level if only one significance test is conducted and the decision rules are formulated before looking at the numerical results in the sample.

In actual practice, researchers often depart from these ideal procedures in a number of ways. When researchers depart from the rules for NHST, the actual risk of Type I error is often much higher than the nominal alpha level that is chosen during the NHST process and used to look up critical values.

The goal of NHST is to be able to make a reject/do not reject decision about a null hypothesis, with a known risk of committing a Type I error. The risk of Type I error should correspond to the nominal alpha level when the assumptions for NHST are satisfied and the rules are followed. Ideally, the researcher should be able to identify the population of interest, should be able to identify the members of that population, and should be able to obtain a truly random sample from the population of interest. The researcher should state the null and alternative hypotheses and select an alpha level and formulate a decision rule (the ranges of values of *z* for which H_0 will be rejected) *before* the researcher looks at the value of the sample mean, *M*, and a corresponding test statistic such as *z*. The researcher should conduct only one statistical significance test or only a limited number of statistical significance tests.

In practice, researchers often violate one or more of these assumptions and rules. When assumptions are violated and these rules are not followed, the actual risk of Type I error is usually much higher than the nominal $\alpha = .05$ level that the researcher sets up as the standard.

3.4.1 ♦ Use of Convenience Samples

One common violation of the rules for NHST involves sampling. Ideally, the sample should be selected randomly from a well-defined population. In laboratory research, investigators often work with convenience samples (as discussed in Chapter 1). The consequence of using convenience samples is that the convenience sample may not be representative of any specific real-world population. A convenience sample may not provide adequate information to test hypotheses about population means for any well-defined real-world population.

3.4.2 ♦ Modification of Decision Rules After the Initial Decision

Second, researchers occasionally change their decisions about the alpha level and/or whether to perform a two-tailed/nondirectional test, versus a one-tailed/directional test, after examining the obtained values of *M* and *z*. For example, a researcher who obtains a sample mean that corresponds to a *z* value of +1.88 would find that this *z* value does not lie in the range of values for the decision to reject H_0 if the criterion for significance that is used is $\alpha = .05$, two-tailed. However, if the researcher subsequently changes the alpha level to $\alpha = .10$, two-tailed, or to $\alpha = .05$, one-tailed (after examining the values of M and z), this would redefine the decision rules in a way that would make the outcome of the study "significant." This is not a legitimate practice; just about any statistical outcome can be judged significant after the fact if the researcher is willing to increase alpha value in order to make that judgment. Another way of redefining the decision rules after the fact would be to change H_0; for example, if the researcher cannot reject H_0: $\mu = 100$, and wants to reject H_0, the researcher might change the null hypothesis to H_0: $\mu = 90$. It should be clear that reverse engineering the decision rules to reject H_0 based on the obtained value of M for a sample of data is not a legitimate practice.

3.4.3 ♦ Conducting Large Numbers of Significance Tests

Another common deviation from the rules for NHST occurs in exploratory research. In some exploratory studies, researchers run a large number of tests; for example, a researcher might want to evaluate whether each of 200 Pearson correlations differs significantly from hypothesized correlations of 0. One way of thinking about the nominal alpha level is as a prediction about the number of Type I errors that can be anticipated when large numbers of significance tests are performed. If we set up a decision rule that leads to a Type I error 5% of the time when H_0 is actually correct, it follows that—even if we generated all the scores in our study using a random number generator, and in the population, all the variables in our study had correlations of 0 with each other—we would expect that if we ran 100 correlations and tested each one for statistical significance, we would find approximately 5 out of these 100 correlations statistically significant. In other words, we would expect to find about 5 significant correlations that are statistically significant even if we generate the data using a process for which the real population correlations are 0 among all pairs of variables. When a researcher reports a large number of significance tests, the likelihood that at least *one* of the outcomes is an instance of a Type I error increases as the number of tests increases.

The increased risk of Type I error that arises when a large number of statistical significance tests are performed is called an **inflated risk of Type I error**. When a researcher reports significance tests for 100 correlations using $\alpha = .05$ as the nominal criterion for the significance of each individual correlation, the true risk of committing at least one Type I error in the set of 100 statistical significance tests is typically substantially greater than the nominal alpha level of .05.

3.4.4 ♦ Impact of Violations of Assumptions on Risk of Type I Error

It is extremely important for researchers to understand that an exact p value, or a statement that the obtained p value is smaller than some preselected alpha value such as $\alpha = .05$, involves making a theoretical estimate of the risk of committing a Type I error that is based on a large number of assumptions and conditions. An estimated risk of Type I error can be expected to be accurate only when a large number of assumptions about procedure are met. *If* the researcher begins with a well-defined actual population; and *if* the scores on the variable of interest are quantitative and reasonably close to interval/ratio level of measurement; and *if* the researcher sets up H_0, H_1, and α prior to examining the data; and *if* the researcher obtains a random sample of independent observations from the population of interest; and *if* the researcher performs only one statistical significance test; *then* the preselected alpha level theoretically corresponds to the risk of Type I error. When one or more of these assumptions are not satisfied, then the actual risk of Type I error may be considerably higher than the nominal alpha level. Unfortunately, in many real-life research situations, one or more of these conditions are frequently not satisfied.

3.5 ♦ Strategies to Limit Risk of Type I Error

3.5.1 ♦ Use of Random and Representative Samples

For all types of research, both experimental and nonexperimental, samples should be representative of the populations about which the researcher wants to make inferences. When the population of interest can be clearly defined, all members can be identified, and a representative sample can be obtained by using random selection methods (possibly combined with systematic sampling procedures such as stratification), then the application of NHST logic to sample means provides a reasonable way of making inferences about corresponding population means. However, in some research domains, the use of convenience samples is common. As discussed in Chapter 1, some convenience samples may not correspond to any well-defined real-life population. This lack of correspondence between the sample in the study and any well-defined real-world population can make the use of NHST procedures invalid.

3.5.2 ♦ Adherence to the Rules for NHST

It is dishonest to change the decision standards (the alpha level and the choice of a directional vs. nondirectional research hypothesis) after examining the outcome of a study. In the extreme, if a researcher is willing to raise the alpha level high enough, just about any outcome can be judged statistically significant.

3.5.3 ♦ Limit the Number of Significance Tests

A simple way of limiting the risk of Type I error is to conduct only one significance test or a limited number of significance tests. It is generally easier to limit the number of tests in experimental studies that involve manipulating just one or two variables and measuring just one or two outcomes; the number of possible analyses is limited by the small number of variables.

Researchers often find it more difficult to limit the number of significance tests in nonexperimental, exploratory studies that include measures of large numbers of variables. It is relatively common to report dozens, or even hundreds, of correlations in large-scale exploratory studies. One possible way for researchers to reduce the risk of Type I error in exploratory studies is to decide ahead of time on a limited number of analyses (instead of running a correlation of every variable with every other variable). Another strategy is to make specific predictions about the pattern of outcomes that is expected (e.g., what pairs of variables are expected to have correlations that are large positive, large negative, or close to 0?) and then to assess how closely the set of obtained correlations matches the predicted pattern (Westen & Rosenthal, 2003).

When researchers set out to do exploratory nonexperimental research, however, they often do not have enough theoretical or empirical background to make such detailed predictions about the pattern of outcomes. The next few sections describe other ways of trying to limit the risk of Type I error in studies that include large numbers of analyses.

3.5.4 ♦ Bonferroni-Corrected Per-Comparison Alpha Levels

A commonly used method in limiting the risk of Type I error when multiple significance tests are performed in either experimental or exploratory studies is the Bonferroni correction. Suppose that the researcher wants to conduct $k = 3$ different significance tests—for example, significance tests about the means on three variables such as intelligence, blood pressure, and depression for the nursing home population. The researcher wants the overall "experiment-wise" risk of Type I error for this entire set of $k = 3$ tests to be limited to .05. The overall or **experiment-wise α** is denoted by $\mathbf{EW_\alpha} = .05$. (The term *experiment-wise* is commonly used to describe an entire set of significance tests even when the study is not, strictly speaking, an experiment.) To limit the size of the experiment-wise Type I error risk, the researcher may decide to use a more conservative "corrected" alpha level for each test. The Bonferroni correction is quite simple. The per-comparison alpha level (PC_α) is given as follows:

$$PC_\alpha = EW_\alpha / k, \tag{3.8}$$

where

EW_α is the experiment-wise α, often set at $EW_\alpha = .05$,

and

k is the number of significance tests performed in the entire experiment or study.

For example, if the researcher wanted to set up three different z tests (one each to test hypotheses about the population means for intelligence, blood pressure, and depression for the nursing home population) and the researcher wanted to keep the experiment-wise overall risk of Type I error limited to .05, the researcher would use a **per-comparison alpha** level of $PC_\alpha = EW_\alpha / k = .05/3 = .017$ as the criterion for statistical significance for each of the individual z tests.

The advantage of the Bonferroni correction procedure is its simplicity; it can be used in a wide range of different situations. The disadvantage is that when the number of tests (k) becomes very large, the per-comparison alpha levels become so small that very few outcomes can be judged statistically significant. Relative to many other methods of trying to control the risk of Type I error, the **Bonferroni procedure** is very conservative.

3.5.5 ♦ Replication of Outcome in New Samples

A crucial consideration in both experimental and nonexperimental research is whether a statistically significant finding from a single study can be replicated in later studies. Even in a carefully controlled experiment that includes only one statistical significance test, a decision to reject H_0 may be an instance of Type I error. Because of sampling error, a single study cannot provide conclusive evidence for or against the null hypothesis; in addition, any single study may have methodological flaws. When a statistically significant difference is found between a sample mean and a population mean using the one-sample *z* test, it is important to replicate this finding across new samples. If the statistically significant finding is actually an instance of Type I error, it is not likely to occur repeatedly across new samples. If successive samples or successive studies replicate the difference, then each replication should gradually increase the researcher's confidence that the outcome is not attributable to Type I error.

3.5.6 ♦ Cross-Validation

Cross-validation is a method that is related to replication. In a cross-validation study, the researcher typically begins with a large sample. The cases in the sample are randomly divided into two separate datasets (there is a data sampling procedure in SPSS that can be used to obtain a random sample, either a specific number of cases or a percentage of cases from the entire dataset). The first dataset may be subjected to extensive exploratory analyses; for example, many one-sample *z* tests may be run on different variables, or many correlations may be run. After running many exploratory analyses, the researcher chooses a limited number of analyses that are theoretically interesting. The researcher then runs that small number of analyses on the second half of the data, to assess whether significant *z* tests or correlations can be replicated in this "new" set of data. A cross-validation study is a useful way of trying to reduce the risk of Type I error. If a pair of variables *X* and *Y* are significantly correlated with each other when the researcher runs a hundred correlations on the first batch of data but are not significantly correlated in the second batch of data, it is reasonable to conclude that the first significant correlation may have been an instance of Type I error. If the *X*, *Y* correlation remains significant when this analysis is performed on a second, new set of data, it is less probable that this correlation is a Type I error.

To summarize, in experiments, the most commonly used method of controlling Type I error is to limit the number of significance tests that are conducted. If a large number of follow-up tests such as comparisons of many pairs of group means are performed, researchers may use **"protected" tests** such as the Bonferroni-corrected PC_α levels to control the risk of Type I error. In nonexperimental or exploratory studies that include measurements of large numbers of variables, researchers often do not limit the number of significance tests. However, they may try to limit the risk of Type I error by using Bonferroni-corrected alpha levels or by running cross-validation analyses. For both experimental and nonexperimental research, the replication of statistically significant outcomes across new samples and new studies is extremely important; the results of a single study should not be viewed as conclusive. The results of any one study could be due to Type I error or methodological flaws; numerous successful replications of a finding gradually increase researcher confidence that the finding is not just an instance of Type I error.

3.6 ♦ Interpretation of Results

3.6.1 ♦ Interpretation of Null Results

Most authorities agree that when a study yields a nonsignificant result, it is not correct to conclude that we should "Accept H_0." There are many possible explanations for a nonsignificant outcome. It is inappropriate to interpret the outcome of an individual study as evidence that H_0 is correct unless these other explanations can be ruled out; and in practice, it is very difficult to rule out many of these possible alternative explanations for a nonsignificant outcome. A significance test may yield nonsignificant results when the null hypothesis is correct for numerous reasons. We can only make a case for the possible inference that the null hypothesis might be correct if we can rule out all the following alternative explanations for a nonsignificant outcome; and in practice, it is not possible to rule out all these alternative explanations completely.

1. The effect size that the researcher is trying to detect (e.g., the magnitude of the difference between μ and μ_{hyp}) is very small.
2. The number of cases in the study (N) may be too small to provide adequate statistical power for the significance test. Sample sizes that are too small to have sufficient statistical power are fairly common (Maxwell, 2004).
3. The measure of the outcome variable may be unreliable or not valid or sensitive to the effects of an intervention.
4. In experiments that involve comparisons of outcomes for groups that receive different dosage levels or types of treatment, the manipulation of the independent variable may be weak, not implemented consistently, or not a valid manipulation of the theoretical construct of interest.
5. The relationship between variables is of a type that the analysis cannot detect (e.g., a Pearson r is not appropriate for detecting curvilinear relationships between variables).
6. A nonsignificant result can arise due to sampling error.

For example, suppose that a developmental psychologist would like to show that the cognitive outcomes for children who are cared for at home (Group 1) do not differ from the outcomes for children who spend at least 20 hr a week in day care (Group 2). If a t test is performed to compare mean cognitive test scores between these two groups and the result is nonsignificant, this result cannot be interpreted as proof that day care has no effect on cognitive outcomes. It is possible that the nonsignificant outcome of the study was due to a small effect size, small sample sizes, variations in the way day care and home care were delivered, unreliable or invalid outcome measures, failure to include the outcome measures that would reflect differences in outcome, and a number of other limitations of the study.

A researcher can present evidence to try to discount each of these alternative explanations. For example, if a study has an N of 10,000 and used an outcome measure that is generally viewed as appropriate, reliable, and valid, these design factors strengthen the possibility that nonsignificant results might be evidence of a lack of difference in the population. However, the results of one study are not conclusive proof of the null hypothesis. If a nonsignificant difference in cognitive test score outcomes for day care versus home care groups can be replicated across many studies with large samples and good quality outcome measures, then as evidence accumulates across repeated studies, the degree of belief that there may be no difference between the populations may gradually become stronger.

Usually, researchers try to avoid setting up studies that predict a nonsignificant outcome. Most researchers feel that a statistically significant outcome represents a "success"; and a review of publication bias by Hubbard and Armstrong (1992) concluded that many journal editors are reluctant to publish results that are not statistically significant. If both researchers and journal editors tend to regard nonsignificant research outcomes as "failures," then it seems likely that Type I errors may be overrepresented among published studies and Type II errors may be overrepresented among unpublished studies.

3.6.2 ♦ Interpretation of Statistically Significant Results

The interpretation of a statistically significant outcome must be made as a carefully qualified statement. A study may yield a statistically significant outcome when there is really no effect in the population for a variety of reasons, such as the following:

1. A statistically significant outcome may arise due to sampling error. That is, even when the null hypothesis H_0: $\mu = \mu_{hyp}$ is correct, a value of the sample mean, M, that is quite different from μ_{hyp} can arise just due to sampling error or chance. By definition, when the nominal alpha level is set at .05, values of M that are far enough away from μ_{hyp} to meet the criterion for the decision to reject H_0 do occur about 5% of the time when the null hypothesis is actually correct. NHST procedures involve a (theoretically) known risk of Type I error; this theoretical level of risk is determined by the selection of an α level before the data are analyzed. However, the actual risk of Type I error corresponds to the nominal alpha level only when all the assumptions for NHST are met and the rules for NHST (such as conducting a limited number of significance tests) are followed.

2. Statistically significant outcomes sometimes occur due to experimenter expectancy effects. There is substantial evidence that when researchers have a preferred outcome for a study, or an idea about a likely outcome for the study, they communicate these expectations to research participants and this in turn influences research participants so that they tend to behave in the ways that the researcher expects (Rosenthal, 1966; Rosenthal & Rosnow, 1980). Experimenter expectancy effects occur in research with human participants, and they are even stronger in research with nonhuman animal subjects. In addition, errors in data entry and computation tend to be in the direction of the researcher's hypothesis.

3. Statistically significant outcomes sometimes occur because of unrecognized confounds; that is, the treatment co-occurs with some other variable, and it is that other variable that influences the outcome.

If any of these problems are present, then drawing the conclusion that the variables are related in some broader population, based on the results of a study, is incorrect. The accuracy of an obtained p value as an estimate of the risk of Type I error is conditional; the p value reported on the computer printout is only an accurate estimate of the true risk of Type I error if all the assumptions involved in null hypothesis significance testing are satisfied. Unfortunately, in many studies, one or more of these assumptions are violated. When the assumptions involved in NHST are seriously violated, the nominal p values reported by SPSS or other computer programs usually underestimate the true risk of Type I error; sometimes this underestimation of risk is substantial. To evaluate whether a p value provides accurate information about the magnitude of risk of Type I error, researchers need to understand the logic and the assumptions of NHST; to recognize how the procedures in their studies may violate those assumptions; and, finally, to realize how violations of assumptions may make their nominal p values inaccurate estimates of the true risk of Type I error.

3.7 ♦ When Is a *t* Test Used Instead of a *z* Test?

When σ is not known, and we have to use $SE_M = s/\sqrt{N}$ to estimate sampling error, we evaluate the distance between a sample mean, M, and the hypothesized population mean, μ, by setting up a t ratio. When we use SE_M to estimate σ_M, the size of the resulting t ratio is evaluated by looking up values in a t distribution with degrees of freedom $(df) = N - 1$, where N is the number of scores in the sample used to estimate M and s. When σ is not known and we use the sample standard deviation, s, to estimate the amount of sampling error, the test statistic for H_0: $\mu = \mu_{\text{hyp}}$ is as follows:

$$t = \frac{M - \mu_{\text{hyp}}}{SE_M}. \tag{3.9}$$

As discussed in Chapter 2, when we use SE_M in place of σ_M, this increases the magnitude of sampling error. We evaluate the size of a t ratio by looking at a t distribution with $N - 1$ *df*, where N is the number of scores in the sample used to estimate M and s.

Note the consistency in the logic across procedures. Back in Chapter 2, when we wanted to evaluate the location of an individual X score relative to a distribution of other individual X scores, we set up a z ratio to evaluate the distance of an individual X score from the mean of a distribution of other individual X scores, in number of population standard deviation units ($z = (X - M)/\sigma$). The t ratio in Equation 3.9 provides similar information about the location of an individual sample mean, M, relative to a theoretical distribution of possible outcomes for M that is assumed to have a mean of μ_{hyp} and a standard error of SE_M.

In many research situations, the population standard deviation, σ, is not known. When σ is not known, we have to make several changes in the test procedure. First, we replace σ_M with SE_M, an estimate of sampling error based on the sample standard deviation, s. Second, we use a t distribution with $N - 1$ *df* (rather than the standard normal distribution) to look up areas that correspond to distances from the mean and to decide whether the sample mean, M, is far away from the hypothesized value, μ_{hyp}. A numerical example of a one-sample t test (using SPSS) appears in Section 3.10.

3.8 ♦ Effect Size

There are several different ways of describing effect size; two of these are discussed here. For the one-sample *z* test or *t* test, one way of describing the effect size is to simply look at the magnitude of the obtained difference between *M*, the sample mean, and μ_{hyp}, the hypothesized population mean, in the units that are used to measure the outcome variable of interest.

3.8.1 ♦ Evaluation of "Practical" (vs. Statistical) Significance

It is useful to distinguish between "statistical significance" and **"practical"** or **"clinical significance"** (Kirk, 1996). A result that is statistically significant may be too small to have much real-world value. A difference between *M* and μ_{hyp} can be statistically significant and yet be too small in actual units to be of much practical or clinical significance. Consider research that has been done to compare the mean IQ for twins with the mean IQ for "singletons"—that is, individual birth children. Suppose that a researcher obtains a sample mean IQ of $M = 98.1$ for $N = 400$ twins (similar to results reported by Record, McKeown, & Edwards, 1970). This sample mean value of $M = 98.1$ can be evaluated relative to the mean IQ for the general population, using H_0: $\mu_{hyp} = 100$, $\sigma = 15$, and $N = 400$ and $\alpha = .05$, two-tailed. Given $\sigma = 15$ and $N = 400$, the value of $\sigma_M = .75$. For this hypothetical example $z = 98.1 - 100/.75 = 2.53$; this obtained value $z = 2.53$ exceeds the critical value of $z = +1.96$, and so this difference would be judged to be statistically significant. However, the actual difference in IQ between twins and singletons was only $98.1 - 100 = -1.9$ points. Is a difference of less than two IQ points large enough to be of any practical or clinical importance? Most researchers do not think this difference is large enough to cause concern. (Note that if *N* had been smaller, for example, if we recalculated the *z* test above using $N = 100$, this 1.9-IQ-point difference would not be judged statistically significant.)

Statistical significance alone is not a guarantee of practical significance or usefulness (Vacha-Haase, 2001). For example, a meta-analysis of the results of studies on the effect of short-term coaching on Scholastic Aptitude Test scores (Powers & Rock, 1993) suggested that the average difference in scores between the coached and control groups was of the order of 15 to 25 points. While this difference could be judged statistically significant in a study with a large *N*, the improvement in scores may be too small to be of much practical value to students. A 15-point improvement would not be likely to improve a student's chance of being accepted by a highly selective university.

On the other hand, when the *N* values are very small, a real effect of a treatment variable may not be detected even when it is relatively strong. For a *z* or *t* ratio to be large when the *N* values are small, it is necessary to have a very large difference between group means (a big difference in dosage levels) and/or very small variances within groups (good control over extraneous variables) in order to have a reasonable chance the *t* ratio will turn out to be large.

There are trade-offs among the design decisions that can affect the size of *z* or *t*. If a researcher knows ahead of time that the effect size that he or she wants to detect may be quite small, then the researcher may want to make the sample large. The next few sections of this chapter provide more specific guidance for decisions about sample size, based on beliefs about the magnitude of the effect that the researcher wants to be able to detect.

3.8.2 ♦ Formal Effect Size Index: Cohen's Little *d*

The problem with $(M - \mu_{hyp})$ as an effect size index is that the size of the difference depends on the units of measurement, and this difference is not always easily interpretable. We may be able to obtain a more interpretable index of effect size by taking another piece of information into account—that is, the standard deviation of the scores on the variable of interest. For many purposes, such as comparing outcomes across different variables or across different studies, it may be useful to have an index of effect size that is "unit free" or standardized—that is, not dependent on the units in which the original measurements were made. Cohen's little d is one possible effect size index; it describes the difference between two means in terms of number of standard deviations as shown in the following equation:

$$d = \frac{\mu - \mu_{hyp}}{\sigma} \text{ or } \frac{M - \mu_{hyp}}{\sigma}, \tag{3.10}$$

where μ is the unknown mean X score of the population about which inferences are to be made (based on a random sample drawn from that population); μ_{hyp} the hypothesized population mean for the variable of interest, X; σ the population standard deviation of scores on the variable of interest, X; and M the mean of the X scores in the sample.

In the previous example, the sample mean IQ for twins was $M = 98.1$ IQ points, the hypothesized population mean (based on the mean IQ for the general population) was $\mu_{hyp} = 100$, and the standard deviation for IQ test scores in the population was $\sigma = 15$. The corresponding d effect size is $d = (98.1 - 100)/15 = -.13$; that is, the sample mean for twins was .13 standard deviations below the mean for the general population. When the population standard deviation, σ, is not known, the sample standard deviation, s, can be used to estimate it, giving the following equation for **Cohen's *d***:

$$d = (M - \mu_{hyp})/s, \tag{3.11}$$

where M is the sample mean; μ_{hyp} is the hypothesized value of the population mean—that is, the specific numerical value in H_0, the null hypothesis; and s is the sample standard deviation of individual X scores.

It is useful to have an effect size index (such as Cohen's d) that is independent of the size of N, the number of cases in the sample; it is also independent of the units in which the X variable is measured (these units of measurement are removed by dividing the difference between means by σ or s). There are three ways we can use an effect size index. First of all, it is useful and informative to report effect size information along with the results of a statistical significance test. Guidelines for reporting research results in some disciplines such as psychology now call for the inclusion of effect size information along with statistical significance tests (APA, 2001). Second, effect size information can be useful when making decisions about the minimum sample size that is needed to have adequate statistical power (as discussed in Section 3.9). Third, effect size indexes that are unit free and independent of N provide information that can be summarized across studies using meta-analysis.

In later chapters, Cohen's d will also be used to index the magnitude of the difference between the means of two different populations, as illustrated in the following example concerning the mean height for males versus the mean height for females:

$$d = \frac{\mu_1 - \mu_2}{s_{\text{pooled}}}, \tag{3.12}$$

where μ_1 is the mean for Population 1 (e.g., mean height for males); μ_2 is the mean for Population 2 (e.g., mean height for females); and s_{pooled} is the overall standard deviation for the variable of interest, averaged or pooled across the two populations. This is typically estimated by averaging the sample variances as shown below:

$$s_{\text{pooled}} = \sqrt{\frac{(n_1 - 1)s_1^2 + (n_2 - 1)s_2^2}{(n_1 - 1) + (n_2 - 1)}}, \tag{3.13}$$

where n_1 is the number of scores in the sample drawn from Population 1, s_1^2 the sample variance in scores of the sample drawn from Population 1, n_2 the number of scores in the sample drawn from Population 2, and s_2^2 the sample variance for the sample drawn from Population 2.

Cohen's *d* can be illustrated graphically; the distance between the means of two normal distributions is related to the amount of overlap between the distributions. A larger value of *d* corresponds to a larger and therefore more easily detectable difference between two distributions of scores. For example, Kling, Hyde, Showers, and Buswell (1999) conducted a review of studies that examined differences in self-esteem between women and men; they found that across a large number of studies, the mean size of the Cohen's *d* effect size was approximately $d = .20$, which corresponds to a rather small effect size. Figure 3.4 illustrates the overlap in the distribution of self-esteem scores for male and female populations, with an effect size of $d = .22$; that is, the mean self-esteem for the male population is about two tenths of a standard deviation higher than the mean self-esteem for the female population. This difference is small; there is substantial overlap in the scores on self-esteem for males and females. On the other hand, there is a much larger gender difference for height. Data reported at http://en.wikipedia.org/wiki/Effect_size suggest that the Cohen's *d* value for gender differences in height is about $d = 2.00$; this is a very large effect size. Figure 3.5 illustrates the amount of overlap in the distributions of female and male heights, given a Cohen's *d* value of 2.00; that is, the population mean for male height is about two standard deviations higher than the population mean for female height.

When the magnitude of a difference indexed by Cohen's *d* is small, the difference between means may be relatively difficult to detect. When the real magnitude of the difference between population means is relatively small, as in Figure 3.4, a researcher typically needs to have a relatively large sample in order to obtain a *t* or *z* value large enough to reject H_0 (the null hypothesis that the male and female population means are equal). On the other hand, when the magnitude of the difference between population means is large, for example, when Cohen's $d = 2.00$ as in Figure 3.5, a researcher may be able to obtain a *t* or *z* value large enough to reject H_0 even when the sample size is relatively small.

Suggested verbal labels for small, medium, and large values of Cohen's *d* in behavioral and social science research are summarized in Table 3.2.

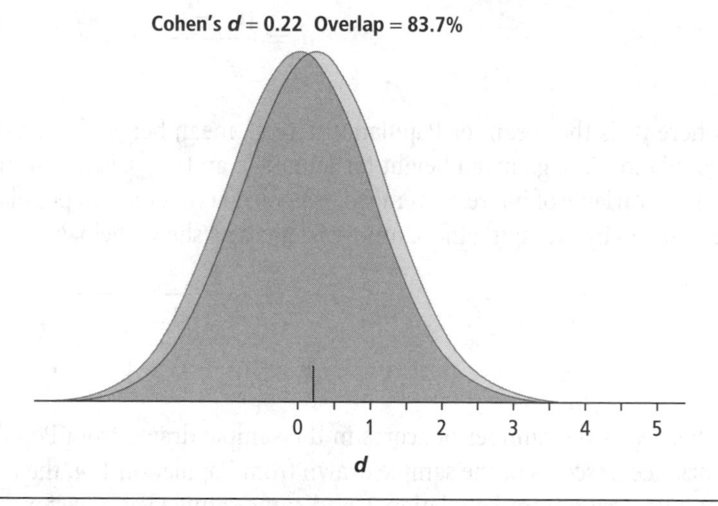

Figure 3.4 ♦ Example of a Small Effect Size: Cohen's $d = .22$

SOURCE: Kling et al. (1999).

NOTE: Across numerous studies, the average difference in self-esteem between male and female samples is estimated to be about .22; the mean self-esteem for males is typically about two tenths of a standard deviation higher than the mean self-esteem for females.

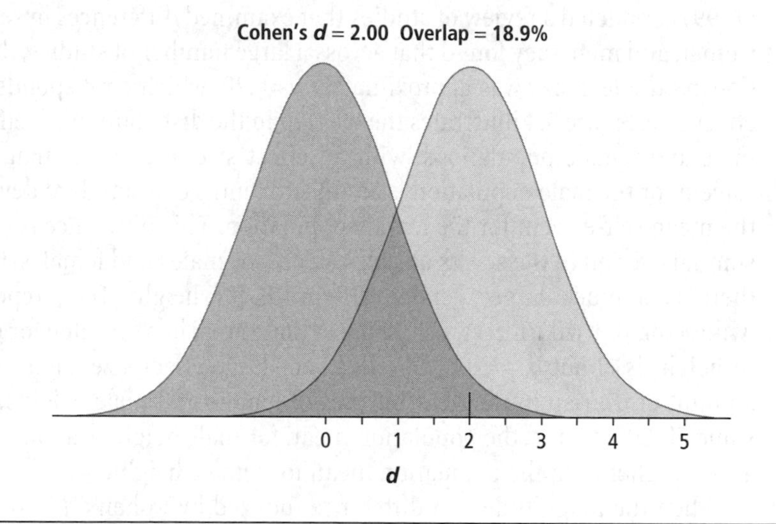

Figure 3.5 ♦ Example of a Large Effect Size: Cohen's $d = 2.00$

SOURCE: http://en.wikipedia.org/wiki/Effect_size.

NOTE: From samples of men and women in the United Kingdom, mean height for males = 1,754 mm; mean height for females = 1,620 mm. The standard deviation for height = 67.5 mm. Therefore, Cohen's $d = (M_{\text{male}} - M_{\text{female}})/s = (1{,}754 - 1{,}620)/67.5 \approx 2.00$.

3.9 ♦ Statistical Power Analysis

Statistical power is defined as the probability of obtaining a value of z or t that is large enough to reject H_0 when H_0 is actually false. In most (although not all) applications of NHST, researchers hope to reject H_0. In many experimental research situations, H_0

Table 3.2 ♦ Suggested Verbal Labels for Cohen's *d* Effect Size Index in Behavioral and Social Science Research

Verbal Label	*Magnitude of d*
Small	$d \leq .20$
Medium	*d* of the order of .5 (e.g., *d* between .20 and .79)
Large	$d \geq .80$

NOTE: Population value of Cohen's $d = (\mu - \mu_{hyp})/\sigma$, where μ is the mean for the population from which the study sample was drawn, μ_{hyp} is the hypothesized value of the population mean, and σ is the standard deviation of scores in the population from which the sample was drawn. In other words, *d* is the distance in number of standard deviation units (σ) between the actual population mean (μ) and the hypothesized population mean stated in the null hypothesis (μ_{hyp}).

corresponds to an assumption that the treatment has no effect on the outcome variable; in many nonexperimental research situations, H_0 corresponds to an assumption that scores on a *Y* outcome variable are not predictable from scores on an *X* independent variable; researchers often hope to demonstrate that a treatment does have an effect on some outcome variable and that they can predict scores on an outcome variable. In other words, researchers often hope to be able to reject a null hypothesis that states that there is no treatment effect or no relationship between variables.

Refer back to Table 3.1 to see the four possible outcomes when decisions are made whether to reject or not reject a null hypothesis. The outcome of interest, at this point, is the one in the upper right-hand corner of the table: the probability of correctly rejecting H_0 when H_0 is false, which is called *statistical power*. Note that within each column of Table 3.1 the probabilities of the two different possible outcomes sum to 100%. Statistical power corresponds to $(1 - \beta)$, where β is the risk of committing a Type I error. The factors that reduce the probability of committing a Type II error (β) also increase statistical power $(1 - \beta)$. Researchers want statistical power to be reasonably high; often, statistical power of .80 is suggested as a reasonable goal.

The risk of committing a Type I error (α) is, in theory, established by the choice of a nominal alpha level. The risk of committing a Type II error (β), on the other hand, depends on several factors, including the nominal alpha level, the magnitude of the true population effect size (such as Cohen's *d*), and the sample size, *N*. Because statistical power $(1 - \beta)$ is the complement of risk of committing a Type II error (β), factors that decrease the risk of Type II error also increase statistical power.

We will begin with some general qualitative statements about how these factors (α, effect size, and *N*) are related to power and then give an empirical example to illustrate how statistical power varies as a function of these factors. In each case, when a statement is made about the effect of changes in one factor (such as *N*, sample size), we assume that the other factors are held constant. Here is a qualitative summary about factors that influence statistical power and the risk of committing a Type II error:

1. Assuming that the population effect size, *d*, and the sample size, *N*, remain constant, statistical power increases (and risk of Type II error decreases) as the value of the nominal alpha level is increased. Usually researchers set the nominal alpha level at .05. In theory, statistical power can be increased by raising the alpha level to $\alpha = .10$

or $\alpha = .20$. However, most researchers are unwilling to accept such high levels of risk of Type I error, and therefore, they prefer changing other features of the research situation (rather than increasing α) to improve statistical power.

2. To see how the other two factors (the effect size, Cohen's *d*; the sample size, *N*) influence statistical power, it is useful to reexamine the equations that are used to compute the *t* ratio. Recall that the one-sample *t* test is calculated as follows: $t = (M - \mu_{\text{hyp}})/(s/\sqrt{N})$. We can "factor out" the term involving the square root of *N* to show that the size of *t* depends on two things: the magnitude of *d* and the magnitude of the square root of *N*.

   Recall that $t = (M - \mu_{\text{hyp}})/(s/\sqrt{N})$; this can be rearranged as follows:

$$\frac{M - \mu_{\text{hyp}}}{s} \times \frac{1}{1/\sqrt{N}}.$$

   Also, recall that $(M - \mu_{\text{hyp}})/s$ is Cohen's *d*. And note that $1/(1/\sqrt{N})$ can be simplified to $\sqrt{N}$. Thus, we have

$$t = d \times \sqrt{N}. \tag{3.14}$$

   In other words, the size of the *t* ratio is related to both the magnitude of the population effect size, Cohen's *d*, and the sample size, *N*. (This equation is similar to equations presented by Rosenthal & Rosnow, 1991, which show how the overall independent samples *t* test can be factored into two parts: one term, *d* in this example, to represent effect size and the other term, square root of *N* in this example, to represent sample size.) If the sample size, *N*, is held constant, *t* increases as *d* increases. If the effect size, *d*, is held constant, *t* increases as $\sqrt{N}$ increases. Therefore, assuming that the sample size, *N*, remains constant, as the population effect size represented by Cohen's *d* increases, we expect that the size of the sample *t* ratio will also tend to increase. The implication of this is that we are more likely to obtain a large value of the *t* ratio (large enough to reject H_0) when the population effect size indexed by *d* is large.

   We can summarize by saying that, other factors being equal, as the magnitude of the population effect size, *d*, increases, statistical power tends to increase and the risk of committing a Type II error (β) tends to decrease.

3. It also follows from the previous argument that, if all other factors in the equation for *t* remain constant, as *N* increases, the size of the obtained *t* ratio will tend to be larger. Therefore, other factors being equal, as the size of *N* increases, statistical power tends to increase (and the risk of committing a Type II error, β, decreases).

The problem with the preceding qualitative statements about the connection between statistical power and the values of α, *d*, and *N* is that they do not take sampling error into account. We can evaluate statistical power more precisely by setting up a graph of the distributions of outcomes of *t* that are expected if H_0 is true and a separate graph of the distribution of outcomes of t that would be expected for a specific population effect size, d. In the following example, let's continue to consider testing hypotheses about intelligence scores. Suppose that the null hypothesis is

$$H_0: \mu = 100,$$

the sample standard deviation $s = 15$, and the sample size is $N = 10$ (therefore $df = 9$).

$$SE_M = 15/\sqrt{N} = 15/\sqrt{10} = 15/3.162 = 4.74.$$

Now let's suppose that the actual population mean is 115. This would make the value of Cohen's $d = [\mu - \mu_{hyp}]/s = [115 - 100]/15 = 1.00$.

From the table of critical values for the t distribution, which appears in Appendix B, the critical values of t for $\alpha = .05$, two-tailed, and $df = 9$ are $t = +2.262$ and $t = -2.262$. Based on Equation 3.7, the critical values of M would therefore be

$$100 - 2.262 \times 4.74 = 89.28$$

and

$$100 + 2.262 \times 4.74 = 110.72.$$

In other words, we would reject H_0: $\mu = 100$ if we obtain a sample mean, M, that is less than 89.28 or greater than 110.72.

To evaluate statistical power, we need to think about two different possible distributions of outcomes for M, the sample mean. The first is the distribution of outcomes that would be expected if H_0 were true; the "reject regions" for the statistical significance test are based on this first distribution. The second is the distribution of outcomes for M that we would expect to see if the effect size $d = 1.00$, that is, if the real population mean (115) were one standard deviation about the hypothesized population mean of 100 points.

The upper panel of Figure 3.6 shows the expected distribution of outcome values of t given H_0: $\mu = 100$, H_1: $\mu \neq 100$, $df = 9$, and $\alpha = .05$ (two-tailed). Using the fact that $N = 10$ and $df = 9$, we can find the critical values of t from the table of the t distribution in Appendix B. For $\alpha = .05$ (two-tailed) with 9 df, we would reject H_0 for values of $t > +2.262$ and for values of $t < -2.262$.

The lower panel of Figure 3.6 shows how these critical values of t correspond to values of M. Using Equation 3.7 along with knowledge of H_0 and SE_M, we can convert each critical value of t into a corresponding critical value of M. For example, a t value of +2.262 corresponds to a sample mean, M, of 110.72. The reject regions for H_0 can be given in terms of obtained values of M. We would reject H_0: $\mu = 100$ for values of $M > 110.72$ and for values of $M < 89.28$.

The preceding discussion shows how the distribution of outcomes for M that is (theoretically) expected when H_0 is assumed to be true is used to figure out the reject regions for H_0 (in terms of values of t or of M).

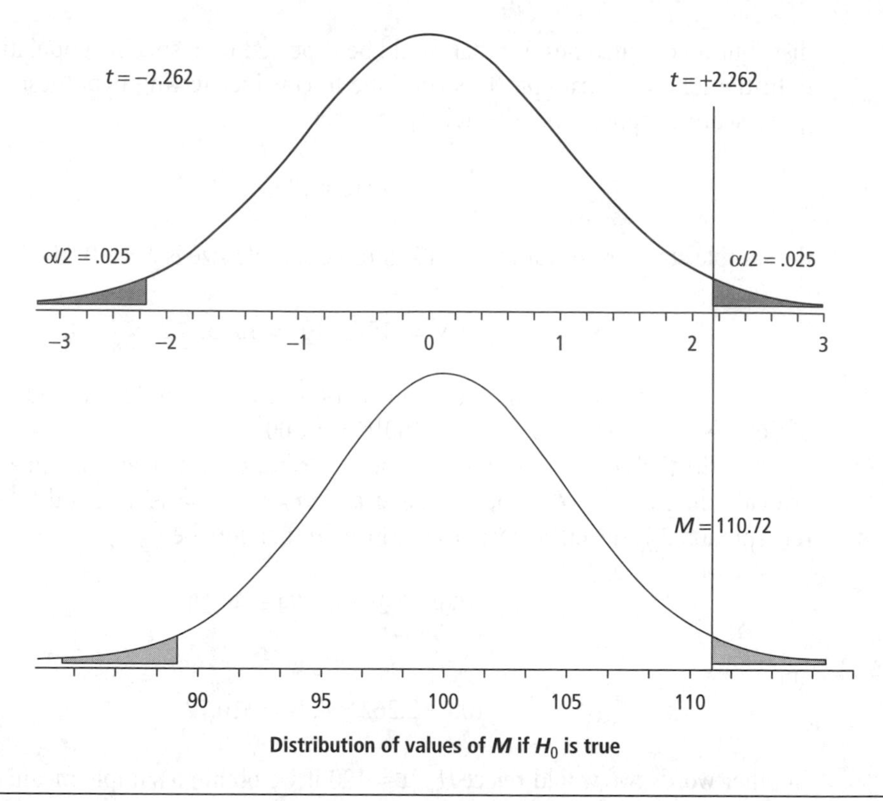

Figure 3.6 ♦ Diagram Showing Reject Regions for the Following Null Hypothesis: H_0: $\mu = 100$, $\sigma_M = 3$, $\alpha = .05$, Two-Tailed

NOTES: Reject regions (in terms of z values) are as follows: Reject H_0 for $z > +1.96$ and for $z < -1.96$. Reject regions (in terms of values of M) are as follows: Reject H_0 for $M < 89.28$ and for $M > 110.72$.

The next step is to ask what values of M would be expected to occur if H_0 is false (in fact, μ is actually equal to 115). An actual population mean of $\mu = 115$ corresponds to a Cohen's d effect size of +1 (i.e., the actual population mean $\mu = 115$ is one standard deviation higher than the value of $\mu_{hyp} = 100$ given in the null hypothesis).

The lower panel of Figure 3.7 illustrates the theoretical sampling distribution of M if the population mean is really equal to 115. We would expect most values of M to be fairly close to 115 if the real population mean is 115; and we can use SE_M to predict the amount of sampling error that is expected to arise for values of M.

The final step involves asking this question: Based on the distribution of outcomes for M that would be expected if μ is really equal to 115 (as shown in the bottom panel of Figure 3.7), how often would we expect to obtain values of the sample mean, M, that are larger than the critical value of $M = 110.72$ (as shown in the upper panel of Figure 3.7)? Note that values of M below the lower critical value of $M = 89.28$ would occur so rarely when μ really is equal to 115 that we can ignore this set of possible outcomes.

To figure out the probability of obtaining sample means, M, greater than 110.72 when $\mu = 115$, we find the t ratio that tells us the distance between the "real" population mean, $\mu = 115$, and the critical value of $M = 110.72$. This t ratio is $t = (M - \mu)/SE_M = (110.72 - 115)/4.74 = -.90$. The likelihood that we will obtain a sample value for M that is large

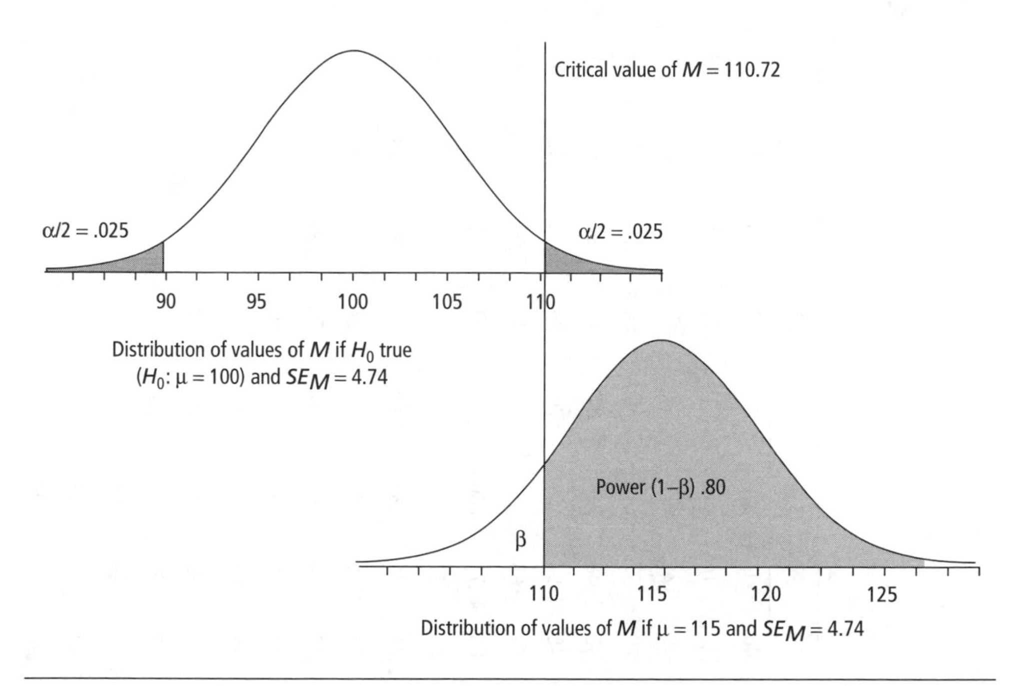

Figure 3.7 ◆ Statistical Power: Probability of Obtaining a Sample Value of *M* That Exceeds the Critical Value of *M* (for H_0: $\mu = 100$ and $\alpha = .05$, Two-Tailed)

NOTES: The upper distribution shows how values of *M* are expected to be distributed if H_0 is true and $\mu = 100$. The shaded regions in the upper and lower tails of the upper distribution correspond to the reject regions for this test. The lower distribution shows how values of *M* would actually be distributed if the population mean, μ, is actually 115; based on this distribution, we see that if μ is really 115, then about .80 or 80% of the outcomes for *M* would be expected to exceed the critical value of *M* (110.72).

enough to be judged statistically significant given the decision rule developed previously (i.e., reject H_0 for $M > 110.72$) can now be evaluated by finding the proportion of the area in a *t* distribution with 9 *df* that lies to the right of $t = -.90$. Tables of the *t* distribution, such as the one in Appendix B, do not provide this information; however, it is easy to find Java applets on the Web that calculate exact tail areas for any specific value of *t* and *df*. Using one such applet, the proportion of the area to the left of $M = 110.72$ and $z = -.90$ (for the distribution centered at $\mu = 115$) was found to be .20 and the proportion of area to the right of $M = 110.72$ and $z = -.90$ was found to be .80. The shaded region on the right-hand side of the distribution in the lower part of Figure 3.7 corresponds to statistical power in this specific situation; that is, *if* we test the null hypothesis H_0: $\mu = 100$, using $\alpha = .05$, two-tailed, and a *t* test with $df = 9$, and *if* the real value of Cohen's $d = +1.00$ (i.e., the real population mean is equal to $\mu = 115$), *then* there is an 80% chance that we will obtain a value of *M* (and therefore a value of *t*) that is large enough to reject H_0.

We can use this logic to figure out what sample size, *N*, is required to achieve a specific desired level of statistical power (usually the desired level of power is at least 80%) when we are planning a study. Tables have been published that map out the minimum *N* needed to achieve various levels of statistical power as a function of the alpha level and the population effect size Cohen's *d*. An example of a statistical power table appears in Table 3.3.

Table 3.3 ♦ Power Tables for the One-Sample t Test Using $\alpha = .05$, Two Tailed

	Cohen's *d*																			
n	*.20*	*.25*	*.30*	*.35*	*.40*	*.45*	*.50*	*.55*	*.60*	*.65*	*.70*	*.75*	*.80*	*.90*	*1.00*	*1.10*	*1.20*	*1.30*	*1.40*	*1.50*
3	.05	.05	.05	.05	.05	.05	.06	.06	.06	.06	.06	.06	.06	.07	.08	.08	.09	.10	.11	.13
4	.05	.05	.06	.06	.06	.07	.07	.08	.08	.09	.10	.10	.11	.14	.17	.21	.25	.31	.37	.44
5	.06	.06	.06	.07	.08	.09	.10	.11	.12	.13	.15	.17	.19	.25	.31	.39	.47	.55	.63	.70
6	.06	.07	.07	.08	.09	.11	.12	.14	.17	.19	.22	.25	.29	.37	.46	.55	.64	.72	.79	.84
7	.06	.07	.08	.10	.11	.13	.16	.18	.21	.25	.29	.33	.38	.48	.58	.67	.75	.82	.87	.91
8	.07	.08	.09	.11	.13	.16	.19	.23	.27	.31	.36	.41	.46	.57	.67	.76	.83	.88	.92	.95
9	.07	.09	.10	.13	.15	.19	.22	.27	.31	.37	.42	.48	.54	.65	.75	.83	.88	.93	.95	.97
10	.08	.09	.12	.14	.18	.21	.26	.31	.36	.42	.48	.54	.60	.71	.80	.87	.92	.95	.97	.98
11	.08	.10	.13	.16	.20	.24	.29	.35	.41	.47	.54	.60	.66	.77	.85	.91	.94	.97	.98	.99
12	.09	.11	.14	.17	.22	.27	.33	.39	.45	.52	.59	.65	.71	.81	.88	.93	.96	.98	.99	.99
13	.09	.12	.15	.19	.24	.30	.36	.42	.49	.56	.63	.70	.75	.85	.91	.95	.97	.99	.99	
14	.10	.13	.16	.21	.26	.32	.39	.46	.53	.61	.67	.74	.79	.88	.93	.96	.98	.99		
15	.10	.13	.17	.22	.28	.35	.42	.49	.57	.64	.71	.77	.82	.90	.95	.97	.99	.99		
16	.11	.14	.19	.24	.30	.37	.45	.53	.60	.68	.74	.80	.85	.92	.96	.98	.99			
17	.11	.15	.20	.26	.32	.40	.48	.56	.64	.71	.77	.83	.87	.93	.97	.99	.99			
18	.12	.16	.21	.27	.34	.42	.50	.59	.67	.74	.80	.85	.89	.95	.98	.99				
19	.12	.17	.22	.29	.36	.45	.53	.62	.69	.76	.82	.87	.91	.96	.98	.99				
20	.13	.17	.23	.30	.38	.47	.56	.64	.72	.79	.84	.89	.92	.97	.99	.99				
21	.13	.18	.24	.32	.40	.49	.58	.67	.74	.81	.86	.90	.94	.97	.99					
22	.14	.19	.26	.33	.42	.51	.60	.69	.76	.83	.88	.92	.95	.98	.99					

n	.20	.25	.30	.35	.40	.45	.50	.55	.60	.65	.70	.75	.80	.90	1.00	1.10	1.20	1.30	1.40	1.50
	Cohen's d																			
23	.14	.20	.27	.35	.44	.53	.63	.71	.78	.85	.89	.93	.95	.98	.99					
24	.15	.21	.28	.36	.46	.55	.65	.73	.80	.86	.91	.94	.96	.99						
30	.18	.25	.35	.45	.56	.66	.75	.83	.89	.93	.96	.98	.99							
40	.23	.33	.45	.58	.69	.79	.87	.92	.96	.98	.99									
50	.28	.41	.54	.68	.79	.88	.93	.97	.98	.99										
60	.33	.47	.63	.76	.86	.93	.97	.99	.99											
70	.37	.54	.70	.82	.91	.96	.98	.99												
80	.42	.60	.75	.87	.94	.98	.99													
90	.46	.65	.80	.91	.96	.99														
100	.51	.70	.84	.93	.98	.99														

SOURCE Reprinted with permission from Dr. Victor Bissonnette.

Table 3.3 can be used to look up the statistical power that corresponds to the situation in the previous example. The previous example involves a single sample t test with an effect size $d = +1.00$, a sample size $N = 10$, and $\alpha = .05$, two-tailed. Table 3.3 provides estimates of statistical power for a one-sample t test with $\alpha = .05$, two-tailed. If you look up the value of d across the top of the table and find the column for $d = +1.00$, and look up the value of N in the rows of the table and find the row that corresponds to $N = 10$, the table entry that corresponds to the estimated power is .80. This agrees with the power estimate that was based on an examination of the distributions of the values of M shown in Figures 3.6 and 3.7.

The power table can be used to look up the minimum N value required to achieve a desired level of power. For example, suppose that a researcher believes that the magnitude of difference that he or she is trying to detect using a one-sample t test corresponds to Cohen's $d = +.50$ and the researcher plans to use $\alpha = .05$, two-tailed. The researcher can read down the column of values for estimated power under the column headed $d = +.50$. Based on the values in Table 3.3, the value of N required to have a statistical power of about .80 to detect an effect size of $d = +.5$ in a one-sample t test with $\alpha = .05$ two-tailed is about $N = 35$.

The true strength of the effect that we are trying to detect—for example, the degree to which the actual population mean, μ, differs from the hypothesized value, μ_{hyp}, as indexed by the population value of Cohen's d—is usually not known. The sample size needed for adequate statistical power can only be approximated by making an educated guess about the true magnitude of the effect as indexed by d. If that guess about the population effect size, d, is wrong, then the estimate of β based on that guess will also be wrong. Information from past studies can often be used to make at least approximate estimates of population effect size.

When is statistical power analysis needed? It is a good idea whenever a researcher plans to conduct a study to think about whether the expected effect size, alpha level, and sample size are likely to be adequate to obtain reasonable statistical power. People who write proposals to compete for research funds from government grant agencies are generally required to include a rationale for decisions about planned sample size that takes statistical power into account.

In summary, a researcher can do several things to try to reduce the magnitude of β and increase statistical power. The researcher can set a higher alpha level (but this trade-off, which involves increasing the risk of Type I error, is usually not considered acceptable). The researcher can increase the size of N, the sample size. Another way of reducing β is by increasing the size of the difference that the researcher is trying to detect—that is, increasing the effect size that corresponds to Cohen's d. One way of increasing the effect size is by increasing the difference $\mu - \mu_{hyp}$, that is, the difference between the actual population mean, μ, and the hypothesized value of the population mean, μ_{hyp}, given in the null hypothesis. Other factors being equal, as this difference $\mu - \mu_{hyp}$ increases, the likelihood of obtaining values of M and z large enough to reject H_0 also increases. Another way of increasing the effect size given by Cohen's d is through reduction of the within-group standard deviation, σ or s. Typically, s can be made smaller by selecting a homogeneous sample of participants and standardizing procedures.

3.10 ♦ Numerical Results for a One-Sample *t* Test Obtained From SPSS

A one-sample *t* test (to test a null hypothesis about the value of the mean of one population) can be performed using SPSS. The data for this empirical example are in the SPSS file named wais.sav; this file contains hypothetical IQ scores for a sample of $N = 36$ randomly sampled residents from a nursing home population. In the following example, we will assume that the researcher does not know the value of σ, the population standard deviation, and must therefore use *s*, the sample standard deviation. Using the data from this sample, the researcher can do the following things:

1. Set up a CI for μ based on the sample values of *M* and *s* (using procedures reviewed in Chapter 2).
2. Test a specific null hypothesis about the value of μ, such as H_0: $\mu = 100$ (using a preselected alpha level such as .05 and a specific alternative hypothesis such as H_1: $\mu \neq 100$). When σ is known, a *z* test may be used (as described in section 3.1). In the following example, we will assume that σ is not known and that we use the sample standard deviation *s* to set up a one sample *t* test.
3. Calculate a sample estimate of Cohen's *d* as an index of effect size.

The menu selections that are needed to run a one-sample *t* test appear in Figure 3.8; from the top-level menu bar, choose <Analyze>, and then from the pull-down menus, click on <Compare Means> and <One-Sample T Test>. These menu selections open up the SPSS dialog window for the One-Sample T Test, which appears in Figure 3.9. Initially, the names of all the variables in the dataset appear in the left-hand window; the variable for the one-sample *t* test is selected from this list and moved into the right-hand window (here, the variable is WAIS IQ score for each person in the sample). The hypothesized value of μ that is specified in the null hypothesis should be typed into the window that has the heading Test Value. In this example, H_0: $\mu = 100$, and therefore the test value is 100.

The *p* value that is reported by SPSS for the one-sample *t* test is a two-tailed or nondirectional *p* value. The Options button opens up a new dialog window where the user can specify the width of the CI; the default choice is a 95% CI (see Figure 3.10).

The results for the one-sample *t* test appear in Figure 3.11. The first panel reports the sample descriptive statistics; that is, the mean IQ for this sample was $M = 103.83$. This sample mean, *M*, is almost four points higher than the value of μ predicted in the null hypothesis, H_0: $\mu = 100$. The obtained *t* ratio was $t(35) = 2.46$, $p = .019$, two-tailed.

If the population mean, μ, really were equal to 100, we would expect that the means for 95% of all samples would lie rather close to 100. Our definition of "rather close" depends on the selected alpha level ($\alpha = .05$, two-tailed, in this example) and on the magnitude of the sampling error, which is given by the sample value of $SE_M = 1.556$. For a *t* distribution with $df = N - 1 = 36 - 1 = 35$, the critical values of *t* that correspond to the top 2.5% and the bottom 2.5% of the area of the *t* distribution can be found in the table of critical values for the *t* distribution in Appendix B. This table shows values of *t* for $df = 30$ and $df = 40$. We can figure out the value of *t* for $df = 35$ by taking the midpoint between the critical values of *t* that appear for $df = 30$ ($t_{\text{critical}} = 2.042$) and for $df = 40$ ($t_{\text{critical}} = 2.021$) in the column for $\alpha = .05$, two-tailed. The

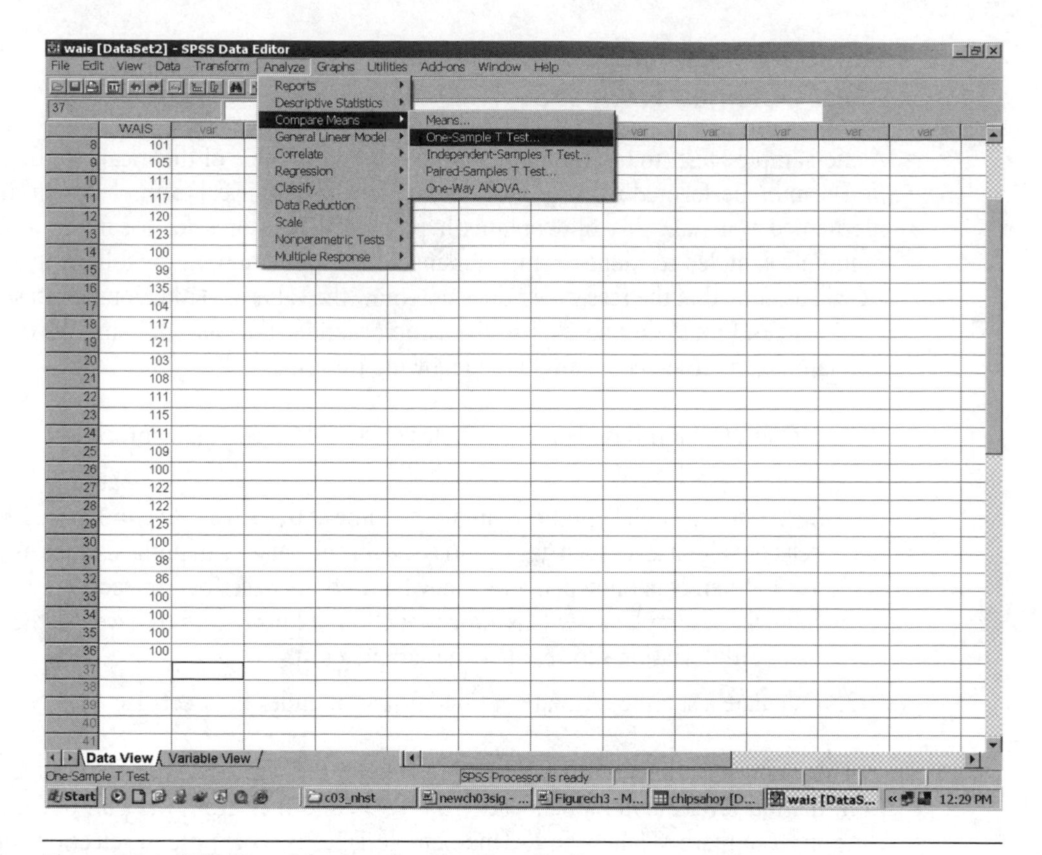

Figure 3.8 ♦ SPSS Menu Selections for a One-Sample t Test

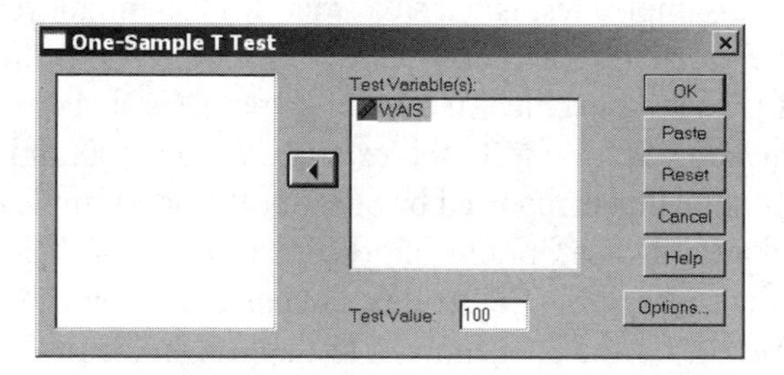

Figure 3.9 ♦ SPSS Dialog Window for the One-Sample t-Test Procedure

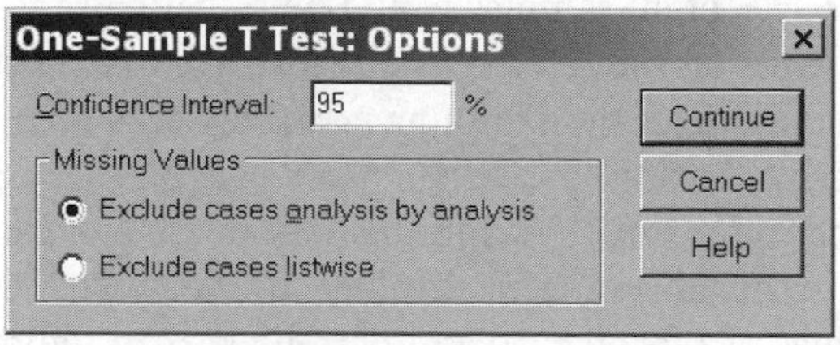

One-Sample T Test: Options

Confidence Interval: 95 %

Missing Values

Exclude cases analysis by analysis

Exclude cases listwise

Continue

Cancel

Help

Figure 3.10 ♦ Options for the One-Sample t Test: Width of the CI

T-Test

One-Sample Statistics

WAIS	N	Mean	Std. Deviation	Std. Error Mean
	36	103.83	9.337	1.556

One-Sample Test

	Test Value = 100					
					95% Confidence Interval of the Difference	
	t	df	Sig. (2-tailed)	Mean Difference	Lower	Upper
WAIS	2.463	35	.019	3.833	.67	6.99

Figure 3.11 ♦ SPSS Output for One-Sample *t* Test on Nursing Home Sample IQ Data

midpoint of these values is $(2.042 + 2.021)/2 = 2.031$. Thus, for a *t* distribution with 35 *df*, the critical values of *t* that correspond to the bottom 2.5% and top 2.5% of the area are as follows: The bottom 2.5% of the outcomes correspond to $t < -2.031$, and the top 2.5% of the outcomes correspond to $t > +2.031$.

Given all this information, we can predict what the distribution of outcomes of sample values of *M* would look like across hundreds or thousands of different samples when we use the sample value *s* (because we do not know the population value of σ). The distribution of values of *M* are expected to have the following characteristics:

1. A mean of 100
2. A shape described by *t* with $df = 35$ (in other words, we use values from a *t* distribution with 35 *df* to figure out how distances from the mean are related to tail areas or proportions)
3. A standard deviation or standard error, SE_M, that depends on *s* and *N*

If all the conditions stated so far are correct, then we would expect that 95% of the outcomes for the sample mean, *M*, would lie between $\mu - 2.031 \times SE_M$ and $\mu + 2.031 \times SE_M$, that is, between $[100 - 2.031 \times 1.556]$ and $[100 + 2.031 \times 1.556]$ in this example, between 96.84 and 103.16. (Conversely, 2.5% of the values of *M* would be less than 96.84, and 2.5% of the values of *M* would be greater than 103.16.)

In other words, our criterion for "values of *M* that would be unlikely to occur if H_0 were true," in this case, is values of $M < 96.84$ and values of $M > 103.16$.

If H_0 is correct ($\mu = 100$) and if $SE_M = 1.556$, it would be quite unusual to see sample means less than 96.84 or greater than 103.16 for samples of size $N = 36$. When we use $\alpha = .05$, two-tailed, as our criterion for statistical significance, "quite unusual" corresponds to the most extreme 5% of outcomes (the bottom 2.5% and the top 2.5% of the distribution of likely values for *M*).

The obtained sample mean $M = 103.83$ is large enough in this example for it to fall within the top 2.5% of the distribution of predicted outcomes for *M* (for the null hypothesis H_0: $\mu = 100$). The obtained *t* ratio, $t = 2.463$, tells us the distance (in number of standard errors) of the sample mean, M, from the hypothesized value $\mu =$ 100. The exact p value given on the SPSS printout ($p = .019$, two-tailed) corresponds to the theoretical probability of obtaining a sample mean that is more than 2.463 standard errors above or below the hypothesized population mean, if H_0 is true and all our other assumptions are correct. If μ really were equal to 100 for the entire nursing home population, the probability of obtaining a sample mean this far away from μ just due to sampling error is $p = .019$—that is, less than 2%. In other words, this outcome $M = 103.83$ is an outcome that is very unlikely to occur if H_0 is true. We can, therefore, reject H_0: $\mu = 100$ (because the outcome of the study, $M = 103.83$, would be an unlikely outcome if this null hypothesis were correct).

A CI was set up for the value of the difference between the unknown population mean, m, and the hypothesized value $\mu = 100$; this 95% CI has a lower boundary of .67 and an upper boundary of 6.99. In other words, given the size of the sample mean IQ, the difference between the unknown population mean IQ, m, for the entire population of nursing home residents and the hypothesized IQ of 100 probably lies between .67 and 6.99 points. As discussed in Chapter 2, in the long run, 95% of the CIs set up using the value of μ, the critical values of t, and the sample value of SE_M should actually contain the true population mean, μ. It seems reasonable in this case to conclude that the true magnitude of the difference between the mean IQ for the nursing home population and the value of 100 stated in the null hypothesis is probably between .67 points and 6.99 points; the nursing home population IQ is probably a few points higher than the value of 100 points given in the null hypothesis.

Finally, we can also report the outcome for this study using Cohen's d to index the effect size. In this example, we have $M = 103.83$, $\mu_{hyp} = 100$, and $s = 9.337$; $d = (103.83 - 100)/9.337 = .41$. Using the verbal labels for effect size in Table 3.2, this would be judged a moderate effect size.

3.11 ♦ Guidelines for Reporting Results

An APA Task Force report (Wilkinson & Task Force on Statistical Inference, 1999) concluded that reports of significance tests alone are not sufficient; additional information about the outcomes of studies should be reported. The current APA (2001) guidelines for reporting results of significance tests, such as the independent samples t test, are as follows:

> When reporting inferential statistics (e.g., t tests, F tests, and chi-square), include information about the obtained magnitude or value of the test statistic, the degrees of freedom, the probability of obtaining a value as extreme as or more extreme than the one obtained, and the direction of the effect. Be sure to include sufficient descriptive statistics (e.g., per-cell sample size, means, correlations, standard deviations) so that the nature of the effect being reported can be understood by the reader and for future meta-analysis. This information is important, even if no significant effect is reported. When point estimates are provided, always include an associated measure of variability (precision) specifying its nature (e.g., the standard error). (p. 22)

The APA guidelines also state that the reporting of CIs is strongly recommended. Note that certain types of information discussed in this chapter are usually not included; for example, it is uncommon to see statements of a formal null hypothesis in a journal article.

3.12 ♦ Summary

3.12.1 ♦ Logical Problems With NHST

The reasons why NHST procedures are problematic are complex; some of the potential problems with NHST are mentioned here. To begin with, the basic logic involved in using the proposition that "the results in the sample would be unlikely to occur if H_0 is true" to make the decision "Reject H_0" is controversial; Sir Ronald Fisher argued in favor of making a decision to reject H_0 in this situation, but other major figures such as Karl Pearson argued against this reasoning.

Even though NHST procedures involve looking for disconfirmatory evidence, they do not involve the kind of theory testing that Karl Popper advocated when he said that theories need to be potentially falsifiable (see Meehl, 1978). Usually researchers want to reject H_0, but that is not usually equivalent to falsifying a theory. In fact, the decision to reject H_0 is often interpreted as (indirect) support for a theory. The researcher's theory more often corresponds to the prediction made in the alternate hypothesis than to the prediction made in the null hypothesis. Some theorists argue that in a sense H_0 is virtually always false (see Krueger, 2001, for a discussion of this point). If that is the case, then results of NHST will always turn out to be statistically significant if sample sizes are made large enough.

In practice, many of the assumptions for NHST are frequently violated; for example, samples are often not randomly selected from any real population, and researchers often report large numbers of significance tests. The desire to obtain statistically significant results can tempt researchers to engage in "data fishing"; researchers may "massage" their data (e.g., by running many different analyses or by deleting extreme scores) until they manage to obtain statistically significant results. When any of these violations of assumptions are present, researchers should explain that their reported p values do not accurately represent the true risk of incorrectly rejecting H_0.

Misunderstandings about the meaning of a statistically significant result are fairly common. A few people mistakenly think that p is the probability that the null hypothesis is true or that $1 - p$ is the probability that the alternative hypothesis is true or that $1 - p$ represents some sort of probability that the results of a study are replicable.[5] None of these interpretations are correct. These misunderstandings occur partly because statistical significance tests do not tell us what we really want to know. As Cohen (1994) said,

> What we want to know is "Given these data, what is the probability that H_0 is true?" But as most of us know, what [NHST] tells us is, "Given that H_0 is true, what is the probability of these (or more extreme) data?" These are not the same, as has been pointed out many times over the years. (p. 997)

In spite of potential logical and practical problems with NHST, however, the APA Task Force did not recommend that this approach to evaluation of data should be entirely abandoned. NHST can help researchers to evaluate whether chance or sampling error is a likely explanation for an observed outcome of a study, and at a minimum, researchers should be able to rule out chance as an explanation for their results before they begin to suggest other interpretations. Instead of abandoning NHST, which can provide useful information about the expected magnitude of sampling error when it is judiciously applied and cautiously and appropriately interpreted, the APA Task Force advocated more complete reporting of other information about the outcomes of studies, including CIs and effect size information.

3.12.2 ♦ Other Applications of the *t* Ratio

The most general form of the *t* ratio is as follows:

$$t = \frac{\text{Sample statistic} - \text{Hypothesized population parameter}}{SE_{\text{sample statistic}}}. \qquad (3.15)$$

In other words, in its most general form, a *t* test provides information about the magnitude of the difference between a sample statistic (such as *M*) and the corresponding value of the population parameter that is given in the null hypothesis (such as μ) in number of standard errors. The first application of this test that is generally covered in introductory statistics involves using a *t* ratio to evaluate whether the mean of a single sample (*M*) is close to, or far from, a hypothesized value of the population mean, μ. However, this test can be applied to several other common sample statistics.

For example, a very common type of experiment involves random assignment of participants to two groups, administration of different treatments to the two groups, and comparisons of the sample means on the outcome variable for Group 1 versus Group 2 (M_1 vs. M_2) to assess whether the outcomes differ significantly between groups that have received different treatments. An independent samples *t* test is used to assess whether there is a significant difference between the means of two populations; the formal null hypothesis can be stated as H_0: $\mu_1 = \mu_2$ or, equivalently, as H_0: $(\mu_1 - \mu_2) = 0$. To assess this null hypothesis, the researcher needs to obtain two samples; the mean from Sample 1, denoted M_1, is used to estimate μ_1, while the mean from Sample 2, M_2, is used to estimate μ_2. The populations that are being compared may correspond to naturally occurring groups (such as women vs. men or smokers vs. nonsmokers); this type of comparison is nonexperimental. In experiments, groups are often formed by random assignment, and then different treatments are administered to each group; in this case, each group corresponds to a hypothetical population of some broader group of people who might be given the treatments in the study. If Group 1 receives a new antidepressant drug and Group 2 receives a placebo, the hypothetical populations to which the researcher might want to generalize results of the study might be characterized as all depressed patients who take the drug versus all depressed patients who receive only a placebo. In most studies, researchers hope to find evidence that the means in the two samples (and, by inference, the means of the hypothetical populations) are different. If the mean emotional intelligence scores of men and women are different, it provides evidence that gender and emotional intelligence are related. If mean depression scores differ between an experimental group that receives a drug and a placebo control, this outcome provides evidence that the drug may have an effect on depression. In most behavioral and social science studies, researchers hope to find evidence that variables are related, and therefore, in most cases, they hope to reject the null hypothesis that the population means are equal.

The formula for the independent samples t ratio for testing a hypothesis about the equality of the means of two separate populations is as follows (note that this is just a specific case of the general form of t shown in Equation 3.15):

$$t = \frac{(M_1 - M_2) - (\mu_1 - \mu_2)}{SE_{M_1 - M_2}}. \tag{3.16}$$

Because the difference between the means of populations that receive different treatments in an experiment ($\mu_1 - \mu_2$) is usually hypothesized to be 0, this is usually reduced to

$$t = \frac{(M_1 - M_2)}{SE_{M_1 - M_2}}. \tag{3.17}$$

Again, the basic idea is simple. The researcher assesses whether the sample statistic outcome ($M_1 - M_2$) is close to or far from the corresponding hypothesized population parameter (usually 0) by setting up a t ratio. The t ratio is the difference $M_1 - M_2$ in terms of number of standard errors. (Note again the parallel to the evaluation of the location of a single score using a z value, $z = (X - M)/s$. When we looked at a z score for a single X value, we wanted to know how far X was from M in number of standard deviations; when we set up a t ratio, we want to know how far $M_1 - M_2$ is from $\mu_1 - \mu_2$ in number of standard errors.) A large difference $M_1 - M_2$ (and a correspondingly large t ratio) is interpreted as evidence that the results of the study are inconsistent with the null hypothesis. As in the preceding section, the critical value for t is obtained from a table of the appropriate t distribution. For the independent samples t test, the applicable $df = n_1 + n_2 - 2$, where n_1 and n_2 are the numbers of participants in Groups 1 and 2, respectively.

The t ratio for testing a hypothesis about a Pearson correlation between two variables is yet another variation on the general form of the t ratio shown in Equation 2.14:

$$t = \frac{r - \rho_{\text{hyp}}}{SE_r}. \tag{3.18}$$

(*Note:* ρ_{hyp} is the unknown population correlation, which we try to estimate by looking at the value of r in our sample. The test formula in Equation 3.18 works only for $\rho_{\text{hyp}} = 0$, not for other hypothesized values of the population correlation. More information about hypothesis tests for correlations appears in Chapters 8 and 9.)

The t test for assessing whether the (raw score) regression slope in an equation of the form $Y' = b_0 + b_1X$ differs from 0 also has the same basic form:

$$t = \frac{b_1 - b_{\text{hyp}}}{SE_b}, \tag{3.19}$$

where b_1 is the sample estimate of the regression slope, SE_{b1} is the standard error of the slope estimate, and b_{hyp} is the corresponding hypothesized population slope, which is usually hypothesized to be 0.

3.12.3 ♦ What Does It Mean to Say "$p < .05$"?

At this point, let's return to the fundamental question, What does it mean to say "$p < .05$"? An obtained *p* value represents a (theoretical) risk of Type I error; researchers want this risk to be low, and usually that means they want *p* to be less than .05. A small obtained value of *p* (such as $p = .021$) tells us that the obtained outcome such as $M_1 - M_2$ was sufficiently large so that a difference between sample means this large or larger would be expected to occur in only about 2.1% of pairs of samples from this population by chance, when $H_0 : \mu_1 = \mu_2$ is correct.

In most research situations, researchers hope to obtain outcomes (such as large *t* ratios) that are unlikely to occur when H_0 is true because they usually hope to reject H_0. It is important to understand that if one or more of the conditions and assumptions are not satisfied, the risk of Type I error is generally much higher than the nominal *p* value.

In addition to difficulties and disputes about the logic of statistical significance testing, there are additional reasons why the results of a single study should not be interpreted as conclusive evidence that the null hypothesis is either true or false. A study can be flawed in many ways that make the results uninformative, and even when a study is well designed and carefully conducted, statistically significant outcomes sometimes arise just by chance. Therefore, the results of a single study should never be treated as conclusive evidence. To have enough evidence to be confident that we know how variables are related, it is necessary to have many replications of a result.

A very brief answer to the question, What does it mean to say "$p < .05$"? is as follows: *A* p *value is a theoretical estimate of the probability (or risk) of committing a Type I error.* A Type I error occurs when the null hypothesis, H_0, is true and the researcher decides to reject H_0. However, researchers need to be aware that the "exact" *p* values that are given on computer printouts may seriously underestimate the true risk of Type I error when the assumptions for NHST are not met and the rules for carrying out statistical significance tests are not followed.

Notes

1. Although the use of NHST involves setting up a decision where researchers tend to look for "disconfirmatory" evidence, NHST should not be confused with the kind of falsification that Karl Popper advocated as a preferred scientific method. In many applications of NHST, the prediction made by the researcher's theory corresponds to H_1 rather than to H_0. Therefore, a decision to reject H_0 is often interpreted as support for a theory (rather than falsification of the researcher's theory). The logic involved in NHST is problematic even when it is well understood, and it is often misunderstood. See Kline (2004) for a discussion of the logical problems involved in NHST.

2. There is a difference between the conclusion "Do not reject H_0" and the statement "Accept H_0." The latter statement ("Accept H_0") is generally considered an inappropriate interpretation; that is, it is too strong a conclusion. Most textbooks explicitly say that researchers should not report their conclusions as "Accept H_0."

3. Beginning in Chapter 5, the null hypothesis generally corresponds to a statement that scores on a predictor and outcome variable are not related. For example, in experiments, H_0 generally corresponds to a hypothesis that the mean outcome scores are the same across groups that receive different treatments or, in other words, the hypothesis that the manipulated treatment variable has no effect on outcomes. In nonexperimental studies, H_0 generally corresponds to the null hypothesis that scores on an outcome variable *Y* cannot be predicted from scores on an independent variable *X*. In most experiments, researchers hope to obtain evidence that a treatment

does have an effect on outcomes, and in most nonexperimental research, researchers hope to obtain evidence that they can predict scores on outcome variables. Thus, in later applications of NHST, researchers usually hope to reject H_0 (because H_0 corresponds to "no treatment effect" or "no relationship between scores on the predictor and outcome variables."

4. This value was obtained by linear interpolation. The tabled values of z and their corresponding tail areas were as follows:

Value of z	*Value of p, Tail Area to the Right of z*
1.64	.045
1.65	.055

The desired exact tail area of .05 is halfway between the two table entries for the tail area; that is, .05 is (.045 + .055)/2. We can therefore find the corresponding exact value of z by finding the midpoint between the two values of z that appear in the table. By linear interpolation, a z score of $(1.64 + 1.65)/2 = 1.645$ corresponds to a tail area of .05.

5. It is not correct to interpret $1 - p$ as "the probability of replicating a significant outcome in future studies" because the probability of obtaining statistically significant outcomes (statistical power) in future studies depends on other factors such as N (sample size). However, Greenwald et al. (1996) have argued that $1 - p$ is *related* to the probability of future replication.

Comprehension Questions

1. What research decision influences the magnitude of risk of a Type I error?
2. What factors influence the magnitude of risk of a Type II error?
3. How are the risk of a Type II error and the statistical power related?
4. Other factors being equal, which type of significance test requires a value of t that is larger (in absolute value) to reject H_0—a directional or a nondirectional test?
5. In your own words, What does it mean to say "$p < .05$"?
6. Describe at least two potential problems with NHST.
7. What is a null hypothesis? An alternative hypothesis?
8. What is an alpha level? What determines the value of α?
9. What is an "exact" p value?
10. What is the difference between a directional and a nondirectional significance test?
11. Why do reported or "nominal" p values often seriously underestimate the true risk of a Type I error?

12. What is statistical power? What information is needed to decide what sample size is required to obtain some desired level of power (such as 80%)?
13. What recommendations did the APA Task Force make about reporting statistical results? Are significance tests alone sufficient?
14. What conclusions can be drawn from a study with a null result?
15. What conclusions can be drawn from a study with a "statistically significant" result?
16. Briefly discuss: What information do you look at to evaluate whether an effect obtained in an experiment is large enough to have "practical" or "clinical" significance?
17. When a researcher reports a p value, "p" stands for "probability" or risk.
    a. What probability does this p refer to?
    b. Do we typically want p to be large or small?
    c. What is the conventional standard for an "acceptably small" p value?
18. Suppose a researcher writes in a journal article that "the obtained p was $p = .032$; thus, there is only a 3.2% chance that the null hypothesis is correct." Is this a correct or incorrect statement?
19. A p value can be interpreted as a (conditional) risk that a decision to reject H_0 is a Type I error, but the p values reported in research papers are valid indications of the true risk of Type I error *only if* the data meet the assumptions for the test and the researcher has followed the rules that govern the use of significance tests. Identify *one* of the most common researcher behaviors that make the actual risk of Type I error much higher than the "nominal" risk of Type I error that is set by choosing an alpha level.
20. Use Table 3.3: Suppose you are planning to do a study where you will use a one-sample t test. Based on past research, you think that the effect size you are trying to detect may be approximately equal to Cohen's $d = .30$. You plan to use $\alpha = .05$, nondirectional (two-tailed). (a) If you want to have power of .80, what minimum N do you need in the sample? (b) If you can only afford to have $N = 20$ participants in the sample, approximately what is the expected level of statistical power?

CHAPTER 4

Preliminary Data Screening

4.1 ♦ Introduction: Problems in Real Data

Real datasets often contain errors, inconsistencies in responses or measurements, outliers, and **missing values**. Researchers should conduct thorough preliminary data screening to identify and remedy potential problems with their data prior to running the data analyses that are of primary interest. Analyses based on a dataset that contains errors, or data that seriously violate assumptions that are required for the analysis, can yield misleading results.

Some of the potential problems with data are as follows: errors in data coding and data entry, **inconsistent responses**, missing values, extreme outliers, nonnormal distribution shapes, within-group sample sizes that are too small for the intended analysis, and nonlinear relations between quantitative variables. Problems with data should be identified and remedied (as adequately as possible) prior to analysis. A research report should include a summary of problems detected in the data and any remedies that were employed (such as deletion of outliers or data transformations) to address these problems.

Version 14 of SPSS for Windows is used for the data analysis examples in this book. The following discussion assumes that readers are familiar with SPSS and that they know how to open, edit, and save SPSS data files. Readers who do not have prior experience with SPSS will find a good basic introduction in Pavkov (2005) or similar SPSS primers. SPSS is menu-driven; the examples presented here specify the selection of menu options and the use of radio buttons and check boxes to request specific summary statistics. Some SPSS procedures offer additional kinds of output information beyond that which is presented in this textbook. SPSS has online help that answers some of the questions students will encounter in their first few sessions using the program. Most of the statistical procedures in SPSS correspond to fairly similar procedures in other widely used programs (such as SAS, STATA, or SYSTAT). The names of commands and the output formats for statistical results are generally fairly similar across these programs. Users of other programs should find that most of the issues discussed in this book concerning data analysis and interpretation of results are applicable.

4.2 ♦ Quality Control During Data Collection

There are many different possible methods of data collection. A psychologist may collect data on personality or attitudes by asking human participants to answer questions on a questionnaire. A medical researcher may use a computer-controlled blood pressure monitor to assess systolic blood pressure (SBP) or other physiological responses. A researcher may record observations of animal behavior. Physical measurements (such as height or weight) may be taken. Most methods of data collection are susceptible to recording errors, and researchers need to know what kinds of errors are likely to occur.

For example, researchers who use self-report data to do research on personality or attitudes need to be aware of common problems with this type of data. Participants may distort their answers because of **social desirability bias**; they may misunderstand questions; they may not remember the events that they are asked to report about; they may deliberately try to "fake good" or "fake bad"; they may even make random responses without reading the questions. A participant may accidentally skip a question on a survey and, subsequently, use the wrong lines on the answer sheet to enter each response; for example, the response to Question 4 may be filled in as Item 3 on the answer sheet, the response to Question 5 may be filled in as Item 4, and so forth. In addition, research assistants have been known to fill in answer sheets themselves instead of having the participants complete them. Good quality control in the collection of self-report data requires careful consideration of question wording and response format and close supervision of the administration of surveys. Converse and Presser (1999); Robinson, Shaver, and Wrightsman (1991); and Stone, Turkkan, Kurtzman, Bachrach, and Jobe (1999) provide a more detailed discussion of methodological issues in the collection of self-report data.

For observer ratings, it is important to consider issues of **reactivity** (i.e., the presence of an observer may actually change the behavior that is being observed). It is also important to establish good interobserver reliability through thorough training of observers and empirical assessment of interobserver agreement. See Aspland and Gardner (2003), Bakeman (2000), Gottman and Notarius (2002), and Reis and Gable (2000) for a discussion of methodological issues in the collection of observational data.

For physiological measures, it is necessary to screen for artifacts (e.g., when electroencephalogram electrodes are attached near the forehead, they may detect eye blinks as well as brain activity; these eye blink artifacts must be removed from the electroencephalogram signal prior to other processing). See Cacioppo, Tassinary, and Berntson (2000) for methodological issues in the collection of physiological data.

This discussion does not cover all possible types of measurement problems, of course; it only mentions a few of the many possible problems that may arise in data collection. Researchers need to be aware of potential problems or sources of artifact associated with any data collection method that they use, whether they use data from archival sources, experiments with animals, mass media, social statistics, or other methods not mentioned here.

4.3 ♦ Example of an SPSS Data Worksheet

The dataset used to illustrate data-screening procedures in this chapter is named bpstudy.sav. The scores appear in Table 4.1, and an image of the corresponding SPSS worksheet appears in Figure 4.1.

Table 4.1 ♦ Data for the Blood Pressure/Social Stress Study

idnum	*GENDER*	*SMOKE*	*AGE*	*SYS1*	*DIA1*	*HR1*	*WEIGHT*
1	2	1	19	118	69	70	135
2	1	1	23	138	74	78	155
3	1	1	23	160	84	61	197
4	2	1	19	118	71	70	140
5	2	4	20	162	103	89	122
6	1	1	20	173	102	79	200
7	2	1	19	128	99	68	128
8	2	1	18	114	66	63	125
9	1	2	22	121	75	63	165
10	1	1	18	122	67	69	170
11	3	1	21	128	81	77	135
12	1		27	162	99	62	165
13	2	1	19	122	69	61	115
14	2	1	18	109	60	70	113
15	2	1	88	157	75	75	118
16	1	1	20	110	56	81	160
17	2	4	20	108	68	87	120
18	1	1	18	125	80	71	155
19	1	1	20	123	67	88	154
20	2	2	19	105	66	70	130
21	2	1	19	187	103	80	149
22	1	1	21	133	76	89	155
23	1	1	19	132	82	57	185
24	2	1	18	135	76	72	120
25	2	1	19	143	71	62	112
26	1	2	20	129	74	71	172
27	2	1	19	112	59	76	130
28	2	2	18	117	60	74	125
29	2	1	19	113	81	93	103
30	1	4	19	94	53	70	150
31	1	1	19	115	69	65	230
32	1	1	20	109	74	89	130
33	2	1	18	116	63	76	155
34	2	2	18	125	74	79	140
35	1	1	19	120	63	56	180
36	2	1	19	114	67	76	118
37	1	1	20	103	79	82	160
38	2	2	19	131	82	68	115
39	2	1	20	99	60	74	123
40	1	1	19	127	77	75	175
41	2	1	19	113	76	81	115
42	2	1	18	104	63	84	120
43	2	1	21	136	79	69	150
44	1	1	21	109	77	59	170
45	2	1	21	147	76	89	125
46	2	1	19	110	73	76	132

(Continued)

Table 4.1 ♦ (Continued)

idnum	*GENDER*	*SMOKE*	*AGE*	*SYS1*	*DIA1*	*HR1*	*WEIGHT*
47	2	1	19	114	69	84	125
48	1	2	22	143	90	65	190
49	2	1	22	148	78	70	190
50	2	1	20	127	66	71	170
51	2	1	20	110	61	73	104
52	2	1	21	105	62	70	128
53	2	1	20	107	70	82	140
54	1	1	21	166	101	72	170
55	2	1	21	93	71	78	110
56	2	1	20	106	65	81	127
57	1	1	19	145	93	86	175
58	1	1	19	134	80	75	175
59	1	1	18	144	80	66	150
60	1	1	18	142	85	76	145
61	1	1	20	114	68	79	160
62	1	1	19	121	51	83	165
63	2	2	18	125	78	75	115
64	1	2	18	128	85	76	175
65	2	1	18	101	73	63	118

SOURCE: Mooney (1990).

NOTES: 1. idnum = arbitrary, unique identification number for each participant. 2. GENDER was coded 1 = Male, 2 = Female. 3. SMOKE was coded 1 = Nonsmoker, 2 = Light smoker, 3 = Moderate smoker, 4 = Heavy or regular smoker. 4. AGE = age in years. 5. SYS1 = SBP at Time 1/baseline. 6. DIA1 = DBP at Time 1/baseline. 7. HR1 = heart rate at Time 1/baseline. 8. WEIGHT = body weight in pounds.

This file contains selected data from a dissertation that assessed the effects of social stress on blood pressure (Mooney, 1990). The most important features in Figure 4.1 are as follows. Each row in the Data View worksheet corresponds to the data for 1 case or 1 participant. In this example, there are a total of $N = 65$ participants; therefore, the dataset has 65 rows. Each column in the Data View worksheet corresponds to a variable; the SPSS variable names appear along the top of the data worksheet. In Figure 4.1 scores are given for the following SPSS variables: idnum, GENDER, SMOKE (smoking status), AGE, SYS1 (SBP at Time 1), DIA1 (diastolic blood pressure, DBP, at Time 1), HR1 (heart rate at Time 1), and WEIGHT. The numerical values contained in this data file were typed into the SPSS Data View worksheet by hand.

The menu bar across the top of the SPSS Data View worksheet in Figure 4.1 can be used to select menus for different types of procedures. The pull-down menu for <File> includes options such as opening and saving data files. The pull-down menus for <Analyze> and <Graphs> provide access to SPSS procedures for data analysis and graphics, respectively.

There are two tabs near the lower left-hand corner of the Data View of the SPSS worksheet; these can be used to toggle back and forth between the Data View (shown in Figure 4.1) and the Variable View (shown in Figure 4.2) versions of the SPSS data file.

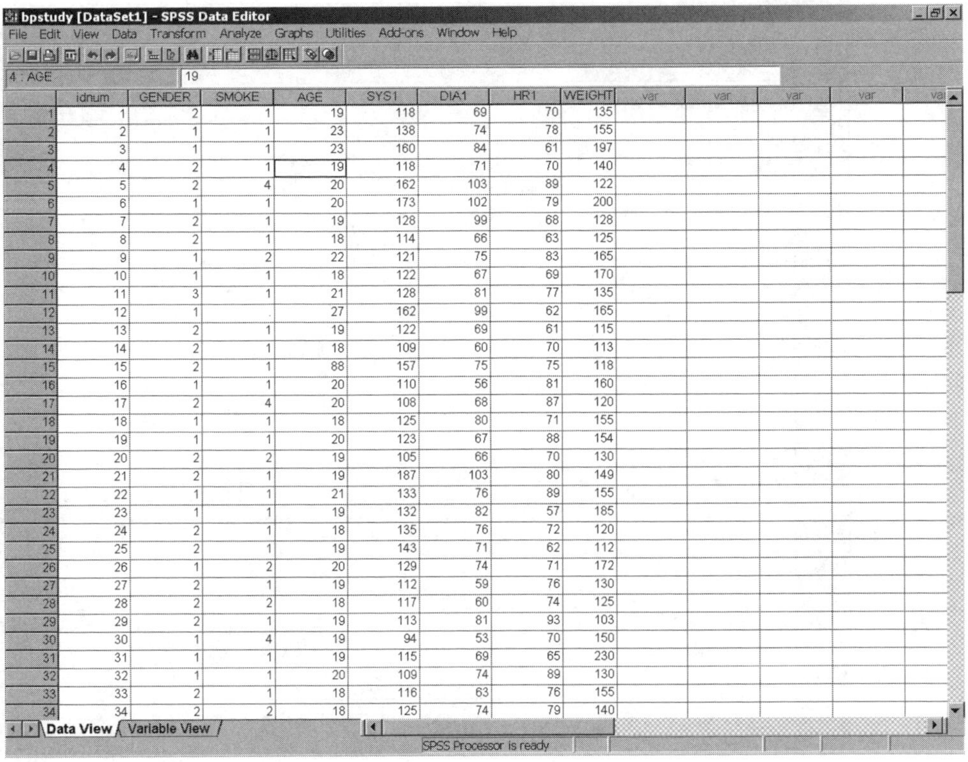

bpstudy [DataSet1] - SPSS Data Editor

File Edit View Data Transform Analyze Graphs Utilities Add-ons Window Help

4 : AGE 19

	idnum	GENDER	SMOKE	AGE	SYS1	DIA1	HR1	WEIGHT	var	var	var	var	var
1	1	2	1	19	118	69	70	135					
2	2	1	1	23	138	74	78	155					
3	3	1	1	23	160	84	61	197					
4	4	2	1	19	118	71	70	140					
5	5	2	4	20	162	103	89	122					
6	6	1	1	20	173	102	79	200					
7	7	2	1	19	128	99	68	128					
8	8	2	1	18	114	66	63	125					
9	9	1	2	22	121	75	83	165					
10	10	1	1	18	122	67	69	170					
11	11	3	1	21	128	81	77	135					
12	12	1	.	27	162	99	62	165					
13	13	2	1	19	122	69	61	115					
14	14	2	1	18	109	60	70	113					
15	15	2	1	88	157	75	75	118					
16	16	1	1	20	110	56	81	160					
17	17	2	4	20	108	68	87	120					
18	18	1	1	18	125	80	71	155					
19	19	1	1	20	123	67	88	154					
20	20	2	2	19	105	66	70	130					
21	21	2	1	19	187	103	80	149					
22	22	1	1	21	133	76	89	155					
23	23	1	1	19	132	82	57	185					
24	24	2	1	18	135	76	72	120					
25	25	2	1	19	143	71	62	112					
26	26	1	2	20	129	74	71	172					
27	27	2	1	19	112	59	76	130					
28	28	2	2	18	117	60	74	125					
29	29	2	1	19	113	81	93	103					
30	30	1	4	19	94	53	70	150					
31	31	1	1	19	115	69	65	230					
32	32	1	1	20	109	74	89	130					
33	33	2	1	18	116	63	76	155					
34	34	2	2	18	125	74	79	140					

Data View / Variable View

SPSS Processor is ready

Figure 4.1 ♦ SPSS Worksheet for the Blood Pressure/Social Stress Study (Data View)

SOURCE: Mooney (1990).

The Variable View of an SPSS worksheet, shown in Figure 4.2, provides a place to document and describe the characteristics of each variable, to supply labels for variables and score values, and to identify missing values.

For example, examine the row of the Variable View worksheet that corresponds to the variable named GENDER. The scores on this variable were numerical; that is, the scores are in the form of numbers (rather than alphabetic characters). Other possible variable types include dates or string variables that consist of alphabetic characters instead of numbers. If the researcher needs to identify a variable as string or date type, the researcher clicks on the cell for Variable Type and selects the appropriate variable type from the pull-down menu list. In the datasets used as examples in this textbook, almost all the variables are numerical.

The Width column indicates how many significant digits the scores on each variable can have. For this example, the variables GENDER and SMOKE were each allowed a one-digit code; the variable AGE was allowed a two-digit code; and the remaining variables (heart rate, blood pressure, and body weight) were each allowed three digits. The Decimals column indicates how many digits are displayed after the decimal point. All the variables in this dataset (such as age in years and body weight in pounds) are given to the nearest integer value, and so all these variables are displayed with "0" digits to the

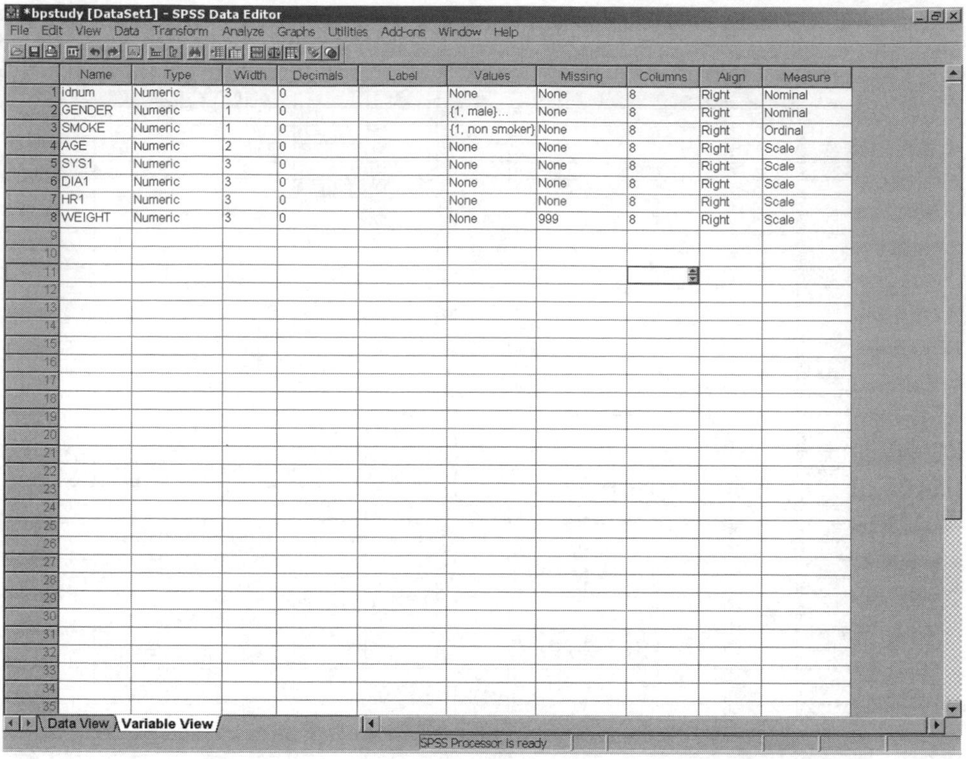

*bpstudy [DataSet1] - SPSS Data Editor

File Edit View Data Transform Analyze Graphs Utilities Add-ons Window Help

	Name	Type	Width	Decimals	Label	Values	Missing	Columns	Align	Measure
1	idnum	Numeric	3	0		None	None	8	Right	Nominal
2	GENDER	Numeric	1	0		{1, male}...	None	8	Right	Nominal
3	SMOKE	Numeric	1	0		{1, non smoker}	None	8	Right	Ordinal
4	AGE	Numeric	2	0		None	None	8	Right	Scale
5	SYS1	Numeric	3	0		None	None	8	Right	Scale
6	DIA1	Numeric	3	0		None	None	8	Right	Scale
7	HR1	Numeric	3	0		None	None	8	Right	Scale
8	WEIGHT	Numeric	3	0		None	999	8	Right	Scale

Data View | Variable View

SPSS Processor is ready

Figure 4.2 ♦ SPSS Worksheet for the Blood Pressure/Social Stress Study (Variable View)

right of the decimal place. If a researcher has a variable, such as grade point average (GPA), that is usually reported to two decimal places (as in GPA = 2.67), then the researcher would select "2" as the number of digits to display after the decimal point.

The next column, Label, provides a place where each variable name can be associated with a longer descriptive label. This is particularly helpful when the brief SPSS variable names are not completely self-explanatory. For example, "body weight in pounds" appears as a label for the variable WEIGHT. The Values column provides a place where labels can be associated with the individual score values of each variable; this is primarily used with nominal or categorical variables. Figure 4.3 shows the dialog window that opens up when the user clicks on the cell for Values for the variable GENDER. To associate each score with a verbal label, the user types in the score (such as "1") and the corresponding verbal label (such as Male) and then clicks the Add button to add this label to the list of value labels. When all the labels have been specified, clicking on OK returns to the main Variable View worksheet. In this example, a score of "1" on GENDER corresponds to Male and a score of "2" on GENDER corresponds to Female.

The column headed Missing provides a place to identify scores as codes for missing values. Consider the following example to illustrate the problem that arises in data analysis when there are missing values. Suppose that 1 participant did not answer the question about body weight. If the data analyst enters a value of "0" for the body weight of this person who did not provide information about body weight and does not identify "0" as a

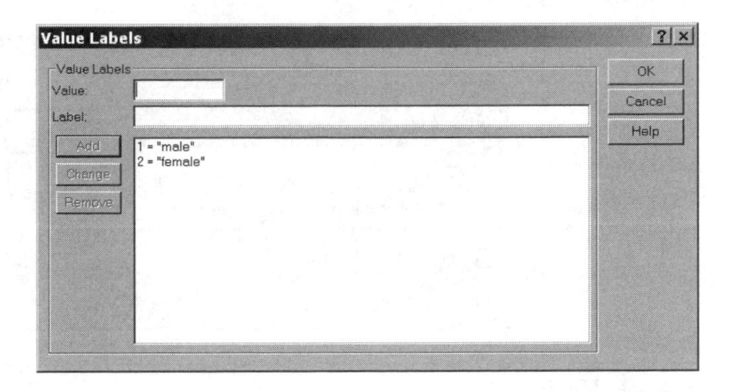

Figure 4.3 ♦ Value Labels

code for a missing value, this value of "0" would be included when SPSS sums the scores on body weight to compute a mean for weight. The sample mean is not robust to outliers; that is, a sample mean for body weight will be substantially lower when a value of "0" is included for 1 participant than it would be if that value of "0" was excluded from the computation of the sample mean. What should the researcher do to make sure that missing values are not included in the computation of sample statistics?

SPSS provides two different ways to handle missing score values. The first option is to leave the cell in the SPSS Data View worksheet that corresponds to the missing score blank. In Figure 4.1, participant number 12 did not answer the question about smoking status; therefore, the cell that corresponds to the response to the variable SMOKE for Participant 12 was left blank. By default, SPSS treats empty cells as "system missing" values. If a table of frequencies is set up for scores on SMOKE, the response for Participant 12 is labeled as a missing value. If a mean is calculated for scores on smoking, the score for Participant 12 is not included in the computation as a value 0; instead, it is omitted from the computation of the sample mean.

A second method is available to handle missing values; it is possible to use different code numbers to represent different types of missing data. For example, a survey question that is a follow-up about the amount and frequency of smoking might be coded "9" if it was not applicable to an individual (because that individual never smoked), "99" if the question was not asked because the interviewer ran out of time, and "88" if the respondent refused to answer the question. For the variable WEIGHT, body weight in pounds, a score of "999" was identified as a missing value by clicking on the cell for Missing and then typing a score of "999" into one of the windows for missing values; see the Missing Values dialog window in Figure 4.4. A score of "999" is defined as a missing value code, and therefore, these scores are not included when statistics are calculated. It is important to avoid using codes for missing values that correspond to possible valid responses. Consider the question, How many children are there in a household? It would not make sense to use a score of "0" or a score of "9" as a code for missing values, because either of these could correspond to the number of children in some households. It would be acceptable to use a code of "99" to represent a missing value for this variable because no single-family household could have such a large number of children.

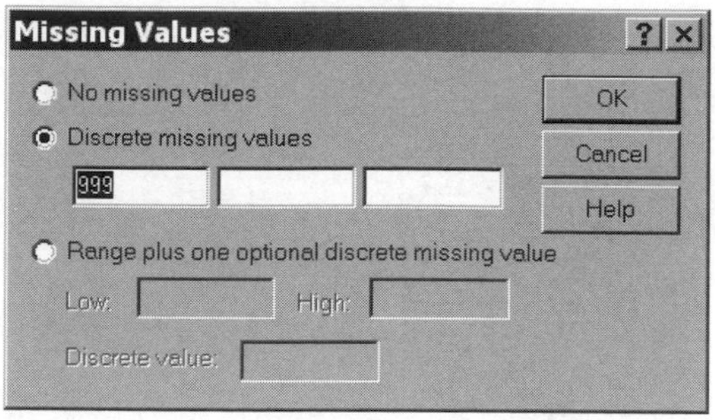

Missing Values

No missing values

Discrete missing values

999

Range plus one optional discrete missing value

Low: High:

Discrete value:

OK

Cancel

Help

Figure 4.4 ♦ Missing Values

The next few columns in the SPSS Variable View worksheet provide control over the way the values are displayed in the Data View worksheet. The column headed Columns indicates the display width of each column in the SPSS Data View worksheet, in number of characters. This was set at eight characters wide for most variables. The column headed Align indicates whether scores will be shown left justified, centered, or (as in this example) right justified in each column of the worksheet.

Finally, the column in the Variable View worksheet that is headed Measure indicates the level of measurement for each variable. SPSS designates each numerical variable as nominal, ordinal, or scale (scale is equivalent to interval/ratio level of measurement, as described in Chapter 1 of this textbook). In this sample dataset, idnum (an arbitrary and unique identification number for each participant) and GENDER were identified as categorical or nominal variables. Smoking status (SPSS variable name SMOKE) was coded on an ordinal scale from 1 to 4, with 1 = Nonsmoker, 2 = Light smoker, 3 = Moderate smoker, and 4 = Heavy smoker. The other variables (heart rate, blood pressure, body weight, and age) are quantitative and interval/ratio, so they were designated as "scale" in level of measurement.

4.4 ♦ Identification of Errors and Inconsistencies

The SPSS data file should be proofread and compared with original data sources (if these are accessible) to correct errors in data coding or data entry. For example, if self-report data are obtained using computer-scorable answer sheets, the correspondence between scores on these answer sheets and scores in the SPSS data file should be verified. This may require proofreading data line by line and comparing the scores with the data on the original answer sheets. It is helpful to have a unique code number associated with each case so that each line in the data file can be matched with the corresponding original data sheet.

Even when line-by-line proofreading has been done, it is useful to run simple exploratory analyses as an additional form of data screening. Rosenthal (cited in Wright, 2003) called this process of exploration "making friends with your data." Subsequent sections of this chapter show how examination of the frequency distribution tables and graphs provides an overview of the characteristics of people in this sample—for example, how many males and females were included in the study, how many nonsmokers versus heavy smokers were included, and the range of scores on physiological responses such as heart rate.

Examining response consistency across questions or measurements is also useful. If a person chooses the response "I have never smoked" to one question and then reports smoking 10 cigarettes on an average day in another question, these responses are inconsistent. If a participant's responses include numerous inconsistencies, the researcher may want to consider removing that participant's data from the data file.

Based on knowledge of the variables and the range of possible response alternatives, a researcher can identify some responses as **"impossible"** or "unlikely." For example, if participants are provided a choice of the following responses to a question about smoking status: 1 = Nonsmoker, 2 = Light smoker, 3 = Moderate smoker, and 4 = Heavy smoker, and a participant marks response number "6" for this question, the value of "6" does not correspond to any of the response alternatives provided for the question. When impossible, unlikely, or inconsistent responses are detected, there are several possible remedies. First, it may be possible to go back to original data sheets or experiment logbooks to locate the correct information and use it to replace an incorrect score value. If that is not possible, the invalid score value can be deleted and replaced with a blank cell entry or a numerical code that represents a missing value. It is also possible to select out (i.e., temporarily or permanently remove) cases that have impossible or unlikely scores.

4.5 ♦ Missing Values

Within the SPSS program, an empty or blank cell in the data worksheet is interpreted as a System Missing value. Alternatively, as described earlier, the Missing Value column in the Variable View worksheet in SPSS can be used to identify some specific numerical codes as missing values and to use different numerical codes to correspond to different types of missing data. For example, for a variable such as verbal Scholastic Aptitude Test (SAT) score, codes such as "888" = "student did not take the SAT" and "999" = "participant refused to answer" could be used to indicate different reasons for the absence of a valid score.

Ideally, a dataset should have few missing values. A systematic pattern of missing observations suggests possible bias in nonresponse. For example, males might be less willing than females to answer questions about negative emotions such as depression; students with very low SAT scores may refuse to provide information about SAT performance more often than students with high SAT scores. To assess whether missing responses on depression are more common among some groups of respondents, or are associated with scores on some other variable, the researcher can set up a variable that is coded "1" (respondent answered a question about depression) versus "0" (respondent did not answer the question about depression). Analyses can then be performed to see whether this variable, which represents missing versus nonmissing data on one variable, is associated with scores on any other variable. If the researcher finds, for example, that a higher proportion of men than women refused to answer a question about depression, it signals possible problems with generalizability of results; for example, conclusions about depression in men can be generalized only to the kinds of men who are willing to answer such questions.

It is useful to assess whether specific individual participants have large numbers of missing scores; if so, data for these participants could simply be deleted. Similarly, it may be useful to see whether certain variables have very high nonresponse rates; it may be necessary to drop these variables from further analysis.

When analyses involving several variables (such as computations of all possible correlations among a set of variables) are performed in SPSS, it is possible to request either **listwise** or **pairwise deletion**. For example, suppose that the researcher wants to use the bivariate correlation procedure in SPSS to run all possible correlations among variables named V_1, V_2, V_3, and V_4. If listwise deletion is chosen, the data for a participant are completely ignored when all these correlations are calculated if the participant has a missing score on any *one* of the variables included in the list. In pairwise deletion, each correlation is computed using data from all the participants who had nonmissing values on that particular pair of variables. For example, suppose that there is one missing score on the variable V_1. If listwise deletion is chosen, then the data for the participant who had a missing score on V_1 is not used to compute any of the correlations (between V_1 and V_2, V_2 and V_3, V_2 and V_4, etc.). On the other hand, if pairwise deletion is chosen, the data for the participant who is missing a score on V_1 cannot be used to calculate any of the correlations that involve V_1 (e.g., V_1 with V_2, V_1 with V_3, V_1 with V_4), but the data from this participant will be used when correlations that don't require information about V_1 are calculated (correlations between V_2 and V_3, V_2 and V_4, and V_3 and V_4). When using listwise deletion, the same number of cases and subset of participants are used to calculate all the correlations for all pairs of variables. When using pairwise deletion, depending on the pattern of missing values, each correlation may be based on a different *N* and a different subset of participants than those used for other correlations.

The default for handling missing data in most SPSS procedures is listwise deletion. The disadvantage of listwise deletion is that it can result in a rather small *N* of participants, and the advantage is that all correlations are calculated using the same set of participants. Pairwise deletion can be selected by the user, and it preserves the maximum possible *N* for the computation of each correlation; however, both the number of participants and the composition of the sample may vary across correlations, and this can introduce inconsistencies in the values of the correlations (as described in more detail by Tabachnick & Fidell, 2007).

When a research report includes a series of analyses and each analysis includes a different set of variables, the *N* of scores that are included may vary across analyses (because different people have missing scores on each variable). This can raise a question in readers' minds: Why do the *N*s change across pages of the research report? When there are large numbers of missing scores, quite different subsets of data may be used in each analysis, and this may make the results not comparable across analyses. To avoid these potential problems, it may be preferable to select out all the cases that have missing values on all the variables that will be used ahead of time, so that the same subset of participants (and the same *N* of scores) are used in all the analyses in a paper.

The default in SPSS is that cases with **system missing values** or scores that are specifically identified as missing values are excluded from computations; but this can result in a substantial reduction in the sample size for some analyses. Another way to deal with missing data is by substitution of a reasonable estimated score value to replace each missing response. Missing value replacement can be done in many different ways; for example, the mean score on a variable can be substituted for all missing values on that variable, or estimated values can be calculated separately for each individual participant using regression methods to predict that person's missing score from her or his scores on other, related variables. This is often called **imputation of missing data**. Procedures for missing value replacement can be rather complex (Schafer, 1997, 1999; Schafer & Olsen, 1998).

Tabachnick and Fidell (2007) summarized their discussion of missing value replacement by saying that the seriousness of the problem of missing values depends on "the pattern of missing data, how much is missing, and why it is missing" (p. 62). They also noted that the decision about how to handle missing data (e.g., deletion of cases or variables, or estimation of scores to replace missing values) is "a choice among several bad alternatives" (p. 63). If some method of imputation or estimation is employed to replace missing values, it is desirable to repeat the analysis with the missing values omitted. Results are more believable, of course, if they are essentially the same with and without the replacement scores.

4.6 ♦ Empirical Example of Data Screening for Individual Variables

In this textbook, variables are treated as either categorical or quantitative (see Chapter 1 for a review of this distinction between two types of variables). Different types of graphs and descriptive statistics are appropriate for use with categorical versus quantitative variables, and for that reason, data screening is discussed separately for categorical and quantitative variables.

4.6.1 ♦ Frequency Distribution Tables

For both categorical (nominal) and quantitative (scale) variables, a table of frequencies can be obtained to assess the number of persons or cases who had each different score value. These frequencies can be converted to proportions or percentages. Examination of a frequency table quickly provides answers to the following questions about each categorical variable: How many groups does this variable represent? What is the number of persons in each group? Are there any groups with *n*s that are too small for the group to be used in analyses that compare groups (e.g., analysis of variance, ANOVA)?

If a group with an *n* of 2 or 3 cases is detected, the researcher needs to decide what to do with the cases in that group. The group could be dropped from all analyses, or if it makes sense to do so, the small *n* group could be combined with one or more of the other groups (by recoding the scores that represent group membership on the categorical variable).

For both categorical and quantitative variables, a frequency distribution also makes it possible to see if there are any "impossible" score values. For instance, if the categorical variable GENDER on a survey has just two response options, 1 = Male and 2 = Female, then scores of "3" and higher are not valid or interpretable responses. Impossible score values should be detected during proofreading, but examination of frequency tables provides another opportunity to see if there are any impossible score values on categorical variables.

SPSS was used to obtain a frequency distribution table for the variables GENDER and AGE. Starting from the data worksheet view (as shown in Figure 4.1), the following menu selections (as shown in Figure 4.5) were made: <Analyze> → <Descriptive Statistics> → <Frequencies>.

The SPSS dialog window for the Frequencies procedure appears in Figure 4.6. To specify which variables are included in the request for frequency tables, the user points to the names of the two variables (GENDER and AGE) and clicks the right-pointing arrow to move these variable names into the right-hand window. Output from this procedure appears in Figure 4.7.

bpstudy [DataSet1] - SPSS Data Editor

File Edit View Data Transform Analyze Graphs Utilities Add-ons Window Help

Analyze menu: Reports; Descriptive Statistics; Compare Means; General Linear Model; Correlate; Regression; Classify; Data Reduction; Scale; Nonparametric Tests; Multiple Response

Descriptive Statistics submenu: Frequencies...; Descriptives...; Explore...; Crosstabs...; Ratio...

4 : AGE 19

	idnum	GENDER					HR1	WEIGHT
1	1	2					70	135
2	2	1					78	155
3	3	1					61	197
4	4	2			118	71	70	140
5	5	2			162	103	89	122
6	6	1			173	102	79	200
7	7	2			128	99	68	128
8	8	2	1	18	114	66	63	125
9	9	1	2	22	121	75	83	165
10	10	1	1	18	122	67	69	170
11	11	3	1	21	128	81	77	135
12	12	1	.	27	162	99	62	165
13	13	2	1	19	122	69	61	115
14	14	2	1	18	109	60	70	113
15	15	2	1	88	157	75	75	118
16	16	1	1	20	110	56	81	160
17	17	2	4	20	108	68	87	120
18	18	1	1	18	125	80	71	155
19	19	1	1	20	123	67	88	154
20	20	2	2	19	105	66	70	130
21	21	2	1	19	187	103	80	149
22	22	1	1	21	133	76	89	155
23	23	1	1	19	132	82	57	185
24	24	2	1	18	135	76	72	120
25	25	2	1	19	143	71	62	112
26	26	1	2	20	129	74	71	172
27	27	2	1	19	112	59	76	130
28	28	2	2	18	117	60	74	125
29	29	2	1	19	113	81	93	103
30	30	1	4	19	94	53	70	150
31	31	1	1	19	115	69	65	230
32	32	1	1	20	109	74	89	130
33	33	2	1	18	116	63	76	155
34	34	2	2	18	125	74	79	140

Data View / Variable View

Frequencies   SPSS Processor is ready

Start | c04_screen | ch04screen - ... | tables4 - Mic... | c04fig - Micro... | bpstudy [D... | 3:14 PM

Figure 4.5 ♦ SPSS Menu Selections: <Analyze> → <Descriptive Statistics> → <Frequencies>

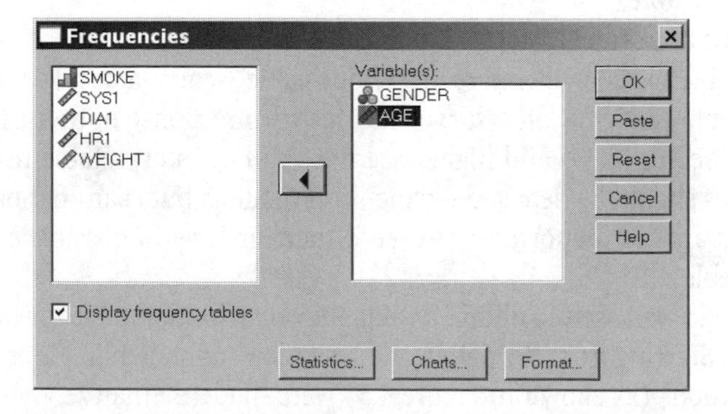

Figure 4.6 ♦ SPSS Dialog Window for the Frequencies Procedure

In the first frequency table in Figure 4.7, there is one "impossible" response for GENDER. The response alternatives provided for the question about gender were 1 = Male and 2 = Female, but a response of 3 appears in the summary table; this does not correspond to a valid response option. In the second frequency table in Figure 4.7, there is an extreme score (88 years) for AGE. This is a possible, but unusual, age for a college student. As will be discussed later in this chapter, scores that are extreme or unusual are often identified as outliers and sometimes removed from the data prior to doing other analyses.

4.6.2 ♦ Removal of Impossible or Extreme Scores

In SPSS, the Select Cases command can be used to remove cases from a data file prior to other analyses. To select out the participant with a score of "3" for GENDER and also the participant with an age of 88, the following SPSS menu selections (see Figure 4.8) would be used: <Data> → <Select Cases>.

Statistics

		GENDER	AGE
N	Valid	65	65
	Missing	0	0

GENDER

		Frequency	Percent	Valid Percent	Cumulative Percent
Valid	1	28	43.1	43.1	43.1
	2	36	55.4	55.4	98.5
	3	1	1.5	1.5	100.0
	Total	65	100.0	100.0	

AGE

		Frequency	Percent	Valid Percent	Cumulative Percent
Valid	18	14	21.5	21.5	21.5
	19	22	33.8	33.8	55.4
	20	14	21.5	21.5	76.9
	21	8	12.3	12.3	89.2
	22	3	4.6	4.6	93.8
	23	2	3.1	3.1	96.9
	27	1	1.5	1.5	98.5
	88	1	1.5	1.5	100.0
	Total	65	100.0	100.0	

Figure 4.7 ♦ Output From the SPSS Frequencies Procedure for GENDER (Prior to Removal of "Impossible" Score Values)

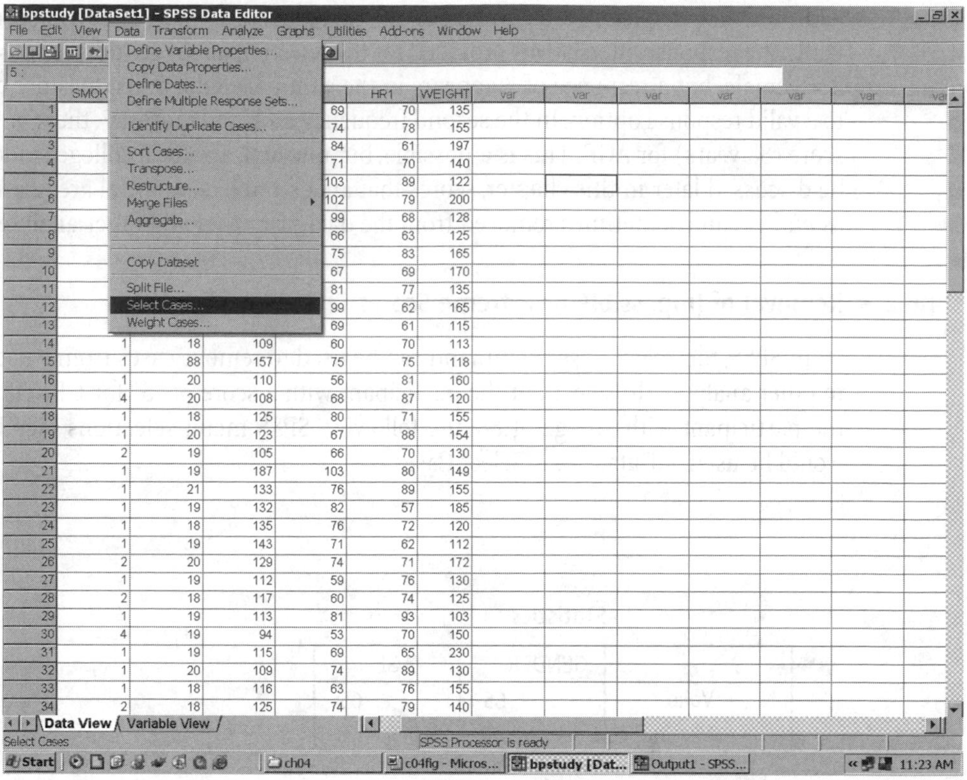

Figure 4.8 ♦ SPSS Menu Selections for <Data> ⟶ <Select Cases> Procedure

The initial SPSS dialog window for Select Cases appears in Figure 4.9.

A logical "If" conditional statement can be used to exclude specific cases. For example, to exclude the data for the person who reported a value of "3" for GENDER, click the radio button for "If condition is satisfied" in the first Select Cases dialog window. Then, in the "Select Cases: If" window, type in the logical condition "GENDER ~= 3." The symbol "~=" represents the logical comparison "not equal to"; thus, this logical "If" statement tells SPSS to include the data for all participants whose scores for GENDER are *not* equal to "3." The entire line of data for the person who reported "3" as a response to GENDER is (temporarily) filtered out or set aside as a result of this logical condition.

It is possible to specify more than one logical condition. For example, to select cases that have valid scores on GENDER and that do not have extremely high scores on AGE, we could set up the logical condition "GENDER ~= 3 and AGE < 70," as shown in Figures 4.10 and 4.11. SPSS evaluates this logical statement for each participant. Any participant with a score of "3" on GENDER and any participant with a score greater than or equal to 70 on AGE is excluded or selected out by this Select If statement.

When a case has been selected out using the Select Cases command, a crosshatch mark appears over the case number for that case (on the far left-hand side of the SPSS data worksheet). Cases that are selected out can be temporarily filtered or permanently deleted. In Figure 4.12, the SPSS data worksheet is shown as it appears after the execution

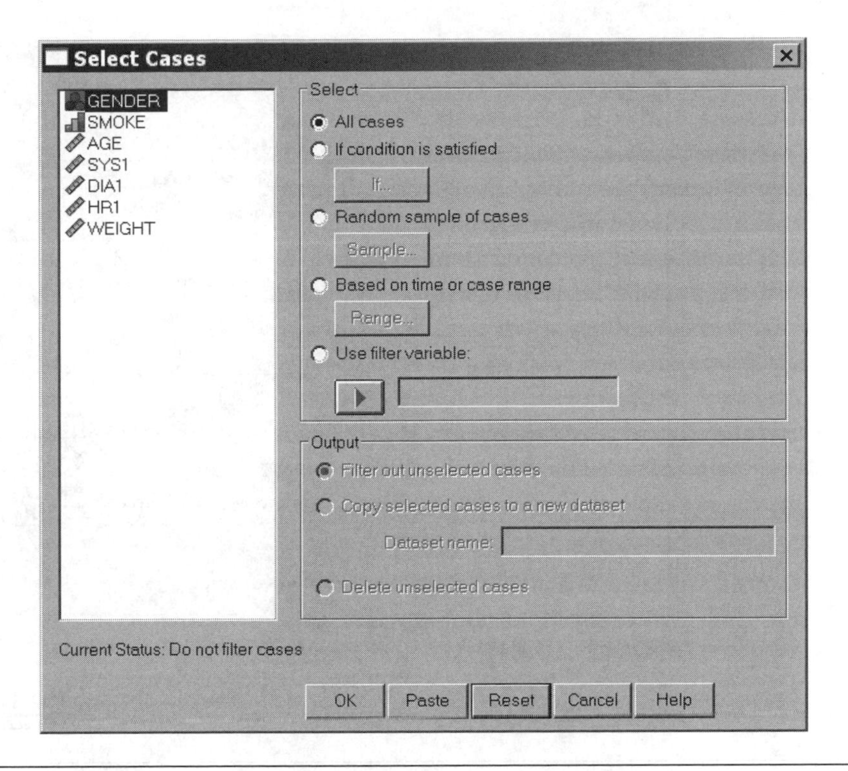

Figure 4.9 ♦ SPSS Dialog Windows for the Select Cases Command

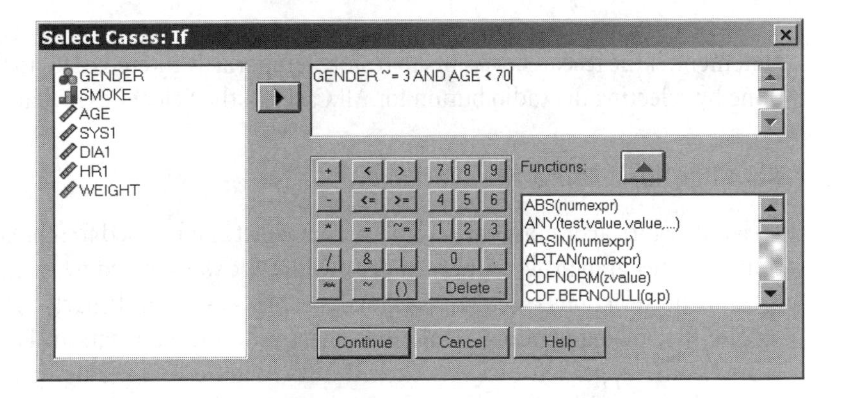

Figure 4.10 ♦ Logical Criterion for Select Cases

NOTE: Include only persons who have a score for GENDER that is not equal to "3" and who have a score for AGE that is less than "70."

of the Data Select If commands just described. Case number 11 (a person who had a score of "3" on GENDER) and case number 15 (a person who had a score of "88" on AGE) are now shown with a crosshatch mark through the case number in the left-hand column of the data worksheet. This crosshatch indicates that unless the Select If condition is explicitly removed, the data for these 2 participants will be excluded from all future analyses. Note that the original *N* of 65 cases has been reduced to an *N* of 63 by this Data Select If

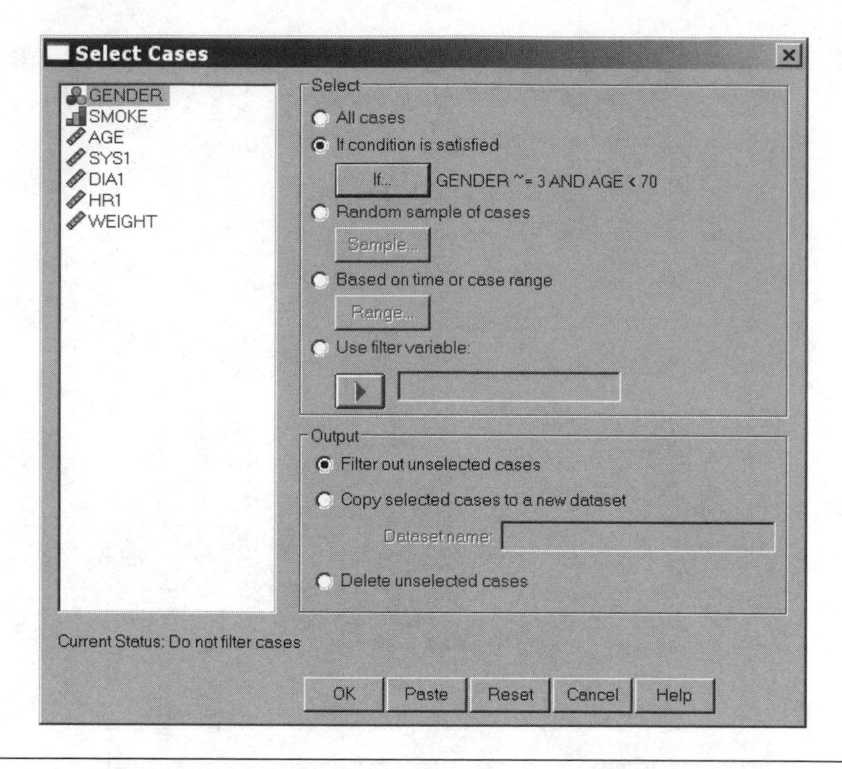

Figure 4.11 ♦ Appearance of the Select Cases Dialog Window After Specification of the Logical "If" Selection Rule

statement. If the researcher wants to restore temporarily filtered cases to the sample, it can done by selecting the radio button for All Cases in the Select Cases dialog window.

4.6.3 ♦ Bar Chart for a Categorical Variable

For categorical or nominal variables, a bar chart can be used to represent the distribution of scores graphically. A bar chart for GENDER was created by making the following SPSS menu selections (see Figure 4.13): <Graphs> → <Bar [Chart] . . . >.

The first dialog window for the bar chart procedure appears in Figure 4.14. In this example, the upper left box was clicked in the Figure 4.14 dialog window to select the "Simple" type of bar chart; the radio button was selected for "Summaries for groups of cases"; then, the Define button was clicked. This opened the second SPSS dialog window, which appears in Figure 4.15.

To specify the form of the bar chart, use the cursor to highlight the name of the variable that you want to graph, and click on the arrow that points to the right to move this variable name into the window under Category Axis. Leave the radio button selection as the default choice, "Bars represent N of cases." This set of menu selections will yield a bar graph with one bar for each group; for GENDER, this is a bar graph with one bar for males and one for females. The height of each bar represents the number of cases in each group. The output from this procedure appears in Figure 4.16. Note that because the invalid score of "3" has been selected out by the prior Select If statement, this score value of "3" is not

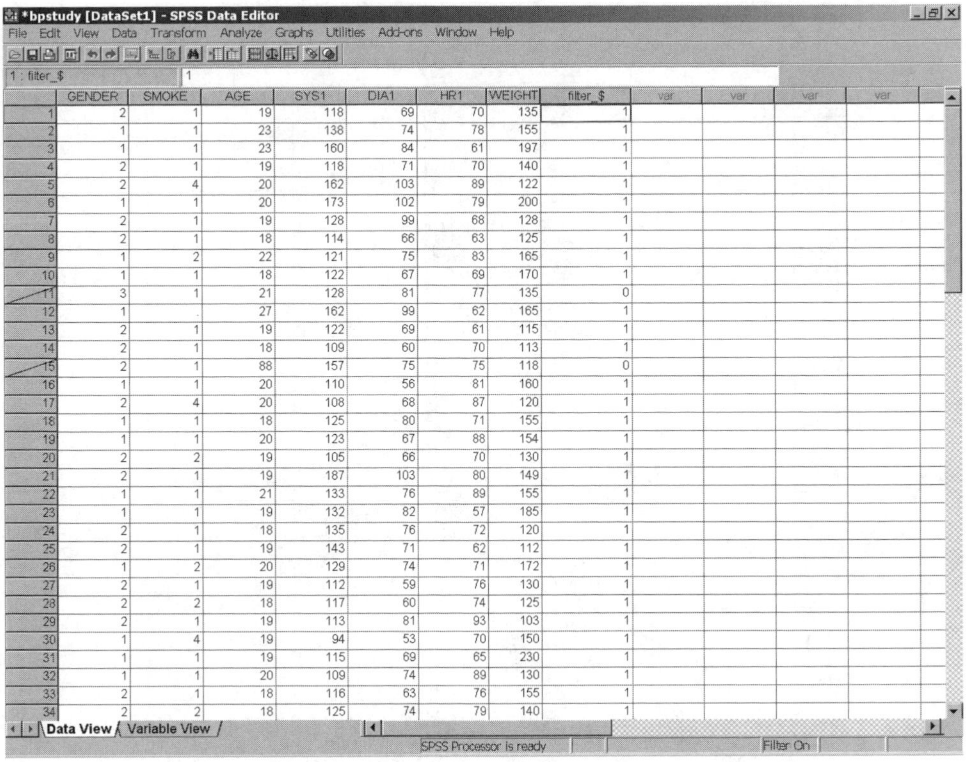

*bpstudy [DataSet1] - SPSS Data Editor

File Edit View Data Transform Analyze Graphs Utilities Add-ons Window Help

1 : filter_$ 1

	GENDER	SMOKE	AGE	SYS1	DIA1	HR1	WEIGHT	filter_$	var	var	var	var
1	2	1	19	118	69	70	135	1				
2	1	1	23	138	74	78	155	1				
3	1	1	23	160	84	61	197	1				
4	2	1	19	118	71	70	140	1				
5	2	4	20	162	103	89	122	1				
6	1	1	20	173	102	79	200	1				
7	2	1	19	128	99	68	128	1				
8	2	1	18	114	66	63	125	1				
9	1	2	22	121	75	83	165	1				
10	1	1	18	122	67	69	170	1				
11	3	1	21	128	81	77	135	0				
12	1	.	27	162	99	62	165	1				
13	2	1	19	122	69	61	115	1				
14	2	1	18	109	60	70	113	1				
15	2	1	88	157	75	75	118	0				
16	1	1	20	110	56	81	160	1				
17	2	4	20	108	68	87	120	1				
18	1	1	18	125	80	71	155	1				
19	1	1	20	123	67	88	154	1				
20	2	2	19	105	66	70	130	1				
21	2	1	19	187	103	80	149	1				
22	1	1	21	133	76	89	155	1				
23	1	1	19	132	82	57	185	1				
24	2	1	18	135	76	72	120	1				
25	2	1	19	143	71	62	112	1				
26	1	2	20	129	74	71	172	1				
27	2	1	19	112	59	76	130	1				
28	2	2	18	117	60	74	125	1				
29	2	1	19	113	81	93	103	1				
30	1	4	19	94	53	70	150	1				
31	1	1	19	115	69	65	230	1				
32	1	1	20	109	74	89	130	1				
33	2	1	18	116	63	76	155	1				
34	2	2	18	125	74	79	140	1				

Data View / Variable View

SPSS Processor is ready Filter On

Figure 4.12 ♦ Appearance of the SPSS Data Worksheet After the Select Cases Procedure

included in the bar graph in Figure 4.16. A visual examination of a set of bar graphs, one for each categorical variable, is a useful way to detect impossible values. The frequency table and bar graphs also provide a quick indication of group size; in this dataset there are $N = 28$ males (score of "1" on GENDER) and $N = 36$ females (score of "2" on GENDER). The bar chart in Figure 4.16, like the frequency table in Figure 4.7, indicates that the male group had fewer participants than the female group.

4.6.4 ♦ Histogram for a Quantitative Variable

For a quantitative variable, a histogram is a useful way to assess the shape of the distribution of scores. As described in Chapter 3, many analyses assume that scores on quantitative variables are at least approximately normally distributed. Visual examination of the histogram is a way to evaluate whether the distribution shape is reasonably close to normal or to identify the shape of a distribution if it is quite different from normal.

In addition, summary statistics can be obtained to provide information about central tendency and dispersion of scores. The mean (M), median, or mode can be used to describe central tendency; the range, standard deviation (s or SD), or variance (s^2) can be used to describe variability or dispersion of scores. A comparison of means, variances, and other descriptive statistics provides the information that a researcher needs to characterize his or her sample and to judge whether the sample is similar enough to some

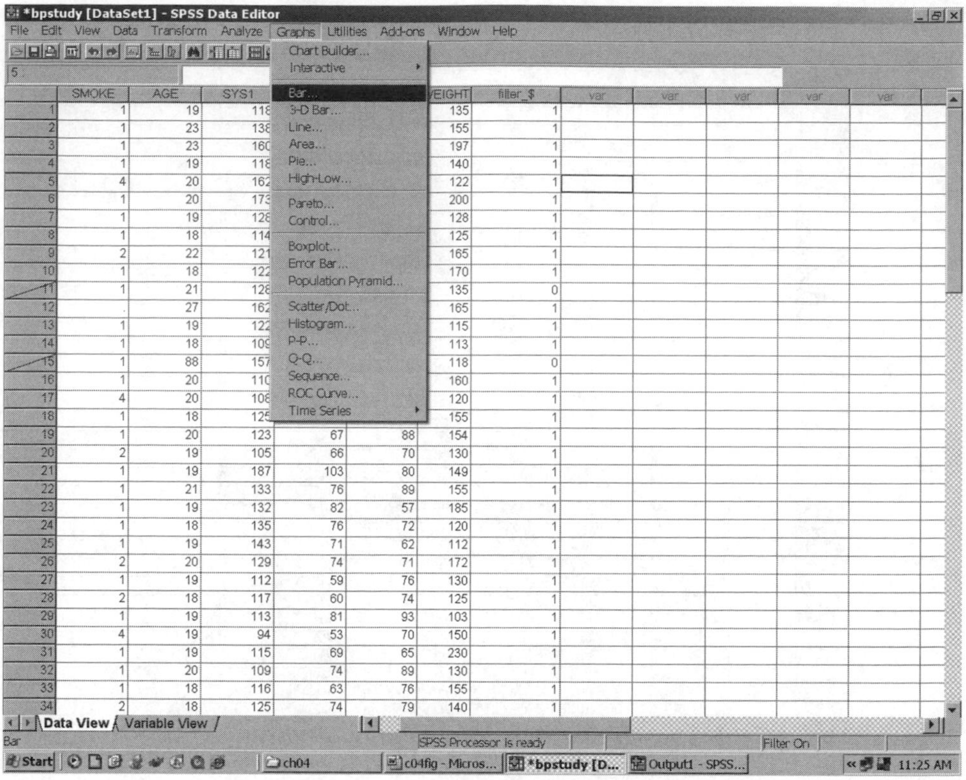

Figure 4.13 ♦ SPSS Menu Selections for the <Graphs> → <Bar [Chart] . . . > Procedure

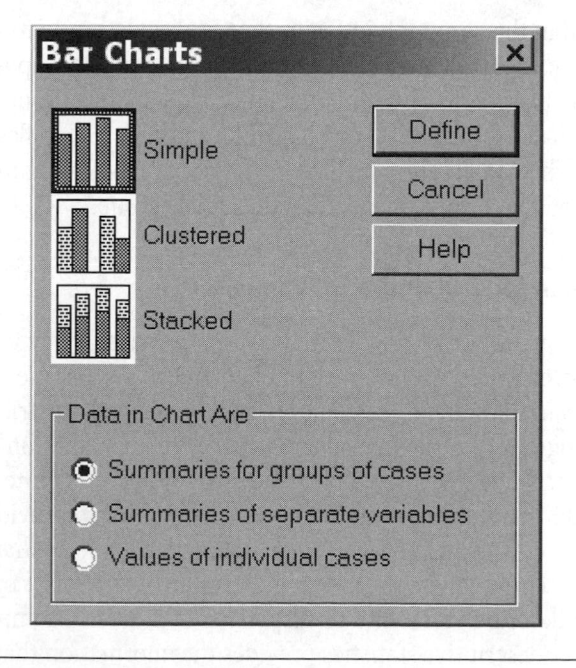

Figure 4.14 ♦ SPSS Bar Charts Dialog Window

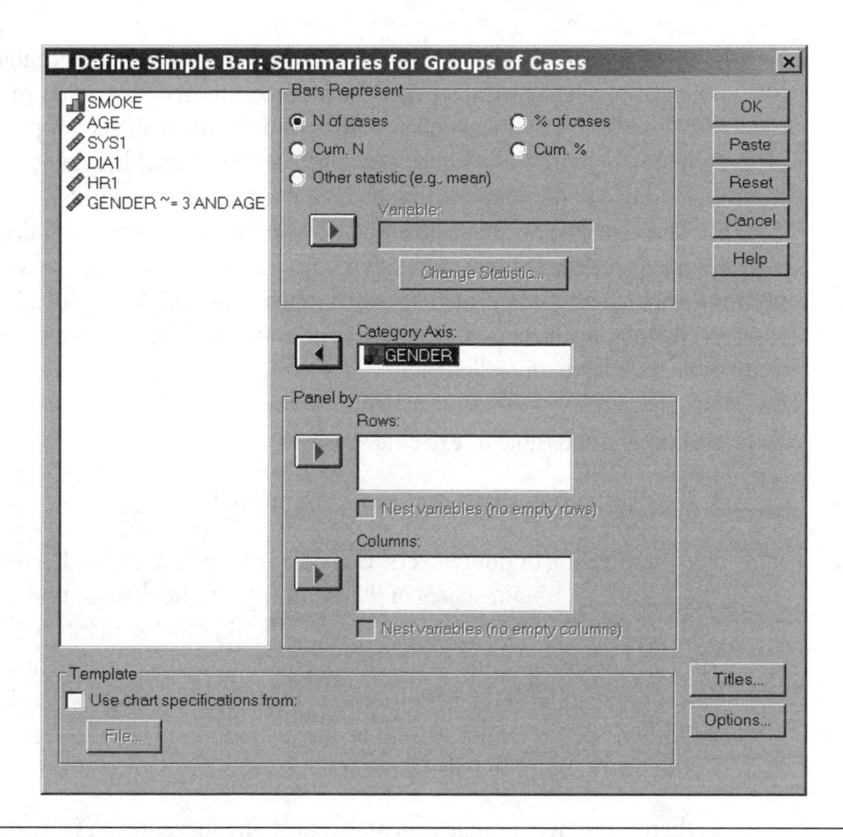

Figure 4.15 ♦ SPSS Define Simple Bar Chart Dialog Window

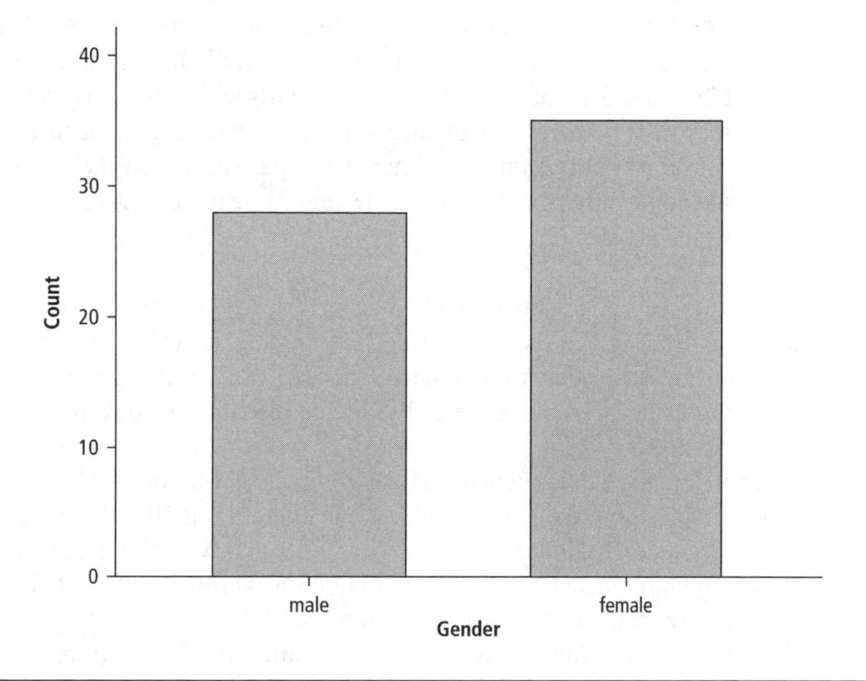

Figure 4.16 ♦ Bar Chart: Frequencies for Each Gender Category

broader population of interest so that results might possibly be generalizable to that broader population (through the principle of "proximal similarity," discussed in Chapter 1). If a researcher conducts a political poll and finds that the range of ages of persons in the sample is from age 18 to 22, for instance, it would not be reasonable to generalize any findings from that sample to populations that include persons above age 22.

When the distribution shape of a quantitative variable is nonnormal, it is preferable to assess central tendency and dispersion of scores using graphic methods that are based on percentiles (such as a **box plot,** also called a **box and whiskers plot**). Issues that can be assessed by looking at frequency tables, histograms, or box and whiskers plots for quantitative scores include the following:

1. Are there impossible or extreme scores?

2. Is the distribution shape normal or nonnormal?

3. Are there ceiling or floor effects? Consider a set of test scores. If a test is too easy and most students obtain scores of 90% and higher, the distribution of scores shows a "ceiling effect"; if the test is much too difficult, most students will obtain scores of 10% and below, and this would be called a "floor effect." Either of these would indicate a problem with the measurement, in particular a lack of sensitivity to individual differences at the upper end of the distribution (when there is a ceiling effect) or the lower end of the distribution (when there is a floor effect).

4. Is there a restricted range of scores? For many measures, researchers know a priori what the minimum and maximum possible scores are, or they have a rough idea of the range of scores. For example, suppose that Verbal SAT scores can range from 250 to 800. If the sample includes scores that range from 550 to 580, the range of Verbal SAT scores in the sample is extremely restricted compared with the range of possible scores. Generally, researchers want a fairly wide range of scores on variables that they want to correlate with other variables. If a researcher wants to "hold a variable constant"—for example, to limit the impact of age on the results of a study by including only persons between 18 and 21 years of age—then a restricted range would actually be preferred.

The procedures for obtaining a frequency table for a quantitative variable are the same as those discussed in the previous section on data screening for categorical variables. Distribution shape for a quantitative variable can be assessed by examining a histogram obtained by making these SPSS menu selections (see Figure 4.17): <Graphs> → <Histogram>.

These menu selections open the Histogram dialog window displayed in Figure 4.18. In this example, the variable selected for the histogram was HR1 (baseline or Time 1 heart rate). Placing a checkmark in the box next to Display Normal Curve requests a superimposed smooth normal distribution function on the histogram plot. To obtain the histogram, after making these selections, click the OK button. The histogram output appears in Figure 4.19. The mean, standard deviation, and *N* for HR1 appear in the legend below the graph.

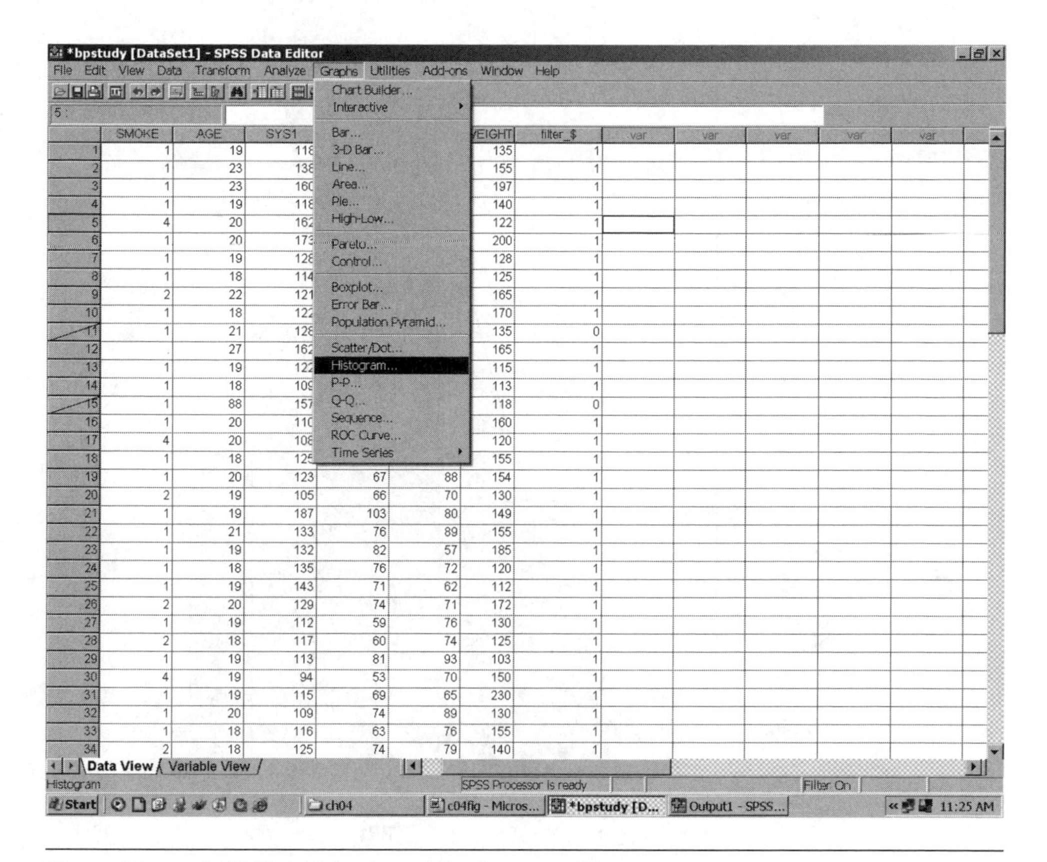

Figure 4.17 ♦ SPSS Menu Selections: <Graphs> → <Histogram>

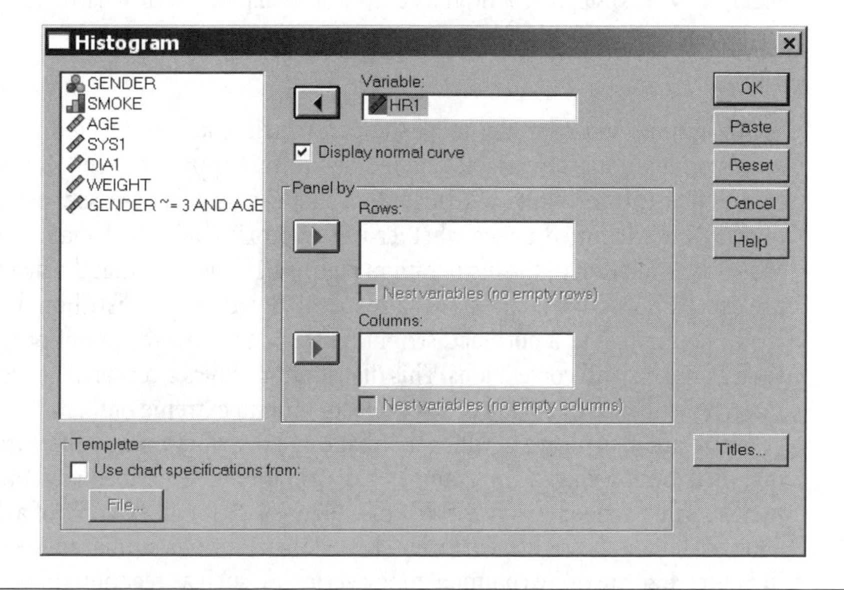

Figure 4.18 ♦ SPSS Dialog Window: Histogram Procedure

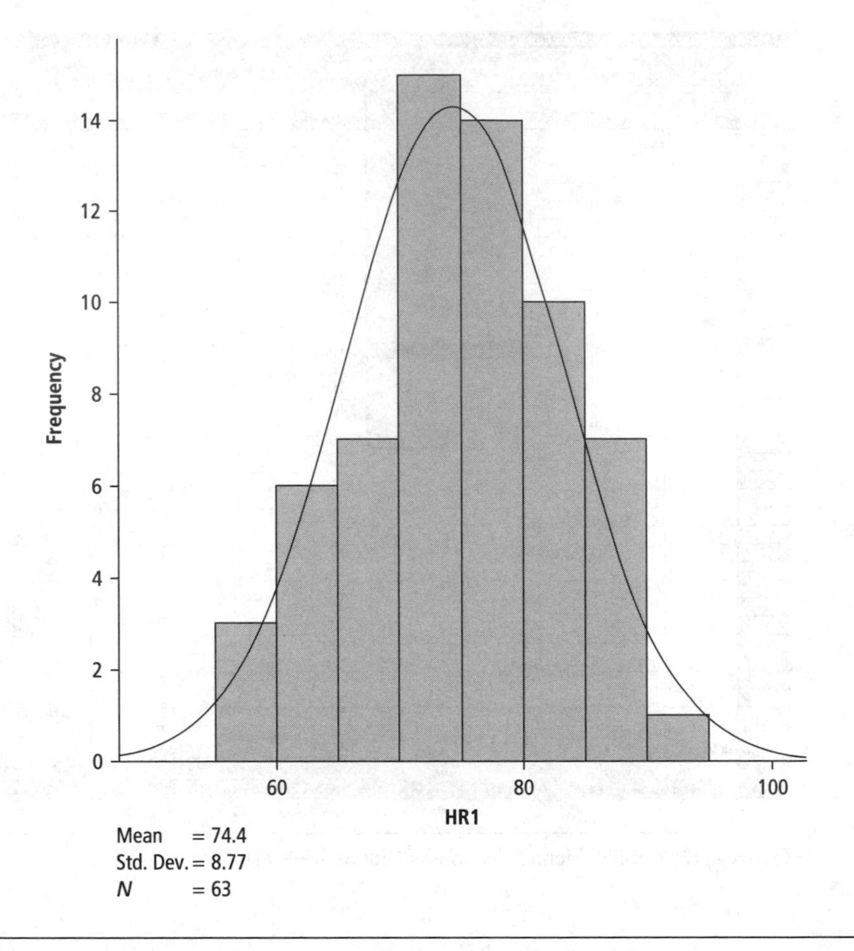

Figure 4.19 ♦ Histogram of Heart Rates With Superimposed Normal Curve

An assumption common to all the parametric analyses covered in this book is that scores on quantitative variables should be (at least approximately) normally distributed. In practice, the normality of distribution shape is usually assessed visually; a histogram of scores is examined to see whether it is approximately "bell shaped" and symmetric. Visual examination of the histogram in Figure 4.19 suggests that the distribution shape is not exactly normal; it is slightly asymmetrical. However, this distribution of sample scores is similar enough to a normal distribution shape to allow the use of parametric statistics such as means and correlations. This distribution shows a reasonably wide range of heart rates, no evidence of ceiling or floor effects, and no extreme outliers.

There are many ways in which the shape of a distribution can differ from an ideal normal distribution shape. For example, a distribution is described as skewed if it is asymmetric, with a longer tail on one side (see Figure 4.20 for an example of a distribution with a longer tail on the right). Positively skewed distributions similar to the one that appears in Figure 4.20 are quite common; many variables, such as reaction time, have a minimum

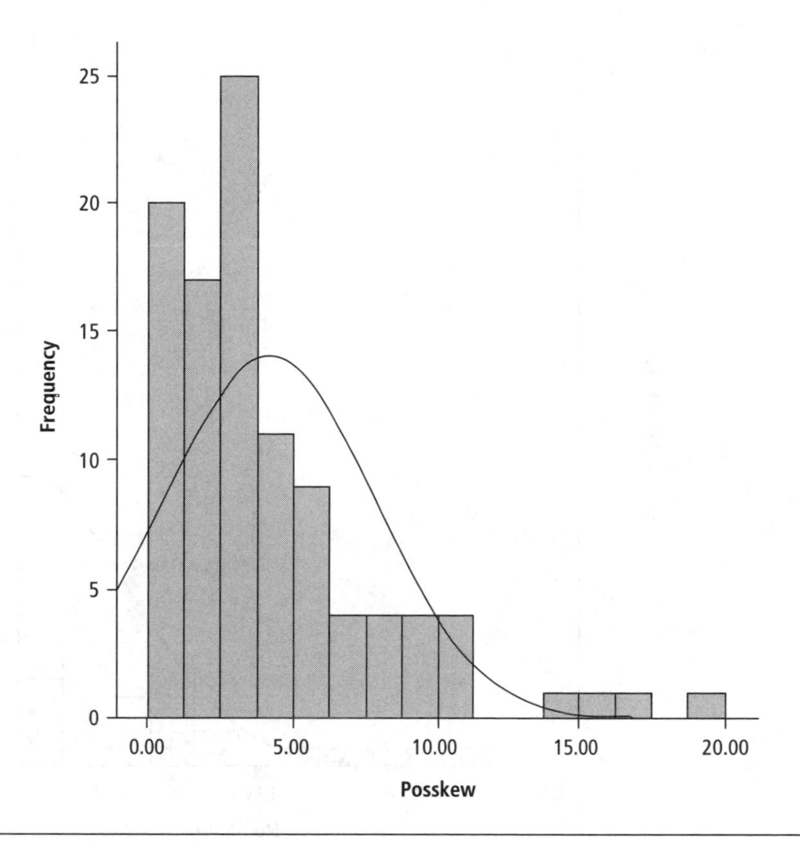

Figure 4.20 ♦ Positively Skewed Distribution

NOTE: Skewness index for this variable is +2.00.

possible value of 0 (which means that that the lower tail of the distribution ends at 0) but do not have a fixed limit at the upper end of the distribution (and therefore the upper tail can be quite long).

A numerical index of **skewness** for a sample set of X scores denoted by $(X_1, X_2, \ldots, X_N)$ can be calculated using the following formula:

$$\text{Skewness} = \frac{\sum(X_i - M_x)^3}{(N - 1)s^3}, \tag{4.1}$$

where M_x is the sample mean of the X scores, s is the sample standard deviation of the X scores, and N is the number of scores in the sample.

For a perfectly normal and symmetrical distribution, skewness has a value of 0. If the skewness statistic is positive, it indicates that there is a longer tail on the right-hand/upper end of the distribution (as in Figure 4.20); if the skewness statistic is negative, it indicates that there is a longer tail on the lower end of the distribution (as in Figure 4.21).

A distribution is described as platykurtic if it is flatter than an ideal normal distribution and leptokurtic if it has a sharper/steeper peak in the center than an ideal normal

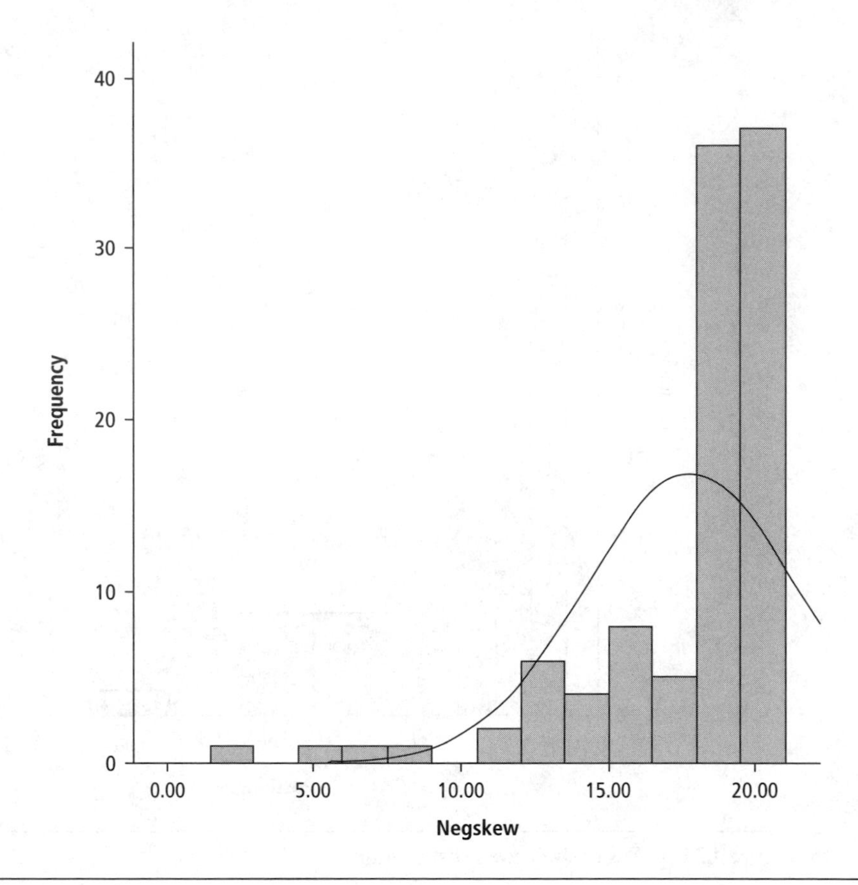

Figure 4.21 ♦ Negatively Skewed Distribution

NOTE: Skewness index for this variable is −2.00.

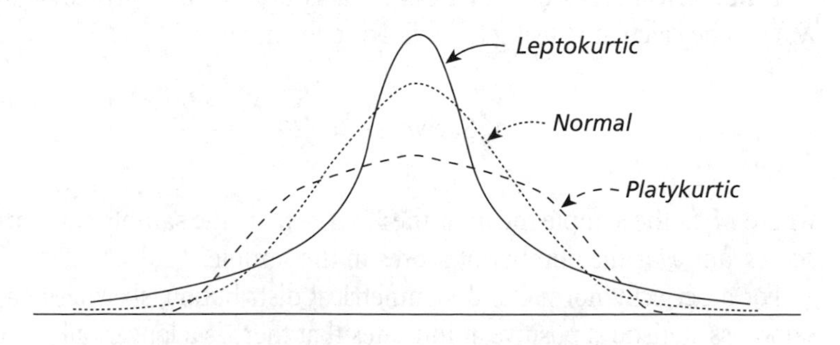

Figure 4.22 ♦ Leptokurtic and Platykurtic Distributions

SOURCE: http://www.murraystate.edu/polcrjlst/p660kurtosis.htm.

distribution (see Figure 4.22). A numerical index of **kurtosis** can be calculated using the following formula:

$$\text{Kurtosis} = \frac{\sum(X - M_x)^4}{(N-1)s^4}, \tag{4.2}$$

where M_x is the sample mean of the X scores, s is the sample standard deviation of the X scores, and N is the number of scores in the sample.

Using Equation 4.2, the kurtosis for a normal distribution corresponds to a value of 3; most computer programs actually report "excess kurtosis"—that is, the degree to which the kurtosis of the scores in a sample differs from the kurtosis expected in a normal distribution. This excess kurtosis is given by the following formula:

$$\text{Excess Kurtosis} = \frac{\sum(X - M_x)^4}{(N-1)s^4} - 3. \tag{4.3}$$

A positive score for excess kurtosis indicates that the distribution of scores in the sample is more sharply peaked than in a normal distribution (this is shown as leptokurtic in Figure 4.22). A negative score for kurtosis indicates that the distribution of scores in a sample is flatter than in a normal distribution (this corresponds to a platykurtic distribution shape in Figure 4.22). The value that SPSS reports as kurtosis corresponds to excess kurtosis (as in Equation 4.3).

A normal distribution is defined as having skewness and (excess) kurtosis of 0. A numerical index of skewness and kurtosis can be obtained for a sample of data to assess the degree of departure from normal distribution shape.

Additional summary statistics for a quantitative variable such as HR1 can be obtained from the SPSS Descriptives procedure by making the following menu selections: <Analyze> → <Descriptive Statistics> → <Descriptives>.

The menu selections shown in Figure 4.23 open the Descriptive Statistics dialog box shown in Figure 4.24. The Options button opens up a dialog box that has a menu with check boxes that offer a selection of descriptive statistics, as shown in Figure 4.25. In addition to the default selections, the boxes for skewness and kurtosis were also checked. The output from this procedure appears in Figure 4.26. The upper panel in Figure 4.26 shows the descriptive statistics for scores on HR1 that appeared in Figure 4.19; skewness and kurtosis for the sample of scores on the variable HR1 were both fairly close to 0. The lower panel shows the descriptive statistics for the artificially generated data that appeared in Figures 4.20 (a set of positively skewed scores) and 4.21 (a set of negatively skewed scores).

It is possible to set up a statistical significance test (in the form of a z ratio) for skewness because SPSS also reports the standard error (SE) for this statistic:

$$z = \frac{\text{Skewness}}{SE_{\text{Skewness}}}. \tag{4.4}$$

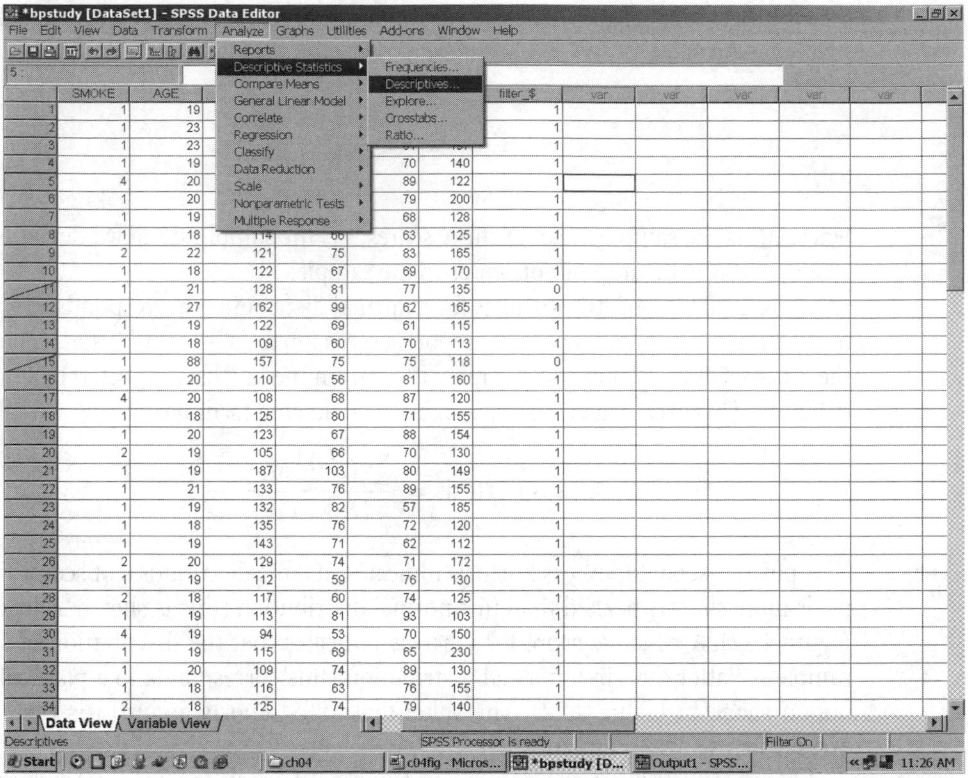

Figure 4.23 ♦ SPSS Menu Selections: <Analyze> → <Descriptive Statistics> → <Descriptives>

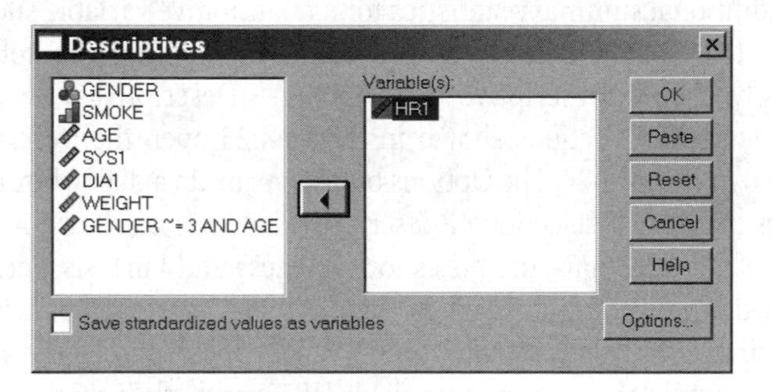

Figure 4.24 ♦ SPSS Descriptive Statistics Procedure

When the N of cases is reasonably large, the resulting z ratio can be evaluated using the standard normal distribution; that is, skewness is statistically significant at the $\alpha = .05$ level (two-tailed) if the z ratio given in Equation 4.4 is greater than 1.96 in absolute value.

A z test can also be set up to test the significance of (excess) kurtosis:

$$z = \frac{\text{Excess kurtosis}}{SE_{\text{Kurtosis}}}. \tag{4.5}$$

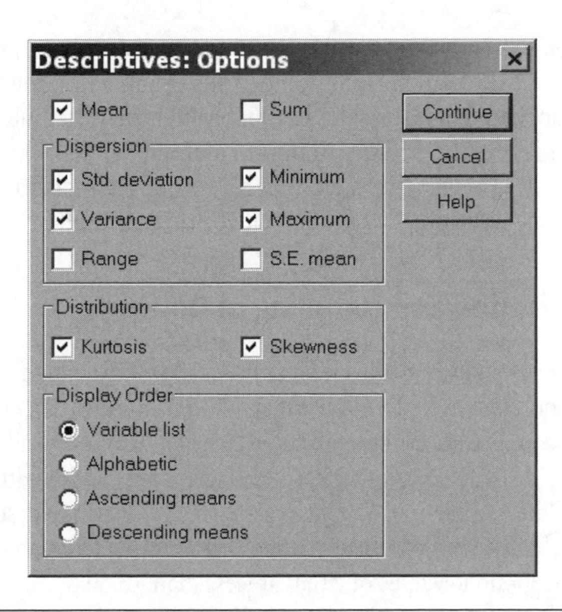

Figure 4.25 ♦ Options for the Descriptive Statistics Procedure

Descriptive Statistics

	N	Minimum	Maximum	Mean	Std.	Variance	Skewness		Kurtosis	
	Statistic	Statistic	Statistic	Statistic	Statistic	Statistic	Statistic	Std. Error	Statistic	Std. Error
HR1	63	56	93	74.40	8.770	76.921	-.005	.302	-.572	.595
Valid N (listwise)	63									

Descriptive Statistics

	N	Minimum	Maximum	Mean	Std.	Variance	Skewness		Kurtosis	
	Statistic	Statistic	Statistic	Statistic	Statistic	Statistic	Statistic	Std. Error	Statistic	Std. Error
posskew	102	1.00	20.00	4.1863	3.62808	13.163	2.029	.239	4.831	.474
negskew	102	2.00	21.00	17.8137	3.62808	13.163	-2.029	.239	4.831	.474
Valid N (listwise)	102									

Figure 4.26 ♦ Output From the SPSS Descriptive Statistics Procedure for Positively Skewed and Negatively Skewed Distributions (as Shown in Figures 4.20 and 4.21)

The tests in Equations 4.4 and 4.5 provide a way to evaluate whether an empirical frequency distribution differs significantly from a normal distribution in skewness or kurtosis. There are formal mathematical tests to evaluate the degree to which an empirical distribution differs from some ideal or theoretical distribution shape (such as the normal curve). If a researcher needs to test whether the overall shape of an empirical frequency distribution differs significantly from normal, it can be done by using the Kolmogorov-Smirnov or Shapiro-Wilk test (both are available in SPSS). In most situations, visual examination of distribution shape is deemed sufficient.

In general, empirical distribution shapes are considered problematic only when they differ dramatically from normal. Some earlier examples of drastically nonnormal distribution shapes appeared in Figures 1.2 (a roughly uniform distribution) and 1.3 (an approximately exponential or J-shaped distribution). Multimodal distributions or very seriously skewed distributions (as in Figure 4.20) may also be judged problematic. A distribution that resembles the one in Figure 4.19 is often judged close enough to normal shape.

4.7 ♦ Identification and Handling of Outliers

An outlier is an extreme score on either the low or the high end of a frequency distribution of a quantitative variable. Many different decision rules can be used to decide whether a particular score is extreme enough to be considered an outlier. When scores are approximately normally distributed, about 99% of the scores should fall within +3 and −3 standard deviations of the sample mean. Thus, for normally distributed scores, z scores can be used to decide which scores to treat as outliers. For example, a researcher might decide to treat scores that correspond to values of z that are less than −3.30 or greater than +3.30 as outliers.

Another method for the detection of outliers uses a graph called a box plot (or a box and whiskers plot). This is a nonparametric exploratory procedure that uses medians and quartiles as information about central tendency and dispersion of scores. The following example uses a box plot of scores on WEIGHT separately for each gender group, as a means of identifying potential outliers on WEIGHT. To set up this box plot for the distribution of weight within each gender group, the following SPSS commands were used: <Graphs> → <Box [Plot] . . . >.

This opens up the first SPSS box plot dialog box, shown in Figure 4.27. For this example, the box marked Simple was clicked, and the radio button for "Summaries for groups of cases" was selected to obtain a box plot for just one variable (WEIGHT) separately for each of two groups (male and female). Clicking on the Define button opened up the second box plot dialog window, as shown in Figure 4.28. The name of the quantitative dependent variable, WEIGHT, was placed in the top window (as the name of the variable); the categorical

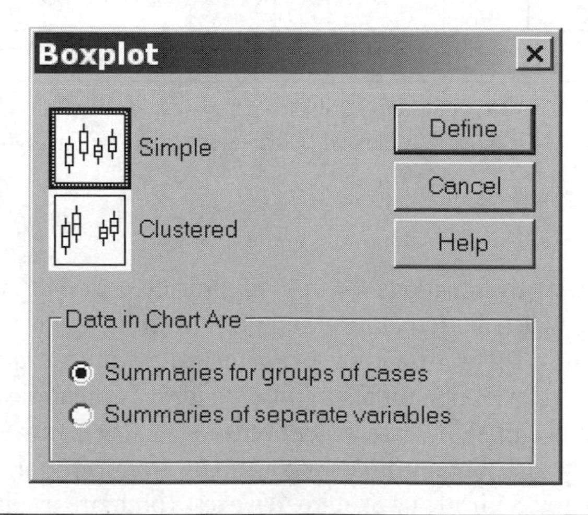

Figure 4.27 ♦ First SPSS Dialog Box for the Box Plot Procedure

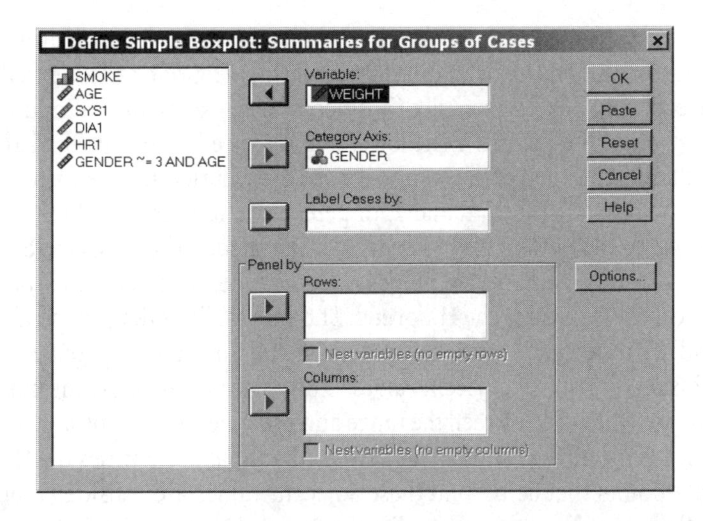

Figure 4.28 ♦ Define Simple Box Plot: Distribution of Weight Separately by Gender

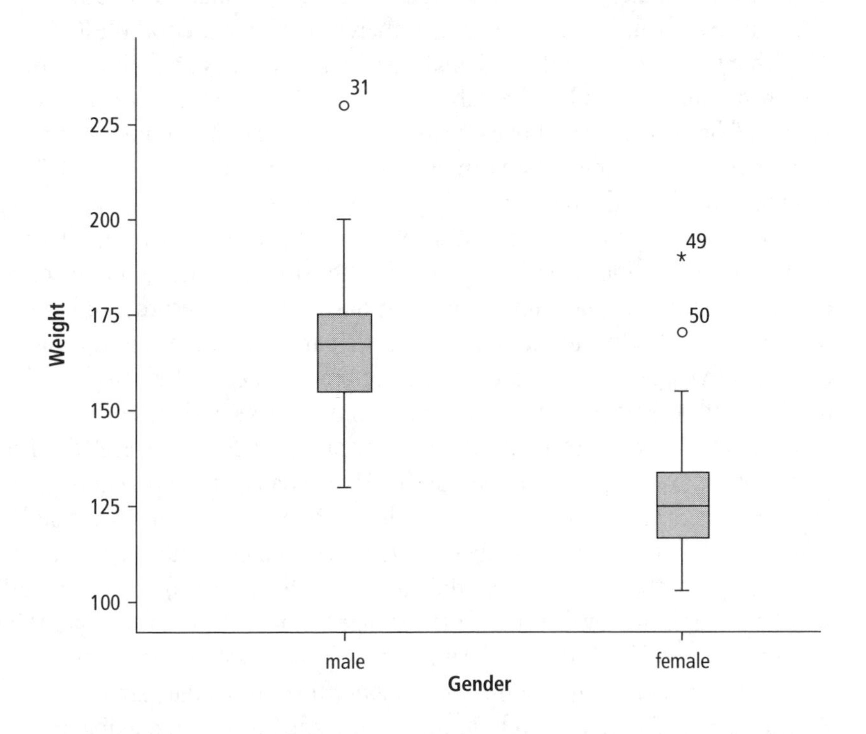

Figure 4.29 ♦ Box Plot of WEIGHT for Each Gender Group

or "grouping" variable (GENDER) was placed in the window for the Category Axis. Clicking the OK button generated the box plot shown in Figure 4.29, with values of WEIGHT shown on the *Y* axis and the categories male and female shown on the *X* axis.

Rosenthal and Rosnow (1991) noted that there are numerous variations of the box plot; the description here is specific to the box plots generated by SPSS and may not correspond exactly to descriptions of box plots given elsewhere. For each group, there is a shaded box that corresponds to the middle 50% of the distribution of scores in that group. The line that bisects this box horizontally (not necessarily exactly in the middle) represents the 50th percentile (the median). The lower and upper edges of this shaded box correspond to the 25th and 75th percentiles of the weight distribution for the corresponding group (labeled on the *X* axis). The 25th and 75th percentiles of each distribution of scores, which correspond to the bottom and top edges of the shaded box, respectively, are called the **hinges**. The distance between the hinges (i.e., the difference between scores at the 75th and 25th percentiles) is called the **H-spread**. The vertical lines that extend above and below the 75th and 25th percentiles are called **"whiskers,"** and the horizontal lines at the ends of the whiskers mark the "adjacent values." The adjacent values are the most extreme scores in the sample which lie between the hinge and the inner fence (not shown on the graph; the inner fence is usually a distance from the median that is 1.5 times the H-spread). Generally, any data points that lie beyond these adjacent values are considered outliers. In the box plot, outliers that lie outside the adjacent values are graphed using small circles. Observations that are extreme outliers are shown as asterisks (*).

Figure 4.29 indicates that the middle 50% of the distribution of body weights for males was between about 160 and 180 lb, and there was one outlier on WEIGHT (case number 31 with a weight of 230 lb) in the male group. For females, the middle 50% of the distribution of weights was between about 115 and 135 lb, and there were two outliers on WEIGHT in the female group; participant number 50 was an outlier (with weight = 170), and participant number 49 was an extreme outlier (with weight = 190). The data record numbers that label the outliers in Figure 4.29 can be used to look up the exact score values for the outliers in the entire listing of data in the SPSS data worksheet or in Table 4.1.

In this dataset, the value of idnum (a variable that provides a unique case number for each participant) was the same as the SPSS line number or record number for all 65 cases. If the researcher wants to exclude the 3 participants who were identified as outliers in the box plot of weight scores for the two gender groups, it could be done by using the following Select If statement: idnum ~= 31 and idnum ~= 49 and idnum ~=50.

Parametric statistics (such as the mean, variance, and Pearson correlation) are not particularly robust to outliers; that is, the value of *M* for a batch of sample data can be quite different when it is calculated with an outlier included than when an outlier is excluded. This raises a problem: Is it preferable to include outliers (recognizing that a single extreme score may have a disproportionate impact on the outcome of the analysis) or to omit outliers (understanding that the removal of scores may change the outcome of the analysis)? It is not possible to state a simple rule that can be uniformly applied to all research situations. Researchers have to make reasonable judgment calls about how to handle extreme scores or outliers. Researchers need to rely on both common sense and honesty in making these judgments.

When the total *N* of participants in the dataset is relatively small, and when there are one or more extreme outliers, the outcomes for statistical analyses that examine the relation between a pair of variables can be quite different when outliers are included versus excluded from an analysis. The best way to find out whether the inclusion of an outlier would make a difference in the outcome of a statistical analysis is to run the analysis both including and excluding the outlier score(s). However, making decisions about how to handle outliers post hoc (after running the analyses of interest) gives rise to a temptation: Researchers may wish to make decisions about outliers based on the way the outliers influence the outcome of statistical analyses. For example, a researcher might find a significant positive correlation between variables *X* and *Y* when outliers are included; but the correlation may become nonsignificant when outliers are removed from the dataset. It would be dishonest to report a significant correlation without also explaining that the correlation becomes nonsignificant when outliers are removed from the data. Conversely, a researcher might also encounter a situation where there is no significant correlation between scores on the *X* and *Y* variables when outliers are included, but the correlation between *X* and *Y* becomes significant when the data are reanalyzed with outliers removed. An honest report of the analysis should explain that outlier scores were detected and removed as part of the data analysis process, and there should be a good rationale for removal of these outliers. The fact that dropping outliers yields the kind of correlation results that the researcher hopes for is not, by itself, a satisfactory justification for dropping outliers. It should be apparent that if researchers arbitrarily drop enough cases from their samples, they can prune their data to fit just about any desired outcome. (Recall the myth of King Procrustes, who cut off the limbs of his guests so that they would fit his bed; we must beware of doing the same thing to our data.)

A less problematic way to handle outliers is to state a priori that the study will be limited to a specific population—that is, to specific ranges of scores on some of the variables. If the population of interest in the blood pressure study is healthy young adults whose blood pressure is within the normal range, this a priori specification of the population of interest would provide a justification for the decision to exclude data for participants with age greater than 30 years and SBP above 140.

Another reasonable approach is to use a standard rule for exclusion of extreme scores (e.g., a researcher might decide at an early stage in data screening to drop all values that correspond to *z* scores in excess of 3.3 in absolute value; this value of 3.3 is an arbitrary standard).

Another method of handling extreme scores (trimming) involves dropping the top and bottom scores (or some percentage of scores, such as the top and bottom 1% of scores) from each group. Winsorizing is yet another method of reducing the impact of outliers: The most extreme score at each end of a distribution is recoded to have the same value as the next highest score.

Another way to reduce the impact of outliers is to apply a **nonlinear transformation** (such as taking the base 10 **logarithm** [log] of the original *X* scores). This type of data transformation can bring outlier values at the high end of a distribution closer to the mean.

Whatever the researcher decides to do with extreme scores (throw them out, Winsorize them, or modify the entire distribution by taking the log of scores), it is a good idea to conduct analyses with the outlier included and with the outlier excluded to see what effect (if any) the decision about outliers has on the outcome of the analysis. If the results are essentially identical no matter what is done to outliers, then either approach could be reported. If the results are substantially different when different things are done with outliers, the researcher needs to make a thoughtful decision about which version of the analysis provides a more accurate and honest description of the situation. In some situations, it may make sense to report both versions of the analysis (with outliers included, and outliers excluded) so that it is clear to the reader how the extreme individual score values influenced the results. None of these choices are ideal solutions; any of these procedures may be questioned by reviewers or editors.

It is preferable to decide on simple exclusion rules for outliers before data are collected and to remove outliers during the preliminary screening stages rather than at later stages in the analysis. It may be preferable to have a consistent rule for exclusion (e.g., excluding all scores that show up as extreme outliers in box plots) rather than to tell a different story to explain why each individual outlier received the specific treatment that it did. The final research report should explain what methods were used to detect outliers, identify scores that were identified as outliers, and make it clear how the outliers were handled (whether extreme scores were removed or modified).

4.8 ♦ Screening Data for Bivariate Analyses

There are three possible combinations of types of variables in bivariate analysis. Both variables may be categorical, both may be quantitative, or one may be categorical and the other quantitative. Separate bivariate data-screening methods are outlined for each of these situations.

4.8.1 ♦ Bivariate Data Screening for Two Categorical Variables

When both variables are categorical, it does not make sense to compute means (the numbers serve only as labels for group memberships); instead, it makes sense to look at the numbers of cases within each group. When two categorical variables are considered jointly, a **cross-tabulation** or contingency table summarizes the number of participants in the groups for all possible combinations of scores. For example, consider GENDER (coded 1 = Male and 2 = Female) and smoking status (coded 1 = Nonsmoker, 2 = Occasional smoker, 3 = Frequent smoker, and 4 = Heavy smoker). A table of cell frequencies for these two categorical variables can be obtained using the SPSS Crosstabs procedure by making the following menu selections: <Analyze> → <Descriptives> → <Crosstabs>.

These menu selections open up the Crosstabs dialog window, which appears in Figure 4.30. The names of the row variable (in this example, GENDER) and the column variable (in this example, SMOKE) are entered into the appropriate boxes. Clicking on the button labeled Cells opens up an additional dialog window, shown in Figure 4.31, where the user specifies the information to be presented in each cell of the contingency table. In this example, both observed (O) and expected (E) frequency counts are shown in each cell (see Chapter 8 in this textbook to see how **expected cell frequencies** are computed from the total number in each row and column of a contingency table). Row percentages were also requested.

The observed cell frequencies in Figure 4.32 show that most of the males and the females were nonsmokers (SMOKE = "1"). In fact, there were very few light smokers and heavy smokers (and no moderate smokers). As a data-screening result, this has two implications: If we wanted to do an analysis (such as a chi-square test of association, as described in Chapter 8) to assess how gender is related to smoking status, the data do not satisfy an assumption about the **minimum expected cell frequencies** required for the chi-square test of association. (For a 2 × 2 table, none of the expected cell frequencies should be less than 5; for larger tables, various sources recommend different standards for minimum expected cell frequencies, but a minimum expected frequency of 5 is recommended here.) In addition, if we wanted to see how gender and smoking status together predict some third variable, such as heart rate, the numbers of participants in most of the groups (such as heavy smoker/females with only $N = 2$ cases) are simply too small.

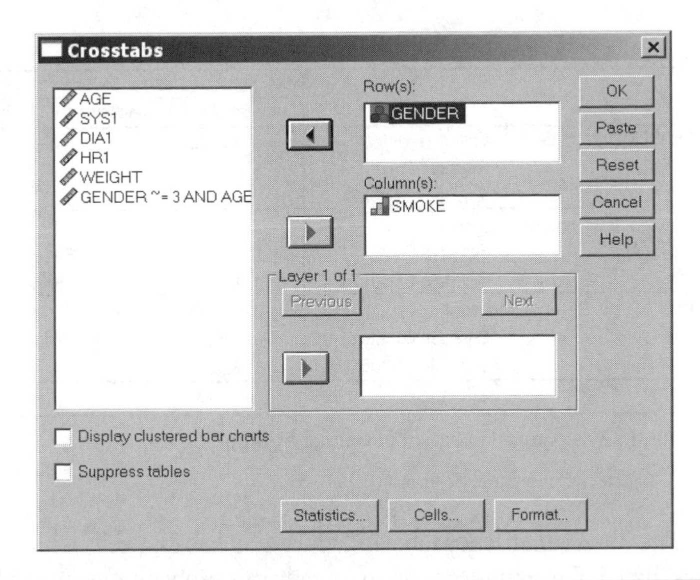

Figure 4.30 ♦ SPSS Crosstabs Dialog Box

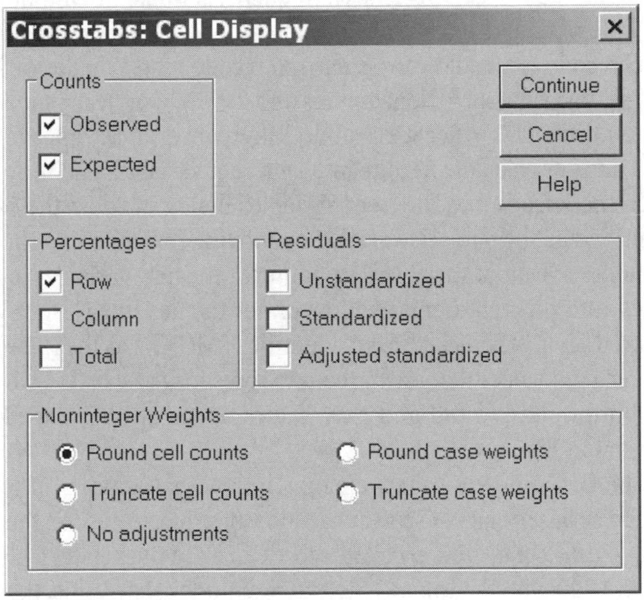

Figure 4.31 ♦ SPSS Crosstabs: Information to Display in Cells

What would we hope to see in preliminary screening for categorical variables? The **marginal frequencies** (e.g., number of males, number of females; number of nonsmokers, light smokers, and heavy smokers) should all be reasonably large. That is clearly not the case in this example: There were so few heavy smokers that we cannot judge whether heavy smoking is associated with gender.

GENDER * SMOKE Crosstabulation

			SMOKE			
			non smoker	light smoker	heavy smoker	Total
GENDER	male	Count	22	4	1	27
		Expected Count	21.8	3.9	1.3	27.0
		% within GENDER	81.5%	14.8%	3.7%	100.0%
	female	Count	28	5	2	35
		Expected Count	28.2	5.1	1.7	35.0
		% within GENDER	80.0%	14.3%	5.7%	100.0%
Total		Count	50	9	3	62
		Expected Count	50.0	9.0	3.0	62.0
		% within GENDER	80.6%	14.5%	4.8%	100.0%

Figure 4.32 ♦ Cross-Tabulation of Gender by Smoking Status

NOTE: Expected cell frequencies less than 10 in three cells.

The 2 × 3 contingency table in Figure 4.32 has four cells with expected cell frequencies less than 5. There are two ways to remedy this problem. One possible solution is to remove groups that have small marginal total *N*s. For example, only 3 people reported that they were "heavy smokers." If this group of 3 people were excluded from the analysis, the two cells with the lowest expected cell frequencies would be eliminated from the table. Another possible remedy is to combine groups (but only if this makes sense). In this example, the SPSS recode command can be used to recode scores on the variable SMOKE so that there are just two values: 1 = Nonsmokers and 2 = Light or heavy smokers. The SPSS menu selections <Compute> <Recode> <Into Different Variable> appear in Figure 4.33; these menu selections open up the Recode into Different Variables dialog box, as shown in Figure 4.34.

In the Recode into Different Variables dialog window, the existing variable SMOKE is identified as the numeric variable by moving its name into the window headed Numeric Variable → Output Variable. The name for the new variable (in this example, SMOKE2) is typed into the right-hand window under the heading Output Variable, and if the button marked Change is clicked, SPSS identifies SMOKE2 as the (new) variable that will contain the recoded values that are based on scores for the existing variable SMOKE. Clicking on the button marked Old and New Values opens up the next SPSS dialog window, which appears in Figure 4.35.

The Old and New Values dialog window that appears in Figure 4.35 can be used to enter a series of pairs of scores that show how old scores (on the existing variable SMOKE) are used to create new recoded scores (on the output variable SMOKE2). For example, under Old Value, the value "1" is entered; under New Value, the value "1" is entered; then, we click the Add button to add this to the list of recode commands. People who have a score of "1" on SMOKE (i.e., they reported themselves as nonsmokers) will also have a score of "1" on SMOKE2 (this will also be interpreted as "nonsmokers"). For the old values "2," "3," and "4" on the existing variable SMOKE, each of these variables is associated with a score of "2" on the new variable SMOKE2. In other words, people who chose responses "2," "3," or "4" on the variable SMOKE (light, moderate, or heavy smokers) will be coded "2" (smokers) on the new variable SMOKE2. Click Continue and then OK to make the recode commands take effect. After the recode command has been executed, a new variable called SMOKE2 will appear in the far right-hand column of the SPSS Data View worksheet; this variable will have scores of "1" (nonsmoker) and "2" (smoker).

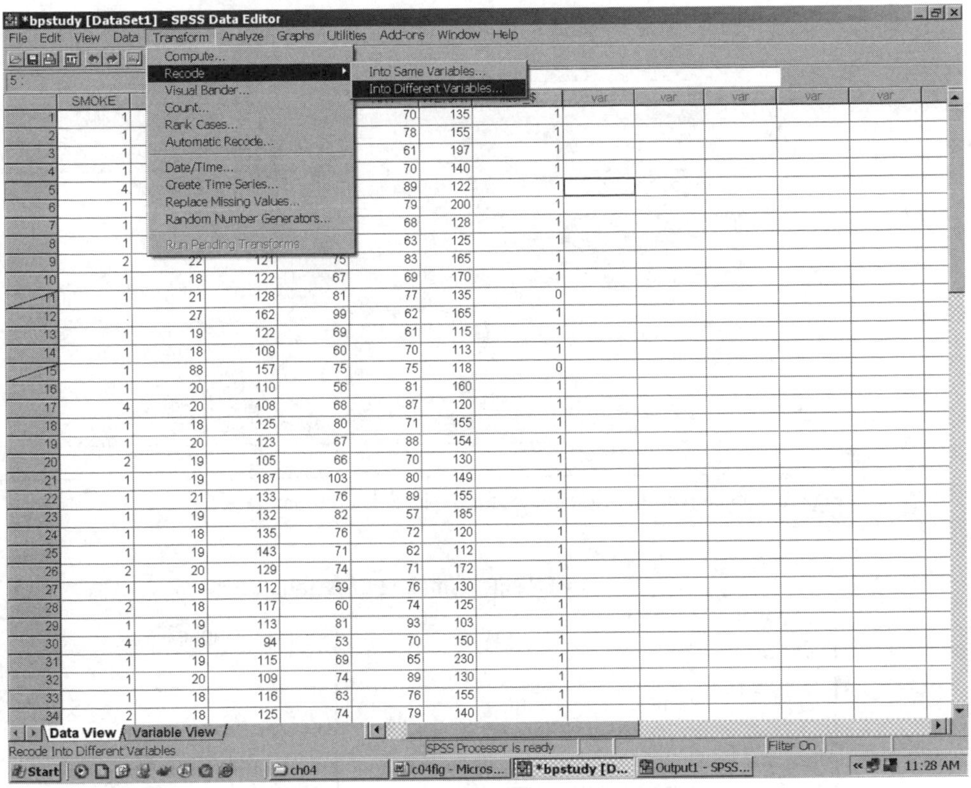

Figure 4.33 ♦ SPSS Menu Selection for the Recode Procedure

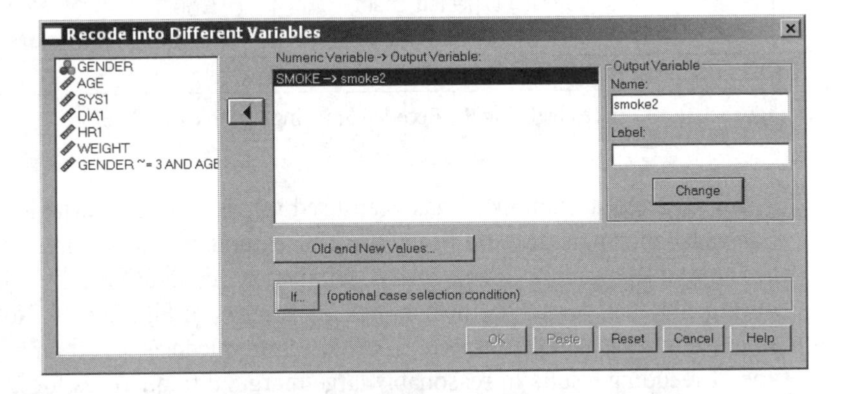

Figure 4.34 ♦ SPSS Recode Into Different Variables Dialog Window

While it is possible to replace the scores on the existing variable SMOKE with recoded values, it is often preferable to put recoded scores into a new output variable. It is easy to lose track of recodes as you continue to work with a data file. It is helpful to retain the variable in its original form so that information remains available.

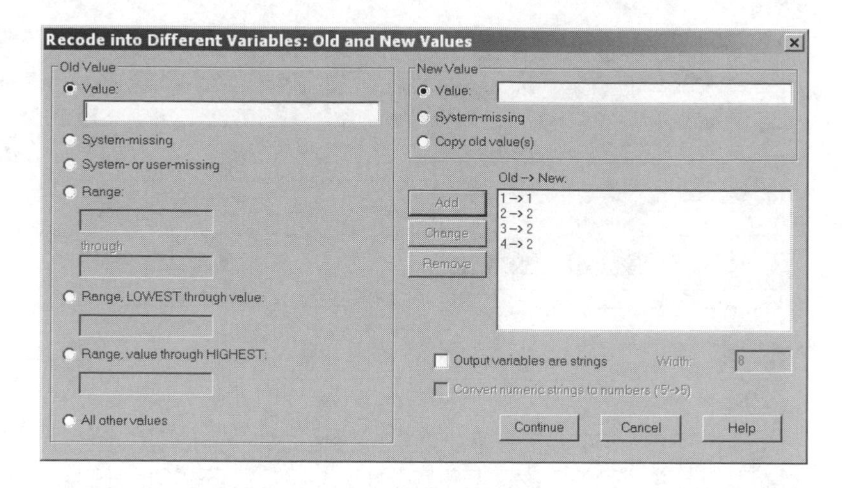

Figure 4.35 ♦ Old and New Values for the Recode Command

GENDER * smoke2 Crosstabulation

			smoke2		
			1	2	Total
GENDER	male	Count	22	5	27
		Expected Count	21.8	5.2	27.0
		% within GENDER	81.5%	18.5%	100.0%
	female	Count	28	7	35
		Expected Count	28.2	6.8	35.0
		% within GENDER	80.0%	20.0%	100.0%
Total		Count	50	12	62
		Expected Count	50.0	12.0	62.0
		% within GENDER	80.6%	19.4%	100.0%

Figure 4.36 ♦ Crosstabs Using the Recoded Smoking Variable (SMOKE2)

After the recode command has been used to create a new variable (SMOKE2), with codes for light, moderate, and heavy smoking combined into a single code for smoking, the Crosstabs procedure can be run using this new version of the smoking variable. The contingency table for GENDER by SMOKE2 appears in Figure 4.36. Note that this new table has no cells with minimum expected cell frequencies less than 5. Sometimes this type of recoding results in reasonably large marginal frequencies for all groups. In this example, however, the total number of smokers in this sample is still too small to conduct further analyses on possible associations between gender and smoking.

4.8.2 ♦ Bivariate Data Screening for One Categorical and One Quantitative Variable

Data analysis methods that compare means of quantitative variables across groups (such as ANOVA) have all the assumptions that are required for univariate parametric statistics:

1. Scores on quantitative variables should be normally distributed.
2. Observations should be independent.

When means on quantitative variables are compared across groups, there is one additional assumption: The variances of the populations (from which the samples are drawn) should be equal. This can be stated as a formal null hypothesis:

$$H_0: \sigma_1^2 = \sigma_2^2.$$

Assessment of possible violations of Assumptions 1 and 2 were described in earlier sections of this chapter. Graphic methods, such as box plots (as described in an earlier section of this chapter), provide a way to see whether groups have similar ranges or variances of scores.

The SPSS *t* test and ANOVA procedures provide a significance test for the null assumption that the population variances are equal (the **Levene test**). Usually, researchers hope that this assumption is not violated, and thus, they usually hope that the ***F* ratio** for the Levene test will be nonsignificant. However, when the *N*s in the groups are equal and reasonably large (approximately $N > 30$ per group), ANOVA is fairly robust to violations of the equal variance assumption (Myers & Well, 1991, 1995).

Small sample sizes create a paradox with respect to the assessment of violations of many assumptions. When *N* is small, significance tests for possible violations of assumptions have low statistical power; and when *N* is small, violations of assumptions are more problematic for the analysis. For example, consider a one-way ANOVA with only 5 participants per group. With such a small *N*, the test for heterogeneity of variance may be significant only when the differences among sample variances are extremely large; however, with such a small *N*, small differences among sample variances might be enough to create problems in the analysis. Conversely, in a one-way ANOVA with 50 participants per group, quite small differences in variance across groups could be judged statistically significant; but with such a large *N*, only fairly large differences in group variances would be a problem. Doing the preliminary test for heterogeneity of variance when *N*s are very large is something like sending out a rowboat to see if the water is safe for the *Queen Mary*. Therefore, it may be reasonable to use very small α levels, such as $\alpha = .001$, for significance tests of violations of assumptions in studies with large sample sizes. On the other hand, researchers may want to set α values that are large (e.g., α of .20 or larger) for preliminary tests of assumptions when *N*s are small.

Tabachnick and Fidell (2007) provide extensive examples of preliminary data screening for comparison of groups. These generally involve repeating the univariate data-screening procedures described earlier (to assess normality of distribution shape and identify outliers) separately for each group and, in addition, assessing whether the **homogeneity of variance assumption** is violated.

It is useful to assess the distribution of quantitative scores within each group and to look for extreme outliers within each group. Refer back to Figure 4.29 to see an example of a box plot that identified outliers on WEIGHT within the gender groups. It might be desirable to remove these outliers or, at least, to consider how strongly they influence the outcome of a *t* test to compare male and female mean weights. The presence of these outlier scores on WEIGHT raises the mean weight for each group; the presence of these outliers also increases the within-group variance for WEIGHT in both groups.

4.8.3 ♦ Bivariate Data Screening for Two Quantitative Variables

Statistics that are part of the **general linear model (GLM)**, such as the Pearson correlation, require several assumptions. Suppose we want to use Pearson r to assess the strength of the relationship between two quantitative variables, X (diastolic blood pressure [DBP]) and Y (systolic blood pressure [SBP]). For this analysis, the data should satisfy the following assumptions:

1. Scores on X and Y should each have a univariate normal distribution shape.
2. The joint distribution of scores on X and Y should have a bivariate normal shape (and there should not be any extreme **bivariate outliers**).
3. X and Y should be linearly related.
4. The variance of Y scores should be the same at each level of X (the homogeneity or **homoscedasticity** of variance assumption).

The first assumption (univariate normality of X and Y) can be evaluated by setting up a histogram for scores on X and Y and by looking at values of skewness as described in Section 4.6.4. The other two assumptions (a bivariate normal distribution shape and a linear relation) can be assessed by examining an X, Y **scatter plot**.

To obtain an X, Y scatter plot, the following menu selections are used: <Graph> → <Scatter>.

From the initial Scatter/Dot dialog box (see Figure 4.37), the Simple Scatter type of scatter plot was selected by clicking on the icon in the upper left part of the Scatter/Dot dialog window. The Define button was used to move on to the next dialog window. In the next dialog window (shown in Figure 4.38), the name of the predictor variable (DBP at Time 1) was placed in the window marked X Axis, and the name of the outcome variable (SBP at Time 1) was placed in the window marked Y Axis. (Generally, if there is a reason to distinguish between the two variables, the predictor or "causal" variable is placed on the X axis in the scatter plot. In this example, either variable could have been designated as the predictor.) The output for the scatter plot showing the relation between scores on DBP

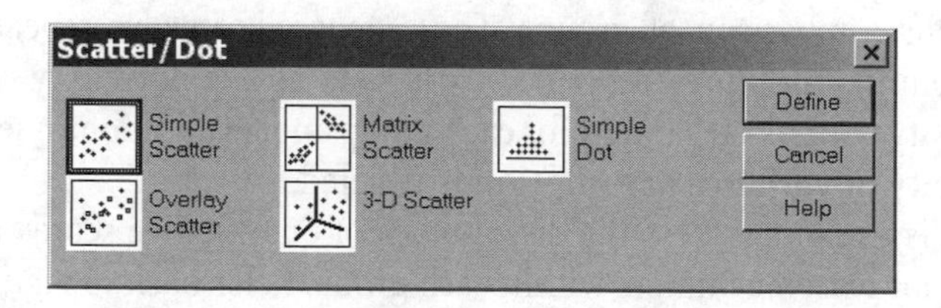

Figure 4.37 ♦ SPSS Dialog Box for the Scatter Plot Procedure

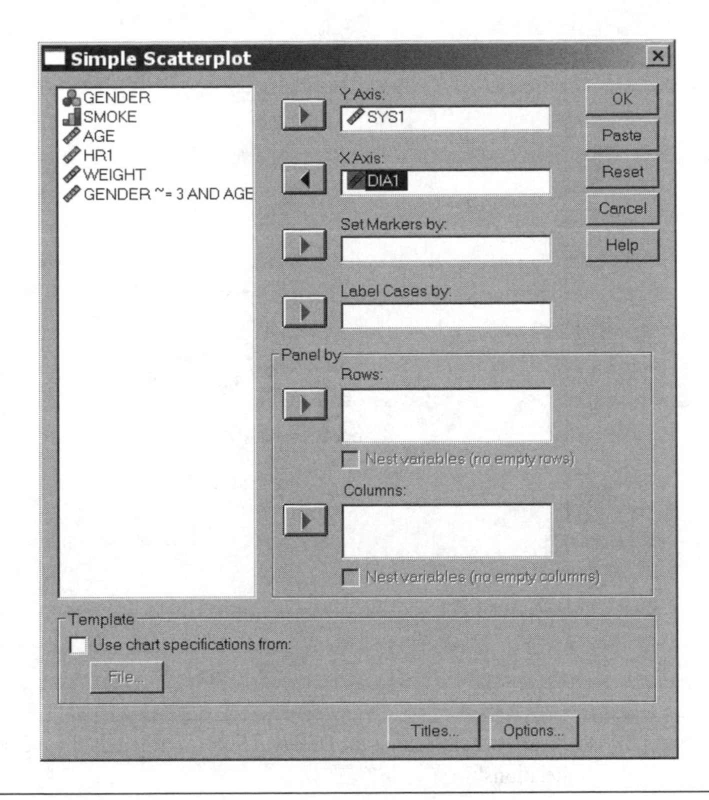

Figure 4.38 ♦ Scatter Plot: Identification of Variables on *X* and *Y* Axes

and SBP in Figure 4.39 shows a strong positive association between DBP and SBP. The relation appears to be fairly linear, and there are no bivariate outliers.

The assumption of bivariate normal distribution is more difficult to evaluate than the assumption of univariate normality, particularly in relatively small samples. Figure 4.40 represents an ideal theoretical bivariate normal distribution. Figure 4.41 is a bar chart that shows the frequencies of scores with specific pairs of *X*, *Y* values; it corresponds approximately to an empirical bivariate normal distribution (note that these figures were not generated using SPSS). *X* and *Y* have a bivariate normal distribution if *Y* scores are normally distributed for each value of *X* (and vice versa). In either graph, if you take any specific value of *X* and look at that cross section of the distribution, the univariate distribution of *Y* should be normal. In practice, even relatively large datasets ($N > 200$) often do not have enough data points to evaluate whether the scores for each pair of variables have a bivariate normal distribution.

There are several problems that may be detectable in a bivariate scatter plot. A bivariate outlier (see Figure 4.42) is a score that falls outside the region in the *X*, *Y* scatter plot where most *X*, *Y* values are located. In Figure 4.42, there is one individual who has a body weight of about 230 lb and SBP of about 110; this combination of score values is "unusual" (in general, persons with higher body weight tended to have higher blood pressure). To be judged

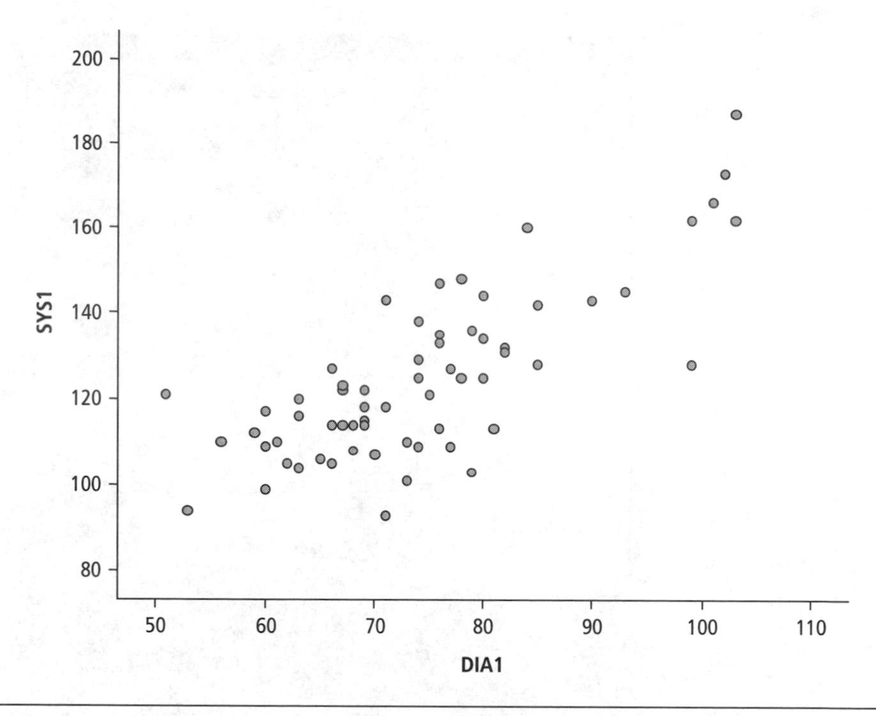

Figure 4.39 ♦ Bivariate Scatter Plot for DBP and SBP (Moderately Strong, Positive, Linear Relationship)

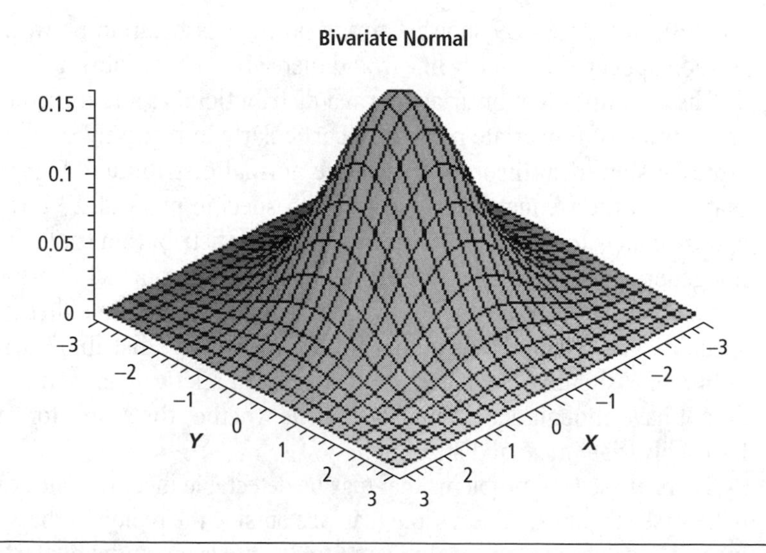

Figure 4.40 ♦ Three-Dimensional Representation of an Ideal Bivariate Normal Distribution

SOURCE: Reprinted with permission from Hartlaub, B., Jones, B. D., & Karian, Z. A., downloaded from www2.kenyon.edu/People/hartlaub/MellonProject/images/bivariate17.gif, supported by the Andrew W. Mellon Foundation.

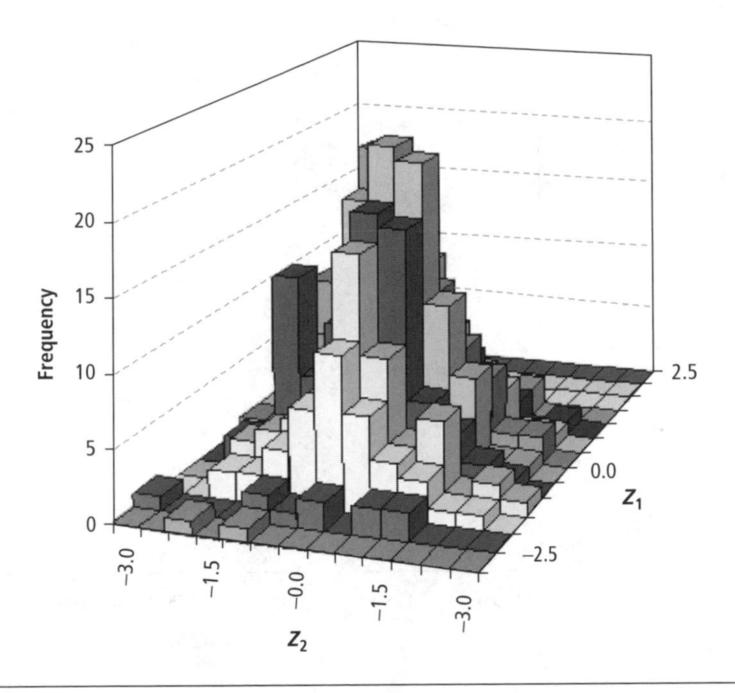

Figure 4.41 ♦ Three-Dimensional Histogram of an Empirical Bivariate Distribution (Approximately Bivariate Normal)

SOURCE: Reprinted with permission from Dr. P. D. M. MacDonald.

NOTE: Z_1 and Z_2 represent scores on the two variables, while the vertical heights of the bars along the "frequency" axis represent the number of cases that have each combination of scores on Z_1 and Z_2. A clear bivariate normal distribution is likely to appear only for datasets with large numbers of observations; this example only approximates bivariate normal.

a bivariate outlier, a score does not have to be a univariate outlier on either *X* or *Y* (although it may be). A bivariate outlier can have a disproportionate impact on the value of Pearson *r* compared with other scores, depending on its location in the scatter plot. Like univariate outliers, bivariate outliers should be identified and examined carefully. It is possible to obtain numerical indexes (such as **Mahalanobis *d***) that provide information about the degree to which individual scores are **multivariate outliers**, but usually the identification of bivariate outliers is done by visual examination of scatter plots. It may make sense in some cases to remove bivariate outliers, but it is preferable to do this early in the data analysis process, with a well-thought-out justification, rather than late in the data analysis process, because the data point does not conform to the preferred linear model.

Heteroscedasticity or heterogeneity of variance refers to a situation where the variance in *Y* scores is greater for some values of *X* than for others. In Figure 4.43, the variance of *Y* scores is much higher for *X* scores near 50 than for *X* values less than 30.

This unequal variance in *Y* across levels of *X* violates the assumption of homoscedasticity of variance; it also indicates that prediction errors for high values of *X* will be systematically larger than prediction errors for low values of *X*. Sometimes a **log transformation** on a *Y* variable that shows heteroscedasticity across levels of *X* can reduce the problem of unequal variance to some degree. However, if this problem cannot be

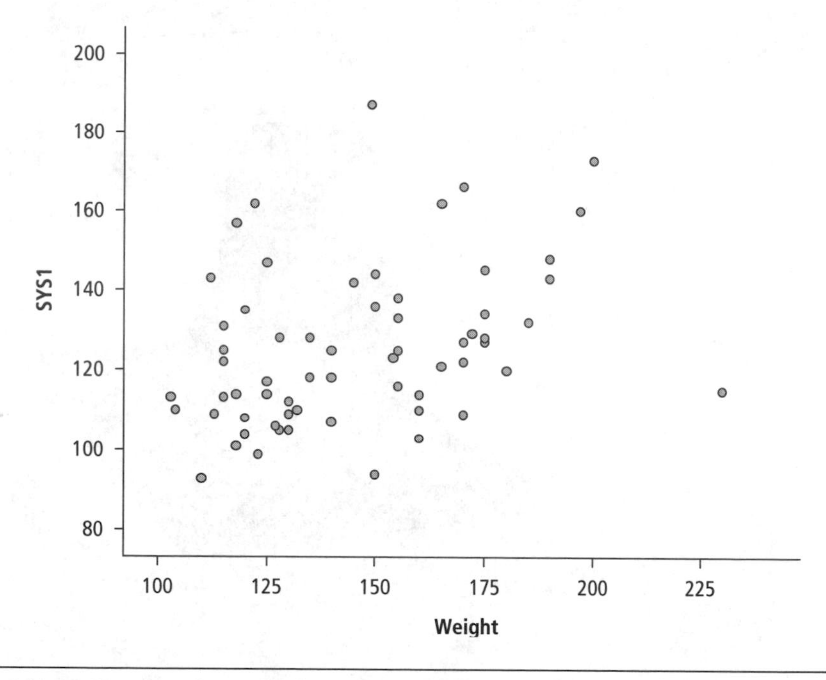

Figure 4.42 ♦ Bivariate Scatter Plot for Weight and SBP

NOTE: Bivariate outlier can be seen in the lower right corner of the graph.

corrected, then the graph that shows the unequal variances should be part of the story that is reported, so that readers understand: It is not just that *Y* tends to increase as *X* increases, in Figure 4.43; the variance of *Y* also tends to increase as *X* increases. Ideally, researchers hope to see reasonably uniform variance in *Y* scores across levels of *X*. In practice, the number of scores at each level of *X* is often too small to evaluate the shape and variance of *Y* values separately for each level of *X*.

4.9 ♦ Nonlinear Relations

Students should be careful to distinguish between these two situations: no relationship between *X* and *Y* versus a nonlinear relationship between *X* and *Y* (i.e., a relationship between *X* and *Y* that is not linear). An example of a scatter plot that shows no relationship of any kind (either linear or curvilinear) between *X* and *Y* appears in Figure 4.44. Note that as the value of *X* increases, the value of *Y* does not either increase or decrease.

In contrast, an example of a curvilinear relationship between *X* and *Y* is shown in Figure 4.45. This shows a strong relationship between *X* and *Y*, but it is not linear; as scores on *X* increase from 0 to 30, scores on *Y* tend to increase, but as scores on *X* increase between 30 and 50, scores on *Y* tend to decrease. An example of a real-world research situation that yields results similar to those shown in Figure 4.45 is a study that examines arousal or level of stimulation (on the *X* axis) as a predictor of task performance (on the *Y* axis). For example, suppose that the score on the *X* axis is a measure of anxiety and the

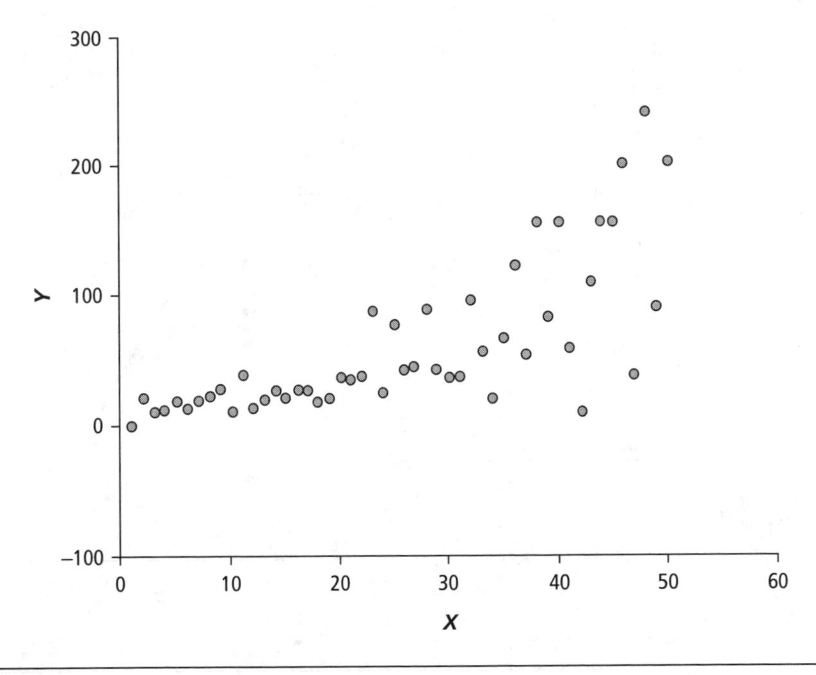

Figure 4.43 ♦ Illustration of Heteroscedasticity of Variance

NOTE: Variance in Y is larger for values of X near 50 than for values of X near 0.

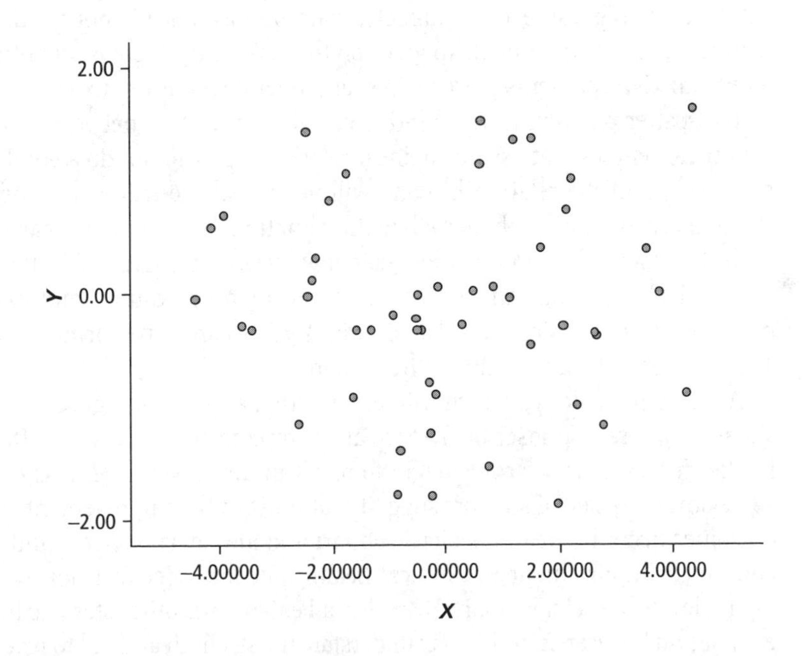

Figure 4.44 ♦ No Relationship Between X and Y

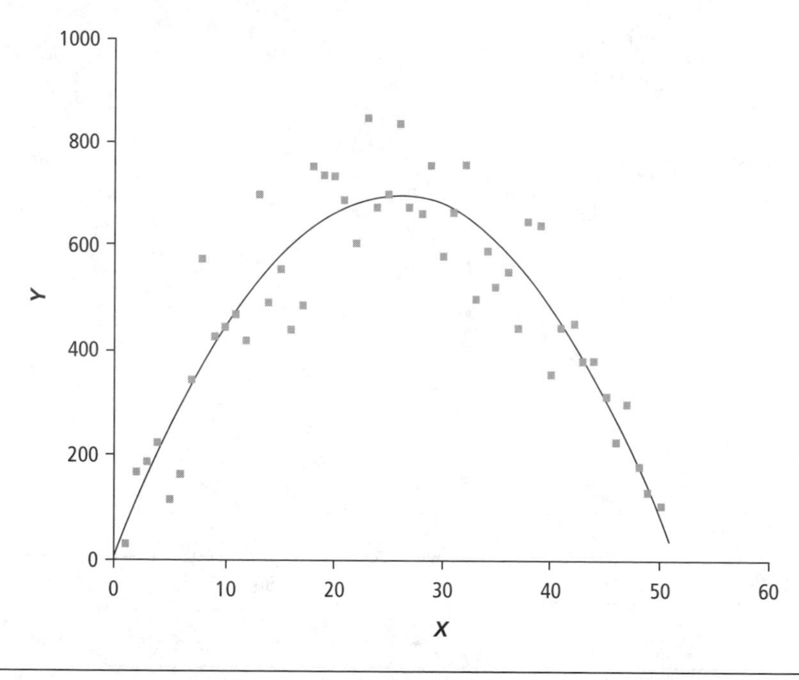

Figure 4.45 ♦ Bivariate Scatter Plot: Inverse U-Shaped Curvilinear Relation Between *X* and *Y*

score on the *Y* axis is a score on an examination. At low levels of anxiety, exam performance is not very good: Students may be sleepy or not motivated enough to study. At moderate levels of anxiety, exam performance is very good: Students are alert and motivated. At the highest levels of anxiety, exam performance is not good: Students may be distracted, upset, and unable to focus on the task. Thus, there is an optimum (moderate) level of anxiety; students perform best at moderate levels of anxiety.

If a scatter plot reveals this kind of curvilinear relation between *X* and *Y*, a Pearson *r* (or other analyses that assume a linear relationship) will not do a good job of describing the strength of the relationship and will not reveal the true nature of the relationship. Other analyses may do a better job in this situation. For example, *Y* can be predicted from both *X* and X^2 (a function that includes an X^2 term is a curve rather than a straight line). Alternatively, students can be separated into high-, medium-, and low-anxiety groups based on their scores on *X*, and a one-way ANOVA can be performed to assess how mean *Y* test scores differ across these three groups.

Another possible type of curvilinear function appears in Figure 4.46. This describes a situation where responses on *Y* reach an asymptote as *X* increases. After a certain point, further increases in *X* scores begin to result in diminishing returns on *Y*. For example, some studies of social support suggest that most of the improvements in physical health outcomes occur between no social support and low social support and that there is little additional improvement in physical health outcomes between low social support and higher levels of social support. Here also, a Pearson *r* or other statistic that assumes a linear relation between *X* and *Y* may understate the strength and fail to reveal the true nature of the association between *X* and *Y*.

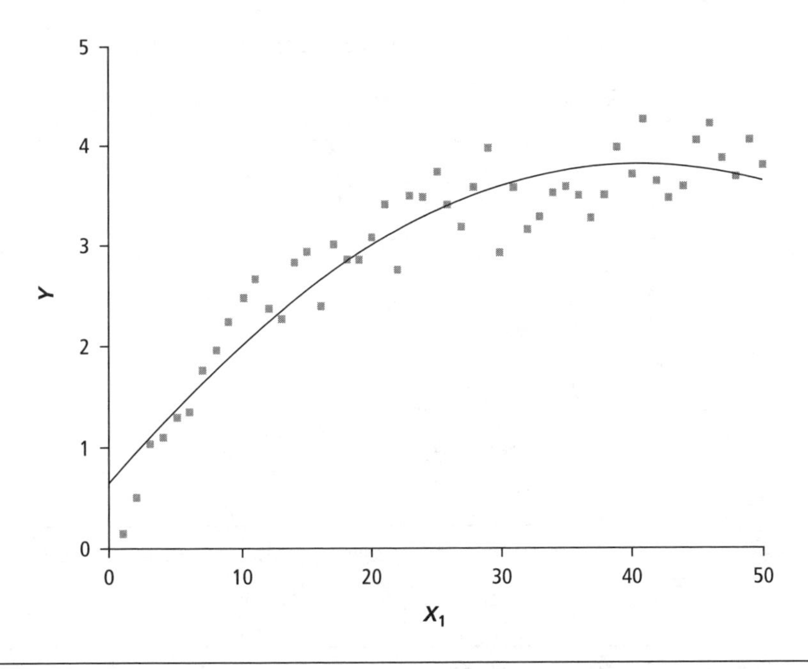

Figure 4.46 ♦ Bivariate Scatter Plot: Curvilinear Relation Between X_1 and Y

If a bivariate scatter plot of scores on two quantitative variables reveals a nonlinear or curvilinear relationship, this nonlinearity must be taken into account in the data analysis. Some nonlinear relations can be turned into linear relations by applying appropriate data transformations; for example, in psychophysical studies, the log of the physical intensity of a stimulus may be linearly related to the log of the perceived magnitude of the stimulus.

4.10 ♦ Data Transformations

A **linear transformation** is one that changes the original X score by applying only simple arithmetic operations (addition, subtraction, multiplication, or division) using constants. If we let b and c represent any two values that are constants within a study, then the arithmetic function $(X - b)/c$ is an example of a linear transformation. The linear transformation that is most often used in statistics is the one that involves the use of M as the constant b and the sample standard deviation s as the constant c: $z = (X - M)/s$. This transformation changes the mean of the scores to 0 and the standard deviation of the scores to 1, but it leaves the shape of the distribution of X scores unchanged.

Sometimes, we want a data transformation that will change the shape of a distribution of scores (or alter the nature of the relationship between a pair of quantitative variables in a scatter plot). Some data transformations for a set of raw X scores (such as the log of X and the log of Y) tend to reduce positive skewness and also to bring extreme outliers at the high end of the distribution closer to the body of the distribution (see Tabachnick &

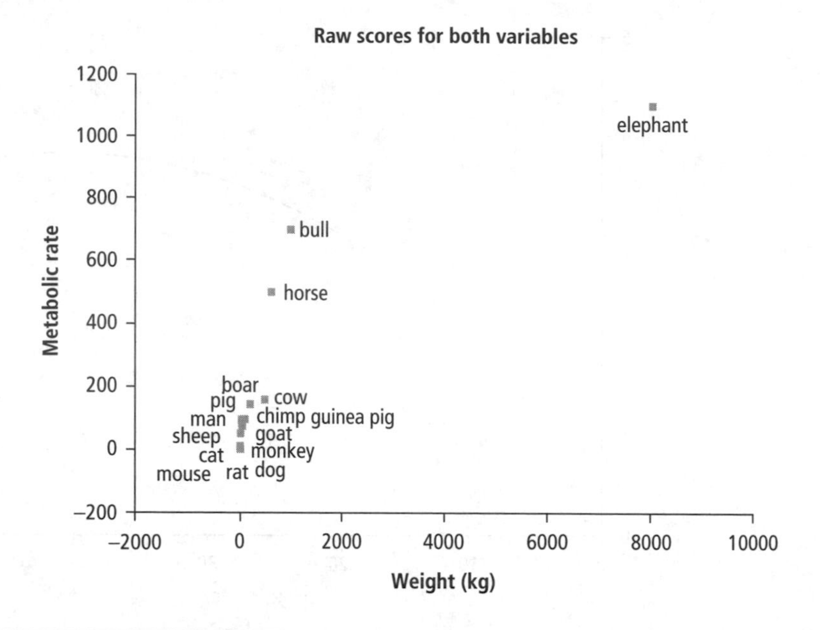

Figure 4.47 ♦ Illustration of the Effect of the Base 10 Log Transformation

NOTE: In Figure 4.47, raw scores for body weight are plotted on the *X* axis; raw scores for metabolic rate are plotted on the *Y* axis. In Figure 4.48, both variables have been transformed using base 10 log. Note that the log plot has more equal spaces among cases (there is more information about the differences among low-body-weight animals, and the outliers have been moved closer to the rest of the scores). Also, when logs are taken for both variables, the relation between them becomes linear. Log transformations do not always create linear relations, of course, but there are some situations where they do.

Fidell, 2007, Chapter 4, for further discussion). Thus, if a distribution is skewed, taking the log or square root of scores sometimes makes the shape of the distribution more nearly normal. For some variables (such as reaction time), it is conventional to do this; log of reaction time is very commonly reported (because reaction times tend to be positively skewed). However, note that changing the scale of a variable (from heart rate to log of heart rate) changes the meaning of the variable and can make interpretation and presentation of results somewhat difficult.

Sometimes a nonlinear transformation of scores on *X* and *Y* can change a nonlinear relation between *X* and *Y* to a linear relation. This is extremely useful, because the analyses included in the family of methods called general linear models usually require linear relations between variables.

A common and useful nonlinear transformation of *X* is the base 10 log of *X*, denoted by $\log_{10}(X)$. When we find the base 10 log of *X*, we find a number *p* such that $10^p = X$. For example, the base 10 log of 1,000 is 3, because $10^3 = 1{,}000$. The *p* exponent indicates order of magnitude.

Consider the graph shown in Figure 4.47. This is a graph of body weight (in kilograms) on the *X* axis with mean metabolic rate on the *Y* axis; each data point represents a mean body weight and a mean metabolic rate for one species. There are some ways in which this graph is difficult to read; for example, all the data points for physically smaller animals are

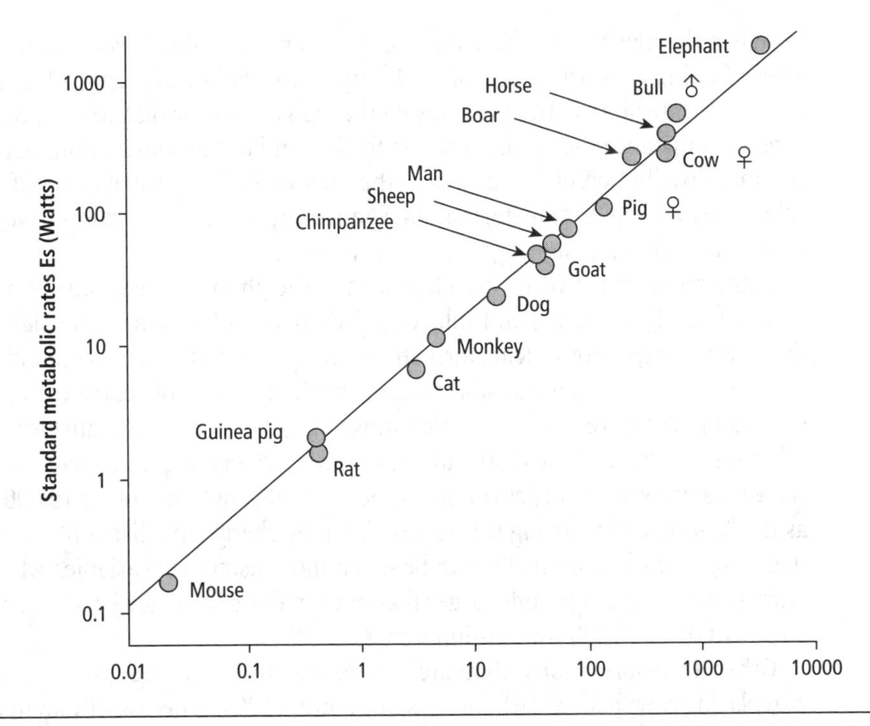

Figure 4.48 ◆ Graph Illustrating That the Relation Between Base 10 Log of Body Weight and Base 10 Log of Metabolic Rate Across Species Is Almost Perfectly Linear

SOURCE: Reprinted with permission from Dr. Tatsuo Motokawa.

crowded together in the lower left-hand corner of the scatter plot. In addition, if you wanted to fit a function to these points, you would need to fit a curve (rather than a straight line).

Figure 4.48 shows the base 10 log of body weight and the base 10 log of metabolic rate for the same set of species as in Figure 4.47. Note that now, it is easy to see the differences among species at the lower end of the body-size scale; and the relation between the logs of these two variables is almost perfectly linear.

In Figure 4.47, the tick marks on the *X* axis represented equal differences in terms of kilograms. In Figure 4.48, the equally spaced points on the *X* axis now correspond to equal spacing between orders of magnitude (e.g., 10^1, 10^2, 10^3, . . .); a one-tick-mark change on the *X* axis in Figure 4.48 represents a change from 10 to 100 kg or 100 to 1,000 kg or 1,000 to 10,000 kg. A cat weighs something like 10 times as much as a dove, a human being weighs something like 10 times as much as a cat, a horse about 10 times as much as a human, and an elephant about 10 times as much as a horse. If we take the log of body weight, these log values ($p = 1, 2, 3$, etc.) represent these orders of magnitude, 10^p (10^1 for a dove, 10^2 for a cat, 10^3 for a human, and so on). If we graphed weights in kilograms using raw scores, we would find a much larger difference between elephants and humans than between humans and cats. The $\log_{10}(X)$ transformation yields a new way of scaling weight in terms of p, the relative orders of magnitude.

When the raw X scores have a range that spans several orders of magnitude (as in the sizes of animals, which vary from < 1 g up to 10,000 kg), applying a log transformation reduces the distance between scores on the high end of the distribution much more than it reduces distances between scores on the low end of the distribution. Depending on the original distribution of X, outliers at the high end of the distribution of X are brought "closer" by the log(X) transformation. Sometimes when raw X scores have a distribution that is skewed to the right, log(X) is nearly normal.

Some relations between variables (such as the physical magnitude of a stimulus, e.g., a weight or a light source) and subjective judgments (of heaviness or brightness) become linear when log or power transformations are applied to the scores on both variables.

Note that when a log transformation is applied to a set of scores with a limited range of possible values (e.g., Likert scale ratings of 1, 2, 3, 4, 5), this transformation has little effect on the shape of the distribution. However, when a log transformation is applied to scores that vary across orders of magnitude (e.g., the highest score is 10,000 times as large as the lowest score), the log transformation may change the distribution shape substantially. Log transformations tend to be much more useful for variables where the highest score is orders of magnitude larger than the smallest score; for example, maximum X is 100 or 1,000 or 10,000 times minimum X.

Other transformations that are commonly used involve power functions—that is, replacing X with X^2, X^c (where c is some power of X, not necessarily an integer value), or $\sqrt{X}$. For specific types of data (such as scores that represent proportions, percentages, or correlations), other types of nonlinear transformations are needed.

Usually, the goals that a researcher hopes to achieve through data transformations include one or more of the following: to make a nonnormal distribution shape more nearly normal, to minimize the impact of outliers by bringing those values closer to other values in the distribution, or to make a nonlinear relationship between variables linear.

One argument against the use of nonlinear transformations has to do with interpretability of the transformed scores. If we take the square root of "number of times a person cries per week," how do we talk about the transformed variable? For some variables, certain transformations are so common that they are expected (e.g., psychophysical data are usually modeled using power functions; measurements of reaction time usually have a log transformation applied to them).

4.11 ♦ Verifying That Remedies Had the Desired Effects

Researchers should not assume that the remedies they use to try to correct problems with their data (such as removal of outliers, or log transformations) are successful in achieving the desired results. For example, after one really extreme outlier is removed, when the frequency distribution is graphed again, other scores may still appear to be relatively extreme outliers. After the scores on an X variable are transformed by taking the **natural log** of X, the distribution of the natural log of X may still be nonnormal. It is important to repeat data screening using the transformed scores in order to make certain that the data transformation had the desired effect. Ideally, the transformed scores will have a nearly normal distribution without extreme outliers, and relations between pairs of transformed variables will be approximately linear.

4.12 ♦ Multivariate Data Screening

Data screening for multivariate analyses (such as **multiple regression** and **multivariate analysis of variance**) begins with screening for each individual variable and bivariate data screening for all possible pairs of variables as described in earlier sections of this chapter. More complex assumptions about data structure will be reviewed as they arise in later chapters. Complete data screening in multivariate studies requires careful examination not just of the distributions of scores for each individual variable but also of the relationships between pairs of variables and subsets of variables. Excellent examples are presented in Tabachnick and Fidell (2007).

4.13 ♦ Reporting Preliminary Data Screening

Many journals in psychology and related fields use the style guidelines published by the American Psychological Association (APA, 2001). This section covers some of the basic guidelines. All APA-style research reports should be double-spaced and single sided with at least 1-in. margins on each page.

A Results section should report data screening and the data analyses that were performed (including results that run counter to predictions). Interpretations and discussion of implications of the results are generally placed in the Discussion section of the paper (except in very brief papers with combined Results/Discussion sections).

Although null hypothesis significance tests are generally reported, the updated fifth edition of the APA *Publication Manual* also calls for the inclusion of effect size information and confidence intervals (CIs), wherever possible, for all major outcomes. Include the basic descriptive statistics that are needed to understand the nature of the results; for example, a report of a one-way ANOVA should include group means and standard deviations as well as *F* values, degrees of freedom, effect size information, and CIs.

Standard abbreviations are used for most statistics—for example, *M* for mean, *SD* for standard deviation (APA, 2001, pp. 140–144). These should be in italic font (APA, 2001, p. 101). Parentheses are often used when these are reported in the context of a sentence, as in, "The average verbal SAT for the sample was 551 ($SD = 135$)."

The sample size (*N*) or the degrees of freedom (*df*) should always be included when reporting statistics. Often the *df* values appear in parentheses immediately following the statistic, as in this example: "There was a significant gender difference in mean score on the Anger In scale, $t(61) = 2.438, p = .018$, two-tailed, with women scoring higher on average than men." Generally, results are rounded to two decimal places, except that *p* values are sometimes given to three decimal places. It is more informative to report exact *p* values than to make directional statements such as $p < .05$. If the printout shows a *p* of .000, it is preferable to report $p < .001$ (the risk of Type I error indicated by *p* is not really zero). When it is possible for *p* values to be either one-tailed or two-tailed (for the independent samples *t* test, for example), this should be stated explicitly.

Tables and figures are often useful ways of summarizing a large amount of information—for example, a list of *t* tests with several dependent variables, a table of correlations among several variables, or the results from multivariate analyses such as multiple regression. See APA (2001, pp. 147–201) for detailed instructions about the preparation of tables and figures. (Tufte, 1983, presents wonderful examples of excellence and awfulness in graphic representations of data.) All tables and figures should be discussed in the text; however, the text should not repeat all the information in a table; it should point out only the highlights. Table and figure headings should be informative enough to be understood on their own. It is common to denote statistical significance using asterisks (e.g., * for $p < .05$, ** for $p < .01$, and *** for $p < .001$), but these should be described by footnotes to the table. Each column and row of the table should have a clear heading. When there is not sufficient space to type out the entire names for variables within the table, numbers or abbreviations may be used in place of variable names, and this should also be explained fully in footnotes to the table.

Horizontal rules or lines should be used sparingly within tables (i.e., not between each row but only in the headings and at the bottom). Vertical lines are not used in tables. Spacing should be sufficient so that the table is readable.

In general, Results sections should include the following information. For specific analyses, additional information may be useful or necessary.

1. The opening sentence of each Results section should state *what analysis was done, with what variables, and to answer what question.* This sounds obvious, but sometimes this information is difficult to find in published articles. An example of this type of opening sentence is, "In order to assess whether there was a significant difference between the mean Anger In scores of men and women, an independent samples t test was performed using the Anger In score as the dependent variable."

2. Next, *describe the data screening that was done to decide whether assumptions were violated, and report any steps that were taken to correct the problems that were detected.* For example, this would include examination of distribution shapes using graphs such as histograms, detection of outliers using box plots, and tests for violations of homogeneity of variance. Remedies might include deletion of outliers, data transformations such as the log, or choice of a statistical test that is more robust to violations of the assumption.

3. The next sentence should report the *test statistic and the associated exact* p *value; also, a statement whether or not it achieved statistical significance*, according to the predetermined alpha level, should be included: "There was a significant gender difference in mean score on the Anger In scale, $t(61) = 2.438$, $p = .018$, two-tailed, with women scoring higher on average than men." The significance level can be given as a range ($p < .05$) or as a specific obtained value ($p = .018$). For nonsignificant results, any of the following methods of reporting may be used: $p > .05$ (i.e., a statement that the p value on the printout was larger than a preselected α level of .05); $p = .38$ (i.e., an exact obtained p value); or just *ns* (an abbreviation for nonsignificant). Recall that the p value is an estimate of the risk of Type I error; in theory, this risk is never zero, although it may be very small. Therefore, when the printout reports a significance or p value of .000, it is more accurate to report it as "$p < .001$" than as "$p = .000$."

4. *Information about the strength of the relationship should be reported.* Most statistics have an accompanying effect size measure. For example, for the independent samples *t* test, Cohen's *d* and η^2 are common effect size indexes. For this example, $\eta^2 = t^2/(t^2 + df) = (2.438)^2/((2.438)^2 + 61) = .09$. Verbal labels may be used to characterize an effect size estimate as small, medium, or large. Reference books such as Cohen's (1988) *Statistical Power Analysis for the Behavioral Sciences* suggest guidelines for the description of effect size.

5. Where possible, *CIs should be reported for estimates.* In this example, the 95% CI for the difference between the sample means was from .048 to .484.

6. *It is important to make a clear statement about the nature of relationships* (e.g., the direction of the difference between group means or the sign of a correlation). In this example, the mean Anger In score for females (M = 2.36, SD = .484) was higher than the mean Anger In score for males (M = 2.10, SD = .353). Descriptive statistics should be included to provide the reader with the most important information. Also, note whether the outcome was consistent with or contrary to predictions; detailed interpretation/discussion should be provided in the Discussion section.

Many published studies report multiple analyses. In these situations, it is important to think about the sequence. Sometimes, basic demographic information is reported in the section about participants in the Methods section of the paper. However, it is also common for the first table in the Results section to provide means and standard deviations for all the quantitative variables and group sizes for all the categorical variables. Preliminary analyses that examine the reliabilities of variables are reported prior to analyses that use those variables. It is helpful to organize the results so that analyses that examine closely related questions are grouped together. It is also helpful to maintain a parallel structure throughout the research paper. That is, a set of questions are outlined in the Introduction, the Methods section describes the variables that are manipulated and/or measured to answer those questions, the Results section reports the statistical analyses that were employed to try to answer each question, and the Discussion section interprets and evaluates the findings relevant to each question. It is helpful to keep the questions in the same order in each section of the paper.

Sometimes, a study has both confirmatory and exploratory components (as discussed in Chapter 1). For example, a study might include an experiment that tests the hypotheses derived from earlier research (confirmatory), but it might also examine the relations among variables to look for patterns that were not predicted (exploratory). It is helpful to make a clear distinction between these two types of results. The confirmatory Results section usually includes a limited number of analyses that directly address questions that were stated in the Introduction; when a limited number of significance tests are presented, there should not be a problem with inflated risk of Type I error. On the other hand, it may also be useful to present the results of other exploratory analyses; however, when many significance tests are performed and no a priori predictions were made, the results should be labeled as exploratory, and the author should state clearly that any *p* values that are reported in this context are likely to underestimate the true risk of Type I error.

Usually, the first part of a Results section reports data screening, reliability assessment, and so on. In some cases, it may be possible to make a general statement such as, "All variables were normally distributed, with no extreme outliers" or "Group variances were not significantly heterogeneous," as a way of indicating that assumptions for the analysis are reasonably well satisfied. Then, follow the analyses that answer the basic research questions. It is helpful to maintain a parallel structure in the Introduction, the Results, and the Discussion; that is, you should raise the questions in the Introduction, report the results for appropriate analyses in the Results, and discuss the implications of your findings in the Discussion—always in the same order.

Put "major questions" first, and if additional exploratory analyses are reported, these may be placed near the end of the Results section, along with information that indicates that they are additional exploratory analyses and not the main point of your study.

An example of a Results section that illustrates some of these points follows. The SPSS printout that yielded these numerical results is not included here; it is provided in the Instructor Supplement materials for this textbook.

Results

An independent samples *t* test was performed to assess whether there was a gender difference in mean Anger In scores. Histograms and box plots indicated that scores on the dependent variable were approximately normally distributed within each group with only one outlier in each group. Because these outliers were not extreme, these scores were retained in the analysis. The Levene test showed a nonsignificant difference between the variances; because the homogeneity of variance assumption did not appear to be violated, the **pooled variances version of the *t* test** was used. The male and female groups had 28 and 37 participants, respectively. The difference in mean Anger In scores was found to be statistically significant, $t(63) = 2.50$, $p = .015$, two-tailed. The mean Anger In score for females ($M = 2.37$, $SD = .482$) was higher than the mean Anger In score for males ($M = 2.10$, $SD = .353$). The effect size, indexed by η^2, was .09. The 95% CI around the difference between these sample means ranged from .05 to .49.

4.14 ♦ Summary and Checklist for Data Screening

The goals of data screening include the following: identification and correction of data errors, detection and decisions about outliers, and evaluation of patterns of missing data and decisions regarding how to deal with missing data. For quantitative variables, the researcher needs to verify that all groups that will be examined in analyses (such as Crosstabs or ANOVA) have a reasonable number of cases. For quantitative variables, it is important to assess the shape of the distribution of scores and to see what information the distribution provides about outliers, ceiling or floor effects, and restricted range. Assumptions specific to the analyses that will be performed (e.g., the assumption of homogeneous population variances for the independent samples *t* test, the assumption of linear relations between variables for Pearson *r*) should be evaluated. Possible remedies for problems with general linear model assumptions that are identified include dropping scores, modifying scores through data transformations, or choosing a different analysis that is more appropriate to the data. After deleting outliers or transforming scores, it is important to check (by rerunning frequency distributions and replotting graphs) in order to make sure that the data modifications actually had the desired effects. A checklist of data-screening procedures is given in Table 4.2.

Preliminary screening also yields information that may be needed to characterize the sample. The Methods section typically reports the numbers of male and female participants, mean and range of age, and other demographic information.

Table 4.2 ♦ Checklist for Data Screening

1. Proofread scores in the SPSS data worksheet against original data sources, if possible.
2. Identify response inconsistencies across variables.
3. During univariate screening of scores on categorical variables,
   a. check for values that do not correspond to valid response alternatives; and
   b. note groups that have *N*s too small to be examined separately in later analyses (decide what to do with small-*N* groups—e.g., combine them with other groups, drop them from the dataset).
4. During univariate screening of scores on quantitative variables, look for
   a. normality of distribution shape (e.g., skewness, kurtosis, other departures from normal shape);
   b. outliers;
   c. scores that do not correspond to valid response alternatives or possible values; and
   d. ceiling or floor effects, restricted range.
5. Consider dropping individual participants or variables that show high levels of incorrect responses or responses that are inconsistent.
6. Note the pattern of "missing" data. If not random, describe how they are patterned.
7. For bivariate analyses involving two categorical variables (e.g., chi-squared),
   a. examine the marginal distributions to see whether the *N*s in each row and column are sufficiently large (if not, consider dropping some categories or combining them with other categories); and
   b. check whether expected values in all cells are greater than 5 (if this is not the case, consider alternatives to χ^2 such as the **Fisher exact test**).

(Continued)

Table 4.2 ♦ (Continued)

8. For bivariate analyses of two continuous variables (e.g., Pearson *r*), examine the scatter plot:
   a. Assess possible violations of **bivariate normality.**
   b. Look for bivariate outliers or **disproportionately influential scores.**
   c. Assess whether the relation between *X* and *Y* is linear. If it is not linear, consider whether to use a different approach to analysis (e.g., divide scores into low, medium, and high groups based on *X* scores and do an ANOVA) or use nonlinear transformations such as log to make the relation more nearly linear.
   d. Assess whether variance in *Y* scores is uniform across levels of *X* (i.e., the assumption of homoscedasticity of variance).
9. For bivariate analyses with one categorical and one continuous variable,
   a. assess the distribution shapes for scores within each group (Are the scores normally distributed?);
   b. look for outliers within each group;
   c. test for possible violations of homogeneity of variance; and
   d. make sure that group sizes are adequate.
10. Verify that any remedies that have been attempted were successful—for example, after removal of outliers, does a distribution of scores on a quantitative variable now appear approximately normal in shape? After taking a log of *X*, is the distribution of *X* more nearly normal, and is the relation of *X* with *Y* more nearly linear?
11. Based on data screening and the success or failure of remedies that were attempted,
   a. Are assumptions for the intended parametric analysis (such as *t* test, ANOVA, or Pearson *r*) sufficiently well met to go ahead and use parametric methods?
   b. If there are problems with these assumptions, should a nonparametric method of data analysis be used?
12. In the report of results, include a description of data-screening procedures and any remedies (such as dropping outliers, imputing values for missing data, or data transformations) that were applied to the data prior to other analyses.

+ − × ÷

Comprehension Questions

1. What are the goals of data screening?
2. What SPSS procedures can be used for data screening of categorical variables?
3. What SPSS procedures can be used for data screening of quantitative variables?
4. What do you need to look for in bivariate screening (for each combination of categorical and quantitative variables)?
5. What potential problems should you look for in the univariate distributions of categorical and quantitative scores?
6. How can a box and whiskers plot (or box plot) be used to look for potential outliers?
7. What assumptions about the data are required for most parametric analyses?
8. How can you identify and remedy the following: errors in data entry, outliers, and missing data?
9. Why is it important to assess whether missing values are randomly distributed throughout the participants and measures? Or in other words, Why is it important to understand what processes lead to missing values?
10. Why are log transformations sometimes applied to scores?
11. Outline the information that should be included in an APA-style Results section.

Data Analysis Project for Univariate and Bivariate Data Screening

Data for this assignment may be provided by your instructor, or use one of the datasets found on the Web site for this textbook. Note that in addition to the variables given in the SPSS file, you can also use variables that are created by compute statements, such as scale scores formed by summing items (e.g., Hostility = H1 + H2 + H3 + H4).

1. Select three variables from the dataset. Choose two of the variables such that they are *good* candidates for correlation/regression and one other variable as a *bad* candidate. Good candidates are variables that meet the assumptions (e.g., normally distributed, reliably measured, interval/ratio level of measurement, etc.). Bad candidates are variables that do not meet assumptions or that have clear problems (restricted range, extreme outliers, gross nonnormality of distribution shape, etc.).
2. For each of the three variables, use the Frequencies procedure to obtain a histogram and all univariate descriptive statistics.

3. For the two "good candidate" variables, obtain a scatter plot. Also, obtain a scatter plot for the "bad candidate" variable with one of the two good variables.

Hand in your printouts for these analyses along with your answers to the following questions (there will be no Results section in this assignment).

1. Explain which variables are good and bad candidates for a correlation analysis, and give your rationale. Comment on the empirical results from your data screening—both the histograms and the scatter plots—as evidence that these variables meet or do not meet the basic assumptions necessary for correlation to be meaningful and "honest." Also, can you think of other information you would want to have about the variables in order to make better informed judgments?
2. Is there anything that could be done (in terms of data transformations, eliminating outliers, etc.) to make your "bad candidate" variable better? If so, what would you recommend?

CHAPTER 5

Comparing Group Means Using the Independent Samples *t* Test

5.1 ♦ Research Situations Where the Independent Samples *t* Test Is Used

One of the simplest research designs involves comparison of mean scores on a quantitative *Y* outcome between two groups; membership in each of the two groups is identified by each person's score on a categorical *X* variable that identifies membership in one of just two groups. The groups that are compared may correspond to naturally occurring groups (for the variable gender, the groups that are compared are male vs. female; for the variable smoking status, the groups that are compared may be smokers vs. nonsmokers). When naturally occurring groups are compared, the study is nonexperimental. Alternatively, in an experimental study, groups are often formed by random assignment and given different treatments. The predictor variable (*X*) in these research situations is a dichotomous group membership variable. A **dichotomous variable** is a categorical variable that has only two values; the values of *X* are labels for group membership. For example, the *X* predictor variable could be dosage level in an experimental drug study (coded 1 = No caffeine, 2 = 150 mg of caffeine), or it could be a code that represents membership in a naturally occurring group such as gender (coded 1 = Male, 2 = Female). The numerical values used to label groups make no difference in the results of the statistical analyses; small integers are usually chosen as values for *X*, for convenience. The outcome variable (*Y*) for an independent samples *t* test is quantitative, for example, anxiety score or heart rate (HR).

Usually, the researcher's goal is to show that there is a significant difference in mean scores on *Y* between the groups. In the context of a well-controlled experimental design, a significant difference in means may be interpreted as evidence that the manipulated independent variable (such as dosage of caffeine) has an effect on the outcome variable (such as HR). However, researchers need to be cautious about making causal inferences. A significant difference between group means may arise due to chance (sampling error); it may be due to some variable that was unintentionally confounded with the treatment variable; or it might be due to artifact, such as experimenter expectancy effects. The results of a single study are not sufficient to justify a causal inference. However, if a result is replicated across many well-controlled experiments and if researchers are careful to rule out possible confounds and artifacts, eventually the accumulated evidence across repeated studies becomes strong enough to warrant cautious inferences about possible causality.

In nonexperimental applications of the *t* test, when the groups being compared are naturally occurring groups (e.g., Americans vs. Japanese, women vs. men, people who meditate vs. nonmeditators), causal inference is not appropriate. This type of comparison can be reported only as a descriptive result and cannot be used to make causal inferences. For example, if meditators score lower on anxiety than nonmeditators, it does not necessarily mean that meditation reduces anxiety; it might be that anxious persons are less likely to choose to meditate. If women score higher on emotional intelligence than men, we can't assume that higher emotional intelligence is due to genetic sex differences; it might have arisen through socialization or other variables that had different effects on women and men.

Occasionally, a researcher hopes to demonstrate that there is no difference between groups; for example, a day care researcher may want to show that there is no difference in cognitive development for children in day care versus home care. The logic of hypothesis testing, described in Chapter 3, does not allow researchers to say that they have "proved" the null hypothesis of no difference between group means. It is necessary to replicate the nonsignificant difference across many studies with large numbers of participants before a case can be made for a lack of difference between populations.

This chapter describes the independent samples *t* test. This test is appropriate when the groups that are compared are between-subjects (between-S) or independent groups (as discussed in Chapter 1). If the data come from a within-subjects (within-S) or repeated measures design, the **paired samples** or **direct difference *t* test** should be used instead of the independent samples *t* test. The paired samples *t* test (for within-S or repeated measures data) is discussed in Chapter 20 of this textbook, along with a more general introduction to methods of analysis for repeated measures data. The independent samples *t* test is a parametric test; that is, the evaluation of the statistical significance of the *t* ratio is based on assumptions that the scores on the outcome variable *Y* are quantitative, interval/ratio, and approximately normally distributed. If the scores on the outcome variable *Y* are ordinal, a nonparametric method of analysis such as the Wilcoxon rank sum test would be more appropriate (as discussed in Chapter 1 of this textbook).

5.2 ♦ A Hypothetical Research Example

As a concrete example, consider the following imaginary experiment. Twenty participants are recruited as a convenience sample; each participant is randomly assigned to one of two groups. The groups are given different dosage levels of caffeine. Group 1 receives 0 mg of caffeine; Group 2 receives 150 mg of caffeine, approximately equivalent to the caffeine in one cup of brewed coffee. Half an hour later, each participant's HR is measured; this is the quantitative *Y* outcome variable. The goal of the study is to assess whether caffeine tends to increase mean HR. In this situation, *X* (the independent variable) is a dichotomous variable (amount of caffeine, 0 vs. 150 mg); *Y*, or HR, is the quantitative dependent

Table 5.1 ♦ Data for Hypothetical Research Example: Effect of Caffeine on HR (0 mg caffeine = "1," 150 mg caffeine = "2")

ID	*Caffeine*	*HR*
1	1	51
2	1	66
3	1	58
4	1	58
5	1	53
6	1	48
7	1	57
8	1	73
9	1	56
10	1	58
11	2	72
12	2	57
13	2	78
14	2	61
15	2	66
16	2	54
17	2	64
18	2	82
19	2	71
20	2	74

variable. See Table 5.1 for a table of scores for these variables. Figure 5.1 shows the corresponding SPSS Data View worksheet. The null hypothesis for the independent samples *t* test is as follows:

$$H_0: \mu_1 = \mu_2 \text{ or } \mu_1 - \mu_2 = 0.$$

In words, this translates into the assumption that the mean HR for a hypothetical population of people who have not recently consumed caffeine (μ_1) is equal to the mean HR for a hypothetical population of people who have recently consumed caffeine equivalent to 150 mg or one cup of coffee (μ_2). In even simpler language, this null hypothesis says that caffeine has no effect on HR (i.e., mean HR is the same whether or not people have recently consumed caffeine). This is an example where the populations of interest are hypothetical, as discussed in Chapter 1; we hope that information about HR for people who do and do not consume caffeine in this convenience sample will provide reasonably good estimates of the mean HR for broader hypothetical populations of all persons who have and have not recently consumed caffeine.

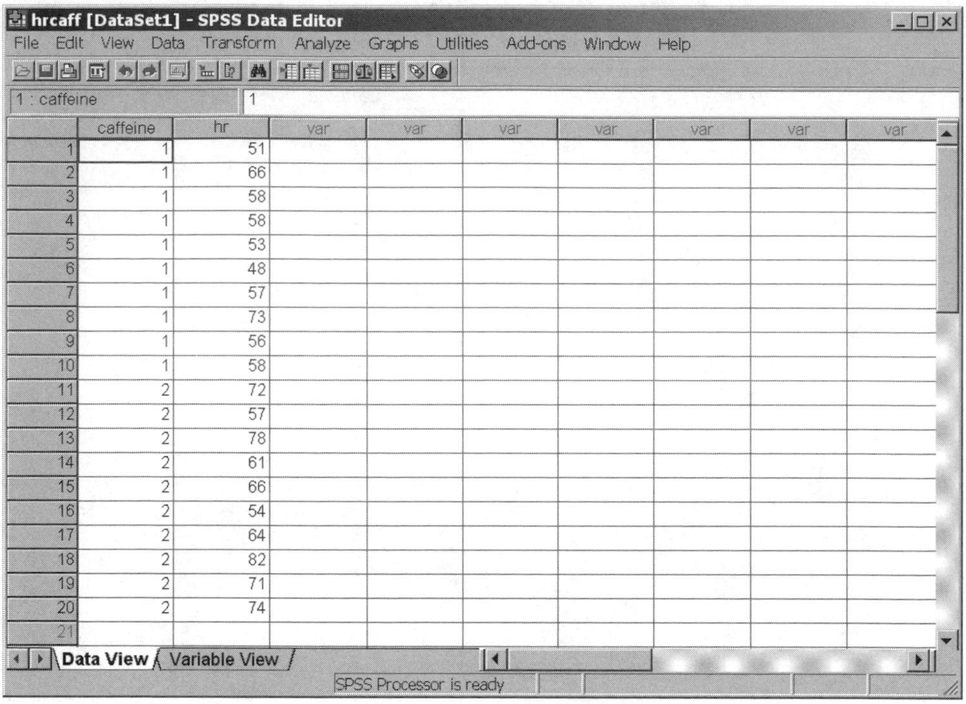

hrcaff [DataSet1] - SPSS Data Editor

File Edit View Data Transform Analyze Graphs Utilities Add-ons Window Help

1 : caffeine | 1

	caffeine	hr	var	var	var	var	var	var	var
1	1	51							
2	1	66							
3	1	58							
4	1	58							
5	1	53							
6	1	48							
7	1	57							
8	1	73							
9	1	56							
10	1	58							
11	2	72							
12	2	57							
13	2	78							
14	2	61							
15	2	66							
16	2	54							
17	2	64							
18	2	82							
19	2	71							
20	2	74							
21									

Data View / Variable View

SPSS Processor is ready

Figure 5.1 ♦ SPSS Data View Worksheet: Caffeine/HR Experiment Data

The researcher is usually interested in evidence that would contradict this hypothesis of no effect. If mean HR in the sample of participants who receive 150 mg of caffeine is much higher than the mean HR in the sample of participants who receive no caffeine, this outcome would lead us to doubt the null hypothesis that caffeine has no effect on HR. How do we decide if one sample mean is "much higher" than another sample mean? We quantify this difference between means by putting it in terms of the number of standard errors and setting up a *t* ratio to assess whether the difference between sample means, $M_1 - M_2$, is large relative to the amount of sampling error represented by $SE_{M_1 - M_2}$. The methods for setting up and evaluating the statistical significance of the difference between means presented in this chapter are an extension of the methods for the one sample *t* test described in Chapter 3.

Researchers usually hope to find a relatively large difference between M_1 and M_2 (in this example, the researcher might hope to find a large difference between mean HR for Groups 1 and 2). That is, when a *t* test ratio is set up to see how large the difference between M_1 and M_2 is relative to the standard error of this difference, $SE_{M_1 - M_2}$, the researcher hopes that this *t* ratio will be "large." When the degrees of freedom (*df*) are greater than 100, and when we set $\alpha = .05$, two-tailed, a *t* ratio greater than 1.96 in absolute value is considered large enough to be judged statistically significant. More generally, we reject H_0 and judge the difference between sample means to be "statistically significant" when the obtained value of *t* calculated for the sample means exceeds the tabled critical value of *t* for the chosen alpha level, the choice of a one- or two-tailed test, and the available degrees of freedom. The table in Appendix B provides critical values of *t* for the most common alpha levels.

5.3 ♦ Assumptions About the Distribution of Scores on the Quantitative Dependent Variable

Ideally, the scores on the *Y* outcome variable should satisfy the following assumptions.

5.3.1 ♦ Quantitative, Approximately Normally Distributed

Scores on the *Y* outcome variable must be quantitative. (It would not make sense to compute a group mean for scores on a *Y* variable that is nominal or categorical level of measurement.) Some statisticians would argue that the scores on *Y* should be at least interval/ratio level of measurement. However, in practice, many researchers require only that these scores are quantitative; many researchers do not insist on true equal interval measurement as a condition for the use of parametric statistics such as the *t* test (see Chapter 1 for further discussion). For example, it is common to compare means on outcome variables that use 5-point (Likert) rating scales, even though the intervals between the scores probably do not correspond to exactly equal differences in favorableness of attitude. Harris (2001) and Tabachnick and Fidell (2007) argue that it is more important that scores on *Y* outcome variables be approximately normally distributed than that they possess true interval/ratio level of measurement properties.

5.3.2 ♦ Equal Variances of Scores Across Groups (the Homogeneity of Variance Assumption)

Ideally, the variance of the *Y* scores should be equal or homogeneous across the two populations that correspond to the samples that are compared in the study. We can write this assumption formally as follows:

$$H_0 : \sigma_1^2 = \sigma_2^2.$$

We can evaluate whether this assumption is violated by computing the sample variances for the scores in the two groups (s_1^2, s_2^2) and setting up an *F* test (such as the Levene test) to assess whether these two sample variances differ significantly. SPSS reports the Levene test as part of the standard SPSS output for the independent samples *t* test. If the *F* ratio for the Levene test is large enough to be statistically significant, it is evidence that the homogeneity of variance assumption is violated. SPSS provides two versions of the independent samples *t* test: one with "equal variances assumed" (also called the pooled variances *t* test) and a second test with "equal variances not assumed" (also called the separate variances *t* test). The Levene test can be used to decide which version of the *t* test to report. If the Levene test shows no significant violation of the homogeneity of variance assumption, the researcher usually reports the "equal variances assumed" version of the *t* test. If the Levene test indicates that the assumption of equal variances is seriously violated, then the researcher may prefer to report the "equal variances not assumed" version of the *t* test; this test has an **adjusted (smaller) degrees of freedom** that provides a more **conservative test**.

5.3.3 ♦ Independent Observations Both Between and Within Groups

The independent samples t test is appropriate (as the name of the test suggests) when the scores in the two samples that are compared are independent of each other; that is, when they are uncorrelated with each other. This assumption is usually satisfied when the design is between-S, that is, the researcher assigns each participant to just one group, and when the scores in Groups 1 and 2 are not matched or paired in any way. This assumption is not satisfied when the design is within-S or repeated measures or where matched or **paired samples** are used (see Chapter 1 for a review of between-S vs. within-S design). When a within-S or repeated measures design is used, the scores are almost always correlated across groups (e.g., if we did a within-S study in which each person's HR was assessed once after 0 mg of caffeine and again after 150 mg of caffeine, there would probably be a positive correlation between the set of HR scores for these two conditions, because individual participants who had relatively high scores on HR without caffeine usually also have relatively high scores on HR with caffeine). When data are obtained from a within-S, repeated measures, or matched samples design, the data analysis must take this correlation into account. Chapter 20 describes the paired samples t test; if the two samples being compared in the study represent repeated measures or matched samples, then the paired samples t test should be used rather than the independent samples t test discussed in this chapter.

A second assumption is that scores are not correlated with other scores within the same group. This type of interdependence among scores can be more difficult to detect (for details, see Kenny & Judd, 1996; Kenny, Mannetti, Pierro, Livi, & Kashy, 2002). Usually, correlations among scores within groups arise when participants are tested in situations where they have opportunities to influence each other's responses. If the research situation provides participants with opportunities to compete, cooperate, imitate, or influence each other's behaviors or judgments through social interaction, discussion, or observation of each other's behavior, it is possible that the responses within groups may not be independent of each other. For example, consider a study in which each group of 12 participants is a mock jury. Each group is given a case to discuss. After group discussion, each individual participant is asked to recommend a length of sentence for the defendant. It is very likely that the individual responses of jurors will not be independent because they have had an opportunity to influence each other's views through discussion. Thus, the 12 juror responses are probably not 12 independent observations. The appropriate unit of analysis in this study may be the jury as a whole (the mean sentence length recommended by each jury of 12). If the assumption that scores within each group are independent of each other is violated, the consequence of this violation may be that the scores are "close together" because of imitation or cooperation; this may mean that the sample variance will underestimate the population variance. A t ratio that is based on scores that are intercorrelated within groups may be larger than it would have been if it had been based on independent observations. Ideally, observations within each group should be obtained in a way such that they are independent of each other.

5.3.4 ♦ Robustness to Violations of Assumptions

The independent samples t test is fairly robust to violations of some of these assumptions (unless sample sizes are very small). A test is "robust" if the rate of Type I error does not increase very much when its assumptions are violated. **Robustness** of a statistical test can be evaluated empirically using Monte Carlo methods.

For example, let's suppose we wanted to do a **Monte Carlo simulation** to assess how much a violation of the equal variances assumption influences the risk of Type I error for an independent samples *t* test. First, we set up a baseline, in which we use populations for which the null hypothesis of no difference between population means is correct and the homogeneity of variance assumption is satisfied. For example, we could create an artificial population of 1,000 normally distributed intelligence quotient (IQ) scores for males (Population 1, $\mu_1 = 100$, $\sigma_1^2 = 225$) and an artificial population of 1,000 normally distributed IQ scores for females (Population 2, $\mu_2 = 100$, $\sigma_2^2 = 225$). What will happen if we use a computer program to draw thousands of pairs of samples of size 30 from each of these two populations and perform a *t* test using $\alpha = .05$ to test whether the sample means differ significantly for each pair of randomly selected samples? We know that for these artificial populations, the null hypothesis (H_0: $\mu_1 = \mu_2$) is actually correct, and we know that the homogeneity of variance assumption (H_0: $\sigma_1^2 = \sigma_2^2$) is satisfied and that the assumptions that population scores are normally distributed is satisfied. If we do thousands of different *t* tests on different pairs of random samples, about 95% of the obtained *t* ratios will be small enough to result in a (correct) decision not to reject H_0; the remaining 5% of the obtained *t* ratios will be large enough to (incorrectly) decide to reject H_0. That is, the empirical outcome of this Monte Carlo simulation is that 5% of the outcomes are instances of Type I error; this corresponds to the risk of Type I error given by the nominal alpha level, $\alpha = .05$.

We can empirically evaluate how much effect different kinds of violations of assumptions have on the risk of Type I error by working with artificial population data that represent different types of violations of assumptions and then running this experiment again. For example, we can ask, Does the risk of Type I error increase above 5% if the assumption of equal population variances is violated? To do this, we use the same values as before, but we change the variance for the hypothetical population of females to a specific value that is different from the variance for males, such as $\sigma_2^2 = 169$. Now, we run thousands of *t* tests on different pairs of samples again and ask, What proportion of these tests yield Type I errors when the equal variance assumption is violated? Myers and Well (1991) reported results for a similar set of Monte Carlo simulations, with the following conclusions:

> If the two sample sizes are equal, there is little distortion in Type I error rate unless *n* is very small and the ratio of the variances is quite large . . . when *n*'s are unequal, whether the Type I error rate is inflated or deflated depends upon the direction of the relation between sample size and population variance. (pp. 69–70)

According to Myers and Well (1991), even when the ratio of σ_1^2/σ_2^2 was as high as 100, *t* tests based on samples of $n = 5$ using $\alpha = .05$ resulted in Type I errors in only 6.6% of the batches of data.

The implication of Monte Carlo simulation results is that the impact of violations of the homogeneity assumption on the risk of committing a Type I error is much greater when the *n*s in the samples are small and when the *n*s in the samples are unequal. When samples sizes are relatively large (e.g., $n > 30$ in each group) and when the *n*s in the two groups are equal, the independent samples *t* test is fairly robust to violations of the homogeneity of variance assumption.

A paradox arises when we conduct statistical significance tests to assess violations of assumptions (such as violations of the homogeneity of variance assumption). When we run a statistical significance test to evaluate whether the difference between variances in two samples is large enough to be statistically significant, and when *n*s in the groups are large, we have good statistical power to detect violations of this assumption, but when *n*s in the groups are large, the independent samples *t* test yields a Type I error rate that is consistent with the nominal alpha level even when the homogeneity of variance assumption is violated. On the other hand, when *n*s in the groups are small, we have poor statistical power to detect violations of the homogeneity of variance assumption; and in small-*n* situations (particularly when *n*s are unequal between groups), violations of the homogeneity of variance assumption are more likely to lead to inflated or deflated risk of Type I error for the independent samples *t* test. When we test the significance of violations of assumptions (such as the homogeneity of variance assumption for the independent samples *t* test), it may be appropriate to use different alpha levels for small versus large samples. For small samples, we can increase the power of the Levene test to detect violations of the homogeneity of variance assumption by setting the alpha level for this test at .10 (instead of .05). For large samples, we can decrease the power of the Levene test to detect violations of the homogeneity of variance assumption by setting the alpha level at .01 or even .001.

To summarize, unless the *n*s in the groups are small and/or extremely unequal, violations of the assumption that scores are drawn from normally distributed populations with equal variances probably do not cause serious problems with risk of Type I error in the independent samples *t* test. If data are extremely nonnormally distributed and/or the populations have extremely different variances and if there are also small *n*s and/or unequal *n*s for the samples, it may be advisable to do one of the following. The equal variances not assumed version of the *t* test may be reported instead of the equal variances assumed test; the equal variances not assumed *t* test, sometimes called Welch's *t*, has a smaller adjusted *df* term that provides a more conservative test. Scores can be converted to ranks, and a nonparametric test such as the Wilcoxon rank test may be performed instead of the independent samples *t* test. Unequal variances and nonnormal distribution shapes are sometimes attributable to a few extreme outliers, and it may make sense to delete these outliers, but this decision should be made prior to conducting the *t* test. These decisions about data analysis should be made based on preliminary consideration of the variances and distribution shapes and should not be made after the fact by choosing the test that leads to the decision to reject H_0 with the smallest estimated *p* value (Myers & Well, 1991, p. 71).

5.4 ♦ Preliminary Data Screening

Preliminary data screening should assess how well these assumptions are satisfied by the data. In this example, we first consider the level of measurement for the scores on the dependent variable HR. HR is quantitative (rather than categorical), and the scores on this variable probably have interval/ratio level of measurement properties. Distribution shape for scores on a quantitative variable can be assessed by examining a histogram of the dependent variable scores. The histogram in Figure 5.2 shows the distribution of HR scores for this sample of 20 participants (data in Table 5.1). The smooth curve superimposed on the graph represents an ideal normal distribution. This distribution is not exactly normal in shape; it is bimodal and somewhat asymmetric. Note that samples as small as $N = 20$ often do not have distributions that conform to normal, even when they are obtained by random selection from normally distributed populations. In this situation, the distribution shape was judged a good enough approximation to normal distribution shape to proceed with an independent samples *t* test.

Additional information can be obtained by setting up a box plot to examine the distribution of HR scores separately for each treatment group, as shown in Figure 5.3. Examination of the box plot in Figure 5.3 provides two kinds of information. First, we can see that most HR scores for Group 2 (which received 150 mg of caffeine) tended to be relatively high, with the middle 50% of scores in this group falling between about 60 and 75 beats per minute (bpm); most HR scores for Group 1 (which received 0 mg caffeine) were lower, with about 50% of these scores between about 54 and 57 bpm. Group 1 included two scores that could be considered outliers (the HR scores for participant numbers 2 and

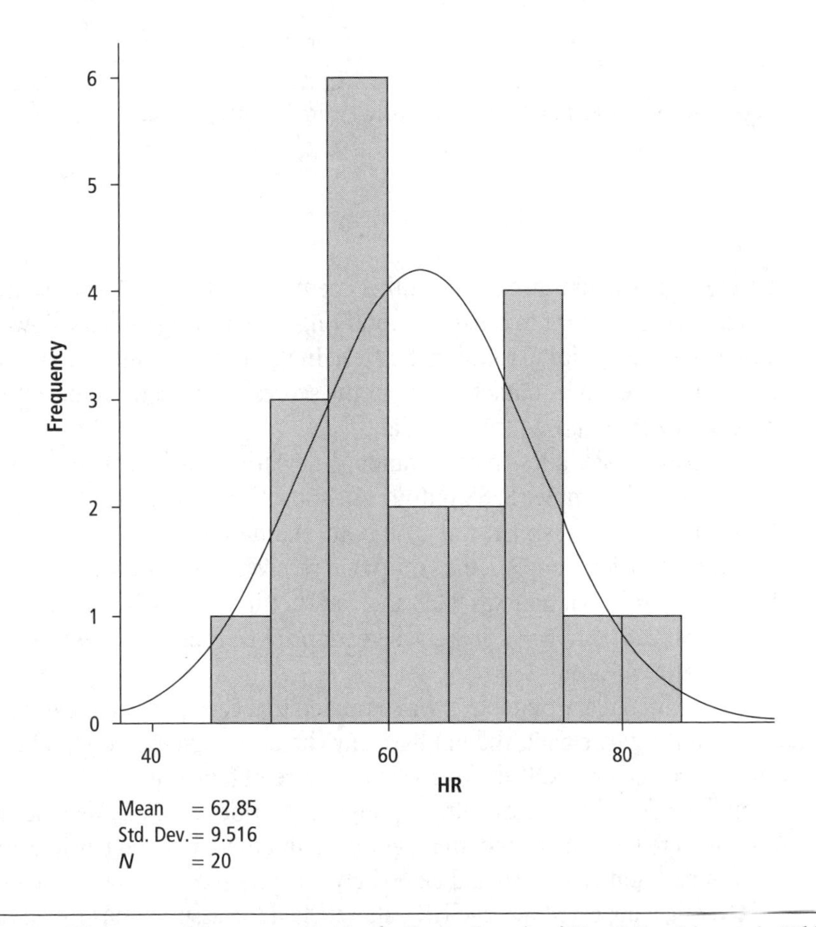

Figure 5.2 ♦ Histogram of Scores on HR (for the Entire Sample of $N = 20$ Participants in Table 5.1)

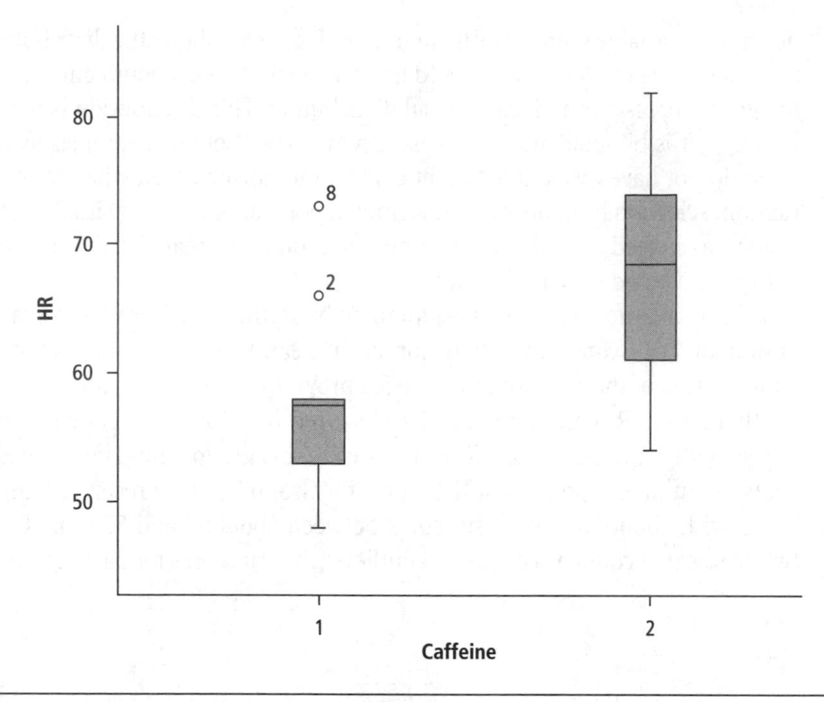

Figure 5.3 ♦ Box Plot for HR Scores Separately by Treatment Group (Caffeine = "1," 0 mg; Caffeine = "2," 150 mg)

8 were higher than those of most other members of Group 1). As discussed in Chapter 4, researchers may want to consider modifying or removing outliers; however, this decision should be made prior to running the *t* test. In this example, although two scores in the first group appeared to be outliers, none of the scores were judged to be extreme outliers, and no scores were removed from the data.

Descriptive statistics (mean, standard deviation, variance) were also obtained separately for each sample (SPSS output not shown). For Sample 1, the group that received 0 mg of caffeine, mean HR was 57.8 bpm, standard deviation was 7.21, and variance in HR was 51.96; for Sample 2, the group that received 150 mg of caffeine, mean HR was 67.9 bpm, standard deviation was 9.09, and variance in HR was 82.54. The ratio between these variances 82.54/51.96 = 1.588, so the variance in Group 2 was less than twice as large as the variance in Group 1.

In this imaginary example, it was assumed that each participant was tested individually and that the participants did not have any chance to influence each other's level of physiological arousal or HR. If the observations were obtained in this manner, the assumption of independence of scores within groups should be satisfied. Also, because independent groups were formed by random assignment and there were no pairings between members of Groups 1 and 2, there should be no between-group correlations of scores on HR.

Overall, it appeared that the HR data satisfied the assumptions for an independent samples *t* test reasonably well. Scores on HR were quantitative and reasonably normally distributed; the variances in HR scores were not drastically unequal across groups; and there was no reason to believe that the scores were correlated with each other, either within or between groups. The only potential problem identified by this preliminary data screening was the presence of two outliers on HR in Group 1. The decision was made, in this case, not to remove these scores. Given the location of the outliers in this case, removing them would slightly increase the difference between the means of the two groups and slightly decrease the variance of scores within Group 1. Thus, we will obtain a more conservative assessment of the outcome by leaving these scores in the sample. If these outliers had been more extreme (e.g., if there had been HR scores above 100 in Group 1), we would need to consider whether those participants fell outside the range of inclusion criteria for the study (we might have decided early on to include only persons with "normal" HR within healthy limits in the study) or whether these extreme scores might have been due to data-recording errors. The sample variances were not significantly different (and thus, there was no indication of violation of the homogeneity of variance assumption); therefore, the equal variances assumed version of the *t* test is the one that will be reported and interpreted.

5.5 ♦ Issues in Designing a Study

There are also some less formal conditions that should be considered. Each group in a study that involves comparisons of group means should have a reasonably large number of scores; in general, an *n* of fewer than 10 participants per group is too small for most research situations. The empirical examples used in early chapters of this textbook include very small numbers of observations to make it easy for readers to duplicate the analyses by hand or to enter the scores into SPSS and reproduce the printouts. However, such small numbers of observations are not intended as a model for research design; larger sample sizes are preferable. Statistical power analysis (see Section 5.9) provides more specific guidance for decisions about sample size.

Researchers should not only ask themselves whether they have a large enough number of participants but also whether they have a large enough number of groups. The use of the *t* test implicitly assumes that two groups are sufficient to assess the effects of the independent variable. If there is any reason to suspect that the relation between the independent variable (or treatment dosage level) and the outcome is curvilinear, then more than two groups need to be included in the study to assess the relationship between variables. For instance, in many studies of environmental psychology, researchers find that optimum task performance occurs at moderate levels of arousal (rather than at high or low levels). Detection of curvilinear relationships between treatment and outcome variables requires at least three groups.

Also, as discussed below, the values of the independent variable (or the dosage levels of the treatment) need to be far enough apart so that there is likely to be a detectable difference in outcomes. If a researcher wants to compare mental processing speeds between "young" and "old" groups, for example, there is a better chance of seeing age-related differences in groups that are 20 versus 70 years old than in groups that are 40 versus 50 years old.

5.6 ♦ Formulas for the Independent Samples *t* Test

In introductory statistics, students are usually required to calculate the *t* test starting from the individual scores in the two groups. This textbook focuses primarily on computational formulas given in terms of intermediate results (mean, *s*, and *n*) for each group rather than individual scores. Note that the subscripts below indicate which Group (1 or 2) a descriptive statistic is based on.

Group 1	*Group 2*
M_1	M_2
s_1	s_2
n_1	n_2

Within each of the two groups, the mean M and standard deviation s are calculated as follows: the subscripts 1 and 2 indicate whether the scores are from Group 1 or Group 2.

Scores on the quantitative outcome variable are denoted by Y_1 when they come from Group 1 and Y_2 when they belong to Group 2.

Calculating the mean for each group,

$$M_1 = \sum Y_1/n_1, \tag{5.1}$$

$$M_2 = \sum Y_2/n_2. \tag{5.2}$$

Calculating the sum of squares (SS) for each group,

$$SS_1 = \sum Y_1^2 - \left(\sum Y_1\right)^2/n_1 = \sum (Y_1 - M_1)^2, \tag{5.3}$$

$$SS_2 = \sum Y_2^2 - \left(\sum Y_2\right)^2/n_2 = \sum (Y_2 - M_2)^2. \tag{5.4}$$

Calculating the variance (s^2) for each group,

$$s_1^2 = SS_1/(n_1 - 1), \tag{5.5}$$

$$s_2^2 = SS_2/(n_2 - 1). \tag{5.6}$$

Calculating s_p^2, the pooled variance,

$$s_p^2 = \frac{[(n_1 - 1)s_1^2 + (n_2 - 1)s_2^2]}{[n_1 + n_2 - 2]}. \tag{5.7}$$

Calculating the standard error of the difference between sample means,

$$SE_{M_1 - M_2} = \sqrt{\frac{s_p^2}{n_1} + \frac{s_p^2}{n_2}}. \tag{5.8}$$

Calculating the independent samples *t* ratio,

$$t = \frac{M_1 - M_2}{SE_{M_1 - M_2}}. \quad (5.9)$$

Calculating the degrees of freedom for the independent samples *t* ratio,

$$df = n_1 + n_2 - 2. \quad (5.10)$$

This discussion focuses on the calculation of *t* tests from the mean and standard deviation for each group instead of in terms of $\sum Y$ and $\sum Y^2$. This version of the computations makes it easier to grasp the conceptual basis of the *t* test and to understand that it compares between-group differences ($M_1 - M_2$) with within-group differences among scores (s_1^2, s_2^2). These versions of the formulas also make it easy for readers to use *t* tests to compare sample means reported in journal articles in situations where the authors do not provide such tests.

A statistical significance test can be used to evaluate whether there is evidence that the homogeneity of variance assumption is violated. The outcome of this significance test is one factor to take into account when deciding which version of the *t* test to use: the equal variances assumed or the equal variances not assumed or the separate variances *t* test. SPSS reports the Levene test to assess whether sample variances differ significantly. Formally, the homogeneity of variance assumption is written as

$$H_0 : \sigma_1^2 = \sigma_2^2. \quad (5.11)$$

Note that in tests of assumptions such as this one, we usually do *not* want to obtain a significant result. For this test, a significant *F* ratio means that an assumption that is required for the use of the pooled variances *t* test has been significantly violated. In the Levene test, each score is converted into an absolute deviation from its group mean, and a one-way analysis of variance is done on these absolute deviations; a significant *F* is interpreted to mean that the variances are not homogeneous; that is, the homogeneity of variance assumption is violated (Rosenthal & Rosnow, 1991, p. 340).

The Levene test is distributed as an *F* ratio with ($k - 1, N - k$) *df*, where *k* is the number of groups and *N* is the total number of scores across all groups. If the Levene test ratio is large (if it exceeds the tabled critical value of *F* for the alpha level set by the researcher), then the researcher should conclude that the group variances differ significantly; this violates an assumption for the pooled variances form of the *t* test. When the Levene test is significant, it may be more appropriate to use the separate variances form of the *t* test.

5.6.1 ♦ The Pooled Variances *t* Test

This is the version of the *t* test usually presented in introductory statistics textbooks. It is preferable in research situations where the variances of the groups being compared are approximately equal. In these situations, it makes sense to pool (average or combine) them into a single estimate of the within-group variability. It is calculated in two steps:

First, we "pool," or average, the two within-group variances (s_1^2 and s_2^2) into a single overall estimate of within-group variance (the pooled variance, s_p^2). Then, we use this pooled variance to compute the standard error term $SE_{M_1-M_2}$ for the t ratio. When we combine these variances, they are weighted by sample size; that is, we multiply each variance by $(n_i - 1)$; this weighting means that a variance that is based on a larger n will be weighted more heavily in the computation of the overall mean than a variance that is based on a smaller n:

$$s_p^2 = [(n_1 - 1)s_1^2 + (n_2 - 1)s_2^2]/[(n_1 - 1) + (n_2 - 1)]. \tag{5.12}$$

If $n_1 = n_2$, this formula reduces to

$$s_p^2 = (s_1^2 + s_2^2)/2. \tag{5.13}$$

The standard error of the $M_1 - M_2$ value, $SE_{M_1-M_2}$, is calculated using this pooled or averaged within-group variance:

$$SE_{M_1-M_2} = \sqrt{\frac{s_p^2}{n_1} + \frac{s_p^2}{n_2}}. \tag{5.14}$$

In general, a t ratio has the form

$$t = \frac{\text{Sample statistic} - \text{Hypothesized parameter}}{SE_{\text{sample statistic}}}. \tag{5.15}$$

The hypothesized population parameter we are usually interested in testing is

$$H_0: \mu_1 = \mu_2 \text{ or } H_0: (\mu_1 - \mu_2) = 0. \tag{5.16}$$

Because we usually assume that this difference equals zero, the term that corresponds to this hypothesized parameter $(\mu_1 - \mu_2)$ is generally set to zero; therefore, the hypothesized parameter term does not appear in the next formula for the t ratio. The sample statistic that estimates the population difference $(\mu_1 - \mu_2)$ is $M_1 - M_2$, and the corresponding t test (the pooled variances version of t) is

$$t = \frac{(M_1 - M_2) - (\mu_1 - \mu_2)}{SE_{M_1-M_2}} = \frac{M_1 - M_2}{SE_{M_1-M_2}}. \tag{5.17}$$

This is a t ratio with $n_1 + n_2 - 2$ df. If the obtained value of t is large enough to exceed the tabled critical values of t for $n_1 + n_2 - 2$ df, the null hypothesis of equal means is rejected, and the researcher concludes that there is a significant difference between the means.

SPSS reports an exact two-tailed p value for the independent samples t test. A two-tailed exact p value corresponds to the combined areas of the upper and lower tails of the t distribution that lie beyond the obtained sample value of t. If we decide to use the conventional $\alpha = .05$ level of significance, an obtained p value less than .05 is interpreted as evidence that the t value is large enough so that it would be unlikely to occur by chance (due to sampling error) if the null hypothesis were true.

5.6.2 ♦ Computation of the Separate Variances *t* Test and Its Adjusted *df*

If the Levene test of the homogeneity of variance assumption turns out to be significant, then a more conservative version of the t test may be used to adjust for the effects of the violation of the equal variance assumption; this version of t is called the separate variances t test. The difference in the computation of the t ratio is that instead of pooling the two within-group variances, the two within-group variances are kept separate when the standard error term $SE_{M_1 - M_2}$ is calculated, as follows:

$$SE_{M_1-M_2} = \sqrt{\frac{s_1^2}{n_1} + \frac{s_2^2}{n_2}}. \tag{5.18}$$

The t ratio for the separate variances or unequal variances situation is calculated using this slightly different version of the standard error. The degrees of freedom for this version of the t test are adjusted to compensate for the violation of the equal variances assumption; the formula for this **adjusted *df***, df', is fairly complex.[1] For this version of t, the degrees of freedom may turn out to be a decimal number (in practice, it is usually reported rounded to the nearest whole number). In general, the adjusted df' for the separate variances version of the t test will be less than the $n_1 + n_2 - 2$ df for the pooled variances t test. The larger the difference in the magnitude of the variances and the ns, the greater the downward adjustment of the degrees of freedom.

5.6.3 ♦ Evaluation of Statistical Significance of a *t* Ratio

As discussed in Chapter 3, the significance of the t ratio for tests about the value of a single-population mean is evaluated by comparing the obtained t ratio with the critical values of t for a t distribution with appropriate degrees of freedom. The critical value of t is obtained by setting the alpha level (usually, this is $\alpha = .05$, two-tailed) and looking up the critical value of t for this alpha level and the degrees of freedom based on the sample size (see the table of critical values for the t distribution in Appendix B). The same logic is used to evaluate the independent samples t ratio. We can either compare the obtained t value with the critical values of t from a table of the t distribution or, given that SPSS reports an exact p value that corresponds to the obtained t value, compare the exact p value on the printout with the predetermined alpha level. If the obtained significance level (labeled "sig" on the SPSS printout and reported as "p" in the Results section) is less than the predetermined alpha level, we can say that the independent samples t value is "statistically significant" and that the sample means differ significantly.

The researcher should decide on an alpha level and whether to use a one-tailed or two-tailed test prior to conducting the independent samples t test. For the following example, the conventional $\alpha = .05$ significance level will be used to evaluate the statistical significance of the t ratio. It would be reasonable to predict that caffeine consumption should be associated with higher HR; therefore, we could reasonably decide to use a one-tailed or directional alternative hypothesis in this research situation:

$$H_1\text{: } \mu_1 < \mu_2 \text{ or } H_1\text{: } (\mu_1 - \mu_2) < 0, \tag{5.19}$$

where μ_1 is the mean HR for a population of persons who have consumed no caffeine and μ_2 is the mean HR for a population of persons who have consumed 150 mg of caffeine. SPSS computes the t ratio as $M_1 - M_2 / SE_{M_1-M_2}$; therefore, if the alternative hypothesis is H_1: $(\mu_1 - \mu_2) < 0$, we would reject H_0 only for large negative values of t (that correspond to negative values of $M_1 - M_2$).

However, the default significance values reported by SPSS for the independent samples t test correspond to a two-tailed or nondirectional test (and this corresponds to the alternative hypothesis H_1: $\mu_1 \neq \mu_2$). When using this nondirectional alternative hypothesis, the researcher rejects H_0 for both large positive and large negative values of t. In the empirical example that follows, the results will be reported using $\alpha = .05$, two-tailed, as the criterion for statistical significance. Researchers who prefer to use a one-tailed test can obtain the one-tailed significance value by dividing the sig or exact p value that appears on the SPSS printout by 2, as will be shown in the following example.

A large t ratio implies that relative to the size of $SE_{M_1-M_2}$, the $M_1 - M_2$ value was large. Usually, the researcher hopes to find a large difference between means as evidence that the treatment variable (or personal factor) that differed across groups was significantly related to the outcome variable.

For an independent samples t test (with equal variances assumed), the appropriate $df = n_1 + n_2 - 2$, where n_1 and n_2 are the number of scores in each sample. For an independent samples t test with unequal variances assumed, the adjusted df reported by SPSS is usually a noninteger value that is less than $n_1 + n_2 - 2$; usually, this is rounded to the nearest integer value when results are reported. This adjusted df value is calculated using the formula that appears in Note 1.

Most computer programs provide an "exact" p value for each test statistic (in a form such as sig = .022 or $p = .013$). By default, SPSS reports two-tailed p or significance values for t tests. The current edition of the *Publication Manual of the American Psychological Association* (APA, 2001) recommends that when an exact p value is available for a test statistic, it should be reported (p. 24). However, when large numbers of tests are reported, such as tables that contain many t tests or correlations, it may be more convenient to use asterisks to denote values that reached some prespecified levels of statistical significance. Conventionally, * indicates $p < .05$, ** indicates $p < .01$, and *** indicates $p < .001$. This is conventional, but this usage should still be explained in footnotes to tables.

To summarize, we reject H_0 for obtained values of t that are larger in absolute value than the critical t values from Appendix B or, equivalently, for obtained p values that are smaller than the predetermined alpha level (usually $\alpha = .05$, two-tailed).

In the preceding empirical example of data from an experiment on the effects of caffeine on HR, $n_1 = 10$ and $n_2 = 10$, therefore $df = n_1 + n_2 - 2 = 18$. If we use $\alpha = .05$, two-tailed, then from the table of critical values of t in Appendix B, the reject regions for this test (given in terms of obtained values of t) would be as follows:

Reject H_0 if obtained $t < -2.101$.

Reject H_0 if obtained $t > +2.101$.

5.6.4 ♦ Confidence Interval Around $M_1 - M_2$

The general formula for a confidence interval (CI) was discussed in Chapter 2. Assuming that the statistic has a sampling distribution shaped like a *t* distribution and that we know how to find the *SE* or standard error of the sample statistic, we can set up a 95% CI for a sample statistic by computing

$$\text{Sample statistic} \pm t_{\text{critical}} \times SE_{\text{sample statistic}},$$

where t_{critical} corresponds to the absolute value of *t* that bounds the middle 95% of the area under the *t* distribution curve with appropriate degrees of freedom. (In other words, $-t_{\text{critical}}$ and $+t_{\text{critical}}$ are the boundaries of the middle 95% of the *t* distribution.) The formulas for the upper and lower bounds of a 95% CI for the difference between the two sample means, $M_1 - M_2$, are as follows:

$$\text{Lower bound of 95\% CI: } (M_1 - M_2) - t_{\text{critical}} \times SE_{M_1-M_2}. \quad (5.20)$$

$$\text{Upper bound of 95\% CI: } (M_1 - M_2) + t_{\text{critical}} \times SE_{M_1-M_2}. \quad (5.21)$$

The value of $SE_{M_1-M_2}$ is obtained from Equation 5.8 above. The degrees of freedom for *t* are calculated as described earlier (for data that satisfy the equal variances assumption, $df = n_1 + n_2 - 2$). Critical values are found in the distribution of *t* with the appropriate *df* value. The value of t_{critical} for a 95% CI is the absolute value of *t* that is the boundary of the middle 95% of the area in the *t* distribution. In the table in Appendix B, the critical values for a 95% CI correspond to the values of *t* that are the boundaries of the middle .95 of the area under the *t* distribution curve. For example, when $df = 18$, for a 95% CI, the value of $t_{\text{critical}} = 2.101$.

5.7 ♦ Conceptual Basis: Factors That Affect the Size of the *t* Ratio

As described in Chapter 2, the numerator and the denominator of the *t* ratio provide information about different sources of variability in the data. Let's look once more at the formula for the pooled variances *t* ratio:

$$t = \frac{M_1 - M_2}{\sqrt{(s_p^2/n_1) + (s_p^2/n_2)}},$$

where M_1 and M_2 represent the means of Groups 1 and 2, s_p^2 is computed from Equation 5.7, and n_1 and n_2 represent the number of scores in Group 1 and Group 2.

This formula includes three kinds of information from the data: the magnitude of the difference between sample means ($M_1 - M_2$); the within-group variability of scores (s_p^2);

and the sample sizes (n_1, n_2). It is useful to think about the way in which increases or decreases in the size of each of these terms influence the outcome value for t when other terms in the equation are held constant. Other factors being equal, an increase in $M_1 - M_2$; an increase in n_1, n_2; or a decrease in s_p^2 will increase the size of the t ratio (see Rosenthal & Rosnow, 1991, p. 303).

Each of these values is influenced by the researcher's decisions. The following sections discuss how researchers can design studies in ways that will tend to increase the size of t and, therefore, increase the likelihood of obtaining a statistically significant difference between group means. Research decisions that affect the size of each of these three components of the t ratio are discussed.

5.7.1 ♦ Design Decisions That Affect the Difference Between Group Means, $M_1 - M_2$

The difference between the means of the two samples, $M_1 - M_2$, estimates the difference between the population means ($\mu_1 - \mu_2$). Imagine a simple two-group experiment in which Group 1 receives no caffeine and Group 2 receives caffeine equivalent to about one cup of strong coffee (150 mg). The outcome measure is HR; mean HR values are computed for these two treatment groups. If $M_1 = 57.8$ bpm and $M_2 = 67.9$ bpm, this increase in HR of -10.1 bpm is (presumably) due partly to the effect of caffeine. Theoretically, the magnitude of this difference in mean HR between groups is also influenced by other extraneous variables that have not been controlled in the study and by any other variables confounded with dosage level of caffeine.

Other factors being equal, the difference between group means in an experimental study tends to be large when the dosage levels of the treatment differ substantially between groups. For instance, assuming that caffeine really does increase the HR, we might reasonably expect the $M_1 - M_2$ value to be larger if we compared 0 and 450 mg of caffeine and smaller if we compared 0 and 50 mg of caffeine. Other factors being equal (i.e., if s_p^2 and n are kept constant), the larger the $M_1 - M_2$ value is, the larger the t ratio will be and the more likely the t ratio is to be judged significant.

A researcher can try to increase the size of $M_1 - M_2$ (and therefore the size of t) through design decisions. In experiments, the dosage levels need to be sufficiently different to create detectable differences in the outcome level. Of course, practical and ethical limitations on dosage levels must be taken into account; we cannot administer 2,000 mg of caffeine to a research participant. In nonexperimental comparisons of groups, subjects need to be chosen so that they are "different enough" to show detectable differences in outcome. For example, a researcher is more likely to see the health effects of smoking when comparing nonsmokers with two-pack-a-day smokers than when comparing nonsmokers with one-pack-a-week smokers; a researcher is more likely to see age-related differences in cognitive processes when comparing groups of 20-year-olds and 70-year-olds than when comparing groups of 20-year-olds and 22-year-olds.

Also, note that if there is a systematic confound of some other variable with the treatment variable of interest, this confound variable may either increase or decrease the size of the observed $M_1 - M_2$ value. For instance, if all the subjects in the group that received caffeine had their HR measured just prior to a final exam and all the subjects that did not receive caffeine had their HR measured on a Friday afternoon at the end of classes, the time of assessment would probably influence their HRs very substantially. The presence of this confound would also make the $M_1 - M_2$ value uninterpretable (even if it is statistically significant), because the researcher would not be able to distinguish how much of the caffeine group's higher average HR was due to caffeine and how much was due to anxiety about the exam. Even in thoughtfully designed experiments, it is possible to have confounds of other factors with the treatment that would compromise the researcher's ability to make causal inferences.

5.7.2 ♦ Design Decisions That Affect Pooled Within-Group Variance, s_p^2

The pooled within-group variability (s_p^2) describes the extent to which participants who were in the same group and received the same treatment exhibited variable or different responses. For example, although the participants who consumed 150 mg of caffeine may have a higher average HR than do participants who did not consume caffeine, there is usually also substantial variability in HR within each group. One person in the 150-mg group may have HR of 82 bpm (due to smoking, taking antihistamines, and anxiety), while another person in the coffee group may have HR of 61 bpm because of good cardiovascular fitness. The deviation of each individual's HR score (Y) from the mean score for group i (M_i), $Y - M_i$, reflects the influence of all other variables that were not controlled in the study on that individual's HR. When these deviations are squared and summed across subjects to estimate s_1^2, s_2^2, and s_p^2, the pooled within-group variance combines the information about the amount of influence of "all other variables" across all the participants in the study.

The s_p^2 term tends to be large when the researcher selects a sample that is very heterogeneous in factors that affect HR (such as age, weight, cardiovascular fitness, gender, anxiety, and substance use). This pooled variance will also tend to be larger when the conditions under which participants are tested are not standardized or constant (e.g., participants are tested at different times in the semester, days of the week, hours of the day; in different settings; or by different researchers whose behavior toward the participants elicits more or less comfort or anxiety).

Other things being equal—that is, for fixed values of $M_1 - M_2$ and n_1, n_2—the larger the value of s_p^2 is, the smaller the *t* ratio will turn out to be, and the less likely it is that the researcher will be able to judge groups significantly different. Therefore, a researcher may want to design the study in ways that tend to reduce s_p^2.

The size of s_p^2 can typically be reduced by

1. recruiting participants who are homogeneous with respect to relevant variables (e.g., age, health, fitness, substance use) (however, note that the use of a very homogeneous sample will limit the generalizability of results; if the sample includes only 20- to 22-year-old healthy men, then the results may not apply to younger or older persons, to women, or to persons in poor health) and

2. keeping experimental and measurement procedures standardized—for instance, by testing all participants on the same day of the week or at the same time of day, training data collectors to behave consistently, and so forth.

However, there is a trade-off involved in this design decision. Selecting a very homogeneous sample helps to reduce error variance due to individual differences among participants, but it also reduces the generalizability of findings. If a researcher decides to include only male participants of age 18 to 22 years who are nonsmokers, the findings of the study may not be generalizable to women, other age groups, and smokers.

5.7.3 ♦ Design Decisions About Sample Sizes, n_1 and n_2

The last element used to calculate the *t* ratio is the sample size information (n_1, n_2). If the values of $M_1 - M_2$ and s^2_p are held constant, then as long as $M_1 - M_2$ is not exactly equal to 0, as n_1 and n_2 increase, the value of *t* will also increase. In fact, as long as the difference between sample means, $M_1 - M_2$, is not exactly equal to zero, even a very small difference between sample means can be judged to be statistically significant if the sample sizes are made sufficiently large.

5.7.4 ♦ Summary: Factors That Influence the Size of *t*

Other things being equal (i.e., assuming that $M_1 - M_2$ is nonzero and is held constant and assuming that s^2_p is constant), the size of the *t* ratio increases as n_1 and n_2 increase. In fact, if the *n*s are large enough, any $M_1 - M_2$ value that is not exactly zero can be judged statistically significant if the *n*s are made sufficiently large. This is potentially a problem because it implies that a determined researcher can make even a difference that is too small to be of any clinical or practical importance turn out to be statistically significant if he or she is persistent enough to collect data on thousands of subjects.

In designing a study, an investigator typically has some control over these factors that influence the size of *t*. The magnitude of the obtained *t* value depends partly upon researcher decisions about the dosage levels of the independent variable (or other factors involving the differences in the groups to be compared); the control of extraneous variables, which affects the within-group variances of scores; and the size of the sample to be included in the study. A large *t* value, one that is large enough to be judged statistically significant, is most likely to occur in a study where the between-group dosage difference, indexed by $M_1 - M_2$, is large; where the within-group variability due to extraneous variables, indexed by s^2_p, is small; and where the *n*s of subjects in the groups are large.

To have the best possible chance of obtaining statistically significant results, a researcher would like to have a strong treatment effect (which produces large differences between sample means), good control over extraneous variables (which makes the within-group variances in scores, s^2_p, small), and large sample sizes. Sometimes it is not possible to achieve all these goals. For instance, a researcher may be able to afford only a small number of participants ($n = 15$) per group. In this situation, unless the researcher can create a large difference in treatment dosage levels and keep within-group variability small, the researcher is unlikely to get a large *t* ratio. In other words, there are trade-offs; it may be possible to compensate for a very small *n* by designing the study in a manner that increases the difference between group means and/or decreases the variance of the scores within groups. On the other hand, if the treatment effect of interest is known to be weak, it may be necessary to run large numbers of subjects to obtain statistically significant results.

A researcher can calculate an approximate estimate of *t* based on the anticipated sizes of the terms involved in a *t* ratio to get a sense of whether the obtained *t* ratio is likely to be large or small given the planned sample size and the anticipated magnitude of effect size. For instance, a researcher who wants to assess the effect of caffeine on HR might look up past research to see what magnitudes of effect sizes other researchers have found using the dosage levels that the researcher has in mind. The researcher can try to guess what values of M_1, M_2, and s_p^2 might be obtained by looking at values obtained in similar past research and then substitute these values into the *t* formula, along with the values of *n* that the researcher plans to use for each group. Of course, the resulting *t* value will not predict the outcome of the planned study precisely. Because of sampling error, these predictions about outcome values of *t* will not be exactly correct; *t* values vary across batches of data. However, the researcher can judge whether this "predicted" *t* value is large or small. If the predicted value of *t* based on the researcher's reasonable guesses about outcomes for the means and variances is extremely small, the researcher should pause and think about whether the planned study makes sense. If the expected *t* is extremely small, then the study is not likely to lead to a conclusion of a statistically significant difference unless the researcher increases the *n*, increases the between-group differences, and/or decreases the within-group differences. In Section 5.9, statistical power analysis tables are described; researchers can use these to make informed judgments about sample size based on anticipated effect size in their research. Section 5.8 provides several quantitative indexes of effect size. Effect size information is usually reported as part of the results of a study, and it is also useful in planning sample size for future studies.

5.8 ♦ Effect Size Indexes for *t*

There are several different effect size indexes that may be used with the independent samples *t* test. Three widely used indexes are described here.

5.8.1 ♦ Eta Squared (η^2)

The **eta squared (η^2)** coefficient is an estimate of the proportion of variance in the *Y*-dependent variable scores that is predictable from group membership. It can be calculated directly from the *t* ratio:

$$\eta^2 = \frac{t^2}{t^2 + df}. \qquad (5.22)$$

Like r^2, η^2 can range in size from 0 (no relationship) to 1.0 (perfect relationship). A related statistic called ω^2 (omega squared) is a downwardly adjusted version of η^2; it provides a more conservative estimate of the proportion of variance in *Y* that may be associated with the group membership variable in the population (Hays, 1994).

It should be emphasized that when we find that we can account for a particular proportion of variance in the sample of scores in a study—for example, $\eta^2 = .40$, or 40%—this obtained proportion of variance is highly dependent on design decisions. It is not necessarily accurate information about the importance of the predictor variable in the real world; it only tells us about the explanatory power of the independent variable relative to other nuisance variables in the somewhat artificial world and the small sample that we are studying. For instance, if we find (in a study of the effects of caffeine on HR) that 40% of the variance in the HR scores is predictable from exposure to caffeine, that does not mean that "out in the world," caffeine is such an important variable that it accounts for 40% of the variability of HR. In fact, out in the real world, there may be a substantial proportion of variation in HR that is associated with other variables such as age, physical health, and body weight that did not contribute to the variance of scores in our experiment because we chose to study only college students who are in good health.

In an experiment, we create an artificial world by manipulating the independent variable and holding other variables constant. In a nonexperimental study, we create an artificial world through the necessarily arbitrary choices that we make when we decide which participants and variables to include and exclude. We have to be cautious when we attempt to generalize beyond the artificial world of our selected samples to a broader, often purely hypothetical population. When we find that a variable is a strong predictor in our study, that result only suggests that it *might* be an important variable outside the laboratory. We need more evidence to evaluate whether it actually is an important factor in other settings than the unique and artificial setting in which we conducted our research. (Of course, in those rare cases where we actually do have random samples of participants taken from well-defined populations, we can be somewhat more confident about making generalizations.)

5.8.2 ♦ Cohen's *d*

For the independent samples *t* test, Cohen's *d* is calculated as follows:

$$d = \frac{M_1 - M_2}{s_p}, \quad (5.23)$$

where s_p is calculated by taking the square root of the value of s_p^2 from Equation 5.7.

In words, *d* indicates how many (pooled) standard deviations apart the group means were in the sample; this can be helpful in visualizing how much overlap there was between the distributions of scores in the two groups. This can be interpreted as a sort of signal-to-noise ratio: The difference in the numerator reflects the influence of the manipulated independent variable (it is also influenced by extraneous variables, and in a poorly designed study, it may also be influenced by confounds). The standard deviation in the denominator reflects the effects of extraneous variables. If the ratio is large (we want it to be large), then it means that the strength of the "signal," such as the effect of caffeine on HR, is large compared with the strength of the "noise" (the effects of all other uncontrolled or extraneous variables on HR).

5.8.3 ♦ Point Biserial *r* (r_{pb})

A *t* ratio can also be converted into a **point biserial correlation** (denoted by r_{pb}); the value of r_{pb} can range from 0 to 1, with 0 indicating no relationship between scores on the dependent variable and group membership:

$$r_{pb} = \sqrt{\frac{t^2}{t^2 + df}}. \quad (5.24)$$

When results of studies that involve comparisons between pairs of group means are combined across many studies using meta-analytic procedures, r_{pb} (sometimes just denoted by r) is often a preferred metric for effect size.

What these three effect size measures have in common is that, unlike the t ratio, they are independent of sample size. A potential problem with the t ratio and other statistical significance tests is that other things being equal, as long as $M_1 - M_2$ is not exactly zero, any nonzero difference between means can be judged to be "statistically significant" if the total N is made large enough. In fact, a significance test (such as a t ratio) can be understood as a product of the size of the effect and the size of the study (Rosenthal & Rosnow, 1991). In general, this relationship is of the form

$$\text{Magnitude of significance test} = \text{Effect size} \times \text{Size of study}. \qquad (5.25)$$

The formula for the independent samples t test can be broken into two separate terms: the effect size t (from Equation 5.9) and the study size $(\sqrt{df}/2)$ (Rosenthal & Rosnow, 1991):

$$t = d\frac{\sqrt{df}}{2}, \qquad (5.26)$$

where d is Cohen's d, as computed in Equation 5.23, and $df = n_1 + n_2 - 2$, where n_1 and n_2 are the numbers of scores in Groups 1 and 2. The term involving the square root of the df value divided by 2 provides information about the sample size of the study.

Table 5.2 provides suggested verbal labels for these effect sizes, and it also shows corresponding values for these three common effect size indexes. These labels for Cohen's d are based on recommendations made by Cohen (1988) for evaluation of effect sizes in social and behavioral research; however, in other research domains, it might make sense to use different labels for effect size (e.g., in psychophysical studies, where the physical magnitudes of stimuli, such as weights of objects, are studied in relation to perceived magnitude, subjectively judged "heaviness," researchers typically expect to find r values on the order of $r = .90$; so in this research domain, an r value that is less than .90 might be considered a "small" or "weak" effect).

5.9 ♦ Statistical Power and Decisions About Sample Size for the Independent Samples *t* Test

Statistical power analysis provides a more formal way to address this question: How does the probability of obtaining a t ratio large enough to reject the null hypothesis (H_0: $\mu_1 = \mu_2$) vary as a function of sample size and effect size? Statistical power is the probability of obtaining a test statistic large enough to reject H_0 when H_0 is false. Researchers generally want to have a reasonably high probability of rejecting the null hypothesis; power of 80% is sometimes used as a reasonable guideline. Cohen (1988) provided tables that can be used to look up power as a function of effect size and n or to look up n as a function of effect size and the desired level of power.

Table 5.2 ♦ Suggested Verbal Labels for Cohen's *d* (Correspondence Between Cohen's *d* and Other Effect Size Indexes)

Verbal Label Suggested by Cohen (1988)	*d*	*r*	r^2 or η^2
	2.0	.707	.500
	1.9	.689	.474
	1.8	.669	.448
	1.7	.648	.419
	1.6	.625	.390
	1.5	.600	.360
	1.4	.573	.329
	1.3	.545	.297
	1.2	.514	.265
	1.1	.482	.232
	1.0	.447	.200
	0.9	.410	.168
Large effect	0.8	.371	.138
	0.7	.330	.109
	0.6	.287	.083
Medium effect	0.5	.243	.059
	0.4	.196	.038
	0.3	.148	.022
Small effect	0.2	.100	.010
	0.1	.050	.002
	0.0	.000	.000

SOURCE: From Cohen, J., *Statistical Power Analysis for the Behavioral Sciences,* copyright © 1988. Reprinted with permission of Lawrence Erlbaum Associates, Inc.

An example of a power table that can be used to look up the minimum required *n* per group to obtain adequate statistical power is given in Table 5.3. This table assumes that the researcher will use the conventional $\alpha = .05$, two-tailed criterion for significance. For other alpha levels, tables can be found in Jaccard and Becker (2002) and Cohen (1988). To use this table, the researcher must first decide on the desired level of power (power of .80 is often taken as a reasonable minimum). Then, the researcher needs to make an educated guess about the population effect size that the study is designed to detect. In an area where similar studies have already been done, the researcher may calculate η^2 values based on the *t* or *F* ratios reported in published studies and then use the average effect size from past research as an estimate of the population effect size. (Recall that η^2 can be calculated by hand from the values of *t* and *df* using Equation 5.22 if the value of η^2 is not actually reported in the journal.) If no similar past studies have been done, the researcher has to make a blind guess; in such situations, it is safer to guess that the effect size in a new research area may be small. Suppose a researcher believes that the population effect size is on the order of $\eta^2 = .20$. Looking at the row that corresponds to power of .80 and the column that corresponds to η^2 of .20, the cell entry of 17 indicates the minimum number of participants required per group to have power of about .80.

Table 5.3 ♦ Sample Size as a Function of Effect Size and Desired Statistical Power

Approximate Sample Sizes Necessary to Achieve Selected Levels of Power for Alpha = –.05, Nondirectional Test, as a Function of Population Values of Eta Squared

	Population Eta Squared									
Power	*.01*	*.03*	*.05*	*.07*	*.10*	*.15*	*.20*	*.25*	*.30*	*.35*
.25	84	28	17	12	8	6	5	3	3	3
.50	193	63	38	27	18	12	9	7	5	5
.60	246	80	48	34	23	15	11	8	7	6
.67	287	93	55	39	27	17	12	10	8	6
.70	310	101	60	42	29	18	13	10	8	7
.75	348	113	67	47	32	21	15	11	9	7
.80	393	128	76	53	36	23	17	13	10	8
.85	450	146	86	61	41	26	19	14	11	9
.90	526	171	101	71	48	31	22	17	13	11
.95	651	211	125	87	60	38	27	21	16	13
.99	920	298	176	123	84	53	38	29	22	18

SOURCE: From *Statistics for the Behavioral Sciences, 4th edition* by Jaccard/Becker, 2002. Reprinted with permission of Wadsworth, a division of Thomson Learning: www.thomsonrights.com. Fax 800-730-2215.

NOTE: All of the tables in this book are based on power estimates given by Cohen (1977).

Given the imprecision of the procedures for estimating the necessary sample sizes, the values contained in this and the other power tables presented in this book are approximate.

Federal agencies that provide research grants now expect statistical power analysis as a part of grant proposals; that is, the researcher has to demonstrate that given reasonable, educated guesses about the effect size, the planned sample size is adequate to provide good statistical power (e.g., at least an 80% chance of judging the effect to be statistically significant). It is not really worth undertaking a study if the researcher knows a priori that the sample size is probably not adequate to detect the effects. In addition, some journal editors now suggest that nonsignificant results be accompanied by information about statistical power. For example, for the obtained effect size, how large an n would have been required for it to be judged significant? This information can be useful in deciding whether the null result represented a lack of any reasonably sized effect or an effect that might have been judged significant given a somewhat larger sample size.

5.10 ♦ Describing the Nature of the Outcome

The nature of the outcome in this simple research situation can be described by the size and sign of the difference between means, $M_1 - M_2$. The researcher should report the direction of the difference; in the caffeine/HR study, which group had a higher HR? The researcher should also comment on the size, in "real" units, of the obtained difference. In the example that follows, the caffeine group's mean HR was approximately 10.1 bpm faster than the no-caffeine group's mean HR. This difference of 10.1 is large enough to be of some clinical or practical importance.

In some research situations, even though the difference that is reported is statistically significant, it is too small to be of much practical or clinical importance. For example, suppose that the mean IQ for individually born people is 100; the mean IQ for people born as twins is 99 points. When this difference is tested in samples of 10,000 or more, it is statistically significant. However, a 1-point difference in IQ is too small to be of much consequence in daily life. As another example, suppose that a researcher evaluates the effect of coaching on Scholastic Aptitude Test (SAT) scores. A coaching program that increases quantitative SAT scores by an average of 10 points could be judged to have a statistically significant effect if the sample size was large; but from the point of view of a student, a 10-point gain in SAT score might be too small to be worth the monetary cost and effort of the program. On the other hand, an improvement of 150 points in SAT score would be large enough to be of some practical importance. One last example: Most dieters would be interested in a program that produced an average weight loss of 30 lb, whereas an average weight loss of 2 lb would be too small to be of interest to most people, even if that small difference were statistically significant.

To evaluate the clinical or practical importance of a difference between means (in beats per minute, SAT scores, pounds, or other units of measurement), the researcher needs to know something about the range of values and the meaning of scores. Some variables in psychological research are measured in arbitrary units, and it can be quite difficult to judge the practical importance of such units; if a person's attraction to another person goes up by 1 point, it may not be possible to describe what that means in terms of any real-world behaviors. For variables that are measured in meaningful units, or for which we have data on population norms (e.g., we know that IQ scores have a mean of 100 points and a standard deviation of about 15 points), it is easier to make judgments about the practical or clinical importance of the difference.

5.11 ♦ SPSS Output and Model Results Section

Prior to doing an independent samples *t* test, the shape of the distribution of scores on the outcome variable should be examined. The SPSS commands to obtain a histogram were covered in Chapter 4. The histogram for the hypothetical study of the effects of caffeine on HR was displayed in Figure 5.2. With small samples ($n < 30$), the distribution shape in the sample is often not a good approximation to normal. In this example, the shape of the distribution of HR scores was somewhat skewed; nevertheless, the distribution shape was judged to be reasonably close to normal.

To obtain an independent samples *t* value using SPSS, the user makes the following menu selections starting from the top level menu above the data worksheet (as shown in Figure 5.4): <Analyze> → <Compare Means> → <Independent-Samples T Test>. This opens up the Independent-Samples T Test dialog box, as shown in Figure 5.5.

The name of the (one or more) dependent variable(s) should be placed in the box marked "Test Variable(s)." For this empirical example, the name of the dependent variable was HR. The name of the grouping variable should be placed in the box headed "Grouping Variable"; for this example, the grouping variable was named "caffeine." In addition, it is necessary to click the Define Groups button; this opens up the Define Groups dialog window that appears in Figure 5.6. Enter the code numbers that identify the groups that are

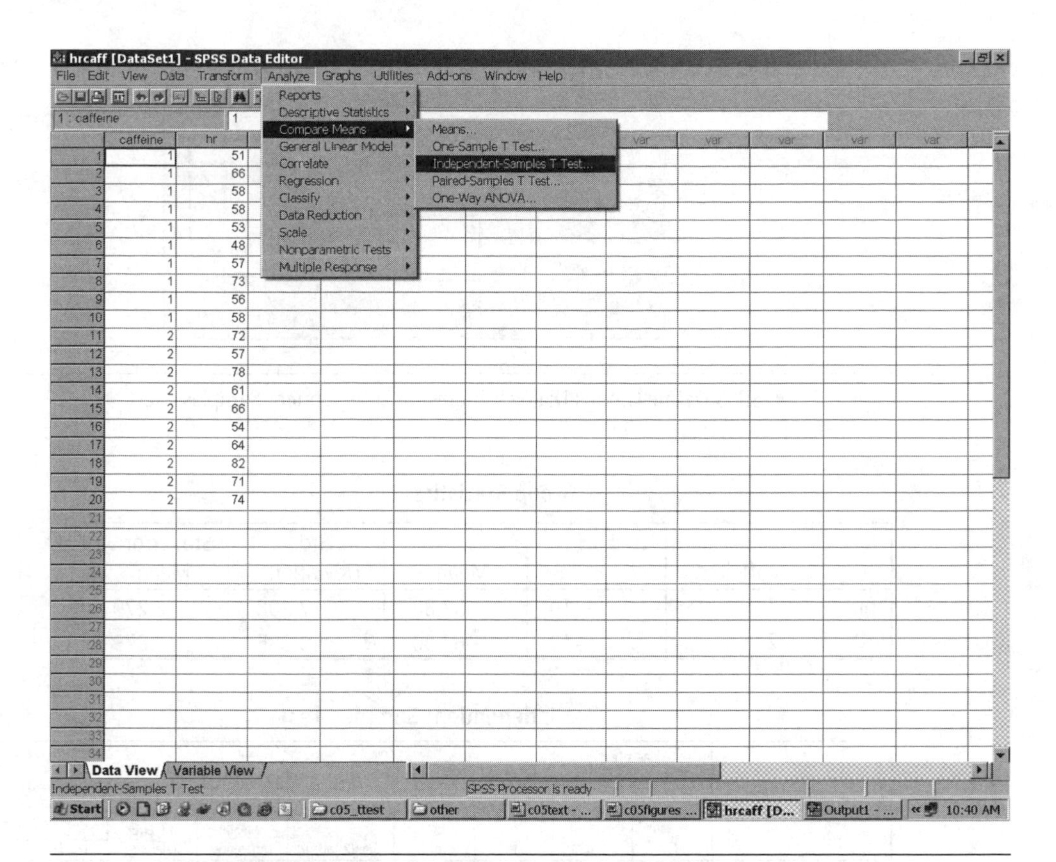

Figure 5.4 ♦ SPSS Menu Selections to Obtain Independent Samples *t* Test

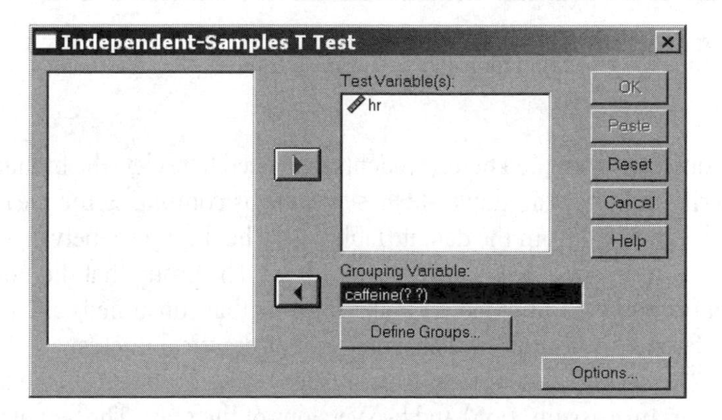

Figure 5.5 ♦ Screen Shot of SPSS Dialog Window for Independent Samples *t* Test

to be compared (in this case, the codes are "1" for the 0-mg-caffeine group and "2" for the 150-mg-caffeine group).Click the OK button to run the specified tests. The output for the independent samples *t* test appears in Figure 5.7.

The first panel, titled "Group statistics," presents the basic descriptive statistics for each group; the *n* of cases, mean, standard deviation, and standard error for each of the two

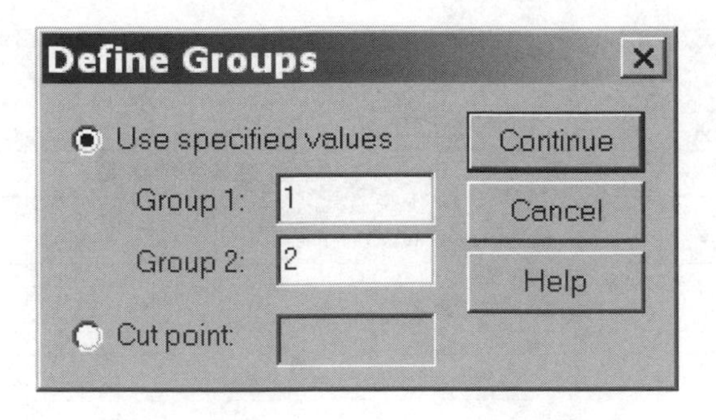

Figure 5.6 ♦ SPSS Define Groups Dialog Window for Independent Samples *t* Test

Group Statistics

	caffeine	N	Mean	Std. Deviation	Std. Error Mean
hr	1	10	57.80	7.208	2.279
	2	10	67.90	9.085	2.873

Independent Samples Test

		Levene's Test for Equality of Variances		t-test for Equality of Means						
									95% Confidence Interval of the Difference	
		F	Sig.	t	df	Sig. (2-tailed)	Mean Difference	Std. Error Difference	Lower	Upper
hr	Equal variances assumed	1.571	.226	−2.754	18	.013	−10.100	3.667	−17.805	−2.395
	Equal variances not assumed			−2.754	17.115	.013	−10.100	3.667	−17.834	−2.366

Figure 5.7 ♦ Output From SPSS Independent Samples *t* Test Procedure

groups are presented here. (Students who need to review the by-hand computation should verify that they can duplicate these results by computing the means and standard deviations by hand from the data in Table 5.1.) The difference between the means, $M_1 - M_2$, in this example was $57.8 - 67.9 = -10.1$ bpm. The group that did not consume caffeine had an HR about 10.1 bpm lower than the group that consumed caffeine; this is a large enough difference, in real units, to be of clinical or practical interest.

The second panel presents the results of the Levene test (the test of the homogeneity of variance assumption) and two versions of the *t* test. The "equal variances assumed" (or pooled variances) *t* is reported on the first line; the separate variances (or equal variances not assumed) version of the *t* test is reported on the second line.

To read this table, look first at the value of *F* for the Levene test and its associated significance (p) value. If p is small ($p < .05$ or $p < .01$), then there is evidence that the homogeneity of variance assumption has been violated. In this case, the researcher may prefer to use the more conservative equal variances not assumed version of *t*. If the Levene test *F* is not significant ($p > .05$), there is no evidence of a problem with this assumption, and the equal variances *t* test can be reported.

In this example, the Levene *F* value is fairly small ($F = 1.571$), and it is not statistically significant ($p = .226$). Thus, it is appropriate in this case to report the equal variances assumed version of the *t* test. The equal variances *t* test result was statistically significant, $t(18) = -2.75$, $p = .013$, two-tailed. Thus, using $\alpha = 0.5$, two-tailed, as the criterion, the 10.1-point difference in HR between the caffeine and no-caffeine groups was statistically significant.

If the researcher had specified a one-tailed test corresponding to the alternative hypothesis H_1: $(\mu_1 - \mu_2) < 0$, this result could be reported as follows: The equal variances *t* test was statistically significant, $t(18) = -2.75$, $p = .0065$, one-tailed. (That is, the one-tailed *p* value is obtained by dividing the two-tailed *p* value by 2.)

SPSS does not provide any of the common effect size indexes for *t*, but these can easily be calculated by hand from the information provided on the printout. Using Equations 5.22, 5.23, and 5.24, the following values can be computed by hand: $\eta^2 = .30$, Cohen's $d = 1.23$, and $r_{pb} = .54$. Using the suggested verbal labels from Table 5.2, these values indicate a large effect size. The most recent edition of the APA publication manual calls for the inclusion of effect size information, so an effect size index should be included in the written Results section.

Results

An independent samples *t* test was performed to assess whether mean HR differed significantly for a group of 10 participants who consumed no caffeine (Group 1) compared with a group of 10 participants who consumed 150 mg of caffeine. Preliminary data screening indicated that scores on HR were multimodal, but the departure from normality was not judged serious enough to require the use of a nonparametric test. The assumption of homogeneity of variance was assessed by the Levene test, $F = 1.57$, $p = .226$; this indicated no significant violation of the equal variance assumption; therefore, the pooled variances version of the *t* test was used. The mean HRs differed significantly, $t(18) = -2.75$, $p = .013$, two-tailed. Mean HR for the no-caffeine group ($M = 57.8$, $SD = 7.2$) was about 10 bpm lower than mean HR for the caffeine group ($M = 67.9$, $SD = 9.1$). The effect size, as indexed by η^2, was .30; this is a large effect. The 95% CI for the difference between sample means, $M_1 - M_2$, had a lower bound of −17.81 and an upper bound of −2.39. This study suggests that consuming 150 mg of caffeine may significantly increase HR, with an increase on the order of 10 bpm.

5.12 ♦ Summary

This chapter reviewed a familiar statistical test (the independent samples *t* test) and provided additional information that may be new to some students about effect size, statistical power, and factors that affect the size of *t*. The *t* test is sometimes used by itself to report results in relatively simple studies that compare group means on a few outcome variables; it is also used as a follow-up in more complex designs that involve larger numbers of groups or outcome variables.

Two important concepts are as follows:

1. Researchers have some control over factors that influence the size of t, in both experimental and nonexperimental research situations.
2. Because the size of t depends to a great extent on our research decisions, we should be cautious about making inferences about the strength of effects in the real world based on the obtained effect sizes in our samples.

Current APA style guidelines require that researchers provide information about effect size when they report statistics; for the independent samples t test, researchers often report one of the following effect size measures: Cohen's d, r_{pb}, or η^2. Past research has not always included effect size information, but it is generally possible to calculate effect sizes from the information in published journal articles.

The t ratio is used to assess the statistical significance for other sample statistics in later chapters of this book. The general form of a t ratio is as follows:

$$t = \frac{\text{Sample statistic} - \text{Hypothesized population parameter}}{SE_{\text{sample statistic}}}.$$

In later chapters, t ratios of a similar form are used to test hypotheses about other sample statistics, including Pearson r and the slopes in regression equations.

Note

1. Computation of the adjusted degrees of freedom (df') for the separate variances t test (from the SPSS algorithms Web page):

$$Z_1 = \left[\frac{s_1^2/N}{s_1^2/n_1 + s_2^2/n_2}\right]\frac{1}{N-1},$$

$$Z_2 = \left[\frac{s_2^2/N}{s_2^2/n_1 + s_2^2/n_2}\right]\frac{1}{N-1},$$

and, finally,

$$df' = \frac{1}{Z_1 + Z_2},$$

where s_1^2 and s_2^2 are the within-group variances of the scores relative to their group means and N is the total number of scores in the two groups combined, $N = n_1 + n_2$.

Comprehension Questions

1. Suppose you read the following in a journal: "The group means did not differ significantly, $t(30.1) = 1.68, p < .05$, two-tailed." You notice that the *ns* in the groups were $n_1 = 40$ and $n_2 = 55$.
   a. What degrees of freedom would you normally expect a *t* test to have when $n_1 = 40$ and $n_2 = 55$?
   b. How do you explain why the degrees of freedom reported here differ from the value you just calculated?
2. What type of effect can a variable that is confounded with the treatment variable in a two-group experimental study have on the obtained value of the *t* ratio?
   a. It always makes *t* larger.
   b. It usually makes *t* smaller.
   c. It can make *t* either larger or smaller or leave it unchanged.
   d. It generally has no effect on the size of the *t* ratio.
3. What is the minimum information that you need to calculate an η^2 effect size from a reported independent samples *t* test?
   a. s_1^2, s_2^2 and n_1, n_2
   b. *t* and *df*
   c. The $M_1 - M_2$ value and s_p
   d. None of the above
4. Aronson and Mills (1959) conducted an experiment to see whether people's liking for a group is influenced by the severity of initiation. They reasoned that when people willingly undergo a severe initiation to become members of a group, they are motivated to think that the group membership must be worthwhile. Otherwise, they would experience cognitive dissonance: Why put up with severe initiation for the sake of a group membership that is worthless? In their experiment, participants were randomly assigned to one of three treatment groups:
   - Group 1 (control) had no initiation.
   - Group 2 (mild) had a mildly embarrassing initiation (reading words related to sex out loud).
   - Group 3 (severe) had a severely embarrassing initiation (reading sexually explicit words and obscene words out loud).

   After the initiation, each person listened to a standard tape-recorded discussion among the group that they would now supposedly be invited to join; this was made to be as dull and banal as possible. Then, they were asked to rate how interesting they thought the discussion was. The researchers expected that people who had undergone the most embarrassing initiation would evaluate the discussion most positively. In the table below, a higher score represents a more positive evaluation.

Experimental Condition

	Control (No Initiation)	Mild Initiation	Severe Initiation
Mean	80.2	81.8	97.6
SD	13.2	21.0	16.6
N	21	21	21

SOURCE: Aronson and Mills (1959).

Results of t Tests Between Group Means	*t*	*p (Two-Tailed)*
Control versus severe	3.66	.001
Mild versus severe	2.62	.02
Control versus mild	.29	NS

a. Were the researcher's predictions upheld? In simple language, what was found?
b. Calculate an effect size (η^2) for each of these three *t* ratios and interpret these.

5. Say whether true or false: The size of η^2 tends to increase as n_1 and n_2 increase.
6. A researcher reports that the η^2 effect size for her study is very large ($\eta^2 = .64$); but the *t* test that she reports is quite small and not statistically significant. What inference can you make about this situation?
   a. The researcher has made a mistake: If η^2 is this large, then *t* must be significant.
   b. The *ns* in the groups are probably rather large.
   c. The *ns* in the groups are probably rather small.
   d. None of the above inferences is correct.
7. A researcher collects data on married couples to see whether men and women differ in their mean levels of marital satisfaction. For each married couple, both the husband and wife fill out a scale that measures their level of marital satisfaction; the researcher carries out an independent samples *t* test to test the null hypothesis that male and female levels of marital satisfaction do not differ. Is this analysis appropriate? Give reasons for your answer.
8. A researcher plans to do a study to see whether people who eat an all-carbohydrate meal have different scores on a mood scale than people who eat an all-protein meal. Thus, the manipulated independent variable is the type of meal (1 = Carbohydrate, 2 = Protein). In past research, the effect sizes that have been reported for the effects of food on mood have been on the order of $\eta^2 = .15$. Based on this and assuming that the researcher plans to use $\alpha = .05$, two-tailed, and wants to have power of about .80, what is the minimum group size that is needed (i.e., how many subjects would be needed in each of the two groups)?

9. Which of the following would be sufficient information for you to calculate an independent samples *t*?
   a. s_1^2, s_2^2; M_1, M_2; and n_1, n_2
   b. *d* and *df*
   c. n_1, n_2, the $M_1 - M_2$ value, and s_p
   d. None of the above
   e. Any of the above (a, b, or c)

10. What changes in research design tend to reduce the magnitude of the within-group variances (s_1^2, s_2^2)? What advantage does a researcher get from decreasing the magnitude of these variances? What disadvantage arises when these variances are reduced?

11. The statistic that is most frequently used to describe the relation between a dichotomous group membership variable and scores on a continuous variable is the independent samples *t*. Name two other statistics that can be used to describe the relationship between these kinds of variables.

12. Suppose that a student conducts a study in which the manipulated independent variable is the level of white noise (60 vs. 65 db). Ten participants are assigned to each level of noise; these participants vary widely in age, hearing acuity, and study habits. The outcome variable is performance on a verbal learning task (how many words on a list of 25 words each participant remembers). The *t* value obtained is not statistically significant. What advice would you give to this student about ways to redesign this study that might improve the chances of detecting an effect of noise on verbal learning recall?

13. In Table 5.4 are listed some of the data obtained in a small experiment run by one of my research methods classes to evaluate the possible effects of food on mood. This was done as a between-subjects study; each participant was randomly assigned to Group 1 (an all-carbohydrate lunch) or Group 2 (an all-protein lunch). One hour after eating lunch, each participant rated his or her moods, with higher scores indicating more agreement with that mood. Select one of the mood outcome variables, enter the data into SPSS, examine a histogram to see if the scores appear to be normally distributed, and conduct an independent samples *t* test to see if mean moods differed significantly between groups. Write up your results in the form of an APA-style Results section, including a statement about effect size. Be certain to state the nature of the relationship between food and mood (i.e., did eating carbohydrates make people more calm or less calm than eating protein)?

    Also, as an exercise, if you wanted to design a better study to assess the possible impact of food on mood, what would you add to this study? What would you change? (For background about research on the possible effects of food on mood, refer to Spring, Chiodo, & Bowen, 1987.)

14. A test of emotional intelligence was given to 241 women and 89 men. The results were as follows: For women, $M = 96.62$, $SD = 10.34$; for men, $M = 89.33$,

Table 5.4 ♦ Data for Problem 13: Hypothetical Study of Effects of Food on Mood

Food[a]	*Calm*	*Anxious*	*Sleepy*	*Alert*
1	4	6	5	5
1	5	4	4	6
1	8	9	3	8
1	3	5	2	5
1	5	5	4	2
1	4	8	2	4
1	6	5	6	6
1	3	5	5	11
1	5	7	7	7
1	5	4	1	6
2	2	6	4	7
2	0	9	0	6
2	3	4	3	12
2	1	6	4	6
2	4	5	0	6
2	2	6	2	3
2	3	4	0	3
2	4	9	3	13
2	2	9	1	12
2	3	4	0	9

NOTE: The moods calm, anxious, sleepy, and alert were rated on a 0- to 15-point scale: 0 = Not at all, 15 = Extremely.

a. Food type was coded 1 = Carbohydrate, 2 = Protein.

SD = 11.61. Was this difference statistically significant ($\alpha = .05$, two-tailed)? How large was the effect, as indexed by Cohen's *d* and by η^2 (For this result, refer to Brackett, Mayer, & Warner, 2004.)

15. What is the null hypothesis for an independent samples *t* test?
16. In what situations should the paired samples *t* test be used rather than the independent samples *t* test?
17. In what situations should the Wilcoxon rank sum test be used rather than the independent samples *t* test?
18. What information does the *F* ratio on the SPSS output for an independent samples *t* test provide; that is, what assumption does it test?
19. Explain briefly why there are two different versions of the *t* test on the SPSS printout and how you decide which one is more appropriate.
20. What is η^2?

CHAPTER 6

One-Way Between-Subjects Analysis of Variance

6.1 ♦ Research Situations Where One-Way Between-Subjects Analysis of Variance (ANOVA) Is Used

A one-way between-subjects (between-S) **ANOVA** is used in research situations where the researcher wants to compare means on a quantitative *Y* outcome variable across two or more groups. Group membership is identified by each participant's score on a categorical *X* predictor variable. ANOVA is a generalization of the *t* test; a *t* test provides information about the distance between the means on a quantitative outcome variable for just two groups, whereas a one-way ANOVA compares means on a quantitative variable across any number of groups. The categorical predictor variable in an ANOVA may represent either naturally occurring groups or groups formed by a researcher and then exposed to different interventions. When the means of naturally occurring groups are compared (e.g., a one-way ANOVA to compare mean scores on a self-report measure of political conservatism across groups based on religious affiliation), the design is nonexperimental. When the groups are formed by the researcher and the researcher administers a different type or amount of treatment to each group while controlling extraneous variables, the design is an experiment.

The term *between-S* (like the term *independent samples*) tells us that each participant is a member of one and only one group and that the members of samples are not matched or paired. When the data for a study consist of repeated measures or paired or matched samples, a repeated measures ANOVA is required (see Chapter 20 for an introduction to the analysis of repeated measures). If there is more than one categorical variable or **factor** included in the study, factorial ANOVA is used (see Chapter 13). When there is just a single factor, textbooks often name this single factor *A*, and if there are additional factors, these are usually designated factors *B*, *C*, *D*, and so forth. If scores on the dependent *Y* variable are in the form of rank or ordinal data, or if the data seriously violate assumptions required for ANOVA, a nonparametric alternative to ANOVA may be preferred.

In ANOVA, the categorical predictor variable is called a *factor*; the groups are called the levels of this factor. In the hypothetical research example introduced in Section 6.2, the factor is called "Types of Stress," and the levels of this factor are as follows: 1, No stress; 2, Cognitive stress from mental arithmetic task; 3, Stressful social role play; and 4, Mock job interview.

Comparisons among several group means could be made by calculating *t* tests for each pairwise comparison among the means of these four treatment groups. However, as described in Chapter 3, doing a large number of significance tests leads to an inflated risk for Type I error. If a study includes *k* groups, there are $k(k - 1)/2$ pairs of means; thus, for a set of four groups, the researcher would need to do $(4 \times 3)/2 = 6$ *t* tests to make all possible pairwise comparisons. If $\alpha = .05$ is used as the criterion for significance for each test, and the researcher conducts six significance tests, the probability that this set of six decisions contains at least one instance of Type I error is greater than .05. Most researchers assume that it is permissible to do a few significance tests (perhaps three or four) without taking special precautions to limit the risk of Type I error. However, if more than a few comparisons are examined, researchers usually want to use one of several methods that offer some protection against inflated risk of Type I error.

One way to limit the risk of Type I error is to perform a single **omnibus test** that examines all the comparisons in the study as a set. The *F* test in a one-way ANOVA provides a single omnibus test of the hypothesis that the means of all *k* populations are equal, in place of many *t* tests for all possible pairs of groups. As a follow-up to a significant overall *F* test, researchers often still want to examine selected pairwise comparisons of group means to obtain more information about the pattern of differences among groups. This chapter reviews the *F* test, which is used to assess differences for a set of more than two group means, and some of the more popular procedures for follow-up comparisons among group means.

The overall null hypothesis for one-way ANOVA is that the means of the *k* populations that correspond to the groups in the study are all equal:

$$H_0: \mu_1 = \mu_2 = \ldots = \mu_k.$$

When each group has been exposed to different types or dosages of a treatment, as in a typical experiment, this null hypothesis corresponds to an assumption that the treatment has no effect on the outcome variable. The alternative hypothesis in this situation is not that all population means are unequal but that there is at least one inequality for one pair of population means in the set.

This chapter reviews the computation and interpretation of the *F* ratio that is used in between-S one-way ANOVA. This chapter also introduces an extremely important concept: the partition of each score into two components—namely, a first component that is predictable from the independent variable that corresponds to group membership and a second component that is due to the collective effects of all other uncontrolled or extraneous variables. Similarly, the total sum of squares can also be partitioned into a sum of squares (between groups) that reflects variability associated with treatment and a sum of squares (within groups) that is due to the influence of extraneous uncontrolled variables. From these sums of squares, we can estimate the proportion of variance in scores that is explained by, accounted for by, or predictable from group membership (or from the different treatments administered to groups). The concept of explained or predicted variance is crucial for understanding ANOVA; it is also important for the comprehension of multivariate analyses described in later chapters.

The best question ever asked by a student in any of my statistics classes was a deceptively simple "Why is there variance?" In the hypothetical research situation described in the following section, the outcome variable is a self-report measure of Anxiety, and the research question is, Why is there variance in anxiety? Why do some persons report more anxiety than other persons? To what extent are the differences in amount of anxiety systematically associated with the independent variable (type of stress) and to what extent are differences in the amount of self-reported anxiety due to other factors (such as trait levels of anxiety, physiological arousal, drug use, gender, other anxiety-arousing events that each participant has experienced on the day of the study, and so forth)? When we do an ANOVA, we identify the part of each individual score that is associated with group membership (and, therefore, with the treatment or intervention, if the study is an experiment) and the part of each individual score that is not associated with group membership and that is, therefore, attributable to a variety of other variables that are not controlled in the study, or "error." In most research situations, researchers usually hope to show that a large part of the scores is associated with, or predictable from, group membership or treatment.

6.2 ♦ Hypothetical Research Example

As an illustration of the application of one-way ANOVA, consider the following imaginary study. Suppose that an experiment is done to compare the effects of four situations: Group 1 is tested in a "no-stress," baseline situation; Group 2 does a mental arithmetic task; Group 3 does a stressful social role play; and Group 4 participants do a mock job interview. For this study, the *X* variable is a categorical variable with codes 1, 2, 3, and 4 that represent which of these four types of stress each participant received. This categorical *X* predictor variable is called a factor; in this case, the factor is called "Type of Stress Intervention"; the four levels of this factor correspond to no stress, mental arithmetic, stressful role play, and mock job interview. At the end of each session, the participants self-report their anxiety on a scale that ranges from 0 = no anxiety to 20 = extremely high anxiety. Scores on anxiety are, therefore, scores on a quantitative *Y* outcome variable. The researcher wants to know whether, overall, anxiety levels differed across these four situations and, if so, which situations elicited the highest anxiety and which treatment conditions differed significantly from baseline and from other types of stress. A convenience sample of 28 participants was obtained; each participant was randomly assigned to one of the four levels of stress. This results in $k = 4$ groups with $n = 7$ participants in each group, for a total of $N = 28$ participants in the entire study. The SPSS Data View worksheet that contains data for this imaginary study appears in Figure 6.1.

6.3 ♦ Assumptions About Scores on the Dependent Variable for One-Way Between-S ANOVA

The assumptions for one-way ANOVA are similar to those described for the independent samples *t* test (see Section 3 of Chapter 5). The scores on the quantitative dependent variable

fakestress [DataSet1] - SPSS Data Editor

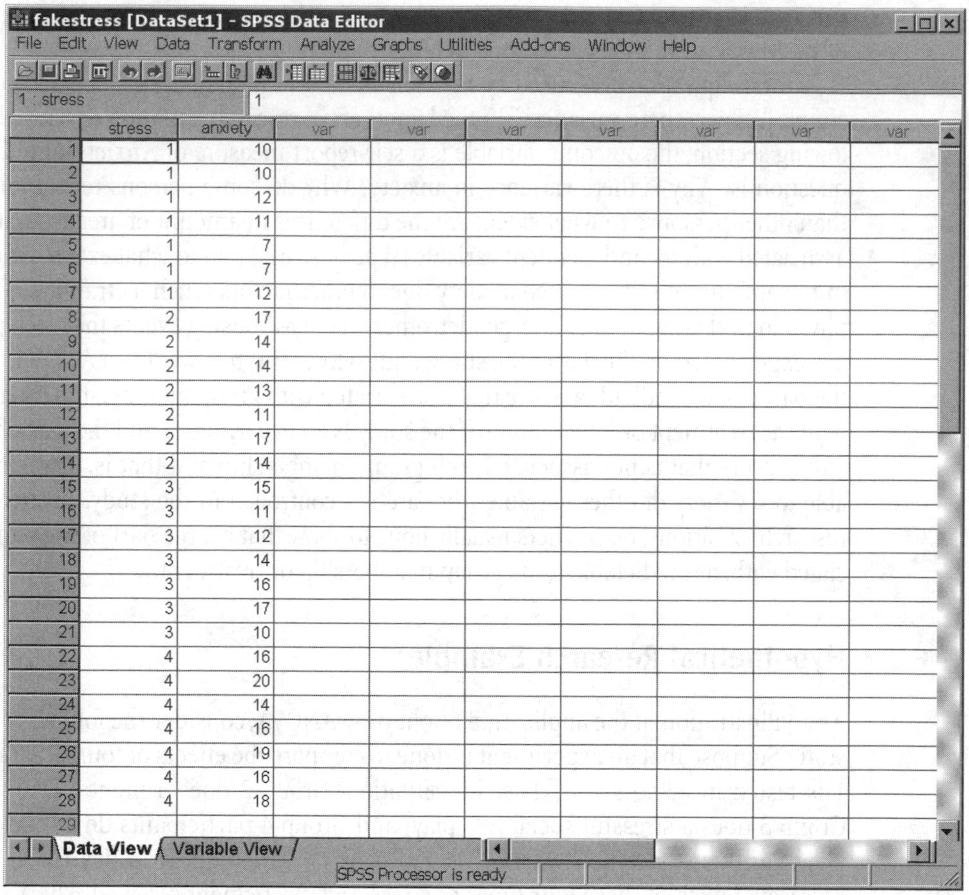

File Edit View Data Transform Analyze Graphs Utilities Add-ons Window Help

1 : stress 1

	stress	anxiety	var	var	var	var	var	var	var
1	1	10							
2	1	10							
3	1	12							
4	1	11							
5	1	7							
6	1	7							
7	1	12							
8	2	17							
9	2	14							
10	2	14							
11	2	13							
12	2	11							
13	2	17							
14	2	14							
15	3	15							
16	3	11							
17	3	12							
18	3	14							
19	3	16							
20	3	17							
21	3	10							
22	4	16							
23	4	20							
24	4	14							
25	4	16							
26	4	19							
27	4	16							
28	4	18							
29									

Data View / Variable View

SPSS Processor is ready

Figure 6.1 ♦ Data View Worksheet for Stress/Anxiety Study

should be quantitative and, at least approximately, interval/ratio level of measurement. The scores should be approximately normally distributed in the entire sample and within each group, with no extreme outliers. The variances of scores should be approximately equal across groups. Observations should be independent of each other, both within groups and between groups. Preliminary screening involves the same procedures as for the *t* test: Histograms of scores should be examined to assess normality of distribution shape, a box and whiskers plot of the scores within each group could be used to identify outliers, and the Levene test (or some other test of homogeneity of variance) could be used to assess whether the homogeneity of variance assumption is violated. If major departures from the normal distribution shape and/or substantial differences among variances are detected, it may be possible to remedy these problems by applying data transformations, such as the base 10 logarithm, or by removing or modifying scores that are outliers, as described in Chapter 4.

6.4 ♦ Issues in Planning a Study

When an experiment is designed to compare treatment groups, the researcher needs to decide how many groups to include, how many participants to include in each group, how to assign participants to groups, and what types or dosages of treatments to administer to each group. In an experiment, the levels of the factor may correspond to different dosage levels of the same treatment variable (such as 0 mg caffeine, 100 mg caffeine, 200 mg caffeine) or to qualitatively different types of interventions (as in the stress study, where the groups received, respectively, no stress, mental arithmetic, role play, or mock job interview stress interventions). In addition, it is necessary to think about issues of experimental control, as described in research methods textbooks; for example, in experimental designs, researchers need to avoid confounds of other variables with the treatment variable. Research methods textbooks (e.g., Cozby, 2004) provide a more detailed discussion of design issues; a few guidelines are listed here.

1. If the treatment variable has a curvilinear relation to the outcome variable, it is necessary to have a sufficient number of groups to describe this relation accurately—usually, at least 3 groups. On the other hand, it may not be practical or affordable to have a very large number of groups; if an absolute minimum *n* of 10 (or better, 20) participants per group are included, then a study that included 15 groups would require a minimum of 150 (or 300) participants.

2. In studies of interventions that may create expectancy effects, it may be necessary to include one or several kinds of placebo and no treatment/control groups for comparison. For instance, studies of the effect of biofeedback on heart rate (HR) sometimes include a group that gets real biofeedback (a tone is turned on when HR increases), a group that gets noncontingent feedback (a tone is turned on and off at random in a way that is unrelated to HR), and a group that receives instructions for relaxation and sits quietly in the lab without any feedback (Burish, 1981).

3. Random assignment of participants to conditions is desirable to try to ensure equivalence of the groups prior to treatment. For example, in a study where HR is the dependent variable, it would be desirable to have participants whose HRs were equal across all groups prior to the administration of any treatments. However, it cannot be assumed that random assignment will always succeed in creating equivalent groups. In studies where equivalence among groups prior to treatment is important, the researcher should collect data on behavior prior to treatment to verify whether the groups are equivalent.[1]

The same factors that affect the size of the *t* ratio (see Chapter 5) also affect the size of *F* ratios: the distances among group means, the amount of variability of scores within each group, and the number, *n*, of participants per group. Other things being equal, an *F* ratio tends to be larger when there are large between-group differences among dosage levels (or participant characteristics). As an example, a study that used three different noise levels in decibels as the treatment variable, for example, 35, 65, and 95 dB, would result in larger differences in group means on arousal and a larger *F* ratio than a study that looked, for example, at these three noise levels, which are closer together: 60, 65, and 70 dB. Similar to the *t* test, the *F* ratio involves a comparison of between-group and within-group variability; the selection of homogeneous participants, standardization of testing conditions, and control over extraneous variables will tend to reduce the magnitude of within-group variability of scores, which in turn tends to produce a larger *F* (or *t*) ratio.

Finally, the F test involves degrees of freedom (df) terms based on number of participants (and also number of groups); the larger the number of participants, other things being equal, the larger the F ratio will tend to be. A researcher can usually increase the size of an F (or t) ratio by increasing the differences in the dosage levels of treatments given to groups and/or reducing the effects of extraneous variables through experimental control, and/or increasing the number of participants. In studies that involve comparisons of naturally occurring groups, such as age groups, a similar principle applies: A researcher is more likely to see age-related changes in mental processing speed in a study that compares ages 20, 50, and 80 than in a study that compares ages 20, 25, and 30.

6.5 ♦ Data Screening

A box and whiskers plot was set up to examine the distribution of anxiety scores within each of the four stress intervention groups; this plot appears in Figure 6.2. Examination of this box and whiskers plot indicates that there are no outlier scores in any of the four groups. Other preliminary data screening that could be done includes a histogram for all the anxiety scores in the entire sample. The Levene test to assess possible violations of the homogeneity of variance assumption is reported in Section 6.11.

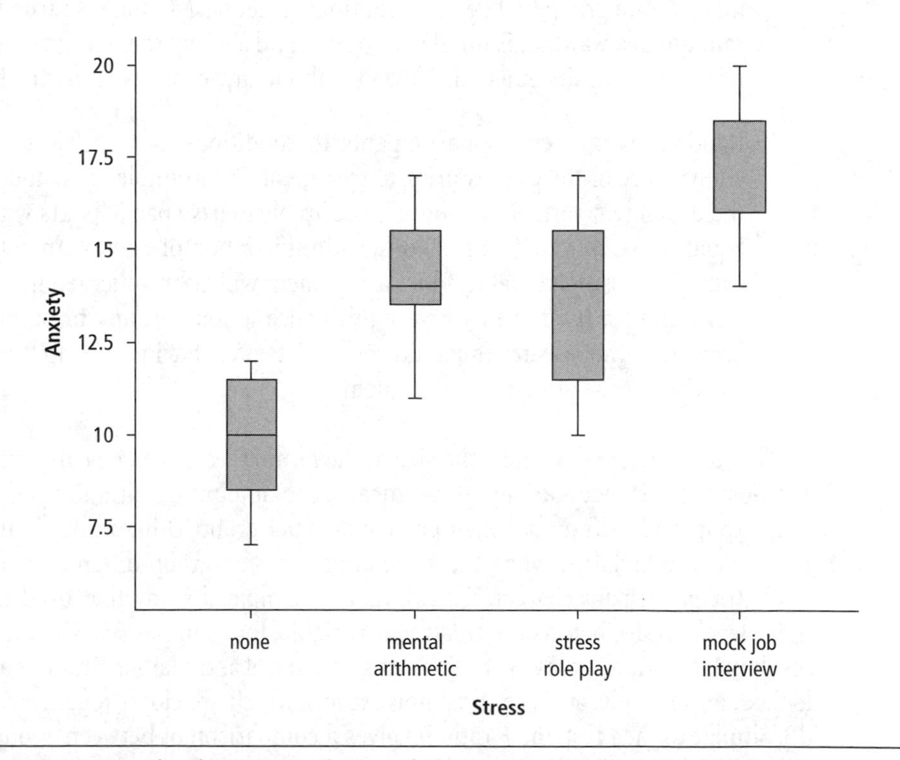

Figure 6.2 ♦ Box Plot Showing Distribution of Anxiety Scores Within Each of the Four Treatment Groups in the Stress/Anxiety Experiment

Table 6.1 ♦ Data for the Partition of HR Scores

Name	*Gender*	*HR*	*Grand Mean*	*Group Mean*	*DevGrand*	*DevGroup*	*Effect*	*Reproduced Score*
Ann	1	81	76.5	86.0	4.5	–5.0	9.5	81.0
Betty	1	85	76.5	86.0	8.5	–1.0	9.5	85.0
Cathy	1	92	76.5	86.0	15.5	6.0	9.5	92.0
John	2	70	76.5	67.0	–6.5	3.0	–9.5	70.0
Ken	2	63	76.5	67.0	–13.5	–4.0	–9.5	63.0
Leo	2	68	76.5	67.0	–8.5	1.0	–9.5	68.0

NOTES: DevGrand = HR – Grand Mean. DevGroup or residual = HR – Group Mean. Effect = Group Mean – Grand Mean. Reproduced Score = Grand Mean + Effect + DevGroup.

6.6 ♦ Partition of Scores Into Components

When we do a one-way ANOVA, the analysis involves partition of each score into two components: a component of the score that is associated with group membership and a component of the score that is not associated with group membership. The sums of squares (*SS*) summarize how much of the variance in scores is associated with group membership and how much is not related to group membership. Examining the proportion of variance that is associated with treatment group membership is one way of approaching the general research question, Why is there variance in the scores on the outcome variable?

To illustrate the partition of scores into components, consider the hypothetical data in Table 6.1. Suppose that we measure HR for each of 6 persons in a small sample and obtain the following set of scores: for the 3 female participants, HR scores of 81, 85, and 92 and for the 3 male participants, HR scores of 70, 63, and 68. In this example, the *X* group membership variable is gender, coded 1 = Female and 2 = Male, and the *Y* quantitative outcome variable is HR. We can partition each HR score into a component that is related to group membership (i.e., related to gender) and a component that is not related to group membership. In a textbook discussion of ANOVA using just two groups, the factor that represents gender might be called Factor *A*.

We will denote the HR score for person *j* in group *i* as Y_{ij}. For example, Cathy's HR of 92 corresponds to Y_{13}, the score for the third person in Group 1. We will denote the **grand mean** (i.e., the mean HR for all $N = 6$ persons in this dataset) as M_Y; for this set of scores, the value of M_Y is 76.5. We will denote the mean for group *i* as M_i. For example, the mean HR for the female group (Group 1) is $M_1 = 86$; the mean HR for the male group (Group 2) is $M_2 = 67$. Once we know the individual scores, the grand mean, and the group means, we can work out a partition of each individual HR score into two components.

The question we are trying to answer is, Why is there variance in HR? In other words, why do some people in this sample have HRs that are higher than the sample mean of 76.5, while others have HRs that are lower than 76.5? Is gender one of the variables that predicts whether a person's HR will be relatively high or low? For each person, we can compute a deviation of that person's individual Y_{ij} score from the grand mean; this tells us how far above (or below) the grand mean of HR each person's score was. Note that in Table 6.1 the value of the grand mean is the same for all 6 participants. The value of the group mean was different for members of Group $i = 1$ (females) than for members of Group $i = 2$ (males). We can use these means to compute the following three deviations from means for each person:

$$\text{Deviation of individual score from grand mean} = \text{DevGrand} = (Y_{ij} - M_Y). \quad (6.1)$$

$$\text{Deviation of individual score from group mean} = \text{DevGroup} = (Y_{ij} - Y_i) = \varepsilon_{ij}. \quad (6.2)$$

$$\text{Deviation of group mean from grand mean} = \text{Effect} = (Y_i - M_Y) = \alpha_i. \quad (6.3)$$

The total deviation of the individual score from the grand mean, DevGrand, can be divided into two components: the deviation of the individual score from the group mean, DevGroup, and the deviation of the group mean from the grand mean, Effect:

$$\begin{array}{ccccc} (Y_{ij} - M_Y) & = & (Y_{ij} - M_i) & + & (M_i - M_Y) \\ \text{Total deviation} & = & \text{Within-group deviation} & + & \text{Between-group difference} \\ & & \text{(residual or error)} & & \text{("effect" for group } i\text{).} \end{array} \quad (6.4)$$

For example, look at the line of data for Cathy in Table 6.1. Cathy's observed HR was 92. Cathy's HR has a total deviation from the grand mean of (92 – 76.5) = +15.5; that is, Cathy's HR was 15.5 beats per minute (bpm) higher than the grand mean for this set of data. Compared with the group mean for the female group, Cathy's HR had a deviation of (92 – 86) = +6; that is, Cathy's HR was 6 bpm higher than the mean HR for the female group. Finally, the "effect" component of Cathy's score is found by subtracting the grand mean (M_Y) from the mean of the group to which Cathy belongs (M_{female} or M_1): (86 – 76.5) = +9.5. This value of 9.5 tells us that the mean HR for the female group (86) was 9.5 bpm higher than the mean HR for the entire sample (and this effect is the same for all members within each group). We could, therefore, say that the "effect" of membership in the female group is to increase the predicted HR score by 9.5. (Note, however, that this example involves comparison of naturally occurring groups, males vs. females, and so we should not interpret a group difference as evidence of causality but merely as a description of group differences.)

Another way to look at these numbers is to set up a predictive equation based on a theoretical model. When we do a one-way ANOVA, we seek to predict each individual score (Y_{ij}) on the quantitative outcome variable from the following theoretical components: the population mean (denoted by μ), the "effect" for group i (often denoted by α_i, because α is the Greek letter that corresponds to A, which is the name usually given to a single factor), and the **residual** associated with the score for person j in group i (denoted by ε_{ij}). Estimates for μ, α_i, and ε_{ij} can be obtained from the sample data as follows:

$$\mu = M_Y, \quad (6.5)$$

$$\alpha_i = M_i - M_Y, \quad (6.6)$$

$$\varepsilon_{ij} = Y_{ij} - M_i. \tag{6.7}$$

Note that the deviations that are calculated as estimates for α_j and ε_{ij} (in Equations 6.6 and 6.7) are the same components that appeared in Equation 6.4. An individual observed HR score Y_{ij} can be represented as a sum of these theoretical components, as follows:

$$Y_{ij} = \mu + \alpha_i + \varepsilon_{ij}. \tag{6.8}$$

In words, Equation 6.8 says that we can predict (or reconstruct) the observed score for person i in group j by taking the grand mean μ, adding the "effect" α_j that is associated with membership in group j, and finally, adding the residual ε_{ij} that tells us how much individual i's score differed from the mean of the group that person i belonged to.

We can now label the terms in Equation 6.4 using this more formal notation:

$(Y_{ij} - M_Y)$	=	α_i	+	ε_{ij},
$(Y_{ij} - M_Y)$	=	$(M_i - M_Y)$	+	$(Y_{ij} - M_i)$
Total deviation for person i in group j	=	"effect" for group i	+	residual for person j. (6.9)

In other words, we can divide the total deviation of each person's score from the grand mean into two components: α_i, the part of the score that is associated with or predictable from group membership (in this case, gender), and ε_{ij}, the part of the score that is not associated with or predictable from group membership (the part of the HR score that is due to all other variables that influence HR, such as smoking, anxiety, drug use, level of fitness, health, and so forth). In words, then, Equation 6.9 says that we can predict each person's HR from the following information: Person j's HR = grand mean + effect of person j's gender on HR + effects of all other variables that influence person j's HR, such as person j's anxiety, health, drug use, fitness, and anxiety.

The collective influence of "all other variables" on scores on the outcome variable, HR in this example, is called the residual, or "error." In most research situations, researchers hope that the components of scores that represent group differences (in this case, gender differences in HR) will be relatively large and that the components of scores that represent within-group variability in HR (in this case, differences among females and among males in HR and thus, differences due to all variables other than gender) will be relatively small.

When we do an ANOVA, we summarize the information about the sizes of these two deviations or components $(Y_{ij} - M_i)$ and $(M_i - M_Y)$ across all the scores in the sample. We cannot summarize information just by summing these deviations; recall from Chapter 2 that the sum of deviations of scores from a sample mean always equals 0, so $\sum (Y_{ij} - M_i) = 0$ and $\sum (M_i - M_Y) = 0$. When we summarized information about distances of scores from a sample mean by computing a sample variance, we avoided this problem by squaring the deviations from the mean prior to summing them. The same strategy is applied in the ANOVA. ANOVA begins by computing the following deviations for each score in the dataset (see Table 6.1 for an empirical example; these terms are denoted DevGrand, DevGroup, and Effect in Table 6.1).

$$(Y_{ij} - M_Y) = (Y_{ij} - M_i) + (M_i - M_Y).$$

To summarize information about the magnitudes of these score components across all the participants in the dataset, we square each term and then sum the squared deviations across all the scores in the entire dataset:

$$\Sigma\,(Y_{ij} - M_Y)^2 = \Sigma\,(Y_{ij} - M_i)^2 + \Sigma\,(M_i - M_Y)^2. \qquad (6.10)$$

$$SS_{\text{total}} = SS_{\text{within}} + SS_{\text{between}},$$
$$SS_{\text{total}} = SS_{\text{residual}} + SS_{\text{treatment}}.$$

These sums of squared deviations are given the following names: SS_{total}, sum of squares total, for the sum of the squared deviations of each score from the grand mean; SS_{within}, for the sum of squared deviations of each score from its group mean, which summarizes information about within-group variability of scores; and SS_{between}, which summarizes information about the distances among (variability among) group means.

Note that, usually, when we square an equation of the form $a = (b + c)$ (as in $[Y_{ij} - M_Y]^2 = ([Y_{ij} - M_i] + [M_i - M_Y])^2$, the result would be of the form $a^2 = b^2 + c^2 + 2bc$. However, the cross-product term (which would correspond to $2bc$) is missing in Equation 6.10. This cross-product term has an expected value of 0 because the $(Y_{ij} - M_i)$ and $(M_i - M_Y)$ deviations are independent of (i.e., uncorrelated with) each other, and therefore, the expected value of the cross product between these terms is 0.

This example of the partition of scores into components and the partition of SS_{total} into SS_{between} and SS_{within} had just two groups, but this type of partition can be done for any number of groups. Researchers usually hope that SS_{between} will be relatively large because this would be evidence that the group means are far apart and that the different participant characteristics or different types or amounts of treatments for each group are, therefore, predictively related to scores on the Y outcome variable.

This concept of variance partitioning is a fundamental and important one. In other analyses later on (such as Pearson correlation and bivariate regression), we will come across a similar type of partition; each Y score is divided into a component that is predictable from the X variable and a component that is not predictable from the X variable.

It may have occurred to you that if you knew each person's scores on physical fitness, drug use, body weight, and anxiety level, and if these variables are predictively related to HR, then you could use information about each person's scores on these variables to make a more accurate prediction of HR. This is exactly the reasoning that is used in analyses such as multiple regression, where we predict each person's score on an outcome variable, such as anxiety, by additively combining the estimated effects of numerous predictor variables. It may have also occurred to you that, even if we know several factors that influence HR, we might be missing some important information; we might not know, for example, that the individual person has a family history of elevated HR. Because some important variables are almost always left out, the predicted score that we generate from a simple additive model (such as grand mean + gender effect + effects of smoking = predicted HR) rarely corresponds exactly to the person's actual HR. The difference between the actual HR and the HR predicted from the model is called a residual or error. In

ANOVA, we predict people's scores by adding (or subtracting) the number of points that correspond to the effect for their treatment group to the grand mean.

Recall that for each subject in the caffeine group in the *t* test example in the preceding chapter, the best prediction for that person's HR was the mean HR for all people in the caffeine group. The error or residual corresponded to the difference between each participant's actual HR and the group mean, and this residual was presumably due to factors that uniquely affected that individual over and above responses to the effects of caffeine.

6.7 ♦ Computations for the One-Way Between-S ANOVA

The formulas for computation of the sums of squares, **mean squares**, and *F* ratio for the one-way between-S ANOVA are presented in this section; they are applied to the data from the hypothetical stress/anxiety study that appear in Figure 6.1.

6.7.1 ♦ Comparison Between the Independent Samples *t* Test and One-Way Between-S ANOVA

The one-way between-S ANOVA is a generalization of the independent samples *t* test. The *t* ratio provides a test of the null hypothesis that two means differ significantly. For the independent samples *t* test, the null hypothesis has the following form:

$$H_0: \mu_1 - \mu_2 = 0 \text{ or } H_0: (\mu_1 - \mu_2) = 0. \quad (6.11)$$

For a one-way ANOVA with *k* groups, the null hypothesis is as follows:

$$H_0: \mu_1 = \mu_2 = \ldots = \mu_k. \quad (6.12)$$

The computation of the independent samples *t* test requires that we find the following for each group: *M*, the sample mean; *s*, the sample standard deviation; and *n*, the number of scores in each group (see Section 6 of Chapter 5). For a one-way ANOVA, the same computations are performed; the only difference is that we have to obtain (and summarize) this information for *k* groups (instead of only two groups as in the *t* test).

Table 6.2 summarizes the information included in the computation of the independent samples *t* test and the one-way ANOVA to make it easier to see how similar these analyses are. For each analysis, we need to obtain information about differences between group means (for the *t* test, we compute $M_1 - M_2$; for the ANOVA, because we have more than two groups, we need to find the variance of the group means $M_1, M_2, \ldots, M_k$). The variance among the *k* group means is called MS_{between}; the formulas to compute this and other intermediate terms in ANOVA appear below. We need to obtain information about the amount of variability of scores within each group; for both the independent samples *t* test and ANOVA, we can begin by computing an *SS* term that summarizes the squared distances of all the scores in each group from their group mean. We then convert that information into a summary about the amount of variability of scores within groups, summarized across all the groups (for the independent samples *t* test, a summary of within-group score variability is provided by s_p^2; for ANOVA, a summary of within-group score variability is called MS_{within}).

Table 6.2 ♦ Comparison Between the Independent Samples *t* Test and One-Way Between-Subjects (Between-S) ANOVA

Type of Statistical Analysis	*Independent Samples t Test (Two Groups)*	*One-Way Between-S ANOVA (k Groups)*
Information about differences between group means (associated with different scores on the *X* predictor, which may correspond to participant characteristics or different types or amount of treatments)	$M_1 - M_2$	$MS_{between}$
Information about variability of scores within groups (which is attributed to error, or the influence of all variables other than *X*)	s_p^2 (computed from SS_1 and SS_2)	MS_{within} (computed from $SS_1, SS_2, \ldots, SS_k$)
Degrees of freedom	$df = n_1 + n_2 - 2$ (or $N - 2$)	Two *df* terms: $df_{between} = k - 1$ $df_{within} = N - k$
Test statistic	$t = \dfrac{M_1 - M_2}{\sqrt{s_p^2/n_1 + (s_p^2/n_2)}}$	$F = \dfrac{MS_{between}}{MS_{within}}$ or $\dfrac{SS_{between}/df_{between}}{SS_{within}/df_{within}}$
Information represented by test statistic	$\dfrac{\text{Between-Group Differences}}{\text{Within-Group Differences}}$	$\dfrac{\text{Between-Group Differences}}{\text{Within-Group Differences}}$

NOTES: The *F* ratio $MS_{between}/MS_{within}$ can be thought of as a ratio of

$$\frac{\text{Between-group differences in scores}}{\text{Within-group differences in scores}};$$

and the *F* ratio can also be interpreted as information about the relative magnitude of

$$\frac{\text{Effects attributed to manipulated treatment variables + effects due to error variables}}{\text{Effects attributed to error variables (all variables other than manipulated variable)}}.$$

For two groups, t^2 is equivalent to *F*.

Any time that we compute a variance (or a mean square), we sum squared deviations of scores from group means. As discussed in Chapter 2, when we look at a set of deviations of *n* scores from their sample mean, only the first $n - 1$ of these deviations are free to vary. Because the sum of all the *n* deviations from a sample mean must equal 0, once we have the first $n - 1$ deviations, the value of the last deviation is not "free to vary"; it must have whatever value is needed to make the sum of the deviations equal to 0. For the *t* ratio, we computed a variance for the denominator, and we need a single *df* term that tells us how many independent deviations from the mean the divisor of the *t* test was based on (for the independent samples *t* test, $df = n_1 + n_2 - 2$, where n_1 and n_2 are the number of scores in Groups 1 and 2. The total number of scores in the entire study is equal to $n_1 + n_2$, so we can also say that the *df* for the independent samples *t* test = $N - 2$).

For an *F* ratio, we compare MS_{between} with MS_{within}; we need to have a separate *df* for each of these mean square terms to indicate how many independent deviations from the mean each *MS* term was based on.

The test statistic for the independent samples *t* test is

$$t = \frac{M_1 - M_2}{\sqrt{(s_p^2/n_1) + (s_p^2/n_2)}}.$$

The test statistic for the one-way between-S ANOVA is $F = MS_{\text{between}}/MS_{\text{within}}$. In each case, we have information about differences between or among group means in the numerator and information about variability of scores within groups in the denominator. In a well-designed experiment, differences between group means are attributed primarily to the effects of the manipulated independent variable *X*, whereas variability of scores within groups is attributed to the effects of all other variables, collectively called error. In most research situations, researchers hope to obtain large enough values of *t* and *F* to be able to conclude that there are statistically significant differences among group means. The method for the computation of these between- and within-group mean squares is given in detail in the following few sections.

The by-hand computation for one-way ANOVA (with *k* groups and a total of *N* observations) involves the following steps. Complete formulas for the *SS* terms are provided in the following sections.

1. Compute $SS_{\text{between groups}}$, $SS_{\text{within groups}}$, and SS_{total}.
2. Compute MS_{between} by dividing SS_{between} by its *df*, $k - 1$.
3. Compute MS_{within} by dividing SS_{within} by its *df*, $N - k$.
4. Compute an *F* ratio: $MS_{\text{between}}/MS_{\text{within}}$.
5. Compare this *F* value obtained with the critical value of *F* from a table of the *F* distribution with $(k - 1)$ and $(N - k)$ *df* (using the table in Appendix C that corresponds to the desired alpha level; for example, the first table provides critical values for $\alpha = .05$). If the *F* value obtained exceeds the tabled critical value of *F* for the predetermined alpha level and the available degrees of freedom, reject the null hypothesis that all the population means are equal.

6.7.2 ♦ Summarizing Information About Distances Between Group Means: Computing MS_{between}

The following notation will be used:

Let *k* be the number of groups in the study.

Let $n_1, n_2, \ldots, n_k$ be the number of scores in Groups $1, 2, \ldots, k$.

Let Y_{ij} be the score of subject number j in group i; $i = 1, 2, \ldots, k$.

Let $M_1, M_2, \ldots, M_k$ be the means of scores in Groups $1, 2, \ldots, k$.

Let N be the total N in the entire study; $N = n_1 + n_2 + \ldots + n_k$.

Let M_Y be the grand mean of all scores in the study (i.e., the total of all the individual scores, divided by N, the total number of scores).

Once we have calculated the means of each individual group ($M_1, M_2, \ldots, M_k$) and the grand mean M_Y, we can summarize information about the distances of the group means, M_j, from the grand mean, M_Y, by computing SS_{between} as follows:

$$SS_{\text{between}} = \sum_{i=1}^{k} n_i (M_i - M_Y)^2 \tag{6.13}$$

$$= n_1(M_1 - M_Y)^2 + n_2(M_2 - M_Y)^2 + \ldots + n_k(M_k - M_Y)^2.$$

For the hypothetical data in Figure 6.1, the mean anxiety scores for Groups 1 through 4 were as follows: $M_1 = 9.86$, $M_2 = 14.29$, $M_3 = 13.57$, and $M_4 = 17.00$. The grand mean on anxiety, M_Y, is 13.68. Each group had $n = 7$ scores. Therefore, for this study,

$$SS_{\text{between}} = 7 \times (9.86 - 13.68)^2 + 7 \times (14.29 - 13.68)^2 + 7 \times (13.57 - 13.68)^2 + 7 \times (17.00 - 13.68)^2$$

$$= 7 \times (-3.82)^2 + 7 \times (.61)^2 + 7 \times (-.11)^2 + 7 \times (+3.32)^2$$

$$= 7 \times 14.5924 + 7 \times .3721 + 7 \times .0121 + 7 \times 11.0224.$$

$SS_{\text{between}} \approx 182$ (this agrees with the value of SS_{between} in the SPSS output that is presented in Section 6.11 except for a small amount of rounding error).

6.7.3 ♦ Summarizing Information About Variability of Scores Within Groups: Computing MS_{within}

To summarize information about the variability of scores within each group we compute MS_{within}. For each group, for groups numbered from $i = 1, 2, \ldots, k$, we first find the sum of squared deviations of scores relative to each group mean, SS_i. The SS for scores within group i is found by taking this sum:

$$SS_i = \sum_{j=1}^{n_k} (Y_{ij} - M_i)^2. \tag{6.14}$$

That is, for each of the k groups, find the deviation of each individual score from the group mean; square and sum these deviations for all the scores in the group.

These within-group *SS* terms for Groups 1, 2, . . . , *k* are summed across the *k* groups to obtain the total SS_{within}:

$$SS_{\text{within}} = \sum_{i=1}^{k} SS_i = SS_1 + SS_2 + \cdots + SS_k. \tag{6.15}$$

For this dataset, we can find the *SS* term for Group 1 (for example) by taking the sum of the squared deviations of each individual score in Group 1 from the mean of Group 1, M_1. The values are shown for by-hand computations; it can be instructive to do this as a spreadsheet, entering the value of the group mean for each participant as a new variable and computing the deviation of each *score* from its group mean and the squared deviation for each participant.

$$SS_1 = (10 - 9.86)^2 + (10 - 9.86)^2 + (12 - 9.86)^2 + (11 - 9.86)^2 + (7 - 9.86)^2 + (7 - 9.86)^2 + (12 - 9.86)^2.$$

$$SS_1 = 26.86.$$

For the four groups of scores in the dataset in Figure 6.1, these are the values of *SS* for each group: $SS_1 = 26.86$, $SS_2 = 27.43$, $SS_3 = 41.71$, $SS_4 = 26.00$.

Thus, the total value of SS_{within} for this set of data is $SS_{\text{within}} = SS_1 + SS_2 + SS_3 + SS_4 = 26.86 + 27.43 + 41.71 + 26.00 = 122.00$.

We can also find SS_{total}; this involves taking the deviation of every individual score from the grand mean, squaring each deviation, and summing the squared deviations across all scores and all groups:

$$SS_{\text{total}} = \sum_{i=1}^{k} \sum_{j=1}^{n_k} (Y_{ij} - M_Y)^2. \tag{6.16}$$

The grand mean $M_Y = 13.68$. This *SS* term includes 28 squared deviations, one for each participant in the dataset, as follows:

$$SS_{\text{total}} = (10 - 13.68)^2 + (10 - 13.68)^2 + (12 - 13.68)^2 + \ldots + (18 - 13.68)^2 = 304.$$

Recall that Equation 6.10 showed that SS_{total} can be partitioned into *SS* terms that represent between- and within-group differences among scores; therefore, the SS_{between} and SS_{within} terms just calculated will sum to SS_{total}:

$$SS_{\text{total}} = SS_{\text{between}} + SS_{\text{within}}. \tag{6.17}$$

For these data, $SS_{\text{total}} = 304$, $SS_{\text{between}} = 182$, and $SS_{\text{within}} = 122$; so the sum of SS_{between} and SS_{within} equals SS_{total} (due to rounding error these values differ slightly from the values that appear on the SPSS output in Section 6.11).

6.7.4 ♦ The *F* Ratio: Comparing MS_{between} With MS_{within}

An *F* ratio is a ratio of two mean squares. A mean square is the ratio of a sum of squares to its degrees of freedom. Note that the formula for a sample variance back in Chapter 2 was just the sum of squares for a sample divided by the degrees of freedom that correspond to the number of independent deviations that were used to compute the *SS*. From Chapter 2, $\text{Var}(X) = SS/df$.

The *df* terms for the two *MS* terms in a one-way between-S ANOVA are based on *k*, the number of groups, and *N*, the total number of scores in the entire study (where $N = n_1 + n_2 + \ldots + n_k$). The between-group *SS* was obtained by summing the deviations of each of the *k* group means from the grand mean; only the first $k - 1$ of these deviations are free to vary, so the between-groups $df = k - 1$, where *k* is the number of groups.

$$df_{\text{between}} = k - 1. \tag{6.18}$$

In ANOVA, the mean square between groups is calculated by dividing SS_{between} by its degrees of freedom:

$$MS_{\text{between}} = \frac{SS_{\text{between}}}{k - 1}. \tag{6.19}$$

For the data in the hypothetical stress/anxiety study, $SS_{\text{between}} = 182$, $df_{\text{between}} = 4 - 1 = 3$, and $MS_{\text{between}} = 182/3 = 60.7$.

The *df* for each *SS* within-group term is given by $n - 1$, where *n* is the number of participants in each group. Thus, in this example, SS_1 had $n - 1$ or $df = 6$. When we form SS_{within}, we add up $SS_1 + SS_2 + \ldots + SS_k$. There are $(n - 1)$ *df* associated with each *SS* term, and there are *k* groups, so the total $df_{\text{within}} = k \times (n - 1)$. This can also be written as

$$df_{\text{within}} = N - k, \tag{6.20}$$

where $N =$ the total number of scores $= n_1 + n_2 + \ldots + n_k$ and *k* is the number of groups.

We obtain MS_{within} by dividing SS_{within} by its corresponding *df*:

$$MS_{\text{within}} = SS_{\text{within}}/(N - k). \tag{6.21}$$

For the hypothetical stress/anxiety data in Figure 6.1, $MS_{\text{within}} = 122/24 = 5.083$.

Finally, we can set up a test statistic for the null hypothesis $H_0\colon \mu_1 = \mu_2 = \ldots = \mu_k$ by taking the ratio of MS_{between} to MS_{within}:

$$F = \frac{MS_{\text{between}}}{MS_{\text{within}}}. \tag{6.22}$$

For the stress/anxiety data, $F = 60.702/5.083 = 11.94$. This *F* ratio is evaluated using the *F* distribution with $(k - 1)$ and $(N - k)$ *df*. For this dataset, $k = 4$ and $N = 28$, so *df* for the *F* ratio are 3 and 24.

An *F* distribution has a shape that differs from the normal or *t* distribution. Because an *F* is a ratio of two mean squares and *MS* cannot be less than 0, the minimum possible value of F is 0. On the other hand, there is no fixed upper limit for the values of F. Therefore, the distribution of F tends to be positively skewed, with a lower limit of 0, as in Figure 6.3. The reject region for significance tests with F ratios consists of only one tail (at the upper end of the distribution). The first table in Appendix C shows the critical values of F for $\alpha = .05$. The second and third tables in Appendix C provide critical values of F for $\alpha = .01$ and $\alpha = .001$. In the hypothetical study of stress and anxiety, the F ratio has df equal to 3 and 24. Using $\alpha = .05$, the critical value of F from the first table in Appendix C with $df = 3$ in the numerator (across the top of the table) and $df = 24$ in the denominator (along the left-hand side of the table) is 3.01. Thus, in this situation, the $\alpha = .05$ decision rule for evaluating statistical significance is to reject H_0 when values of $F > +3.01$ are obtained. A value of 3.01 cuts off the top 5% of the area in the right-hand tail of the F distribution with df equal to 3 and 24, as shown in Figure 6.3. The obtained $F = 11.94$ would therefore be judged statistically significant.

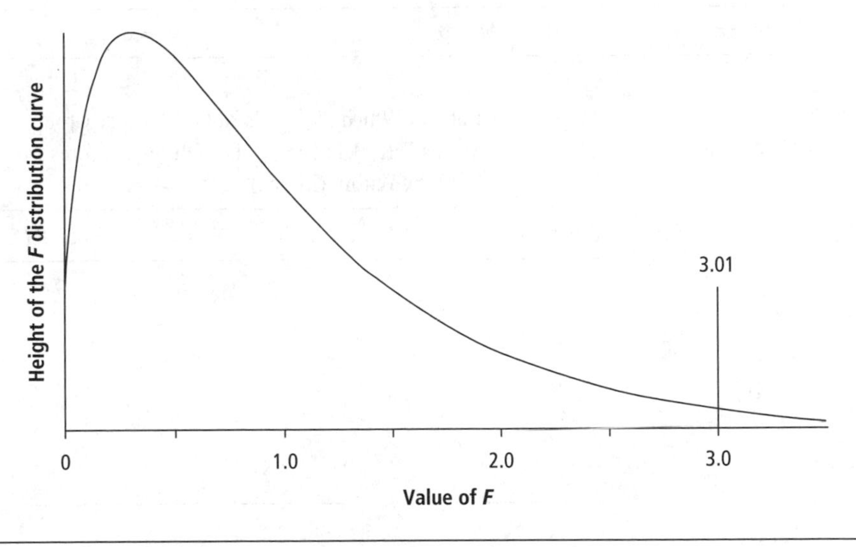

Figure 6.3 ♦ F Distribution With 3 and 24 df

6.7.5 ♦ Patterns of Scores Related to the Magnitudes of MS_{between} and MS_{within}

It is important to understand what information about pattern in the data is contained in these SS and MS terms. SS_{between} is a function of the distances among the group means ($M_1, M_2, \ldots, M_k$); the farther apart these group means are, the larger SS_{between} tends to be. Most researchers hope to find significant differences among groups, and therefore, they want SS_{between} (and F) to be relatively large. SS_{within} is the total of squared within-group deviations of scores from group means. SS_{within} would be 0 in the unlikely event that all scores within each group were equal to each other. The greater the variability of scores within each group, the larger the value of SS_{within}.

Consider the example shown in Table 6.3, which shows hypothetical data for which SS_{between} would be 0 (because all the group means are equal); however, SS_{within} is not 0 (because the scores vary within groups). Table 6.4 shows data for which SS_{between} is not 0 (group means differ) but SS_{within} is 0 (scores do not vary within groups). Table 6.5 shows data for which both $SS_{between}$ and SS_{within} are nonzero. Finally, Table 6.6 shows a pattern of scores for which both $SS_{between}$ and SS_{within} are 0.

Table 6.3 ♦ Data for Which $SS_{between}$ Is 0 (Because All the Group Means Are Equal), but SS_{within} Is Not 0 (Because Scores Vary Within Groups)

Group A1	*Group A2*	*Group A3*	*Group A4*
2	4	5	1
4	5	5	3
6	6	6	6
8	7	7	9
10	8	7	11
$M_1 = 6$	$M_2 = 6$	$M_3 = 6$	$M_4 = 6$

Table 6.4 ♦ Data for Which $SS_{between}$ Is Not 0 (Because Group Means Differ), but SS_{within} Is 0 (Because Scores Do Not Vary Within Groups)

Group B1	*Group B2*	*Group B3*
3	5	7
3	5	7
3	5	7
3	5	7
3	5	7
3	5	7
3	5	7

Table 6.5 ♦ Data for Which $SS_{between}$ and SS_{within} Are Both Nonzero

Group C1	*Group C2*	*Group C3*
1	8	22
3	11	18
5	9	19
3	5	21
1	5	17

Table 6.6 ♦ Data for Which Both SS_{within} and $SS_{between}$ Equal 0

Group D1	*Group D2*	*Group D3*
5	5	5
5	5	5
5	5	5
5	5	5

6.7.6 ◆ Expected Value of F When H_0 Is True

The population variances that are estimated by this ratio of sample mean squares (based on the algebra of expected mean squares[2]) are as follows:

$$F = \frac{MS_{\text{between}}}{MS_{\text{within}}} = \frac{n\sigma_\alpha^2 + \sigma_\varepsilon^2}{\sigma_\varepsilon^2}, \tag{6.23}$$

where σ_α^2 is the population variance of the alpha group effects, that is, the amount of variance in the Y scores that is associated with or predictable from the group membership variable, and σ_ε^2 is the population error variance, that is, the variance in scores that is due to all variables other than the group membership variable or manipulated treatment variable. Earlier, the null hypothesis for a one-way ANOVA was given as

$$H_0\text{: } \mu_1 = \mu_2 = \cdots = \mu_k.$$

An alternative way to state this null hypothesis is that all the α_i effects are equal to 0 (and therefore equal to each other): $\alpha_1 = \alpha_2 = \cdots \alpha_k$. Therefore, another form of the null hypothesis for ANOVA is as follows:

$$H_0\text{: } \sigma_\alpha^2 = 0. \tag{6.24}$$

It follows that if H_0 is true, then the expected value of the F ratio is close to 1. If F is much greater than 1, we have evidence that σ_α^2 may be larger than 0.

How is F distributed across thousands of samples? First of all, note that the MS terms in the F ratio must be positive (MS_{between} can be 0 in rare cases where all the group means are exactly equal; MS_{within} can be 0 in even rarer cases where all the scores within each group are equal; but because these are sums of squared terms neither MS can be negative).

The sums of squared independent normal variables have a chi-square distribution; thus, the distributions of each of the mean squares are chi-square variates. An F distribution is a ratio of two chi-square variates. Like chi-square, the graph of an F distribution has a lower tail that ends at 0, and it tends to be skewed with a long tail off to the right. (See Figure 6.3 for a graph of the distribution of F with $df = 3$ and 24.) We reject H_0 for large F values (which cut off the upper 5% in the right-hand tail). Thus, F is almost always treated as a one-tailed test. (It is possible to look at the lower tail of the F distribution to evaluate whether the obtained sample F is too small for it to be likely to have arisen by chance, but this is rarely done.)

Note that for the two-group situation, F is equivalent to t^2. Both F and t are ratios of an estimate of between-group differences to within-group differences; between-group differences are interpreted as being primarily due to the manipulated independent variable, while within-group variability is due to the effects of extraneous variables. Both t and F are interpretable as signal-to-noise ratios, where the "signal" is the effect of the manipulated independent variable; in most research situations, we hope that this term will be relatively large. The size of the signal is evaluated relative to the magnitude of "noise," the variability due to all other extraneous variables; in most research situations, we hope that this will be relatively small. Thus, in most research situations, we hope for values of t or F that are large enough for the null hypothesis to be rejected. A significant F can be interpreted as evidence that the between-groups independent variable had a detectable effect on the outcome variable. In rare circumstances, researchers hope to affirm the null hypothesis, that is, to demonstrate that the independent variable has no detectable effect, but this claim is actually quite difficult to prove.

If we obtain an F ratio large enough to reject H_0, what can we conclude? The alternative hypothesis is not that *all* group means differ from each other significantly but that there is at least one (and possibly more than one) significant difference between group means:

$$H_1\text{: } \mu_1 \neq \mu_2 \text{ and/or } \mu_2 \neq \mu_3 \text{ and/or } \mu_1 \neq \mu_3 \ldots. \tag{6.25}$$

A significant F tells us that there is at least one significant difference among group means; by itself, it does not tell us where that difference lies. It is necessary to do additional tests to identify the one or more significant differences (see Section 6.10).

6.7.7 ♦ Confidence Intervals (CIs) for Group Means

Once we have information about means and variances for each group, we can set up a CI around the mean for each group (using the formulas from Chapter 2) or a CI around any difference between a pair of group means (as described in Chapter 5). SPSS provides a plot of group means with error bars that represent 95% CIs (the SPSS menu selections to produce this graph appear in Figures 6.12 and 6.13, and the resulting graph appears in Figure 6.14; these appear at the end of the chapter along with the output from the Results section for the one-way ANOVA).

6.8 ♦ Effect-Size Index for One-Way Between-S ANOVA

By comparing the sizes of these *SS* terms that represent variability of scores between and within groups, we can make a summary statement about the comparative size of the effects of the independent and extraneous variables. In fact, the proportion of the total variability (SS_{total}) that is due to between-group differences is given by

$$\eta^2 = \frac{SS_{\text{between}}}{SS_{\text{total}}}. \tag{6.26}$$

In the context of a well-controlled experiment, these between-group differences in scores are presumably, primarily due to the manipulated independent variable; in a nonexperimental study that compares naturally occurring groups, this proportion of variance is reported only to describe the magnitudes of differences between groups, and it is not interpreted as evidence of causality. An eta squared (η^2) is an effect-size index given as a proportion of variance; if $\eta^2 = .50$, then 50% of the variance in the Y_{ij} scores is due to between-group differences. This is the same eta squared that was introduced in the previous chapter as an effect-size index for the independent samples t test; verbal labels that can be used to describe effect sizes are provided in Table 5.2. If the scores in a two-group t test are partitioned into components using the logic just described here and then summarized by creating sums of squares, the η^2 value obtained will be identical to the η^2 that was calculated from the t and df terms.

It is also possible to calculate eta squared from the F ratio and its df; this is useful when reading journal articles that report F tests without providing effect-size information:

$$\eta^2 = \frac{df_{\text{between}} \times F}{df_{\text{between}} \times F + df_{\text{within}}}. \tag{6.27}$$

An eta squared is interpreted as the proportion of variance in scores on the Y outcome variable that is predictable from group membership (i.e., from the score on X, the predictor variable). Suggested verbal labels for eta-squared effect sizes are given in Table 5.2.

An alternative effect-size measure sometimes used in ANOVA is called omega squared (ω^2) (see Hays, 1994). The eta-squared index describes the proportion of variance due to between-group differences in the sample, but it is a biased estimate of the proportion of variance that is theoretically due to differences among the populations. The ω^2 index is essentially a (downwardly) adjusted version of eta squared that provides a more conservative estimate of variance among population means; however, eta squared is more widely used in statistical power analysis and as an effect-size measure in the literature.

6.9 ♦ Statistical Power Analysis for One-Way Between-S ANOVA

Table 6.7 is an example of a statistical power table that can be used to make decisions about sample size when planning a one-way between-S ANOVA with $k = 3$ groups and $\alpha = .05$. Using Table 6.7, given the number of groups, the number of participants, the predetermined alpha level, and the anticipated population effect size estimated by eta squared, the researcher can look up the minimum n of participants per group that is required to obtain various levels of statistical power. The researcher needs to make an educated guess: How large an effect is expected in the planned study? If similar studies have been conducted in the past, the eta-squared values from past research can be used to estimate effect size; if not, the researcher may have to make a guess based on less exact information. The researcher chooses the alpha level (usually .05), calculates df_{between} (which equals $k - 1$, where k is the number of groups in the study), and decides on the desired level of statistical power (usually .80, or 80%). Using this information, the researcher can use the tables in Cohen (1988) or in Jaccard and Becker (2002) to look up the minimum sample size per group that is needed to achieve the power of 80%. For example, using Table 6.7, for an alpha level of .05, a study with three groups and $df_{\text{between}} = 2$, a population eta-squared value of .15, and a desired level of power of .80, the minimum number of participants required per group would be 19.

Java applets are available on the web for statistical power analysis; typically, if the user identifies a Java applet that is appropriate for the specific analysis (such as between-S one-way ANOVA) and enters information about alpha, the number of groups, population effect size, and desired level of power, the applet provides the minimum per-group sample size required to achieve the user-specified level of statistical power.

Table 6.7 ♦ Statistical Power for One-Way Between-S ANOVA With $k = 3$ Groups Using $\alpha = .05$

	Population Eta Squared									
Power	*.01*	*.03*	*.05*	*.07*	*.10*	*.15*	*.20*	*.25*	*.30*	*.35*
.10	22	8	5	4	3	2	2	2	—	—
.50	165	55	32	23	16	10	8	6	5	4
.70	255	84	50	35	24	16	11	9	7	6
.80	319	105	62	44	30	19	14	11	9	7
.90	417	137	81	57	39	25	18	14	11	9
.95	511	168	99	69	47	30	22	16	13	11
.99	708	232	137	96	65	41	29	22	18	14

SOURCE: From *Statistics for the Behavioral Sciences, 4th edition* by Jaccard/Becker, 2002. Reprinted with permission of Wadsworth, a division of Thomson Learning: www.thomsonrights.com. Fax 800-730-2215.

NOTE: Each table entry corresponds to the minimum *n* required in each group to obtain the level of statistical power shown.

6.10 ♦ Nature of Differences Among Group Means

6.10.1 ♦ Planned Contrasts

The idea behind **planned contrasts** is that the researcher identifies a limited number of comparisons between group means before looking at the data. The test statistic that is used for each comparison is essentially identical to a *t* ratio except that the denominator is usually based on the MS_{within} for the entire ANOVA, rather than just the variances for the two groups involved in the comparison. Sometimes an *F* is reported for the significance of each contrast, but *F* is equivalent to t^2 in situations where only two group means are compared or where a contrast has only 1 *df*.

For the means of Groups a and b, the null hypothesis for a simple contrast between M_a and M_b is as follows:

$$H_0\colon \mu_a = \mu_b$$

or

$$H_0\colon \mu_a - \mu_b = 0.$$

The test statistic can be in the form of a *t* test:

$$t = \frac{M_a - M_b}{\sqrt{MS_{within}/n}}, \tag{6.28}$$

where *n* is the number of cases within each group in the ANOVA. (If the *n*s are unequal across groups, then an average value of *n* is used; usually, this is the **harmonic**[3] **mean of the *n*s**.)

Note that this is essentially equivalent to an ordinary *t* test. In a *t* test, the measure of within-group variability is s_p^2; in a one-way ANOVA, information about within-group variability is contained in the term MS_{within}. In cases where an *F* is reported as a significance test for a contrast between a pair of group means, *F* is equivalent to t^2. The *df* for this *t* test equal $N - k$, where *N* is the total number of cases in the entire study and *k* is the number of groups.

When a researcher uses planned contrasts, it is possible to make other kinds of comparisons that may be more complex in form than a simple pairwise comparison of means. For instance, suppose that the researcher has a study in which there are four groups; Group 1 receives a placebo and Groups 2 to 4 all receive different antidepressant drugs. One hypothesis that may be of interest is whether the average depression score combined across the three drug groups is significantly lower than the mean depression score in Group 1, the group that received only a placebo.

The null hypothesis that corresponds to this comparison can be written in any of the following ways:

$$H_0: \mu_1 = \frac{\mu_2 + \mu_3 + \mu_4}{3}$$

or

$$H_0: \mu_1 - \frac{\mu_2 + \mu_3 + \mu_4}{3} = 0.$$

In words, this null hypothesis says that when we combine the means using certain weights (such as +1, −1/3, −1/3, and −1/3), the resulting composite is predicted to have a value of 0. This is equivalent to saying that the mean outcome averaged or combined across Groups 2 to 4 (which received three different types of medication) is equal to the mean outcome in Group 1 (which received no medication). Weights that define a contrast among group means are called contrast coefficients. Usually, contrast coefficients are constrained to sum to 0, and the coefficients themselves are usually given as integers for reasons of simplicity. If we multiply this set of contrast coefficients by 3 (to get rid of the fractions), we obtain the following set of contrast coefficients that can be used to see if the combined mean of Groups 2 to 4 differs from the mean of Group 1 (+3, −1, −1, −1). If we reverse the signs, we obtain the following set (−3, +1, +1, +1), which still corresponds to the same contrast. The *F* test for a contrast detects the magnitude, and not the direction, of differences among group means; therefore, it does not matter if the signs on a set of contrast coefficients are reversed.

In SPSS, users can select an option that allows them to enter a set of contrast coefficients to make many different types of comparisons among group means. To see some possible contrasts, imagine a situation in which there are $k = 5$ groups.

This set of contrast coefficients simply compares the means of Groups 1 and 5 (ignoring all the other groups): (+1, 0, 0, 0, −1).

This set of contrast coefficients compares the combined mean of Groups 1 to 4 with the mean of Group 5: (+1, +1, +1, +1, −4).

Contrast coefficients can be used to test for specific patterns, such as a linear trend (scores on the outcome variable might tend to increase linearly if Groups 1 through 5 correspond to equally spaced dosage levels of a drug): (−2, −1, 0, +1, +2).

A curvilinear trend can also be tested; for instance, the researcher might expect to find that the highest scores on the outcome variable occur at moderate dosage levels of the independent variable. If the five groups received five equally spaced different levels of background noise and the researcher predicts the best task performance at a moderate level of noise, an appropriate set of contrast coefficients would be (−1, 0, +2, 0, −1).

When a user specifies contrast coefficients, it is necessary to have one coefficient for each level or group in the ANOVA; if there are k groups, each contrast that is specified must include k coefficients. A user may specify more than one set of contrasts, although usually the number of contrasts does not exceed $k - 1$ (where k is the number of groups). The following simple guidelines are usually sufficient to understand what comparisons a given set of coefficients makes:

1. Groups with positive coefficients are compared with groups with negative coefficients; groups that have coefficients of 0 are omitted from such comparisons.
2. It does not matter which groups have positive versus negative coefficients; a difference can be detected by the contrast analysis whether or not the coefficients code for it in the direction of the difference.
3. For contrast coefficients that represent trends, if you draw a graph that shows how the contrast coefficients change as a function of group number (X), the line shows pictorially what type of trend the contrast coefficients will detect. Thus, if you plot the coefficients (−2, −1, 0, +1, +2) as a function of the group numbers 1, 2, 3, 4, 5, you can see that these coefficients test for a linear trend. The test will detect a linear trend whether it takes the form of an increase or a decrease in mean Y values across groups.

When a researcher uses more than one set of contrasts, he or she may want to know whether those contrasts are logically independent, uncorrelated, or **orthogonal**. There is an easy way to check whether the contrasts implied by two sets of contrast coefficients are orthogonal or independent. Essentially, to check for orthogonality, you just compute a (shortcut) version of a correlation between the two lists of coefficients. First, you list the coefficients for Contrasts 1 and 2 (make sure that each set of coefficients sums to 0, or this shortcut will not produce valid results).

Contrast 1:	(−2,	−1,	0,	+1,	+2)
Contrast 2:	(+1,	−1,	0,	0,	0)

You cross-multiply each pair of corresponding coefficients (i.e., the coefficients that are applied to the same group) and then sum these cross products. In this example, you get

Contrast 1:	(−2,	−1,	0,	+1,	+2)	
Contrast 2:	(+1,	−1,	0,	0,	0)	
Cross product:	(−2)(1)	(−1)(−1)	(0)(0)	(+1)(0)	(+2)(0)	
Sum of products:	−2	+1	+0	+0	+0	= −1

In this case, the result is −1. This means that the two contrasts above are not independent or orthogonal; some of the information that they contain about differences among means is redundant.

Consider a second example that illustrates a situation in which the two contrasts are orthogonal or independent:

Linear contrast:	−2	−1	0	+1	+2	
Curvilinear contrast:	−1	0	+2	0	−1	
Cross product:	(−2)(−1)	(−1)(0)	(0)(+2)	(+1)(0)	(+2)(−1)	
Sum of cross products:	+2	+0	+0	+0	−2	= 0

In this second example, the curvilinear contrast is orthogonal to the linear trend contrast. In a one-way ANOVA with k groups, it is possible to have up to $(k - 1)$ orthogonal contrasts. The preceding discussion of contrast coefficients assumed that the groups in the one-way ANOVA had equal *n*s. When the *n*s in the groups are unequal, it is necessary to adjust the values of the contrast coefficients so that they take unequal group size into account; this is done automatically in programs such as SPSS.

6.10.2 ♦ Post Hoc or "Protected" Tests

If the researcher wants to make all possible comparisons among groups or does not have a theoretical basis for choosing a limited number of comparisons before looking at the data, it is possible to use test procedures that limit the risk of Type I error by using "protected" tests. Protected tests use a more stringent criterion than would be used for planned contrasts in judging whether any given pair of means differs significantly. One method for setting a more stringent test criterion is the Bonferroni procedure, described in Chapter 3. The Bonferroni procedure requires that the data analyst use a more conservative (smaller) alpha level to judge whether each individual comparison between group means is statistically significant. For instance, in a one-way ANOVA with $k = 5$ groups, there are $k \times (k - 1)/2 = 10$ possible pairwise comparisons of group means. If the researcher wants to limit the overall experiment-wise risk of Type I error (EW_α) for the entire set of 10 comparisons to .05, one possible way to achieve this is to set the per-comparison alpha (PC_α) level for each individual significance test between means at α_{EW}/(number of **post hoc tests** to be performed). For example, if the experimenter wants an experiment-wise α of .05 when doing $k = 10$ post hoc comparisons between groups, the alpha level for each individual test would be set at EW_α/k, or .05/10, or .005 for each individual test. The t test could be calculated using the same formula as for an ordinary t test, but it would be judged significant only if its obtained p value were less than .005. The Bonferroni procedure is extremely conservative, and many researchers prefer less conservative methods of limiting the risk of Type I error.

Dozens of post hoc or protected tests have been developed to make comparisons among means in ANOVA that were not predicted in advance. Some of these procedures are intended for use with a limited number of comparisons; other tests are used to make all possible pairwise comparisons among group means. Some of the better known post hoc tests include the Scheffe, the Newman-Keuls, and the Tukey honestly significant difference (HSD). The Tukey HSD has become popular because it is moderately conservative and easy to apply; it can be used to perform all possible pairwise comparisons of means and is available as an option in widely used computer programs such as SPSS. The menu for the SPSS one-way ANOVA procedure includes the Tukey HSD test as one of many options for post hoc tests; SPSS calls it the Tukey procedure.

The Tukey HSD test (and several similar post hoc tests) uses a different method of limiting the risk of Type I error. Essentially, the Tukey HSD test uses exactly the same formula as a *t* ratio, but the resulting test ratio is labeled "*q*" rather than "*t*," to remind the user that it should be evaluated using a different sampling distribution. The Tukey HSD test and several related post hoc tests use critical values from a distribution called the "Studentized range statistic," and the test ratio is often denoted by the letter *q*:

$$q = \frac{M_a - M_b}{\sqrt{MS_{\text{within}}/n}}. \tag{6.29}$$

Values of the *q* ratio are compared with critical values from tables of the Studentized range statistic (see the table in Appendix F). The Studentized range statistic is essentially a modified version of the *t* distribution. Like *t*, its distribution depends on the numbers of subjects within groups, but the shape of this distribution also depends on *k*, the number of groups. As the number of groups (*k*) increases, the number of pairwise comparisons also increases. To protect against inflated risk of Type I error, larger differences between group means are required for rejection of the null hypothesis as *k* increases. The distribution of the Studentized range statistic is broader and flatter than the *t* distribution and has thicker tails; thus, when it is used to look up critical values of *q* that cut off the most extreme 5% of the area in the upper and lower tails, the critical values of *q* are larger than the corresponding critical values of *t*.

This formula for the Tukey HSD test could be applied by computing a *q* ratio for each pair of sample means and then checking to see if the obtained *q* for each comparison exceeded the critical value of *q* from the table of the Studentized range statistic. However, in practice, a computational shortcut is often preferred. The formula is rearranged so that the cutoff for judging a difference between groups to be statistically significant is given in terms of differences between means rather than in terms of values of a *q* ratio.

$$\text{HSD} = q_{\text{critical}} \times \sqrt{\frac{MS_{\text{within}}}{n}}. \tag{6.30}$$

Then, if the obtained difference between any pair of means (such as $M_a - M_b$) is greater in absolute value than this HSD, this difference between means is judged statistically significant.

An HSD criterion is computed by looking up the appropriate critical value of q, the Studentized range statistic, from a table of this distribution (see the table in Appendix F). The critical q value is a function of both n, the average number of subjects per group, and k, the number of groups in the overall one-way ANOVA. As in other test situations, most researchers use the critical value of q that corresponds to $\alpha = .05$, two-tailed. This critical q value obtained from the table is multiplied by the error term to yield HSD. This HSD is used as the criterion to judge each obtained difference between sample means. The researcher then computes the absolute value of the difference between each pair of group means $(M_1 - M_2)$, $(M_1 - M_3)$, and so forth. If the absolute value of a difference between group means exceeds the HSD value just calculated, then that pair of group means is judged to be significantly different.

When a Tukey HSD test is requested from SPSS, SPSS provides a summary table that shows all possible pairwise comparisons of group means and reports whether each of these comparisons is significant. If the overall F for the one-way ANOVA were statistically significant, it implies that there should be at least one significant contrast among group means. However, it is possible to have situations in which a significant overall F is followed by a set of post hoc tests that do not reveal any significant differences among means. This can happen because protected post hoc tests are somewhat more conservative, and thus require slightly larger between-group differences as a basis for a decision that differences are statistically significant, than the overall one-way ANOVA.

6.11 ♦ SPSS Output and Model Results

To run the one-way between-S ANOVA procedure in SPSS, make the following menu selections from the menu at the top of the data view worksheet, as shown in Figure 6.4: <Analyze> <Compare Means> <One-Way ANOVA>. This opens up the dialog window in Figure 6.5. Enter the name of one (or several) dependent variables into the window called Dependent List; enter the name of the categorical variable that provides group membership information into the window named Factor. For this example, additional windows were accessed by clicking on the buttons marked Post Hoc, Contrasts, and Options. The SPSS screens that correspond to this series of menu selections are shown in Figures 6.6 through 6.8.

From the menu of post hoc tests, this example used the one SPSS calls Tukey (this corresponds to the Tukey HSD test). To define a contrast that compared the mean of Group 1 (no stress) with the mean of the three stress treatment groups combined, these contrast coefficients were entered one at a time: +3, −1, −1, −1. From the list of options, "Descriptive" statistics and "Homogeneity of variance test" were selected by placing checks in the boxes next to the names of these tests.

The output for this one-way ANOVA is reported in Figure 6.9. The first panel provides descriptive information about each of the groups: mean, standard deviation, n, a 95% CI for the mean, and so forth. The second panel shows the results for the Levene test of the assumption of homogeneity of variance; this is an F ratio with $(k - 1)$ and $(N - k)$ df. The obtained F was not significant for this example; there was no evidence that the homogeneity of variance assumption had been violated. The third panel shows the ANOVA source table with the overall F; this was statistically significant, and this implies that there

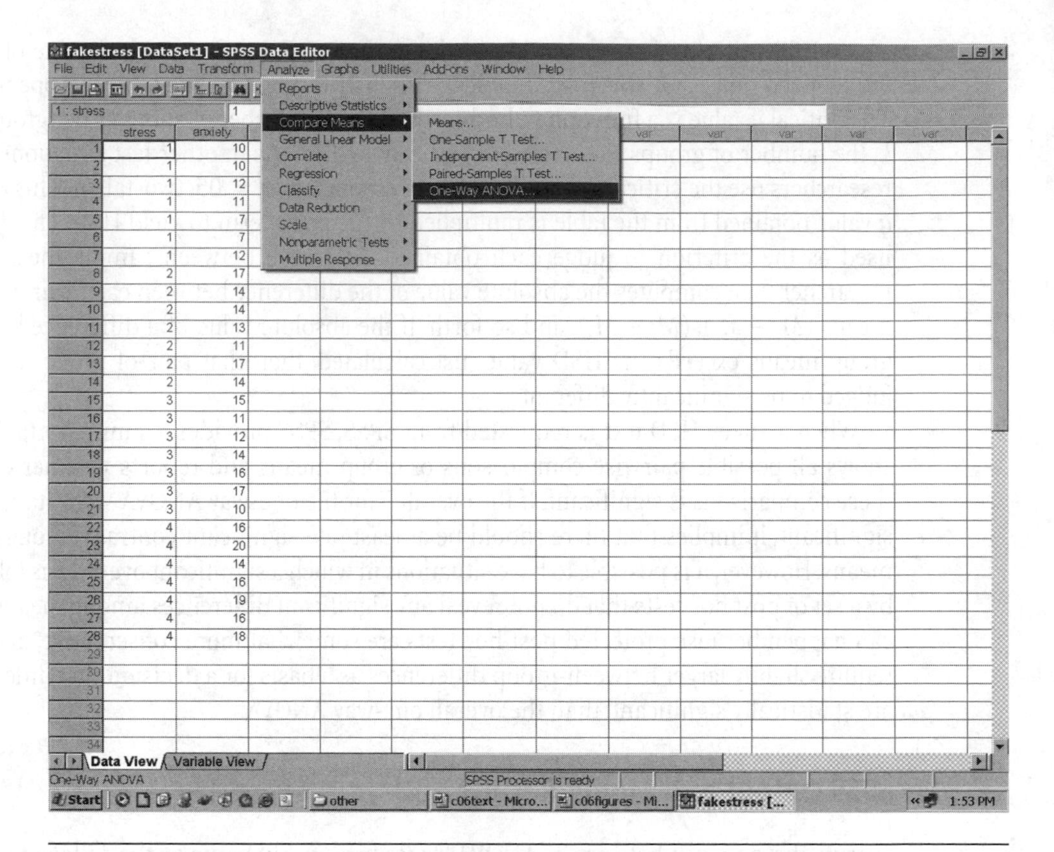

Figure 6.4 ♦ SPSS Menu Selections for One-Way Between-S ANOVA

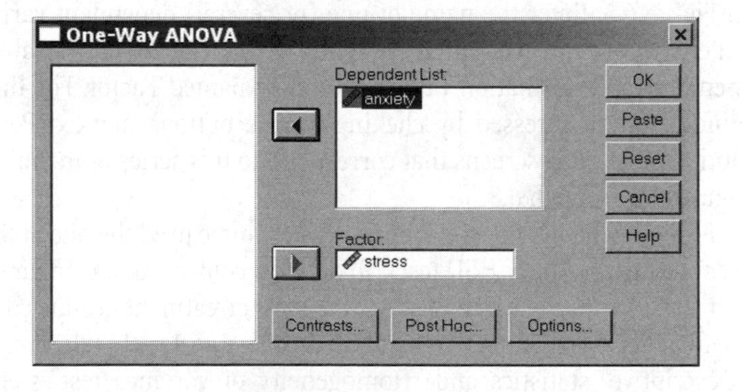

Figure 6.5 ♦ One-Way Between-S ANOVA Dialog Window

was at least one significant contrast between group means. In practice, a researcher would not report both planned contrasts and post hoc tests; however, both were presented for this example to show how they are obtained and reported.

Figure 6.10 shows the output for the planned contrast that was specified by entering these contrast coefficients: (+3, −1, −1, −1). These contrast coefficients correspond to a

One-Way ANOVA: Post Hoc Multiple Comparisons

Equal Variances Assumed

LSD
Bonferroni
Sidak
Scheffe
R-E-G-W F
R-E-G-W Q
S-N-K
Tukey
Tukey's-b
Duncan
Hochberg's GT2
Gabriel
Waller-Duncan
Type I/Type II Error Ratio: 100
Dunnett
Control Category: Last
Test
2-sided < Control > Control

Equal Variances Not Assumed

Tamhane's T2
Dunnett's T3
Games-Howell
Dunnett's C

Significance level: .05

Continue Cancel Help

Figure 6.6 ♦ SPSS Dialog Window, One-Way ANOVA, Post Hoc Test Menu

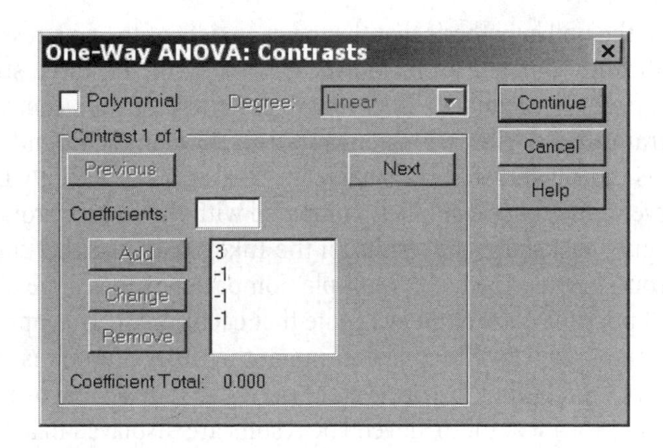

Figure 6.7 ♦ Specification of a Planned Contrast

NOTE: Null hypothesis about weighted linear composite of means that is represented by this set of contrast coefficients:

$$(+3)\,\mu_1 + (-1)\,\mu_2 + (-1)\,\mu_3 + (-1)\,\mu_4 = 0$$

or

$$\mu_1 - \frac{\mu_2 + \mu_3 + \mu_4}{3} = 0$$

or

$$\mu_1 = \frac{\mu_2 + \mu_3 + \mu_4}{3}$$

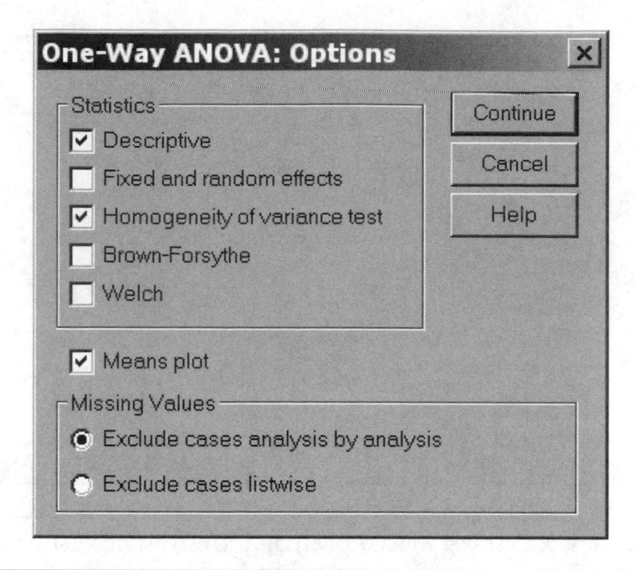

Figure 6.8 ♦ SPSS Dialog Window for One-Way ANOVA: Options

test of the null hypothesis that the mean anxiety of the no stress group (Group 1) was not significantly different from the mean anxiety of the three stress intervention groups (Groups 2–4) combined. SPSS reported a *t* test for this contrast (some textbooks and programs use an *F* test). This *t* was statistically significant, and examination of the group means indicated that the mean anxiety level was significantly higher for the three stress intervention groups combined, compared with the control group.

Figure 6.11 shows the results for the Tukey HSD tests that compared all possible pairs of group means. The table "Multiple Comparisons" gives the difference between means for all possible pairs of means (note that each comparison appears twice; that is, Group a is compared with Group b, and in another line, Group b is compared with Group a). Examination of the "sig" or *p* values indicates that several of the pairwise comparisons were significant at the .05 level. The results are displayed in a more easily readable form in the last panel under the heading "Homogeneous Subsets." Each subset consists of group means that were not significantly different from each other using the Tukey test. The no stress group was in a subset by itself; in other words, it had significantly lower mean anxiety than any of the three stress intervention groups. The second subset consisted of the stress role play and mental arithmetic groups, which did not differ significantly in anxiety. The third subset consisted of the mental arithmetic and mock job interview groups.

Note that it is possible for a group to belong to more than one subset; the anxiety score for the mental arithmetic group was not significantly different from the stress role play or the mock job interview groups. However, because the stress role play group differed significantly from the mock job interview group, these three groups did not form one subset.

Note also that it is possible for all the Tukey HSD comparisons to be nonsignificant even when the overall *F* for the one-way ANOVA is statistically significant. This can

Descriptives

anxiety

	N	Mean	Std. Deviation	Std. Error	95% Confidence Interval for Mean		Minimum	Maximum
					Lower Bound	Upper Bound		
none	7	9.86	2.116	.800	7.90	11.81	7	12
mental arithmetic	7	14.29	2.138	.808	12.31	16.26	11	17
stress role play	7	13.57	2.637	.997	11.13	16.01	10	17
mock job interview	7	17.00	2.082	.787	15.07	18.93	14	20
Total	28	13.68	3.356	.634	12.38	14.98	7	20

Test of Homogeneity of Variances

anxiety

Levene Statistic	df1	df2	Sig.
.453	3	24	.718

ANOVA

anxiety

	Sum of Squares	df	Mean Square	F	Sig.
Between Groups	182.107	3	60.702	11.941	.000
Within Groups	122.000	24	5.083		
Total	304.107	27			

Figure 6.9 ♦ SPSS Output for One-Way ANOVA

Contrast Coefficients

Contrast	STRESS none	mental arithmetic	stress role play	mock job interview
1	3	-1	-1	-1

Contrast Tests

		Contrast	Value of Contrast	Std. Error	t	df	Sig. (2-tailed)
ANXIETY	Assume equal variances	1	-15.29	2.952	-5.178	24	.000
	Does not assume equal	1	-15.29	2.832	-5.397	11.054	.000

Figure 6.10 ♦ SPSS Output for Planned Contrasts

Multiple Comparisons

Dependent Variable: ANXIETY
Tukey HSD

(I) STRESS	(J) STRESS	Mean Difference (I-J)	Std. Error	Sig.	95% Confidence Interval Lower Bound	Upper Bound
none	mental arithmetic	-4.43 *	1.205	.006	-7.75	-1.10
	stress role play	-3.71 *	1.205	.025	-7.04	-.39
	mock job interview	-7.14 *	1.205	.000	-10.47	-3.82
mental arithmetic	none	4.43 *	1.205	.006	1.10	7.75
	stress role play	.71	1.205	.933	-2.61	4.04
	mock job interview	-2.71	1.205	.138	-6.04	.61
stress role play	none	3.71 *	1.205	.025	.39	7.04
	mental arithmetic	-.71	1.205	.933	-4.04	2.61
	mock job interview	-3.43 *	1.205	.042	-6.75	-.10
mock job interview	none	7.14 *	1.205	.000	3.82	10.47
	mental arithmetic	2.71	1.205	.138	-.61	6.04
	stress role play	3.43 *	1.205	.042	.10	6.75

*. The mean difference is significant at the .05 level.

Homogeneous Subsets

ANXIETY

Tukey HSD [a]

STRESS	N	Subset for alpha = .05 1	2	3
none	7	9.86		
stress role play	7		13.57	
mental arithmetic	7		14.29	14.29
mock job interview	7			17.00
Sig.		1.000	.933	.138

Means for groups in homogeneous subsets are displayed.

a. Uses Harmonic Mean Sample Size = 7.000.

Figure 6.11 ♦ SPSS Output for Post Hoc Test (Tukey HSD)

happen because the Tukey HSD test requires a slightly larger difference between means to achieve significance.

In this imaginary example, as in some research studies, the outcome measure (anxiety) is not a standardized test for which we have norms. The numbers by themselves do not tell us whether the mock job interview participants were moderately anxious or twitching, stuttering wrecks. Studies that use standardized measures can make comparisons to test norms to help readers understand whether the group differences were large enough to be of clinical or practical importance. Alternatively, qualitative data about the behavior of participants can also help readers understand how substantial the group differences were.

SPSS one-way ANOVA does not provide an effect-size measure, but this can easily be calculated by hand. In this case, eta squared is found by taking the ratio $SS_{\text{between}}/SS_{\text{total}}$ from the ANOVA source table: $\eta^2 = .60$.

To obtain a graphic summary of the group means that shows the 95% CI for each mean, use the menu selections shown in Figures 6.12 and 6.13 to set up a graph with error bars; this graph appears in Figure 6.14. Generally, the group means are reported either in a summary table (as in Table 6.8) or as a graph (as in Figure 6.14); it is not necessary to include both because they provide the same information.

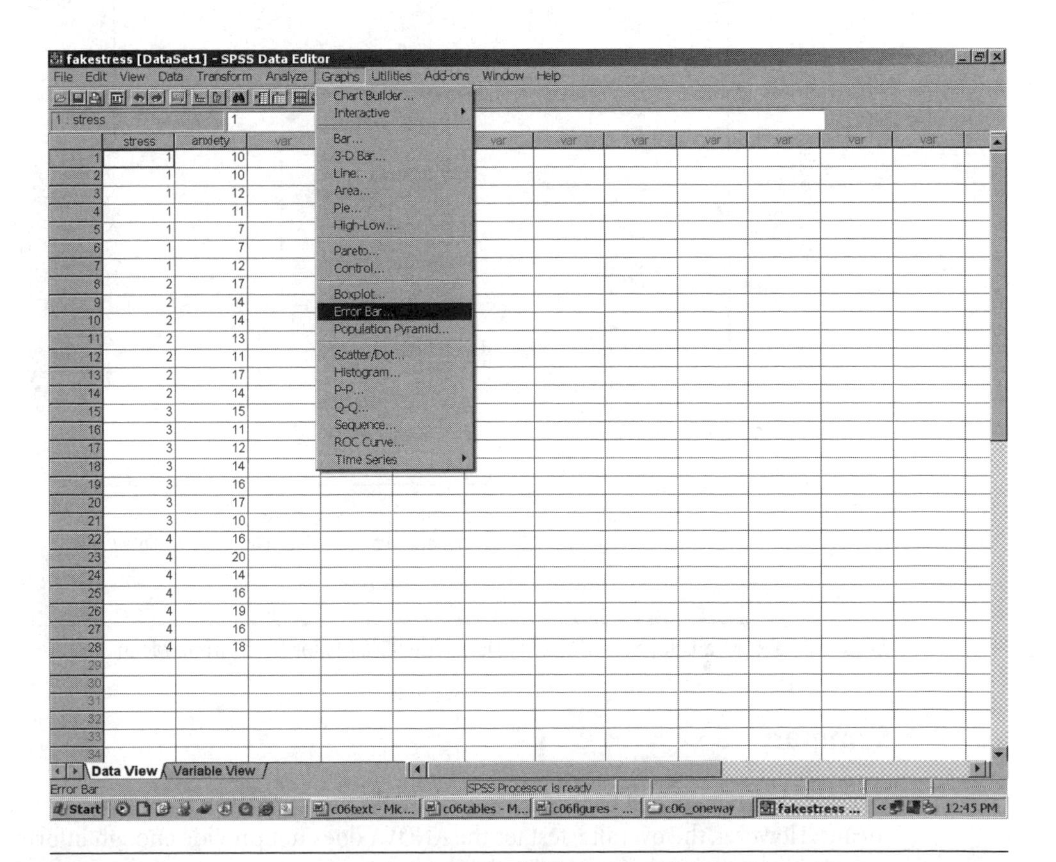

Figure 6.12 ♦ SPSS Menu Selections for Graph of Cell Means With Error Bars or CIs

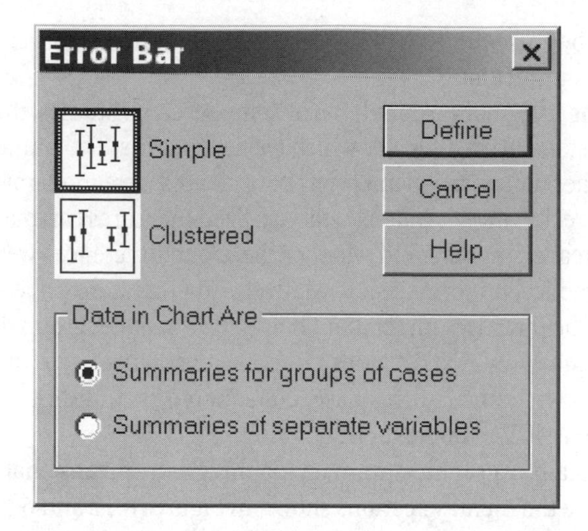

Figure 6.13 ♦ Initial SPSS Dialog Window for Error Bar Graph

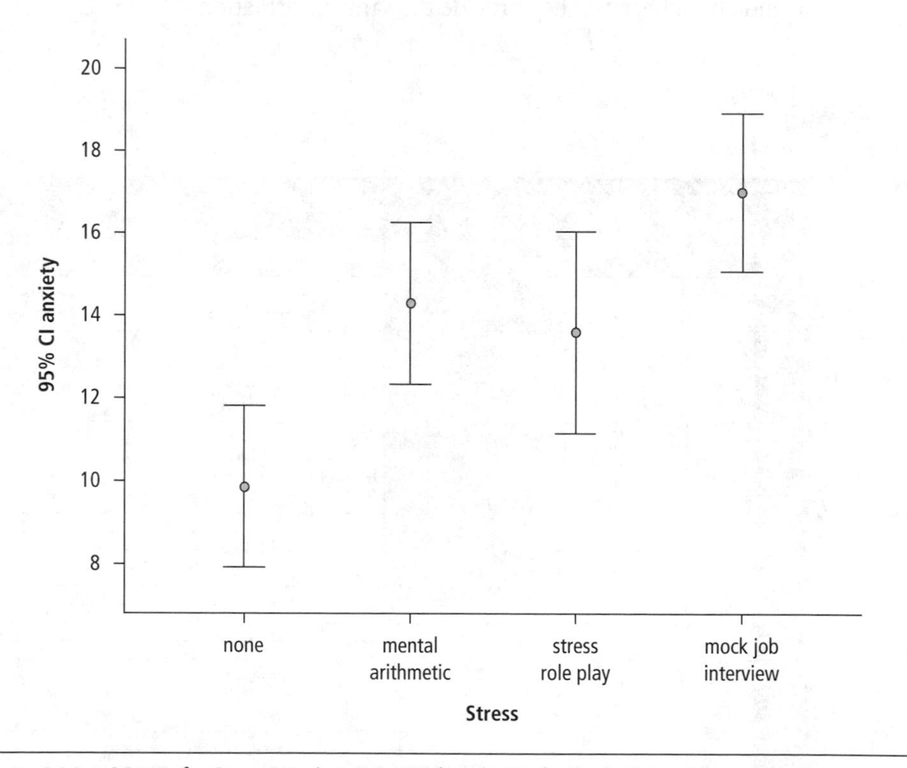

Figure 6.14 ♦ Means for Stress/Anxiety Data With 95% CIs for Each Group Mean

6.12 ♦ Summary

One-way between-S ANOVA provides a method for comparison of more than two group means. However, the overall *F* test for the ANOVA does not provide enough information to completely describe the pattern in the data. It is often necessary to perform additional

Table 6.8 ♦ Mean Anxiety Scores Across Types of Stress

Type of Stress	*Group 1, No Stress*	*Group 2, Role Play*	*Group 3, Arithmetic*	*Group 4, Job Interview*
M	9.86	14.29	13.57	17.00
(*SD*)	(2.12)	(2.14)	(2.64)	(2.08)
n	7	7	7	7

Results

A one-way between-S ANOVA was done to compare the mean scores on an anxiety scale (0 = not at all anxious, 20 = extremely anxious) for participants who were randomly assigned to one of four groups: Group 1 = control group/no stress; Group 2 = mental arithmetic; Group 3 = stressful role play; Group 4 = mock job interview. Examination of a histogram of anxiety scores indicated that the scores were approximately normally distributed with no extreme outliers. Prior to the analysis, the Levene test for homogeneity of variance was used to examine whether there were serious violations of the assumption of homogeneity of variance across groups, but no significant violation was found: $F(3, 24) = .718$, $p = .72$.

The overall F for the one-way ANOVA was statistically significant, $F(3, 24) = 11.94$, $p < .001$. This corresponded to an effect size of $\eta^2 = .60$; that is, about 60% of the variance in anxiety scores was predictable from the type of stress intervention. This is a large effect. The means and standard deviations for the four groups are shown in Table 6.8.

One planned contrast (comparing the mean of Group 1, no stress, to the combined means of Groups 2 to 4, the stress intervention groups) was performed. This contrast was tested using $\alpha = .05$, two-tailed; the t test that assumed equal variances was used because the homogeneity of variance assumption was not violated. For this contrast, $t(24) = -5.18$, $p < .001$. The mean anxiety score for the no stress group ($M = 9.86$) was significantly lower than the mean anxiety score for the three combined stress intervention groups ($M = 14.95$).

In addition, all possible pairwise comparisons were made using the Tukey HSD test. Based on this test (using $\alpha = .05$), it was found that the no stress group scored significantly lower on anxiety than all three stress intervention groups. The stressful role play ($M = 13.57$) was significantly less anxiety producing than the mock job interview ($M = 17.00$). The mental arithmetic task produced a mean level of anxiety ($M = 14.29$) that was intermediate between the other stress conditions, and it did not differ significantly from either the stress role play or the mock job interview. Overall, the mock job interview produced the highest levels of anxiety. Figure 6.14 shows a 95% CI around each group mean.

comparisons among specific group means to provide a complete description of the pattern of differences among group means. These can be **a priori** or **planned contrast comparisons** (if a limited number of differences were predicted in advance). If the researcher did not make a limited number of predictions in advance about differences between group means, then the researcher may use protected or post hoc tests to do any follow-up comparisons; the Bonferroni procedure and the Tukey HSD test were described here, but many other post hoc procedures are available.

The most important concept from this chapter is the idea that a score can be divided into components (one part that is related to group membership or treatment effects and a second part that is due to the effects of all other "extraneous" variables that uniquely influence individual participants). Information about the relative sizes of these components can be summarized across all the participants in a study by computing the sum of squared deviations (*SS*) for the between-group and within-group deviations. Based on the *SS* values, it is possible to compute an effect-size estimate (η^2) that describes the proportion of variance predictable from group membership (or treatment variables) in the study. Researchers usually hope to design their studies in a manner that makes the proportion of explained variance reasonably high and that produces statistically significant differences among group means. However, researchers have to keep in mind that the proportion of variance due to group differences in the artificial world of research may not correspond to the "true" strength of the influence of the variable out in the "real world." In experiments, we create an artificial world by holding some variables constant and by manipulating the treatment variable; in nonexperimental research, we create an artificial world through our selection of participants and measures. Thus, results of our research should be interpreted with caution.

Notes

1. If the groups are not equivalent—that is, if the groups have different means on participant characteristics that may be predictive of the outcome variable—**ANCOVA** (see Chapter 15) may be used to correct for the nonequivalence by statistical control. However, it is preferable to identify and correct any differences in the composition of groups prior to data collection, if possible.

2. When the formal algebra of expected mean squares is applied to work out the population variances that are estimated by the sample values of MS_{within} and $MS_{between}$, the following results are obtained if the group *n*s are equal (Winer, Brown, & Michels, 1991):

$MS_{within} = \sigma^2_\varepsilon$, the population variance due to error (all other extraneous variables)

$MS_{between} = n\,\sigma^2_\alpha + \sigma^2_\varepsilon$, where σ^2_α is the population variance due to the effects of membership in the naturally occurring group or to the manipulated treatment variable. Thus, when we look at *F* ratio estimates, we obtain

$$F = \frac{n\sigma^2_\alpha + \sigma^2_\varepsilon}{\sigma^2_\varepsilon}.$$

The null hypothesis (that all the population means are equal) can be stated in a different form, which says the variance of the population means is zero:

$$H_0\text{: } \sigma^2_\alpha = 0.$$

If H_0 is true and $\sigma^2_\alpha = 0$, then the expected value of the *F* ratio is 1. Values of *F* that are substantially larger than 1, and that exceed the critical value of *F* from tables of the *F* distribution, are taken as evidence that this null hypothesis may be false.

3. *Calculating the harmonic mean of cell or group* ns: When the groups have different *n*s, the harmonic mean *H* of a set of unequal *n*s is obtained by using the following equation:

$$\frac{1}{H} = \frac{1}{k} \times \sum_{i=1}^{k} \frac{1}{n_i},$$

$$H = \frac{1}{(1/k) \times \sum_{i=1}^{k} (1/n_i)}.$$

That is, sum the inverses of the *ns* of the groups, divide this sum by the number of groups, and then take the inverse of that result to obtain the harmonic mean (*H*) of the *ns*.

Comprehension Questions

1. A nonexperimental study was done to assess the impact of the accident at the Three Mile Island (TMI) nuclear power plant on nearby residents (Baum, Gatchel, & Schaeffer, 1983). Data were collected from residents of the following four areas:

   Group 1: Three Mile Island, where a nuclear accident occurred ($n = 38$)

   Group 2: Frederick, with no nuclear power plant nearby ($n = 27$)

   Group 3: Dickerson, with an undamaged coal power plant nearby ($n = 24$)

   Group 4: Oyster Creek, with an undamaged nuclear power plant nearby ($n = 32$)

   Several different measures of stress were taken for people in these four groups. The researchers hypothesized that residents who lived near TMI (Group 1) would score higher on a wide variety of stress measures than people who lived in the other three areas included as comparisons. One-way ANOVA was performed to assess differences among these four groups on each outcome. Selected results are reported below for you to discuss and interpret. Here are results for two of their outcome measures: Stress (total reported stress symptoms) and Depression (score on Beck Depression Inventory). Each cell lists the mean, followed by the standard deviation in parentheses.

Outcome Variable	*TMI Nuclear Plant With Accident*	*No Nuclear Plant*	*Undamaged Coal Plant*	*Undamaged Nuclear Plant*	*F(3, 117)*
Stress symptoms	25.97 (21.0)	14.54 (11.5)	16.63 (11.8)	16.16 (13.5)	3.827
Depression	6.00 (6.5)	3.64 (3.3)	3.54 (3.6)	3.50 (4.2)	2.104

a. Write a Results section in which you report whether or not these overall differences were statistically significant for each of these two outcome variables (using $\alpha = .05$). You will need to look up the critical value for F, and in this instance, you will not be able to include an exact p value. Include an eta-squared effect-size index for each of the F ratios (you can calculate this by hand from the information given in the table). Be sure to state the nature of the differences: Did the TMI group score higher or lower on these stress measures relative to the other groups?
b. Would your conclusions change if you used $\alpha = .01$ instead of $\alpha = .05$ as your criterion for statistical significance?
c. Name a follow-up test that could be done to assess whether all possible pairwise comparisons of group means were significant.
d. Write out the contrast coefficients to test whether the mean for Group 1 (people who lived near TMI) differed from the average for the other three comparison groups.
e. Here is some additional information about scores on the Beck Depression Inventory. For purposes of clinical diagnosis, Beck, Steer, and Brown (1996) suggested the following cutoffs:

   0–13: Minimal depression

   14–19: Mild depression

   20–28: Moderate depression

   29–63: Severe depression

   In light of this additional information, what would you add to your discussion of the outcomes for depression in the TMI group versus groups from other regions? (Did the TMI accident make people severely depressed?)

2. Sigall and Ostrove (1975) did an experiment to assess whether the physical attractiveness of a defendant on trial for a crime had an effect on the severity of the sentence given in mock jury trials. Each of the participants in this study was randomly assigned to one of the following three treatment groups; every participant received a packet that described a burglary and gave background information about the accused person. The three treatment groups differed in the type of information they were given about the accused person's appearance. Members of Group 1 were shown a photograph of an attractive person; members of Group 2 were shown a photograph of an unattractive person; members of Group 3 saw no photograph. Some of their results are described here. Each participant was asked to assign a sentence (in years) to the accused person; the researchers predicted that more attractive persons would receive shorter sentences.
   a. Prior to assessment of the outcome, the researchers did a manipulation check. Members of Groups 1 and 2 rated the attractiveness (on a 1 to 9 scale, with 9 being the most attractive) of the person in the photo. They reported that for the attractive photo, $M = 7.53$; for the unattractive photo, $M = 3.20$, $F(1, 108) = 184.29$. Was this difference statistically significant (using $\alpha = .05$)?
   b. What was the effect size for the difference in (2a)?
   c. Was their attempt to manipulate perceived attractiveness successful?

d. Why does the F ratio in (2a) have just $df=1$ in the numerator?

e. The mean length of sentence given in the three groups was as follows:

   Group 1: Attractive photo, $M = 2.80$

   Group 2: Unattractive photo, $M = 5.20$

   Group 3: No photo, $M = 5.10$

   They did not report a single overall F comparing all three groups; instead, they reported selected pairwise comparisons. For Group 1 versus Group 2, $F(1, 108) = 6.60, p < .025$.

   Was this difference statistically significant? If they had done an overall F to assess the significance of differences of means among all three groups, do you think this overall F would have been statistically significant?

f. Was the difference in mean length of sentence in part (2e) in the predicted direction?

g. Calculate and interpret an effect-size estimate for this obtained F.

h. What additional information would you need about this data to do a Tukey honestly significant difference test to see whether Groups 2 and 3, and 1 and 3, differed significantly?

3. Suppose that a researcher has conducted a simple experiment to assess the effect of background noise level on verbal learning. The manipulated independent level is the level of white noise in the room (Group 1 = low level = 65 dB; Group 2 = high level = 70 dB). (Here are some approximate reference values for decibel noise levels: 45 dB, whispered conversation; 65 dB, normal conversation; 80 dB, vacuum cleaner; 90 dB, chain saw or jack hammer; 120 dB, rock music played very loudly). The outcome measure is number of syllables correctly recalled from a 20-item list of nonsense syllables. Participants ranged in age from 17 to 70 and had widely varying levels of hearing acuity; some of them habitually studied in quiet places and others preferred to study with the television or radio turned on. There were 5 participants in each of the two groups. The researcher found no significant difference in mean recall scores between these groups.

   a. Describe three specific changes to the design of this noise/learning study that would be likely to increase the size of the t ratio (and, therefore, make it more likely that the researcher would find a significant effect).

   b. Also, suppose that the researcher has reason to suspect that there is a curvilinear relation between noise level and task performance. What change would this require in the research design?

4. Suppose that Kim is a participant in a study that compares several coaching methods to see how they affect math Scholastic Aptitude Test (SAT) scores. The grand mean of math SAT scores for all participants (M_Y) in the study is 550. The group that Kim participated in had a mean math SAT score of 565. Kim's individual score on the math SAT was 610.

   a. What was the estimated residual component (ε_{ij}) of Kim's score, that is, the part of Kim's score that was not related to the coaching method? (Both parts of this question call for specific numerical values as answers.)

   b. What was the "effect" (α_i) component of Kim's score?

5. What pattern in grouped data would make $SS_{\text{within}} = 0$? What pattern within data would make $SS_{\text{between}} = 0$?

6. Assuming that a researcher hopes to demonstrate that a treatment or group membership variable makes a significant difference in outcomes, which term does the researcher hope will be larger, MS_{between} or MS_{within}? Why?

7. Explain the following equation:

$$(Y_{ij} - M_Y) = (Y_{ij} - M_j) + (M_j - M_Y).$$

What do we gain by breaking the $(Y_{ij} - M_Y)$ deviation into two separate components, and what do each of these components represent? Which of the terms on the right-hand side of the equation do researchers typically hope will be large, and why?

8. What is H_0 for a one-way ANOVA? If H_0 is rejected, does that imply that each mean is significantly different from every other mean?

9. What information do you need to decide on a sample size that will provide adequate statistical power?

10. In the equation $\alpha_j = (M_j - M_Y)$, what do we call the α_j term?

11. If there is an overall significant F in a one-way ANOVA, can we conclude that the group membership or treatment variable caused the observed differences in the group means? Why or why not?

12. How can eta squared be calculated from the SS values in an ANOVA summary table?

13. Which of these types of tests is more conservative: planned contrasts or post hoc/protected tests?

14. Name two common post hoc procedures for the comparison of means in an ANOVA.

CHAPTER 7

Bivariate Pearson Correlation

7.1 ◆ Research Situations Where Pearson r Is Used

Pearson r is typically used to describe the strength of the linear relationship between two quantitative variables. Often, these two variables are designated X (predictor) and Y (outcome). Pearson r has values that range from −1.00 to +1.00. The sign of r provides information about the direction of the relationship between X and Y. A positive correlation indicates that as scores on X increase, scores on Y also tend to increase; a negative correlation indicates that as scores on X increase, scores on Y tend to decrease; and a correlation near 0 indicates that as scores on X increase, scores on Y neither increase nor decrease in a linear manner. As an example, consider the hypothetical data in Figure 7.1. Suppose that a time-share sales agency pays each employee a base wage of \$10,000 per year and, in addition, a commission of \$1,500 for each sale that the employee completes. An employee who makes zero sales would earn \$10,000; an employee who sold 4 time-shares would make \$10,000 + \$1,500 × 4 = \$16,000. In other words, for each one-unit increase in the number of time-shares sold (X), there is a \$1,500 increase in wages. Figure 7.1 illustrates a perfect linear relationship between number of units sold (X_1) and wages in dollars (Y_1).

The absolute magnitude of Pearson r provides information about the strength of the linear association between scores on X and Y. For values of r close to 0, there is no linear association between X and Y. When $r = +1.00$, there is a **perfect positive linear association**; when $r = -1.00$, there is a **perfect negative linear association**. Intermediate values of r correspond to intermediate strength of the relationship. Figures 7.2 through 7.5 show examples of data for which the correlations are $r = +.75$, $r = +.50$, $r = +.23$, and $r = .00$.

Pearson r is a standardized or unit-free index of the strength of the linear relationship between two variables. No matter what units are used to express the scores on the X and Y variables, the possible values of Pearson r range from −1 (a perfect negative linear relationship) to +1 (a perfect positive linear relationship). Consider, for example, a correlation between height and weight. Height could be measured in inches, centimeters, or feet; weight could be measured in ounces, pounds, or kilograms. When we correlate scores on height and weight for a given sample of people, the correlation has the same value no matter which of these units are used to measure height and weight. This happens because

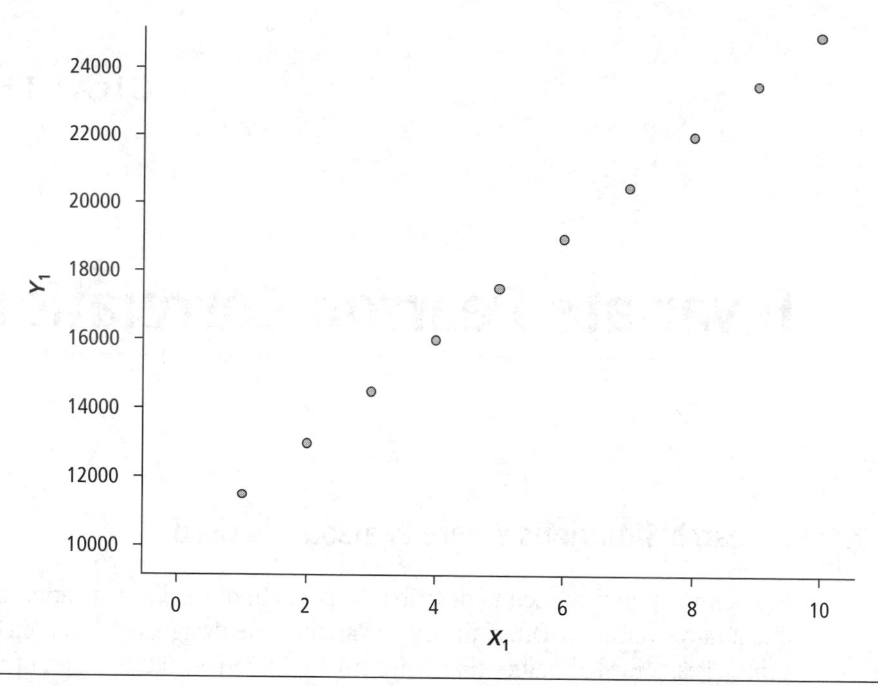

Figure 7.1 ♦ Scatter Plot for a Perfect Linear Relationship, $r = +1.00$ ($Y_1 = 10{,}000 + 1{,}500 \times X_1$; e.g., for $X_1 = 4$, $Y_1 = 16{,}000$)

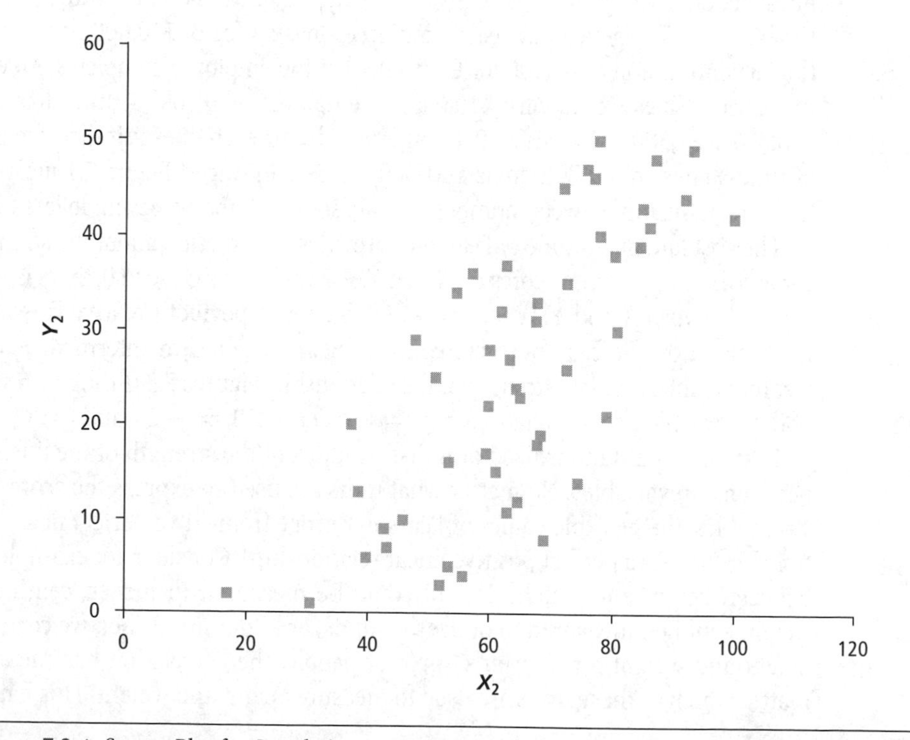

Figure 7.2 ♦ Scatter Plot for Correlation $r = +.75$

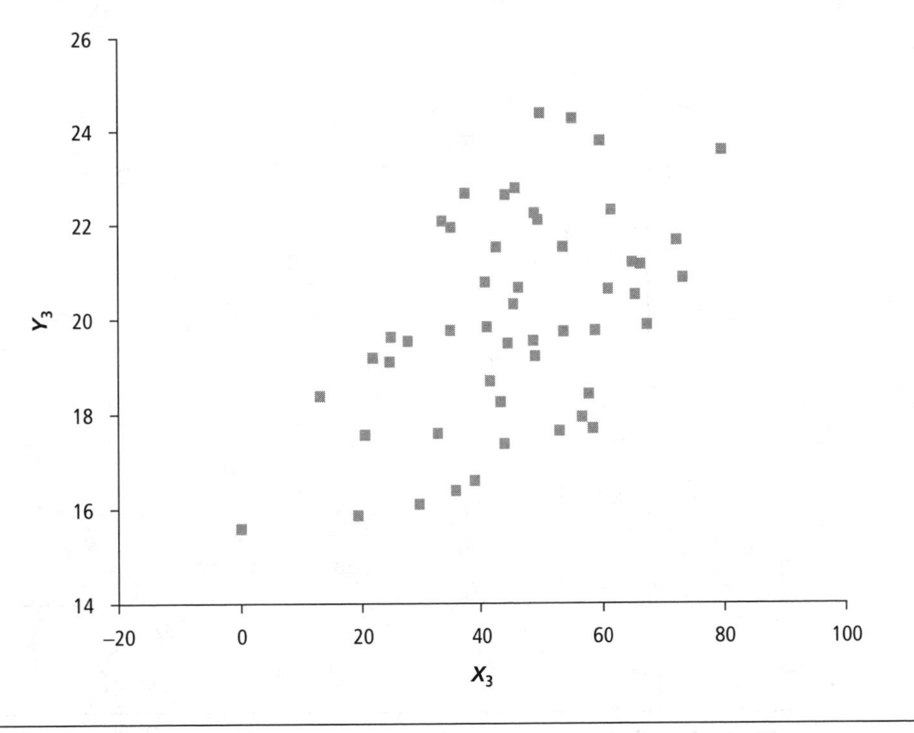

Figure 7.3 ♦ Scatter Plot for Correlation $r = +.50$

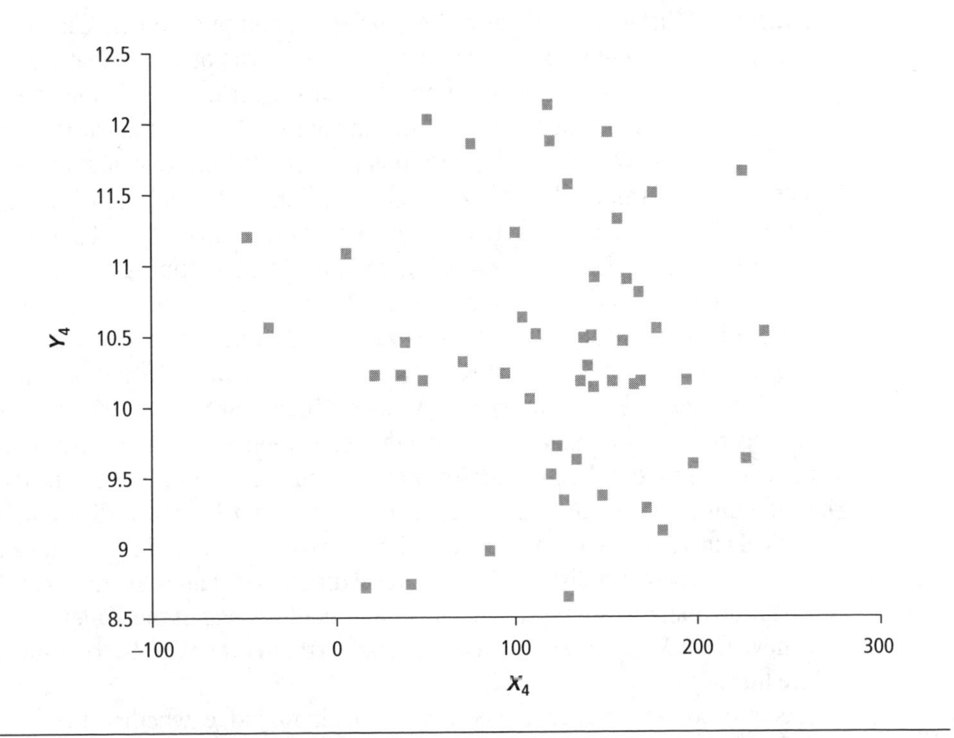

Figure 7.4 ♦ Scatter Plot for Correlation $r = +.23$

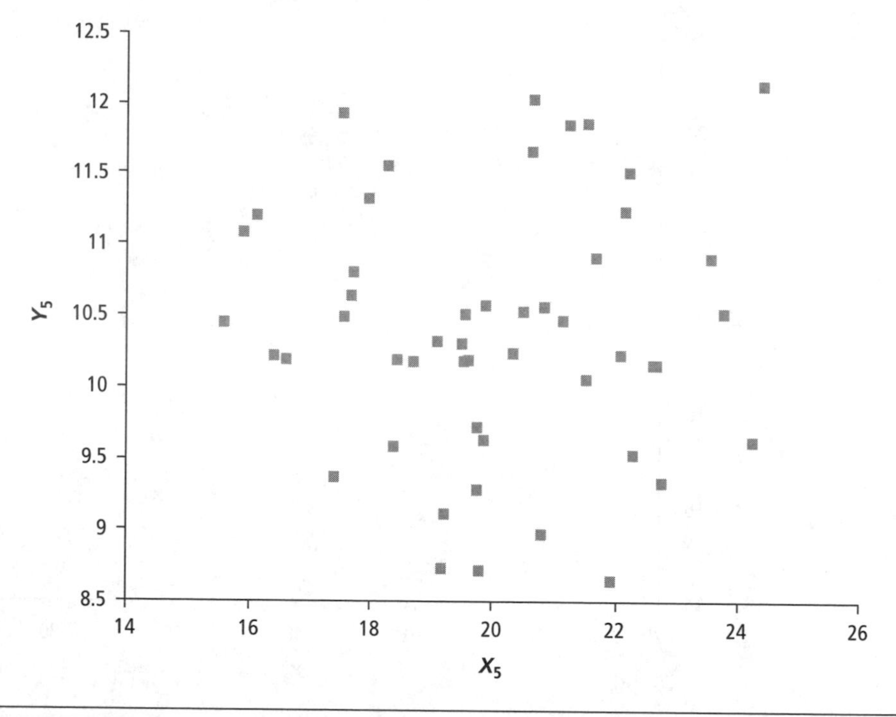

Figure 7.5 ♦ Scatter Plot for Unrelated Variables With Correlation $r = .00$

the scores on X and Y are converted to z scores (i.e., they are converted to unit-free or standardized distances from the mean) during the computation of Pearson r.

Figures 7.1 through 7.5 provide examples of patterns of scores that yield correlations that range from +1 through 0. For X and Y to have a perfect correlation of +1, all the X, Y points must lie on a straight line as shown in Figure 7.1. Perfect linear relations are rarely seen in real data. When the relationship is perfectly linear, we can make an exact statement about how values of Y change as values of X increase; for Figure 7.1, we can say that for a 1-unit increase in X_1, there is exactly a 1,500-unit increase in Y_1. As the strength of the relationship weakens (e.g., $r = +.75$), we can only make approximate statements about how Y changes for each 1-unit increase in X. In Figure 7.2, with $r = +.75$, we can make the (less precise) statement that the mean value of Y_2 tends to increase as the value of X_2 increases. For example, for relatively low X_2 scores (between 30 and 60), the mean score on Y_2 is about 15. For relatively high X_2 scores (between 80 and 100), the mean score of the Y_2 scores is approximately 45. When the correlation is less than 1 in absolute value, we can no longer predict Y scores perfectly from X, but we can predict that the mean of Y will be different for different values of X. In the scatter plot for $r = +.75$, the points form an elliptical cluster (rather than a straight line). If you look at an individual value of X, you can see that there is a distribution of several different Y values for each X. When we do a correlation analysis, we assume that the amount of change in the Y mean is consistent as we move from $X = 1$ to $X = 2$ to $X = 3$, and so forth; in other words, we assume that X and Y are linearly related.

As r becomes smaller, it becomes difficult to judge whether there is any linear relationship simply from visual examination of the scatter plot. The data in Figure 7.4

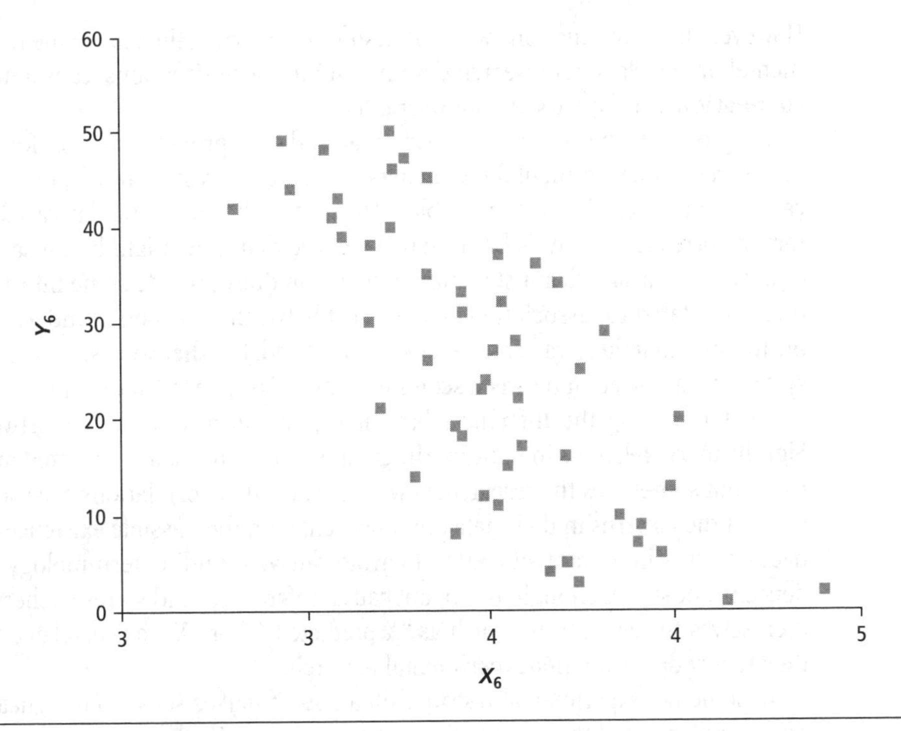

Figure 7.6 ♦ Scatter Plot for Negative Correlation, $r = -.75$

illustrate a weak positive correlation ($r = .23$). In this graph, it is difficult to see an increase in mean values of Y_4 as X_4 increases because the changes in the mean of Y are so small.

Figure 7.5 shows one type of scatter plot for which the correlation is 0; there is no tendency for Y_5 scores to be larger at higher values of X_5. In this example, scores on X are not related to scores on Y, linearly or nonlinearly. Whether X_5 is between 16 and 18, or 18 and 20, or 22 and 24, the mean value of Y is the same (approximately 10). On the other hand, Figure 7.6 illustrates a strong negative linear relationship ($r = -.75$); in this example, as scores on X_6 increase, scores on Y_6 tend to decrease.

Pearson correlation is often applied to data collected in nonexperimental studies; because of this, textbooks often remind students that "correlation does not imply causation." However, it is possible to apply Pearson r to data collected in experimental situations. For example, in a psychophysics experiment, a researcher might manipulate an independent variable (such as the weight of physical objects) and measure a dependent variable (such as the perceived heaviness of the objects). After scores on both variables are transformed (using power or log transformations), a correlation is calculated to show how perceived heaviness is related to the actual physical weight of the objects; in this example, where the weights of objects are varied by the researcher under controlled conditions, it is possible to make a causal inference based on a large Pearson r. The ability to make a causal inference is determined by the nature of the research design, not by the statistic that happens to be used to describe the strength and nature of the relationship between the variables. When data are collected in the context of a carefully controlled experiment, as in the psychophysical research example, a causal inference may be appropriate.

However, in many situations where Pearson *r* is reported, the data come from nonexperimental or correlational research designs, and in those situations, causal inferences from **correlation coefficients** are not warranted.

Despite the inability of nonexperimental studies to provide evidence for making causal inferences, nonexperimental researchers often are interested in the possible existence of causal connections between variables. They often choose particular variables as predictors in correlation analysis because they believe that they might be causal. The presence or absence of a significant statistical association does provide some information: Unless there is a statistical association of some sort between scores on *X* and *Y*, it is not plausible to think that these variables are causally related. In other words, the existence of some systematic association between scores on *X* and *Y* is a necessary (but not sufficient) condition for making the inference that there is a causal association between *X* and *Y*. Significant correlations in nonexperimental research are usually reported merely descriptively, but sometimes the researchers want to show that correlations exist so that they can say that the patterns in their data are consistent with the possible existence of causal connections. It is important, of course, to avoid causal-sounding terminology when the evidence is not strong enough to warrant causal inferences, and so researchers usually limit themselves to saying things such as "*X* predicted *Y*" or "*X* was correlated with *Y*" when they report data from nonexperimental research.

In some nonexperimental research situations, it makes sense to designate one variable as the predictor and the other variable as the outcome. If scores on *X* correspond to events earlier in time than scores on *Y* or if there is reason to think that *X* might cause *Y*, then researchers typically use the scores on *X* as predictors. For example, suppose that *X* is an assessment of mother/infant attachment made when each participant is a few months old, and *Y* is an assessment of adult attachment style made when each participant is 18 years old. It would make sense to predict adult attachment at age 18 from infant attachment; it would not make much sense to predict infant attachment from adult attachment. In many nonexperimental studies, the *X* and *Y* variables are both assessed at the same point in time, and it is unclear whether *X* might cause *Y*, *Y* might cause *X*, or whether both *X* and *Y* might be causally influenced by other variables. For example, suppose a researcher measures grade point average (GPA) and self-esteem for a group of first-year university students. There is no clear justification for designating one of these variables as a predictor; the choice of which variable to designate as the *X* or predictor variable in this situation is arbitrary.

7.2 ♦ Hypothetical Research Example

As a specific example of a question that can be addressed by looking at a Pearson correlation, consider some survey data collected from 118 university students about their heterosexual dating relationships. The variables in this dataset are described in Table 7.1; the scores are in a dataset named love.sav. Only students who were currently involved in a serious dating relationship were included. They provided several kinds of information, including their own gender, partner gender, and a single-item rating of attachment style. They also filled out Sternberg's Triangular Love Measure (Sternberg, 1997). Based on answers to several questions, total scores were calculated for the degree of intimacy, commitment, and passion felt toward the current relationship partner.

Table 7.1 ♦ Description of "Love" Dataset in the File Named love.sav

Variable	*Variable Label*	*Value Labels*
Gender	Gender of student	1 = Female 2 = Male
Genpart	Gender of the student's partner	1 = Female 2 = Male
Length	Length of the dating relationship	1 = Less than 1 month 2 = 1 to 6 months 3 = 6 months to 1 year 4 = 12 years 5 = More than 2 years
Attach	Student's attachment style (single-item measure, from paragraph description)	1 = Secure 2 = Avoidant 3 = Anxious
Times	Number of times participant has been in love	
Intimacy		
Commitment		
Passion		

NOTE: $N = 118$ college student participants (88 female, 30 male).

Later in the chapter, we will use Pearson r to describe the strength of the linear relationship among pairs of these variables and to test whether these correlations are statistically significant. For example, we can ask whether there is a strong positive correlation between scores on intimacy and commitment, and between passion and intimacy.

7.3 ♦ Assumptions for Pearson *r*

The assumptions that need to be met for Pearson r to be an appropriate statistic to describe the relationship between a pair of variables are as follows:

1. *Each score on X should be independent of other X scores (and each score on Y should be independent of other Y scores).* For further discussion of the assumption of independence among observations and the data collection methods that tend to create problems with this assumption, see Chapter 4.
2. *Scores on both X and Y should be quantitative and normally distributed.* Some researchers would state this assumption in an even stronger form: Adherents to strict measurement theory would also require scores on X and Y to be interval/ratio level of measurement. In practice, Pearson r is often applied to data that are not interval/ratio level of measurement; for example, the differences between the scores on 5-point rating scales of attitudes probably do not represent exactly equal differences in

attitude strength, but it is common practice for researchers to apply Pearson r to this type of variable. Harris (2001) summarized arguments about this issue and concluded that it is more important that scores be approximately normally distributed than that the variables satisfy the requirement of true equal interval level of measurement. This does not mean that we should completely ignore issues of level of measurement (see Chapter 1 for further comment on this controversial issue). We can often obtain useful information by applying Pearson r even when the data are obtained by using measurement methods that may fall short of the requirements for true equal interval differences between scores; for example, it is common to apply Pearson correlation to scores obtained using 5-point Likert-type rating scales.

Pearson r can also be applied to data where X or Y (or both) are true dichotomous variables—that is, categorical variables with just two possible values; in this case, it is called a **phi coefficient (ϕ)**. The phi coefficient and other alternate forms of correlation for dichotomous variables are discussed in Chapter 8.

3. *Scores on Y should be linearly related to scores on X.* Pearson r does not effectively detect curvilinear or nonlinear relationships. An example of a curvilinear relationship between X and Y variables that would not be well described by a Pearson r appears in Figure 7.7.

4. *X, Y scores should have a bivariate normal distribution.* Three-dimensional representations of the bivariate normal distribution were shown in Figures 4.40 and 4.41, and the appearance of a bivariate normal distribution in a two-dimensional X, Y scatter plot

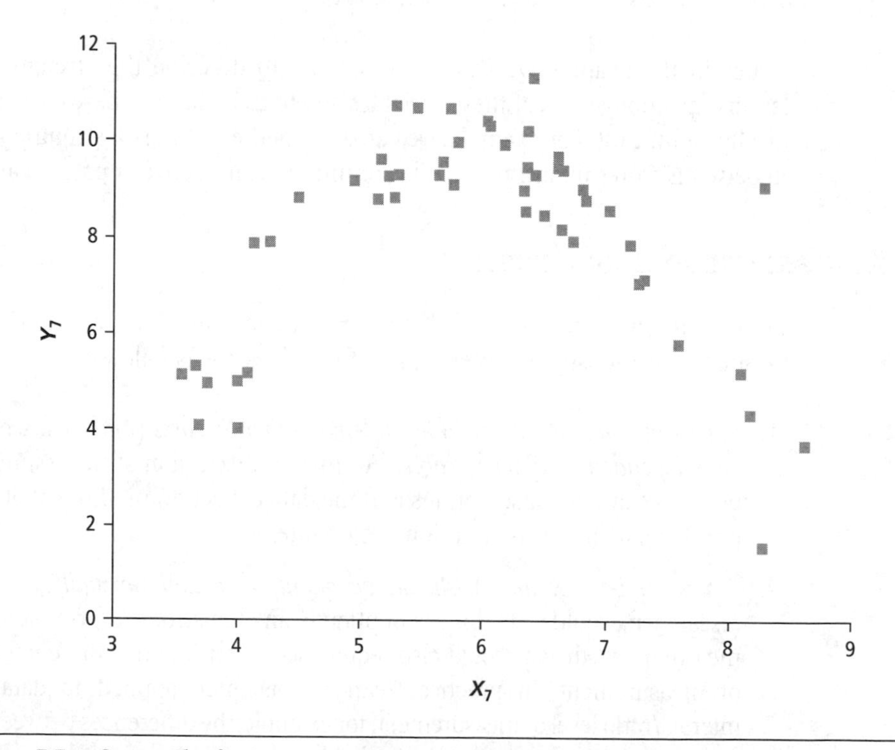

Figure 7.7 ♦ Scatter Plot for Strong Curvilinear Relationship (for These Data, $r = .02$)

appears in Figure 7.8. For each value of X, values of Y should be approximately normally distributed. This assumption also implies that there should not be extreme bivariate outliers. Detection of bivariate outliers is discussed in the next section (on preliminary data-screening methods for correlation).

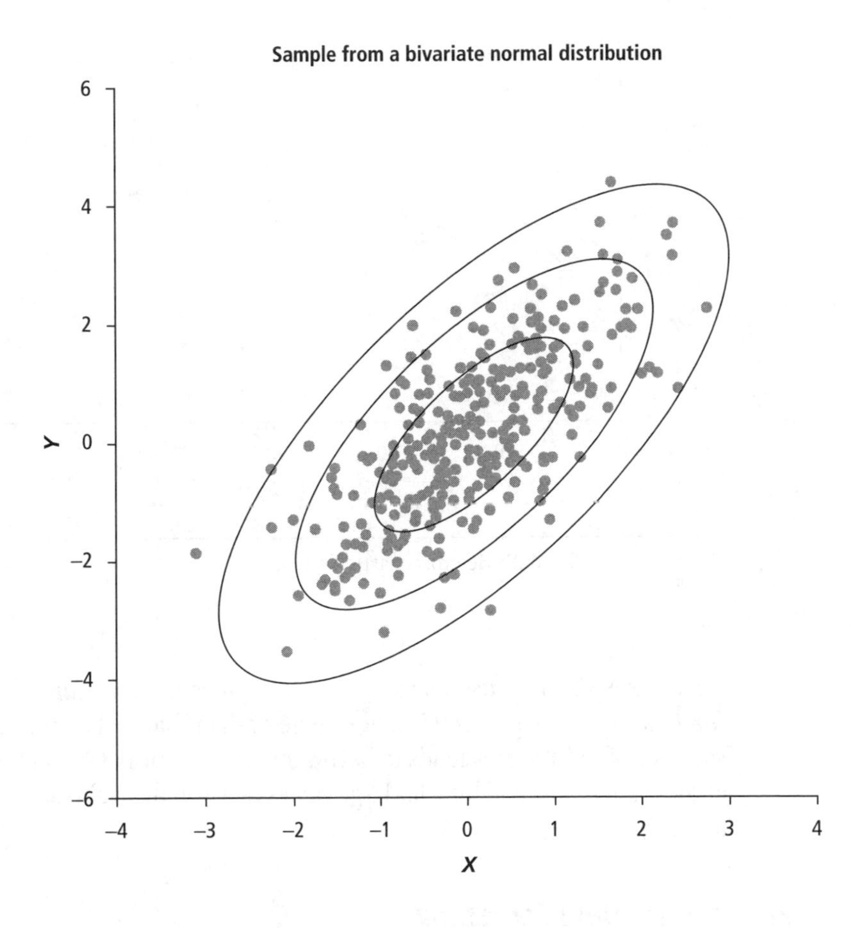

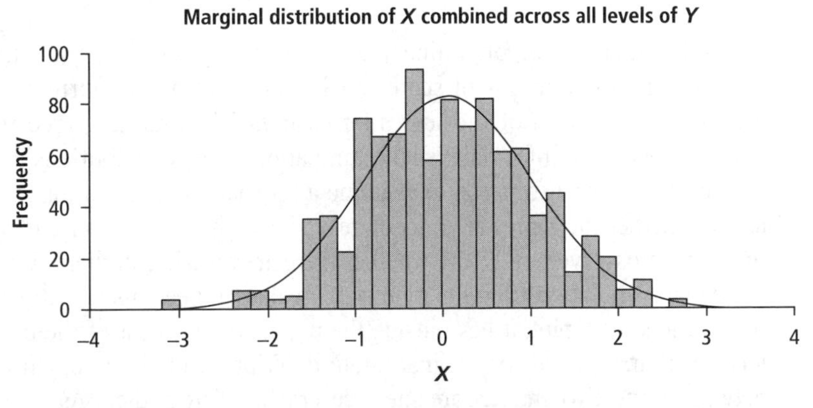

Figure 7.8 ♦ Scatter Plot That Shows a Bivariate Normal Distribution for X and Y

SOURCE: www.survo.fi/gallery/010.html.

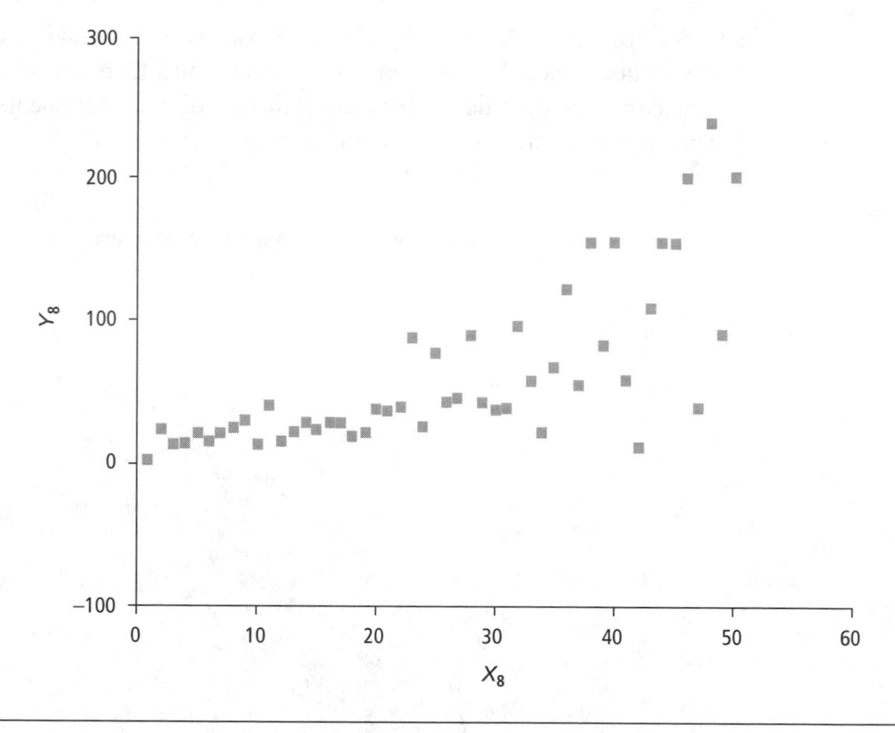

Figure 7.9 ♦ Scatter Plot With Heteroscedastic Variance

5. *Scores on Y should have roughly equal or homogeneous variance across levels of X (and vice versa).* Figure 7.9 is an example of data that violate this assumption; the variance of the *Y* scores tends to be low for small values of *X* (on the left-hand side of the scatter plot) and high for large values of *X* (on the right-hand side of the scatter plot).

7.4 ♦ Preliminary Data Screening

General guidelines for preliminary data screening were given in Chapter 4. To assess whether the distributions of scores on *X* and *Y* are nearly normal, the researcher can examine a histogram of the scores for each variable. As described in Chapter 4, most researchers rely on informal visual examination of the distributions to judge normality.

The researcher also needs to examine a bivariate scatter plot of scores on *X* and *Y* to assess whether the scores are linearly related, whether the variance of *Y* scores is roughly uniform across levels of *X*, and whether there are bivariate outliers. A bivariate outlier is a score that is an unusual combination of *X*, *Y* values; it need not be extreme on either *X* or *Y*, but in the scatter plot, it lies outside the region where most of the other *X*, *Y* points are located. Pearson *r* can be an inaccurate description of the strength of the relationship between *X* and *Y* when there are one or several bivariate outliers. As discussed in Chapter 4, researchers should take note of outliers and make thoughtful decisions about whether to retain, modify, or remove them from the data. Figure 7.10 shows an example of a set of

$N = 50$ data points; when the extreme bivariate outlier is included (as in the upper panel), the correlation between X and Y is +.64; when the correlation is recalculated with this outlier removed (as shown in the scatter plot in the lower panel), the correlation changes to $r = -.10$. Figure 7.11 shows data for which a single bivariate outlier deflates the value of Pearson r; when the circled data point is included, $r = +.53$; when it is omitted, $r = +.86$. It is not desirable to have the outcome of a study depend on the behavior represented by a single data point; the existence of this outlier makes it difficult to evaluate the relationship between the X and Y variables. It would be misleading to report a correlation of $r = +.64$ for the data that appear in Figure 7.10 without including the information that this large positive correlation would be substantially reduced if one bivariate outlier was omitted. In some cases, it may be more appropriate to report the correlation with the outlier omitted.

It is important to examine a scatter plot of the X, Y scores when interpreting a value of r. A scatter plot makes it possible to assess whether there are violations of assumptions of r that make the Pearson r value a poor index of relationship; for instance, the scatter plot can reveal a nonlinear relation between X and Y or extreme outliers that have a disproportionate impact on the obtained value of r. When Pearson correlation is close to zero, it can mean that there is no relationship between X and Y; but correlations close to zero can also occur when there is a nonlinear relationship.

For this example, histograms of scores on the two variables (commitment and intimacy) were obtained by selecting the <Graph> → <Histogram> procedure; SPSS menu selections for this were outlined in Chapter 4. An optional box in the Histogram dialog window can be checked to obtain a normal curve superimposed on the histogram; this can be helpful in assessment of the distribution shape.

The histograms for commitment and intimacy (shown in Figures 7.12 and 7.13) do not show perfect normal distribution shapes; both distributions were skewed. Possible scores on these variables ranged from 15 to 75; most people rated their relationships near the maximum value of 75 points. Thus, there was a ceiling effect such that scores were compressed at the upper end of the distribution. Only a few people rated their relationships low on commitment and intimacy, and these few low scores were clearly separate from the body of the distributions. As described in Chapter 4, researchers need to take note of outliers and decide whether they should be removed from the data or recoded. However, these are always judgment calls. Some researchers prefer to screen out and remove unusually high or low scores, as these can have a disproportionate influence on the size of the correlation (particularly in small samples). Some researchers (e.g., Tabachnick & Fidell, 2007) routinely recommend the use of transformations (such as logs) to make nonnormal distribution shapes more nearly normal. (It can be informative to "experiment" with the data and see whether the obtained correlation changes very much when outliers are dropped or transformations are applied.) For the analysis presented here, no transformations were applied to make the distribution shapes more nearly normal; the r value was calculated with the outliers included and also with the outliers excluded.

The bivariate scatter plot for self-reported intimacy and commitment (in Figure 7.14) shows a positive, linear, and moderate to strong association between scores; that is, persons who reported higher scores on intimacy also reported higher scores on commitment. Although the pattern of data points in Figure 7.14 does not conform perfectly to the ideal bivariate normal distribution shape, this scatter plot does not show any serious

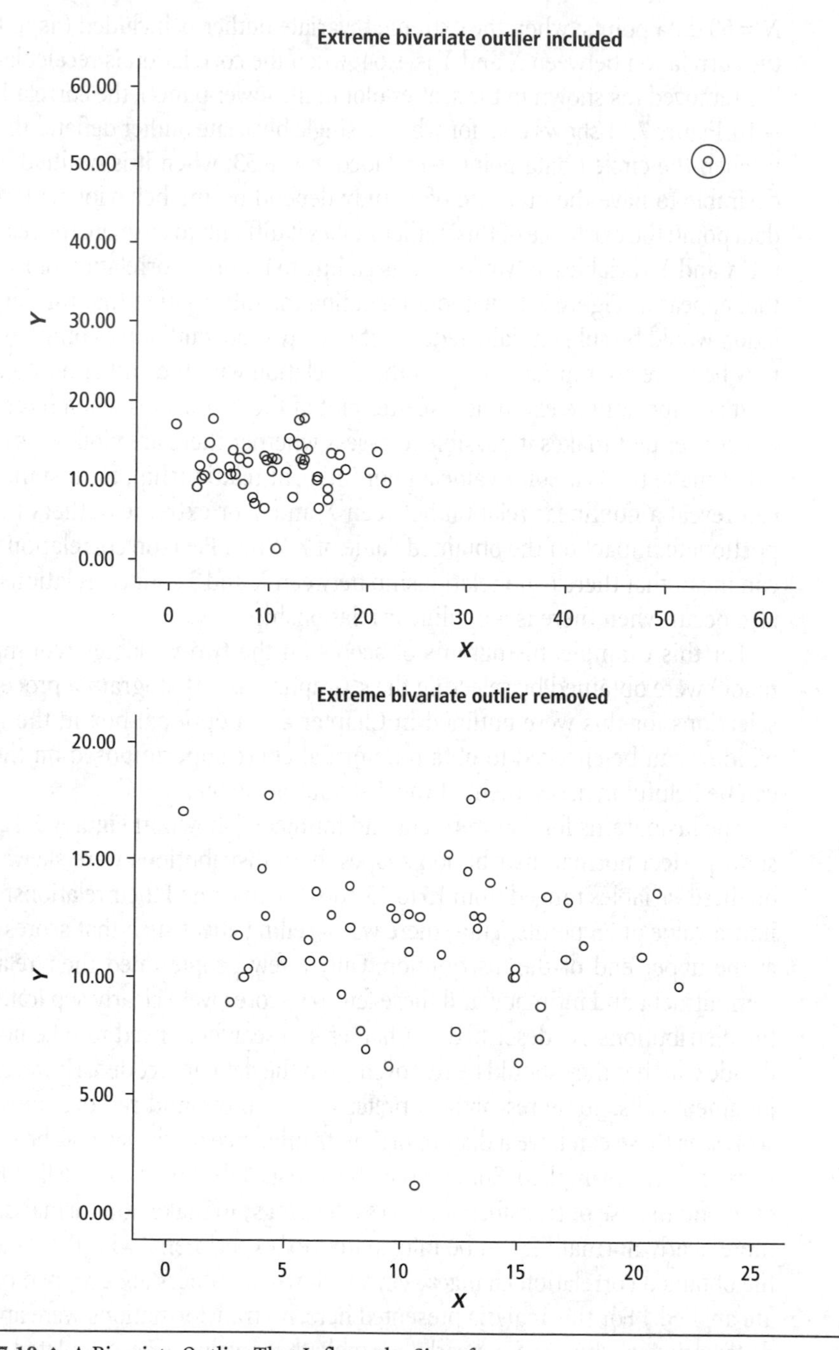

Figure 7.10 ♦ A Bivariate Outlier That Inflates the Size of r

NOTE: With bivariate outlier included, Pearson $r(48) = +.64, p < .001$; with bivariate outlier removed, Pearson $r(47) = -.10$, not significant.

problems. X and Y are approximately linearly related; their bivariate distribution is not extremely different from bivariate normal; there are no extreme bivariate outliers; and while the variance of Y is somewhat larger at low values of X than at high values of X, the differences in variance are not large.

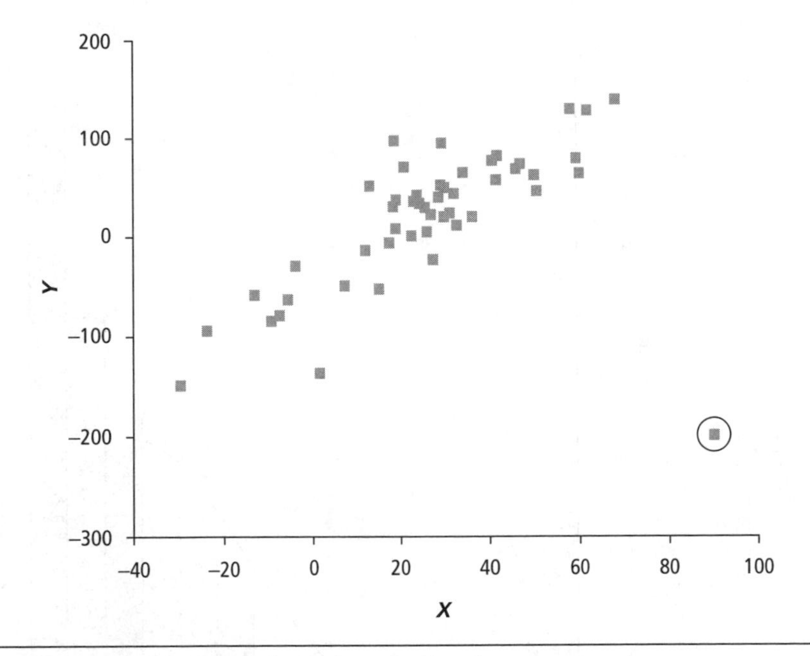

Figure 7.11 ♦ A Bivariate Outlier That Deflates the Size of *r*

NOTE: With bivariate outlier included, Pearson $r(48) = +.532, p < .001$; with bivariate outlier removed, Pearson $r(47) = +.86, p < .001$.

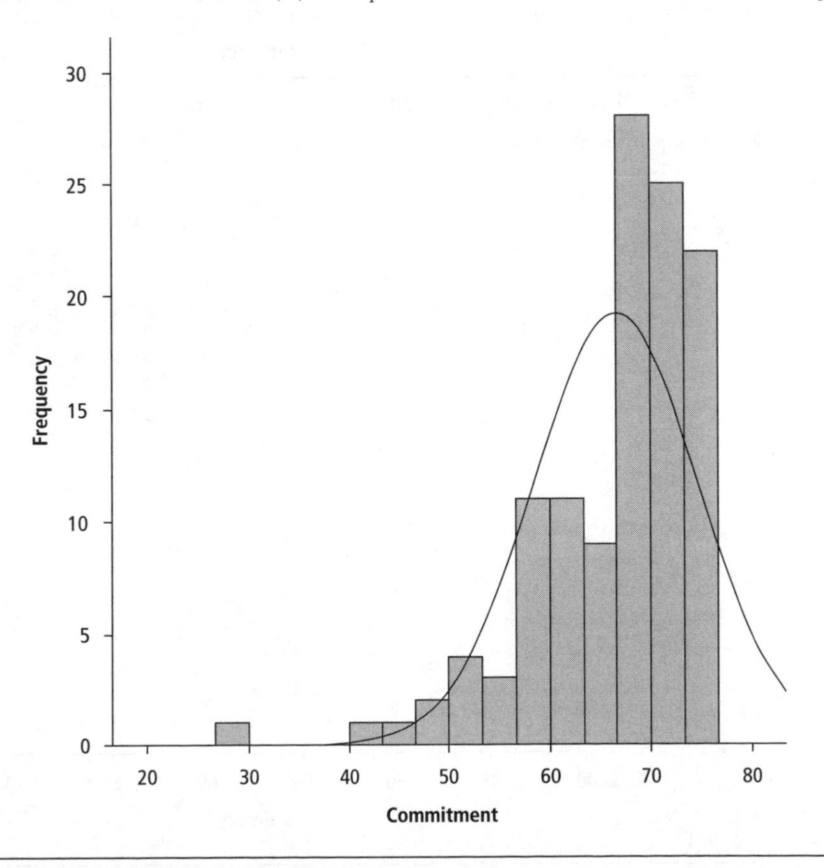

Figure 7.12 ♦ Data Screening: Histogram of Scores for Commitment

NOTE: Descriptive statistics: Mean = 66.63, $SD = 8.16$, $N = 118$.

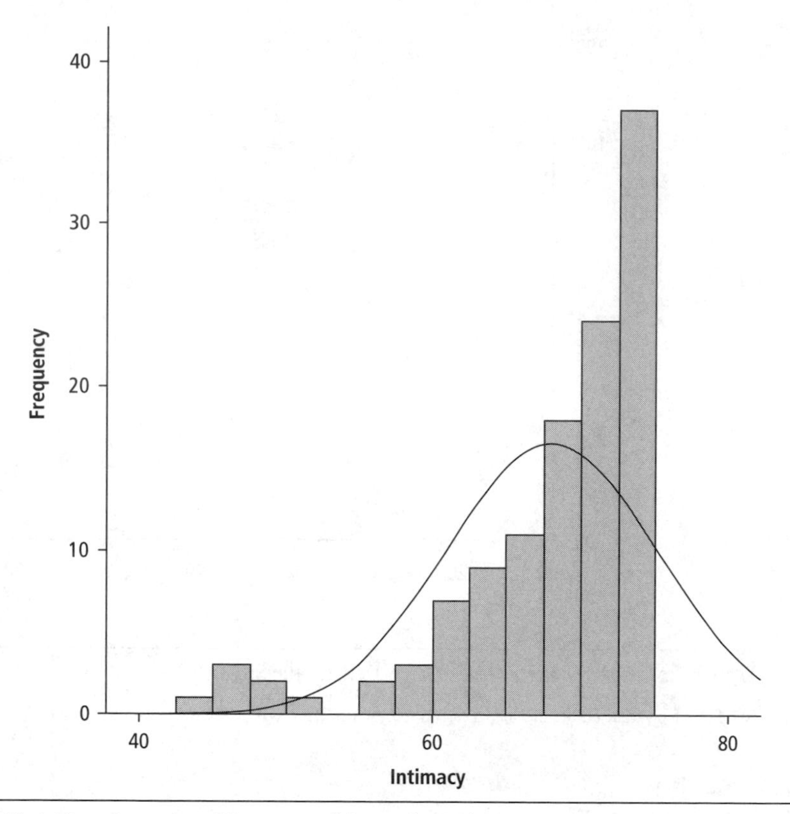

Figure 7.13 ♦ Data Screening: Histogram of Scores for Intimacy

NOTE: Descriptive statistics: Mean = 68.04, *SD* = 7.12, *N* = 118.

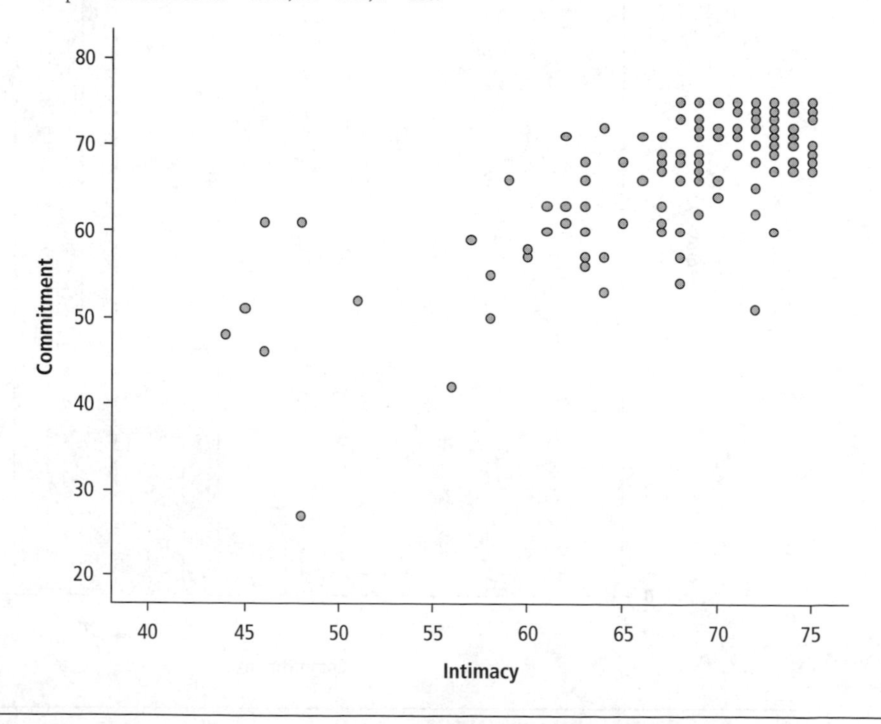

Figure 7.14 ♦ Scatter Plot for Prediction of Commitment From Intimacy

7.5 ♦ Design Issues in Planning Correlation Research

Several of the problems at the end of this chapter use data with very small numbers of cases so that students can easily calculate Pearson r by hand or enter the data into SPSS. However, in general, studies that report Pearson r should be based on fairly large samples. Pearson r is not robust to the effect of extreme outliers, and the impact of outliers is greater when the N of the sample is small. Values of Pearson r show relatively large amounts of sampling error across different batches of data, and correlations obtained from small samples often do not replicate well. In addition, fairly large sample sizes are required so that there is adequate statistical power for the detection of differences between different correlations. Because of sampling error, it is not realistic to expect sample correlations to be a good indication of the strength of the relationship between variables in samples smaller than $N = 30$. When N is less than 30, the size of the correlation can be greatly influenced by just one or two extreme scores. In addition, researchers often want to choose sample sizes large enough to provide adequate statistical power (see Section 7.1). It is advisable to have an N of at least 100 for any study where correlations are reported.

It is extremely important to have a reasonably wide range of scores on both the predictor and the outcome variables. In particular, the scores should cover the range of behaviors to which the researcher wishes to generalize. For example, in a study that predicts verbal Scholastic Aptitude Test (VSAT) scores from GPA, one would ideally want to include a wide range of scores on both variables, with VSAT scores ranging from 250 to 800 and GPAs that range from very poor to excellent marks.

A report of a single correlation is not usually regarded as sufficient to be the basis of a thesis or a publishable paper (American Psychological Association, 2001, p. 5). Studies that use Pearson r generally include correlations among many variables and may include other analyses. Sometimes researchers report correlations among all possible pairs of variables; this often results in reporting hundreds of correlations in a single paper. This leads to an inflated risk of Type I error. A more thoughtful and systematic approach involving the examination of selected correlations is usually preferable (as discussed in Chapter 1). In exploratory studies, statistically significant correlations that are detected by examining dozens or hundreds of tests need to be replicated through cross-validation or new data collection before they can be treated as "findings."

7.6 ♦ Computation of Pearson *r*

The version of the formula for the computation of Pearson r[1] that is easiest to understand conceptually is as follows:

$$r = \sum (z_X \times z_Y)/(N), \tag{7.1}$$

where $z_X = (X - M_X)/s_X$, $z_Y = (Y - M_Y)/s_Y$, and N = number of cases (number of X, Y pairs of observations).

There are alternative versions of this formula that are easier to use and give less rounding error when Pearson r is calculated by hand. The version of the formula above is more helpful in understanding how the Pearson r value can provide information about the spatial distribution of X, Y data points in a scatter plot. This conceptual formula can be used for by-hand computation; it corresponds to the following operations. First, each X and Y score is converted to a standard score or z score; then, for each participant, z_X is multiplied by z_Y; these products are summed across all participants; and finally, this sum is divided by the number of participants. The resulting value of r falls within the range $-1.00 \leq r \leq +1.00$.

Because we convert X and Y to standardized or z scores, the value of r does not depend on the units that were used to measure these variables. If we take a group of subjects and express their heights in both inches (X_1) and centimeters (X_2) and their weights in pounds (Y_1) and kilograms (Y_2), the correlation between X_1, Y_1 and between X_2, Y_2 will be identical. In both cases, once we convert height to a z score, we are expressing the individual's height in terms of a unit-free distance from the mean. A person's z score for height will be the same whether we work with height in inches, feet, or centimeters.

Another formula for Pearson r is based on the **covariance** between X and Y:

$$\text{Cov}(X, Y) = \frac{\sum[(X - M_X)(Y - M_Y)]}{N}, \tag{7.2}$$

where M_X is the mean of the X scores, M_Y is the mean of the Y scores, and N is the number of X, Y pairs of scores.

Note that the variance of X is equivalent to the covariance of X with itself:

$$s_X^2 = \sum (X - M_X)^2/(N - 1) = \sum (X - M_X)(X - M_X)/(N - 1). \tag{7.3}$$

Pearson r can be calculated from the covariance of X with Y as follows:

$$r = \text{Cov}(X, Y)/(s_X s_Y). \tag{7.4}$$

A covariance,[2] like a variance, is an arbitrarily large number; its size depends on the units used to measure the X and Y variables. For example, suppose a researcher wants to assess the relation between height (X) and weight (Y). These can be measured in many different units: Height can be given in inches, feet, meters, or centimeters, and weight can be given in terms of pounds, ounces, kilograms, or grams. If height is stated in inches and weight in ounces, the numerical scores given to most people will be large and the covariance will turn out to be very large. However, if heights are given in feet and weights in pounds, both the scores and the covariances between scores will be smaller values. Covariance, thus, depends on the units of measurement the researcher happened to use. This can make interpretation of covariance difficult, particularly in situations where the units of measurement are arbitrary.

Pearson correlation can be understood as a standardized covariance: The values of r fall within a fixed range from -1 to $+1$, and the size of r does not depend on the units of measurement the researcher happened to use for the variables. Whether height was measured in inches, feet, or meters, when the height scores are converted to standard or z scores, information about the units of measurement is lost. Because correlation is standardized, it is easier to interpret, and it is possible to set up some verbal guidelines to describe the sizes of correlations.

Table 7.2 ♦ Computation of Pearson *r* for a Set of Scores on HR and Self-Reported Tension

HR	*Tension*	z_{HR}	$z_{tension}$	$z_X \times z_Y$
69	10	−.60062	.16010	−.10
58	5	−1.37790	−1.17409	1.62
80	8	.17665	−.37357	−.07
89	15	.81261	1.49429	1.21
101	15	1.66055	1.49429	2.48
67	7	−.74195	−.64041	.48
67	9	−.74195	−.10674	.08
68	6	−.67129	−.90725	.61
80	6	.17665	−.90725	−.16
96	13	1.30724	.96062	1.26

NOTE: $\Sigma z_X \times z_Y = 7.41$, Pearson $r = \Sigma z_X \times z_Y /(N{-}1) = 7.41/9 = +.82$.

Here is a numerical example that shows the computation of Pearson *r* for a small dataset that contains $N = 10$ pairs of scores on heart rate (HR) and self-reported tension (see Table 7.2).

The first two columns of this table contain the original scores for the variables HR and tension. The next two columns contain the *z* scores for each variable, z_{HR} and z_{tension} (these *z* scores or standard scores can be saved as output from the SPSS Descriptive Statistics procedure). For this example, HR is the *Y* variable and tension is the *X* variable. The final column contains the product of z_X and z_Y for each case, with $\sum z_X z_Y$ at the bottom. Finally, Pearson *r* was obtained by taking $\sum z_X z_Y / (N - 1) = +7.41/9 = .823$. (The values of *r* reported by SPSS use $N - 1$ in the divisor rather than *N* as in most textbook formulas for Pearson *r*. When *N* is large, for example, *N* greater than 100, the results do not differ much whether *N* or $N - 1$ is used as the divisor.) This value of *r* agrees with the value obtained by running the SPSS bivariate correlation/Pearson *r* procedure on the data that appear in Table 7.2.

7.7 ♦ Statistical Significance Tests for Pearson *r*

7.7.1 ♦ Testing the Hypothesis that $\rho_{XY} = 0$

The most common statistical significance test is for the statistical significance of an individual correlation. The population value of the correlation between *X* and *Y* is denoted by the Greek letter ***rho*** **(ρ)**. Given an obtained sample *r* between *X* and *Y*, we can test the null hypothesis that ρ_{XY} in the population equals 0. The formal null hypothesis that corresponds to the lack of a (linear) relationship between *X* and *Y* is

$$H_0\text{: } \rho_{XY} = 0. \qquad (7.5)$$

When the population correlation ρ_{XY} is 0, the sampling distribution of *r*s is shaped like a normal distribution (for large *N*) or a *t* distribution with $N - 2$ *df* (for small *N*), except that the tails are not infinite (the tails end at +1 and −1); see the top panel in Figure 7.15.

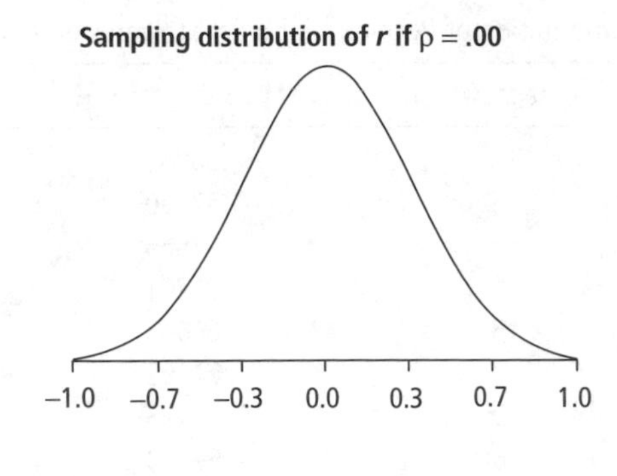

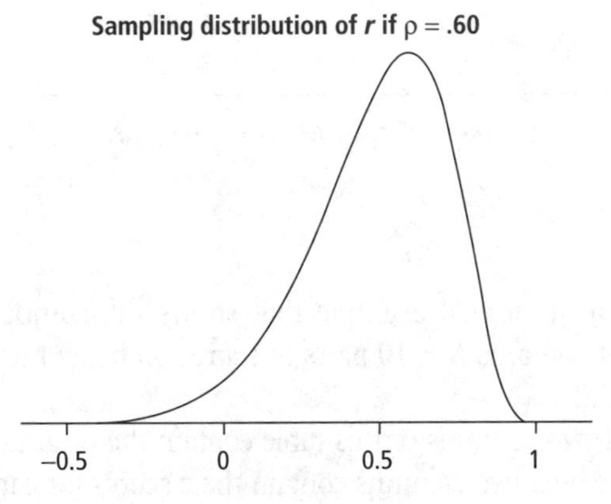

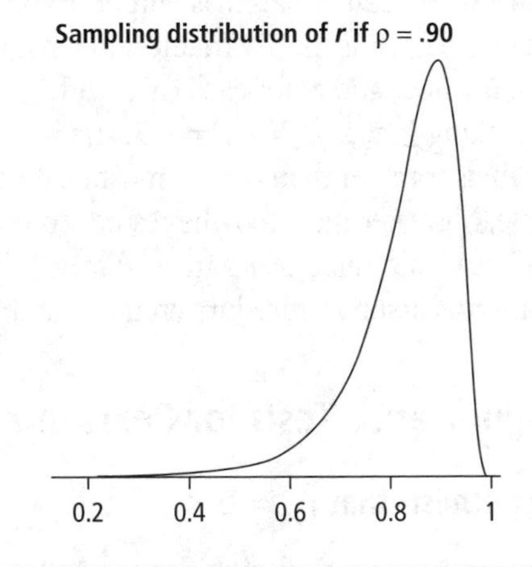

Figure 7.15 ♦ Sampling Distributions for r With $N = 12$

That is, when the true population correlation ρ is actually 0, most sample rs tend to be close to 0; and the sample rs tend to be normally distributed, but the tails of this distribution are not infinite (as they are for a true normal distribution), because sample correlations cannot be outside the range of −1 to +1. Because the sampling distribution for this situation is roughly that of a normal or t distribution, a t ratio to test this null hypothesis can be set up as follows:

$$t = \frac{r - \rho_0}{SE_r}. \quad (7.6)$$

The value of SE_r is given by the following equation:

$$SE_r = \frac{\sqrt{1 - r^2}}{\sqrt{N - 2}}. \quad (7.7)$$

Substituting this value of SE_r from Equation 7.7 into Equation 7.6 and rearranging the terms yields the most widely used formula for a t test for the significance of a sample r value; this t test has $N - 2$ degrees of freedom (*df*); the hypothesized value of ρ_0 is 0.

$$t = \frac{r - \rho_0}{SE_r} = \frac{r}{\sqrt{1 - r^2}/\sqrt{N - 2}} = \frac{r \times \sqrt{N - 2}}{\sqrt{1 - r^2}}. \quad (7.8)$$

It is also possible to set up an F ratio, with $(1, N - 2)$ *df*, to test the significance of sample r. This F is equivalent to t^2; it has the following form:

$$F = \frac{r^2 \times (N - 2)}{(1 - r^2)}. \quad (7.9)$$

Programs such as SPSS provide an exact p value for each sample correlation (a two-tailed p value by default; a one-tailed p value can be requested). Critical values of the t and F distributions are provided in Appendixes B and C. It is also possible to look up whether r is statistically significant as a function of degrees of freedom and the r value itself directly in the table in Appendix E (without having to calculate t or F).

7.7.2 ♦ Testing Other Hypotheses About ρ_{XY}

It is uncommon to test null hypotheses about other specific hypothesized values of ρ_{XY} (such as H_0: $\rho_{XY} = .90$). For this type of null hypothesis, the sampling distribution of r is not symmetrical and therefore cannot be approximated by a t distribution. For example, if $\rho_{XY} = .90$, then most sample rs will be close to .90; and sample rs will be limited to the range from −1 to +1, so the sampling distribution will be extremely skewed (see the bottom panel in Figure 7.15). To correct for this nonnormal distribution shape, a data transformation is applied to r before testing hypotheses about nonzero hypothesized values of ρ. The **Fisher *r* to *Z* transformation** rescales sample rs in a way that yields a more nearly normal distribution shape, which can be used for hypothesis testing. (Note that in this book, lowercase z always refers to a standard score; uppercase Z refers to the Fisher Z transformation based on r. Some books label the Fisher Z using a lowercase z or z'.) The r to Fisher Z transformation is shown in Table 7.3 (for reference, it is also included in Appendix G at the end of this book).

The value of Fisher Z that corresponds to a sample Pearson r is usually obtained by table lookup, although Fisher Z can be obtained from this formula:

$$\text{Fisher } Z = \tfrac{1}{2}[\ln(1 + r) - \ln(1 - r)] = \tanh^{-1} r. \quad (7.10)$$

Table 7.3 ♦ Transformation of Pearson r to Fisher Z'

r	Z'	r	Z'
		(Continued)	
0.0000	0.0000	0.5000	0.5493
0.0100	0.0100	0.5100	0.5627
0.0200	0.0200	0.5200	0.5763
0.0300	0.0300	0.5300	0.5901
0.0400	0.0400	0.5400	0.6042
0.0500	0.0500	0.5500	0.6184
0.0600	0.0601	0.5600	0.6328
0.0700	0.0701	0.5700	0.6475
0.0800	0.0802	0.5800	0.6625
0.0900	0.0902	0.5900	0.6777
0.1000	0.1003	0.6000	0.6931
0.1100	0.1104	0.6100	0.7089
0.1200	0.1206	0.6200	0.7250
0.1300	0.1307	0.6300	0.7414
0.1400	0.1409	0.6400	0.7582
0.1500	0.1511	0.6500	0.7753
0.1600	0.1614	0.6600	0.7928
0.1700	0.1717	0.6700	0.8107
0.1800	0.1820	0.6800	0.8291
0.1900	0.1923	0.6900	0.8480
0.2000	0.2027	0.7000	0.8673
0.2100	0.2132	0.7100	0.8872
0.2200	0.2237	0.7200	0.9076
0.2300	0.2342	0.7300	0.9287
0.2400	0.2448	0.7400	0.9505
0.2500	0.2554	0.7500	0.9730
0.2600	0.2661	0.7600	0.9962
0.2700	0.2769	0.7700	1.0203
0.2800	0.2877	0.7800	1.0454
0.2900	0.2986	0.7900	1.0714
0.3000	0.3095	0.8000	1.0986
0.3100	0.3205	0.8100	1.1270
0.3200	0.3316	0.8200	1.1568
0.3300	0.3428	0.8300	1.1881
0.3400	0.3541	0.8400	1.2212
0.3500	0.3654	0.8500	1.2562
0.3600	0.3769	0.8600	1.2933
0.3700	0.3884	0.8700	1.3331
0.3800	0.4001	0.8800	1.3758
0.3900	0.4118	0.8900	1.4219
0.4000	0.4236	0.9000	1.4722
0.4100	0.4356	0.9100	1.5275
0.4200	0.4477	0.9200	1.5890
0.4300	0.4599	0.9300	1.6584
0.4400	0.4722	0.9400	1.7380
0.4500	0.4847	0.9500	1.8318
0.4600	0.4973	0.9600	1.9459
0.4700	0.5101	0.9700	2.0923
0.4800	0.5230	0.9800	2.2976
0.4900	0.5361	0.9900	2.6467

SOURCE: From Instructor's Edition: *Hyperstat,* 2nd Edition by Lane, D. M., copyright © 2001. Reprinted with permission of Custom Publishing, a division of Thomson Learning: www.thomsonrights.com.

A Fisher Z value can also be converted back into an r value by using Table 7.3.

For the Fisher Z, the standard error (SE) does not depend on ρ but only on N; the sampling distribution of Fisher Z scores has this standard error:

$$SE_Z \cong \frac{1}{\sqrt{N-3}}. \tag{7.11}$$

Thus, to test the null hypothesis,

$$H_0: \rho_{hyp} = .9. \tag{7.12}$$

With $N = 28$ and an observed sample r of .8, the researcher needs to do the following:

1. Convert ρ_{hyp} to a corresponding Fisher Z value, Z_{hyp}, by looking up the Z value in Table 7.3. For $\rho_{hyp} = .90, Z_{hyp} = 1.472$.
2. Convert the observed sample r (r_{sample}) to a Fisher Z value (Z_{sample}) by looking up the corresponding Fisher Z value in Table 7.3. For an observed r of .80, $Z_{sample} = 1.099$.
3. Calculate SE_Z from Equation 7.11:

$$SE_Z = 1/\sqrt{N-3} = 1/\sqrt{28-3} = 1/5 = .20.$$

4. Compute the z ratio as follows:

$$z = (Z_{samp} - Z_{hyp})/SE_Z = (1.472 - 1.099)/.20 = 1.865.$$

5. For $\alpha = .05$, two-tailed, the reject region for a z test is $z > 1.96$ and $z < -1.96$; therefore, do not reject the null hypothesis that $\rho = .90$.

The Fisher Z transformation is also used when testing the null hypothesis that the value of ρ is equal between two different populations, $H_0: \rho_1 = \rho_2$. For example, we can test whether the correlation between X and Y is significantly different for women versus men. Fisher Z is also used to set up confidence intervals (CIs) for correlation estimates.

7.7.3 ♦ Assessing Differences Between Correlations

It can be quite problematic to compare correlations that are based on different samples or populations, or that involve different variables, because there are so many factors that can artifactually influence the size of Pearson r (many of these factors are discussed in Section 7.9). For example, suppose a researcher wants to evaluate whether the correlation between emotional intelligence (EI) and drug use is stronger for males than for females. If the scores on drug use have a much more restricted range in the female sample than in the male sample, this restricted range of scores in the female sample might make the correlation between these variables smaller for females. If the measurement of drug use has lower reliability for females than for males, this difference in reliability could also artifactually reduce the magnitude of the correlation between EI and drug use in the female sample. If two correlations differ significantly, this difference might arise due to artifact (such as a narrower range of scores used to compute one r) rather than because of a difference in the true strength of the relationship. Researchers have to be very cautious when comparing correlations, and they should acknowledge possible artifacts that might have led to different r values (sources of artifacts are discussed in Section 7.9).

It is useful to have statistical significance tests for comparison of correlations; these at least help to answer the question whether the difference between a pair of correlations is so small that it could very likely be due to sampling error. Obtaining statistical significance is a necessary, but not a sufficient, condition for concluding that a genuine difference in the strength of relationship is present. Two types of comparisons between correlations are described here.

In the first case, the test compares the strength of the correlation between the same two variables in two different groups or populations. Suppose that the same set of variables (such as X = EI and Y = drug abuse or DA) are correlated in two different groups of participants (Group 1 = Males, Group 2 = Females). We might ask whether the correlation between EI and DA is significantly different for men versus women. The corresponding null hypothesis is

$$H_0: \rho_1 = \rho_2. \tag{7.13}$$

To test this hypothesis, the Fisher Z transformation has to be applied to both sample r values. Let r_1 be the sample correlation between EI and DA for males and r_2 the sample correlation between EI and DA for females; N_1 and N_2 are the numbers of participants in the male and female groups, respectively.

First, using Table 7.3, look up the Z_1 value that corresponds to r_1 and the Z_2 value that corresponds to r_2.

Next, apply the following formula:

$$z = \frac{Z_1 - Z_2}{\sqrt{1/(N_1 - 3) + 1/(N_2 - 3)}}. \tag{7.14}$$

The test statistic z is evaluated using the standard normal distribution; if the obtained z ratio is greater than +1.96 or less than −1.96, then the correlations r_1 and r_2 are judged significantly different using $\alpha = .05$, two-tailed. This test should only be used when the N in each sample is fairly large, preferably $N > 100$.

A second situation of interest involves comparison of two different predictor variables. Suppose the researcher wants to know whether the correlation of X with Z is significantly different from the correlation of Y with Z. The corresponding null hypothesis is

$$H_0: \rho_{XZ} = \rho_{YZ}. \tag{7.15}$$

This test does not involve the use of Fisher Z transformations. Instead, we need to have all three possible bivariate correlations (r_{XZ}, r_{YZ}, and r_{XY}); N = total number of participants. The test statistic (from Lindeman, Merenda, & Gold, 1980) is a t ratio of this form:

$$t = (r_{XZ} - r_{YZ})\sqrt{\frac{(N-3)(1 + r_{XY})}{2 \times (1 - r_{XY}^2 - r_{XZ}^2 - r_{YZ}^2 + 2r_{XY}r_{XZ}r_{YZ})}}. \tag{7.16}$$

The resulting t value is evaluated using critical values from the t distribution with $(N - 3)$ df. Even if a pair of correlations is judged to be statistically significantly different using these tests, the researcher should be very cautious about interpreting this result. Different size correlations could arise because of differences across populations or across predictors in factors that affect the size of r discussed in Section 7.9, such as range of scores, reliability of measurement, outliers, and so forth.

7.7.4 ♦ Reporting Many Correlations: Need to Control Inflated Risk of Type I Error

Journal articles rarely report just a single Pearson r; in fact, the *Publication Manual of the American Psychological Association* (American Psychological Association, 2001) states that this is not sufficient for a reportable study. Unfortunately, however, many studies report such a large number of correlations that evaluation of statistical significance becomes problematic. Suppose that $k = 20$ variables are measured in a nonexperimental study. If the researcher does all possible bivariate correlations, there will be $k \times (k - 1)/2$ different correlations, in this case $(20 \times 19)/2 = 190$ correlations. When we set our risk of Type I error at $\alpha = .05$, this implies that out of 100 statistical significance tests that are done (on data from populations that really have no relationship between the X and Y variables), about 5 tests will be instances of Type I error (rejection of H_0 when H_0 is true). When a journal article reports 200 correlations, for example, one would expect that about 5% of these (10 correlations) should be statistically significant using the $\alpha = .05$ significance level, even if the data were completely random. Thus, of the 200 correlations, it is very likely that at least some of the significance tests (on the order of 9 or 10) will be instances of Type I error. If the researcher runs 200 correlations and finds that the majority of them (let's say, 150 out of 200) are significant, then it seems likely that at least some of these correlations are not merely artifacts of chance. However, if a researcher reports 200 correlations and only 10 are significant, then it's quite possible that the researcher has found nothing beyond the expected number of Type I errors. It is even more problematic for the reader when it's not clear how many correlations were run; if a researcher runs 200 correlations, hand selects 10 statistically significant rs after the fact, and then reports only the 10 rs that were judged to be significant, it is extremely misleading to the reader, who is no longer able to evaluate the true magnitude of the risk of Type I error.

In general, it is common in exploratory nonexperimental research to run large numbers of significance tests; this inevitably leads to an inflated risk of Type I error. That is, the probability that the entire research report contains at least one instance of Type I error is much higher than the nominal risk of $\alpha = .05$ that is used for any single significance test. There are several possible ways to deal with this problem of inflated risk of Type I error.

7.7.4.1 ♦ *Limiting the Number of Correlations*

One approach is to limit the number of correlations that will be examined at the outset, before looking at the data, based on theoretical assumptions about which predictive relations are of interest. The possible drawback of this approach is that it may preclude serendipitous discoveries. Sometimes, unexpected observed correlations do point to relationships among variables that were not anticipated from theory but that can be confirmed in subsequent replications.

7.7.4.2 ♦ *Cross-Validation of Correlations*

A second approach is cross-validation. In a cross-validation study, the researcher randomly divides the data into two batches; thus, if the entire study had data for $N = 500$ participants, each batch would contain 250 cases. The researcher then does extensive exploratory analysis on the first batch of data and decides on a limited number of correlations or predictive equations that seem to be interesting and useful. Then, the researcher reruns this small set of correlations on the second half of the data. If the relations between variables remain significant in this fresh batch of data, it is less likely that these relationships were just instances of Type I error. The main problem with this approach is that researchers often don't have large enough numbers of cases to make this possible.

7.7.4.3 ♦ *Bonferroni Procedure: A More Conservative Alpha Level for Tests of Individual Correlations*

A third approach is the Bonferroni procedure. Suppose that the researcher plans to do $k = 10$ correlations and wants to have an experiment-wise alpha (EW_α) of .05. To keep the risk of obtaining at least one Type I error as low as 5% for a set of $k = 10$ significance tests, it is necessary to set the per-comparison alpha (PC_α) level lower for each individual test. Using the Bonferroni procedure, the PC_α used to test the significance of each individual r value is set at EW_α/k; for example, if $EW_\alpha = .05$ and $k = 10$, each individual correlation has to have an observed p value less than .05/10, or .005, to be judged statistically significant. The main drawback of this approach is that it is quite conservative. Sometimes the number of correlations that are tested in exploratory studies is quite large (100 or 200 correlations have been reported in some recent papers). If the error rate were adjusted by dividing .05 by 100, the resulting PC_α would be so low that it would almost never be possible to judge individual correlations significant. Sometimes the experiment-wise α for the Bonferroni test is set higher than .05; for example, $EW_\alpha = .10$ or .20.

If the researcher does not try to limit the risk of Type I error in any of these three ways (by limiting the number of significance tests, by doing a cross-validation, or by using a Bonferroni-type correction for alpha levels), then at the very least, the researcher should explain in the write-up that the p values that are reported probably underestimate the true overall risk of Type I error. In these situations, the Discussion section of the paper should reiterate that the study is exploratory, that relationships detected by running large numbers of significance tests are likely to include large numbers of Type I errors, and that replications of the correlations with new samples are needed before researchers can be confident that the relationships are not simply due to chance or sampling error.

7.8 ♦ Setting Up CIs for Correlations

If the researcher wants to set up a CI using a sample correlation, it is necessary to use the Fisher Z transformation (in cases where r is not equal to 0). The upper and lower bounds of the 95% CI can be calculated by applying the usual formula for a CI to the Fisher Z values that correspond to the sample r.

The general formula for a CI is as follows:

$$\text{Lower bound of 95\% CI} = \text{Sample statistic} - t_{\text{crit}} \times SE_{\text{sample statistic}}, \tag{7.17}$$

$$\text{Upper bound of 95\% CI} = \text{Sample statistic} + t_{\text{crit}} \times SE_{\text{sample statistic}}. \tag{7.18}$$

To set up a CI around a sample r value (let $r = .50$, with $N = 43$, for example), first, look up the Fisher Z value that corresponds to $r = .50$; from Table 7.3, this is Fisher $Z = .549$. For $N = 43$ and $df = N - 3 = 40$,

$$SE_Z = \frac{1}{\sqrt{N-3}} = \frac{1}{\sqrt{43-3}} = \frac{1}{\sqrt{40}} = .158.$$

For $N = 43$ and a 95% CI, t_{crit} is approximately equal to +2.02 for the top 2.5% of the distribution. The Fisher Z values and the critical values of t are substituted into the equations for the lower and upper bounds:

$$\text{Lower bound of 95\% CI} = \text{Fisher } Z - 2.02 \times SE_Z = .549 - 2.02 \times .157 = .232,$$
$$\text{Upper bound of 95\% CI} = \text{Fisher } Z + 2.02 \times SE_Z = .549 + 2.02 \times .157 = .866.$$

Table 7.3 is used to transform these boundaries given in terms of Fisher Z back into estimated correlation values:

$$\text{Fisher } Z = .232 \text{ is equivalent to } r = .23,$$
$$\text{Fisher } Z = .866 \text{ is equivalent to } r = .70.$$

Thus, if a researcher obtains a sample r of .50 with $N = 43$, the 95% CI is from .23 to .70. If this CI does not include 0, then the sample r would be judged statistically significant using $\alpha = .05$, two-tailed. The SPSS Bivariate Correlation procedure does not provide CIs for r values, but these can easily be calculated by hand.

7.9 ♦ Factors that Influence the Magnitude and Sign of Pearson *r*

7.9.1 ♦ Pattern of Data Points in the *X*, *Y* Scatter Plot

To understand how the formula for correlation can provide information about the location of points in a scatter plot and how it detects a tendency for high scores on Y to co-occur with high or low scores on X, it is helpful to look at the arrangement of points in an X, Y scatter plot (see Figure 7.16). Consider what happens when the scatter plot is divided into four quadrants or regions: scores that are above and below the mean on X and scores that are above and below the mean on Y.

The data points in Regions II and III are cases that are **"concordant"**; these are cases for which high X scores were associated with high Y scores or low X scores were paired with low Y scores. In Region II, both z_X and z_Y are positive, and their product is also positive; in Region III, both z_X and z_Y are negative, so their product is also positive. If most of the data

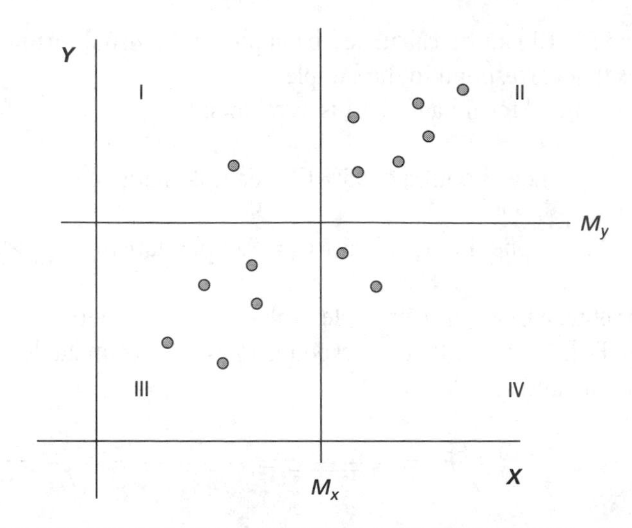

Figure 7.16 ♦ *X, Y* Scatter Plot Divided Into Quadrants (Above and Below the Means on *X* and *Y*)

points fall in Regions II and/or III, it follows that most of the contributions to the $\sum (z_X \times z_Y)$ sum of products will be positive, and the correlation will tend to be large and positive.

The data points in Regions I and IV are **"discordant"** because these are cases where high *X* went with low *Y* and/or low *X* went with high *Y*. In Region I, z_X is negative and z_Y is positive; in Region IV, z_X is positive and z_Y is negative. This means that the product of z_X and z_Y for each point that falls in Regions I or IV is negative. If there are a large number of data points in Regions I and/or IV, then most of the contributions to $\sum (z_X \times z_Y)$ will be negative, and *r* will tend to be negative.

If the data points are about evenly distributed among the four regions, then positive and negative values of $z_X \times z_Y$ will be about equally common, they will tend to cancel each other out when summed, and the overall correlation will be close to zero. This can happen because *X* and *Y* are unrelated (as in Figure 7.5) or in situations where there is a strongly curvilinear relationship (as in Figure 7.7). In either of these situations, high *X* scores are associated with high *Y* scores about as often as high *X* scores are associated with low *Y* scores.

Note that any time a statistical formula includes a product between variables of the form $\sum (X \times Y)$ or $\sum z_X \times z_Y$, the computation provides information about correlation or covariance. These products summarize information about the spatial arrangement of *X*, *Y* data points in the scatter plot; these summed products tend to be large and positive when most of the data points are in the upper right and lower left (concordant) areas of the scatter plot. In general, formulas that include sums such as $\sum X$ or $\sum Y$ provide information about the means of variables (just divide by *N* to get the mean). Terms that involve $\sum X^2$ or $\sum Y^2$ provide information about variability. Awareness about the information that these terms provide makes it possible to decode the kinds of information that are included in more complex computational formulas. Any time a $\sum (X \times Y)$ term appears, one of the elements of information included in the computation is covariance or correlation between *X* and *Y*.

Correlations provide imperfect information about the "true" strength of predictive relationships between variables. Many characteristics of the data, such as restricted ranges of scores, nonnormal distribution shape, outliers, and low reliability, can lead to over- or underestimation of the correlations between variables. Correlations and covariances provide the basic information for many other multivariate analyses (such as multiple regression and multivariate analysis of variance). It follows that artifacts that influence the values of sample correlations and covariances will also affect the results of other multivariate analyses. It is therefore extremely important for researchers to understand how characteristics of the data, such as restricted range, outliers, or measurement unreliability, influence the size of Pearson *r*, for these aspects of the data also influence the sizes of regression coefficients, **factor loadings**, and other coefficients used in multivariate models.

7.9.2 ♦ Biased Sample Selection: Restricted Range or Extreme Groups

The ranges of scores on the *X* and *Y* variables can influence the size of the sample correlation. If the research goal is to estimate the true strength of the correlation between *X* and *Y* variables for some population of interest, then the ideal sample should be randomly selected from the population of interest and should have distributions of scores on both *X* and *Y* that are representative of, or similar to, the population of interest. That is, the mean, variance, and distribution shape of scores in the sample should be similar to the population mean, variance, and distribution shape.

Suppose that the researcher wants to assess the correlation between GPA and VSAT scores. If data are obtained for a random sample of many students from a large high school with a wide range of student abilities, scores on GPA and VSAT are likely to have wide ranges (GPA from about 0 to 4.0, VSAT from about 250 to 800). See Figure 7.17 for hypothetical data that show a wide range of scores on both variables. In this example, when a wide range of scores are included, the sample correlation between VSAT and GPA is fairly high ($r = +.61$).

However, samples are sometimes not representative of the population of interest; because of accidentally biased or intentionally selective recruitment of participants, the distribution of scores in a sample may differ from the distribution of scores in the population of interest. Some sampling methods result in a restricted range of scores (on *X* or *Y* or both variables). Suppose that the researcher obtains a convenience sample by using scores for a class of honors students. Within this subgroup, the range of scores on GPA may be quite restricted (3.3–4.0), and the range of scores on VSAT may also be rather restricted (640–800). Within this subgroup, the correlation between GPA and VSAT scores will tend to be smaller than the correlation in the entire high school, as an artifact of restricted range. Figure 7.18 shows the subset of scores from Figure 7.17 that includes only cases with GPAs greater than 3.3 and VSAT scores greater than 640. For this group, which has a restricted range of scores on both variables, the correlation between GPA and VSAT scores drops to +.34. It is more difficult to predict a 40- to 60-point difference in VSAT scores from a .2- or .3-point difference in GPA for the relatively homogeneous group of honors students whose data are shown in Figure 7.18 than to predict the 300- to 400-point differences in VSAT scores from 2- or 3-point differences in GPA in the more diverse sample shown in Figure 7.17.

In general, when planning studies, researchers should try to include a reasonably wide range of scores on both predictor and outcome variables. They should also try to include the entire range of scores about which they want to be able to make inferences because it

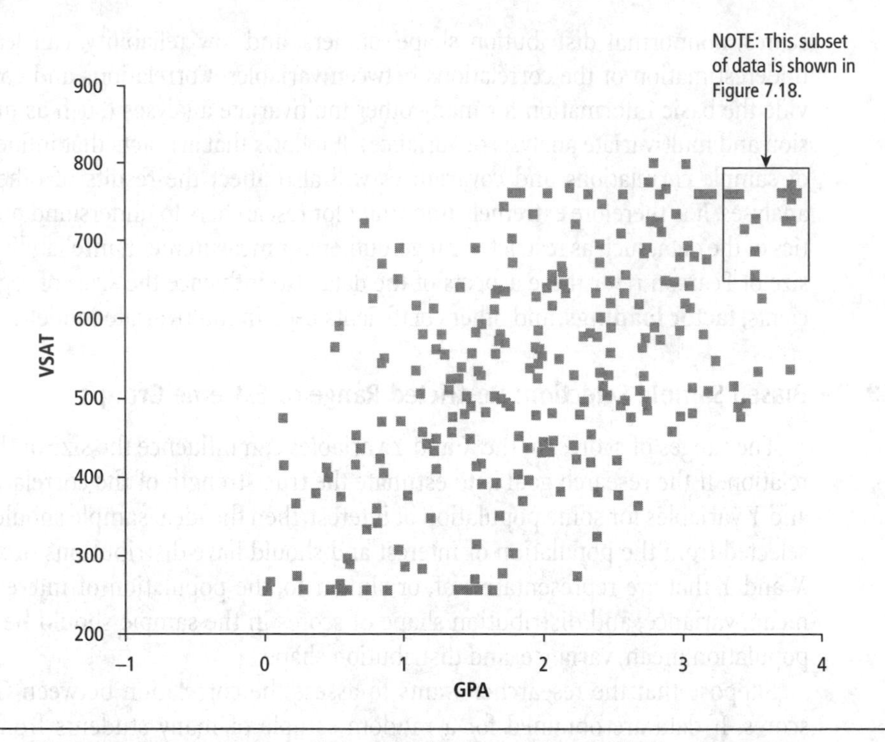

Figure 7.17 ♦ Correlation Between GPA and VSAT Scores in Data With Unrestricted Range ($r = +.61$)

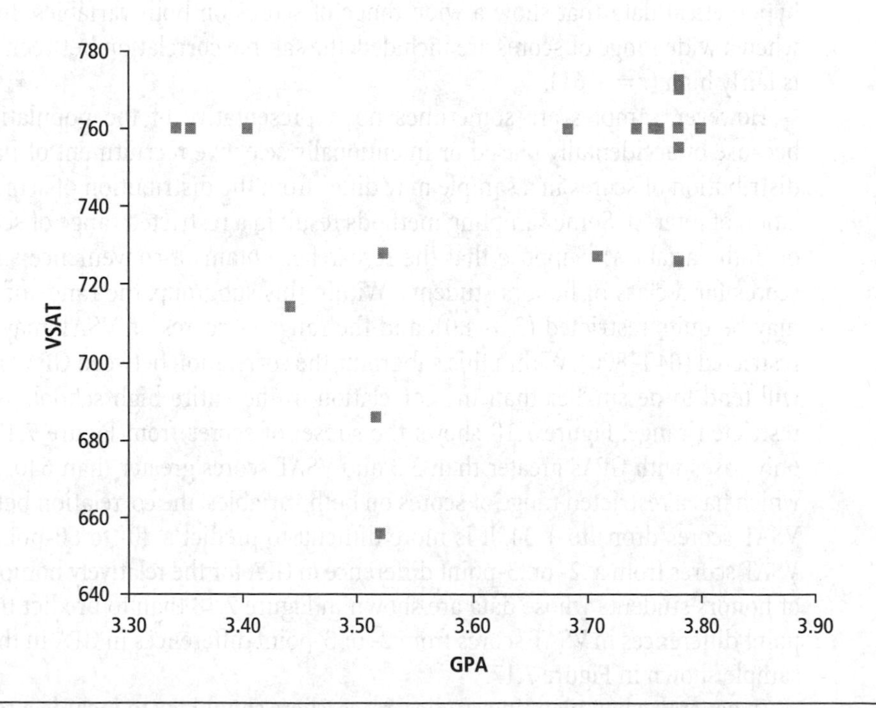

Figure 7.18 ♦ Correlation Between GPA and VSAT Scores in a Subset of Data With Restricted Range (Pearson $r = +.34$)

NOTE: This is the subset of the data in Figure 7.17 for which GPA > 3.3 and VSAT > 640.

is risky to extrapolate correlations beyond the range of scores for which you have data. For example, if you show that there is only a small correlation between age and blood pressure for a sample of participants with ages up to 40 years, you cannot safely assume that the association between age and blood pressure remains weak for ages of 50, 60, and 80 (for which you have no data). Even if you find a strong linear relation between two variables in your sample, you cannot assume that this relation can be extrapolated beyond the range of X and Y scores for which you have data (or, for that matter, to different types of research participants).

A different type of bias in correlation estimates occurs when a researcher purposefully selects groups that are extreme on both X and Y variables. This is sometimes done in early stages of research in an attempt to ensure that a relationship can be detected. Figure 7.19 illustrates the data for GPA and VSAT for two extreme groups selected from the larger batch of data in Figure 7.17 (honors students vs. failing students). The correlation between GPA and VSAT scores for this sample that comprises two extreme groups was $r = +.93$. The Pearson r obtained for samples that are formed by looking only at extreme groups tends to be much higher than the correlation for the entire range of scores. When extreme groups are used, the researcher should note that the correlation for this type of data typically overestimates the correlation that would be found in a sample that included the entire range of possible scores. Examination of extreme groups can be legitimate in early stages of research, as long as researchers understand that the correlations obtained from such samples do not describe the strength of relationship for the entire range of scores.

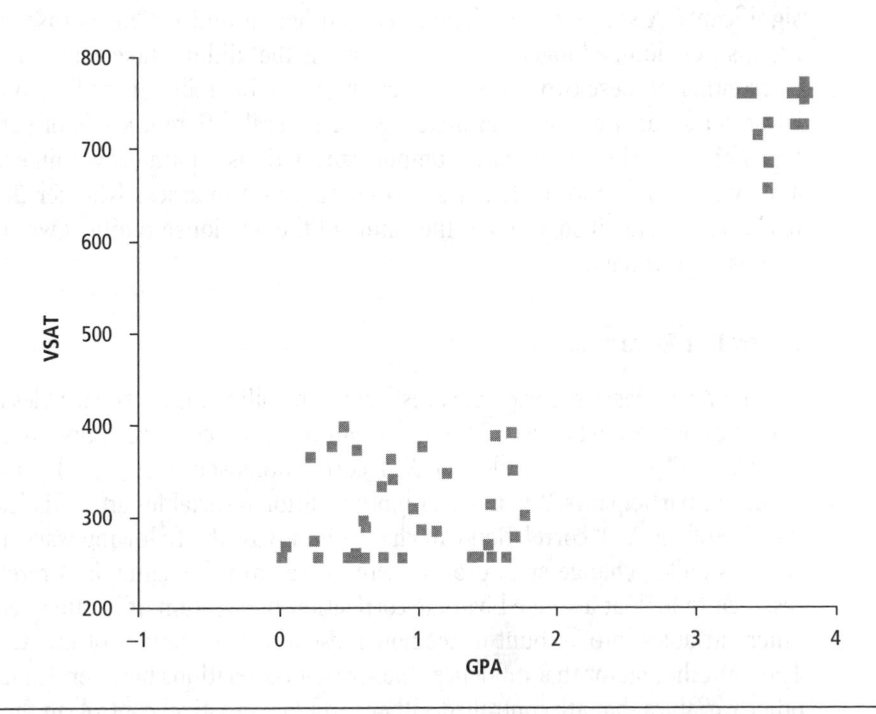

Figure 7.19 ♦ Correlation Between GPA and VSAT Scores Based on Extreme Groups (Pearson $r = +.93$)

NOTE: Two subsets of the data in Figure 7.17 (low group, GPA < 1.8 and VSAT < 400; high group, GPA > 3.3 and VSAT > 640).

7.9.3 ♦ Correlations for Samples That Combine Groups

It is important to realize that a correlation between two variables (for instance, X = EI and Y = drug use) may be different for different types of people. For example, Brackett, Mayer, and Warner (2004) found that EI was significantly predictive of illicit drug use behavior for males but not for females (men with higher EI engaged in less drug use). The scatter plot (of hypothetical data) in Figure 7.20 illustrates a similar but stronger **interaction effect**—"different slopes for different folks." In Figure 7.20, there is a fairly strong negative correlation between EI and drug use for males; scores for males appear as triangular markers in Figure 7.20. In other words, there was a tendency for males with higher EI to use drugs less. For women (data points shown as circular markers in Figure 7.20), drug use and EI were not significantly correlated. The gender differences shown in this graph are somewhat exaggerated (compared with the actual gender differences Brackett et al., 2004, found in their data), to make it clear that there were differences in the correlation between EI and drugs for these two groups (women vs. men).

A **spurious correlation** can also arise as an artifact of between-group differences. The hypothetical data shown in Figure 7.21 show a positive correlation between height and violent behavior for a sample that includes both male and female participants ($r = +.687$). Note that the overall negative correlation between height and violence occurred because women were low on both height and violence compared with men; the apparent correlation between height and violence is an artifact that arises because of gender differences on both the variables. Within the male and female groups, there was no significant correlation between height and violence ($r = -.045$ for males, $r = -.066$ for females, both not significant). A spurious correlation between height and violence arose when these two groups were lumped together into one analysis that did not take gender into account.

In either of these two research situations, it can be quite misleading to look at a correlation for a batch of data that mixes two or several different kinds of participants' data together. It may be necessary to compute correlations separately within each group (separately for males and for females, in this example) to assess whether the variables are really related and, if so, whether the nature of the relationship differs within various subgroups in your data.

7.9.4 ♦ Control of Extraneous Variables

Chapter 10 describes ways of statistically controlling for other variables that may influence the correlation between an X and Y pair of variables. For example, one simple way to "control for" gender is to calculate X, Y correlations separately for the male and female groups of participants. When one or more additional variables are statistically controlled, the size of the X, Y correlation can change in any of the following ways: It may become larger, smaller, change sign, drop to zero, or remain the same. It is rarely sufficient in research to look at a single bivariate correlation in isolation; it is often necessary to take other variables into account to see how these affect the nature of the X, Y relationship. Thus, another factor that influences the sizes of correlations between X and Y is the set of other variables that are controlled, either through statistical control (in the data analysis) or through experimental control (by holding some variables constant, for example).

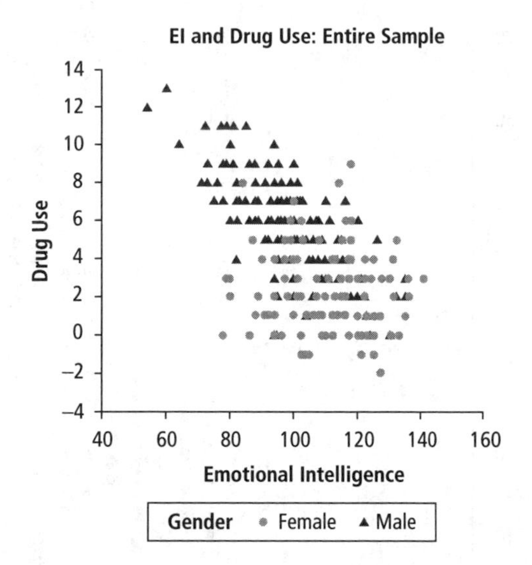

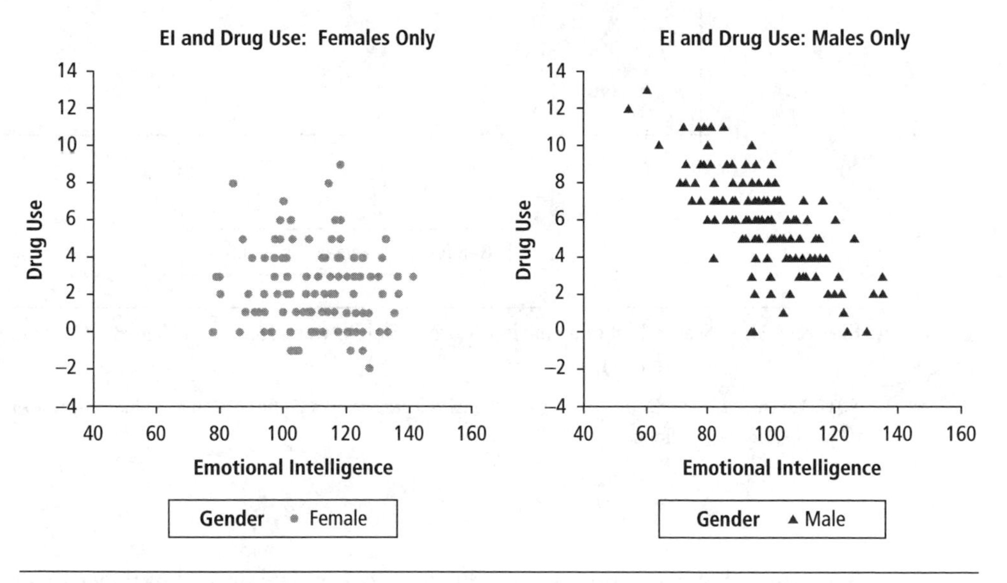

Figure 7.20 ♦ Scatter Plot for Interaction Between Gender and EI as Predictors of Drug Use: "Different Slopes for Different Folks"

NOTE: Correlation between EI and drug use for entire sample is $r(248) = -.60, p < .001$; correlation within female subgroup (circular markers) is $r(112) = -.11$, not significant; correlation within male subgroup (triangular markers) is $r(134) = -.73$, $p < .001$.

7.9.5 ♦ Disproportionate Influence by Bivariate Outliers

Like the sample mean (M), Pearson r is not robust against the influence of outliers. A single bivariate outlier can lead to either gross overestimation or gross underestimation of the value of Pearson r (refer back to Figures 7.10 and 7.11 for visual examples).

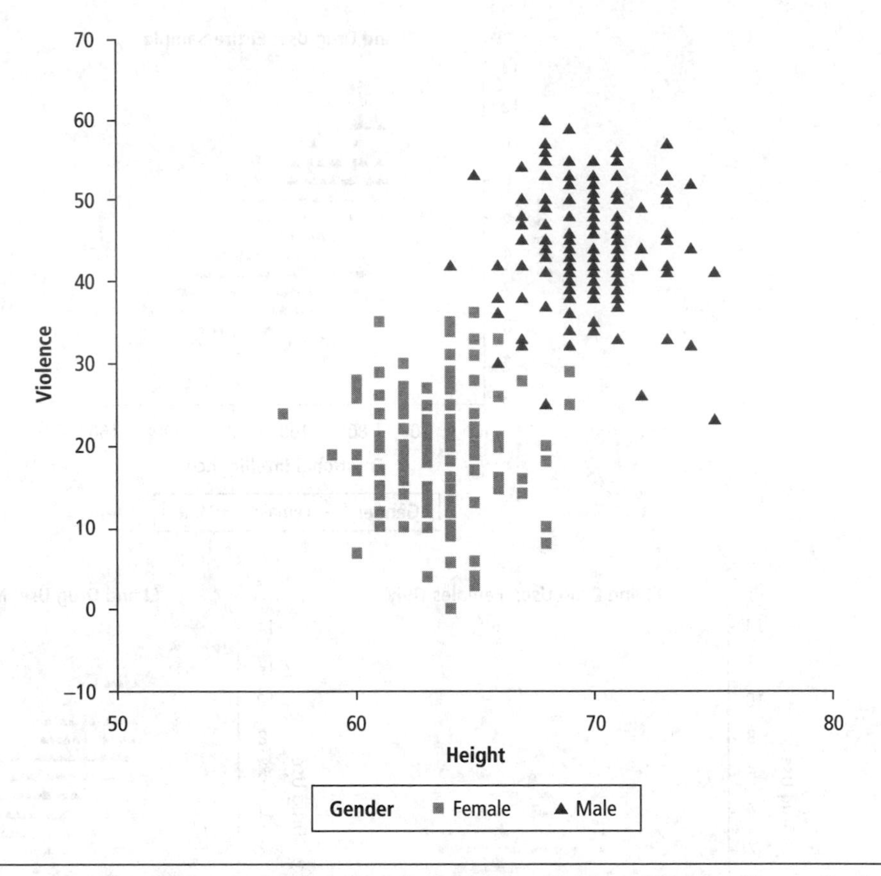

Figure 7.21 ♦ Scatter Plot of a Spurious Correlation Between Height and Violence (Due to Gender Differences)

NOTE: For entire sample, $r(248) = +.687, p < .001$; male subgroup only, $r(134) = -.045$, not significant; female subgroup only, $r(112) = -.066$, not significant.

Sometimes bivariate outliers arise due to errors in data entry, but they can also be valid scores that are unusual combinations of values (it would be unusual to find a person with a height of 72 in. and a weight of 100 lb, for example). Particularly in relatively small samples, a single unusual data value can have a disproportionate impact on the estimate of the correlation. For example, in Figure 7.10, if the outlier in the upper right-hand corner of the scatter plot is included, $r = +.64$; if that outlier is deleted, r drops to $-.10$. It is not desirable to have the outcome of a study hinge on the scores of just one or a few unusual participants. An outlier can either inflate the size of the sample correlation (as in Figure 7.10) or deflate it (as in Figure 7.11). For Figure 7.11, the r is $+.532$ if the outlier in the lower right-hand corner is included in the computation of r; the r value increases to $+.86$ if this bivariate outlier is omitted.

The impact of a single extreme outlier is much more problematic when the total number of cases is small ($N < 30$, for example). Researchers need to make thoughtful judgments about whether to retain, omit, or modify these unusual scores (see Chapter 4 for more discussion on the identification and treatment of outliers). If outliers are modified or deleted, the decision rules for doing this should be explained, and it should be made clear how many cases were involved. Sometimes it is useful to report the correlation results with and without the outlier, so that readers can judge the impact of the bivariate outlier for themselves.

As samples become very small, the impact of individual outliers becomes greater. Also, as the degrees of freedom become very small, we get "overfitting." In a sample that contains only 4 or 5 data points, a straight line is likely to fit rather well, and Pearson r is likely to be large in absolute value, even if the underlying relation between variables in the population is not strongly linear.

7.9.6 ♦ Shapes of Distributions of *X* and *Y*

Pearson r can be used as a standardized regression coefficient to predict the standard score on Y from the standard score on X (or vice versa).

This equation is as follows:

$$z'_Y = r \times z_X \text{ or } z'_Y = \beta \times z_X. \tag{7.19}$$

Correlation is a symmetric index of the relationship between variables; that is, the correlation between X and Y is the same as the correlation between Y and X. (There are some relationship indices that are nonsymmetrical; for instance, the raw score **slope coefficient** to predict scores on Y from scores on X is not generally the same as the raw score coefficient to predict scores on X from scores on Y.) One interpretation of the Pearson r is that it predicts the position of the score on the Y variable (in standard score or z score units) from the position of the score on the X variable in z score units. For instance, one study found a correlation of about .32 between the height of wives (X) and the height of husbands (Y). The equation to predict husband height (in z score units) from wife height (in z score units) is as follows:

$$z'_Y = r \times z_X;$$

this equation predicts standard scores on Y from standard scores on X. Because r is a symmetrical index of relationship, it is also possible to use it to predict z_X from z_Y:

$$z'_X = r \times z_Y.$$

In words, Equation 7.19 tells us that one interpretation of correlation is in terms of relative distance from the mean. That is, if X is 1 *SD* from its mean, we predict that Y will be r *SD* units from its mean. If husband and wife heights are correlated +.32, then a woman who is 1 *SD* above the mean in the distribution of female heights is predicted to have a husband who is about 1/3 *SD* above the mean in the distribution of male heights (and vice versa).

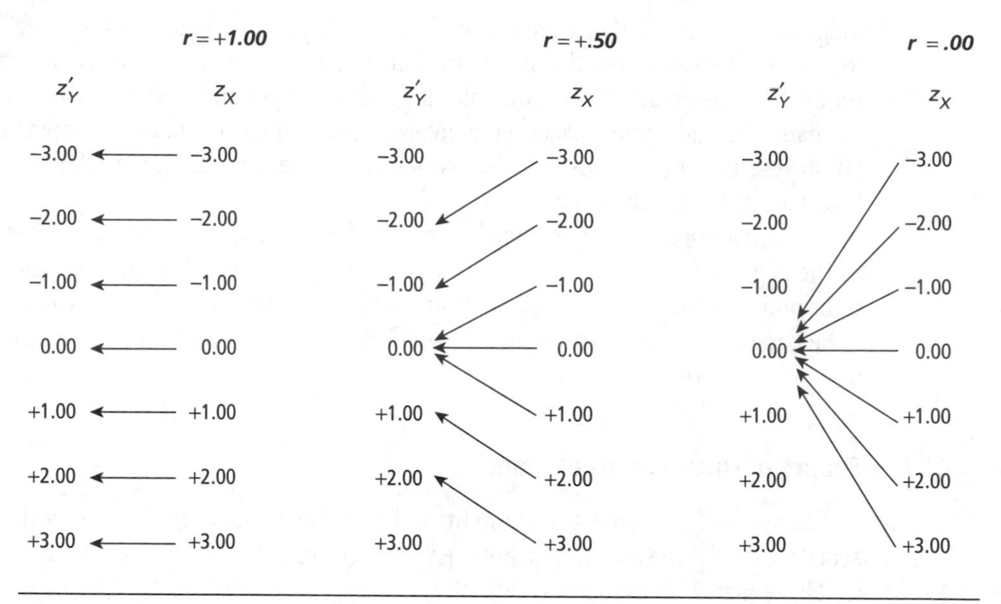

Figure 7.22 ◆ Mapping of Standardized Scores From z_X to z'_Y for Three Values of Pearson r

A correlation of +1.0 implies an exact one-to-one mapping of locations of Y scores from locations of X scores. An exact one-to-one correspondence of score locations relative to the mean can occur only when the distribution shapes of the X and Y variables are identical. For instance, if both X and Y are normally distributed, it is possible to map Y scores one-to-one on corresponding X scores. If r happens to equal +1, Equation 7.19 for the prediction of z'_Y from z_X implies that there must be an exact one-to-one mapping of score locations. That is, when $r = +1$, each person's score on Y is predicted to be exactly the same distance from the mean of Y as their score on X was from the mean of X. If the value of r is less than 1, we find that the predicted score on Y is always somewhat closer to the mean than the score on X. See Figure 7.22 for a representation of the mapping from z_X to z'_Y for three different values of r: $r = +1$, $r = +.5$, and $r = .00$.

This phenomenon, that the predicted score on the dependent variable is closer to the mean than the score on the independent variable whenever the correlation is less than 1, is known as **regression toward the mean**. (We call the equation to predict Y from scores on one or more X variables a "regression equation" for this reason.) When X and Y are completely uncorrelated ($r = 0$), the predicted score for all participants corresponds to $z_Y = 0$ regardless of their scores on X. That is, when X and Y are not correlated, our best guess at any individual's score on Y is just the mean of Y (M_Y). If X and Y are positively correlated (but r is less than 1), then for participants who score above the mean on X, we predict scores that are above the mean (but not as far above the mean as the X scores) on Y.

Note that we can only obtain a perfect one-to-one mapping of z score locations on the X and Y scores when X and Y both have exactly the same distribution shape. That is, we

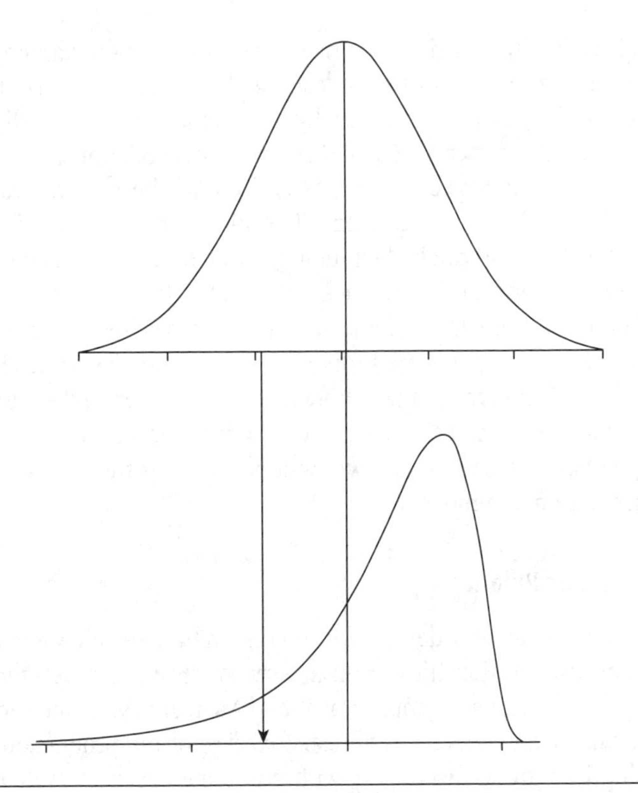

Figure 7.23 ♦ Failure of One-to-One Mapping of Score Location for Different Distribution Shapes

NOTE: For example, there are more scores located about 1 *SD* below the mean in the upper/normal distribution than in the lower/skewed distribution, and there are more scores located about 1 *SD* above the mean in the lower/skewed distribution than in the upper/normal distribution. When the shapes of distributions for the *X* and *Y* scores are very different, this difference in distribution shape makes it impossible to have an exact one-to-one mapping of score locations for *X* and *Y*; this in turn makes it impossible to obtain a sample correlation close to +1.00 between scores on *X* and *Y*.

can only obtain a Pearson r of +1 if X and Y both have the same shape. As an example, consider a situation where X has a normal distribution shape and Y has a skewed distribution shape (see Figure 7.23). It is not possible to make a one-to-one mapping of the scores with z_X values greater than $z_X = +1$ to corresponding z_Y values greater than +1; the Y distribution has far more scores with z_Y values greater than +1 than the X distribution in this example.

This example illustrates that when we correlate scores on two quantitative variables that have different distribution shapes, the maximum possible correlation that can be obtained will artifactually be less than 1 in absolute value. Perfect one-to-one mapping (that corresponds to $r = 1$) can only arise when the distribution shapes for the X and Y variables are the same.

It is desirable for all the quantitative variables in a multivariable study to have nearly normal distribution shapes. However, if you want to compare X_1 and X_2 as predictors of Y and if the distribution shapes of X_1 and X_2 are different, then the variable with a distribution less similar to the distribution shape of Y will artifactually tend to have a lower correlation with Y. This is one reason why comparisons among correlations for different predictor variables can be misleading. The effect of distribution shape on the size of Pearson r is one of the reasons why it is important to assess distribution shape for all the variables before interpreting and comparing correlations.

Thus, correlations can be artifactually small because the variables that are being correlated have drastically different distribution shapes. Sometimes nonlinear transformations (such as log) can be helpful in making a skewed or exponential distribution appear more nearly normal. Tabachnick and Fidell (2007) recommend using such transformations and/or removing extreme outlier scores to achieve more normal distribution shapes when scores are nonnormally distributed. Some analysts (e.g., Harris, 2001; Tabachnick & Fidell, 2007) suggest that it is more important for researchers to screen their data to make sure that the distribution shapes of quantitative variables are approximately normal in shape than for researchers to worry about whether the scores have true interval/ratio level of measurement properties.

7.9.7 ♦ Curvilinear Relations

Pearson r can only detect a *linear* relation between an X and a Y variable. Various types of nonlinear relationships result in very small r values even though the variables may be strongly related; for example, Hayes and Meltzer (1972) showed that favorable evaluation of speakers varied as a curvilinear function of the proportion of time the person spent talking in a group discussion, with moderate speakers evaluated most positively (refer back to Figure 7.7 for an example of this type of curvilinear relation). An r of 0 does not necessarily mean that two variables are unrelated, only that they are not *linearly* related. When nonlinear or curvilinear relations appear in a scatter plot (for an example of one type of curvilinear relation, see Figure 7.7), Pearson r is not an appropriate statistic to describe the strength of the relationship between variables. Other approaches can be used when the relation between X and Y is nonlinear. One approach is to apply a data transformation to one or both variables (e.g., replace X by $\log(X)$). Sometimes a relationship that is not linear in the original units of measurement becomes linear under the right choice of data transformation. Another approach is to recode the scores on the X predictor variable so that instead of having continuous scores, you have a group membership variable ($X = 1$, low; $X = 2$, medium; $X = 3$, high); an analysis of variance (ANOVA) to compare means on Y across these groups can detect a nonlinear relationship such that Y has the highest scores for medium values of X. Another approach is to predict scores on Y from scores on X^2 as well as X.

7.9.8 ♦ Transformations of Data

Linear transformations of X and Y that involve addition, subtraction, multiplication, or division by constants do not change the magnitude of r (except in extreme cases where very large or very small numbers can result in important information getting lost in rounding error). Nonlinear transformations of X and Y (such as log, square root, X^2, etc.) can change correlations between variables substantially, particularly if there are extreme outliers or if there are several log units separating the minimum and maximum data values. Transformations may be used to make nonlinear relations linear, to minimize the impact of extreme outliers, or to make nonnormal distributions more nearly normal.

For example, refer back to the data that appear in Figure 7.10. There was one extreme outlier (circled). One way to deal with this outlier is to select it out of the dataset prior to calculating Pearson *r*. Another way to minimize the impact of this bivariate outlier is to take the base 10 logarithms of both *X* and *Y* and correlate the log-transformed scores. Taking the logarithm reduces the value of the highest scores, so that they lie closer to the body of the distribution.

An example of data that show a strong linear relationship only after a log transformation is applied to both variables was presented in Chapter 4, Figures 4.47 and 4.48. Raw scores on body mass and metabolic rate showed a curvilinear relation in Figure 4.47; when the base 10 log transformation was applied to both variables, the log-transformed scores were linearly related.

Finally, consider a positively skewed distribution (refer back to Figure 4.20). If you take either the base 10 or natural log of the scores in Figure 4.20 and do a histogram of the logs, the extreme scores on the upper end of the distribution appear to be closer to the body of the distribution, and the distribution shape can appear to be closer to normal when a log transformation is applied. However, a log transformation changes the shape of a sample distribution of scores substantially only when the maximum score on *X* is 10 or 100 times as high as the minimum score on *X*; if scores on *X* are ratings on a 1 to 5 point scale, for example, a log transformation applied to these scores will not substantially change the shape of the distribution.

7.9.9 ♦ Attenuation of Correlation Due to Unreliability of Measurement

Other things being equal, when the *X* and *Y* variables have low measurement reliability, this low reliability tends to decrease or attenuate their observed correlations with other variables. A reliability coefficient for an *X* variable is often denoted by r_{XX}. One way to estimate a reliability coefficient for a quantitative *X* variable would be to measure the same group of participants on two occasions and correlate the scores at Time 1 and Time 2 (a test-retest correlation). (Values of reliability coefficients r_{XX} range from 0 to 1.) The magnitude of this **attenuation** of observed correlation as an artifact of unreliability is given by the following formula:

$$r_{XY} = \rho_{XY}\sqrt{r_{XX} \times r_{YY}}, \tag{7.20}$$

where r_{XY} is the observed correlation between *X* and *Y*, ρ_{XY} is the "real" correlation between *X* and *Y* that would be obtained if both variables were measured without error, and r_{XX} and r_{YY} are the test-retest reliabilities of the variables.

Because reliabilities are usually less than perfect ($r_{XX} < 1$), Equation 7.20 implies that the observed r_{XY} will generally be smaller than the "true" population correlation ρ_{XY}. The lower the reliabilities, the greater the predicted attenuation or reduction in magnitude of the observed sample correlation.

It is theoretically possible to correct for attenuation and estimate the true correlation ρ_{XY}, given obtained values of r_{XY}, r_{XX}, and r_{YY}; but note that if the reliability estimates themselves are inaccurate, this estimated true correlation may be quite misleading. Equation 7.21 can be used to generate attenuation-corrected estimates of correlations; however, keep in mind that this correction will be inaccurate if the reliabilities of the measurements are not precisely known:

$$\widehat{\rho}_{XY} = \frac{r_{XY}}{\sqrt{r_{XX} \times r_{YY}}}. \tag{7.21}$$

7.9.10 ♦ Part-Whole Correlations

If you create a new variable that is a function of one or more existing variables (as in $X = Y + Z$, or $X = Y - Z$), then the new variable X will be correlated with its component parts Y and Z. Thus, it is an artifact that the total Wechsler Adult Intelligence Scale (WAIS) score (which is the sum of WAIS verbal + WAIS quantitative) is correlated to the WAIS verbal subscale. Part-whole correlations can also occur as a consequence of item overlap: If two psychological tests have identical or very similar items contained in them, they will correlate artifactually because of item overlap. For example, many depression measures include questions about fatigue, sleep disturbance, and appetite disturbance. Many physical illness symptom checklists also include questions about fatigue and sleep and appetite disturbance. If a depression score is used to predict a physical symptom checklist score, a large part of the correlation between these could be due to duplication of items.

7.9.11 ♦ Aggregated Data

Correlations can turn out to be quite different when they are computed on individual participant data versus **aggregated data** where the units of analysis correspond to groups of participants. It can be misleading to make inferences about individuals based on aggregated data; sociologists call this the "ecological fallacy." Sometimes relationships appear much stronger when data are presented in aggregated form (e.g., when each data point represents a mean, median, or rate of occurrence for a geographical region). Keys (1980) collected data on serum cholesterol and on coronary heart disease outcomes for $N = 12{,}763$ men who came from 19 different geographical regions around the world. He found a correlation near zero between serum cholesterol and coronary heart disease for individual men; however, when he aggregated data for each of the 19 regions and looked at median serum cholesterol for each region as a predictor of the rate of coronary heart disease, the r for these aggregated data (in Figure 7.24) went up to +.80.

7.10 ♦ Pearson r and r^2 as Effect Size Indexes

The indexes that are used to describe the effect size or strength of linear relationship in studies that report Pearson r values are usually either just r itself or r^2, which estimates the proportion of variance in Y that can be predicted from X (or equivalently, the proportion of variance in X that is predictable from Y). The proportion of explained variance (r^2) can be diagramed by using overlapping circles, as shown in Figure 7.25. Each circle represents the unit variance of z scores on each variable, and the area of overlap is proportional to r^2, the shared or explained variance. The remaining area of each circle corresponds to $1 - r^2$, for example, the proportion of variance in Y that is not explained by X. In meta-analysis research, where results are combined across many studies, it is more common to use r itself as the effect size indicator. Cohen (1988) suggested the following verbal labels for sizes of r: r of about .10 or less ($r^2 < .01$) is small, r of about .30 ($r^2 = .09$) is medium, and r greater than .50 ($r^2 > .25$) is large. (Refer back to Table 5.2 for a summary of suggested verbal labels for r and r^2.)

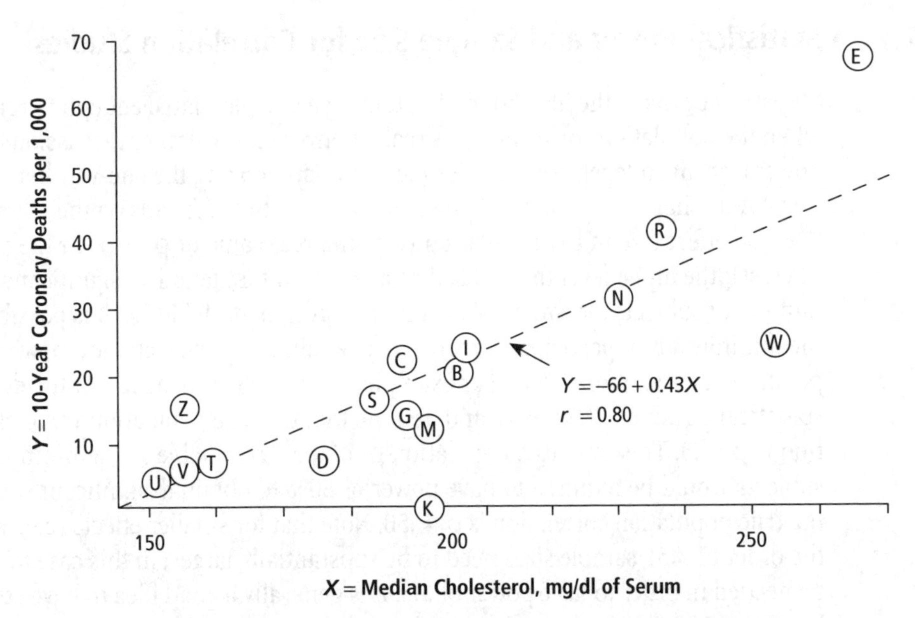

Figure 7.24 ♦ A Study in Which the Correlation Between Serum Cholesterol and Cardiovascular Disease Outcomes Was $r = +.80$ for Aggregated Scores (Each Data Point Summarizes Information for One of the 19 Geographic Regions in the Study). In Contrast, the Correlation Between These Variables for Individual-Level Scores Was Close to Zero

SOURCE: Reprinted by permission of the publisher from *Seven Countries: A Multivariate Analysis of Death and Coronary Heart Disease* by Ancel Keys, p. 122, Harvard University Press, copyright © 1980 by the President and Fellows of Harvard College.

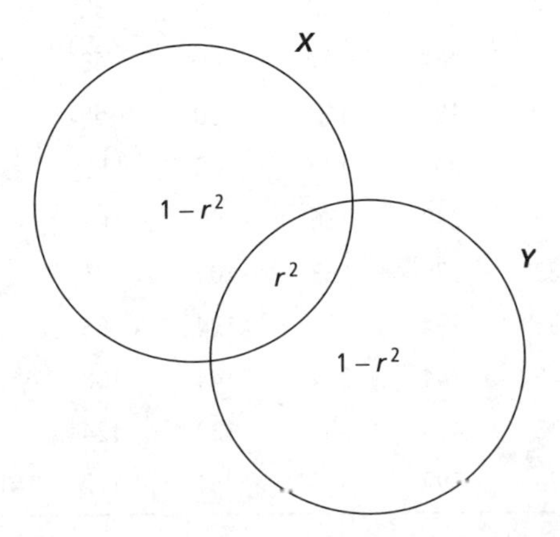

Figure 7.25 ♦ Proportion of Shared Variance Between X and Y Corresponds to r^2, the Proportion of Overlap Between the Circles

7.11 ♦ Statistical Power and Sample Size for Correlation Studies

Statistical power is the likelihood of obtaining a sample *r* large enough to reject H_0: $\rho = 0$ when the population correlation ρ is really nonzero. As in earlier discussions of statistical power (for the independent samples *t* test in Chapter 5 and the one-way between-subjects ANOVA in Chapter 6), statistical power for a correlation depends on the following factors: the true effect size in the population (e.g., the real value of ρ or ρ^2 in the population of interest); the alpha level that is used as a criterion for statistical significance; and *N*, the number of subjects for whom we have *X*, *Y* scores. Using Table 7.4, it is possible to look up the minimum *N* of participants required to obtain adequate statistical power for different population correlation values. For example, let $\alpha = .05$, two-tailed; set the desired level of statistical power at .80 or 80%; and assume that the true population value of the correlation is $\rho = .5$. This implies a population ρ^2 of .25. From Table 7.4, a minimum of $N = 28$ subjects would be required to have power of 80% to obtain a significant sample result if the true population correlation is $\rho = .50$. Note that for smaller effects (e.g., a ρ^2 value on the order of .05), sample sizes need to be substantially larger; in this case, $N = 153$ would be needed in order to have power of .80. It is generally a good idea to have studies with at least $N = 100$ cases where correlations are reported, partly to have adequate statistical power but also to avoid situations where there is not enough information to evaluate whether assumptions (such as bivariate normality and linearity) are satisfied and to avoid situations where one or two extreme outliers can have a large effect on the size of the sample *r*.

Table 7.4 ♦ Statistical Power for Pearson *r*

Approximate Sample Sizes Necessary to Achieve Selected Levels of Power for Alpha = .05, Nondirectional Test, as a Function of the Population Correlation Coefficient Squared

	Population Correlation Coefficient Squared									
Power	*.01*	*.03*	*.05*	*.07*	*.10*	*.15*	*.20*	*.25*	*.30*	*.35*
.25	166	56	34	25	17	12	9	8	6	6
.50	384	127	76	54	38	25	19	15	12	10
.60	489	162	97	69	48	31	23	18	15	13
.67	570	188	112	80	55	36	27	21	17	14
.70	616	203	121	86	59	39	29	23	18	15
.75	692	228	136	96	67	43	32	25	20	17
.80	783	258	153	109	75	49	36	28	23	19
.85	895	294	175	124	85	56	41	32	26	21
.90	1,046	344	204	144	100	65	47	37	30	25
.95	1,308	429	255	180	124	80	58	46	37	30
.99	1,828	599	355	251	172	111	81	63	50	42

SOURCE: From *Statistics for the Behavioral Sciences, 4th edition* by Jaccard/Becker, 2002. Reprinted with permission of Wadsworth, a division of Thomson Learning: www.thomsonrights.com. Fax 800-730-2215.

7.12 ♦ Interpretation of Outcomes for Pearson *r*

7.12.1 ♦ "Correlation Does Not Necessarily Imply Causation" (So What Does It Imply?)

Introductory statistics students are generally told, "Correlation does not imply causation." This can be more precisely stated as "Correlational (or nonexperimental) design does not imply causation." When data are obtained from correlational or nonexperimental research designs, we cannot make causal inferences. In the somewhat uncommon situations where Pearson r is applied to data from well-controlled experiments, tentative causal inferences may be appropriate. As stated earlier, it is the nature of the research design, not the statistic that happens to be used to analyze the data, that determines whether causal inferences might be appropriate.

Many researchers who conduct nonexperimental studies and apply Pearson r to variables do so with causal theories implicitly in mind. When a researcher measures social stress and blood pressure in natural settings, and correlates these measures, the researcher expects to see a positive correlation, in part because the researcher suspects that social stress may "cause" an increase in blood pressure. Very often, predictor variables are selected for use in correlational research because the researcher believes that they may be "causal." If the researcher finds no statistical association between X and Y (a Pearson r that is close to 0 and no evidence of any other type of statistical association, such as a curvilinear relation), this finding of no statistical relationship is inconsistent with the belief that X causes Y. If X did cause Y, then increases in X should be statistically associated with changes in Y. If we find a significant correlation, it is consistent with the idea that X might cause Y, but it is not sufficient evidence to prove that X causes Y. Scores on X and Y can be correlated for many reasons; a causal link between X and Y is only one of many situations that tends to create a correlation between X and Y. A statistical association between X and Y (such as a significant Pearson r) is a necessary, but not sufficient, condition for the conclusion that X might cause Y.

Many other situations (apart from "X causes Y") may give rise to correlations between X and Y. Here is a list of some of the possible reasons for an X, Y correlation. We can conclude that X causes Y only if we can rule out *all* these other possible explanations. (And in practice, it is almost never possible to rule out all these other possible explanations in nonexperimental research situations; well-controlled experiments come closer to accomplishing the goal of ruling out rival explanations, although a single experiment is not a sufficient basis for a strong causal conclusion.)

Reasons why X and Y may be correlated include the following:

1. X may be a cause of Y.
2. Y may be a cause of X.
3. X might cause Z, and in turn, Z might cause Y. In causal sequences like this, we say that the Z variable "mediates" the effect of X on Y or that Z is a **"mediator"** or **"mediating" variable.**
4. X is confounded with some other variable Z, and Z predicts or causes Y. In a well-controlled experiment, we try to artificially arrange the situation so that no other

variables are confounded systematically with our X intervention variable. In nonexperimental research, we often find that our X variable is correlated with or confounded with many other variables that might be the "real" cause of Y. In most nonexperimental studies, there are potentially dozens or hundreds of potential nuisance (or confounded) Z variables. For instance, we might try to predict student GPA from family structure (single parent vs. two parent); but if students with single-parent families have lower GPAs than students from two-parent families, this might be because single-parent families have lower incomes and lower-income neighborhoods may have poor-quality schools that lead to poor academic performance.

5. X and Y might actually be measures of the same thing, instead of two separate variables that could be viewed as cause and effect. For example, X could be a depression measure that consists of questions about both physical and mental symptoms; Y could be a checklist of physical health symptoms. If X predicts Y, it might be because the depression measure and the health measure both included the same questions (about fatigue, sleep disturbance, low energy level, appetite disturbance, etc.).
6. Both X and Y might be causally influenced by some third variable Z. For instance, both ice cream sales (X) and homicides (Y) tend to increase when temperatures (Z) go up. When X and Y are both caused or predicted by some third variable and X has no direct causal influence on Y, the X, Y correlation is called spurious.
7. Sometimes a large X, Y correlation arises just due to chance and sampling error; it just happens that the participants in the sample tended to show a strong correlation because of the luck of the draw, even though the variables are not correlated in the entire population. (When we use statistical significance tests, we are trying to rule out this possible explanation, but we can never be certain that a significant correlation was not simply due to chance.)

This list does not exhaust the possibilities; apart from the single Z variable mentioned in these examples, there could be numerous other variables involved in the X, Y relationship.

In well-controlled experiments, a researcher tries to arrange the situation so that no other Z variable is systematically confounded with the independent variable X. Experimental control makes it possible, in theory, to rule out many of these possible rival explanations. However, in nonexperimental research situations, when a large correlation is found, all these possible alternative explanations have to be considered and ruled out before we can make a case for the interpretation that "X causes Y"; and in practice, it is not possible to rule out all of these alternative reasons why X and Y are correlated in nonexperimental data. It is possible to make the case for a causal interpretation of a correlation somewhat stronger by statistically controlling for some of the Z variables that you know are likely to be confounded with X, but it is never possible to identify and control for all the possible confounds. That's why we have to keep in mind that "correlational *design* does not imply causation."

7.12.2 ♦ Interpretation of Significant Pearson r Values

Pearson r describes the strength and direction of the linear predictive relationship between variables. The sign of r indicates the direction of the relationship; for a positive r, as scores on X increase, scores on Y also tend to increase; for a negative r, as scores on X increase, scores on Y tend to decrease. The r value indicates the magnitude of change in z'_Y for a one-SD change in z_X. For example, if $r = +.5$, for a one-SD increase in z_X, we predict a .5-SD increase in z_Y. When researchers limit themselves to descriptive statements about predictive relationships or statistical associations between variables, it is sufficient to describe this in terms of the direction and magnitude of the change in z_Y associated with change in z_X.

Thus, a significant Pearson correlation can be interpreted as information about the degree to which scores on X and Y are linearly related, or the degree to which Y is predictable from X. Researchers often examine correlations among variables as a way of evaluating whether variables might possibly be causally related. In many studies, researchers present a significant correlation between X and Y as evidence that X and Y might *possibly* be causally related. However, researchers should not interpret correlations that are based on nonexperimental research designs as evidence for causal connections. Experiments in which X is manipulated and other variables are controlled and the Y outcome is measured provide stronger evidence for causal inference.

7.12.3 ♦ Interpretation of a Nonsignificant Pearson *r* Value

A Pearson r near 0 does not always indicate a complete lack of relationship between variables. A correlation that is not significantly different from zero might be due to a true lack of any relationship between X and Y (as in Figure 7.5); but correlations near zero also arise when there is a strong but curvilinear relation between X and Y (as in Figure 7.7) or when one or a few bivariate outliers are not consistent with the pattern of relationship suggested by the majority of the data points (as in Figure 7.11). For this reason, it is important to examine the scatter plot before concluding that X and Y are not related.

If examination of the scatter plot suggests a nonlinear relation, a different analysis may be needed to describe the X, Y association. For example, using the multiple regression methods described in a later chapter, Y may be predicted from X, X^2, X^3, and other powers of X. Other nonlinear transformations (such as log of X) may also convert scores into a form where a linear relationship emerges. Alternatively, if the X variable is recoded to yield three or more categories (e.g., if income in dollars is recoded into low-, medium-, and high-income groups), a one-way ANOVA comparing scores among these three groups may reveal differences.

If a sample correlation is relatively large but not statistically significant due to small sample size, it is possible that the correlation might be significant in a study with a larger sample. Only additional research with larger samples can tell us whether this is the case. When a nonsignificant correlation is obtained, the researcher should not conclude that the study proves the absence of a relation between X and Y. It would be more accurate to say that the study did not provide evidence of a linear relationship between X and Y.

7.13 ♦ SPSS Output and Model Results Write-Up

To obtain a Pearson correlation using SPSS, the menu selections (from the main menu above the data worksheet) are <Analyze> → <Correlate> → <Bivariate>, as shown in Figure 7.26. These menu selections open up the SPSS Correlations dialog window, shown in Figure 7.27.

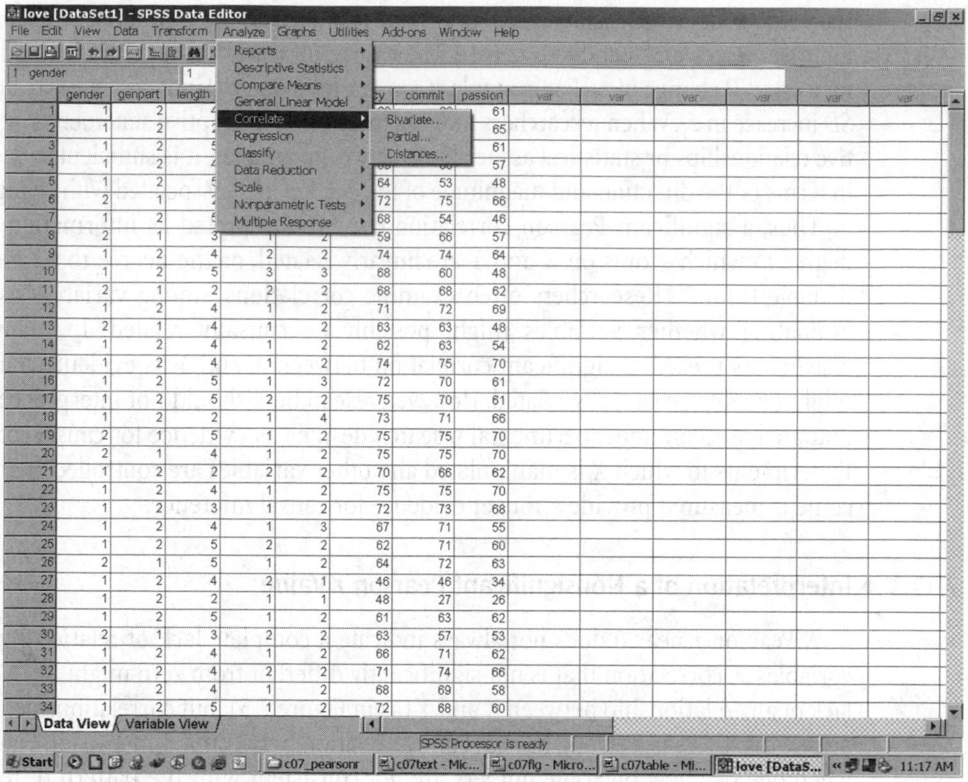

Figure 7.26 ♦ Menu Selections for the Bivariate Correlation Procedure

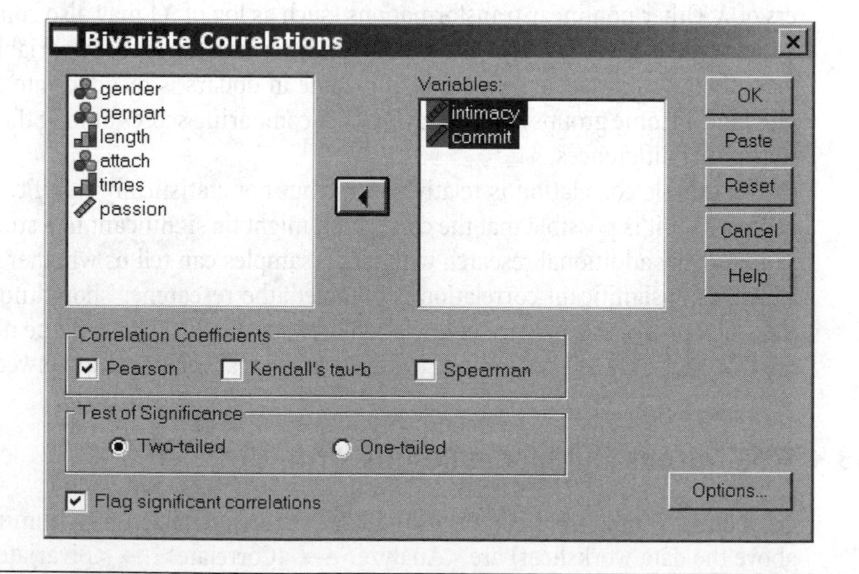

Figure 7.27 ♦ SPSS Correlations Procedure Dialog Window

Correlations

		intimacy	commit
intimacy	Pearson Correlation	1	.745**
	Sig. (2-tailed)		.000
	N	118	118
commit	Pearson Correlation	.745**	1
	Sig. (2-tailed)	.000	
	N	118	118

**. Correlation is significant at the 0.01 level

Figure 7.28 ♦ SPSS Output for the Bivariate Correlation Procedure

The data analyst uses the cursor to highlight the names of the variables in the left-hand window (which lists all the variables in the active data file) that are to be intercorrelated. Then, the user clicks on the arrow button to move two or more variable names into the list of variables to be analyzed. In this example, the variables to be correlated are named commit (commitment) and intimacy. Other boxes can be checked to determine whether or not significance tests are to be displayed, and whether two-tailed or one-tailed p values are desired. To run the analyses, click the OK button. The output from this procedure is displayed in Figure 7.28, which shows the value of the Pearson correlation ($r = +.745$), the p value (which would be reported as $p < .001$, two-tailed),[3] and the N of data pairs the correlation was based on ($N = 118$). The degrees of freedom for this correlation is given by $N - 2$, so in this example, the correlation has 116 *df.*

It is possible to run correlations among many pairs of variables. The SPSS Correlations dialog window that appears in Figure 7.29 includes a list of five variables: intimacy, commit, passion, length (of relationship), and times (the number of times the person has been in love). If the data analyst enters a list of five variables as shown in this example, SPSS runs the bivariate correlations among all possible pairs of these five variables (this results in $(5 \times 4)/2 = 10$ different correlations), and these correlations are reported in a summary table as shown in Figure 7.30. Note that because correlation is "symmetrical" (i.e., the correlation between X and Y is the same as the correlation between Y and X), the correlations that appear in the upper right-hand corner of the table in Figure 7.30 are the same as those that appear in the lower left-hand corner. When such tables are presented in journal articles, usually only the correlations in the upper right-hand corner are shown.

Sometimes a data analyst wants to obtain summary information about the correlations between a set of predictor variables X_1 and X_2 (length and times) and a set of outcome variables Y_1, Y_2, and Y_3 (intimacy, commitment, and passion). To do this, we need to paste and edit SPSS Syntax. Look again at the SPSS dialogue window in Figure 7.29; there is a button labeled Paste just below OK. Clicking the Paste button opens up a new window, called a Syntax window, and pastes the SPSS commands (or syntax) for correlation that were generated by the user's menu selections into this window; the initial SPSS Syntax window appears in Figure 7.31. This syntax can be saved, printed, or edited. In this example, we will edit the syntax; in Figure 7.32, the SPSS keyword WITH has been inserted into the list of variable names so that the list of variables in the CORRELATIONS

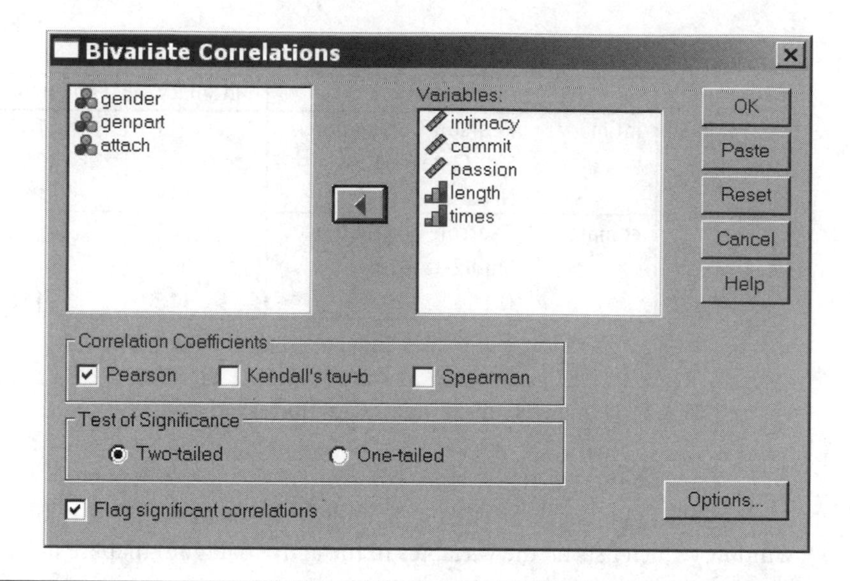

Figure 7.29 ♦ Bivariate Correlations Among All Possible Pairs of Variables in a List

Correlations

		intimacy	commit	passion	length	times
intimacy	Pearson Correlation	1	.745**	.735**	.175	-.008
	Sig. (2-tailed)		.000	.000	.058	.934
	N	118	118	114	118	115
commit	Pearson Correlation	.745**	1	.835**	.197*	.008
	Sig. (2-tailed)	.000		.000	.033	.929
	N	118	118	114	118	115
passion	Pearson Correlation	.735**	.835**	1	.074	-.041
	Sig. (2-tailed)	.000	.000		.432	.670
	N	114	114	114	114	111
length	Pearson Correlation	.175	.197*	.074	1	-.090
	Sig. (2-tailed)	.058	.033	.432		.340
	N	118	118	114	118	115
times	Pearson Correlation	-.008	.008	-.041	-.090	1
	Sig. (2-tailed)	.934	.929	.670	.340	
	N	115	115	111	115	115

**. Correlation is significant at the 0.01 level (2-tailed).

*. Correlation is significant at the 0.05 level (2-tailed).

Figure 7.30 ♦ SPSS Output: Bivariate Correlations Among All Possible Pairs of Variables

command now reads, "intimacy commit passion WITH length times." (It does not matter whether the SPSS commands are in uppercase or lowercase; the word WITH appears in uppercase characters in this example to make it easy to see the change in the command.) For a correlation command that includes the key word WITH, SPSS computes correlations for all pairs of variables on the lists that come before and after WITH. In this example,

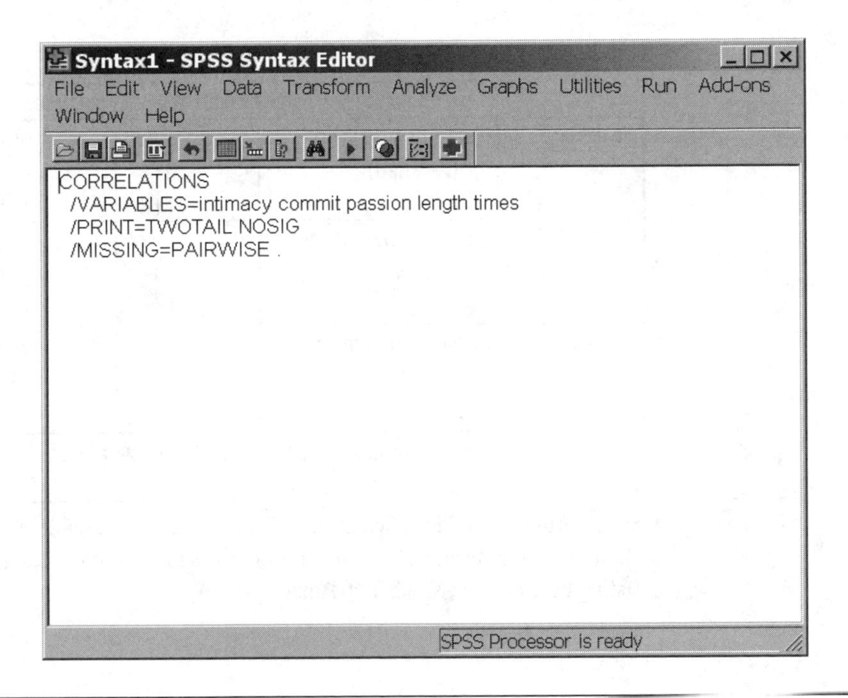

Figure 7.31 ♦ Pasting SPSS Syntax Into a Syntax Window and Editing Syntax

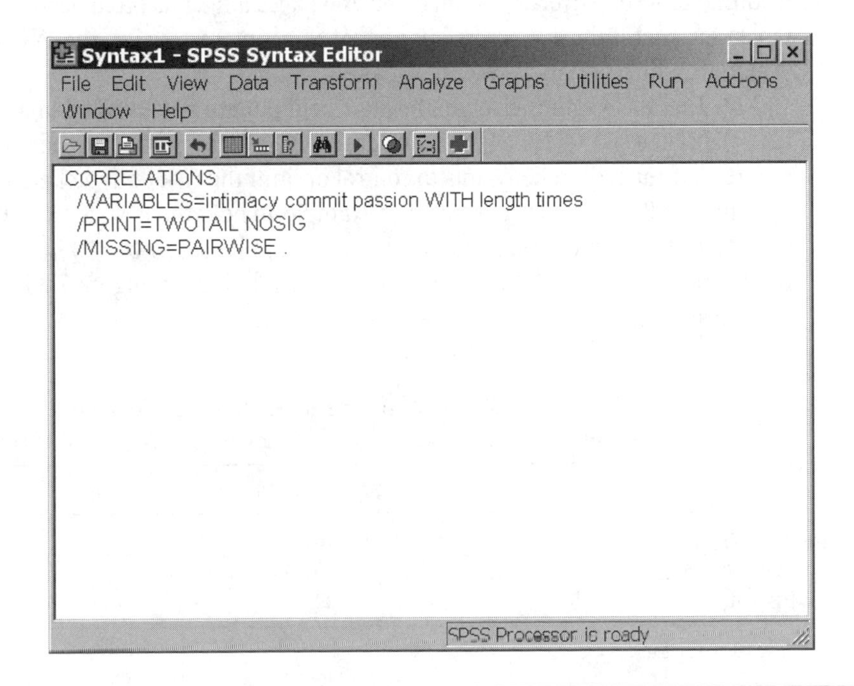

Figure 7.32 ♦ Edited SPSS Syntax

Correlations

		length	times
intimacy	Pearson Correlation	.175	-.008
	Sig. (2-tailed)	.058	.934
	N	118	115
commit	Pearson Correlation	.197*	.008
	Sig. (2-tailed)	.033	.929
	N	118	115
passion	Pearson Correlation	.074	-.041
	Sig. (2-tailed)	.432	.670
	N	114	111

*. Correlation is significant at the 0.05 level (2-tailed).

Figure 7.33 ♦ SPSS Output From Edited Syntax: Correlations Between Variables in the First List (Intimacy, Commitment, Passion) And Variables in the Second List (Length of Present Dating Relationship, Number of Times in Love)

each variable in the first list (intimacy, commit, and passion) is correlated with each variable in the second list (length and times). This results in a table of six correlations, as shown in Figure 7.33.

If many variables are included in the list for the bivariate correlation procedure, the resulting table of correlations can be several pages long. It is often useful to set up smaller tables for each subset of correlations that is of interest, using the WITH command to designate which variables should be paired.

Note that the *p* values given in the SPSS printout are not adjusted in any way to correct for the inflated risk of Type I error that arises when large numbers of significance tests are reported. If the researcher wants to control or limit the risk of Type I error, it can be done by using the Bonferroni procedure. For example, to hold the EW_α level to .05, the PC_α level used to test the six correlations in Table 7.5 could be set to $\alpha = .05/6 = .008$. Using this Bonferroni-corrected PC_α level, none of the correlations in Figure 7.33 would be judged statistically significant.

Table 7.5 ♦ Correlations Between Sternberg Love Scales (Intimacy, Commitment, and Passion) and Length of Relationship and Number of Times in Love ($N = 118$ Participants)

	Length of Relationship	*Times in Love*
Intimacy	.18	–.01
Commitment	.20*	.01
Passion	.07	–.04

$*p < .05$, two-tailed.

NOTE: The *p* values for these correlations are reported for each single significance test; the *p* values have not been adjusted to correct for inflation of Type I error that may arise when multiple significance tests are reported.

Results: One Correlation

A Pearson correlation was performed to assess whether levels of intimacy in dating relationships could be predicted from levels of commitment on a self-report survey administered to 118 college students currently involved in dating relationships. Commitment and intimacy scores were obtained by summing items on two of the scales from Sternberg's Triangular Love Scale; the range of possible scores was from 15 (low levels of commitment or intimacy) to 75 (high levels of commitment or intimacy). Examination of histograms indicated that the distribution shapes were not close to normal for either variable. Both distributions were skewed. For both variables, most scores were near the high end of the scale, which indicated the existence of ceiling effects; and there were a few isolated outliers at the low end of the scale. However, the skewness was not judged severe enough to require data transformation or removal of outliers. The scatter plot of intimacy with commitment suggested a positive linear relationship. The correlation between intimacy and commitment was statistically significant, $r(116) = +.75$, $p < .001$ (two-tailed).[2] The r^2 was .56; thus, about 56% of the variance in intimacy could be predicted from levels of commitment. This relationship remained strong and statistically significant, $r(108) = +.64$, $p < .001$) even when outliers with scores less than 56 on intimacy and 49 on commitment were removed from the sample.

Results: Several Correlations

Pearson correlations were performed to assess whether levels of self-reported intimacy, commitment, and passion in dating relationships could be predicted from the length of the dating relationship and the number of times the participant has been in love, based on a self-report survey administered to 118 college students currently involved in dating relationships. Intimacy, commitment, and passion scores were obtained by summing items on scales from Sternberg's Triangular Love Scale; the range of possible scores was from 15 to 75 on each of the three scales. Examination of histograms indicated that the distribution shapes were not close to normal for any of these variables; distributions of scores were negatively skewed for intimacy, commitment, and passion. Most scores were near the high end of the scale, which indicated the existence of ceiling effects; and there were a few isolated outliers at the low end of the scale. However, the skewness was not judged severe enough to require data transformation or removal of outliers. The scatter plots suggested that when there were relationships between pairs of variables, these relationships were (weakly) linear. The six Pearson correlations are reported in Table 7.5. Only the correlation between commitment and length of relationship was statistically significant, $r(116) = +.20$,

(Continued)

(Continued)

$p < .05$ (two-tailed). The r^2 was .04; thus, only about 4% of the variance in commitment scores could be predicted from length of the relationship; this is a weak positive relationship. There was a tendency for participants who reported longer relationship duration to report higher levels of commitment.

If Bonferroni-corrected PC_α levels are used to control for the inflated risk of Type I error that occurs when multiple significance tests are performed, the PC_α level is .05/6 = .008. Using this more conservative criterion for statistical significance, none of the six correlations in Table 7.5 would be judged statistically significant.

7.14 ♦ Summary

This chapter described the use of Pearson *r* to describe the strength and direction of the linear association between a pair of quantitative variables. The conceptual formula for *r* (given in terms of z_X and z_Y) was discussed. Computing products between z_X and z_Y provides information about the spatial distribution of points in an *X*, *Y* scatter plot and the tendency for high scores on *X* to be systematically associated with high (or low) scores on *Y*. Fisher *Z* was introduced because it is needed for the construction of confidence intervals around sample *r* values and in some of the significance tests for *r*. Pearson *r* itself can be interpreted as an effect size index; sometimes, r^2 is also reported to describe the strength of relationship in terms of the proportion of variance in *Y* that is predictable from *X*. Pearson *r* is symmetric: That is, the correlation between *X* and *Y* is identical to the correlation between *Y* and *X*. Many of the analyses introduced in later chapters (such as **partial correlation** and multiple regression) are based on Pearson *r*; factors that artifactually influence the magnitude of Pearson *r* (such as restricted range, unreliability, and bivariate outliers) can also influence the magnitude of regression coefficients. Pearson *r* is often applied to *X* and *Y* variables that are both quantitative; however, it is also possible to use special forms of Pearson *r* (such as the point biserial correlation) when one or both of the variables are dichotomous (categorical, with only two categories), as discussed in Chapter 8.

Notes

1. One version of the formula to calculate Pearson *r* from the raw scores on *X* and *Y* is as follows:

$$r_{XY} = \frac{\sum(XY) - \frac{(\sum X)^2 \times (\sum Y)^2}{N}}{\sqrt{\left[\sum X^2 - \frac{(\sum X)^2}{N}\right] \times \left[\sum Y^2 - \frac{(\sum Y)^2}{N}\right]}}.$$

This statistic has $N - 2$ *df*, where *N* is the number of (*X*, *Y*) pairs of observations.

The value of Pearson r reported by SPSS is calculated using $N - 1$ as the divisor:

$$r = \frac{\sum z_X z_Y}{(N-1)}.$$

2. In more advanced statistical methods such as structural equation modeling, covariances rather than correlation are used as the basis for estimation of model parameters and evaluation of model fit.

3. Note that SPSS printouts sometimes report "$p = .000$" or "Sig = .000." (Sig is an SPSS abbreviation for significance, and p means probability or risk of Type I error; you want the risk of error to be small, so you usually hope to see p values that are small, i.e., less than .05.) These terms, p and sig, represent the theoretical risk of Type I error. This risk can never be exactly zero, but it becomes smaller and smaller as r increases and/or as N increases. It would be technically incorrect to report an exact p value as $p = .000$ in your write-up. Instead, when SPSS gives p as .000, you should write, "$p < .001$," to indicate that the risk of Type I error was estimated to be very small.

Comprehension Questions

1. A meta-analysis (Anderson & Bushman, 2001) reported that the average correlation between time spent playing violent video games (X) and engaging in aggressive behavior (Y) in a set of 21 well-controlled experimental studies was .19. This correlation was judged to be statistically significant. In your own words, what can you say about the nature of the relationship?

2. Harker and Keltner (2001) examined whether emotional well-being in later life could be predicted from the facial expressions of 141 women in their college yearbook photos. The predictor variable of greatest interest was the "positivity of emotional expression" in the college yearbook photo. They also had these photographs rated on physical attractiveness. They contacted the same women for follow-up psychological assessments at age 52 (and at other ages, data not shown here). Here are the correlations of these two predictors (based on ratings of the yearbook photo) with several of their self-reported social and emotional outcomes at age 52:

	In College Photo	
At Age 52	*Physical Attractiveness*	*Positivity of Facial Expression*
Negative emotionality	.04	−.27
Nurturance	−.06	.22
Well-being	.03	.27

a. Which of the six correlations above are statistically significant (i) if you test each correlation using $\alpha = .05$, two-tailed, and (ii) if you set $EW_\alpha = .05$ and use Bonferroni-corrected tests?
b. How would you interpret their results?
c. Can you make any causal inferences from this study? Give reasons.
d. Would it be appropriate for the researchers to generalize these findings to other groups, such as men?
e. What additional information would be available to you if you were able to see the scatter plots for these variables?

3. Are there ever any circumstances when a correlation such as Pearson *r* can be interpreted as evidence for a causal connection between two variables? If yes, what circumstances?

4. For a correlation of −.64 between *X* and *Y*, each one-*SD* change in z_X corresponds to a predicted change of _____ *SD* in z_Y.

5. A researcher says that 50% of the variance in blood pressure can be predicted from HR and that blood pressure is positively associated with HR. What is the correlation between blood pressure and HR?

6. Suppose that you want to do statistical significance tests for four correlations and you want your EW_α to be .05. What PC_α would you use if you apply the Bonferroni procedure?

7. Suppose that you have two different predictor variables (*X* and *Z*) that you use to predict scores on *Y*. What formula would you need to use to assess whether their correlations with *Y* differ significantly? What information would you need to have to do this test? Which of these is the appropriate null hypothesis?

   $H_0: \rho = 0$
   $H_0: \rho = .9$
   $H_0: \rho_1 = \rho_2$
   $H_0: \rho_{XY} = \rho_{ZY}$

   What test statistic should be used for each of these null hypotheses?

8. Why should researchers be very cautious about comparison of correlations that involve different variables?

9. How are *r* and r^2 interpreted?

10. Draw a diagram to show r^2 as an overlap between circles.

11. If "correlation does not imply causation," what does it imply?

12. What are some of the possible reasons for large correlations between a pair of variables, *X* and *Y*?

13. What does it mean to say that *r* is a symmetrical measure of a relationship?

14. Suppose that two raters (Rater A and Rater B) each assign physical attractiveness scores (0 = Not at all attractive to 10 = Extremely attractive) to a set of seven facial

photographs. Pearson r is a common index of **interrater reliability** or agreement on quantitative ratings. A correlation of +1 would indicate perfect rank order agreement between raters, while an r of 0 would indicate no agreement about judgments of relative attractiveness. Generally rs of .8 to .9 are considered desirable when reliability is assessed. The attractiveness ratings are as follows:

Photo	*Rater A*	*Rater B*
1	3	5
2	5	5
3	8	9
4	7	8
5	6	4
6	10	9
7	5	4

a. Compute the Pearson correlation between the Rater A/Rater B attractiveness ratings. What is the obtained r value?
b. Is your obtained r statistically significant? (Unless otherwise specified, use $\alpha = .05$, two-tailed, for all significance tests.)
c. Are the Rater A and Rater B scores "reliable"? Is there good or poor agreement between raters?

15. From a review of Chapters 5 and 6, what other analyses could you do with the variables in the SPSS dataset love.sav (variables described in Table 7.1)? Give examples of pairs of variables for which you could do t tests or one-way ANOVA. Your teacher may ask you to run these analyses and write them up.

16. Explain how the formula $r = \sum z_X z_Y / N$ is related to the pattern of points in a scatter plot (i.e., the numbers of concordant/discordant pairs).

17. What assumptions are required for a correlation to be a valid description of the relation between X and Y?

18. What is a bivariate normal distribution? Sketch the three-dimensional appearance of a bivariate normal distribution of scores.

19. When $r = 0$, does it necessarily mean that X and Y are completely unrelated?

20. Discuss how nonlinear relations may result in small rs.

21. Sketch the sampling distribution of r when $\rho = 0$ and the sampling distribution of ρ when $r = .80$. Which of these two distributions is nonnormal? What do we do to correct for this nonnormality when we set up significance tests?

22. What is a Fisher Z, how is it obtained, and what is it used for?
23. In words, what does the equation $z'_Y = r \times z_X$ say?
24. Take one of the following datasets: the SPSS file love.sav or some other dataset provided by your teacher, data obtained from your own research, or data downloaded from the Web. From your chosen dataset, select a pair of variables that would be appropriate for Pearson r. Examine histograms and a scatter plot to screen for possible violations of assumptions; report any problems and any steps you took to remedy problems, such as removal of outliers (see Chapter 4). Write up a brief Results section reporting the correlation, its statistical significance, and r^2. Alternatively, your teacher may ask you to run a list or group of correlations and present them in table form. If you are reporting many correlations, you may want to use Bonferroni-protected tests.

CHAPTER 8

Alternative Correlation Coefficients

8.1 ♦ Correlations for Different Types of Variables

Pearson correlation is generally introduced as a method to evaluate the strength of linear association between scores on two quantitative variables, an *X* predictor and a *Y* outcome variable. If the scores on *X* and *Y* are at least interval level of measurement and if the other assumptions for Pearson *r* are satisfied (e.g., *X* and *Y* are linearly related, *X* and *Y* have a bivariate normal distribution), then Pearson *r* is generally used to describe the strength of the linear relationship between variables. However, we also need to have indexes of correlation for pairs of variables that are not quantitative or that fall short of having equal-interval level of measurement properties or that have joint distributions that are not bivariate normal. This chapter discusses some of the more widely used alternative bivariate statistics that describe strength of association or correlation for different types of variables.

When deciding which index of association (or which type of correlation coefficient) to use, it is useful to begin by identifying the type of measurement for the *X* and *Y* variables. The *X* or independent variable and the *Y* or dependent variable may each be any of the following types of measurement:

1. Quantitative with interval/ratio measurement properties
2. Quantitative but only ordinal or rank level of measurement
3. Nominal or categorical with more than two categories
4. Nominal with just two categories
   a. A true **dichotomy**
   b. An artificial dichotomy

Table 8.1 presents a few of the most widely used correlation statistics. For example, if both *X* and *Y* are quantitative and interval/ratio (and if the other assumptions for Pearson *r* are satisfied), the **Pearson product-moment correlation** is often used to describe the strength of linear association between scores on X and Y. If the scores come in the form of rank or ordinal data or if it is necessary to convert scores into ranks to get rid of problems such as severely nonnormal distribution shapes or outliers, then Spearman r or Kendall's tau (τ) may be used. If scores on X correspond to a true dichotomy and scores on Y are interval/ratio level of measurement, the point biserial correlation may be used. If scores on X and Y both correspond to true dichotomies, the phi coefficient (ϕ) can be reported. Details about computation and interpretation of these various types of correlation coefficients appear in the following sections of this chapter.

Some of the correlation indexes listed in Table 8.1, including Spearman r, point biserial r, and the phi coefficient, are actually equivalent to Pearson r. For example, a Spearman r can be obtained by converting scores on X and Y into ranks (if they are not already in the form of ranks) and then computing a Pearson r for the ranked scores. A point biserial r can be obtained by computing a Pearson r for scores on a truly dichotomous X variable that has only two values (e.g., gender, coded 1 = Female, 2 = Male) and scores on a quantitative Y variable (such as heart rate, HR). A phi coefficient can be obtained by computing a Pearson r between scores on two true dichotomies (e.g., Does the person take a specific drug, 1 = No, 2 = Yes; Does the person have a heart attack within 1 year, 1 = No, 2 = Yes). Alternative computational formulas are available for Spearman r, point biserial r, and the phi coefficient, but the same numerical results can be obtained by applying the formula for Pearson r. Thus, Spearman r, point biserial r, and phi coefficient are equivalent to Pearson r. Within SPSS, you obtain the same results when you use the Pearson r procedure to compute a correlation between drug use and death as when you request a phi coefficient between drug use and death in the Crosstabs procedure. On the other hand, some of the other correlation statistics listed in Table 8.1 (such as the **tetrachoric correlation r_{tet}**, biserial r, and Kendall's tau) are *not* equivalent to Pearson r.

For many combinations of variables shown in Table 8.1, several different statistics can be reported as an index of association. For example, for two truly dichotomous variables, such as drug use and death, Table 8.1 lists the phi coefficient as an index of association, but it is also possible to report other statistics such as chi-square and **Cramer's V**, described in this chapter, or **log odds ratios**, described in Chapter 21 on binary logistic regression.

Later chapters in this textbook cover statistical methods that are implicitly or explicitly based on Pearson r values and covariances. For example, in multiple regression (in Chapters 11 and 14), the slope coefficients for regression equations can be computed based on sums of squares and **sums of cross products** based on the X and Y scores, or from the Pearson correlations among variables and the means and standard deviations of variables. For example, we could predict a person's HR from that person's scores on several different X predictors (X_1 = gender, coded 1 = Female, 2 = Male; X_2 = age in years; X_3 = body weight in pounds):

$$Y = b_0 + b_1 X_1 + b_2 X_2 + b_3 X_3. \tag{8.1}$$

$$\text{HR} = 64 + (-2.5) \times \text{Gender} + 1.2 \times \text{Years} + 1.7 \times \text{Weight}.$$

When researchers use dichotomous variables (such as gender) as predictors in multiple regression, they implicitly assume that it makes sense to use Pearson r to index the

Table 8.1 ♦ Widely Used Correlations for Various Types of Independent Variables (X) and Dependent Variables (Y) (Assuming That Groups Are Between Subjects or Independent)

	Y Is Quantitative, Interval/Ratio	*Y Is Rank or Ordinal*	*Y Is Categorical With More Than Two Levels*	*Y Is a True Dichotomy*	*Y Is an Artificial Dichotomy*
X Is Quantitative, Interval/ Ratio Level of Measurement	Pearson r (Chapter 7)	[a]	Eta (η) or eta squared (η^2)	Point biserial r (r_{pb}) or η or η^2	Biserial r (r_b)
X Scores Are Obtained as Rank or Ordinal Data or Converted Into Ranks	[a]	Spearman r (r_s) or Kendall's tau (τ)	Epsilon squared (effect size for Kruskal-Wallis test)	Glass rank biserial correlation (effect size for Wilcoxon rank sum test)	[b]
X Is Categorical With More Than Two Levels	Eta (η) or eta squared (η^2)	Epsilon squared (effect size for Kruskal-Wallis test)	Cramer's V	Cramer's V	[b]
X Is a True Dichotomy	Point biserial r (r_{pb}) or η or η^2	Glass rank biserial correlation (effect size for Wilcoxon rank sum test)	Cramer's V	Phi coefficient (ϕ)	[b]
X Is an Artificial Dichotomy	Biserial r (r_b)	[b]	[b]	[b]	Tetrachoric r (r_{tet})

a. There may be no purpose-designed statistic for some combinations of types of variables, but it is usually possible to downgrade your assessment of the level of measurement of one or both variables. For example, if you have an X variable that is interval/ratio and a Y variable that is ordinal, you could convert scores on X to ranks and use Spearman r. It might also be reasonable to apply Pearson r in this situation.

b. In practice, researchers do not always pay attention to the existence of artificial dichotomies when they select statistics. Tetrachoric r and biserial r are rarely reported. I do not know of any tests specifically designed for other situations involving artificially dichotomized variables.

strength of relationship between scores on gender and scores on the outcome variable HR. In this chapter, we will examine the point biserial r and the phi coefficient and verify that they are equivalent to Pearson r. An important implication of this equivalence is that we can use true dichotomous variables as predictors in regression analysis.

However, some problems can arise when we include dichotomous predictors in correlation-based analyses such as regression. In Chapter 7, for example, it was pointed out that the maximum value of Pearson r, $r = +1.00$, can occur only when the scores on X and Y have identical distribution shapes. This condition is not met when we correlate scores on a dichotomous X variable such as gender with scores on a quantitative variable such as HR.

Many of the statistics included in Table 8.1, such as Kendall's tau, will not be mentioned again in later chapters of this textbook. However, they are included because you might encounter data that require these alternative forms of correlation analysis and they are occasionally reported in journal articles. The understanding of Pearson r, point biserial correlation, and phi coefficient that is developed in this chapter is necessary background for later topics (such as multiple regression). When dichotomous (or dummy) predictor variables are included in multiple regression, you need to understand what information a Pearson r can provide in such situations.

8.2 ♦ Two Research Examples

To illustrate some of these alternative forms of correlation, two small datasets will be used. The first dataset, which appears in Table 8.2 and Figure 8.1, consists of a hypothetical set of scores on a true dichotomous variable (gender) and a quantititave variable that has interval/ratio level of measurement properties (height). The relationship between gender and height can be assessed by doing an independent samples t test to compare means on height across the two gender groups (as described in Chapter 5). However, an alternative way to describe the strength of association between gender and height is to calculate a point biserial correlation, r_{pb}, as shown in this chapter.

The second set of data come from an actual study (Friedmann, Katcher, Lynch, & Thomas, 1980) in which 92 men who had a first heart attack were asked whether or not they owned a dog. Dog ownership is a true dichotomous variable, coded 0 = No and 1 = Yes; this was used as a predictor variable. At the end of a 1-year follow-up, the researchers recorded whether each man had survived; this was the outcome or dependent variable. Thus, the outcome variable, survival status, was also a true dichotomous variable, coded 0 = No, 1 = Yes. The question in this study was whether survival status was predictable from dog ownership. The strength of association between these two true dichotomous variables can be indexed by several different statistics, including the phi coefficient; a test of statistical significance of the association between two nominal variables can be obtained by performing a **chi-square (χ^2) test** of association. The data from the Friedmann et al. study appear in the form of a data file in Table 8.3 and as a summary table of observed cell frequencies in Table 8.4.

The correlation index that you are likely to encounter most often in reading journals is the Pearson correlation and its equivalents (such as phi). However, you may also encounter (or need to use) other types of correlation; therefore, it is useful to consider some alternative forms of correlation indexes.

Table 8.2 ♦ Data for the Point Biserial r Example: Gender and Height (1 = Male, 2 = Female)

Gender	*Height*	*Gender*	*Height*
		(Continued)	
1	70	2	63
1	68	2	63
1	68	2	62
1	68	2	66
1	74	2	63
1	68	2	68
1	67	2	69
1	67	2	63
1	70	2	62
1	71	2	61
1	67	2	62
1	65	2	62
1	68	2	61
1	71	2	64
1	68	2	64
1	68	2	65
1	70	2	65
1	68	2	60
1	69	2	64
1	68	2	65
1	71	2	64
1	69	2	68
1	68	2	59
1	69	2	62
1	69	2	69
1	69	2	64
1	71	2	64
1	71	2	63
1	69	2	64
1	70	2	65
1	72	2	61
1	73	2	66
1	67	2	66
1	66	2	65

Similarities among the indexes of association (correlation indexes) covered in this chapter include the following:

1. The size of r (its absolute magnitude) provides information about the strength of association between X and Y. In principle, the range of possible values for the Pearson correlation is $-1 \leq r \leq +1$; however, in practice, the maximum possible values of r may be limited to a narrower range. Perfect correlation (either $r = +1$ or $r = -1$) is possible only when the X, Y scores have identical distribution shapes.

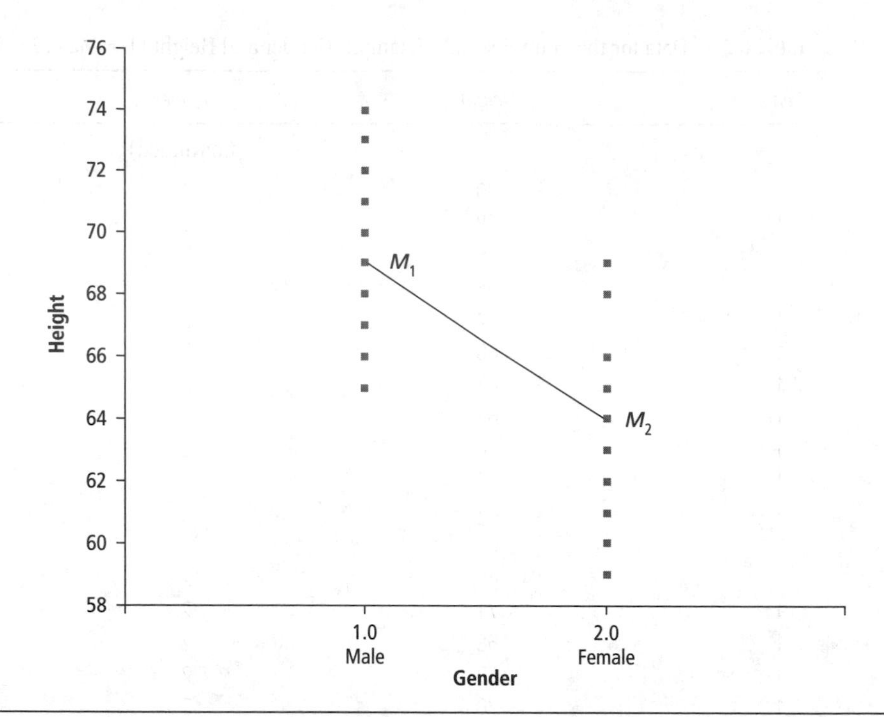

Figure 8.1 ◆ Scatter Plot for Relation Between a True Dichotomous Predictor (Gender) and a Quantitative Dependent Variable (Height)

NOTE: M_1 = mean male height; M_2 = mean female height.

When distribution shapes for X and Y differ, the maximum possible correlation between X and Y is often somewhat less than 1 in absolute value. For the phi coefficient, we can calculate the maximum possible value of phi given the marginal distributions of X and Y. Some indexes of association covered later in the textbook, such as the log odds ratios in Chapter 21, are scaled quite differently and are not limited to values between −1 and +1.

2. For correlations that can have a plus or a minus sign, the sign of r provides information about the direction of association between scores on X and scores on Y. However, in many situations, the assignment of lower versus higher scores is arbitrary (e.g., gender, coded 1 = Female, 2 = Male), and in such situations, researchers need to be careful to pay attention to the codes that were used for categories when they interpret the sign of a correlation. Some types of correlation (such as η and Cramer's V) have a range from 0 to +1—that is, they are always positive.

3. Some (but not all) of the indexes of association discussed in this chapter are equivalent to Pearson r.

Ways in which the indexes of association may differ:

1. The interpretation of the meaning of these correlations varies. Chapter 7 described two useful interpretations of Pearson r. One involves the "mapping" of scores from z_X to z_Y, or the prediction of a z'_Y score from a z_X score for each individual

Table 8.3 ♦ Dog Ownership/Survival Data

Dog Ownership	*Survival Status*	*Dog Ownership*	*Survival Status*
		(Continued)	
1	1	1	1
1	1	1	1
1	1	1	1
1	1	1	1
1	1	1	0
1	1	1	0
1	1	1	0
1	1	0	1
1	1	0	1
1	1	0	1
1	1	0	1
1	1	0	1
1	1	0	1
1	1	0	1
1	1	0	1
1	1	0	1
1	1	0	1
1	1	0	1
1	1	0	1
1	1	0	1
1	1	0	1
1	1	0	1
1	1	0	1
1	1	0	1
1	1	0	1
1	1	0	1
1	1	0	1
1	1	0	1
1	1	0	1
1	1	0	1
1	1	0	1
1	1	0	1
1	1	0	1
1	1	0	1
1	1	0	1
1	1	0	0
1	1	0	0
1	1	0	0
1	1	0	0
1	1	0	0
1	1	0	0
1	1	0	0
1	1	0	0
1	1	0	0
1	1	0	0
1	1	0	0

SOURCE: Friedmann et al. (1980).

NOTE: Prediction of survival status (true dichotomous variable) from dog ownership (true dichotomous variable): Dog ownership: 0 = No, 1 = Yes. Survival status: 0 = No, 1 = Yes.

Table 8.4 ♦ Dog Ownership and Survival Status 1 Year After the First Heart Attack

	Survival Status			
Dog Ownership	*No*	*Yes*	*Row*	*Total*
No	$b = 11$	$a = 28$	39	
Yes	$d = 3$	$c = 50$	53	
Column total	14	78		92

SOURCE: Friedmann et al. (1980).

NOTE: The table shows the observed frequencies for outcomes in a survey study of $N = 92$ men who have had a first heart attack. The frequencies in the cells denoted by b and c represent concordant outcomes (b indicates answer "No" for both variables, c indicates answer "Yes" for both variables). The frequencies denoted by a and d represent discordant outcomes (i.e., an answer of "Yes" for one variable and "No" for other variable). When calculating a phi coefficient by hand from the cell frequencies in a 2×2 table, information about the frequencies of concordant and discordant outcomes is used.

participant. A Pearson r of 1 can occur only when there is an exact one-to-one correspondence between distances from the mean on X and distances from the mean on Y and that in turn can happen only when X and Y have identical distribution shapes. A second useful interpretation of Pearson r was based on the squared correlation (r^2). A squared Pearson correlation can be interpreted as "the proportion of variance in Y scores that is linearly predictable from X," and vice versa. However, some of the other correlation indexes—even though they are scaled so that they have the same range from -1 to $+1$ like Pearson r—have different interpretations.

2. Some of the indexes of association summarized in this chapter are applicable only to very specific situations (such as 2×2 tables), while other indexes of association (such as the chi-square test of association) can be applied in a wide variety of situations.

3. Most of the indexes of association discussed in this chapter are symmetrical. For example, Pearson r is symmetrical because the correlation between X and Y is the same as the correlation between Y and X. However, there are some **asymmetrical indexes of association** (such as **lambda** and **Somers's *d***). There are some situations where the ability to make predictions is asymmetrical; for example, consider a study about gender and pregnancy. If you know that an individual is pregnant, you can predict gender (the person must be female) perfectly. However, if you know that an individual is female, you cannot assume that she is pregnant. For further discussion of asymmetrical indexes of association, see Everitt (1977).

4. Some of the indexes of association for ordinal data are appropriate when there are large numbers of tied ranks; others are appropriate only when there are not many tied ranks. For example, problems can arise when computing Spearman r using the formula that is based on differences between ranks. Furthermore, indexes that describe strength of association between categorical variables differ in the way they handle tied scores; some statistics subtract the number of ties when they evaluate the numbers of concordant and discordant pairs (e.g., Kendall's tau), while other statistics ignore cases with tied scores.

8.3 ♦ Correlations for Rank or Ordinal Scores

Spearman r is applied in situations where the scores on X and Y are both in the form of ranks or in situations where the researcher finds it necessary or useful to convert X and Y scores into ranks to get rid of problems such as extreme outliers or extremely nonnormal distribution shapes. One way to obtain Spearman r, in by-hand computation, is as follows. First, convert scores on X into ranks. Then, convert scores on Y into ranks. If there are ties, assign the mean of the ranks for the tied scores to each tied score. For example, consider this set of X scores:

X	*Rank of X:* R_X
30	1
28	2
25	(3 + 4 + 5)/3 = 4
25	(3 + 4 + 5)/3 = 4
25	(3 + 4 + 5)/3 = 4
24	6
20	7
12	8

For each participant, let d_i be the difference between ranks on the X and Y variables. The value of Spearman r, denoted by r_s, can be found in either of two ways:

1. Compute the Pearson correlation between R_X (rank on the X scores) and R_Y (rank on the Y scores).

2. Use the formula below to compute Spearman r (r_s) from the differences in ranks:

$$\text{Spearman } r = r_s = \frac{1 - 6\sum (d_i)^2}{n(n^2 - 1)}, \tag{8.2}$$

where d_i = the difference between ranks = $(R_X - R_Y)$ and n = the number of pairs of (X, Y) scores or the number of d_i differences.

If $r_s = +1$, there is perfect agreement between the ranks on X and Y; if $r_s = -1$, the rank orders on X and Y are perfectly inversely related (e.g., the person with the highest score on X has the lowest score on Y).

Another index of association that can be used in situations where X and Y are either obtained as ranks or converted into ranks is Kendall's tau; there are two variants of this called **Kendall's tau-b** and **Kendall's tau-c**. In most cases, the values of Kendall's τ and Spearman r lead to the same conclusion about the nature of the relationship between X and Y. See Liebetrau (1983) for further discussion.

8.4 ♦ Correlations for True Dichotomies

Most introductory statistics books only show Pearson r applied to pairs of quantitative variables. Generally, it does not make sense to apply Pearson r in situations where X and/or Y are categorical variables with more than two categories. For example, it would not make sense to compute a Pearson correlation to assess whether the categorical variable Political Party Membership (coded 1= Democrat, 2 = Republican, 3 = Independent, 4 = Socialist, and so forth) is related to income level. The numbers used to indicate party membership serve only as labels and do not convey any quantitative information about differences among parties. The mean income level could go up, go down, or remain the same as the X scores change from 1 to 2, 2 to 3, and so on; there is no reason to expect a consistent linear increase (or decrease) in income as the value of the code for political party membership increases.

However, when a categorical variable has only two possible values (such as gender, coded 1 = Male, 2 = Female, or survival status, coded 1 = Alive, 0 = Dead), we can use the Pearson correlation and related correlation indexes to relate them to other variables. To see why this is so, consider this example: X is gender (coded 1= Male, 2 = Female); Y is height, a quantitative variable (hypothetical data appear in Table 8.2, and a graph of these scores is shown in Figure 8.1). Recall that Pearson r is an index of the linear relationship between scores on two variables. When X is dichotomous, the only possible relation it can have with scores on a continuous Y variable is linear. That is, as we move from Group 1 to Group 2 on the X variable, scores on Y may increase, decrease, or remain the same. In any of these cases, we can depict the X, Y relationship by drawing a straight line to show how the mean Y score for $X = 1$ differs from the mean Y score for $X = 2$.

See Figure 8.1 for a scatter plot that shows how height (Y) is related to gender (X); clearly, mean height is greater for males (Group 1) than for females (Group 2). We can describe the relationship between height (Y) and gender (X) by doing an independent samples t test to compare mean Y values across the two groups identified by the X variable; or we can compute a correlation (either a Pearson r or a point biserial r) to describe how these variables are related. We shall see that the results of these two analyses provide equivalent information. By extension, it is possible to include dichotomous variables in some of the multivariable analyses covered in later chapters of this book. For example, when dichotomous variables are included as predictors in a multiple regression, they are usually called **"dummy" variables.** First, however, we need to consider one minor complication.

Pearson r can be applied to dichotomous variables when they represent true dichotomies—that is, naturally occurring groups with just two possible outcomes. One common example of a true dichotomous variable is gender (coded 1= Male, 2 = Female); another is survival status in a follow-up study of medical treatment (1 = Patient survives; 0 = Patient dies). However, sometimes we encounter artificial dichotomies. For instance, when we take a set of quantitative exam scores that range from 15 to 82 and impose an arbitrary cutoff (scores less than 65 are fail, scores greater than 65 are pass), this type of dichotomy is "artificial." The researcher has lost some of the information about variability of scores by artificially converting them to a dichotomous group membership variable.

When a dichotomous variable is an artificially created dichotomy, there are special types of correlation; their computational formulas involve terms that attempt to correct for the information about variability that was lost in the artificial dichotomization. The correlation of an artificial dichotomy with a quantitative variable is called a **biserial** r (r_b); the correlation between two artificial dichotomies is called a tetrachoric r (r_{tet}). These are not examples of Pearson r; they use quite different computational procedures and they are rarely used.

When a dichotomous variable represents a naturally occurring group membership or true dichotomy, it can be calculated using the usual formulas for Pearson r. There are some alternative formulas that may be used, but the results are equivalent to the ones obtained by application of the ordinary formula for Pearson r. It does not matter whether the group memberships are coded as 1, 2 or 0, 1. The name for the correlation between a true dichotomy and a quantitative variable is point biserial r (r_{pb}); the name for the correlation between two true dichotomous variables is phi (ϕ). These are described in more detail in the following sections.

8.4.1 ♦ Point Biserial r (r_{pb})

If a researcher has data on a true dichotomous variable (such as gender) and a continuous variable (such as emotional intelligence, EI), the relationship between these two variables can be assessed by calculating a t test to assess the difference in mean EI for the male versus female groups or by calculating a point biserial r to describe the increase in EI scores in relation to scores on gender. The values of t and r_{pb} are related, and each can easily be converted into the other. The t value can be compared with critical values of t to assess statistical significance. The r_{pb} value can be interpreted as a standardized index of effect size, or the strength of the relationship between group membership and scores on the outcome variable:

$$t = \frac{r_{pb}\sqrt{N-2}}{\sqrt{1-r_{pb}^2}} \tag{8.3}$$

and

$$r_{pb} = \sqrt{\frac{t^2}{t^2+df}}. \tag{8.4}$$

In this equation, $df = N - 2$; N = total number of subjects. The sign of r_{pb} can be determined by looking at the direction of change in Y across levels of X. This conversion between r_{pb} and t is useful because t can be used to assess the statistical significance of r_{pb}, and r_{pb} or r_{pb}^2 can be used as an index of the effect size associated with t.

To illustrate the correspondence between r_{pb} and t, SPSS was used to run two different analyses on the hypothetical data shown in Figure 8.1. The predictor or X variable is a true dichotomy (X = gender, coded 1= Male, 2 = Female); the Y outcome variable, height, is quantitative. First, the independent samples t test was run to assess the significance of the difference of mean height for the male versus female groups (the procedures for running an independent samples t test using SPSS were presented in Chapter 5). The results are shown in the top two panels of Figure 8.2. The difference in mean height for males (M_1 = 69.03) and females (M_2 = 63.88) was statistically

Group Statistics

	GENDER	N	Mean	Std. Deviation	Std. Error Mean
HEIGHT	male	34	69.03	1.946	.334
	female	34	63.88	2.409	.413

Independent Samples Test

		Levene's Test for Equality of Variances		t-test for Equality of Means						
									95% Confidence Interval of the Difference	
		F	Sig.	t	df	Sig. (2-tailed)	Mean Difference	Std. Error Difference	Lower	Upper
HEIGHT	Equal variances assumed	.975	.327	9.691	66	.000	5.15	.531	4.087	6.207
	Equal variances not assumed			9.691	63.205	.000	5.15	.531	4.086	6.208

Correlations

		GENDER	HEIGHT
GENDER	Pearson Correlation	1	-.766**
	Sig. (2-tailed)	.	.000
	N	68	68
HEIGHT	Pearson Correlation	-.766**	1
	Sig. (2-tailed)	.000	.
	N	68	68

**. Correlation is significant at the 0.01 level (2-tailed).

Figure 8.2 ♦ SPSS Output: Independent Samples *t* Test (*Top*) and Pearson *r* (*Bottom*) for Data in Figure 8.1

significant, $t(66) = 9.69$, $p < .001$. The mean height for females was about 5 in. lower than the mean height for males. Second, a point biserial correlation between height and gender was obtained by using the Pearson correlation procedure in SPSS: <Analyze> → <Correlate> → <Bivariate>. The results for this analysis are shown in the bottom panel of Figure 8.2; the correlation between gender and height was statistically significant, $r_{pb}(66) = -.77$, $p < .001$. The nature of the relationship was that having a higher score on gender (i.e., being female) was associated with a lower score on height. The reader may wish to verify that when these values are substituted into Equations 8.3 and 8.4, the r_{pb} value can be reproduced from the t value and the t value can be obtained from the value of r_{pb}. Also, note that when η^2 is calculated from the value of t as discussed in Chapter 5, η^2 is equivalent to r^2_{pb}.

This demonstration is one of the many places in the book where readers will see that analyses that were introduced in different chapters in most introductory statistics textbooks turn out to be equivalent. This occurs because most of the statistics that we use in the behavioral sciences are special cases of a larger data analysis system called the general linear model. In the most general case, the general linear model may include multiple predictor and multiple outcome variables, and it can include quantitative and dichotomous variables on both the predictor and the outcome side of the analysis. Thus, when we predict a quantitative Y from a quantitative X variable, or a quantitative Y from a categorical X variable, these are special cases of the general linear model where we limit the number and type of variables on one or both sides of the analysis (the predictor and the dependent variable).

8.4.2 ♦ Phi Coefficient (ϕ)

The phi coefficient (ϕ) is the version of Pearson r that is used when both X and Y are true dichotomous variables. It can be calculated from the formulas given earlier for the general Pearson r using score values of 0 and 1, or 1 and 2, for the group membership variables; the exact numerical value codes that are used do not matter, although 0, 1 is the most conventional representation. Alternatively, phi can be computed from the cell frequencies in a 2 × 2 table that summarizes the number of cases for each combination of X and Y scores. Table 8.5 shows the way the frequencies of cases in the four cells of a 2 × 2 table are labeled to compute phi from the cell frequencies. Assuming that the cell frequencies a through d are as shown in Table 8.5 (i.e., a and d correspond to "discordant" outcomes and b and c correspond to "concordant" outcomes), here is a formula that may be used to compute phi directly from the cell frequencies:

$$\phi = \frac{bc - ad}{\sqrt{(a+b)\times(a+c)\times(b+d)\times(c+d)}}, \tag{8.5}$$

where b and c are the number of cases in the concordant cells of a 2 × 2 table and a and d are the number of cases in the discordant cells of a 2 × 2 table.

In Chapter 7, you saw that the Pearson correlation turned out to be large and positive when most of the points fell into the concordant regions of the X, Y scatter plot that appeared in Figure 7.16 (high values of X paired with high values of Y and low values of X paired with low values of Y). Calculating products of z scores was a way to summarize the information about score locations in the scatter plot and to assess whether most cases were concordant or discordant on X and Y. The same logic is evident in the formula to

Table 8.5 ♦ Labels for Cell Frequencies in a 2 × 2 Contingency Table (a) as Shown in Most Textbooks and (b) as Shown in Crosstab Tables From SPSS

	X = *Low (0)*	*X* = *High (1)*
(a) In most textbooks		
Y = High (1)	*a*	*b*
Y = Low (0)	*c*	*d*
(b) In SPSS Crosstabs output		
Y = Low (0)	*b*	*a*
Y = High (1)	*d*	*c*

NOTES: Cases are called concordant if they have high scores on both X and Y or low scores on both X and Y. Cases are called discordant if they have low scores on one variable and high scores on the other variable. a = Number of cases with X low and Y high (discordant), b = number of cases with X high and Y high (concordant), c = number of cases with Y low and X low (concordant), and d = number of cases with X low and Y high (discordant). In textbook presentations of the phi coefficient, the 2 × 2 table is usually oriented so that values of X increase from left to right and values of Y increase from bottom to top (as they would in an X, Y scatter plot). However, in the Crosstabs tables produced by SPSS, the arrangement of the rows is different (values of Y increase as you read *down* the rows in an SPSS table). If you want to calculate a phi coefficient by hand from the cell frequencies that appear in the SPSS Crosstabs output, you need to be careful to look at the correct cells for information about concordant and discordant cases. In most textbooks, as shown in this table, the concordant cells b and c are in the major diagonal of the 2 × 2 table—that is, the diagonal that runs from lower left to upper right. In SPSS Crosstabs output, the concordant cells b and c are in the minor diagonal—that is, the diagonal that runs from upper left to lower right.

calculate the phi coefficient. The $b \times c$ product is large when there are many concordant cases; the $a \times d$ product is large when there are many discordant cases. The phi coefficient takes on its maximum value of +1 when all the cases are concordant (i.e., when the a and d cells have frequencies of 0). The ϕ coefficient is 0 when $b \times c = a \times d$—that is, when there are as many concordant as discordant cases.

A formal significance test for phi can be obtained by converting it into a chi-square; in the following equation, N represents the total number of scores in the contingency table:

$$\chi^2 = N \times \phi. \tag{8.6}$$

This is a chi-square statistic with 1 degree of freedom (*df*). Those who are familiar with chi-square from other statistics courses will recognize it as one of the many possible statistics to describe relationships between categorical variables based on tables of cell frequencies. For χ^2 with 1 *df* and $\alpha = .05$, the critical value of χ^2 is 3.84; thus, if the obtained χ^2 exceeds 3.84, then phi is statistically significant at the .05 level.

When quantitative X and Y variables have different distribution shapes, it limits the maximum possible size of the correlation between them because a perfect one-to-one mapping of score location is not possible when the distribution shapes differ. This issue of distribution shape also applies to the phi coefficient. If the proportions of yes/no or 0/1 codes on the X and Y variables do not match (i.e., if p_1, the probability of a yes code on X, does not equal p_2 the probability of a yes code on Y), then the maximum obtainable size of the phi coefficient may be much less than 1 in absolute value. This limitation on the magnitude of phi occurs because unequal marginal frequencies make it impossible to have 0s in one of the diagonals of the table (i.e., in a and d, or in b and c).

For example, consider the hypothetical research situation that is illustrated in Table 8.6. Let's assume that the participants in a study include 5 dead and 95 live subjects and 40 Type B and 60 Type A personalities and then try to see if it is possible to arrange the 100 cases into the four cells in a manner that results in a diagonal pair of cells with 0s in it. You will discover that it can't be done. You may also notice, as you experiment with arranging the cases in the cells of Table 8.6, that that there are only six possible outcomes for the study—depending on the way the 5 dead people are divided between Type A and Type B personalities; you can have 0, 1, 2, 3, 4, or 5 Type A/dead cases, and the rest of the cell frequencies are not free to vary once you know the number of cases in the Type A/dead cell.[1]

It is possible to calculate the maximum obtainable size of phi as a function of the marginal distributions of X and Y scores and to use this as a point of reference in evaluating whether the obtained phi coefficient was relatively large or small. The formula for ϕ_{max} is as follows:

$$\phi_{\max} = \sqrt{\frac{p_i q_j}{q_i p_j}}, \quad p_j \geq p_i, \quad i \neq j. \tag{8.7}$$

That is, use the larger of the values p_1 and p_2 as p_j in the formula above. For instance, if we correlate an X variable (coronary-prone personality, coded Type A = 1, Type B = 0, with a 60%/40% split) with a Y variable (death from heart attack, coded 1= Dead, 0 = Alive, with a 5%/95% split), the maximum possible ϕ that can be obtained in this situation is about .187 (see Table 8.6).

Table 8.6 ♦ Computation of ϕ_{max} for a Table With Unequal Marginals

Death From Coronary Heart Disease	*Coronary-Prone Personality* Type B = 0	Type A = 1	*Proportions*
Dead = 1	a	b	$p_i = (a + b)/n = .05$
Alive = 0	c	d	$q_i = (c + d)/n = .95$
Proportions	$q_j = (a + c)/n$ $q_j = .40$	$p_j = (b + d)/n$ $p_j = .60$	Total $n = a + b + c + d$

NOTE: To determine what the maximum possible value of ϕ is given these marginal probabilities, apply Equation 8.7:

$$\phi_{max} = \sqrt{\frac{p_i q_j}{q_i p_j}}, \quad p_j \geq p_i, \quad i \neq j.$$

Because p_1 (.60) > p_2 (.05), we let $p_j = .60, q_j = .40; p_i = .05, q_i = .95$:

$$\phi_{max} = \sqrt{\frac{(.05)(.40)}{(.95)(.60)}} = \sqrt{\frac{.02}{.57}} = \sqrt{.0351} = .1873.$$

Essentially, ϕ_{max} is small when the marginal frequencies are unequal because there is no way to arrange the cases in the cells that would make the frequencies in both of the discordant cells equal zero. One reason why correlations between measures of personality and disease outcomes are typically quite low is that, in most studies, the proportions of persons who die, or have heart attacks, or have other specific disease outcomes of interest are quite small. If a predictor variable (such as gender) has a 50/50 split, the maximum possible correlation between variables such as gender and heart attack may be quite small because the marginal frequency distributions for the variables are so different. This limitation is one reason why many researchers now prefer other ways of describing the strength of association, such as the **odds ratios** that can be obtained using binary logistic regression (see Chapter 21).

8.5 ♦ Correlations for Artificially Dichotomized Variables

Artificial dichotomies arise when researchers impose an arbitrary cutoff point on continuous scores to obtain groups; for instance, students may obtain a continuous score on an exam ranging from 1 to 100, and the teacher may impose a cutoff to determine pass/fail status (1 = Pass, for scores of 70 and above; 0 = Fail, for scores of 69 and below). There are special forms of correlation that may be used for artificial dichotomous scores (biserial r, usually denoted by r_b, and tetrachoric r, usually denoted by r_{tet}). These are rarely used; they are discussed only briefly here.

8.5.1 ♦ Biserial r (r_b)

Suppose that the artificially dichotomous Y variable corresponds to a "pass" or "fail" decision. Let M_{Xp} be the mean of the quantitative X scores for the pass group and p be the proportion of people who passed; let M_{Xq} be the mean of the X scores for the fail group and q be the proportion of people who failed. Let h be the height of the normal distribution at the point where the pass/fail cutoff was set for the distribution of Y. Let s_X be the standard deviation of all the X scores. Then,

$$r_b = \frac{(M_{Xp} - M_{Xq})}{s_X} \times \left(\frac{pq}{h}\right) \quad (8.8)$$

(from Lindeman et al., 1980, p. 74).

8.5.2 ♦ Tetrachoric r (r_{tet})

Tetrachoric r is a correlation between two artificial dichotomies. The trigonometric functions included in this formula provide an approximate adjustment for the information about the variability of scores, which is lost when variables are artificially dichotomized; these two formulas are only approximations; the exact formula involves an infinite series.

The cell frequencies are given in the following table:

	$X = 0$	$X = 1$
$Y = 1$	a	b
$Y = 0$	c	d

where b and c are the concordant cases (the participant has a high score on X and a high score on Y, or a low score on X and a low score on Y); and a and d are the discordant cases (the participant has a low score on X and a high score on Y, or a high score on X and a low score on Y); and n = the total number of scores, $n = a + b + c + d$.

If there is a 50/50 split between the number of 0s and the number of 1s on both the X and the Y variables (this would occur if the artificial dichotomies were based on median splits)—that is, if $(a + b) = (c + d)$ and $(a + c) = (b + d)$, then an exact formula for tetrachoric r is as follows:

$$r_{tet} = \sin\left[90^\circ \left(\frac{b + c - a - d}{n}\right)\right]. \quad (8.9)$$

However, if the split between the 0/1 groups is not made by a median split on one or both variables, a different formula provides an approximation for tetrachoric r that is a better approximation for this situation:

$$r_{tet} = \cos\left(\frac{180^\circ}{1 + \sqrt{bc/ad}}\right) \quad (8.10)$$

(from Lindeman et al., 1980, p. 79).

8.6 ♦ Assumptions and Data Screening for Dichotomous Variables

For a dichotomous variable, the closest approximation to a normal distribution would be a 50/50 split (i.e., half zeros and half ones). Situations where the group sizes are extremely unequal (e.g., 95% zeros and 5% ones) should be avoided for two reasons. First, when the absolute number of subjects in the smaller group is very low, the outcome of the analysis may be greatly influenced by scores for one or a few cases. For example, in a 2 × 2 contingency table, when a row has a very small total number of cases (such as five), then the set of cell frequencies in the entire overall 2 × 2 table will be entirely determined by the way the five cases in that row are divided between the two column categories. It is undesirable to have the results of your study depend on the behavior of just a few scores. When chi-square is applied to contingency tables, the usual rule is that no cell should have an expected cell frequency less than 5. A more appropriate analysis for tables where some rows or columns have very small *N*s and some cells have expected frequencies less than 5 is the Fisher exact test. Second, the maximum possible value of the phi coefficient is constrained to be much smaller than +1 or −1 when the proportions of ones for the *X* and *Y* variables are far from equal.

8.7 ♦ Analysis of Data: Dog Ownership and Survival After a Heart Attack

Friedmann et al. (1980) reported results from a survey of patients who had a first heart attack. The key outcome of interest was whether or not the patient survived at least 1 year (coded 0 = No, 1 = Yes). One of the variables they assessed was dog ownership (0 = No, 1 = Yes). The results for this sample of 92 patients are shown in Table 8.4. Three statistics will be computed for this table, to assess the relationship between pet ownership and survival: a phi coefficient computed from the cell frequencies in this table; a Pearson correlation calculated from the 0,1 scores; and a chi-square test of significance. Using the formula in Equation 8.5, the phi coefficient for the data in Table 8.4 is .310. The corresponding chi-square, calculated from Equation 8.6, is 8.85. This chi-square exceeds the critical value of chi-square for a 1-*df* table (χ^2 critical = 3.84), so we can conclude that there is a significant association between pet ownership and survival. Note that the phi coefficient is just a special case of Pearson *r*, so the value of the obtained correlation between pet ownership and survival will be the same whether it is obtained from the SPSS bivariate Pearson correlation procedure or as a ϕ coefficient from the SPSS Crosstabs procedure.

Although it is possible to calculate chi-square by using the values of ϕ and *N*, it is also instructive to consider another method for the computation of chi-square, a method based on the sizes of the discrepancies between observed frequencies and expected frequencies, based on a null hypothesis that the row and column variables are not related. We will reanalyze the data in Table 8.4 and compute chi-square directly from the cell frequencies. Our notation for the observed frequency of scores in each cell will be *O*; the expected cell frequency for each cell is denoted by *E*. The expected cell frequency is the number of observations that are expected to fall in each cell under the null hypothesis that the row and column variables are independent. These expected values for *E* are generated from a simple model that tells us what cell frequencies we would expect to see if the row and column variables were independent.

First, we need to define independence between events *A* (such as owning a dog) and *B* (surviving 1 year after a heart attack). If Pr(*A*) = Pr(*A*|B)—that is, if the **unconditional probability** of *A* is the same as the **conditional probability** of *A* given *B*, then *A* and *B* are independent. Let's look again at the observed frequencies given in Table 8.4 for pet ownership and coronary disease patient survival. The unconditional probability that any patient in the study will be alive at the end of 1 year is denoted by Pr(alive at the end of 1 year) and is obtained by dividing the number of persons alive by the total N in the sample; this yields 78/92 or .85. In the absence of any other information, we would predict that any randomly selected patient has about an 85% chance of survival. Here are two of the conditional probabilities that can be obtained from this table. The conditional probability of surviving 1 year for dog owners is denoted by Pr(survived 1 year|owner of dog); it is calculated by taking the number of dog nonowners who survived and dividing by the total number of dog owners, 50/53, which yields .94. This is interpreted as a 94% chance of survival for dog owners. The conditional probability of survival for nonowners is denoted by Pr(survived 1 year|nonowner of dog); it is calculated by taking the number of dog nonowners who survived and dividing by the total number of nonowners of dogs; this gives 28/39 or .72—that is, a 72% chance of survival for nonowners of dogs. If survival were independent of dog ownership, then these three probabilities should all be equal: Pr(alive|owner) = Pr(alive|nonowner) = Pr(alive). For this set of data, these three probabilities are *not* equal. In fact, the probability of surviving for dog owners is higher than for nonowners and higher than the probability of surviving in general for all persons in the sample. We need a statistic to help us evaluate whether this difference between the conditional and unconditional probabilities is statistically significant or whether it is small enough to be reasonably attributed to sampling error. In this case, we can evaluate significance by setting up a model of the expected frequencies we should see in the cells if ownership and survival were independent.

For each cell, the expected frequency, E—the number of cases that would be in that cell if group membership on the row and column variables were independent—is obtained by taking (Row total × Column total)/Table total N. For instance, for the dog owner/alive cell, the expected frequency E = (Number of dog owners × Number of survivors)/Total N in the table. Another way to look at this computation for E is that E = Column total × (Row total/Total N); that is, we take the total number of cases in a column and divide it so that the proportion of cases in Row 1 equals the proportion of cases in Row 2. For instance, the expected number of dog owners who survive 1 year if survival is independent of ownership = Total number of dog owners × Proportion of all people who survive 1 year = 53 × (78/92) = 44.9. That is, we take the 53 dog owners and divide them into the same proportions of survivors and nonsurvivors as in the overall table. These expected frequencies, E, for the dog ownership data are summarized in Table 8.7.

Note that the Es (expected cell frequencies if H_0 is true and the variables are not related) in Table 8.7 sum to the same marginal frequencies as the original data in Table 8.4. All we have done is reapportion the frequencies into the cells in such a way that Pr(A) = Pr($A|B$). That is, for this table of Es,

Pr(survived 1 year) = 78/92 = .85,

Pr(survived 1 year|owner) = 44.9/53 = .85, and

Pr(survived 1 year|nonowner) = 33.1/39 = .85.

In other words, if survival is *not* related to ownership of a dog, then the probability of survival should be the same in the dog owner and non-dog-owner groups, and we have

Table 8.7 ♦ Expected Cell Frequencies (If Dog Ownership and Survival Status Are Independent) for the Data in Tables 8.3 and 8.4

	Survival Status		
Dog Ownership	*No*	*Yes*	*Row Total, N*
(a) Expected cell frequencies (E)			
No	$E = 5.9$	$E = 33.1$	39
Yes	$E = 8.1$	$E = 44.9$	53
Column total *N*	13	78	92
(b) Observed – expected cell frequencies (O – E)			
No	$(11 - 5.9) = +5.1$	$(28 - 33.1) = -5.1$	0
Yes	$(3 - 8.1) = -5.1$	$(50 - 44.9) = +5.1$	0
Column total *N*	0	0	

figured out what the cell frequencies would have to be in order to make those probabilities or proportions equal.

Next we compare the *E*s (the frequencies we would expect if owning a dog and survival are independent) and *O*s (the frequencies we actually obtained in our sample). We want to know if our actually observed frequencies are close to the ones we would expect if H_0 were true; if so, it would be reasonable to conclude that these variables are independent. If *O*s are very far from *E*s, then we can reject H_0 and conclude that there is some relationship between these variables. We summarize the differences between *O*s and *E*s across cells by computing the following statistic:

$$\chi^2 = \sum (O - E)^2/E. \tag{8.11}$$

Note that the $(O–E)$ deviations sum to zero within each row and column, which means that once you know the $(O–E)$ deviation for the first cell, the other three $(O–E)$ values in this 2×2 table are not free to vary. In general, for a table with r rows and c columns, the number of independent deviations $(O–E) = (r–1)(c–1)$, and this is the *df* for the chi-square. For a 2×2 table, $df = 1$.

In this example,

$$\begin{aligned}\chi^2 &= (50 - 44.9)^2/44.9 + (28 - 33.1)^2/33.1 + (3 - 8.1)^2/8.1 + (11 - 5.9)^2/5.9\\ &= (5.1)2/44.9 + (5.1)2/33.1 + (5.1)2/8.1 + (5.1)2/5.9\\ &= 8.85.\end{aligned}$$

This obtained value agrees with the value of chi-square that we computed earlier from the phi coefficient, and it exceeds the critical value of chi-square that cuts off 5% in the right-hand tail for the 1-*df* distribution of χ^2 (critical value = 3.84). Therefore, we conclude that there is a statistically significant relation between these variables, and the nature of the relationship is that dog owners have a significantly higher probability of surviving 1 year after a heart attack (about 94%) than nonowners of dogs (72%). Survival is not independent of pet ownership; in fact, in this sample, there is a significantly higher *rate* of survival for dog owners.

The most widely reported effect size for the chi-square test of association is Cramer's V. Cramer's V can be calculated for contingency tables with any number of rows and columns. For a 2×2 table, Cramer's V is equal to the absolute value of phi. Values of Cramer's V range from 0 to 1 regardless of table size (but only if the row marginal totals equal the column marginal totals). Values close to 0 indicate no association; values close to 1 indicate strong association:

$$V = \sqrt{\frac{\chi^2}{n \times m}}, \tag{8.12}$$

where chi-square is computed from Equation 8.11, n is the total number of scores in the sample, and m is the minimum of (Number of rows – 1), (Number of columns – 1).

The statistical significance of Cramer's V can be assessed by looking at the associated chi-square; Cramer's V can be reported as effect size information for a chi-square analysis. Cramer's V is a **symmetrical index of association**; that is, it does not matter which is the independent variable.

Chi-square **goodness-of-fit** tests can be applied to 2×2 tables (as in this example); they can also be applied to contingency tables with more than two rows or columns. Chi-square also has numerous applications later in statistics as a generalized goodness-of-fit test.[2] Although chi-square is commonly referred to as a "goodness-of-fit" test, note that the higher the chi-square value, in general, the worse the agreement between the model used to generate expected values that correspond to some model and the observed data. When chi-square is applied to contingency tables, the expected frequencies generated by the model correspond to the null hypothesis that the row and column variables are independent. Therefore, a chi-square large enough to be judged statistically significant is a basis for rejection of the null hypothesis that group membership on the row variable is unrelated to group membership on the column variable.

When chi-square results are reported, the write-up should include the following:

1. A table that shows the observed cell frequencies and either row or column percentages (or both).

2. The obtained value of chi-square, its *df*, and whether it is statistically significant.

3. An assessment of the effect size; this can be phi, for a 2×2 table; other effect sizes such as Cramer's V are used for larger tables.

4. A statement about the nature of the relationship, stated in terms of differences in proportions or of probabilities. For instance, in the pet ownership example, the researchers could say that the probability of surviving for 1 year is much higher for owners of dogs than for people who do not own dogs.

8.8 ♦ Chi-Square Test of Association (Computational Methods for Tables of Any Size)

The method of computation for chi-square described in the preceding section can be generalized to contingency tables with more than two rows and columns. Suppose that the table has r rows and c columns. For each cell, the expected frequency, E, is computed by multiplying the corresponding row and column total Ns and dividing this product by the N of cases in the entire table. For each cell, the deviation between O (observed) and E (expected) frequencies is calculated, squared, and divided by E (expected frequency for that cell). These terms are then summed across the $r \times c$ cells. For a 2×3 table, for example, there are six cells and six terms included in the computation of chi-square. The df for the chi-square test on an $r \times c$ table is calculated as follows:

$$df = (r - 1) \times (c - 1). \tag{8.13}$$

Thus, for instance, the degrees of freedom for a 2×3 table is $(2 - 1) \times (3 - 1) = 2$ df. Only the first two $(O - E)$ differences between observed and expected frequencies in a 2×3 table are free to vary. Once the first two deviations are known, the remaining deviations are determined because of the requirement that $(O - E)$ sum to zero down each column and across each row of the table. Critical values of chi-square for any df value can be found in the table in Appendix D.

8.9 ♦ Other Measures of Association for Contingency Tables

Until about 10 years ago, the most widely reported statistic for the association between categorical variables in contingency tables was the chi-square test of association (sometimes accompanied by ϕ or Cramer's V as effect size information). The chi-square test of association is still fairly widely reported. However, many research situations involve prediction of outcomes that have low base rates (e.g., fewer than 100 out of 10,000 patients in a medical study may die of coronary heart disease). The effect size indexes most commonly reported for chi-square, such as phi, are constrained to be less than +1.00 when the marginal distribution for the predictor variable differs from the marginal distribution of the outcome variable; for instance, in some studies, about 60% of patients have the Type A coronary-prone personality, but only about 5% to 10% of the patients develop heart disease. Because the marginal distributions (60/40 split on the personality predictor variable vs. 90/10 or 95/5 split on the outcome variable) are so different, the maximum possible value of phi or Pearson r is restricted; even if there is a strong association between personality and disease, phi cannot take on values close to +1 when the marginal distributions of the X and Y variables are greatly different. In such situations, effect size measures, such as phi, that are **marginal dependent** may give an impression of effect size that is unduly pessimistic.

Partly for this reason, different descriptions of association are often preferred in clinical studies; in recent years, odds ratios have become the most popular index of the strength of association between a risk factor (such as smoking) and a disease outcome (such as lung cancer) or between a treatment and an outcome (such as survival). Odds ratios are usually obtained as part of a binary logistic regression analysis. A brief definition of odds ratios is provided in the glossary, and a more extensive explanation of this increasingly popular approach to summarizing information from 2 × 2 tables that correlate risk and outcome (or intervention and outcome) is provided in Chapter 21.

In addition, there are dozens of other statistics that may be used to describe the patterns of scores in contingency tables. Some of these statistics are applicable only to tables that are 2 × 2; others can be used for tables with any number of rows and columns. Some of these statistics are marginal dependent, while others are not dependent on the marginal distributions of the row and column variables. Some of these are symmetric indexes, while others (such as lambda and Somers's *d*) are asymmetric; that is, they show a different reduction in uncertainty for prediction of *Y* from *X* than for prediction of *X* from *Y*. The **McNemar test** is used when a contingency table corresponds to repeated measures—for example, participant responses on a binary outcome variable before versus after an intervention. A full review of these many contingency table statistics is beyond the scope of this book; see Everitt (1977) and Liebetrau (1983) for more comprehensive discussion of contingency table analysis.

8.10 ♦ SPSS Output and Model Results Write-Up

Two SPSS programs were run on the data in Table 8.4 to verify that the numerical results obtained by hand earlier were correct. The SPSS Crosstabs procedure was used to compute phi and chi-square (this program also reports numerous other statistics for contingency tables). The SPSS bivariate correlation procedure (as described earlier in Chapter 7) was also applied to these data to obtain a Pearson *r* value.

To enter the dog owner/survival data into SPSS, one column was used to represent each person's score on dog ownership (coded 0 = Did not own dog, 1 = Owned dog), and a second column was used to enter each person's score on survival (0 = Did not survive for 1 year after heart attack, 1 = Survived for at least 1 year). The number of lines with scores of 1, 1 in this dataset corresponds to the number of survivors who owned dogs. The complete set of data for this SPSS example appears in Table 8.3.

The SPSS menu selections to run the Crosstabs procedure were as follows (from the top-level menu, make these menu selections, as shown in Figure 8.3): <Analyze> → <Descriptive Statistics> → <Crosstabs>.

This opens the SPSS dialog window for the Crosstabs procedure, shown in Figure 8.4. The names of the row and column variables were placed in the appropriate windows. In this example, the row variable corresponds to the score on the predictor variable (dog ownership), and the column variable corresponds to the score on the outcome variable (survival status). The Statistics button was clicked to access the menu of optional statistics to describe the pattern of association in this table, as shown in Figure 8.5. The optional statistics selected included chi-square, phi, and Cramer's *V*. In addition, the Cells button in the main Crosstabs dialog window was used to open up the Crosstabs Cell Display menu, which appears in Figure 8.6. In addition to the observed frequency for each cell, the expected frequencies for each cell and row percentages were requested.

The output from the Crosstabs procedure for these data appears in Figure 8.7. The first panel shows the contingency table with observed and expected cell frequencies and row

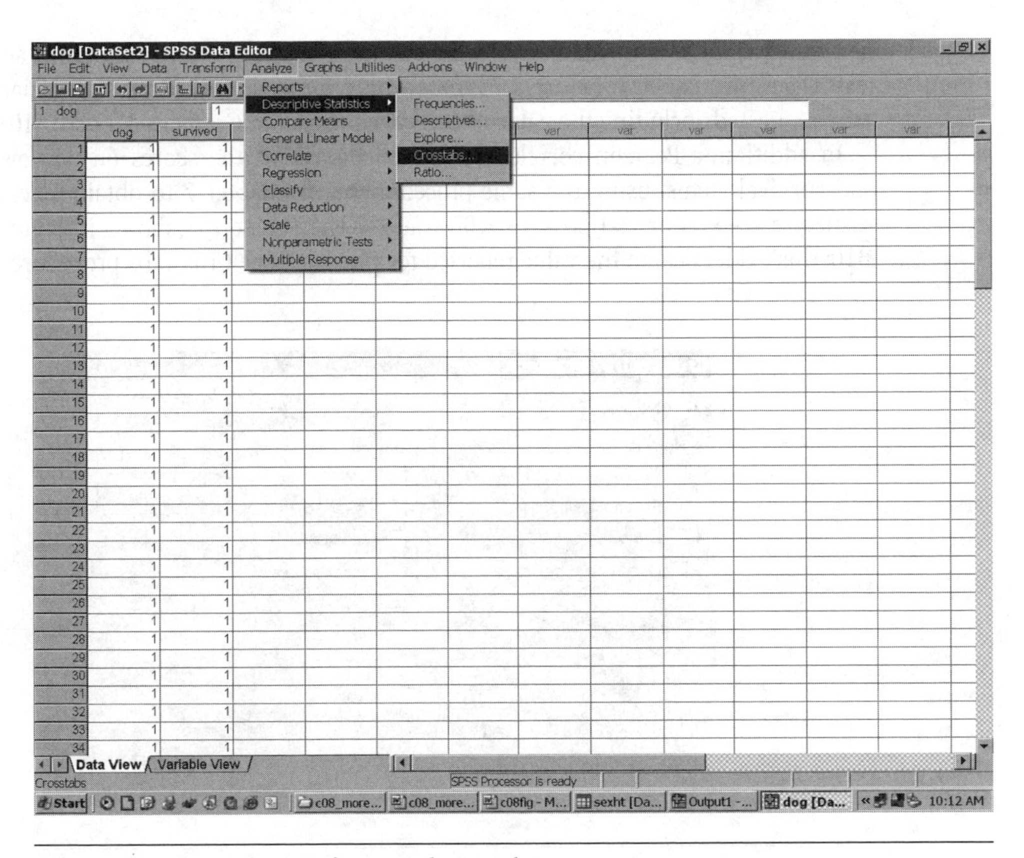

Figure 8.3 ♦ Menu Selections for Crosstabs Procedure

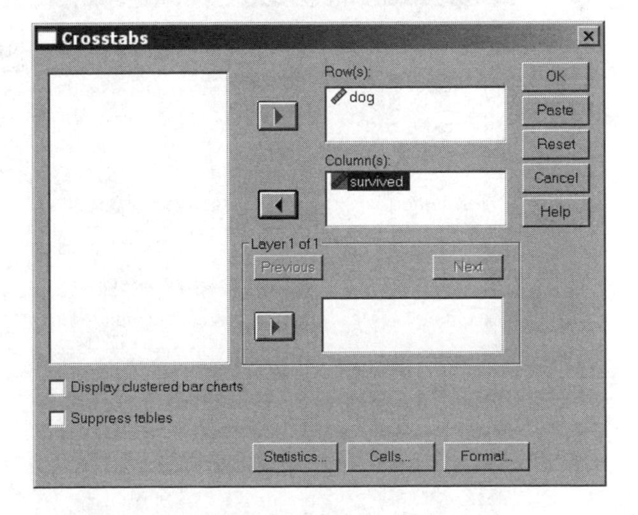

Figure 8.4 ♦ SPSS Crosstabs Main Dialog Window

percentages. The second panel reports the obtained value of χ^2 (8.85) and some additional tests. The third panel in Figure 8.7 reports the symmetric measures of association that were requested, including the value of ϕ (.310) and that of Cramer's V (also .310).

In addition, a Pearson correlation was calculated for the scores on dog ownership and survival status, using the same procedure as in Chapter 7 to obtain a correlation: <Analyze> → <Correlation> → <Bivariate>. The Pearson r (shown in Figure 8.8) is .310; this is identical to the value reported for phi using the Crosstabs procedure above.

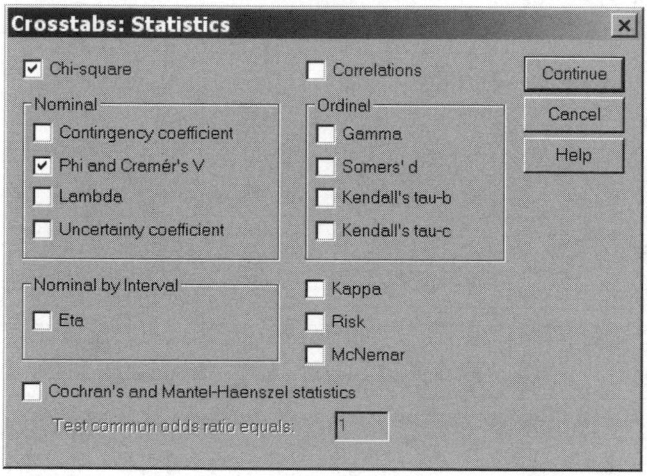

Crosstabs: Statistics

Chi-square
Correlations
Continue
Cancel
Help
Nominal
Contingency coefficient
Phi and Cramér's V
Lambda
Uncertainty coefficient
Ordinal
Gamma
Somers' d
Kendall's tau-b
Kendall's tau-c
Nominal by Interval
Eta
Kappa
Risk
McNemar
Cochran's and Mantel-Haenszel statistics
Test common odds ratio equals: 1

Figure 8.5 ♦ SPSS Crosstabs Statistics Dialog Window

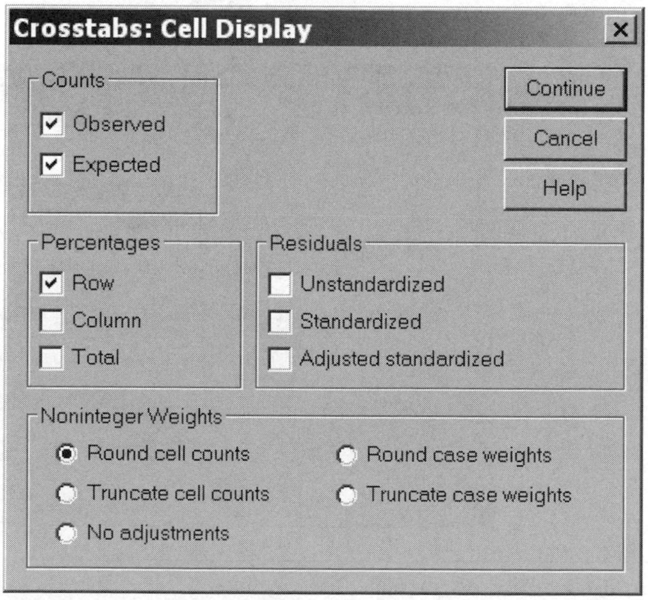

Crosstabs: Cell Display

Counts
Observed
Expected
Continue
Cancel
Help
Percentages
Row
Column
Total
Residuals
Unstandardized
Standardized
Adjusted standardized
Noninteger Weights
Round cell counts
Round case weights
Truncate cell counts
Truncate case weights
No adjustments

Figure 8.6 ♦ SPSS Crosstabs Cell Display Dialog Window

dog * survived Crosstabulation

			survived		
			Did Not Survive	Survived	Total
dog	Does Not Own a Dog	Count	11	28	39
		Expected Count	5.9	33.1	39.0
		% within dog	28.2%	71.8%	100.0%
	Owns a Dog	Count	3	50	53
		Expected Count	8.1	44.9	53.0
		% within dog	5.7%	94.3%	100.0%
Total		Count	14	78	92
		Expected Count	14.0	78.0	92.0
		% within dog	15.2%	84.8%	100.0%

Chi-Square Tests

	Value	df	Asymp. Sig. (2-sided)	Exact Sig. (2-sided)	Exact Sig. (1-sided)
Pearson Chi-Square	8.851[b]	1	.003		
Continuity Correction[a]	7.190	1	.007		
Likelihood Ratio	9.011	1	.003		
Fisher's Exact Test				.006	.004
Linear-by-Linear Association	8.755	1	.003		
N of Valid Cases	92				

a. Computed only for a 2x2 table

b. 0 cells (.0%) have expected count less than 5. The minimum expected count is 5.93.

Symmetric Measures

		Value	Approx. Sig.
Nominal by Nominal	Phi	.310	.003
	Cramer's V	.310	.003
N of Valid Cases		92	

a. Not assuming the null hypothesis.

b. Using the asymptotic standard error assuming the null hypothesis.

Figure 8.7 ♦ SPSS Output From Crosstabs Procedure for Dog/Survival Status Data in Tables 8.3 and 8.4

Correlations

		DOG	SURVIVED
DOG	Pearson Correlation	1	.310**
	Sig. (2-tailed)	.	.003
	N	92	92
SURVIVED	Pearson Correlation	.310**	1
	Sig. (2-tailed)	.003	.
	N	92	92

**. Correlation is significant at the 0.01 level (2-tailed).

Figure 8.8 ♦ SPSS Output From Pearson Correlation Procedure for Dog/Survival Status Data in Tables 8.3 and 8.4

Results

A survey was done to assess numerous variables that might predict survival for 1 year after a first heart attack; there were 92 patients in the study. Only one predictor variable is reported here: dog ownership. Expected cell frequencies were examined to see whether there were any expected frequencies less than 5; the smallest expected cell frequency was 5.9. (If there were one or more cells with expected frequencies less than 5, it would be preferable to report the Fisher exact test rather than chi-square.) Table 8.4 shows the observed cell frequencies for dog ownership and survival status. Of the 53 dog owners, 3 did not survive; of the 39 nonowners of dogs, 11 did not survive. A phi coefficient was calculated to assess the strength of this relationship: $\phi = .310$. This corresponds to a medium-size effect. This was a **statistically significant association**: $\chi^2(1) = 8.85$, $p < .05$. This result was also statistically significant by the Fisher exact test, $p = .006$. The nature of the relationship was that dog owners had a significantly higher proportion of survivors (94%) than nondog owners (72%). Because this study was not experimental, it is not possible to make a causal inference.

8.11 ♦ Summary

This chapter provided information about different forms of correlation that are appropriate when X and Y are rank/ordinal or when one or both of these variables are dichotomous. This chapter demonstrated that Pearson r can be applied in research situations where one or both of the variables are true dichotomies. This is important because it means that true dichotomous variables may be used in many other multivariate analyses that build on variance partitioning and use covariance and correlation as information about the way variables are interrelated.

The chi-square test of association for contingency tables was presented in this chapter as a significance test that can be used to evaluate the statistical significance of the phi correlation coefficient. However, chi-square tests have other applications, and it is useful for students to understand the chi-square as a general goodness-of-fit test; for example, chi-square is used as one of the numerous goodness-of-fit tests for structural equation models.

This chapter described only a few widely used statistics that can be applied to contingency tables. There are many other possible measures of association for contingency tables; for further discussion, see Everitt (1977) or Liebetrau (1983). Students who anticipate that they will do a substantial amount of research using dichotomous outcome variables should refer to Chapter 21 in this book for an introductory discussion of binary logistic regression; **logistic regression** is presently the most widely used analysis for this type of data. For categorical outcome variables with more than two categories, polytomous logistic regression can be used (Menard, 2001). In research situations where there are several categorical predictor variables and one categorical outcome variable, log linear analysis is often reported.

Notes

1. In other words, conclusions about the outcome of this study depend entirely on the outcomes for these five individuals, regardless of the size of the total *N* for the table (and it is undesirable to have a study where a change in outcome for just one or two participants can greatly change the nature of the outcome).

2. There are other applications of chi-square apart from its use to evaluate the association between row and column variables in contingency tables. For example, in structural equation modeling (SEM), chi-square tests are performed to assess how much the **variance/covariance matrix** that is reconstructed from SEM parameters differs from the original variance/covariance matrix calculated from the scores. A large chi-square for an SEM model is usually interpreted as evidence that the model is a poor fit—that is, the model does not do a good job of reconstructing the variances and covariances.

Comprehension Questions

1. How are point biserial r (r_{pb}) and the phi coefficient different from Pearson r?
2. How are biserial r (r_b) and tetrachoric r (r_{tet}) different from Pearson r?
3. Is high blood pressure diagnosis (defined as high blood pressure = 1 = Systolic pressure equal to or greater than 140 mm of mercury, low blood pressure = 0 = Systolic pressure less than 140mm of mercury) a true dichotomy or an artificial dichotomy?
4. The data in the table below were collected in a famous social-psychological field experiment. The researchers examined a common source of frustration for drivers: a car stopped at a traffic light that fails to move when the light turns green. The variable they manipulated was the status of the frustrating car (1 = High status, expensive, new; 0 = Low status, inexpensive, old). They ran repeated trials in which they stopped at a red light, waited for the light to turn green, and then did not move the car; they observed whether the driver in the car behind them honked or not (1 = Honked, 0 = Did not honk). They predicted that people would be more likely to honk at low-status cars than at high-status cars (Doob & Gross, 1968).

   This table reports part of their results:

	Car Status	
Response	*Low*	*High*
Honked	32	18
Did not honk	6	18

   a. Calculate phi and chi-square by hand for the table above, and write up a Results section that describes your findings and notes whether the researchers' prediction was upheld.

b. Enter the data for this table into SPSS. To do this, create one variable in the SPSS worksheet that contains scores of 0 or 1 for the variable status and another variable in the SPSS worksheet that contains scores of 0 or 1 for the variable honking (e.g., because there were 18 people who honked at a high-status car, you will enter 18 lines with scores of 1 on the first variable and 1 on the second variable).
c. Using SPSS, do the following: Run the Crosstabs procedure and obtain both phi and chi-square; also, run a bivariate correlation (and note how the obtained bivariate correlation compares with your obtained phi).
d. In this situation, given the marginal frequencies, what is the maximum possible value of phi?
e. The researchers manipulated the independent variable (status of the car) and were careful to control for extraneous variables. Can they make a causal inference from these results? Give reasons for your answer.

5. When one or both of the variables are dichotomous, the Pearson r has specific names; for example, when a true dichotomy is correlated with a quantitative variable, what is this correlation called? When two true dichotomous variables are correlated, what is this correlation called?

6. What information should be included in the report of a chi-square test of contingency?

7. The table below gives the percentage of people who were saved (vs. lost) when the *Titanic* sank. The table provides information divided into groups by class (first class, second class, third class, and crew) and by gender and age (children, women, men).

Titanic Disaster—Official Casualty Figures

Passenger Category	*Percent Saved*	*Percent Lost*	*Number Saved*	*Number Lost*	*Number Aboard*
Children, First Class	100.00	0.00	6	0	6
Children, Second Class	100.00	0.00	24	0	24
Women, First Class	97.22	2.78	140	4	144
Women, Crew	86.96	13.04	20	3	23
Women, Second Class	86.02	13.98	80	13	93
Women, Third Class	46.06	53.94	76	89	165
Children, Third Class	34.18	65.82	27	52	79
Men, First Class	32.57	67.43	57	118	175
Men, Crew	21.69	78.31	192	693	885
Men, Third Class	16.23	83.77	75	387	462
Men, Second Class	8.33	91.67	14	154	168
Total	**31.97**	**68.08**	**711**	**1,513**	**2,224**

SOURCE: British Parliamentary Papers, Shipping Casualties (Loss of the Steamship 'Titanic"). 1912, cmd. 6352, 'Report of a Formal Investigation into the circumstances attending the foundering on the 15th April, 1912, of the British Steamship 'Titanic," of Liverpool, after striking ice in or near Latitude 41= 46' N., Longitude 50 =14' W., North Atlantic Ocean, whereby loss of life ensued.' (London: His Majesty's Stationery Office, 1912), page 42.

The information in the table is sufficient to set up some simple chi-square tests. For example, let's ask the question, Was there a difference in the probability of being saved for women passengers in first class versus women passengers in third class? There were a total of 309 women in first and third class. The relevant numbers from the table on page 336 appear in the table below.

	Women Passengers		
	First Class	*Third Class*	*Total N*
Saved	140	76	216
Lost	4	89	93
Total *N*	144	165	309

Compute a phi coefficient using the observed cell frequencies in the table above. Also, compute a chi-square statistic for the observed frequencies in the table above. Write up your results in paragraph form.

Was there a statistically significant association between being in first class and being saved when we look at the passenger survival data from the *Titanic*? How strong was the association between class and outcome (e.g., how much more likely were first class passengers to be saved than were third class passengers)?

CHAPTER 9

Bivariate Regression

9.1 ♦ Research Situations Where Bivariate Regression Is Used

A bivariate regression analysis provides an equation that predicts raw scores on a quantitative Y variable from raw scores on an X variable; in addition, it also provides an equation to predict z or **standardized scores** on Y from standardized scores on X. The predictor or X variable is usually also quantitative, but it can be a dichotomous variable (as described in Chapter 8). Bivariate regression analysis is closely related to Pearson r. Like Pearson r, bivariate regression assumes that the relation between Y and X is linear; this implies that scores on Y can be predicted as a linear function of scores on X using an equation of the following form:

$$Y' = b_0 + bX, \tag{9.1}$$

where Y' is the predicted score on the outcome (Y) variable, $\boldsymbol{b_0}$ is the **intercept** or constant term, and $\boldsymbol{b}$ is the slope. The intercept term in this equation, b_0, is the predicted Y score when X equals 0. The nature of the relation between X and Y is described by the slope coefficient, b, which can be interpreted as the number of units of change in the raw score on Y that is predicted for a one-unit increase in the raw score on X. Figure 9.1 shows the graph of a line with an intercept $b_0 = 30$ and a slope $b = 10$. The line crosses the Y axis at $Y = 30$, and for each one-unit increase in X, Y increases by 10 points.

The magnitude of Pearson r provides information about how well predicted Y' scores generated from Equation 9.1 will match the actual Y scores. For example, if Pearson $r = 1$, predicted Y' scores will be identical to the observed Y scores for every subject. As Pearson r decreases in absolute magnitude, the distances between predicted Y' scores and actual Y scores increase. Pearson r tells us how well we can expect the predictions from our linear regression equation to fit the data; the larger r is (in absolute value), the closer the correspondence between the predictions from our regression equation and the actual Y scores.

The b_0 and b (intercept and slope) coefficients in Equation 9.1 are the coefficients that provide the best possible predictions for Y. They are obtained using ordinary least squares (OLS)[1] estimation procedures; that is, the formulas to compute b_0 and b provide the prediction equation for which the sum of the squared differences between actual Y and predicted Y', that is, $\sum (Y - Y')^2$, is minimized.

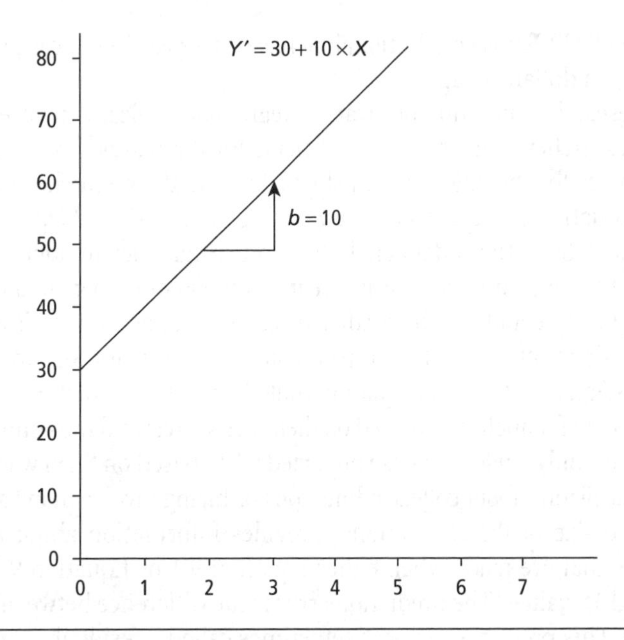

Figure 9.1 ◆ Bivariate Regression Line With Intercept $b_0 = 30$ and Slope $b = 10$

NOTE: $Y' = b_0 + bX$
$= 30 + 10X$.

The value of b_0, the intercept, can have a meaningful interpretation in situations where the *X* predictor variable can take on the value of 0, as in the example above. However, in situations where *X* never equals 0 (e.g., in a prediction of salary from IQ), the intercept is just the point where the regression line crosses the *Y* axis; it is a prediction for a value of *X* that is far outside the range of actual values of *X*.

In Chapter 7, an equation was given to predict standard or *z* scores on *Y* from *z* scores on *X*:

$$z'_Y = r \times z_X \text{ or } z'_Y = \beta \times z_X. \tag{9.2}$$

Thus, we have two different forms of the prediction equation for the variables *X* and *Y*. In Equation 9.1, predictions for raw scores on *Y* are given the terms of raw scores for *X* (in the original units of measurement). In Equation 9.2, predictions for z_Y are made from values of z_X. Equation 9.2 is a "unit-free" or standardized version of the prediction equation.[2] Both forms of the equation are useful, and they provide slightly different information. The standard score or *z* score form (Equation 9.2) is useful when researchers want to describe the strength of the relationship between *X* and *Y* in standardized, unit-free terms. The raw score form of the predictive equation (Equation 9.1) is useful when the units used to measure both *X* and *Y* correspond to "real" quantities such as dollars or years or points on standardized tests for which we have normative data. When *X* and *Y* are given in meaningful units, the value of *b* provides useful information about predicted magnitudes of change. For example, for each hour of Scholastic Aptitude Test (SAT) preparation or coaching given to a student, how many additional points will that student probably score on the SAT? For each additional increase of 1 year in education, how much does starting salary in dollars go up?

Research application of bivariate regression typically involves two or three steps. First, the researcher estimates the coefficients for the regression equation (i.e., finds numerical estimates for the intercept and slope). Second, the researcher assesses how well or poorly this equation predicts scores on Y (using both statistical significance tests and effect size indexes). Sometimes data analysts who use statistics in applied settings go on to a third step: They may use the bivariate regression equation to generate predicted scores for individual people and use the predicted scores as a basis for decisions. For example, a college admissions officer might use past data on a student's SAT to estimate coefficients for a regression equation; that equation could be used to predict future college grade point average (GPA) for applicants, based on their SAT scores; and the admissions officer might decide to admit only students whose predicted GPA (based on SAT) was greater than 2.0. (In practice, decisions about college admissions or hiring involve many additional factors.)

The size of the correlation r provides information about the size of the prediction errors that are made when either Equation 9.1 or Equation 9.2 is used to generate predicted Y_i' values. The prediction error is the difference between Y_i and Y_i' for each participant. This prediction error is sometimes called a residual.

$$Y_i - Y_i' = \text{prediction error or residual for participant } i. \tag{9.3}$$

When $r = +1$ or -1, there is a perfect linear relation between X and Y, and the value of the $Y_i - Y_i'$ prediction error is 0 for every case; there is perfect prediction. As the value of r decreases in absolute value, the sizes of these prediction errors increase.

Pearson r is used to describe the strength of the linear relationship between one predictor and one outcome variable. However, regression analysis can be generalized to include more than one predictor. The correlation between the actual Y scores and the predicted Y' scores in a regression analysis is called a **multiple *R*** (or sometimes just ***R***) because the prediction may be based on multiple predictor variables. When we run a bivariate regression in SPSS, using just one predictor variable, the output will include a multiple R. When we have just one predictor, R is equivalent to r. When we begin to look at analyses with more than one predictor, R will provide information about the predictive usefulness of an entire set of several predictors, while r is used to denote the relationship between just one predictor variable and an outcome variable.

9.2 ♦ A Research Example: Prediction of Salary From Years of Job Experience

A small set of hypothetical data for the empirical example is given in Table 9.1. The predictor, X, is the number of years employed at a company; the outcome, Y, is the annual salary in dollars. The research question is whether salary changes in a systematic (linear) way as years of job experience or job seniority increases. In other words, how many dollars more salary can an individual expect to earn for each additional year of employment?

Table 9.1 ♦ Data for Salary/Years of Job Experience: SPSS Analysis

Case	*Years*	*Salary ($)*
1	1	33,470
2	1	14,841
3	1	27,314
4	1	38,837
5	1	39,212
6	2	27,493
7	2	46,372
8	2	49,651
9	2	35,832
10	2	39,652
11	3	45,720
12	3	39,141
13	3	39,964
14	3	61,903
15	3	45,791
16	4	30,674
17	4	31,660
18	4	61,870
19	4	44,082
20	4	20,890
21	4	54,679
22	4	45,284
23	5	50,086
24	5	38,183
25	5	46,498
26	5	49,140
27	6	43,685
28	6	42,755
29	6	29,751
30	7	55,918
31	7	55,937
32	7	46,955
33	8	56,501
34	8	38,640
35	9	56,600
36	9	52,769
37	9	43,708
38	10	78,096
39	10	65,049
40	12	67,432
41	12	64,543
42	13	82,911
43	14	85,162
44	14	79,647
45	14	66,159
46	15	54,180
47	17	82,461
48	19	69,694
49	20	95,735
50	22	94,320

9.3 ♦ Assumptions and Data Screening

Preliminary data screening for bivariate regression is the same as for Pearson correlation, with one minor additional requirement. When we examine a scatter plot in a regression analysis, the X or predictor variable is conventionally graphed on the X or horizontal axis of the scatter plot and the Y or outcome variable is graphed on the vertical axis. Preliminary examination of the univariate distributions of X and Y tells the researcher whether these distributions are reasonably normal, whether there are univariate outliers, and whether the range of scores on both X and Y is wide enough to avoid problems with restricted range. Preliminary examination of the scatter plot enables the researcher to assess whether the X, Y relation is linear, whether the variance of Y scores is fairly uniform across levels of X, and whether there are bivariate outliers that require attention.

In applications of Pearson r, both X and Y variables could be either quantitative or dichotomous. For bivariate regression, the Y (dependent) variable is assumed to be quantitative; the X (predictor) variable can be either quantitative or dichotomous.

Figure 9.2 shows the distributions of scores on the quantitative predictor variable years of job experience. This distribution is positively skewed (i.e., there is a longer tail on the upper end of the distribution). When a distribution is positively skewed, sometimes a log transformation of the scores has a more nearly normal distribution shape. However, for these data, the natural log of years of job experience (shown in Figure 9.3) also had a rather nonnormal distribution shape. Therefore, the raw scores on years (rather than the log-transformed scores) were used in subsequent analyses. The distribution of scores on salary, which appears in Figure 9.4, was somewhat more normal in shape.

The scatter plot that shows how values of the dependent or outcome variable salary (plotted on the vertical or Y axis) were associated with values of the predictor variable years of job experience (plotted on the horizontal or X axis) appears in Figure 9.5. The relation appears to be positive and reasonably linear; there were no bivariate outliers. There were not enough data points in this small dataset to make a good evaluation of whether Y scores are normally distributed at each level of X, but it appears that the assumption of bivariate normality was reasonably well satisfied.

9.4 ♦ Issues in Planning a Bivariate Regression Study

Unlike Pearson r, bivariate regression analysis requires the researcher to make a distinction between the predictor and outcome variables, and the slope coefficient for an equation to predict Y from X is not the same as the slope coefficient for an equation to predict X from Y. (Pearson r, on the other hand, was symmetrical: The correlation of X with Y was identical to the correlation of Y with X.)

In experimental research situations, it is clear which variable should be treated as the independent variable; the independent variable is the one that has been manipulated by the researcher. In nonexperimental research situations, it is often not clear which variable should be treated as the predictor. In some nonexperimental research, the decision to treat one variable as the predictor and the other as the outcome variable may be completely arbitrary. However, there are several considerations that may be helpful in deciding which of two variables (A or B) to designate as the predictor or X variable in a bivariate regression analysis.

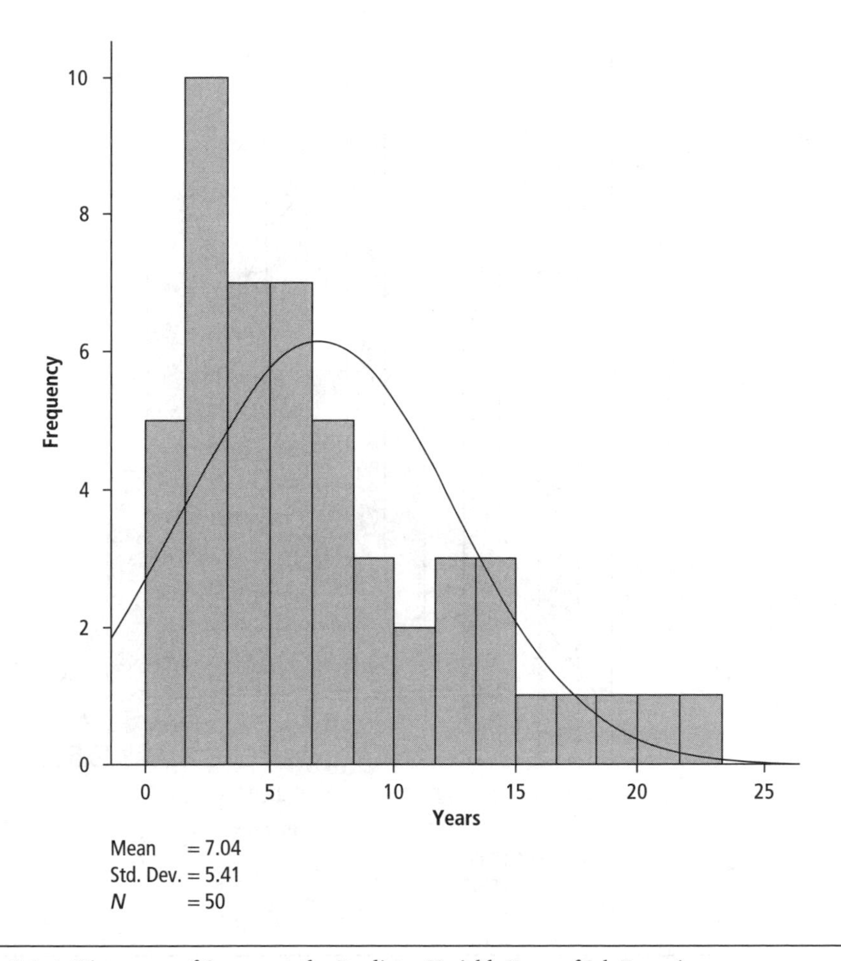

Figure 9.2 ♦ Histogram of Scores on the Predictor Variable Years of Job Experience

If *A* is measured earlier than *B* in time or if *A* assesses some characteristic of the subject that probably existed before *B* occurred, then it makes sense to use *A* to predict *B* (rather than use *B* to predict *A*). Examples: If you have high school GPA (*A*) and college GPA (*B*) as variables, it makes sense to use high school GPA to predict college GPA. For variables such as gender (*A*) and emotional intelligence (*B*), it makes sense to predict emotional intelligence from gender, because gender is established early in life and emotional intelligence probably develops later.

If *A* might be a cause of *B*, then it makes sense to use *A* to predict *B*. If *A* is hours spent studying, and *B* is exam grade, it makes more sense to predict exam grade (*B*) from hours of studying (*A*) than to predict hours of studying from exam grade. If caffeine consumption is *A* and heart rate is *B*, it makes more sense to predict *B* from *A*. If *A* is a strong predictor of *B* in a regression analysis, it does not prove that *A* causes *B*. However, if we have reason to suspect that *A* is a cause of *B*, or that *A* has some influence on *B*, it makes more sense to give the *A* variable the role of predictor in the statistical analysis.

There are some research situations where there is no strong a priori reason to assume causality or unidirectional influence; in these situations, it can still be useful to set up a

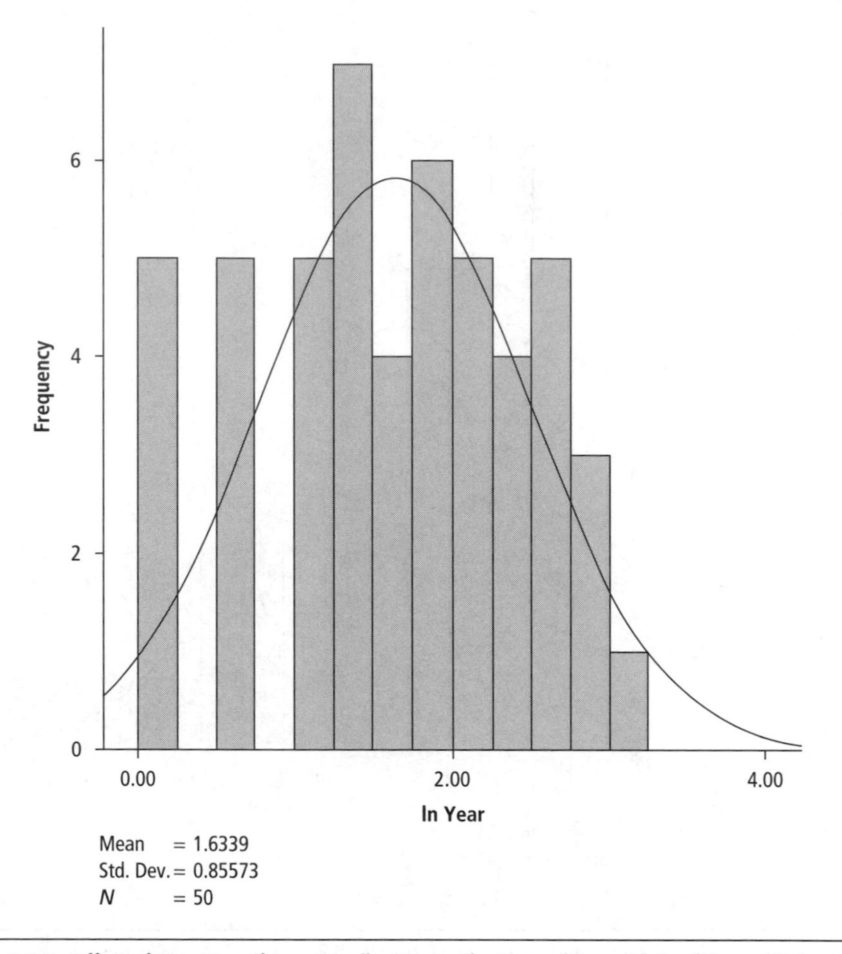

Figure 9.3 ♦ Effect of Data Transformation (ln Year = the Natural Logarithm of Years of Job Experience)

bivariate regression equation, but the decision of which variable to use as predictor may be arbitrary. For example, if A is self-esteem and B is GPA, it may be just as reasonable to predict GPA from self-esteem as to predict self-esteem from GPA.

Apart from the need to distinguish between a predictor and an outcome variable, the same issues apply to the design of a bivariate regression study as for a bivariate correlation study; see Section 7.5 for details. It is important to avoid restricted scores on both X and Y because a restricted range tends to reduce the size of the correlation, and that, in turn, implies that a linear prediction equation will tend to have large prediction errors. Many of the factors that affect the size of Pearson r, described in Chapter 7, also affect the size of b, the regression slope coefficient (because b is essentially a rescaled version of r).

9.5 ♦ Formulas for Bivariate Regression

The best estimate for the slope b for the predictive Equation 9.1 (the OLS estimate that minimizes the sum of squared prediction errors) can be calculated as follows:

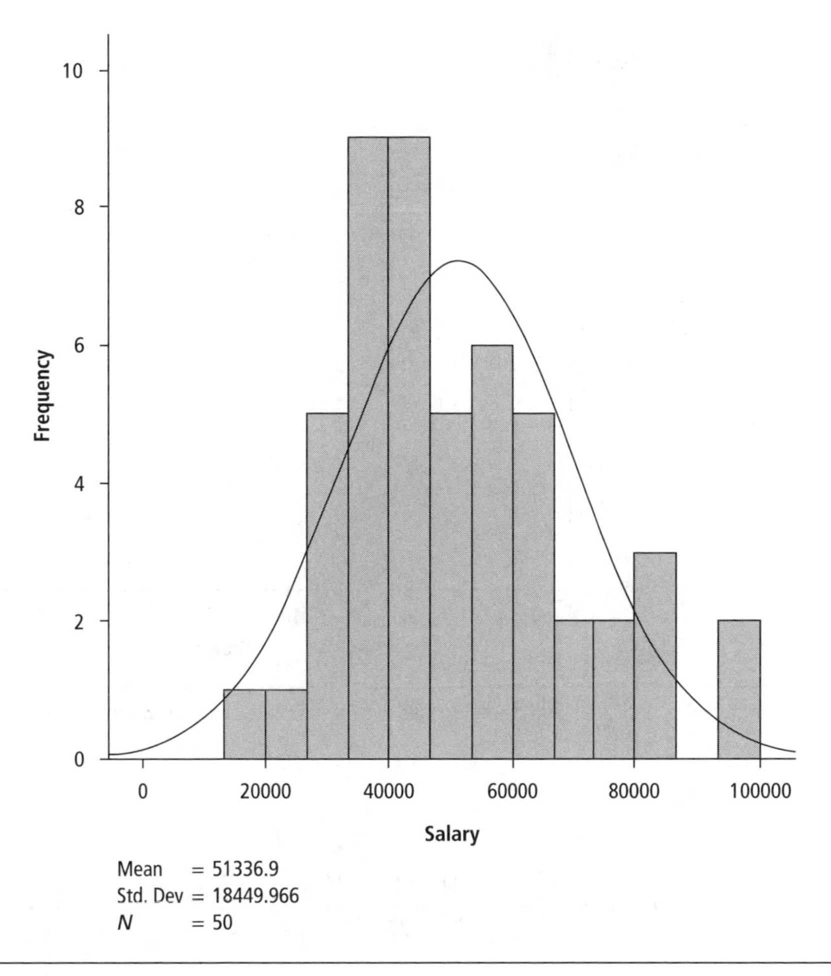

Figure 9.4 ♦ Histogram of Scores on the Outcome Variable Salary

$$b = r\frac{s_Y}{s_X}. \tag{9.4}$$

This equation makes it clear that b is essentially a rescaled version of r. The information necessary to rescale the correlation r so that it can be used to generate a predicted score for Y in terms of the original units used to measure Y is the standard deviation (*SD*) of Y (s_Y) and the standard deviation of X (s_X). Notice several implications of Equation 9.4: If r equals 0, then b also equals 0; the sign of b is determined by the sign of r; other factors being equal, as r increases, b increases, and as s_Y (the standard deviation of the outcome variable) increases, b increases. On the other hand, as s_X (the standard deviation of the predictor variable) increases, if other terms remain constant, then b decreases.

A computationally efficient (but less conceptually clear) version of the formula for b is as follows:

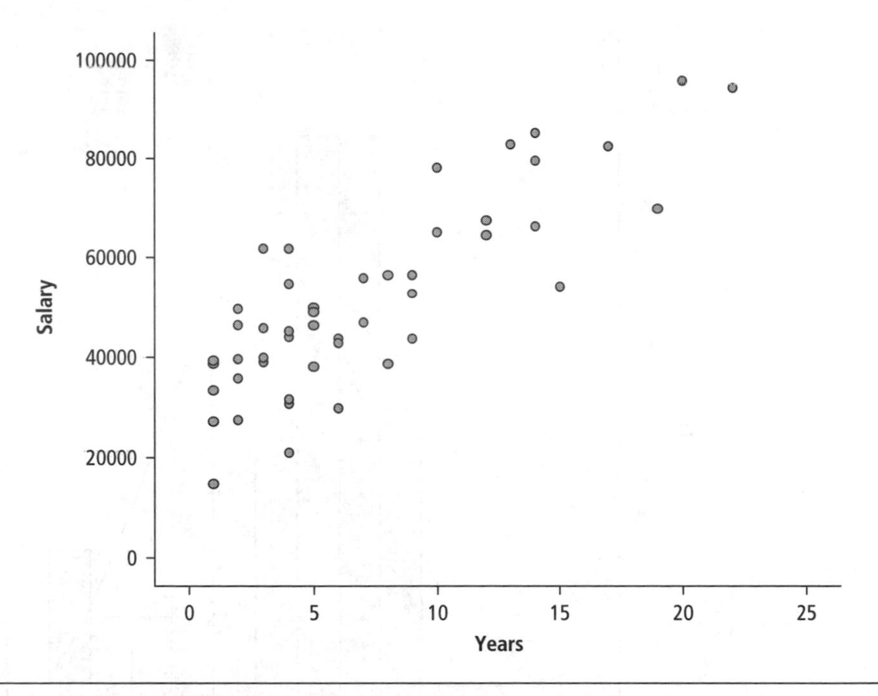

Figure 9.5 ♦ Scatter Plot Showing Salary in Relation to Years of Job Experience

$$b = \frac{\sum (X - M_X)(Y - M_Y)}{\sum (X - M_X)^2} = \frac{\sum XY - (\sum X \sum Y/N)}{\sum X^2 - \left((\sum X)^2 / N \right)}. \tag{9.5}$$

The terms included in Equation 9.5 involve sums of squared *X*s (thus, the variance of *X* is one kind of information that is taken into account in a regression analysis). Other terms in Equation 9.5 involve products between *X* and *Y* (thus, covariance or correlation between *X* and *Y* is another kind of information that is taken into account in a bivariate regression). As the equations for multivariate statistics become more complicated in future chapters, it is still possible to recognize meaningful "chunks" of information and, therefore, to be aware of the kinds of information about the data that are included in the computation.

The intercept b_0 (the predicted value of *Y* when $X = 0$, or the point on the graph where the regression line crosses the *Y* axis) can be computed from the means of *X* and *Y* as follows:

$$b_0 = M_Y - bM_X, \tag{9.6}$$

where M_Y is the mean of the *Y* scores in the sample and M_X is the mean of the *X* scores in the sample.

9.6 ♦ Statistical Significance Tests for Bivariate Regression

The null hypothesis for a regression that uses one predictor variable can be stated in three different ways. In bivariate regression, these three null hypotheses are all logically equivalent:

H_0: $\rho = 0$,

H_0: $b = 0$,

H_0: $R = 0$.

In words, there are three different ways to formulate the null hypothesis that Y is unrelated to X. The first null hypothesis says that the population correlation between X and Y, ρ, is 0; the second null hypothesis says that b, the predicted increase in Y' for a one-unit increase in X, is 0; and the third null hypothesis says that the multiple R correlation between actual Y_i and the predicted Y_i' computed from the score on X_i equals 0.

In future chapters, when we look at multiple regressions that have two or more predictor variables, the null hypothesis that the slope for each individual predictor variable equals 0 is different from the null hypothesis that the Y_i' value constructed from the entire set of predictor variables has a multiple R correlation of 0 with the actual Y values. In future chapters, we will use tests of the significance of a slope coefficient b for each of several X predictor variables to assess the significance of contribution for each individual predictor variable and the test of R to assess the significance of the entire group of predictor variables considered as a set.

We have already covered procedures to test H_0: $\rho = 0$ in Chapter 7. Significance tests for H_0: $b = 0$ and H_0: $R = 0$ are introduced here.

SPSS provides an estimate of the standard error for the coefficient b (SE_b). This is used to set up a t ratio to test the null hypothesis that the population value of $b = 0$. The formula to compute SE_b is as follows:

$$SE_b = \sqrt{\frac{\sum (Y - Y')^2/(N-2)}{\sum (X - M_X)^2}}. \tag{9.7}$$

This standard error estimate for b can be used to set up a t ratio as follows:

$$t = \frac{\text{Sample statistic} - \text{Hypothesized parameter}}{SE_{\text{sample statistic}}} = \frac{b - 0}{SE_b}. \tag{9.8}$$

This t ratio is evaluated relative to a t distribution with $N - 2$ degrees of freedom (df), where N is the number of cases. In practice, when regression uses only one predictor variable, this test is logically equivalent to the test that Pearson r between X and Y is equal to 0 (H_0: $r = 0$; see Chapter 7) or that the multiple R for the equation to predict Y from a single X predictor variable equals 0.

The null hypothesis H_0: $R = 0$ is tested using an F ratio. When the regression has just one predictor variable, this F test can be calculated from the Pearson r between X and Y as follows:

$$F = \frac{r^2}{(1 - r^2)/(N - 2)} = \frac{r^2(N - 2)}{1 - r^2}. \tag{9.9}$$

Critical values from the F distribution with $(k, N - k - 1)$ df are used to judge whether this F ratio is significant (N = total number of participants; k = number of predictor variables = 1, in this situation). An F ratio can also be used to test the significance of the overall regression equation in situations where there is more than one predictor variable.

For each participant, a predicted Y_i' value can be obtained by applying Equation 9.1 to that participant's X_i score, and a prediction error or residual can be computed by taking the difference between the actual observed Y_i value and Y_i', where Y_i' is the value of Y predicted from X for participant i; thus, we have $Y_i - Y_i'$ = actual Y_i − predicted Y_i = the residual or prediction error for participant i.

In terms of the scatter plot shown in Figure 9.6, $Y_i - Y_i'$ is the vertical distance between an actual (X_i, Y_i) observation and the Y_i' value on the regression line that corresponds to the predicted score for Y based on a linear relationship of Y with X. We want $Y_i - Y_i'$, the prediction errors, to be as small as possible. We can obtain the variance for the prediction errors by squaring and summing these residuals. We can obtain the standard error of regression estimates by taking the square root of the error variance. The standard error of the regression estimate (often denoted in textbooks as $s_{y.x}$) is similar to a standard deviation; it tells us something about the typical distance between Y' and Y. SPSS denotes this term as $\boldsymbol{SE}_{\textbf{est}}$.[3] Researchers usually hope that this standard error, SE_{est}, will be relatively small:

$$SE_{\text{est}} = \sqrt{\frac{\sum (Y_i - Y_i')^2}{N - 2}}. \tag{9.10}$$

The **standard error of the estimate** is an indication of the typical size of these residuals or prediction errors. An alternate way to calculate SE_{est} shows that it is a function of the variability of scores on Y, the dependent variable, and the strength of the relationship between the predictor and outcome variables given by r^2:

$$SE_{\text{est}} = \sqrt{(1 - r^2)} \times s_Y. \tag{9.11}$$

Note that if $r = 0$, SE_{est} takes on its maximum possible value of s_Y. When $r = 0$, the best predicted value of Y is M_Y no matter what the X score happens to be; and the typical size of a prediction error in this situation is s_Y, the standard deviation of the Y scores. If $r = +1$ or -1, SE_{est} takes on its minimum possible value of 0, and the best prediction of Y is (exactly) $Y = b_0 + bX$. That is, when $r = 1$ in absolute value, all the actual Y values fall on the regression line (as in Figure 7.1). The size of SE_{est} provides the researcher with a sense of the typical size of prediction errors; when SE_{est} is close to 0, the prediction errors are small. As SE_{est} approaches s_Y, this tells us that information about a participant's X score does not improve the ability to predict the participant's score on Y very much.

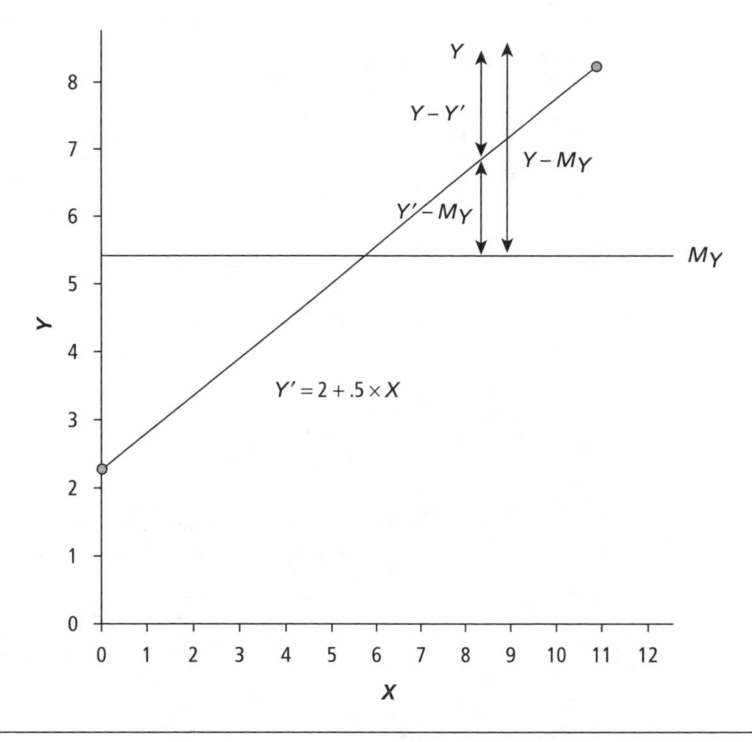

Figure 9.6 ♦ Components of the Deviation of *Y* From the Grand Mean of *Y*

In the idealized situation represented in Figure 9.7, SE_{est} corresponds to the standard deviation of the distribution of actual *Y* scores around the regression line at each value of *X*. Recall that an assumption of Pearson *r* (and therefore of bivariate regression) was that the variances of *Y* should be homoscedastic (i.e., the variance of *Y* should be roughly equal across levels of *X*). For an example of data that violate this assumption, see Figure 7.9. SE_{est} serves as a reasonable estimate of the size of prediction errors across all values of *X* only when the *Y* scores are homoscedastic. Figure 9.7 illustrates the meaning of SE_{est}. We imagine that actual *Y* scores are normally distributed at each value of *X*. The mean of the actual *Y* scores at each value of *X* corresponds to Y', the predicted *Y* score for each value of *X*. The standard deviation or standard error that describes the distance of the actual *Y* values from Y' is given by SE_{est}. SE_{est}, an index of the variability of the distributions of *Y* scores separately for each *X* value, is assumed to be uniform across levels of *X*. In actual datasets, there are usually too few observations to assess the normality of the distribution of *Y* at each level of *X* and to judge whether the variances of *Y* are uniform across levels of *X*.

SE_{est} can be interpreted as the typical size of $Y_i - Y_i'$ prediction errors or residuals that are obtained when the $Y' = b_0 + bX$ equation is used to generate predicted Y' scores. Theoretically, if normality assumptions are satisfied, about 95% of the actual *Y* scores should lie within the range bounded by $Y' + 2 \times SE_{est}$ and $Y' - 2 \times SE_{est}$. However, this is a summary value, which assumes that the variance of *Y* is homogeneous across levels of *X* and that *Y* is normally distributed at each value of *X*; these assumptions are often violated in practice.

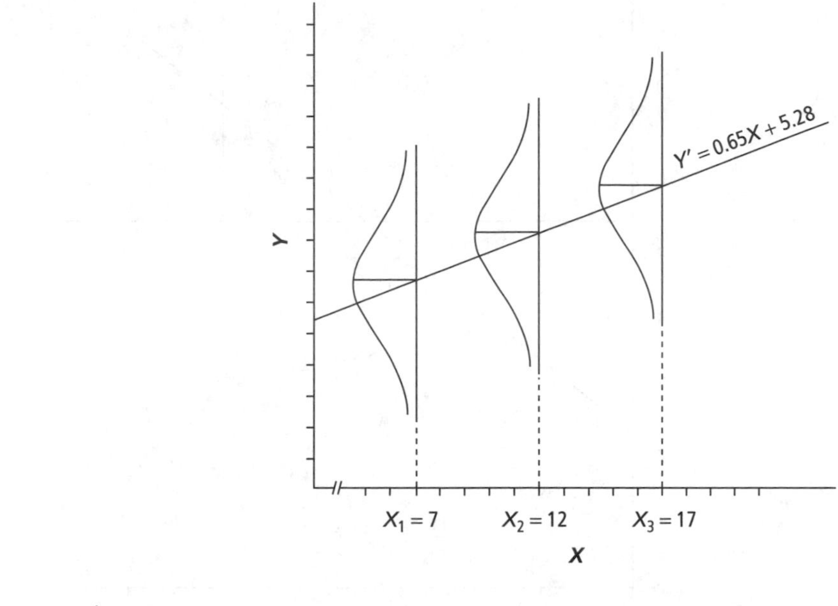

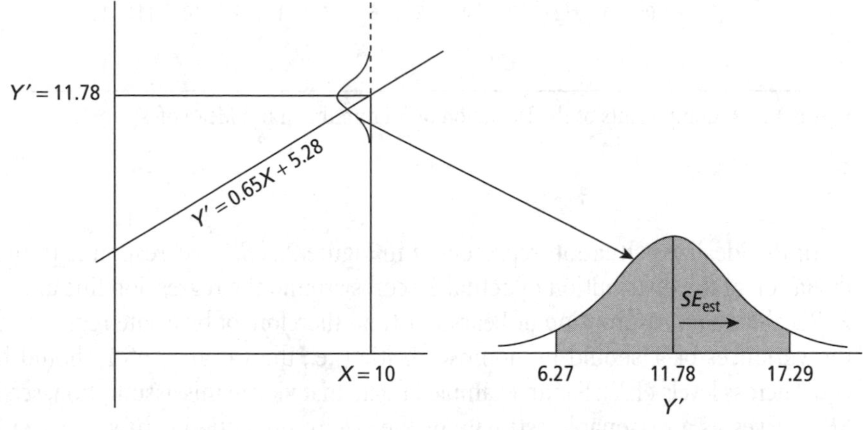

Figure 9.7 ♦ Normal Distribution of Values of Y at Each Level or Score on X

Detail: Distribution of Values of Y for $X = 10$

SOURCE: From Hinkle, D. E., Wiersma, W., & Jurs, S. G., *Applied Statistics for the Behavioral Sciences, 3rd Edition,* copyright © 1994 by the Houghton Mifflin Company. Used with permission.

NOTE: Values of Y for this value of X are normally distributed with a mean that equals the predicted value of Y for $X = 10$ ($Y' = 11.78$) and a theoretical standard deviation given by $SE_{est} = 2.81$.

9.7 ♦ Setting Up Confidence Intervals Around Regression Coefficients

Using SE_b, the upper and lower limits of the 95% confidence interval (CI) around b can be calculated using the usual formula:

$$\text{Lower limit: } b - t_{crit} \times SE_b, \quad (9.12)$$

$$\text{Upper limit: } b + t_{crit} \times SE_b, \quad (9.13)$$

where t_{crit} is the critical value of t that cuts off the top 2.5% of the area in a t distribution with $N - 2$ *df*.

9.8 ♦ Factors That Influence the Magnitude and Sign of *b*

The *b* coefficient essentially combines two kinds of information. First, it includes information about the strength of the relationship between *X* and *Y* (i.e., the correlation *r*). Second, it adds information about the units in which the raw scores on *Y* and *X* are measured.

Here is another version of the regression equation to predict *Y* from *X*:

$$(Y' - M_Y) = r\frac{s_Y}{s_X}(X - M_X). \quad (9.14)$$

To predict Y_i' from X_i, we go through the following steps:

1. Subtract M_X from *X* to find out how far the *X* score was from the *X* mean.
2. Divide this $(X - M_X)$ distance by s_X to convert the distance from the *X* mean into unit-free, standard score terms (this results in z_X).
3. Multiply this *z* score for *X* by Pearson *r* to predict z'_Y (as in Equation 9.2).
4. Multiply this z'_Y predicted value for *Y* in standard score terms by s_Y to restore the predicted *Y* score to a distance from the *Y* mean, in the units originally used to measure *Y*.
5. Add M_Y to adjust the predicted *Y* score: This gives us a value for Y'.

By minor rearrangement of the preceding equation, we see that

$$Y' = M_Y + b(X - M_X).$$

If we multiply $b(X - M_X)$ and combine terms, we find that

$$Y' = [M_Y - b(M_X)] + bX.$$

We can replace the constant term $(M_Y - bM_X)$ with b_0; b_0 is the "constant" or intercept that must be added to all predictions to adjust for the means of *X* and *Y*. This results in Equation 9.1: $(Y'_i = b_0 + bX_i)$.

This example demonstrates that the *b* slope coefficient in Equation 9.1 is just a rescaled version of the *r* or β slope in Equation 9.2. The *b* coefficient is scaled so that the information about raw *X* scores can be used to predict the raw score on *Y*.

Thus, we have two different forms of the prediction equation: an equation to predict standard scores on *Y* from standard scores on *X* ($z'_Y = rz_x$ or βz_X, from the previous chapter) and an unstandardized equation to predict raw scores on *Y* from raw scores on

X ($Y' = b_0 + bX$). If the raw score units for X and Y are meaningful, it may be preferable to report the unstandardized equation; when the units of measurement are meaningful, the b coefficient can be interpreted as information about the size of changes in Y that are predicted for a one-unit change in X. For instance, if we have a regression equation to predict starting salary in dollars from years of post-high-school education, readers may want to know by how many dollars income is predicted to increase for each 1-year increment in education. If there is only a $300 increase in income for each year of education, this increase has less practical significance or importance than if there is a $5,000 increase in income for each additional year of education.

In situations where the units used to measure X and Y are entirely arbitrary, or where an estimate of the strength of the relationship that is independent of the units of measurement is needed, it may be preferable to report the standardized form of the equation. For instance, suppose a researcher finds that liking (rated on a scale from 0 to 36) is predictable from attitude similarity (scored on a scale from 0 to 52). It is not particularly informative to talk about the number of points of change in liking as a function of a one-point increase in attitude similarity.

Of course, the researcher can report both the raw score and the standardized forms of the regression equation. This distinction between two different models—one in terms of z scores and one in terms of raw scores—will occur again for almost every analysis that we cover.

9.8.1 ♦ Factors That Affect the Size of the *b* Coefficient

Note that the sign of b is determined by the sign of r (if r is positive, then b must also be positive). The size of b increases as r and s_Y increase; it decreases as s_X increases. This should make intuitive sense. If we want to predict from an X variable that has a small variance (years of education) to a Y variable that has a large variance (salary in dollars), the b coefficient will tend to be large; we would expect to see an increase of hundreds or thousands of dollars for each 1-year increase in education. On the other hand, if we make predictions to an X variable that has a small variance (life satisfaction on a 1 to 10 scale) from a Y variable that has a large variance (gross national product [GNP] per capita in thousands of dollars), the b coefficient tends to be extremely small; a $1.00 increase in GNP per capita "buys" only a small fraction of a point in life satisfaction. Note also that factors that artifactually increase or decrease the size of r (see Chapter 7) will also influence the size of b; sources of artifact such as unreliability can make values of b difficult to interpret or compare.

9.8.2 ♦ Comparison of Coefficients for Different Predictors or for Different Groups

Sometimes researchers want to make comparisons of the predictive strength of two or more different predictor variables (let's call these X_1, X_2, and X_3). It is necessary to keep in mind that the size of b slopes in the raw score form of the predictive equation (Equation 9.1) depends on the units used to measure X and Y (unlike the size of r or β, the coefficient in the raw score form of the predictive equation, which is standardized or unit free). The correlation between height and weight is the same whether height is given in inches, feet, or meters and weight in ounces, pounds, or kilograms. On the other hand, the *b* coefficient to predict weight from height will change depending on what units are used to measure both height and weight.

If a researcher has several different *X* predictor variables (let's call them X_1, X_2, X_3), and the researcher wants to compare them to see which is the best predictor of *Y*, the researcher may compare their correlations with *Y*, r_{1Y}, r_{2Y}, r_{3Y} (where r_{iY} is the correlation of X_i with *Y*), to see which *X* variable has the strongest statistical association with *Y*. The *b* coefficient in Equation 9.1 cannot be compared across different predictor variables such as X_1 and X_2 if those variables have different units of measurement; the magnitude of a *b* slope coefficient depends on the units used to measure the predictor and outcome variables, as well as the strength of direction of the relationship between the variables.

However, comparisons of correlations or slopes should be made very cautiously. Chapter 7 described many factors, apart from the strength of the relationship between *X* and *Y*, that can artifactually affect the size of the correlation between them; this included factors such as unreliability of measurement, difference in distribution shapes, and restricted range. If X_1 is a better predictor of *Y* than X_2, this might be because X_1 was more reliably measured than X_2 or because X_2 had a very restricted range of scores. Researchers should be aware of these problems and should try to rule them out (e.g., demonstrate that X_1 and X_2 have similar reliabilities) if possible. Because it is never really possible to rule out all possible sources of artifacts, comparisons of the predictive strength of variables should always be made cautiously and reported with an appropriate discussion of the problems inherent in these comparisons.

9.9 ♦ Effect Size/Partition of Variance in Bivariate Regression

The effect size index for a bivariate regression is simply the Pearson *r* (or r^2) between the predictor and outcome variables; see Table 5.2 for suggested verbal labels for *r* as an effect size. Earlier, it was stated (without much explanation) that r^2 could be interpreted as the proportion of variance in *Y* that is predictable from *X* (or vice versa). Let's look at the way in which regression analysis partitions the *Y* score into two parts: a component that is predictable from, explained by, or related to *X* and a component that is not predictable from or related to *X*. We can work out a partition of the sums of squares by squaring and summing these explained and not-explained score components across subjects in regression, just as we squared and summed the between-group/explained-by-treatment and within-group/not-explained-by-treatment components of scores in one-way analysis of variance (ANOVA) (see Chapter 6).

We can take the total deviation of an individual participant's Y_i score from the *Y* mean and divide it into two components: the part of the score that *is* predictable from *X* and the part that is *not* predictable from *X* (refer back to Figure 9.6 to see these deviations for one specific *Y* value):

$$\underset{\text{Total deviation}}{Y_i - M_Y} = \underset{\substack{\text{Part of } Y \text{ score that is not} \\ \text{related to } X \text{ (residual)}}}{Y_i - Y_i'} + \underset{\substack{\text{Part of } Y \text{ score that is} \\ \text{predictable from } X \\ \text{(predicted part).}}}{Y_i' - M_Y} \tag{9.15}$$

In the above equation, Y_i is the observed score for an individual participant *i*; M_Y is the grand mean of scores on *Y*; and $Y_i' = b_0 + bX_i$, the predicted *Y* score generated by applying the regression equation to that same person's score on *X*. Note that this division of the score into components is conceptually similar to the partition of scores into components by one-way ANOVA that was described in Chapter 6:

$$\underset{\text{Total deviation}}{(Y_{ij} - M_Y)} = \underset{\substack{\text{Within-group deviation due} \\ \text{to extraneous variables}}}{(Y_{ij} - M_i)} + \underset{\substack{\text{Between-group deviation} \\ \text{predictable from treatment} \\ \text{variable,}}}{(M_i - M_Y)} \qquad (9.16)$$

where Y_{ij} is the score for participant *j* in group *i* and M_i is the mean of the *Y* scores in treatment group *i*. In the one-way ANOVA, the predicted score for each participant is the mean of a treatment group that a person belongs to, whereas in bivariate regression, the predicted score for each participant is a linear function of that person's *X* score. In both cases, the underlying logic is the same: We divide the total deviation (which tells us whether an individual participant's score is above or below the grand mean) into two components: a part that is *not* explained by the predictor variable and a part that *is* explained by the predictor variable. In both analyses, the researcher typically hopes that the explained part will be large relative to the unexplained part of the scores.

In ANOVA, we wanted $Y_{ij} - M_i$, the within-group deviations, to be small (because this meant that the amount of variation due to extraneous variables would be small). Similarly, in bivariate regression, we want $Y_i - Y_i'$, the prediction errors (or distance of the actual *Y* data points from the regression prediction line), to be as small as possible. The $Y_i' - M_Y$ part of the score in regression represents the improvement or change in prediction of *Y* that we obtain when we use Y_i' to predict a person's score instead of using M_Y, the grand mean of *Y*. We want this component of the score, the part that is predicted from the regression line, to be relatively large. Regression, like ANOVA, involves looking at a signal-to-noise ratio: how much of the scores is predictable from the analysis (regression line) versus not predicted by the analysis.

As in ANOVA, we can summarize the information about the relative sizes of the predicted and not-predicted components of scores by squaring these components and summing them across all scores. This gives the following equation for the partition of the sums of squares (*SS*) in regression:

$$\sum (Y_i' - M_Y)^2 = \sum (Y_i - Y_i')^2 + \sum (Y_i' - M_Y)^2. \qquad (9.17)$$

The terms in Equation 9.17 can be given names and the equation rewritten as

$$SS_{\text{total}} = SS_{\text{residual}} + SS_{\text{regression}}. \qquad (9.18)$$

Note that this is the same kind of equation that we had for a partition of sums of squares in one-way ANOVA, where we had $SS_{\text{total}} = SS_{\text{within}} + SS_{\text{between}}$. The residual and within-group *SS* terms both correspond to variance not explained by the predictor variable (and therefore due to extraneous variables); the regression and between-group *SS* terms correspond to variance that is predictable from the score on X (in regression) or from group membership (in ANOVA), respectively. In ANOVA, the proportion of explained variance was assessed by looking at the ratio of $SS_{between}$ to SS_{total}; this ratio was called eta squared and interpreted as the proportion of variance due to between-group differences. In regression, this ratio is called R^2, but it has a similar interpretation: the proportion of variance in Y that is predictable from X.

$$R^2 = SS_{regression}/SS_{total} = \text{Proportion of variance in } Y \text{ predicted by } X. \quad (9.19)$$

R^2 is essentially equivalent to η^2, except that R^2 assumes a linear relationship between Y and X. An eta squared does not assume that Y scores increase linearly as a function of increases in X scores. As in ANOVA, we can test the significance of the model fit by looking at a ratio of mean squares (MS) terms and calling this ratio an F. In the more general case of multiple regression, where we can use more than one predictor variable to generate predicted Y scores, the fit of the regression is assessed by calculating a correlation between the actual Y scores and the predicted Y' scores (which are calculated from X scores); this type of correlation is called a multiple R (because there can be multiple predictor X variables). For a bivariate regression with just a single predictor variable, multiple R is equal to the Pearson r between X and Y. The null hypothesis that is used in significance tests of bivariate or multivariate regressions states that the multiple R equals 0:

$$H_0: R = 0.$$

The degrees of freedom for the SS terms are $df = k$ for $SS_{regression}$ (k = number of predictors = 1, in this instance) and $df = N - k - 1$ for $SS_{residual}$, where N is the number of cases:

$$MS_{regression} = SS_{regression}/k, \quad (9.20)$$

$$MS_{residual} = SS_{residual}/(N - k - 1), \quad (9.21)$$

$$F = MS_{regression}/MS_{residual}, \text{ with } (k, N - k - 1)\ df. \quad (9.22)$$

If the obtained F exceeds the critical value of F from the table of critical values for F in Appendix C, we can reject the null hypothesis and conclude that the overall regression is statistically significant. This version of the F test can be used (and it is reported by SPSS) for a bivariate regression. It will also be used in situations where the regression includes multiple predictor variables.

The F ratio can also be calculated directly from the multiple R:

$$F = \frac{R^2/k}{(1 - R^2)/(N - k - 1)}, \quad (9.23)$$

where N is the number of cases and k is the number of predictor variables; F has $(k, N - k - 1)\ df$.

Note that whether F is calculated as a ratio of mean squares or as a ratio of R^2 to $1 - R^2$, it provides information about the relative sizes of "explained" variance versus "unexplained" variance, or signal versus noise, similar to F ratios obtained in ANOVA. If F is large enough to exceed the tabled critical value of F for $(k, N - k - 1)$ df, we can reject H_0 and conclude that our model, using X as a predictor, predicts a statistically significant portion of the variance in Y.

In bivariate regression (i.e., a regression with only two variables, one outcome variable Y and one predictor variable X), these three tests (that b, r, and R equal 0) are logically equivalent. Multiple R is the same as Pearson r (when there is just one predictor), and $b = r(s_Y/s_X)$. In bivariate regression, if $r = 0$, then both R and b must also equal 0. SPSS presents all three of these significance tests even when only one predictor variable is used in the regression analysis. This discussion should make it clear that the three test results are equivalent when there is just one predictor.

In a later chapter where the use of multiple predictor variables regression is introduced, it becomes necessary to distinguish between the fit of the overall multiple regression (which is assessed by multiple R) and the significance of individual variables (which can be assessed by testing the b slope coefficients for each individual predictor); these terms are not identical, and provide quite distinct information, when we have more than one predictor.

9.10 ◆ Statistical Power

For a bivariate regression, the assessment of statistical power can be done using the same tables as those used for Pearson r (see Table 7.4). In general, if the researcher assumes that the strength of the squared correlation between X and Y in the population is weak, the number of cases required to have power of .80 or higher is rather large. Tabachnick and Fidell (2007) suggested that the ratio of cases (N) to number of predictor variables (k) should be on the order of $N > 50 + 8k$ or $N > 104 + k$ (whichever is larger) for regression analysis. This implies that N should be at least 105 even with one predictor variable and that N needs to be even larger when we begin to add more predictors to our regression analyses. They also noted that more cases are needed if the distribution of scores on Y is skewed, if the magnitude of the population correlation ρ is believed to be small, or if measurement reliability for some variables in the analysis is poor.

9.11 ◆ Raw Score Versus Standard Score Versions of the Regression Equation

On reading a computer printout of a bivariate regression, is it usually easy to tell which set of coefficients applies to z scores and which to raw scores. First, the **beta coefficient (β)** that is used to predict z scores on Y from z scores on X is not accompanied by an intercept value; there is no need to adjust for the mean of Y or X when the regression is given in terms of z scores because z scores always have means of 0. Second, beta coefficients usually tend to be less than 1 in absolute value, although in multiple regressions, beta coefficients sometimes fall outside this range. Third, raw score slope coefficients (b coefficients used to predict raw scores on Y from raw scores on X) can have virtually any value, but their size tends to be dominated by the ratio s_Y/s_X. Thus, if the *Y* variable scores have a much larger range than *X*, *b* tends to be a large number.

Which form of the equation is more useful? A beta coefficient (equivalent to *r* in the one-predictor-variable case) tells us about the strength of the relationship between *X* and *Y* in unit-free, standardized terms. For a one-*SD* increase in z_X, it tells us how much change (in standard deviation units) we can expect in z_Y. When we have more than one predictor variable, comparing their beta weights will give us some information about their relative predictive usefulness, independent of the sizes of the units of measurement. When we are interested primarily in theory testing, and particularly if the units of measurement of *X* and *Y* are arbitrary and not meaningful, the beta coefficient may be the most useful information.

On the other hand, a *b* or raw score slope coefficient tells us, for a one-unit increase in *X*, how many units of change we can predict in *Y*, in terms of the original units of measurement. For instance, for each additional year of education, how many additional dollars of income do we predict? For each additional pound of body weight, how large a change in blood pressure (in millimeters of mercury) do we predict? If our interest is applied or clinical, and if the units of measurement of both variables are meaningful, knowing the effects of treatment in real units may be very important. Sometimes, of course, people report and interpret both forms of the slope (β and *b*).

Many of the factors that artifactually affect the size of Pearson *r* can also make the values of the *b* slope coefficient potentially misleading. For this reason, we should be extremely cautious about making claims that one predictor has a stronger predictive relation with *Y* than another predictor. For instance, suppose that a researcher does two separate bivariate regressions: In one, salary (*Y*) is predicted from intelligence quotient (IQ) (X_1); in the other, salary is predicted from need for achievement (X_2). If the *b* coefficient is larger in the equation that uses X_1 as a predictor than in the equation that uses X_2 as a predictor, can the researcher argue that IQ is more strongly related to salary than need for achievement? Such a claim would have to be made very cautiously. First, the researcher would want to check that the difference in magnitudes of the correlations for these two predictors was statistically significant. In addition, the researcher would want to consider the possible influence of other factors (besides the strength of the relationship between variables) that might have created an artifactual difference; for instance, if IQ has a much higher reliability than need for achievement or if need for achievement had a very restricted range of scores, either of these factors might have decreased the size of the correlation between need for achievement and salary and made that variable appear to be more weakly related to salary.

9.12 ♦ Removing the Influence of *X* From the *Y* Variable by Looking at Residuals From Bivariate Regression

In bivariate regression, each individual $Y_i - M_Y$ deviation is partitioned into two components: one component that is related to X_i ($Y_i' - M_Y$) and another component that is not related to X_i ($Y_i - Y_i'$). This partition is very useful in some research situations. Imagine the following situation. Suppose we have a *Y* variable that is a paper-and-pencil test of mountain-climbing skills. Unfortunately, whether or not people do well on this test depends rather heavily on verbal skills (X) that have little to do with actually surviving out in the mountains. It would be useful if we could remove the influence of verbal ability (X) from our mountain-climbing skills test scores (Y). This can be accomplished by doing a bivariate regression to predict scores on the mountain-climbing test (Y) from verbal ability test scores (X) and calculating the residuals from this regression ($Y - Y'$). In theory, the $Y - Y'$ residual corresponds to a part of the Y test score that is *unrelated* to verbal ability. Thus, $Y - Y'$ should be a better measure of mountain-climbing ability; this residual corresponds to a part of the Y score that is not related to verbal ability. When we use the residual from this regression as a variable in subsequent analyses, we can say that we have "partialled out" or "controlled for" or "removed the influence of" the X variable.

9.13 ♦ Empirical Example Using SPSS

To run a bivariate linear regression, make the following menu selections: <Analyze> → <Regression> → <Linear> (as shown in Figure 9.8); this opens up the main dialog window for the SPSS Linear Regression procedure, which appears in Figure 9.9. The name of the dependent variable (salary) and the name of the predictor variable (years) were moved into the windows marked Dependent and Independent in this main dialog window. It is possible at this point to click OK and run a regression analysis; however, for this example, the following additional selections were made. The Statistics button was clicked to open up the Statistics dialog window; a check box in this window was marked to request values of the 95% CI for b, the slope to predict raw scores on Y from raw scores on X (Figure 9.10). The Plots button was used to open the dialog window to request diagnostic plots of residuals (Figure 9.11). SPSS creates temporary variables that contain values for the standardized predicted scores from the regression (***ZPRED**) and the standardized residuals from the regression (***ZRESID**), as well as other versions of the predicted and residual values for each case. In this example, *ZRESID was plotted (as the Y variable) against scores on *ZPRED (which was designated as the X variable for the residuals plot). Finally, the Save button was used to request that SPSS save values for unstandardized predicted scores and residuals into the SPSS worksheet; saved scores on these specific variables were requested by placing checkmarks in the check boxes for these variables in the dialog window for Save, which appears in Figure 9.12.

The regression output that appears in Figure 9.13 begins with a panel that indicates what variable(s) were entered as predictors; in this example, only one predictor variable (years) was used. The second panel (under the heading Model Summary) provides information about overall model fit: How well did the regression equation including all predictors (just one predictor in this case) predict scores on the outcome variable? The multiple R of .83 that appears in the Model Summary table is the correlation between actual salaries and predicted salaries; because the regression equation has just one predictor, multiple R is equivalent to the Pearson r between these two variables. R^2 is the proportion of variance in salary that is predictable from years of job experience; $R^2 = .69$—that is, about 69% of the variance in salary was predictable from years of job experience. **Adjusted R^2** (sometimes called **"shrunken R^2"**) is a downwardly adjusted version of R^2 that takes into account both the N of cases and k, the number of predictor variables; it provides a more conservative estimate of the true proportion of explained variance. The

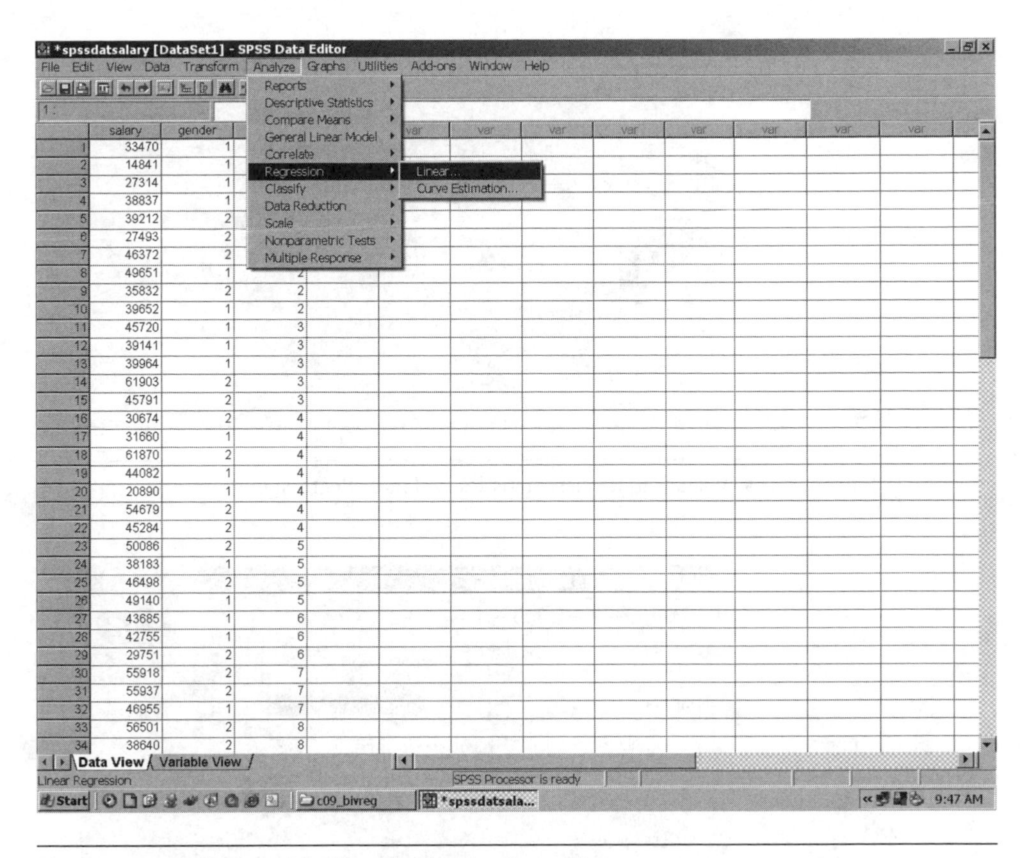

Figure 9.8 ♦ SPSS Menu Selections for Bivariate Regression

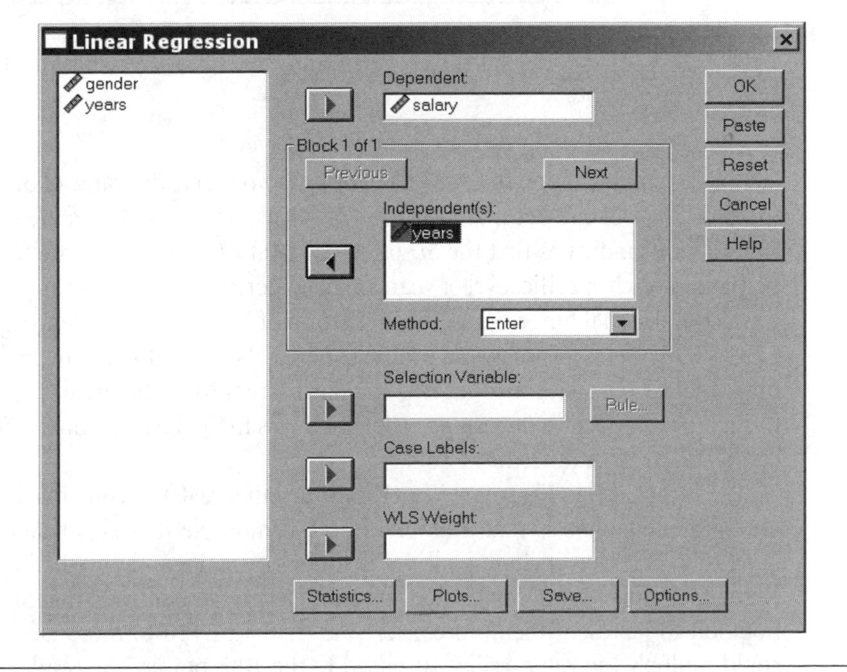

Figure 9.9 ♦ Main Dialog Window for SPSS Linear Regression

Linear Regression: Statistics

Regression Coefficients
Estimates
Confidence intervals
Covariance matrix
Model fit
R squared change
Descriptives
Part and partial correlations
Collinearity diagnostics
Continue
Cancel
Help
Residuals
Durbin-Watson
Casewise diagnostics
Outliers outside: 3 standard deviations
All cases

Figure 9.10 ♦ Window to Request Statistics for Linear Regression

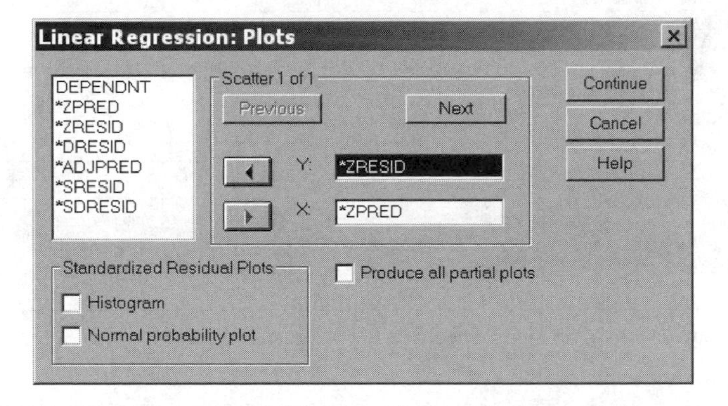

Figure 9.11 ♦ Window to Request Plot of Standardized Residuals by Standardized Predicted Salaries

last piece of information in the Model Summary panel is the value of SE_{est}, with the heading Std. Error of the Estimate. For this empirical example, the value of SE_{est} was about \$10,407; this indicates that the *SD* of actual salary (relative to predicted salary) for participants at each specific level of years of job experience was on the order of \$10,000; thus, it was not unusual for errors in the prediction of individual salaries to be on the order of \$10,000. Most employees would probably have salaries within about ±\$20,000 of the predicted salary (i.e., within about two standard errors from the prediction line). Thus, even though the proportion of variance explained was high, the individual errors of prediction could still be large in terms of dollars.

The third panel in Figure 9.13 (under the heading ANOVA) contains the significance test for the null hypothesis that multiple $R = 0$; if multiple R is significant, then the overall regression equation including all predictors was significantly predictive of the outcome. Note that in this instance, the values of the *SS* terms were so large that SPSS used scientific notation to represent them; to convert the term 1.5E+10 to more familiar notation, one could multiply the value 1.5 by 10 raised to the 10th power. The result that is of primary interest here is the *F* ratio: $F(1, 48) = 105.99$, $p < .001$. Because p is less than .05, this *F* ratio

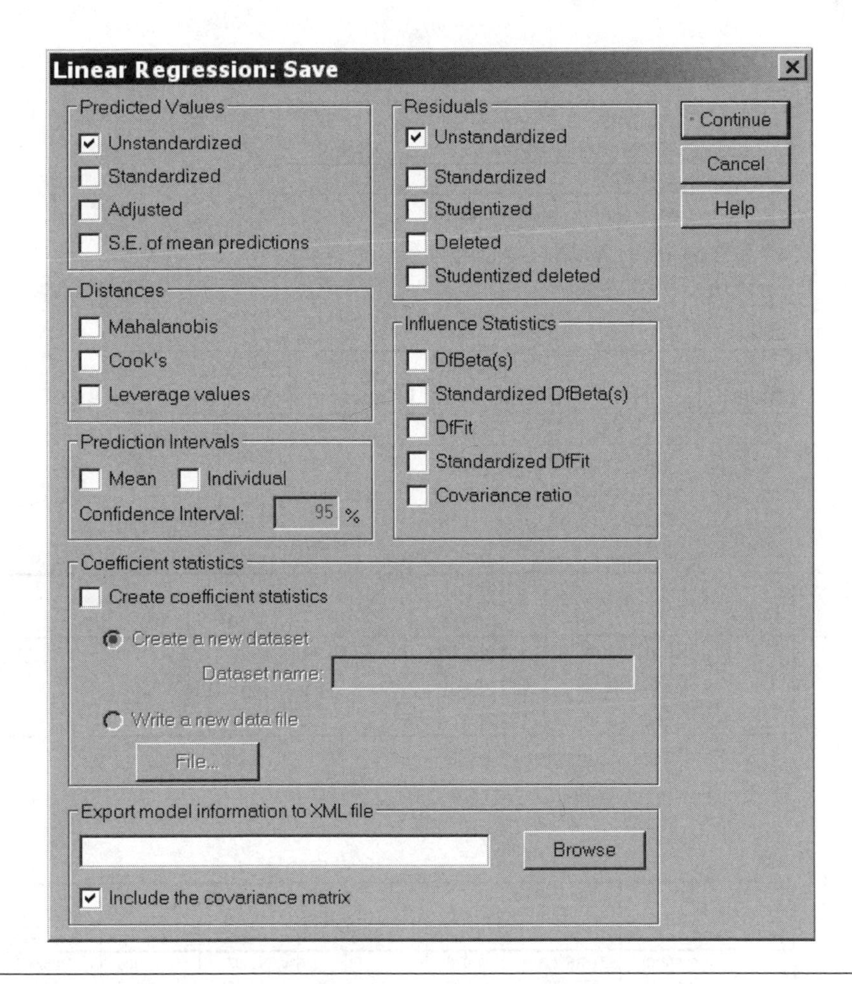

Figure 9.12 ♦ Window to Save Unstandardized Predicted Scores and Unstandardized Residuals From Regression

can be judged statistically significant at the conventional .05 level. The overall multiple *R*, and thus the overall regression equation, is judged significantly predictive of salary.

The last panel in Figure 9.13 shows the coefficients for both the raw score and the standard score version of the regression equation. The row labeled Constant provides the estimated value of b_0, the intercept ($b_0 = 31416.72$) and a *t* test to evaluate whether this differed significantly from 0. The intercept b_0 was significantly different from 0; $t(48) = 12.92$, $p < .001$. In this example, the intercept has a meaningful interpretation; it would represent the predicted salary for a person with 0 years of job experience (entry-level salary). In some research situations, scores of $X = 0$ are not within the range of possible values for *X*. Usually, the researcher is primarily interested in the significance of the slope coefficient (*b*); the significance test for b_0 may not be of interest and is often omitted from research reports when the intercept is not meaningful as a prediction.

The row labeled "years" provides the information about the predictor variable years. From left to right, the second column (headed B) contains the raw score coefficient, usually denoted by *b*, which can be used to predict salary in dollars from years of job

Variables Entered/Removed[b]

Model	Variables Entered	Variables Removed	Method
1	YEARS[a]	.	Enter

a. All requested variables entered.

b. Dependent Variable: SALARY

Model Summary

Model	R	R Square	Adjusted R Square	Std. Error of the Estimate
1	.830[a]	.688	.682	10407.343

a. Predictors: (Constant), YEARS

ANOVA[b]

Model		Sum of Squares	df	Mean Square	F	Sig.
1	Regression	1.15E+10	1	1.148E+10	105.995	.000[a]
	Residual	5.20E+09	48	108312781.8		
	Total	1.67E+10	49			

a. Predictors: (Constant), YEARS

b. Dependent Variable: SALARY

Coefficients[a]

Model		Unstandardized Coefficients		Standardized Coefficients	t	Sig.	95% Confidence Interval for B	
		B	Std. Error	Beta			Lower Bound	Upper Bound
1	(Constant)	31416.72	2431.038		12.923	.000	26528.786	36304.646
	years	2829.572	274.838	.830	10.295	.000	2276.972	3382.171

a. Dependent Variable: salary

Figure 9.13 ♦ SPSS Bivariate Regression Output: Prediction of Salary From Years

experience. Together with the intercept (in the same column), these two values can be used to write out the raw score regression equation:

$$Y' = b_0 + bX$$

$$\text{Salary}' = \$31416.72 + \$2829.57 \times \text{Years}.$$

In words, the typical predicted salary for a new employee with 0 years of job experience is $31,416.72; and for each year of experience, predicted salary increases by $2,829.57. The next column, headed Std. Error, provides the standard error associated with this regression coefficient (denoted by SE_b). The t test to assess whether b differs significantly from 0 was obtained by dividing b by SE_b; the t value is given on the SPSS printout in a column headed t. The reader may wish to verify that the t ratio 10.295 is the ratio of the values in the B (2829.572) and Std. Error (274.838) columns.

The fourth column in the Coefficients table, which appears at the bottom of Figure 9.13, headed Standardized Coefficients/Beta, contains the coefficient used to predict z_Y from z_X. No intercept or constant is needed because *z* scores have a mean of 0. Thus, the equation to predict standardized/*z* score salary from standardized years of job experience is

$$z'_Y = \beta z_X$$

$$z'_{\text{salary}} = .83\ z_X.$$

When there is just one predictor variable, β = Pearson *r*. When there are multiple predictors, as in the upcoming discussion of multiple regression in Chapters 11 and 14, the βs are calculated from the *r*s, but they are not simply equal to the *r*s in most research situations. In words, this equation says that for a one-*SD* increase in years of job experience, the predicted increase in salary is .83 *SD*s; this is a strong relationship.

Additional results produced by the SPSS regression procedure appear in Figure 9.14; this shows the SPSS Data View worksheet with the two new variables that were created and saved (raw or unstandardized scores for predicted salary, denoted by PRE_1 to indicate that these are predicted scores from the first regression analysis, and raw or unstandardized residuals, denoted by RES_1). For example, the participant whose data are given in Line 2 of the SPSS data file had a RES_1 value of −19405.67; in words, this person's actual salary was $19,405.67 lower than the salary that was predicted for this person by substituting this person's years of job experience (1) into the regression equation to generate a predicted salary. Figure 9.15 shows that, if you obtain a Pearson correlation between the new saved variable PRE_1 and the actual salary, this correlation is .83; this is equivalent to the multiple *R* value for the regression equation, which was also .83.

The graph of standardized residuals requested as part of the regression analysis appears in Figure 9.16. When the assumptions for regression are satisfied by the data, the points in this plot should appear within a fairly uniform band from left to right and most standardized residuals should be between −3 and +3. This graph shows that the assumptions for regression appear to be reasonably well satisfied. Figure 9.17 (page 366) provides standards for comparison of plots of residuals. Ideally, when residuals (either standardized or unstandardized) are graphed against predicted scores, the graph should look like the one in Figure 9.17a. Figures 9.17b, c, and d show what a graph of residuals might look like in the presence of three different kinds of violations of assumptions for linear regression. Figure 9.17b shows what the residuals might look like if scores on the *Y* outcome variable (and also on the residuals) are skewed rather than normally distributed. Figure 9.17c shows how the residuals might look if *X* and *Y* had a strong curvilinear relationship. Figure 9.17d shows how the residuals might look if the prediction errors or residuals have greater variance for high predicted *Y* scores than for low predicted *Y* scores—that is, if the data violate the assumption of homoscedasticity. Because the residual plot obtained for the salary/years of job experience data most closely resembles the graph in Figure 9.17a, the residual plot does not show evidence that the assumptions of normality, linearity, and homoscedasticity were violated in the data presented for this empirical example.

It is useful to show the fitted regression line superimposed on the original *X*, *Y* scatter plot. This can be done using the SPSS Chart Editor to add a regression line to the SPSS scatter plot obtained earlier. Position the mouse so that the cursor points to the scatter plot you

*spssdatsalary [DataSet1] - SPSS Data Editor

2 : years 1

	salary	gender	years	PRE_1	RES_1
1	33470	1	1	34246.28792	-776.56249
2	14841	1	1	34246.28792	-19405.66886
3	27314	1	1	34246.28792	-6931.92485
4	38837	1	1	34246.28792	4590.58440
5	39212	2	1	34246.28792	4965.45425
6	27493	2	2	37075.85973	-9583.20399
7	46372	2	2	37075.85973	9296.57472
8	49651	1	2	37075.85973	12575.63271
9	35832	2	2	37075.85973	-1243.44331
10	39652	1	2	37075.85973	2576.07169
11	45720	1	3	39905.43154	5814.73481
12	39141	1	3	39905.43154	-764.92228
13	39964	1	3	39905.43154	58.81115
14	61903	2	3	39905.43154	21997.17200
15	45791	2	3	39905.43154	5885.19178
16	30674	2	4	42735.00334	-12061.46866
17	31660	1	4	42735.00334	-11074.82165
18	61870	2	4	42735.00334	19135.22186
19	44082	1	4	42735.00334	1346.78006
20	20890	1	4	42735.00334	-21845.48672
21	54679	2	4	42735.00334	11944.42943
22	45284	2	4	42735.00334	2548.58407
23	50086	2	5	45564.57515	4521.39127
24	38183	1	5	45564.57515	-7381.63862
25	46498	2	5	45564.57515	933.56663
26	49140	1	5	45564.57515	3575.14341
27	43685	1	6	48394.14696	-4709.15550
28	42755	1	6	48394.14696	-5639.54381
29	29751	2	6	48394.14696	-18643.08938
30	55918	2	7	51223.71877	4693.97929
31	55937	2	7	51223.71877	4713.22113
32	46955	1	7	51223.71877	-4269.18322
33	56501	2	8	54053.29058	2447.47843
34	38640	2	8	54053.29058	-15413.06654

Data View / Variable View

SPSS Processor is ready

Figure 9.14 ♦ SPSS Data Worksheet With Saved Variables, PRE_1 (Predicted Salary) and RES_1 (Residual or Prediction Error), for Each Participant

Correlations

		Unstandardized Predicted Value	salary
Unstandardized Predicted Value	Pearson Correlation	1	.830**
	Sig. (2-tailed)		.000
	N	50	50
salary	Pearson Correlation	.830**	1
	Sig. (2-tailed)	.000	
	N	50	50

**. Correlation is significant at the 0.01 level (2-tailed).

Figure 9.15 ♦ Correlation Between Actual Salary and Predicted Salary Scores Saved From the Bivariate Regression Procedure

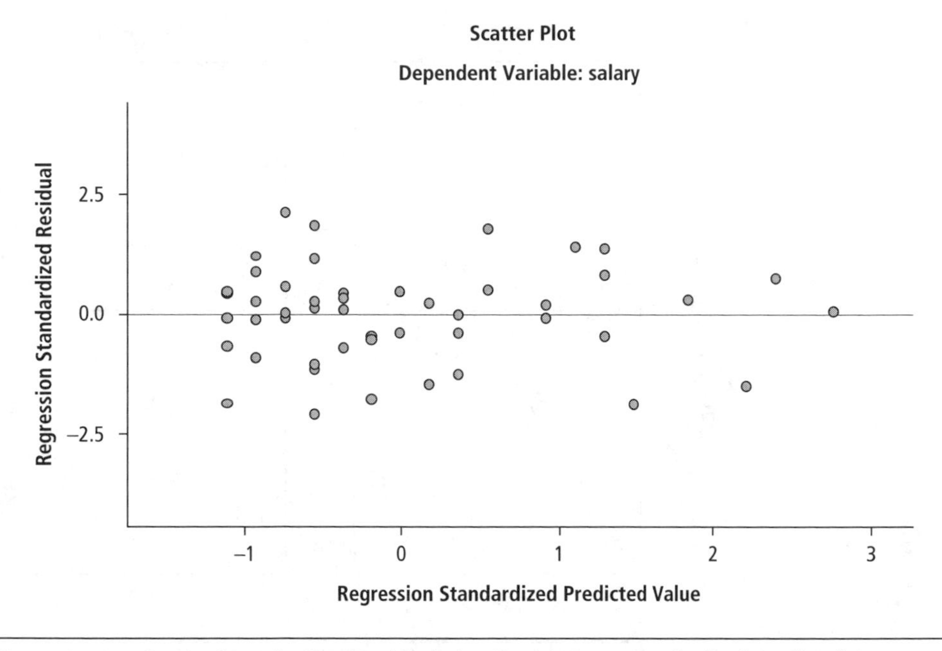

Figure 9.16 ♦ Graph of Standardized Residuals (on *Y* axis) Versus Standardized Predicted Scores on Dependent Variable (Salary)

want to edit, and click the right mouse button; a menu appears, as shown in Figure 9.18 (page 367); from this menu, select <SPSS Chart Object> and <Open> to open the chart for editing. Figure 9.19 (page 368) shows the Chart Editor main dialog window. To add a fitted line, click on <Elements> in the top-level menu bar of the Chart Editor, then select and click on <Fit Line at Total> as shown in Figure 9.20 (page 369). This opens the dialog window for Properties of the fit line, which appears in Figure 9.21 (page 370). In this example, the radio button for Linear was selected to request the best-fitting linear function. In addition, the radio button for Individual CIs was also selected to request the boundaries for a 95% CI around the regression line. Click Close to exit from the Properties window and exit from the Chart Editor. The resulting plot appears in Figure 9.22 (page 371). The middle line represents the best-fitting linear regression line that corresponds to the equation reported earlier (predicted Salary = 31,416.72 + 2829.57 × Years). The lines above and below the regression line correspond to the upper and lower boundaries of the 95% CI for the regression. As expected, most of the actual *Y* values lie within these bounds.

The box below shows a brief sample write-up for a bivariate regression. You can use either Pearson *r* or multiple *R* to describe the relationship when you just have one predictor variable; you can either compute an *F* ratio or just look up the critical values of Pearson *r* to evaluate significance.

9.13.1 ♦ Information to Report From a Bivariate Regression

A significance test should be done (you can report an *F* or a direct table lookup of the significance of Pearson *r* or base this judgment on the exact *p* value that is given for the *F* ratio on the SPSS printout) to assess whether *Y* and *X* are significantly related:

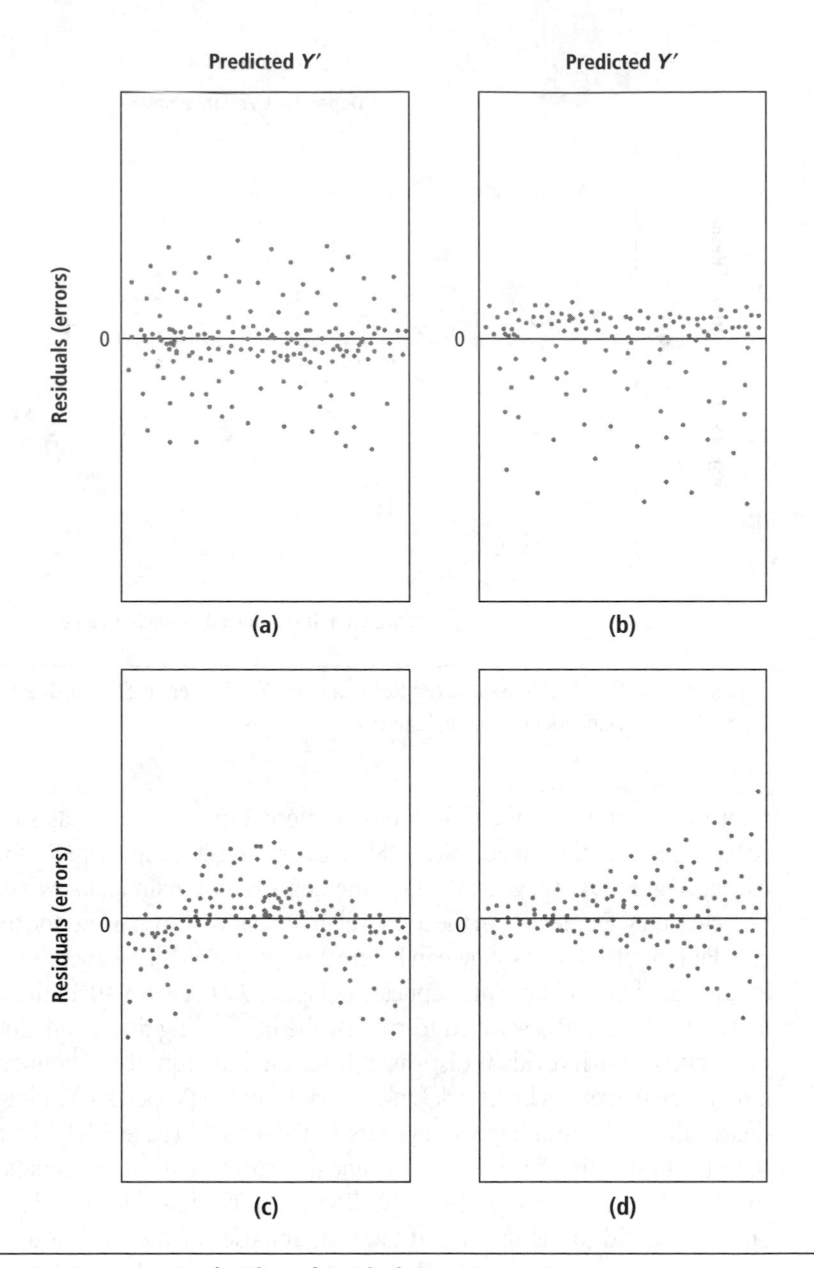

Figure 9.17 ♦ Interpretation for Plots of Residuals From Regression

SOURCE: Tabachnick and Fidell (2007, p. 126).

NOTES:

a. Residuals that meet the assumptions: They are normally distributed around 0, have the same variance for all values of predicted Y', and do not show any systematic pattern.

b. Residuals that are not normally distributed around 0 (evidence that the assumption of normality is violated).

c. Residuals that show a curvilinear pattern (evidence that the assumption that the relation between X and Y is linear has been violated).

d. Residuals that show heteroscedasticity: That is, residuals have higher variance for high scores on predicted Y than for low Y'.

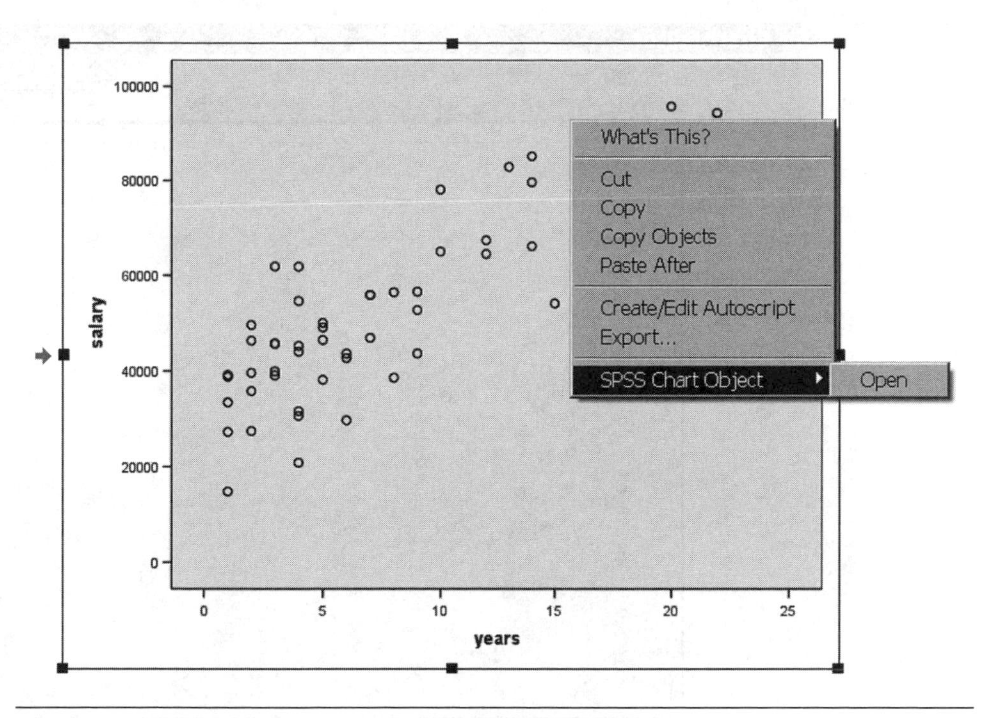

Figure 9.18 ♦ SPSS Menu Commands to Open and Edit an *X*, *Y* Scatter Plot

If *X* and *Y* are not significantly related and the scatter plot for *Y* and *X* does not suggest a curvilinear relationship, then typically, the Results section would end with the statement that there is no significant (linear) relationship between the variables. If there is a strong curvilinear relationship, you might try a different analysis, such as converting *X* to a categorical variable with high/medium/low groups and doing a one-way ANOVA to see if scores on *Y* are significantly different across these three or more groups defined by the *X* variable.

If *Y* and *X* are significantly linearly related, then you should go on to report the following:

1. An equation to predict a raw score on *Y* from a raw score on *X* (and/or an equation to predict z_Y from z_X)
2. A statement about the nature of the relationship between *X* and *Y*—that is, whether *Y* tends to increase or decrease as *X* increases
3. An assessment of the strength of the relationship—for example, R^2 or r^2, the percentage of variance in *Y* that is predictable from *X*
4. A scatter plot showing how *Y* and *X* are related (this is particularly important if there are outliers and/or any indications of curvilinearity)
5. A CI for estimate of the slope *b*

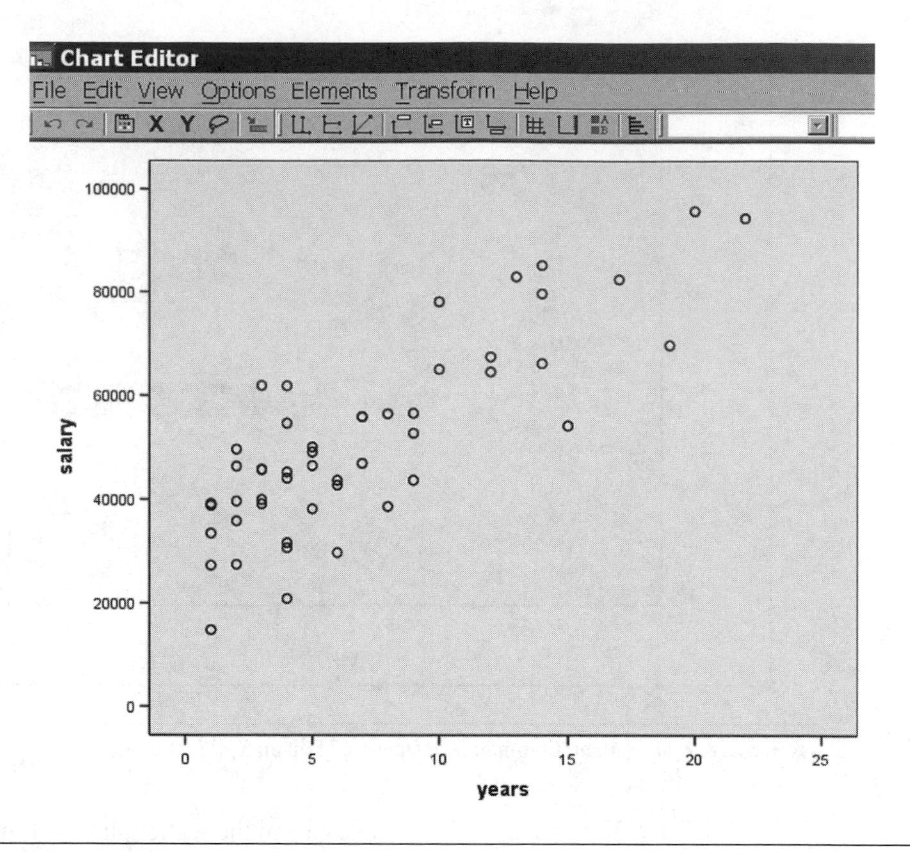

Figure 9.19 ♦ SPSS Chart Editor Main Dialog Window

Results

A bivariate regression was performed to evaluate how well salary (in dollars) could be predicted from years of job experience. Preliminary data screening indicated that the scores on salary were reasonably normally distributed. Scores on years were positively skewed, but because a log transformation did not make the shape of the distribution closer to normal, raw scores were used rather than log-transformed scores for years of job experience in subsequent analyses. A scatter plot indicated that the relation between X and Y was positive and reasonably linear and there were no bivariate outliers. The correlation between years of job experience and salary was statistically significant, $r(48) = .83$, $p < .001$. The regression equation for predicting starting salary from years of job experience was found to be $Y' = 31416.72 + 2829.57 \times X$. The r^2 for this equation was .69; that is, 69% of the variance in salary was predictable from years of job experience. This is a very strong relationship; increases in years of job experience tended to be associated with increases in salary. The 95% CI for the slope to predict salary in dollars from years of job experience ranged from 2,277 to 3,382; thus, for each 1 year of increase in years of job experience, the predicted salary increased by about $2,300 to $3,400. A scatter plot of the data with a 95% CI around the fitted regression line appears in Figure 9.22.

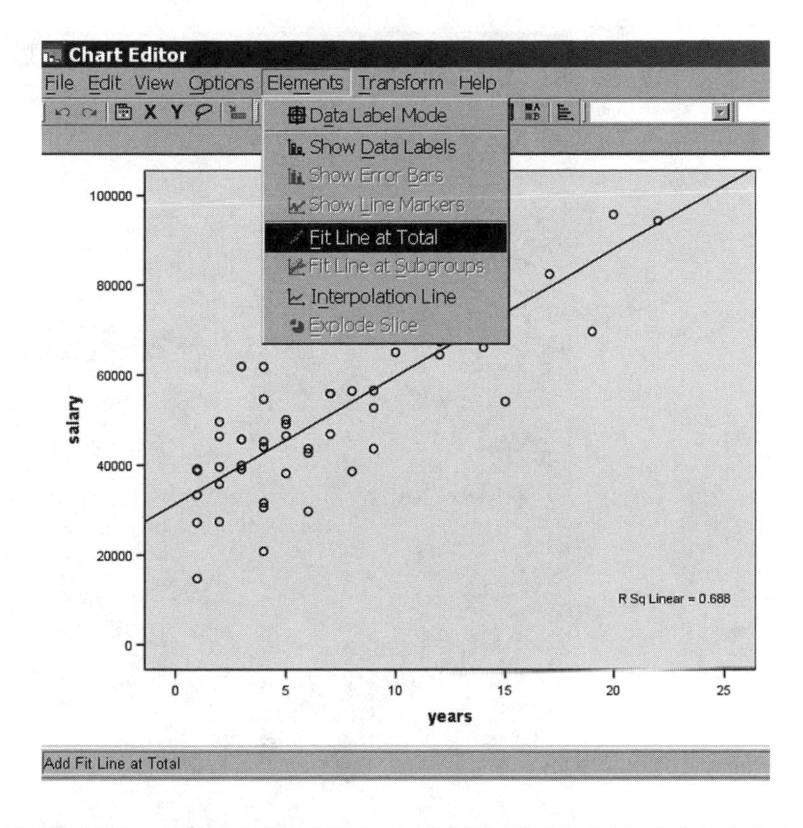

Figure 9.20 ♦ SPSS Menu Selections to Add Fit Line (Regression Line) to Scatter Plot

9.14 ♦ Summary

Assuming that X and Y are linearly related, scores on a quantitative Y variable can be predicted from scores on a quantitative (or dichotomous) X variable by using a prediction equation in either a raw score form,

$$Y' = b_0 + bX,$$

or a standardized, z score form,

$$z'_Y = \beta z_X.$$

The b coefficient can be interpreted in terms of "real" units; for example, for each additional hour of exercise (X), how many pounds of weight (Y) can a person expect to lose? The beta coefficient is interpretable as a unit-free or standardized index of the strength of the linear relationship: For each one-*SD* increase in z_X, how many standard deviations do we predict that z_Y will change?

The significance of a regression prediction equation can be tested by using a t test (to test whether the raw score slope b differs significantly from 0) or an F test (to assess whether the r or R differs significantly from 0). In bivariate regression, where there is only one predictor variable, these two types of tests are equivalent. In multiple regression,

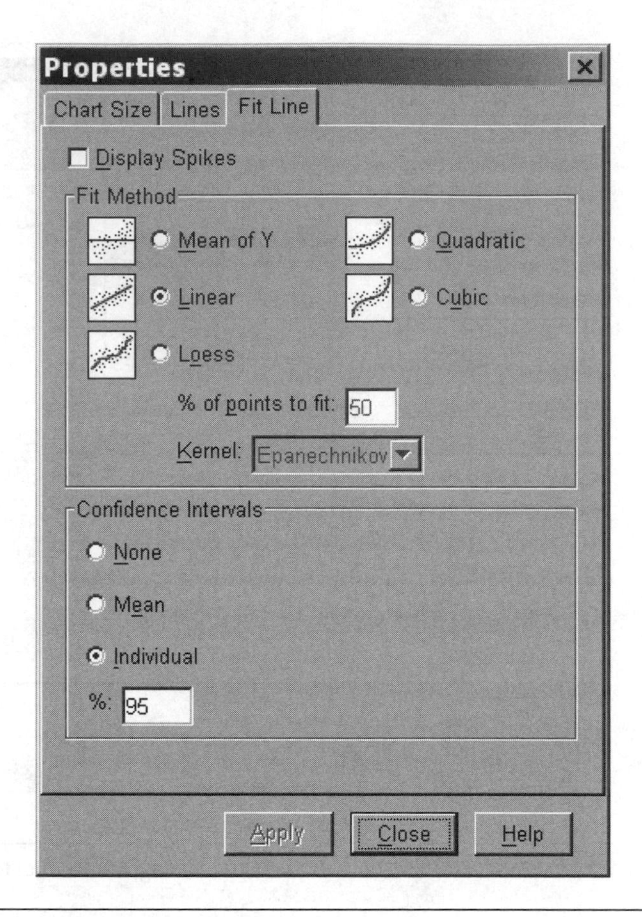

Figure 9.21 ♦ SPSS Dialog Window for Properties of Fit Line

where there are several predictor variables, tests of each *b* slope coefficient provide information about the significance of the contribution of each individual predictor, while a test of *R* tests the significance of the entire equation that includes all the predictors.

Like ANOVA, regression involves a partition of scores into components: a part that can be predicted from *X* and a part that cannot be predicted from *X*. Squaring and summing these components gives us a partition of SS_{total} into $SS_{\text{regression}} + SS_{\text{residual}}$, which is analogous to the partition of sums of squares that we saw in ANOVA ($SS_{\text{total}} = SS_{\text{between}} + SS_{\text{within}}$).

Researchers often choose to use certain variables as predictors in regression because they believe that these variables may possibly cause or influence the dependent variable. However, finding that an *X* variable is significantly predictive of a *Y* variable in a regression does not prove that *X* caused *Y* (unless *X* and *Y* represent data from a well-controlled experiment). The most we can say when we find that an *X* variable is predictive of *Y* in correlational research is that this finding is consistent with theories that *X* might be a cause of *Y*.

In later chapters, we will go beyond the one-predictor-variable situation,

$$Y' = b_0 + bX,$$

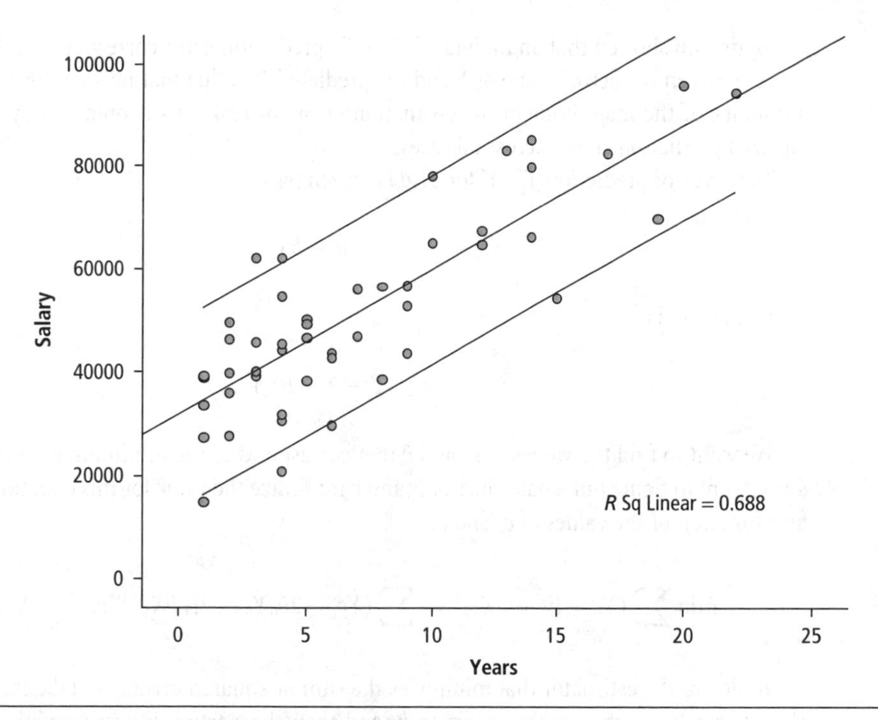

Figure 9.22 ♦ Scatter Plot With Fitted Regression Line (and 95% CI)

to consider equations that include multiple predictor variables, such as

$$Y' = b_0 + b_1X_1 + b_2X_2 + \ldots + b_kX_k.$$

Notes

1. Students who have enough background in calculus to understand the concept of a partial derivative may find it useful to see how the partial derivative was used to figure out the formulas that produce the "best" estimates of b_0 and b—that is, the estimates of the regression intercept and slope for which the sum of squared prediction errors is minimized. The bivariate regression to predict Y_i from X_i is denoted as

$$Y_i' = b_0 + bX_i,$$

where b_0 is the constant or intercept and b is the slope for the regression line.

The OLS estimate of b, the slope for the regression line to predict Y from X, is the value of b for which the sum of squared errors (SSE), the sum of the squared deviations, is minimized.

For the empirical example that follows, this regression line had the following parameters:

$$Y_i' = 4.53 + 1.30X.$$

Figure 9.6 showed that an individual ($Y - Y'_i$) prediction error corresponds to the vertical distance between the actual value of Y and the predicted Y' value that falls on the regression line. A summary of the magnitude of the estimation errors or residuals is obtained by summing these squared prediction errors across all cases.

The error of prediction $Y_i - Y'_i$ for case i is given by

$$Y_i - b_0 - bX_i.$$

So the *SSE* is

$$\Sigma (Y_i - b_0 - bX_i)^2.$$

We want to find the values of b_0 and b that correspond to the minimum possible value of *SSE*, so we want to figure out what values of b_0 and b minimize the value for this equation that gives *SSE* as a function of the values of b_0 and b:

$$\min \sum (Y_i - b_0 - bX_i)^2 = \sum (Y_i^2 - 2b_0 Y_i - 2b_0 bX_i + b_0^2 + b^2 X_i^2) = Q.$$

To derive the estimator that minimizes the sum of squared errors, we take the partial derivative of this function Q with respect to b_0 and then the partial derivative of this function with respect to b.

First, we take the partial derivative of the function Q with respect to b_0; the value of the intercept

$$\frac{\partial Q}{\partial b_0} = \sum_n (-2Y_i + 2b_0 + 2bX_i)$$

Setting this derivative to 0 yields the following equation for the intercept b_0:

$$nb_0 = \sum_n (Y_i - bX_i)$$

$$b_0 = \frac{\sum Y_i}{n} - \frac{b \sum X_i}{n}$$

or

$$b_0 = M_Y - bM_X,$$

where M_X and M_Y are the sample means for X and Y, respectively.

To obtain the formula for the OLS value of the slope coefficient b, we take the derivative of this function with respect to b:

$$\frac{\partial Q}{\partial b} = \sum_n (-2X_i Y_i + 2bX_i^2 + 2b_0 X_i).$$

The next step is to set this derivative to 0 and solve for b; this equation is equivalent to Equation 9.5, although it appears in a slightly different form here:

$$b = \frac{\sum X_i Y_i - b_0 \sum X_i}{\sum X_i^2}.$$

When we set the derivative equal to 0, we find the place where the function for *SSE* has the minimum possible value, as shown in Figure 9.23, where the minimum value of *SSE* was obtained when the value of the slope estimate b was equal to 1.30.

When models are more complex, it is not always possible to solve for partial derivatives and to obtain exact analytic solutions for the values of parameters such as the intercept b_0 and slope b. In situations where OLS solutions are not available (e.g., the estimation of path coefficients in structural equation models), brute force computational methods can be used. In effect, this involves a grid search; the program systematically calculates the value of *SSE* for different possible values of b (e.g., $b = 1.00$, $b = 1.10$, $b = 1.20$, $b = 1.30$, $b = 1.40$, $b = 1.50$). The value of b for which *SSE* has the minimum value is determined empirically by grid search—that is, by computing the value of *SSE* for different candidate values of b until the value of b for which *SSE* is minimized is found. This method is called maximum likelihood estimation; for a simple regression, maximum likelihood estimates yield the same values of b_0 and b as the OLS estimates.

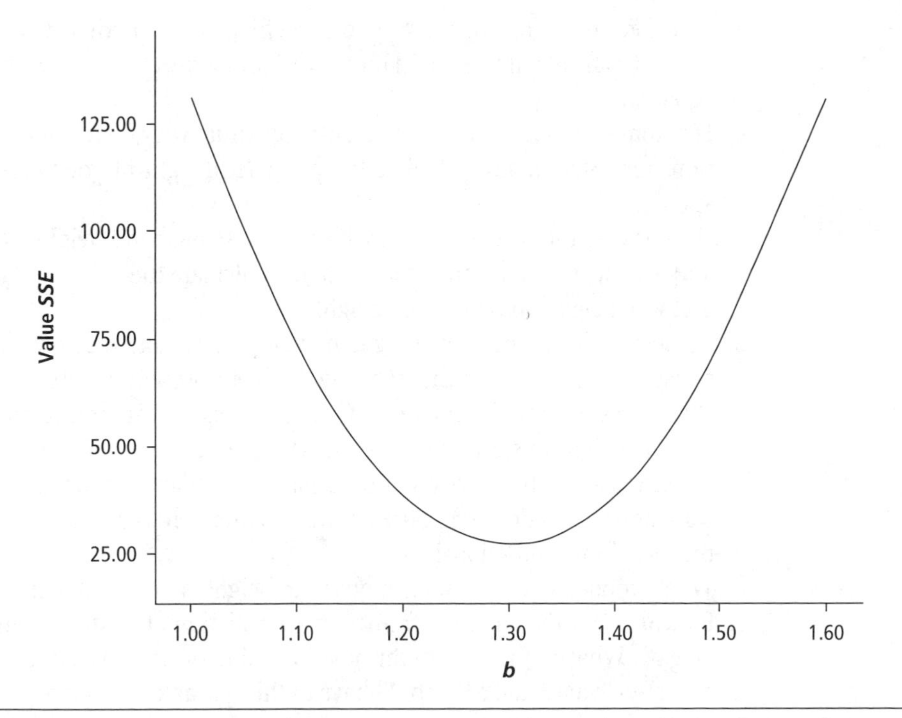

Figure 9.23 ♦ Graph of Values of *SSE* as a Function of Different Values of b, the Slope Coefficient

NOTE: For this dataset, the minimum possible value of *SSE* (27.23) is obtained for $b = 1.30$; for any other value of b, *SSE* is greater than 27.23; thus, $b = 1.30$ is the value of b that has the minimum possible *SSE*.

2. Pearson r is a symmetrical index of linear relationship. It can be used to predict z score locations on Y from z score locations on X ($z'_Y = rz_X$); it can also be used to predict z score locations on X from z score locations on Y ($z'_X = rz_Y$). In the examples presented here, X denotes the predictor and Y denotes the outcome variable.

3. In some textbooks, the standard error of the estimate is denoted by $s_{Y.X}$. A subscript of the form $Y.X$ is read as follows: "Y predicted from X."

Comprehension Questions

1. In a survey that included assessment of husband and wife heights, Hodges, Krech, and Crutchfield (1975) reported the following results. Let's treat wife height as the predictor (X) variable and husband height as the outcome (Y) variable:

$$r_{XY} = .32$$

Wife height: $M_X = 64.42$, $s_X = 2.56$
Husband height: $M_Y = 70.46$, $s_Y = 2.87$
$N = 1,296$

   a. Calculate the values of the b_0 and b coefficients to predict husband height in inches (Y) from wife height in inches (X), and write out this raw score predictive equation.
   b. For women: What is your own height? Substitute your own height into the equation from Step a, and calculate the predicted height of your present or future spouse.
   c. Now reverse the roles of the variables (i.e., use husband height as the predictor and wife height as the outcome variable). Calculate the values of b_0 and b to predict wife height from husband height.
   d. For men: What is your own height in inches? Substitute your own height into the equation in Step c to predict the height of your present or future spouse.
   e. What is the equation to predict z_Y (husband height in standard score units) from z_X (wife height in standard score units)? Include the numerical value of the coefficient used in this version of the equation. In your own words, what does this standardized version of the prediction equation tell us about "regression toward the mean" for predictions?
   f. What proportion of variance in husband height is predictable from wife height?
   g. Find SE_{est} (for the equation from Step a, prediction of husband height from wife height). What is the minimum possible value of SE_{est}? What is the maximum numerical value that SE_{est} could have in this research situation?

2. Table 9.2 displays a small set of (real) data that show GNP per capita (X) and mean life satisfaction for a set of 19 nations (Y).

Table 9.2 ♦ Data for 19 Nations on GNP per Capita and Life Satisfaction

Country	*Life Satisfaction*	*GNP per Capita (in Dollars)*
Denmark	8.03	8,470
Sweden	8.02	10,071
Switzerland	7.98	9,439
Norway	7.90	8,762
Netherlands	7.77	7,057
N. Ireland	7.77	3,560
Ireland	7.76	3,533
Finland	7.73	5,814
Luxembourg	7.64	7,754
United States	7.57	10,765
Britain	7.52	4,972
Belgium	7.33	7,978
W. Germany	7.23	9,507
Austria	7.14	6,311
France	6.63	8,619
Spain	6.60	2,830
Italy	6.58	4,191
Japan	6.39	7,244
Greece	5.85	2,881

a. Enter the data into SPSS.

b. Run the <Analyze> → <Descriptive Statistics> → <Descriptives> procedure; obtain the mean and standard deviation for each variable.

c. Do appropriate preliminary data screening, then run the bivariate regression to predict life satisfaction from GNP. Write up the results in paragraph form, including statistical significance, effect size, and nature of relationship. Also, use the <Save> command to save the unstandardized residuals as a new variable in your SPSS worksheet. Include effect size and nature of relationship even if the regression is not statistically significant.

d. Using the <Transform> → <Compute> command, create a new variable that tells you the squared unstandardized residual for each case. Using the <Descriptives> command, obtain the sum of the squared residuals across all 19 cases. Note that this sum should correspond to $SS_{residual}$ in the ANOVA table in your SPSS regression printout.

e. Show that you can use the *SS* values in the ANOVA table in the SPSS printout to reproduce the value of R^2 on the SPSS printout.

3. Given that 1 meter (m) = 39.37 in., answer the following:

   a. Write an equation to convert height given in inches (X) to height expressed in meters (Y). What is the value of the intercept for this equation? What is the value of the slope?

   b. Use this equation to convert your own height from inches to meters.

c. What would you expect the correlation between a person's height in inches (X) and corresponding heights for the same person in meters (Y) to be?

4. Let's revisit some data that we looked at in Chapter 8, Table 8.2. Let X = gender, coded 1 = Male, 2 = Female. Let Y = height. Using SPSS, run a bivariate regression to predict height from gender. If you do not still have your output from analyses you ran in Chapter 8, also run the Pearson correlation between gender and height and the independent samples t test comparing mean heights for male and female groups.
   a. Compare the F for your bivariate regression with the t from your independent samples t test. How are these related?
   b. Compare the multiple R from your bivariate regression with the r from your bivariate correlation; compare the R^2 from the regression with an eta squared effect size computed by hand from your t test. How are these related?
   c. What do you conclude regarding these three ways of analyzing the data?

5. Discuss the two forms of the regression equation, raw score and z score, and their verbal interpretations. When is each form of the analysis more useful?

6. How are r, b, and β related to each other?

7. Discuss bivariate regression as a partition of each score into components and as a partition of SS_{total} into $SS_{residual}$ and $SS_{regression}$ (note the analogy to the partitions of scores in one-way ANOVA).

8. What is a multiple R? How is multiple R^2 interpreted?

9. What information should be included in a report of a bivariate regression?

Data Analysis Project for Bivariate Regression

Choose two quantitative variables and designate one as the predictor. Conduct data screening using appropriate SPSS procedures (including a scatter plot), and report any problems you found and what you did to correct them. Obtain a Pearson r, and do a bivariate regression. Hand in the following:

a. Excerpts from your SPSS printouts (if you refer to them in your writing, please indicate clearly what page of the SPSS printout you are talking about)
b. An APA-style Results section that includes a (more thorough than usual) discussion of preliminary data screening and the results for your correlation and regression analysis
c. Answers to the list of questions below:
   i. Write out the raw score prediction equation. In words and/or equations, explain how the b_0 or intercept term was calculated and what it means. Explain how the b or slope coefficient can be calculated from the values of r, s_Y, and s_X and describe the nature of the predictive relationship between X and Y in terms of the sign and magnitude of the b slope coefficient.

ii. Identify a particular score on the scatter plot and read the X, Y values (as nearly as you can) from the graph. Calculate the predicted value of Y' for this value of X. Now find the three deviations: Y – the mean of Y, $Y - Y'$, and Y' – the mean of Y. Draw these deviations on your scatter plot and label them (like the graph in Figure 9.6). In words, what is $Y - Y'$? What is Y' – the mean of Y? Which of these two deviations do we hope will be larger and why? *Hint:* if you have trouble with this, try choosing a data point that is *not* sandwiched in between the regression line and the line that represents the mean of Y. It will be less confusing if you choose a Y value such that Y – the mean of Y is the largest of the three deviations.

iii. Write out the standardized score prediction equation—that is, the equation to predict from z score on the independent variable to z score on the dependent variable.

iv. How does the slope in the standardized score prediction equation in Step ii differ from the slope in the raw score equation in Step i? In what situations would you be more interested in the raw score versus the standardized score slope as a way of describing how the variables are related?

v. By hand, do a significance test on the correlation (use either F or t) and verify that your results are consistent with the p value given on the printout.

vi. What percentage of variance is explained by the independent variable? Is this a relatively strong or weak effect, in your opinion?

vii. Consider the value of SE_{est} (also known as $s_{Y.X}$). In words, what information does SE_{est} provide? What is the smallest value this could have had? What is the largest value this could have had, for this particular dependent variable? Relative to this range of possible values, do you think that SE_{est} is large? What does this tell you about how well or how poorly your regression equation predicts outcome scores?

viii. Consider the factors that influence the size of r. Do any of these factors make you doubt whether this correlation is a good indication of the real strength of the relationship between these variables? If so, which factors are you most concerned about?

ix. Can you think of any modifications you would want to make if you were reanalyzing these data (or collecting new data) to get a better description of how these two variables are related?

CHAPTER 10

Adding a Third Variable

Preliminary Exploratory Analyses

10.1 ♦ Three-Variable Research Situations

In previous chapters, we reviewed the bivariate correlation (Pearson r) as an index of the strength of the linear relationship between one independent variable (X) and one dependent variable (Y). This chapter moves beyond the two-variable research situation to ask, "Does our understanding of the nature and strength of the predictive relationship between a predictor variable, X_1, and a dependent variable, Y, *change* when we take a third variable, X_2, into account in our analysis, and if so, how does it change?" In this chapter, X_1 denotes a predictor variable, Y denotes an outcome variable, and X_2 denotes a third variable that may be involved in some manner in the X_1, Y relationship. For example, we will examine whether age (X_1) is predictive of systolic blood pressure (SBP) (Y) when body weight (X_2) is statistically controlled.

We will examine two preliminary exploratory analyses that make it possible to statistically control for scores on the X_2 variable; these analyses make it possible to assess whether controlling for X_2 changes our understanding about whether and how X_1 and Y are related. First, we can split the data file into separate groups based on participant scores on the X_2 control variable and then compute Pearson correlations or bivariate regressions to assess how X_1 and Y are related separately within each group. Although this exploratory procedure is quite simple, it can be very informative. Second, if the assumptions for partial correlation are satisfied, we can compute a partial correlation to describe how X_1 and Y are correlated when scores on X_2 are statistically controlled. The concept of statistical control that is introduced in this chapter continues to be important in later chapters that discuss analyses that include multiple predictor variables.

Partial correlations are sometimes reported as the primary analysis in a journal article. In this textbook, partial correlation analysis is introduced primarily to explain the concept of statistical control. The data analysis methods for the three-variable research situation that are presented in this chapter are suggested as preliminary exploratory analyses that can help a data analyst evaluate what kinds of relationships among variables should be taken into account in later, more complex analyses.

For partial correlation to provide accurate information about the relationship between variables, the following assumptions about the scores on X_1, X_2, and Y must be reasonably well satisfied. Procedures for data screening to identify problems with these assumptions were reviewed in detail in Chapters 7 and 9, and detailed examples of data screening are not repeated here.

1. The scores on X_1, X_2, and Y should be quantitative. It is also acceptable to have predictor and control variables that are dichotomous. Most of the control variables (X_2) that are used as examples in this chapter have a small number of possible score values because this limitation makes it easier to work out in detail the manner in which X_1 and Y are related at each level or score value of X_2. However, the methods described here can be generalized to situations where the X_2 control variable has a large number of score values, as long as X_2 meets the other assumptions for Pearson correlation.
2. Scores on X_1, X_2, and Y should be reasonably normally distributed. For dichotomous predictor or control variables, the closest approximation to a normal distribution occurs when the two groups have an equal number of cases.
3. For each pair of variables (X_1 and X_2, X_1 and Y, and X_2 and Y), the joint distribution of scores should be bivariate normal, and the relation between each pair of variables should be linear. The assumption of linearity is extremely important. If X_1 and Y are nonlinearly related, then Pearson r does not provide a good description of the strength of the association between them.
4. Variances in scores on each variable should be approximately the same across scores on other variables (the homogeneity of variance assumption).
5. The slope that relates X_1 to Y should be equal or homogeneous across levels of the X_2 control variable. (That is, the overall partial correlation does not provide an adequate description of the nature of the X_1, Y relationship when the nature of the X_1, Y relationship differs across levels of X_2 or, in other words, when there is an **interaction** between X_1 and X_2 as predictors of Y.)

When interpreting partial and **semipartial correlations (*sr*)**, factors that can artifactually influence the magnitude of Pearson correlations must be considered. For example, if X_1 and Y both have low measurement reliability, the correlation between X_1 and Y will be attenuated or reduced, and any partial correlation that is calculated using r_{1Y} may also be inaccurate.

We can compute three bivariate (or **zero-order**) Pearson correlations for a set of variables that includes X_1, X_2, and Y. When we say that a correlation is "zero-order," we mean that the answer to the question, "How many other variables were statistically controlled or partialled out when calculating this correlation?" is 0 or none.

r_{Y1} or r_{1Y} denotes the zero-order bivariate Pearson correlation between Y and X_1.

r_{Y2} or r_{2Y} denotes the zero-order correlation between Y and X_2.

r_{12} or r_{21} denotes the zero-order correlation between X_1 and X_2.

Note that the order of the subscripts in the zero-order correlation between X_1 and Y can be changed: The correlation can be either r_{1Y} or r_{Y1}. A zero-order Pearson correlation is symmetrical; that is, the correlation between X_1 and Y is identical to the correlation between Y and X_1. We can compute a partial correlation between X_1 and Y, controlling for one or more variables (such as X_2, X_3, and so forth). In a **first-order partial correlation** between X_1 and Y, controlling for X_2, the term *first order* tells us that only one variable (X_2) was statistically controlled when assessing how X_1 and Y are related. In a **second-order partial correlation,** the association between X_1 and Y is assessed while statistically controlling for two variables; for example, $r_{Y1.23}$ would be read as "the partial correlation between Y and X_1, statistically controlling for X_2 and X_3." In a kth order partial correlation, there are k control variables. This chapter examines first-order partial correlation in detail; the conceptual issues involved in the interpretation of higher-order partial correlations are similar.

The three zero-order correlations listed above (r_{1Y}, r_{2Y}, and r_{12}) provide part of the information that we need to work out the answer to our question, "When we control for, or take into account, a third variable called X_2, how does that change our description of the relation between X_1 and Y?" However, examination of separate scatter plots that show how X_1 and Y are related separately for each level of the X_2 variable provides additional, important information.

In all the following examples, a distinction is made among three variables: an independent or predictor variable (denoted by X_1), a dependent or outcome variable (Y), and a control variable (X_2). The preliminary analyses in this chapter provide ways of exploring whether the nature of the relationship between X_1 and Y changes when you remove, partial out, or statistically control for the X_2 variable.

Associations or correlations of X_2 with X_1 and Y can make it difficult to evaluate the true nature of the relationship between X_1 and Y. This chapter describes numerous ways in which including a single control variable (X_2) in the data analysis may change our understanding of the nature and strength of the association between an X_1 predictor and a Y outcome variable. Terms that describe possible roles for an X_2 or control variable (confound, **mediation,** suppressor, interaction) are explained using hypothetical research examples.

10.2 ♦ First Research Example

Suppose that X_1 = height, Y = vocabulary test score, and X_2 = grade level (Grade 1, 5, or 9). In other words, measures of height and vocabulary are obtained for groups of school children who are in Grades 1, 5, and 9. A scatter plot for hypothetical data, with vocabulary scores plotted in relation to height, appears in Figure 10.1. Case markers are used to identify group membership on the control variable—namely, grade level; that is, scores for first graders appear as "1," scores for fifth graders appear as "5," and scores for ninth graders appear as "9" in this scatter plot. This scatter plot shows an example in which there are three clearly separated groups of scores. Students in Grades 1, 5, and 9 differ so much in both height and vocabulary scores that there is no overlap between the groups on these variables; both height and vocabulary increase across grade levels. Real data can show a similar pattern; this first example shows a much clearer separation between groups than would typically appear in real data.

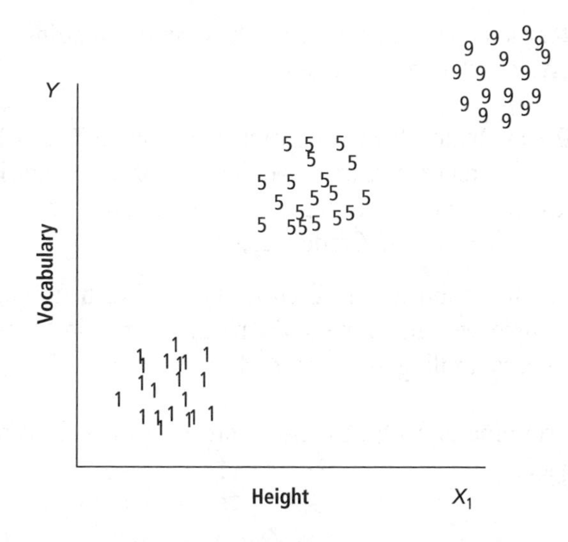

Figure 10.1 ♦ Hypothetical Data That Show a Spurious Correlation Between Y (Vocabulary) and X_1 (Height)

NOTE: Height and vocabulary are correlated only because scores on both variables increase across grade level (Grade Levels are 1, 5, and 9).

When all the data points in Figure 10.1 are examined (ignoring membership in grade-level groups), it appears that the correlation between X_1 (height) and Y (vocabulary) is large and positive ($r_{1Y} \approx +.8$). On the other hand, if you examine the X_1,Y scores within any one grade (e.g., Grade 1 children, whose scores are denoted by "1" in the scatter plot), the r_{1Y} correlation between height and vocabulary scores within each level of X_2 appears to be close to 0. This is an example of one type of situation where the apparent correlation between X_1 and Y is different when you ignore the X_2 variable (grade level) than when you take the X_2 variable into account. When you ignore grade level and compute a Pearson correlation between height and vocabulary for all the scores, there appears to be a strong positive linear association between height and vocabulary. If you take grade level into account by examining the X_1, Y (height, vocabulary) relationship separately within each grade level, there appears to be no correlation between X_1 and Y. Later in this chapter, we will see that one reasonable interpretation for this outcome would be that the correlation between height and vocabulary is "spurious." That is, height and vocabulary are not *directly* associated with each other; however, both height and vocabulary tend to increase across grade levels. The increase in height and vocabulary that occurs from Grades 1 to 5 to 9 creates a pattern in the data such that when all grade levels are treated as one group, height and vocabulary appear to be positively correlated.

10.3 ♦ Exploratory Statistical Analyses for Three-Variable Research Situations

There are several different possible ways to analyze data when we introduce a third variable X_2 into an analysis of variables X_1 and Y. This chapter describes two preliminary or exploratory analyses that can be helpful in understanding what happens when a control variable X_2 is included in an analysis:

1. We can obtain a bivariate correlation between X_1 and Y separately for subjects who have different scores on X_2. For instance, if X_2 is grade level (coded "1," "5," and "9"), we can compute three separate r_{1Y} correlations between height and vocabulary for the students within Grades 1, 5, and 9.

2. We can obtain a partial correlation between X_1 and Y, controlling for X_2. For example, we can compute the partial correlation between height and vocabulary score, controlling for grade level.

We will examine each of these two approaches to analysis to see what information they can provide.

10.4 ♦ Separate Analysis of X_1, Y Relationship for Each Level of the Control Variable X_2

One simple way to take a third variable X_2 into account when analyzing the relation between X_1 and Y is to divide the dataset into groups based on scores on the X_2 or control variable and then obtain a Pearson correlation or bivariate regression to assess the nature of the relationship between X_1 and Y separately within each of these groups. Suppose that X_2, the variable you want to control for, is grade level, whereas X_1 and Y are continuous, interval/ratio variables (height and vocabulary). Table 10.1 shows a small hypothetical dataset for which the pattern of scores is similar to the pattern in Figure 10.1; that is, both height and vocabulary tend to increase across grade levels.

A first step in the exploratory analysis of these data is to generate an X_1, Y scatter plot with case markers that identify grade level for each child. To do this, make the following SPSS menu selections: <Graph> → <Scatter/Dot>, then choose the "Simple" type of scatter plot. This set of menu selections opens up the SPSS dialog window for the simple scatter plot procedure, which appears in Figure 10.2. As in earlier examples of scatter plots, the names of the variables for the X and Y axes are identified. In addition, the variable grade is moved into the "Set Markers by" window. When the "Set Markers by" command is included, different types of markers are used to identify cases for Grades 1, 5, and 9. (Case markers can be modified using the SPSS Chart Editor to make them more distinctive in size or color.)

The scatter plot generated by these SPSS menu selections appears in Figure 10.3. These data show a pattern similar to the simpler example in Figure 10.1; that is, there are three distinct clusters of scores (all the Grade 1s have low scores on both height and vocabulary, the Grade 5s have intermediate scores on both height and vocabulary, and the Grade 9s have high scores on both height and vocabulary). A visual examination of the scatter plot in Figure 10.3 suggests that height and vocabulary are positively correlated when a correlation is calculated using all the data points, but the correlation between height and vocabulary is approximately 0 within each grade level.

We can obtain bivariate correlations for each pair of variables (height and vocabulary, height and grade, and vocabulary and grade) by making the SPSS menu selections

Table 10.1 ♦ Hypothetical Data for a Research Example Involving a Spurious Correlation Between Height and Vocabulary

Grade	*Height*	*Vocabulary*
1	46	43
1	48	45
1	48	30
1	48	18
1	46	50
1	47	53
1	48	46
1	53	33
1	49	47
1	53	54
1	49	28
1	49	51
1	51	37
1	47	33
1	49	42
1	52	50
5	54	51
5	54	79
5	54	49
5	52	50
5	58	66
5	52	65
5	53	44
5	55	73
5	52	51
5	55	47
5	54	65
5	55	52
5	53	50
5	58	48
5	59	53
5	52	59
9	57	73
9	63	62
9	62	67
9	62	80
9	61	79
9	58	67
9	57	85
9	60	64
9	61	75
9	62	62
9	62	72
9	66	85
9	62	69
9	61	75
9	64	60
9	62	58

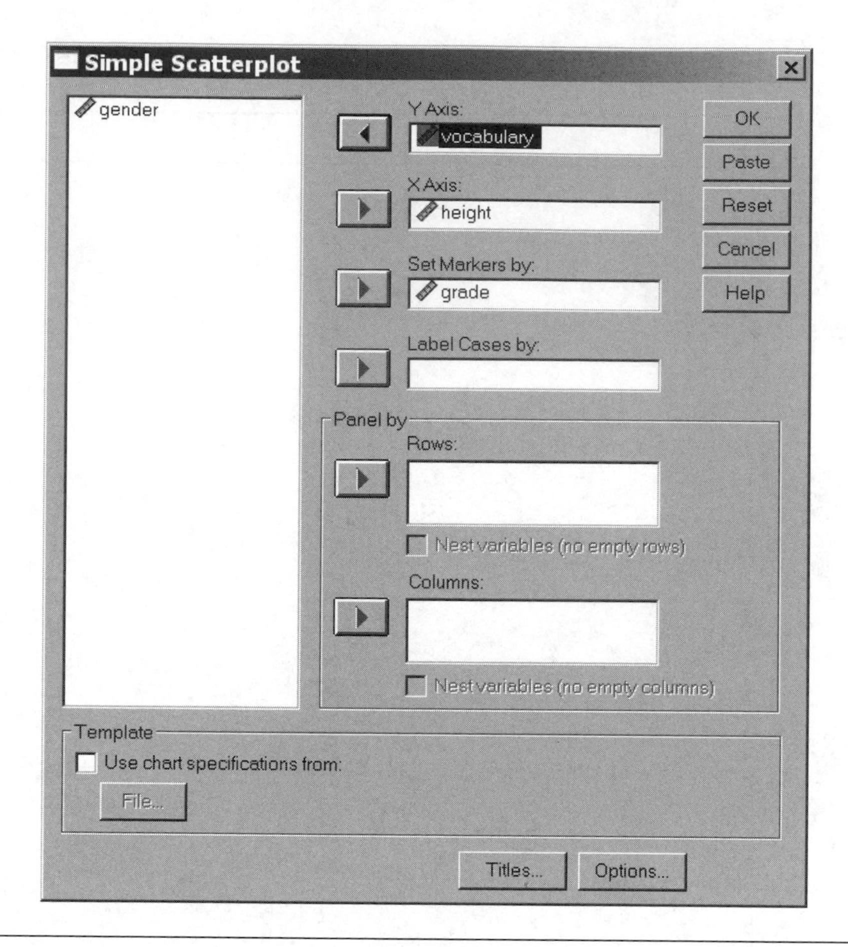

Figure 10.2 ♦ SPSS Dialog Window for Scatter Plot With Case Markers

<Analyze> → <Correlate> → <Bivariate> and entering the three variable names. These three bivariate correlations appear in Figure 10.4.

In this example, all three zero-order bivariate correlations are large and positive. This is consistent with what we see in the scatter plot in Figure 10.3. Height increases across grade levels; vocabulary increases across grade levels; and if we ignore grade and compute a Pearson correlation between height and vocabulary for all students in all three grades, that correlation is also positive.

Does the nature of the relationship between height and vocabulary appear to be different when we statistically control for grade level? One way to answer this question is to obtain the correlation between height and vocabulary separately for each of the three grade levels. We can do this conveniently using the <Split File> command in SPSS.

To obtain the correlations separately for each grade level, the following SPSS menu selections are made (as shown in Figure 10.5): <Data> → <Split File>. The SPSS dialog window for the Split File procedure appears in Figure 10.6. To split the file into separate groups based on scores on the variable grade, the user clicks the radio button for "Organize output by groups" and then enters the name of the grouping variable (in this case, the control variable grade) in the window for "Groups Based on," as shown in Figure 10.6. Once the <Split File>

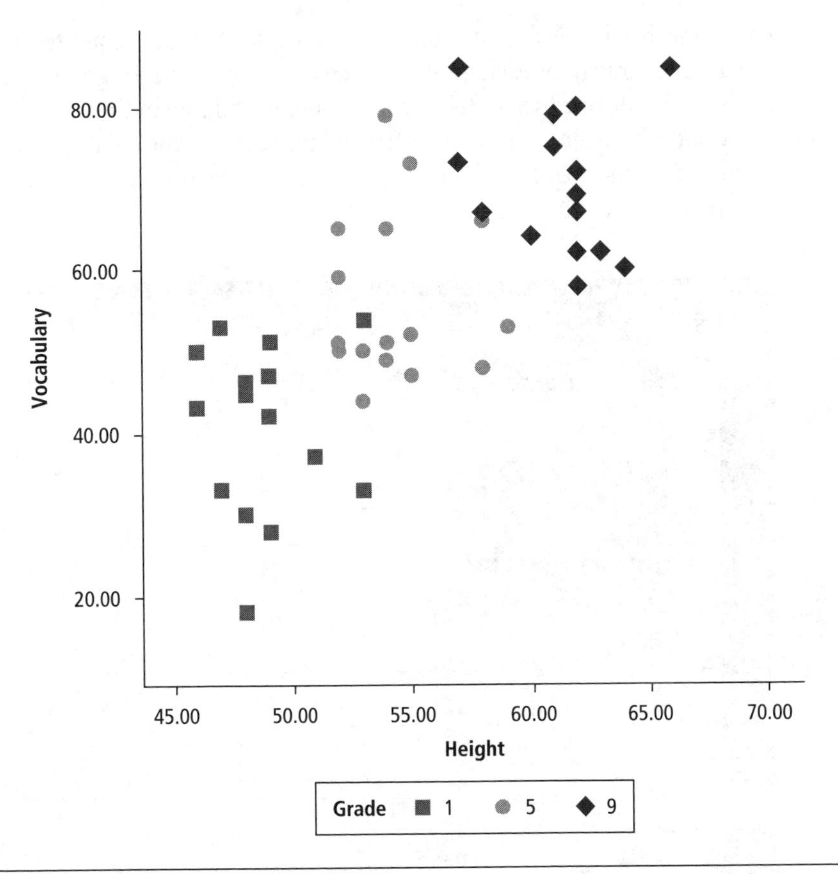

Figure 10.3 ♦ A Bivariate Scatter Plot: Vocabulary (on *Y* Axis) Against Height (on *X* Axis)

NOTE: Case markers identify grade level for each data point (Grades 1, 5, 9).

Correlations

		height	vocabulary	grade
height	Pearson Correlation	1	.716**	.913**
	Sig. (2-tailed)		.000	.000
	N	48	48	48
vocabulary	Pearson Correlation	.716**	1	.787**
	Sig. (2-tailed)	.000		.000
	N	48	48	48
grade	Pearson Correlation	.913**	.787**	1
	Sig. (2-tailed)	.000	.000	
	N	48	48	48

**. Correlation is significant at the 0.01 level (2-tailed).

Figure 10.4 ♦ The Bivariate Zero-Order Pearson Correlations Between All Pairs of Variables: X_1 (Height), Y (Vocabulary), and X_2 (Grade Level)

command has been run, any subsequent analyses that are requested are performed and reported separately for each group based on the score value for grade; in this example, each analysis is performed separately for the groups of children in Grades 1, 5, and 9. Note that to carry out subsequent analyses that treat all the scores in the dataset as one group, you need to go back to the Split File dialog window shown in Figure 10.6 and select the radio button for "Analyze all cases, do not create groups."

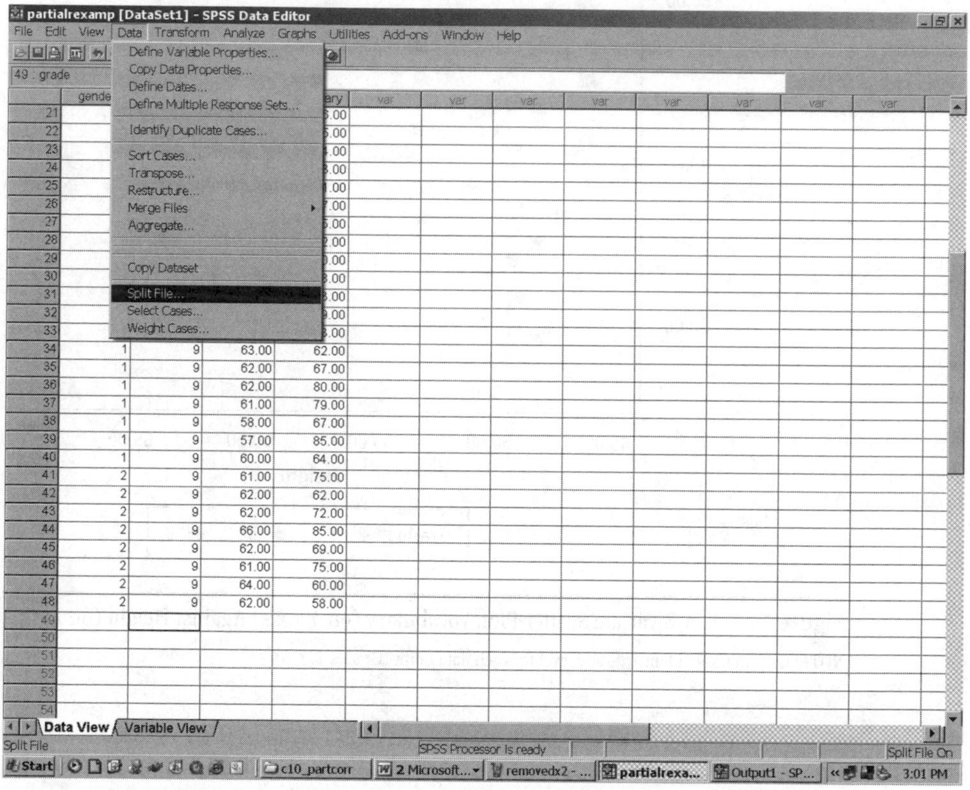

Figure 10.5 ♦ SPSS Menu Selections for the Split File Command

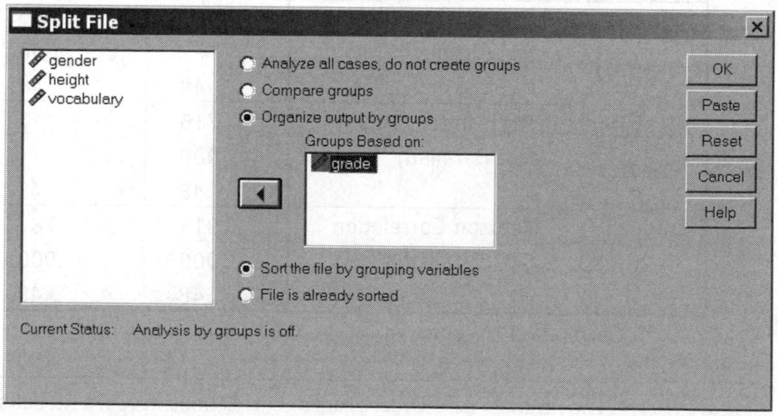

Figure 10.6 ♦ SPSS Dialog Window for Split File: Grade Level X_2 Is the "Control" Variable in This Situation

To obtain the Pearson correlation between height and vocabulary separately within each grade, the user next makes the menu selections <Analyze> → <Correlate> → <Bivariate> and enters the names of the variables (height and vocabulary) in the dialog window for the bivariate correlation procedure. The corresponding output appears in Figure 10.7. Within each grade level, the correlation between height and vocabulary did not differ significantly from 0. To summarize, the overall Pearson correlation between height and vocabulary (combining the scores for students in all three grade levels) was positive and statistically significant (from Figure 10.4, the correlation between height and vocabulary was +.72). After we statistically control for grade level by calculating the height, vocabulary correlation separately for students within each grade level, we find that the correlation between these variables within each grade level was close to 0 (from Figure 10.7, $r = .07, .03$, and $-.14$, for students in Grade 1, Grade 5, and Grade 9, respectively). Scatter plots are not shown for each grade, but they can provide valuable additional information when sample sizes are larger than in this example.

In this situation, our understanding of the nature of the relationship between height (X_1) and vocabulary (Y) is quite different, depending on whether or not we statistically control for grade (X_2). If we ignore grade, height and vocabulary appear to be positively correlated; if we statistically control for grade, height and vocabulary appear to be uncorrelated.

Another possible outcome when the X_1, Y relationship is examined separately for each value of X_2 is that the slope or correlation between X_1 and Y may differ across levels of X_2. This outcome suggests that there may be an interaction between X_1 and X_2 as predictors of Y. If there is evidence that X_1 and Y have significantly different correlations or slopes across levels of X_2, then it would be misleading to report a single overall partial correlation between X_1 and Y, controlling for X_2. Later chapters in this textbook describe ways to set up analyses that include an interaction between X_1 and X_2 and test the statistical significance of interactions (see Chapters 12 and 13).

10.5 ♦ Partial Correlation Between X_1 and Y, Controlling for X_2

Another way to evaluate the nature of the relationship between X_1 (height) and Y (vocabulary) while statistically controlling for X_2 (grade) is to compute a partial correlation between X_1 and Y, controlling for or partialling out X_2. The following notation is used to denote the partial correlation between Y and X_1, controlling for X_2: $r_{Y1.2}$. The subscript 1 in $r_{Y1.2}$ refers to the predictor variable X_1, and the subscript 2 refers to the control variable X_2. When the subscript is read, pay attention to the position in which each variable is mentioned relative to the "." in the subscript. The period within the subscript divides the subscripted variables into two sets. The variable or variables to the right of the period in the subscript are used as predictors in a regression analysis; these are the variables that are statistically controlled or partialled out. The variable or variables to the left of the period in the subscript are the variables for which the partial correlation is assessed while taking one or more control variables into account. Thus, in $r_{Y1.2}$, the subscript *Y1.2* denotes the partial correlation between X_1 and Y, controlling for X_2.

In the partial correlation, the order in which the variables to the left of the period in the subscript are listed does not signify any difference in the treatment of variables; we could read either $r_{Y1.2}$ or $r_{1Y.2}$ as "the partial correlation between X_1 and Y, controlling for X_2." However, changes in the position of variables (before versus after the period) do reflect

Correlations

grade = 1

Correlations[a]

		height	vocabulary
height	Pearson Correlation	1	.067
	Sig. (2-tailed)		.806
	N	16	16
vocabulary	Pearson Correlation	.067	1
	Sig. (2-tailed)	.806	
	N	16	16

a. grade = 1

grade = 5

Correlations[a]

		height	vocabulary
height	Pearson Correlation	1	.031
	Sig. (2-tailed)		.909
	N	16	16
vocabulary	Pearson Correlation	.031	1
	Sig. (2-tailed)	.909	
	N	16	16

a. grade = 5

grade = 9

Correlations[a]

		height	vocabulary
height	Pearson Correlation	1	-.141
	Sig. (2-tailed)		.603
	N	16	16
vocabulary	Pearson Correlation	-.141	1
	Sig. (2-tailed)	.603	
	N	16	16

a. grade = 9

Figure 10.7 ♦ Height (X_1) and Vocabulary (Y) Correlations for Each Level of Grade (X_2)

a difference in their treatment. For example, we would read $r_{Y2.1}$ as "the partial correlation between X_2 and Y, controlling for X_1."

Another common notation for partial correlation is pr_1. The subscript 1 associated with pr_1 tells us that the partial correlation is for the predictor variable X_1. In this notation, it is implicit that the dependent variable is Y and that other predictor variables, such as X_2, are statistically controlled. Thus, pr_1 is the partial correlation that describes the predictive relation of X_1 to Y when X_2 (and possibly other additional variables) is controlled for.

10.6 ♦ Understanding Partial Correlation as the Use of Bivariate Regression to Remove Variance Predictable by X_2 From Both X_1 and Y

One way to calculate the partial r between X_1 and Y, controlling for X_2, is to carry out the following series of simple and familiar analyses. First, use bivariate regression to obtain the residuals for the prediction of X_1 from X_2; these residuals (X_1^*) represent the parts of the X_1 scores that are not predictable from X_2 or are not correlated with X_2. Second, use bivariate regression to obtain the residuals for the prediction of Y from X_2; these residuals (Y^*) represent the parts of the Y scores that are not predictable from or correlated with X_2. Third, compute a Pearson correlation between the X_1^* and Y^* residuals. This Pearson correlation is equivalent to the partial correlation $r_{Y1.2}$; it tells us how strongly X_1 is correlated with Y when the variance that is predictable from the control variable X_2 has been removed from both X_1 and Y. This method using correlations of regression residuals is not the most convenient computational method, but the use of this approach provides some insight into what it means to statistically control for X_2.

X_1 and Y are the variables of interest. X_2 is the variable we want to statistically control for; we want to remove the variance that is associated with or predictable from X_2 from both the X_1 and the Y variables.

First, we perform a simple bivariate regression to predict X_1 from X_2: X_1' is the predicted score for X_1 based on X_2; that is, X_1' is the part of X_1 that is related to or predictable from X_2:

$$X_1' = b_0 + bX_2. \tag{10.1}$$

The residuals from this regression, denoted by X_1^*, are calculated by finding the difference between the actual value of X_1 and the predicted value of X_1' for each case: $X_1^* = (X_1 - X_1')$. The X_1^* residual is the part of the X_1 score that is *not* predictable from or related to X_2.

Next, we perform a similar regression to predict Y from X_2:

$$Y' = b_0 + bX_2. \tag{10.2}$$

Then, we take the residuals Y^*, where $Y^* = (Y - Y')$. Y^* gives us the part of Y that is *not* related to or predictable from X_2. (Note that the b_0 and b coefficients will have different numerical values for the regressions in Equations 10.1 and 10.2.)

This method can be used to compute the partial correlation between height and vocabulary, controlling for grade level, for the data in the previous research example, where X_1 = height, Y = vocabulary, and X_2 is the control variable grade.

As described in Chapter 9, we can run the SPSS regression procedure and use the <Save> command to save computational results, such as the unstandardized residuals for each case; these appear as new variables in the SPSS worksheet. Figure 10.8 shows the SPSS Regression dialog window to run the regression that is specified in Equation 10.1 (to predict X_1 from X_2—in this example, height from grade). Figure 10.9 shows the SPSS Data View worksheet after performing the regressions in Equations 10.1 (predicting height from grade) and 10.2 (predicting vocabulary score from grade).The residuals from these two separate regressions were saved as new variables and renamed. RES_1, renamed Resid_Height, refers to the part of the scores on the X_1 variable, height, that were not predictable from the control or X_2 variable, grade; RES_2, renamed Resid_Voc, refers to the part of the scores on the Y variable, vocabulary, that were not predictable from the control variable, grade. Resid_Height corresponds to X_1^*, and Resid_Voc corresponds to Y^* in the previous general description of this analysis (Figure 10.10). Finally, we can obtain the bivariate Pearson correlation between these two new variables, Resid_Height and Resid_Voc (X_1^* and Y^*). The correlation between these residuals, $r = -.012$ in Figure 10.11, corresponds to the value of the partial correlation between X_1 and Y, controlling for or partialling out X_2. Note that X_2 is partialled out or removed from both variables. This partial $r = -.012$ tells us that X_1 (height) is not significantly correlated with Y (vocabulary) when variance that is predictable from grade level (X_2) has been removed from or partialled out of *both* the X_1 and the Y variables.

10.7 ♦ Computation of Partial *r* From Bivariate Pearson Correlations

There is a simpler direct method for the computation of the partial r between X_1 and Y, controlling for X_2, based on the values of the three bivariate correlations

r_{Y1}, the correlation between Y and X_1,

r_{Y2}, the correlation between Y and X_2, and

r_{12}, the correlation between X_1 and X_2.

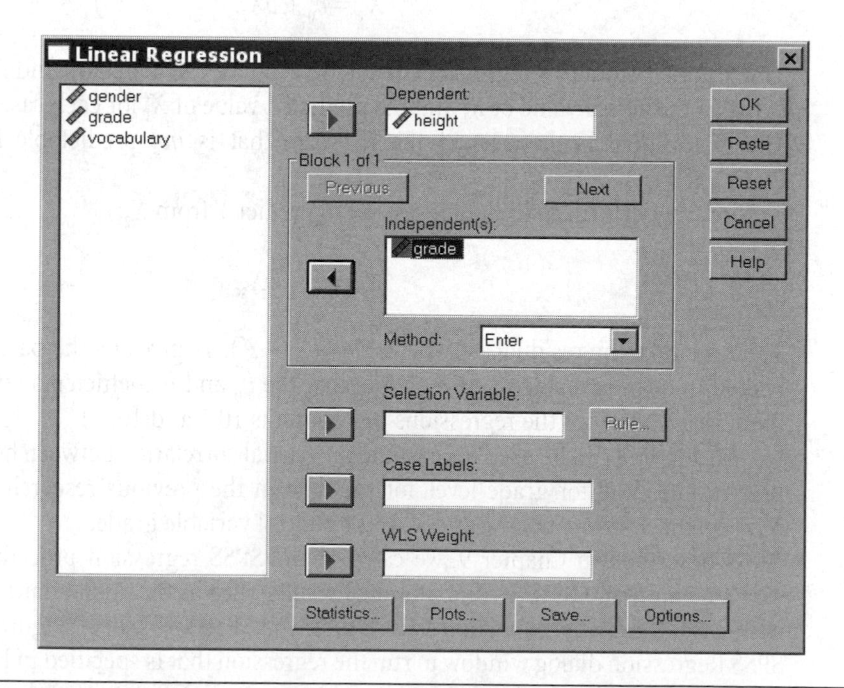

Figure 10.8 ♦ Bivariate Regression to Predict Height (X_1) From Grade in School (X_2)

NOTE: The unstandardized residuals from this regression were saved as RES_1 and renamed Resid_Height. A bivariate regression was also performed to predict vocabulary from grade; the residuals from this regression were saved as RES_2 and then renamed Resid_Voc.

partialrexamp [DataSet1] - SPSS Data Editor

	gender	grade	height	vocabulary	RES_1	RES_2
1	1	1	46.00	43.00	-2.69792	1.63542
2	1	1	48.00	45.00	-.69792	3.63542
3	1	1	48.00	30.00	-.69792	-11.36458
4	1	1	48.00	18.00	-.69792	-23.36458
5	1	1	46.00	50.00	-2.69792	8.63542
6	1	1	47.00	53.00	-1.69792	11.63542
7	1	1	48.00	46.00	-.69792	4.63542
8	1	1	53.00	33.00	4.30208	-8.36458
9	2	1	49.00	47.00	.30208	5.63542
10	2	1	53.00	54.00	4.30208	12.63542
11	2	1	49.00	28.00	.30208	-13.36458
12	2	1	49.00	51.00	.30208	9.63542
13	2	1	51.00	37.00	2.30208	-4.36458
14	2	1	47.00	33.00	-1.69792	-8.36458
15	2	1	49.00	42.00	.30208	.63542
16	2	1	52.00	50.00	3.30208	8.63542
17	1	5	54.00	51.00	-.85417	-5.14583
18	1	5	54.00	79.00	-.85417	22.85417
19	1	5	54.00	49.00	-.85417	-7.14583
20	1	5	52.00	50.00	-2.85417	-6.14583
21	1	5	58.00	66.00	3.14583	9.85417
22	1	5	52.00	65.00	-2.85417	8.85417
23	1	5	53.00	44.00	-1.85417	-12.14583
24	1	5	55.00	73.00	.14583	16.85417
25	2	5	52.00	51.00	-2.85417	-5.14583
26	2	5	55.00	47.00	.14583	-9.14583
27	2	5	54.00	65.00	-.85417	8.85417
28	2	5	55.00	52.00	.14583	-4.14583
29	2	5	53.00	50.00	-1.85417	-6.14583
30	2	5	58.00	48.00	3.14583	-8.14583
31	2	5	59.00	53.00	4.14583	-3.14583
32	2	5	52.00	59.00	-2.85417	2.85417
33	1	9	57.00	73.00	-4.01042	2.07292
34	1	9	63.00	62.00	1.98958	-8.92708

Figure 10.9 ♦ SPSS Data View Worksheet

NOTE: RES_1 and RES_2 are the saved unstandardized residuals for the prediction of height from grade and vocabulary from grade; these were renamed Resid_Height and Resid_Voc.

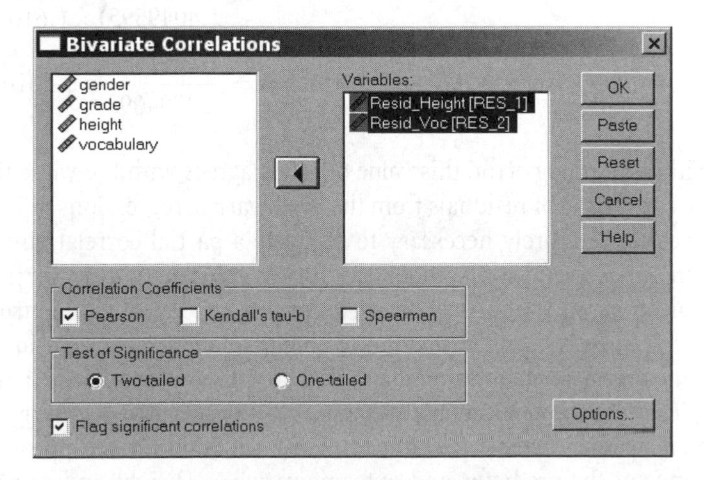

Figure 10.10 ♦ Correlation Between Residuals for Prediction of Height From Grade and Residuals for Prediction of Vocabulary From Grade

Correlations

		Resid_Height	Resid_Voc
Resid_Height	Pearson Correlation	1	-.012
	Sig. (2-tailed)		.937
	N	48	48
Resid_Voc	Pearson Correlation	-.012	1
	Sig. (2-tailed)	.937	
	N	48	48

Figure 10.11 ♦ Correlations Between Residuals of Regressions in Which the Control Variable Grade (X_2) Was Used to Predict Scores on the Other Variables (X_1, Height and Y, Vocabulary)

NOTE: The variable Resid_Height contains the residuals from the bivariate regression to predict height (X_1) from grade (X_2). The variable Resid_Voc contains the residuals from the bivariate regression to predict vocabulary (Y) from grade (X_2). These residuals correspond to the parts of the X_1 and Y scores that are not related to or not predictable from grade (X_2).

The formula to calculate the partial r between X_1 and Y, controlling for X_2, directly from the Pearson correlations is as follows:

$$pr_1 = r_{Y1.2} = \frac{r_{1Y} - (r_{12} \times r_{2Y})}{\sqrt{1 - r_{12}^2}\sqrt{1 - r_{2Y}^2}}. \quad (10.3)$$

In the preceding example, where X_1 = height, Y = vocabulary, and X_2 = grade, the corresponding bivariate correlations were $r_{1Y} = +.716$, $r_{2Y} = +.787$, and $r_{12} = +.913$. If these values are substituted into Equation 10.3, the partial correlation $r_{Y1.2}$ is as follows:

$$\frac{+.716 - (.913 \times .787)}{\sqrt{1 - .913^2}\sqrt{1 - .787^2}} = \frac{.716 - .71853}{\sqrt{.166431}\sqrt{.380631}}$$

$$= \frac{-.00253}{(.4049595) \times (.6169529)}$$

$$= \frac{-.00253}{.2498409} \approx -.010.$$

Within rounding error, this value of −.010 agrees with the value that was obtained from the correlation of residuals from the two bivariate regressions reported in Figure 10.11. In practice, it is rarely necessary to calculate a partial correlation by hand. However, it is sometimes useful to use Equation 10.3 to calculate partial correlations as a secondary analysis based on tables of bivariate correlations reported in journal articles.

The most convenient method of obtaining a partial correlation when you have access to the original data is using the Partial Correlations procedure in SPSS.

The SPSS menu selections, <Analyze> → <Correlate> → <Partial>, shown in Figure 10.12 open up the Partial Correlations dialog window, which appears in Figure 10.13. The names of the predictor and outcome variables (height and vocabulary) are entered in the window that is headed Variables. The name of the control variable, grade, is entered in the window under the heading "Controlling for." (Note that more than one variable can be placed in this window; that is, we can include more than one control variable.) The output for this procedure appears in Figure 10.14, where the value of the partial correlation between height and vocabulary, controlling for grade, is given as $r_{1Y.2} = -.012$; this partial correlation is not significantly different from 0 (and is identical to the correlation between Resid_Height and Resid_Voc reported in Figure 10.11).

*partialrexamp [DataSet1] - SPSS Data Editor

File Edit View Data Transform Analyze Graphs Utilities Add-ons Window Help

Analyze: Reports ▸ / Descriptive Statistics ▸ / Compare Means ▸ / General Linear Model ▸ / Correlate ▸ / Regression ▸ / Classify ▸ / Data Reduction ▸ / Scale ▸ / Nonparametric Tests ▸ / Multiple Response ▸

Correlate: Bivariate... / Partial... / Distances...

1 : gender 1

	gender	grade		
1	1	1		
2	1	1		
3	1	1		
4	1	1		
5	1	1		
6	1	1		
7	1	1		
8	1	1	[illegible]	[illegible]
9	2	1	49.00	47.00
10	2	1	53.00	54.00
11	2	1	49.00	28.00
12	2	1	49.00	51.00
13	2	1	51.00	37.00
14	2	1	47.00	33.00
15	2	1	49.00	42.00
16	2	1	52.00	50.00
17	1	2	54.00	51.00
18	1	2	54.00	79.00
19	1	2	54.00	49.00
20	1	2	52.00	50.00
21	1	2	58.00	66.00
22	1	2	52.00	65.00
23	1	2	53.00	44.00
24	1	2	55.00	73.00
25	2	2	52.00	51.00
26	2	2	55.00	47.00
27	2	2	54.00	65.00
28	2	2	55.00	52.00
29	2	2	53.00	50.00
30	2	2	58.00	48.00
31	2	2	59.00	53.00
32	2	2	52.00	59.00
33	1	3	57.00	73.00
34	1	3	63.00	62.00

Data View / Variable View

Partial — SPSS Processor is ready

Start | c10_partcorr | *partialrex... | Output1 - SP... | Chart Editor | ch10figures -... | 1:55 PM

Figure 10.12 ♦ SPSS Data View Worksheet With Menu Selections for Partial Correlation

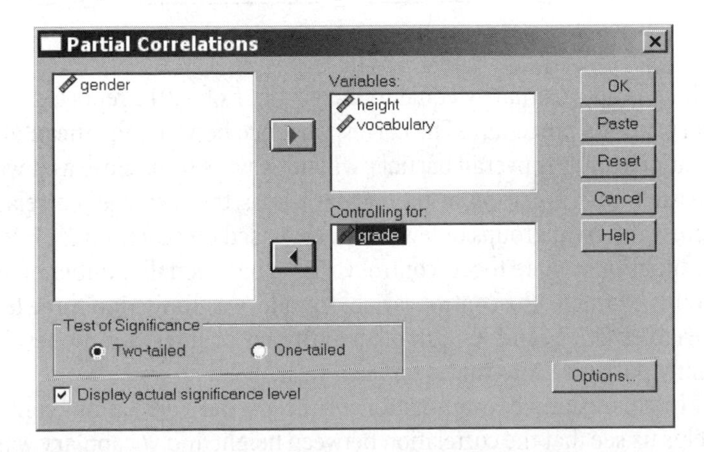

Figure 10.13 ♦ SPSS Dialog Window for the Partial Correlations Procedure

Correlations

Control Variables			height	vocabulary
grade	height	Correlation	1.000	-.012
		Significance (2-tailed)	.	.938
		df	0	45
	vocabulary	Correlation	-.012	1.000
		Significance (2-tailed)	.938	.
		df	45	0

Figure 10.14 ♦ Output From the SPSS Partial Correlations Procedure: First-Order Partial Correlation Between Height (X_1) and Vocabulary (Y), Controlling for Grade Level (X_2)

NOTE: The partial correlation between height and vocabulary, controlling for grade ($r = -.012$), is identical to the correlation between Resid_Height and Resid_Voc ($r = -.012$), which appeared in Figure 10.11.

10.8 ♦ Intuitive Approach to Understanding Partial *r*

In Chapter 7, we saw how the value of a bivariate correlation r_{1Y} is related to the pattern of points in the X_1, Y scatter plot. In some three-variable research situations, we can understand the outcome for partial correlation by looking at the pattern of points in the X_1, Y scatter plot. The partial correlation between X_1 and Y, controlling for X_2, is *approximately* the average of the r_{1Y} values obtained by correlating X_1 and Y separately within each group defined by scores on the X_2 variable. (This is not an exact computational method for partial correlation; however, thinking about partial correlation in this way sometimes helps us understand the way the three variables are interrelated.) In Figure 10.7, the correlations between height and vocabulary for Grades 1, 5, and 9 were $r_1 = .067$, $r_5 = .031$, and $r_9 = -.141$, respectively. If you compute the average of these three within-group correlations, the mean of the within-group correlations *approximately* equals the partial correlation reported earlier for these data:

$$\frac{r_1 + r_5 + r_9}{3} = \frac{.067 + .031 - .141}{3} = -.014,$$

which is approximately equal to the partial r of $-.012$ reported by SPSS from the Partial Correlations procedure. The correspondence between the mean of the within-group correlations and the overall partial r will not always be as close as it was in this example. One possible interpretation of the partial r is as the "average" correlation between X_1 and Y across different groups or levels that are based on scores on X_2. It is easier to visualize this in situations where the X_2 control variable has a small number of levels or groups, and so, in this example, the control variable, grade, was limited to three levels. However, all three variables (X_1, Y, and X_2) can be quantitative or interval/ratio level of measurement, with many possible score values for each variable.

In this example, examination of the partial correlation, controlling for grade level, helps us see that the correlation between height and vocabulary was completely due to the fact that both height and vocabulary increased from Grade 1 to 5 to 9. Height and vocabulary increase with age, and the observed zero-order correlation between height and vocabulary arose only because each of these variables is related to the third variable X_2 (grade level or age). When we statistically control for grade level, the X_1, Y relationship disappears: That is, for children who are within the same grade level, there is no relation between height and vocabulary. In this situation, we could say that the X_1, Y correlation was "completely accounted for" or "completely explained away" by the X_2 variable. The apparent association between height and vocabulary is completely explained away by the association of these two variables with grade level. We could also say that the X_1, Y or height, vocabulary correlation was "completely spurious." A correlation between two variables is said to be spurious if there is no direct association between X_1 and Y and if X_1 and Y are correlated only because both these variables are caused by or are correlated with some other variable (X_2).

In the example above, partial correlation provides an understanding of the way X_1 and Y are related that is quite different from that obtained using the simple zero-order correlation. If we control for X_2, and the partial correlation between X_1 and Y, controlling for X_2, $r_{Y1.2}$, is equal to or close to 0, we may conclude that X_1 and Y are not directly related.

10.9 ♦ Significance Tests, Confidence Intervals, and Statistical Power for Partial Correlations

10.9.1 ♦ Statistical Significance of Partial *r*

The null hypothesis that a partial correlation equals 0 can be tested by setting up a *t* ratio that is similar to the test for the statistical significance of an individual zero-order Pearson correlation. The SPSS Partial Correlations procedure provides this statistical significance test; SPSS reports an exact *p* value for the statistical significance of partial *r*. The degrees of freedom (*df*) for a partial correlation are $N - k$, where k is the total number of variables that are involved in the partial correlation and N is the number of cases or participants.

10.9.2 ♦ Confidence Intervals for Partial *r*

Most textbooks do not present detailed formulas for standard errors or confidence intervals for partial correlations. Olkin and Finn (1995) provided formulas for computation of the standard error for partial correlations; however, the formulas are complicated and not easy to work with by hand. SPSS does not provide standard errors or confidence interval estimates for partial correlations. SPSS add-on programs, such as ZumaStat (www.zumastat.com), can be used to obtain confidence intervals for partial correlations.

10.9.3 ♦ Effect Size, Statistical Power, and Sample Size Guidelines for Partial *r*

Like Pearson r (and r^2), the partial correlation $r_{Y1.2}$ and squared partial correlation $r^2_{Y1.2}$ can be interpreted directly as information about effect size or strength of association between variables. The effect-size labels for the values of Pearson r and r^2 that appeared in Table 5.2 can reasonably be used to describe the effect sizes that correspond to partial correlations.

In terms of statistical power and sample size, the guidelines about sample size needed to detect zero-order Pearson correlations provided in Chapter 7 can be used to set lower limits for the sample size requirements for partial correlation. However, when a researcher wants to explore the association between X_1 and Y separately for groups that have different scores on an X_2 control variable, the minimum N that might be required to do a good job may be much higher. Table 7.4 provides approximate sample sizes needed to achieve various levels of statistical power for different population effect sizes; the population effect size in this table is given in terms of ρ^2. For example, based on Table 7.5, a researcher who believes that the population effect size (squared correlation) for a pair of variables X_1 and Y is of the order of $\rho^2 = .25$ would need a sample size of $N = 28$ to have statistical power of about .80 for a zero-order Pearson correlation between X_1 and Y.

Assessment of the relationship between X_1 and Y that takes a control variable X_2 into account should probably use a larger N than the minimum value suggested in Table 7.5. If the X_2 control variable is SES and the X_2 variable has three different values, the population effect size for the X_1, Y relationship is of the order of $\rho^2_{Y1.2} = .25$, and the researcher wants to obtain good quality information about the nature of the relationship between X_1 and Y separately within each level of SES, it would be helpful to have a minimum sample size of $N = 28$ within *each* of the three levels of SES: in other words, a total N of $3 \times 28 = 84$. Using a larger value of N (which takes the number of levels of the X_2 variable into account) may help the researcher obtain a reasonably good assessment of the nature and significance of the relationship between X_1 and Y within each group, based on scores on the X_2 variable. However, when the X_2 control variable has dozens or hundreds of possible values, it may not be possible to have a sufficiently large N for each possible value of X_2 to provide an accurate description of the nature of the relationship between X_1 and Y at each level of X_2. This issue should be kept in mind when interpreting results from studies where the control variable has a very large number of possible score values.

10.10 ♦ Interpretation of Various Outcomes for $r_{Y1.2}$ and r_{Y1}

When we compare the size and sign of the zero-order correlation between X_1 and Y with the size and sign of the partial correlation between X_1 and Y, controlling for X_2, several different outcomes are possible. The following sections identify some possible interpretations for each of these outcomes. The value of r_{1Y}, the zero-order correlation between X_1 and Y, can range from −1 to +1. The value of $r_{1Y.2}$, the partial correlation between X_1 and Y, controlling for X_2, can also range from −1 to +1, and in principle, any combination of values of r_{1Y} and $r_{1Y.2}$ can occur (although some outcomes are much more common than others). The diagram in Figure 10.15 shows the square that corresponds to all possible combinations of values for r_{1Y} and $r_{1Y.2}$; this is divided into regions labeled *a*, *b*, *c*, *d*, and *e*. The combinations of r_{1Y} and $r_{1Y.2}$ values that correspond to each of these regions in Figure 10.15 have different possible interpretations, as discussed in the next few sections.

For each region that appears in Figure 10.15, the corresponding values of the zero-order correlation and partial correlation and the nature of possible interpretations are as follows:

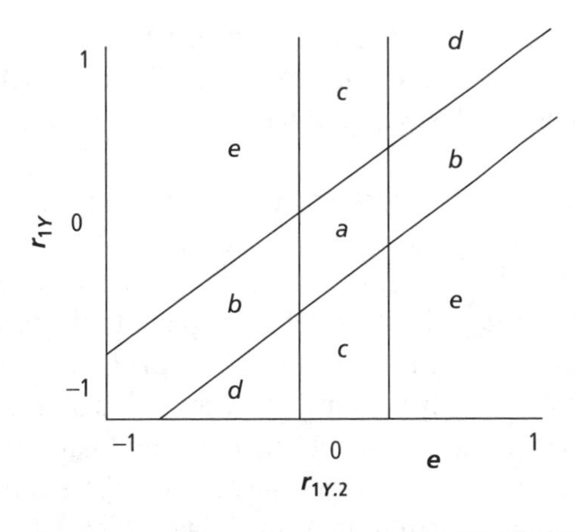

Figure 10.15 ♦ Possible Values for the Partial Correlation $r_{1Y.2}$ (on the Horizontal Axis) and the Zero-Order Pearson Correlation r_{1Y} (on the Vertical Axis)

SOURCE: Adapted from Davis (1971).

Area a. This corresponds to outcomes where the zero-order r_{Y1} is approximately equal to 0 and the partial $r_{Y1.2}$ is also approximately equal to 0. Based on this outcome, we might conclude that X_1 is not (linearly) related to Y whether X_2 is controlled or not. (Note that if the X_1, Y scatter plot shows an indication of a curvilinear relation, a different statistic may be needed to describe the corresponding relationship.)

Area b. This corresponds to outcomes where the value of the partial $r_{Y1.2}$ is approximately equal to the value of the zero-order r_{1Y}, and both the partial $r_{Y1.2}$ and the zero-order correlation are not equal to 0. Based on this outcome, we might conclude that controlling for X_2 does not change the apparent strength and nature of the relationship of X_1 with Y; we might say that X_2 is "irrelevant" to understanding the X_1, Y relation.

Area c. This corresponds to the outcome where the zero-order correlation r_{Y1} is significantly different from 0 but the partial correlation $r_{Y1.2}$ is not significantly different from 0. In other words, X_1 and Y appear to be related if you ignore X_2, but their relationship disappears when X_2 is statistically controlled. There are at least two different interpretations for this outcome. One possible interpretation is that the correlation between X_1 and Y is spurious; both X_1 and Y might have been caused by X_2 or might be correlated with X_2. In this situation, we might say that the X_2 variable completely "accounts for" or "explains away" the apparent association between X_1 and Y. Alternatively, it is possible that X_2 mediates a causal relationship between X_1 and Y. That is, there could be a causal chain such that first X_1 causes X_2 and then X_2 causes Y. If X_1 affects Y *only* through its intermediate effect on X_2, we might say that its effect on Y is completely mediated by X_2. Unfortunately, correlation analysis alone cannot determine which of these two interpretations for Area *c* outcomes is more appropriate. Unless there is additional evidence to support the plausibility of a mediated causal sequence, it is more conservative to interpret this outcome as evidence of a spurious correlation between X_1 and Y. (Chapter 11 will present methods that provide a more rigorous test for the mediated **causal model**, but even these methods will not lead to conclusive proof that the variables are causally associated.)

Area d. This outcome occurs when the partial correlation, $r_{Y1.2}$, is smaller than the zero-order correlation, r_{1Y} and has the same sign as r_{1Y}, but the partial correlation is significantly different from 0. This is a very common outcome. (Note that the likelihood of outcomes *a, b, c, d,* and *e* does not correspond to the proportion of areas that these outcomes occupy in the graph in Figure 10.15.) For outcomes that correspond to Area *d*, we might say that X_2 "partly explains" the relationship of X_1 to Y; when X_2 is controlled, X_1 and Y are still related, but the relationship is weaker than when X_2 is ignored. These values for partial and zero-order correlation suggest that there could be both a direct path from X_1 to Y and an indirect path from X_1 to Y via X_2. Alternatively, under certain conditions, we might also interpret this outcome as being consistent with the theory that X_2 "partly mediates" a causal connection between X_1 and Y.

Area e. This outcome occurs when the partial correlation $r_{Y1.2}$ is opposite to the zero-order r_{1Y} in sign and/or when $r_{Y1.2}$ is larger than r_{1Y} in absolute value. In other words, controlling for X_2 either makes the X_1, Y relationship stronger or changes the direction of the relationship. When this happens, we often say that X_2 is a **suppressor variable**; that is, the effect of X_2 (if it is ignored) is to suppress or alter the apparent relationship between X_1 and Y; only when X_2 is controlled do you see the "true" relationship. Although Area *e* corresponds to a large part of the diagram in Figure 10.15, this is not a very common outcome, and the interpretation of this type of outcome can be difficult.

The preceding discussion mentions some possible interpretations for outcomes for partial *r* that involve noncausal associations between variables and other interpretations for values of partial *r* that involve hypothesized causal associations between variables. Of course, the existence of a significant zero-order or partial correlation between a pair of variables is not conclusive evidence that the variables are causally connected. However, researchers sometimes do have causal hypotheses in mind when they select the variables to measure in nonexperimental studies, and the outcomes from correlational analyses can be interpreted as (limited) information about the plausibility of some possible causal hypotheses. Readers who do not expect to work with causal models in any form may want to skip Sections 10.11 through 10.13. However, those who plan to study more advanced methods, such as structural equation modeling, will find the introduction to causal models presented in the next few sections useful background for these more advanced analytic methods.

The term *model* can have many different meanings in different contexts. In this textbook, the term *model* generally refers to either (a) a theory about possible causal and noncausal associations among variables, often presented in the form of path diagrams, or (b) an equation to predict Y from scores on one or more X predictor variables; for example, a regression to predict Y from scores on X_1 can be called a regression model. The choice of variables to include as predictors in regression equations is sometimes (although not always) guided by an implicit causal model. The following sections briefly explain the nature of the "causal" models that can be hypothesized to describe the relationships between two variables or among three variables.

10.11 ♦ Two-Variable Causal Models

One possible model for the association between an X_1 predictor and a Y outcome variable is that the variables are not associated, either causally or noncausally. The first row in Table 10.2 illustrates this situation. If r_{1Y} is equal to or nearly equal to 0, we do not need to include a **causal** or **noncausal path** between X_1 and Y in a path model that shows how X_1 and Y are related. If X_1 and Y are not systematically related; for example, if X_1 and Y have a correlation of 0, they probably are not related either causally or noncausally, and the causal model does not need to include a direct path between X_1 and Y.[1]

What statistical result would be evidence consistent with a causal model that assumes no relation between X_1 and Y? Let's assume that X_1 and Y are both quantitative variables and that all the assumptions for Pearson r are satisfied. In this situation, we can use Pearson r to evaluate whether scores on X_1 and Y are systematically related. If we obtain a correlation between X_1 and Y of $r_{1Y} = 0$ or $r_{1Y} \approx 0$, we would tentatively conclude that there is no association (either causal or noncausal) between X_1 and Y.

There are three possible ways in which X_1 and Y could be associated. First, X_1 and Y could be noncausally associated; they could be correlated or confounded but with neither variable being a cause of the other. The second row in Table 10.2 illustrates the path model for this situation; when X_1 and Y are noncausally associated, this noncausal association is represented by a bidirectional arrow. In some textbooks, the arrow that represents a noncausal association is straight rather than curved (as shown here). It does not matter whether the arrow is straight or curved; the key thing to note is whether the arrow is bidirectional or unidirectional. Second, we might hypothesize that X_1 causes Y; this causal hypothesis is represented by a unidirectional arrow or path that points away from the cause (X_1) toward the outcome (Y), as shown in the third row of Table 10.2. Third, we might hypothesize that Y causes X_1; this causal hypothesis is represented by a unidirectional arrow or path that leads from Y to X_1, as shown in the last row of Table 10.2.

Any hypothesized theoretical association between X_1 and Y (whether it is noncausal, whether X_1 causes Y, or whether Y causes X_1) should lead to the existence of a systematic statistical association (such as a Pearson r significantly different from 0) between X_1 and Y. In this example, where Pearson r is assumed to be an appropriate index of association, variables that are either causally or noncausally associated with each other should have correlations that are significantly different from 0. However, when we find a nonzero correlation between X_1 and Y, it is difficult to decide which one of these three models (X_1 and Y are noncausally related, X_1 causes Y, or Y causes X_1) provides the correct explanation. To further complicate matters, the relationship between X_1 and Y may involve additional variables; X_1 and Y may be correlated with each other because they are both causally influenced by (or noncausally correlated with) some third variable, X_2, for example.

Table 10.2 ♦ Four Possible Hypothesized Paths Between Two Variables (X_1 and Y)

Verbal Description of the Relationship Between X_1 and Y	*Path Model for X_1 and Y*	*Comment on Path Model*	*Corresponding Value of the r_{1Y} Correlation*
X_1 and Y are not associated in any way (either causally or noncausally)	X_1 Y	No arrow or path between X_1 and Y.	$r_{1Y} = 0$ or $r_{1Y} \approx 0$
X_1 and Y are associated but not in a causal way. X_1 and Y co-occur, or are confounded, but neither variable is the cause of the other	$X_1 \leftrightarrow Y$ (curved bidirectional arrow)	Bidirectional arrow or path between X_1 and Y. We use a bidirectional path when our theory says that X_1 is predictive of Y, or is correlated with Y, but X_1 is not a cause of Y.	$r_{1Y} \neq 0$
X_1 is a cause of Y	$X_1 \rightarrow Y$	Unidirectional arrow that points from the cause (X_1) toward the outcome or effect (Y) or path from X_1 to Y. We use this unidirectional "causal" path when our theory involves the hypothesis that X_1 causes Y.	$r_{1Y} \neq 0$
Y is a cause of X_1	$Y \rightarrow X_1$	Unidirectional arrow that points from the cause (Y) toward the outcome or effect (X_1). We use this unidirectional causal path when our theory involves the hypothesis that Y causes X_1.	$r_{1Y} \neq 0$

Although we can propose theoretical models that involve hypothesized causal connections, the data that we collect in typical nonexperimental studies only yield information about correlations. Correlations can be judged consistent or inconsistent with hypothetical causal models; however, finding correlations that are consistent with a specific causal model does not constitute proof of that particular model. Usually, there are several causal models that are equally consistent with an observed set of correlations.

If we find that the correlation r_{1Y} is not significantly different from 0 and/or is too small to be of any practical or theoretical importance, we would interpret that as evidence that that particular situation is more consistent with a model that has no path between X_1 and Y (as in the first row of Table 10.2) than with any of the other three models that do have paths between X_1 and Y. However, obtaining a value of r_{1Y} that is too small to be statistically significant does not conclusively rule out models that involve causal connections between X_1 and Y. We can obtain a value of r_{1Y} that is too small to be statistically significant in situations where X_1 really does influence Y, but this small correlation could be either due to sampling error or due to artifacts such as attenuation of correlation due to unreliability of the measurement or due to a nonlinear association between X_1 and Y.

On the other hand, if we find that the r_{1Y} correlation is statistically significant and large enough to be of some practical or theoretical importance, we might tentatively decide that this outcome is more consistent with one of the three models that include a path between X_1 and Y (as shown in Rows 2–4 of Table 10.2) than with the model that has no path between X_1 and Y (as in Row 1 of Table 10.2). A significant r_{1Y} correlation suggests that there may be a direct (or an indirect) path between X_1 and Y; however, we would need additional information to decide which hypothesis is the most plausible: that X_1 causes Y, that Y causes X_1, that X_1 and Y are noncausally associated, or that X_1 and Y are connected through their relationships with other variables.

In this chapter, we shall also see that when we take a third variable (X_2) into account, the correlation between X_1 and Y can change in many different ways. For a correlation to provide accurate information about the nature and strength of the association between X_1 and Y, we must have a correctly specified model. A model is correctly specified if it includes all the variables that need to be taken into account (because they are involved in causal or noncausal associations between X_1 and Y) and, also, if it does not include any variables that should not be taken into account. A good theory can be helpful in identifying the variables that should be included in a model; however, we can never be certain that the hypothetical causal model that we are using to select variables is correctly specified; it is always possible that we have omitted a variable that should have been taken into account.

For all these reasons, we cannot interpret correlations or partial correlations as either conclusive proof or conclusive disproof of a specific causal model. We can only evaluate whether correlation and partial correlation values are consistent with the results we would expect to obtain given different causal models. Experimental research designs provide more rigorous means to test causal hypotheses than tests that involve correlational analysis of nonexperimental data.

10.12 ♦ Three-Variable Models: Some Possible Patterns of Association Among X_1, Y, and X_2

When there were just two variables (X_1 and Y) in the model, there were four different possible hypotheses about the nature of the relationship between X_1 and Y—namely, X_1 and Y are not directly related either causally or noncausally, X_1 and Y are noncausally associated or confounded, X_1 causes Y, or Y causes X_1 (as shown in Table 10.2). When we expand a model to include a third variable, the number of possible models that can be considered becomes much larger. There are three pairs of variables

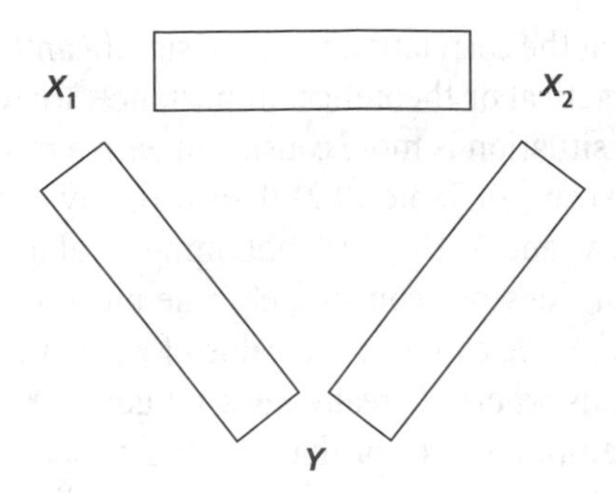

Figure 10.16 ♦ The Set of All Logically Possible Hypothesized "Causal" Models for a Set of Three Variables (X_1, X_2, and Y).

NOTE: The set can be obtained by filling in each of the rectangles with one of the four types of paths described in Table 10.2

(X_1 and X_2, X_1 and Y, and X_2 and Y), and each pair of variables can be related in any of the four ways that were described in Table 10.2. Each rectangle in Figure 10.16 can be filled in with any of the four possible types of path (no relation, noncausal association, or two different directions of cause). The next few sections of this chapter describe some of the different types of causal models that might be proposed as hypotheses for the relationships among three variables.

Using data from a nonexperimental study, we cannot prove or disprove any of these models. However, we can interpret some outcomes for correlation and partial correlation as being either consistent with or not consistent with some of the logically possible models. This may make it possible, in some research situations, to reduce the set of models that could be considered as plausible explanations for the relationships among variables.

The causal models that might be considered reasonable candidates for various possible combinations of values of r_{Y1} and $r_{Y1.2}$ are discussed in greater detail in the following sections.

10.12.1 ♦ X_1 and Y Are Not Related Whether You Control for X_2 or Not

One possible hypothetical model is that none of the three variables (X_1, X_2, and Y) is either causally or noncausally related to the others (see Figure 10.17). If we obtain Pearson r values for r_{12}, r_{1Y}, and r_{2Y} that are not significantly different from 0 (and all the three correlations are too small to be of any practical or theoretical importance), those correlations would be consistent with a model that has no paths among any of the three pairs of variables, as shown in Figure 10.17. The partial correlation between X_1 and Y, controlling for X_2, would also be 0 or very close to 0 in this situation. A researcher who

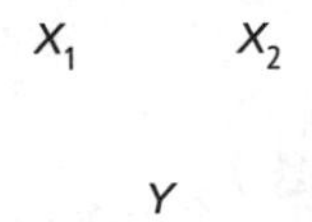

X_1 X_2

Y

Figure 10.17 ♦ No Direct Paths (Either Causal or Noncausal) Between Any Pairs of Variables

obtains values close to 0 for all the bivariate (and partial) correlations would probably conclude that none of the variables is related to the others either causally or noncausally. (Of course, this conclusion could be incorrect if the variables are related nonlinearly, because Pearson r is not appropriate to assess the strength of nonlinear relations.)

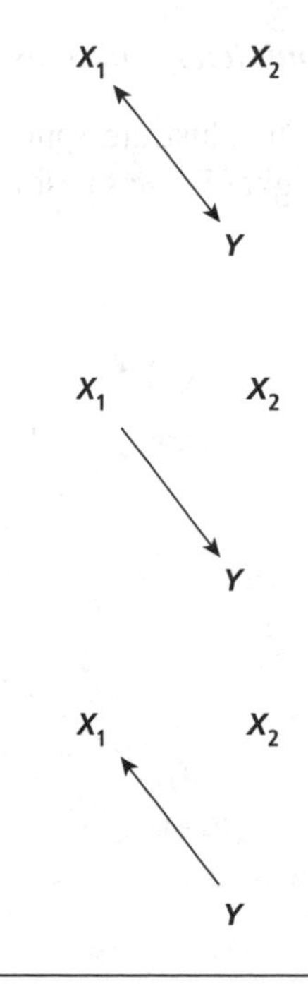

Figure 10.18 ♦ A Theoretical Causal Model in Which X_1 Is Either Causally or Noncausally Related to Y and X_2 Is Not Related to Either X_1 or Y

NOTE: Possible interpretation: X_2 is "irrelevant" to the X_1, Y relationship. It would be reasonable to hypothesize that there is some direct path between X_1 and Y, but we would need information beyond the existence of a significant correlation to evaluate whether the correlation occurred because X_1 and Y are noncausally associated, because X_1 causes Y, or because Y causes X_1. For the pattern of associations shown in these three-path models, we would expect that $r_{1Y} \neq 0$, $r_{12} \approx 0$, $r_{2Y} \approx 0$, and $r_{1Y.2} \approx r_{1Y}$.

10.12.2 ♦ X_2 Is Irrelevant to the X_1, Y Relationship

A second possible theoretical model is that X_1 is either causally or noncausally related to Y and that the X_2 variable is "irrelevant" to the X_1, Y relationship. If this model is correct, then we should obtain a statistically significant correlation between X_1 and Y (large enough to be of some practical or theoretical importance). The correlations of X_1 and Y with X_2 should be 0 or close to 0. The partial correlation between X_1 and Y, controlling for X_2, should be approximately equal to the zero-order correlation between X_1 and Y; that is, $r_{1Y.2} \approx r_{1Y}$. Figure 10.18 shows three different hypothetical causal models that would be logically consistent with this set of correlation values.

10.12.3 ♦ When You Control for X_2, the X_1, Y Correlation Drops to 0 or Close to 0

As mentioned earlier, there are two quite different possible explanations for this outcome. The causal models that are consistent with this outcome do not need to include a direct path between the X_1 and Y variables. However, there are many different possible models that include causal and/or noncausal paths between X_1 and X_2 and X_2 and Y, and these could point to two quite different interpretations. One possible explanation is that the X_1, Y correlation may be completely accounted for by X_2 (or completely spurious). Another possible explanation is that there may be a causal association between X_1 and Y that is "completely mediated" by X_2. However, several additional conditions should be met before we consider the mediated causal model a likely explanation, as described by Baron and Kenny (1986) and discussed in Chapter 11 of this book.

10.12.3.1 ♦ *Completely Spurious Correlation*

To illustrate spurious correlations, consider the previous research example where height (X_1) was positively correlated with vocabulary (Y). It does not make sense to think that there is some direct connection (causal or otherwise) between height and vocabulary. However, both these variables are related to grade (X_2). We could argue that there is a noncausal correlation or confound between height and grade level and between vocabulary and grade level. Alternatively, we could propose that a maturation process that occurs from Grade 1 to 5 to 9 "causes" increases in both height and vocabulary. The theoretical causal models that appear in Figure 10.19 illustrate these two hypotheses. In the top diagram, there is no direct path between height and vocabulary (but these variables are correlated with each other in the observed data because both these variables are positively correlated with grade). In the bottom diagram, the unidirectional arrows represent the hypothesis that grade level or the associated physical and cognitive maturation causes increases in both height and vocabulary. In either case, the model suggests that the correlation between height and vocabulary is entirely "accounted for" by their relationship with the X_2 variable. If all the variance associated with the X_2 variable is removed (through a partial correlation $r_{Y1.2}$, for example), the association between height and vocabulary disappears. It is, therefore, reasonable in this case to say that the correlation between height and vocabulary was completely accounted for or explained by grade level or that the correlation between height and vocabulary was completely spurious (the variables were correlated with each other only because they were both associated with grade level).

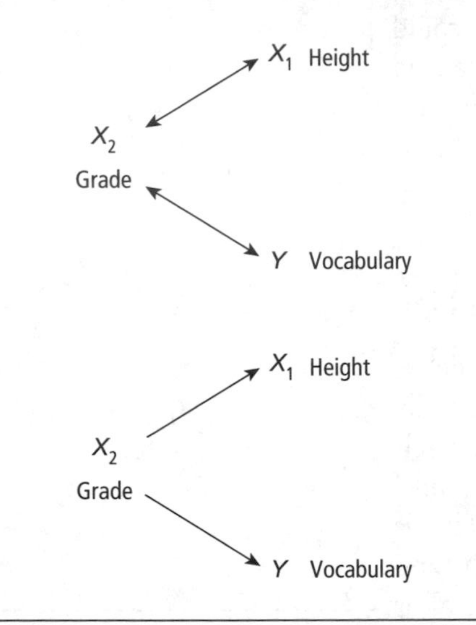

Figure 10.19 ♦ Two Possible Theoretical Models for the Relationships Among X_1 (Height), Y (Vocabulary Score), and X_2 (Grade Level) That Are Consistent With the Observed Correlations and Partial Correlations Obtained for These Variables

NOTE: In both these models, there is no direct path between X_1 and Y. Any observed correlation between X_1 and Y occurs because both X_1 and Y are noncausally associated with X_2 (as shown in the top diagram) or because both X_1 and Y are caused by X_2 (as shown in the bottom diagram). In this example, any correlation between X_1 and Y is spurious; X_1 and Y are correlated only because they are both correlated with X_2.

Some examples of spurious correlation intentionally involve foolish or improbable variables. For example, ice cream sales may increase as temperatures rise; homicide rates may also increase as temperatures rise. If we control for temperature, the correlation between ice cream sales and homicide rates drops to 0, so we would conclude that there is no direct relationship between ice cream sales and homicide but that the association of each of these variables with outdoor temperature (X_2) creates the spurious or misleading appearance of a connection between ice cream consumption and homicide.

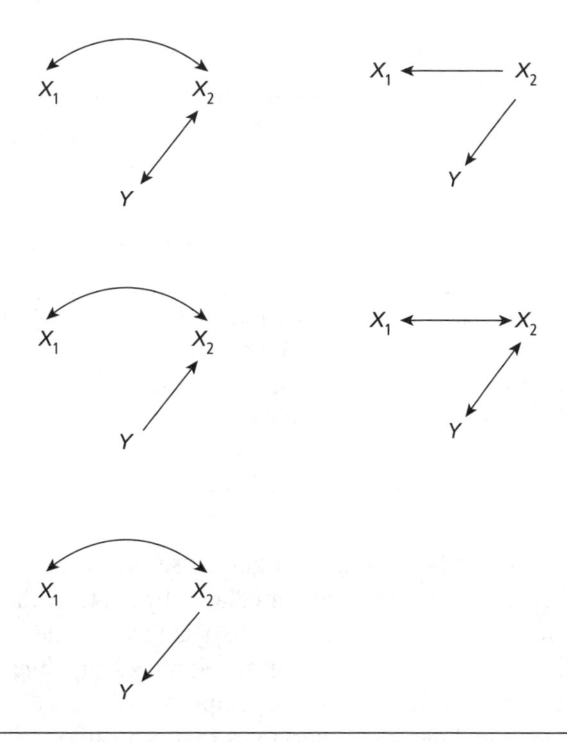

Figure 10.20 ♦ A Theoretical Causal Model in Which X_1 Has No Direct Association With Y, but Because X_1 Is Correlated With X_2 (and X_2 Is Either Causally or Noncausally Associated With Y), X_1 Is Also Correlated With Y

NOTE: In this situation we might say that the observed correlation between X_1 and Y is spurious; it is entirely due to the association of X_1 with X_2 and the association of X_2 with Y. Another possible interpretation for the path model in the lower left-hand corner of the figureis that X_1 is completely redundant with X_2 as a predictor of Y; X_1 provides no predictive information about Y that is not already available in the X_2 variable. A pattern of correlations that would be consistent with these path models is as follows: $r_{1Y} \neq 0$, but $r_{1Y.2} \approx 0$. (However, the mediated causal models in Figure 10.20 would also be consistent with these correlation values.)

There are actually many different theoretical or causal models that would be equally consistent with the outcome values $r_{Y1} \neq 0$ and $r_{Y1.2} = 0$. Some of these models appear in Figure 10.20. The characteristic that all these models for the $r_{Y1} \neq 0$ and $r_{Y1.2} = 0$ situation have in common is that they do *not* include a direct path between X_1 and Y (either causal or noncausal). In every path model in Figure 10.20, the observed correlation between X_1 and Y is due to an association (either causal or noncausal) between X_1 and X_2 and between X_2 and Y. Figure 10.20 omits the special case of models that involve mediated causal sequences, such as $X_1 \rightarrow X_2 \rightarrow Y$; the next section provides a brief preliminary discussion of mediated causal models.

10.12.3.2 ♦ *Completely Mediated Association Between X_1 and Y*

There are some research situations in which it makes sense to hypothesize that there is a causal sequence such that first X_1 causes X_2 and then X_2 causes Y. A possible example is the following: Increases in age (X_1) might cause increases in body weight

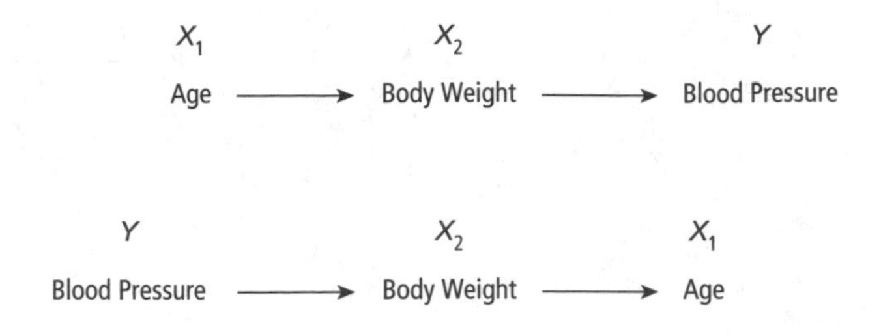

Figure 10.21 ♦ Completely Mediated Causal Models (in Which X_2 Completely Mediates a Causal Connection Between X_1 and Y)

NOTE: Top panel: a plausible mediated model in which age causes an increase in body weight, then body weight causes an increase in blood pressure. Bottom panel: an implausible mediated model.

(X_2); increases in body weight might cause increases in SBP (Y). If the relationship between X_1 and Y is completely mediated by X_2, then there is no direct path leading from X_1 to Y; the only path from X_1 to Y is through the mediating variable, X_2. Figure 10.21 illustrates two possible causal sequences in which X_2 is a mediating variable between X_1 and Y. Because of the temporal ordering of variables and our commonsense understandings about variables that can causally influence other variables, the first model ($X_1 \rightarrow X_2 \rightarrow Y$, i.e., age influences weight, and weight influences blood pressure) seems reasonably plausible. The second model ($Y \rightarrow X_2 \rightarrow X_1$, which says that blood pressure influences weight and then weight influences age) does not seem plausible.

What pattern of correlations and partial correlations would be consistent with these completely mediated causal models? If we find that $r_{Y1.2} = 0$ and $r_{Y1} \neq 0$, this outcome is logically consistent with the entire set of models that include paths (either causal or noncausal) between X_1 and X_2 and X_2 and Y but that do *not* include a direct path (either causal or noncausal) between X_1 and Y. Examples of these models appear in Figures 10.19, 10.20, and 10.21. However, models that do not involve mediated causal sequences (which appear in Figures 10.19 and 10.20) and mediated causal models (which appear in Figure 10.21) are equally consistent with the empirical outcome where $r_{1Y} \neq 0$ and $r_{1Y} = 0$. The finding that $r_{1Y} \neq 0$ and $r_{1Y} = 0$ suggests that we do not need to include a direct path between X_1 and Y in the model, but this empirical outcome does not tell us which among the several models that do not include a direct path between X_1 and Y is the "correct" model to describe the relationships among the variables.

We can make a reasonable case for a completely mediated causal model only in situations where it makes logical and theoretical sense to think that perhaps X_1 causes X_2 and then X_2 causes Y; where there is appropriate temporal precedence, such that X_1 happens before X_2 and X_2 happens before Y; and where additional statistical analyses yield results that are consistent with the mediation hypothesis (e.g., Baron & Kenny, 1986; also, see Chapter 11 in this textbook). When a mediated causal sequence does not make sense, it is more appropriate to invoke the less informative explanation that X_2 accounts for the X_1, Y correlation (this explanation was discussed in the previous section).

10.12.4 ♦ When You Control for X_2, the Correlation Between X_1 and Y Becomes Smaller (but Does Not Drop to 0 and Does Not Change Sign)

This may be one of the most common outcomes when partial correlations are compared with zero-order correlations. The implication of this outcome is that the association between X_1 and Y can be only partly accounted for by a (causal or noncausal) path via X_2. A direct path (either causal or noncausal) between X_1 and Y is needed in the model, even when X_2 is included in the analysis.

If most of the paths are thought to be noncausal, then the explanation for this outcome is that the relationship between X_1 and Y is "partly accounted for" or "partly explained by" X_2. If there is theoretical and empirical support for a possible mediated causal model, an interpretation of this outcome might be that X_2 "partly mediates" the effects of X_1 on Y, as discussed below. Figure 10.22 provides examples of general models in which an X_2 variable partly accounts for the X_1, Y relationship. Figure 10.23 provides examples of models in which the X_2 variable partly mediates the X_1, Y relationship.

10.12.4.1 ♦ X_2 *Partly Accounts for the* X_1, Y *Association, or* X_1 *and* X_2 *Are Correlated Predictors of* Y

Hansell, Sparacino, and Ronchi (1982) found a negative correlation between facial attractiveness (X_1) of high-school-aged women and their blood pressure (Y); that is, less attractive young women tended to have higher blood pressure. This correlation did not occur for young men. They tentatively interpreted this correlation as evidence of social stress; they reasoned that for high school girls, being relatively unattractive is a source of stress that may lead to high blood pressure. This correlation could be spurious, due to a causal or noncausal association of both attractiveness ratings and blood pressure with a third variable, such as body weight (X_2). If heavier people are rated as less attractive and tend to have higher blood pressure, then the apparent link between attractiveness and blood pressure might be due to the associations of both these variables with body weight. When Hansell et al. obtained a partial correlation between attractiveness and blood pressure, controlling for weight, they still saw a sizeable relationship between attractiveness and blood pressure. In other words, the correlation between facial attractiveness and blood pressure was apparently not completely accounted for by body weight. Even when weight was statistically controlled for, less attractive people tended to have higher blood pressure. The models in Figure 10.22 are consistent with their results.

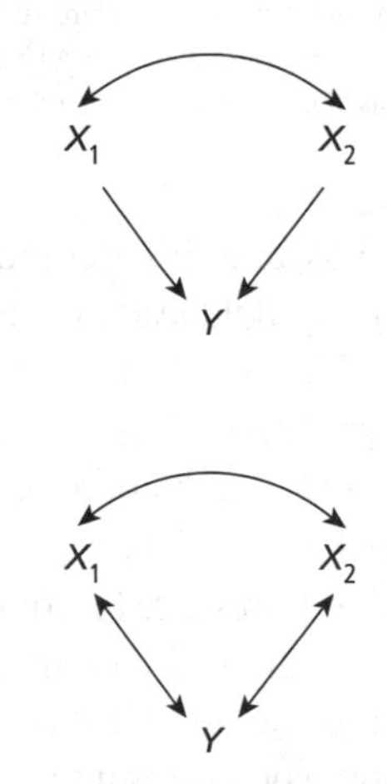

Figure 10.22 ♦ Two Possible Models in Which X_2 Partly Explains or Partly Accounts for the X_1, Y Relationship

NOTE: Top: X_1 and X_2 are correlated or confounded causes of Y (or X_1 and X_2 are partly redundant as predictors of Y). Bottom: All three variables are noncausally associated with each other. The pattern of correlations that would be consistent with this model is as follows: $r_{Y1} \neq 0$ and $r_{Y1.2} < r_{Y1}$ but with $r_{Y1.2}$ significantly greater than 0.

10.12.4.2 ♦ X_2 *Partly Mediates the* X_1, Y *Relationship*

An example of a research situation where a "partial mediation" hypothesis might make sense is the following. Suppose a researcher conducts a nonexperimental study and measures the following three variables: the X_1 predictor variable, age in years; the Y outcome variable, SBP (given in millimeters of mercury); and the X_2 control variable, body weight (measured in pounds or kilograms or other units). Because none of the variables has been manipulated by the researcher, the data cannot be used to prove causal connections among variables. However, it is conceivable that as people grow older, normal aging causes an increase in body weight. It is also conceivable that as people become heavier, this increase in body weight causes an increase in blood pressure. It is possible that body weight completely mediates the association between age and blood pressure (if this were the case, then people who do not gain weight as they age should not show any age-related increases in blood pressure). It is also conceivable that increase in age causes increases in blood pressure through other pathways and not solely through weight gain. For example, as people age, their arteries may become clogged with deposits of fats or lipids, and this accumulation of fats in the arteries may be another pathway through which aging could lead to increases in blood pressure.

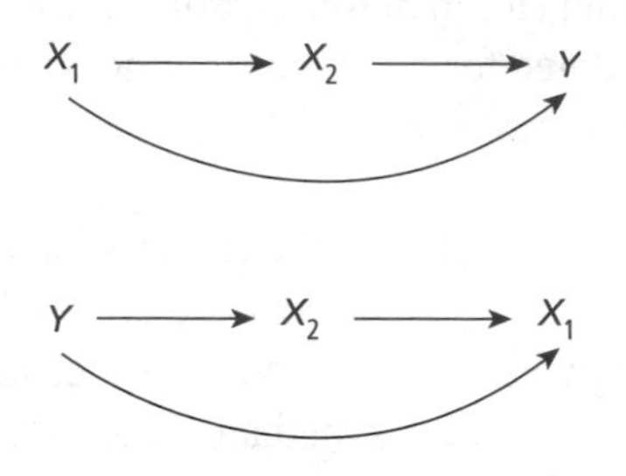

Figure 10.23 ♦ Two Possible Models for Partly Mediated Causation Between X_1 and Y (X_2 Is the Mediator)

NOTE: Top: X_1 causes or influences X_2, and then X_2 causes or influences Y; in addition, X_1 has effects on Y that are not mediated by X_2. Bottom: Y causes or influences X_2, and then X_2 causes or influences X_1; in addition, Y has effects on X_1 that are not mediated by X_2. The pattern of correlations that would be consistent with this model is as follows: $r_{Y1} \neq 0$ and $r_{Y1.2} < r_{1Y}$ but with $r_{Y1.2}$ significantly greater than 0. (Because this is the same pattern of correlations that is expected for the models in Figure 10.22, we cannot determine whether one of the partly mediated models in Figure 10.23 or the "partly redundant predictor" model in Figure 10.22 is a better explanation, based on the values of correlations alone.)

Figure 10.24 shows two hypothetical causal models that correspond to these two hypotheses. The first model in Figure 10.24 represents the hypothesis that the effects of age (X_1) on blood pressure (Y) are completely mediated by weight (X_2). The second model in Figure 10.24 represents the competing hypothesis that age (X_1) also has effects on blood pressure (Y) that are not mediated by weight (X_2).

In this example, the X_1 variable is age, the X_2 mediating variable is body weight, and the Y outcome variable is SBP. If preliminary data analysis reveals that the correlation between age and blood pressure $r_{1Y} = +.8$ but that the partial correlation between age and blood pressure, controlling for weight, $r_{1Y.2} = .00$, we may tentatively conclude that the relationship between age and blood pressure is completely mediated by body weight. In other words, as people age, increase in age causes an increase in SBP but only if age causes an increase in body weight.

We will consider a hypothetical example of data for which it appears that the effects of age on blood pressure may be only partly mediated by or accounted for by weight. Data on age, weight, and blood pressure for $N = 30$ appear in Table 10.3. Pearson correlations were performed between all pairs of variables; these bivariate zero-order correlations appear in Figure 10.25. All three pairs of variables were significantly

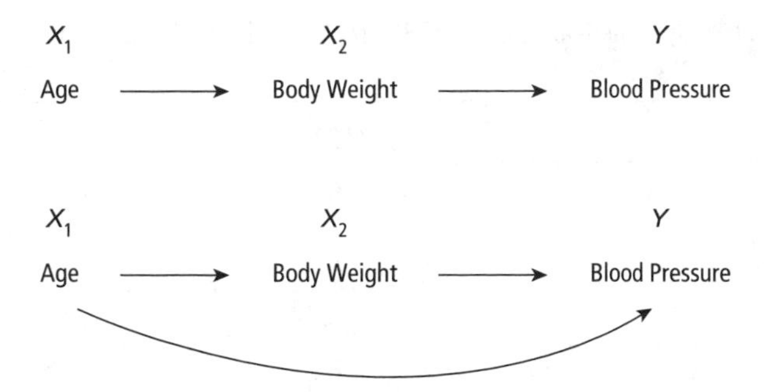

Figure 10.24 ♦ Hypothetical Mediated Models for Age, Weight, and Blood Pressure
Hypothetical Model Showing a Fully Mediated Relationship
Hypothetical Model Showing a Partially Mediated Relationship

NOTES: Preliminary evidence consistent with a fully mediated model is as follows: Partial correlation between age and blood pressure, controlling for weight ($r_{1Y.2}$), is less than the zero-order correlation between age and blood pressure (r_{1Y}), and $r_{1Y.2}$ is not significantly greater than 0.

Preliminary evidence consistent with a partial mediation model is as follows: Partial correlation between age and blood pressure, controlling for weight ($r_{1Y.2}$) is less than the zero-order correlation between age and blood pressure (r_{1Y}), but $r_{1Y.2}$ is significantly greater than 0.

See Baron and Kenny (1986) for a discussion of additional conditions that should be met before a mediated model is proposed as an explanation for correlational results.

positively correlated. A partial correlation was calculated between age and blood pressure, controlling for weight (the SPSS partial correlation output appears in Figure 10.26). The partial correlation between age and blood pressure, controlling for weight, was $r_{Y1.2} = +.659$, $p = .005$, two-tailed. The zero-order correlation between age and blood pressure (not controlling for weight) was $r_{1Y} = +.782$, $p < .001$. Because the partial correlation was smaller than the zero-order correlation between age and blood pressure but was still substantially greater than 0, a partially mediated model (Figure 10.24) could reasonably be considered. It is possible that age has some influence on blood pressure through its effects on weight and the effects of weight on blood pressure, but the partial correlation results suggest that age may also have effects on blood pressure that are not mediated by weight gain. For a researcher who wants to test mediated causal models, an examination of partial correlation provides only a preliminary test. Additional analyses of these data will be discussed in Chapter 11, to evaluate whether other conditions necessary for mediation are satisfied (Baron & Kenny, 1986).

10.12.5 ♦ When You Control for X_2, the X_1, Y Correlation Becomes Larger Than r_{1Y} or Becomes Opposite in Sign Relative to r_{1Y}

When $r_{Y1.2}$ is larger than r_{Y1} or when $r_{Y1.2}$ is opposite in sign relative to r_{1Y}, X_2 is described as a "suppressor" variable. In other words, the true strength or true sign of the X_1, Y association is "suppressed" by the X_2 variable; the true strength or sign of the X_1, Y correlation becomes apparent only when the variance associated with the X_2 suppressor variable is removed by partialling X_2 out. This corresponds to the regions marked Area *e* in Figure 10.15. In spite of the fact that Area *e* occupies a large amount of the space in

Table 10.3 ♦ Hypothetical Data for an Example of a Partly Mediated Relationship: Age, Weight, and Blood Pressure

Age	*Weight*	*Blood Pressure*
25	82	89
30	142	151
33	66	37
35	113	127
37	123	65
40	147	96
41	115	103
43	178	194
44	115	176
48	116	74
52	181	228
55	164	158
57	189	177
58	133	169
60	188	184
61	192	195
63	207	201
65	219	197
67	177	202
66	244	257
69	199	187
71	158	128
73	200	247
72	219	246
75	187	231
78	195	225
80	214	214
82	154	288
84	118	266
85	125	206

NOTE: $N = 30$ cases. Age in years, body weight in pounds, and SBP in millimeters of mercury.

Figure 10.15, suppression is a relatively uncommon outcome. The following sections describe two relatively simple forms of suppression. For a discussion of other forms that suppression can take, see Cohen, Cohen, West, and Aiken (2003).

10.12.5.1 ♦ *Suppression of Error Variance in a Predictor Variable*

Consider the following hypothetical situation. A researcher develops a paper-and-pencil test of "mountain survival skills." The score on this paper-and-pencil test is the X_1 predictor variable. The researcher wants to demonstrate that scores on this paper-and-pencil test (X_1) can predict performance in an actual mountain survival situation (the score for this survival test is the Y outcome variable). The researcher knows that, to some extent, performance on the paper-and-pencil test depends on the level of verbal ability (X_2). However, verbal ability is completely uncorrelated with performance in the actual mountain survival situation.

Correlations

		Age	Weight	Blood Pressure
Age	Pearson Correlation	1	.563**	.782**
	Sig. (2-tailed)		.001	.000
	N	30	30	30
Weight	Pearson Correlation	.563**	1	.672**
	Sig. (2-tailed)	.001		.000
	N	30	30	30
BloodPressure	Pearson Correlation	.782**	.672**	1
	Sig. (2-tailed)	.000	.000	
	N	30	30	30

**. Correlation is significant at the 0.01 level (2-tailed).

Figure 10.25 ♦ Zero-Order Correlations Among Age, Weight, and Blood Pressure (for the Hypothetical Data From Table 10.3)

Correlations

Control Variables			Age	Blood Pressure
Weight	Age	Correlation	1.000	.659
		Significance (2-tailed)	.	.000
		df	0	27
	BloodPressure	Correlation	.659	1.000
		Significance (2-tailed)	.000	.
		df	27	0

Figure 10.26 ♦ Partial Correlation Between Age and Blood Pressure, Controlling for Weight (for the Hypothetical Data From Table 10.3)

The hypothetical data in the file named papertest_suppress.sav show what can happen in this type of situation. The correlation between the score on the paper-and-pencil test (X_1) and actual mountain survival skills (Y) is $r = +.50$. The correlation between verbal ability (X_2) and actual mountain survival skills is not significantly different from 0 ($r = -.049$). We, thus, have a situation where the predictor variable of interest, X_1, is reasonably predictive of the outcome variable, Y. Note that the control variable, X_2 (verbal ability), is highly correlated with the predictor variable of interest, X_1 ($r_{12} = +.625$), but X_2 has a correlation close to 0 with the outcome or Y variable.

The partial correlation between X_1 and Y, controlling for X_2, is calculated (i.e., the partial correlation between performance on the paper-and-pencil test and actual mountain survival skills, controlling for verbal ability) and is found to be $r_{1Y.2} = +.75$. That is, when the variance in the paper-and-pencil test scores (X_1) due to verbal ability (X_2) is partialled out or removed, the remaining part of the X_1 scores become more predictive of actual mountain survival skills.

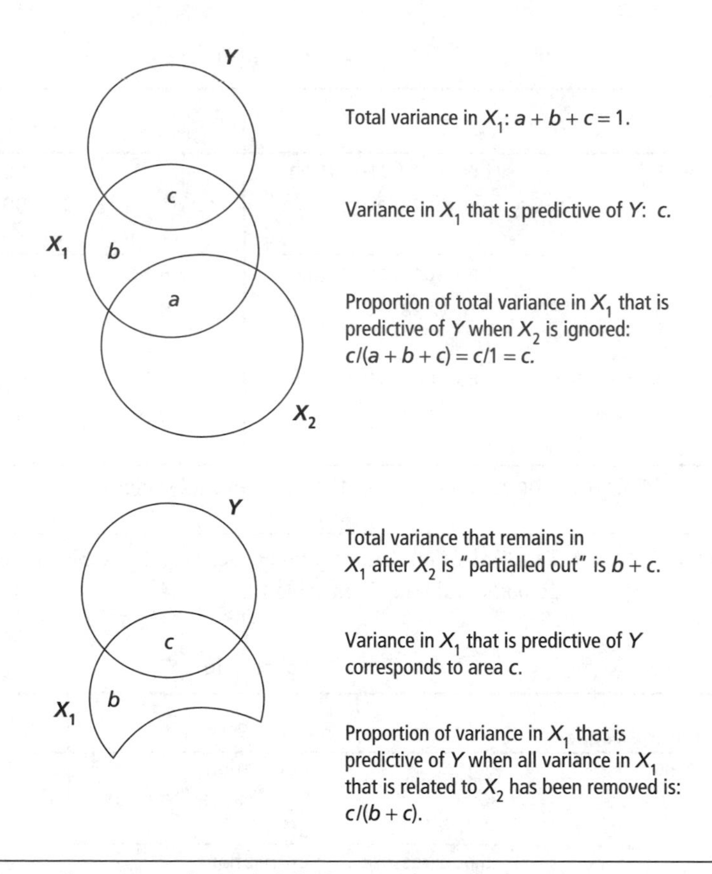
Y
Total variance in X_1: $a + b + c = 1$.
c
Variance in X_1 that is predictive of Y: c.
X_1
b
a
Proportion of total variance in X_1 that is predictive of Y when X_2 is ignored: $c/(a + b + c) = c/1 = c$.
X_2
Y
Total variance that remains in X_1 after X_2 is "partialled out" is $b + c$.
c
Variance in X_1 that is predictive of Y corresponds to area c.
X_1
b
Proportion of variance in X_1 that is predictive of Y when all variance in X_1 that is related to X_2 has been removed is: $c/(b + c)$.

Figure 10.27 ♦ X_2 (a Measure of Verbal Ability) Is a Suppressor of Error Variance in the X_1 Predictor (a Paper-and-Pencil Test of Mountain Survival Skills); Y Is a Measure of Actual Mountain Survival Skills

The overlapping circle diagrams that appear in Figure 10.27 can help us understand what might happen in this situation. The top diagram shows that X_1 is correlated with Y and X_2 is correlated with X_1; however, X_2 is not correlated with Y (the circles that represent the variance of Y and the variance of X_2 do not overlap). If we ignore the X_2 variable, the squared correlation between X_1 and Y (r^2_{1Y}) corresponds to Area c in Figure 10.27. The total variance in X_1 is given by the sum of areas $a + b + c$. In these circle diagrams, the total area equals 1.00; therefore, the sum $a + b + c = 1$. The proportion of variance in X_1 that is predictive of Y (when we do not partial out the variance associated with X_2) is equivalent to $c/(a + b + c) = c/1 = c$.

When we statistically control for X_2, we remove all the variance that is predictable from X_2 from the X_1 variable, as shown in the bottom diagram in Figure 10.27. The second diagram shows that after the variance associated with X_2 is removed, the remaining variance in X_1 corresponds to the sum of areas $b + c$. The variance in X_1 that is predictive of Y corresponds to Area c. The proportion of the variance in X_1 that is predictive of Y after we partial out or remove the variance associated with X_2 now corresponds to $c/(b + c)$. Because $(b + c)$ is less than 1, the proportion of variance in X_1 that is associated with Y after removal of the variance associated with X_2 (i.e., $r^2_{Y1.2}$) is actually higher than the original proportion of variance in Y that was predictable from X_1 when X_2 was not controlled (i.e.,r^2_{Y1}). In this example, $r_{Y1.2} = +.75$, whereas $r_{Y1} = +.50$. In other words, when you control for verbal ability, the score on the paper-and-pencil test accounts for $.75^2 = .56$ or 56% of the variance in Y. When you do *not* control for verbal ability, the score on the paper-and-pencil test only predicted $.50^2 = .25$ or 25% of the variance in Y.

In this situation, the X_2 control variable suppresses irrelevant or **error variance** in the X_1 predictor variable. When X_2 is statistically controlled, X_1 is more predictive of Y than when X_2 is not statistically controlled. It is not common to find a suppressor variable that makes some other predictor variable a better predictor of Y in actual research. However, sometimes a researcher can identify a factor that influences scores on the X_1 predictor and that is not related to or predictive of the scores on the outcome variable Y. In this example, verbal ability was one factor that influenced scores on the paper-and-pencil test, but it was almost completely unrelated to actual mountain survival skills. Controlling for verbal ability (i.e., removing the variance associated with verbal ability from the scores on the paper-and-pencil test) made the paper-and-pencil test a better predictor of mountain survival skills.

10.12.5.2 ♦ *A Second Type of Suppression*

Another possible form of suppression occurs when the sign of $r_{Y1.2}$ is opposite to the sign of r_{Y1}. In the example we are going to discuss now, r_{1Y}, the zero-order correlation between crowding (X_1) and crime rate (Y_2) across neighborhoods, is large and positive. However, when you control for X_2 (level of neighborhood socioeconomic status or SES), the sign of the partial correlation between X_1 and Y, controlling for X_2, $r_{Y1.2}$, becomes negative. A hypothetical situation where this could occur is shown in Figure 10.28.

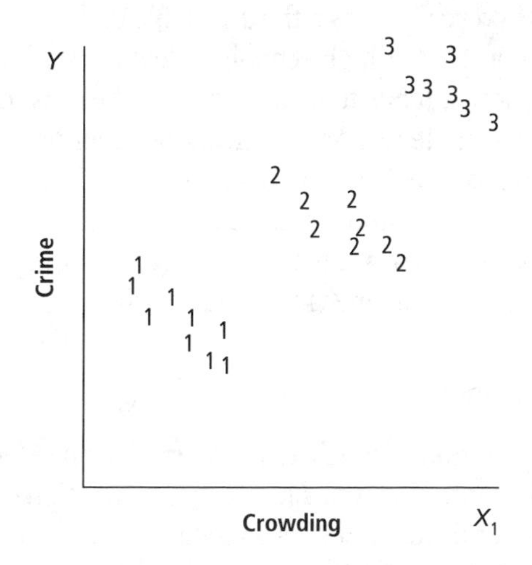

Figure 10.28 ♦ Example of One Type of Suppression

NOTE: On this graph, cases are marked by SES level of the neighborhood (1 = High SES, 2= Middle SES, 3 = Low SES). When SES is ignored, there is a large positive correlation between X_1 (neighborhood crowding) and Y (neighborhood crime). When the X_1, Y correlation is assessed separately within each level of SES, the relationship between X_1 and Y becomes negative. The X_2 variable (SES) suppresses the true relationship between X_1 (crowding) and Y (crime). Crowding and crime appear to be positively correlated when we ignore SES; when we statistically control for SES, it becomes clear that within SES levels, crowding and crime are actually negatively related.

In this hypothetical example, the unit of analysis or case is "neighborhood"; for each neighborhood, X_1 is a measure of crowding, Y is a measure of crime rate, and X_2 is a measure of income level (SES). X_2 (SES) is coded as follows: 1 = Upper class, 2 = Middle class, 3 = Lower class. The pattern in this graph represents the following hypothetical situation. This example was suggested by correlations reported by Freedman (1975), but it illustrates a much stronger form of suppression than Freedman found in his data. For the hypothetical data in Figure 10.28, if you ignore SES and obtain the zero-order correlation between crowding and crime, you would obtain a large positive correlation, suggesting that crowding predicts crime. However, there are two confounds present: Crowding tends to be greater in lower-SES neighborhoods (3 = Low SES), and the incidence of crime also tends to be greater in lower-SES neighborhoods.

Once you look separately at the plot of crime versus crowding within each SES category, however, the relationship becomes quite different. Within the lowest SES neighborhoods (SES code 3), crime is negatively associated with crowding (i.e., more crime takes place in "deserted" areas than in areas where there are many potential witnesses out on the streets). Freedman (1975) suggested that crowding, per se, does not "cause" crime; it just happens to be correlated with something else that is predictive of crime—namely, poverty or low SES. In fact, within neighborhoods matched in SES, Freedman reported that higher population density was predictive of *lower* crime rates.

Another example of this type of suppression, where the apparent direction of relationship reverses when a control variable is taken into account, was reported by Guber (1999). The unit of analysis in this study was each of the 50 states in the United States. The X_1 predictor variable was per-student annual state spending on education; the Y outcome variable was the mean Scholastic Aptitude Test (SAT) score for students in each state. The zero-order bivariate correlation, r_{Y1}, was large and negative; in other words, states that spent the largest amount of money per student on education actually tended to have lower SAT scores. However, there is a third variable that must be taken into account in this situation: the proportion of high school students in each state who take the SAT. When the proportion is low, it tends to mean that only the most capable students took the SAT and the mean SAT score tends to be high; on the other hand, in states where the proportion of students who take the SAT is high, the mean SAT score tends to be lower. When the proportion of high school students in each state who took the SAT (X_2) was statistically controlled, the partial correlation between state spending on education per pupil and mean SAT scores (controlling for X_2) became positive.

10.12.6 ♦ "None of the Above"

The interpretations described above provide an extensive, but not exhaustive, description of possible interpretations for partial correlation outcomes. A partial correlation can be misleading or difficult to interpret when assumptions such as linearity are violated or when the value of the partial r depends disproportionately on the outcomes for a few cases that have outlier scores.

A preliminary examination of scatter plots for the X_1, Y relationship separately within each value of X_2 can be helpful in evaluating whether correlation is or is not a good description of the association between X_1 and Y at each level of X_2. One possible outcome is that the slope or correlation for the X_1, Y relationship differs across levels of the X_2

control variable. If this occurs, the overall partial correlation $r_{Y1.2}$ can be misleading; a better way to describe this outcome would be to report different slopes or different correlations between X_1 and Y for each level of X_2. In this situation, we can say that there is an interaction between X_1 and X_2 as predictors of Y or that the X_2 variable moderates the predictive relationship between X_1 and Y. When a preliminary analysis of three-variable research situations suggests the presence of an interaction, data analyses that take interaction into account (as discussed in Chapters 12 and 13) are needed.

10.13 ♦ Mediation Versus Moderation

A classic paper by Baron and Kenny (1986) distinguished between mediating and moderating variables. X_2 is a *mediating* variable if it represents an intermediate step in a causal sequence, such that X_1 causes X_2 and then X_2 causes Y. Mediation should not be confused with **moderation.** X_2 is a *moderating* variable if the nature of the X_1, Y relationship is different for people who have different score values on the X_2 moderator variable; or, to say this in another way, X_2 is a moderator of the X_1, Y relationship if there is an interaction between X_1 and X_2 as predictors of Y. When moderation occurs, there are "different slopes for different folks"; the slope that predicts Y from X_1 differs depending on the score on the X_2 control variable. The analyses described in this chapter provide only preliminary information about the possible existence of mediation or moderation. If a preliminary examination of data suggests that mediation or moderation is present, then later analyses (such as multiple regression) need to take these issues into account.

10.13.1 ♦ Preliminary Analysis to Identify Possible Moderation

For example, suppose that X_2 corresponds to gender, coded as follows: 1 = Male, 2 = Female; X_1 is the need for power, and Y is job performance evaluations. Figure 10.29 shows hypothetical data for a sample of male and female managers. Each manager took a test to assess his or her level of need for power; this was the X_1 or predictor variable. After 6 months in a management position, each manager's employees completed self-report evaluations of the manager's job performance (Y). Let's suppose that back in the 1970s, employees responded more favorably to male managers who scored high on need for power but that employees responded less favorably to female managers who scored high on need for power. We can set up a scatter plot to show how evaluations of job performance (Y) are related to managers' need for power (X_1); different types of markers identify which scores belong to male managers (X_2 = m) and which scores belong to female managers (X_2 = f). Gender of the manager is the X_2 or "controlled for" variable in this example.

If we look only at the scores for male managers (each case for a male is denoted by "m" in Figure 10.29), there is a positive correlation between need for power (X_1) and job performance evaluations (Y). If we look only at the scores for female managers (denoted by "f" in Figure 10.29), there is a negative correlation between need for power (X_1) and job performance evaluations (Y). We can test whether the r_{1Y} correlation for the male subgroup and the r_{1Y} correlation for the female subgroups are significantly different, using significance tests from Chapter 7. Later chapters show how interaction terms can be included in regression analyses to provide different estimates of the slope for the prediction of Y from X_1 for subgroups defined by an X_2 control variable.

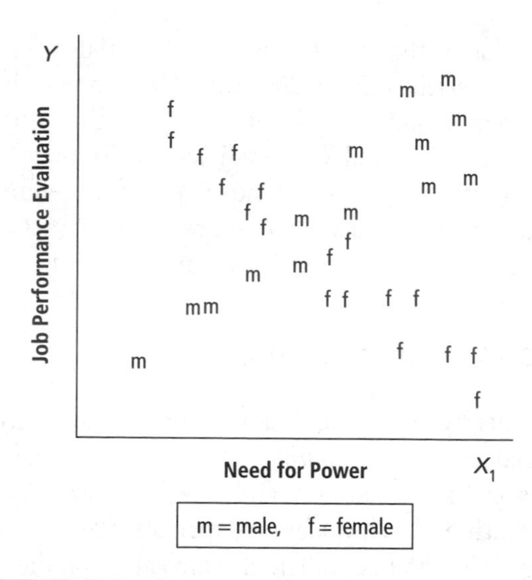

Figure 10.29 ♦ Interaction Between Gender (X_2) and Need for Power (X_1) as Predictors of Job Performance Evaluation (Y): For the Male Group, X_1 and Y Are Positively Correlated; Within the Female Group, X_1 and Y Are Negatively Correlated

In this example, we could say that gender and need for power interact as predictors of job performance evaluation; more specifically, for male managers, their job performance evaluations increase as their need for power scores increase, whereas for female managers, their job performance evaluations decrease as their need for power scores increase. We could also say that gender "moderates" the relationship between need for power and job performance evaluation.

In this example, the slopes for the subgroups (male and female) had opposite signs. Moderation or interaction effects do not have to be this extreme. We can say that gender moderates the effect of X_1 on Y if the slopes to predict Y from X_1 are significantly different for males and females. The slopes do not actually have to be opposite in sign for an interaction to be present. Another type of interaction occurs when there is no correlation between X_1 and Y for females and a strong positive correlation between X_1 and Y for males. Yet another kind of interaction is seen when the b slope coefficient to predict Y from X_1 is positive for both females and males but is significantly larger in magnitude for males than for females.

Note that partial correlations do not provide information about the existence of moderation or interaction, and in fact, the values of partial correlations can be quite misleading when an interaction or moderator variable is present. When we compute a partial correlation, it is essentially the mean of the within-group slopes between Y and X_1 when these slopes are calculated separately for each group of scores on the X_2 variable. If the slope to predict Y from X_1 differs across values of X_2, it is misleading to average these slopes together; to provide an accurate description of the relations among variables, you need to provide information about the way in which the X_1, Y relationship differs across levels or scores on the X_2 variable. For the scatter plot that appears in Figure 10.29, the zero-order correlation, r_{1Y}, between need for power and job performance evaluation, ignoring gender, would be close to 0. The partial r between need for power and job performance evaluation, controlling for gender, would be (approximately, not exactly) the mean of the r between these variables for men and the r between these variables for women; $r_{Y1.2}$ would also be close to 0. Neither the zero-order r nor the partial r would tell us that there is an interaction present. To detect the interaction, we need to look at a scatter plot or compute correlations between these variables separately for each gender (using the SPSS Split File command to examine male and female cases separately).

10.13.2 ♦ Preliminary Analysis to Detect Possible Mediation

One possible causal model that might be applicable in a three-variable research situation involves a mediated causal process. For example, consider a situation in which the X_1 variable first causes X_2, and then the X_2 variable causes Y. If the relationship between X_1 and Y is completely mediated by X_2, then the only path that leads from X_1 to Y is a path via X_2. If the relationship between X_1 and Y is only partly mediated by X_2, there may be two paths that lead from X_1 to Y. One path involves the indirect or **mediated relationship** via X_2, as shown above, but there is an additional direct path from X_1 to Y.

What evidence would be consistent with (but not proof of) a mediated causal process model? The first requirement for consideration of a mediated causal model is that it must be consistent with a theory that predicts that there could be a causal sequence (or at least with common sense), such that first X_1 causes X_2 and then X_2 causes Y. Temporal priority among variables should also be considered when asking whether a particular mediation model makes sense. Causes must precede effects in time. If we first measure X_1, then measure X_2, and then measure Y, the temporal order of measurements is consistent with a causal theory in which X_1 causes X_2, which then causes Y. Of course, obtaining measurements in this temporal sequence does not prove that the events actually occur in this temporal order. In many nonexperimental studies, researchers measure all three variables (X_1, X_2, and Y) at the same point in time. When all variables are measured at one point in time, it is difficult to rule out other possible sequences (such as Y causes X_2, which then causes X_1).

If a theory includes the hypothesis that X_2 mediates a causal sequence from X_1 to Y, then examining the partial correlation $r_{Y1.2}$ provides preliminary evidence as to whether any mediation that is present is complete or only partial. If $r_{Y1.2}$ is not significantly different from 0, it is possible that the X_1, Y relationship is completely mediated by X_2. On the other hand, if $r_{Y1.2} < r_{1Y}$, and $r_{Y1.2}$ is significantly different from 0, then perhaps the X_1, Y relationship is only partly mediated by X_2. However, several additional conditions should be satisfied before a researcher interprets an outcome as evidence of a mediated causal process (Baron & Kenny, 1986). Chapter 11 in this textbook will demonstrate statistical procedures (including a regression equation in which Y is predicted from both X_1 and X_2) to evaluate whether these additional conditions are satisfied. However, correlation analysis of data obtained in nonexperimental research situations does not provide a strong basis for making causal inferences. If a researcher has a hypothesis about a mediated causal sequence, there may be ways to test this model using experimental methods.

10.13.3 ♦ Experimental Tests for Mediation Models

The best evidence to support a mediated causal model would come from an experiment in which the researcher can manipulate the presumed causal variable X_1, control the mediating variable X_2 (either by experimental means or statistically), control other variables that might influence the variables in the study, and measure the outcome variable Y. For example, Levine, Gordon, and Fields (1978) theorized that placebo pain relief is mediated by the release of endorphins (naturally occurring chemicals in the body that mimic the effects of narcotic drugs). They tested this mediated process model experimentally. The pain induction involved medically necessary dental surgery. After surgery, half the patients received an intravenous dose of naloxone (a drug that blocks the pain-relieving effects of both artificial narcotic drugs and endogenous opioids). The other half of the patients received an intravenous dose of an inert substance. Subsequently, all patients were given a placebo pain reliever that they were told should reduce their pain. Patients who had received naloxone (which blocks the pain-relieving effects of opiates and endorphins) did not report any pain relief from the placebo injection. In contrast, many of the patients who had not received naloxone reported substantial pain relief in response to the placebo. The theoretical model was as follows: Administration of a placebo (X_1) causes the release of endorphins (X_2) in at least some patients; the release of endorphins (X_2) causes a reduction in pain (Y). Note that in most studies, only about 30% to 35% of patients who receive placebos experience pain relief.

Administration of placebo → Release of endorphins → Pain relief

By administering naloxone to one group of patients, these investigators were able to show that placebos had no effect on pain when the effects of endorphins were blocked. This outcome was consistent with the proposed mediated model. It is possible that the effects of placebos on pain are mediated by the release of endorphins.

10.14 ♦ Model Results

The first research example introduced early in the chapter examined whether height (X_1) and vocabulary (Y) are related when grade level (X_2) is statistically controlled. The results presented in Figures 10.3 through 10.14 can be summarized briefly.

Results

The relation between height and vocabulary score was assessed for $N = 48$ students in three different grades in school: Grade 1, Grade 5, and Grade 9. The zero-order Pearson r between height and vocabulary was statistically significant: $r(46) = .72$, $p < .001$, two-tailed. A scatter plot of vocabulary scores by height (with individual points labeled by grade level) suggested that both vocabulary and height tended to increase with grade level (see Figure 10.3). It seemed likely that the correlation between vocabulary and height was spurious—that is, entirely attributable to the tendency of both these variables to increase with grade level.

To assess this possibility, the relation between vocabulary and height was assessed controlling for grade. Grade was controlled for in two different ways. A first-order partial correlation was computed for vocabulary and height, controlling for grade. This partial *r* was not statistically significant: $r(45) = -.01$, $p = .938$. In addition, the correlation between Height and Vocabulary was computed separately for each of the three grade levels. For Grade = 1, $r = .067$; for Grade = 5, $r = .031$; for Grade = 9, $r = -.141$. None of these correlations was statistically significant, and the differences among these three correlations were not large enough to suggest the presence of an interaction effect (i.e., there was no evidence that the nature of the relationship between vocabulary and height differed substantially across grades).

When grade was controlled for, either by partial correlation or by computing Pearson *r* separately for each grade level, the correlation between vocabulary and height became very small and was not statistically significant. This is consistent with the explanation that the original correlation was spurious. Vocabulary and height are correlated only because both variables increase across grade levels (and not because of any direct causal or noncausal association between height and vocabulary).

10.15 ♦ Summary

Partial correlation can be used to provide preliminary exploratory information that may help the researcher understand relations among variables. In this chapter, we have seen that when we take a third variable, X_2, into account, our understanding of the nature and strength of the association between X_1 and Y can change in several different ways.

This chapter outlines two methods to evaluate how taking X_2 into account as a control variable may modify our understanding of the way in which an X_1 predictor variable is related to a Y outcome variable. The first method involved dividing the dataset into separate groups, based on scores on the X_2 control variable (using the Split File command in SPSS), and then examining scatter plots and correlations between X_1 and Y separately for each group. In the examples in this chapter, the X_2 control variables had a small number of possible score values (e.g., when gender was used as a control variable, it had just two values, male and female; when grade level in school and SES were used as control variables, they had just three score values). The number of score values on X_2 variables was kept small in these examples to make it easy to understand the examples. However, the methods outlined here are applicable in situations where the X_2 variable has a larger number of possible score values, as long as the assumptions for Pearson correlation and partial correlation are reasonably well met. Note, however, that if the X_2 variable has 40 possible different score values, and the total number of cases in a dataset is only $N = 50$, it is quite likely that when any one score is selected (e.g., $X_2 = 33$), there may be only one or two cases with that value of X_2. When the *ns* within groups based on the value of X_2 become very small, it becomes impossible to evaluate assumptions such as linearity and normality within the subgroups, and estimates of the strength of association between X_1 and Y that are based on extremely small groups are not likely to be very reliable. The minimum sample sizes that were suggested for Pearson correlation and bivariate regression were on the order of $N = 100$. Sample sizes should be even larger for studies where an X_2 control variable is taken into account, particularly in situations where the researcher suspects the presence of an interaction or moderating variable; in these situations, the researcher needs to estimate a different slope to predict Y from X_1 for each score value of X_2.

We can use partial correlation to statistically control for an X_2 variable that may be involved in the association between X_1 and Y as a rival explanatory variable, a confound, a mediator, a suppressor, or in some other role. However, statistical control is generally a less effective method for dealing with extraneous variables than experimental control. Some methods of experimental control (such as random assignment of participants to treatment groups) are, at least in principle, able to make the groups equivalent with respect to hundreds of different participant characteristic variables. However, when we measure and statistically control for one specific X_2 variable in a nonexperimental study, we have controlled for only one of many possible rival explanatory variables. In a nonexperimental study, there may be dozens or hundreds of other variables that are relevant to the research question and whose influence is not under the researcher's control; when we use partial correlation and similar methods of statistical control, we are able to control statistically for only a few of these variables.

The next three chapters continue to examine the three-variable research situation, but they do so using slightly different approaches. In Chapter 11, X_1 and X_2 are both used as predictors in a multiple regression to predict Y. Chapter 12 discusses the use of dummy-coded or dichotomous variables as predictors in regression and demonstrates the use of product terms in regression to represent interaction or moderation. Chapter 13 reviews two-way factorial ANOVA; this is a version of the three-variable situation where both the X_1 and X_2 predictor variables are categorical and the outcome variable Y is quantitative. From Chapter 14, we begin to consider multivariate analyses that may include more than three variables. Multivariate analyses can include multiple predictors and/or multiple outcomes and/or multiple control variables or **covariates**.

In this chapter, many questions were presented in the context of a three-variable research situation. For example, is X_1 confounded with X_2 as a predictor? When you control for X_2, does the partial correlation between X_1 and Y drop to 0? In multivariate analyses, we often take several additional variables into account when we assess each X_1, Y predictive relationship. However, the same issues that were introduced here in the context of three-variable research situations continue to be relevant for studies that include more than three variables.

Note

1. This assumes that Pearson r is an appropriate statistical analysis to describe the strength of the relationship between X_1 and Y; if the assumptions required for Pearson r are violated; for example, if the relation between X_1 and Y is nonlinear, then Pearson r is not an appropriate analysis to describe the strength of the association between these variables. In this case, data transformations may be applied to scores on one or both variables to make the relationship between them more linear, or an entirely different statistical analysis may be required to describe the strength of the association between X_1 and Y.

+ − × ÷

Comprehension Questions

1. When we assess X_1 as a predictor of Y, there are several ways in which we can add a third variable (X_2) and several "stories" that may describe the relations among variables. Explain what information can be obtained from the following two analyses:
   I. Assess the X_1, Y relation separately for each group on the X_2 variable.
   II. Obtain the partial correlation (partial r of Y with X_1, controlling for X_2).
   a. Which of these analyses (I or II) makes it possible to detect an interaction between X_1 and X_2? Which analysis assumes that there is no interaction?
   b. If there is an interaction between X_1 and X_2 as predictors of Y, what pattern would you see in the scatter plots in Analysis I?
2. Discuss each of the following as a means of illustrating the partial correlation between X_1 and Y, controlling for X_2. What can each analysis tell you about the strength and the nature of this relationship?
   I. Scatter plots showing Y versus X_1 (with X_2 scores marked in plot)
   II. Partial r as the correlation between the residuals when X_1 and Y are predicted from X_2
3. Explain how you might interpret the following outcomes for partial r:
   a. $r_{1Y} = .70$ and $r_{1Y.2} = .69$
   b. $r_{1Y} = .70$ and $r_{1Y.2} = .02$
   c. $r_{1Y} = .70$ and $r_{1Y.2} = -.54$
   d. $r_{1Y} = .70$ and $r_{1Y.2} = .48$
4. What does the term *partial* mean when it is used in connection with correlations?
5. Suppose you correlate age with SBP and find a strong positive correlation. Compute a first-order partial correlation between age and SBP, controlling for weight (body weight).
   a. What would you conclude if this partial r were almost exactly 0?
   b. What would you conclude if this partial r were a little smaller than the zero-order r between age and SBP but still substantially larger than 0?
   c. What would you conclude if the zero-order r and the partial r for these variables were essentially identical?
6. Suppose you want to predict job performance by a police officer (Y) from scores on the Police Skills Inventory, a paper-and-pencil test (X_1). You also have a measure of general verbal ability (X_2). Is it possible that $r_{1Y.2}$ could actually be larger than r_{1Y}? Give reasons for your answer.
7. Give a plausible example of a three-variable research problem in which partial correlation would be a useful analysis; make sure that you indicate which of your three variables is the "controlled for" variable. What results might you expect to obtain for this partial correlation, and how would you interpret your results?

8. Describe two (of the many possible) problems that would make a partial correlation "nonsense" or uninterpretable.

9. Choose three variables. The predictor and outcome variables should both be quantitative. The control variable for this exercise should be dichotomous (although, in general, the control variable can be quantitative, that type of variable doesn't lend itself to the analysis of separate groups). For this set of three variables, do the following. Run Pearson *r*s among all three variables. Make a scatter plot of scores on the predictor and outcome variables (with cases labeled by group membership on the control variable). Determine the partial correlation.

CHAPTER 11

Multiple Regression With Two Predictor Variables

11.1 ♦ Research Situations Involving Regression With Two Predictor Variables

Until Chapter 10, we considered analyses that used only one predictor variable to predict scores on a single outcome variable. For example, in Chapter 9, bivariate regression was used to predict salary in dollars (Y) from years of job experience (X). However, it is natural to ask whether we could make a better prediction based on information about two predictor variables (we will denote these predictors X_1 and X_2). In this chapter, we will examine regression equations that use two predictor variables. The notation for a raw score regression equation to predict the score on a quantitative Y outcome variable from scores on two X variables is as follows:

$$Y' = b_0 + b_1X_1 + b_2X_2. \tag{11.1}$$

As in bivariate regression, there is also a standardized form of this predictive equation:

$$z'_Y = \beta_1 z_{X_1} + \beta_2 z_{X_2}. \tag{11.2}$$

A regression analysis that includes more than one predictor variable can provide answers to several different kinds of questions. First of all, we can do an omnibus test to assess how well scores on Y can be predicted when we use the entire set of predictor variables (i.e., X_1 and X_2 combined). Second, as a follow up to a significant overall regression analysis, we can also assess how much variance is predicted uniquely by each individual predictor variable when other predictor variables are statistically controlled (e.g., what proportion of the variance in Y is uniquely predictable by X_1 when X_2 is statistically controlled). We can make comparisons to evaluate whether the X_1 is more or less strongly predictive of Y than the X_2 predictor variable; however, such comparisons must be made with caution, because the sizes of regression slope coefficients (like the sizes of Pearson correlations) can be artifactually influenced by differences in the range, reliability, distribution shape, and other characteristics of the X_1 and X_2 predictors (as discussed in Chapter 7). In addition, regression results provide additional information to help evaluate hypotheses about the "causal" models described in Chapter 10—for example, models in which the relationship between X_1 and Y may be mediated fully or partly by X_2.

In Chapter 9, we saw that to compute the coefficients for the equation $Y' = b_0 + bX$, we needed the correlation between X and Y (r_{XY}), and the mean and standard deviation of X and Y. In regression analysis with two predictor variables, we need the means and standard deviations of Y, X_1, and X_2 and the correlation between each predictor variable and the outcome variable Y (r_{1Y} and r_{2Y}). However, we also have to take into account (and adjust for) the correlation between the predictor variables (r_{12}).

The discussion of partial correlation in Chapter 10 demonstrated how to calculate an adjusted or "partial" correlation between an X_1 predictor variable and a Y outcome variable that statistically controls for a third variable (X_2). A multiple regression that includes both X_1 and X_2 as predictors uses similar methods to statistically control for other variables when assessing the individual contribution of each predictor variable (note that linear regression and correlation only control for *linear* associations between predictors). A regression analysis that uses both X_1 and X_2 as predictors of Y provides information about how X_1 is related to Y while controlling for X_2 and, conversely, how X_2 is related to Y while controlling for X_1. In this chapter, we will see that the regression slope b_1 in Equation 11.1 and the partial correlation (pr_1, described in Chapter 10) provide similar information about the nature of the predictive relationship of X_1 with Y when X_2 is controlled.

The b_1 and b_2 regression coefficients in Equation 11.1 represent *partial* slopes. That is, b_1 represents the number of units of change in Y that is predicted for each one-unit increase in X_1 when X_2 is statistically controlled. Why do we need to statistically control for X_2 when we use both X_1 and X_2 as predictors? Chapter 10 described a number of different ways in which an X_2 variable could modify the relationship between two other variables, X_1 and Y. Any of the situations described for partial correlations in Chapter 10 (such as suppression) may also arise in regression analysis. In many research situations, X_1 and X_2 are partly redundant (or correlated) predictors of Y; in such situations, we need to control for, or partial out, the part of X_1 that is correlated with or predictable from X_2 in order to avoid "double counting" the information that is contained in both the X_1 and X_2 variables. To understand why this is so, consider a trivial prediction problem. Suppose that you want to predict people's total height in inches (Y) from two measurements that you make using a yardstick: distance from hip to top of head (X_1) and distance from waist to floor (X_2). You cannot predict Y by summing X_1 and X_2, because X_1 and X_2 contain some duplicate information (the distance from waist to hip). The $X_1 + X_2$ sum would overestimate Y because it includes the waist-to-hip distance twice. When you perform a multiple regression of the form shown in Equation 11.1, the b coefficients are adjusted so that when the X_1 and X_2 variables are correlated or contain redundant information, this information is not double counted. Each variable's contribution to the prediction of Y is estimated using computations that partial out other predictor variables; this corrects for, or removes, any information in the X_1 score that is predictable from the X_2 score (and vice versa).

However, regression using X_1 and X_2 to predict Y can also be used to assess other types of situations described in Chapter 10—for example, situations where the association between X_1 and Y is positive when X_1 is not taken into account and negative when X_1 is taken into account.

11.2 ♦ Hypothetical Research Example

As a concrete example of a situation in which two predictors might be assessed in combination, consider the data used in Chapter 10. A researcher measures age (X_1) and weight (X_2) and uses these two variables to predict blood pressure (Y) (data for this example appeared in Table 10.3). In this situation, it would be reasonable to expect that the predictor variables would be correlated with each other to some extent (e.g., as people get older, they often tend to gain weight). It is plausible that both predictor variables might contribute unique information toward the prediction of blood pressure. For example, weight might directly cause increases in blood pressure, but in addition, there might be other mechanisms through which age causes increases in blood pressure; for example, age-related increases in artery blockage might also contribute to increases in blood pressure. In this analysis, we might expect to find that the two variables together are strongly predictive of blood pressure and that each predictor variable contributes significant unique predictive information. Also, we would expect that both coefficients would be positive (i.e., as age and weight increase, blood pressure should also tend to increase).

Many outcomes are possible when two variables are used as predictors in a multiple regression. The overall regression analysis can either be significant or not significant, and each predictor variable may or may not make a statistically significant unique contribution. As we saw in the discussion of partial correlation (in Chapter 10), the assessment of the contribution of an individual predictor variable controlling for another variable can lead to the conclusion that a predictor provides useful information even when another variable is statistically controlled or, conversely, that a predictor becomes nonsignificant when another variable is statistically controlled. The same types of interpretations (e.g., spuriousness, possible mediated relationships, and so forth) that were described for partial correlation outcomes in Chapter 10 can be considered as possible explanations for multiple regression results. In this chapter, we will examine the two-predictor-variable situation in detail; comprehension of the two-predictor situation will be helpful in understanding regression analyses with more than two predictors in later chapters.

To summarize, when we include two (or more) predictor variables in a regression, we sometimes choose one or more of the predictor variables because we hypothesize that they might be causes of the Y variable or at least useful predictors of Y. On the other hand, sometimes rival predictor variables are included in a regression because they are correlated with, confounded with, or redundant with a primary explanatory variable; in some situations, researchers hope to demonstrate that a rival variable completely "accounts for" the apparent correlation between the primary variable of interest and Y, while in other situations, researchers hope to show that rival variables do not completely account for any correlation of the primary predictor variable with the Y outcome variable. Sometimes a well-chosen X_2 control variable can be used to partial out sources of measurement error in another X_1 predictor variable (e.g., verbal ability is a common source of measurement error when paper-and-pencil tests are used to assess skills that are largely nonverbal, such as playing tennis or mountain survival). An X_2 variable may also be included as a predictor because the researcher suspects that the X_2 variable may "suppress" the relationship of another X_1 predictor variable with the Y outcome variable. Chapter 10 described how partial correlation and scatter plots could be used for preliminary examination of these types of outcomes in three-variable research situations. This chapter shows that regression analysis using X_1 and X_2 as predictors of Y provides additional information about three-variable research situations.

11.3 ♦ Graphic Representation of Regression Plane

In Chapter 9, a two-dimensional graph was used to diagram the scatter plot of Y values for each value of X. The regression prediction equation $Y' = b_0 + bX$ corresponded to a line on this graph. If the regression fits the data well, most of the actual Y scores fall relatively close to this line. The b coefficient represented the slope of this line (for a one-unit increase in X, the regression equation predicted a b-unit increase in Y').

When we add a second predictor variable, X_2, we need to use a three-dimensional graph to represent the pattern on scores for three variables. Imagine a cube with X_1, X_2, and Y dimensions; the data points form a cluster in this three-dimensional space. The best-fitting regression equation, $Y' = b_0 + b_1X_1 + b_2X_2$, can be represented as a plane that intersects this space; for a good fit, we need a **regression plane** that has points clustered close to it in this three-dimensional space. See Figure 11.1 for a graphic representation of a regression plane.

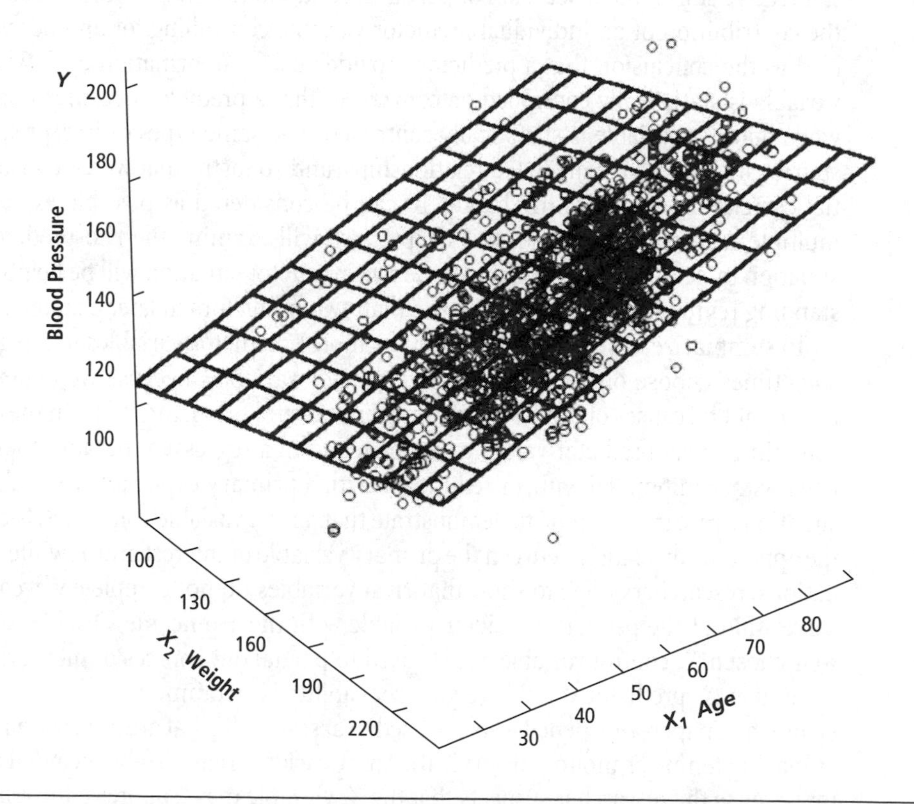

Figure 11.1 ♦ Three-Dimensional Graph of Multiple Regression Plane with X_1 and X_2 as Predictors of Y

SOURCE: Reprinted with permission from Palmer, M., http://ordination.okstate.edu/plane.jpg.

A more concrete way to visualize this situation is to imagine the X_1, X_2 points as locations on a tabletop (where X_1 represents the location of a point relative to the longer side of the table and X_2 represents the location along the shorter side). You could draw a grid on the top of the table to show the location of each subject's X_1, X_2 pair of scores on the flat plane represented by the tabletop. When you add a third variable, Y, you need to add a third dimension to show the location of the Y score that corresponds to each particular pair of X_1, X_2 score values; the Y values can be represented by points that float in space above the top of the table. For example, X_1 can be age, X_2 can be weight, and Y can be blood pressure. The regression plane can then be represented by a piece of paper held above the tabletop, oriented so that it is centered within the cluster of data points that float in space above the table. The b_1 slope represents the degree of tilt in the paper in the X_1 direction, parallel to the width of the table (i.e., the slope to predict blood pressure from age for a specific weight). The b_2 slope represents the slope of the paper in the X_2 direction, parallel to the length of the table (i.e., the slope to predict blood pressure from weight at some specific age).

Thus, the partial slopes b_1 and b_2, described earlier, can be understood in terms of this graph. The b_1 partial slope (in the regression equation $Y' = b_0 + b_1X_1 + b_2X_2$) has the following verbal interpretation: For a one-unit increase in scores on X_1, the best-fitting regression equation makes a b_1-point increase in the predicted Y' score (controlling for or partialling out any changes associated with the other predictor variable, X_2).

11.4 ♦ Semipartial (or "Part") Correlation

Chapter 10 described how to calculate and interpret a partial correlation between X_1 and Y, controlling for X_2. One method that can be used to obtain $r_{Y1.2}$ (the partial correlation between X_1 and Y, controlling for X_2) is to perform a simple bivariate regression to predict X_1 from X_2, run another regression to predict Y from X_2, and then correlate the residuals from these two regressions (X_1^* and Y^*). This correlation was denoted by $r_{1Y.2}$, which is read as "the partial correlation between X_1 and Y, controlling for X_2." This partial r tells us how X_1 is related to Y when X_2 has been removed from or partialled out of both the X_1 and the Y variables. The squared partial r correlation, $r^2_{Y1.2}$, can be interpreted as the proportion of variance in Y that can be predicted from X_1 when all the variance that is linearly associated with X_2 is removed from both the X_1 and the Y variables.

Partial correlations are sometimes reported in studies where the researcher wants to assess the strength and nature of the X_1, Y relationship with the variance that is linearly associated with X_2 completely removed from both variables. This chapter introduces a slightly different statistic (the semipartial or **part correlation**) that provides information about the **partition of variance** between predictor variables X_1 and X_2 in regression in a more convenient form. A semipartial correlation is calculated and interpreted slightly differently from the partial correlation, and a different notation is used. The semipartial (or "part") correlation between X_1 and Y, controlling for X_2, is denoted by $r_{Y(1.2)}$. Another common notation for the semipartial correlation is sr_i, where X_i is the predictor variable. In this notation for semipartial correlation, it is implicit that the outcome variable is Y; the predictive association between X_i and Y is assessed while removing the variance from X_i that is shared with any other predictor variables in the regression equation.

To obtain this semipartial correlation, we remove the variance that is associated with X_2 from only the X_1 predictor (and not from the Y outcome variable). For example, to obtain the semipartial correlation $r_{Y(1.2)}$, the semipartial correlation that describes the strength of the association between Y and X_1 when X_2 is partialled out of X_1, do the following:

1. First, run a simple bivariate regression to predict X_1 from X_2. Obtain the residuals (X_1^*) from this regression. X_1^* represents the part of the X_1 scores that is not predictable from or correlated with X_2.
2. Then, correlate X_1^* with Y to obtain the semipartial correlation between X_1 and Y, controlling for X_2. Note that X_2 has been partialled out of, or removed from, only the other predictor variable X_1; the variance associated with X_2 has not been partialled out of or removed from Y, the outcome variable.

This is called a semipartial correlation because the variance associated with X_2 is removed from only one of the two variables (and not removed entirely from both X_1 and Y as in the partial correlation presented in Chapter 10).

It is also possible to compute the semipartial correlation, $r_{Y(1.2)}$, directly from the three bivariate correlations (r_{12}, r_{1Y}, and r_{2Y}):

$$r_{Y(1.2)} = \frac{r_{1Y} - r_{2Y} \times r_{12}}{\sqrt{1 - r_{12}^2}}. \tag{11.3}$$

In most research situations, the partial and semipartial correlations (between X_1 and Y, controlling for X_2) yield similar values. The squared semipartial correlation has a simpler interpretation than the squared partial correlation when we want to describe the partitioning of variance among predictor variables in a multiple regression. The squared semipartial correlation between X_1 and Y, controlling for X_2—that is, $r^2_{Y(1.2)}$ or sr_1^2—is equivalent to the proportion of the total variance of Y that is predictable from X_1 when the variance that is shared with X_2 has been partialled out of X_1. It is more convenient to report squared semipartial correlations (instead of squared partial correlations) as part of the results of regression analysis because squared semipartial correlations correspond to proportions of the total variance in Y, the outcome variable, that are associated uniquely with each individual predictor variable.

11.5 ♦ Graphic Representation of Partition of Variance in Regression With Two Predictors

In multiple regression analysis, one goal is to obtain a partition of variance for the dependent variable Y (blood pressure) into variance that can be accounted for or predicted by each of the predictor variables, X_1 (age) and X_2 (weight), taking into account the overlap or correlation between the predictors. Overlapping circles can be used to represent the proportion of shared variance (r^2) for each pair of variables in this situation. Each circle has a total area of 1 (this represents the total variance of z_Y, for example). For each pair of variables, such as X_1

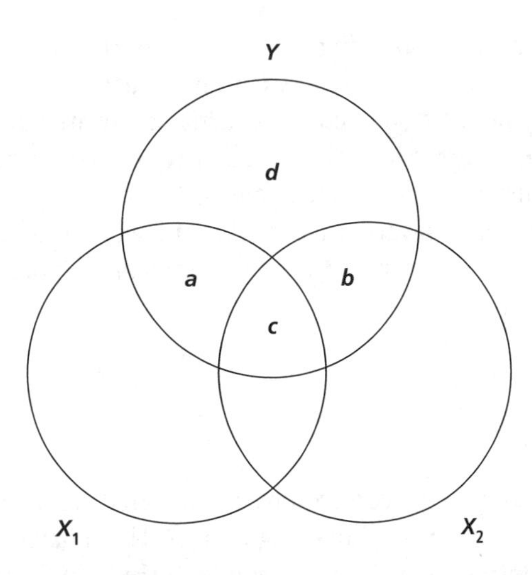

Figure 11.2 ♦ Partition of Variance of Y in a Regression With Two Predictor Variables X_1 and X_2

NOTE: The areas a, b, c, and d correspond to the following proportions of variance in Y, the outcome variable:

Area a	sr_1^2	The proportion of variance in Y that is predictable uniquely from X_1 when X_2 is statistically controlled or partialled out
Area b	sr_2^2	The proportion of variance in Y that is predictable uniquely from X_2 when X_1 is statistically controlled or partialled out
Area c		The proportion of variance in Y that could be explained by either X_1 or X_2; Area c can be obtained by subtraction, e.g., $c = 1 - (a + b + d)$
Area $a + b + c$	$R^2_{Y.12}$	The overall proportion of variance in Y predictable from X_1 and X_2 combined
Area d	$1 - R^2_{Y.12}$	The proportion of variance in Y that is not predictable from either X_1 or X_2

and Y, the squared correlation between X_1 and Y (i.e., r^2_{Y1}) corresponds to the proportion of the total variance of Y that overlaps with X_1, as shown in Figure 11.2.

The total variance of the outcome variable (such as Y, blood pressure) corresponds to the circle in Figure 11.2 with sections that are labeled a, b, c, and d. We will assume that the total area of this circle corresponds to the total variance of Y and that Y is given in z score units, so the total variance or total area $a + b + c + d$ in this diagram corresponds to a value of 1.0. As in earlier examples, overlap between circles that represent different variables corresponds to squared correlation; the total area of overlap between X_1 and Y (which corresponds to the sum of Areas a and c) is equal to r^2_{1Y}, the squared correlation between X_1 and Y. One goal of multiple regression is to obtain information about the partition of variance in the outcome variable into the following components. Area d in the diagram corresponds to the proportion of variance in Y that is not predictable from either X_1 or X_2. Area a in this diagram corresponds to the proportion of variance in Y that is uniquely predictable from X_1 (controlling for or partialling out any variance in X_1 that is shared with X_2). Area b corresponds to the proportion of variance in Y that is uniquely predictable from X_2 (controlling for or partialling out any variance in X_2 that is shared with the other predictor, X_1). Area c corresponds to a proportion of variance in Y that can be predicted by either X_1 or X_2. We can use results from a multiple regression analysis that predicts Y from X_1 and X_2 to deduce the proportions of variance that correspond to each of these areas, labeled a, b, c, and d, in this diagram.

We can interpret squared semipartial correlations as information about variance partitioning in regression. We can calculate zero-order correlations among all these variables by running Pearson correlations of X_1 with Y, X_2 with Y, and X_1 with X_2. The overall squared zero-order bivariate correlations between X_1 and Y and between X_2 and Y correspond to the areas that show the total overlap of each predictor variable with Y as follows:

$$a + c = r_{Y1}^2,$$

$$b + c = r_{Y2}^2.$$

The squared partial correlations and squared semipartial rs can also be expressed in terms of areas in the diagram in Figure 11.2. The squared semipartial correlation between X_1 and Y, controlling for X_2, corresponds to Area a in Figure 11.2; the squared semipartial correlation sr_1^2 can be interpreted as "the proportion of the total variance of Y that is uniquely predictable from X_1." In other words, sr_1^2 (or $r_{Y(1.2)}^2$) corresponds to Area a in Figure 11.2.

The squared partial correlation has a somewhat less convenient interpretation; it corresponds to a ratio of areas in the diagram in Figure 11.2. When a partial correlation is calculated, the variance that is linearly predictable from X_2 is removed from the Y outcome variable, and therefore, the proportion of variance that remains in Y after controlling for X_2 corresponds to the sum of Areas a and d. The part of this remaining variance in Y that is uniquely predictable from X_1 corresponds to Area a; therefore, the squared partial correlation between X_1 and Y, controlling for X, corresponds to the ratio $a/(a + d)$. In other words, pr_1^2 (or $r_{Y1.2}^2$) corresponds to a ratio of areas, $a/(a + d)$.

We can "reconstruct" the total variance of Y, the outcome variable, by summing Areas a, b, c, and d in Figure 11.2. Because Areas a and b correspond to the squared semipartial correlations of X_1 and X_2 with Y, it is more convenient to report squared semipartial correlations (instead of squared partial correlations) as effect size information for a multiple regression. Area c represents variance that could be explained equally well by either X_1 or X_2.

In multiple regression, we seek to partition the variance of Y into components that are uniquely predictable from individual variables (Areas a and b) and areas that are explainable by more than one variable (Area c). We will see that there is more than one way to interpret the variance represented by Area c. The most conservative strategy is not to give either X_1 or X_2 credit for explaining the variance that corresponds to Area c in Figure 11.2. Areas a, b, c, and d in Figure 11.2 correspond to proportions of the total variance of Y, the outcome variable, as given in the table below the overlapping circles diagram.

In words, then, we can divide the total variance of scores on the Y outcome variable into four components when we have two predictors: the proportion of variance in Y that is uniquely predictable from X_1 (Area a, sr_1^2); the proportion of variance in Y that is uniquely predictable from X_2 (Area b, sr_2^2); the proportion of variance in Y that could be predicted from either X_1 or X_2 (Area c, obtained by subtraction); and the proportion of variance in Y that cannot be predicted from either X_1 or X_2 (Area d, $1 - R_{Y.12}^2$).

Note that the sum of the proportions for these four areas, $a + b + c + d$, equals 1 because the circle corresponds to the total variance of Y (an area of 1.00). In this chapter, we will see that information obtained from the multiple regression analysis that predicts scores on Y from X_1 and X_2 can be used to calculate the proportions that correspond to each of these four areas (a, b, c, and d). When we write up results, we can comment on whether the two variables combined explained a large or a small proportion of variance in Y; we can also note how much of the variance was predicted uniquely by each predictor variable.

If X_1 and X_2 are uncorrelated with each other, then there is no overlap between the circles that correspond to the X_1 and X_2 variables in this diagram and Area c is 0. However, in most applications of multiple regression, X_1 and X_2 are correlated with each other to some degree; this is represented by an overlap between the circles that represent the variances of X_1 and X_2. When some types of suppression are present, the value obtained for Area c by taking 1.0 – Area a – Area b – Area d can actually be a negative value; in such situations, the overlapping circle diagram may not be the most useful way to think about variance partitioning. The partition of variance that can be made using multiple regression allows us to assess the total predictive power of X_1 and X_2 when these predictors are used together and also to assess their unique contributions, so that each predictor is assessed while statistically controlling for the other predictor variable.

In regression, as in many other multivariable analyses, the researcher can evaluate results in relation to several different questions. The first question is, Are the two predictor variables together significantly predictive of Y? Formally, this corresponds to the following null hypothesis:

$$H_0: R_{Y.12} = 0. \tag{11.4}$$

In Equation 11.4, an explicit notation is used for R (with subscripts that specifically indicate the dependent and independent variables). That is, $R_{Y.12}$ denotes the multiple R for a regression equation in which Y is predicted from X_1 and X_2. In this subscript notation, the variable to the left of the period in the subscript is the outcome or dependent variable; the numbers to the right of the period represent the subscripts for each of the predictor variables (in this example, X_1 and X_2). This explicit notation is used when it is needed to make it clear exactly which outcome and predictor variables are included in the regression.

In most reports of multiple regression, these subscripts are omitted; and it is understood from the context that $\boldsymbol{R^2}$ stands for the proportion of variance explained by the entire set of predictor variables that are included in the analysis. Subscripts on R and R^2 are generally used only when it is necessary to remove possible ambiguity. Thus, the formal null hypothesis for the overall multiple regression can be written more simply as follows:

$$H_0: R = 0. \tag{11.5}$$

Recall that multiple R refers to the correlation between Y and Y' (i.e., the correlation between observed scores on Y and the predicted Y' scores that are formed by summing the weighted scores on X_1 and X_2, $Y' = b_0 + b_1X_1 + b_2X_2$).

A second set of questions that can be addressed using multiple regression involves the unique contribution of each individual predictor. Sometimes, data analysts do not test the significance of individual predictors unless the F for the overall regression is statistically significant. Requiring a significant F for the overall regression before testing the significance of individual predictor variables used to be recommended as a way to limit the increased risk of Type I error that arises when many predictors are assessed; however, the requirement of a significant overall *F* for the regression model as a condition for conducting significance tests on individual predictor variables probably does not provide much protection against Type I error in practice.

For each predictor variable in the regression—for instance, for X_i—the null hypothesis can be set up as follows:

$$H_0: b_i = 0, \tag{11.6}$$

where b_i represents the unknown population raw score slope[1] that is estimated by the sample slope. If the b_i coefficient for predictor X_i is statistically significant, then there is a significant increase in predicted *Y* values that is uniquely associated with X_i (and not attributable to other predictor variables).

It is also possible to ask whether X_1 is more strongly predictive of *Y* than X_2 (by comparing β_1 and β_2). However, comparisons between regression coefficients must be interpreted very cautiously; factors that artifactually influence the magnitude of correlations (discussed in Chapter 7) can also artifactually increase or decrease the magnitude of slopes.

11.6 ♦ Assumptions for Regression With Two Predictors

For the simplest possible multiple regression with two predictors, as given in Equation 11.1, the assumptions that should be satisfied are basically the same as the assumptions described in the earlier chapters on Pearson correlation and bivariate regression. Ideally, all the following conditions should hold:

1. The *Y* outcome variable should be a quantitative variable with scores that are approximately normally distributed. Possible violations of this assumption can be assessed by looking at the univariate distributions of scores on *Y*. The X_1 and X_2 predictor variables should be normally distributed and quantitative, or one or both of the predictor variables can be dichotomous (or dummy) variables (as will be discussed in Chapter 12). If the outcome variable, *Y*, is dichotomous, then a different form of analysis (binary logistic regression) should be used (see Chapter 21).

2. The relations among all pairs of variables (X_1, X_2), (X_1, Y), and (X_2, Y) should be linear. This assumption of linearity can be assessed by examining bivariate scatter plots for all possible pairs of these variables. Also, as discussed in Chapter 4 on data screening, these plots should not have any extreme bivariate outliers.

3. There should be no interactions between variables, such that the slope that predicts *Y* from X_1 differs across groups that are formed based on scores on X_2. An alternative way to state this assumption is that the regressions to predict *Y* from X_1 should be homogeneous across levels of X_2. This can be qualitatively assessed by grouping subjects based on scores on the X_2 variable and running a separate X_1, *Y* scatter plot or bivariate regression for each group; the slopes should be similar across groups. If this assumption is violated and if the slope relating *Y* to X_1 differs across levels of X_2, then it would not be possible to use a flat plane to represent the relation among

the variables as in Figure 11.1. Instead, you would need a more complex surface that has different slopes to show how Y is related to X_1 for different values of X_2. (Chapter 12 demonstrates how to include interaction terms in regression models and how to test for the statistical significance of interactions between predictors.)

4. Variance in Y scores should be homogeneous across levels of X_1 (and levels of X_2); this assumption of homogeneous variance can be assessed in a qualitative way by examining bivariate scatter plots to see whether the range or variance of Y scores varies across levels of X. For an example of a scatter plot of hypothetical data that violate this homogeneity of variance assumption, see Figure 7.9. Formal tests of homogeneity of variance are possible, but they are rarely used in regression analysis. In many real-life research situations, researchers do not have a sufficiently large number of scores for each specific value of X to set up a test to verify whether the variance of Y is homogeneous across values of X.

As in earlier analyses, possible violations of these assumptions can generally be assessed reasonably well by examining the univariate frequency distribution for each variable and the bivariate scatter plots for all pairs of variables. Many of these problems can also be identified by graphing the standardized residuals from regression—that is, the $z_Y - z'_Y$ prediction errors. Chapter 9 discussed the problems that can be detected by the examination of plots of residuals in bivariate regression; the same issues should be considered when examining plots of residuals for regression analyses that include multiple predictors. That is, the mean and variance of these residuals should be fairly uniform across levels of z'_Y, and there should be no pattern in the residuals (there should not be a linear or curvilinear trend). Also, there should not be extreme outliers in the plot of standardized residuals. Some of the problems that are detectable through visual examination of residuals can also be noted in univariate and bivariate data screening; however, examination of residuals may be uniquely valuable as a tool for the discovery of multivariate outliers. A multivariate outlier is a case that has an unusual combination of values of scores for variables such as X_1, X_2, and Y (even though the scores on the individual variables may not, by themselves, be outliers). A more extensive discussion of the use of residuals for assessment of violations of assumptions and the detection and possible removal of multivariate outliers is provided in Chapter 4 of Tabachnick and Fidell (2007). Multivariate or bivariate outliers can have a disproportionate impact on estimates of b or β slope coefficients (just as they can have a disproportionate impact on estimates of r, as described in Chapter 7). That is, sometimes omitting a few extreme outliers results in drastic changes in the size of b or β coefficients. It is undesirable to have the results of a regression analysis depend to a great extent on the values of a few extreme or unusual data points.

If extreme bivariate or multivariate outliers are identified in preliminary data screening, it is necessary to decide whether the analysis is more believable with these outliers included, with the outliers excluded, or using a data transformation (such as log of X) to reduce the impact of outliers on slope estimates. See Chapter 4 in this textbook for further discussion of issues to consider in decisions on the handling of outliers. If outliers are identified and removed, the rationale and decision rules for the handling of these cases should be clearly explained in the write-up of results.

The hypothetical data for this example consist of data for 30 cases on three variables (see Table 10.3): blood pressure (Y), age (X_1), and weight (X_2). Before running the multiple regression, scatter plots for all pairs of variables were examined, descriptive statistics were obtained for each variable, and zero-order correlations were computed for all pairs of variables using the methods described in previous chapters. It is also a good idea to examine histograms of the distribution of scores on each variable, as discussed in Chapter 4, to assess whether scores on continuous predictor variables are reasonably normally distributed without extreme outliers.

A matrix of scatter plots for all possible pairs of variables was obtained through the SPSS menu sequence <Graph> → <Scatter/Dot>, followed by clicking on the Matrix Scatter icon, shown in Figure 11.3. The names of all three variables (age, weight, and blood pressure) were entered in the dialog box for matrix scatter plots, which appears in Figure 11.4. The SPSS output shown in Figure 11.5 (page 436) shows the matrix scatter plots for all pairs of variables: X_1 with Y, X_2 with Y, and X_1 with X_2. Examination of these scatter plots suggested that relations between all pairs of variables were reasonably linear and there were no bivariate outliers. Variance of blood pressure appeared to be reasonably homogenous across levels of the predictor variables. The bivariate Pearson correlations for all pairs of variables appear in Figure 11.6 (page 436).

Based on preliminary data screening (including histograms of scores on age, weight, and blood pressure that are not shown here), it was judged that scores were reasonably normally distributed, relations between variables were reasonably linear, and there were no outliers extreme enough to have a disproportionate impact on the results. Therefore, it seemed appropriate to perform a multiple regression analysis on these data; no cases were dropped, and no data transformations were applied.

If there appear to be curvilinear relations between any variables, then the analysis needs to be modified to take this into account. For example, if Y shows a curvilinear pattern across levels of X_1, one way to deal with this is to recode scores on X_1 into group membership codes (e.g., if X_1 represents income in dollars, this could be recoded as three groups: low, middle, and high income levels); then, an analysis of variance (ANOVA) can be used to see whether means on Y differ across these groups (based on low, medium, or high X scores). Another possible way to incorporate nonlinearity into a regression analysis is to include X^2 (and perhaps higher powers of X, such as X^3) as a predictor of Y in a regression equation of the following form:

$$Y' = b_0 + b_1 X^1 + b_2 X^2 + b_3 X^3 + \cdots. \quad (11.7)$$

In practice, it is rare to encounter situations where powers of X higher than X^2, such as X^3 or X^4 terms, are needed. Curvilinear relations that correspond to a U-shaped or inverse U-shaped graph (in which Y is a function of X and X^2) are more common.

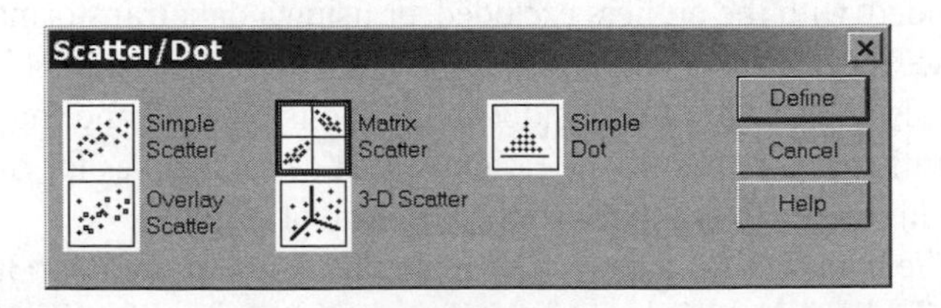

Figure 11.3 ♦ SPSS Dialog Window for Request of Matrix Scatter Plots

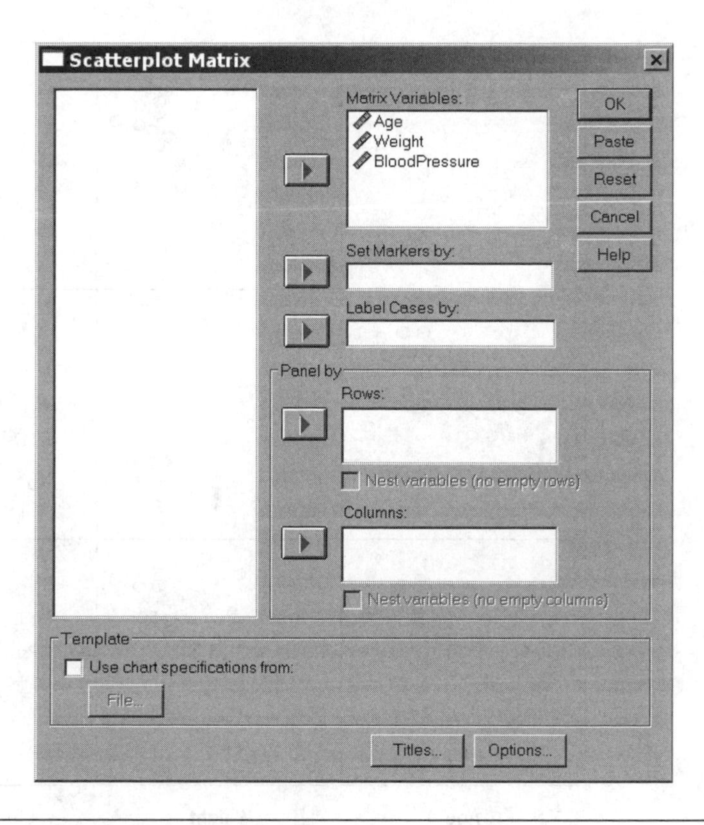

Figure 11.4 ♦ SPSS Scatterplot Matrix Dialog Window

NOTE: This generates a matrix of all possible scatter plots between pairs of listed variables (e.g., age with weight, age with blood pressure, and weight with blood pressure).

Finally, if an interaction between X_1 and X_2 is detected, it is possible to incorporate one or more interaction terms into the regression equation using methods that will be described in Chapter 12. A regression equation that does not incorporate an interaction term when there is in fact an interaction between predictors can produce misleading results. When we do an ANOVA, most programs automatically generate interaction terms to represent interactions among all possible pairs of predictors. However, when we do regression analyses, interaction terms are not generated automatically; if we want to include interactions in our models, we have to add them explicitly. The existence of possible interactions among predictors is therefore easy to overlook when regression analysis is used.

11.7 ♦ Formulas for Regression Coefficients, Significance Tests, and Confidence Intervals

11.7.1 ♦ Formulas for Standard Score Beta Coefficients

The coefficients to predict z'_Y from z_{X_1}, z_{X_2} ($z'_Y = \beta_1 z_{X_1} + \beta_2 z_{X_2}$) can be calculated directly from the zero-order Pearson rs among the three variables Y, X_1, and X_2, as shown in Equations 11.8 and 11.9. In a subsequent section, a simple path model is used to show how these formulas were derived:

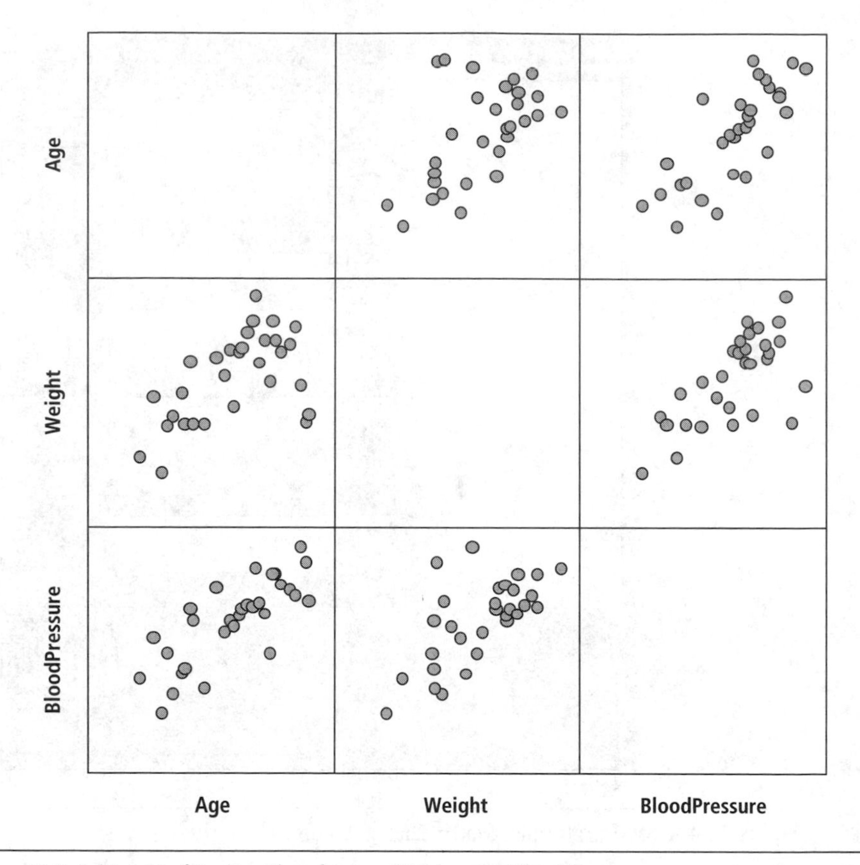

Figure 11.5 ♦ Matrix of Scatter Plots for Age, Weight, and Blood Pressure

Correlations

		Age	Weight	Blood Pressure
Age	Pearson Correlation	1	.563**	.782**
	Sig. (2-tailed)		.001	.000
	N	30	30	30
Weight	Pearson Correlation	.563**	1	.672**
	Sig. (2-tailed)	.001		.000
	N	30	30	30
BloodPressure	Pearson Correlation	.782**	.672**	1
	Sig. (2-tailed)	.000	.000	
	N	30	30	30

**. Correlation is significant at the 0.01 level (2-tailed).

Figure 11.6 ♦ Bivariate Correlations Among Age, Weight, and Blood Pressure

$$\beta_1 = \frac{r_{Y1} - r_{12}r_{Y2}}{1 - r_{12}^2} \qquad (11.8)$$

and

$$\beta_2 = \frac{r_{Y2} - r_{12}r_{Y1}}{1 - r_{12}^2}. \qquad (11.9)$$

11.7.2 ♦ Formulas for Raw Score (*b*) Coefficients

Given the beta coefficients and the means (M_Y, M_{X_1}, and M_{X_2}) and standard deviations (SD_Y, SD_{X_1}, and SD_{X_2}) of Y, X_1, and X_2, respectively, it is possible to calculate the *b* coefficients for the raw score prediction equation shown in Equation 11.1 as follows:

$$b_1 = \frac{SD_Y}{SD_{X_1}} \times \beta_1 \qquad (11.10)$$

and

$$b_2 = \frac{SD_Y}{SD_{X_2}} \times \beta_2. \qquad (11.11)$$

Note that these equations are analogous to Equation 9.4 for the computation of *b* from *r* (or β) in a bivariate regression, where $b = (SD_Y / SD_X)r_{XY}$. To obtain *b* from β, we needed to restore the information about the scales in which *Y* and the predictor variable were measured (information that is not contained in the unit-free beta coefficient). As in the bivariate regression described in Chapter 9, a *b* coefficient is a rescaled version of β—that is, rescaled so that the coefficient can be used to make predictions from raw scores rather than *z* scores.

Once we have estimates of the b_1 and b_2 coefficients, we can compute the intercept b_0:

$$b_0 = M_Y - b_1 M_{X_1} - b_2 M_{X_2}. \qquad (11.12)$$

This is analogous to the way the intercept was computed for a bivariate regression in Chapter 9, $b_0 = M_Y - bM_X$. There are other by-hand computational formulas to compute *b* from the sums of squares and sums of cross products for the variables; however, the formulas shown in the preceding equations make it clear how the *b* and β coefficients are related to each other and to the correlations among variables. In a later section of this chapter, you will see how the formulas to estimate the beta coefficients can be deduced from the correlations among the variables, using a simple path model for the regression. The computational formulas for the beta coefficients, given in Equations 11.8 and 11.9, can be understood conceptually: They are not just instructions for computation. These equations tell us that the values of the beta coefficients are influenced not only by the correlation between each *X* predictor variable and *Y* but also by the correlations between the *X* predictor variables.

11.7.3 ♦ Formulas for Multiple *R* and Multiple R^2

The multiple R can be calculated by hand. First of all, you could generate a predicted Y' score for each case by substituting the X_1 and X_2 raw scores into the equation and computing Y' for each case. Then, you could compute the Pearson r between Y (the actual Y score) and Y' (the predicted score generated by applying the regression equation to X_1 and X_2). Squaring this Pearson correlation yields R^2, the multiple R squared; this tells you what proportion of the total variance in Y is predictable from X_1 and X_2 combined.

Another approach is to examine the ANOVA source table for the regression (part of the SPSS output). As in the bivariate regression, SPSS partitions SS_{total} for Y into $SS_{\text{regression}}$ + SS_{residual}. Multiple R^2 can be computed from these sums of squares:

$$R^2 = \frac{SS_{\text{regression}}}{SS_{\text{total}}}. \tag{11.13}$$

There is a slightly different version of this overall goodness-of-fit index called the "adjusted" or "shrunken" R^2. This is adjusted for the effects of sample size (N) and number of predictors. There are several formulas for adjusted R^2; Tabachnick and Fidell (2007) provided this example:

$$R^2_{\text{adj}} = 1 - (1 - R^2)\left(\frac{N-1}{N-k-1}\right), \tag{11.14}$$

where N is the number of cases, k is the number of predictor variables, and R^2 is the squared multiple correlation given in Equation 11.13. R^2_{adj} tends to be smaller than R^2; it is much smaller than R^2 when N is relatively small and k is relatively large. In some research situations where the sample size N is very small relative to the number of variables k, the value reported for R^2_{adj} is actually negative; in these cases, it should be reported as 0. For computations involving the partition of variance (as shown in Figure 11.14 on page 459), the unadjusted R^2 was used rather than the adjusted R^2.

11.7.4 ♦ Test of Significance for Overall Regression: Overall *F* Test for H_0: $R = 0$

As in bivariate regression, an ANOVA can be performed to obtain sums of squares that represent the proportion of variance in Y that is and is not predictable from the regression, the sums of squares can be used to calculate mean squares (MS), and the ratio $MS_{\text{regression}}/MS_{\text{residual}}$ provides the significance test for R. N stands for the number of cases, and k is the number of predictor variables. For the regression examples in this chapter, the number of predictor variables, k, equals 2. (Chapter 14 shows how these procedures can be generalized to handle situations where there are more than two predictor variables.)

$$F = \frac{SS_{\text{regression}}/k}{SS_{\text{residual}}/(N-k-1)}, \tag{11.15}$$

with ($k, N - k - 1$) degrees of freedom (df).

If the obtained F ratio exceeds the tabled critical value of F for the predetermined alpha level (usually $\alpha = .05$), then the overall multiple R is judged statistically significant.

11.7.5 ♦ Test of Significance for Each Individual Predictor: *t* Test for H_0: $b_i = 0$

Recall from Chapter 2 that many sample statistics can be tested for significance by examining a *t* ratio of the following form:

$$t = \frac{\text{Sample statistic} - \text{Hypothesized population parameter}}{SE_{\text{sample statistic}}}.$$

The printout from SPSS includes an estimated standard error (SE_b) associated with each raw score slope coefficient (*b*). This standard error term can be calculated by hand in the following way. First, you need to know SE_{est}, the standard error of the estimate, defined earlier in Chapter 9, which can be computed as

$$SE_{\text{est}} = SD_Y\sqrt{(1 - R^2)} \times \sqrt{\frac{N}{N-2}}. \quad (11.16)$$

As described in Chapter 9 on bivariate regression, SE_{est} describes the variability of the observed or actual *Y* values around the regression prediction at each specific value of the predictor variables. In other words, it gives us some idea of the typical magnitude of a prediction error when the regression equation is used to generate a Y' predicted value. Using SE_{est}, it is possible to compute an SE_b term for each *b* coefficient, to describe the theoretical sampling distribution of the slope coefficient. For predictor X_i, the equation for SE_{b_i} is as follows:

$$SE_{b_i} = \frac{SE_{\text{est}}}{\sqrt{\sum (X_i - M_{X_i})^2}}. \quad (11.17)$$

The hypothesized value of each *b* slope coefficient is 0. Thus, the significance test for each raw score b_i coefficient is obtained by the calculation of a *t* ratio, b_i divided by its corresponding *SE* term:

$$t_i = \frac{b_i}{SE_{b_i}} \quad \text{with } (N - k - 1)\ df. \quad (11.18)$$

If the *t* ratio for a particular slope coefficient, such as b_1, exceeds the tabled critical value of *t* for $N - k - 1$ *df*, then that slope coefficient can be judged statistically significant. Generally, a two-tailed or nondirectional test is used.

Some multiple regression programs provide an *F* test (with 1 and $N - k - 1$ *df*) rather than a *t* test as the significance test for each *b* coefficient. Recall that when the numerator has only 1 *df*, *F* is equivalent to t^2.

11.7.6 ♦ Confidence Interval for Each *b* Slope Coefficient

A confidence interval (CI) can be set up around each sample b_i coefficient, using SE_{bi}. To set up a 95% CI, for example, use the *t* distribution table to look up the critical value of *t* for $N - k - 1$ *df* that cuts off the top 2.5% of the area, t_{crit}:

$$\text{Upper bound of 95\% CI} = b_i + t_{\text{crit}} \times SE_{b_i}. \quad (11.19)$$

$$\text{Lower bound of 95\% CI} = b_i - t_{\text{crit}} \times SE_{b_i}. \quad (11.20)$$

11.8 ◆ SPSS Regression Results

To run the SPSS linear regression procedure and to save the predicted Y' scores and the unstandardized residuals from the regression analysis, the following menu selections were made: <Analyze> → <Regression> → <Linear>.

In the SPSS linear regression dialog window (which appears in Figure 11.7), the name of the dependent variable (blood pressure) was entered in the box labeled Dependent; the names of both predictor variables were entered in the box labeled Independent. CIs for the *b* slope coefficients and values of the part and partial correlations were requested in addition to the default output by clicking the button Statistics and checking the boxes for CIs and for part and partial correlations, as shown in the previous linear regression example in Chapter 9. Note that the value that SPSS calls a "part" correlation is called the "semipartial" correlation by most textbook authors. The part correlations are needed to calculate the squared part or semipartial correlation for each predictor variable and to work out the partition of variance for blood pressure. Finally, as in the regression example that was presented in Chapter 9, the Plots button was clicked, and a graph of standardized residuals against standardized predicted scores was requested to evaluate whether assumptions for regression were violated. The resulting SPSS syntax was copied into the Syntax window by clicking the Paste button; this syntax appears in Figure 11.8.

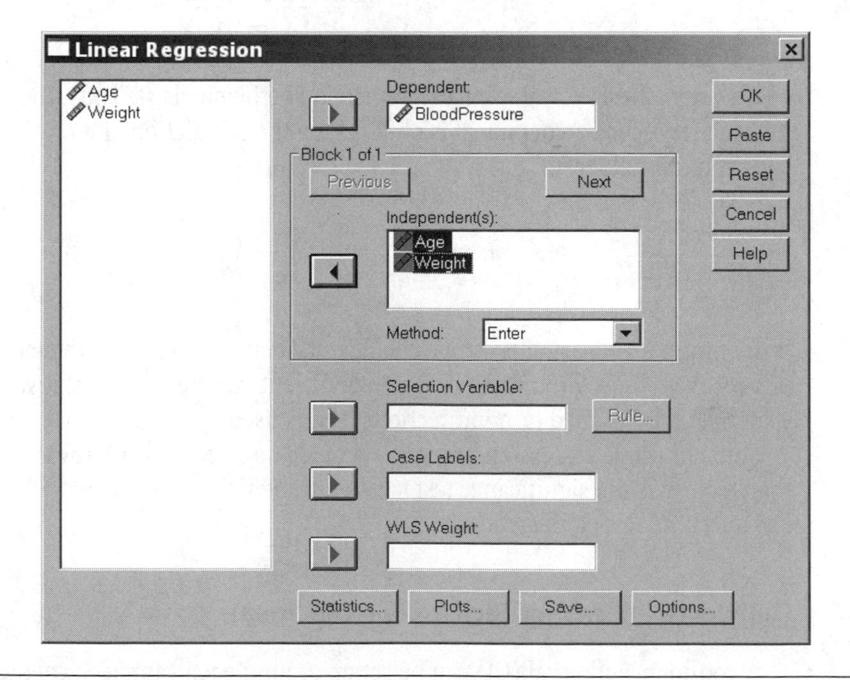

Figure 11.7 ◆ SPSS Linear Regression Dialog Window for a Regression to Predict Blood Pressure From Age and Weight

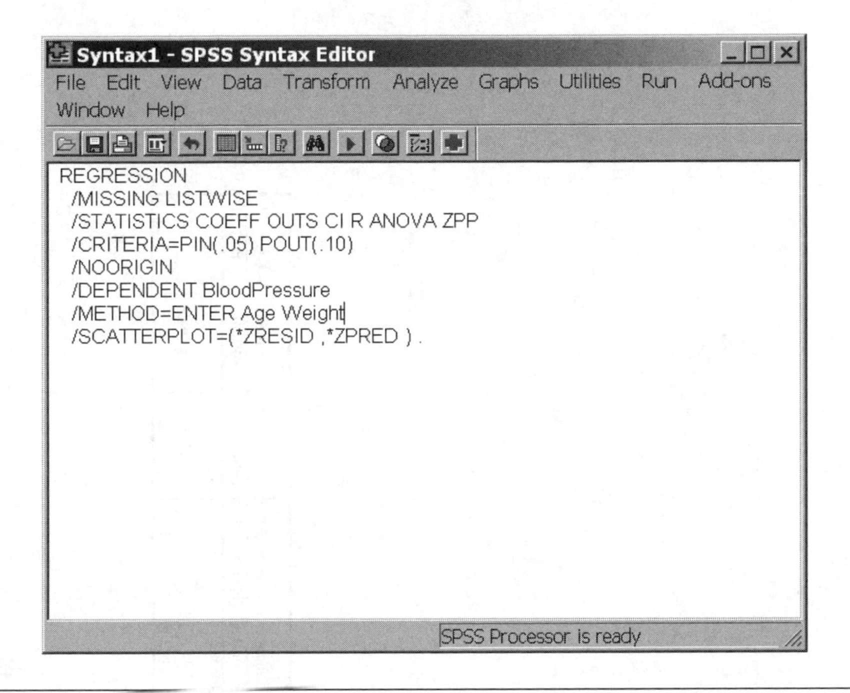

Figure 11.8 ♦ Syntax for the Regression to Predict Blood Pressure From Age and Weight (Including Part and Partial Correlations and a Plot of Standardized Residuals)

The resulting output for the regression to predict blood pressure from both age and weight appears in Figure 11.9, and the plot of the standardized residuals for this regression appears in Figure 11.10. The overall regression was statistically significant: $R = .83$, $F(2, 27) = 30.04$, $p < .001$. Thus, blood pressure could be predicted at levels significantly above chance from scores on age and weight combined. In addition, each of the individual predictor variables made a statistically significant contribution. For the predictor variable age, the raw score regression coefficient b was 2.16, and this b slope coefficient differed significantly from 0, based on a t value of 4.55 with $p < .001$. The corresponding effect size for the proportion of variance in blood pressure uniquely predictable from age was obtained by squaring the value of the part correlation of age with blood pressure to yield $sr^2_{\text{age}} = .24$. For the predictor variable weight, the raw score slope $b = .50$ was statistically significant: $t = 2.62$, $p = .014$; the corresponding effect size was obtained by squaring the part correlation for weight, $sr^2_{\text{weight}} = .08$. The pattern of residuals that is shown in Figure 11.10 does not indicate any problems with the assumptions (refer back to Chapter 9 for a discussion of the evaluation of pattern in residuals from regression). These regression results are discussed and interpreted more extensively in the model Results section that appears near the end of this chapter.

11.9 ♦ Conceptual Basis: Factors That Affect the Magnitude and Sign of β and *b* Coefficients in Multiple Regression With Two Predictors

It may be intuitively obvious that the predictive slope of X_1 depends, in part, on the value of the zero-order Pearson correlation of X_1 with Y. It may be less obvious, but the value of

Variables Entered/Removed[b]

Model	Variables Entered	Variables Removed	Method
1	Weight, Age[a]	.	Enter

a. All requested variables entered.

b. Dependent Variable: BloodPressure

Model Summary[b]

Model	R	R Square	Adjusted R Square	Std. Error of the Estimate
1	.831[a]	.690	.667	36.692

a. Predictors: (Constant), Weight, Age

b. Dependent Variable: BloodPressure

ANOVA[b]

Model		Sum of Squares	df	Mean Square	F	Sig.
1	Regression	80882.13	2	40441.066	30.039	.000[a]
	Residual	36349.73	27	1346.286		
	Total	117231.9	29			

a. Predictors: (Constant), Weight, Age

b. Dependent Variable: BloodPressure

Coefficients[a]

Model		Unstandardized Coefficients		Standardized Coefficients	t	Sig.	95% Confidence Interval for B		Correlations		
		B	Std. Error	Beta			Lower Bound	Upper Bound	Zero-order	Partial	Part
1	(Constant)	-28.046	27.985		-1.002	.325	-85.466	29.373			
	Age	2.161	.475	.590	4.551	.000	1.187	3.135	.782	.659	.488
	Weight	.490	.187	.340	2.623	.014	.107	.873	.672	.451	.281

a. Dependent Variable: BloodPressure

Residuals Statistics[a]

	Minimum	Maximum	Mean	Std. Deviation	N
Predicted Value	66.13	249.62	177.27	52.811	30
Residual	-74.752	63.436	.000	35.404	30
Std. Predicted Value	-2.104	1.370	.000	1.000	30
Std. Residual	-2.037	1.729	.000	.965	30

a.Dependent Variable: BloodPressure

Figure 11.9 ♦ Output From SPSS Linear Regression to Predict Blood Pressure From Age and Weight

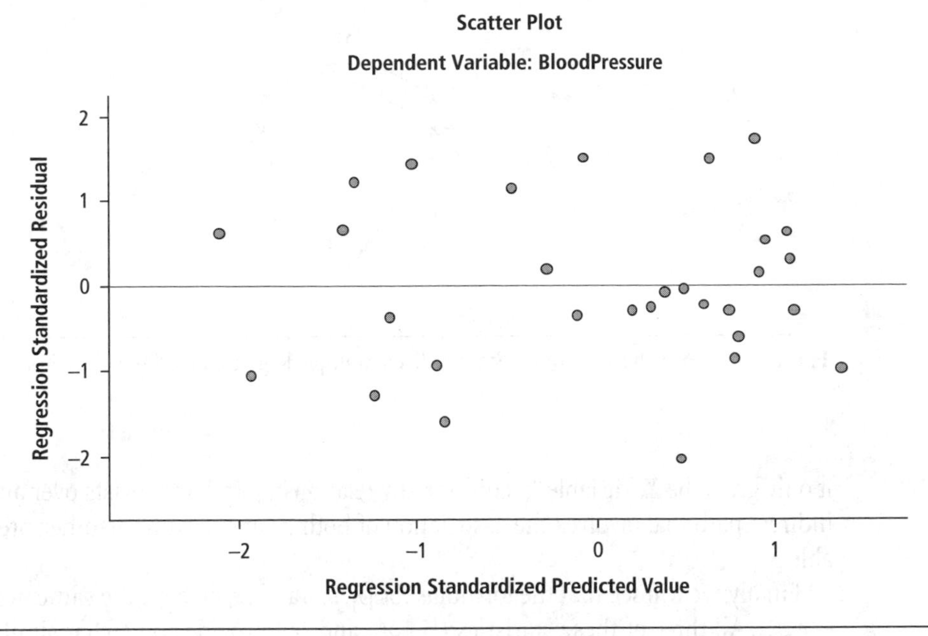

Figure 11.10 ♦ Plot of Standardized Residuals From Linear Regression to Predict Blood Pressure From Age and Weight

the slope coefficient for each predictor is also influenced by the correlation of X_1 with other predictors, as you can see in Equations 11.8 and 11.9. Often, but not always, we will find that an X_1 variable that has a large correlation with Y also tends to have a large beta coefficient; the sign of beta is often, but not always, the same as the sign of the zero-order Pearson r. However, depending on the magnitudes and signs of the r_{12} and r_{2Y} correlations, a beta coefficient (like a partial correlation) can be larger, smaller, or even opposite in sign compared with the zero-order Pearson r_{1Y}. The magnitude of a β_1 coefficient, like the magnitude of a partial correlation pr_1, is influenced by the size and sign of the correlation between X_1 and Y; it is also affected by the size and sign of the correlation(s) of the X_1 variable with other variables that are statistically controlled in the analysis.

In this section, we will examine a path diagram model of a two-predictor multiple regression to see how estimates of the beta coefficients are found from the correlations among all three pairs of variables involved in the model: r_{12}, r_{Y_1}, and r_{Y_2}. This analysis will make several things clear. First, it will show how the sign and magnitude of the standard score coefficient β_i for each X_i variable is related to the size of r_{Yi}, the correlation of that particular predictor with Y, and also to the size of the correlation of X_i and all other predictor variables included in the regression (at this point, this is the single correlation r_{12}).

Second, it will explain why the numerator for the formula to calculate β_1 in Equation 11.8 above has the form $r_{Y_1} - r_{12}r_{Y_2}$. In effect, we begin with the "overall" relationship between X_1 and Y, represented by r_{Y_1}; we subtract from this the product $r_{12} \times r_{Y_2}$, which represents an indirect path from X_1 to Y via X_2. Thus, the estimate of the β_1 coefficient is adjusted so that

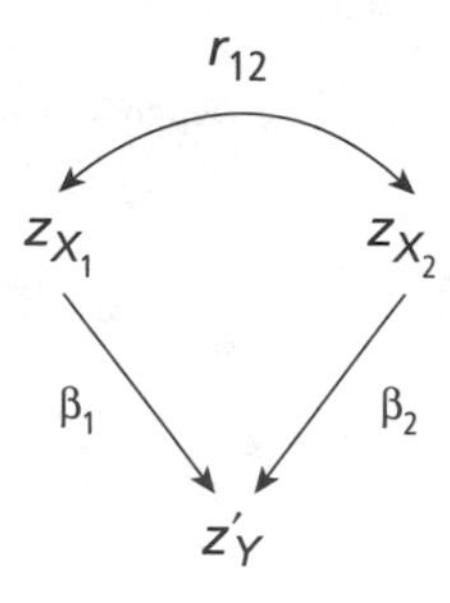

Figure 11.11 ♦ Path Diagram for Standardized Multiple Regression to Predict z'_Y From z_{X_1} and z_{X_2}

it only gives the X_1 variable "credit" for any relationship to Y that exists over and above the indirect path that involves the association of both X_1 and Y with the other predictor variable X_2.

Finally, we will see that the formulas for β_1, pr_1, and sr_1 all have the same numerator: $r_{Y_1} - r_{12}r_{Y2}$. All three of these statistics (β_1, pr_1, and sr_1) provide somewhat similar information about the nature and strength of the relation between X_1 and Y, controlling for X_2, but they are scaled slightly differently (by using different divisors) so that they can be interpreted and used in different ways.

Consider the regression problem in which you are predicting z scores on y from z scores on two independent variables X_1 and X_2. We can set up a path diagram to represent how two predictor variables are related to one outcome variable (Figure 11.11).

The path diagram in Figure 11.11 corresponds to this regression equation:

$$z'_Y = \beta_1 z_{X_1} + \beta_2 z_{X_2}. \tag{11.21}$$

Path diagrams represent hypothetical models (often called "causal" models, although we cannot prove causality from correlational analyses) that represent our hypotheses about the nature of the relations between variables. In this example, the path model is given in terms of z scores (rather than raw X scores) because this makes it easier to see how we arrive at estimates of the beta coefficients. When two variables in a path model diagram are connected by a double-headed arrow, it represents a hypothesis that the two variables are correlated or confounded (but there is no hypothesized causal connection between the variables). The Pearson r between these predictors indexes the strength of this confounding or correlation. A single-headed arrow ($X \rightarrow Y$) indicates a theorized causal relationship (such that X causes Y), or at least a directional predictive association between the variables. The "path coefficient" or regression coefficient (i.e., a beta coefficient) associated with it indicates the estimated strength of the predictive relationship through this direct path. If there is no arrow connecting a pair of variables, it indicates a lack of any direct association between the pair of variables, although the variables may be connected through indirect paths.

The path diagram that is usually implicit in a multiple regression analysis has the following general form: Each of the predictor (X) variables has a unidirectional arrow pointing from X to Y, the outcome variable. All pairs of X predictor variables are connected to each other by double-headed arrows that indicate correlation or confounding, but no presumed causal linkage, among the predictors. Figure 11.11 shows the path diagram for the standardized (z score) variables in a regression with two correlated predictor variables z_{X_1} and z_{X_2}. This model corresponds to a causal model in which z_{X_1} and z_{X_2} are represented as "partially redundant" or correlated causes or predictors of z_Y (as discussed in Chapter 10).

Our problem is to deduce the unknown path coefficients or standardized regression coefficients associated with the direct (or causal) path from each of the z_X predictors, β_1 and β_2, in terms of the known correlations r_{12}, r_{Y1}, and r_{Y2}. This is done by applying the tracing rule, as described in the following section.

11.10 ♦ Tracing Rules for Causal Model Path Diagrams

The idea behind path models is that an adequate model should allow us to reconstruct the observed correlation between any pair of variables (e.g., r_{Y1}), by tracing the paths that lead from X_1 to Y through the path system, calculating the strength of the relationship for each path, and then summing the contributions of all possible paths from X_1 to Y.

Kenny (1979) provided a clear and relatively simple statement about the way in which the paths in this causal model can be used to reproduce the overall correlation between each pair of variables: "The correlation between X_i and X_j equals the sum of the product of all the path coefficients [these are the beta weights from a multiple regression] obtained from each of the possible tracings between X_i and X_j. The set of tracings includes all possible routes from X_i to X_j given that (a) the same variable is not entered twice and (b) a variable is not entered through an arrowhead and left through an arrowhead" (p. 30). In general, the traced paths that lead from one variable, such as z_{X_1}, to another variable, such as z'_Y, may include one direct path and also one or more indirect paths.

We can use the tracing rule to reconstruct exactly[2] the observed correlation between any two variables from a path model from correlations and the beta coefficients for each path. Initially, we will treat β_1 and β_2 as unknowns; later, we will be able to solve for the betas in terms of the correlations.

Now, let's look in more detail at the multiple regression model with two independent variables (represented by the diagram in Figure 11.11). The path from z_{X_1} to z_{X_2} is simply r_{12}, the observed correlation between these variables. We will use the labels β_1 and β_2 for the coefficients that describe the strength of the direct, or unique, relationship of X_1 and X_2, respectively, to Y. β_1 indicates how strongly X_1 is related to Y after we have taken into account, or partialled out, the indirect relationship of X_1 to Y involving the path via X_2. β_1 is a partial slope: the number of standard deviation units of change in z_Y we predict for a one-*SD* change in z_{X_1} when we have taken into account, or partialled out, the influence of z_{X_2}. If z_{X_1} and z_{X_2} are correlated, we must somehow correct for the redundance of information they provide when we construct our prediction of Y; we don't want to double count information that is included in both z_{X_1} and z_{X_2}. That is why we need to correct for the correlation of z_{X_1} with z_{X_2} (i.e., take into account the indirect path from z_{X_1} to z_Y via z_{X_2}) in order to get a clear picture of how much predictive value z_{X_1} has that is *unique* to z_{X_1} and not somehow related to z_{X_2}.

For each pair of variables (z_{X_1} and z_Y, z_{X_2} and z_Y), we need to work out all possible paths from z_{X_i} to z_Y; if the path has multiple steps, the coefficients along that path are multiplied with each other. After we have calculated the strength of association for each path, we sum the contributions across paths. For the path from z_{X_1} to z'_Y, in the diagram above, there is one direct path from z_{X_1} to z'_Y, with a coefficient of β_1. There is also one indirect path from z_{X_1} to z'_Y via z_{X_2}, with two coefficients enroute (r_{12} and β_2); these are multiplied to give the strength of association represented by the indirect path, $r_{12} \times \beta_2$. Finally, we should be able to reconstruct the entire observed correlation between z_{X_1} and z_Y (r_{Y_1}) by summing the contributions of all possible paths from z_{X_1} to z'_Y in this path model. This reasoning based on the tracing rule yields the equation below:

$$\begin{array}{ccccc} \text{Total correlation} & = & \text{Direct path} & + & \text{Indirect path} \\ r_{Y_1} & = & \beta_1 & + & r_{12} \times \beta_2. \end{array} \tag{11.22}$$

Applying the same reasoning to the paths that lead from z_{X_2} to z'_Y, we arrive at a second equation of this form:

$$r_{Y_2} = \beta_2 + r_{12} \times \beta_1. \tag{11.23}$$

Equations 11.22 and 11.23 are called the normal equations for multiple regression; they show how the observed correlations (r_{Y_1} and r_{Y_2}) can be perfectly reconstructed from the regression model and its parameter estimates β_1 and β_2. We can solve these equations for values of β_1 and β_2 in terms of the known correlations r_{12}, r_{Y_1}, and r_{Y_2} (these equations appeared earlier as Equations 11.8 and 11.9):

$$\beta_1 = \frac{r_{Y_1} - r_{12} r_{Y_2}}{1 - r_{12}^2}$$

and

$$\beta_2 = \frac{r_{Y_2} - r_{12} r_{Y_1}}{1 - r_{12}^2}.$$

The numerator for the betas is the same as the numerator of the partial correlation, which we have seen in Chapter 10. Essentially, we take the overall correlation between X_1 and Y and subtract the correlation we would predict between X_1 and Y due to the relationship through the indirect path via X_2; whatever is left, we then attribute to the direct or unique influence of X_1. In effect, we "explain" as much of the association between X_1 and Y as we can by first looking at the indirect path via X_2 and only attributing to X_1 any additional relationship it has with Y that is above and beyond that indirect relationship. We then divide by a denominator that scales the result (as a partial slope or beta coefficient, in these two equations, or as a partial correlation, as in Chapter 10).

Note that if the value of β_1 is zero, we can interpret it to mean that we do not need to include a direct path from X_1 to Y in our model. If $\beta_1 = 0$, then any statistical relationship or correlation that exists between X_1 and Y can be entirely explained by the indirect path involving X_2. Possible explanations for this pattern of results include the following: that X_2 causes both X_1 and Y and the X_1,Y correlation is spurious, or that X_2 is a mediating variable and X_1 only influences Y through its influence on X_2. This is the basic idea that underlies path analysis or so-called "causal" modeling: If we find that we do not need to include a direct path between X_1 and Y, then we can simplify the model by dropping a path. We will not be able to prove causality from path analysis; we can only decide whether a causal or theoretical model that has certain paths omitted is sufficient to reproduce the observed correlations and, therefore, is "consistent" with the observed pattern of correlations.

11.11 ♦ Comparison of Equations for β, *b*, *pr*, and *sr*

By now, you may have recognized that β, *b*, *pr*, and *sr* are all slightly different indexes of how strongly X_1 predicts Y when X_2 is controlled. Note that the (partial) standardized slope or β coefficient, the partial *r*, and the semipartial *r* all have the same term in the numerator: They are scaled differently by dividing by different terms, to make them interpretable in slightly different ways, but generally, they are similar in magnitude. The numerators for partial *r* (*pr*), semipartial *r* (*sr*), and beta (β) are identical. The denominators differ slightly because they are scaled to be interpreted in slightly different ways (squared partial *r* as a proportion of variance in *Y* when X_2 has been partialled out of *Y*; squared semipartial *r* as a proportion of the total variance of *Y*; and beta as a partial slope, the number of standard deviation units of change in *Y* for a one-unit *SD* change in X_1). It should be obvious from looking at the formulas that *sr*, *pr*, and β tend to be similar in magnitude and must have the same sign. (These equations are all repetitions of equations given earlier, and therefore, they are not given new numbers here.)

Standard score slope coefficient β:

$$\beta_1 = \frac{r_{Y1} - r_{Y2}r_{12}}{(1 - r_{12}^2)}.$$

Raw score slope coefficient *b* (a rescaled version of the β coefficient):

$$b_1 = \beta_1 \times \frac{SD_Y}{SD_{X_1}}.$$

Partial correlation to predict *Y* from X_1, controlling for X_2 (removing X_2 completely from both X_1 and *Y*, as explained in Chapter 10):

$$pr_1 \text{ or } r_{Y1.2} = \frac{r_{Y1} - r_{Y2}r_{12}}{\sqrt{(1 - r_{Y2}^2) \times (1 - r_{12}^2)}}.$$

Semipartial (or part) correlation to predict *Y* from X_1, controlling for X_2 (removing X_2 only from X_1, as explained in this chapter):

$$sr \text{ or } r_{Y(1.2)} = \frac{r_{Y1} - r_{Y2}r_{12}}{\sqrt{(1 - r_{12}^2)}}.$$

Because these equations all have the same numerator (and they differ only in that the different divisors scale the information so that it can be interpreted and used in slightly different ways), it follows that your conclusions about how X_1 is related to Y when you control for X_2 tend to be fairly similar no matter which of these four statistics (b, β, pr, or sr) you use to describe the relationship. If any one of these four statistics exactly equals 0, then the other three also equal 0, and all these statistics must have the same sign. They are scaled or sized slightly differently so that they can be used in different situations (to make predictions from raw vs. standard scores and to estimate the proportion of variance accounted for relative to the total variance in Y or only the variance in Y that isn't related to X_2).

The difference among the four statistics above is subtle: β_1 is a partial slope (how much change in z_Y is predicted for a one-*SD* change in z_{X_1} if z_{X_2} is held constant). The partial r describes how X_1 and Y are related if X_2 is removed from both variables. The semipartial r describes how X_1 and Y are related if X_2 is removed only from X_1. In the context of multiple regression, the squared semipartial r (sr^2) provides the most convenient way to estimate effect size and variance partitioning. In some research situations, analysts prefer to report the b (raw score slope) coefficients as indexes of the strength of the relationship among variables. In other situations, standardized or unit-free indexes of strength of relationship (such as β, sr, or pr) are preferred.

11.12 ♦ Nature of Predictive Relationships

When reporting regression, it is important to note the signs of b and β coefficients, as well as their size, and to state whether these signs indicate relations that are in the predicted direction. Researchers sometimes want to know whether a pair of b or β coefficients differ significantly from each other. This can be a question about the size of b in two different groups of subjects: For instance, is the β slope coefficient to predict salary from years of job experience significantly different for male versus female subjects? Alternatively, it could be a question about the size of b or β for two different predictor variables in the same group of subjects (e.g., Which variable has a stronger predictive relation to blood pressure: age or weight?).

It is important to understand how problematic such comparisons usually are. Our estimates of β and b coefficients are derived from correlations; thus, any factors that artifactually influence the sizes of correlations (as described in Chapter 7), such that the correlations are either inflated or deflated estimates of the real strength of the association between variables, can also potentially affect our estimates of β and b. Thus, if women have a restricted range in scores on drug use (relative to men), a difference in the Pearson r and the beta coefficient to predict drug use for women versus men might be artifactually due to a difference in the range of scores on the outcome variable for the two groups. Similarly, a difference in the reliability of measures for the two groups could create an artifactual difference in the size of Pearson r and regression coefficient estimates. It is probably never possible to rule out all possible sources of artifact that might explain the different sizes of r and β coefficients (in different samples or for different predictors). If a researcher wants to interpret a difference between slope coefficients as evidence for a difference in the strength of the association between variables, the researcher should demonstrate that the two groups do not differ in range of scores, distribution shape of scores, reliability of measurement, existence of outliers, or other factors that may affect the size of correlations (as described in Chapter 7). However, no matter how many possible sources of artifact are taken into account, comparison of slopes and correlations remains problematic. Chapter 12 provides a method (using dummy variables and interaction terms) to test whether two groups, such as women versus men, have significantly different slopes for the prediction of Y from some X_i variable. More sophisticated methods that can be used to test equality of specific model parameters, whether they involve comparisons across groups or across different predictor variables, are available within the context of structural equation modeling (SEM) analysis using programs such as AMOS.

11.13 ♦ Effect Size Information in Regression With Two Predictors

11.13.1 ♦ Effect Size for Overall Model

The effect size for the overall model—that is, the proportion of variance in Y that is predictable from X_1 and X_2 combined—is estimated by computation of an R^2. This R^2 is shown on the SPSS printout; it can be obtained either by computing the correlation between observed Y and predicted Y' scores and squaring this correlation or by taking the ratio $SS_{\text{regression}}/SS_{\text{total}}$:

$$R^2 = \frac{SS_{\text{regression}}}{SS_{\text{total}}}. \tag{11.24}$$

Note that this formula for the computation of R^2 is analogous to the formulas given in earlier chapters for eta squared ($\eta^2 = SS_{\text{between}}/SS_{\text{total}}$ for an ANOVA; $R^2 = SS_{\text{regression}}/SS_{\text{total}}$ for multiple regression). R^2 differs from η^2 in that R^2 assumes a *linear* relation between scores on Y and scores on the predictors. On the other hand, η^2 detects differences in mean values of Y across different values of X, but these changes in the value of Y do not need to be a linear function of scores on X. Both R^2 and η^2 are estimates of the proportion of variance in Y scores that can be predicted from independent variables. However, R^2 (as described in this chapter) is an index of the strength of *linear* relationship, while η^2 detects patterns of association that need not be linear.

For some statistical power computations, such as those presented by Green (1991), a different effect size for the overall regression equation, called f^2, is used:

$$f^2 = R^2/(1 - R^2). \tag{11.25}$$

11.13.2 ♦ Effect Size for Individual Predictor Variables

The most convenient effect size to describe the proportion of variance in Y that is uniquely predictable from X_i is the squared semipartial correlation between X_i and Y, controlling for all other predictors. This semipartial (also called the part) correlation between each predictor and Y can be obtained from the SPSS regression procedure by checking the box for the "part and partial" correlations in the optional statistics dialog box. The semipartial or part correlation (sr) from the SPSS output can be squared by hand to yield an estimate of the proportion of uniquely explained variance for each predictor variable (sr^2).

If the part (also called semipartial) correlation is not requested, it can be calculated from the t statistic associated with the significance test of the b slope coefficient. It is useful to know how to calculate this by hand so that you can generate this effect size measure for published regression studies that don't happen to include this information:

$$sr_i^2 = \frac{t_i^2}{df_{\text{residual}}}(1 - R^2), \quad (11.26)$$

where t_i is the ratio b_i/SE_{b_i} for the X_i predictor variable, the df residual $= N - k - 1$, and R^2 is the multiple R^2 for the entire regression equation. The verbal interpretation of sr_i^2 is the proportion of variance in Y that is uniquely predictable from X_i (when the variance due to other predictors is partialled out of X_i).

Some multiple regression programs do not provide the part or semipartial correlation for each predictor, and they report an F ratio for the significance of each b coefficient; this F ratio may be used in place of t_i^2 to calculate the effect size estimate:

$$sr_i^2 = \frac{F}{df_{\text{residual}}}(1 - R^2). \quad (11.27)$$

11.14 ♦ Statistical Power

Tabachnick and Fidell (2007) discussed a number of issues that need to be considered in decisions about sample size; these include alpha level, desired statistical power, number of predictors in the regression equation, and anticipated effect sizes. They suggest the following simple guidelines. Let k be the number of predictor variables in the regression (in this chapter, $k = 2$). The effect size index used by Green (1991) was f^2, where $f^2 = R^2/(1 - R^2)$; $f^2 = .15$ is considered a medium effect size. Assuming medium effect size and $\alpha = .05$, the minimum desirable N for testing the significance of multiple R is $N > 50 + 8k$, and the minimum desirable N for testing the significance of individual predictors is $N > 104 + k$. Tabachnick and Fidell recommended that the data analyst choose the larger number of cases required by these two decision rules. Thus, for the regression analysis with two predictor variables described in this chapter, assuming that the researcher wants to detect medium-size effects, a desirable minimum sample size would be $N = 106$. (Smaller Ns are used in many of the demonstrations and examples in this textbook, however.) If there are substantial violations of assumptions (e.g., skewed rather than normal distribution shapes) or low measurement reliability, then the minimum N should be substantially larger; see Green (1991) for more detailed instructions. If N is extremely large (e.g., $N > 5{,}000$), researchers may find that even associations that are too weak to be of any practical or clinical importance turn out to be statistically significant.

To summarize, then, the guidelines described above suggest that a minimum N of about 106 should be used for multiple regression with two predictor variables to have reasonable power to detect the overall model fit that corresponds to approximately medium-size R^2 values. If more precise estimates of required sample size are desired, the guidelines given by Green (1991) may be used. In general, it is preferable to have sample sizes that are somewhat larger than the minimum values suggested by these decision rules. In addition to having a large enough sample size to have reasonable statistical power, researchers should also have samples large enough so that the CIs around the estimates of slope coefficients are reasonably narrow. In other words, we should try to have sample sizes that are large enough to provide reasonably precise estimates of slopes and not just samples that are large enough to yield "statistically significant" results.

11.15 ♦ Issues in Planning a Study

11.15.1 ♦ Sample Size

A minimum N of at least 100 cases is desirable for a multiple regression with two predictor variables (the rationale for this recommended minimum sample size is given in Section 11.14 on statistical power). The examples presented in this chapter use fewer cases, so that readers who want to enter data by hand or perform computations by hand or in an Excel spreadsheet can replicate the analyses shown.

11.15.2 ♦ Selection of Predictor Variables

The researcher should have some theoretical rationale for the choice of independent variables. Often, the X_1, X_2 predictors are chosen because one or both of them are implicitly believed to be "causes" of Y (although a significant regression does not provide evidence of causality). In some cases, the researcher may want to assess the combined predictive usefulness of two variables or to judge the relative importance of two predictors (e.g., How well do age and weight in combination predict blood pressure? Is age a stronger predictor of blood pressure than weight?). In some research situations, one or more of the variables used as predictors in a regression analysis serve as control variables that are included to control for competing causal explanations or to control for sources of contamination in the measurement of other predictor variables.

There are several variables that are often used to control for contamination in the measurement of predictor variables. For example, many personality test scores are related to social desirability; if the researcher includes a good measure of social desirability response bias as a predictor in the regression model, the regression may yield a better description of the predictive usefulness of the personality measure. Alternatively, of course, controlling for social desirability could make the predictive contribution of the personality measure drop to zero. If this occurred, the researcher might conclude that any apparent predictive usefulness of that personality measure was due entirely to its social desirability component.

After making a thoughtful choice of predictors, the researcher should try to anticipate the possible different outcomes and the various possible interpretations to which these would lead. Selection of predictor variables based on "data fishing"—that is, choosing predictors because they happen to have high correlations with the Y outcome variable in the sample of data in hand—is not recommended. Regression analyses that are set up in this way are likely to report "significant" predictive relationships that are instances of Type I error. It is preferable to base the choice of predictor variables on past research and theory rather than on sizes of correlations. (Of course, it is possible that a large correlation that turns up unexpectedly may represent a serendipitous finding; however, replication of the correlation with new samples should be obtained.)

11.15.3 ♦ Multicollinearity Among Predictors

Although multiple regression can be a useful tool for separating the unique predictive contributions of correlated predictor variables, it does not work well when predictor variables are extremely highly correlated (in the case of multiple predictors, high correlations among many predictors is referred to as **multicollinearity**). In the extreme case, if two predictors are perfectly correlated, it is impossible to distinguish their predictive contributions; and, in fact, regression coefficients cannot be calculated in this situation.

To understand the nature of this problem, consider the partition of variance illustrated in Figure 11.12 for two predictors, X_1 and X_2, that are highly correlated with each other. When there is a strong correlation between X_1 and X_2, most of the explained variance cannot be attributed uniquely to either predictor variable; in this situation, even if the overall multiple R is statistically significant, neither predictor may be judged statistically significant. The area (denoted by "Area *c*" in Figure 11.12) that corresponds to the variance in Y that could be predicted from either X_1 or X_2 tends to be quite large when the predictors are highly intercorrelated, whereas Areas *a* and *b*, which represent the proportions of variance in Y that can be uniquely predicted from X_1 and X_2, respectively, tend to be quite small.

Extremely high correlations between predictors (in excess of .9 in absolute value) may suggest that the two variables are actually measures of the same underlying construct (Berry, 1993). In such cases, it may be preferable to drop one of the variables from the predictive equation. Alternatively, sometimes it makes sense to combine the scores on two or more highly correlated predictor variables into a single index by summing or averaging them; for example, if income and occupational prestige are highly correlated predictors, it

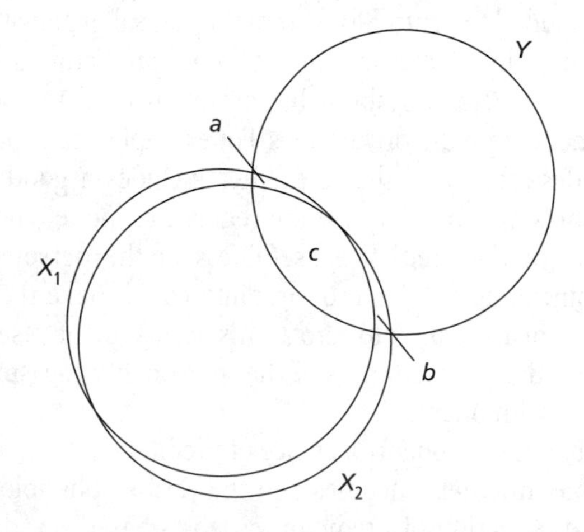

Figure 11.12 ♦ Diagram of Partition of Variance With Highly Correlated (Multicollinear) Predictors

NOTE: Area *c* becomes very large and Areas *a* and *b* become very small when there is a large correlation between X_1 and X_2.

may make sense to combine these into a single index of socioeconomic status, which can then be used as a predictor variable. (See Chapter 19 for a discussion of issues to consider when combining scores across items or variables to form scales.)

Stevens (1992) identified three problems that arise when the predictor variables in a regression are highly intercorrelated (as shown in Figure 11.12). First, a high level of correlation between predictors can limit the size of multiple *R*, because the predictors are "going after much of the same variance" in *Y*. Second, as noted above, it makes assessment of the unique contributions of predictors difficult. When predictors are highly correlated with each other, Areas *a* and *b*, which represent their unique contributions, tend to be quite small. Finally, the error variances associated with each *b* slope coefficient (SE_b) tend to be large when the predictors are highly intercorrelated; this means that the CIs around estimates of *b* are wider and, also, that power for statistical significance tests is lower.

11.15.4 ♦ Range of Scores

As in correlation analyses, there should be a sufficient range of scores on both the predictor and the outcome variables to make it possible to detect relations between them. This, in turn, requires that the sample be drawn from a population in which the variables of interest show a reasonably wide range. It would be difficult, for example, to demonstrate strong age-related changes in blood pressure in a sample with ages that ranged only from 18 to 25 years; the relation between blood pressure and age would probably be stronger and easier to detect in a sample with a much wider range in ages (e.g., from 18 to 75).

11.16 ♦ Use of Regression With Two Predictors to Test Mediated Causal Models

Chapter 10 discussed several different hypothetical causal models that correspond to different types of associations among variables. For a set of three variables, such as age (X_1), weight (X_2), and blood pressure (Y), one candidate model that might be considered is a model in which X_1 and X_2 are correlated or confounded or redundant predictors of Y—namely, the model that is usually implicit when multiple regression is performed to predict Y from both X_1 and X_2. However, a different model can also be considered in some research situations—namely, a model in which the X_2 variable "mediates" a causal sequence that leads from X_1 to X_2 to Y. It is conceivable that increasing age causes weight gain and that weight gain causes increases in blood pressure; effects of age on blood pressure might be either partly or fully mediated by weight increases.

Chapter 10 showed that examination of a partial correlation between age and blood pressure, controlling for weight, was one preliminary way to assess whether the age/blood pressure association might be partly or fully mediated by weight. A more thorough test of the mediation hypothesis requires additional information. A mediated causal model cannot be proved to be correct using data from a nonexperimental study; however, we can examine the pattern of correlations and partial correlations in nonexperimental data to evaluate whether a partly or fully mediated model is a plausible hypothesis.

Baron and Kenny (1986) identified several conditions that should be met as evidence that there could potentially be a mediated causal process in which X_1 causes Y (partly or completely) through its effects on X_2. In the following example, the hypothetical "fully mediated" model is as follows: age → weight → blood pressure. In words, this model says that the effects of age on blood pressure are entirely mediated by weight. Figure 11.13 (based on the discussion in Baron & Kenny, 1986) illustrates several tests that should be performed to test hypotheses about mediated causal models.

To test a theoretical mediated causal model in which $X_1 \rightarrow X_2 \rightarrow Y$ (e.g., age → weight → blood pressure), the first piece of information that is needed is the strength of the association between X_1 (the initial cause) and Y (the ultimate outcome), which represents the "total" relationship between X_1 and Y. In this example, all indexes of strength of association that are reported are correlations or unit-free beta coefficients, but note that it is also possible (and in some situations, it may be preferable) to assess the strength of the associations between variables by using raw score regression coefficients.

The first condition required for the mediated causal model ($X_1 \rightarrow X_2 \rightarrow Y$) to be plausible is that the r_{1Y} correlation should be statistically significant and large enough to be of some practical or theoretical importance. We can obtain this correlation by requesting the Pearson correlation between X_1 and Y, age and blood pressure. From Figure 11.6, the overall correlation between age and blood pressure was $r = +.78$, $p < .001$, two-tailed. Therefore, the first condition for mediation is satisfied. (An alternative approach would be to show that the b raw score slope in a regression that predicts Y from X_1 is statistically significant and reasonably large). The r_{1Y} correlation corresponds to the path labeled c in the top diagram in Figure 11.13; this correlation represents the "total" strength of the association between X_1 and Y, including both direct and indirect paths that lead from X_1 to Y.

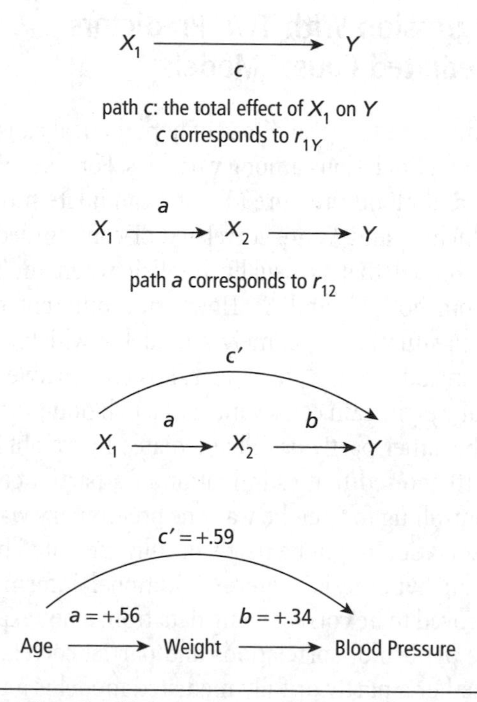

Figure 11.13 ♦ Conditions That Should Be Met for a Hypothetical Mediated Model in Which X_2 Mediates the Effect of X_1 on Y

SOURCE: Adapted from Baron and Kenny (1986).

The second condition required for mediation to be a reasonable hypothesis is that the path that leads from X_1 to X_2 should correspond to a relationship that is statistically significant and large enough to be of some theoretical or practical importance. This can be assessed by examining the Pearson correlation between X_1 and X_2, age and weight ($r = +.56, p < .001$, two-tailed) or by examining the b slope coefficient in a regression that predicts X_2 from X_1. This correlation corresponds to the path denoted by a in the second path diagram in Figure 11.13.

Two additional paths need to be evaluated; the coefficients for these paths can be obtained by performing a regression to predict Y from both X_1 and X_2 (in this example, to predict blood pressure from both age and weight). The path denoted by b in the third path diagram in Figure 11.13 represents the strength of the predictive relationship between X_2 and Y when X_1 is statistically controlled. From the standardized regression equation $z'_Y = \beta_1 z_{X_1} + \beta_2 z_{X_2}$, we can use β_2 to represent the strength of the relationship for the path from X_2 to Y in a unit-free index of strength of association. From the SPSS regression output that appears in Figure 11.9, the beta coefficient to predict standardized blood pressure from standardized weight while statistically controlling for age was $\beta = .34$. Finally, the c' path in the last two path diagrams in Figure 11.13 represents the strength of the predictive relationship between X_1 (age) and Y (blood pressure) when X_2 (weight) is statistically controlled. The strength of this c' relationship is given in unit-free terms by the beta coefficient for age as a predictor of blood pressure in a regression that includes weight as the other predictor. In this example (from Figure 11.9), the value of beta that corresponds to the c' path is +.59.

A fully mediated model (in which the effects of age on blood pressure occur entirely through the effect of age on weight) would lead us to expect that the beta standard score (or b raw score) regression coefficient that corresponds to the c' path should not be significantly different from 0.

A "partly mediated model" (in which age has part of its effects on blood pressure through weight but age also has additional effects on blood pressure that are not mediated by weight) would lead us to expect that the c' path coefficient should be significantly greater than 0 and, also, that the c' path coefficient is significantly smaller than the c path coefficient. In addition, the strength of the "mediated relationship" between X_1 and Y, age and blood pressure, is estimated by multiplying the a and b coefficients for the path that leads from age to weight and then from weight to blood pressure: In this example, the product $a \times b = +.56 \times +.34 = +.20$. We would also want this product, $a \times b$, to be significantly larger than 0 and large enough to be of some practical or theoretical importance before we would conclude that there is a mediated causal path from X_1 to Y.

In this particular example, all the path coefficients were reported in terms of unit-free beta weights (it is also possible to report path coefficients in terms of raw score regression coefficients). Because the c' path corresponds to a statistically significant b coefficient and other conditions for a mediation model have been met, the data appear to be consistent with the hypothesis that the effects of age on blood pressure are "partly" mediated by weight. For a one-*SD* increase in the z score for age, we would predict a +.59 increase in the z score for blood pressure due to the "direct" effects of age on blood pressure (or, more precisely, the effects of age on blood pressure that are not mediated by weight). For a one-*SD* increase in the z score for age, we would also predict a .20 increase in the z score for blood pressure due to the indirect path via weight. The total association between age and blood pressure, given by the Pearson correlation between age (X_1) and blood pressure (Y), $r_{1Y} = +.78$, can be reproduced by summing the strength of the relationship for the "direct" path c' (+.59) and the strength of the association via the indirect path that involves X_2 (weight) (.19); +.59 + .20 ≈ +.79.

Procedures are available to test the hypothesis that c' is significantly less than c and to test whether the $a \times b$ product, which represents the overall strength of the mediated relationship, is significantly different from 0. For datasets with large *N*s, significance tests for partial and full mediation models can be performed using the procedures suggested by Sobel (1982). If only bivariate correlations and simple regression results are available, online calculators to test for mediation are available at this Web site: http://www.psych.ku.edu/preacher/sobel/sobel.htm; however, the use of these procedures is not recommended for datasets with small *N*s. For a more detailed discussion of mediation, see MacKinnon, Lockwood, Hoffman, West, and Sheets (2002).

11.17 ♦ Results

The results of an SPSS regression analysis to predict blood pressure from both age and weight (for the data in Table 10.3) are shown in Figure 11.9. A summary table was created based on a format suggested by Tabachnick and Fidell (2007); see Table 11.1. If a large number of different regression analyses are performed with different sets of predictors and/or different outcome variables, other table formats that report less complete information are sometimes used. This table provides a fairly complete summary, including bivariate correlations among all the predictor and outcome variables; mean and standard deviation for each variable involved in the analysis; information about the overall fit of the regression model (multiple R and R^2 and the associated F test); the b coefficients for the raw score regression equation, along with an indication whether each b coefficient differs significantly from zero; the beta coefficients for the standard score regression equation; and a squared part or semipartial correlation (sr^2) for each predictor that represents the proportion of variance in the Y outcome variable that can be predicted uniquely from each predictor variable, controlling for all other predictors in the regression equation. The example given in the Results below discusses age and weight as correlated or partly redundant predictor variables, because this is the most common implicit model when regression is applied. Alternatively, these results could be reported as evidence consistent with a partly mediated model (as shown in Figure 11.13).

Table 11.1 ♦ Results of Standard Multiple Regression to Predict Blood Pressure (Y) from Age (X_1) and Weight (X_2)

Variables	*Blood Pressure*	*Age*	*Weight*	*b*	*β*	sr^2_{unique}
Age	+.78***			+2.161***	+.59	+.24
Weight	+.67***	+.56		+.490*	+.34	+.08
			Intercept =	−28.05		
Means	177.3	58.3	162.0			
SD	63.6	17.4	44.2			
						R^2 = .690
						R^2_{adj} = .667
						R = .831***

***$p < .001$; **$p < .01$; *$p < .05$.

Results

Initial examination of blood pressure data for a sample of $N = 30$ participants indicated that there were positive correlations between all pairs of variables. However, the correlation between the predictor variables age and weight, $r = +.56$, did not indicate extremely high multicollinearity.

For the overall multiple regression to predict blood pressure from age and weight, $R = .83$ and $R^2 = .69$. That is, when both age and weight were used as predictors, about 69% of the variance in blood pressure could be predicted. The adjusted R^2 was .67. The overall regression was statistically significant, $F(2, 27) = 30.04$, $p < .001$. Complete results for the multiple regression are presented in Table 11.1.

Age was significantly predictive of blood pressure when the variable weight was statistically controlled: $t(27) = 4.55$, $p < .001$. The positive slope for age as a predictor of blood pressure indicated that there was about a 2-mmHg increase in blood pressure for each 1-year increase in age, controlling for weight. The squared semipartial that estimated how much variance in blood pressure was uniquely predictable from age was $sr^2 = .24$. About 24% of the variance in blood pressure was uniquely predictable from age (when weight was statistically controlled).

Weight was also significantly predictive of blood pressure when age was statistically controlled: $t(27) = 2.62$, $p = .014$. The slope to predict blood pressure from weight was approximately $b = +.49$; in other words, there was about a half-point increase in blood pressure in millimeters of mercury for each 1-lb increase in body weight. The sr^2 for weight (controlling for age) was .08. Thus, weight uniquely predicted about 8% of the variance in blood pressure when age was statistically controlled.

The conclusion from this analysis is that the original zero-order correlation between age and blood pressure ($r = .78$ or $r^2 = .61$) was partly (but not entirely) accounted for by weight. When weight was statistically controlled, age still uniquely predicted 24% of the variance in blood pressure. One possible interpretation of this outcome is that age and weight are partly redundant as predictors of blood pressure; to the extent that age and weight are correlated with each other, they compete to explain some of the same variance in blood pressure. However, each predictor was significantly associated with blood pressure even when the other predictor variable was significantly controlled; both age and weight contribute uniquely useful predictive information about blood pressure in this research situation.

The predictive equations were as follows:

Raw score version:

$$\text{Blood pressure}' = -28.05 + 2.16 \times \text{Age} + .49 \times \text{Weight}.$$

Standard score version:

$$z_{\text{blood pressure}}' = .59 \times z_{\text{age}} + .34 \times z_{\text{weight}}.$$

Although residuals are rarely discussed in the results sections of journal articles, examination of plots of residuals can be helpful in detecting violations of assumptions or multivariate outliers; either of these problems would make the regression analysis less credible. The graph of standardized residuals against standardized predicted scores (in Figure 11.10) did not suggest any problem with the residuals. If all the assumptions for regression analysis are satisfied, the mean value of the standardized residuals should be 0 for all values of the predicted score, the variance of residuals should be uniform across values of the predicted score, the residuals should show no evidence of linear or curvilinear trend, and there should be no extreme outliers.

Although this is not usually reported in a journal article, it is useful to diagram the obtained partition of variance so that you understand exactly how the variance in *Y* was divided. Figure 11.14 shows the specific numerical values that correspond to the variance components that were identified in Figure 11.2. Area *d* was calculated by finding $1 - R^2 = 1 - .69 = .31$, using the R^2 value of .69 from the SPSS printout. Note that the unadjusted R^2 was used rather than the adjusted R^2; the adjusted R^2 can actually be negative in some instances. Numerical estimates for the proportions of unique variance predictable from each variable represented by Areas *a* and *b* were obtained by squaring the part correlations (also called the semipartial correlation) for each predictor. For age, the part or semipartial correlation on the SPSS printout was $sr_{age} = .488$; the value of sr^2_{age} obtained by squaring this value was about .24. For weight, the part or semipartial correlation reported by SPSS was $sr_{weight} = .281$; therefore, $sr^2_{weight} = .08$. Because the sum of all four areas $(a + b + c + d)$ equals 1, once the values for Areas *a*, *b*, and *d* are known, a numerical value for Area *c* can be obtained by subtraction $(c = 1 - a - b - d)$.

In this example, 69% of the variance in blood pressure was predictable from age and weight in combination (i.e., $R^2 = .69$). This meant that 31% $(1 - R^2 = 1 - .69 = .31)$ of the variance in salaries could not be predicted from these two variables. Twenty-four percent of the variance in blood pressure was uniquely predictable from age (the part correlation for age was .488, so $sr^2_{age} = .24$). Another 8% of the variance in blood pressure was uniquely predictable from weight (the part correlation for weight was .281, so the squared part correlation for weight was about $sr^2_{weight} = .08$). Area *c* was obtained by subtraction of Areas *a*, *b*, and *d* from 1: $1 - .24 - .08 - .31 = .37$. Thus, the remaining 37% of the variance in blood pressure could be predicted equally well by age or weight (because these two predictors were confounded or redundant to some extent).

Note that it is possible, although unusual, for Area *c* to turn out to be a negative number; this can occur when one (or both) of the semipartial *r*s for the predictor variables are larger in absolute value than their zero-order correlations with *Y*. When Area *c* is large, it indicates that the predictor variables are fairly highly correlated with each other and therefore "compete" to explain the same variance. If Area *c* turns out to be negative, then the overlapping circles diagram shown in Figure 11.2 may not be the best way to think about what is happening; a negative value for Area *c* suggests that some kind of suppression is present, and suppressor variables can be difficult to interpret.

11.18 ♦ Summary

Regression with two predictor variables can provide a fairly complete description of the predictive usefulness of the X_1 and X_2 variables, although it is important to keep in mind

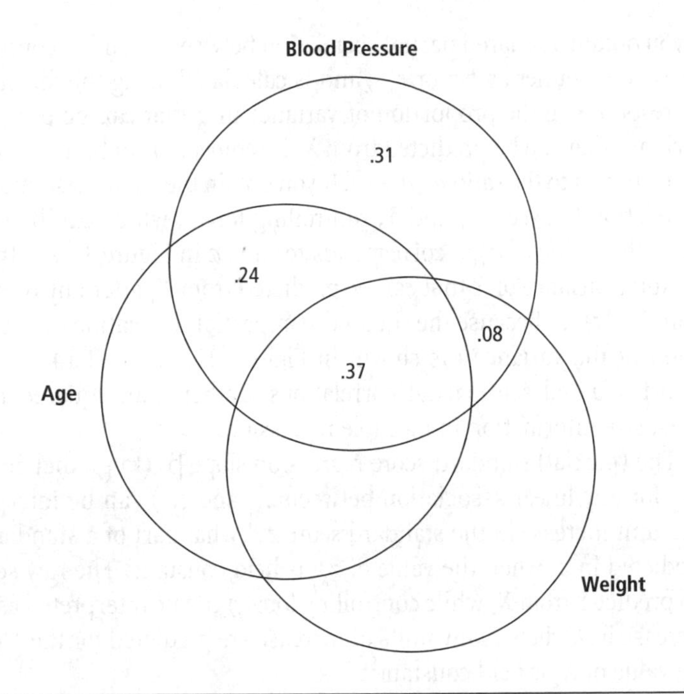

Figure 11.14 ♦ Diagram of Partition of Variance for Prediction of Blood Pressure (Y) From Age (X_1) and Weight (X_2)

NOTES:

Proportion of variance in blood pressure not predictable from age or weight = $1 - R^2 = .31$ = Area *d*.

Proportion of variance in blood pressure uniquely predictable from age, controlling for weight = $sr^2_{\text{age}} = .24$ = Area *a*.

Proportion of variance in blood pressure uniquely predictable from weight, controlling for age = $sr^2_{\text{weight}} = .08$ = Area *b*.

Proportion of variance in blood pressure predictable by either age or weight = $1 - a - b - d = .37$ = Area *c*.

Areas in the diagram do not correspond exactly to the proportions; the diagram is only approximate.

that serious violations of the assumptions (such as nonlinear relations between any pair of variables and/or an interaction between X_1 and X_2) can invalidate the results of this simple analysis. Violations of these assumptions can often be detected by preliminary data screening that includes all bivariate scatter plots (e.g., X_1 vs. X_2, Y vs. X_1, and Y vs. X_2) and scatter plots that show the X_1, Y relationship separately for groups with different scores on X_2. Examination of residuals from the regression can also be a useful tool for identification of violation of assumptions.

Note that the regression coefficient b to predict a raw score on Y from X_1 while controlling for X_2 only "partials out" or removes or controls for the part of the X_1 scores that is *linearly* related to X_2. If there are nonlinear associations between the X_1 and X_2 predictors, then linear regression methods are not an effective way to describe the unique contributions of the predictor variables.

So far, you have learned several different analyses that can be used to evaluate whether X_1 is (linearly) predictive of Y. If you obtain a squared Pearson correlation between X_1 and Y, r^2_{1Y}, the value of r^2_{1Y} estimates the proportion of variance in Y that is predictable from X_1 when you do *not* statistically control for or partial out any variance associated with other predictor variables such as X_2. In Figure 11.2, r^2_{1Y} corresponds to the sum of Areas *a* and *c*.

If you obtain a squared partial correlation between X_1 and Y, controlling for X_2 (which can be denoted either as pr_1^2 or $r^2_{Y1.2}$ and is calculated using the formulas in Chapter 10), $r^2_{Y1.2}$ corresponds to the proportion of variance in Y that can be predicted from X_1 when the variance that can be predicted from X_2 is removed from both Y and X_1; in Figure 11.2, $r^2_{Y1.2}$ corresponds to the ratio $a/(a + d)$. If you obtain the squared semipartial (or squared part) correlation between X_1 and Y, controlling for X_2, which can be denoted by either sr_1^2 or $r^2_{Y(1.2)}$, this value of $r^2_{Y(1.2)}$ corresponds to Area a in Figure 11.2—that is, the proportion of the total variance of Y that can be predicted from X_1 after any overlap with X_2 is removed from (only) X_1. Because the squared semipartial correlations can be used to deduce a partition of the variance (as shown in Figures 11.2 and 11.14), data analysts more often report squared semipartial correlations (rather than squared partial correlations) as effect size information in multiple regression.

The (partial) standard score regression slope β_1 (to predict z_Y from z_{X1} while controlling for any linear association between z_{X_1} and z_{X_2}) can be interpreted as follows: For a one-unit increase in the standard score z_{X_1}, what part of a standard deviation increase is predicted in z_Y when the value of z_{X_2} is held constant? The raw score regression slope b_1 (to predict Y from X_1 while controlling for X_2) can be interpreted as follows: For a one-unit increase in X_1, how many units of increase are predicted for the Y outcome variable when the value of X_2 is held constant?

In some circumstances, data analysts find it more useful to report information about the strength of predictive relationships using unit-free or standardized indexes (such as β or $r^2_{Y(1.2)}$). This may be particularly appropriate when the units of measurement for the X_1, X_2, and Y variables are all arbitrary; or when a researcher wants to try to compare the predictive usefulness of an X_1 variable with the predictive usefulness of an X_2 variable that has completely different units of measurement than X_1. (However, such comparisons should be made very cautiously because differences in the sizes of correlations, semipartial correlations, and beta coefficients may be partly due to differences in the ranges or distribution shapes of X_1 and X_2 or the reliabilities of X_1 and X_2 or other factors that can artifactually influence the magnitude of correlations that were discussed in Chapter 7.)

In other research situations, it may be more useful to report the strength of predictive relationships by using raw score regression slopes. These may be particularly useful when the units of measurement of the variables have some "real" meaning—for example, when we ask how much blood pressure increases for each 1-year increase in age.

Later chapters show how regression analysis can be extended in several different ways. Chapter 12 describes the use of dichotomous or dummy variables (such as gender) as predictors in regression. When dummy variables are used as predictors, regression analysis can be used to obtain results that are equivalent to those from more familiar methods (such as one-way ANOVA). Chapter 12 also discusses the inclusion of interaction terms in regression models. Chapter 13 demonstrates that interaction terms can also be examined using factorial ANOVA. Chapter 14 discusses the generalization of regression analysis to situations with more than two predictor variables. In the most general case, a multiple regression equation can have k predictors:

$$Y = b_0 + b_1 X_1 + b_2 X_2 + b_3 X_3 + \ldots + b_k X_k. \tag{11.28}$$

The predictive contribution of each variable (such as X_1) can be assessed while controlling for all other predictors in the equation (e.g., $X_2, X_3, \ldots, X_k$). When we use this approach to variance partitioning—that is, each predictor is assessed controlling for all other predictors in the regression equation, the method of variance partitioning is often called **"standard"** or **"simultaneous" multiple regression** (Tabachnick & Fidell, 2007). When variance was partitioned between two predictor variables X_1 and X_2 in this chapter, the "standard" method of partitioning was used; that is, sr_1^2 was interpreted as the proportion of variance in Y that was uniquely predictable by X_1 when X_2 was statistically controlled, and sr_2^2 was interpreted as the proportion of variance in Y that was uniquely predictable from X_2 when X_1 is statistically controlled. In Chapter 14, we will see that it is possible to use other approaches to variance partitioning (in which the overlap that corresponds to Area c in Figures 11.2 and 11.14 may be arbitrarily attributed to one of the predictor variables in the regression). In this chapter, a more conservative approach was used in variance partitioning; any variance that could be predicted by either X_1 or X_2 (Area c) was not attributed to either of the individual predictor variables.

Notes

1. Formal treatments of statistics use β to represent the population slope parameter in this equation for the null hypothesis. This notation is avoided in this textbook because it is easily confused with the more common use of β as the sample value of the standardized slope coefficient—that is, the slope to predict z'_Y from z_{X1}. In this textbook, β always refers to the sample estimate of a standard score regression slope.

2. It is possible to reconstruct the correlations (r_{Y_1}, r_{Y_2}) exactly from the model coefficients (β_1, β_2) in this example, because this regression model is "just identified"; that is, the number of parameters being estimated (r_{12}, β_1, and β_2) equals the number of correlations used as input data. In advanced applications of path model logic such as SEM, researchers generally constrain some of the model parameters (e.g., path coefficients) to fixed values, so that the model is "overidentified." For instance, if a researcher assumes that $\beta_1 = 0$, the direct path from z_{X_1} to z_Y is omitted from the model. When constraints on parameter estimates are imposed, it is generally not possible to reproduce the observed correlations perfectly from the constrained model. In SEM, the adequacy of a model is assessed by checking to see how well the reproduced correlations (or reproduced variances and covariances) implied by the paths in the overidentified SEM model agree with the observed correlations (or covariances). The tracing rule described here can be applied to standardized SEM models to see approximately how well the SEM model reconstructs the observed correlations among all pairs of variables. The formal goodness-of-fit statistics reported by SEM programs are based on goodness of fit of the observed variances and covariances rather than correlations.

+ − × ÷

Comprehension Questions

1. Consider the following hypothetical data (in the file named pr1_mr2iv.sav):

Anxiety	*Weight*	*SBP*
7	90	137
10	130	158
35	30	163
19	151	133
17	170	106
11	190	128
18	210	168
36	91	145
19	90	149
22	95	143
28	130	145
25	150	123
4	110	145
25	150	143
18	110	113
7	150	119
16	230	139
25	315	173
28	250	176
10	210	178
34	271	174
22	250	176
17	230	156
16	185	144
9	201	154

The research question is this: How well can systolic blood pressure (SBP) be predicted from anxiety and weight combined? Also, how much variance in blood pressure is uniquely explained by each of these two predictor variables?

a. As preliminary data screening, generate a histogram of scores on each of these three variables, and do a bivariate scatter plot for each pair of variables. Do you see evidence of violations of assumptions? For example, do any variables have nonnormal distribution shapes? Are any pairs of variables related in a way that is not linear? Are there bivariate outliers?

b. Run a regression analysis to predict SBP from weight and anxiety. As in the example presented in the chapter, make sure that you request the part and partial correlation statistics and a graph of the standardized residuals (ZRESID) against the standardized predicted values (ZPRED).

c. Write up a Results section. What can you conclude about the predictive usefulness of these two variables, individually and combined?
d. Does examination of the plot of standardized residuals indicate any serious violation of assumptions? Explain.
e. Why is the b coefficient associated with the variable weight so much smaller than the b coefficient associated with the variable anxiety (even though weight accounted for a larger unique share of the variance in SBP)?
f. Set up a table (similar to the one shown as Table 11.1) to summarize the results of your regression analysis.
g. Draw a diagram (similar to the one in Figure 11.14) to show how the total variance of SBP is partitioned into variance that is uniquely explained by each predictor, variance that can be explained by either predictor, and variance that cannot be explained, and fill in the numerical values that represent the proportions of variance in this case.
h. Were the predictors highly correlated with each other? Did they compete to explain the same variance? (How do you know?)
i. Ideally, how many cases should you have to do a regression analysis with two predictor variables?

2. What is the null hypothesis for the overall multiple regression?
3. What null hypothesis is used to test the significance of each individual predictor variable in a multiple regression?
4. Which value on your SPSS printout gives the correlation between the observed Y and predicted Y values?
5. Which value on your SPSS printout gives the proportion of variance in Y (the dependent variable) that is predictable from X_1 and X_2 as a set?
6. Which value on your SPSS printout gives the proportion of variance in Y that is uniquely predictable from X_1, controlling for or partialling out X_2?
7. Explain how the normal equations for a two-predictor multiple regression can be obtained from a path diagram that shows z_{X_1} and z_{X_2} as correlated predictors of z_Y, by applying the tracing rule.
8. The normal equations show the overall correlation between each predictor and Y broken down into two components—for example, $r_{1Y} = \beta_1 + r_{12}\beta_2$. Which of these components represents a direct (or unique) contribution of X_1 as a predictor of Y, and which one shows an indirect relationship?
9. For a regression (to predict Y from X_1 and X_2), is it possible to have a significant R but nonsignificant b coefficients for both X_1 and X_2? If so, under what circumstances would this be likely to occur?
10. Draw three overlapping circles to represent the variance and shared variance among X_1, X_2, and Y, and label each of the following: sr_1^2, sr_2^2, and $1 - R^2$. What interpretation is given to the three-way overlap? How do you deduce the area of

the three-way overlap from the information on your SPSS printout? (Can this area ever turn out to be negative, and if so, how does this come about?)

11. What is multicollinearity in multiple regression, and why is it a problem?
12. How do you report effect size and significance test information for the entire regression analysis?
13. In words, what is this null hypothesis: $H_0: b = 0$?
14. How can you report the effect size and significance test for each individual predictor variable?
15. How are the values of b and β similar? How are they different?
16. If $r_{12} = 0$, then what values would β_1 and β_2 have?
17. What does the term *partial* mean when it is used in the term *partial slope*?

CHAPTER 12

Dummy Predictor Variables and Interaction Terms in Multiple Regression

12.1 ♦ Research Situations Where Dummy Predictor Variables Can Be Used

Previous examples of regression analysis have used scores on quantitative X variables to predict scores on a quantitative Y variable. However, it is possible to include group membership or categorical predictor variables as predictors in regression analysis. This can be done by creating dummy (dichotomous) predictor variables to represent information about group membership. A dummy or dichotomous predictor variable provides Yes/No information for questions about group membership. For example, a simple dummy variable to represent gender corresponds to the following question: Is the participant female ("0") or male ("1")? Gender is an example of a two-group categorical variable that can be represented by a single dummy variable.

When we have more than two groups, we can use a set of dummy variables to provide information about group membership. For example, suppose that a study includes members of $k = 3$ political party groups. The categorical variable political party has the following scores: 1 = Democrat, 2 = Republican, and 3 = Independent. We might want to find out whether mean scores on a quantitative measure of political conservatism (Y) differ across these three groups. One way to answer this question is to perform a one-way analysis of variance (ANOVA) that compares mean conservatism (Y) across the three political party groups. In this chapter, we will see that we can also use regression analysis to evaluate how political party membership is related to scores on political conservatism.

However, we should not set up a regression to predict scores on conservatism from the *multiple-group* categorical variable political party, with party membership coded 1= Democrat, 2 = Republican, and 3 = Independent. Multiple-group categorical variables usually do not work well as predictors in regression, because scores on a quantitative outcome variable, such as "conservatism," will not necessarily increase linearly with the score on the categorical variable that provides information about political party. The score values that represent political party membership may not be rank ordered in a way that is

monotonically associated with changes in conservatism; as we move from Group 1 = Democrat to Group 2 = Republican, scores on conservatism may increase, but as we move from Group 2 = Republican to Group 3 = Independent, conservatism may decrease. Even if the scores that represent political party membership are rank ordered in a way that is monotonically associated with level of conservatism, the amount of change in conservatism between Groups 1 and 2 may not be equal to the amount of change in conservatism between Groups 2 and 3. In other words, scores on a multiple-group categorical predictor variable (such as political party coded 1 = Democrat, 2 = Republican, and 3 = Independent) are not necessarily *linearly* related to scores on quantitative variables.

If we want to use the categorical variable political party to predict scores on a quantitative variable such as conservatism, we need to represent the information about political party membership in a different way. Instead of using one categorical predictor variable with codes 1 = Democrat, 2 = Republican, and 3 = Independent, we can create two dummy or dichotomous predictor variables to represent information about political party membership, and we can then use these two dummy variables as predictors of conservatism scores in a regression. Political party membership can be assessed by creating dummy variables (denoted by D_1 and D_2) that correspond to two Yes/No questions. In this example, the first dummy variable D_1 corresponds to the following question: Is the participant a member of the Democratic party? Coded 1 = Yes, 0 = No. The second dummy variable D_2 corresponds to the following question: Is the participant a member of the Republican Party? Coded 1 = Yes, 0 = No. We assume that group memberships for individuals are mutually exclusive and exhaustive, that is, each case belongs to only one of the three groups, and every case belongs to one of the three groups identified by the categorical variable. When these conditions are met, a third dummy variable is not needed to identify the members of the third group because, for Independents, the answers to the first two questions that correspond to the dummy variables D_1 and D_2 would be "No." In general, when we have k groups or categories, a set of $k - 1$ dummy variables is sufficient to provide complete information about group membership. Once we have represented political party group membership by creating scores on two dummy variables, we can set up a regression to predict scores on conservatism (Y) from the scores on the two dummy predictor variables D_1 and D_2:

$$Y' = b_0 + b_1D_1 + b_2D_2. \tag{12.1}$$

In this chapter, we will see that the information about the association between group membership (represented by D_1 and D_2) and scores on the quantitative Y variable that can be obtained from the regression analysis in Equation 12.1 is equivalent to the information that can be obtained from a one-way ANOVA that compares means on Y across groups or categories. It is acceptable to use *dichotomous* predictor variables in regression and correlation analysis. This works because (as discussed in Chapter 8) a dichotomous categorical variable has only two possible score values, and the only possible relationship between scores on a dichotomous predictor variable and a quantitative outcome variable is a linear one. That is, as you move from a score of 0 = Female on gender to a score of 1 = Male on gender, mean height or mean annual salary may either increase, decrease, or stay the same; any change that can be observed across just two groups can be represented as linear. Similarly, if we represent political party membership using two dummy variables, each dummy variable represents a contrast between the means of two groups; for example, the D_1 dummy variable can represent the difference in mean conservatism between Democrats and Independents, and the D_2 dummy variable can represent the mean difference in conservatism between Republicans and Independents.

This chapter uses empirical examples to demonstrate that regression analyses that use dummy predictor variables (similar to Equation 12.1) provide information that is equivalent to the results of more familiar analyses for comparison of group means (such as ANOVA). There are several reasons why it is useful to consider dummy predictor variables as predictors in regression analysis. First, the use of dummy variables as predictors in regression provides a simple demonstration of the fundamental equivalence between ANOVA and multiple regression; ANOVA and regression are both special cases of a more general analysis called the general linear model (GLM). Second, researchers often want to include group membership variables (such as gender) along with other predictors in a multiple regression. Therefore, it is useful to examine examples of regression that include dummy variables along with quantitative predictors.

This chapter also introduces the use of products between predictor variables as a way of incorporating one or more interaction terms in a regression analysis. A regression analysis can include an interaction between two dummy-coded variables, an interaction between a dummy variable and a quantitative predictor variable, or an interaction between two quantitative predictor variables. This chapter discusses how to set up and interpret interaction terms in regression models. Interactions that involve one or two dummy variables are usually relatively easy to interpret. Interactions between two quantitative predictors can be more difficult to evaluate.

The computational procedures for regression remain the same when we include one or more dummy predictor variables. The most striking difference between dummy variables and quantitative variables is that the scores on dummy variables usually have small integer values (such as "1," "0," and "–1"). The use of small integers as codes simplifies the interpretation of the regression coefficients associated with dummy variables. When dummy variables are used as predictors in a multiple regression, the b raw score slope coefficients provide information about differences between group means. The specific group means that are compared differ depending on the method of coding that is used for dummy variables, as explained in the following sections. Except for this difference in the interpretation of regression coefficients, regression analysis remains essentially the same when dummy predictor variables are included.

12.2 ♦ Empirical Example

The hypothetical data for this example are provided by a study of predictors of annual salary in dollars for a group of $N = 50$ college faculty members; the complete data appear in Table 12.1. Predictor variables include the following: gender, coded 0 = Female and 1 = Male; years of job experience; college, coded 1 = Liberal Arts, 2 = Sciences, 3 = Business; and an overall merit evaluation. Additional columns in the SPSS data worksheet in Figure 12.1, such as D_1, D_2, E_1, and E_2, represent alternative ways of coding group membership, which are discussed later in this chapter. All subsequent analyses in this chapter (except for the graph in Figure 12.16) are based on the data in Table 12.1.

Table 12.1 ♦ Hypothetical Data for Salary and Predictors of Salary for $N = 50$ College Faculty

Salary	*Years*	*Gender*	*College*	*Merit*
31	0	0	1	30
32	0	0	3	10
33	1	0	1	15
34	1	0	2	29
33	2	0	1	30
40	2	0	3	59
39	3	0	2	55
51	3	0	3	30
54	3	0	3	17
38	4	0	1	11
39	4	0	2	45
43	4	0	2	40
37	5	0	1	55
39	5	0	1	30
41	6	0	1	31
41	6	0	1	30
42	7	0	2	50
46	9	0	2	18
49	12	0	1	35
54	15	0	1	30
30	1	1	2	7
34	2	1	1	7
42	2	1	1	30
36	2	1	2	30
43	3	1	1	28
44	3	1	3	24
46	4	1	1	7
43	4	1	2	51
45	4	1	2	22
47	4	1	2	35
44	5	1	1	34
34	5	1	2	57
46	6	1	1	43
51	7	1	1	35
47	7	1	2	21
50	8	1	1	30
51	8	1	1	15
51	9	1	2	24
51	9	1	3	30
54	10	1	3	56
63	10	1	3	55
56	12	1	3	19
58	13	1	3	44
58	14	1	1	44
59	14	1	1	38
58	14	1	2	38
66	17	1	1	50
67	19	1	2	37
59	20	1	2	40
64	22	1	3	44

NOTES: Salary, annual salary in thousands of dollars; years, years of job experience; gender, dummy-coded gender (0 = Female and 1 = Male); college, membership coded 1 = Liberal Arts, 2 = Sciences, 3 = Business; merit, overall merit evaluation based on publications, teaching, and service.

	salary	years	gender	genyears	geneff	college	d1	d2	e1	e2	merit	Zmerit	Zyears	zyearmerit	Zsalary
1	31	0	0	0	-1	1	1	0	1	0	30	-.20458	-1.28239	.26	-1.56520
2	32	0	0	0	-1	3	0	0	-1	-1	10	-1.61545	-1.28239	2.07	-1.46263
3	33	1	0	0	-1	1	1	0	1	0	15	-1.26273	-1.09919	1.39	-1.36006
4	34	1	0	0	-1	2	0	1	0	1	29	-.27512	-1.09919	.30	-1.25750
5	33	2	0	0	-1	1	1	0	1	0	30	-.20458	-.91599	.19	-1.36006
6	40	2	0	0	-1	3	0	0	-1	-1	59	1.84119	-.91599	-1.69	-.64208
7	39	3	0	0	-1	2	0	1	0	1	55	1.55901	-.73279	-1.14	-.74465
8	51	3	0	0	-1	3	0	0	-1	-1	30	-.20458	-.73279	.15	.48618
9	54	3	0	0	-1	3	0	0	-1	-1	17	-1.12164	-.73279	.82	.79388
10	38	4	0	0	-1	1	1	0	1	0	11	-1.54490	-.54960	.85	-.84722
11	39	4	0	0	-1	2	0	1	0	1	45	.85358	-.54960	-.47	-.74465
12	43	4	0	0	-1	2	0	1	0	1	40	.50086	-.54960	-.28	-.33437
13	37	5	0	0	-1	1	1	0	1	0	55	1.55901	-.36640	-.57	-.94979
14	39	5	0	0	-1	1	1	0	1	0	30	-.20458	-.36640	.07	-.74465
15	41	6	0	0	-1	1	1	0	1	0	31	-.13403	-.18320	.02	-.53951
16	41	6	0	0	-1	1	1	0	1	0	30	-.20458	-.18320	.04	-.53951
17	42	7	0	0	-1	2	0	1	0	1	50	1.20629	.00000	.00	-.43694
18	46	9	0	0	-1	2	0	1	0	1	18	-1.05110	.36640	-.39	-.02667
19	49	12	0	0	-1	1	1	0	1	0	35	.14814	.91599	.14	.28104
20	54	15	0	0	-1	1	1	0	1	0	30	-.20458	1.46559	-.30	.79388
21	30	1	1	1	1	2	0	1	0	1	7	-1.82708	-1.09919	2.01	-1.66777
22	34	2	1	2	1	1	1	0	1	0	7	-1.82708	-.91599	1.67	-1.25750
23	36	2	1	2	1	2	0	1	0	1	30	-.20458	-.91599	.19	-1.05236
24	42	2	1	2	1	1	1	0	1	0	30	-.20458	-.91599	.19	-.43694
25	43	3	1	3	1	1	1	0	1	0	28	-.34566	-.73279	.25	-.33437
26	44	3	1	3	1	3	0	0	-1	-1	24	-.62784	-.73279	.46	-.23181
27	43	4	1	4	1	2	0	1	0	1	51	1.27684	-.54960	-.70	-.33437
28	45	4	1	4	1	2	0	1	0	1	22	-.76892	-.54960	.42	-.12924
29	46	4	1	4	1	1	1	0	1	0	7	-1.82708	-.54960	1.00	-.02667
30	47	4	1	4	1	2	0	1	0	1	36	.14014	-.54960	-.08	.07590
31	34	5	1	5	1	2	0	1	0	1	57	1.70010	-.36640	-.62	-1.25750
32	44	5	1	5	1	1	1	0	1	0	34	.07760	-.36640	-.03	-.23181
33	46	6	1	6	1	1	1	0	1	0	43	.71249	-.18320	-.13	-.02667
34	47	7	1	7	1	2	0	1	0	1	21	-.83947	.00000	.00	.07590

Figure 12.1 ♦ SPSS Data Worksheet for Hypothetical Faculty Salary Study

The first research question that can be asked using these data is whether there is a significant difference in mean salary between males and females (ignoring all other predictor variables). This question could be addressed by conducting an independent samples *t* test to compare male and female means on salary. In this chapter, a one-way ANOVA is performed to compare mean salary for male versus female faculty; then, salary is predicted from gender by doing a regression analysis to predict salary scores from a dummy variable that represents gender. The examples presented in this chapter demonstrate that ANOVA and regression analysis provide equivalent information about gender differences in mean salary. Examples or demonstrations such as the ones presented in this chapter do not constitute formal mathematical proof. Mathematical statistics textbooks provide formal mathematical proof of the equivalence of ANOVA and regression analysis.

The second question that will be addressed is whether there are significant differences in salary across the three colleges. This question will be addressed by doing a one-way ANOVA to compare mean salary across the three college groups and by using dummy variables that represent college group membership as predictors in a regression. This example demonstrates that membership in k groups can be represented by a set of $(k - 1)$ dummy variables.

The third question that will be addressed involves assessment of a possible interaction between gender and years in position as predictors of salary. A regression will be used to predict salary from gender and years of experience. An interaction term will be added to the multiple regression equation to represent a possible interaction between gender and years of job experience (i.e., to assess whether there is a significant difference between the slopes that predict salary from years of experience for males versus females).

An analysis that includes an interaction term between two quantitative predictor variables (years of job experience and merit) can also be included. These analyses do not exhaust all the possible combinations of predictors for this set of hypothetical salary data; however, they serve to introduce the use of dummy predictors and interaction terms in regression analysis.

12.3 ♦ Screening for Violations of Assumptions

When we use one or more dummy variables as predictors in regression, the assumptions are essentially the same as for any other regression analysis (and the assumptions for a one-way ANOVA). As in other applications of ANOVA and regression, scores on the outcome variable *Y* should be quantitative and approximately normally distributed. If the *Y* outcome variable is categorical, logistic regression analysis should be used instead of linear regression; a brief introduction to binary logistic regression is presented in Chapter 21. Potential violations of the assumption of an approximately normal distribution shape for the *Y* outcome variable can be assessed by examining a histogram of scores on *Y*; the shape of this distribution should be reasonably close to normal. As described in Chapter 4, if there are extreme outliers or if the distribution shape is drastically different from normal, it may be appropriate to drop a few extreme scores, modify the value of a few extreme scores, or by using a data transformation such as the logarithm of *Y*, make the distribution of *Y* more nearly normal.

The variance of *Y* scores should be fairly homogenous across groups—that is, across levels of the dummy variables. The *F* tests used in ANOVA are fairly robust to violations of this assumption, unless the numbers of scores in the groups are small and/or unequal. When comparisons of means on *Y* across multiple groups are made using the SPSS *t* test procedure, one-way, or GLM, the Levene test can be requested to assess whether the homogeneity of variance is seriously violated. SPSS multiple regression does not provide a formal test of the assumption of homogeneity of variance across groups. It is helpful to examine a graph of the distribution of scores within each group (such as a box plot) to assess visually whether the scores of the *Y* outcome variable appear to have fairly homogeneous variances across levels of each *X* dummy predictor variable.

The issue of group size should also be considered in preliminary data screening (i.e., How many people are there in each of the groups represented by codes on the dummy variables?). For optimum statistical power and greater robustness to violations of assumptions, such as the homogeneity of variance assumption, it is preferred that there are equal numbers of scores in each group. The minimum number of scores within each group should be large enough to provide a reasonably accurate estimate of group means. For any groups that include fewer than 10 or 20 scores, estimates of the group means may have confidence intervals that are quite wide. The guidelines about the minimum number of scores per group from Chapter 5 (on the independent samples *t* test) and Chapter 6 (between subjects [between-S] one-way ANOVA) can be used to judge whether the numbers in each group that correspond to a dummy variable, such as gender, are sufficiently large to yield reasonable statistical power and reasonable robustness against violations of assumptions.

For the hypothetical faculty salary data in Table 12.1, the numbers of cases within the groups are (barely) adequate. There were 20 female and 30 male faculty in the sample, of whom 22 were Liberal Arts faculty, 17 were Sciences faculty, and 11 were Business faculty. Larger group sizes are desirable in real-world applications of dummy variable analysis; relatively small numbers of cases were used in this example to make it easy for students to verify computations, such as group means, by hand.

All the SPSS procedures that are used in this chapter, including box plot, scatter plot, Pearson correlation, one-way ANOVA, and linear regression, have been introduced and discussed in more detail in earlier chapters. Only the output from these procedures appears in this chapter. For a review of the menu selections and the SPSS dialog windows for any of these procedures, refer back to the chapter in which each analysis was first introduced. The only new material in this chapter concerns the use of dummy variables and products between predictor variables that represent interactions as predictors in regression.

To assess possible violations of assumptions, the following preliminary data screening was performed. A histogram was set up to assess whether scores on the quantitative outcome variable salary were reasonably normally distributed; this histogram appears in Figure 12.2. Although the distribution of salary values was multimodal, the salary scores did not show a distribution shape that was drastically different from a normal distribution. In real-life research situations, salary distributions are often positively skewed, such that there is a long tail at the upper end of the distribution (because there is usually no fixed upper limit for salary) and a truncated tail at the lower end of the distribution (because salary values cannot be lower than 0). Skewed distributions of scores can sometimes be made more nearly normal in shape by taking the base 10 or natural log of the salary scores. Logarithmic transformation was judged unnecessary for the artificial salary data that are used in the example in this chapter. Also, there were no extreme outliers in salary, as can be seen in Figure 12.2.

A box plot of salary scores was set up to examine the distribution of outcomes for the female and male groups (Figure 12.3). A visual examination of this box plot did not reveal any extreme outliers within either group; median salary appeared to be lower for females than for males. The box plot suggested that the variance of salary scores might be larger for males than for females. This is a possible violation of the homogeneity of variance assumption; in the one-way ANOVA presented in a later section, the Levene test was requested to evaluate whether this difference between salary variances for the female and male groups was statistically significant (and the Levene test was not significant).

Figure 12.4 shows a scatter plot of salary (on the *Y* axis) as a function of years of job experience (on the *X* axis) with different case markers for female and male faculty. The relation between salary and years appears to be linear, and the slope that predicts salary from years appears to be similar for the female and male groups, although there appears to be a tendency for females to have slightly lower salary scores than males at each level of years of job experience; also, the range for years of experience was smaller for females (0–15) than for males (0–22). Furthermore, the variance of salary appears to be reasonably homogeneous across years in the scatter plot that appears in Figure 12.4, and there

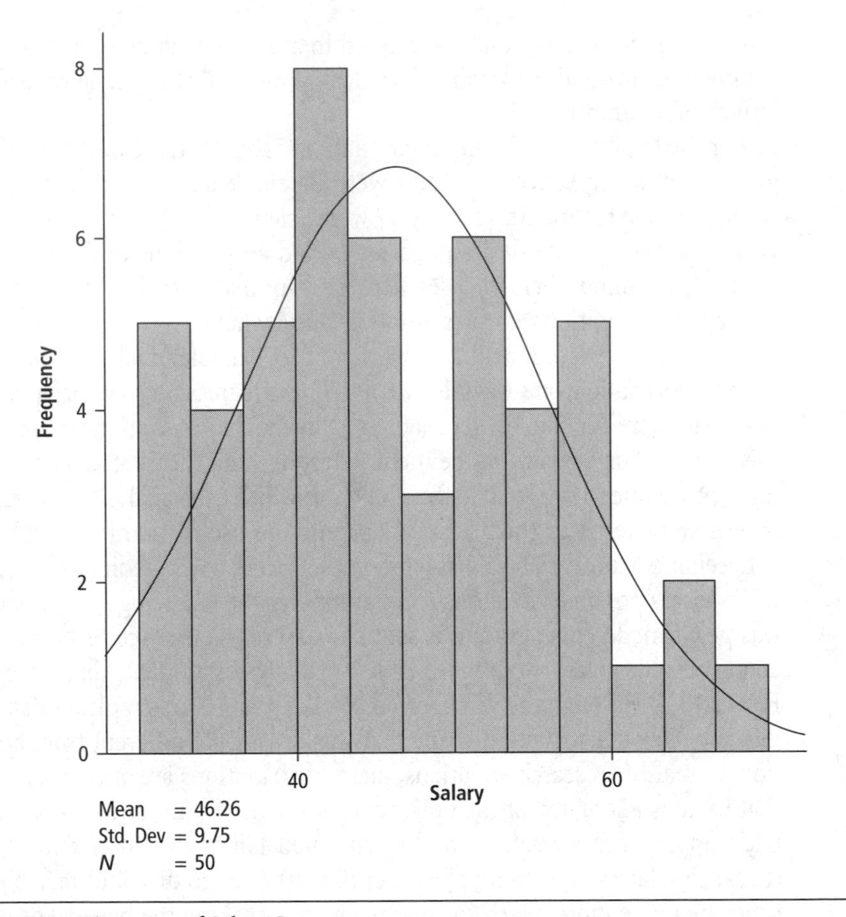

Figure 12.2 ♦ Histogram of Salary Scores

were no extreme bivariate outliers. Subsequent regression analyses will provide us with more specific information about gender and years as predictors of salary. We will use a regression analysis that includes the following predictors: years, a dummy variable to represent gender, and a product of gender and years. The results of this regression will help answer the following questions: Is there a significant increase in predicted salary associated with years of experience (when gender is statistically controlled)? Is there a significant difference in predicted salary between males and females (when years of experience is statistically controlled)? Is there an interaction between gender and years such that the predicted increase in salary for each 1-year increment in years of experience is different for males as compared with females?

12.4 ♦ Issues in Planning a Study

Essentially, when we use a dummy variable as a predictor in a regression, we have the same research situation as when we do a *t* test or ANOVA (both analyses predict scores on the *Y* outcome variable for two groups). When we use several dummy variables as predictors in a regression, we have the same research situation as in a one-way ANOVA (both analyses compare means across several groups). Therefore, the issues reviewed in

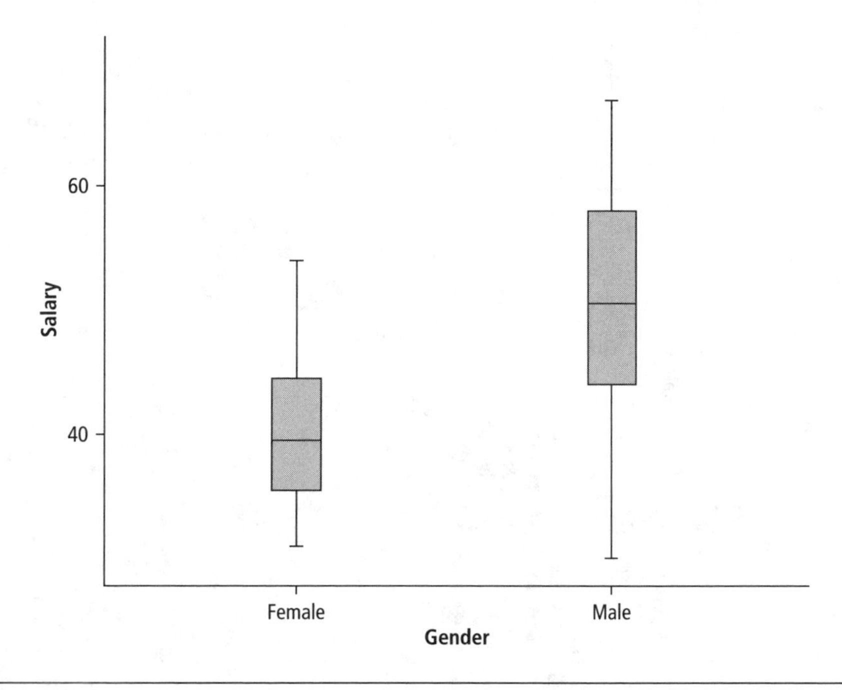

Figure 12.3 ♦ Box Plot of Salary Scores for Female and Male Groups

planning studies that use *t* tests (in Chapter 5) and studies that use one-way ANOVA (in Chapter 6) are also relevant when we use a regression analysis as the method of data analysis. To put it briefly, if the groups that are compared received different "dosage" levels of some treatment variable, the dosage levels need to be far enough apart to produce detectable differences in outcomes. If the groups are formed on the basis of participant characteristics (such as age), the groups need to be far enough apart on these characteristics to yield detectable differences in outcome. Other variables that might create within-group variability in scores may need to be experimentally or statistically controlled to reduce the magnitude of error variance, as described in Chapters 5 and 6.

It is important to check that every group has a reasonable minimum number of cases. If any group has fewer than 10 cases, the researcher may decide to combine that group with one or more other groups (if it makes sense to do so) or exclude that group from the analysis. The statistical power tables that appeared in Chapters 5 and 6 can be used to assess the minimum sample size needed per group to achieve reasonable levels of statistical power.

12.5 ♦ Parameter Estimates and Significance Tests for Regressions With Dummy Variables

The use of one or more dummy predictor variables in regression analysis does not change any of the computations for multiple regression described in Chapter 11. The estimates of *b* coefficients, the *t* tests for the significance of individual *b* coefficients, and the overall *F* test for the significance of the entire regression equation are all computed using the methods described in Chapter 11. Similarly, sr^2 (the squared semipartial correlation), an estimate of the proportion of variance uniquely predictable from each dummy predictor

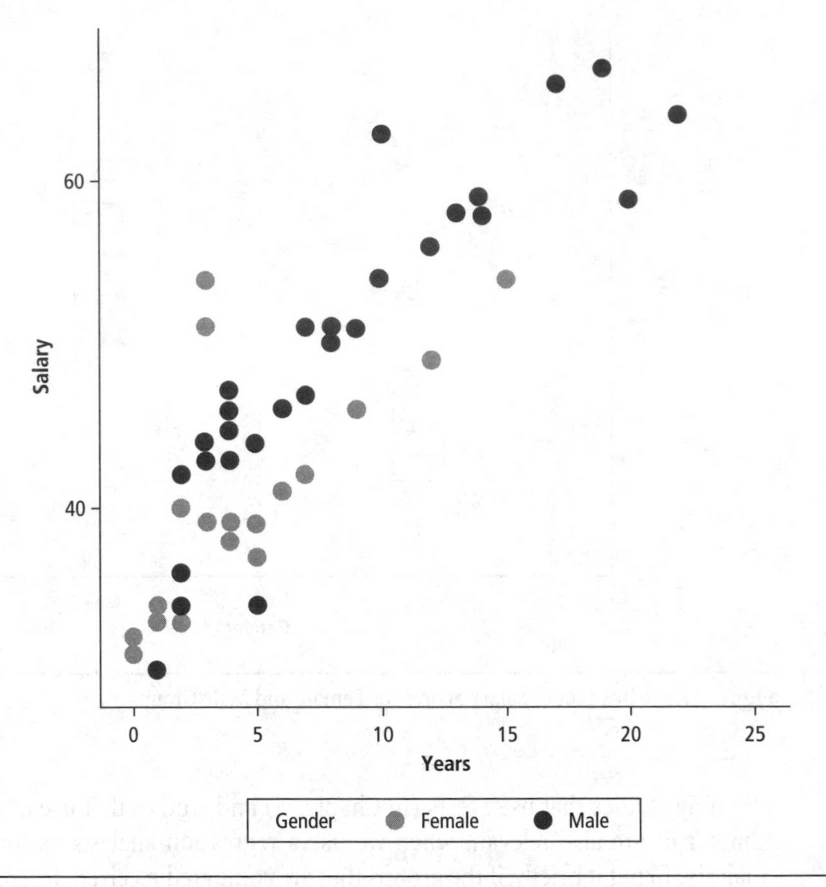

Figure 12.4 ♦ Scatter Plot of Salary by Years With Case Markers for Gender

variable, can be calculated and interpreted for dummy predictors in a manner similar to that described in Chapter 11. Confidence intervals for each b coefficient are obtained using the methods described in Chapter 11.

However, the *interpretation* of b coefficients when they are associated with dummy-coded variables is slightly different from the interpretation when they are associated with continuous predictor variables. Depending on the method of coding that is used, the b_0 (intercept) coefficient may correspond to the mean of one of the groups or to the grand mean of Y. The b_i coefficients for each dummy-coded predictor variable may correspond to contrasts between group means or to differences between group means and the grand mean.

12.6 ♦ Group Mean Comparisons Using One-Way Between-S ANOVA

12.6.1 ♦ Gender Differences in Mean Salary

A one-way between-S ANOVA was performed to assess whether mean salary differed significantly between female and male faculty. No other variables were taken into account in this analysis. A test of the homogeneity of variance assumption was requested (the Levene test).

The Levene test was not statistically significant: $F(1, 48) = 2.81, p = .1$. Thus, there was no statistically significant difference between the variances of salary for the female and male groups; the homogeneity of variance assumption was not violated. The mean salary for males (M_{male}) was 49.9 thousand dollars per year; the mean salary for females (M_{female}) was 40.8 thousand dollars per year. The difference between mean annual salary for males and females was statistically significant at the conventional $\alpha = .05$ level: $F(1, 48) = 13.02$, $p = .001$. The difference between the means ($M_{male} - M_{female}$) was +9.1 thousand dollars (i.e., on average, male faculty earned about 9.1 thousand dollars more per year than female faculty). The eta-squared effect size for this gender difference was .21; in other words, about 21% of the variance in salaries could be predicted from gender. This corresponds to a large effect size (refer back to Table 5.2 for suggested verbal labels for values of η^2 and R^2).

This initial finding of a gender difference in mean salary is not necessarily evidence of gender bias in salary levels. Within this sample, there was a tendency for female faculty to have fewer years of experience and for male faculty to have more years of experience, and salary increases as a function of years of experience. It is possible that the gender difference we see in this ANOVA is partly or completely accounted for by differences between males and females in years of experience. Subsequent analyses will address this by examining whether gender still predicts different levels of salary when "years of experience" is statistically controlled by including it as a second predictor of salary.

Note that when the numbers of cases in the groups are unequal, there are two different ways in which a grand mean can be calculated. An unweighted grand mean of salary can be obtained by simply averaging male and female mean salary, ignoring sample size. The unweighted grand mean is $(40.8 + 49.9)/2 = 45.35$. However, in many statistical analyses, the estimate of the grand mean is weighted by sample size. The weighted grand mean in this example is found as follows:

$$[(n_{male} \times M_{male}) + (n_{female} \times M_{female})]/(n_{male} + n_{female})$$
$$= [(20 \times 40.8) + (30 \times 49.9)]/(20+30)$$
$$= 46.26.$$

The grand mean for salary reported in the one-way ANOVA in Figure 12.5 corresponds to the weighted grand mean. (This weighted grand mean is equivalent to the sum of the 50 individual salary scores divided by the number of scores in the sample.) In the regression analyses reported later in this chapter, the version of the grand mean that appears in the results corresponds to the unweighted grand mean.

12.6.2 ♦ College Differences in Mean Salary

In a research situation that involves a categorical predictor variable with more than two levels or groups and a quantitative outcome variable, the most familiar approach to data

Descriptives

salary

	N	Mean	Std. Deviation	Std. Error	95% Confidence Interval for Mean Lower Bound	95% Confidence Interval for Mean Upper Bound	Minimum	Maximum
Female	20	40.80	6.986	1.562	37.53	44.07	31	54
Male	30	49.90	9.714	1.774	46.27	53.53	30	67
Total	50	46.26	9.750	1.379	43.49	49.03	30	67

Test of Homogeneity of Variances

salary

Levene Statistic	df1	df2	Sig.
2.809	1	48	.100

ANOVA

salary

	Sum of Squares	df	Mean Square	F	Sig.
Between Groups	993.720	1	993.720	13.019	.001
Within Groups	3663.900	48	76.331		
Total	4657.620	49			

Figure 12.5 ♦ One-Way Between-S ANOVA: Mean Salary for Females and Males

analysis is a one-way ANOVA. To evaluate whether mean salary level differs for faculty across the three different colleges in the hypothetical dataset in Table 12.1, we can conduct a between-S one-way ANOVA using the SPSS ONEWAY procedure. The variable college is coded 1 = Liberal Arts, 2 = Sciences, and 3 = Business. The outcome variable, as in previous analyses, is annual salary in thousands of dollars. Orthogonal contrasts between colleges were also requested by entering custom contrast coefficients. The results of this one-way ANOVA appear in Figure 12.6.

The means on salary were as follows: 44.8 for faculty in Liberal Arts, 44.7 in Sciences, and 51.6 in Business. The overall F for this one-way ANOVA was not statistically significant: $F(2, 47) = 2.18, p = .125$. The effect size, eta squared, was obtained by taking the ratio $SS_{between}/SS_{total}$; for this ANOVA, $\eta^2 = .085$. This is a medium effect. In this situation, if we want to use this sample to make inferences about some larger population of faculty, we would not have evidence that the proportion of variance in salary that is predictable from college is significantly different from 0. However, if we just want to describe the strength of the association between college and salary within this sample, we could say that, for this sample, about 8.5% of the variance in salary was predictable from college. The orthogonal contrasts that were requested made the following comparison. For Contrast 1, the custom contrast coefficients were "+1," "−1," and "0"; this corresponds to a comparison of the mean salaries between College 1 (Liberal Arts) and College 2 (Sciences); this contrast was not statistically significant: $t(47) = .037, p = .97$. For Contrast 2, the custom contrast coefficients were "+1," "+1," and "−2"; this corresponds to a comparison of the

salary

Descriptives

	N	Mean	Std. Deviation	Std. Error	95% Confidence Interval for Mean		Minimum	Maximum
					Lower Bound	Upper Bound		
Liberal Arts	22	44.82	9.261	1.975	40.71	48.92	31	66
Sciences	17	44.71	9.777	2.371	39.68	49.73	30	67
Business	11	51.55	9.658	2.912	45.06	58.03	32	64
Total	50	46.26	9.750	1.379	43.49	49.03	30	67

Test of Homogeneity of Variances

salary

Levene Statistic	df1	df2	Sig.
.005	2	47	.995

ANOVA

salary

	Sum of Squares	df	Mean Square	F	Sig.
Between Groups	394.091	2	197.045	2.172	.125
Within Groups	4263.529	47	90.713		
Total	4657.620	49			

Contrast Coefficients

Contrast	college		
	Liberal Arts	Sciences	Business
1	1	-1	0
2	1	1	-2

Contrast Tests

		Contrast	Value of Contrast	Std. Error	t	df	Sig. (2-tailed)
salary	Assume equal variances	1	.11	3.076	.037	47	.971
		2	-13.57	6.515	-2.082	47	.043
	Does not assume equal variances	1	.11	3.086	.036	33.580	.971
		2	-13.57	6.591	-2.058	16.027	.056

Figure 12.6 ♦ One-Way Between-S ANOVA: Mean Salary Across Colleges

mean salary for Liberal Arts and Sciences faculty combined, compared with the Business faculty. This contrast was statistically significant: $t(47) = -2.058, p = .036$. Business faculty had a significantly higher mean salary than the two other colleges (Liberal Arts and Sciences) combined.

The next sections show that regression analyses with dummy predictor variables can be used to obtain the same information about the differences between group means. Dummy variables that represent group membership (such as female and male, or Liberal Arts, Sciences, and Business colleges) can be used to predict salary in regression analyses. We will see that the information about differences among group means that can be obtained by doing regression with dummy predictor variables is equivalent to the information that we can obtain from one-way ANOVA. In future chapters, both dummy variables and quantitative variables are used as predictors in regression.

12.7 ♦ Three Methods of Coding for Dummy Variables

The three coding methods that are most often used for dummy variables are given below (the details regarding these methods are presented in subsequent sections along with empirical examples):

1. Dummy coding of dummy variables
2. Effect coding of dummy variables
3. Orthogonal coding of dummy variables

In general, when we have k groups, we need only $(k - 1)$ dummy variables to represent information about group membership. Most dummy variable codes can be understood as answers to Yes/No questions about group membership. For example, to represent college group membership, we might include a dummy variable D_1 that corresponds to the question, Is this faculty member in Liberal Arts? and code the responses as 1 = Yes, 0 = No. If there are k groups, a set of $(k - 1)$ Yes/No questions provides complete information about group membership. For example, if a faculty member reports responses of "No" to the questions, Are you in Liberal Arts? and Are you in Science? and there are only three groups, then that person must belong to the third group (in this example, Group 3 is the Business college).

The difference between dummy coding and effect coding is in the way in which codes are assigned for members of the last group—that is, the group that does not correspond to an explicit Yes/No question about group membership. In dummy coding of dummy variables, members of the last group receive a score of "0" on all the dummy variables. (In effect coding of dummy variables, members of the last group are assigned scores of "–1" on all the dummy variables.) This difference in codes results in slightly different interpretations of the b coefficients in the regression equation, as described in the subsequent sections of this chapter.

12.7.1 ♦ Regression With Dummy-Coded Dummy Predictor Variables

12.7.1.1 ♦ *Two-Group Example With a Dummy-Coded Dummy Variable*

Suppose we want to predict salary (Y) from gender; gender is a **dummy-coded dummy variable** with codes of 0 for female and 1 for male participants in the study. In a previous section, this difference was evaluated by doing a one-way between-S ANOVA to compare mean salary across female and male groups. We can obtain equivalent information about the magnitude of gender differences in salary from a regression analysis that uses a dummy-coded variable to predict salary. For this simple two-group case (prediction of salary from dummy-coded gender), we can write a regression equation using gender to predict salary (Y) as follows:

$$\text{Salary}' \text{ or } Y' = b_0 + b_1 \times \text{Gender}. \tag{12.2}$$

From Equation 12.2, we can work out two separate prediction equations: one that makes predictions of Y for females and one that makes predictions of Y for males. To do this, we substitute the values of "0" (for females) and "1" (for males) into Equation 12.2 and simplify the expression to obtain these two different equations:

$$Y' = b_0 \text{ (for females, gender = "0")}, \tag{12.3}$$

$$Y' = b_0 + b_1 \text{ (for males, gender = "1")}. \tag{12.4}$$

These two equations tell us that the constant value b_0 is the best prediction of salary for females, and the constant value ($b_0 + b_1$) is the best prediction of salary for males. This implies that b_0 = mean salary for females, $b_0 + b_1$ = mean salary for males, and b_1 = the difference between mean salary for the male versus female groups. The slope coefficient b_1 corresponds to the difference in mean salary for males and females. If the b_1 slope is significantly different from 0, it implies that there is a statistically significant difference in mean salary for males and females.

The results of the regression in Figure 12.7 provide the numerical estimates for the raw score regression coefficients (b_0 and b_1) for this set of data:

$$\text{Salary}' = 40.8 + 9.1 \times \text{Gender}.$$

For females, with gender = "0," the predicted mean salary given by this equation is $40.8 + 9.1 \times 0 = 40.8$. Note that this is the same as the mean salary for females in the one-way ANOVA output in Figure 12.5. For males, with gender= "1," the predicted salary given by this equation is $40.8 + 9.1 = 49.9$. Note that this value is equal to the mean salary for males in the one-way ANOVA in Figure 12.5.

The b_1 coefficient in this regression was statistically significant: $t(48) = 3.61, p = .001$. The F test reported in Figure 12.6 is equivalent to the square of the t test value for the null hypothesis that $b_1 = 0$ in Figure 12.7 ($t = +3.608, t^2 = 13.02$). Note also that the eta-squared effect size associated with the ANOVA ($\eta^2 = .21$) and the R^2 effect size associated with the regression were equal; in both analyses, about 21% of the variance in salaries was predictable from gender.

When we use a dummy variable with codes of "0" and "1" to represent membership in two groups, the value of the b_0 intercept term in the regression equation is equivalent to the mean of the group for which the dummy variable has a value of "0." The b_1 "slope" coefficient represents the difference (or contrast) between the means of the two groups. The slope, in this case, represents the change in the mean level of Y when you move from a code of "0" (female) to a code of "1" (male) on the dummy predictor variable.

Figure 12.8 shows a scatter plot of salary scores (on the Y axis) as a function of gender code (on the X axis). In this graph, the intercept b_0 corresponds to the mean on the dependent variable (salary) for the group that had a dummy variable score of "0" (females). The slope, b_1, corresponds to the difference between the means for the two groups—that is, the change in the predicted mean when you move from a code of "0" to a code of "1" on the dummy variable. That is, $b_1 = M_{\text{male}} - M_{\text{female}}$, the change in salary when you move from

Variables Entered/Removed[b]

Model	Variables Entered	Variables Removed	Method
1	gender[a]		Enter

a. All requested variables entered.

b. Dependent Variable: salary

Model Summary

Model	R	R Square	Adjusted R Square	Std. Error of the Estimate
1	.462[a]	.213	.197	8.737

a. Predictors: (Constant), gender

ANOVA[b]

Model		Sum of Squares	df	Mean Square	F	Sig.
1	Regression	993.720	1	993.720	13.019	.001[a]
	Residual	3663.900	48	76.331		
	Total	4657.620	49			

a. Predictors: (Constant), gender

b. Dependent Variable: salary

Coefficients[a]

Model		Unstandardized Coefficients		Standardized Coefficients	t	Sig.	Correlations		
		B	Std. Error	Beta			Zero-order	Partial	Part
1	(Constant)	40.800	1.954		20.884	.000			
	gender	9.100	2.522	.462	3.608	.001	.462	.462	.462

a. Dependent Variable: salary

Figure 12.7 ♦ Regression to Predict Salary From Dummy-Coded Gender

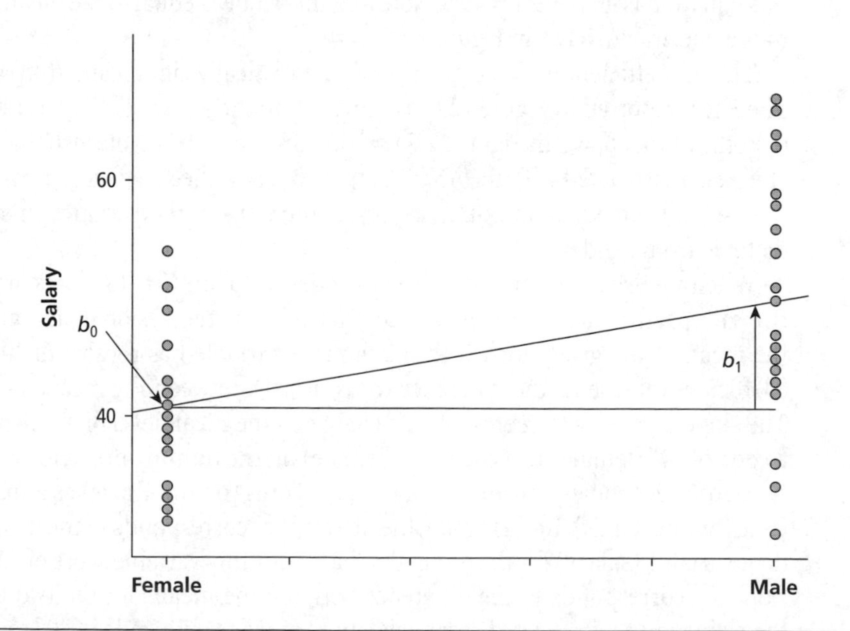

Figure 12.8 ♦ Graph for Regression to Predict Salary From Dummy-Coded Gender (0 = Female, 1 = Male)

a score of "0" (female) to a score of "1" (male). The test of the statistical significance of b_1 is equivalent to the t test of the difference between the mean Y values for the two groups represented by the dummy variable in the regression.

12.7.1.2 ♦ *Multiple Group Example With Dummy-Coded Dummy Variables*

When there are multiple groups (number of groups = k), group membership can be represented by scores on a set of $(k - 1)$ dummy variables. Each dummy variable essentially represents a Yes/No question about group membership. In the preceding example, there are $k = 3$ college groups in the faculty data. In this example, we will use two dummy variables denoted by D_1 and D_2 to represent information about college membership in a regression analysis. D_1 corresponds to the following question: Is the faculty member from the Liberal Arts College? 1 = Yes, 0 = No. D_2 corresponds to the following question: Is the faculty member from the Sciences College? 1 = Yes, 0 = No. For dummy coding, members of the last group receive a score of "0" on all the dummy variables. In this example, faculty from the Business College received scores of "0" on both the D_1 and D_2 dummy-coded dummy variable. For the set of three college groups, the dummy-coded dummy variables that provide information about college group membership were coded as follows:

	D_1	D_2
Liberal Arts	1	0
Sciences	0	1
Business	0	0

Now that we have created dummy variables that represent information about membership in the three college groups as scores on a set of dummy variables, mean salary can be predicted from college groups by a regression analysis that uses the dummy-coded dummy variables shown above as predictors:

$$Y' = b_0 + b_1D_1 + b_2D_2. \qquad (12.5)$$

The results of the regression (using dummy-coded dummy variables to represent career group membership) are shown in Figure 12.9. Note that the overall F reported for the regression analysis in Figure 12.9 is identical to the overall F reported for the one-way ANOVA in Figure 12.6 ($F[2, 47] = 2.17, p = .125$) and that the η^2 for the one-way ANOVA is identical to the R^2 for the regression ($\eta^2 = R^2 = .085$). Note also that the b_0 coefficient in the regression results in Figure 12.9 (b_0 = constant = 51.55) equals the mean salary for the group that was assigned score values of 0 on all the dummy variables (the mean salary for the Business faculty was 51.55). Note also that the b_i coefficients for each of the two dummy variables represent the difference between the mean of the corresponding group and the mean of the comparison group whose codes were all 0; for example, $b_1 = -6.73$, which corresponds to the difference between the mean salary of Group 1, Liberal Arts (M_1 = 44.82) and mean salary of the comparison group, Business (M_3 = 51.55); the value of the b_2 coefficient ($b_2 = -6.84$) corresponds to the difference between mean salary for the Science (M = 44.71) and Business (M = 51.55) groups. This regression analysis with

Variables Entered/Removed[b]

Model	Variables Entered	Variables Removed	Method
1	d2, d1 [a]		Enter

a. All requested variables entered.

b. Dependent Variable: salary

Model Summary

Model	R	R Square	Adjusted R Square	Std. Error of the Estimate
1	.291 [a]	.085	.046	9.524

a. Predictors: (Constant), d2, d1

ANOVA[b]

Model		Sum of Squares	df	Mean Square	F	Sig.
1	Regression	394.091	2	197.045	2.172	.125 [a]
	Residual	4263.529	47	90.713		
	Total	4657.620	49			

a. Predictors: (Constant), d2, d1

b. Dependent Variable: salary

Coefficients[a]

Model		Unstandardized Coefficients		Standardized Coefficients			Correlations		
		B	Std. Error	Beta	t	Sig.	Zero-order	Partial	Part
1	(Constant)	51.545	2.872		17.949	.000			
	d1	-6.727	3.517	-.346	-1.913	.062	-.132	-.269	-.267
	d2	-6.840	3.685	-.336	-1.856	.070	-.116	-.261	-.259

a. Dependent Variable: salary

Figure 12.9 ♦ Regression to Predict Salary From Dummy-Coded College Membership

dummy variables to represent college membership provided information equivalent to a one-way ANOVA to predict salary from college.

12.7.2 ♦ Regression With Effect-Coded Dummy Predictor Variables

12.7.2.1 ♦ *Two-Group Example With an Effect-Coded Dummy Variable*

We will now code the scores for gender slightly differently, using a method called "effect coding of dummy variables." In effect coding of dummy variables, a score value of "1" is used to represent a "Yes" answer to a question about group membership; membership in the group that does not correspond to a "Yes" answer on any of the group membership questions is represented by a score of "–1." In the following example, the **effect-coded dummy variable** "geneff" (Is the participant male, Yes or No?) is coded "+1" for males and "–1" for females. The variable geneff is called an effect-coded dummy variable because we used "–1" (rather than "0") as the value that represents membership in the last group. Our overall model for the prediction of salary (*Y*) from gender, represented by the effect-coded dummy variable geneff, can be written as follows:

$$Y' = b_0 + b_1 \times \text{Geneff}. \tag{12.6}$$

Substituting the values of "+1" for males and "–1" for females, the predictive equations for males and females become

$$Y' = b_0 + b_1 \text{ (for males, with geneff coded "+1")}, \tag{12.7}$$

$$Y' = b_0 - b_1 \text{ (for females, with geneff coded "–1")}. \tag{12.8}$$

From earlier discussions on *t* tests and ANOVA, we know that the best predicted value of *Y* for males is equivalent to the mean on *Y* for males, M_{male}; similarly, the best predicted value of *Y* for females is equal to the mean on *Y* for females, M_{female}. The two equations above, therefore, tell us that $M_{\text{male}} = b_0 + b_1$ and $M_{\text{female}} = b_0 - b_1$. What does this imply for the values of b_0 and b_1? The mean for males is b_1 units above b_0; the mean for females is b_1 units below b_0. With a little thought, you will see that the intercept b_0 must equal the grand mean on salary for both genders combined.

Note that when we calculate a grand mean by combining group means, there are two different possible ways to calculate the grand mean. If the groups have the same numbers of scores, these two methods yield the same result, but when the groups have unequal numbers of cases, these two methods for computation of the grand mean yield different results. Whenever you do analyses with unequal numbers in the groups, you need to decide whether the unweighted or the weighted grand mean is a more appropriate value to report. In some situations, it may not be clear what default decision a computer program uses (i.e., whether the program reports the weighted or the unweighted grand mean), but it is possible to calculate both the weighted and unweighted grand means by hand from the group means; when you do this, you will be able to determine which version of the grand mean was reported on the SPSS printout.

The unweighted grand mean for salary for males and females is obtained by ignoring the number of cases in the groups and averaging the group means together for males and females. For the male and female salary data that appeared in Table 12.1, the **unweighted mean** is $(M_{\text{male}} + M_{\text{female}})/2 = (40.80 + 49.90)/2 = 45.35$. Note that the b_0 constant or intercept term in the regression in Figure 12.10 that uses effect-coded gender to predict salary corresponds to this unweighted grand mean of 45.35. When you run a regression to predict scores on a quantitative outcome variable from effect-coded dummy predictor variables, and the default methods of computation are used in SPSS, the b_0 coefficient in the regression equation corresponds to the unweighted grand mean, and effects (or differences between group means and grand means) are reported relative to this unweighted grand mean as a reference point. This differs slightly from the one-way ANOVA output in Figure 12.5, which reported the weighted grand mean for salary (46.26).

When effect-coded dummy predictor variables are used, the slope coefficient b_1 corresponds to the "effect" of gender; that is, $+b_1$ is the distance between the male mean and the grand mean, and $-b_1$ is the distance between the female mean and the grand mean. In Chapter 6 on one-way ANOVA, the terminology used for these distances (group mean minus grand mean) was *effect*. The effect of membership in Group *i* in a one-way ANOVA is represented by α_i, where $\alpha_i = M_i - M_Y$, the mean of Group *i* minus the grand mean of *Y* across all groups.

This method of coding ("+1" versus "−1") is called "effect coding," because the intercept b_0 in Equation 12.5 equals the unweighted grand mean for salary, M_Y, and the slope coefficient b_i for each effect-coded dummy variable E represents the effect for the group that has a code of "1" on that variable—that is, the difference between that group's mean on Y and the unweighted grand mean. Thus, when effect-coded dummy variables are used to represent group membership, the b_0 intercept term equals the grand mean for Y, the outcome variable, and each b_i coefficient represents a contrast between the mean of one group versus the unweighted grand mean (or the "effect" for that group). The significance of b_1 for the effect-coded variable geneff is a test of the significance of the difference between the mean of the corresponding group (in this example, males) and the unweighted grand mean. Given that geneff is coded "−1" for females and "+1" for males and given that the value of b_1 is significant and positive, the mean salary of males is significantly higher than the grand mean (and the mean salary for females is significantly lower than the grand mean).

Note that we do not have to use a code of "+1" for the group with the higher mean and a code of "−1" for the group with the lower mean on Y. The sign of the b coefficient can be either positive or negative; it is the combination of signs (on the code for the dummy variable and the b coefficient) that tells us which group had a mean that was lower than the grand mean.

The overall F result for the regression analysis that predicts salary from effect-coded gender (geneff) in Figure 12.10 is identical to the F value in the earlier analyses of gender and salary reported in Figures 12.5 and 12.7: $F(1, 48) = 13.02$, $p = .001$. The effect size given by η^2 and R^2 is also identical across these three analyses ($R^2 = .21$). The only difference between the regression that uses a dummy-coded dummy variable (Figure 12.7) and the regression that uses an effect-coded dummy variable to represent gender (in Figure 12.10) is in the way in which the b_0 and b_1 coefficients are related to the grand mean and group means.

12.7.2.2 ♦ *Multiple Group Example With Effect-Coded Dummy Variables*

If we used effect coding instead of dummy coding to represent membership in the three college groups used as an example earlier, group membership could be coded as follows:

	E_1	E_2
Liberal Arts	1	0
Sciences	0	1
Business	−1	−1

That is, E_1 and E_2 still represent Yes/No questions about group membership. E_1 corresponds to the following question: Is the faculty member in Liberal Arts? Coded 1 = Yes, 0 = No. E_2 corresponds to the following question: Is the faculty member in Sciences? Coded 1 = Yes, 0 = No. The only change when effect coding (instead of dummy coding) is used is that members of the Business college (the one group that does not correspond directly to a Yes/No question) now receive codes of "−1" on both the variables E_1 and E_2.

Variables Entered/Removed[b]

Model	Variables Entered	Variables Removed	Method
1	geneff [a]		Enter

a. All requested variables entered.

b. Dependent Variable: salary

Model Summary

Model	R	R Square	Adjusted R Square	Std. Error of the Estimate
1	.462 [a]	.213	.197	8.737

a. Predictors: (Constant), geneff

ANOVA[b]

Model		Sum of Squares	df	Mean Square	F	Sig.
1	Regression	993.720	1	993.720	13.019	.001 [a]
	Residual	3663.900	48	76.331		
	Total	4657.620	49			

a. Predictors: (Constant), geneff

b. Dependent Variable: salary

Coefficients[a]

Model		Unstandardized Coefficients		Standardized Coefficients	t	Sig.	Correlations		
		B	Std. Error	Beta			Zero-order	Partial	Part
1	(Constant)	45.350	1.261		35.962	.000			
	geneff	4.550	1.261	.462	3.608	.001	.462	.462	.462

a. Dependent Variable: salary

Figure 12.10 ♦ Regression to Predict Salary From Effect-Coded Gender

We can run a regression to predict scores on salary from the two effect-coded dummy variables E_1 and E_2:

$$\text{Salary}' \text{ or } Y' = b_0 + b_1E_1 + b_2E_2. \quad (12.9)$$

The results of this regression analysis are shown in Figure 12.11.

When effect coding is used, the intercept or b_0 coefficient is interpreted as an estimate of the (unweighted) grand mean for the Y outcome variable, and each b_i coefficient represents the effect for one of the groups—that is, the contrast between a particular group mean and the grand mean. (Recall that when dummy coding was used, the intercept b_0 was interpreted as the mean of the "last" group, namely, the group that did not correspond to a "Yes" answer on any of the dummy variables, and each b_i coefficient corresponded to the difference between one of the group means and the mean of the "last" group, the group that is used as the reference group for all comparisons.) In Figure 12.11, the overall F value and the overall R^2 are the same as in the two previous analyses that compared mean

Variables Entered/Removed[b]

Model	Variables Entered	Variables Removed	Method
1	e2, e1 [a]	.	Enter

a. All requested variables entered.

b. Dependent Variable: salary

Model Summary

Model	R	R Square	Adjusted R Square	Std. Error of the Estimate
1	.291 [a]	.085	.046	9.524

a. Predictors: (Constant), e2, e1

ANOVA[b]

Model		Sum of Squares	df	Mean Square	F	Sig.
1	Regression	394.091	2	197.045	2.172	.125 [a]
	Residual	4263.529	47	90.713		
	Total	4657.620	49			

a. Predictors: (Constant), e2, e1

b. Dependent Variable: salary

Coefficients[b]

Model		Unstandardized Coefficients		Standardized Coefficients	t	Sig.	Correlations		
		B	Std. Error	Beta			Zero-order	Partial	Part
1	(Constant)	47.023	1.403		33.525	.000			
	e1	-2.205	1.828	-.179	-1.206	.234	-.238	-.173	-.168
	e2	-2.317	1.935	-.177	-1.197	.237	-.237	-.172	-.167

a. Dependent Variable: salary

Figure 12.11 ♦ Regression to Predict Salary From Effect-Coded College Membership

salary across college (in Figures 12.6 and 12.9): $F(2, 47) = 2.12, p = .125; R^2 = .085$. The b coefficients for Equation 12.9, from Figure 12.11, are as follows:

$$\text{Salary}' = 47.03 - 2.205 \times E_1 - 2.317 \times E_2.$$

The interpretation is as follows: The (unweighted) grand mean of salary is 47.03 thousand dollars per year. Members of the Liberal Arts faculty have a predicted annual salary that is 2.205 thousand dollars less than this grand mean; members of the Sciences faculty have a predicted salary that is 2.317 thousand dollars less than this grand mean. Neither of these differences between a group mean and the grand mean is statistically significant at the $\alpha = .05$ level.

Because members of the Business faculty have scores of "−1" on both E_1 and E_2, the predicted mean salary for Business faculty is

$$\text{Salary}' = 47.03 + 2.205 + 2.317 = 51.5 \text{ thousand dollars per year.}$$

We do not have a significance test to evaluate whether the mean salary for Business faculty is significantly higher than the grand mean. If we wanted to include a significance test for this contrast, we could do so by rearranging the dummy variable codes associated with group membership, such that membership in the Business College group corresponded to an answer of "Yes" on either E_1 or E_2.

12.7.3 ♦ Orthogonal Coding of Dummy Predictor Variables

We can set up contrasts among group means in such a way that the former are orthogonal (the term *orthogonal* is equivalent to uncorrelated or independent). One method of creating orthogonal contrasts is to set up one contrast that compares Group 1 versus Group 2 and a second contrast that compares Groups 1 and 2 combined versus Group 3, as in the example below:

	Group		
	1 (Liberal Arts)	*2 (Sciences)*	*3 (Business)*
O_1	+1	−1	0
O_2	+1	+1	−2

The codes across each row should sum to 0. For each **orthogonally coded dummy variable,** the groups for which the code has a positive sign are contrasted with the groups for which the code has a negative sign; groups with a code of "0" are ignored. Thus, O_1 compares the mean of Group 1 (Liberal Arts) with the mean of Group 2 (Sciences).

To figure out which formal null hypothesis is tested by each contrast, we form a weighted linear composite that uses these codes. That is, we multiply the population mean μ_k for Group k by the contrast coefficient for Group k and sum these products across the k groups; we set that weighted linear combination of population means equal to 0 as our null hypothesis.

In this instance, the null hypotheses that correspond to the contrast specified by the two O_i orthogonally coded dummy variables are as follows:

$$H_0 \text{ for } O_1\text{: } (+1)\mu_1 + (-1)\mu_2 + (0)\mu_3 = 0.$$

That is,

$$H_0 \text{ for } O_1\text{: } \mu_1 - \mu_2 = 0 \text{ (or } \mu_1 = \mu_2\text{)}.$$

The O_2 effect-coded dummy variable compares the average of the first two group means (i.e., the mean for Liberal Arts and Sciences combined) with the mean for the third group (Business):

$$H_0 \text{ for } O_2\text{: } (+1)\mu_1 + (+1)\mu_2 + (-2)\mu_3,$$
$$H_0 \text{ for } O_2\text{: } (\mu_1 + \mu_2) - 2\mu_3 = 0,$$

or

$$\frac{\mu_1 + \mu_2}{2} - \mu_3 = 0 \quad \text{or} \quad \frac{\mu_1 + \mu_2}{2} = \mu_3.$$

We can assess whether the contrasts are orthogonal by taking the cross products and summing the corresponding coefficients. Recall that products between sets of scores provide information about covariation or correlation; see Chapter 7 for details. Because each of the two variables O_1 and O_2 has a sum (and, therefore, a mean) of 0, each code represents a deviation from a mean. When we compute the sum of cross products between corresponding values of these two variables, we are, in effect, calculating the numerator of the correlation between O_1 and O_2. In this example, we can assess whether O_1 and O_2 are orthogonal or uncorrelated by calculating the following sum of cross products. For O_1 and O_2, the sum of cross products of the corresponding coefficients is

$$(+1)(+1) + (-1)(+1) + (0)(-2) = 0.$$

Because this sum of the products of corresponding coefficients is 0, we know that the contrasts specified by O_1 and O_2 are orthogonal.

Of course, as an alternate way to see whether the O_1 and O_2 predictor variables are orthogonal or uncorrelated, SPSS can also be used to calculate a correlation between O_1 and O_2; if the contrasts are orthogonal, the Pearson r between O_1 and O_2 will equal 0 (provided that the numbers in the groups are equal).

For each contrast specified by a set of codes (e.g., the O_1 set of codes), any group with a 0 coefficient is ignored, and groups with opposite signs are contrasted. The direction of the signs for these codes does not matter; the contrast represented by the codes ("+1," "−1," and "0") represents the same comparison as the contrast represented by ("−1," "+1," "0"). The *b* coefficients obtained for these two sets of codes would be opposite in sign, but the significance of the difference between the means of Group 1 and Group 2 would be the same whether Group 1 was assigned a code of "+1" or "−1."

Figure 12.12 shows the results of a regression in which the dummy predictor variables O_1 and O_2 are coded to represent the same orthogonal contrasts. Note that the *t* tests for significance of each contrast are the same in both Figure 12.6, where the contrasts were requested as optional output from the ONEWAY procedure, and Figure 12.12, where the contrasts were obtained by using orthogonally coded dummy variables as predictors of salary. Only one of the two contrasts, the contrast that compares the mean salary for Liberal Arts and Sciences faculty with the mean salary for Business faculty, was statistically significant.

Orthogonal coding of dummy variables can also be used to perform a trend analysis—for example, when the groups being compared represent equally spaced dosage levels along a continuum. The following example shows orthogonal coding to represent linear versus quadratic trends for a study in which the groups receive three different dosage levels of caffeine.

Variables Entered/Removed[b]

Model	Variables Entered	Variables Removed	Method
1	o2, o1[a]		Enter

a. All requested variables entered.

b. Dependent Variable: salary

Model Summary

Model	R	R Square	Adjusted R Square	Std. Error of the Estimate
1	.291 [a]	.085	.046	9.524

a. Predictors: (Constant), o2, o1

ANOVA[b]

Model		Sum of Squares	df	Mean Square	F	Sig.
1	Regression	394.091	2	197.045	2.172	.125 [a]
	Residual	4263.529	47	90.713		
	Total	4657.620	49			

a. Predictors: (Constant), o2, o1

b. Dependent Variable: salary

Coefficients[a]

Model		Unstandardized Coefficients		Standardized Coefficients	t	Sig.	Correlations		
		B	Std. Error	Beta			Zero-order	Partial	Part
1	(Constant)	47.023	1.403		33.525	.000			
	o1	.056	1.538	.005	.037	.971	-.013	.005	.005
	o2	-2.261	1.086	-.291	-2.082	.043	-.291	-.291	-.291

a. Dependent Variable: salary

Figure 12.12 ♦ Regression to Predict Salary From Dummy Variables That Represent Orthogonal Contrasts

	0 mg	*150 mg*	*300 mg*
O_1	−1	0	+1
O_2	+1	−2	+1

Note that the sum of cross products is again 0 [(−1)(+1) + (0)(−2) + (+1)(+1)], so these contrasts are orthogonal. A simple way to understand what type of trend is represented by each line of codes is to visualize the list of codes for each dummy variable as a template or graph. If you place values of "−1," "0," and "+1" from left to right on a graph, it is clear that these codes represent a linear trend. The set of coefficients "+1," "−2," and "+1" or, equivalently, "−1," "+2," and "−1" represent a quadratic trend. So, if b_1 (the coefficient for O_1) is significant with this set of codes, the linear trend is significant and b_1 is the amount of change in the dependent variable Y from 0 to 150 mg or from 150 to 300 mg. If b_2 is significant, there is a quadratic (curvilinear) trend.

12.8 ♦ Regression Models That Include Both Dummy and Quantitative Predictor Variables

We can do a regression analysis that includes one (or more) dummy variables and one (or more) continuous predictors, as in the following example:

$$Y' = b_0 + b_1D + b_2X. \tag{12.10}$$

How is the b_1 coefficient for the dummy variable, D, interpreted in the context of this multiple regression with another predictor variable? The b_1 coefficient still represents the estimated difference between the means of the two groups; if gender was coded "0," "1" as in the first example, b_1 still represents the difference between mean salary Y for males and females. However, in the context of this regression analysis, this difference between means on the Y outcome variable for the two groups is assessed while statistically controlling for any differences in the quantitative X variable (such as years of job experience).

We could set up a model to evaluate whether salary (Y) is predictable from dummy-coded gender and from years of experience on the job: salary$' = b_0 + b_1 \times$ gender $+ b_2 \times$ years. In this equation, the b_1 coefficient would be interpreted as an estimate of the gender difference in predicted salary when "years of experience" is held constant or statistically controlled. However, it is possible that there could be an interaction between gender and years; that is, the annual salary increase for each additional year of job experience might be different for females and males. The next section shows how to include a term in the regression equation to represent this interaction between gender and years.

12.9 ♦ Tests for Interaction (or Moderation)

By creating a new variable that is the product of X (a continuous predictor variable) and D (a dummy-coded predictor), we can add a term that represents the potential interaction between D and X to this regression equation. For example, suppose we want to predict salary (Y) from gender, represented by a dummy-coded dummy variable denoted by D, and years of experience. It is possible that the slope that represents annual salary increase that occurs with each additional year of job experience is different for males and females; therefore, we might want to include an interaction term in the analysis. We can create an interaction term, $D \times X$, by forming the product of the effect-coded dummy variable and the continuous predictor variable. (The SPSS Transform/Compute procedure can be used to create a new variable that corresponds to the product of gender and years; in the SPSS worksheet that appears in Figure 12.1, the variable genyears corresponds to the product of gender and years, denoted $D \times X$ in the equation below.) The full regression model has the following form:

$$Y' = b_0 + b_1D + b_2X + b_3(D \times X). \tag{12.11}$$

For the example that involves the prediction of salary from dummy-coded gender and years of experience, the equation becomes

$$\text{Salary}' = b_0 + b_1 \times \text{Gender} + b_2 \times \text{Years} + b_3 \times \text{Genyears}.$$

Equation 12.11 predicts scores on Y (salary) from gender; from years of experience; and from the gender by years interaction ("genyears"). To see what predictions Equation 12.11 makes for the male and female groups, as in prior sections, substitute the "0," "1" codes for the variable gender into the equation, simplify the expressions, and write down separate equations for males and females:

$$Y' = b_0 + b_2X \qquad (12.12)$$

for females, and

$$Y' = (b_0 + b_1) + (b_2 + b_3)X \qquad (12.13)$$

for males.

These two equations (12.12 and 12.13) represent lines that can have different intercepts and different slopes for the best-fitting regression line to predict Y from X. The intercept for females is b_0; the intercept for males is $(b_0 + b_1)$. If the b_1 coefficient is statistically significant, then the intercepts for the regression lines for males versus females are significantly different. These two equations also represent prediction lines that may have different slopes for the prediction of Y from X for males and females; for females, the slope to predict Y from X is b_2; for males, the slope to predict Y from X is $(b_2 + b_3)$. If b_3 is statistically significant, then, the slopes to predict Y from X are significantly different for males as compared with females. Or, to phrase it another way, if b_3 is significant, there is a significant interaction between years and gender as predictors of Y ("different slopes for different folks," to use an expression from Robert Rosenthal).

From the regression output shown in Figure 12.13, the coefficients for Equation 12.11 would be as given below.

Overall

$$\text{Salary}' = b_0 + b_1 \times \text{Gender} + b_2 \times \text{Years} + b_3 \times \text{Genyears},$$
$$\text{Salary}' = 35.12 + 1.94 \times \text{Gender} + 1.24 \times \text{Years} + .26 \times \text{Genyears}.$$

For females (gender = 0 and genyears = 0) this prediction equation becomes

$$\begin{aligned}\text{Salary}' &= b_0 + b_1 \times 0 + b_2 \times \text{Years} + b_3 \times (0 \times \text{Years})\\ &= b_0 + b_2 \times \text{Years}\\ &= 35.12 + 1.24 \times \text{Years}.\end{aligned}$$

That is, for females, the predicted "starting salary" for a person with 0 years of experience is 35.12, and each one point increase in years of experience predicts a 1.24 increase in annual salary (in thousands of dollars).

For males (gender = 1 and genyears = years), this prediction equation becomes

$$\begin{aligned}\text{Salary}' &= b_0 + (b_1 \times 1) + b_2 \times \text{Years} + b_3 \times (1 \times \text{Years})\\ &= b_0 + b_1 + (b_2 + b_3) \times \text{Years}\\ &= 35.12 + 1.94 + (1.24 + .26) \times \text{Years}\\ &= 37.06 + 1.50 \times \text{Years}.\end{aligned}$$

Variables Entered/Removed[b]

Model	Variables Entered	Variables Removed	Method
1	genyears, gender, years[a]		Enter

a. All requested variables entered.
b. Dependent Variable: salary

Model Summary

Model	R	R Square	Adjusted R Square	Std. Error of the Estimate
1	.882 [a]	.778	.764	4.740

a. Predictors: (Constant), genyears, gender, years

ANOVA[b]

Model		Sum of Squares	df	Mean Square	F	Sig.
1	Regression	3624.141	3	1208.047	53.770	.000 [a]
	Residual	1033.479	46	22.467		
	Total	4657.620	49			

a. Predictors: (Constant), genyears, gender, years
b. Dependent Variable: salary

Coefficients[a]

Model		Unstandardized Coefficients		Standardized Coefficients	t	Sig.	Correlations		
		B	Std. Error	Beta			Zero-order	Partial	Part
1	(Constant)	35.117	1.675		20.970	.000			
	years	1.236	.282	.692	4.383	.000	.866	.543	.304
	gender	1.936	2.289	.098	.846	.402	.462	.124	.059
	genyears	.258	.320	.164	.808	.423	.816	.118	.056

a. Dependent Variable: salary

Figure 12.13 ♦ Regression to Predict Salary From Gender, Years, and Gender × Years Interaction

We need to decide whether an interaction term (between gender and years) should be included in the regression model. The slope coefficient associated with the interaction term genyears was not statistically significant; from Figure 12.13, the raw score slope for the interaction was $b = .258$, $t(46) = .808$, $p > .05$. Therefore, it is not useful and not necessary to use separate equations with different slopes to predict salary from years of experience for females versus males. If we drop the nonsignificant interaction term genyears from the equation and run a new regression to predict salary from only the **main effects** for gender and years, that is, $\text{Salary}' = b_0 + b_1 \times \text{Years} + b_2 \times \text{Gender}$, we obtain the results that appear in Figure 12.14.

Based on the regression analysis in Figure 12.14, we would say that both gender and years are significantly predictive of salary. The coefficient to predict salary from gender (controlling for years of experience) was $b_2 = 3.36$, with $t(47) = 2.29$, $p = .026$. The corresponding squared semipartial (or part) correlation for gender as a predictor of salary was $sr^2 = (.159)^2 = .03$. The coefficient to predict salary from years of experience, controlling

Variables Entered/Removed[b]

Model	Variables Entered	Variables Removed	Method
1	gender, years[a]		Enter

a. All requested variables entered.

b. Dependent Variable: salary

Model Summary

Model	R	R Square	Adjusted R Square	Std. Error of the Estimate
1	.880[a]	.775	.765	4.722

a. Predictors: (Constant), gender, years

ANOVA[b]

Model		Sum of Squares	df	Mean Square	F	Sig.
1	Regression	3609.469	2	1804.734	80.926	.000[a]
	Residual	1048.151	47	22.301		
	Total	4657.620	49			

a. Predictors: (Constant), gender, years

b. Dependent Variable: salary

Coefficients[a]

Model		Unstandardized Coefficients		Standardized Coefficients			Correlations		
		B	Std. Error	Beta	t	Sig.	Zero-order	Partial	Part
1	(Constant)	34.193	1.220		28.038	.000			
	years	1.436	.133	.804	10.830	.000	.866	.845	.749
	gender	3.355	1.463	.170	2.293	.026	.462	.317	.159

a. Dependent Variable: salary

Figure 12.14♦ Regression to Predict Salary From Gender and Years (Without an Interaction Term)

for gender, was $b = 1.44$, $t(47) = 10.83$, $p < .001$; the corresponding sr^2 effect size for years was $.749^2 = .56$. This analysis suggests that controlling for years of experience partly accounts for the observed gender differences in salary, but it does not completely account for gender differences; even after years of experience is taken into account, males still have an average salary that is about 3.36 thousand dollars higher than females at each level of years of experience. However, this gender difference is relatively small in terms of the proportion of explained variance. About 56% of the variance in salaries is uniquely predictable from years of experience (this is a strong effect). About 3% of the variance in salary is predictable from gender (this is a medium-sized effect). Within this sample, for each level of years of experience, females are paid about 3.35 thousand dollars less than males who have the same number of years of experience; this is evidence of possible gender bias. Of course, it is possible that this remaining gender difference in salary might be accounted for by other variables. Perhaps more women are in the College of Liberal Arts, which has lower salaries, and more men are in the Business and Science Colleges, which

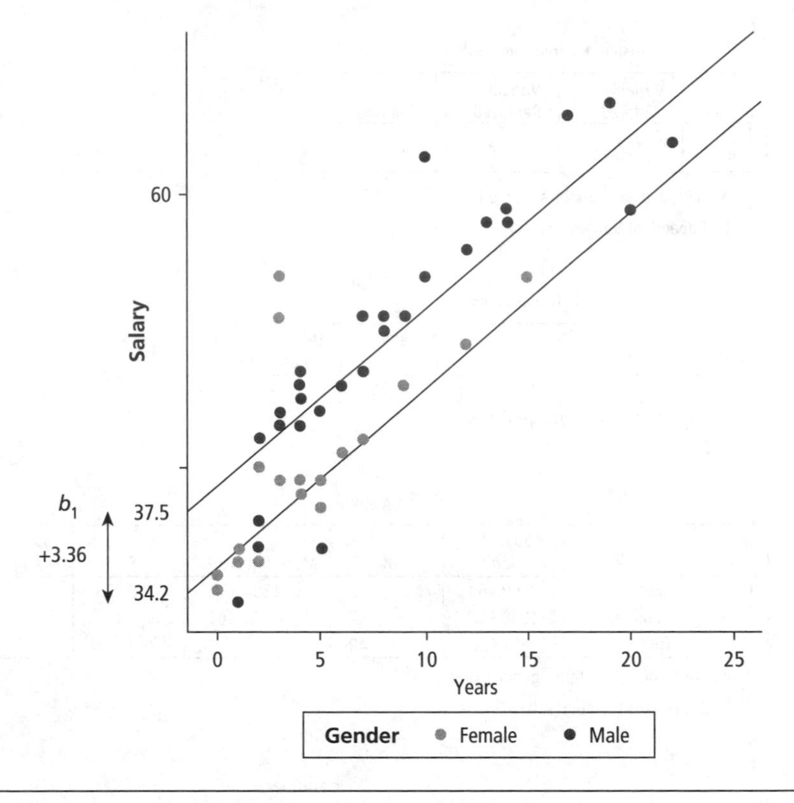

Figure 12.15 ♦ Graph of the Regression Lines to Predict Salary From Years of Experience Separately for Males and Females

NOTE: This is based on the regression analysis in Figure 12.14: Salary′ = 34.2 + 3.36 × Gender + 1.44 × Years; gender coded 0 = Female and 1 = Male.

tend to receive higher salaries. Controlling for other variables such as college might help us to account for part of the gender difference in mean salary levels.

The last two regression analyses suggested that there was no significant gender by years interaction; in other words, there was no gender difference in the amount of salary increase for each 1-year increment in years of experience. The slope to predict salary from years is not significantly different for males and females. However, after the interaction term was dropped from the analysis, a regression that predicted salary from years of experience and gender indicated that there was a significant difference between mean salary for males and that for females, overall. This corresponds to a situation where the regression line to predict salary from years of experience has the same slope but a different intercept for males and females, as shown in Figure 12.15.

The raw score *b* coefficient associated with gender in the regression in Figure 12.14 had a value of $b = 3.355$; this corresponds to the difference between the intercepts of the regression lines for males and females in Figure 12.15. For this set of data, it appears that gender differences in salary may be due to a difference in starting salaries (i.e., the salaries paid to faculty with 0 years of experience) and not due to differences between the annual raises in salary given to male and female faculty. The model Results section at the end of the chapter provides a more detailed discussion of the results that appear in Figures 12.14 and 12.15.

The best interpretation for the salary data in Table 12.1, based on the analyses that have been performed so far, appears to be the following: Salary significantly increases as a function of years of experience; there is also a gender difference in salary (such that males are paid significantly higher salaries than females) even when the effect of years of experience is statistically controlled. However, keep in mind that statistically controlling for additional predictor variables (such as college and merit) in later analyses could substantially change the apparent magnitude of gender differences in salary.

What would an interaction between years and gender as predictors of salary look like? Figure 12.16 shows a graph for a different set of hypothetical faculty salary data (these are different from the scores in Table 12.1). In this graph, the intercepts for the regression lines for male and female faculty are approximately equal, but the slope to predict salary from years of experience is much steeper for male faculty than for female faculty. Regression coefficients for the data displayed in Figure 12.16 appear below the graph. If male and female faculty have similar starting salaries (at years of experience = 0), but male faculty receive an average annual salary increase of 2.767 + 4.434 = 7.2 thousand dollars per year while female faculty receive an annual average salary increase of only 2.767 thousand dollars per year, the regression line to predict salary from years would have a significantly steeper slope for male faculty than for female faculty.

Let us summarize Sections 12.8 and 12.9. We have seen that categorical and quantitative independent variables can both be used as predictors in regression, provided that the group memberships are represented by dichotomous dummy variables. When we examine interactions by creating products of variables in a regression (a product between two dummy variables or a product between a dummy variable and a continuous variable), it is relatively easy to describe the nature of the interaction (we can figure out "different slopes for different folks"; for example, we can graph the regression line to predict salary from years separately for males and females). If one or both of the variables that are used to create a product term to represent an interaction in a regression equation are dummy variables, then the interpretation of the nature of the interaction is usually relatively simple. We can plot the values of the regression line for each possible code on the dummy variable and then compare the regression lines for groups that correspond to different codes on each dummy variable.

Sometimes, researchers want to examine potential interactions between pairs of quantitative predictor variables. Interactions between quantitative predictor variables are typically much more difficult to analyze and interpret than interactions that involve dummy variables.

12.10 ♦ Interaction Terms That Involve Two Quantitative Predictors

It is possible to create an interaction term to represent an interaction between two quantitative predictor variables. For example, the dataset in Table 12.1 included two quantitative predictor variables: years of job experience and merit. We could set up a regression to predict salary from just the main effects of these two quantitative predictors:

$$\text{Salary}' = b_0 + b_1 \times \text{Years} + b_2 \times \text{Merit}. \tag{12.14}$$

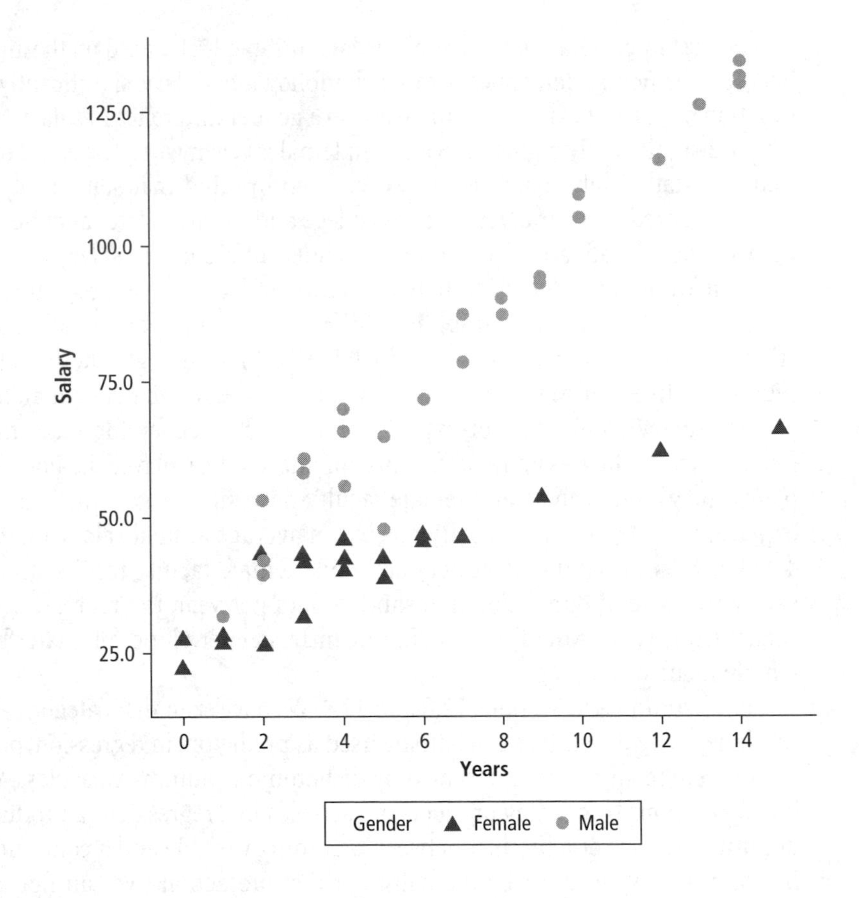

Coefficients[a]

Model		Unstandardized Coefficients		Standardized Coefficients		
		B	Std. Error	Beta	t	Sig.
1	(Constant)	28.278	1.957		14.449	.000
	years	2.767	.329	.367	8.400	.000
	gender	2.969	2.912	.048	1.020	.314
	genyears	4.434	.425	.660	10.426	.000

a. Dependent Variable: salary

Figure 12.16 ♦ An Alternative Example: Statistically Significant Years × Gender Interaction in Prediction of Salary

It is conceivable that there could be an interaction between merit and years; for example, the annual salary increase for each additional year of experience might be higher for faculty with high merit evaluations than for faculty with low merit evaluations.

When trying to solve the problem of how to represent interaction, we might be tempted to create a new variable that is the product of the raw score on years multiplied by the raw score on merit; but this approach can be problematic. Jaccard, Wan, and Turrisi (1990) suggested that in many research situations it may be preferable to convert quantitative variables into z scores before forming a product term to represent an interaction. If this conversion to z scores is not made, then the apparent magnitude of the interaction, and the magnitude of the product term that represents an interaction with the main effect predictors, varies depending on the units of measurement for each of the quantitative variables; this is not desirable. Furthermore, it is easier to interpret the nature of interactions when z scores are used for both main effect and interaction terms, because the equation can be evaluated for representative values of z such as $z = -2, -1, 0, +1$, and $+2$ for each of the quantitative predictor variables.

For the regression example that appears in Figure 12.17, all three variables were converted to z scores (z_{salary}, z_{years}, and z_{merit}); the product between z_{years} and z_{merit} was also included as a predictor variable to represent a possible interaction between years and merit as predictors of salary.

$$z'_{\text{salary}} = \beta_1 z_{\text{years}} + \beta_2 z_{\text{merit}} + \beta_3 z_{\text{years}} z_{\text{merit}}. \quad (12.15)$$

The significance test for the β_3 coefficient associated with this product term provides a test of the significance of the interaction between merit and years as predictors of salary. In this empirical example, the interaction between merit and years was not statistically significant. We could say that "there is no significant interaction between merit and years as predictors of salary" or that "merit is not a significant moderator of the effect of 'years' on 'salary'." The interpretation of the nature of a significant interaction between two quantitative predictors can be difficult, and the nature of the conclusions about the interaction may differ depending on how the interaction term is created. For a more extensive discussion of these complex issues, see Jaccard et al. (1990).

12.11 ♦ Effect Size and Statistical Power

As discussed in Chapter 11, we can represent the effect size for the regression as a whole (i.e., the proportion of variance in Y is predictable from a set of variables that may include dummy-coded and/or continuous variables) by reporting multiple R and multiple R^2 as our overall effect size measures. We can represent the strength of the unique predictive contribution of any particular variable by reporting the estimate of sr^2_{unique} for each predictor variable, as in Chapter 11. The proportion of variance given by sr^2 can be used to describe the proportion of variance uniquely predictable from the contrast specified by a dummy variable, in the same manner in which it describes the proportion of variance uniquely predictable from a continuous predictor variable.

When dummy variables are included in regression analysis, the regression is essentially equivalent to a one-way ANOVA (or a t test, if only two groups are being compared). Therefore, the tables presented in Chapter 5 (independent samples t test) and Chapter 6 (one-way ANOVA) can be used to look up reasonable minimum sample sizes per group for anticipated effect sizes that are small, medium, or large. Whether the method used to make predictions and compare means across groups is ANOVA or regression, none of the groups should have a very small n. If $n < 20$ per group, nonparametric analyses may be more appropriate.

Variables Entered/Removed[b]

Model	Variables Entered	Variables Removed	Method
1	zyearmerit, Zscore(years), Zscore(merit)[a]		Enter

a. All requested variables entered.

b. Dependent Variable: Zscore(salary)

Model Summary

Model	R	R Square	Adjusted R Square	Std. Error of the Estimate
1	.866[a]	.750	.734	.51606285

a. Predictors: (Constant), zyearmerit, Zscore(years), Zscore(merit)

ANOVA[b]

Model		Sum of Squares	df	Mean Square	F	Sig.
1	Regression	36.749	3	12.250	45.996	.000[a]
	Residual	12.251	46	.266		
	Total	49.000	49			

a. Predictors: (Constant), zyearmerit, Zscore(years), Zscore(merit)

b. Dependent Variable: Zscore(salary)

Coefficients[a]

Model		Unstandardized Coefficients		Standardized Coefficients			Correlations		
		B	Std. Error	Beta	t	Sig.	Zero-order	Partial	Part
1	(Constant)	.005	.079		.060	.952			
	Zscore(merit)	.004	.087	.004	.042	.966	.280	.006	.003
	Zscore(years)	.868	.084	.868	10.365	.000	.866	.837	.764
	zyearmerit	-.016	.102	-.013	-.153	.879	.183	-.023	-.011

a. Dependent Variable: Zscore(salary)

Figure 12.17 ♦ Regression to Predict *z* Score on Salary From *z* Scores on Years, Merit, and a Years × Merit Interaction Term

12.12 ♦ Nature of the Relationship and/or Follow-Up Tests

When dummy-coded group membership predictors are included in a regression analysis, the information that individual coefficients provide is equivalent to the information obtained from planned contrasts between group means in an ANOVA. The choice of the method of coding (dummy, effect, orthogonal) and the decision as to which group to code as the "last" group determine which set of contrasts the regression will include.

Whether we compare multiple groups by performing a one-way ANOVA or by using a regression equation with dummy-coded group membership variables as predictors, the written results should include the following: means and standard deviations for scores in each group, confidence intervals for each group mean, an overall *F* test to report whether there were significant differences in group means, planned contrasts or post hoc tests to identify which specific pairs of group means differed significantly, and a discussion of the direction of differences between group means. Effect size information about the proportion of explained variance (in the form of an η^2 or sr^2) should also be included.

12.13 ♦ Results

The hypothetical data showing salary scores for faculty (in Table 12.1) were analyzed in several different ways to demonstrate the equivalence between ANOVA and regression with dummy variables, to illustrate the interpretation of *b* coefficients for dummy variables in regression, and to show how product terms can be added to regression equations to represent interactions. The text below reports the regression analysis for prediction of salary from years of experience and gender (as shown in Figure 12.14).

Results

To assess whether gender and years of experience significantly predict faculty salary, a regression analysis was performed to predict faculty annual salary in thousands of dollars from gender (dummy-coded 0 = Female, 1 = Male) and years of experience. (Previous regression analyses that are not reported here indicated that there was no statistically significant interaction between gender and years as predictors of salary.) The distribution of salary was roughly normal, the variances of salary scores were not significantly different for males and females, and scatter plots did not indicate nonlinear relations or bivariate outliers. No data transformations were applied to scores on salary and years, and all 50 cases were included in the regression analysis.

The results of this regression analysis (SPSS output in Figure 12.14) indicated that the overall regression equation was significantly predictive of salary; $R = .88$, $R^2 = .78$, adjusted $R^2 = .77$, $F(2, 47) = 80.93$, $p < .001$. Salary could be predicted almost perfectly from gender and years of job experience. Each of the two individual predictor variables was statistically significant. The raw score coefficients for the predictive equation were as follows:

$$\text{Salary}' = 34.19 + 3.36 \times \text{Gender} + 1.44 \times \text{Years}.$$

When controlling for the effect of years of experience on salary, the magnitude of the gender difference in salary was 3.36 thousand dollars. That is, at each level of years of experience, male annual salary was about 3.36 thousand dollars higher than female salary. This difference was statistically significant: $t(47) = 2.29$, $p = .026$.

For each 1-year increase in experience, the salary increase was approximately 1.44 thousand dollars for both females and males. This slope for the prediction of salary from years of experience was statistically significant: $t(47) = 10.83$, $p < .001$. The graph in Figure 12.15 illustrates the regression lines to predict salary for males and females separately. The intercept (i.e., predicted salary for 0 years of experience) was significantly higher for males than for females; however, the slope (i.e., the amount of increase in salary for each 1-year increase in experience) was not significantly higher for males as compared with that for females.

(Continued)

(Continued)

The squared semipartial correlation for years as a predictor of salary was $sr^2 = .56$; thus, years of experience uniquely predicted about 56% of the variance in salary (when gender was statistically controlled). The squared semipartial correlation for gender as a predictor of salary was $sr^2 = .03$; thus, gender uniquely predicted about 3% of the variance in salary (when years of experience was statistically controlled). The results of this analysis suggest that there was a systematic difference between salaries for male and female faculty and that this difference was approximately the same at all levels of years of experience. Statistically controlling for years of job experience, by including it as a predictor of salary in a regression that also used gender to predict salary, yielded results that suggest that the overall gender difference in mean salary was partly, but not completely, accounted for by gender differences in years of job experience. (Note that the remaining gender difference in salary might be accounted for partly or entirely by other variables related to both gender and salary, for example, whether each faculty member teaches in the Liberal Arts, Sciences, or Business college.)

12.14 ♦ Summary

This chapter presented examples that demonstrated the equivalence of ANOVA and regression analyses that use dummy variables to represent membership in multiple groups. This discussion has presented demonstrations and examples rather than formal proofs; mathematical statistics textbooks provide formal proofs of equivalence between ANOVA and regression. ANOVA and regression are different special cases of the GLM.

If duplication of ANOVA using regression were the only application of dummy variables, it might not be worth spending so much time on them. However, dummy variables have important practical applications. Researchers often want to include group membership variables (such as gender) among the predictors that they use in multiple regression, and it is important to understand how the coefficients for dummy variables are interpreted.

This chapter also demonstrated that products between predictor variables can be used to test for the presence of interaction effects in regression analyses. When one or both of the variables involved in an interaction are dummy variables, the interpretation of the nature of the interaction is relatively simple. However, if both the variables involved in an interaction are quantitative, evaluation of possible interaction effects becomes more complex.

The advantage of choosing ANOVA as the method for comparing group means is that the SPSS procedures provide a wider range of options for follow-up analysis—for example, post hoc protected tests. Also, when ANOVA is used, interaction terms are generated automatically for all pairs of (categorical) predictor variables or factors, so it is less likely that a researcher will fail to notice an interaction when the analysis is performed as a factorial ANOVA (as discussed in Chapter 13) than when a comparison of group means is performed using dummy variables as predictors in a regression. ANOVA does not assume a linear relationship between scores on categorical predictor variables and scores on quantitative outcome variables. A quantitative predictor can be added to an ANOVA model (this type of analysis, called analysis of covariance or ANCOVA, is discussed in Chapter 15).

On the other hand, an advantage of choosing regression as the method for comparing group means is that it is easy to use quantitative predictor variables along with group membership predictor variables to predict scores on a quantitative outcome variable. Regression analysis yields equations that can be used to generate different predicted scores for cases with different score values on both categorical and dummy predictor variables. A possible disadvantage of the regression approach is that interaction terms are not automatically included in a regression; the data analyst must specifically create a new variable (the product of the two variables involved in the interaction) and add that new variable as a predictor. Thus, unless they specifically include interaction terms in their models, data analysts who use regression analysis may fail to notice interactions between predictors. A data analyst who is careless may also set up a regression model that is "nonsense"; for example, it would not make sense to predict political conservatism (Y) from scores on a categorical X_1 predictor variable that has codes "1" for Democrat, "2" for Republican, and "3" for Independent. Regression assumes a linear relationship between predictor and outcome variables; political party represented by just one categorical variable with three possible score values probably would not be linearly related to an outcome variable such as conservatism. To compare group means using regression in situations where there are more than two groups, the data analyst needs to create dummy variables to represent information about group membership. In some situations, it may be less convenient to create new dummy variables (and run a regression) than to run an ANOVA.

Ultimately, however, ANOVA and regression with dummy predictor variables yield essentially the same information about predicted scores for different groups. In many research situations, ANOVA may be a more convenient method to assess differences among group means. However, regression with dummy variables provides a viable alternative, and in some research situations (where predictor variables include both categorical and quantitative variables), a regression analysis may be a more convenient way of setting up the analysis.

+ − × ÷

Comprehension Questions

1. Suppose that a researcher wants to do a study to assess how scores on the dependent variable HR differ across groups that have been exposed to various types of stress. Stress group membership was coded as follows:

   Group 1, no stress/baseline

   Group 2, mental arithmetic

   Group 3, pain induction

   Group 4, stressful social role play

   The basic research questions are whether these four types of stress elicited significantly different HRs overall and which specific pairs of groups differed significantly.

   a. Set up dummy-coded dummy variables that could be used to predict HR in a regression.

      Note that it might make more sense to use the "no stress" group as the one that all other group means are compared with, rather than the group that happens to be listed last in the list above (stressful role play). Before working out the contrast coefficients, it may be helpful to list the groups in a different order:

      Group 1, mental arithmetic

      Group 2, pain induction

      Group 3, stressful social role play

      Group 4, no stress/baseline

      Set up dummy-coded dummy variables to predict scores on HR from group membership for this set of four groups.

      Write out in words which contrast between group means each dummy variable that you have created represents.

   b. Set up effect-coded dummy variables that could be used to predict HR in a regression.

      Describe how the numerical results for these effect-coded dummy variables (in 1b) differ from the numerical results obtained using dummy-coded dummy variables (in 1a). Which parts of the numerical results will be the same for these two analyses?

   c. Set up the coding for orthogonally coded dummy variables that would represent these orthogonal contrasts:

      Group 1 versus 2

      Groups 1 and 2 versus 3

      Groups 1, 2, and 3 versus 4

2. Suppose that a researcher does a study to see how level of anxiety (A_1 = Low, A_2 = Medium, A_3 = High) is used to predict exam performance (Y). Here are hypothetical data for this research situation. Each column represents scores on Y (exam scores).

Comprehension Questions

A_1, *Low Anxiety*	A_2, *Medium Anxiety*	A_3, *High Anxiety*
72	86	65
81	93	79
54	81	74
66	80	80
71	92	74

a. Would it be appropriate to do a Pearson correlation (and/or linear regression) between anxiety, coded "1," "2," "3" for (low, medium, high), and exam score? Justify your answer.
b. Set up orthogonally coded dummy variables (O_1, O_2) to represent linear and quadratic trends, and run a regression analysis to predict exam scores from O_1 and O_2. What conclusions can you draw about the nature of the relationship between anxiety and exam performance?
c. Set up dummy-coded dummy variables to contrast each of the other groups with Group 2, Medium anxiety; run a regression to predict exam performance (Y) from these dummy-coded dummy variables.
d. Run a one-way ANOVA on these scores; request contrasts between Group 2, Medium anxiety, and each of the other groups. Do a point-by-point comparison of the numerical results for your ANOVA printout with the numerical results for the regression in (2c), pointing out where the results are equivalent.

3. Why is it acceptable to use a dichotomous predictor variable in a regression when it is not usually acceptable to use a categorical variable that has more than two values as a predictor in regression?
4. Why are values such as "+1," "0," and "−1" generally used to code dummy variables?
5. How does the interpretation of regression coefficients differ for dummy coding of dummy variables versus effect coding of dummy variables (hint: in one type of coding, b_0 corresponds to the grand mean; in the other, b_0 corresponds to the mean of one of the groups).
6. If you have k groups, why do you only need $k - 1$ dummy variables to represent group membership? Why is it impossible to include k dummy variables as predictors in a regression when you have k groups?
7. How does orthogonal coding of dummy variables differ from dummy and effect coding?
8. Write out equations to show how regression can be used to duplicate a t test or a one-way ANOVA.
9. How is an interaction between variables represented in a regression analysis?

CHAPTER 13

Factorial Analysis of Variance

13.1 ♦ Research Situations and Research Questions

Factorial analysis of variance is used in research situations where two or more group membership variables (called "factors") are used to predict scores on one quantitative outcome variable (such as the number of symptoms on a physical symptom checklist or the score on a self-report test of anxiety). This is a generalization of one-way analysis of variance (ANOVA) (described in Chapter 6).

In a factorial ANOVA, one or more of the factors may represent membership in treatment groups in an experiment (such as dosage levels of caffeine, presentation of different stimuli, or different types of therapy). In addition, one (or more) of the factors may represent membership in naturally occurring groups (such as gender, personality type, or diagnostic group). Although factorial ANOVA is most frequently used with experimental data, it can also be applied in nonexperimental research situations—that is, research situations where all the factors correspond to naturally occurring groups rather than to different treatments administered by the researcher.

A common notation to describe **factorial designs** is as follows. Each factor is designated by an uppercase letter (the first factor is typically named A, the second factor is named B, and so forth). The levels or groups within each factor are designated by numbers; thus, if the A factor corresponds to two different levels of social support, the groups that correspond to low and high levels of social support could be labeled A_1 and A_2. The number of levels (or groups) within each factor is denoted by a corresponding lowercase letter (e.g., if the A factor has two levels, then lowercase $a = 2$ represents the number of levels of Factor A).

This chapter reviews factorial ANOVA in which the factors are crossed (i.e., every level of the A factor is paired with every level of the B factor). For example, if the A factor has three levels and the B factor has four levels, then the fully crossed design involves 12 groups. Each of the three levels of A are paired with each of the four levels of B; the number of groups or cells for an $A \times B$ factorial is given by the product $a \times b$.

A factor in ANOVA can be either "fixed" or "random." This distinction is based on the nature of the correspondence between the levels of the factor that are actually included in the study compared with all possible levels of the factor that might exist in a hypothetical "real world." A factor is treated as fixed if all the levels that exist in the real world are represented in the study. For example, gender is often treated as a **fixed factor**; there are two possible levels for this factor (male and female), and a design that includes male and female levels of gender as a factor includes all the possible levels for this factor. (It is possible that we could add other categories such as transgender to the factor gender, but most past studies have not included this level of gender.) A factor is also treated as fixed when the levels of the factor are systematically selected to cover the entire range of interest; for example, there are potentially thousands of different possible dosage levels of caffeine, but if a researcher systematically administered dosages of 0, 100, 200, 300, and 400 mg, the researcher would probably treat the caffeine dosage factor as fixed because these five levels systematically cover the range of dosage levels that are of interest.

A factor is random when the levels included in the sample represent a very small percentage of the possible levels for that factor. For example, suppose that a researcher uses a sample of six randomly selected facial photographs in a study of person perception. There are more than 6 billion people in the world; the six randomly selected faces included in the study represent an extremely small fraction of all the possible faces in the world. In this situation, we would call "face" a **random factor** with six levels. Random factors arise most commonly in research where stimulus materials are randomly sampled from large domains—for example, when we choose 10 words from the more than 50,000 words in English. When we include "subjects" as a factor, levels of that factor are almost always treated as a random factor, because the 20 participants included in a study represent an extremely small fraction of all the persons in the world. Most of the ANOVA examples in this textbook are limited to fixed factors.

The *F* ratio that is set up to test the significance of any factor needs to use a different error term when it is crossed with a fixed factor than when it is crossed with a random factor. Typically, when we test effects for a factor that is crossed with some random factor, the *F* ratio that is set up compares mean of squares effect with a mean of squares for an interaction instead of a mean of squares within groups. In SPSS univariate factorial ANOVA, we can identify factors as fixed or random by entering them into the lists for fixed versus random factors in the main dialog window for general linear modeling (GLM), and SPSS should generate an appropriate *F* ratio for each effect. The logic behind the choice of error terms models that include random factors is complex; for a detailed discussion, see Myers and Well (1995).

This chapter covers only between-subject (between-S) designs; that is, each participant contributes a score in only one cell (each cell represents one combination of the *A* and *B* treatment levels). For now, we will assume that the number of participants in each cell or group (n) is equal for all cells. The number of scores within each cell is denoted by n; thus, the total number of scores in the entire dataset, N, is found by computing the product $n \times a \times b$. Unequal numbers in the cells or groups require special treatment when we compute row and column means and when we partition the SS_{total} into sum of squares (SS) terms for the main effects and interactions, as described in the appendix to this chapter.

A hypothetical example of a 2 × 2 factorial design is used throughout this chapter, based on the data in Table 13.1. In this imaginary study, Factor *A* is level of social support (A_1 = Low social support, A_2 = High social support). Factor *B* is level of stress (B_1 = Low stress, B_2 = High stress). The quantitative outcome variable (*Y*) is the number of physical illness symptoms for each participant. There are $n = 5$ scores in each of the four groups, and thus, the overall number of participants in the entire dataset $N = a \times b \times n = 2 \times 2 \times 5 = 20$ scores. Table 13.2 reports the means for the groups included in this 2 × 2 factorial design. The research questions focus on the pattern of means. Are there substantial differences in mean level of symptoms between people with low versus high social support? Are there substantial differences in mean level of symptoms between groups that report low versus high levels of stress? Is there a particular combination of circumstances (e.g., low social support and high stress) that predicts a much higher level of symptoms? As in one-way ANOVA, *F* ratios will be used to assess whether differences among group means are statistically significant. Because the factors in this hypothetical study correspond to naturally occurring group memberships rather than experimentally administered treatments, this is a nonexperimental or correlational design, and results cannot be interpreted as evidence of causality.

It is generally useful to ask how each new analysis compares with earlier, simpler analyses. What information do we obtain from a factorial ANOVA that we cannot obtain from a simpler one-way ANOVA? Figure 13.1 presents results for a one-way ANOVA for the data in Table 13.1 that tests whether mean level of symptoms differ across levels of social

Table 13.1 ♦ Data for a Hypothetical 2 × 2 Factorial ANOVA: Prediction of Symptoms From Social Support, Factor *A* (Socsup_*A*: 1 = Low, 2 = High), and Stress, Factor *B* (Stress_*B*: 1 = Low, 2 = High)

Socsup_A	*Stress_B*	*Symptom*
1	1	2
1	1	3
1	1	5
1	1	4
1	1	5
1	2	11
1	2	13
1	2	10
1	2	16
1	2	14
2	1	2
2	1	4
2	1	6
2	1	5
2	1	3
2	2	6
2	2	7
2	2	4
2	2	6
2	2	7

Table 13.2 ♦ Table of Means on Number of Symptoms for the 2 × 2 Factorial ANOVA Data in Table 13.1

	B_1 *(Low Stress)*	B_2 *(High Stress)*	
A_1 (low social support)	$M_{AB_{11}} = 3.80$	$M_{AB_{12}} = 12.80$	$M_{A_1} = 8.30$
A_2 (high social support)	$M_{AB_{21}} = 4.00$	$M_{AB_{22}} = 6.00$	$M_{A_2} = 5.00$
	$M_{B_1} = 3.90$	$M_{B_2} = 9.40$	$M_Y = 6.65$

Descriptives

SYMPTOM

	N	Mean	Std. Deviation	Std. Error	95% Confidence Interval for Mean		Minimum	Maximum
					Lower Bound	Upper Bound		
Low Social Support	10	8.30	5.078	1.606	4.67	11.93	2	16
High Social Support	10	5.00	1.700	.537	3.78	6.22	2	7
Total	20	6.65	4.056	.907	4.75	8.55	2	16

ANOVA

SYMPTOM

	Sum of Squares	df	Mean Square	F	Sig.
Between Groups	54.450	1	54.450	3.797	.067
Within Groups	258.100	18	14.339		
Total	312.550	19			

Figure 13.1 ♦ One-Way ANOVA Results: Symptoms for Low- Versus High-Social-Support Groups

NOTE: Data from Table 13.1.

support. The difference in symptoms in this one-way ANOVA, a mean of 8.3 for the low-social-support group versus a mean of 5.00 for the high-social-support group, was not statistically significant at the conventional $\alpha = .05$ level: $F(1, 18) = 3.80, p = .067$.

In this chapter, we will see that a factorial ANOVA of the data in Table 13.1 provides three significance tests: a significance test for the main effect of Factor *A* (social support), a significance test for the main effect of Factor *B* (stress), and a test for an interaction between the *A* and *B* factors (social support by stress). The ability to detect potential interactions between predictors is one of the major advantages of factorial ANOVA compared with one-way ANOVA designs.

However, adding a second factor to ANOVA can be helpful in another way. Sometimes the error term (SS_{within}) that is used to compute the divisor for *F* ratios decreases substantially when one or more additional factors are taken into account in the analysis. Later in this chapter, we will see that the main effect for social support (which was not significant in the one-way ANOVA reported in Figure 13.1) becomes significant when we do a factorial ANOVA on these data.

To assess the pattern of outcomes in a two-way factorial ANOVA, we will test three separate null hypotheses.

13.1.1 ♦ First Null Hypothesis: Test of Main Effect for Factor *A*

$$H_0 : \mu_{A1} = \mu_{A2}.$$

The first null hypothesis (main effect for Factor A) is that the population means on the quantitative Y outcome variable are equal across all levels of A. We will obtain information relevant to this null hypothesis by computing an SS term that is based on the observed sample means for Y for the A_1 and A_2 groups. (Of course, the A factor may include more than two groups.) If these sample means are "far apart" relative to within-group variability in scores, we conclude that there is a statistically significant difference for the A factor. As in one-way ANOVA (Chapter 6), we evaluate whether sample means are far apart by examining an F ratio that compares MS_{between} (in this case, MS_A, a term that tells us how far apart the means of Y are for the A_1 and A_2 groups) with the mean square (MS) that summarizes the amount of variability of scores within groups, MS_{within}. To test the null hypothesis of no main effect for the A factor, we will compute an F ratio, F_A:

$$F_A = MS_A/MS_{\text{within}}, \text{ with } (a - 1, N - ab) \text{ degrees of freedom } (df).$$

13.1.2 ♦ Second Null Hypothesis: Test of Main Effect for Factor *B*

$$H_0 : \mu_{B1} = \mu_{B2}.$$

For all the levels of B that are included in your factorial design, the second null hypothesis (main effect for Factor B) is that the population means on Y are equal across all levels of B. We will obtain information relevant to this null hypothesis in this empirical example by looking at the observed sample means for Y in the B_1 and B_2 groups. We can compute an F ratio to test whether observed differences in means across levels of B are statistically significant:

$$F_B = MS_B/MS_{\text{within}}, \text{ with } (b - 1, N - ab)\ df.$$

13.1.3 ♦ Third Null Hypothesis: Test of the *A* × *B* Interaction

This null hypothesis can be written algebraically;[1] for the moment, it is easier to state it verbally:

$$H_0\text{: No } A \times B \text{ interaction.}$$

When there is no $A \times B$ interaction, the lines in a graph of the cell means are parallel, as shown in Figure 13.2a. When an interaction is present, the lines in the graph of cell means are not parallel, as in Figure 13.2b. Robert Rosenthal has aptly described interaction as "different slopes for different folks." Interaction can also be called "moderation," as discussed in Chapter 10. If the lines in a graph of cell means are not parallel (as in Figure 13.2b) and if the F ratio that corresponds to the interaction is statistically significant, we can say that Factor A (social support) interacts significantly with Factor B (stress) to predict different levels of symptoms; or we can say that social support moderates the association between stress and symptoms.

The graphs in Figure 13.2 represent two different possible outcomes for the hypothetical study of social support and stress as predictors of symptoms. If there is no interaction between social support and stress, then the difference in mean symptoms between the high-stress (B_2) and low-stress (B_1) groups is the same for the A_1 (low social support) group as it is for the A_2 (high social support) group, as illustrated in Figure 13.2a. In this graph, both

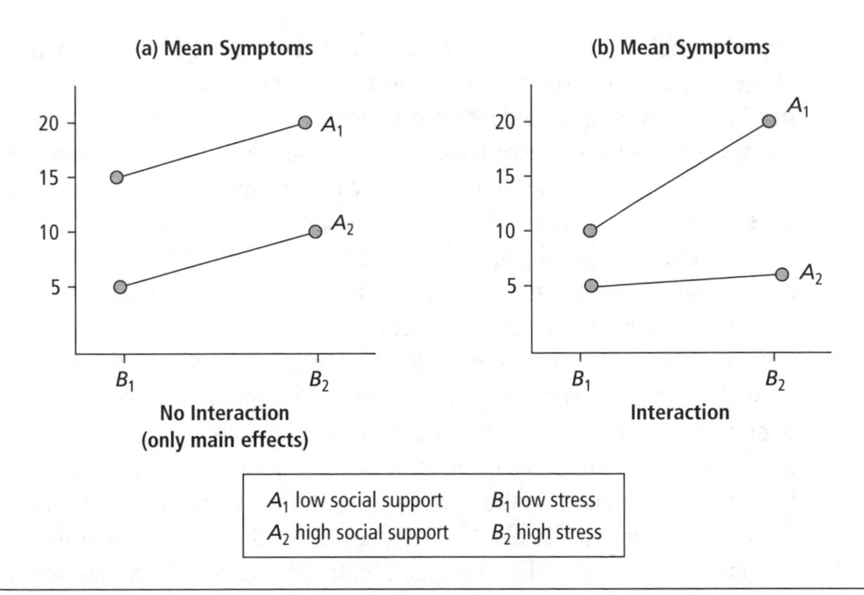

Figure 13.2 ♦ Patterns of Cell Means Predicted by the No Interaction or Direct Effects Hypothesis (a) and the Interaction or Buffering Hypothesis (b)

NOTE: These graphs are not based on the data in Table 13.1; they represent different hypothetical outcomes.

the A_1 and A_2 groups showed the same difference in mean symptoms between the low- and high-stress conditions; that is, mean symptom score was 5 points higher in the high-stress condition (B_2) than in the low-stress condition (B_1). Thus, the difference in mean symptoms across levels of stress was the same for both low- and high-social-support groups in Figure 13.2a. (If we saw this pattern of means in data in an experimental factorial design, where the *B* factor represented a stress intervention administered by the experimenter, we might say that the effect of an experimentally manipulated increase in stress was the same for the A_1 and A_2 groups.) The 5-point difference between means corresponds to the slope of the line that we obtain by connecting the points that represent cell means.

On the other hand, the graph in Figure 13.2b shows a possible interaction between social support and stress as predictors of symptoms. For the low-social-support group (A_1), there was a 10-point difference in mean symptoms between the low- and high-stress conditions. For the high-social-support group (A_2), there was only a 1-point difference between mean symptoms in low- versus high-stress conditions. Thus, the difference in mean symptoms across levels of stress was not the same for the two social support groups. Factorial ANOVA provides a significance test (an *F* ratio for the interaction) that is used to judge whether the observed difference in slopes in the graph of cell means is large enough to be statistically significant. When a substantial interaction effect is present, the interpretation of results often focuses primarily on the interaction rather than on main effects.

13.2 ♦ Screening for Violations of Assumptions

The assumptions for a factorial ANOVA are essentially the same as those for a one-way ANOVA. First, we assume that scores on the outcome (*Y*) variable are quantitative and approximately normally distributed. This can be assessed by examination of a histogram of the frequency distribution for *Y*. We also assume that the scores are obtained in a manner that leads to independent observations, as discussed in Chapter 4. We assume that the variances of scores are reasonably homogeneous across groups; SPSS provides an optional test (the Levene test) to assess whether this assumption of homogeneity of variance is violated. Factorial ANOVA is fairly robust to violations of the normality assumption and the homogeneity of variance assumption unless the numbers of cases in the cells are very small and/or unequal. Often, when the Levene test indicates significant violation of the homogeneity of variance assumption, this is due to outliers in a few cells; see Chapter 4 for issues to consider in the identification and handling of outliers.

We assume in this chapter that the design is completely between-S (i.e., each participant provides data for only one observation, in one group or cell of the design). When data involve repeated measures or matched samples, we need to use repeated measures analytic methods that take the correlations among scores into account, as discussed in Chapter 20 of this textbook. We also assume, for the moment, that the number of observations is equal in all cells. The appendix to this chapter discusses possible ways to handle the problems that arise when the numbers of scores in cells are unequal.

13.3 ♦ Issues in Planning a Study

First of all, the number of levels and the dosage levels included for each factor should be adequate to cover the range of dosage levels of interest and to detect nonlinear changes as a function of dosage level if these may exist. Dosage levels should be far enough apart to yield detectable differences, whether the factor represents an experimentally manipulated treatment such as amount of caffeine administered or a preexisting participant characteristic such as age. Issues about the number and spacing of dosage levels in ANOVA were discussed in Chapter 6.

The total number of groups ($a \times b$) should not be excessively large. Even if only 10 subjects are run in each group, it can be quite difficult to fill up the cells in a 4×6 design, for example.[2] The number of observations in each cell should be large enough to provide adequate statistical power to detect group differences. In practice, this usually means a minimum number of 10 to 20 cases per cell; see Section 13.8 on statistical power later in this chapter for more detailed guidelines about the minimum number of scores per group needed to detect differences that correspond to small, medium, or large effect sizes. In addition, the number of cases within each cell should be large enough to make the confidence intervals for estimates of group means reasonably narrow.

If possible, in experimental situations, the number of participants per cell should be made equal. This simplifies the analysis and provides the maximum power possible for a given overall *N*. If random assignment of participants to cells or groups is possible, it can enhance the internal validity of the study by minimizing the likelihood of confounds between preexisting group differences and exposure to different levels of treatment factors.

In nonexperimental situations where factorial designs are used, it may be difficult or impossible to obtain equal numbers in the cells due to naturally occurring confounds among variables in nonexperimental research. For instance, suppose that a researcher wants to predict mean age at death from two factors that correspond to naturally occurring participant characteristics: gender (male, female) and smoking status (cigar smoker, nonsmoker of cigars). Interpretation of the results of this study would be complicated by the confound between gender and smoking status (most cigar smokers tend to be male, and most females tend not to smoke cigars). An earlier mean age at death for cigar smokers than for nonsmokers of cigars may, therefore, be due partly to the higher percentage of men in the cigar-smoking group; mean life expectancy for men is typically lower than for women. When equal numbers of scores in the groups are not obtainable or when the unequal numbers represent a real confound between variables and not just under sampling within certain cells, it may be necessary to use statistical methods to distinguish how much variance is uniquely predictable from each factor when the other is statistically controlled; factorial ANOVA with unequal numbers in the cells uses the same kinds of procedures as multiple regression to partial out shared variance among predictors. This is discussed further in the appendix to this chapter.

13.4 ♦ Empirical Example: Description of Hypothetical Data

The data shown in Table 13.1 are hypothetical results of a survey on social support, stress, and symptoms. There are scores on two predictor variables: Factor *A*, level of social support (1 = Low, 2 = High) and Factor *B*, level of stress (1 = Low, 2 = High). The dependent variable was number of symptoms of physical illness. Because the numbers of scores in the cells are equal across all four groups in this study (each group has $n = 5$ participants), this study does not raise problems of confounds between factors; it is an **orthogonal factorial ANOVA.** Based on past research on social support, there are two different hypotheses about the combined effects of social support and stress on physical illness symptoms. One hypothesis corresponds to a direct- or main-effects-only model; this model suggests that stress and social support each independently influence symptoms, but there is no interaction between these predictors. The other hypothesis is a buffering effect or interaction model. Some researchers hypothesize that people who have high levels of social support are protected against or buffered from the effects of stress; in this model, we expect to see high levels of physical symptoms associated with higher levels of stress only for persons who have low levels of social support and not for persons who have high levels of social support. The patterns of cell means that correspond to these two different hypotheses appear in Figure 13.2. The direct-effects model predicts that symptoms decrease as a function of increasing social support and symptoms increase as a function of increases in stress. However, the direct-effects model does not include an interaction between social support and stress; the direct-effects model predicts that no matter what level of social support a person has, the difference in physical symptoms between people with low and high stress should be about the same. The pattern of cell means that would be consistent with the predictions of this direct-effects (no interaction) hypothesis is shown in Figure 13.2a.

The buffering hypothesis about the effects of social support includes an interaction effect. The buffering hypothesis predicts that people with low levels of social support should show more symptoms under high stress than under low stress. However, some theories suggest that a high level of social support buffers (or protects against) the impact of stress, and therefore, people with high levels of social support are predicted to show little or no difference in symptoms as a function of level of stress. This is an example of "different slopes for different folks," a phrase suggested by Robert Rosenthal. For the low-social-support group, there is a large increase in symptoms when you contrast the high versus low stress levels. For the high-social-support group, a much smaller change in symptoms is predicted between the low- and high-stress conditions. A pattern of cell means that would be consistent with this buffering hypothesis is shown in Figure 13.2b.

A factorial ANOVA of the data provides the information needed to judge whether there is a statistically significant interaction between stress and social support as predictors of symptoms and, also, whether there are significant main effects of stress and social support on symptoms.

13.5 ♦ Computations for Between-S Factorial ANOVA

How can we summarize the information about the pattern of scores on *Y* in the sample data in this research situation? To begin, we need to find the grand mean of *Y*, the *Y* mean for each level of *A*, the *Y* mean for each level of *B*, and the *Y* mean for each of the cells. All subsequent analyses are based on assessment of the differences among these means (compared with the variances of scores within cells).

13.5.1 ♦ Notation for Sample Statistics That Estimate Score Components in Factorial ANOVA

Let Y_{ijk} denote the score for participant *k* within the group that corresponds to level *i* of the *A* factor and level *j* of the *B* factor.

Let M_Y equal the grand mean of the *Y* scores for the entire study. It is an estimate of μ_Y, the population grand mean of *Y* scores, obtained by summing all the *Y* scores and dividing by the total *N*.

Let M_{Ai} equal the mean of *Y* scores for level *i* of Factor *A*. Each M_{Ai} is an estimate of μ_{Ai}, the population mean for *Y* in each level of *A*. If each row of the data table represents a level of the *A* factor, as in Table 13.1, then we find the mean for level *i* of Factor *A* by summing all the scores within row *i* and dividing by the number of participants in that row. Alternatively, when the number of scores is equal across all groups or cells, we can simply average the cell means across each row of the design.

Let M_{Bj} equal the mean of *Y* scores for level *j* of Factor *B*. Each M_{Bj} is an estimate of μ_{Bj}, the population mean for *Y* in each level of *B*. If each column of the data table represents a level of the *B* factor, as in Table 13.1, then we find the mean for level *j* of Factor *B* by summing all the scores within column *j* and dividing that sum by the number of participants in column *j*.

Let M_{ABij} equal the observed sample mean of *Y* in the cell that represents the combined effects of level *i* of the *A* factor and level *j* of the *B* factor. Each cell mean was obtained by summing the scores within that cell and dividing that sum by *n*, the number of scores in the cell.

$Y_{ijk} - M_{ABij}$ denotes the residual or prediction error for each participant, or the within-groups deviation of an individual score from the mean of the group to which the score belongs.

$Y_{ijk} - M_Y$ is the "total deviation" for each participant—that is, the difference between the individual score and the grand mean or Y for the entire study.

13.5.2 ♦ Notation for Theoretical Effect Terms (or Unknown Population Parameters) in Factorial ANOVA

The theoretical model that we use when we do an ANOVA is an equation that represents each individual score as a sum of effects of all the theoretical components in the model. In the one-way ANOVA described in Chapter 6, the model for the score of person j in treatment Group A_i was represented as follows:

$$Y_{ij} = \mu_Y + \alpha_i + \varepsilon_{ij},$$

where μ_Y is the grand mean of Y; α_i represents the "effect" of the ith level of the A factor on people's scores; and ε_{ij} represents the residual or error, which captures any unique factors that influenced each person's score. (See Chapter 6 for a review of these concepts.) The μ_Y term is estimated by the sample grand mean M_Y; the α_i term is estimated by $M_{Ai} - M_Y$, the distance of each sample group mean from the sample grand mean, and the residual ε_{ij} is estimated by the difference between the individual Y_{ij} score and the mean for the ith level of Factor A—that is, $\varepsilon_{ij} = Y_{ij} - M_{Ai}$.

In a two-way factorial ANOVA, we need to add a term to this model to represent the main effect for a second factor (B) and, also, a term to represent a possible interaction between the A and B factors. The following effect components can be estimated for each participant once we have calculated the grand mean and all the cell, row, and column means.

Let α_i be the effect of level i for Factor A:

$$\alpha_i = M_{A_i} - M_Y. \tag{13.1}$$

Let β_j be the effect of level j for Factor B:

$$\beta_j = M_{B_j} - M_Y. \tag{13.2}$$

Let $\alpha\beta_{ij}$ be the interaction effect for the i, j cell:

$$\alpha\beta_{ij} = M_{AB_{ij}} - \mu_Y - \alpha_i - \beta_j. \tag{13.3}$$

Let ε_{ijk} be the residual, or unexplained part, of each individual score:

$$\varepsilon_{ijk} = Y_{ijk} - M_{AB_{ij}}. \tag{13.4}$$

For each observation in an $A \times B$ factorial model, each individual observed Y_{ijk} score corresponds to an additive combination of these theoretical effects:

$$Y_{ijk} = \mu_Y + \alpha_i + \beta_j + \alpha\beta_{ij} + \varepsilon_{ijk}. \tag{13.5}$$

The "no interaction" null hypothesis is equivalent to the assumption that for all cells, this $\alpha\beta_{ij}$ term is equal to or close to zero or, in other words, that scores can be adequately predicted from the reduced (no interaction) model:

$$Y_{ijk} = \mu_Y + \alpha_i + \beta_j + \varepsilon_{ijk}. \tag{13.6}$$

The equation for this reduced (no interaction) model says that the Y score for each person and/or the mean of Y for each cell can be predicted from just the additive main effects of the A and B factors. When there is no interaction, we do not need to add an adjustment factor ($\alpha\beta$) to predict the mean for each cell. The $\alpha\beta$ effect represents something "different" that happens for particular combinations of levels of A with levels of B, which cannot be anticipated simply by summing their main effects. Thus, the null hypothesis of "no interaction" can be written algebraically: H_0: $\alpha\beta_{ij} = 0$, for all i and j.

These theoretical terms (the α, β, and $\alpha\beta$ effects) are easier to comprehend when you see how the observed scores can be separated into these components. That is, we can

Table 13.3 ♦ Summary of Score Components in Factorial ANOVA

Name of Component	*Population Parameter*	*Sample Estimate*	*Name in Excel Spreadsheet*
Grand mean of Y	μ_Y	M_Y	G_MEAN
Effect of Level i of Factor A	α_i	$M_{A_i} - M_Y$	A_EFF
Effect of Level j of Factor B	β_j	$M_{B_j} - M_Y$	B_EFF
Interaction effect for the combination of Level i of A with Level j of B	$\alpha\beta_{ij}$	$M_{AB_{ij}} - (M_Y + M_{A_i} + M_{B_j})$	AB_EFF
Residual or error for each participant	ε_{ijk}	$Y_{ijk} - M_{AB_{ij}}$	RESID
Deviation from grand mean for each participant	$Y_{ijk} - \mu_Y$	$Y_{ijk} - M_Y$	DEV
Squared A effects	α_i^2	$(M_{A_i} - M_Y)^2$	AEFFSQ
Squared B effects	β_j^2	$(M_{B_j} - M_Y)^2$	BEFFSQ
Squared interaction effects	$\alpha\beta_{ij}^2$	$[M_{AB_{ij}} - (M_Y + M_{A_i} + M_{B_j})]^2$	ABEFFSQ
Squared residuals	ε_{ijk}^2	$(Y_{ijk} - M_{AB_{ij}})^2$	RESIDSQ
Squared total deviations	$(Y_{ijk} - \mu_Y)^2$	$(Y_{ijk} - M_Y)^2$	DEV_SQ

obtain a numerical estimate for each effect for each participant. Table 13.3 summarizes the set of components of scores in a factorial ANOVA (and shows the correspondence between the theoretical effect components in Equation 13.5, the sample estimates, and the names of these terms in the Excel spreadsheet).

The Excel spreadsheet shown in Table 13.4 illustrates that each score in Table 13.1 can be "taken apart" into these components. The column headed "G_MEAN" in the spreadsheet represents the grand mean (μ_Y or M_Y). This is obtained by summing all the scores and dividing by the total N. For all participants, this value is the same: G_MEAN = 6.65

The effect estimates (α_i) for the A factor are calculated for each group, A_1 and A_2; for each group, the sample estimate of this effect (denoted by A_EFF in the Excel spreadsheet) is calculated by subtracting the grand mean from the A_i group mean. The row, column, and cell means needed to compute these effect components are shown in Table 13.2. For all members of the A_1 group, A_EFF = $M_{A1} - M_Y = 8.30 - 6.65 = + 1.65$; for all members of the A_2 group, A_EFF = $M_{A2} - M_Y = 5.00 - 6.65 = - 1.65$.

The effect estimates (β_j) for the B factor can be calculated separately for the B_1 and B_2 groups; for each group, the sample estimate of this effect (denoted by B_EFF in the Excel spreadsheet) is found by subtracting the grand mean from the B_j group mean. For all members of the B_1 group, B_EFF = $M_{B1} - M_Y = 3.90 - 6.65 = - 2.75$; for all members of the B_2 group, B_EFF = $(M_{B2} - M_Y) = (9.40 - 6.65) = + 2.75$.

The sample estimate of the interaction effect for each cell or A, B treatment combination, $\alpha\beta$, is found by applying the formula in Equation 13.3. That is, for each group, we begin with the cell mean and subtract the grand mean and the corresponding A_EFF and B_EFF terms; in the Excel spreadsheet this estimated interaction component is called AB_EFF. For example, for all members of the AB_{11} group, AB_EFF = $M_{AB11} - M_Y -$ A_EFF $-$ B_EFF $= 3.80 - 6.65 - 1.65 - (-2.75) = -1.75$. The sample estimates of these effects (α_i, β_j, and $\alpha\beta_{ij}$) are summarized in Table 13.4.

The sum of these effects across all the scores in the study (and for each row and/or column in the design) must be 0. This is the case because effect estimates are deviations from a mean, and by definition, the sum of deviations from means equals 0. Therefore, we can't summarize information about the sizes of these effect components simply by summing them for all the participants; these sums would just turn out to be 0. If we square these components and then sum them, however, it provides a way to summarize information about the relative sizes of these terms across all scores in our sample. Each of the *SS* terms obtained in a factorial ANOVA is just the sum of squared effects across all the scores in the study; for example, SS_A is obtained by summing the squared A_EFF terms across all scores.

The residual (in the column headed RESID in the Excel spreadsheet in Table 13.4) for each individual score was calculated by subtracting the group mean from each individual score. The sum of these residuals must also equal 0 (within each group and across the entire study). We can verify that these estimated effect components can be used to reconstruct the original observed scores; you should take the data from one line of the spreadsheet in Table 13.4 and verify that the original observed *Y* score can be reproduced by summing the theoretical effect components into which we have divided the scores, using the logic of ANOVA:

$$Y = \text{G\_MEAN} + \text{A\_EFF} + \text{B\_EFF} + \text{AB\_EFF} + \text{RESID}.$$

If we look at the sizes of these score components for an individual participant—for example, the first subject in the AB_{11} group (the first line of data in Table 13.4), we can

Table 13.4 ♦ Excel Spreadsheet for Data in Table 13.1

A	*B*	*Y*	*G_MEAN*	*CELLMEAN*	*A_EFF*	*B_EFF*	*AB_EFF*	*RESID*	*RESIDSQ*	*AEFFSQ*	*BEFFSQ*	*ABEFFSQ*	*DEV*	*DEV_SQ*
1	1	2	6.65	3.80	1.65	−2.75	−1.75	−1.80	3.24	2.72	7.56	3.06	−4.65	21.62
1	1	3	6.65	3.80	1.65	−2.75	−1.75	−.80	.64	2.72	7.56	3.06	−3.65	13.32
1	1	5	6.65	3.80	1.65	−2.75	−1.75	1.20	1.44	2.72	7.56	3.06	−1.65	2.72
1	1	4	6.65	3.80	1.65	−2.75	−1.75	.20	.04	2.72	7.56	3.06	−2.65	7.02
1	1	5	6.65	3.80	1.65	−2.75	−1.75	1.20	1.44	2.72	7.56	3.06	−1.65	2.72
1	2	11	6.65	12.80	1.65	2.75	1.75	−1.80	3.24	2.72	7.56	3.06	4.35	18.92
1	2	13	6.65	12.80	1.65	2.75	1.75	.20	.04	2.72	7.56	3.06	6.35	40.32
1	2	10	6.65	12.80	1.65	2.75	1.75	−2.80	7.84	2.72	7.56	3.06	3.35	11.22
1	2	16	6.65	12.80	1.65	2.75	1.75	3.20	10.24	2.72	7.56	3.06	9.35	87.42
1	2	14	6.65	12.80	1.65	2.75	1.75	1.20	1.44	2.72	7.56	3.06	7.35	54.02
2	1	2	6.65	4.00	−1.65	−2.75	1.75	−2.00	4.00	2.72	7.56	3.06	−4.65	21.62
2	1	4	6.65	4.00	−1.65	−2.75	1.75	.00	.00	2.72	7.56	3.06	−2.65	7.02
2	1	6	6.65	4.00	−1.65	−2.75	1.75	2.00	4.00	2.72	7.56	3.06	−.65	.42
2	1	5	6.65	4.00	−1.65	−2.75	1.75	1.00	1.00	2.72	7.56	3.06	−1.65	2.72
2	1	3	6.65	4.00	−1.65	−2.75	1.75	−1.00	1.00	2.72	7.56	3.06	−3.65	13.32
2	2	6	6.65	6.00	−1.65	2.75	−1.75	.00	.00	2.72	7.56	3.06	−.65	.42
2	2	7	6.65	6.00	−1.65	2.75	−1.75	1.00	1.00	2.72	7.56	3.06	.35	.12
2	2	4	6.65	6.00	−1.65	2.75	−1.75	−2.00	4.00	2.72	7.56	3.06	−2.65	7.02
2	2	6	6.65	6.00	−1.65	2.75	−1.75	.00	.00	2.72	7.56	3.06	−.65	.42
2	2	7	6.65	6.00	−1.65	2.75	−1.75	1.00	1.00	2.72	7.56	3.06	.35	.12
Sum					.00	.00	.00	.00	45.60	54.45	151.25	61.25	.00	312.55
									SS_{within}	SS_A	SS_B	SS_{AxB}		SS_{total}

NOTES: Computation of sums of squares for factorial ANOVA from score components in an Excel spreadsheet:

Each Y score can be represented as the sum of the following components:	G_MEAN	+	A_EFF	+	B_EFF	+	AB_EFF	+	RESID
These correspond to the following theoretical score components:	μ_Y	+	α_i	+	β_j	+	$\alpha\beta_{ij}$	+	ε_{ijk}

To compute the sums of squares for the factorial ANOVA, square the columns that contain these effect components (A_EFF, B_EFF, AB_EFF, and RESID), and place these squared components in columns that are named AEFFSQ, BEFFSQ, ABEFFSQ, and RESIDSQ. Sum each column of squared components. The sum of AEFFSQ across all N participants is SS_A. The sum of BEFFSQ across all N participants is SS_B. The sum of ABEFFSQ across all N participants is $SS_{A \times B}$. The sum of RESIDSQ across all N participants equals SS_{within}. The sum of DEV_SQ across all N participants equals SS_{total}. To check, verify that $SS_{\text{total}} = SS_A + SS_B + SS_{A \times B} + SS_{\text{within}}$.

provide a verbal interpretation of how that participant's score is predicted from the effects in our 2 × 2 factorial ANOVA. The observed value of the score Y_{111} was 2. ($Y_{111} = 2 =$ the number of symptoms for participant 1 in the A_1, low-social-support group, and B_1, low-stress-level group.) The corresponding numerical estimates for the effect components for this individual score, from the first line in Table 13.3, were as follows (from the columns headed G_MEAN, A_EFF, B_EFF, AB_EFF, and RESID):

$$\begin{array}{ccccccccccc} Y_{111} & = & \text{G\_MEAN} & + & \text{A\_EFF} & + & \text{B\_EFF} & + & \text{AB\_EFF} & + & \text{RESID} \\ 2.00 & = & 6.65 & + & 1.65 & + & (-2.75) & + & (-1.75) & + & (-1.80). \end{array}$$

Recall that the sample estimates of effects in the equation above correspond to the terms in the theoretical model for a two-way factorial ANOVA shown earlier in Equation 13.5: $Y_{ijk} = \mu_Y + \alpha_i + \beta_j + \alpha\beta_{ij} + \varepsilon_{ijk}$.

When we translate this equation into a sentence, the story about the prediction of symptoms from social support and stress is as follows. The observed number of symptoms for Participant Y_{111} can be predicted by starting with the grand mean, 6.65; adding the "effect" of being in the low-social-support group (which predicts an increase of 1.65 in number of symptoms); adding the effect of being in the low-stress-level group (which predicts a decrease of 2.75 symptoms); and adding an effect term to represent the nature of the interaction (the group that was low in social support and low in stress showed 1.75 fewer symptoms than would be expected from an additive combination of the effects of social support and stress). Finally, we add the residual (−1.80) to indicate that this individual participant scored 1.8 units lower in symptoms than the average for other participants in this group.

Note that the term *effect* throughout this discussion of ANOVA refers to theoretical estimated effect components (such as α_i) in the ANOVA model. The use of the term *effect* does not necessarily imply that we can make a causal inference. It is not appropriate to make causal inferences from nonexperimental research designs; we can only interpret ANOVA effects as possible evidence of causal connections when the effects are found in the context of well-controlled experimental designs.

The goal of the factorial ANOVA is to summarize the sizes of these effect components (the α, β, and $\alpha\beta$ effects) across all scores, so that we can assess which of these effects tend to be relatively large. Because these effects are deviations from means, we cannot just sum them (the sum of deviations from a mean equals 0 by definition). Instead, we will sum these squared effects (across all participants) to obtain a sum of squares for each effect.

13.5.3 ♦ Formulas for Sums of Squares and Degrees of Freedom

The formulas for by-hand computations of a two-way factorial ANOVA are often presented in a form that minimizes the number of arithmetic operations needed, but in that form, it is often not very clear how the patterns in the data are related to the numbers you obtain. A spreadsheet approach to the computation of sums of squares makes it clearer how each *SS* term provides information about the different (theoretical) components that make up the scores. The operations described here yield the same numerical results as the formulas presented in introductory statistics textbooks.

We will need three subscripts to denote each individual score. The first subscript (i) tells you which level of Factor A the score belongs to, the second subscript (j) tells you which level of Factor *B* the score belongs to, and the third subscript (*k*) tells you which participant within that group the score belongs to. In general, an individual score is denoted by Y_{ijk}. For example, Y_{215} would indicate a score from Level 2 of Factor *A*, Level 1 of Factor *B*, and the fifth subject within that group or cell.

The model, or equation, that describes the components that predict each Y_{ijk} score was given earlier (as Equation 13.5):

$$Y_{ijk} = \mu_Y + \alpha_i + \beta_j + \alpha\beta_{ij} + \varepsilon_{ijk}.$$

Suppose, for example, that *Y* is the number of physical illness symptoms, the *A* factor (which corresponds to the α-effect parameter) is level of social support; and the *B* factor (which corresponds to the β-effect parameter) is level of stress. The αβ parameter represents potential interaction effects between social support and stress. Finally, the ε parameter represents all the unique factors that influence the symptoms of subject *k* in Groups $A = i$ and $B = j$ or, in other words, all other variables (such as depression, poverty, diet, and so forth) that might influence a person's symptoms. In our data, we can use sample statistics to estimate each of these theoretical components of the score.

The Excel spreadsheet in Table 13.4 shows these individual effect components for each individual score. To obtain the sums of squares that will summarize the sizes of these components across all the data in the study, we need to square the estimates of α_i, β_j, $\alpha\beta_{ij}$, and ε_{ijk} for each score and then sum these squared terms, as shown in the spreadsheet in Table 13.4. Equations that summarize these operations in more familiar notation are given below.

Spreadsheet

$$SS_{\text{total}} = \Sigma_{ijk} (Y_{ijk} - M_Y)^2 \qquad \Sigma\ (\text{DEV\_SQ}) \qquad (13.7)$$

$$SS_A = \Sigma_{ijk} (M_{Ai} - M_Y)^2 \text{ or } \Sigma_{ijk} \alpha^2_i \qquad \Sigma\ (\text{AEFFSQ}) \qquad (13.8)$$

$$SS_B = \Sigma_{ijk} (M_{Bj} - M_Y)^2 \text{ or } \Sigma_{ijk} \beta^2_j \qquad \Sigma\ (\text{BEFFSQ}) \qquad (13.9)$$

$$SS_{A\times B} = \Sigma_{ijk} [M_{ABij} - (M_Y + M_{Ai} + M_{Bj})]^2 \text{ or } \Sigma_{ijk} \alpha\beta^2_{ij} \qquad \Sigma\ (\text{ABEFFSQ}) \qquad (13.10)$$

$$SS_{\text{within}} = \Sigma_{ijk} (Y_{ijk} - M_{ABij})^2 \qquad \Sigma\ (\text{RESIDSQ}) \qquad (13.11)$$

You can check the computations of these *SS* terms by verifying that this equality holds:

$$SS_{\text{total}} = SS_A + SS_B + SS_{A\times B} + SS_{\text{within}}.$$

To find the degrees of freedom that correspond to each sum of squares, we need to use the information about the number of levels of the *A* factor, *a*; the number of levels of the *B* factor, *b*; the number of cases in each cell of the design, *n*; and the total number of scores, *N*:

$$df_A = a - 1, \tag{13.12}$$

$$df_B = b - 1, \tag{13.13}$$

$$df_{A \times B} = (a - 1)(b - 1), \tag{13.14}$$

$$df_{\text{within}} = ab(n - 1), \tag{13.15}$$

$$df_{\text{total}} = N - 1. \tag{13.16}$$

The degrees of freedom for a factorial ANOVA are also additive, so you should check that $N - 1 = (a - 1) + (b - 1) + (a - 1)(b - 1) + ab(n - 1)$.

To find the mean square for each effect, divide the sum of squares by its corresponding degrees of freedom; this is done for all terms, except that (conventionally) a mean square is not reported for SS_{total}:

$$MS_A = SS_A/df_A, \tag{13.17}$$

$$MS_B = SS_B/df_B, \tag{13.18}$$

$$MS_{A \times B} = SS_{A \times B}/df_{A \times B}, \tag{13.19}$$

$$MS_{\text{within}} = SS_{\text{within}}/df_{\text{within}}. \tag{13.20}$$

The F for each effect (A, B and interaction) is found by dividing the mean square for that effect by mean square error.

$$F_A = MS_A/MS_{\text{within}}, \tag{13.21}$$

$$F_B = MS_B/MS_{\text{within}}, \tag{13.22}$$

$$F_{A \times B} = MS_{A \times B}/MS_{\text{within}}. \tag{13.23}$$

The following numerical results were obtained for the data in Table 13.1 (based on the spreadsheet in Table 13.4).

$$SS_A = \Sigma\,(\text{AEFFSQ}) = 54.45,$$

$$SS_B = \Sigma\,(\text{BEFFSQ}) = 151.25,$$

$$SS_{A \times B} = \Sigma\,(\text{ABEFFSQ}) = 61.25,$$

$$SS_{\text{within}} = \Sigma\,(\text{RESIDSQ}) = 45.60.$$

For this example, $a = 2$ and $b = 2$; there were $n = 5$ participants in each A, B group, or a total of $a \times b \times n = N = 20$ scores. From these values of a, b, and n, we can work out the degrees of freedom.

$$df_A = a - 1 = 2 - 1 = 1,$$

$$df_B = b - 1 = 2 - 1 = 1,$$

$$df_{A \times B} = (a - 1)(b - 1) = (2 - 1)(2 - 1) = 1,$$

$$df_{\text{within}} = N - ab = 20 - (2 \times 2) = 16.$$

For each sum of squares, we calculate the corresponding mean square by dividing by the corresponding degrees of freedom:

$$MS_A = SS_A/df_A = 54.45/1 = 54.45,$$

$$MS_B = SS_B/df_B = 151.25/1 = 151.25,$$

$$MS_{A \times B} = SS_{A \times B}/df_{A \times B} = 61.25/1 = 61.25,$$

$$MS_{\text{within}} = SS_{\text{within}}/df_{\text{within}} = 45.60/16 = 2.85.$$

Finally, we obtain an F ratio to test each of the null hypotheses by dividing the MS for each of the three effects (A, B, and $A \times B$) by MS_{within}:

$$F_A = MS_A/MS_{\text{within}} = 54.45/2.85 = 19.10,$$

$$F_B = MS_B/MS_{\text{within}} = 151.25/2.85 = 53.07,$$

$$F_{A \times B} = MS_{A \times B}/MS_{\text{within}} = 61.25/2.85 = 21.49.$$

These computations for a factorial ANOVA can be summarized in table form as follows. Most journal articles no longer include intermediate results such as sums of squares and mean squares. Instead, for each effect (A and B main effect and the $A \times B$ interaction), journal articles typically include the following information: the F ratio and its associated df and p values; an effect size estimate such as η^2; and tables or graphs of cell means or row and column means, sometimes with confidence intervals for each estimated mean.

Source	*SS*	*df*	*MS*	*F*
Social support (*A*)	SS_A	df_A	MS_A	F_A
Stress (*B*)	SS_B	df_B	MS_B	F_B
$A \times B$	$SS_{A \times B}$	$df_{A \times B}$	$MS_{A \times B}$	$F_{A \times B}$
Within group	SS_{within}	df_{within}	MS_{within}	
Total	SS_{total}	df_{total}		

The source tables produced by the GLM procedure in SPSS contain additional lines. For example, SPSS reports a sum of squares for the combined effects of *A*, *B*, and the interaction of $A \times B$; this combined test (both main effects and the interaction effect) is rarely reported in journal articles. There is a difference in the way SPSS (and some other programs) labels SS_{total}, compared with the notation used here and in most other statistics textbooks. The term that is generally called SS_{total} (here and in most other statistics textbooks) is labeled Corrected Total on the SPSS printout. The term that SPSS labels "SS_{total}" was calculated by taking $\Sigma(Y_{ijk} - 0)^2$—that is, the sum of the squared deviations of all scores from 0; this sum of squares can be used to test the null hypothesis that the grand mean for the entire study equals 0; this term is usually not of interest. When you read the source table from SPSS GLM, you can generally ignore the lines that are labeled Corrected Model, Intercept, and Total. You will use the *SS* value in the line that SPSS labels Corrected Total as your value for SS_{total} if you do a by hand computation of η^2 effect size estimates.

13.6 ♦ Conceptual Basis: Factors That Affect the Size of Sums of Squares and *F* Ratios in Factorial ANOVA

Just as in previous chapters on the independent samples *t* test and one-way ANOVA, in factorial ANOVA, we want to assess whether group means are far apart relative to the within-group variability of scores. The same factors that affected the size of *t* and *F* in these earlier analyses are relevant in factorial ANOVA.

13.6.1 ♦ Distances Between Group Means (Magnitude of the α and β Effects)

Other factors being equal, group means tend to be farther apart and $SS_{between}$ and the *F* ratio for an effect tend to be larger, when the dosage levels of treatments administered to a group are different enough to produce detectable differences. For instance, if a researcher compares groups that receive 0 versus 300 mg of caffeine, the effects of caffeine on heart rate will probably be larger than if the researcher compares 0 versus 30 mg of caffeine. For comparison of preexisting groups that differ on participant characteristics, differences between group means tend to be larger (and *F* ratios tend to be larger) when the groups are chosen so that these differences are substantial; for example, a researcher has a better chance of finding age-related differences in blood pressure when comparing groups that are age 20 versus age 70 than when comparing groups that are age 20 versus age 25.

13.6.2 ♦ Number of Scores (*n*) Within Each Group or Cell

Assuming a constant value of *SS*, MS_{within} tends to become smaller when the *n* of participants within groups is increased (because $MS = SS/df$, and *df* for the within-group *SS* term increases as *N* increases). This, in turn, implies that the value of an *F* ratio often tends to be higher as the total sample size *N* increases, assuming that all other aspects of the data (such as the distances between group means and the within-cell variation of scores) remain the same. This corresponds to a commonsense intuition. Most of the statistical significance test statistics that you have encountered so far (such as the independent samples *t* test and the *F* ratio in ANOVA) tend to yield larger values as the sample size *N* is increased, assuming that other terms involved in the computation of *F* (such as the distances between group means and the variability of scores within groups) remain the same and that the sample means are not exactly equal across groups.

13.6.3 ♦ Variability of Scores Within Groups or Cells (Magnitude of MS_{within})

Other factors being equal, the variance of scores within groups (MS_{within}) tends to be smaller, and therefore, F ratios tend to be larger, when sources of error variance within groups can be controlled (e.g., through selection of homogeneous participants, through standardization of testing or observation methods, by holding extraneous variables constant so that they do not create variations in performance within groups). In this chapter, we will see that including a blocking factor can be another way to reduce the variability of scores within groups.

In addition to the ability to detect potential interaction effects, a factorial ANOVA design also offers researchers a possible way to reduce SS_{within}, the variability of scores within groups, by *blocking* on subject characteristics. For example, suppose that a researcher wants to study the effects of social support on symptoms, but another potential predictor variable, stress, is also related to symptoms. The researcher can use stress as a "blocking factor" in the analysis. In this hypothetical example involving a 2 × 2 factorial, the factor that represents social support (A_1 = Low, A_2 = High) can be crossed with a second blocking factor, stress (B_1 = Low, B_2 = High). We would say that the participants have been "blocked on level of stress."

Just as in studies that use t tests and one-way ANOVA, researchers who use factorial ANOVA designs can try to maximize the size of F ratios by increasing the differences between the treatment dosage levels or participant characteristics that differentiate groups, by controlling extraneous sources of variance through experimental or statistical control, or by increasing the number of cases within groups.

13.7 ♦ Effect Size Estimates for Factorial ANOVA

For each of the sources of variance in a two-way factorial ANOVA, simple η^2 effect size estimates can be computed either from the sums of squares or from the F ratio and its degrees of freedom.

$$\eta_A^2 = SS_A/SS_{\text{total}} = \frac{df_A \times F_A}{df_A \times F_A + df_{\text{within}}}, \tag{13.24}$$

$$\eta_B^2 = SS_B/SS_{\text{total}} = \frac{df_B \times F_B}{df_B \times F_B + df_{\text{within}}}, \tag{13.25}$$

$$\eta_{A\times B}^2 = SS_{A\times B}/SS_{\text{total}} = \frac{df_{A\times B} \times F_{A\times B}}{df_{A\times B} \times F_{A\times B} + df_{\text{within}}}. \tag{13.26}$$

For the data in Table 13.1, the effect size η^2 for each of the three effects can also be computed by taking the ratio of SS_{effect} to SS_{total}:

$$\eta_A^2 = SS_A / SS_{\text{total}} = 54.50/312.55 = .17,$$

$$\eta_B^2 = SS_B / SS_{\text{total}} = 151.25/312.55 = .48,$$

$$\eta_{A\times B}^2 = SS_{A\times B} / SS_{\text{total}} = 61.25/312.55 = .20.$$

Note that the η^2 values above correspond to unusually large effect sizes. These hypothetical data were intentionally constructed so that the effects would be large.

When effect sizes are requested from the SPSS GLM program, a **partial η^2** is reported for each main effect and interaction instead of the simple η^2 effect sizes that are defined by Equations 13.24 to 13.26. For a partial η^2 for the effect of the A factor, the divisor is $SS_A + SS_{\text{within}}$ instead of SS_{total}; that is, the variance that can be accounted for by the main effect of B and the $A \times B$ interaction is removed when the partial η^2 effect size is calculated. The partial η^2 that describes the proportion of variance that can be predicted from A (social support) when the effects of B (stress) and the $A \times B$ interaction are statistically controlled would be calculated as follows:

$$\text{Partial } \eta_A^2 = \frac{SS_A}{SS_A + SS_{\text{within}}} = \frac{54.50}{54.50 + 45.60} = .54. \quad (13.27)$$

Note that the simple η^2 tells us what proportion of the total variance in scores on the Y outcome variable is predictable from each factor in the model, such as Factor A. The partial η^2 tells us what proportion of the remaining variance in Y outcome variable scores is predictable from the A factor after the variance associated with other predictors in the analysis (such as main effect for B and the $A \times B$ interaction) has been removed. These partial η^2 values are typically larger than simple η^2 values. Note that eta squared is only a description of the proportion of variance in the Y outcome scores that is predictable from a factor, such as Factor A, in the sample. It can be used to describe the strength of the association between variables in a sample, but eta squared tends to overestimate the proportion of variance in Y that is predictable from Factor A in some broader population. Other effect size estimates such as omega squared (ω^2, described in Hays, 1994) may provide estimates of the population effect size. However, eta squared is more widely reported than omega squared as a sample effect size in journal articles, and eta squared is more widely used in statistical power tables.

13.8 ♦ Statistical Power

As in other ANOVA designs, the basic issue is the minimum number of cases required in each cell (to test the significance of the interaction) or in each row or column (to test the A and B main effects) to have adequate statistical power. If interactions are not predicted or of interest, then the researcher may be more concerned with the number of cases in each level of the A and B factors rather than with cell size. To make a reasonable judgment about the minimum number of participants required to have adequate statistical power, the researcher needs to have some idea of the population effect size. If comparable past research reports F ratios, these can be used to compute simple estimated effect sizes (η^2), using the formulas provided in Section 13.7.

Table 13.5 (from Jaccard & Becker, 1997) can be used to decide on a reasonable minimum number of scores per group using eta squared as the index of population effect size. These tables are for significance tests using an α level of .05. (Jaccard and Becker also provided statistical power tables for other levels of alpha.) The usual minimum level of power desired is .80. To use the statistical power table, the researcher needs to know the degrees of freedom for the effect to be tested and needs to assume a population eta-squared value. Suppose that a researcher wants to test the significance of dosage level of caffeine with three dosage levels using $\alpha = .05$. The *df* for this main effect would be 2. If past research has reported effect sizes on the order of .15, then from Table 13.5, the researcher would locate the portion of the table for designs with $df = 2$, the column in the table for $\eta^2 = .15$, and the row in the table for power = .80. The *n* given by the table for this combination of *df*, η^2, and power is $n = 19$. Thus, to have an 80% chance of detecting an effect that accounts for 15% of the variance in scores for a factor that compares three groups (2 *df*), a minimum of 19 participants are needed in each group (each dosage level of caffeine). Note that a larger number of cases may be needed to obtain reasonably narrow confidence intervals for each estimated group mean; it is often desirable to have sample sizes larger than the minimum numbers suggested by power tables.

13.9 ♦ Nature of the Relationships, Follow-Up Tests, and Information to Include in the Results

A factorial ANOVA provides information about three questions:

1. Is the $A \times B$ interaction statistically significant? If so, how strong is the interaction, and what is the nature of the interaction? Often, when there is a statistically significant interaction, the description of outcomes of the study focuses primarily on the nature of the interaction.

2. Is the *A* main effect statistically significant? If so, how strong is the effect, and what is the nature of the differences in means across levels of *A*?

3. Is the *B* main effect statistically significant? If so, how strong is the effect, and what is the nature of the differences in means across levels of *B*?

In an orthogonal factorial design, the outcomes of these three questions are independent; that is, you can have any combination of yes/no answers to the three questions about significance of the $A \times B$ interaction and the *A* and *B* main effects.

Each of the three questions listed above requires essentially the same kind of presentation of information that was described in Chapter 6 for one-way ANOVA. For each effect, it is customary to present an *F* test (to assess statistical significance). If *F* is nonsignificant, usually there is no further discussion. If *F* is significant, then some indication of effect size (such as η^2) should be presented, and the magnitudes and directions of the differences among group means should be discussed and interpreted. In particular, were the obtained differences between group means in the predicted direction?

It is increasingly common for journal articles to present a series of many analyses. When there are no significant main effects or interactions, it might be considered a waste

Table 13.5 ♦ Statistical Power Tables for Factorial ANOVA

	Degrees of Freedom Between or Effect = 1, Alpha = .05 POPULATION ETA SQUARED																		
Power	.01	.03	.05	.07	.10	.15	.20	.25	.30	.35	.40	.45	.50	.55	.60	.65	.70	.75	.80
.10	22	8	5	4	3	2	2	2	—	—	—	—	—	—	—	—	—	—	—
.50	193	63	38	27	18	12	9	7	5	5	4	3	3	3	2	2	2	2	—
.70	310	101	60	42	29	18	13	10	8	7	6	5	4	4	3	3	2	2	2
.80	393	128	76	53	36	23	17	13	10	8	7	6	5	4	4	3	3	2	2
.90	526	171	101	71	48	31	22	17	13	11	9	7	6	5	5	4	3	3	2
.95	651	211	125	87	60	39	27	21	16	13	11	9	8	6	5	5	4	3	3
.99	920	298	176	123	84	53	38	29	22	18	15	12	10	9	7	6	5	4	3

	Degrees of Freedom Between or Effect = 2, Alpha = .05 POPULATION ETA SQUARED																		
Power	.01	.03	.05	.07	.10	.15	.20	.25	.30	.35	.40	.45	.50	.55	.60	.65	.70	.75	.80
.10	22	8	5	4	3	2	2	2	—	—	—	—	—	—	—	—	—	—	—
.50	165	55	32	23	16	10	8	6	5	4	3	3	3	2	2	2	2	2	—
.70	255	84	50	35	24	16	11	9	7	6	5	4	4	3	3	2	2	2	2
.80	319	105	62	44	30	19	14	11	9	7	6	5	4	4	3	3	2	2	2
.90	417	137	81	57	39	25	18	14	11	9	7	6	5	4	4	3	3	2	2
.95	511	168	99	69	47	30	22	16	13	11	9	7	6	5	4	4	3	3	2
.99	708	232	137	96	65	41	29	22	18	14	12	10	8	7	6	5	4	3	3

SOURCE: From Jaccard and Becker (1997).

of space to provide detailed information about each factorial ANOVA. Even when factorial ANOVAs yield significant results, it is rare for researchers to report intermediate numerical results (such as sums of squares or mean squares). It is preferable to include information that is useful to the reader. For instance, if a researcher reports a set of 10 ANOVAs and only 6 of them yielded significant results, the researcher might report the results of the 4 nonsignificant ANOVAs in a few sentences and only report information about row, column, or cell outcomes for effects that were judged significant. When results are reported for groups, it is useful to report the number of cases and the standard deviation as well as the mean for each group that is included. The *Publication Manual of the American Psychological Association* (APA, 2001) provides useful guidelines for reporting the results of factorial ANOVA.

13.9.1 ♦ Nature of a Two-Way Interaction

One of the most common ways to illustrate a significant interaction in a two-way factorial ANOVA is to graph the cell means (as in Figure 13.2) or to present a table of cell means (as in Table 13.2). Visual examination of a graph makes it possible to "tell a story" about the nature of the outcomes. The potential problem is that the researcher may fail to notice that the location of cell means on such a graph is due, at least in part, to any main effects that are present. Rosenthal and Rosnow (1991) argue that this common practice of examining observed cell means can be misleading and that it is more informative to present a table of the interaction effects (here, the interaction effect for each cell is denoted by $\alpha\beta_{ij}$); these interaction effects have the main effects for the A and B factors (and the grand mean) subtracted from them. For the data in this chapter, the effects (α_i, β_j, and $\alpha\beta_{ij}$) are summarized in Table 13.6. A verbal interpretation of the nature of the interaction, based on this table, might focus on the $\alpha\beta_{22}$ effect estimate of −1.75: Persons in the high-social-support group who were exposed to high levels of stress experienced 1.75 fewer symptoms on average than would be predicted just from an additive combination of the effects of stress and social support; high social support may have reduced the impact of high stress on physical symptoms.

In spite of Rosenthal and Rosnow's (1991) advice, it is common for researchers to illustrate the nature of an interaction by presenting a graph or a table of cell means and to test the significance of differences among cell means. Given a 2 × 2 ANOVA in which Factor A represents two levels of social support and Factor B represents two levels of stress, a researcher might follow up a significant two-way interaction with an "analysis of simple main effects." For example, the researcher could ask whether within each level of B, the means for cells that represent two different levels of A differ significantly (or, conversely, the researcher could ask whether separately within each level of A, the two B group means differ significantly). If these comparisons are made based on a priori theoretical predictions, it may be appropriate to do planned comparisons; the error term used for these could be the MS_{within} for the overall two-way factorial ANOVA. If comparisons among cell means are made post hoc (without any prior theoretically based predictions), then it may be more appropriate to use protected tests (such as the Tukey honestly significant difference, described in Chapter 6), to make comparisons among means. The APA Task Force (Wilkinson &

Table 13.6 ♦ Effects (α, β, and $\alpha\beta$) for the Social Support, by Stress Factorial Data Presented in Table 13.1

	B_1 *(Low Stress)*	B_2 *(High Stress)*	α_i
A_1 (low social support)	$\alpha\beta_{11} = -1.75$	$\alpha\beta_{21} = +1.75$	+1.65
A_2 (high social support)	$\alpha\beta_{21} = +1.75$	$\alpha\beta_{22} = -1.75$	−1.65
β_j	−2.75	+2.75	Grand mean 6.65

NOTES: Note that within each row and column, the sum of the $\alpha\beta$ effects must be 0. The interaction effect has $(a - 1)(b - 1)$ *df*; in this example, *df* for the $A \times B$ interaction = 1. In a 2 × 2 table, only one of the $\alpha\beta$ terms in the table is "free to vary." The constraint that requires these $\alpha\beta$ terms to sum to 0 implies that once we know the first $\alpha\beta$ term, the values of all the remaining terms are determined. Across all levels of each factor, the sum of effects must equal 0; thus, $\alpha_1 + \alpha_2 = 0$, and $\beta_1 + \beta_2 = 0$. Only $a - 1$—in this case, 1—of the alpha effects is free to vary; and only $b - 1$—that is, 1—of the beta effects is free to vary. As soon as we know the numerical value of the first alpha effect and the first beta effect, the values of the remaining terms are determined (because the effects must sum to 0).

Task Force on Statistical Inference, 1999) suggested that it is more appropriate to specify a limited number of planned contrasts in advance rather than to make all possible comparisons among cell means post hoc.

Sometimes, particularly when the computer program does not include post hoc tests as an option or when there are problems with homogeneity of variance assumptions, it is convenient to do independent samples *t* tests to compare pairs of cell means and to use the Bonferroni procedure to control overall risk of Type I error. The Bonferroni procedure, described in earlier chapters, is very simple: If a researcher wants to perform *k* number of *t* tests, with an overall experiment-wise error rate (EW_α) of .05, then the per-comparison alpha (PC_α) that would be used to assess the significance of each individual *t* test would be set at $PC_\alpha = EW_\alpha/k$. Usually, EW_α is set at .05 (although higher levels such as EW_α = .10 or .20 may also be reasonable). Thus, if a researcher wanted to do three *t* tests among cell means as a follow-up to a significant interaction, the researcher might require a *p* value less than .05/3 = .0167 for each *t* test as the criterion for statistical significance.

13.9.2 ♦ Nature of Main Effect Differences

When a factor has only two levels or groups, and if the *F* for the main effect for that factor is statistically significant, no further tests are necessary to understand the nature of the differences between group means. However, when a factor has more than two levels or groups, the researcher may want to follow up a significant main effect for the factor with either planned contrasts or post hoc tests to compare means for different levels of that factor (as described in Chapter 6). It is important, of course, to note whether significant differences among means were in the predicted direction.

13.10 ♦ Factorial ANOVA Using the SPSS GLM Procedure

The hypothetical data in Table 13.1 represent scores for a nonexperimental study; the two factors were as follows: Factor A, amount of social support (A_1 = Low, A_2 = High) and

Factor B, stress (B_1 = Low, B_2 = High). The outcome variable was number of physical symptoms reported on a checklist.

To obtain the factorial ANOVA, the **SPSS GLM procedure** was used, and the following menu selections were made: <Analyze> → <General Linear Model> → <Univariate>.

The selection <Univariate> is made because there is only one outcome variable in this example. (In later chapters, we will see that multiple outcome variables can be included in a GLM analysis.) This set of menu selections brings up the main GLM Dialog window, shown in Figure 13.3; the names of the factors (both treated as fixed factors in this example) were entered in the Fixed Factor(s) window. The name of the single outcome variable (symptom) was entered in the Dependent Variable window. To obtain printouts of the means, it is necessary to use the Options button in order to access the dialog window for GLM options; this Options dialog window is shown in Figure 13.4. The default method for computation of sums of squares (**SS Type III**) was used. For these data, because the numbers of scores are equal across all groups, the results do not differ if different methods of computation for sums of squares are selected. The appendix to this chapter provides a brief description of the differences between SS Type III and other methods of computation of sums of squares in GLM. These differences are only important when the design is **nonorthogonal**—that is, when the numbers of scores are not equal or balanced across the cells in a factorial design.

To request a printout of the means, highlight the list of effects in the left-hand window, and move the entire list into the right-hand window that is headed "Display Means for," as shown in Figure 13.4. This results in printouts of the grand mean, the mean for each row, each column, and each cell in this 2 × 2 factorial design. If a test of homogeneity of

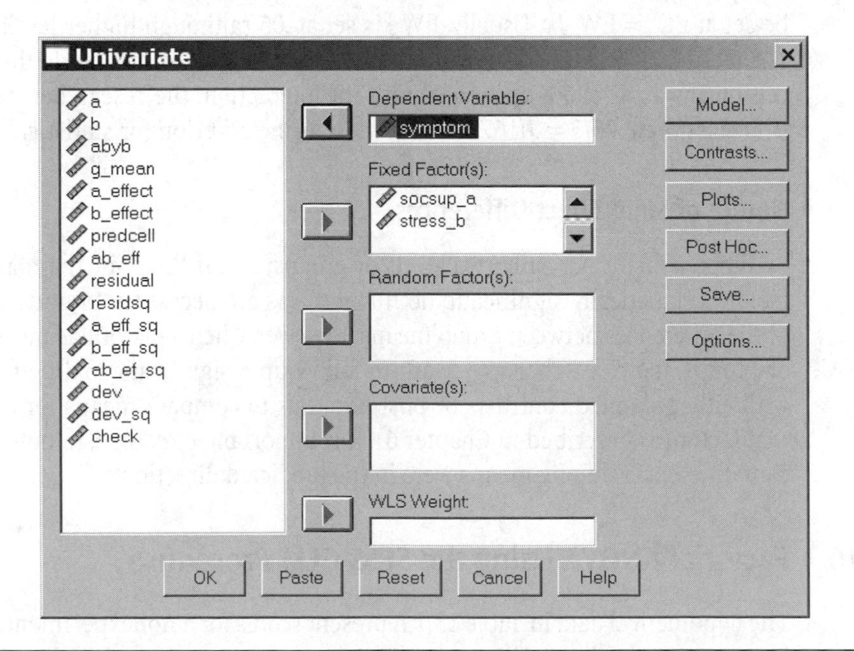

Figure 13.3 ♦ Main Dialog Window for the SPSS GLM Procedure

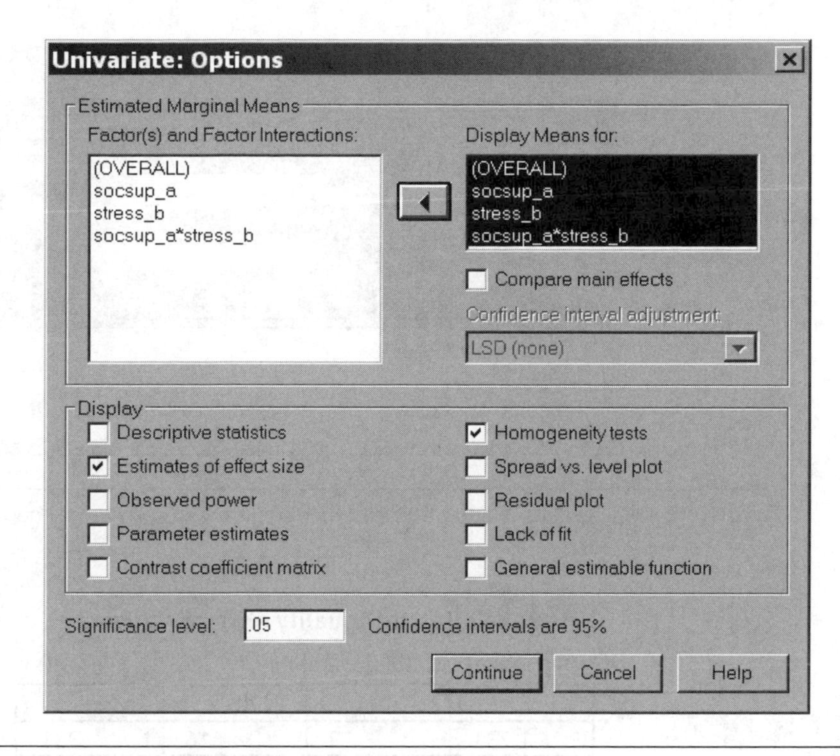

Figure 13.4 ♦ Options Window for the GLM Procedure

variance (the Levene test) is desired, check the box for Homogeneity tests. Close this window by clicking OK.

It is also useful to generate a graph of cell means; to do this, click the Plots button in the main GLM dialog window. This opens up the Plots dialog window, shown in Figure 13.5. To place the stress factor on the horizontal axis of the plot, move this factor name into the Horizontal Axis box; to define a separate line for each level of the social support factor, move this factor name into the Separate Lines box; then, click Add to request this plot.

In addition, a follow-up analysis was performed to compare mean symptoms between the low- versus high-stress groups, separately for the low-social-support group and the high-social-support group. An analysis that compares means on one factor (such as stress) within the groups that are defined by a second factor (such as level of social support) is called an analysis of simple main effects. There are several ways to conduct an analysis of simple main effects. For the following example, the <Data> → <Split File> command was used, and social support was identified as the variable for which separate analyses were requested for each group. An independent samples *t* test was performed to evaluate whether mean symptoms differed between the low- versus high-stress conditions, separately within the low- and high-social-support groups.

Results of the Levene test for the homogeneity of variance assumption are displayed in Figure 13.6. This test did not indicate a significant violation of the homogeneity of variance assumption: $F(3, 16) = 1.23, p = .33$.

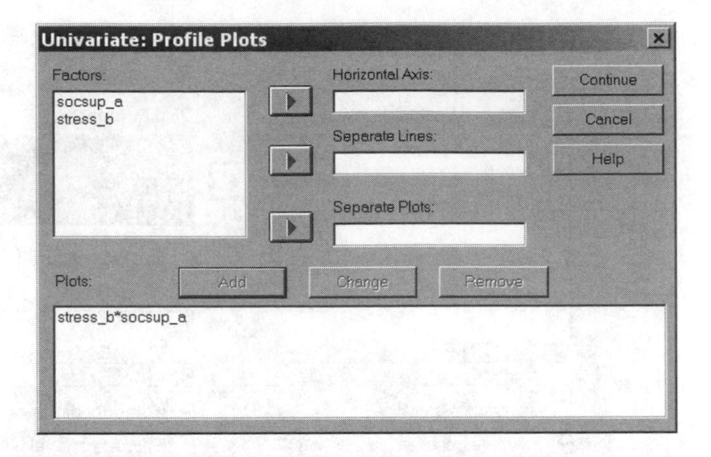

Figure 13.5 ♦ Profile Plots Window for the GLM Procedure

Levene's Test of Equality of Error Variances[a]

Dependent Variable: symptom

F	df1	df2	Sig.
1.231	3	16	.331

Tests the null hypothesis that the error variance of the dependent variable is equal across groups.

a. Design: Intercept+socsup_a+stress_b+socsup_a * stress_b

Figure 13.6 ♦ Levene Test for Homogeneity of Variance Assumption

NOTE: Output from the GLM procedure applied to data in Table 13.1.

The GLM source table that summarizes the factorial ANOVA to predict symptoms from social support and stress appears in Figure 13.7. The line labeled Corrected Model represents an "omnibus" test: This test tells us whether the combined effects of A, B, and $A \times B$ are statistically significant. This omnibus F is rarely reported. The line labeled Intercept tests the null hypothesis that the grand mean for the outcome variable equals 0; this is rarely of interest. (If the outcome variable represented a **change score**, it might be useful to ask whether the mean change score for the overall study differed significantly from 0.) The lines that are labeled with the names of the factors (socsup_A, stress_B, and socsup_A × stress_B) represent the main effects and the interaction; note that the values (of SS, df, MS, and F) reported on each of these lines of the table correspond to the values that were calculated using an Excel spreadsheet earlier in this chapter (refer back to Table 13.4). The line labeled Error corresponds to the term that is usually called "within groups" in textbook descriptions of ANOVA. The SS_{total} that is usually reported in textbook examples, and is used to calculate estimates of effect size, corresponds to the line of the table that SPSS labeled Corrected

Tests of Between-Subjects Effects

Dependent Variable: symptom

Source	Type III Sum of Squares	df	Mean Square	F	Sig.	Partial Eta Squared
Corrected Model	266.950[a]	3	88.983	31.222	.000	.854
Intercept	884.450	1	884.450	310.333	.000	.951
socsup_a	54.450	1	54.450	19.105	.000	.544
stress_b	151.250	1	151.250	53.070	.000	.768
socsup_a * stress_b	61.250	1	61.250	21.491	.000	.573
Error	45.600	16	2.850			
Total	1197.000	20				
Corrected Total	312.550	19				

a. R Squared = .854 (Adjusted R Squared = .827)

Figure 13.7 ♦ Source Table for Factorial ANOVA Output From the SPSS GLM Procedure

NOTE: Data from Table 13.1.

Total (it is called "corrected" because the grand mean was "corrected for" or subtracted from each score before the terms were squared and summed). Finally, the line that SPSS designated Total corresponds to the sum of the squared scores on the outcome variable ($\sum Y^2$); this term is not usually of any interest (unless the researcher wants to test the null hypothesis that the grand mean of scores on the outcome variable Y equals 0).

Tables of means (grand mean, row means, column means, and cell means) appear in Figure 13.8. A graph of cell means (with separate lines to illustrate the outcomes for the low-social-support group, A_1, and the high-social-support group, A_2) is given in Figure 13.9. For this hypothetical dataset, the obtained pattern of cell means was similar to the pattern predicted by the buffering hypothesis (illustrated in Figure 13.2b). An increase in stress (from low to high) was associated with an increase in reported symptoms only for the low-social-support group (A_1). The high-social-support group (A_2) showed very little increase in symptoms as stress increased; this outcome could be interpreted as consistent with the buffering hypothesis illustrated by the graph of cell means in Figure 13.2b. It appears that high social support buffered or protected people against the effects of stress.

In some situations the researcher may want to conduct an analysis of simple main effects—that is, to ask whether the mean level of symptoms differs between the high- and low-stress groups, separately for an analysis of data in the high-social-support and low-social-support groups. The analysis reported in Figure 13.10 provides this information; independent samples t tests were performed to compare cell means, separately for the high-social-support group and the low-social-support group.

In the Results section that follows, simple η^2 effect sizes that were calculated by hand from the sums of squares using Equations 13.24 to 13.26 are reported. Note that these simple η^2 effects are smaller than the partial η^2 effect sizes that appear in the SPSS GLM output.

1. Grand Mean

Dependent Variable: symptom

Mean	Std. Error	95% Confidence Interval	
		Lower Bound	Upper Bound
6.650	.377	5.850	7.450

2. socsup_a

Dependent Variable: symptom

socsup_a	Mean	Std. Error	95% Confidence Interval	
			Lower Bound	Upper Bound
Low Social Support	8.300	.534	7.168	9.432
High Social Support	5.000	.534	3.868	6.132

3. stress_b

Dependent Variable: symptom

stress_b	Mean	Std. Error	95% Confidence Interval	
			Lower Bound	Upper Bound
Low Stress	3.900	.534	2.768	5.032
High Stress	9.400	.534	8.268	10.532

4. socsup_a * stress_b

Dependent Variable: symptom

socsup_a	stress_b	Mean	Std. Error	95% Confidence Interval	
				Lower Bound	Upper Bound
Low Social Support	Low Stress	3.800	.755	2.200	5.400
	High Stress	12.800	.755	11.200	14.400
High Social Support	Low Stress	4.000	.755	2.400	5.600
	High Stress	6.000	.755	4.400	7.600

Figure 13.8 ♦ Tables of Means From the SPSS GLM Procedure

NOTE: Data from Table 13.1.

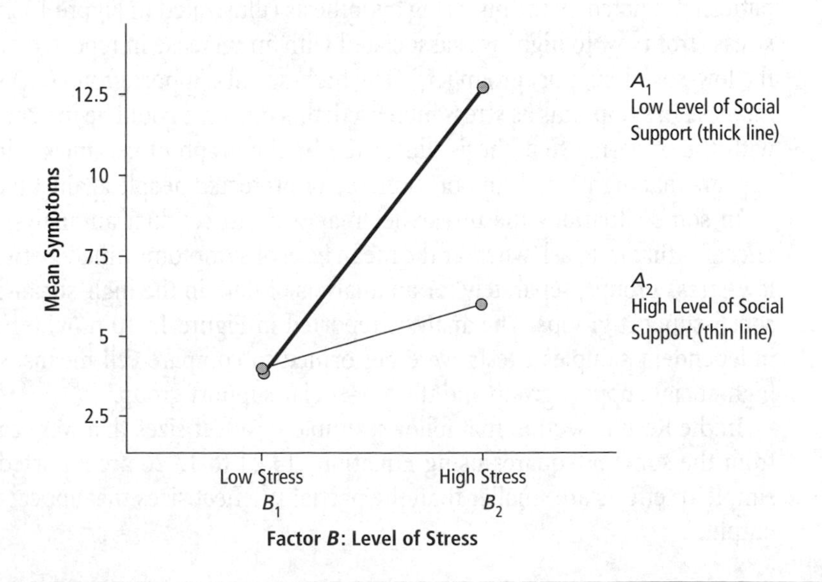

Figure 13.9 ♦ Graph of Cell Means (Profile Plots) From the SPSS GLM Procedure

NOTE: Data from Table 13.1.

SOCSUP_A = low

Group Statistics[a]

	STRESS_B	N	Mean	Std. Deviation	Std. Error Mean
SYMPTOM	low	5	3.80	1.304	.583
	high	5	12.80	2.387	1.068

a. SOCSUP_A = low

Independent Samples Test[a]

		Levene's Test for Equality of Variances		t-test for Equality of Means						
									95% Confidence Interval of the Difference	
		F	Sig.	t	df	Sig. (2-tailed)	Mean Difference	Std. Error Difference	Lower	Upper
SYMPTOM	Equal variances assumed	1.762	.221	-7.398	8	.000	-9.00	1.217	-11.805	-6.195
	Equal variances not assumed			-7.398	6.191	.000	-9.00	1.217	-11.955	-6.045

a. SOCSUP_A = low

SOCSUP_A = high

Group Statistics[a]

	STRESS_B	N	Mean	Std. Deviation	Std. Error Mean
SYMPTOM	low	5	4.00	1.581	.707
	high	5	6.00	1.225	.548

a. SOCSUP_A = high

Independent Samples Test[a]

		Levene's Test for Equality of Variances		t-test for Equality of Means						
									95% Confidence Interval of the Difference	
		F	Sig.	t	df	Sig. (2-tailed)	Mean Difference	Std. Error Difference	Lower	Upper
SYMPTOM	Equal variances assumed	.571	.471	-2.236	8	.056	-2.00	.894	-4.063	.063
	Equal variances not assumed			-2.236	7.529	.058	-2.00	.894	-4.085	.085

a. SOCSUP_A = high

Figure 13.10 ♦ Contrasts Between Cell Means Through Independent Samples *t* Tests

NOTE: The SPSS Split File command was used to obtain *t* tests separately for the low- and high-social-support groups.

Results

A 2 × 2 factorial ANOVA was performed using SPSS GLM to assess whether number of reported symptoms (*Y*) could be predicted from level of social support (A_1 = Low, A_2 = High), level of stress (B_1 = Low, B_2 = High), and the interaction between social support and stress. Based on the buffering hypothesis, it was expected that the high-social-support group (A_2) would show little or no increase in symptoms at higher levels of stress, whereas the low-social-support group (A_1) was predicted to show substantially higher levels of symptoms under high-stress conditions than under low-stress conditions. This was an orthogonal factorial design; each of the four cells had the same number of participants ($n = 5$).

Preliminary data screening was done to assess whether the assumptions for ANOVA were seriously violated. Examination of a histogram of scores on the outcome variable suggested that the symptom scores had a skewed distribution; however, no data transformation was applied. The Levene test indicated no significant violation of the homogeneity of variance assumption.

As predicted, there was a statistically significant social support by stress interaction: $F_{A \times B}$ (1, 16) = 21.49, $p < .001$. The corresponding effect size estimate ($\eta^2 = .20$) indicated a strong effect. The graph of cell means (in Figure 13.9) indicated that the low-social-support/high-stress group had a much higher level of mean symptoms than the other three groups. The pattern of cell means was consistent with the pattern that would be predicted from the buffering hypothesis. For the low-social-support group, the mean number of symptoms was greater in the high-stress condition ($M = 12.8$) than in the low-stress condition ($M = 3.8$). On the other hand, for persons high in social support, symptoms were not much greater in the high-stress condition ($M = 6.0$) than in the low-stress condition ($M = 4.0$).

Planned contrasts were done to assess whether these differences (between low- and high-stress groups) were significant within the low- and high-social-support groups. For the low-social-support group, the difference between means for the low- versus high-stress groups was statistically significant: $t(8) = -7.40$, $p < .001$. For the high-social-support group, the difference between means for the low- versus high-stress groups was not significant: $t(8) = -2.24$, $p = .056$. The nature of the obtained interaction was consistent with the prediction based on the buffering hypothesis. Participants with high levels of social support did not report significantly higher symptoms under high stress; participants with low levels of social support did report significantly higher symptoms under high stress.

There were also significant main effects: for social support, $F_A(1, 16) = 19.10$, $p < .001$, with an associated η^2 effect size estimate of .17; for stress, $F_B(1, 16) = 53.07$, $p < .001$, with an estimated effect size of $\eta^2 = .48$ (an extremely large effect size).

Table 13.6 shows the main effects for social support (α), stress (β), and the social support by stress interaction ($\alpha\beta$). Although the interaction between social support and stress was statistically significant, as predicted by the buffering hypothesis, there were also strong main effects.

13.10.1 ♦ Further Discussion of Results: Comparison of the Factorial ANOVA (in Figures 13.7 and 13.8) With the One-Way ANOVA (in Figure 13.1)

If a researcher does a one-way ANOVA to compare mean symptoms across two levels of social support (as shown earlier in Figure 13.1), and these two groups include participants who are both high and low on social stress, then the differences in symptoms that are associated with stress will tend to make the SS_{within} term relatively large. This, in turn, tends to reduce the F ratio for the effects of social stress. When the researcher blocks on stress by including stress as a factor (as in the factorial ANOVA carried out using the Excel spreadsheet in Table 13.4 and in the corresponding SPSS output in Figure 13.7), each cell of the factorial design is homogeneous on level of stress. The SS_{within} is smaller in the factorial ANOVA than in the one-way ANOVA because in the factorial ANOVA, SS_{within} no longer includes differences in symptoms that are associated with different levels of stress or any variability in symptoms associated with an interaction between stress and social support.

Recall that $MS_{within} = SS_{within}/df_{within}$. When we add a blocking factor (such as stress), SS_{within} may decrease; however, df_{within} will also decrease, because we take a few degrees of freedom away from the within-groups term when we add a main effect and an interaction to our analysis. The net effect of adding a blocking factor to a design can be either an increase or a decrease in MS_{within}, depending on whether the decrease in the SS_{within} is large enough to offset the reduction in df_{within}. If the **blocking variable** is well chosen—that is, if it is responsible for a lot of the within-group variability, then the net effect of blocking is usually a decrease in MS_{within} and an increase in the F ratio for the detection of main effects. In the social support/stress example presented here, the F ratio for the effects of social stress is larger in a factorial ANOVA (where effects of stress are controlled for, or removed from SS_{within}, by blocking) than in the one-way ANOVA on social support (which included the effects of stress-related variations in symptoms as part of SS_{within}).

When we compare the results of the factorial analysis of the data in Table 13.1 with the results of the one-way ANOVA (presented in Figure 13.1), several differences are apparent. In the one-way ANOVA (predicting symptoms from social support, ignoring levels of stress), SS_{within} was 258.1, df_{within} was 18, and MS_{within} was 14.34. The overall F for the effect of social support was $F = 3.80$, and this was not statistically significant.

In the factorial ANOVA (predicting symptoms from social support, stress, and interaction between social support and stress), SS_{within} was 45.60 (much smaller than the SS_{within} in the one-way analysis because the main effect of stress and the effect of the interaction were removed from SS_{within}). The df_{within} was reduced to 16 (one df was taken away by the main effect of stress, and one df was used for the interaction). In the factorial ANOVA, the MS_{within} value was 2.85; this was much smaller than the MS_{within} in the one-way ANOVA. Finally, the factorial ANOVA yielded significant effects for both social support and stress, as well as the interaction between them, whereas the one-way ANOVA did not yield a statistically significant effect for social support. In this example, the effect of adding a second factor (stress) was to reduce the error term MS_{within}, which was used to compute F ratios, as well as to provide new information about the predictive value of stress and an interaction between stress and social support.

This example illustrates one of the potential benefits of a factorial design (and, indeed, of any statistical analysis that includes additional predictor variables). Sometimes the inclusion of an additional factor or predictor variable substantially reduces the size of

SS_{within} (or SS_{error} or $SS_{residual}$, in a regression analysis). The inclusion of additional predictor variables can lead to a reduction of MS_{error} or MS_{within}, unless the reduction of degrees of freedom in the divisor for MS_{within} is large enough to outweigh the reduction in the *SS* term. Thus, the inclusion of new predictor variables sometimes (but not always) results in larger *F* ratios for the tests of the predictive usefulness of other variables in the analysis.

13.11 ♦ Summary

This chapter demonstrated that adding a second factor to ANOVA (setting up a factorial analysis of variance) can provide more information than a one-way ANOVA analysis. When a second factor is added, factorial ANOVA provides information about main effects for two factors and, in addition, information about potential interactions between factors. Also, blocking on a second factor (such as level of stress) can reduce the variability of scores within cells (SS_{within}); this, in turn, can lead to larger *F* ratios for the tests of main effects. Note that the *F* test for the main effect of social support was statistically significant in the factorial ANOVA reported in the Results section, even though it was not statistically significant in the one-way ANOVA reported at the beginning of the chapter. The addition of the stress factor and the interaction term resulted in a smaller value of SS_{within} in the factorial design (relative to the one-way ANOVA). The reduction in degrees of freedom for the error term (*df* was reduced by 2) did not outweigh this reduction in SS_{within}, so in this situation, MS_{within} was smaller in the factorial ANOVA than in the one-way ANOVA, and the *F* ratio for the main effect of social support was larger in the factorial ANOVA than in the one-way ANOVA.

This chapter described the meaning of an interaction in a 2 × 2 factorial design. When an interaction is present, the observed cell means differ from the cell means that would be predicted by simply summing the estimates of the grand mean μ, row effect α_i, and column effect β_j for each cell. The significance test for the interaction allows the researcher to judge whether departures from the pattern of cell means that would be predicted by a simple additive combination of row and column effects are large enough to be judged statistically significant. A major goal of this chapter was to make it clear exactly how the pattern of cell means is related to the presence or absence of an interaction. Description of the nature of an interaction usually focuses on the pattern of cell means (as summarized either in a graph or a table). However, as pointed out by Rosenthal and Rosnow (1991), the observed cell means actually represent a combination of effects (grand mean + row + column + interaction, or $\mu_Y + \alpha_i + \beta_j + \alpha\beta_{ij}$). If the researcher wants to isolate the interaction effect, these row and column effects should be subtracted from the cell means and the remaining $\alpha\beta$ interaction effect terms should be presented (as in Table 13.6). Comparisons among cell means should be interpreted very cautiously because these differences are often partly due to main effects.

We can expand on the basic factorial ANOVA design described in this chapter in many different ways. First of all, we can include more than two levels on any factor; for example, we could run a 2 × 3 or a 4 × 6 factorial ANOVA. If significant *F*s are obtained for main effects on factors that have more than two levels, post hoc comparisons (or planned contrasts) may be used to assess which particular levels or groups differed significantly. We can combine or cross more than two factors. When more than two factors are included in a factorial ANOVA, each one is typically designated by an uppercase letter (e.g., A, B, C, D). A fully crossed three-way factorial $A \times B \times C$ ANOVA combines all levels of A with all levels of B and C. A possible disadvantage of three-way and higher-order factorials is that theories rarely predict three-way (or higher order) interactions and three- or four-way interactions can be difficult to interpret. Furthermore, it may be difficult to fill all the cells (in a $3 \times 2 \times 5$ factorial, e.g., we would need enough participants to fill 30 cells or groups).

Factors can be combined in ways that do not involve crossing. For example, suppose that a researcher wants to show three photographs of female targets and three photographs of male targets in a person perception study. The six individual photos can be included as factors, but in this example, photos are **"nested"** in gender. Photos 1, 2, and 3 belong to the female target group, whereas Photos 4, 5, and 6 belong to the male target group. In this case, we would say that the factor "photo" is nested in the factor target "gender."

We can obtain repeated measures on one or more factors; for example, in a study that assesses the effect of different dosage levels of caffeine on male and female participants, we can expose each participant to every dosage level of the drug. Chapter 20 describes issues in the analysis of repeated measures designs.

We can combine group membership predictors (or factors) with continuous predictors (usually called "covariates") to see whether group means differ when we statistically control for scores on covariates; this type of analysis is called analysis of covariance, or ANCOVA. This is discussed in Chapter 15.

We can measure multiple outcome variables and ask whether the patterns of means on a set of several outcome variables differ in ways that suggest main effects and/or interaction effects; when we include multiple outcome measures, our analysis is called a multivariate analysis of variance, or **MANOVA.**

A complete treatment of all these forms of ANOVA (e.g., nested designs, multiple factors, **mixed models** that include both between-S and repeated measures factors) is beyond the scope of this textbook. More advanced textbooks provide detailed coverage of these various forms of ANOVA.

Appendix: Nonorthogonal Factorial ANOVA (ANOVA With Unbalanced Numbers of Cases in the Cells or Groups)

There are three different possible patterns of *n*s (numbers of cases in groups) in factorial designs, as shown in Table 13.7. Unequal *n*s raise two questions. First, how do the unequal *n*s affect the partition of the variance among effects? (We will see that unequal *n*s can imply some degree of confounding between treatments or group memberships.) And second, how do we handle the unequal *n*s when we combine means across cells? Does it make more sense to use weighted or unweighted combinations of cell means?

First, let's review three possible group size situations in factorial ANOVA:

1. *All cells have equal* ns: This is the situation assumed in most introductory presentations of factorial ANOVA, and it does not create any special problems in data analysis. The data analyst does not need to worry about whether or not to weight group means by cell *n*s and does not need to worry about possible confounds

Table 13.7 ♦ Factorial ANOVAs With Different Arrangements of Cell *ns* in a Study of Gender, by Type of Drug

	$Drug_1$	$Drug_2$
(a) Equal *ns* (orthogonal factorial between gender and drug)		
Male	$n = 50$	$n = 50$
Female	$n = 50$	$n = 50$
(b) Unequal but balanced *ns* (orthogonal factorial)		
Male	$n = 40$	$n = 40$
Female	$n = 60$	$n = 60$
(c) Unequal and not balanced cell *ns* (nonorthogonal factorial)		
Male	$n = 70$	$n = 30$
Female	$n = 30$	$n = 70$

between factors. Table 13.7a shows an example of a factorial design with equal *ns* in all cells.

2. *Cell* ns *are not equal, but they are balanced:* Table 13.7b shows unequal cell *ns*; however, the *ns* are balanced. When we say that *ns* are balanced in a factorial ANOVA, we mean that the proportion of cases in the B_1 and B_2 groups is the same for the A_1 group as it is for the A_2 group. For example, the two groups (Drug 1 vs. Drug 2) in Table 13.7b each have the same proportional membership on the other factor, gender; within each drug group, 60% of the scores are from female, and 40% are from male participants. When cell *ns* are balanced, there is no confound between gender and type of drug; we do not have a situation where people in the Drug 2 group are more likely to be male than people in the Drug 1 group.

   We can index the degree of confounding between *A* and *B* by performing a chi-square test on the $A \times B$ table of cell *ns* in Table 13.7b. Because the proportion of males is the same for both drug groups, the χ^2 for this table of cell *ns* is 0, which indicates that there was no confound or association between group memberships on drug and gender.

   However, even though there is no confound between gender and drug in Table 13.7b, the unequal *ns* of males versus females in this design still pose a potential problem. If we want to estimate the overall population mean response to each drug (for a population with 50% male/50% female members), we may want to adjust or correct for the overrepresentation of female participants in the study. A simple way to do this would be to add the mean for males and mean for females and divide by 2, to obtain an "unweighted" overall mean for each drug. (See the section below on computation of weighted vs. unweighted means.)

3. *Cell* ns *are unequal and also not balanced:* Table 13.7c shows a situation in which the cell *ns* are not balanced; that is, the proportion of males in Drug Group 1 is not the same as the proportion of males in Drug Group 2. In this example, 70% of the people who received Drug 1 were male, and only 30% of the people who received

Drug 2 were male. The proportion of males is not balanced (or equal) across levels of drug; there is a confound between drug and gender (i.e., most of the people who got Drug 1 are male, whereas most of the people who received Drug 2 are female).

If we average the responses of all participants who received Drug 1, the overall mean for Drug 1 will disproportionately reflect the response of males (because they are a majority of the cases in this drug group). The data analyst needs to recognize that a confound between factors in a factorial design creates the same kind of problem in variance partitioning in factorial ANOVA as that discussed in Chapter 11 for multiple regression. When factors are not orthogonal, the group membership predictor variables are correlated, and therefore, to some extent, they compete to explain the same variance. When we have a nonorthogonal factorial ANOVA (as in Table 13.7c), we need to adjust our estimates of the *SS* or η^2 for each effect (e.g., the *A* effect) to correct for confounding or correlation between factors.

In most experimental research situations, of course, researchers design their studies so that the cells have equal *n*s of cases. In the laboratory, researchers can artificially arrange group memberships so that there is no confound or correlation between the independent variables or factors.

Partition of Variance in Orthogonal Factorial ANOVA

If the *n*s in the groups are equal (or balanced), we have an orthogonal ANOVA; the two independent variables *A* and *B* are uncorrelated—that is, not confounded with each other; in addition, the main effects of *A* and *B* are orthogonal to, or uncorrelated with, the $A \times B$ interaction term. In an orthogonal factorial ANOVA, we can diagram the partitioning of SS_{total} for *Y* as shown in Figure 13.11.

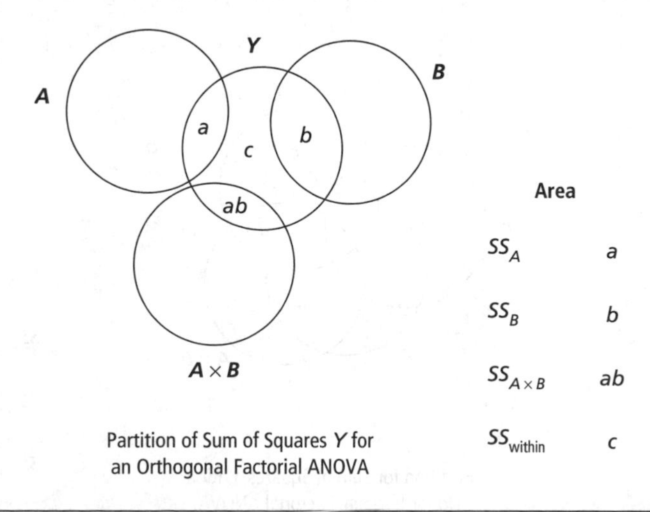

Figure 13.11 ♦ Diagram Showing Partition of Variance Among *A*, *B*, and $A \times B$ Effects in an Orthogonal Factorial ANOVA

NOTE: Equal or balanced *n*s in the cells.

The total area of the circle that represents Y corresponds to SS_{total}, and the overlap between the circles that represent Factor A and Y is equivalent to SS_A. The proportion of explained variance (represented by the overlap between Y and A) is equivalent to η^2_A ($\eta^2_A = SS_A/SS_{total}$).

When we have an orthogonal factorial design, there is no confound between the group memberships on the A and B factors; the predictive Factors A and B do not compete to explain the same variance. The $A \times B$ interaction is also orthogonal to (uncorrelated with) the main effects of the A and B factors. Therefore, when we diagram the variance-partitioning situation for a factorial ANOVA (as in Figure 13.11), there is no overlap between the circles that represent A, B, and the $A \times B$ interaction. Because these group membership variables are independent of each other in an orthogonal design, there is no need to statistically control for any correlation between predictors.

Partition of Variance in Nonorthogonal Factorial ANOVA

When the *ns* in the cells are not balanced, it implies that group memberships (on the A and B factors) are not independent; in such situations, there is a confound or correlation between the A and B factors, and probably also between the main effects and the $A \times B$ interaction. In a nonorthogonal factorial ANOVA, the predictor variables (factors) are correlated; they compete to explain some of the same variance. The variance-partitioning problem in a nonorthogonal factorial ANOVA is illustrated by the diagram in Figure 13.12.

Recall that in the earlier discussion (in Chapter 11) of multiple regression, we had examined a similar problem. When we wanted to predict scores on Y from intercorrelated continuous predictor variables X_1 and X_2, we needed to take into account the overlap in variance that could be explained by these variables. In fact, we can use variance-partitioning strategies similar to those that can be used in multiple regression to compute the *SS* terms in a nonorthogonal factorial ANOVA. Chapter 14 will show that there are several ways to handle the problem of variance partitioning in regression analysis, and the same logic can be used to evaluate variance partitioning in nonorthogonal factorial ANOVA.

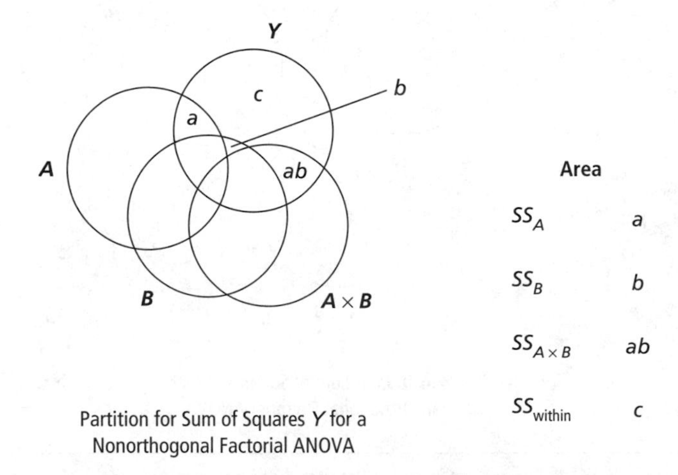

Figure 13.12 ♦ Diagram Showing Partition of Variance Among A, B, and $A \times B$ Effects in a Nonorthogonal Factorial ANOVA (With Unbalanced Cell *ns*) When Type III SS Computations Are Used

Briefly, we can use one of these two strategies for variance partitioning. We can assess the sum of squares or proportion of variance explained by each predictor variable while controlling for *all* other predictor variables and interaction terms in the model. To obtain this type of variance partitioning, we do a **"standard"** or **"simultaneous" regression;** that is, we enter all predictor variables into the regression in one step. In GLM, to obtain the type of variance partitioning that corresponds to a standard or simultaneous regression, we request the SS Type III method of computation. Alternatively, we can enter predictors into the predictive model in steps (either one variable or a group of variables per step); and we can assess the predictive contribution of each independent variable while controlling for only variables that were entered into the analysis in the same step and in previous steps. This corresponds to regression analysis called hierarchical or statistical (or less precisely, "stepwise"). To obtain this type of variance partitioning in SPSS GLM, we request the **SS Type I** method of computation; this is explained in Chapter 14.

When we request the SS Type III computation in GLM, we obtain an *SS* value for Factor *A* that has any variance that Factor *A* shares with other predictors such as Factor *B* or the $A \times B$ interaction partialled out or removed from SS_A. When we request the SS Type I computation in GLM, the independent variables in the model are entered in a user-specified order, and each predictor is assessed while controlling only for predictors that were entered in earlier steps. If the factorial design is orthogonal, these two methods produce identical results. By default, SPSS GLM uses the SS Type III computation; you click on the **Custom Models** button in the main GLM dialog box to select other types of computation methods for sums of squares, such as SS Type I, and to specify the order of entry of predictors and indicate whether some terms such as interactions should be included or excluded from the model.

Other optional methods of computing sums of squares can be requested from the SPSS GLM Custom Model (SS Types II and IV). These are more complicated and are rarely used.

Weighted Versus Unweighted Estimates of Row, Column, and Grand Means

Unequal *n*s (in the cells or within the groups defined by each factor) can create another potential problem for data analysts: When cell *n*s are unequal, there is more than one possible way to combine means across groups. For example, when we compute a row mean by combining data from the cells in one row of a table and the cell means are based on different *n*s, there are three ways we could compute an overall mean for the row of the table:

1. We can sum all the individual scores within a row and divide by the total *n* of cases in that row. This results in a row mean that is "weighted" by sample size; that is, the overall row mean will be closer to the mean of the cell that had a larger *n*.

2. We can multiply each cell mean by its corresponding *n* (this just reproduces the sum of the scores in the cell), add these products across cells, and divide by the total *n* of subjects in all the cells that are being combined. This also produces a "weighted" estimate of the mean. With a little thought, you should realize that this method is actually identical to Method 1. Whether we use Method 1 or 2 to compute a **weighted mean**, we are "weighting" each cell mean by its corresponding *n*; thus, the results from cells with larger *n*s are counted or weighted more heavily in the computation of the overall row mean.

Table 13.8 ♦ Cell Means and Cell *ns* to Illustrate Different Outcomes for Weighted Versus Unweighted Means

	B_1 (Nonsmokers)	B_2 (Pipe Smokers)
A_1 (males)	$M_{AB_{11}} = 77$ $n_{AB_{11}} = 40$	$M_{AB_{12}} = 67$ $n_{AB_{12}} = 20$
A_2 (females)	$M_{AB_{21}} = 84$ $n_{AB_{21}} = 48$	$M_{AB_{22}} = 74$ $n_{AB_{22}} = 2$

NOTE: Dependent variable: mean age at death.

3. We can add the cell means in a row, and divide by the number of means. This yields an "unweighted" (or "equally weighted") estimate of the row mean (Table 13.8). This version of the row mean gives the same weight to the cell mean from each group regardless of the *ns* in the groups.

When the groups in a factorial ANOVA have unequal *ns*, the estimates of row and column means that are obtained using weighted versus unweighted means can differ substantially. Consider the following example based on a hypothetical table of means with unequal *ns* in the cells. Mean age at death is reported in a study where the row factor was gender (A_1 = Male, A_2 = Female) and the column factor was smoking behavior (B_1 = Nonsmoker, B_2 = Pipe smoker). Because there are few female pipe smokers, the *n* for the AB_{22} cell was much lower than the *ns* for the other three cells in this 2×2 design; this is an example of a nonorthogonal factorial design. The *weighted* mean age at death for all pipe smokers is computed as follows:

$$M_{B_2} = \frac{n_{AB_{12}} \times M_{AB_{12}} + n_{AB_{22}} \times M_{AB_{22}}}{n_{AB_{12}} + n_{AB_{22}}} = \frac{(20 \times 67) + (2 \times 74)}{20 + 2} = 67.6. \quad (13.28)$$

Note that when we weight for the size of cell *ns*, the overall column mean is closer to the cell mean that was associated with a larger *n*; in this case, the unweighted mean for all pipe smokers (67.6) was much closer to the mean for male pipe smokers ($n = 20, M = 67$) than to the mean for female pipe smokers ($n = 2, M = 74$).

The *unweighted* mean age at death for all pipe smokers is computed by just averaging the cell means:

$$M_{B_1} = \frac{M_{AB_{12}} + M_{AB_{22}}}{2} = \frac{67 + 74}{2} = 70.5. \quad (13.29)$$

Note that this unweighted mean is halfway between the means for the male and female pipe-smoking groups; that is, information from male participants was not given greater weight or importance in computing this version of the mean. If unequal cell *ns* arise because there really *is* a confound between the *A* and *B* factors in the natural environment (as in this example involving gender and pipe smoking), then it can be misleading to ignore this natural confound. It might be preferable to report a weighted average in this example, because then the weighted mean age at death for pipe smokers in our study would be consistent with the fact that most pipe smokers (out in the real world) are also male. On the other hand, we can't easily separate the effects of gender and smoking in this example. The low mean age at death for pipe smokers is at least partly due to gender effect (most pipe smokers were male, and men tend to die younger than women).

When we calculate an unweighted mean across cells, the unweighted row mean is midway between the two cell means. We "unconfound" the effects of confounded factors when we look at an unweighted mean. The unweighted mean provides an estimate of what the column mean would have been if the *ns* in the male and female groups had been equal (i.e., if there were no confound between gender and pipe smoking). When unequal *ns* arise accidentally, as in laboratory experiments with missing data, it may be preferable to use unweighted means when we combine information across groups.

Should we use weighted or unweighted cell means? There is no easy answer to this question. If the difference in the cell *ns* occurred "by accident" and does not correspond to some real confound between the *A* and *B* factors out in the real world, then it may make sense to ignore the unequal *ns* and use unweighted means. On the other hand, when the confound in our data corresponds to a confound that exists in the real world, we may prefer to report weighted means.

Beginning data analysts may prefer to avoid these difficulties by collecting data in ways that result in equal *ns* in the cells. Once we begin to work with unequal *ns* and nonorthogonal designs, we have to think carefully whether weighted or unweighted means make more sense. Also, when we work with nonorthogonal factorial designs, we need to be aware that the SS_A, SS_B, and $SS_{A \times B}$ terms need to be adjusted for possible correlations among predictors. SPSS uses a default method of computation for *SS* terms (called Type III SS) that automatically controls for any confounds among factors. Type III SS involves a method of variance partitioning that is equivalent to a standard or simultaneous multiple regression (see Chapter 14 in this book); each effect is tested while statistically controlling for the other effects in the model. For example, the estimate of SS_A does not include any variance that could be explained by SS_B or $SS_{A \times B}$.

Notes

1. Here is an algebraic statement of the null hypothesis that there is no $A \times B$ interaction:

$$H_0 = \alpha\beta_{ij}, \text{ for all values of } i \text{ and } j.$$

If $\alpha\beta_{ij} = 0$ for all cells, it implies that scores on *Y* can be reproduced reasonably well by a model that includes only additive main effects for the *A* and *B* factors: $Y_{ij} = \mu_Y + \alpha_i + \beta_j + \varepsilon_{ij}$.

Cell means generated by this type of additive (no interaction) model yield parallel lines when cell means are plotted.

2. It may seem painfully obvious that the total number of cells should not be too large; but a student once asked me why SPSS generated error messages for a $2 \times 3 \times 2 \times 5 \times 3 \times 2 \times 2$ factorial ANOVA. Of course, the *ns* in the cells were unequal, the number of cases in most cells was very small, and in fact, many cells were empty. (Even if this analysis had run, it would have been impossible to interpret the four-, five-, six-, and seven-way interactions.)

+ − × ÷

Comprehension Questions

1. Consider the following actual data from a study by Lyon and Greenberg (1991). The first factor in their factorial ANOVA was family background; female participants were classified into two groups (Group 1: codependent, women with an alcoholic parent; Group 2: non-codependent, women with nonalcoholic parents). Members of these two groups were randomly assigned to one of two conditions; they were asked to donate time to help a man who was described to them as either Mr. Wrong (exploitative, selfish, and dishonest) or Mr. Right (nurturant, helpful). The researchers predicted that women from a non-codependent/nonalcoholic family background would be more helpful to a person described as nurturant and helpful, whereas women from a codependent/alcoholic family background would be more helpful to a person described as needy, exploitative, and selfish.

   The table of means below represents the amount of time donated in minutes in each of the four cells of this 2 × 2 factorial design. In each cell, the first entry is the mean, and the standard deviation is given in parentheses. The *n* in each cell was 12.

	B_1, *Mr. Wrong*	B_2, *Mr. Right*
A_1 (codependent family background)	133.84 (54.24)	12.50 (29.88)
A_2 (non-codependent family background)	0.00 (0.00)	60.00 (70.06)

   The reported *F* ratios were as follows:

   $F_A(1, 44) = 9.89, p < .003.$

   $F_B(1, 44) = 4.99, p < .03.$

   $F_{A \times B}(1, 44) = 43.64, p < .0001.$

   a. Calculate an η^2 effect size for each of these effects (*A* and *B* main effects and the $A \times B$ interaction). (Recall that $\eta^2 = df_{\text{between}} \times F/[df_{\text{between}} \times F + df_{\text{within}}]$.)
   b. Calculate the row means, column means, and grand mean from these cell means.
   c. Calculate the α_i and β_j effects for each level of *A* and *B*.
   d. For each cell, calculate the $\alpha\beta_{ij}$ interaction effect.
   e. Set up a table that summarizes these effects (similar to Table 13.6).
   f. Write up a Results section that presents these findings and provides an interpretation of the results.
   g. What were the values of the 12 individual scores in the A_2/B_1 group? How did you arrive at them?

Comprehension Questions

2. Do the following analyses using the hypothetical data below. In this imaginary experiment, participants were randomly assigned to receive either no caffeine ("1") or 150 mg of caffeine ("2") and to a no-exercise condition ("1") or half an hour of exercise on a treadmill ("2"). The dependent variable was heart rate in beats per minute.

Factorial Design to Assess Effects of Caffeine and Exercise on Heart Rate

Caffeine	*Exercise*	*Heart Rate*
1	1	65
1	1	70
1	1	75
1	1	60
1	2	75
1	2	80
1	2	85
1	2	70
2	1	80
2	1	85
2	1	90
2	1	75
2	2	90
2	2	95
2	2	100
2	2	85

a. Graph the distribution of heart rate scores (as a histogram). Is this distribution reasonably normal? Are there any outliers?
b. Compute the row, column, and grand means by hand. Set up a table that shows the mean and *n* for each group in this design.
c. Calculate the α, β, and $\alpha\beta$ effect estimates, and report these in a table (similar to Table 13.6).
d. Comment on the pattern of effects in the table you reported in (c). Do you see any evidence of possible main effects and/or interactions?
e. Set up an Excel or SPSS worksheet to break each score down into components, similar to the spreadsheet shown in Table 13.4.
f. Sum the squared α, β, and $\alpha\beta$ effects in the spreadsheet you just created to find your estimates for SS_A (main effects of caffeine), SS_B (main effects of exercise), and $SS_{A \times B}$ (interaction between caffeine and exercise).
g. Run a factorial ANOVA using the SPSS GLM procedure. Verify that the values of the *SS* terms in the SPSS GLM printout agree with the *SS* values you obtained from your spreadsheet. Make sure that you request cell means, a test of homogeneity of variance, and a plot of cell means (as in the example in this chapter).

h. What null hypothesis is tested by the Levene statistic? What does this test tell you about possible violations of an assumption for ANOVA?

i. Write up a Results section. What conclusions would you reach about the possible effects of caffeine and exercise on heart rate? Is there any indication of an interaction?

3. Write out an equation to represent a 2 × 2 factorial ANOVA (make up your own variable names); include an interaction term in the equation.

4. Consider these two tables of cell means. Which one shows a possible interaction, and which one does not show any evidence of an interaction?

Table 1

	B_1	B_2
A_1	10	15
A_2	20	25

Table 2

	B_1	B_2
A_1	10	25
A_2	18	18

Based on these two tables, show the following:

a. When an interaction is not present, a cell mean is just the linear combination of the row effect for *A* and the column effect for *B*.

b. When an interaction is present, a cell mean is something other than just the linear combination of these effects.

5. What information do you need to judge the statistical significance and the effect size for each of these sources of variance: *A*, *B*, and *A* × *B*?

6. If there is a statistically significant interaction, how can you remove the main effects of *A* and *B* from the cell means to see the interaction effects alone, more clearly?

The following questions can be answered only if you have read the appendix to this chapter:

7. Consider the following three tables; the value in each cell is the *N* of observations for that group. Which one of these tables represents a nonorthogonal factorial ANOVA?

Table 1

	B_1	B_2
A_1	10	10
A_2	10	10

Table 2

	B_1	B_2
A_1	15	30
A_2	30	60

Table 3

	B_1	B_2
A_1	20	40
A_2	60	10

8. How is a weighted average of group means different from an unweighted average of group means? What is the weighting factor? Under what circumstances would each type of mean be preferred?
9. Using overlapping circles to represent shared variance, show how the partition of variance differs between these two situations:
   a. An orthogonal factorial ANOVA
   b. A nonorthogonal factorial ANOVA
   (Which of these, *a* or *b*, is more similar to the usual situation in a multiple regression analysis using nonexperimental data?)
10. How does Type III SS differ from Type I SS in GLM?

CHAPTER 14

Multiple Regression With More Than Two Predictors

14.1 ♦ Research Questions

The extension of multiple regression to situations in which there are more than two predictor variables is relatively straightforward. The raw score version of a two-predictor multiple regression equation (as described in Chapter 11) was written as follows:

$$Y' = b_0 + b_1X_1 + b_2X_2. \tag{14.1}$$

The raw score version of a regression equation with k predictor variables is written as follows:

$$Y' = b_0 + b_1 X_1 + b_2 X_2 + \ldots + b_k X_k. \tag{14.2}$$

In Equation 14.1, the b_1 slope represents the predicted change in Y for a one-unit increase in X_1, controlling for X_2. When there are more than two predictors in the regression, the slope for each individual predictor is calculated controlling for *all* other predictors; thus, in Equation 14.2, b_1 represents the predicted change in Y for a one-unit increase in X_1, controlling for $X_2, X_3, \ldots, X_k$ (i.e., controlling for all other predictor variables included in the regression analysis). For example, a researcher might predict first-year medical school grade point average (Y) from a set of several predictor variables such as college grade point average (X_1), Medical College Admissions Test (MCAT) physics score (X_2), MCAT biology score (X_3), quantitative evaluation of the personal goals statement on the application (X_4), a score on a self-reported empathy scale (X_5), and so forth. One goal of the analysis may be to evaluate whether this entire set of variables is sufficient information to predict medical school performance; another goal of the analysis may be to identify which of these variables are most strongly predictive of performance in medical school.

The standard score version of a regression equation with k predictors is represented as follows:

$$z'_Y = \beta_1 z_{X_1} + \beta_2 z_{X_2} + \cdots + \beta_k z_{X_k}. \tag{14.3}$$

The beta coefficients in the standard score version of the regression can be compared across variables to assess which of the predictor variables are more strongly related to the *Y* outcome variable when all the variables are represented in *z* score form. (This comparison must be interpreted with caution, for reasons discussed in Chapter 11; beta coefficients, like correlations, may be influenced by many types of artifacts such as unreliability of measurement and restricted range of scores in the sample.)

We can conduct an overall or omnibus significance test to assess whether the entire set of all *k* predictor variables significantly predicts scores on *Y*; we can also test the significance of the slopes, b_i, for each individual predictor to assess whether each X_i predictor variable is significantly predictive of *Y* when all other predictors are statistically controlled.

In addition, the inclusion of more than two predictor variables in a multiple regression can serve the following purposes (based on a suggestion from Dr. Michael D. Biderman):

1. A regression that includes several predictor variables can be used to evaluate theories that include several variables that—according to theory—predict or influence scores on the outcome variable.

2. In a regression with more than two predictors, it is possible to assess the predictive usefulness of an X_i variable that is of primary interest while statistically controlling for more than one extraneous variable. As seen in Chapter 10, when we control for "other" variables, the apparent nature of the relation between X_i and *Y* can change in many different ways.

3. Sometimes a better prediction of scores on the *Y* outcome variable can be obtained by using more than two predictor variables. However, we should beware the "kitchen sink" approach to selection of predictors. It is not a good idea to run a regression that includes 10 or 20 predictor variables that happen to be strongly correlated with the outcome variable in the sample data; this approach increases the risk of Type I error. It is preferable to have a rationale for the inclusion of each predictor; each variable should be included (a) because a well-specified theory says it could be a "causal influence" on *Y*, (b) because it is known to be a useful predictor of *Y*, or (c) because it is important to control for the specific variable when assessing the predictive usefulness of other variables, because the variable is confounded with or interacts with other variables, for example.

4. When we use dummy predictor variables to represent group membership (as in Chapter 12), and the categorical variable has more than four levels, we need to include more than two dummy predictor variables to represent group membership.

5. In a regression with more than two predictor variables, we can use X^2 and X^3 (as well as *X*) to predict scores on a *Y* outcome variable; this provides us with a way to test for curvilinear associations between *X* and *Y*.

Two new issues are addressed in this chapter. First, when we expand multiple regression to include *k* predictors, we need a general method for the computation of β and *b* coefficients that works for any number of predictor variables. These computations can be represented using matrix algebra; however, the reader does not need a background in matrix algebra to understand the concepts involved in the application of multiple regression. This chapter provides an intuitive description of the computation of regression coefficients; students who want to understand these computations in more detail will find a brief introduction to the matrix algebra for multiple regression in Appendix 14.A at the end of this chapter. Second, there are several different methods for entry of predictor variables into multiple regression. These methods use different logic to partition the variance in the Y outcome variable among the individual predictor variables. Subsequent sections of this chapter describe these three major forms of order of entry in detail:

1. *Simultaneous or standard regression:* All the X predictor variables are entered in one step.
2. *Hierarchical or* ***sequential*** *or* ***user-determined regression:*** X predictor variables are entered in a series of steps, with the order of entry determined by the data analyst.
3. ***Statistical*** *or* ***data-driven regression:*** The order of entry is based on the predictive usefulness of the individual X variables.

Both (2) and (3) are sometimes called "stepwise" regression. However, in this chapter, the term ***stepwise*** will be used in a much narrower sense, to identify one of the options for statistical or data-driven regression that is available in the SPSS regression program.

In this chapter, all three of these approaches to regression will be applied to the same data analysis problem. In general, the simultaneous approach to regression is preferable: It is easier to understand, and all the predictor variables are given equal treatment. In standard or simultaneous regression, when we ask the question, "What other variables were statistically controlled while assessing the predictive usefulness of the X_i predictor?" the answer is always "All the other X predictor variables." In a sense, then, all predictor variables are treated equally; the predictive usefulness of each X_i predictor variable is assessed controlling for *all* other predictors.

On the other hand, when we use hierarchical or statistical regression analysis, which involves running a series of regression equations with one or more predictor variables added at each step, the answer to the question, "What other variables were statistically controlled while assessing the predictive usefulness of the X_i predictor variable?" is "Only the other predictor variables entered in the same step or in previous steps." Thus, the set of "statistically controlled variables" differs across the X_i predictor variables in hierarchical or statistical regression (analyses in which a series of regression analyses are performed). Predictor variables in sequential or statistical regression are treated "differently" or "unequally"; that is, the contributions for some of the X_i predictor variables are assessed controlling for none or few other predictors, while the predictive contributions of other variables are assessed controlling for most, or all, of the other predictor variables. Sometimes it is possible to justify this "unequal" treatment of variables based on theory or temporal priority of the variables, but sometimes the decisions about order of entry are arbitrary.

Direct or standard or simultaneous regression (i.e., a regression analysis in which all predictor variables are entered in one step) usually, but not always, provides a more conservative assessment of the contribution made by each individual predictor. That is, usually the proportion of variance that is attributed to an X_i predictor variable is smaller when that variable is assessed in the context of a direct or standard or simultaneous regression, controlling for *all* the other predictor variables, than when the X_i predictor variable is entered in an early step in a hierarchical or statistical method of regression (and therefore is assessed controlling for only a subset of the other predictor variables).

The statistical or data-driven method of entry is not recommended because this approach to order of entry often results in inflated risk of Type I error (variables that happen to have large correlations with the Y outcome variable in the sample, due to sampling error, tend to be selected earliest as predictors). This method is included here primarily because it is sometimes reported in journal articles. Statistical or data-driven methods of entry yield the largest possible R^2 using the smallest number of predictor variables within a specific sample, but they often yield analyses that are not useful for theory evaluation (or even for prediction of individual scores in different samples).

14.2 ♦ Empirical Example

The hypothetical research problem for this chapter involves prediction of scores on a physics achievement test from the following predictors: intelligence quotient (IQ), emotional intelligence (EI), verbal Scholastic Ability Test (VSAT), math SAT (MSAT), and gender (coded 1 = Male, 2 = Female). The first question is, How well are scores on physics predicted when this entire set of five predictor variables is included? The second question is, How much variance does each of these predictor variables uniquely account for? This second question can be approached in three different ways, using standard, hierarchical, or statistical methods of entry. Table 14.1 shows hypothetical data for 200 participants (100 male, 100 female) on these six variables—that is, the five predictor variables and the score on the dependent variable. (Note that because five subjects have missing data on physics, the actual N in the regression analyses that follow is $N = 195$.)

14.3 ♦ Screening for Violations of Assumptions

As the number of variables in analyses increases, it becomes increasingly time-consuming to do a thorough job of preliminary data screening. Detailed data screening will no longer be presented for the empirical examples from this point onward because it would require a great deal of space; instead, there is only a brief description of the types of analyses that should be conducted for preliminary data screening.

First, for each predictor variable (and the outcome variable), you need to set up a histogram to examine the shape of the distribution of scores. Ideally, all quantitative variables (and particularly the Y outcome variable) should have approximately normal distribution shapes. If there are extreme outliers, the researcher should make a thoughtful decision whether to remove or modify these scores (see Chapter 4 for discussion of outliers). If there are dummy-coded predictors, the two groups should ideally have approximately equal *ns*, and in any case, no group should have fewer than 10 cases.

Second, a scatter plot should be obtained for every pair of quantitative variables. The scatter plots should show a linear relation between variables, homogeneous variance (for the variable plotted on the vertical axis) at different score values (of the variable plotted on the horizontal axis), and no extreme bivariate outliers.

Table 14.1 ♦ Data for SPSS Regression Examples in Chapter 14

ID	*Age*	*Gender*	*EI*	*IQ*	*VSAT*	*MSAT*	*Create*	*Abstract*	*Physics*
1	17	1	96	102	392	487	8	10	93
2	18	1	106	112	593	538	15	12	103
3	18	1	92	91	364	458	7	13	88
4	18	1	102	126	680	550	15	13	92
5	17	1	95	120	616	505	12	12	84
6	17	1	111	84	522	467	4	8	75
7	19	1	94	91	450	457	9	9	75
8	17	1	113	134	740	569	18	10	103
9	17	1	103	102	558	515	6	11	83
10	18	1	83	112	519	467	14	15	76
11	17	1	109	91	516	477	8	9	64
12	17	1	116	83	471	442	3	7	66
13	17	1	106	100	485	464	9	7	78
14	18	1	105	99	491	542	12	10	108
15	18	1	89	104	505	511	16	8	97
16	18	1	112	86	551	463	6	11	71
17	17	1	92	103	495	508	11	12	89
18	17	1	102	104	561	552	15	10	117
19	18	1	102	85	366	441	5	2	80
20	17	1	109	78	354	442	9	9	91
21	18	1	100	79	522	439	4	4	64
22	17	1	85	112	531	481	8	6	90
23	17	1	130	93	578	457	7	10	55
24	19	1	90	88	444	486	6	2	102
25	18	1	97	75	314	446	6	4	83
26	18	1	98	88	433	492	7	9	86
27	17	1	124	107	712	546	16	11	97
28	17	1	97	105	587	521	6	11	87
29	17	1	109	105	600	558	11	10	101
30	18	1	96	83	534	455	9	7	77
31	17	1	116	99	566	496	9	4	93
32	18	1	111	95	496	500	7	5	86
33	17	1	103	82	428	475	15	8	87
34	17	1	121	108	624	527	8	13	75
35	18	1	84	90	406	465	7	12	76
36	18	1	106	90	451	469	7	5	86
37	18	1	109	101	541	498	11	10	86
38	17	1	121	83	562	481	7	6	75
39	19	1	130	104	580	452	7	12	62
40	18	1	98	86	471	482	5	9	.
41	17	1	90	109	479	511	16	8	110
42	17	1	115	75	427	443	11	2	83
43	17	1	106	119	497	564	14	10	119
44	19	1	102	72	392	461	5	2	76
45	17	1	106	92	463	513	5	12	89
46	16	1	103	98	490	466	9	9	65
47	17	1	102	104	446	502	9	11	112
48	17	1	109	121	688	553	15	15	98
49	17	1	100	107	609	512	13	9	66
50	17	1	96	101	449	500	9	8	81

ID	*Age*	*Gender*	*EI*	*IQ*	*VSAT*	*MSAT*	*Create*	*Abstract*	*Physics*
51	18	1	110	128	655	594	17	15	117
52	18	1	98	90	469	444	13	10	84
53	16	1	87	96	441	480	9	9	97
54	18	1	109	104	518	485	6	12	66
55	18	1	114	90	588	421	9	7	56
56	17	1	114	93	536	500	10	10	82
57	18	1	91	84	397	490	7	10	88
58	18	1	103	98	518	491	14	9	81
59	18	1	106	119	594	583	13	13	111
60	16	1	115	99	654	461	10	7	73
61	17	1	102	121	590	579	14	8	113
62	18	1	86	102	352	534	5	12	119
63	17	1	83	96	358	458	13	11	76
64	17	1	107	81	455	492	8	11	83
65	18	1	113	103	518	499	12	10	96
66	18	1	109	101	607	492	16	8	84
67	19	1	86	68	322	444	1	1	91
68	18	1	75	119	484	550	16	12	118
69	18	1	114	114	579	519	10	10	97
70	17	1	107	94	557	475	10	7	66
71	18	1	101	115	523	573	10	12	117
72	17	1	106	99	561	466	10	7	83
73	18	1	100	79	354	487	3	7	92
74	18	1	96	86	411	517	2	7	95
75	17	1	81	114	508	502	15	11	79
76	17	1	108	83	570	455	8	4	65
77	17	1	115	100	712	474	13	9	54
78	17	1	103	86	415	450	8	8	80
79	17	1	.	.	.	476	9	6	.
80	17	1	104	94	435	492	9	10	95
81	17	1	87	109	507	548	18	10	102
82	17	1	103	108	676	597	16	12	102
83	18	1	98	125	558	561	11	14	118
84	17	1	93	108	518	448	11	14	61
85	18	1	108	94	444	471	9	13	69
86	18	1	117	77	555	446	7	9	54
87	17	1	82	107	473	491	10	8	79
88	18	1	85	114	503	473	9	7	63
89	17	1	114	110	733	544	11	13	94
90	18	1	85	96	395	498	8	11	98
91	18	1	79	92	374	497	13	10	96
92	17	1	87	118	485	474	15	11	82
93	18	1	96	82	459	449	8	11	62
94	18	1	98	80	425	444	7	9	78
95	17	1	86	94	423	457	4	9	73
96	17	1	116	88	658	513	12	11	73
97	17	1	102	109	673	516	8	12	79
98	17	1	104	57	339	444	3	2	83
99	18	1	105	126	626	587	16	11	118
100	17	1	86	112	463	508	14	14	93

(Continued)

Table 14.1 ♦ (Continued)

ID	*Age*	*Gender*	*EI*	*IQ*	*VSAT*	*MSAT*	*Create*	*Abstract*	*Physics*
101	16	2	109	105	474	450	6	9	87
102	18	2	121	107	639	481	9	6	53
103	18	2	120	101	554	504	14	11	78
104	17	2	92	132	562	574	10	14	102
105	17	2	120	88	492	494	13	5	72
106	17	2	117	98	582	522	9	11	82
107	19	2	118	85	471	451	6	5	62
108	18	2	108	103	653	534	8	12	74
109	19	2	115	107	605	.	8	12	.
110	18	2	101	69	320	473	4	8	91
111	17	2	95	97	354	532	10	11	98
112	17	2	105	81	399	416	6	9	56
113	17	2	125	109	532	563	13	12	98
114	18	2	115	83	501	453	6	6	61
115	17	2	127	88	599	507	14	9	61
116	16	2	108	91	558	427	9	12	57
117	17	2	100	95	479	496	8	9	74
118	17	2	106	121	588	570	13	13	106
119	16	2	111	107	618	478	8	12	79
120	17	2	109	95	548	465	9	9	59
121	17	2	114	115	672	532	13	11	87
122	17	2	101	97	438	433	11	9	75
123	17	2	112	81	524	453	3	5	53
124	16	2	111	129	724	542	13	11	72
125	17	2	108	94	550	464	12	7	69
126	17	2	109	89	566	440	5	6	54
127	17	2	126	87	436	474	5	11	83
128	17	2	99	93	463	478	5	9	98
129	17	2	108	108	637	535	11	7	90
130	16	2	134	101	440	506	9	15	77
131	17	2	102	77	372	452	3	3	70
132	17	2	110	118	618	499	13	14	85
133	17	2	123	133	736	597	15	13	103
134	18	2	120	105	584	486	10	10	69
135	18	2	116	95	503	448	2	14	63
136	17	2	103	113	533	501	12	9	76
137	17	2	113	124	653	541	17	12	87
138	17	2	127	122	724	574	15	14	84
139	18	2	100	86	437	479	6	10	71
140	17	2	120	117	615	510	15	11	64
141	17	2	107	92	478	471	9	9	74
142	17	2	125	122	777	564	10	13	91
143	17	2	122	123	685	550	11	9	82
144	17	2	119	98	614	476	8	12	62
145	16	2	108	90	498	452	12	9	60
146	17	2	113	60	397	447	4	4	65
147	17	2	102	109	353	507	11	8	97
148	17	2	105	105	472	513	4	9	89
149	17	2	115	117	599	540	13	12	94
150	17	2	116	97	567	513	12	10	73

ID	*Age*	*Gender*	*EI*	*IQ*	*VSAT*	*MSAT*	*Create*	*Abstract*	*Physics*
151	17	2	119	98	536	464	4	7	72
152	17	2	113	114	665	570	13	14	91
153	17	2	120	88	540	430	8	8	44
154	17	2	115	91	562	447	9	1	50
155	18	2	134	103	712	564	6	11	90
156	17	2	115	119	640	524	9	12	79
157	17	2	117	100	526	506	11	12	74
158	17	2	114	107	583	516	7	8	85
159	17	2	123	103	561	485	10	9	61
160	18	2	119	98	602	485	14	9	57
161	18	2	110	96	560	458	9	8	67
162	16	2	108	88	514	364	8	9	41
163	17	2	100	91	486	469	4	8	84
164	16	2	127	110	686	558	13	9	96
165	18	2	118	80	436	457	4	4	86
166	17	2	129	121	793	481	8	12	44
167	18	2	115	118	666	538	9	10	72
168	17	2	112	97	469	483	9	6	.
169	17	2	110	109	630	473	11	8	51
170	17	2	101	100	446	464	5	10	76
171	17	2	110	85	387	439	11	7	75
172	18	2	108	130	694	549	15	13	81
173	17	2	139	106	722	518	11	9	52
174	18	2	112	116	487	436	12	7	63
175	17	2	120	100	598	523	11	15	84
176	19	2	124	93	552	460	8	10	58
177	17	2	115	100	616	451	2	8	54
178	18	2	118	94	438	467	12	11	85
179	16	2	111	77	394	412	3	9	60
180	17	2	93	119	502	497	10	13	79
181	18	2	107	118	562	527	16	11	79
182	18	2	106	88	476	451	7	13	61
183	17	2	106	111	644	541	17	13	87
184	18	2	131	102	717	570	10	11	80
185	17	2	124	103	623	436	7	10	39
186	17	2	100	122	629	533	12	11	83
187	17	2	98	129	681	601	19	14	114
188	18	2	97	96	485	489	8	10	75
189	18	2	108	92	440	476	8	4	.
190	17	2	105	107	462	495	7	10	82
191	17	2	130	88	598	495	8	12	65
192	17	2	125	118	731	489	13	7	58
193	17	2	114	81	497	478	11	14	69
194	17	2	104	91	485	519	9	10	81
195	17	2	126	107	753	446	9	8	29
196	17	2	117	102	564	480	9	9	69
197	17	2	128	104	801	455	4	0	36
198	17	2	128	109	657	478	8	7	69
199	17	2	104	89	456	474	7	8	69
200	17	2	108	92	383	440	6	6	67

NOTE: The corresponding SPSS file for this table is ch14predictphysics.sav.

Detection of possible multivariate outliers is most easily handled by examination of plots of residuals from the multiple regression, and/or examination of information about individual cases (such as Mahalanobis *D* or **leverage** statistics) that can be requested and saved into the SPSS worksheet from the regression program. See Tabachnick and Fidell (2007, pp. 73–76) for further discussion of methods for detection and handling of multivariate outliers.

14.4 ♦ Issues in Planning a Study

Usually, regression analysis is used in nonexperimental research situations, in which the researcher has manipulated none of the variables. In the absence of an experimental design, causal inferences cannot be made. However, researchers often select at least some of the predictor variables for regression analysis because they believe that these might be "causes" of the outcome variable. If an X_i variable that is theorized to be a "cause" of Y fails to account for a significant amount of variance in the Y variable in the regression analysis, this outcome may weaken the researcher's belief that the X_i variable has a causal connection with Y. On the other hand, if an X_i variable that is thought to be "causal" does uniquely predict a significant proportion of variance in Y even when confounded variables or competing causal variables are statistically controlled, this outcome may be interpreted as consistent with the possibility of causality. Of course, neither outcome provides proof for or against causality. An X_i variable may fail to be a statistically significant predictor of Y in a regression (even if it really is a cause) for many reasons: poor measurement reliability, restricted range, Type II error, a relation that is not linear, an improperly specified model, and so forth. On the other hand, an X_i variable that is not a cause of Y may significantly predict variance in Y because of some artifact; for instance, X_i may be correlated or confounded with some other variable that causes Y, or we may have an instance of Type I error.

As discussed in Chapter 11, the proportion of variance uniquely accounted for by X_i in a multiple regression, sr_i^2, is calculated in a way that adjusts for the correlation of X_i with all other predictors in the regression equation. We can obtain an accurate assessment of the proportion of variance attributable to X_i only if we have a correctly specified model—that is, a regression model that includes all the predictors that should be included and that does not include any predictors that should not be included. A good theory provides guidance about the set of variables that should be taken into account when trying to explain people's scores on a particular outcome variable. However, in general, we can never be sure that we have a correctly specified model.

What should be included in a correctly specified model? First, all the relevant "causal variables" that are believed to influence or predict scores on the outcome variable should be included. This would, in principle, make it possible to sort out the unique contributions of causes that may well be confounded or correlated with each other. In addition, if our predictor variables are "contaminated" by sources of measurement bias (such as general verbal ability or social desirability), measures of these sources of bias should also be included as predictors. In practice, it is not possible to be certain that we have a complete list of causes or a complete assessment of sources of bias. Thus, we can never be certain that we have a correctly specified model. In addition, a correctly specified model should include any moderator variables.

Usually, when we fail to include competing causes as predictor variables, the X_i variables that we do include in the equation may appear to be stronger predictors than they really are. For example, when we fail to include measures of bias (e.g., a measure of social desirability), this may lead to either over- or underestimation of the importance of individual X predictor variables. Finally, if we include irrelevant predictor variables in our regression, sometimes these take explained variance away from other predictors.

We must be careful, therefore, to qualify or limit our interpretations of regression results. The proportion of variance explained by a particular X_i predictor variable is specific to the sample of data and to the type of participants in the study; it is also specific to the context of the other variables that are included in the regression analysis. When predictor variables are added to (or dropped from) a regression model, the sr_i^2 that indexes the unique variance explained by a particular X_i variable can either increase or decrease; the β_i that represents the partial slope for z_{X_i} can become larger or smaller (Kenny, 1979, called this "bouncing betas"). Thus, our judgment about the apparent predictive usefulness of an individual X_i variable is context dependent in at least three ways: it may be unique to the peculiarities of the particular sample of data, it may be limited to the types of participants included in the study, and it varies as a function of the other predictor variables that are included in (and excluded from) the regression analysis.

Past research (and well-developed theory) can be extremely helpful in deciding what variables ought to be included in a regression analysis, in addition to any variables whose possible causal usefulness a researcher wants to explore. Earlier chapters described various roles that variables can play. Regression predictors may be included because they are of interest as possible causes; however, predictors may also be included in a regression analysis because they represent competing causes that need to be controlled for, confounds that need to be corrected for, sources of measurement error that need to be adjusted for, moderators, or extraneous variables that are associated with additional random error.

The strongest conclusion that a researcher is justified in drawing when a regression analysis is performed on data from a nonexperimental study is that a particular X_i variable is (or is not) significantly predictive of Y when a specific set of other X variables (that represent competing explanations, confounds, sources of measurement bias, or other extraneous variables) is controlled. If a particular X_i variable is still significantly predictive of Y when a well-chosen set of other predictor variables is statistically controlled, the researcher has a slightly stronger case for the possibility that X_i might be a cause of Y than if the only evidence is a significant zero-order Pearson correlation between X_i and Y. However, it is by no means proof of causality; it is merely a demonstration that, after we control for the most likely competing causes that we can think of, X_i continues to account uniquely for a share of the variance in Y.

Several other design issues are crucial, in addition to the appropriate selection of predictor variables. It is important to have a reasonably wide range of scores on the Y outcome variable and on the X predictor variables. As discussed in Chapter 7, a restricted range can artifactually reduce the magnitude of correlations, and restricted ranges or scores can also reduce the size of regression slope coefficients. Furthermore, we cannot assume that the linear regression equation will make accurate predictions for scores on X predictor variables that lie outside the range of X values in the sample. We need to have a sample size that is sufficiently large to provide adequate statistical power (see Section 14.12) and also large enough to provide reasonably narrow confidence intervals for the estimates of b slope coefficients; the larger the number of predictor variables (k), the larger the required sample size.

14.5 ♦ Computation of Regression Coefficients With *k* Predictor Variables

In Chapter 11, we worked out the normal equations for a standardized multiple regression with two predictors using a path diagram to represent z_{X_1} and z_{X_2} as correlated predictors of z_Y. This resulted in the following equations to describe the way that the overall correlation between a specific predictor z_{X_1} and the outcome z_Y (i.e., r_{1Y}) can be "deconstructed" into two components, a direct path from z_{X_1} to z_Y (the strength of this direct or unique predictive relationship is represented by β_1) and an indirect path from z_{X_1} to z_Y via z_{X_2} (represented by $r_{12}\beta_2$):

$$r_{Y1} = \beta_1 + r_{12}\beta_2, \tag{14.4}$$

$$r_{Y2} = r_{12}\beta_1 + \beta_2. \tag{14.5}$$

We are now ready to generalize the procedures for the computation of β and b regression coefficients to regression equations that include more than two predictor variables. On a conceptual level, when we set up a regression that includes k predictor variables (as shown in Equations 14.1 and 14.2), we need to calculate the β_i and b_i partial slope coefficients that make the best possible prediction of Y from each X_i predictor variable (the beta coefficients are applied to z scores on the variables, while the b coefficients are applied to the raw scores in the original units of measurement). These partial slopes must control for or partial out any redundancy or linear correlation of X_i with all the other predictor variables in the equation (i.e., $X_1, X_2, \ldots, X_k$). When we had only two predictors, z_{X_1} and z_{X_2}, we needed to control for or partial out the part of the predictive relationship of z_{X_1} with z_Y that could be accounted for by the path through the correlated predictor variable z_{X_2}. More generally, the path model for a regression with several correlated predictor variables has the form shown in Figure 14.1.

When we compute a standardized partial slope β_1, which represents the unique predictive contribution of z_{X_1}, we must "partial out" or remove all the indirect paths from z_{X_1} to z_Y via each of the other predictor variables ($z_{X_2}, z_{X_3}, \ldots, z_{X_k}$). A formula to calculate an estimate of β_1 from the bivariate correlations among all the other predictors, and the other estimates of $\beta_2, \beta_3, \ldots, \beta_k$, is obtained by subtracting the indirect paths from the overall r_{1Y} correlation (as in Equation 14.3, except that in the more general case with k predictor variables, multiple indirect paths must be "subtracted out" when we assess the unique predictive relationship of each z_{X_i} variable with z_Y). In addition, the divisor for each beta slope coefficient takes into account the correlation between each predictor and Y and the correlations between all pairs of predictor variables. The point to understand is that the calculation of a β_i coefficient includes information about the magnitude of the Pearson correlation between X_i and Y; but the magnitude of the β_i coefficient is also adjusted for the correlations of X_i with all other X predictor variables (and the association of the other X predictors with the Y outcome variable). Because of this adjustment, a β_i slope coefficient can differ in size and/or in sign from the zero-order Pearson correlation of X_i with Y. This issue was discussed more extensively in Chapters 10 and 11, and the same conceptual issues are applicable here. Controlling for other predictors can greatly

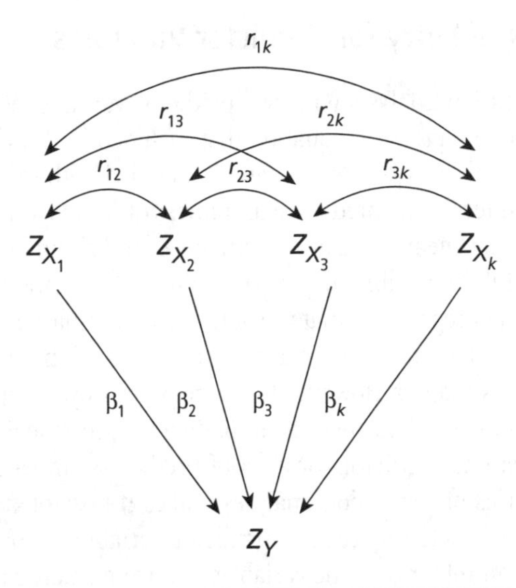

Figure 14.1 ♦ Path Model for Multiple Regression With k Predictor Variables ($X_1, X_2, \ldots, X_k$) as Correlated Predictors of Y

change our understanding of the strength and direction of the association between an individual X_i predictor variable and Y. For example, controlling for a highly correlated competing predictor variable may greatly reduce the apparent strength of the association between X_i and Y. On the other hand, controlling for a suppressor variable can actually make the predictive association between X_i and Y (represented by β_i) stronger than the zero-order correlation r_{1Y} or different in sign (see Chapter 10 for a discussion of suppressor variables).

Students who want to understand the computational procedures for multiple regression with more than two predictors in greater detail should see Appendix 14.A at the end of this chapter for a brief introduction to matrix algebra and an explanation of the matrix algebra computation of the b and β slope coefficients for multiple regression with k predictor variables.

Once we have obtained estimates of the beta coefficients, we can obtain the corresponding b coefficients (to predict raw scores on Y from raw scores on the X predictor variables) by rescaling the slopes to take information about the units of measurement of the predictor and outcome variable into account, as we did earlier in Chapter 11:

$$b_i = \beta_i(SD_y/SD_i), \tag{14.6}$$

where SD_i is the standard deviation of X_i, the ith independent variable. The intercept b_0 is calculated from the means of the Xs and their b coefficients using the following equation:

$$b_0 = M_Y - b_1M_{X1} - b_2M_{X2} - \ldots - b_kM_{Xk}. \tag{14.7}$$

14.6 ♦ Methods of Entry for Predictor Variables

When regression analysis with two predictor variables was introduced in Chapter 11, we calculated one regression equation that included both predictors. In Chapter 11, only one regression analysis was reported, and it included both the X_1 and X_2 variables as predictors of Y. In Chapter 11, the predictive usefulness of X_1 was assessed while controlling for or partialling out any linear association between X_2 and X_1, and the predictive usefulness of X_2 was assessed while controlling for or partialling out any linear association between X_1 and X_2. The method of regression that was introduced in Chapter 11 (with all predictors entered at the same time) is equivalent to a method of regression that is called "standard" or "simultaneous" in this chapter. However, there are other ways to approach an analysis that involves multiple predictors. It is possible to conduct a regression analysis as a series of analyses and to enter just one predictor (or a set of predictors) in each step in this series of analyses. Doing a series of regression analyses makes it possible to evaluate how much additional variance is predicted by each X_i predictor variable (or by each set of predictor variables) when you control for only the variables that were entered in prior steps.

Unfortunately, the nomenclature that is used for various methods of entry of predictors into regression varies across textbooks and journal articles. In this textbook, three major approaches to method of entry are discussed. These are listed here and then discussed in more detail in later sections of this chapter:

1. *Standard or simultaneous or direct regression:* In this type of regression analysis, only one regression equation is estimated, all the X_i predictor variables are added at the same time, and the predictive usefulness of each X_i predictor is assessed while statistically controlling for any linear association of X_i with all other predictor variables in the equation.

2. *Sequential or hierarchical regression* (*user-determined order of entry*): In this type of regression analysis, the data analyst decides on an order of entry for the predictor variables based on some theoretical rationale. A series of regression equations are estimated. In each step, either one X_i predictor variable or a set of several X_i predictor variables are added to the regression equation.

3. *Statistical regression* (*data-driven order of entry*): In this type of regression analysis, the order of entry of predictor variables is determined by statistical criteria. In Step 1, the single predictor variable that has the largest squared correlation with Y is entered into the equation; in each subsequent step, the variable that is entered into the equation is the one that produces the largest possible increase in the magnitude of R^2.

Unfortunately, the term *stepwise* regression is sometimes used in a nonspecific manner to refer to any regression analysis that involves a series of steps with one or more additional variables entered at each step in the analysis (i.e., to either Method 2 or 3 described above). It is sometimes unclear whether authors who label an analysis as a stepwise regression are referring to a hierarchical/sequential regression (user-determined) or to a statistical regression (data-driven) when they describe their analysis as stepwise. In this chapter, the term *stepwise* is defined in a narrow and specific manner; stepwise refers to one of the specific methods that the SPSS program uses for the entry of predictor variables in statistical regression. To avoid confusion, it is preferable to state as clearly as possible in simple language how the analysis was set up—that is, to make an explicit statement about order of entry of predictors (e.g., see Section 14.15).

14.6.1 ♦ Standard or Simultaneous Method of Entry

The method of regression that was described in Chapter 11 corresponds to *standard* multiple regression (this is the term used by Tabachnick & Fidell, 2007). This method is also widely referred to as simultaneous or direct regression. In standard or simultaneous multiple regression, all the predictor variables are entered into the analysis in one step, and coefficients are calculated for just one regression equation that includes the entire set of predictors. The effect size that describes the unique predictive contribution of each X variable, sr^2_{unique}, is adjusted to partial out or control for any linear association of X_i with all the other predictor variables. This standard or simultaneous approach to multiple regression usually provides the most conservative assessment of the unique predictive contribution of each X_i variable. That is, usually (but not always) the proportion of variance in Y that is attributed to a specific X_i predictor variable is smaller in a standard regression analysis than when the X_i predictor is entered in an early step in a sequential or statistical series of regression equations. The standard method of regression is usually the simplest version of multiple regression to run and report.

14.6.2 ♦ Sequential or Hierarchical (User-Determined) Method of Entry

Another widely used method of regression involves running a series of regression analyses; at each step, one X_i predictor (or a set, group, or block of X_i predictors) selected by the data analyst for theoretical reasons is added to the regression analysis. The key issue is that the order of entry of predictors is determined by the data analyst (rather than by the sizes of the correlations among variables in the sample data). Tabachnick and Fidell (2007) called this method of entry (in which the data analyst decides on the order of entry of predictors) sequential or hierarchical regression. Sequential regression involves running a *series* of multiple regression analyses. In each step, one or more predictor variables are added to the model, and the predictive usefulness of each X_i variable (or set of X_i variables) is assessed by asking how much the R^2 for the regression model increases in the step when each predictor variable (or set of predictor variables) is first added to the model. When just one predictor variable is added in each step, the **increment in R^2, R^2_{inc},** is equivalent to **sr^2_{inc},** the squared part correlation for the predictor variable in the step when it first enters the analysis.

14.6.3 ♦ Statistical (Data-Driven) Order of Entry

In a statistical regression, the order of entry for predictor variables is based on statistical criteria. SPSS offers several different options for statistical regression. In forward regression, the analysis begins without any predictor variables included in the regression equation; in each step, the X_i predictor variable that produces the largest increase in R^2 is added to the regression equation. In backward regression, the analysis begins with all predictor variables included in the equation; in each step, the X_i variable is dropped, which leads to the smallest reduction in the overall R^2 for the regression equation. SPSS stepwise is a combination of **forward** and **backward methods;** the analysis begins with no predictor variables in the model; in each step, the X_i predictor that adds the largest amount to the R^2 for the equation is added to the model; but if any X_i predictor variable no longer makes a significant contribution to R^2, that variable is dropped from the model. Thus, in an SPSS stepwise statistical regression, variables are added in each step, but variables can also be dropped from the model if they are no longer significant (after the addition of other predictors). The application of these three different methods of statistical regression (forward, backward, and stepwise) may or may not result in the same set of X_i predictors in the final model. As noted earlier, many writers use the term *stepwise* in a very broad sense to refer to any regression analysis where a series of regression equations are estimated with predictors added to the model in each step. In this chapter, stepwise is used to refer specifically to one of the types of variable entry that SPSS provides for statistical regression. Table 14.2 summarizes the preceding discussion about types of regression and nomenclature for these methods.

14.7 ♦ Variance Partitioning in Regression for Standard or Simultaneous Regression Versus Regressions That Involve a Series of Steps

The three methods of entry of predictor variables (standard/simultaneous, sequential/hierarchical, and statistical) handle the problem of partitioning-explained variance among predictor variables somewhat differently. Figure 14.2 illustrates this difference in variance partitioning.

In a standard or simultaneous entry multiple regression, each predictor is assessed controlling for all other predictors in the model; each X_i predictor variable gets credit only for variance that it shares uniquely with Y and not with any other X predictors (as shown in Figure 14.2a). In sequential or statistical regression, each X_i variable's contribution is assessed controlling only for the predictors that enter in the same or earlier steps; variables that are entered in later steps are not taken into account (as shown in Figure 14.2b). In the example shown in Figure 14.2a, the standard regression, X_1 is assessed controlling for X_2 and X_3, X_2 is assessed controlling for X_1 and X_3, and X_3 is assessed controlling for X_1 and X_2. (Note that, to keep notation simple in the following discussion, the number subscripts for predictor variables in the following discussion indicate the step in which each variable is entered.)

On the other hand, in sequential or statistical regression, X_1 (the variable that enters in Step 1) is assessed controlling for none of the other predictors; X_2, the variable that enters in Step 2, is assessed only for variables that entered in prior steps (in this case, X_1); and X_3, the variable that enters in Step 3, is assessed controlling for all variables that entered in prior steps (X_1 and X_2) (see Figure 14.2b). In this example, the X_1 variable would get credit for a much smaller proportion of variance in a standard regression (shown in Figure 14.2a) than in sequential or statistical regression (shown in Figure 14.2b). Sequential or statistical regression essentially makes an arbitrary decision to give the variable that entered in an earlier step (X_1) credit for variance that could be explained just

Table 14.2 ♦ Summary of Nomenclature for Various Types of Regression

Common Names for the Procedure	*What the Procedure Involves*
Standard or simultaneous or direct regression	A single regression analysis is performed to predict Y from $X_1, X_2, \ldots, X_k$. The predictive contribution of each X_i predictor is assessed while statistically controlling for linear associations with all the other X predictor variables.
Sequential or hierarchical regression[a] (user-determined order of entry)	A series of regression analyses are performed. On each step, one or several X_i predictor variables are entered into the equation. The order of entry is determined by the data analyst based on a theoretical rationale. The predictive usefulness of each X_i predictor variable is assessed while controlling for any linear association of X_i with other predictor variables that enter on the same step or on previous steps.
Statistical (or data-driven) regression[a]	A series of regression analyses are performed. X_i predictor variables are added to and/or dropped from the regression model at each step. An X_i predictor is added if it provides the maximum increase in R^2 (while controlling for predictors that are already in the model). An X_i predictor is dropped if removing it results in a nonsignificant reduction in R^2. Within SPSS, there are three types of statistical regression: forward, backward, and stepwise.

a. Some authors use the term *stepwise* to refer to either sequential or statistical regression. That use is avoided here because it introduces ambiguity and is inconsistent with the specific definition of stepwise that is used by the SPSS regression program.

as well by variables that entered in later steps (X_2 or X_3). The decision about order of entry is sometimes arbitrary. Unless there are strong theoretical justifications, or the variables were measured at different points in time, it can be difficult to defend the decision to enter a particular predictor in an early step.

Usually, although not always, a predictor variable tends to explain more variance when it is entered in an early step in a series of regression analyses, because a variable that is entered by itself in the first step is assessed while statistically controlling for no other predictors. A variable that is entered by itself in the second step is assessed while statistically controlling only for the one or more variables entered in the first step. However, there are some situations where controlling for an X_2 suppressor variable makes X_1 appear to be a stronger predictor of Y (as discussed in Chapter 10).

When a sequential (or hierarchical or user-determined) method of variable entry is used, it may be possible to justify order of entry based on the times when the scores were obtained for predictors and/or the roles that the predictor variables play in the theory. If the X_1, X_2, and X_3 predictors were measured a year, a month, and a week prior to the

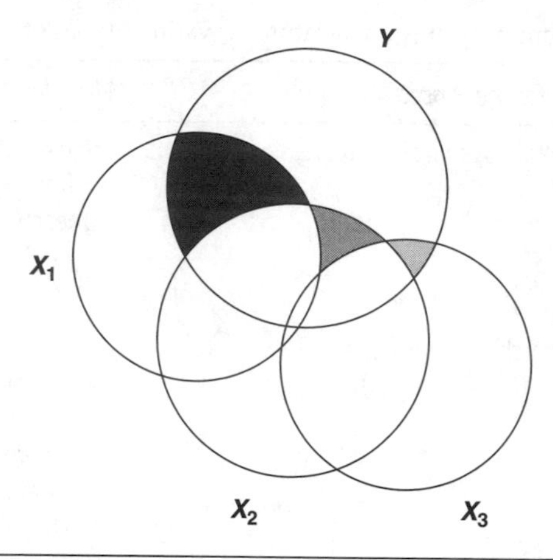

$Y' = b_0 + b_1X_1 + b_2X_2 + b_3X_3$

NOTE: Figure 14.2a shows partition of variance among predictor variables in a standard regression. The contribution of each predictor is assessed controlling for all other predictors. The shaded areas correspond to the squared semipartial (or part) correlation of each predictor with Y (i.e., sr^2_{unique}).

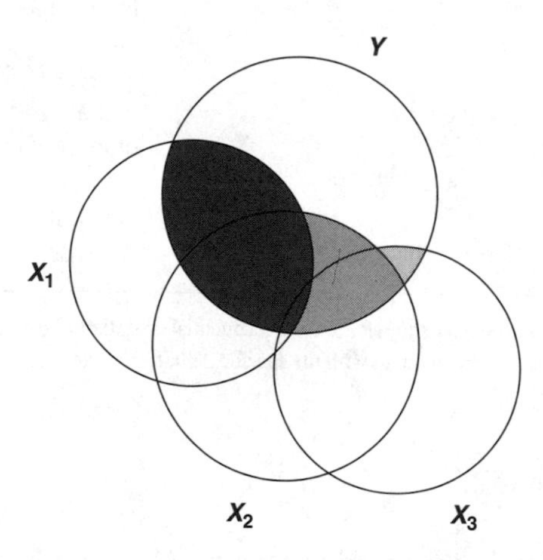

Figure 14.2 ♦ Partition of Variance Among Three Predictor Variables (X_1, X_2, and X_3)

NOTE: Figure 14.2b shows partition of variance among predictor variables in sequential or statistical regression. In this example, the predictor variables are added one at a time and three separate regressions are performed.

Step 1: $Y' = b_0 + b_1X_1$

Proportion of variance attributed to $X_1 = R^2$ for Step 1 regression (the black area in the diagram).

Step 2: $Y' = b_0 + b_1X_1 + b_2X_2$ (Step 2 model)

Proportion of variance attributed to X_2 = increment in R^2 when X_2 is added to the model $= R^2_{\text{Step2}} - R^2_{\text{Step1}}$ = proportion of variance in Y that is uniquely predictable from X_2 (controlling for X_1) (dark gray area in the diagram).

Step 3: $Y' = b_0 + b_1X_1 + b_2X_2 + b_3X_3$ (Step 3 model)

Proportion of variance attributed to X_3 = increment in R^2 when X_3 is added to the model $= R^2_{\text{Step3}} - R^2_{\text{Step2}}$ = proportion of variance in Y that is uniquely predictable from X_3, controlling for X_1 and X_2 (light gray area in the diagram).

assessment of Y, respectively, then it might make sense to control for the X_1 variable before assessing what additional adjustments in the prediction of Y should be made based on the later values of X_2 and X_3. Usually, our theories include some "preferred" predictor whose usefulness the researcher wants to demonstrate. When this is the case, the researcher can make a stronger case for the usefulness of her or his preferred predictor if she or he demonstrates that the preferred variable is still significantly predictive of Y when control or nuisance variables, or competing explanatory variables, have been controlled for or taken into account. It makes sense, in general, to include "control," "nuisance," or "competing" variables in the sequential regression in early steps and to include the predictors that the researcher wants to subject to the most stringent test and to make the strongest case for in later steps. (Unfortunately, researchers sometimes do the opposite—that is, enter their "favorite" variables in early steps so that their preferred variables will get credit for larger shares of the variance.)

When statistical (data-driven) methods of variable entry are used (i.e., when predictor variables are entered into the model in order of their predictive usefulness), it is difficult to defend the resulting partition of variance. It is quite likely that this method of variable selection for predictors will result in an inflated risk of Type I error—that is, variables whose sample correlations overestimate their true population correlations with Y are likely to be included as predictors when statistical methods of variable entry are used. Tabachnick and Fidell (2007) pointed out that the significance tests on SPSS printouts are not adjusted to correct for the inflated risk of Type I error that arises when statistical regression methods are used.

It is usually simpler to use standard regression methods rather than sequential regression methods. The use of statistical regression is not recommended under any circumstances; this is a data-fishing technique that produces the largest R^2 possible from the minimum number of predictors, but it is likely to capitalize on chance, to result in a model that makes little sense, and to include predictors whose significance is due to Type I error. If the researcher cannot resist the temptation to use statistical regression in spite of this warning, then at least the researcher should use the modified test procedures suggested by Wilkinson and Dallal (1981) to assess the statistical significance of the overall regression, as described later in the chapter.

14.8 ♦ Significance Test for an Overall Regression Model

In interpreting the results of a multiple regression analysis with k predictor variables, two questions are considered. First, is the overall multiple regression significantly predictive of Y? This corresponds to the null hypothesis that the multiple R (between the Y' calculated from the Xs and the observed Y) equals 0. The test statistic for this omnibus test is the same as described in Chapter 11; it is an F ratio with k, $N - k - 1$ degrees of freedom (df), where N is the number of participants or cases and k is the number of predictor variables.

The null hypothesis for the overall test of the regression is

$$H_0: R = 0. \tag{14.8}$$

The F ratio that tests this null hypothesis can be calculated either from the sums of squares (SS) or from the overall R^2 for regression:

$$F = \frac{SS_{\text{regression}}/k}{SS_{\text{residual}}/(N - k - 1)} \tag{14.9}$$

or, equivalently,

$$F = \frac{R^2/k}{(1 - R^2)/(N - k - 1)}. \tag{14.10}$$

SPSS provides an exact p value for the F ratio for the overall regression. If the obtained p value is smaller than the preselected alpha level, then the null hypothesis is rejected; the researcher concludes that Y scores can be predicted significantly better than chance when the entire set of predictor variables (X_1 through X_k) is used to calculate the predicted Y score. There is disagreement whether a statistically significant omnibus F test should be required before doing follow-up tests to assess the predictive contribution of individual predictor variables. Requiring a statistically significant overall F for the model might provide some protection against inflated risk of Type I error; however, it does not provide a guarantee of protection. When the omnibus F is significant, the researcher usually goes on to assess the predictive usefulness of each predictor variable (or sometimes the significance of sets or blocks of predictors considered as groups).

The omnibus test for the overall model is done the same way for standard (simultaneous) and sequential (user-determined) methods of entry. The researcher examines the F for the standard regression equation or for the equation in the final step of the sequential regression. If the p or significance value associated with this F ratio (on the SPSS printout) is less than the predetermined alpha level (usually .05), then the overall multiple regression is judged to be statistically significant. For these cases, the significance levels reported by SPSS may be accurate but only if the researcher has run a single regression analysis. If the researcher has run a dozen different variations of the regression before deciding on a "best" model, then the p value on the SPSS printout may seriously underestimate the real risk of Type I error that arises when a researcher goes "data fishing," in search of a combination of predictors that yields a large R^2 value.

In statistical methods of regression, the p values given on the SPSS printout generally underestimate the true risk of Type I error. If the backward or stepwise methods of entry are used, there is no easy way to correct for this inflated risk of Type I error. If "method = forward" is used, then the tables provided by Wilkinson and Dallal (1981), reproduced in Appendix 14.B at the end of this chapter, can be used to look up appropriate critical values for multiple R^2. The value of critical R depends on the following factors: the desired alpha level (usually $\alpha = .05$); the number of "candidate" predictor variables, k (variables are counted whether or not they actually are entered into the analysis); the residual df, $N - k - 1$; and the **F-to-enter** that the user tells SPSS to use as a criterion in deciding whether to enter potential predictors into the regression equation. The table provides critical values of R^2 for F-to-enter values of 2, 3, and 4. For example, using the table in Appendix 14.B at the end of this chapter, if there are $k = 20$ candidate predictor variables, $N = 221$ subjects, $N - k - 1 = 200$ df, $\alpha = .05$, and F-to-enter $= 3.00$, then the critical value from the table is $R^2 = .09$ (decimal points are omitted in the table). That is, the final regression equation described in the preceding example can be judged statistically significant at the .05 level if its multiple R^2 exceeds .09.

Note that the risk of Type I error for the overall regression model will be close to the nominal alpha level only if the researcher ran just one multiple regression and conducted just one omnibus F test. In many cases, researchers run numerous regression models before they decide which one to report; when this has been done, it would be more honest to make it clear to the reader that the real risk of Type I error is greater than the nominal Type I error or p value on the SPSS printout.

For statistical multiple regression, the problem of inflated risk of Type I error can be even greater, because the researcher essentially gives the computer program a list of candidate predictor variables and asks the computer to "fish" for the few predictors that explain the highest proportions of variance in Y in the particular sample of data. The larger the number of candidate predictor variables, the higher the likelihood that the final multiple regression equation will have an inflated risk of Type I error.

14.9 ♦ Significance Tests for Individual Predictors in Multiple Regression

The assessment of the predictive contribution of each individual X variable is handled differently in standard or simultaneous regression (in contrast to the sequential and statistical approaches). For standard regression, the researcher examines the t ratio (with $N - k - 1$ df) that assesses whether the b_i partial slope coefficient is statistically significant for each X_i predictor, for the printout of the one regression model that is reported.

When you run either sequential or statistical regression, you actually obtain a series of regression equations. For now, let's assume that you have added just one predictor variable in each step (it is also possible to add groups or blocks of variables in each step). To keep the notation simple, let's also suppose that the variable designated X_1 happens to enter first, X_2 second, and so forth. The order of entry may either be determined arbitrarily by the researcher (the researcher tells SPSS which variable to enter in Step 1, Step 2, and so forth) or be determined statistically; that is, the program checks at each step to see which variable would increase the R^2 the most if it were added to the regression and adds that variable to the regression equation.

In either sequential or statistical regression, then, a series of regression analyses is performed as follows:

Step 1:

Add X_1 to the model.

Predict Y from X_1 only.

Step 1 model: $Y' = b_0 + b_1X_1$.

X_1 gets credit for $R^2_{\text{incremental}}$ in Step 1.

$R^2_{\text{inc}} = R^2_{\text{Step1}} - R^2_{\text{Step0}}$ (assume that R^2 in Step 0 was 0).

Note that sr_1^2 is identical to R^2 for the Step 1 model and r_{1Y}^2, because at this point no other variable is controlled for when assessing the predictive usefulness of X_1. Thus, X_1 is assessed controlling for none of the other predictors.

R^2_{inc} or sr^2_{inc} for Step 1 corresponds to the black area in Figure 14.2b.

Step 2:

Add X_2 to the model.

Predict Y from both X_1 and X_2.

Step 2 model: $Y' = b_0 + b_1X_1 + b_2X_2$.

(Note that b_1 in the Step 2 equation must be reestimated controlling for X_2; so it is not, in general, equal to b_1 in Step 1.)

X_2 gets credit for the increase in R^2 that occurs in the step when X_2 is added to the model:

$R^2_{inc} = R^2_{Step2} - R^2_{Step1}$.

Note that R^2_{inc} is equivalent to sr_2^2, the squared part correlation for X_2 in Step 2, and these terms correspond to the medium gray area in Figure 14.2b.

Step 3:

Add X_3 to the model.

Predict Y from X_1, X_2, and X_3.

Step 3 model: $Y' = b_0 + b_1X_1 + b_2X_2 + b_3X_3$.

X_3 gets credit for the proportion of incremental variance $R^2_{Step3} - R^2_{Step2}$ or sr_3^2 (the squared part correlation associated with X_3 in the step when it enters). This corresponds to the light gray area in Figure 14.2b.

Researchers do not generally report complete information about the regression analyses in all these steps. Usually, researchers report the b and beta coefficients for the equation in the *final* step and the multiple R and overall F ratio for the equation in the *final* step with all predictors included. To assess the statistical significance of each X_i individual predictor, the researcher looks at the t test associated with the b_i slope coefficient associated with X_i in the step when X_i first enters the model. If this t ratio is significant *in the step when X_i first enters the model,* this implies that the X_i variable added significantly to the explained variance, in the step when it first entered the model, controlling for all the predictors that entered in earlier steps. The R^2_{inc} for each predictor variable X_i on the step when X_i first enters the analysis provides effect size information (the estimated proportion of variance in Y that is predictable from X_i, statistically controlling for all other predictor variables included in this step).

It is also possible to report the overall F ratio for the multiple R at Step 1, Step 2, Step 3, and so forth, but this information is not always included.

Notice that it would not make sense to report a set of b coefficients such that you took your b_1 value from Step 1, b_2 from Step 2, and b_3 from Step 3; this mixed set of slope coefficients could not be used to make accurate predictions of Y. Note also that the value of b_1

(the slope to predict changes in Y from increases in X_1) is likely to change when you add new predictors in each step, and sometimes this value changes dramatically. As additional predictors are added to the model, b_1 usually decreases in absolute magnitude, but it can increase in magnitude or even change sign when other variables are controlled (see Chapter 10 to review discussion of ways that the apparent relationship of X_1 to Y can change when you control for other variables).

When reporting results from a standard or simultaneous regression, it does not matter what order the predictors are listed in your regression summary table. However, when reporting results from a sequential or statistical regression, predictors should be listed in the order in which they entered the model; readers expect to see the first entered variable on line 1, the second entered variable on line 2, and so forth.

The interpretation of $b_1, b_2, b_3, \ldots, b_k$ in a k predictor multiple regression equation is similar to the interpretation of regression slope coefficients described in Chapter 11. For example, b_1 represents the number of units of change predicted in the raw Y score for a one-unit change in X_1, controlling for or partialling out all the other predictors ($X_2, X_3, \ldots, X_k$). Similarly, the interpretation of β_1 is the number of standard deviations of change in predicted z_Y, for a 1 *SD* increase in z_{X_1}, controlling for or partialling out any linear association between z_{X_1} and all the other predictors.

For standard multiple regression, the null hypothesis of interest for each X_i predictor is

$$H_0: b_i = 0. \tag{14.11}$$

That is, we want to know whether the raw score slope coefficient b_i associated with X_i differs significantly from 0. As in Chapter 11, the usual test statistic for this situation is a t ratio:

$$t = \frac{b_i}{SE_{b_i}}, \quad \text{with } N - k - 1 \; df. \tag{14.12}$$

SPSS and other programs provide exact (two-tailed) p values for each t test. If the obtained p value is less than the predetermined alpha level (which is usually set at $\alpha = .05$, two-tailed), the partial slope b_i for X_i is judged statistically significant. Recall (from Chapters 10 and 11) that if b_i is 0, then b_i, sr_i, and pr_i also equal 0. Thus, this t test can also be used to judge whether the proportion of variance that is uniquely predictable from X_i, sr_i^2, is statistically significant.

In sequential and statistical regression, the contribution of each X_i predictor variable is assessed in the step when X_i first enters the regression model. The null hypothesis is

$$H_0 : R^2_{\text{inc}} = 0. \tag{14.13}$$

In words, this is the null hypothesis that the *increment* in multiple R^2 in the step when X_i enters the model equals 0. Another way to state the null hypothesis about the incremental amount of variance that can be predicted when X_i is added to the model is

$$H_0 : sr_i^2 = 0. \tag{14.14}$$

In words, this is the null hypothesis that the squared part correlation associated with X_i in the step when X_i first enters the regression model equals 0. When b, β, sr, and pr are calculated (see Chapters 10 and 11), they all have the same terms in the numerator. They are scaled differently (using different divisors), but if sr is 0, then b must also equal 0. Thus, we can use the t ratio associated with the b_i slope coefficient to test the null hypothesis $H_0 : sr_i^2 = 0$ (or, equivalently, $H_0 : R^2_{\text{inc}} = 0$).

There is also an F test that can be used to assess the significance of the change in R^2 in a sequential or statistical regression, from one step to the next, for any number of added variables. In the example at the end of this chapter, it happens that just one predictor variable is added in each step. However, we can add a set or group of predictors in each step. To test the null hypothesis $H_0 : R^2_{\text{inc}} = 0$ for the general case where m variables are added to a model that included k variables in the prior step, the following F ratio can be used:

Let $R^2_{\text{wo}} = R^2$ for the reduced model with only k predictors.

Let $R^2_{\text{with}} = R^2$ for the full model that includes k predictors and m additional predictors.

Let $N =$ number of participants or cases.

Let $R^2_{\text{inc}} = R^2_{\text{with}} - R^2_{\text{wo}}$ (note that R^2_{with} must be equal to or greater than R^2_{wo}).

The test statistic for $H_0 : R^2_{\text{inc}} = 0$ is an F ratio with $m, N - k - m - 1$ df:

$$F_{\text{inc}} = \frac{R^2_{\text{inc}}/m}{(1 - R^2_{\text{with}})/(N - k - m - 1)}. \tag{14.15}$$

When you enter just one new predictor X_i in a particular step, the F_{inc} for X_i equals the squared t ratio associated with X_i.

14.10 ♦ Effect Size

14.10.1 ♦ Effect Size for Overall Regression (Multiple *R*)

For all three methods of regression (standard, sequential, and statistical), the effect size for the overall regression model that includes all the predictors is indexed by multiple R, multiple R^2, and adjusted multiple R^2 (for a review of these, see Chapter 11). For standard regression, because there is just one regression equation, it is easy to locate this overall multiple R and R^2. For sequential and statistical regression, researchers always report multiple R and R^2 in the final step. Occasionally, they also report R and R^2 for every individual step.

14.10.2 ♦ Effect Sizes for Individual Predictor Variables (sr^2)

For standard or simultaneous regression, the most common effect size index for each individual predictor variable is sr_i^2; this is the squared part correlation for X_i. SPSS regression can report the part correlation (if requested); this value is squared by hand to provide an estimate of the unique proportion of variance predictable from each X_i variable. We will call this sr^2_{unique} to indicate that it estimates the proportion of variance that each X predictor uniquely explains (i.e., variance that is not shared with *any* of the other predictors).

For either sequential or statistical regression, the effect size that is reported for each individual predictor is labeled either sr^2_{inc} or R^2_{inc} (i.e., the increase in R^2 in the step when that predictor variable first enters the model). When just one new predictor variable enters in a step, R^2_{inc} is equivalent to the sr^2_i value associated with X_i in the step when X_i first enters the model. If you request "R square change" as one of the statistics from SPSS, you obtain a summary table that shows the total R^2 for the regression model at each step and also the R^2 increment at each step of the analysis.

To see how the partition of variance among individual predictors differs when you compare standard regression with sequential/statistical regression, reexamine Figure 14.2. The sr^2_{unique} for variables X_1, X_2, and X_3 in a standard regression correspond to the black, dark gray, and light gray areas in Figure 14.2a, respectively. The $sr^2_{incremental}$ (or R^2_{inc}) terms for X_1, X_2, and X_3 in either a sequential or a statistical regression correspond to the black, dark gray, and light gray areas in Figure 14.2b, respectively. Note that in a standard regression, when predictors "compete" to explain the same variance in Y, none of the predictors get credit for the explained variance that can be explained by other predictors. By contrast, in the sequential and statistical regressions, the contribution of each predictor is assessed controlling only for predictors that entered in earlier steps. As a consequence, when there is "competition" between variables to explain the same variance in Y, the variable that enters in an earlier step gets credit for explaining that shared variance. In sequential and statistical regressions, the total R^2 for the final model can be reconstructed by summing sr^2_1, sr^2_2, sr^2_3, and so forth. In standard or simultaneous regression, the sum of the sr^2_{unique} contributions for the entire set of X_is is usually less than the overall R^2 for the entire set of predictors.

14.11 ♦ Changes in *F* and *R* as Additional Predictors Are Added to a Model in Sequential or Statistical Regression

Notice that as you add predictors to a regression model, an added predictor variable may produce a 0 or positive change in R^2; it cannot decrease R^2. However, the adjusted R^2, defined in Chapter 11, takes the relative sizes of k, number of predictor variables, and N, number of participants, into account; adjusted R^2 may go down as additional variables are added to an equation. The F ratio for the entire regression equation may either increase or decrease as additional variables are added to the model. Recall that

$$F = [R^2/df_{regression}]/[(1 - R^2)/df_{residual}],$$
$$\text{where } df_{regression} = k \text{ and } df_{residual} = N - K - 1. \tag{14.16}$$

As additional predictors are added to a regression equation, R^2 may increase (or remain the same), $df_{regression}$ increases, and $df_{residual}$ decreases. If R^2 goes up substantially, this increase may more than offset the change in $df_{regression}$, and if so, the net effect of adding an additional predictor variable is an increase in F. However, if you add a variable that produces little or no increase in R^2, the F for the overall regression may go down, because the loss of degrees of freedom for the residual term may outweigh any small increase in R^2. In general, F goes up if the added variables contribute a large increase in R^2; but if you add "garbage" predictor variables that use up a degree of freedom without substantially increasing R^2, the overall F for the regression can go down as predictor variables are added.

14.12 ♦ Statistical Power

According to Tabachnick and Fidell (2007), the ratio of N (number of cases) to k (number of predictors) has to be "substantial" for a regression analysis to give believable results. Based on work by Green (1991), they recommended a minimum $N > 50 + 8k$ for tests of multiple R and a minimum of $N > 104 + k$ for tests of significance of individual predictors. The larger of these two minimum Ns should be used to decide how many cases are needed. Thus, for a multiple regression with $k = 5$ predictors, the first rule gives $N > 75$ and the second rule gives $N > 109$; at least 109 cases should be used.

This decision rule should provide adequate statistical power to detect medium effect sizes; however, if the researcher wants to be able to detect weak effect sizes, or if there are violations of assumptions such as nonnormal distribution shapes, or if measurements have poor reliability, larger Ns are needed. If statistical regression methods (such as stepwise entry) are used, even larger Ns should be used. Larger sample sizes than these minimum sample sizes based on statistical power analysis are required to make the confidence intervals around estimates of b slope coefficients reasonably narrow. Note also that the higher the correlations among predictors, the larger the sample size that will be needed to obtain reasonably narrow confidence intervals for slope estimates.

On the other hand, in research situations where the overall sample size N is very large (e.g., $N > 10{,}000$), researchers may find that even effects that are too small to be of any practical or clinical importance may turn out to be statistically significant. For this reason, it is important to include information about effect size along with statistical significance tests.

Notice also that if a case has missing values on any of the variables included in the regression, the effective N is decreased. SPSS provides choices about handling data with missing observations. In pairwise deletion, each correlation is calculated using all the data available for that particular pair of variables. Pairwise deletion can result in quite different Ns (and different subsets of cases) used for each correlation, and this inconsistency is undesirable. In listwise deletion, a case is entirely omitted from the regression if it is missing a value on any one variable. This provides consistency in the set of data used to estimate all correlations, but if there are many missing observations on numerous variables, listwise deletion of missing data can lead to a very small overall N.

14.13 ♦ Nature of the Relationship Between Each *X* Predictor and *Y* (Controlling for Other Predictors)

It is important to pay attention to the sign associated with each b or β coefficient and to ask whether the direction of the relation it implies is consistent with expectations. It's also useful to ask how the partial and semipartial *r*s and *b* coefficients associated with a particular X_i variable compare to its zero-order correlation with *Y*, in both size and sign. As described in Chapter 10, when one or more other variables are statistically controlled, the apparent relation between X_i and *Y* can become stronger or weaker, become nonsignificant, or even change sign. The same is true of the partial slopes (and semipartial correlations) associated with individual predictors in multiple regression. The "story" about prediction of *Y* from several *X* variables may need to include discussion of the ways in which controlling for some *X*s changes the apparent importance of other predictors.

It is important to include a matrix of correlations as part of the results of a multiple regression (not only correlations of each X_i with *Y* but also correlations among all the *X* predictors). The correlations between the *X*s and *Y* provide a baseline against which to evaluate whether including each X_i in a regression with other variables statistically controlled has made a difference in the apparent nature of the relation between X_i and *Y*. The correlations among the *X*s should be examined to see whether there were strong correlations among predictors (also called strong multicollinearity). When predictors are highly correlated with each other, they may compete to explain much of the same variance; also, when predictors are highly correlated, the researcher may find that none of the individual b_i slope coefficients are significant even when the overall *R* for the entire regression is significant. If predictors X_1 and X_2 are very highly correlated, the researcher may want to consider whether they are, in fact, both measures of the same thing; if so, it may be better to combine them (perhaps by averaging X_1 and X_2 or z_{X_1} and z_{X_2}) or drop one of the variables.

One type of information provided about multicollinearity among predictors is **"tolerance."** The tolerance for a candidate predictor variable X_i is the proportion of variance in X_i that is not predictable from other *X* predictor variables that are already included in the regression equation. For example, suppose that a researcher has a regression equation to predict *Y′* from scores on X_1, X_2, and X_3: $Y' = b_0 + b_1X_1 + b_2X_2 + b_3X_3$. Suppose that the researcher is considering whether to add predictor variable X_4 to this regression. Several kinds of information are useful in deciding whether X_4 might possibly provide additional useful predictive information. One thing the researcher wants to know is, How much of the variance in X_4 is not already explained by (or accounted for) the other predictor variables already in the equation? To estimate the proportion of variance in X_4 that is not predictable from, or shared with, the predictor variables already included in the analysis—that is, X_1 through X_3—we could set up a regression to predict scores on the candidate variable X_4 from variables X_1 through X_3; the tolerance of candidate predictor variable X_4 is given by $1 - R^2$ for the equation that predicts X_4 from the other predictors X_1 through X_3. The minimum possible value of tolerance is 0; tolerance of 0 indicates that the candidate X_4 variable contains no additional variance or information that is not already present in predictor variables X_1 through X_3 (and that therefore X_4 cannot provide any "new" predictive information that is not already provided by X_1 through X_3). The maximum possible value of tolerance is 1.0; this represents a situation in which the predictor variable X_4 is completely uncorrelated with the other set of predictor variables already included in the model. If we are interested in adding the predictor variable X_4 to a regression analysis, we typically hope that it will have a tolerance that is not close to 0; tolerance that is substantially larger than 0 is evidence that X_4 provides new information not already provided by the other predictor variables.

If the researcher believes that there may be interactions between predictors, it is possible to include interaction terms in multiple regression equations (as described in Chapter 12 on dummy variables). For example, suppose that a researcher suspects that there is an interaction between gender (coded 0 = Male, 1 = Female) and EI (X) as predictors of drug use (Y). Thus, EI may be strongly predictive of drug use for males but not for females. To test the significance of a possible gender-by-EI interaction, the researcher computes a new variable that is the product of gender times EI (this could be called GenByEI). The multiple regression equation that includes main effects for gender and EI as well as a term to represent a possible interaction would be as follows:

$$Y' = b_0 + b_1 \times \text{Gender} + b_2 \times \text{EI} + b_3 \times \text{GenByEI}. \quad (14.17)$$

If the b_3 slope coefficient associated with GenByEI is statistically significant, this can be interpreted as evidence of a significant interaction ("different slopes for different folks," in the words of Robert Rosenthal; specifically, a different slope relating Y to EI for males than for females).

If the researcher thinks that there could be a nonlinear relation between an X predictor variable and Y, this can be represented by adding an X^2 term to the equation (and possibly also X^3)—for example, $Y' = b_0 + b_1X + b_2X^2$. If b_2 is statistically significant, this is evidence of a quadratic trend (such that Y varies as a U-shaped or an inverse U-shaped function of X).

14.14 ♦ Assessment of Multivariate Outliers in Regression

Examination of a histogram makes it possible to detect scores that are extreme univariate outliers; examination of bivariate scatter plots makes it possible to identify observations that are bivariate outliers (i.e., they represent unusual combinations of scores on X and Y, even though they may not be extreme on either X or Y alone); these are scores that lie outside the "cloud" that includes most of the data points in the scatter plot. It can be more difficult to detect multivariate outliers, as graphs that involve multiple dimensions are complex. Regression analysis offers several kinds of information about individual cases that can be used to identify multivariate outliers. For a more complete discussion, see Tabachnick and Fidell (2007, chap. 4, pp. 73–76; chap. 5, pp. 161–167).

First, SPSS can provide graphs of residuals (actual Y – predicted Y scores) against other values, such as Y' predicted scores. In a graph of standardized (z score) residuals, about 99% of the values should lie between –3 and +3; any observations with standardized residual z score values >3 in absolute value represent cases for which the regression made an unusually poor prediction. We should not necessarily automatically discard such cases, but it may be informative to examine these cases carefully to answer questions such as: Is the poor fit due to data entry errors? Was there something unique about this case that might explain why the regression prediction was poor for this participant? In addition, SPSS can provide saved scores on numerous case-specific diagnostic values such as "leverage" (slopes change by a large amount when cases with large leverage index values are dropped from the analysis; thus, such cases are inordinately influential). Mahalanobis *D* is another index available in SPSS that indicates the degree to which observations are multivariate outliers.

14.15 ♦ SPSS Example and Results

As an illustration of the issues in this chapter, the data in Table 14.1 are analyzed using three different methods of multiple regression. The first analysis is standard or simultaneous regression; all five predictor variables are entered in one step. The second analysis is sequential regression; the five predictor variables are entered in a user-determined sequence, one in each step, in an order that was specified by using SPSS menu command selections. The third analysis is a statistical (data-driven) regression using "method = forward." In this analysis, a statistical criterion was used to decide the order of entry of variables. At each step, the predictor variable that would produce the largest increment in R^2 was added to the model; when adding another variable would not produce a statistically significant increment in R^2, no further variables were entered. Note that to use the Wilkinson and Dallal (1981) table (reproduced as Appendix 14.B at the end of this chapter) to assess statistical significance of the overall final model obtained through forward regression, the user must specify a required minimum *F*-to-enter that matches one of the *F* values included in the Wilkinson and Dallal table (i.e., *F*-to-enter = 2.00, 3.00, or 4.00).

Details of data screening for these analyses were omitted. See Chapter 4 for guidelines on data screening. Prior to doing a multiple regression, the following preliminary screening should be done:

1. Histogram of scores on each predictor variable and *Y*: Check to see that the distribution shape is reasonably normal and that there are no extreme outliers or "impossible" score values.

2. Scatter plots between every pair of variables (e.g., all pairs of *X* variables and each *X* with *Y*). The scatter plots should show linear relations, homoscedastic variance, and no extreme bivariate outliers.

Three different multiple regressions were performed using the same data. Ordinarily, only one method is reported in a journal article; results from all three methods are reported here to illustrate how the nature of conclusions about the relative importance of predictors may differ depending on the method of entry that is chosen. Note that gender could have been dummy coded "+1" and "−1" or "+1" and "0" (as in the examples of dummy variables that were presented in Chapter 12). However, the proportion of variance that is predictable from gender in a multiple regression is the same whether gender is coded "+1", "0" or "+1", "+2" as in the examples that follow.

14.15.1 ♦ SPSS Screen Shots, Output, and Results for Standard Regression

Results for a Standard or Simultaneous Multiple Regression

Scores on a physics achievement test were predicted from the following variables: gender (coded 1 = Male, 2 = Female), emotional intelligence (EI), intelligence quotient (IQ), verbal SAT (VSAT), and math SAT (MSAT). The total *N* for this sample was 200; five cases were dropped due to missing data on at least one variable and, therefore, for this analysis *N* = 195. Preliminary data screening included examination of histograms of scores on all six variables and examination of scatter plots for all pairs of variables. Univariate distributions were reasonably normal with no extreme outliers; bivariate relations were fairly linear, all slopes had the expected signs, and there were no bivariate outliers.

Standard multiple regression was performed; that is, all predictor variables were entered in one step. Zero-order, part, and partial correlations of each predictor with physics were requested in addition to the default statistics. Results for this standard multiple regression are summarized in Table 14.3. See Figures 14.3 through 14.7 for SPSS menu selections and syntax and Figure 14.8 for the SPSS output.

Table 14.3 ♦ Results of Standard Multiple Regression to Predict Physics (*Y*) From Gender, IQ, EI, VSAT, and MSAT

	Physics	*Gender*	*IQ*	*EI*	*VSAT*	*MSAT*		*b*	*b*	sr^2_{unique}
Gender	−.37							−7.45***	-.21	.03
IQ	.34	.11						.13	.11	<.01
EI	−.39	.48	.04					−.03	-.02	<.01
VSAT	−.13	.23	.64	.55				−.09***	-.56	.10
MSAT	.69	-.03	.70	.04	.48			.36***	.88	.38
							Intercept =	−45.20***		
Mean	79.6	—[a]	100.1	107.6	535.2	493.7				
SD	17.6	—[a]	14.7	12.3	105.9	43.4				
								R^2 = .805		
								R^2_{adj} = .800		
								R = .897***		

***$p < .001$.

a. Because gender was a dummy variable (coded 1 = Male, 2 = Female), mean and standard deviation were not reported. The sample included $n = 100$ males and $n = 100$ females.

(Continued on page 580)

Predicting Physics Scores [DataSet1] - SPSS Data Editor

File Edit View Data Transform Analyze Graphs Utilities Add-ons Window Help

Reports
Descriptive Statistics
Compare Means
General Linear Model
Mixed Models
Correlate
Regression
Loglinear
Classify
Data Reduction
Scale
Nonparametric Tests
Survival
Multiple Response

Linear...
Curve Estimation...
Binary Logistic...
Multinomial Logistic...
Ordinal...
Probit...
Nonlinear...
Weight Estimation...
2-Stage Least Squares...

1 : idno 135

	idno	age			iq	vsat	msat	create	mech	abstract	physics	var
1	135	17			102	392	487	8	13	10	93	
2	136	18			112	593	538	15	71	12	103	
3	137	18					458	7	46	13	88	
4	138	18					550	15	53	13	92	
5	139	17					505	12	39	12	84	
6	140	17					467	4	18	8	75	
7	141	19					457	9	31	9	75	
8	142	17					569	18	71	10	103	
9	143	17					515	6	43	11	83	
10	144	18					467	14	76	15	76	
11	145	17	1	109			477	8	31	9	64	
12	146	17	1	116			442	3	23	7	66	
13	147	17	1	106	100	485	464	9	24	7	78	
14	148	18	1	105	99	491	542	12	49	10	108	
15	149	18	1	89	104	505	511	16	78	8	97	
16	150	18	1	112	86	551	463	6	27	11	71	
17	151	17	1	92	103	495	508	11	48	12	89	
18	152	17	1	102	104	561	552	15	78	10	117	
19	153	18	1	102	85	366	441	5	28	2	80	
20	154	17	1	109	78	354	442	9	49	9	91	
21	155	18	1	100	79	522	439	4	14	4	64	
22	156	17	1	85	112	531	481	8	66	6	90	
23	157	17	1	130	93	578	457	7	36	10	55	
24	158	19	1	90	88	444	486	6	37	2	102	
25	159	18	1	97	75	314	446	6	5	4	83	
26	160	18	1	98	88	433	492	7	47	9	86	
27	161	17	1	124	107	712	546	16	52	11	97	
28	162	17	1	97	105	587	521	6	23	11	87	
29	163	17	1	109	105	600	558	11	58	10	101	
30	164	18	1	96	83	534	455	9	68	7	77	
31	165	17	1	116	99	566	496	9	34	4	93	
32	166	18	1	111	95	496	500	7	5	5	86	
33	167	17	1	[illegible]	[illegible]	[illegible]	[illegible]	16	[illegible]	0	[illegible]	
34	168	17	1	121	108	624	527	8	62	13	75	

Data View / Variable View

Linear Regression SPSS Processor is ready

Start ch14mr_3 Permissions Lo... ch14figure - Mi... Predicting P... 11:29 AM

Figure 14.3 ♦ Menu Selections for SPSS Linear Regression Procedure

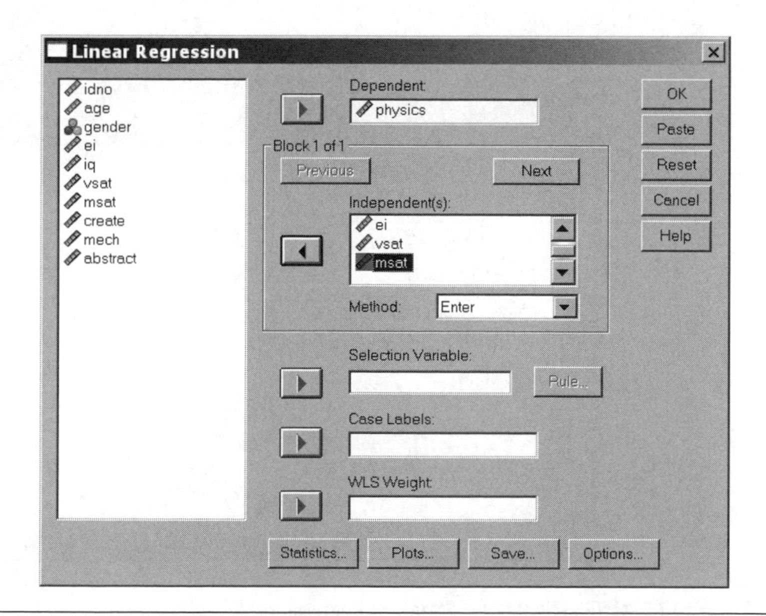

Figure 14.4 ♦ Regression Main Dialog Window for Standard or Simultaneous Linear Regression in SPSS (Method = Enter)

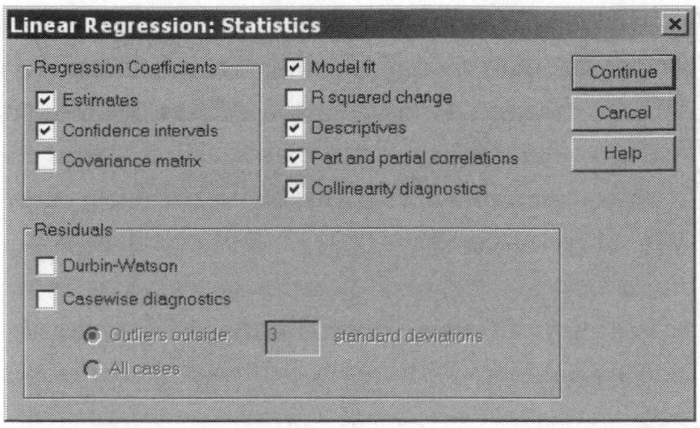

Linear Regression: Statistics

Regression Coefficients
- [x] Estimates
- [x] Confidence intervals
- [ ] Covariance matrix

- [x] Model fit
- [ ] R squared change
- [x] Descriptives
- [x] Part and partial correlations
- [x] Collinearity diagnostics

Continue
Cancel
Help

Residuals
- [ ] Durbin-Watson
- [ ] Casewise diagnostics

Outliers outside: 3 standard deviations
All cases

Figure 14.5 ♦ Statistics Requested for Standard Linear Regression in SPSS

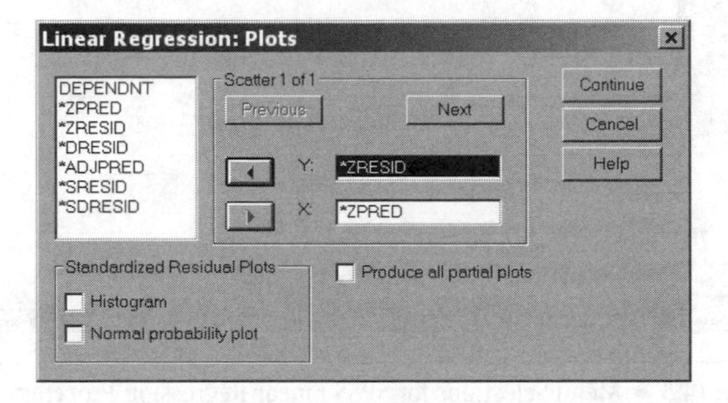

Figure 14.6 ♦ Request for Plot of Standardized Residuals From Linear Regression

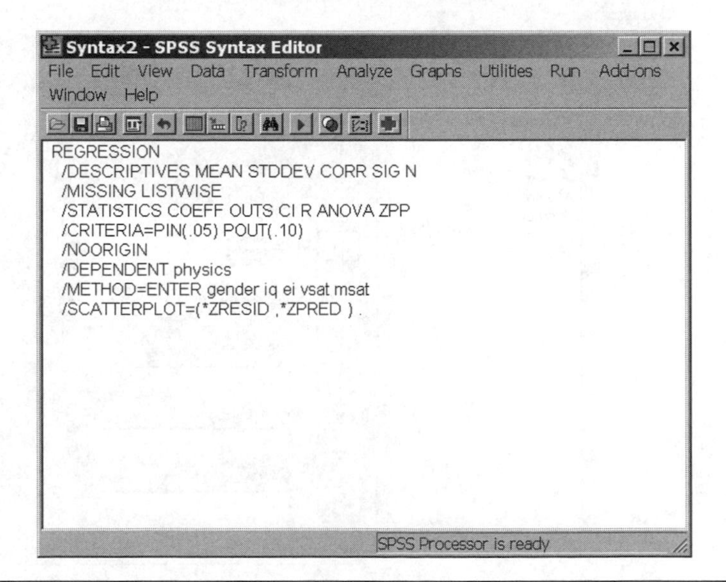

Figure 14.7 ♦ SPSS Syntax for Standard Regression

NOTE: Standard regression obtained by clicking Paste button after making all the menu selections.

Descriptive Statistics

	Mean	Std. Deviation	N
physics	79.57	17.582	195
gender	1.50	.501	195
iq	100.09	14.687	195
ei	107.57	12.257	195
vsat	535.24	105.912	195
msat	493.67	43.367	195

Correlations

		physics	gender	iq	ei	vsat	msat
Pearson Correlation	physics	1.000	-.368	.344	-.394	-.129	.690
	gender	-.368	1.000	.108	.483	.234	-.028
	iq	.344	.108	1.000	.036	.641	.704
	ei	-.394	.483	.036	1.000	.548	.040
	vsat	-.129	.234	.641	.548	1.000	.484
	msat	.690	-.028	.704	.040	.484	1.000
Sig. (1-tailed)	physics	.	.000	.000	.000	.036	.000
	gender	.000	.	.066	.000	.001	.350
	iq	.000	.066	.	.308	.000	.000
	ei	.000	.000	.308	.	.000	.289
	vsat	.036	.001	.000	.000	.	.000
	msat	.000	.350	.000	.289	.000	.
N	physics	195	195	195	195	195	195
	gender	195	195	195	195	195	195
	iq	195	195	195	195	195	195
	ei	195	195	195	195	195	195
	vsat	195	195	195	195	195	195
	msat	195	195	195	195	195	195

Variables Entered/Removed[b]

Model	Variables Entered	Variables Removed	Method
1	msat, gender, ei, vsat, iq[a]	.	Enter

a. All requested variables entered.

b. Dependent Variable: physics

Model Summary[b]

Model	R	R Square	Adjusted R Square	Std. Error of the Estimate
1	.897[a]	.805	.800	7.870

a. Predictors: (Constant), msat, gender, ei, vsat, iq

b. Dependent Variable: physics

ANOVA[b]

Model		Sum of Squares	df	Mean Square	F	Sig.
1	Regression	48262.46	5	9652.492	155.833	.000[a]
	Residual	11706.91	189	61.941		
	Total	59969.37	194			

a. Predictors: (Constant), msat, gender, ei, vsat, iq

b. Dependent Variable: physics

Coefficients[a]

Model		Unstandardized Coefficients		Standardized Coefficients			95% Confidence Interval for B		Correlations		
		B	Std. Error	Beta	t	Sig.	Lower Bound	Upper Bound	Zero-order	Partial	Part
1	(Constant)	-45.201	9.125		-4.954	.000	-63.200	-27.202			
	gender	-7.453	1.335	-.212	-5.583	.000	-10.086	-4.819	-.368	-.376	-.179
	iq	.130	.070	.109	1.860	.064	-.008	.268	.344	.134	.060
	ei	-.032	.072	-.022	-.443	.658	-.173	.110	-.394	-.032	-.014
	vsat	-.093	.010	-.563	-9.643	.000	-.113	-.074	-.129	-.574	-.310
	msat	.357	.019	.881	19.155	.000	.320	.394	.690	.812	.616

a. Dependent Variable: physics

Figure 14.8 ♦ SPSS Output for Standard or Simultaneous Regression: Prediction of Physics From Gender, IQ, EI, VSAT, and MSAT

NOTE: All the predictors entered in one step.

(Results continued)

To assess whether there were any multivariate outliers, the standardized residuals from this regression were plotted against the standardized predicted values (see the last panel in Figure 14.9). There was no indication of pattern, trend, or heteroscedasticity in this graph of residuals, nor were there any outliers; thus, it appears that the assumptions required for multiple regression were reasonably well met.

The overall regression, including all five predictors, was statistically significant, $R = .90$, $R^2 = .81$, adjusted $R^2 = .80$, $F(5, 189) = 155.83$, $p < .001$. Physics scores could be predicted quite well from this set of five variables, with approximately 80% of the variance in physics scores accounted for by the regression.

To assess the contributions of individual predictors, the *t* ratios for the individual regression slopes were examined. Three of the five predictors were significantly predictive of physics scores; these included gender, $t(189) = -5.58$, $p < .001$; VSAT, $t(189) = -9.64$, $p < .001$; and MSAT, $t(189) = 19.15$, $p < .001$. The nature of the predictive relation of gender was as expected; the negative sign for the slope for gender indicated that higher scores on gender (i.e., being female) predicted lower scores on physics. The predictive relation of MSAT to physics was also as predicted; higher scores on MSATs predicted higher scores on physics. However, scores on the VSAT were negatively related to physics; that is, higher

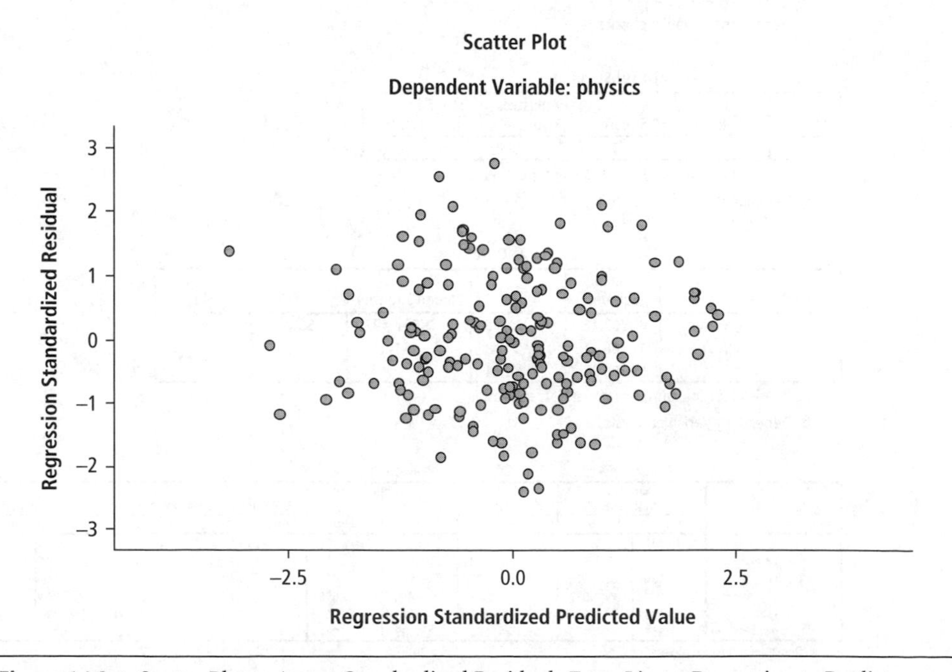

Figure 14.9 ♦ Scatter Plot to Assess Standardized Residuals From Linear Regression to Predict Physics Scores From Gender, IQ, EI, VSAT, and MSAT

VSAT scores predicted lower scores on physics, which was contrary to expectations. The negative partial *r* for the prediction of physics from VSAT controlling for the other four predictors ($pr = -.57$) was stronger than the zero-order Pearson *r* for the prediction of physics from VSAT without controlling for other variables ($r = -.13$), an indication of possible suppression effects—that is, it appears that the part of VSAT that was unrelated to MSAT and IQ was strongly predictive of poorer performance on physics. (See Chapter 10 for a review of suppression.) The proportions of variance uniquely explained by each of these predictors (sr^2_{unique}, obtained by squaring the part correlation from the SPSS printout) were as follows: $sr^2 = .03$ for gender, $sr^2 = .096$ for VSAT, and $sr^2 = .38$ for MSAT. Thus, in this sample and in the context of this set of predictors, MSAT was the strongest predictor of physics.

The other two predictor variables (EI and IQ) were not significantly related to physics when other predictors were statistically controlled; their partial slopes were not significant. Overall, physics scores were highly predictable from this set of predictors; the strongest unique predictive contributions were from MSAT and VSAT, with a smaller contribution from gender. Neither EI nor IQ were significantly predictive of physics scores in this regression, even though these two variables had significant zero-order correlations with physics; apparently, the information that they contributed to the regression was redundant with other predictors.

14.15.2 ♦ SPSS Screen Shots, Output, and Results for Sequential Regression

Results for Sequential or Hierarchical Regression (User-Determined Order of Entry)

Scores on a physics achievement test were predicted from the following variables: gender (coded 1 = Male, 2 = Female), emotional intelligence (EI), intelligence quotient (IQ), verbal SAT (VSAT), and math SAT (MSAT). The total *N* for this sample was 200; five cases were dropped due to missing data on at least one variable and, therefore, for this analysis $N = 195$. Preliminary data screening included examination of histograms of scores on all six variables and examination of scatter plots for all pairs of variables. Univariate distributions were reasonably normal with no extreme outliers; bivariate relations were fairly linear, all slopes had the expected signs, and there were no bivariate outliers.

Sequential multiple regression was performed; that is, each predictor variable was entered in one step in an order that was determined by the researcher, as follows: Step 1, gender; Step 2, IQ; Step 3, EI; Step 4, VSAT; and Step 5, MSAT. The rationale for this order of entry was that factors that emerge earlier in development were entered in earlier steps; VSAT was entered prior to MSAT arbitrarily. Zero-order, part, and partial correlations of each predictor with physics were requested in addition to the default statistics. Results for this sequential multiple regression are summarized in Tables 14.4 and 14.5. (See Figures 14.10 through 14.12 for SPSS commands and Figure 14.13 for SPSS output.)

(Continued)

(Continued)

Table 14.4 ◆ Results of Sequential Multiple Regression to Predict Physics (*Y*) From IQ, EI, VSAT, MSAT, and Gender

	Physics	*Gender*	*IQ*	*EI*	*VSAT*	*MSAT*	*b*	*b*	$sr^2_{incremental}$
Gender	–.37						–7.45***	–.21	.135***
IQ	.34	.11					.13	.11	.149***
EI	–.39	.48	.04				–.03	–.02	.058***
VSAT	–.13	.23	.64	.55			–.09***	–.56	.084***
MSAT	.69	–.03	.70	.04	.48		.36***	.88	.379***
						Intercept =	–45.20***		
Mean	79.6	—[a]	100.1	107.6	535.2	493.7			
SD	17.6	—[a]	14.7	12.3	105.9	43.4			
							R^2 = .805		
							R^2_{adj} = .800		
							R = .897***		

NOTE: One predictor entered in each step in a user-determined sequence.

$^{***}p < .001$.

a. Because gender was a dummy variable (coded 1 = Male, 2 = Female), mean and standard deviation were not reported. The sample included $n = 100$ males and $n = 100$ females.

The overall regression, including all five predictors, was statistically significant, $R = .90$, $R^2 = .81$, adjusted $R^2 = .80$, $F(5, 189) = 155.83$, $p < .001$. Physics scores could be predicted quite well from this set of five variables, with approximately 80% of the variance in physics scores accounted for by the regression.

Table 14.5 ◆ Summary of R^2 Values and R^2 Changes at Each Step in the Sequential Regression in Table 14.4

	Predictors Included	*R^2 for Model*	*F for Model*	*R^2 Change*	*F for R^2 Change*
1	Gender	.135	F(1, 193) = 30.14***	.135	F(1, 193) = 30.14***
2	Gender, IQ	.284	F(2, 192) = 38.10***	.149	F(1, 192) = 39.97***
3	Gender, IQ, EI	.342	F(3, 191) = 33.08***	.058	F(1, 191) = 16.78***
4	Gender, IQ, EI, VSAT	.426	F(4, 190) = 35.23**	.084	F(1, 190) = 27.76***
5	Gender, IQ, EI, VSAT, MSAT	.805	F(5, 189) = 155.83***	.379	F(1, 189) = 366.90***

$^{***}p < .001$.

(Continued on page 586)

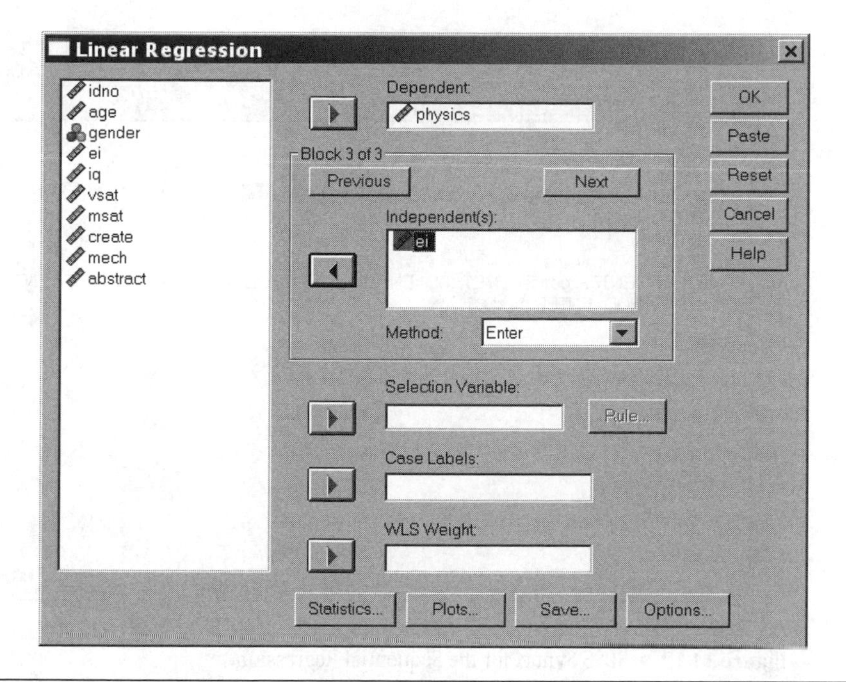

Figure 14.10 ♦ SPSS Dialog Box for Sequential or Hierarchical Regression

NOTE: One variable entered in each step in a user-determined sequence. To enter gender in Step 1, place the name gender in the box, then click the Next button to go on to the next step. In this example, only one variable was entered in each step; it is possible to enter more than one variable in each step. The dialog window is shown only for the step in which EI is entered as the third predictor.

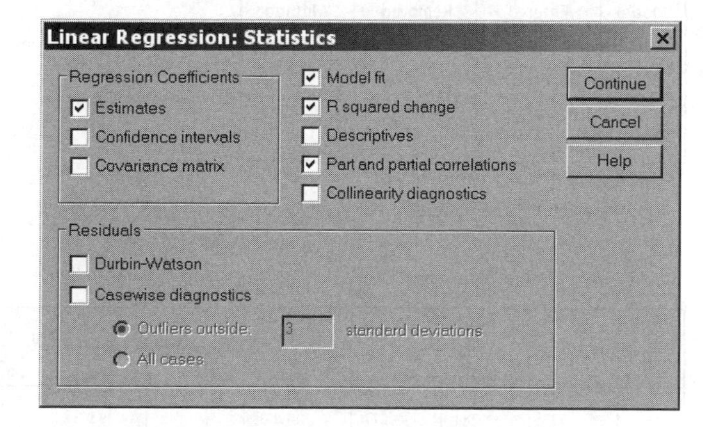

Figure 14.11 ♦ Additional Statistics Requested for Sequential or Hierarchical Regression: *R* Squared Change

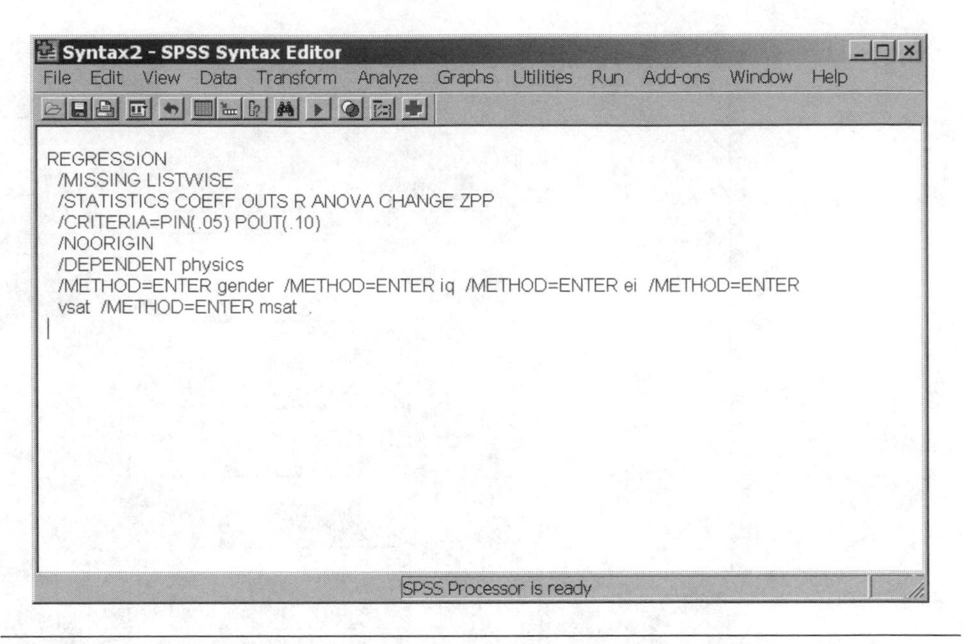

Figure 14.12 ♦ SPSS Syntax for the Sequential Regression

NOTE: Prediction of physics from five variables each entered in one step: Step 1, gender; Step 2, IQ; Step 3, EI; Step 4, VSAT; Step 5, MSAT.

Variables Entered/Removed[b]

Model	Variables Entered	Variables Removed	Method
1	gender [a]	.	Enter
2	iq [a]	.	Enter
3	ei [a]	.	Enter
4	vsat [a]	.	Enter
5	msat [a]	.	Enter

a. All requested variables entered.

b. Dependent Variable: physics

Model Summary

					Change Statistics				
Model	R	R Square	Adjusted R Square	Std. Error of the Estimate	R Square Change	F Change	df1	df2	Sig. F Change
1	.368[a]	.135	.131	16.394	.135	30.143	1	193	.000
2	.533[b]	.284	.277	14.953	.149	39.975	1	192	.000
3	.585[c]	.342	.332	14.374	.058	16.778	1	191	.000
4	.653[d]	.426	.414	13.462	.084	27.759	1	190	.000
5	.897[e]	.805	.800	7.870	.379	366.897	1	189	.000

a. Predictors: (Constant), gender

b. Predictors: (Constant), gender, iq

c. Predictors: (Constant), gender, iq, ei

d. Predictors: (Constant), gender, iq, ei, vsat

e. Predictors: (Constant), gender, iq, ei, vsat, msat

Figure 14.13 ♦ Selected Output From SPSS Sequential Regression

Model		Sum of Squares	df	Mean Square	F	Sig.
1	Regression	8100.807	1	8100.807	30.143	.000[a]
	Residual	51868.56	193	268.749		
	Total	59969.37	194			
2	Regression	17039.07	2	8519.535	38.102	.000[b]
	Residual	42930.30	192	223.595		
	Total	59969.37	194			
3	Regression	20505.76	3	6835.253	33.082	.000[c]
	Residual	39463.61	191	206.616		
	Total	59969.37	194			
4	Regression	25536.37	4	6384.094	35.227	.000[d]
	Residual	34433.00	190	181.226		
	Total	59969.37	194			
5	Regression	48262.46	5	9652.492	155.833	.000[e]
	Residual	11706.91	189	61.941		
	Total	59969.37	194			

a. Predictors: (Constant), gender
b. Predictors: (Constant), gender, iq
c. Predictors: (Constant), gender, iq, ei
d. Predictors: (Constant), gender, iq, ei, vsat
e. Predictors: (Constant), gender, iq, ei, vsat, msat
f. Dependent Variable: physics

Coefficients[a]

		Unstandardized Coefficients		Standardized Coefficients			Correlations		
Model		B	Std. Error	Beta	t	Sig.	Zero-order	Partial	Part
1	(Constant)	98.876	3.707		26.675	.000			
	gender	-12.891	2.348	-.368	-5.490	.000	-.368	-.368	-.368
2	(Constant)	54.553	7.783		7.009	.000			
	gender	-14.365	2.154	-.410	-6.668	.000	-.368	-.434	-.407
	iq	.465	.074	.388	6.323	.000	.344	.415	.386
3	(Constant)	90.472	11.527		7.849	.000			
	gender	-9.700	2.363	-.277	-4.104	.000	-.368	-.285	-.241
	iq	.460	.071	.384	6.500	.000	.344	.426	.382
	ei	-.394	.096	-.275	-4.096	.000	-.394	-.284	-.240
4	(Constant)	53.387	12.887		4.143	.000			
	gender	-11.822	2.250	-.337	-5.255	.000	-.368	-.356	-.289
	iq	.858	.100	.716	8.537	.000	.344	.527	.469
	ei	.044	.123	.031	.363	.717	-.394	.026	.020
	vsat	-.087	.017	-.526	-5.269	.000	-.129	-.357	-.290
5	(Constant)	-45.201	9.125		-4.954	.000			
	gender	-7.453	1.335	-.212	-5.583	.000	-.368	-.376	-.179
	iq	.130	.070	.109	1.860	.064	.344	.134	.060
	ei	-.032	.072	-.022	-.443	.658	-.394	-.032	-.014
	vsat	-.093	.010	-.563	-9.643	.000	-.129	-.574	-.310
	msat	.357	.019	.881	19.155	.000	.690	.812	.616

a. Dependent Variable: physics

Excluded Variables[e]

Model		Beta In	t	Sig.	Partial Correlation	Collinearity Statistics Tolerance
1	iq	.388[a]	6.323	.000	.415	.988
	ei	-.283[a]	-3.826	.000	-.266	.767
	vsat	-.045[a]	-.656	.512	-.047	.945
	msat	.680[a]	14.858	.000	.731	.999
2	ei	-.275[b]	-4.096	.000	-.284	.767
	vsat	-.501[b]	-6.853	.000	-.444	.562
	msat	.821[b]	12.880	.000	.682	.494
3	vsat	-.526[c]	-5.269	.000	-.357	.303
	msat	.866[c]	15.469	.000	.747	.489
4	msat	.881[d]	19.155	.000	.812	.488

a. Predictors in the Model: (Constant), gender
b. Predictors in the Model: (Constant), gender, iq
c. Predictors in the Model: (Constant), gender, iq, ei
d. Predictors in the Model: (Constant), gender, iq, ei, vsat
e. Dependent Variable: physics

NOTE: Prediction of physics from five variables each entered in one step in a user-determined sequence: Step 1, gender; Step 2, IQ; Step 3, EI; Step 4, VSAT; Step 5, MSAT (descriptive statistics and correlations were omitted because they are the same as for the standard regression).

(Results continued)

To assess the contributions of individual predictors, the *t* ratios for the individual regression slopes were examined for each variable in the step when it first entered the analysis. In Step 1, gender was statistically significant, $t(193) = -5.49$, $p < .001$; $R^2_{increment}$ (which is equivalent to sr^2_{inc}) was .135. The nature of the relation of gender to physics was as expected; the negative sign for the slope for gender indicated that higher scores on gender (i.e., being female) predicted lower scores on physics. IQ significantly increased the R^2 when it was entered in Step 2, $t(192) = 6.32$, $R^2_{inc} = .149$. (Note that the contribution of IQ, which is assessed in this analysis without controlling for EI, VSAT, and MSAT, appears to be much stronger than in the standard regression, where the contribution of IQ was assessed by controlling for all other predictors.) EI significantly increased the R^2 when it was entered in Step 3, $t(191) = -4.10$, $p < .001$, $R^2_{inc} = .058$. VSAT significantly increased the R^2 when it was entered in Step 4, $t(190) = -5.27$, $p < .001$, $R^2_{inc} = .084$. MSAT significantly increased the R^2 when it was entered in the fifth and final step, $t(189) = 19.16$, $p < .001$, $R^2_{inc} = .379$. Except for VSAT and EI, which were negatively related to physics, the slopes of all predictors had the expected signs.

Overall, physics scores were highly predictable from this set of predictors; the strongest unique predictive contribution was from MSAT (even though this variable was entered in the last step). All five predictors significantly increased the R^2 in the step when they first entered.

14.15.3 ♦ SPSS Screen Shots, Output, and Results for Statistical Regression

Results of Statistical Regression Using Method = Forward to Select Predictor Variables for Which the Increment in R^2 in Each Step Is Maximized

A statistical regression was performed to predict scores on physics from the following candidate predictor variables: gender (coded 1 = Male, 2 = Female), emotional intelligence (EI), intelligence quotient (IQ), verbal SAT (VSAT), and math SAT (MSAT). The total *N* for this sample was 200; five cases were dropped due to missing data on at least one variable and, therefore, for this analysis $N = 195$. Preliminary data screening included examination of histograms of scores on all six variables and examination of scatter plots for all pairs of variables. Univariate distributions were reasonably normal with no extreme outliers; bivariate relations were fairly linear, all slopes had the expected signs, and there were no bivariate outliers.

Statistical multiple regression was performed using Method = Forward with the F-to-enter criterion value set at $F = 3.00$. That is, in each step, SPSS entered the one predictor variable that would produce the largest increase in R^2. When the F ratio for the R^2 increase due to additional variables fell below $F = 3.00$, no further variables were added to the model. This resulted in the following order of entry: Step 1, MSAT; Step 2, VSAT; Step 3, gender, and Step 4, IQ. EI did not enter the equation. See Figures 14.14 through 14.16 for SPSS commands and Figure 14.17 for SPSS output. Results of this sequential regression are summarized in Table 14.6.

(Continued on page 590)

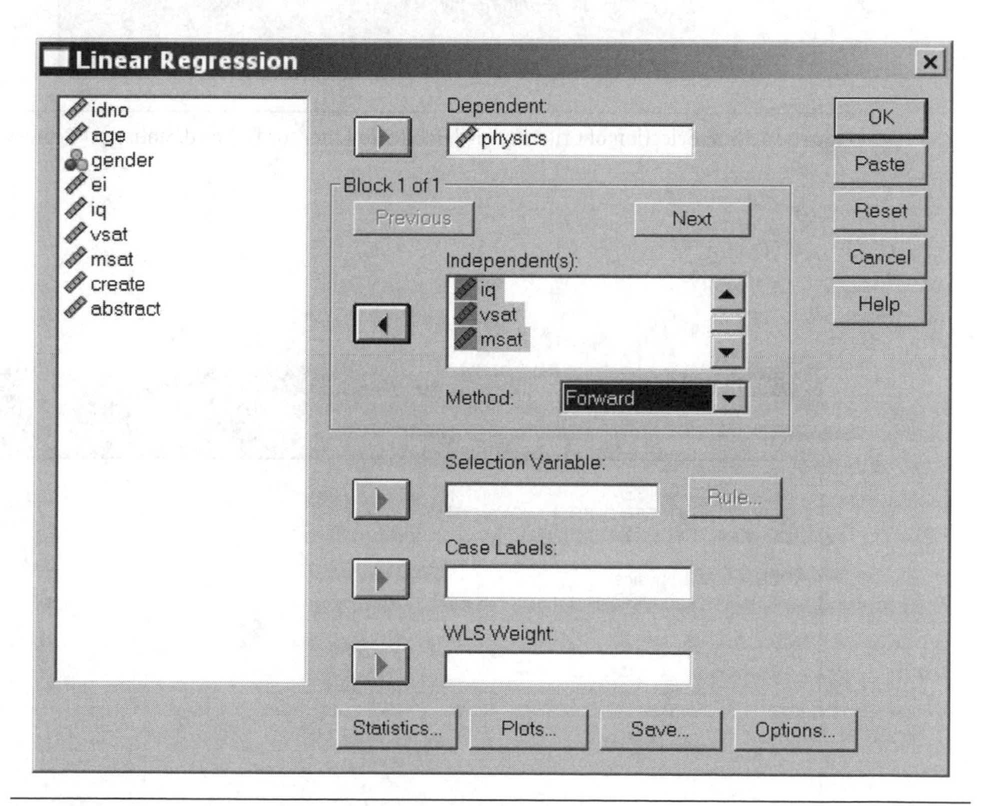

Figure 14.14 ♦ SPSS Menu Selections for Statistical Regression (Method = Forward)

NOTE: Variables are entered by data-driven order of entry.

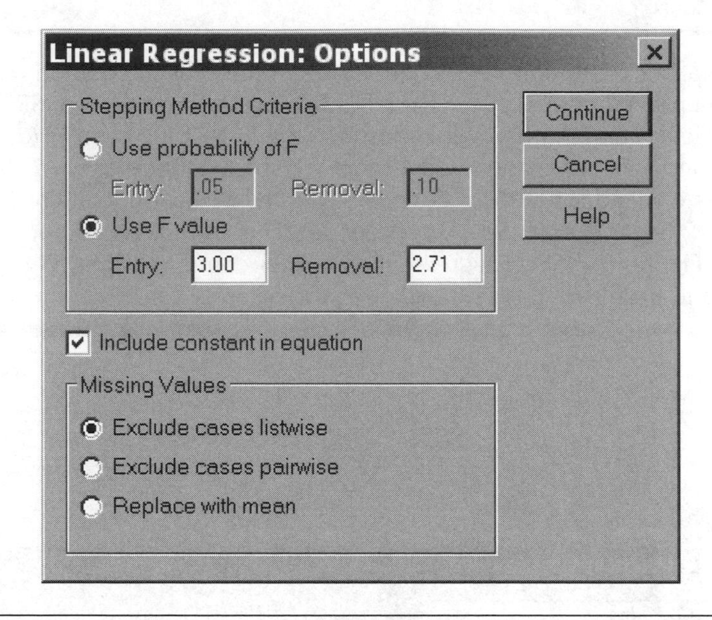

Figure 14.15 ♦ Selection of Criterion for $F = 3.00$ to Enter for Forward/Statistical Regression

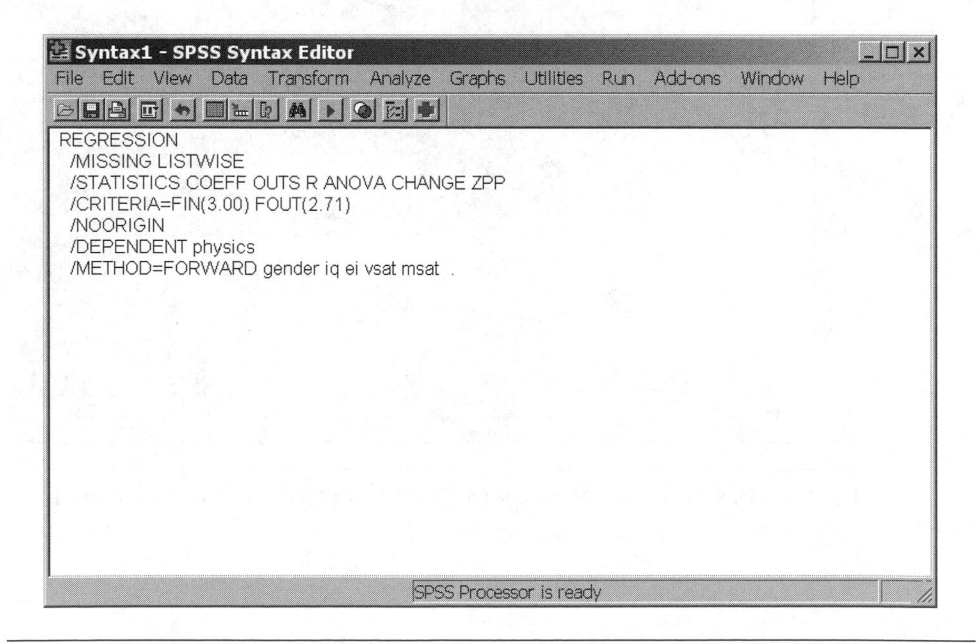

Figure 14.16 ♦ SPSS Syntax for Statistical Regression (Method = Forward, F-to-Enter Set at $F = 3.00$)

Model Summary

Model	R	R Square	Adjusted R Square	Std. Error of the Estimate	Change Statistics				
					R Square Change	F Change	df1	df2	Sig. F Change
1	.690[a]	.476	.473	12.757	.476	175.468	1	193	.000
2	.869[b]	.756	.753	8.732	.280	219.988	1	192	.000
3	.894[c]	.799	.796	7.945	.043	40.901	1	191	.000
4	.897[d]	.805	.800	7.854	.006	5.477	1	190	.020

a. Predictors: (Constant), msat

b. Predictors: (Constant), msat, vsat

c. Predictors: (Constant), msat, vsat, gender

d. Predictors: (Constant), msat, vsat, gender, iq

ANOVA[e]

Model		Sum of Squares	df	Mean Square	F	Sig.
1	Regression	28557.95	1	28557.952	175.468	.000[a]
	Residual	31411.42	193	162.753		
	Total	59969.37	194			
2	Regression	45330.62	2	22665.311	297.275	.000[b]
	Residual	14638.75	192	76.243		
	Total	59969.37	194			
3	Regression	47912.47	3	15970.825	253.003	.000[c]
	Residual	12056.90	191	63.125		
	Total	59969.37	194			
4	Regression	48250.29	4	12062.572	195.569	.000[d]
	Residual	11719.08	190	61.679		
	Total	59969.37	194			

a. Predictors: (Constant), msat

b. Predictors: (Constant), msat, vsat

c. Predictors: (Constant), msat, vsat, gender

d. Predictors: (Constant), msat, vsat, gender, iq

e. Dependent Variable: physics

Coefficients[a]

Model		Unstandardized Coefficients		Standardized Coefficients	t	Sig.	Correlations		
		B	Std. Error	Beta			Zero-order	Partial	Part
1	(Constant)	-58.543	10.467		-5.593	.000			
	msat	.280	.021	.690	13.246	.000	.690	.690	.690
2	(Constant)	-63.401	7.171		-8.841	.000			
	msat	.398	.017	.983	24.115	.000	.690	.867	.860
	vsat	-.100	.007	-.604	-14.832	.000	-.129	-.731	-.529
3	(Constant)	-50.271	6.841		-7.349	.000			
	msat	.382	.015	.943	25.077	.000	.690	.876	.814
	vsat	-.089	.006	-.535	-13.828	.000	-.129	-.707	-.449
	gender	-7.590	1.187	-.216	-6.395	.000	-.368	-.420	-.207
4	(Constant)	-47.871	6.839		-7.000	.000			
	msat	.357	.019	.880	19.200	.000	.690	.812	.616
	vsat	-.096	.007	-.581	-13.509	.000	-.129	-.700	-.433
	gender	-7.732	1.175	-.220	-6.582	.000	-.368	-.431	-.211
	iq	.145	.062	.121	2.340	.020	.344	.167	.075

a. Dependent Variable: physics

Figure 14.17 ♦ Selected Output From SPSS Statistical Regression (Method = Forward) *(Continued)*

Excluded Variables[e]

Model		Beta In	t	Sig.	Partial Correlation	Collinearity Statistics Tolerance
1	gender	-.349[a]	-7.613	.000	-.482	.999
	iq	-.280[a]	-3.964	.000	-.275	.505
	ei	-.423[a]	-9.952	.000	-.583	.998
	vsat	-.604[a]	-14.832	.000	-.731	.766
2	gender	-.216[b]	-6.395	.000	-.420	.919
	iq	.103[b]	1.812	.072	.130	.387
	ei	-.161[b]	-3.722	.000	-.260	.633
3	iq	.121[c]	2.340	.020	.167	.386
	ei	-.066[c]	-1.470	.143	-.106	.527
4	ei	-.022[d]	-.443	.658	-.032	.412

a. Predictors in the Model: (Constant), msat

b. Predictors in the Model: (Constant), msat, vsat

c. Predictors in the Model: (Constant), msat, vsat, gender

d. Predictors in the Model: (Constant), msat, vsat, gender, iq

e. Dependent Variable: physics

Figure 14.17 ♦ (Continued)

NOTE: Variables are entered by data-driven order of entry.

(Results continued)

Table 14.6 ♦ Results of Statistical (Method = Forward) Multiple Regression to Predict Physics (Y) From IQ, EI, VSAT, MSAT, and Gender

	Physics	*MSAT*	*VSAT*	*Gender*	*IQ*	b^a	b	$sr^2_{\text{incremental}}$
MSAT	.69					.36***	.88	.476
VSAT	−.13	.48				−.09***	−.58	.280
Gender	−.37	−.03	.23			−7.73***	−.22	.043
IQ	.34	.70	.64	.11		.14*	.12	.006
EI (not entered)	−.39	.04	.55	.48	.04	—	—	—
						Intercept = −47.87***		
Mean[b]	79.6	493.7	535.2	—	100.1			
SD	17.6	43.4	105.9	—	14.7			
						R^2 = .805[c]		
						R^2_{adj} = .800		
						R^a = .897***		

*p < .05, ***p < .001.

a. The significance values given on SPSS printout for all individual b coefficients were all <.001; however, forward regression leads to an inflated risk of Type I error and, therefore, these p values may underestimate the true risk of Type I error in this situation. The significance of the multiple R for the final model was assessed using critical values that are appropriate for forward regression from Wilkinson and Dallal (1981) rather than the p values on the SPSS printout.

b. Because gender was a dummy variable (coded 1 = Male, 2 = Female), mean and standard deviation were not reported. The sample included n = 100 males and n = 100 females.

c. In general, a multiple R would be smaller for a subset of four predictors than for five predictors. In this example, after rounding, the multiple R for four predictors was the same as the multiple R for five predictors, in Table 14.4, in the first three decimal places.

The overall regression, including four of the five candidate predictors, was statistically significant, $R = .90$, $R^2 = .81$, adjusted $R^2 = .80$, $F(4, 190) = 195.57$, $p < .001$. Physics scores could be predicted quite well from this set of four variables, with approximately 80% of the variance in physics scores accounted for by the regression. (Note that the F test on the SPSS printout does not accurately reflect the true risk of Type I error in this situation. The critical values of R^2 from the Wilkinson and Dallal table in Appendix 14.B at the end of this chapter provide a more conservative test of significance for the overall regression. For $k = 5$ candidate predictor variables, $df_{\text{residual}} = N - k - 1$ = between 150 and 200, F-to-enter set at $F = 3.00$, and $\alpha = .01$. The critical value of R^2 from the second page of the Wilkinson and Dallal (1981) table in Appendix 14.B at the end of this chapter was R^2 between .05 and .04; thus, using the more conservative Wilkinson and Dallal test, this overall multiple regression would still be judged statistically significant.)

To assess the statistical significance of the contributions of individual predictors, the F ratio for R^2 increment was examined for each variable in the step when it first entered the analysis. In Step 1, MSAT was entered; it produced an R^2 increment of .473, $F(1, 193) = 175.47$. In Step 2, VSAT was entered; it produced an R^2 increment of .280, $F(1, 192) = 219.99$, $p < .001$. In Step 3, gender was entered; it produced an R^2 increment of .043, $F(1, 191) = 40.90$, $p < .001$. In Step 4, IQ was entered; it produced an R^2 increment of .006, $F(1, 190) = 5.48$, $p = .020$. (EI was not entered because its F ratio was below the criterion set for F-to-enter using forward regression.) Except for VSAT, which was negatively related to physics, the slopes of all predictors had signs that were in the predicted direction.

Overall, physics scores were highly predictable from this set of predictors; the strongest unique predictive contribution was from MSAT (when this variable was entered in the first step). All four predictors significantly increased the R^2 in the step when they first entered.

14.16 ♦ Summary

When more than two predictor variables are included in a regression, the basic logic remains similar to the logic in regression with two predictors: The slope and proportion of variance associated with each predictor variable is assessed controlling for other predictor variables. With k predictors, it becomes possible to assess several competing causal variables, along with nuisance or confounded variables; it also become possible to include additional terms to represent interactions.

There are three major methods of variable entry: standard, sequential, and statistical. In most applications, standard or simultaneous entry (all predictor variables entered in a single step) provides the simplest and most conservative approach for assessment of the contributions of individual predictors. When there is temporal priority among predictor variables (X_1 is measured earlier in time than X_2, for example), or when there is a strong theoretical rationale for order of entry, it can be useful to set up a regression as a sequential or hierarchical analysis in which the order of entry of predictor variables is determined by the data analyst. The statistical or data-driven methods of entry can be used in situations where the only research goal is to identify the smallest possible set of predictor variables that will generate the largest possible R^2 value, but the statistical approach to regression is generally not a good method of testing theoretical hypotheses about the importance of predictor variables; in data driven on statistical series of regression equations, variables enter because they are strongly predictive of Y in the sample, but the variables that enter the regression may not have any meaningful relationship to the Y outcome variable or with other predictors. In addition, they may not provide accurate predictions for scores on Y outcome variables in different samples.

Appendix 14.A: A Review of Matrix Algebra Notation and Operations and Application of Matrix Algebra to Estimation of Slope Coefficients for Regression With More Than *k* Predictor Variables

A matrix is a table of numbers or variables. An entire matrix of scores on X predictor variables is usually denoted by a bold type symbol ($\mathbf{X}$), while individual values or elements are denoted by nonbold characters with subscripts to indicate which row and which column they come from. This discussion is limited to matrices that are two dimensional; that is, they correspond to a set of rows and a set of columns. The matrix that is usually of interest in statistics is the matrix of data values; this corresponds to the SPSS worksheet. Each row of the matrix contains scores that belong to one participant or case. Each column of the matrix contains scores that correspond to values of one variable. The element X_{ij} is the score in Row i and Column j; it is the score for participant number i on variable X_j. Suppose that N is the number of participants, and k is the number of variables. The matrix that contains data for N participants on k variables is described as an N by k matrix. The example below shows the notation for a matrix with k equals to three variables and four participants. The first subscript tells you what row of the matrix the element is in; the second subscript tells you the column. Thus, X_{41} is the score for Participant 4 on variable X_1:

$$\mathbf{X} = \begin{array}{c} \\ P_1 \\ P_2 \\ P_3 \\ P_4 \end{array} \begin{array}{c} \begin{array}{ccc} X_1 & X_2 & X_3 \end{array} \\ \begin{bmatrix} X_{11} & X_{12} & X_{13} \\ X_{21} & X_{22} & X_{23} \\ X_{31} & X_{32} & X_{33} \\ X_{41} & X_{42} & X_{43} \end{bmatrix} \end{array}.$$

Several other matrices are common in statistics. One is a correlation matrix usually denoted by $\mathbf{R}$; it contains the correlations among all possible pairs of k variables. For example, based on the scores above for the matrix $\mathbf{X}$, we could compute correlations between all possible pairs of variables (X_1 and X_2, X_1 and X_3, and X_2 and X_3) and then summarize the entire set of correlations in an $\mathbf{R}$ matrix:

$$\mathbf{R} = \begin{array}{c} \\ X_1 \\ X_2 \\ X_3 \end{array} \begin{array}{c} \begin{array}{ccc} X_1 & X_2 & X_3 \end{array} \\ \begin{bmatrix} 1.0 & r_{21} & r_{13} \\ r_{21} & 1.0 & r_{23} \\ r_{31} & r_{32} & 1.0 \end{bmatrix} \end{array}.$$

Note several characteristics of this matrix. All the diagonal elements equal 1 (because the correlation of a variable with itself is, by definition, 1.0). The matrix is "symmetric" because each element below the diagonal equals one corresponding element above the diagonal—for example, $r_{12} = r_{21}$, $r_{13} = r_{31}$, and $r_{23} = r_{32}$. When we write a correlation matrix, we do not need to fill in the entire table. It is sufficient to provide just the elements below the diagonal to provide complete information about correlations among the variables. In general, if we have k predictor variables, we have $k \times (k - 1)/2$ correlations. In this example, with $k = 3$, we have $3 \times (2 - 1)/2 = 3$ correlations in the matrix **R**. For a set of k variables, **R** is a $k \times k$ matrix.

Another frequently encountered matrix is called a variance/covariance matrix; the population parameter version is usually denoted by **Σ**, while the sample estimate version is generally denoted by **S**. This matrix contains the variance for each of the k variables in the diagonal, and the covariances for all possible pairs of variables are the off-diagonal elements. Like **R**, **S** is symmetrical; that is, the covariance between X_1 and X_2 is the same as the covariance between X_2 and X_1:

$$\mathbf{S} = \begin{array}{c} \\ X_1 \\ X_2 \\ X_3 \end{array} \begin{array}{c} \begin{array}{ccc} X_1 & X_2 & X_3 \end{array} \\ \begin{bmatrix} \text{Var}(X_1) & \text{Cov}(X_1, X_2) & \text{Cov}(X_1, X_3) \\ \text{Cov}(X_2, X_1) & \text{Var}(X_2) & \text{Cov}(X_2, X_3) \\ \text{Cov}(X_3, X_1) & \text{Cov}(X_3, X_2) & \text{Var}(X_3) \end{bmatrix} \end{array}.$$

Recall that a correlation is essentially a standardized covariance, that is,

$$r_{XY} = \text{Cov}(X, Y)/(SD_X \times SD_Y).$$

Thus, if we have **S**, it can easily be converted to an **R** matrix.

One additional matrix that is useful is a matrix that contains the sums of squares (SS) and the sums of cross products (SCP) for all the predictor variables:

$$SS(X) = \Sigma\,(X - M_X)^2,$$

$$\text{SCP}(X, Y) = \Sigma\,(X - M_X) \times (Y - M_Y).$$

The SS for a variable is equivalent to the SCP of that variable with itself. Thus, the **SCP** matrix for a set of variables X_1, X_2, and X_3 would have elements as follows:

$$\mathbf{SCP} = \begin{array}{c} \\ X_1 \\ X_2 \\ X_3 \end{array} \begin{array}{c} \begin{array}{ccc} X_1 & X_2 & X_3 \end{array} \\ \begin{bmatrix} \text{SS}(X_1) & \text{SCP}(X_1, X_2) & \text{SCP}(X_1, X_3) \\ \text{SCP}(X_2, X_1) & \text{SS}(X_2) & \text{SCP}(X_2, X_3) \\ \text{SCP}(X_3, X_1) & \text{SCP}(X_3, X_2) & \text{SS}(X_3) \end{bmatrix} \end{array}.$$

The covariance between two variables is just the SCP for those variables divided by N:

$$\text{Cov}(X, Y) = \text{SCP}(X, Y)/N.$$

Thus, we can easily convert an **SCP** matrix into an **S** matrix by dividing every element of the **SCP** matrix by N or df.

Another useful matrix is called the identity matrix, and it is denoted by **I**. The diagonal elements of an identity matrix are all 1s, and the off-diagonal elements are all 0s, as in the following example of a 3×3 identity matrix:

$$\mathbf{I} = \begin{bmatrix} 1 & 0 & 0 \\ 0 & 1 & 0 \\ 0 & 0 & 1 \end{bmatrix}.$$

There are several special cases of matrix form. For example, a row vector is a matrix that has just one row and any number of columns. A column vector is a matrix that has just one column and any number of rows. A commonly encountered column vector is the set of scores on Y, the dependent variable in a regression. The score in each row of this vector corresponds to the Y score for one participant. If $N = 4$, then there will be four scores; in general, the column vector Y is a matrix with N rows and one column:

$$Y = \begin{bmatrix} Y_1 \\ Y_2 \\ Y_3 \\ Y_4 \end{bmatrix}.$$

A matrix of size 1×1 (i.e., just one row and one column) is called a scalar; it corresponds to a single number (such as N).

Matrix Addition and Subtraction

It is possible to do arithmetic operations with matrices. For example, to add matrix **A** to matrix **B**, you add the corresponding elements (the elements in **A** that have the same combination of subscripts as the corresponding elements in **B**). Two matrices can be added (or subtracted) only if they are of the same size (i.e., they have the same number of rows and the same number of columns).

Here are three small matrices that will be used for practice in matrix operations:

$$\mathbf{A} = \begin{bmatrix} 5 & 3 \\ 2 & 1 \end{bmatrix}, \quad \mathbf{B} = \begin{bmatrix} 2 & 1 \\ 5 & 8 \end{bmatrix}, \quad \mathbf{C} = \begin{bmatrix} 4 & 2 & 6 \\ 7 & 3 & 9 \end{bmatrix}.$$

To add **B** to **A**, you simply add the corresponding elements:

$$\mathbf{A} + \mathbf{B} = \mathbf{D} = \begin{bmatrix} (5+2) & (3+1) \\ (2+5) & (1+8) \end{bmatrix} = \begin{bmatrix} 7 & 4 \\ 7 & 9 \end{bmatrix}.$$

To subtract matrix **B** from **A** (**A** − **B** = **E**), you need to subtract the corresponding elements, as follows:

$$\mathbf{A} + \mathbf{B} = \mathbf{E} = \begin{bmatrix} (5-2) & (3-1) \\ (2-5) & (1-8) \end{bmatrix} = \begin{bmatrix} 3 & 2 \\ -3 & -7 \end{bmatrix}.$$

It is not possible to add **A** to **C**, or subtract **C** from **A**, because the **A** and **C** matrices do not have the same dimensions; **C** has three columns, while **A** has only two columns.

Matrix Multiplication

Matrix multiplication is somewhat more complex than addition because it involves both multiplying and summing elements. It is only possible to multiply a **Q** matrix by a **W** matrix if the number of columns of **Q** equals the number of rows in **W**. In general, if you multiply an $m \times n$ matrix by a $q \times w$ matrix, n must equal q, and the product matrix will be of dimensions $m \times w$. Note also that the product **A** × **B** is not generally equal to the product **B** × **A**. The order in which the terms are listed makes a difference in how the elements are combined.

The method of matrix multiplication I learned from David Kenny makes the process easy. To multiply **A** × **B**, set up the computations as follows: Write the first matrix, **A**, on the left; write the second matrix, **B**, at the top of the page; put the product matrix (which we will call **D**) between them. Notice that we are multiplying a 2 × 2 matrix by another 2 × 2 matrix. The multiplication is possible because the number of columns in the first matrix (**A**) equals the number of rows in the second matrix (**B**). When we multiply an $m \times n$ matrix by a $q \times w$ matrix, the inner dimensions (n and q) must be equal, and these terms disappear; the product matrix **D** will be of size $m \times w$. Initially, the elements of the product matrix **D** are unknown (d_{11}, d_{12}, etc.):

$$B = \begin{bmatrix} 2 & 1 \\ 5 & 8 \end{bmatrix}$$

$$A = \begin{bmatrix} 5 & 3 \\ 2 & 1 \end{bmatrix} \begin{bmatrix} d_{11} & d_{12} \\ d_{21} & d_{22} \end{bmatrix}.$$

To find each element of **D**, you take the vector product of the row immediately to the left and the column immediately above. That is, you cross multiply corresponding elements for each row combined with each column, and then sum these products. For the problem above, the four elements of the **D** matrix are found as follows:

$$d_{11} = a_{11} \times b_{11} + a_{12} \times b_{21}$$
$$(5 \times 2) + (3 \times 5)$$
$$25$$

$$d_{12} = a_{11} \times b_{12} + a_{12} \times b_{22}$$
$$(5 \times 1) + (3 \times 8)$$
$$29$$

$$d_{21} = a_{21} \times b_{11} + a_{22} \times b_{21}$$
$$(2 \times 2) + (1 \times 5)$$
$$9$$

$$d_{22} = a_{21} \times b_{12} + a_{22} \times b_{22}$$
$$(2 \times 1) + (1 \times 8)$$
$$10.$$

Note that matrix multiplication is not commutative: **A** × **B** does not usually yield the same result as **B** × **A**.

To multiply two matrices, the number of columns in the first matrix must equal the number of rows in the second matrix; if this is not the case, then you cannot match the elements up one to one to form the vector product. In the sample matrices given above, matrix **A** was 2 × 2 and **C** was 2 × 3. Thus, we can multiply **A** × **C**, and obtain a 2 × 3 matrix as the product; however, we cannot multiply **C** × **A**.

When you multiply any matrix by the identity matrix **I**, it is the matrix algebra equivalent of multiplication by 1. That is, **A** × **I** = **A**. As an example,

$$I = \begin{bmatrix} 1 & 0 \\ 0 & 1 \end{bmatrix}$$

$$A = \begin{bmatrix} 5 & 3 \\ 2 & 1 \end{bmatrix} \begin{bmatrix} (5 * 1) + (3 * 0) = 5 & (5 * 0) + (1 * 3) = 3 \\ (2 * 1) + (1 * 0) = 2 & (2 * 0) + (1 * 1) = 1 \end{bmatrix}.$$

Matrix Inverse

This is the matrix operation that is equivalent to division. In general, dividing by a constant c is equivalent to multiplying by $1/c$ or c^{-1}. Both $1/c$ and c^{-1} are notations for the inverse of c. The matrix equivalent of 1 is the identity matrix, which is a square matrix with 1s in the diagonal and all other elements 0.

Suppose that we need to find the inverse of matrix **F**, and the elements of this **F** matrix are as follows:

$$\mathbf{F} = \begin{bmatrix} 2 & 6 \\ 4 & 1 \end{bmatrix}.$$

The inverse of **F**, denoted by $\mathbf{F}^{-1}$, is the matrix that yields **I** when we form the product $\mathbf{F} \times \mathbf{F}^{-1}$. Thus, one way to find the elements of the inverse matrix $\mathbf{F}^{-1}$ is to set up the following multiplication problem and solve for the elements of $\mathbf{F}^{-1}$:

$$F^{-1} = \begin{bmatrix} f_{11} & f_{12} \\ f_{21} & f_{22} \end{bmatrix}$$

$$F = \begin{bmatrix} 2 & 6 \\ 1 & 4 \end{bmatrix} \begin{bmatrix} 1 & 0 \\ 0 & 1 \end{bmatrix}.$$

When we carry out this multiplication, we find the following set of four equations in four unknowns:

$$2f_{11} + 6f_{21} = 1, 2f_{12} + 6f_{22} = 0,$$

$$1f_{11} + 4f_{21} = 0, 1f_{12} + 4f_{22} = 1.$$

We can solve for the elements of **F** by substitution:

$$\text{From } 1f_{11} + 4f_{21} = 0, \text{ we get } f_{11} = -4f_{21}.$$

Substituting this for f_{11} in the first equation, we get

$$2f_{11} + 6f_{21} = 1,$$

$$2(-4f_{21}) + 6f_{21} = 1,$$

$$-2f_{21} = 1,$$

$$f_{21} = -1/2.$$

Substituting this value for f_{21} back into the equation we had for f_{11} in terms of f_{21} yields

$$f_{11} = -4f_{21} = -4(-1/2) = +2.$$

Similarly, we can solve the other pair of equations:

$$2f_{12} + 6f_{22} = 0$$

can be rearranged to give

$$6f_{22} = -2f_{12},$$

$$f_{22} = -1/3 \times f_{12}.$$

Substituting this value for f_{22} (in terms of f_{12}) into the last equation, we have

$$1f_{12} + 4f_{22} = 1,$$

$$1f_{12} + 4(-1/3 \times f_{12}) = 1,$$

$$-1/3f_{12} = 1,$$

$$f_{12} = -3.$$

We can now find f_{22} by using this known value for f_{12}:

$$f_{22} = -1/3f_{12} = -1/3 \times (-3) = +1.$$

The inverse matrix $\mathbf{F}^{-1}$ is thus

$$\begin{bmatrix} 2 & -3 \\ -\frac{1}{2} & 1 \end{bmatrix}.$$

To check, we multiply $\mathbf{FF}^{-1}$ to make sure that the product is $\mathbf{I}$:

$$\begin{bmatrix} 2 & -3 \\ 1/2 & 1 \end{bmatrix}$$

$$\begin{bmatrix} 2 & 6 \\ 1 & 4 \end{bmatrix} \begin{bmatrix} 4-3=1 & -6+6=0 \\ 2-2=0 & -3+4=1 \end{bmatrix}.$$

There are computational shortcuts and algorithms that can be used to compute inverses for larger matrices. The method used here should make it clear that the inverse of $\mathbf{F}$, $\mathbf{F}^{-1}$, is the matrix for which the $\mathbf{F} \times \mathbf{F}^{-1}$ product equals the identity matrix $\mathbf{I}$. Returning to our original definition of the inverse of c as $1/c$, it should be clear that just as multiplying by $(1/c)$ is equivalent to dividing by c, multiplying a matrix by $\mathbf{F}^{-1}$ is essentially equivalent to "dividing" by the $\mathbf{F}$ matrix.

Matrix Transpose

The transpose operation involves interchanging the rows and columns of a matrix. Consider the $\mathbf{C}$ matrix given earlier:

$$\mathbf{C} = \begin{bmatrix} 4 & 2 & 6 \\ 7 & 3 & 9 \end{bmatrix}.$$

The transpose of $\mathbf{C}$, denoted by $\mathbf{C}'$, is found by turning Column 1 into Row 1, Column 2 into Row 2, and so forth. Thus,

$$\mathbf{C}' = \begin{bmatrix} 4 & 7 \\ 2 & 3 \\ 6 & 9 \end{bmatrix}.$$

If $\mathbf{C}$ is a 2×3 matrix, then $\mathbf{C}'$ will be a 3×2 matrix.

The transpose operation is useful because we sometimes need to "square" a matrix—that is, multiply a matrix by itself. For example, the **X** matrix is an $N \times k$ matrix that contains scores for N participants on k variables. If we need to obtain sums of squared elements, we cannot multiply $\mathbf{X} \times \mathbf{X}$ (the number of columns in the first matrix must equal the number of rows in the second matrix). However, we can multiply $\mathbf{X'X}$; this will be a $k \times N$ matrix multiplied by an $N \times k$ matrix, and it will yield a $k \times k$ matrix in which the terms are sums of squared scores and sums of cross products of scores; these are the building blocks needed to obtain the sum of cross products (**SCP**) or variance/covariance matrix (**S**) for the k variables or the correlation matrix **R**.

Determinant

The **determinant of a matrix**, such as **A**, is denoted by $|\mathbf{A}|$. For a 2 × 2 matrix, the determinant is found by subtracting the product of the elements on the minor diagonal (from upper right to lower left corner) from the product of the elements on the major diagonal (from upper left to lower right):

$$\left\| \begin{bmatrix} a & b \\ c & d \end{bmatrix} \right\| = ad - bc.$$

The computation of determinants for larger matrices is more complex but essentially involves similar operations—that is, forming products of elements along the major diagonals (upper left to lower right) and summing these products; and forming products of elements along minor diagonals (from lower left to upper right) and subtracting these products.

The determinant has a useful property: It tells us something about the amount of *nonshared* or nonredundant variance among the rows and/or columns of a matrix. If the determinant is 0, then at least one row is perfectly predictable from some linear combination of one or more other rows in the matrix. This means that the matrix is singular: One predictor variable in the data matrix is perfectly predictable from other predictor variables.

For example, consider the correlation matrix **R** for X_1 and X_2:

$$\mathbf{R} = \begin{bmatrix} 1 & r_{12} \\ r_{12} & 1 \end{bmatrix}.$$

The determinant of $\mathbf{R} = (1)(1) - (r_{12} \times r_{12}) = 1 - r_{12}^2$. Here, $1 - r_{12}^2$ is the proportion of variance in X_1 that is not shared with X_2 and vice versa. If $r_{12} = 1$, then the determinant of **R** will be 0; this would tell us that X_1 is perfectly predictable from X_2. When you have an **R** matrix for k variables, the determinant provides information about multicollinearities that are less obvious; for example, if $X_3 = X_1 + X_2$, the determinant of **R** for a matrix of correlations among X_1, X_2, and X_3 will be 0.

When a matrix has a determinant of 0 you *cannot* calculate an inverse for the matrix (because computation of the inverse, in most algorithms, involves dividing by the determinant).

Here is an example of a matrix **G**; let Column 1 = X_1 and Column 2 = X_2; note that $X_2 = 3 \times X_1$; that is, X_2 is perfectly predictable from X_1:

$$\mathbf{G} = \begin{bmatrix} 2 & 6 \\ 4 & 12 \end{bmatrix}.$$

The determinant of $\mathbf{G} = (2 \times 12) - (4 \times 6) = 24 - 24 = 0$.

Thus, it would not be possible to compute an inverse for this **G** matrix. When we do a multiple regression, the program computes a determinant for the matrices (such as **X′X** and **R**) that it calculates from the data. If a determinant of exactly 0 is found, this means that there is a perfect correlation between two predictors, or that one predictor variable is perfectly predictable from a weighted linear combination of other predictor variables. When this happens, a regression analysis cannot be performed. A zero determinant is reported using several different kinds of error message: determinant = 0, singular matrix, or matrix not of full rank.

Before you attempt to compute an inverse for a matrix, you should ask two questions. First, is the matrix square (i.e., number of rows equal to number of columns)? You can only compute an inverse for a square matrix. Second, does the matrix have a determinant of 0? If the determinant equals 0, you cannot calculate an inverse.

In running multiple regression programs, information about the determinant of the **X** and **R** matrices is helpful in the assessment of multicollinearity—that is, the strength of correlation among predictors. Perfect multicollinearity means that at least one *X* variable can be predicted perfectly from one or more of the other *X* variables. A determinant near 0 indicates strong, but not perfect, multicollinearity; a near-zero determinant suggests that predictors are highly correlated. For reasons discussed in Chapter 11 (on bivariate regression), it is better to avoid situations in which the predictors are very highly correlated with each other. When there are correlations in excess of .9 among predictors, the *X* variables "compete" to explain the same variance, and sometimes no single variable is significant. Furthermore, as correlations among predictors increase, the width of the confidence intervals around estimates of *b* slope coefficients increases.

These matrix operations can be applied to the problem of finding *b* coefficient estimates in regression with *k* predictor variables.

Recall (from Chapter 11) that the normal equations for a regression to predict z_Y from z_{X_1} and z_{X_2} were as follows:

$$r_{Y_1} = \beta_1 + r_{12}\beta_2,$$

$$r_{Y_2} = r_{12}\beta_1 + \beta_2.$$

These equations were obtained by applying the tracing rule to a path diagram that represented z_{X_1} and z_{X_2} as correlated predictors of z_Y; these equations show how the observed correlation between each predictor variable and the *Y* outcome variable could be reconstructed from the direct and indirect paths that lead from each predictor variable to *Y*.

The next step is to rewrite these two equations using matrix notation. Once we have set up these equations in matrix algebra form, the computations can be represented using simpler notation, and the computational procedures can easily be generalized to situations with any number of predictors. The matrix notation that will be used is as follows:

$\mathbf{R}_{iY}$ is a column vector containing the correlation of each X predictor variable with Y; for the case with two independent variables, the elements of $\mathbf{R}_{iY}$ are

$$\begin{bmatrix} r_{Y_1} \\ r_{Y_2} \end{bmatrix}.$$

In general, $\mathbf{R}_{iY}$ is a column vector with k elements, each of which corresponds to the correlation of one X predictor variable with Y.

β_i is a column vector containing the beta coefficient, or standardized slope coefficient, for each of the predictors. For two predictors, the elements of β_i are as follows:

$$\begin{bmatrix} \beta_1 \\ \beta_2 \end{bmatrix}.$$

In general, the beta vector has k elements; each one is the beta coefficient for one of the z_{Xi} predictor variables.

$\mathbf{R}_{ii}$ is a matrix of correlations among all possible pairs of the predictors:

$$\mathbf{R}_{ii} = \begin{bmatrix} 1 & r_{12} \\ r_{12} & 1 \end{bmatrix}.$$

In the case of two predictors, $\mathbf{R}_{ii}$ is a 2×2 matrix; in the more general case of k predictors, this is a $k \times k$ matrix that includes the correlations among all possible pairs of predictor variables. The normal equations above (Equations 14a and 14b) can be written more compactly in matrix notation as

$$\mathbf{R}_{iY} = \mathbf{R}_{ii} \times \beta_i.$$

To verify that the matrix algebra equation (Equation 14.5) is equivalent to the set of two normal equations given earlier, multiply the **R** matrix by the beta vector:

$$\begin{bmatrix} \beta_1 \\ \beta_2 \end{bmatrix}$$

$$\begin{bmatrix} 1 & r_{12} \\ r_{12} & 1 \end{bmatrix} \begin{bmatrix} \beta_1 & + r_{12}\beta_2 \\ r_{12}\beta_1 + & \beta_2 \end{bmatrix}.$$

When we had just two predictor variables, it was relatively easy to solve this set of two equations in two unknowns by hand using substitution methods (see Chapter 11). However, we can write a solution for this problem using matrix algebra operations; this matrix algebra version provides a solution for the more general case with k predictor variables.

To solve this equation—that is, to solve for the vector of beta coefficients in terms of the correlations among predictors and the correlations between predictors and Y, we just multiply both sides of this matrix equation by the inverse of $\mathbf{R}_{ii}$. Because $\mathbf{R}_{ii}\mathbf{R}_{ii}^{-1} = \mathbf{I}$, the identity matrix, this term disappears, leaving us with the following equation:

$$\boldsymbol{\beta} = \mathbf{R}^{-1}{}_{ii}\,\mathbf{R}_{iY}.$$

This matrix equation thus provides a way to calculate the set of beta coefficients from the correlations; when k, the number of predictor variables, equals 2, this is equivalent to the operations we went through to get the betas from the rs when we solved the normal equations by hand using substitution methods. This matrix notation generalizes to matrices of any size—that is, situations involving any number of predictors. It is not difficult to find the beta coefficients by hand when you only have two predictors, but the computations become cumbersome as the number of predictor variables increases. Matrix algebra gives us a convenient and compact notation that works no matter how many predictor variables there are.

Thus, no matter how many predictors we have in a multiple regression, the betas are found by multiplying the column vector of correlations between predictors and Y by the inverse of the correlation matrix among the predictors. Computer programs use algorithms for matrix inverse that are much more computationally efficient than the method of computing matrix inverse shown earlier in this appendix.

Note that if the determinant of $\mathbf{R}_{ii}$ is 0—that is, if any predictor is perfectly predictable from one or more other predictors, then we cannot do this computation (it would be equivalent to dividing by 0).

Using the Raw Score Data Matrices for *X* and *Y* to Calculate *b* Coefficients

Another matrix representation of multiple regression involves calculating the b coefficients directly from the raw scores. Let $\mathbf{X}$ be the independent variable data matrix (each row = one subject's raw scores; each column = scores on one of the independent variables). We will assume that prior to all the other computations in this section, each X variable and the Y variable are converted into deviations from their means. This X matrix is usually augmented by adding a column of 1s; this column of 1s will provide the information needed to estimate the intercept term, a. Let $\mathbf{Y}$ be the vector of raw scores of each subject on the dependent variable. To keep this very simple, let's just use two predictors and four subjects, but the method applies to any number of variables and subjects. The matrix computation that gives us the vector of b coefficients is

$$\mathbf{B} = (\mathbf{X}'\mathbf{X})^{-1}\,\mathbf{X}'\mathbf{Y},$$

where the $\mathbf{B}$ vector corresponds to the list of raw score regression coefficients, including b_0, the intercept:

$$\begin{bmatrix} b_0 \\ b_1 \\ b_2 \\ \vdots \\ b_k \end{bmatrix}.$$

$\mathbf{B} = \mathbf{X'X}^{-1}\mathbf{X'Y}$ includes terms that are analogous to those in the earlier standard score equation, $\beta = \mathbf{R}^{-1}{}_{ii}\,\mathbf{R}_{ii'}\ \mathbf{X'X}$, like $\mathbf{R}_{ii'}$, contains information about covariances (or correlations) among the predictors. $\mathbf{X'Y}$, like $\mathbf{R}_{iY}$, contains information about the covariation (or correlation) between each predictor and Y. Each x_{ij} element is the deviation from the mean of the score of subject i on variable j; X_i is the vector of scores of all subjects on variable i. Let's look at this product $\mathbf{X'X}$ (with an added column of 1s in $\mathbf{X}$):

$$\begin{array}{c} \\ \begin{bmatrix} 1 & 1 & 1 & 1 \\ X_{11} & X_{21} & X_{31} & X_{41} \\ X_{12} & X_{22} & X_{32} & X_{42} \end{bmatrix} \end{array} \begin{array}{c} \begin{bmatrix} 1 & X_{11} & X_{12} \\ 1 & X_{21} & X_{22} \\ 1 & X_{31} & X_{32} \\ 1 & X_{41} & X_{42} \end{bmatrix} \\ \begin{bmatrix} N & \sum X_1 & \sum X_2 \\ \sum X_1 & X_{11}^2 + X_{21}^2 + X_{31}^2 + X_{41}^2 & X_{11}X_{12} + X_{21}X_{22} + X_{31}X_{32} + X_{41}X_{42} \\ \sum x_2 & X_{11}X_{12} + X_{21}X_{22} + X_{31}X_{32} + X_{41}X_{42} & X_{12}^2 + X_{22}^2 + X_{32}^2 + X_{42}^2 \end{bmatrix} \end{array}.$$

The elements of this product matrix correspond to the following familiar terms:

$$\mathbf{X'X} = \begin{bmatrix} N & \sum X_1 & \sum X_2 \\ \sum X_1 & \sum X_1^2 & \sum X_1 X_2 \\ \sum X_2 & \sum X_1 X_2 & \sum X_2^2 \end{bmatrix}.$$

The diagonal elements of $\mathbf{X'X}$ contain the squared scores on each of the predictors or the basic information about the variance of each predictor. The off-diagonal elements of $\mathbf{X'X}$ contain the cross products of all possible pairs of predictors or the basic information about the covariance of each pair of predictors. The first row and column just contain the sums of the scores on each predictor, and the first element is N, the number of subjects. This is all the information that is needed to calculate the vector of raw score regression coefficients, $\mathbf{B} = (b_0, b_1, b_2, \ldots, b_k)$. We can obtain the variance/covariance matrix $\mathbf{S}$ for this set of predictors by making a few minor changes in the way we compute $\mathbf{X'X}$. If we replace the scores on each X variable with deviations from the scores on the means of each variable, omit the added column of 1s, form $\mathbf{X'X}$, and then divide each element of $\mathbf{X'X}$ by N, the number of scores, we obtain the variance/covariance matrix. The sample value of this matrix is usually denoted by $\mathbf{S}$; the corresponding population parameter matrix is $\mathbf{\Sigma}$:

$$\mathbf{S} = \begin{bmatrix} \mathrm{Var}(X_1) & \mathrm{Cov}(X_1, X_2) \\ \mathrm{Cov}(X_1, X_2) & \mathrm{Var}(X_2) \end{bmatrix}.$$

Similarly, $\mathbf{X'Y}$ involves cross multiplying scores on the predictors with scores on the dependent variable (Y) to obtain a sum of cross products term for each predictor or X variable with Y:

$$\begin{bmatrix} 1 & 1 & 1 & 1 \\ x_{11} & x_{21} & x_{31} & x_{41} \\ x_{12} & x_{22} & x_{32} & x_{42} \end{bmatrix} \begin{bmatrix} y_1 \\ y_2 \\ y_3 \\ y_4 \end{bmatrix} = \begin{bmatrix} y_1 + y_2 + y_3 + y_4 \\ x_{11}y_1 + x_{21}y_2 + x_{31}y_3 + x_{41}y_4 \\ x_{12}y_1 + x_{22}y_2 + x_{32}y_3 + x_{42}y_4 \end{bmatrix}.$$

The elements of this product matrix correspond to

$$\mathbf{X'Y} = \begin{bmatrix} \sum Y \\ \sum X_1 Y \\ \sum X_2 Y \end{bmatrix}.$$

This $\mathbf{X'Y}$ product vector contains information that is related to the SCPs and could be used to calculate the SCP for each predictor variable with Y.

We can convert the variance/covariance matrix $\mathbf{S}$ to the correlation matrix $\mathbf{R}_{ii}$ in a few easy steps. Correlation is essentially a standardized covariance—that is, the X, Y covariance divided by SD_Y and SD_X. If we divide each row and each column through by the standard deviation for the corresponding variables, we can easily obtain $\mathbf{R}_{ii}$ from $\mathbf{S}$.

$\mathbf{X'X}$ and $\mathbf{X'Y}$ are the basic building blocks or "chunks" that are included in the computation of many multivariate statistics. Understanding the information that these matrices contain will help you to recognize what is going on as we look at additional multivariate techniques. Examination of the matrix algebra makes it clear that the values of β or b coefficients are influenced by the correlations among predictors as well as by the correlations between predictors and Y.

In practice, most computer programs find the b coefficients from $\mathbf{X'X}$ and $\mathbf{X'Y}$ and then derive other regression statistics from these, but it should be clear from this discussion that either the raw scores on the Xs and Y or the correlations among all the variables along with means and standard deviations for all variables are sufficient information to do a multiple regression analysis.

To summarize, computer programs typically calculate the raw score regression coefficients, contained in the vector **B**, with elements ($b_0, b_1, b_2, \ldots, b_k$) by performing the following matrix algebra operation: $\mathbf{B} = \mathbf{X'X}^{-1}\mathbf{X'Y}$. This computation cannot be performed if $\mathbf{X'X}$ has a zero determinant; a zero determinant indicates that at least one X variable is perfectly predictable from one or more other X variables. Once we have the **B** vector, we can write a raw score prediction equation as follows:

$$Y' = b_0 + b_1 X_1 + b_2 X_2 + \ldots + b_k X_k.$$

Each b_i coefficient has an associated standard error (SE_{bi}). This SE term may be used to set up a confidence interval for each b_i and to set up a t ratio to test the null hypothesis that each b_i slope equals 0. This is done using the same formulas reported earlier in Chapter 11.

The standardized slopes or beta coefficients can also be calculated for any number of predictors by doing the following matrix computation: $\boldsymbol{\beta} = \mathbf{R}_{ii}^{-1}\, \mathbf{R}_{iY}$. Once we have this beta vector, we can set up the equation to predict standard scores on Y from standardized or z scores on the predictors as follows:

$$z'_Y = \beta_1 z_{X_1} + \beta_2 z_{X_2} + \cdots + \beta_k z_{X_k}.$$

Once we have computed the βs we can convert them to bs (or vice versa).

It would be tedious to do the matrix algebra by hand (in particular, the computation of the inverse of $\mathbf{X'X}$ is time-consuming). You will generally obtain estimates of slope coefficients and other regression results from the computer program. Even though we will not do by-hand computations of coefficients from this point onward, it is still potentially useful to understand the matrix algebra formulas. First of all, when you understand the $(\mathbf{X'X})^{-1}\mathbf{X'Y}$ equation "as a sentence," you can see what information is taken into account in the computation of the b coefficients. In words, $(\mathbf{X'X})^{-1}\mathbf{X'Y}$ tells us to form all the sums of cross products between each pair of X predictor variables and Y (i.e., calculate $\mathbf{X'Y}$) and then divide this by the matrix that contains information about all the cross products for all pairs of X predictor variables (i.e., $X'X$). The expression $(\mathbf{X'X})^{-1}\mathbf{X'Y}$ is the matrix algebra generalization of the computation for the single b slope coefficient in a regression of the form $Y' = b_0 + b_1X_1$. For a one-predictor regression equation, the estimate of the raw score slope $b = \text{SCP}/SS_X$, where SCP is the sum of cross products between X and Y, $\text{SCP} = \sum[(X - M_X) \times (Y - M_Y)]$, and SS_X is the sum of squares for X, $SS_X = \sum (X - M_X)^2$. The information contained in the matrix $\mathbf{X'X}$ is the multivariate generalization of SS_X. If we base all our computations on deviations of raw scores from the appropriate means, the $X'X$ matrix includes the sum of squares for each individual X_i predictor and also the cross products for all pairs of X predictor variables; these cross products provide the information we need to adjust for correlation or redundancy among predictors. The information contained in the matrix $\mathbf{X'Y}$ is the multivariate generalization of SCP; these sums of cross products provide information that can be used to obtain a correlation between each X_i predictor and Y. Regression slope coefficients are obtained by dividing SCP by SS_X (or by premultiplying $\mathbf{X'Y}$ by the inverse of $\mathbf{X'X}$, which is the matrix equivalent of dividing by $\mathbf{X'X}$).

To compute b or β coefficients, we need to know about the covariance or correlation between each X predictor and Y, but we also need to take into account the covariances or correlations among all the X predictors. Thus, adding or dropping an X_i predictor can change the slope coefficient estimates for all the other X variables. The beta slope coefficient for a predictor can change (sometimes dramatically) as other predictors are added to (or dropped from) the regression equation.

Understanding the matrix algebra makes some of the error messages from compute programs intelligible. When SPSS or some other program reports "zero determinant," "singular matrix," or "matrix not of full rank," this tells you that there is a problem with the **X′X** matrix; specifically, it tells you that at least one X predictor variable is perfectly predictable from other X predictors. We could also say that there is perfect multicollinearity among predictors when the determinant of **X′X** is 0. For example, suppose that X_1 = VSAT, X_2 = MSAT, and X_3 = total SAT (verbal + math). If you try to predict college grade point average from X_1, X_2, and X_3, the **X** data matrix will have a determinant of 0 (because "total SAT" X_3 is perfectly predictable from X_1 and X_2). To get rid of this problem, you would have to drop the X_3 variable from your set of predictors. Another situation in which perfect multicollinearity may arise involves the use of dummy variables as predictors (refer back to Chapter 12). If you have k groups and you use k dummy variables, the score on the last dummy variable will be perfectly predictable from the scores on the first $k - 1$ dummy variables, and you will have a zero determinant for **X′X**. To get rid of this problem, you have to drop one dummy variable.

Another reason why it may be useful to understand the matrix algebra for multiple regression is that once you understand what information is contained in meaningful chunks of matrix algebra (such as **X′X**, which contains information about the variances and covariances of the X predictors), you will recognize these same terms again in the matrix algebra for other multivariate procedures.

Appendix 14.B: Tables for Wilkinson and Dallal (1981) Test of Significance of Multiple R^2 in Method = Forward Statistical Regression

Critical Values for Squared Multiple Correlation (R^2) in Forward Stepwise Selection
$\alpha = .05$

		N – k – 1															
k	*F*	*10*	*12*	*14*	*16*	*18*	*20*	*25*	*30*	*35*	*40*	*50*	*60*	*80*	*100*	*150*	*200*
2	2	43	38	33	30	27	24	20	16	14	13	10	8	6	5	3	2
2	3	40	36	31	27	24	22	18	15	13	11	9	7	5	4	2	2
2	4	38	33	29	26	23	21	17	14	12	10	8	7	5	4	3	2
3	2	49	43	39	35	32	29	24	21	18	16	12	10	8	7	4	2
3	3	45	40	36	32	29	26	22	19	17	15	11	9	7	6	4	3
3	4	42	36	33	29	27	25	20	17	15	13	11	9	7	5	4	3
4	2	54	48	44	39	35	33	27	23	20	18	15	12	10	8	5	4
4	3	49	43	39	36	33	30	25	22	19	17	14	11	8	7	5	4
4	4	45	39	35	32	29	27	22	19	17	15	12	10	8	6	5	3
5	2	58	52	47	43	39	36	31	26	23	21	17	14	11	9	6	5
5	3	52	46	42	38	35	32	27	24	21	19	16	13	9	8	5	4
5	4	46	41	38	35	52	29	24	21	18	16	13	11	9	7	5	4
6	2	60	54	50	46	41	39	33	29	25	23	19	16	12	10	7	5
6	3	54	48	44	40	37	34	29	25	22	20	17	14	10	8	6	5
6	4	48	43	39	36	33	30	26	23	20	17	14	12	9	7	5	4
7	2	61	56	51	48	44	41	35	30	27	24	20	17	13	11	7	5
7	3	59	50	46	42	39	36	31	26	23	21	18	15	11	9	7	5
7	4	50	45	41	38	35	32	27	24	21	18	15	13	10	8	6	4
8	2	62	58	53	49	46	43	37	31	28	26	21	18	14	11	8	6
8	3	57	52	47	43	40	37	32	28	24	22	19	16	12	10	7	5
8	4	51	46	42	39	36	33	28	25	22	19	16	14	11	9	7	5
9	2	63	59	54	51	47	44	38	33	30	27	22	19	15	12	9	6
9	3	58	53	49	44	41	38	33	29	25	23	20	16	12	10	7	6
9	4	52	46	43	40	37	34	29	25	23	20	17	14	11	10	7	6
10	2	64	60	55	52	49	46	39	34	31	28	23	20	16	13	10	7
10	3	59	54	50	45	42	39	34	30	26	24	20	17	13	11	8	6
10	4	52	47	44	41	38	35	30	26	24	21	18	15	12	10	8	6
12	2	66	62	57	54	51	48	42	37	33	30	25	22	17	14	10	8
12	3	60	55	52	47	44	41	36	31	28	25	22	19	14	12	9	7
12	4	53	48	45	41	39	36	31	27	25	22	19	16	13	11	9	7
14	2	68	64	60	56	53	50	44	39	35	32	27	24	18	15	11	8
14	3	61	57	53	49	46	43	37	32	29	27	23	20	15	13	10	8
14	4	43	49	46	42	40	37	32	29	26	23	20	17	13	11	9	7
16	2	69	65	61	58	55	53	46	41	37	34	29	25	20	17	12	9
16	3	61	58	54	50	47	44	38	34	31	28	24	21	17	14	11	8
16	4	53	50	46	43	40	38	33	30	27	24	21	18	14	12	10	8
18	2	70	67	63	60	57	55	49	44	40	36	31	27	21	18	13	9
18	3	62	59	55	51	49	46	40	35	32	30	26	23	18	15	12	9
18	4	54	50	46	44	41	38	34	31	28	25	22	19	15	13	11	8
20	2	72	68	64	62	59	56	50	46	42	38	33	28	22	19	14	10
20	3	62	60	56	52	50	47	42	37	34	31	27	24	19	16	12	9
20	4	54	50	46	44	41	37	35	32	29	26	23	20	16	14	11	8

(Continued)

(Continued)

		N − k − 1															
k	*F*	*10*	*12*	*14*	*16*	*18*	*20*	*25*	*30*	*35*	*40*	*50*	*60*	*80*	*100*	*150*	*200*
2	2	59	53	48	43	40	36	30	26	23	20	17	14	11	9	7	5
2	3	58	52	46	42	38	35	30	25	22	19	16	13	10	8	6	4
2	4	57	49	44	39	36	32	26	22	19	16	13	11	8	7	5	4
3	2	67	60	55	50	46	42	35	30	27	24	20	17	13	11	7	5
3	3	63	58	52	47	43	40	34	29	25	22	19	16	12	10	7	5
3	4	61	54	48	44	40	37	31	26	23	20	16	14	11	9	6	5
4	2	70	64	58	53	49	46	39	34	30	27	23	19	15	12	8	6
4	3	67	62	56	51	47	44	37	32	28	25	21	18	14	11	8	6
4	4	64	58	52	47	43	40	34	29	26	23	19	16	13	11	7	6
5	2	73	67	61	57	52	49	42	37	32	29	25	21	16	13	9	7
5	3	70	65	59	54	50	46	39	34	30	27	23	19	15	12	9	7
5	4	65	60	55	50	46	43	36	31	28	25	20	17	14	12	8	6
6	2	74	69	63	59	55	51	44	39	34	31	26	23	18	14	10	8
6	3	72	67	61	56	51	48	41	36	32	28	24	20	16	13	10	7
6	4	66	61	56	52	48	45	38	33	29	26	22	19	15	13	9	7
7	2	76	70	65	60	56	53	46	40	36	33	28	25	19	15	11	9
7	3	73	68	62	57	53	50	42	37	33	30	25	21	17	14	10	8
7	4	67	62	58	54	49	46	40	35	31	28	23	20	16	14	10	8
8	2	77	72	66	62	58	55	48	42	38	34	29	26	20	16	12	9
8	3	74	69	63	58	54	51	44	39	34	31	26	22	18	15	11	9
8	4	67	63	59	55	50	47	41	36	32	29	24	21	17	15	11	9
9	2	78	73	67	63	60	56	49	43	39	36	31	27	21	17	12	10
9	3	74	69	64	59	56	52	45	40	35	32	27	23	19	16	12	9
9	4	68	63	60	56	51	48	42	37	33	30	25	22	18	16	12	9
10	2	79	74	68	65	61	58	51	45	40	37	32	28	22	18	13	10
10	3	74	69	65	50	57	53	47	41	37	33	28	24	20	17	13	10
10	4	68	64	61	56	52	49	43	38	34	31	26	23	19	17	13	9
12	2	80	75	70	66	63	60	53	48	43	39	34	30	24	20	14	11
12	3	74	70	66	62	58	55	48	43	39	35	30	26	21	18	14	10
12	4	69	65	61	57	53	50	44	40	35	32	27	24	20	18	13	10
14	2	81	76	71	68	65	62	55	50	45	41	36	32	25	21	15	11
14	3	74	70	67	63	60	56	50	45	41	37	31	27	22	19	15	11
14	4	69	65	61	57	54	52	45	41	36	33	28	25	21	19	14	10
16	2	82	77	72	69	66	63	57	52	47	43	38	34	27	22	16	12
16	3	74	70	67	64	61	58	52	47	42	39	33	29	23	20	15	11
16	4	70	66	62	58	55	52	46	42	37	34	29	26	22	20	14	11
18	2	82	78	73	70	67	65	59	54	49	45	39	35	28	23	17	12
18	3	74	70	67	65	62	59	53	48	44	41	35	30	24	21	16	12
18	4	70	65	62	58	55	53	47	43	38	35	30	27	23	20	15	11
20	2	82	78	74	71	68	66	60	55	50	46	41	36	29	24	18	13
20	3	74	70	67	65	62	60	55	60	46	42	36	32	26	22	17	12
20	4	70	66	62	58	55	53	47	43	39	36	31	28	24	21	16	11

SOURCE: Reprinted with permission from Leland Wilkinson and Jerry Dallal.

NOTE: Decimals are omitted; k = number of candidate predictors; N = sample size; F = criterion F-to-enter.

+ − × ÷

Comprehension Questions

1. Describe a hypothetical study for which multiple regression with more than two predictor variables would be an appropriate analysis. Your description should include one dependent variable and three or more predictors.
   a. For each variable, provide specific information about how the variable would be measured and whether or not it is quantitative, normally distributed, and so forth.
   b. Assuming that you are trying to detect a medium size R^2, what is the minimum desirable N for your study (based on the guidelines in this chapter)?
   c. What regression method (e.g., standard, sequential, or statistical) would you prefer to use for your study, and why? (If sequential, indicate the order of entry or predictors that you would use and the rationale for the order of entry.)
   d. What pattern of results would you predict in the regression; that is, would you expect the overall multiple R to be significant? Which predictor(s) would you expect to make a significant unique contributions, and why?
2. Sketch the scatter plot for a graph of $z_{\text{residuals}}$ by $z_{\text{predicted}}$ scores that illustrates each of the following outcomes:
   a. Assumptions are met: no trend or pattern in residuals, homogeneous variance of z_{residual} across levels of $z_{\text{predicted}}$, no extreme outliers.
   b. Same as (2a) except for one extreme outlier residual.
   c. Same as (2a) but with heteroscedasticity of residuals.
3. What types of research situations often make use of multiple regression analysis with more than two predictors?
4. What is the general path diagram for multiple regression?
5. What are the different methods of entry for predictors in multiple regression?
6. What kind of reasoning would justify a decision to enter some variables earlier than others in a hierarchical regression analysis?
7. Suppose a researcher runs a "forward entry" statistical multiple regression with a group of 20 candidate predictor variables; the final model includes five predictors. Why are the p values shown on the SPSS printout not a good indication of the true risk of Type I error in this situation? What correction can be used when you test the significance of the overall multiple R for a forward regression?
8. How can we obtain significance tests for sets or blocks of predictor variables?
9. What information do we need to assess possible multicollinearity among predictors?

10. What information should be included in the report of a standard or simultaneous multiple regression (all predictors entered in one step)?
11. What information should be included in the report of a hierarchical or statistical multiple regression (i.e., a series of regressions with one variable or a group of variables entered in each step)?
12. What parts of the information that is reported for standard versus hierarchical regression are identical? What parts are different?
13. What is an R^2_{inc}? How is the R^2_{inc} for a single variable that enters in a particular step related to the sr^2_{inc} for that same single variable in the same step?
14. What can the residuals from a regression tell us about possible violations of assumptions for regression?
15. How can we identify disproportionately influential scores and/or multivariate outliers?
16. What is the correlation between Y and $(b_0 + b_1X_1 + b_2X_2 + \ldots + b_kX_k)$?
17. What is "tolerance?" What is the range of possible values for tolerance? Do we usually want the tolerance for a candidate predictor variable to be low or high?

Suggested Data Analysis Project for Multiple Regression

1. Select four independent variables and one dependent variable. They can be interval/ratio level of measurement (quantitative), or you may also include dummy variables to represent group membership.
2. Run a standard or simultaneous regression (i.e., all variables are entered in one step) with this set of variables.
3. Then run a hierarchical regression in which you arbitrarily specify an order of entry for the four variables.
4. Write up a Results section separately for each of these analyses using the model results sections and the tables in this chapter as a guide. Answer the following questions about your analyses:

   How does the equation in the last step of your hierarchical analysis (with all four variables entered) compare with your standard regression?

   Draw overlapping circle diagrams to illustrate how the variance is partitioned among the X predictor variables in each of these two analyses; indicate (by giving a numerical value) what percentage of variance is attributed (uniquely) to each independent variable. In other words, how does the variance partitioning in these two analyses differ? Which is the more conservative approach?

   Look at the equations for Steps 1 and 2 of your hierarchical analysis.

a. Calculate the difference between the R^2 values for these two equations; show that this equals one of the squared part correlations (which one?) on your printout.
b. Evaluate the statistical significance of this change in R^2 between Steps 1 and 2. How does this F compare to the t statistic for the slope of the predictor variable that entered the model at Step 2?

Consider the predictor variable that you entered in the first step of your hierarchical analysis. How does your evaluation of the variance due to this variable differ when you look at it in the hierarchical analysis compared with the way you look at it in the standard analysis? In which analysis does this variable look more "important"—that is, appears to explain more variance? Why?

Now consider the predictor variable that you entered in the last (fourth) step of your hierarchical analysis and compare your assessment of this variable in the hierarchical analysis (in terms of proportions of explained variance) with your assessment of this variable in the standard analysis.

Look at the values of R and F for the overall model as they change from Steps 1 through 4 in the hierarchical analysis. How do R and F change in this case as additional variables are added in? In general, does R tend to increase or decrease as additional variables are added to a model? In general, under what conditions does F tend to increase or decrease as variables are added to a model?

Suppose you had done a statistical (data-driven) regression (using the forward method of entry) with this set of four predictor variables. Which (if any) of the four predictor variables would have entered the equation and which one would have entered first? Why?

If you had done a statistical regression (using a method of entry such as forward) rather than a hierarchical (user-determined order of entry), how would you change your significance testing procedures? Why?

CHAPTER 15

Analysis of Covariance

15.1 ♦ Research Situations and Research Questions

Previous chapters have examined research situations where the outcome variable of interest is a quantitative *Y* variable. The group means on *Y* scores can be compared by performing analysis of variance (ANOVA; Chapter 6). Scores on a quantitative *Y* outcome variable can also be predicted from a regression model that includes several predictor variables (Chapters 11 and 14). Analysis of covariance (ANCOVA) combines one or more categorical predictor variables and one or more quantitative predictor variables, called covariates, to predict scores on a quantitative *Y* outcome variable. ANCOVA is, in a sense, a combination of ANOVA and regression; it can also be understood as a specific type of regression analysis in which at least one predictor variable represents group membership, usually membership in some sort of treatment group. In ANCOVA, the typical research situation is as follows. A researcher has a set of treatment groups that correspond to **levels of a factor** (Factor *A* in the example in this chapter corresponds to three types of teaching method). The researcher has pre-intervention quantitative measures of one or more individual participant characteristics, such as ability (X_c). The goal of the analysis is to assess whether scores on a quantitative *Y* outcome variable (such as scores on a final exam) differ significantly across levels of *A* when we statistically control for the individual differences among participants that are measured by the X_c covariate. The type of statistical control used in classic or traditional ANCOVA requires the absence of any interaction between the *A* treatment and the X_c covariate. We can thus think of ANCOVA as ANOVA with one or more added quantitative predictors called covariates. Because we can represent group membership predictor variables using dummy-coded predictor variables in regression (as discussed in Chapter 12), we can also understand ANCOVA as a special case of regression analysis in which the goal is to assess differences among treatment groups while controlling for scores on one or more covariates.

ANCOVA is often used in research situations where mean scores on a quantitative outcome variable are compared across groups that may not be equivalent in terms of participant characteristics (such as age or baseline level of mood). Comparison groups in a study are nonequivalent if the groups differ (prior to treatment) on one or more participant characteristics, such as age, motivation, or ability, that might possibly influence the outcome variable. **Nonequivalent comparison groups** are often encountered in quasi-experimental research. A quasi experiment is a study that resembles an experiment (up to a point); quasi-experimental designs typically involve comparison of behavior for two groups that do versus do not receive some intervention or multiple groups that receive different types of treatment or intervention. However, in a quasi experiment, the researcher generally has less control over the research situation than in a true experiment. Quasi experiments are often conducted in field settings (such as schools) where researchers may have to administer interventions to already existing groups such as classrooms of students. In most quasi experiments, the researcher does not have the ability to randomly assign individual participants to different treatment conditions. At best, the researcher may be able to randomly assign groups of students in different classrooms to different treatment conditions; but such quasi-experimental designs often result in a situation where prior to the treatment or intervention, the groups that are compared differ in participant characteristics, such as age, motivation, or ability. This nonequivalence can lead to a confound between the average participant characteristics for each group (such as motivation) and the type of treatment received.

One reason to use ANCOVA is that ANCOVA provides a way to assess whether mean outcome scores differ across treatment groups when a statistical adjustment is made to control for (or try to remove the effects of) different participant characteristics across groups. In most ANCOVA situations, the variable that we designate as the covariate X_c is a measure of some participant characteristic that differs across groups prior to treatment. In an ANCOVA, we want to evaluate whether scores on Y, the outcome variable of interest, differ significantly across groups when the Y scores are adjusted for group differences on the X_c covariate. Of course, this adjustment works well only if the assumptions for ANCOVA are not violated. The statistical adjustment for differences in participant characteristics between groups that can be made using ANCOVA is not typically as effective as experimental methods of controlling the composition of groups (such as random assignment of individuals to groups, creating matched samples, and so forth).

Even in a well-controlled laboratory experiment, a researcher sometimes discovers that groups that were formed by random assignment or more systematic matching techniques are not closely equivalent. If this nonequivalence is discovered early in the research process (e.g., before the intervention, the researcher notices that the mean age for participants is higher in the A_1 treatment group than in the A_2 treatment group), it may be possible to re-randomize and reassign participants to groups to get rid of this systematic difference between groups. Sometimes, even in a true experiment, the groups that are formed by random assignment of individual participants to conditions turn out not exactly equivalent with respect to characteristics such as age due to unlucky randomization. If this problem is not corrected before the intervention is administered, even a true experiment can end up with a comparison of nonequivalent groups. ANCOVA is usually applied to data from quasi-experimental studies, but it may also be used to try to correct for nonequivalence of participant characteristics in the analysis of data from true experiments. However, when it is possible, it is preferable to correct the problem by redoing the assignment of participants to conditions to get rid of the nonequivalence between groups through experimental control; statistical control for nonequivalence, using ANCOVA, should be used only when the problem of nonequivalent groups cannot be avoided through design.

In ANCOVA, the null hypothesis that is of primary interest is whether means on the quantitative Y outcome variable differ significantly across groups after adjustment for scores on one or more covariates. For example, consider a simple experiment to assess the possible effects of food on mood. Some past research of possible drug-like effects on mood (Spring et al., 1987) suggests that the consumption of an all-carbohydrate meal has a calming effect on mood, while an all-protein meal may increase alertness. This hypothesis can be tested by doing an experiment. Suppose that participants are randomly assigned to Group $A = 1$ (an all-carbohydrate lunch) or Group $A = 2$ (an all-protein lunch). We will consider two different possible covariates. The first is a measurement of a preexisting participant characteristic that may be linearly predictive of scores on the Y outcome variable. Before lunch, each participant rates "calmness" on a scale from "0" (not at all calm) to "10" (extremely calm); this score is X_c, the quantitative covariate. One hour after lunch, participants rate calmness again using the same scale; this second rating is Y, the outcome variable.

If we do a *t* test or an ANOVA to compare means on Y (calmness) across levels of A (carbohydrate vs. protein meal), the research question would be whether mean Y (calmness) is significantly higher in the carbohydrate group than in the protein group (when we *do not* take into account the individual differences in calmness before lunch or at baseline). When we do an ANCOVA to predict Y (calmness after lunch) from both A (type of food) and X_c (pretest level of calmness before lunch), the research question becomes whether mean calmness after lunch is significantly higher for the carbohydrate group than for the protein group when we control for or partial out any part of Y (the after-lunch calmness rating) that is linearly predictable from X_c (the before-lunch or baseline calmness rating).

The covariate in the preceding example (X_c) was a pretest measure of the same mood (calmness) that was used as the outcome or dependent variable. However, a covariate can be a measure of some other variable that is relevant in the research situation, either because it is confounded with treatment group membership or because it is linearly predictive of scores on the Y outcome variable, or both. For example, the covariate measured prior to the intervention in the food/mood study might have been a measure of self-reported caffeine consumption during the 2-hour period prior to the study. Even if random assignment is used to place participants in the two different food groups, it is possible that due to "unlucky randomization," one group might have a higher mean level of self-reported caffeine consumption prior to the lunch intervention. It is also possible that caffeine consumption is negatively correlated with self-reported calmness (i.e., participants may tend to report lower calmness when they have consumed large amounts of caffeine). Statistically controlling for caffeine consumption may give us a clearer idea of the magnitude of the effects of different types of food on mood for two reasons. First of all, statistically controlling for caffeine consumption as a covariate may help to correct for any preexisting group differences or confounds between the amount of caffeine consumed and the type of food. Second, whether or not there is any confound between caffeine consumption and type of food, partialling out the part of the calmness scores that can be linearly predicted from caffeine consumption may reduce the error variance (MS_{within} in ANOVA or MS_{residual} if the analysis is a regression) and increase statistical power.

The regression equation for this simple ANCOVA is as follows: Let $A = 0$ for the protein group and $A = 1$ for the carbohydrate group; that is, A is a dummy-coded dummy variable (see Chapter 12). Let X_c be calmness ratings before lunch and let Y be calmness ratings after lunch. The regression equation that corresponds to an ANCOVA to predict Y from A and X_c is as follows:

$$Y' = b_0 + b_1 X_c + b_2 A. \quad (15.1)$$

Controlling for an X_c covariate variable can be helpful in two different ways. First of all, it is possible that X_c is confounded with A. When we randomly assign participants to conditions in an experiment, we hope that random assignment will result in groups that are equivalent on all relevant participant characteristics. For example, in the food/mood study, if we create treatment groups through random assignment, we hope that the resulting groups will be similar in gender composition, smoking behavior, level of stress, and any other factors that might influence calmness. In particular, if we have measured X_c, calmness before lunch, we can check to see whether the groups that we obtained are approximately equal on this variable. In addition, when we calculate the regression coefficient for the dummy variable for group membership, we can statistically control for any confound of X_c with A; that is, we can assess the relation of A to Y, controlling for or partialling out X_c. If A is still related to Y when we control for X_c, then we know that the group differences in mean Y are not entirely explained due to a confound with the covariate.

Including a covariate can also be helpful in research situations where the covariate is not confounded with the treatment variable A. Consider a regression analysis in which we use only A to predict Y:

$$Y' = b_0 + b_1 A. \quad (15.2)$$

The $SS_{residual}$ term for Equation 15.1 will be equal to or less than the $SS_{residual}$ for Equation 15.2. Therefore, the F ratio that is set up to evaluate the significance of the A variable for the analysis in Equation 15.1 (which includes X_c as a predictor) may have a smaller $MS_{residual}$ value than the corresponding F ratio for the analysis in Equation 15.2 (which does not include X_c as a predictor). If $SS_{residual}$ decreases more than $df_{residual}$ when the covariate is added to the analysis, then $MS_{residual}$ will be smaller for the ANCOVA in Equation 15.1 than for the ANOVA in Equation 15.2, and the F ratio for the main effect of A will tend to be larger for the analysis in Equation 15.1 than for the analysis in Equation 15.2. When we obtain a substantially smaller $SS_{residual}$ term by including a covariate in our analysis (as in Equation 15.1), we can say that the covariate X_c acts as a **"noise suppressor"** or an "error variance suppressor."

When we add a covariate X_c to an analysis that compares Y means across levels of A, taking the X_c covariate into account while assessing the strength and nature of the association between Y and A can change our understanding of that relationship in two ways. Controlling for X_c can, to some degree, correct for a confound between X_c and type of A treatment; this correction may either increase or decrease the apparent magnitude of the differences between Y outcomes across levels of A. In addition, controlling for X_c can reduce the amount of error variance, and this can increase the power for the F test to assess differences in Y means across levels of A.

We will prefer ANCOVA over ANOVA in situations where

1. the covariate, X_c, is confounded with the group membership variable A, and we can only assess the unique effects of A when we statistically control for or partial out X_c and/or

2. the covariate, X_c, is strongly predictive of Y, the outcome variable, and therefore, including X_c as a predictor along with A (as in Equation 15.1) gives us a smaller error term and a larger F ratio for assessment of the main effect of A.

We can use ANCOVA only in situations where the assumptions for ANCOVA are met. All the usual assumptions for ANOVA and regression must be met for ANCOVA to be applied (i.e., scores on the Y outcome variable should be approximately normally distributed with homogeneous variances across levels of A, and scores on Y must be linearly related to scores for X_c). Additional assumptions are required for the use of ANCOVA, including the following. The covariate (X_c) must be measured prior to the administration of the A treatments, and there must be no treatment by covariate ($A \times X_c$) interaction. In addition, factors that cause problems in the estimation of regression models generally (such as unreliability of measurement, restricted ranges of scores, and so forth; see Chapter 7) will also create problems in ANCOVA.

15.2 ♦ Empirical Example

The hypothetical example for ANCOVA will be an imaginary quasi-experimental study to compare the effectiveness of three different teaching methods in a statistics course (Factor A). Group A_1 receives self-paced instruction (SPI), Group A_2 participates in a seminar, and Group A_3 receives lecture instruction. We will assume that the groups were not formed by random assignment; instead, each teaching method was carried out in a different preexisting classroom of students. The scores used in this example appear in Table 15.1.

The X_c covariate in this hypothetical research example corresponds to scores on a pretest of math ability. As we shall see, there are small differences among the A_1, A_2, and A_3 groups on this math pretest X_c variable. Thus, there is some confound between the mean level of math ability prior to the intervention in each group and the type of teaching method received by each group. ANCOVA can help to adjust or correct for these differences in preexisting participant characteristics. The Y outcome variable in this hypothetical study corresponds to the score on a standardized final exam for the statistics course. If all the assumptions of ANCOVA are met, an ANCOVA in which we examine **adjusted means** on the Y outcome variable (i.e., Y scores that have been statistically adjusted to remove linear association with the X_c covariate) may provide us with clearer information about the nature of the effects of the three different teaching methods included in the study.

For this example, there are $n = 7$ participants in each of the three groups, for a total N of 21 (in real-life research situations, of course, larger numbers of scores in each group would be preferable). The covariate, X_c, is a math ability pretest. The outcome, Y, is the score on a standardized final exam. The scores for A, X_c, and Y and alternative forms of dummy-coded variables that represent treatment level appear in Table 15.1.

The research question is this: Do the three teaching methods (A) differ in mean final exam scores (Y) when pretest scores on math ability (X_c) are statistically controlled by doing an ANCOVA? What is the nature of the differences among group means? (That is, which pairs of groups differed significantly, and which teaching methods were associated with the highest mean exam scores?) As part of the interpretation, we will also want to consider the nature of the confound (if any) between A and X_c. (That is, did the groups differ in math ability?) Also, how did controlling for X_c change our understanding of the pattern of Y means? (That is, how did the ANCOVA results differ from a one-way ANOVA comparing Y means across levels of A?)

Table 15.1 ♦ Data for Hypothetical ANCOVA Study

	A	X_c	Y	D_1	D_2	Int_1	Int_2
1	1	12	19	1	0	12	0
2	1	7	13	1	0	7	0
3	1	10	10	1	0	10	0
4	1	8	9	1	0	8	0
5	1	8	12	1	0	8	0
6	1	5	6	1	0	5	0
7	1	13	18	1	0	13	0
8	2	12	22	0	1	0	12
9	2	15	28	0	1	0	15
10	2	14	24	0	1	0	14
11	2	12	27	0	1	0	12
12	2	10	24	0	1	0	10
13	2	8	16	0	1	0	8
14	2	5	21	0	1	0	5
15	3	11	18	−1	−1	−11	−11
16	3	6	21	−1	−1	−6	−6
17	3	9	24	−1	−1	−9	−9
18	3	10	21	−1	−1	−10	−10
19	3	15	28	−1	−1	−15	−15
20	3	7	15	−1	−1	−7	−7
21	3	13	21	−1	−1	−13	−13

SOURCE: Horton (1978, p. 175).

NOTES: Factor A is a categorical variable that represents which of three teaching methods each student received; Y is the final exam score, and X_c is the pretest on ability. Using an ANOVA program such as the GLM procedure in SPSS, the categorical variable A would be used to provide information about group membership, while the Y and X_c scores would be the dependent variable and covariate, respectively. It is also possible to run an ANCOVA using a linear regression program, but in this situation, we need to use dummy-coded variables to represent information about treatment group membership. Dummy-coded variables D_1 and D_2 provide an alternate way to represent treatment group membership. The variables Int_1 and Int_2 are the products of D_1 and D_2 with X_c, the covariate. To run the ANCOVA using linear regression, we could predict scores on Y from scores on D_1, D_2, and X_c. To test for the significance of a treatment by covariate interaction, we could assess whether the additional terms Int_1 and Int_2 are associated with a significant increase in R^2 for this regression analysis.

It is usually more convenient to use the SPSS general linear model (GLM) procedure (which was introduced in Chapter 13) to perform ANCOVA. The ANCOVA example in this chapter was performed using SPSS GLM. However, if group membership is represented by dummy variables (such as D_1 and D_2), ANCOVA can also be performed using the SPSS Regression procedure. The regression version of ANCOVA is not presented in this chapter.

15.3 ♦ Screening for Violations of Assumptions

ANCOVA refers to a predictive model that includes a mix of both categorical and continuous predictor variables; however, there are several additional conventional limitations that are set on an analysis before we can call it an ANCOVA:

1. As in other linear correlation and regression analyses, we assume that the X_c and Y quantitative variables have distribution shapes that are reasonably close to normal, that the relation between X_c and Y is linear, and that the variance of Y is reasonably homogeneous across groups on the A treatment variable.
2. The covariate X_c should not be influenced by the treatment or intervention; to ensure this, X_c should be measured or observed before we administer treatments to the groups. (If the covariate is measured after the treatment, then the covariate may also have been influenced by the treatment, and we cannot distinguish the effects of the intervention from the covariate.) In the hypothetical research example that involves comparison of teaching methods, the X_c covariate is measured prior to the teaching intervention, and therefore, the X_c covariate cannot be influenced by the teaching intervention.
3. In practice, we cannot expect ANCOVA adjustments to be really effective in separating the effects of A and X_c if they are very strongly confounded. ANCOVA makes sense only in situations where the correlation or confound between A and X_c is weak to moderate. Caution is required in interpreting ANCOVA results where the rank ordering of Y means across treatment groups changes drastically or where the distributions of X_c scores differ greatly across the A groups.
4. We assume that there is no treatment by covariate or $A \times X_c$ interaction; that is, we assume that the slope for the regression prediction of Y from X_c is the same within each group included in the A factor. Another term for this assumption of no **treatment by covariate interaction** is ***homogeneity of regression.*** Procedures to test whether there is a significant interaction are discussed below.
5. Measurement of the covariate should have high reliability.

If the assumption of no treatment by covariate interaction is violated, the data analyst may choose to report the results of an analysis that does include a treatment by covariate interaction; however, this analysis would no longer be called an ANCOVA (it might be described as a regression model that includes group membership, covariate, and group membership by covariate interaction terms).

Potential violations of these assumptions can be evaluated by preliminary data screening. At a minimum, preliminary data screening for ANCOVA should include the following:

1. *Examination of histograms for* Y *and each of the* X_c *covariates; all distributions should be approximately normal in shape with no extreme outliers:* The histograms in Figure 15.1 and the scatter plot in Figure 15.2 indicate that for this small hypothetical dataset, these assumptions do not appear to be severely violated.
2. *Examination of scatter plots between* Y *and each* X_c *covariate and between all pairs of* X_c *covariates:* All relations between pairs of variables should be approximately linear with no extreme bivariate outliers.
3. *Evaluation of the homogeneity of variance assumption for* Y *and assessment of the degree to which the* X_c *covariate is confounded with (i.e., systematically different across) levels of the* A *treatment factor:* Both of these can be evaluated by running a

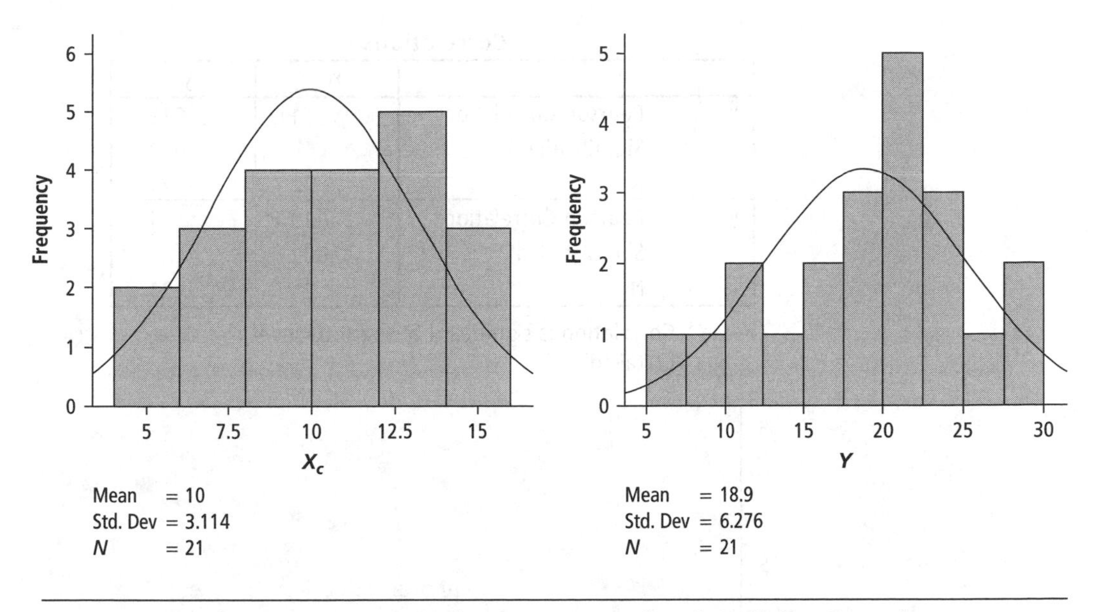

Figure 15.1 ♦ Preliminary Data Screening for ANCOVA: Normality of Distribution Shapes for Y (Quantitative Outcome Variable) and X_c (Quantitative Covariate)

simple one-way between-subjects ANOVA using A as the group membership variable and using X_c and Y as separate outcome variables. The use of the SPSS ONEWAY procedure was described in greater detail in Chapter 6. Here, only selected results are reported (see Figure 15.3). Examination of the results in Figure 15.3 indicates that there is no statistically significant violation of the assumption of homogeneity of variance for scores on Y.

This preliminary one-way ANOVA provides information about the nature of any confound between X_c and A. The mean values of X_c did not differ significantly across levels of A: $F(2, 18) = .61, p = .55, \eta^2 = .06$. (This eta-squared effect size was obtained by dividing SS_{between} by SS_{total} for the one-way ANOVA to predict X_c from A.) Thus, the confound between level of ability (X_c) and type of teaching method (A) experienced by each of the groups was relatively weak.

The one-way ANOVA to predict Y scores from type of teaching method (A) indicated that when the scores on the covariate are not taken into account, there are statistically significant differences in mean final exam scores across the three treatment groups: $F(2, 18) = 12.26, p < .001, \eta^2 = .58$. The nature of the differences in final exam scores across types of teaching group was as follows: Mean exam scores were highest for the seminar group, intermediate for the lecture group, and lowest for the SPI group.

This one-way ANOVA to assess differences in Y means across A groups (without the X_c covariate) provides us with a basis for comparison. We can evaluate whether or how the inclusion of an X_c covariate in an ANCOVA model changes our understanding of the way in which mean Y scores differ across treatment groups, in comparison with the simple one-way ANOVA reported in Figure 15.3 (in which the X_c

Correlations

		xc	y
xc	Pearson Correlation	1	.629**
	Sig. (2-tailed)		.002
	N	21	21
y	Pearson Correlation	.629**	1
	Sig. (2-tailed)	.002	
	N	21	21

**. Correlation is significant at the 0.01 level (2-tailed).

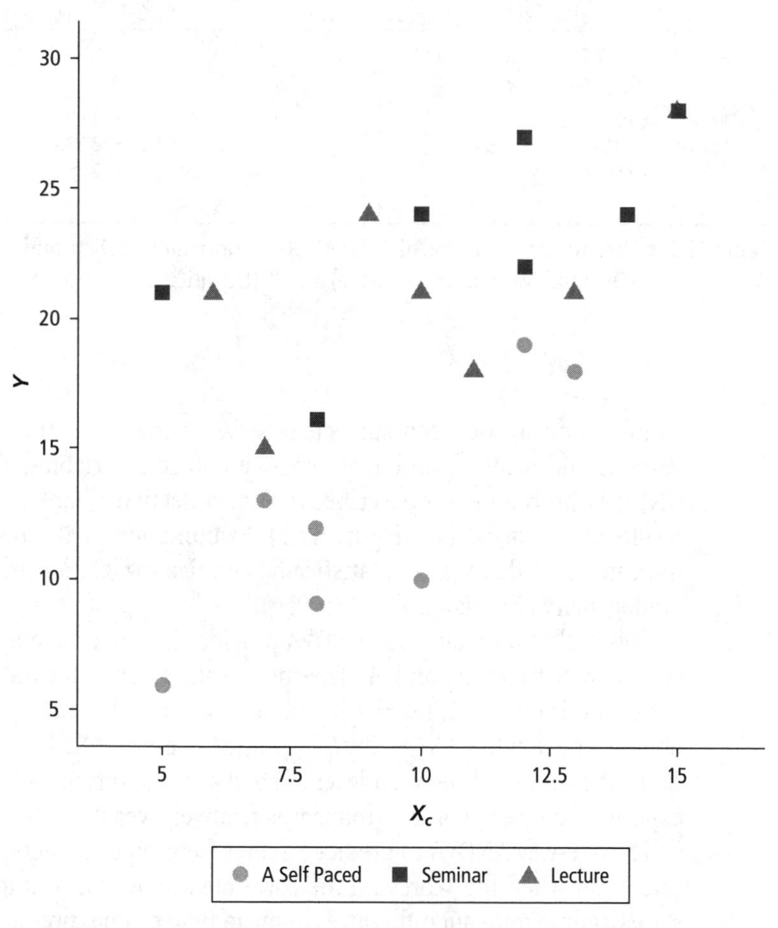

Figure 15.2 ♦ Preliminary Data Screening for ANCOVA: Relationship Between Y and X_c

covariate is not statistically controlled). When we control for the X_c covariate in an ANCOVA reported later in this chapter, the apparent association between type of teaching method (A) and score on the final exam (Y) could become weaker or stronger; it is even possible that the rank ordering of adjusted means on Y can be different from the rank ordering of the original Y means.

Descriptives

		N	Mean	Std. Deviation	Std. Error	95% Confidence Interval for Mean Lower Bound	95% Confidence Interval for Mean Upper Bound	Minimum	Maximum
xc	Self Paced	7	9.00	2.828	1.069	6.38	11.62	5	13
	Seminar	7	10.86	3.485	1.317	7.63	14.08	5	15
	Lecture	7	10.14	3.185	1.204	7.20	13.09	6	15
	Total	21	10.00	3.114	.680	8.58	11.42	5	15
y	Self Paced	7	12.43	4.721	1.784	8.06	16.79	6	19
	Seminar	7	23.14	4.018	1.519	19.43	26.86	16	28
	Lecture	7	21.14	4.140	1.565	17.31	24.97	15	28
	Total	21	18.90	6.276	1.370	16.05	21.76	6	28

Test of Homogeneity of Variances

	Levene Statistic	df1	df2	Sig.
xc	.128	2	18	.881
y	.202	2	18	.819

ANOVA

		Sum of Squares	df	Mean Square	F	Sig.
xc	Between Groups	12.286	2	6.143	.608	.555
	Within Groups	181.714	18	10.095		
	Total	194.000	20			
y	Between Groups	454.381	2	227.190	12.265	.000
	Within Groups	333.429	18	18.524		
	Total	787.810	20			

Figure 15.3 ♦ One-Way ANOVA to Evaluate Differences in Group Means Across Types of Teaching Methods (A) for the Covariate X_c and for the Unadjusted Outcome Scores on Y

4. *Assessment of possible treatment by covariate interactions for each covariate:* This can be requested in SPSS GLM by requesting a custom model that includes interaction terms. Researchers usually hope that the test of a treatment by covariate interaction in an ANCOVA research situation will not be statistically significant. If this interaction term is significant, it is evidence that an important assumption of ANCOVA is violated. If the treatment by covariate or $A \times X_c$ interaction term is nonsignificant, then this interaction term is dropped from the model and an ANCOVA is performed without an interaction term between treatment and covariate. The ANCOVA that is used to generate adjusted group means (in the final Results section) generally does not include a treatment by covariate interaction. This simple ANCOVA corresponds to the type of ANCOVA model represented in Equation 15.1.

15.4 ♦ Variance Partitioning in ANCOVA

The most common approach to variance partitioning in ANCOVA is similar to the hierarchical method of variance partitioning described in Chapter 14. In a typical ANCOVA, the researcher enters one (or more) X covariates as predictors of the Y outcome variable in Step 1. In Step 2, categorical predictors that represent treatment group membership

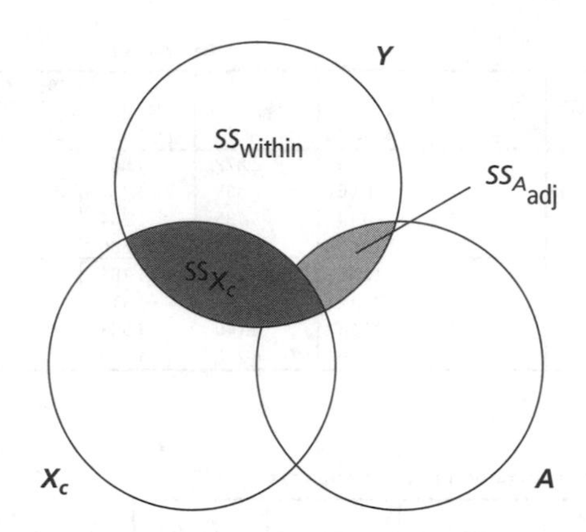

Type I Method of Sum of Squares in GLM

Figure 15.4a ♦ Variance Partitioning With One Treatment (A) and One Covariate (X_c) Using Type I SS (Hierarchical Approach)

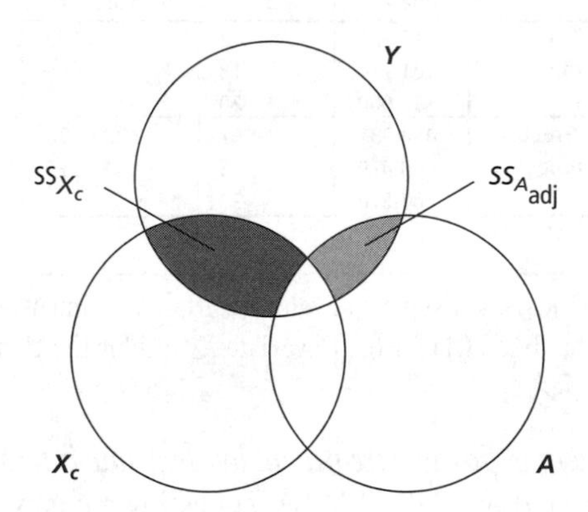

Type III Method of Sum of Squares in GLM
Each predictor assessed controlling for all other predictors

Figure 15.4b ♦ Variance Partitioning Using Type III SS (Standard Regression Approach)

Figure 15.4 ♦ Variance Partitioning in an ANCOVA

NOTE: In both figures, the dark gray area corresponds to SS_{X_c}, the light gray area corresponds to $SS_{A_{adj}}$, and the sum of squares represents the remaining A main effect after statistically controlling for X_c, the covariate.

are added, and the data analysts examine the results to see whether the categorical predictor variable(s) contribute significantly to our ability to predict scores on the Y outcome variable when scores on one or more covariates are statistically controlled.

We can use overlapping circles to represent the partition of SS_{total} for Y into sum of squares components for the covariate and for the treatment factor (just as we used overlapping circles in Chapter 14 to illustrate the partition of variance for Y, the outcome variable in a regression analysis, among correlated predictor variables). In SPSS GLM, we can obtain this type of variance partitioning (enter the covariate in Step 1, enter the treatment group membership variable in Step 2) by using the Custom Models window in GLM to request the SS Type I method of computation for sums of squares and by listing the predictor variables in the Custom Models window in the sequence in which they should enter the model. The variance partition that results from these analysis specifications is depicted by the overlapping circles in Figure 15.4a. We obtain a sum of squares for X_c; if we request SS Type I computation, this sum of squares for the covariate X_c is not adjusted for *A*, the treatment group variable. Then, we calculate an adjusted *SS* for *A* that tells us how strongly the *A* treatment variable is predictively related to scores on the *Y* outcome variable when the covariate X_c has been statistically controlled or partialled out. Figure 15.4a illustrates the partition of *SS* that is obtained when the SS Type I computation method is chosen; this is the most conventional approach to ANCOVA.

If you use the default method for computation of sums of squares in GLM, SS Type III, the *SS* term for each predictor is calculated by controlling for or adjusting for all the other predictors in the model. If the SS Type III method is used to compute the sums of squares in this situation, SS_A is adjusted for X_c and SS_{X_c} is adjusted for *A*, as shown in Figure 15.4b. Note that the $SS_{A_{adj}}$ term is the same whether you use Type I or Type III SS; in either instance, $SS_{A_{adj}}$ is adjusted for X_c. However, the *SS* term for the X_c covariate is estimated differently depending on whether Type I or Type III sum of squares is requested. Conventionally, when reporting ANCOVA, readers usually expect to see the Type I approach to variance partitioning, and therefore, the use of Type I SS is recommended. However, the primary interest in ANCOVA is usually the differences in group means rather than the proportion of variance that is predictable from the covariate, and the conclusion about that question is the same whether the data analyst selects Type I or Type III SS. More detailed information about the different methods of computation for sums of squares can be obtained by clicking the Help button while in the SPSS GLM dialog window.

An ANCOVA model can include multiple covariates. An ANCOVA model can include more than one factor or treatment group variable, and in such cases, it will generally include interactions among factors. If you use the SS Type III method in GLM, then every predictor variable in the analysis is assessed controlling for every other predictor. If you use the SS Type I method (or sequential method of entry), the most conventional order of entry is as follows: all covariates first, all main effects next, all two-way interactions between factors after that, and so forth. When SS Type I is selected, predictor variables are entered into the analysis in the sequence in which they appear in the list provided by the data analyst. Note that each predictor is assessed by controlling for every other predictor that is entered in the same step or in previous steps.

15.5 ♦ Issues in Planning a Study

When a researcher plans a study that involves comparisons of groups that are exposed to different treatments or interventions and if the researcher wants to make causal inferences, the preferred design is a well-controlled experiment. Random assignment of participants to groups and/or creation of matched samples are methods of experimental control for participant characteristics that are usually, although not always, effective in ruling out confounds between participant characteristics and type of treatment. Even in a well-designed experiment (in which participant characteristics are not confounded with treatments), it can be useful to measure one or more participant characteristics and to include them in the analysis as covariates so that the magnitude of the SS_{within} or SS_{residual} term used in the F test can be reduced.

In quasi experiments, and sometimes even in randomized groups experiments, we may find that one or more participant characteristics are confounded with treatments. When such confounds are present, they make group differences difficult to interpret; we need to statistically control for or partial out the participant characteristics.

A common choice of covariate is a pretest measure on the same behavior that is assessed as the outcome. Thus, ANCOVA may be used to assess pretest/posttest differences across multiple treatment groups. (See the appendix to this chapter for a brief discussion of several alternative methods of analysis for pretest/posttest data.) However, any participant characteristic that has a strong correlation with Y may be useful as a noise suppressor in the data analysis; and any participant characteristic that is known to be confounded with treatment group membership, and thus is a rival explanation for any observed differences in outcomes, should be controlled as a covariate when comparing group means.

The statistical control of confounds using ANCOVA is not as powerful a method for dealing with participant characteristics as true experimental control (through random assignment of participants to conditions and/or holding participant characteristics constant by sampling homogeneous participants). Note that if we control for a covariate that is an inappropriate choice, an invalid measure of the construct that we want to control, or unreliably measured or if the basic assumptions for ANCOVA are badly violated, then the results of ANCOVA can be misleading.

15.6 ♦ Formulas for ANCOVA

ANCOVA can be understood as an ANOVA that compares adjusted Y means across groups (the adjustment involves regression of Y on one or more X_c covariates and subtracting any part of Y that is related to covariates). Note, however, that the regression slope that is used to make this adjustment is a partial or within-group slope (the slope for prediction of Y from X_c when you control for A, the treatment group variable).

It is somewhat easier to describe this adjustment if we describe ANCOVA as a regression analysis to predict Y from one or more covariates and from dummy variables that represent membership in treatment groups. An equation for regression analysis that is equivalent to an ANCOVA for the teaching method study described in Section 15.4 would look like this:

$$Y' = b_0 + b_1 D_1 + b_2 D_2 + b_3 (X_c - M_{Xc}), \qquad (15.3)$$

where Y is the outcome variable, final statistics exam score; X_c is the covariate, pretest on math ability; M_{X_c} is the grand mean of the covariate; and D_1, D_2 are the dummy variables that represent membership in the three teaching method groups. The b_3 regression coefficient is the partial regression slope that predicts Y from X_c while controlling for A group membership. The assumption of no treatment by covariate interaction is the assumption that the b_3 slope to predict Y from $X_c - M_{X_c}$ is the same within each of the A groups; the b_3 partial slope is, essentially, the average of the three within-group slopes to predict Y from $X_c - M_{X_c}$ (see Chapter 11 for a more complete description of the concept of a partial slope).

We can rewrite the previous equation as follows:

$$Y' - b_3 (X_c - M_{X_c}) = b_0 + b_1 D_1 + b_2 D_2. \quad (15.4)$$

The new term on the left-hand side of Equation 15.4, $Y - b_3 Y' - b_3 (X_c - M_{X_c})$, corresponds to the adjusted Y score (Y^*). When we subtract $b_3 (X_c - M_{X_c})$ from Y', we are removing from Y' the part of each Y score that is linearly predictable from X_c, to obtain an adjusted predicted Y^* score that is uncorrelated with X_c. The resulting equation,

$$Y^* = b_0 + b_1 D_1 + b_2 D_2, \quad (15.5)$$

corresponds to a one-way ANOVA to assess whether the adjusted Y^* means differ significantly across groups. (Refer back to Chapter 12 for a discussion of the use of dummy variables to represent a one-way ANOVA as a regression equation.) The partition of the total variance of Y into the proportion of variance that is predictable from X_c and the proportion of variance that is predictable from A, statistically controlling for X_c, can be assessed by running appropriate regression analyses.[1]

The idea that students need to keep in mind is that ANCOVA is only a special case of regression in which one or more quantitative predictors are treated as covariates and one or more categorical variables represent membership in treatment groups. Using the matrix algebra for multiple regression (presented in Appendix 14.A in Chapter 14) with dummy variables to represent treatment group membership (in Chapter 12) will yield parameter (slope) estimates for the ANCOVA model.

Matrix algebra is required to represent the computations of the various SS terms in ANCOVA (see Tabachnick & Fidell, 2007, pp. 203–208, for the matrix algebra). Only a conceptual description of the terms is presented here. In this simple example with only one covariate and one treatment factor, we need to obtain SS_{X_c}; this represents the part of SS_{total} for Y that is predictable from X_c, a covariate. We need to obtain an SS for A, usually called an adjusted SS_A, to remind us that we are adjusting for or partialling out the covariate; $SS_{A_{\text{adj}}}$ represents the part of SS_{total} for Y that is predictable from membership in the groups on Factor A when the X_c covariate is statistically controlled. Finally, we will also have SS_{within} or SS_{residual}, the part of the SS_{total} for Y that is not explained by either X_c or A.

The primary goal in ANCOVA is generally to assess whether the A main effects remain significant when one or more covariates are controlled. However, an F ratio may also be set up to assess the significance of each covariate, as follows:

$$F_{X_c} = [SS_{X_c} / df_{X_c}] / [SS_{\text{residual}} / df_{\text{residual}}]. \quad (15.6)$$

The null hypothesis that is usually of primary interest in ANCOVA is that the adjusted Y^* means are equal across levels of A:

$$H_0 : \mu^*_{A1} = \ldots = \mu^*_{Ak}. \quad (15.7)$$

In words, Equation 15.7 corresponds to the null hypothesis that the adjusted Y^* means in the population are equal across levels of A; the means are adjusted by subtracting the parts of Y scores that are predictable from one or more X_c covariates. The corresponding F ratio to test the null hypothesis in Equation 15.7 is as follows:

$$F_A = [SS_{A_{adj}} / df_A] / [SS_{residual} / df_{residual}]. \quad (15.8)$$

15.7 ♦ Computation of Adjusted Effects and Adjusted Y^* Means

The unadjusted effect (α_i) for level or Group i of an A factor in a one-way ANOVA was introduced in Chapter 6. The sample estimate of α_1 for Group 1 is obtained by finding the difference between the A_1 group mean on Y and the grand mean for Y. The unadjusted effects in the ANCOVA are calculated as follows for each group:

$$\alpha_i = Y_{A_i} - Y_{grand}, \quad (15.9)$$

where Y_{A_i} is the mean of Y in Group A_i and Y_{grand} is the grand mean of Y across all groups included in the study. To do an ANCOVA, we need to compute adjusted effects, α^*_i, which have the X_c covariate partialled out:

$$\alpha^*_i = (Y_{A_i} - Y_{grand}) - b_3 \times (X_{cA_i} - X_{cgrand}), \quad (15.10)$$

where X_{cA_i} is the mean of the covariate X_c in Group A_i, X_{cgrand} is the mean of the X_c covariate across all groups in the study, and b_3 is the partial slope to predict Y from X_c in a regression that controls for A group membership by including dummy predictors, as in Equation 15.3.

Once we have adjusted effects, α^*_i, we can compute an estimated adjusted Y mean for each A_i group ($Y^*_{A_i}$) by adding the grand mean (Y_{grand}) to each adjusted effect estimate (α^*_i):

$$Y^*_{A_i} = \alpha^*_i + Y_{grand}. \quad (15.11)$$

15.8 ♦ Conceptual Basis: Factors That Affect the Magnitude of $SS_{A_{adj}}$ and $SS_{residual}$ and the Pattern of Adjusted Group Means

When we control for a covariate X_c, several outcomes are possible for our assessment of the effects of the treatment group membership variable A. These outcomes are analogous to the outcomes described for partial correlation analysis in Chapter 10:

1. When we control for X_c, the main effect of A (which was significant in an ANOVA) may become nonsignificant (in the ANCOVA), and/or the effect size for A may decrease.

2. When we control for X_c, the main effect of A (which was nonsignificant in an ANOVA) may become significant (in the ANCOVA), and/or the effect size for A may increase.

3. When we control for X_c, the conclusion about the statistical significance and effect size for A may not change.

4. When we control for X_c and calculate the adjusted group means using an ANCOVA, the rank ordering of Y means across levels of the A treatment group factor may change; in

such cases, our conclusions about the nature of the relationship of *A* and *Y* may be quite different when we control for X_c than when we do not. This outcome is not common, but it can occur. Data analysts should be skeptical of adjusted means when adjustment for covariates drastically changes the rank ordering of treatment group means; when adjustments are this dramatic, there may be rather serious confounds between X_c and *A*, and in such situations, ANCOVA adjustments may not be believable.

15.9 ♦ Effect Size

When SPSS GLM is used to do an ANCOVA, effect sizes for predictor variables can be indexed by eta-squared values, as in previous chapters in this book. Note that there are two versions of η^2 that can be calculated for an ANCOVA. In ANCOVA, the *SS* term for a main effect, adjusted for the covariate, is called an adjusted *SS*. In an ANCOVA using SS Type I (or sequential regression), with one *A* factor and one X_c covariate, SS_{total} for *Y* is partitioned into the following sum of squares components:

$$SS_{total} = SS_{X_c} + SS_{A_{adj}} + SS_{residual}. \tag{15.12}$$

We can compute a "simple" η^2 ratio to assess the proportion of variance accounted for by the *A* main effect relative to the *total* variance in *Y*; this is the version of eta squared that was discussed in the review of one-way ANOVA (see Chapter 6).

$$\eta_A^2 = SS_{A_{adj}} / SS_{total} = \frac{SS_{A_{adj}}}{SS_{A_{adj}} + SS_{X_c} + SS_{residual}}. \tag{15.13}$$

Alternatively, we can set up a "partial" η^2 to assess the proportion of variance in *Y* that is predictable from *A* when the variance associated with other predictors (such as the X_c covariates) has been partialled out or removed. When effect size information is requested, SPSS GLM reports a partial η^2 for each effect. For the simple ANCOVA example in this chapter, this corresponds to the partial η^2 in Equation 15.14:

$$\text{Partial } \eta_A^2 = SS_{A_{adj}} / [SS_{A_{adj}} + SS_{residual}]. \tag{15.14}$$

This partial η^2 tells us what proportion of the *Y* variance that remains after we statistically control for X_c is predictable from *A*. Note that the partial η^2 given by Equation 15.14 is usually larger than the simple η^2 in Equation 15.13 because SS_{X_c} is not included in the divisor for partial η^2.

More generally, partial η^2 for an effect in a GLM model is given by the following equation:

$$\text{Partial } \eta_{effect}^2 = SS_{effect} / [S_{effect} + SS_{residual}]. \tag{15.15}$$

15.10 ♦ Statistical Power

Usually, researchers are not particularly concerned about statistical power for assessment of the significance of the X_c covariate(s). The covariates are usually included primarily because we want to adjust for them when we assess whether Y means differ across groups (rather than because of interest in the predictive usefulness of the covariates). We are more concerned with statistical power for assessment of main effects and interactions. To make an informed decision about the minimum n per group, you can use the same power tables as the ones provided in Chapter 13 on factorial ANOVA.

Note that including one or more covariates can increase statistical power for tests of main effects by reducing the size of SS_{residual} or SS_{within} and, therefore, increasing the size of F; but an increase in power is obtained only if the covariates are fairly strongly predictive of Y. Note also that if covariates are confounded with group membership, adjustment for covariates can move the group means farther apart (which will increase $SS_{A_{\text{adj}}}$ and, thus, increase F_A), or adjustment for the covariate X_c can move the group means closer together (which will decrease $SS_{A_{\text{adj}}}$ and reduce F_A). Thus, when one or more covariates are added to an analysis, main effects for group differences can become either stronger or weaker. It is also possible that controlling for covariates will change the rank ordering of Y means across levels of A, although this is not a common outcome. When a covariate is an effective noise suppressor, its inclusion in ANCOVA can improve statistical power. However, under some conditions, including a covariate decreases the apparent magnitude of between-group differences or fails to reduce within-group or error variance; in these situations, ANCOVA may be less powerful than ANOVA.

15.11 ♦ Nature of the Relationship and Follow-Up Tests: Information to Include in the Results Section

First, you should describe what variables were included (one or more covariates, one or more factors, and the name of the Y outcome variable). You should specify which method of variance partitioning you have used (e.g., Type I vs. Type III SS) and the order of entry of the predictors. Second, you should describe the preliminary data screening and state whether there are violations of assumptions. Most of these assumptions (e.g., normally distributed scores on Y, linear relation between X_c and Y) are similar to those for regression. You should check to see whether there are homogeneous variances for Y across groups. In addition, you need to assess possible treatment by covariate interaction(s); this can be done by using GLM (as in the example at the end of this chapter) or by adding an interaction to a regression model. If there are no significant interactions, rerun the ANCOVA with the treatment by covariate interaction terms omitted and report the results for that model. If there are significant interactions, blocking on groups based on X_c scores may be preferable to ANCOVA.

In addition to checking for possible violations of assumptions, I also recommend running ANOVAs to see how Y varies as a function of treatment groups (not controlling for X_c); this gives you a basis for comparison, so that you can comment on how controlling for X_c changes your understanding of the pattern of Y means across groups. Also, run an ANOVA to see how X_c means vary across groups; this provides information about the nature of any confounds between group memberships and covariates.

Your Results section will typically include F ratios, df values, and effect size estimates for each predictor; effect sizes are typically indexed by partial η^2 when you use GLM. If you include a source table for your ANCOVA, it is conventional to list the variables in the following order in the source table: first, the covariate(s); then, main effects; then, if ANCOVA involves a factorial design, two-way interactions and higher-order interactions.

To describe the nature of the pattern of Y means across groups, you should include a table of both unadjusted and adjusted means for the Y outcome variable. The adjusted Y means are corrected to remove any association with the covariate(s) or X_c variables; they are, in effect, estimates of what the means of Y might have been if the treatment groups had been exactly equal on the X_c covariates.

It is important to notice exactly how the ANCOVA (which assesses the pattern of Y means, controlling for the X_c covariates) differs from the corresponding ANOVA (which assessed the pattern of Y means without making adjustment for covariates). The overall F for main effects can become larger or smaller when you control for covariates depending on whether the adjustment to the Y means that is made using the covariates moves the Y means closer together, or farther apart, and whether $SS_{residual}$ or SS_{within} is much smaller in the ANCOVA than it was in the ANOVA.

If you have a significant main effect for a factor in your ANCOVA and the factor has more than two levels, you may want to do post hoc tests or planned contrasts to assess which pairs of group means differ significantly. If you have a significant interaction between factors in a factorial ANCOVA, you may want to present a graph of cell means, a table of cell means, or a table of adjusted effects (as in Chapter 13 on factorial ANOVA) to provide information about the nature of the interaction.

15.12 ♦ SPSS Analysis and Model Results

To understand how Y, X_c, and A are interrelated, the following analyses were performed. First, a one-way ANOVA was performed to see whether Y differed across levels of A. This analysis provided an answer to the question, If you do *not* control for X_c, how does Y differ across levels of A, if at all? A one-way ANOVA was run to see whether X_c differed across levels of A. This analysis answered the question, To what extent did the A treatment groups begin with preexisting differences on X_c, the covariate, and what was the nature of any confound between X_c and A? These results were obtained using the SPSS ONEWAY program (described in Chapter 6). The results for these ANOVAs (shown in Figure 15.3) indicated that the means on the covariate did not differ significantly across the three treatment groups; thus, there was a very weak confound between treatment and covariate (and that's good). To the extent that there was a confound between X_c (ability) and A (type of teaching method), students in Groups 2 and 3 (seminar and lecture) had slightly higher pretest math ability than students in Group 1 (the SPI group). Mean scores on Y, the final exam score, differed significantly across groups: Group 2 (seminar) and Group 3 (lecture) had higher mean final exam scores than Group 1 (SPI). We want to see whether controlling for the slight confound between teaching method and pretest math ability changes the Y differences across teaching groups.

To assess whether X_c and Y are linearly related, a scatter plot and a bivariate Pearson correlation between X_c and A were obtained. ANCOVA assumes that covariates are linearly related to the outcome variable Y. Unless there is a nonzero correlation between X_c and Y and/or some association between X_c and A, it is not useful to include X_c as a covariate. The scatter plot and Pearson r were obtained using commands that were introduced in earlier chapters. The relation between X_c and Y appeared to be fairly linear (see the scatter plot in Figure 15.2 on page 620). The correlation between X_c and Y was strong and positive: $r(19) = .63, p = .02$. Thus, X_c seemed to be sufficiently correlated to Y to justify its inclusion as a covariate. The preceding analyses provide two kinds of information: information about possible violations of assumptions (such as homogeneity of variance and linearity of association between X_c and Y) and information about all the bivariate associations between pairs of variables in this research situation. That is, How does the mean level of X_c differ across groups based on Factor A? Is Y, the outcome variable, strongly linearly related to X_c, the covariate? How do mean Y scores differ across levels of A when the covariate X_c is not statistically controlled? It is useful to compare the final ANCOVA results with the original ANOVA to predict Y from A, so that we can see whether and how controlling for X_c has changed our understanding of the nature of the association between Y and A.

Another assumption of ANCOVA is "no treatment by covariate interaction"; that is, the slope relating Y to X_c should be the same within each of the A_1, A_2, and A_3 groups. In the scatter plot in Figure 15.2 on page 620, the points were labeled with different markers for members of each group. Visual examination of the scatter plots does not suggest much difference among groups in the slope of the function to predict Y from X_c. Of course, it would be desirable to have much larger numbers of scores in each of the three groups when trying to assess the nature of the association between X_c and Y separately within each level of A. To examine the Y by X_c regression separately within each group, you could use the SPSS <Data> → <Split File> command to run a separate regression to predict Y from X_c for each level of the A factor—that is, for groups A_1, A_2, and A_3; results of separate regressions are not reported here.

To test for a possible violation of the assumption that there is no treatment by covariate ($A \times X_c$) interaction, a preliminary GLM analysis was conducted with an interaction term included in the model. Figure 15.5 shows the SPSS menu selections for the GLM procedure. From the top level menu in the SPSS Data View worksheet, the following menu selections were made: <Analyze> → <General Linear Model> → <Univariate>. (This ANCOVA is "univariate" in the sense that there is only one Y outcome variable.) Figure 15.6 shows the main dialog window for SPSS Univariate GLM. The names of the dependent variable (Y), treatment variable or factor (A), and covariate (X_c) were entered in the appropriate boxes in the main dialog window.

To add an interaction term to the ANCOVA model, it is necessary to request a custom model. Interactions between factors and covariates are not included by default; however, they can be specifically requested in the Model window. Clicking the button Model in the main GLM dialog window, seen in Figure 15.6, opens up the Model dialog window, which appears in Figure 15.7.

To specify a preliminary ANCOVA that included an $X_c \times A$ interaction term, the following selections were made within the Model window. Both X_c and A were highlighted in the left-hand window, which included all possible factors and covariates, and the term *Interaction* was chosen from the pull-down menu in the Build Term(s) window. Clicking the right arrow created an $A \times X_c$ interaction term and placed it in the list of terms to be included in the Custom Model in the right-hand window in Figure 15.7. Note that it was also necessary to move the terms that represented the main effect of A and the effect of X_c into this list in order to test for the interaction while controlling for main effects. The window that is labeled Sum of Squares has a pull-down menu that offers a choice of four methods of computation for sums of squares (labeled Types I, II, III, and IV). For this preliminary analysis, select Type III; this ensures that the interaction will be assessed while statistically controlling for both X_c and A. Click the Continue button to return to the main GLM dialog window, then OK to run the analysis. (Additional statistics such as adjusted group means and model parameter estimates were not requested for this preliminary analysis.)

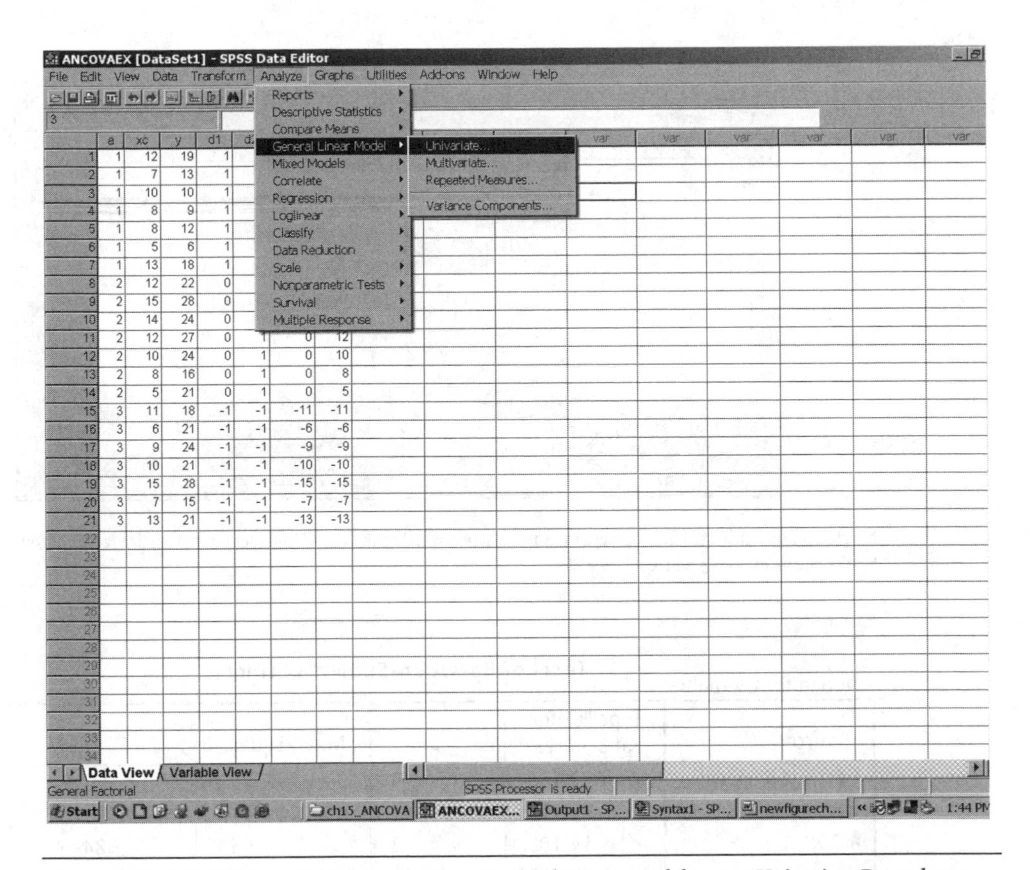

Figure 15.5 ♦ SPSS Menu Selections for the <General Linear Model> → <Univariate Procedure>

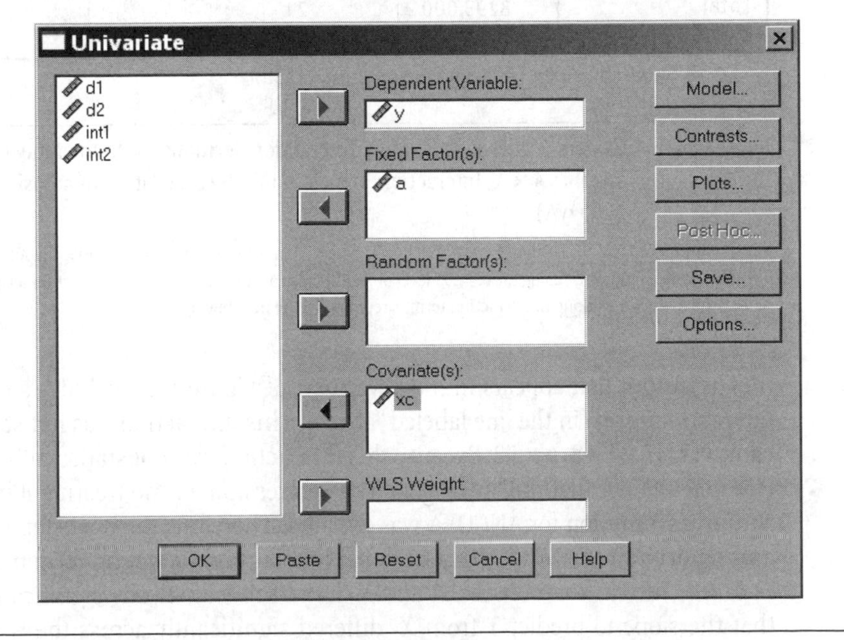

Figure 15.6 ♦ SPSS GLM Univariate Main Dialog Window

NOTE: A is the grouping variable, Y the outcome variable, and X_c the covariate.

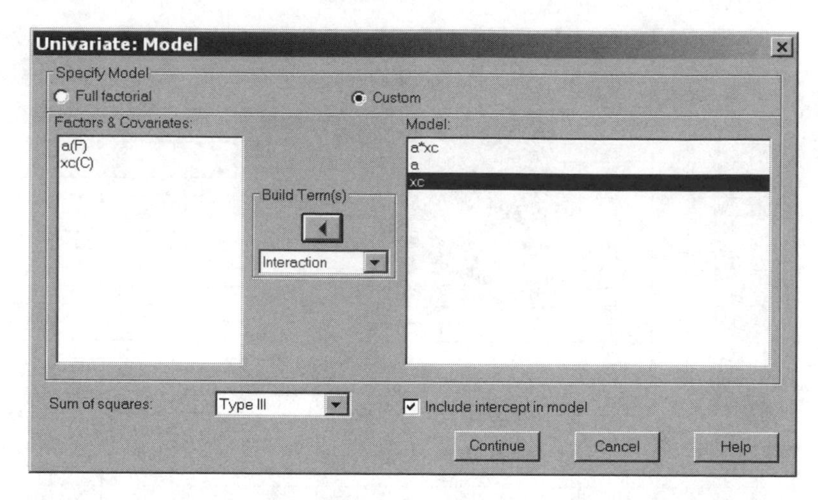

NOTE: SS Type III is used here because we want to assess whether the $A \times X_c$ interaction is statistically significant, controlling for the effects of both A and X_c.

Tests of Between-Subjects Effects

Dependent Variable: y

Source	Type III Sum of Squares	df	Mean Square	F	Sig.
Corrected Model	633.192 [a]	5	126.638	12.286	.000
Intercept	139.316	1	139.316	13.516	.002
a * xc	14.108	2	7.054	.684	.520
a	74.671	2	37.335	3.622	.052
xc	174.702	1	174.702	16.949	.001
Error	154.617	15	10.308		
Total	8293.000	21			
Corrected Total	787.810	20			

a. R Squared = .804 (Adjusted R Squared = .738)

Figure 15.7 ♦ Custom Model Specification: Interaction Term Added to the Model (to Test for Possible $A \times X_c$ Interaction, Which Would Be a Violation of a Basic Assumption for ANCOVA)

NOTE: Output for GLM that includes A, X_c, and $A \times X_c$ interaction term in the model. Because the $A \times X_c$ interaction term was not statistically significant (using the $\alpha = .05$ level of significance as the criterion), this interaction term is dropped from the model; the ANCOVA without an interaction term is reported in the next few figures.

The output that appears in the lower part of Figure 15.7 includes a test for the $A \times X_c$ interaction term (in the line labeled "a*xc"). This interaction was not statistically significant: $F(2, 15) = .68$, $p = .52$. Because this interaction was not statistically significant, there was no evidence that the "homogeneity of regression" or "no treatment by covariate interaction" assumption for ANCOVA was violated. Therefore, the next step will be to conduct and report a final ANCOVA that does *not* include an interaction term between A and X_c.

If this interaction between A and X_c had been statistically significant, it would mean that the slope to predict Y from X_c differed significantly across the A groups and that, therefore, it would be misleading to use the same regression slope to create adjusted group means for all the *A* groups (which is what ANCOVA does, in effect). At this point, the data analyst can report a model that includes an interaction term between treatment and covariate, but that analysis would not be called an ANCOVA. Horton (1978) mentioned the possibility of "ANCOVA with heterogeneous slopes"; however, most authorities require the absence of treatment by covariate interaction before they will call the analysis an ANCOVA.

Given that, for these hypothetical data, there was no significant interaction between treatment and covariate, the next step is to run a new GLM analysis that does not include this interaction term. To be certain that this interaction term is excluded from this new model, it is a good idea to click the Reset button for GLM so that all the menu selections made in the earlier analysis are cleared. We identify the variables in the main Univariate GLM dialog window (as shown in Figure 15.6). To request additional descriptive statistics, click the Statistics button in the main dialog window; the Statistics dialog window appears in Figure 15.8. In the following example, these were the requested statistics: First, the terms that represent the OVERALL (grand) mean and the mean for each level of Factor *A* (a) were highlighted and moved into the window under the heading Display Means for. Check boxes were used to request "Estimates of effect size" (i.e., partial η^2 values) and "Parameter estimates" (i.e., coefficients for the regression model that predicts *Y* from scores on X_c and dummy variables to represent group membership). The box to Compare main effects was checked to request selected comparisons between pairs of group means on the *A* factor. From the pull-down menu directly below this, the option for Bonferroni-corrected adjustments was selected. The Continue button was clicked to return to the main GLM Univariate dialog window.

To request the use of Type I SS computational methods, and to make sure that X_c (the covariate) was entered before *A* (the treatment factor), the Model button was clicked. Within the Model dialog window, which appears in Figure 15.9, SS Type I was selected from the pull-down menu for type of sums of squares. The Custom Specify Model radio button was clicked, then the terms X_c and *A* were highlighted and moved into the Model window. The order in which the terms appear in this list is the order in which the predictors will enter the model. That is, for the analysis specified in Figure 15.9, the X_c covariate will be entered in the first step, and then an adjusted *SS* for *A* will be calculated controlling for, or adjusting for, X_c. Clicking Continue returns us to the main dialog window. The Paste button was used to place the SPSS syntax that was generated by these menu selections into an SPSS Syntax window, which appears in Figure 15.10.

These menu selections correspond to an ANCOVA to assess whether mean scores on *Y* differ significantly across levels of *A* when the X_c covariate is statistically controlled. Results for this analysis appear in Figures 15.11 and 15.12. The result that is typically of primary interest in an ANCOVA is the *F* test for the *A* main effect, controlling for the covariate X_c. This appears in the Tests of Between-Subjects Effects panel in Figure 15.11, in the row for "a." The mean *Y* scores differed significantly across levels of the *A* factor when the covariate X_c was statistically controlled: $F(2, 17) = 15.47$, $p < .001$. The "adjusted" *Y* means (i.e., adjusted to remove the association with the covariate X_c) are given in Figure 15.12, in the panel labeled Estimates. Table 15.2 summarizes the original unadjusted *Y* means (from the one-way ANOVA that was reported in Figure 15.3) along with the adjusted *Y* means obtained from the GLM ANCOVA reported in Figures 15.11 and 15.12.

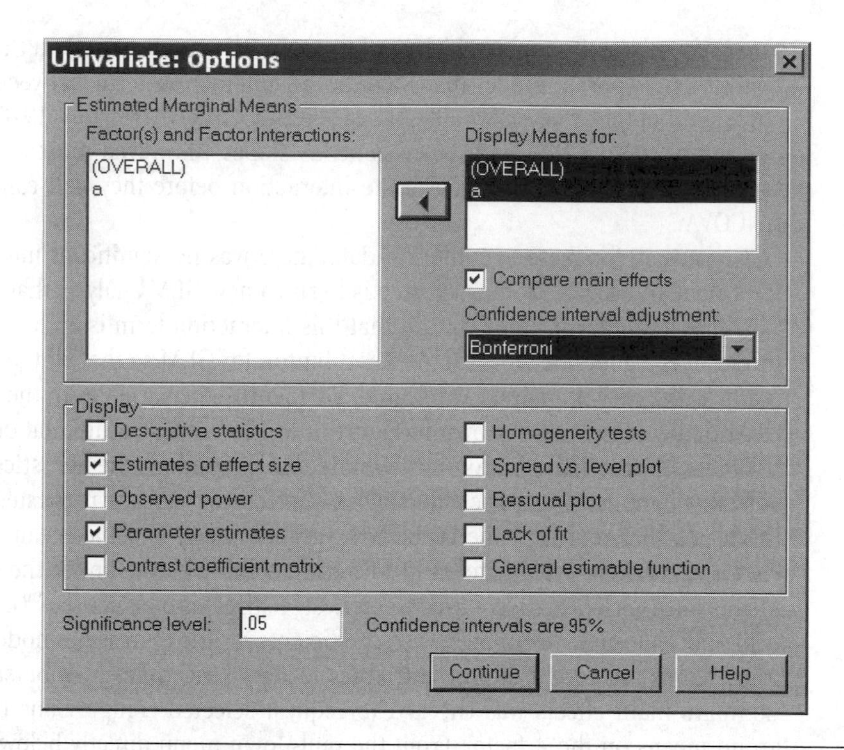

Figure 15.8 ♦ Options Selected for the Final ANCOVA Model: Display Means for Each Level of *A* (Type of Teaching Method)

NOTE: Do Bonferroni-corrected contrasts between group means. Report estimates of effect size (partial η^2) and parameter estimates for the regression version of the ANCOVA.

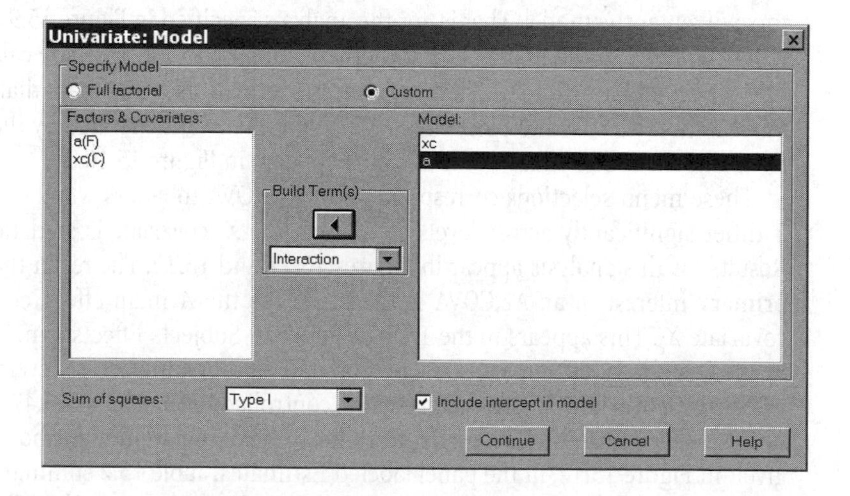

Figure 15.9 ♦ Custom Model for Final Version of ANCOVA

NOTE: Select Type I method of computation of *SS*. List variables in the order of entry (first X_c, then *A*) so that the *SS* for *A* is calculated while partialling out or controlling for X_c.

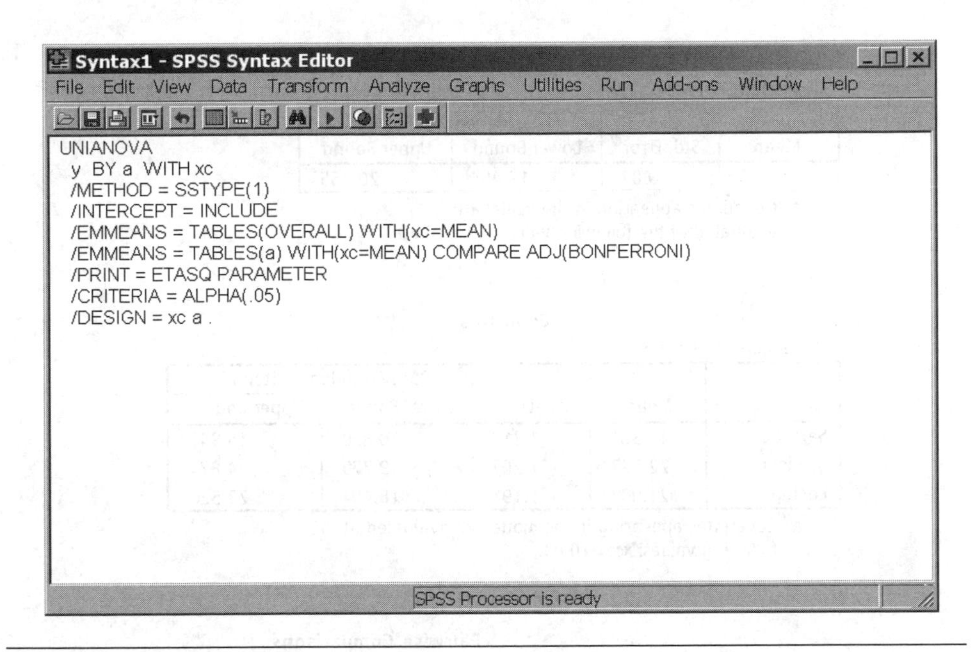
Syntax1 - SPSS Syntax Editor

```
UNIANOVA
 y BY a WITH xc
 /METHOD = SSTYPE(1)
 /INTERCEPT = INCLUDE
 /EMMEANS = TABLES(OVERALL) WITH(xc=MEAN)
 /EMMEANS = TABLES(a) WITH(xc=MEAN) COMPARE ADJ(BONFERRONI)
 /PRINT = ETASQ PARAMETER
 /CRITERIA = ALPHA(.05)
 /DESIGN = xc a .
```


Figure 15.10 ♦ SPSS Syntax for the Final ANCOVA Analysis Specified by the Menu Selections Above

Between-Subjects Factors

a	Value Label	N
1	Self Paced	7
2	Seminar	7
3	Lecture	7

Tests of Between-Subjects Effects

Dependent Variable: y

Source	Type I Sum of Squares	df	Mean Square	F	Sig.	Partial Eta Squared
Corrected Model	619.085 [a]	3	206.362	20.792	.000	.786
Intercept	7505.190	1	7505.190	756.191	.000	.978
xc	311.938	1	311.938	31.430	.000	.649
a	307.146	2	153.573	15.473	.000	.645
Error	168.725	17	9.925			
Total	8293.000	21				
Corrected Total	787.810	20				

a. R Squared = .786 (Adjusted R Squared = .748)

Parameter Estimates

Dependent Variable: y

Parameter	B	Std. Error	t	Sig.	95% Confidence Interval		Partial Eta Squared
					Lower Bound	Upper Bound	
Intercept	11.486	2.653	4.330	.000	5.890	17.083	.524
xc	.952	.234	4.074	.001	.459	1.445	.494
[a=1]	-7.626	1.705	-4.473	.000	-11.223	-4.029	.541
[a=2]	1.320	1.692	.780	.446	-2.250	4.890	.035
[a=3]	0[a]	.	.	.	.	.	.

a. This parameter is set to zero because it is redundant.

Figure 15.11 ♦ Output From the Final ANCOVA Analysis

1. Grand Mean

Dependent Variable: y

Mean	Std. Error	95% Confidence Interval	
		Lower Bound	Upper Bound
18.905[a]	.687	17.454	20.355

a. Covariates appearing in the model are evaluated at the following values: xc = 10.00.

Estimates

Dependent Variable: y

a	Mean	Std. Error	95% Confidence Interval	
			Lower Bound	Upper Bound
Self Paced	13.381[a]	1.213	10.820	15.941
Seminar	22.327[a]	1.207	19.779	24.874
Lecture	21.007[a]	1.191	18.494	23.520

a. Covariates appearing in the model are evaluated at the following values: xc = 10.00.

Pairwise Comparisons

Dependent Variable: y

(I) a	(J) a	Mean Difference (I-J)	Std. Error	Sig.[a]	95% Confidence Interval for Difference[a]	
					Lower Bound	Upper Bound
Self Paced	Seminar	-8.946 *	1.739	.000	-13.563	-4.329
	Lecture	-7.626 *	1.705	.001	-12.153	-3.099
Seminar	Self Paced	8.946 *	1.739	.000	4.329	13.563
	Lecture	1.320	1.692	1.000	-3.173	5.813
Lecture	Self Paced	7.626 *	1.705	.001	3.099	12.153
	Seminar	-1.320	1.692	1.000	-5.813	3.173

Based on estimated marginal means

*. The mean difference is significant at the .05 level.

a. Adjustment for multiple comparisons: Bonferroni.

Figure 15.12 ♦ Estimated (i.e., Adjusted for the Covariate X_c) Means on the *Y* Outcome Variable Across Levels of the *A* Teaching Method Factor

Table 15.2 ♦ Adjusted and Unadjusted Group Means for Final Statistics Exam Scores

		Unadjusted (c)	*Adjusted for X_c (Y_i^*)*	*Ability Pretest*
Group 1	Self-paced instruction	12.43	13.38	
Group 2	Seminar	23.14	22.32	
Group 3	Lecture	21.14	21.00	

Results

A quasi-experimental study was performed to assess whether three different teaching methods produced different levels of performance on a final exam in statistics. Group A_1 was given SPI, Group A_2 participated in a seminar, and Group A_3 received lecture instruction. Because the assignment of participants to groups was not random, there was a possibility that there might be preexisting differences in math ability; to assess this, students were given a pretest on math ability (X_c), and this score was used as a covariate. The dependent variable, Y, was the score on the final exam.

Preliminary data screening was done; scores on X_c and Y were reasonably normally distributed with no extreme outliers. The scatter plot for X_c and Y showed a linear relation with no bivariate outliers. Scores on the math pretest (X_c) did not differ significantly across groups: $F(2, 18) = .608$, $p = .55$; however, the SPI group (A_1) scored slightly lower on the math pretest than the other two groups.

To assess whether there was an interaction between treatment and covariate, a preliminary ANCOVA was run using SPSS GLM with a custom model that included an $X_c \times A$ interaction term. This interaction was not statistically significant: $F(2, 15) = .684$, $p = .52$. This indicated no significant violation of the homogeneity of regression (or no treatment by covariate interaction) assumption. (Examination of the X_c by Y association within each of the A groups, as marked in Figure 15.2 on page 620, suggested that the X_c by Y relation was reasonably linear within each group; however, the sample size within each group, $n = 7$, was extremely small in this example. Much better assessments could be made for linearity and possible treatment by covariate interaction if the numbers of scores were much larger for each group.) The assumption of no treatment by covariate interaction that is required for ANCOVA appeared to be satisfied; the final ANCOVA reported, therefore, does not include this interaction term.

When math ability at pretest (X_c) was not statistically controlled, the difference in final statistics exam scores (Y) was statistically significant: $F(2, 18) = 12.26$, $p < .001$. The main effect for type of instruction in the final ANCOVA using math ability scores (X_c) as a covariate was also statistically significant: $F(2, 17) = 15.47$, $p < .001$. The strength of the association between teaching method and final exam score was $\eta^2 = .58$ when math scores were not used as a covariate versus partial $\eta^2 = .645$ when math scores were used as a covariate. The effects of teaching method on final exam score appeared to be somewhat stronger when baseline levels of math ability (X_c) were statistically controlled.

Adjusted and unadjusted group means for the final exam scores are given in Table 15.2. The rank ordering of the group means was not changed by adjustment for the covariate; however, after adjustment, the means were slightly closer together. Nevertheless, the F for the main effect of instruction in the ANCOVA was larger than the F when X_c was not statistically controlled, because controlling for the variance associated with pretest math ability substantially reduced the size of the within-group error variance.

(Continued)

(Continued)

After adjustment for pretest math scores, students who received either lecture or seminar forms of instruction scored higher ($M_3 = 21.00$ and $M_2 = 22.32$) than students who received SPI ($M_1 = 13.38$). The parameter estimates in Figure 15.11 that correspond to the slope coefficients for the dummy variable for the $A = 1$ group and the $A = 2$ group provide information about contrasts between the adjusted Y means of the A_1 versus A_3 group and the adjusted Y means of the A_2 versus A_3 group, respectively. After adjustment for the X_c covariate, the difference between mean final exam scores for the A_1 group was 7.626 points lower than for the A_3 group, and this difference was statistically significant: $t(17) = -4.473$, $p < .001$. After adjustment for the X_c ability covariate, the difference between the mean final exam scores for the A_2 group was only 1.32 points lower than for the A_3 group, and this difference was not statistically significant: $t(17) = .780$, $p = .446$.

To summarize, after adjustment for the scores on the X_c ability pretest, mean final exam scores were significantly lower for the SPI group than for the lecture group, while adjusted mean final exam scores did not differ significantly between the seminar and lecture groups. (Additional post hoc tests, not reported here, could be run to make other comparisons, such as SPI vs. seminar group means.)

15.13 ♦ Additional Discussion of ANCOVA Results

When SPSS GLM is used to run ANCOVA, it reports both unadjusted and adjusted Y means. We can compute the "unadjusted" effect for Y for each group by hand by subtracting the grand mean of Y (18.90) from each Y group mean. This gives the following:

Unadjusted α_i Effect for Group

General Case	$Y_{A_i} - Y_{grand} = \alpha_{Y_i}$
Group 1	12.43 – 18.90 = –6.47
Group 2	23.14 – 18.90 = 4.24
Group 3	21.14 – 18.90 = 2.24

Part of these unadjusted effects may be associated with preexisting group differences on X_c, the covariate. To see what the magnitude of these preexisting differences on X_c are, we compute the α_{X_c} effect for each group.

General Case	$X_{c_i} - X_{c_{grand}} = \alpha_{x_{c_i}}$
Group 1	9.00 – 10.00 = –1.00
Group 2	10.86 – 10.00 = +.86
Group 3	10.14 – 10.00 = +.14

Note that the computation of effects becomes a little more complicated when the numbers of scores in the groups are unequal, because we then need to choose whether to use the weighted or unweighted grand mean as the basis for computing deviations.

Now, we will need to know the raw score slope that relates scores on Y to scores on X_c while controlling for/taking into account membership on the A grouping variable. SPSS GLM provides this slope if "Parameter estimates" are requested as part of the GLM output; see the last panel in Figure 15.11. The raw score partial slope relating scores on Y to scores on X_c, controlling for A, is $b = +.952$. (It can be demonstrated that this partial slope approximately equals the average of the three within-group regression slopes to predict Y from X_c; however, the three separate within-group regressions to predict Y from X_c are not presented here.)

Now let's calculate the adjusted $\alpha^*_{Y_i}$ effect for each of the three groups:

	Adjusted Y Deviation or Effect (Removing Part of the Y Mean That Is Related to X_c)
General Case	$\alpha_{Y_i} - \quad b(\alpha_{X_{c_i}}) = \quad \alpha^*_{Y_i}$
Group 1	−6.48 − .952 (−1.00) = −5.528
Group 2	4.24 − .952 (+.86) = 3.421
Group 3	2.24 − .952 (+.14) = 2.107

Notice that these adjusted effects are essentially predictions of how far each Y group mean would have been from the grand mean if parts of the Y score that are linearly associated with X_c were completely removed or partialled out.

To construct the predicted or adjusted group means on Y, Y^*_i, adjusted for X_c, the covariate, you add the grand mean of Y to each of these adjusted effects. This gives us the following:

	Adjusted Effect + Grand Mean = Adjusted Mean for Group A_i
General Case	$\alpha^*_i + \ Y_{grand} = \quad Y^*_{A_i}$
Group 1	−5.528 + 18.90 = 13.372
Group 2	3.421 + 18.90 = 22.321
Group 3	2.107 + 18.90 = 21.007

Within rounding error, these agree with the values given for the adjusted or estimated means in the second panel headed Estimates in Figure 15.12.

Finally, the partitions of sums of squares for the ANOVA results (prediction of Y from A) and the ANCOVA results (prediction of Y from A, controlling for X_c) are graphically illustrated in Figures 15.13 and 15.14. The inclusion of the X_c covariate resulted in the following changes in the values of the SS terms. Before adjustment for the covariate, the value of SS_A was 454.38; this numerical value comes from the one-way ANOVA to predict Y from A that was represented in Figure 15.3, and the partition of SS for Y in this one-way ANOVA is shown graphically in Figure 15.13. After controlling for the X_c covariate, the adjusted $SS_{A_{adj}}$ was reduced to 311.94; this adjusted SS for A is obtained from the ANCOVA results in Figure 15.11, and the new partition of SS in the ANCOVA is shown graphically in

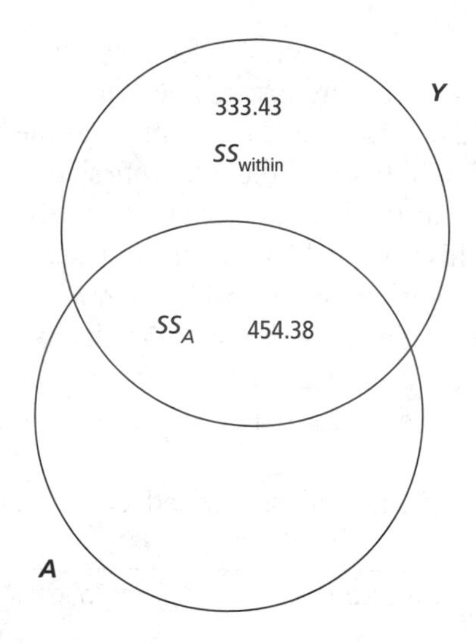

Figure 15.13 ♦ Partition of *SS* in ANOVA to Predict *Y* From *A* for Data in Table 15.1 (Ignoring the Covariate X_c)

NOTE: Areas in the diagrams are not exactly proportional to the magnitudes of *SS* terms. $SS_{total} = SS_A + SS_{within} = 454.38 + 333.43 = 787.81$.

Figure 15.14. By comparing the partition of the total *SS* for *Y* between the one-way ANOVA (as shown in Figure 15.13) and the ANCOVA (as shown in Figure 15.14), we can evaluate how statistically controlling for the X_c covariate has changed the partition of *SS*. The slight reduction in the *SS* for *A* when X_c was statistically controlled is consistent with the observation that after adjustment for X_c, the adjusted *Y* means were slightly closer together than the original (unadjusted) *Y* means. See Table 15.2 for a side-by-side comparison of the unadjusted and adjusted *Y* means. However, the inclusion of X_c in the analysis also had a substantial impact on the magnitude of SS_{within}. For the one-way ANOVA to predict *Y* from *A*, SS_{within} was 333.43. For the ANCOVA to predict *Y* from *A* while controlling for X_c, SS_{within} was (substantially) reduced to 168.73. Thus, X_c acted primarily as a noise suppressor in this situation; controlling for X_c reduced the magnitude of SS_A by a small amount; however, controlling for X_c resulted in a much greater reduction in the magnitude of SS_{within}. As a consequence, the *F* ratio for the main effect of *A* was larger in this ANCOVA than in the simple ANOVA; and the partial η^2 effect size (.645), which represents the strength of the association between *A* and *Y*, was larger in the ANCOVA than the simple η^2 (.577) in the ANOVA. Note, however, that including an X_c covariate can result in quite a different outcome, such as a substantial reduction in SS_A and not much reduction in SS_{within}.

15.14 ♦ Summary

ANCOVA is likely to be used in the following research situations.

1. It is used to correct for nonequivalence in subject characteristics (either on a pretest measure or on another variable related to *Y*). In these situations, the ANCOVA

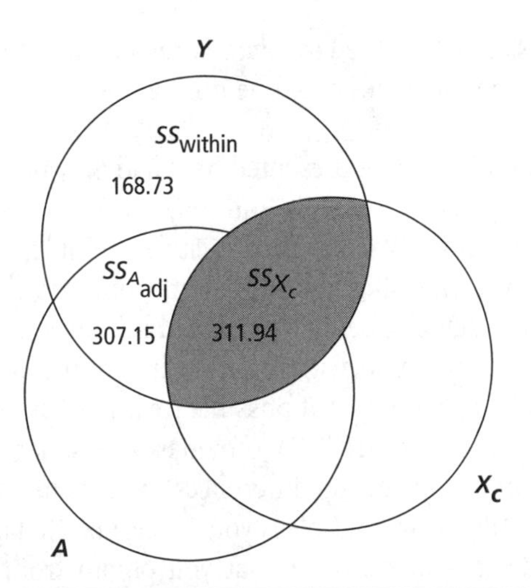

Figure 15.14 ♦ Partition of *SS* in ANCOVA to Predict *Y* From *A* and X_c Using Type I SS, With X_c Entered Prior to *A*

NOTE: $SS_{\text{total}} = SS_{X_c} + SS_{A_{\text{adj}}} + SS_{\text{within}} = 311.94 + 307.15 + 168.75 = 787.81.$

"adjusts" the group means for the *Y* outcome variable to the values they might have had if means on the covariate (X_c) had been equal across groups. When considering this application of ANCOVA, relevant questions include the following. How far apart were the groups on the covariate? Are the adjusted group means closer together, or farther apart, than the unadjusted group means? Researchers should interpret large adjustments to *Y* means cautiously, particularly if the adjustments result in larger group differences or a different rank ordering of groups.

2. It can be used to analyze pretest/posttest data. Several other methods that can be used to analyze pretest/posttest scores are briefly discussed in the appendix to this chapter.

3. It may be used to reduce error variance due to individual differences among subjects (i.e., a covariate may be included as noise suppressor).

Conventional restrictions on ANCOVA include the following:

1. The covariate X_c is a normally distributed quantitative variable, and it must be linearly related to *Y*.

2. There should be no treatment by covariate or $A \times X_c$ interaction (you need to test for this; if there is a significant interaction, you should not call your analysis an ANCOVA).

3. The covariate should not be influenced by the treatment (this is not usually a problem if the covariate is measured *before* treatment is administered but often is a problem if the covariate is measured after the treatment).

4. We assume that the covariate is reliably measured; otherwise, adjustments made using the covariates could be misleading.

The ANCOVA example presented here had only one factor and one covariate. ANCOVA may have more than one factor and may include two-way and higher-order interactions among factors. ANCOVA may also include multiple covariates and/or multiple outcome variables. However, it should be apparent from this discussion that keeping track of how the variance is actually being partitioned becomes more complex as you add more variables to your design. Also, with multiple factors and/or multiple covariates, it is necessary to check for a large number of possible treatment by covariate interactions.

In some situations, ANCOVA provides a reasonable method for trying to remove or control for preexisting group differences in participant characteristics that are measured as covariates. But be wary: Unless you are certain that all the assumptions of ANCOVA are met, the adjusted scores on *Y* that you obtain from ANCOVA may be misleading. As Tabachnick and Fidell (2007) stated, "In all uses of ANCOVA, however, adjusted means must be interpreted with some caution. The mean *Y* score after adjustment for covariates may not correspond to any situation in the real world" (p. 200).

If you expect to do a great deal of quasi-experimental research involving **nonequivalent control groups,** Cook and Campbell's (1979) book *Quasi-Experimentation* is an extremely valuable resource on design and analysis.

Appendix: Alternative Methods for the Analysis of Pretest/Posttest Data

Often it is desirable to obtain pretest observations on the outcome behavior (e.g., in research on the effects of stress on blood pressure, we nearly always include a prestress or baseline measure of blood pressure). Although **pretest/posttest designs** are extremely common, there is continuing controversy about the best way to analyze such data. Here are the most common methods for the analysis of data from a simple pretest/posttest design (e.g., blood pressure measures for *N* subjects at two points in time, before and after treatment):

1. *Analysis of gain scores or difference scores or change scores:* For each of the *N* subjects, simply compute the change or **gain** or **difference score** (posttest blood pressure − pretest blood pressure); use this change score in any further analysis. For instance, the **correlated samples** or direct difference ***t*** **test** (discussed in Chapter 20) assesses whether the mean change differs significantly from 0; ANOVAs may be performed to see if change scores differ across groups such as male/female; correlational or regression analyses may be done to see if change scores are predictable from continuous variables such as weight, IQ, trait measures, etc.
2. *ANCOVA using the posttest score as the outcome or* Y *variable and the pretest score as the covariate:* Differences among groups are assessed statistically controlling for, or partialling out, the pretest scores; in effect, the adjusted *Y* scores are estimates of the outcomes we might have obtained if the groups had been equal on the pretest or covariate variable.

3. *Repeated measures ANOVA in which the pretest/posttest measures are represented as two levels of a within-subjects trials factor:* Interactions between this within-subjects trials factor and other between-subjects factors such as sex may be used to assess whether between-subjects factors predict different changes in response across time.

4. *Multiple regression, in which the* X_c *pretest variable is entered as a predictor of* Y *and pretest levels on* X_c *are statistically controlled when assessing whether other* X_i *predictors are significantly related to* Y: In effect, the researcher controls for or partials out the part of Y that can be predicted from pretest scores on X_c and treats the remaining part of Y as an index of change; partial slopes for other X_i predictors in the multiple regression are interpreted as predictors of change in Y. For instance, if X_c is depression at Time 1, Y is depression at Time 2, and X_1, X_2, and X_3 correspond to social support, coping skills, and stress, respectively, we would set up a regression to predict Y from X_c, X_1, X_2, and X_3 and interpret the b coefficients associated with X_1, X_2, and X_3 as predictors of *change* in level of depression from Time 1 to Time 2.

5. *Growth curve analysis:* This is a relatively new method for analysis of repeated measures (usually, studies that use growth curve analysis involve more than two repeated measures, but two measures can be done as a special case). It has become particularly popular as a way of assessing individual participant patterns of change over time in developmental psychology research. Typically, participants have two or several measures on a particular cognitive or emotional test per subject. If there are only two scores, all you can do to assess change is compute a change score (equivalent to a linear slope); X_c, the pretest score, is the intercept for the growth function and d, the difference scores, is its slope, and with only two points, the function must be linear. If there are more than two scores, you can begin to fit curves and estimate parameters such as intercept and slope for various types of growth curves; these growth curve parameters can be used as dependent variables in a multivariate analysis. Thus, growth curve analysis represents yet another way to look at pretest/posttest data.

Potential Problems With Gain or Change Scores

There is a potential problem with the use of change scores or gain scores, and so ANCOVA is often viewed as a better means of controlling for pretest differences.

Here is the issue:

Let Y = posttest and X_c = pretest.

If we set up a bivariate regression to predict Y from X (as we implicitly do in an ANCOVA), we will obtain an equation like this: $Y' = b_0 + b_{X_c}$. In situations where X_c is a pretest on the same measurement as Y, the slope b is generally less than 1.

To obtain the part of the Y scores that is not related to or predictable from X_c—that is, an estimate of the change from pretest to posttest that is not correlated with the pretest score X_c, we could use the residuals from this simple bivariate regression:

$$\text{Adjusted change score } Y^* = Y - Y' \text{ or } Y - (b_0 + bX).$$

Note that Y^* or $Y - Y'$ is uncorrelated with X_c because of the way it is constructed, by using regression to divide each Y score into two components, one that is related to $X_c (Y')$ and the other that is not related to X_c $(Y - Y')$.

On the other hand, when we compute a simple gain or change score $d = Y - X_c$, we are calculating something like a regression residual, except that we essentially assume that $b = 1$. By construction, $Y - Y'$ is uncorrelated with X_c, and this estimate of Y' usually involves a value of b that is less than 1 (because pretest and posttest scores usually have similar standard deviations and are less than perfectly correlated). Subtracting $b \times X_c$, as in the regression or ANCOVA, will yield an index of change that is uncorrelated with X_c, by construction. When we subtract $1 \times X_c$ (as in the computation of a simple difference score), it is often an "overcorrection." When Y and X_c are repeated measures on the same variable, b is almost always less than 1. Often, although not always, simple change scores or gain scores that are computed by taking the $Y - X_c$ difference are artifactually negatively correlated with pretest levels of X_c. We can, of course, run a Pearson correlation between d and X_c to assess empirically whether this artifactual negative correlation is actually present.

Researchers often use simple gain scores or change scores even though they may be artifactually correlated with pretest levels; in many research areas (such as studies of cardiovascular reactivity), it has become conventional to do so. However, editors and reviewers sometimes point out potential problems with simple change scores; they may recommend ANCOVA or repeated measures ANOVA as alternative methods of analysis for pretest/posttest data.

Apart from the possible artifactual correlation between d, a simple change score, and X_c, the pretest, there is another issue to consider in the analysis of pretest/posttest data. Sometimes the nature of the differences between groups in a pretest/posttest design is quite different depending on the type of analysis that is used. For example, whether or not the between-group differences are statistically significant may depend on whether the researcher did an ANOVA on gain scores or an ANCOVA, controlling for pretest scores as a covariate. The rank ordering of groups on outcomes may even differ depending on which analysis was used; this phenomenon is called **Lord's paradox** (Lord, 1967). While it is not common for the nature of the effects in the study to appear entirely different for ANOVA versus ANCOVA, the researcher needs to be aware of this as a possibility.

It may be informative to analyze pretest/posttest data in several different ways (e.g., an ANOVA on change scores, an ANCOVA using X_c as the covariate, and a repeated measures ANOVA). If all the different analyses yield essentially the same results, the researcher can then report the analysis that seems the clearest and mention (perhaps in a footnote) that alternate analyses lead to the same conclusions. However, if the analyses yield radically different outcomes (e.g., group differences being significant in one analysis and not in the other or the apparent rank ordering of outcome changes across analyses), then the researcher faces a dilemma. Expert advice may be needed to decide which of the analyses provides the best assessment of change; otherwise, the results of the study may have to be judged inconclusive.

If you use change scores, it is a good idea to run a correlation between the change scores and pretest scores (X_c) in order to see if there is a negative correlation. Ideally, you might want d and X_c to be unrelated. Sometimes, there is no negative correlation between simple change scores and pretest; and in this case, you can dismiss one of the potential problems by showing empirically that no artifactual negative correlation between X_c and d exists. On the other hand, it is possible that there is a relationship between initial status on the pretest variable and the amount of change; and if this makes sense, then generating an adjusted change score that is forced to be uncorrelated with pretest level may not make sense.

Each of the various methods for the analysis of pretest/posttest data has both advantages and potential drawbacks. There is no single method that is always the correct choice in all situations. Researchers need to think carefully about the choice among these options, particularly because the different analyses do not necessarily lead to the same conclusions.

Note

1. The partitioning of variance in this situation can be worked out by representing the ANCOVA as a multiple regression problem (following Keppel & Zedeck, 1989, pp. 458–461) in which scores on Y are predicted from scores on both A (a treatment group variable) and X_c (a covariate). First, we need to figure out the proportion of variance in Y that is predictable from X_c, the covariate. This can be done by finding R^2_{YX} for the following regression analysis: $Y = b_0 + bX_c$. Next, as in any other sequential or hierarchical regression analysis, we want to assess how much additional variance in Y is predictable from A when we control for X_c. To assess this R^2 increment, we can run a regression to obtain R^2_{YXA}, $Y' = a + b_1 X_c + b_2 D_1 + b_2 D_2$, where D_1 and D_2 are dummy-coded variables that represent membership in the A treatment groups. The proportion of variance uniquely predictable from A, adjusting for or controlling for $X_c(R^2_{A_{adj}})$, corresponds to the difference between these two R^2 terms: $R^2_{A_{adj}} = R^2_{YXA} - R^2_{YX}$. Finally, to obtain the adjusted-error variance or within-group variance, we calculate $1 - R^2_{YXA}$. Values for SS_{X_c}, $SS_{A_{adj}}$, and $SS_{residual}$ can be obtained by multiplying the SS_{total} for Y by these three proportions. When the SS terms are obtained in this manner, they should sum to $SS_{total} = SS_{X_c} + SS_{A_{adj}} + SS_{residual}$.

+ − × ÷

Comprehension Questions

1. What is a quasi-experimental design? What is a nonequivalent control group?
2. Why is ANCOVA often used in situations that involve nonequivalent control groups?
3. Is the statistical correction for nonequivalence among groups likely to be as effective a means of controlling for the covariate as experimental methods of control (such as creating equivalent groups by random assignment)?
4. What is a covariate?
5. What information is needed to judge whether scores on the covariate (X_c) differ across groups? What information is needed to assess whether scores on the covariate X are strongly linearly related to Y (the dependent variable)?
6. Is there any point in controlling for X as a covariate if X is not related to group membership and also not related to Y?
7. The usual or conventional order of entry in an ANCOVA is as follows:

   Step 1: one or more covariates

   Step 2: one or more group membership variables

   (if the design is factorial, interactions are often entered after main effects).

   Why is this order of entry preferred to group variables in Step 1 and covariates in Step 2?
8. Running an ANCOVA is almost (but not exactly) like running the following two-step analysis:

   First, predict Y (the dependent variable) from X_1 (the covariate), and take the residuals from this regression; this residual is denoted by X^*.

   Second, run an ANOVA on X^* to see if the mean on X^* differs across levels of A, the group membership variable.

   How is the two-step analysis described here different from what happens when you run an ANCOVA?
9. When we say that the group means in an ANCOVA are "adjusted," what are they adjusted for?
10. When SPSS reports "predicted" group means, what does this mean?
11. What is the homogeneity of regression assumption? Why is a violation of this assumption problematic? How can this assumption be tested?
12. When adjusted group means are compared with unadjusted group means, which of the following can occur? (Answer yes or no to each.)

- The adjusted group means can be closer together than the unadjusted group means.
- The adjusted group means can be different in rank order across groups from the unadjusted means.
- The adjusted group means can be farther apart than the unadjusted group means.

13. Draw two overlapping circles to represent the shared variance between *A* (the treatment or group membership variable) and *Y* (the outcome variable). Now draw three overlapping circles to represent the shared variance among *Y* (a quantitative outcome variable), *X* (a quantitative covariate), and *A* (a group membership variable).

    Use different versions of these diagrams to illustrate that the effect of including the covariate can be primarily to reduce the SS_{error}, or primarily to reduce SS_A.

Data Analysis Project for ANCOVA

For this assignment, use the small batch of data provided below. Enter the data into an SPSS worksheet—one column for each variable (group membership, *A*; dependent variable, *Y*; and covariate, X_c).

To answer all the following questions you will need to do these analyses:

1. Run a one-way ANOVA to compare mean levels of *Y* across groups on the *A* factor (without the covariate) and also a one-way ANOVA to compare mean levels of X_c, the covariate, across groups on the *A* factor (to see if scores on the covariate differ across treatment groups).

2. Create histograms for scores on X_c and *Y* and a scatter plot and bivariate Pearson *r* to assess the association between X_c and *Y*.

3. Run a preliminary ANCOVA using the GLM procedure, with *A* as the group membership factor, *Y* as the dependent variable, and X_c as the covariate. In addition, use the Custom Model dialog box to create an interaction term between treatment (*A*) and covariate (X_c). The purpose of this preliminary ANCOVA is to test whether the assumption of no treatment by covariate interaction is violated. This is not the version of the analysis that is generally reported as the final result.

4. Run your final ANCOVA using the GLM procedure. This is the same as in Part 2, except that you drop the $A \times X_c$ interaction term. Assuming that you have no significant treatment by covariate interaction, this is the ANCOVA for which you would report results. Use Type I SS, and make sure that the covariate precedes the group membership predictor variable in the list of predictors in the Custom Model dialog window in GLM.

5. Optionally, at your instructor's discretion, you may also want to use the Regression procedure to run the same ANCOVA as in Part 3 above. Use dummy coding for the two dummy variables that represent treatment group membership. Enter the

covariate in Step 1 of a hierarchical analysis; enter the two dummy variables in Step 2. You should find that the results of this regression duplicate the results of your ANCOVA in Part 3.

Data for the Preceding Questions

These are hypothetical data. We will imagine that a three-group quasi-experimental study was done to compare the effects of three treatments on the aggressive behavior of male children. X_c, the covariate, is a pretest measure of aggressiveness: the number of aggressive behaviors emitted by each child when the child is first placed in a neutral playroom situation. This measure was done prior to exposure to the treatment. Children could not be randomly assigned to treatment groups, and so the groups did not start out exactly equivalent on aggressiveness. The dependent variable, Y, is a posttest measure: the number of aggressive behaviors emitted by each child after exposure to one of the three treatments. Treatment A consisted of three different films. The A_1 group saw a cartoon animal behaving aggressively. The A_2 group saw a human female model behaving aggressively. The A_3 group saw a human male model behaving aggressively. The question is whether these three models elicited different amounts of aggressive behavior when you do (and do not) control for individual differences in baseline aggressiveness. The scores are given on the top of the next page.

Let's further assume that there was very good interrater reliability on these frequency counts of behaviors and that they are interval/ratio level of measure, normally distributed, and independent observations.

A1		A2		A3	
X_c	Y	X_c	Y	X_c	Y
3	8	13	18	14	26
7	12	17	22	10	22
10	16	10	16	8	20
8	14	9	14	4	16
15	20	4	8	2	14
8	12	6	10	8	18
15	18	2	6	6	12

Write up your Results using APA style; hand in your printouts. In addition, answer the following questions about your results:

1. Using overlapping circles, show how SS_{total} was partitioned in the ANOVA (to predict Y from A) versus the ANCOVA (to predict Y from A, controlling for X_c). Indicate what SS values correspond to each slice of the circles in these diagrams.

2. Compare the unadjusted SS terms from your ANOVA (for error and the effect of treatment) with the adjusted SS terms from your ANCOVA. Was the effect of including the covariate primarily to decrease error variance, primarily to take variance

away from the treatment variable, or both? Does the effect of treatment look stronger or weaker when you control for the covariate?

3. Note the pattern of differences across treatment groups on X_c, the covariate or pretest. To what extent were there large differences in aggressiveness before the treatment was even administered? What was the nature of the confound: Did the most aggressive boys get a treatment that was highly effective or ineffective at eliciting aggression?
4. Look at the adjusted and unadjusted deviations that you got when you ran ANCOVA using the ANOVA program. Find and report the within-group slope that relates Y to X_c (for this, you will have to look at your regression printout). Show how one of the adjusted means can be calculated.
5. After adjustment, were the group means closer together or farther apart? Is there any change in the rank order of group means when you compare adjusted versus unadjusted means? Do you think these adjustments are believable?
6. We make a number of assumptions to do ANCOVA: that the treatment does not affect the covariate, that there is no treatment by covariate interaction, and that the covariate is reliably measured. How well do you think these assumptions were met in this case? Why would it be a problem if these assumptions were not met?
7. Only if you have covered the appendix material that discusses other possible methods of analysis for pretest/posttest designs, briefly describe three other ways you could analyze these data.

CHAPTER 16

Discriminant Analysis

16.1 ♦ Research Situations and Research Questions

Up to this point, the analyses that have been discussed have involved the prediction of a score on a quantitative *Y* outcome variable. For example, in multiple regression (MR) (Chapter 14), we set up an equation to predict the score on a quantitative *Y* outcome variable (such as heart rate) from a combination of scores on several predictor variables (such as gender, anxiety, smoking, caffeine consumption, and level of physical fitness). However, in some research situations, the *Y* outcome variable of primary interest is categorical. For example, in medical research, we may want to predict whether a patient is more likely to die (outcome Group 1) or survive (outcome Group 2) based on the patient's scores on several relevant predictor variables. This chapter discusses discriminant analysis (DA) as one possible way to approach the problem of predicting group membership on a categorical *Y* outcome variable from several *X* predictors. As in MR, these *X* predictors are often quantitative variables; the *X* predictors in DA may also include dummy-coded variables that represent information about group membership.

Like MR, DA involves finding optimal weighted linear combinations of scores on several *X* predictor variables—that is, sums or composites of *X* scores that make the best possible predictions for scores on a *Y* outcome variable. The difference between MR and DA is in the type of outcome variable. For MR, the *Y* outcome variable is quantitative; for DA, the *Y* outcome variable is categorical. Because DA involves a different type of outcome variable, different types of information are needed to evaluate the adequacy of prediction. In MR, we assess the accuracy of prediction of scores on a quantitative *Y* outcome variable by looking at the magnitude of multiple *R* (the correlation between actual *Y* and predicted *Y′* scores) and the magnitude of $SS_{residual}$ (residual sum of squares). For DA, we will examine other types of information, such as the percentage of cases for whom the predicted group membership agrees with the actual group membership.

In both MR and DA, the goal is to find an "optimal weighted linear combination" of scores on a set of *X* predictor variables—that is, a combination of raw scores or *z* scores for the *X* variables that makes the best possible prediction of scores on the outcome variable. A "weighted linear combination" of scores on a set of *X* variables is a sum of scores on several *X* variables. A combination of scores on *X* variables is linear as long as we use only addition and subtraction in forming the composites; if we use nonlinear arithmetic operations, such as division, multiplication, X^2, $\sqrt{X}$, and so forth, then the resulting composite is a nonlinear combination of scores on the X variables. In MR, the weight associated with each X predictor variable corresponds to its regression slope or regression coefficient. The weights for an optimal weighted linear combination in MR can be given in raw score form; that is, b slope coefficients or weights are applied to raw scores on X variables to estimate the raw score on a Y variable. Alternatively, the optimal predictive weights may be reported in standardized form as beta coefficients that are applied to z scores on X variables to generate predicted z scores on a Y outcome variable. In the MR analyses described in Chapter 14, the optimal weighted linear combination of scores on X can be obtained from a raw score version of the MR equation, as shown in Equation 16.1:

$$Y' = b_0 + b_1 X_1 + b_2 X_2 + \ldots + b_k X_k. \tag{16.1}$$

In the standardized version of MR, we set up an optimal weighted linear combination of standardized or z scores on the X predictor variables:

$$z'_Y = \beta_1 z_{X_1} + \beta_2 z_{X_2} + \ldots + \beta_k z_{X_k}. \tag{16.2}$$

The magnitude and sign of the β_i coefficient for each z_{X_i} predictor variable indicate how an increase in scores on each z_{X_i} variable that is included in the composites influences the overall composite predicted score (z'_Y). In other words, each β_i coefficient tells us how much "weight" is given to z_{X_i} when forming the sum of the z scores across several predictors.

Let's consider a different example involving a weighted linear composite. In this example with the specific weighted linear composite U given in Equation 16.3,

$$U = +.8 \times z_{X_1} - .4 \times z_{X_2} + .03 \times z_{X_3}, \tag{16.3}$$

the score on the z_{X_1} variable has a stronger association with the value of the composite U than the score on the z_{X_3} variable, because z_{X_1} has a much larger beta value or weight associated with it. High scores on the linear composite U, described by Equation 16.3, will be strongly and positively associated with high scores on z_{X_1} (because z_{X_1} has a large positive weight of +.80). High scores on U will be moderately associated with low scores on z_{X_2}, because z_{X_2} has a moderate negative weight (−.40) in the formula used to compute the composite variable U. Scores on the new linear composite function U will have a correlation close to 0 with scores on z_{X_3} because the weight associated with the z_{X_3} term in the computation of U (.03) is extremely small.

In MR, the goal of the analysis is to find regression slopes or weights that yield an **optimal weighted linear composite** of scores on the X predictor variables; this composite is "optimal" in the sense that it has the maximum possible correlation with the quantitative Y outcome variable in MR and the minimum possible value of SS_{residual}. By analogy, in DA, the goal of the analysis is to find the DA function coefficients that yield the optimal weighted linear composites of scores on the X predictor variables; the optimal weighted linear composites obtained in DA, which will be called "**discriminant functions**," are optimal in the sense that they have the largest possible SS_{between} and the smallest possible

SS_{within} for the groups defined by the *Y* categorical outcome variable. Another sense in which the discriminant function scores obtained from DA may be optimal is that these scores can be used to classify cases into groups with the highest possible percentage of correct classifications into groups.

DA can be performed to answer one or both of the following types of questions (Huberty, 1994). First of all, DA can be used to make *predictions about group membership outcomes.* For example, DA could be used to predict whether an individual patient is more likely to be a member of Group 1 (patients who are having heart attacks) or Group 2 (patients whose chest pain is not due to heart attacks) based on each individual patients' scores on variables that can be assessed during the initial intake interview and lab tests, such as age, gender, high-density lipoprotein (HDL) and low-density lipoprotein (LDL) cholesterol, triglyceride levels, smoking, intensity of reported chest pain, and systolic blood pressure (SBP). In the initial study to develop the predictive model, we need data that include actual group membership (e.g., based on later more definitive tests, "Is the patient a member of Group 1, heart attack, or Group 2, chest pain that is not due to heart attack?"). We can use a sample of patients for whom the correct diagnosis is known (based on subsequent diagnostic work that takes longer to obtain), to develop a statistical formula that makes the best possible prediction of ultimate diagnosis using the information available during the initial intake interview. Once we have this statistical formula, we may be able to use this formula to try to predict the ultimate diagnosis for new groups of incoming patients for whom we have only the initial intake information. Similarly, an admissions officer at a university might do an initial model development study to see what combination of variables (verbal Scholastic Aptitude Test [SAT], math SAT, high school grade point average, high school class rank, and so forth) best predicts whether applicants to a college subsequently succeed (graduate from the college) or fail (drop out or fail). Once a statistical prediction model has been developed using data for students whose college outcomes are known, this model can be applied to the information obtained from the applications of new incoming students to try to predict the group outcome (succeed or fail) for these new individuals whose actual group outcomes are not yet known.

Second, *DA can be used to describe the nature of differences between groups.* If we can identify a group of patients who have had heart attacks (Group 1) and compare them with a group of patients who have not had heart attacks (Group 2), we can determine what pattern of scores on variables such as age, HDL and LDL cholesterol, triglyceride levels, smoking, anxiety, hostility, and SBP best describes the differences between these 2 groups of patients. In studies that involve DA, researchers may be interested in the accurate prediction of group membership or a description of the nature of differences between the groups, or both kinds of information.

DA is one of several analyses that may be used to predict group membership on a categorical outcome variable. When all the variables involved in an analysis are categorical (both the outcome and all the predictor variables), log-linear analysis is more appropriate. Another approach to the prediction of group membership outcomes that has become increasingly popular in recent years is called logistic regression. Logistic regression involves evaluating the **odds** of different group outcomes; for example, if a patient is five times as likely to die as to survive, the odds of death are described as 5:1. Logistic regression is sometimes preferred to DA because it requires less restrictive assumptions about distributions of scores. In addition, results derived from logistic regression analysis can be interpreted as information about the changes in the likelihood of negative outcomes as scores on each predictor variable increase. A brief introduction to binary logistic regression (which is limited to categorical outcome variables that have only two possible outcomes) is provided in Chapter 21 of this textbook.

DA can provide useful information about the nature of group differences. Discriminant functions can be used to describe the dimensions (i.e., combinations of scores on variables) on which groups differ. In DA, research questions that can be addressed include the following: Can we predict group membership at levels above chance, based on participants' scores on the discriminating variables? What pattern of scores on the predictor variables predicts membership in each group? Which predictor variables provide unique information about group membership when correlations with other predictor variables are statistically controlled?

In some applications of DA, different types of **classification errors** have different consequences or costs. For instance, when a patient who is at high risk for heart disease is erroneously predicted to be a member of the low-risk group, the patient may not receive the necessary treatment. Different costs and consequences arise when a healthy patient is incorrectly predicted to belong to a high-risk category (which may lead to unnecessary diagnostic tests or treatment). One type of information that can be obtained as part of DA is a table that shows the numbers and types of classification errors (i.e., participants for whom the predicted group membership is not the same as the actual group membership).

In DA, as in MR, it is assumed that the predictor (or discriminating) variables are likely to be correlated with each other to some degree. Thus, it is necessary to statistically control for these intercorrelations when we form optimal weighted linear combinations of the discriminating variables (i.e., discriminant functions).

A discriminant function is an optimal weighted linear combination of scores on the X predictor variables. The equation for a discriminant function may be given in terms of the raw scores on the X predictor variables or in terms of z scores or standard scores. In a situation where the Y outcome variable corresponds to just 2 groups, the equation for a single standardized discriminant function D_1 that is used to predict group membership is written as follows (where p is the number of discriminating variables):

$$D_1 = d_{11} z_1 + d_{12} z_2 + \ldots + d_{1p} z_p, \tag{16.4}$$

where $z_1, z_2, \ldots, z_p$ are the z score or standard score versions of the predictor variables X_1, $X_2, \ldots, X_p$. These Xs are usually quantitative predictor variables, although it is possible for one or more of these predictors to be dichotomous or dummy variables. The equation for the discriminant function D_1 in Equation 16.4 is analogous to the standardized regression in Equation 16.2, except that the outcome variable for DA corresponds to a prediction of group membership, whereas the outcome variable in regression was a score on a quantitative (Y) variable. Later, we will see that when the Y outcome variable corresponds to k groups, we may be able to calculate up to $(k - 1)$ different discriminant functions. (The number of discriminant functions that can be obtained is also limited by the number of predictor variables, p, or more formally, by the number of predictor variables that are not completely predictable from other predictor variables.)

As a hypothetical example, suppose that we want to predict membership in one of the following 3 diagnostic groups (Group 1 = No diagnosis; Group 2 = Neurotic; Group 3 = Psychotic). Suppose that for each patient, we have scores on the following quantitative measures: X_1, anxiety; X_2, depression; X_3, delusions; and X_4, hallucinations. In our initial study, we have patients whose diagnoses (none, neurotic, psychotic) are known; we want to develop formulas that tell us how well these group memberships can be predicted from scores on the Xs. If we can develop predictive equations that work well to predict group membership, the equations may help us characterize how the members of the 3 groups differ; also, in future clinical work, if we encounter a patient whose diagnosis is unknown, for whom we have scores on X_1 through X_4, we may be able to use the equations to predict a diagnosis for that patient.

If the categorical variable that is predicted in the DA has just 2 groups, only one discriminant function is calculated. In general, if there are k groups and p discriminating variables, the number of different discriminant functions that will be obtained is the minimum of these two values: $(k - 1)$ = (number of groups − 1) and p, the number of discriminating variables. The units or the specific values of scores on the discriminant function D do not matter; the D values are generally scaled to have a mean of 0 and a within-group variance of 1, similar to standardized z scores. What does matter is the relation of discriminant scores to group membership. When we have just 2 groups on the Y outcome variable, the goal of DA is to find the weighted linear combination of scores on the standardized predictor variables (D_1, in Equation 16.4) that has the largest possible variance between groups and the smallest possible variance within groups. If we take the new variate or function called D_1 (where the value of D_1 is given by Equation 16.4) as the outcome variable and perform a one-way ANOVA on these scores across groups on the Y variable, then DA provides an optimal set of weights—that is, the set of weights that yields values of D_1 scores for which the F ratio in this ANOVA is maximized.

If there are more than 2 groups, for example if $k = 3$, more than one discriminant function may be needed to differentiate between group members; for instance, in a DA with 3 groups, the first discriminant function might distinguish members of Group 1 from Groups 2 and 3, while the second discriminant function might distinguish members of Group 3 from Groups 1 and 2.

Boundary decision rules can be applied to the values of discriminant functions to make predictions about group membership. For example, if there are just 2 groups, there will be only one discriminant function, D_1. The boundary that divides the D_1 values for most Group 1 members from those of most members of Group 2 might fall at $D_1 = 0$, with lower scores on D_1 for members of Group 1. In that case, our decision rule would be to classify persons with D_1 scores less than 0 as members of Group 1 and to classify persons with D_1 scores greater than 0 as members of Group 2. When there are two or more discriminant functions, the decision rule depends jointly on the values of several discriminant functions. For two discriminant functions, the decision rule can be shown graphically in the form of a "**territorial map**" with D_1 (scores on the first discriminant function) on the X axis and D_2 (scores on the second discriminant function) on the Y axis. Each case can be plotted in this two-dimensional map based on its score on D_1 (along the horizontal axis) and its score on D_2 (along the vertical axis). If the DA is successful in creating discriminant functions that have scores that differ clearly across groups, then the cases that belong to each group will tend to fall close together in the territorial map and members of different groups will tend to occupy different regions in the space given by this map. We can set up boundaries to represent how the members of Groups 1, 2, and 3 are spatially distributed in this map (based on their combinations of scores on D_1 and D_2). When DA is successful, we can draw boundaries to separate this map into a separate spatial region that corresponds to each group. When a participant has scores on D_1 and D_2 that place him or her in the part of the territorial map that is occupied mostly by members of Group 1, then we will classify that individual as a member of Group 1, as in the example that follows.

Figure 16.1 shows a hypothetical territorial map in which the members of 3 groups (no diagnosis, neurotic, and psychotic) can be differentiated quite well by using scores on two discriminant functions. In this imaginary example, suppose that the first discriminant function, D_1, is a composite of scores on the variables in which there are large positive **discriminant function coefficients** for the variables anxiety and depression (and near-zero coefficients for delusions and hallucinations). From Figure 16.1 we can see that members of the "neurotic" group scored high on D_1; if D_1 is a function that involves large positive weights or coefficients for the predictor variables anxiety and depression, then the diagram in Figure 16.1 tells us that the neurotic group scored high on a composite of anxiety and depression compared with the other 2 groups. High scores on this first discriminant function D_1 correspond to high scores on anxiety and depression, and they are not related to people's scores on delusions or hallucinations; therefore, we might interpret D_1 as a dimension that represents how much each person exhibits "everyday psychopathologies." On the other hand, suppose that the second discriminant function, D_2, has large coefficients for delusions and hallucinations. Also, suppose that the psychotic group scored high on D_2 (and thus, high on a composite score for delusions and hallucinations) relative to the other 2 groups. We might interpret the combination of variables that is represented by D_2 as an index of "major psychopathologies" or "delusional disorders." In Figure 16.1, members of the three groups have D_1 and D_2 scores that correspond to clearly separated regions in the territorial map. In contrast, Figure 16.2 shows a second hypothetical territorial map in which the groups cannot be well discriminated because the discriminant function scores overlap substantially among all 3 groups.

Sometimes a discriminant function that combines information about scores on two or more variables can accurately distinguish between members of 2 groups even though the groups do not differ significantly on either of the two individual variables. Consider the hypothetical data shown in Figure 16.3, which illustrate a problem in physical anthropology. Suppose that we want to classify a set of skulls as belonging to apes (A) or humans (H) based on measurements such as jaw width (X_1) and forehead height (X_2). Scores on X_1 and X_2 are shown as a scatter plot in Figure 16.3. The 2 groups (whose scores are marked in the scatter plot using A for ape and H for human) do not differ substantially on their scores on either X_1 or X_2 alone. However, a weighted combination of scores on X_1 and X_2 can be used to create a new function; this new discriminant function D_1 is represented by the dotted line in Figure 16.3. The boundary between the A and H groups appears as a solid line in Figure 16.3. This boundary corresponds to a decision rule that is based on scores on the discriminant function. In this example, if the score on the discriminant function D_1, which is represented by the dotted line, is greater than 0, we would classify the case as a member of group H (human); if the score on D_1 is less than 0, we would classify the case as a member of group A (ape); this decision rule would provide perfectly

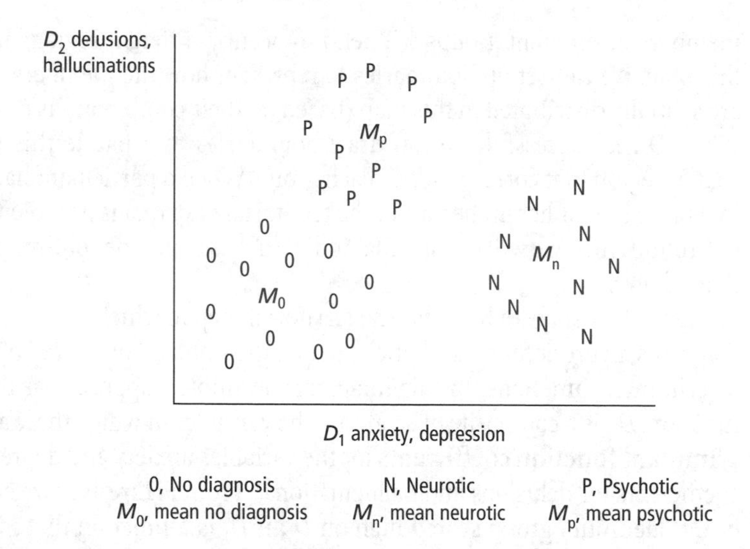

Figure 16.1 ♦ Hypothetical Example of a Discriminant Analysis Study

NOTE: Three diagnostic groups: 0 = No diagnosis, N = Neurotic, and P = Psychotic; four discriminating variables (X_1, anxiety; X_2, depression; X_3, delusions; and X_4, hallucinations).

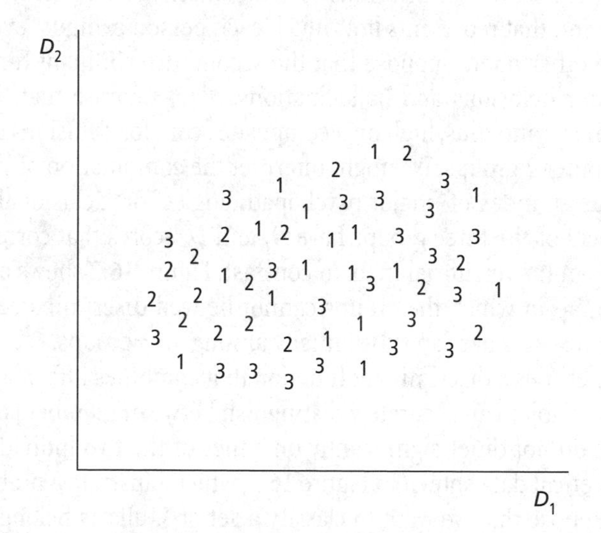

Figure 16.2 ♦ Hypothetical Example: Discriminant Analysis With Substantial Overlap in Values of Both D_1 and D_2 Among Groups (and, Therefore, Poor Prediction of Group Membership From Discriminant Function Scores)

accurate group classification. An alternate view of this graph appears in Figure 16.4. Suppose that we use the following decision rule to predict group membership based on D_1 scores: for cases with scores on $D_1 > 0$, classify as H; for scores on $D_1 < 0$, classify the case as A. It is apparent from the graph that appears in Figure 16.4 that scores on this new function D_1 differentiate the 2 groups quite well.

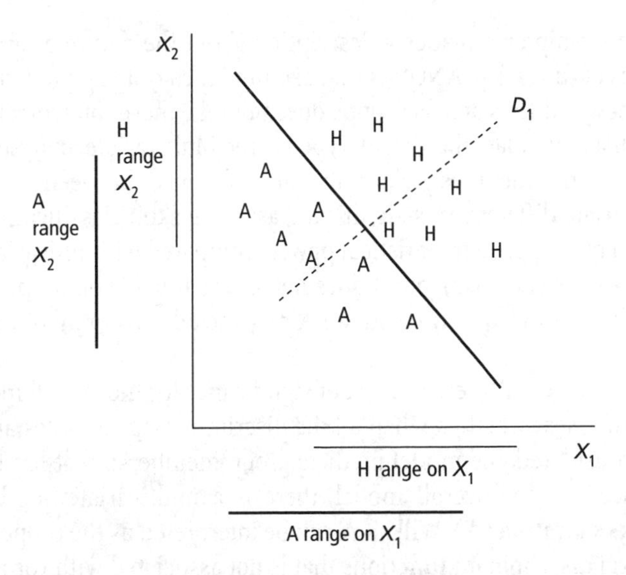

Figure 16.3 ♦ Hypothetical Example of a Situation in Which Discrimination Between Groups A and H Is Better for a Combination of Scores on Two Variables (X_1, X_2) Than for Either Variable Alone

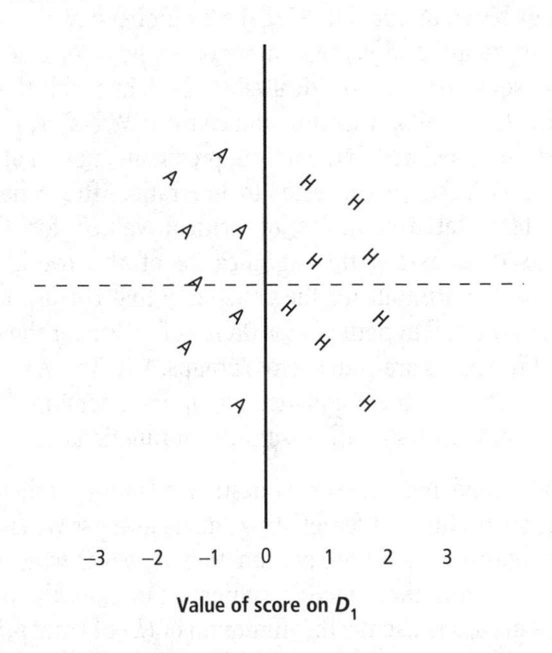

Figure 16.4 ♦ Group Classification Decision Rule for Data in Figure 16.3

NOTE: D_1 (dotted line) is the score on one discriminant function. If the score on $D_1 > 0$, classify as H. If the score on $D_1 < 0$, classify as A.

In some research situations, DA provides a similar description of the way in which individual variables differ across groups compared with a series of univariate ANOVAs on the individual variables (Huberty & Morris, 1989). However, a DA takes the intercorrelations among the p predictor variables into account when it makes predictions for group membership or provides a description about the nature of group differences, whereas a series of univariate ANOVAs that examine each of the *p* predictor variables in isolation to see how it differs across groups does not take these intercorrelations into account, and it will usually inflate the risk of Type I error. Multivariate analyses that consider patterns of scores can sometimes yield significant differences between groups even when none of the univariate differences is significant, as in the skull classification example in Figure 16.3. DA may have greater statistical power (compared with univariate ANOVAs for each of the *p* predictor variables), but it does not always provide better power.

The research questions for a DA may include any or all of the following:

1. First of all, there is a test of significance for the overall model—that is, an omnibus test to assess whether, when all the discriminating variables and discriminant functions are considered, the model predicts group membership at levels significantly better than chance. For the overall model, there is a multivariate goodness-of-fit statistic called **Wilks's lambda (Λ)**. Wilks's Λ can be interpreted as the proportion of variance in one or several discriminant functions that is not associated with (or predictive of) group membership. Thus, Wilks's Λ might be better thought of as a "badness-of-fit" measure, because it is an estimate of the proportion of unexplained variance; thus, it is analogous to $(1 - \eta^2)$ in an ANOVA or $(1 - R^2)$ in a regression. There are three different *df* terms associated with a Wilks's Λ based on the following three numbers: *N*, the overall number of cases; *k*, the number of groups; and *p*, the number of predictor or discriminating variables. It would be cumbersome to look up critical values for Λ in a table that needs to be indexed by three different *df* terms. SPSS Discriminant converts Wilks's Λ to a χ^2. This conversion from Λ to χ^2 and the associated *df* are only approximate; note that in some programs, including SPSS GLM, Wilks's Λ is converted to an *F* ratio. After Λ has been converted into a more familiar test statistic, tables of critical values for these familiar distributions can be used to assess the significance of the overall DA model. SPSS provides *p* values on the printout for the chi-square that corresponds to Wilks's Λ. One way to understand the null hypothesis for DA is as follows: for the population, means on the discriminant functions are equal across groups. Another way to understand the null hypothesis is to say that, in the population, group membership is not related to (i.e., cannot be predicted from) scores on the discriminant functions.

2. The second major research question in DA is the following: How many discriminant functions are useful in differentiating among groups? When we have only $k = 2$ groups, we can only obtain one discriminant function. However, when there are more than 2 groups, there may be more than one discriminant function; the number of discriminant functions for *k* groups is usually the minimum of $(k - 1)$ and *p*, where *p* is the number of predictor variables. (This assumes that the correlation matrix for the *p* variables does not have a determinant of 0; that is, none of the *p* predictor variables is perfectly predictable from one or more other predictor variables.) The discriminant functions are extracted in such a way that scores on the first discriminant function, D_1, have the largest possible between-group differences. Scores on the second discriminant function, D_2, must be uncorrelated with scores on D_1, and D_2 scores must predict the largest possible between-group variance that is not explained by D_1.

The discriminant functions are numbered (1, 2, 3, etc.). They are rank ordered from the most predictive to the least predictive of group membership. Scores on the discriminant functions D_1, D_2, D_3, and so forth are uncorrelated with each other. If we have $k = 5$ groups and at least four predictor variables, for example, we will obtain $k - 1 = 4$ different discriminant functions. Our research question will be the following: How many of these four discriminant functions are useful in predicting group membership?

To answer this question, we conduct a "**dimension reduction analysis**"; that is, we evaluate how many discriminant functions are needed to achieve reasonably accurate predictions of group membership. SPSS DA actually tests the significance of *sets* of discriminant functions rather than the significance of each individual discriminant function; the tests are set up in a manner that may seem counterintuitive at first. For instance, if we have four discriminant functions (denoted by D_1 through D_4), SPSS will provide significance tests for the following sets of functions:

Set 1: D_1, D_2, D_3, D_4
Set 2: D_2, D_3, D_4 (i.e., all discriminant functions after, or not including, the first)
Set 3: D_3, D_4
Set 4: D_4

The proportion of between-groups variance associated with each set of discriminant functions typically decreases as we reduce the number of discriminant functions included in the sets. If Set 2 is significant, then Set 1 is usually also significant because Set 1 includes the same functions as Set 2. In practice, it is helpful to read this list from the bottom up. If Set 4 is not significant, then we can drop D_4 from our interpretation. If Set 3 is also not significant, then we can drop D_3 also from our interpretation. If Set 2 is significant, then we may decide that we need to include the first two discriminant functions in our interpretation and discussion of group differences. Usually, we hope that we can account for between-group differences using a relatively small set of discriminant functions. Each discriminant function can be thought of as a "dimension" along which groups may differ. When we make the decision to include only the first one or two discriminant functions from a larger set of discriminant functions in our interpretation of DA, it is called "dimension reduction"; that is, we decide that we only need the information about group differences that is provided by scores on the first one or two discriminant functions (and that later functions do not add much additional information about the nature of group differences or do not lead to much improvement in the prediction of group membership). Sometimes, but not always, the discriminant functions can be given meaningful interpretations or labels as descriptors of underlying "dimensions" on which groups differ.

3. Another possible question is related to the evaluation of the contribution of individual predictors or discriminating variables: Which variables are useful in discriminating among groups or predicting group membership? SPSS Discriminant (unlike SPSS regression) does not provide direct statistical significance tests for the contribution of individual predictor variables. DA provides three pieces of information about the contributions of individual predictors; together, these can be used to decide which predictor variables provide important and/or unique information about group differences.

First of all, as described above, the researcher decides how many of the discriminant functions should be interpreted. Then, only for the discriminant functions that are retained for interpretation, the researcher asks the following question: Which predictor variable(s) had "large" correlations (in absolute value) with each function? The magnitudes of correlations between predictor variables and discriminant functions are judged relative to some arbitrary cutoff value; for example, a correlation in excess of .5 (in absolute value) could be interpreted as an indication that a predictor is strongly related to a discriminant function. (The choice of a cutoff value such as .50 is arbitrary.)

Second, the standardized discriminant function coefficient for each variable can be examined to assess which predictor variables were given the greatest weight in computing each discriminant function score when intercorrelations among the discriminating variables are taken into account. The standardized discriminant function coefficients (denoted by d in Equation 16.4) are analogous to beta weights or coefficients in an MR. To see how scores on a particular predictor are related to group membership, we need to consider two things: First of all, is a high score on X_i associated with a high or low score on D? Second, does each group tend to score high or low on discriminant function D? Together, these two pieces of information help us understand how X_i scores differ across groups in the context of this multivariate analysis.

Third, the univariate ANOVAs show how scores on each individual discriminating variable differ across groups when each variable's correlations with other predictors are *not* taken into account. For each discriminating variable, there is an F ratio that indicates whether means differed significantly across groups; note, however, that the p values that SPSS reports for these univariate Fs are not corrected for inflated risk of Type I error.

If all these three pieces of information provide a consistent story about the strength and nature of the relationship between group membership and X_i scores, then it is relatively easy to write up a description of how groups differ on the Xs and/or what pattern of X scores leads us to predict that an individual is a member of each group. However, you may recall (from Chapter 10) that the zero-order correlation of an individual X_i with Y and the partial slope (β_i) to predict z_Y from z_{X_i}, while controlling for other Xs, can be quite different. The same kind of difference can arise in DA; that is, the discriminant function coefficient d_i for z_{X_i} can have a sign that implies a different association between group membership and X_i scores than found in a simple one-way ANOVA. When the results of the DA (in which each X_i predictor variable's association with group membership is assessed while controlling for all other X predictors) differ qualitatively from the results of univariate ANOVAs (in which means on each X_i variable across groups are assessed separately for each X variable, without taking intercorrelations among the Xs into account), the researcher needs to think very carefully about descriptions of group differences.

4. Which groups can (or cannot) be discriminated? As in an ANOVA, if there are more than 2 groups in a DA, our null hypothesis is essentially that the means on the discriminant functions are equal across all groups. If we find that the overall DA is significant, and there are more than 2 groups, follow-up tests are needed to assess which of the pairwise contrasts between groups are statistically significant. A significant overall DA usually implies that at least 1 group differs significantly from the others (just as a significant one-way ANOVA usually implies at least one significant contrast between group means). However, a significant overall DA analysis does not necessarily imply that all groups can

be distinguished. The SPSS DA program provides several kinds of information about group differences. The contingency table at the end of the DA output summarizes information about the numbers of persons who were correctly classified as members of each group and also indicates the numbers of misclassifications. This table makes it possible to see whether there were certain groups whose members were frequently misclassified into other groups. It is also possible to examine the territorial map (described later) to see if the clusters of scores on Discriminant Functions 1 and 2 were clearly separated in the territorial map (as in Figure 16.1 or Figure 16.3) or mixed together (as in Figure 16.2).

5. What meaning (if any) can we attach to each discriminant function? (This question is optional and is likely to be of greater interest in more theoretical studies.) It is sometimes useful to name or label a discriminant function; this name should be based on a careful consideration of which *X* variables had high correlations with this discriminant function and which groups show the largest differences on the discriminant function. The names given to discriminant functions are analogous to the names given to factors or components in a factor analysis (see Chapter 18). It may make sense, particularly in theory-driven studies, to think of the discriminant functions as "dimensions" along which the groups differ. In the hypothetical study for which results are shown in Figure 16.1, we might call D_1 (which is highly correlated with anxiety and depression) the "everyday psychological distress" dimension; the "neurotic" group had high scores on this discriminant function, and thus, high scores on anxiety and depression, compared with both the "no-diagnosis" and "psychotic" groups. The second discriminant function D_2 was highly correlated with delusions and hallucinations; this might be called a "major psychopathology" or "delusional disorder" dimension; on this dimension, the psychotic group scored higher than either the no-diagnosis or neurotic group.

However, sometimes in practice, the variables that are given most weight in forming the discriminant function cannot be easily interpreted. If it does not make sense to think of the measures that make the strongest contribution to a particular discriminant function as different measures of the same underlying construct or dimension, an interpretation should not be forced. In research where the selection of predictor variables is guided by theory, it is more likely that the discriminant functions will be interpretable as meaningful "dimensions" along which groups differ.

6. What types of classification errors are most common? A classification error occurs when the predicted group membership for an individual case does not correspond to the actual group membership. Do we need to modify decision rules to reduce the occurrence of certain types of costly prediction errors? (This question is optional and is most likely to be of interest in clinical prediction applications.) For example, DA may be used as a statistical method of making medical diagnoses (based on past records of initial symptom patterns and final diagnosis or medical outcome) or in hiring or admission decisions (based on the past history of applicant credential information and success/failure in the job or academic program). In such applied situations, a goal of research may be to generate a formula that can be used to make predictions about future individual cases for which the ultimate medical, job, or educational outcome is not yet known.

Consider the problem of deciding whether a patient is having a heart attack (Group 1) or not having a heart attack (Group 2), based on initial information such as the patient's age, gender, blood pressure, serum cholesterol, description of location and intensity of pain, and so forth. A physician needs to make an initial guess about diagnosis based on incomplete preliminary information (a definitive diagnosis of heart attack requires more invasive tests that take more time). There are two possible kinds of classification errors: a false positive (the physician thinks the patient is having a heart attack when the patient is not really having a heart attack) and a false negative (the physician thinks the patient is not having a heart attack when the patient really is having a heart attack). In clinical situations, these two types of errors have different types of costs and consequences associated with them. A false positive results in unnecessarily alarming the patient, running unnecessary tests, or perhaps administering unnecessary treatment; there may be some discomfort and even risk, as well as financial cost, associated with this type of error. Usually, in medicine, a false-negative error is seen as more costly; if a patient who really has a life-threatening condition is mistakenly judged to be healthy, necessary care may not be given, and the patient may get worse or even die (and the physician may possibly be sued for negligence). Thus, in medical decisions, there may be a much greater concern with avoiding false negatives than avoiding false positives.

There are a number of ways in which we may be able to reduce the occurrence of false negatives—for example, by finding more useful predictor variables. However, an alternative way of reducing false negatives involves changing the cutoff boundaries for classification into groups. For example, if we decide to classify people as hypertensive when they have SBP (systolic blood pressure) > 130 (instead of SBP > 145), lowering the cutoff to 130 will decrease the false-negative rate, but it will also increase the false-positive rate. The classification of individual cases and the costs of different types of classification errors are not usually major concerns in basic or theory-driven research, but these are important issues when DA is used to guide practical or clinical decision making.

16.2 ♦ Introduction of an Empirical Example

As an example for DA, consider the data shown in Table 16.1. This is a subset of cases and variables from the Project Talent study (Lohnes, 1966); high school students took aptitude and ability tests, and they were asked to indicate their choice of future career. For this analysis, students who selected the following types of future careers were selected: Group 1, business; Group 2, medicine; and Group 3, teaching/social work. Six of the achievement and aptitude test scores were used as predictors of career choice: abstract (reasoning), reading, English, math, mechanic (mechanical aptitude), and office (skills). Later in this chapter, the results of DAs are presented for the data in Table 16.1; all six predictors were entered in one step (this is analogous to a standard or simultaneous MR). In addition, one-way ANOVA was performed to show how the means on the saved scores on Discriminant function 1 differed across the 3 groups. A simple unit-weighted composite variable was also created (raw scores on office minus raw scores on math); one-way ANOVA was performed to assess how the means of the simple unit composite variable (the difference between office and math scores) varied across groups.

16.3 ♦ Screening for Violations of Assumptions

It may be apparent from the previous section that DA involves some of the same issues as ANOVA (it assesses differences in means on the discriminant functions across groups)

Table 16.1 ♦ Data for the Discriminant Analysis Example in This Chapter

Cargroup	*Gender*	*English*	*Reading*	*Mechanic*	*Abstract*	*Math*	*Office*
1	1	90	28	12	9	26	6
1	1	92	43	15	12	37	6
2	1	96	44	15	10	43	11
1	1	83	33	15	11	27	23
1	1	85	39	12	12	36	23
3	1	47	24	8	4	11	13
2	1	94	45	15	14	46	14
1	1	98	44	17	9	31	13
1	1	87	41	16	13	40	24
3	1	82	28	15	7	27	26
2	1	76	39	18	10	36	17
1	1	77	27	14	9	20	0
1	1	89	40	7	10	24	26
2	1	95	45	13	12	28	24
1	1	90	35	9	7	20	19
1	1	86	24	9	11	12	13
1	1	71	17	13	9	15	23
1	1	85	35	11	10	22	0
1	1	70	12	8	2	11	10
2	1	92	46	14	10	38	6
1	1	98	35	18	9	25	17
1	1	83	31	14	10	22	27
2	1	68	4	19	12	30	7
3	1	81	41	13	11	16	7
1	1	96	43	19	12	37	16
3	1	89	40	16	13	39	23
2	2	94	39	4	9	15	15
2	2	92	37	18	11	29	20
2	2	93	46	5	11	28	36
3	2	89	42	9	12	32	23
1	2	86	23	4	8	15	37
3	2	95	40	9	11	35	29
1	2	84	37	6	9	11	15
1	2	101	43	15	12	40	11
1	2	87	28	6	6	7	36
3	2	98	45	9	9	20	26
1	2	95	41	13	12	22	34
1	2	81	43	7	9	15	13
3	2	88	40	9	11	28	26
1	2	88	17	2	5	12	30
1	2	84	25	10	7	18	30
2	2	85	24	8	11	19	17

(Continued)

Table 16.1 ♦ (Continued)

Cargroup	*Gender*	*English*	*Reading*	*Mechanic*	*Abstract*	*Math*	*Office*
1	2	81	32	7	9	16	37
3	2	93	32	4	15	24	17
1	2	83	32	4	3	13	37
1	2	99	39	12	10	25	37
2	2	91	43	8	9	29	3
3	2	107	42	13	12	39	1
3	2	94	41	10	14	41	30
1	2	87	28	10	10	20	27
1	2	92	30	12	9	19	30
3	2	104	42	13	9	33	21
2	2	79	28	10	12	27	19
1	2	72	14	4	9	7	39
1	2	63	15	8	4	9	27
1	2	94	32	11	8	22	20
3	2	84	31	5	9	27	11
1	2	80	21	11	7	18	30
1	2	82	32	8	8	5	33
1	2	74	27	7	1	13	34
1	2	102	44	11	12	27	33
1	2	96	44	9	12	30	31
1	2	93	44	14	9	25	16
1	2	84	42	17	8	16	31
1	2	93	33	10	9	17	27
2	2	81	28	4	8	18	29
2	2	95	42	21	9	35	26
2	2	66	17	6	4	14	21
3	2	103	37	9	12	24	10
3	2	102	42	11	10	34	20
1	2	86	30	8	6	21	26
1	2	91	42	7	8	17	24
1	2	103	33	9	10	19	39
3	2	111	46	14	13	47	24
1	2	88	35	13	7	11	19
2	2	87	36	12	15	26	23
1	2	77	35	6	10	18	15
2	2	71	16	2	9	7	0
1	2	109	45	14	13	33	33
1	2	95	31	4	10	16	24
2	2	93	37	9	11	44	10
1	2	84	27	8	10	26	29
1	2	79	38	6	12	19	33
1	2	100	40	8	7	23	24
2	2	88	34	8	9	21	21

Cargroup	*Gender*	*English*	*Reading*	*Mechanic*	*Abstract*	*Math*	*Office*
3	2	75	37	11	0	11	24
1	2	87	41	12	7	16	39
1	2	85	30	7	8	20	23
1	2	82	22	4	6	13	36
1	2	78	42	12	9	20	37
3	2	102	48	11	13	41	9
3	2	96	45	6	9	22	23
3	2	76	40	8	8	20	26
3	2	75	29	2	4	10	24
2	2	86	35	6	7	22	7
1	2	81	24	10	11	14	34
1	2	100	46	12	11	23	23
2	2	96	0	0	0	36	17
1	2	84	30	13	8	27	36
1	2	81	31	11	12	19	31
1	2	99	38	8	13	24	40
1	2	91	36	10	9	16	30
1	2	92	37	14	9	23	39
1	2	87	31	2	9	23	30
3	2	98	33	13	10	26	21
1	2	76	26	7	10	20	33
1	2	88	21	6	7	11	36
3	2	86	36	8	10	20	14
3	2	86	35	6	3	28	3
1	2	80	24	7	10	19	30
1	2	90	21	6	9	23	34
3	2	90	33	10	9	20	27
1	2	90	38	8	9	25	34
3	2	90	39	12	10	25	16
1	2	89	37	12	13	16	27
1	2	99	29	10	13	23	33
1	2	100	42	9	14	21	39
1	2	94	42	12	12	20	33
1	2	81	19	7	7	11	19
3	2	88	27	7	12	16	33
3	2	93	25	9	9	25	36
1	2	88	18	2	8	16	40
2	2	105	46	11	11	34	9
1	2	91	25	7	7	18	40
2	2	94	39	9	10	29	13
3	2	67	14	7	7	17	31
1	2	107	38	6	9	33	37
2	2	74	17	4	4	6	26

(Continued)

Table 16.1 ♦ (Continued)

Cargroup	*Gender*	*English*	*Reading*	*Mechanic*	*Abstract*	*Math*	*Office*
1	2	90	33	10	10	17	40
1	2	90	30	3	7	18	39
1	2	86	41	9	7	27	37
1	2	109	40	8	12	31	29
3	2	72	33	8	13	21	14
1	2	102	30	8	9	20	23
3	2	89	28	8	7	19	27
1	2	90	28	6	5	20	37
3	2	87	41	6	8	31	26
1	2	91	24	7	11	16	37
3	2	100	31	8	13	25	23
3	2	58	14	8	3	20	24
3	2	90	19	9	3	17	16
1	2	98	40	9	11	18	30
2	2	78	36	12	7	22	29
1	2	83	23	4	5	13	23
1	2	89	34	11	7	23	29
3	2	73	20	10	7	14	34
1	2	85	37	12	13	20	36
2	2	80	29	7	6	18	6
1	2	91	37	6	10	22	26
3	2	86	38	10	10	24	11
3	2	107	43	10	14	34	29
2	2	83	22	1	8	14	23
1	2	102	44	8	7	23	24
1	2	89	43	7	10	29	39
1	2	86	27	4	11	17	31
1	2	93	23	5	7	15	37
2	2	93	36	13	12	24	30
1	2	88	37	9	12	17	31
1	2	70	18	13	8	26	23
3	2	84	29	12	11	19	19
1	2	90	30	7	10	10	29
1	2	98	45	13	11	24	31
1	2	85	32	8	10	19	33
1	2	88	39	7	7	15	30
2	2	95	28	6	10	17	34
1	2	84	27	8	8	8	29
3	2	100	48	15	13	27	11
1	2	72	20	7	9	17	26
1	2	87	22	6	8	18	37

NOTE: Categorical variables: cargroup, coded 1 = Business, 2 = Medicine, 3 = Teaching; gender, coded 1 = Male, 2 = Female. Quantitative predictor variables: English, reading, mechanic, abstract, math, office.

and some issues similar to those in MR (it takes intercorrelations among predictors into account when constructing the optimal weighted composite for each discriminant function). The assumptions about data structure required for DA are similar to those for ANOVA and regression but also include some new requirements. Just as in regression, the quantitative X predictor variables should be approximately normally distributed, with no extreme outliers, and all pairs of X predictor variables should be linearly related, with no extreme outliers (this can be assessed by an examination of scatter plots). Just as in ANOVA, the numbers in the groups should exceed some minimum size. (If the number of scores within a group is smaller than the number of predictor variables, for example, it is not possible to calculate independent estimates of all the values of within-group variances and covariances. Larger sample sizes are desirable for better statistical power.)

One additional assumption is made, which is a multivariate extension of a more familiar assumption. When we conduct a DA, we need to compute a sum of cross products (SCP). Let $\mathbf{SCP}_i$ stand for the sample **sum of cross products matrix** for the scores within group i. The elements of the sample **SCP** matrix are given below; an **SCP** matrix is computed separately within each group, and also, for the entire dataset with all groups combined.

$$\mathbf{SCP} = \begin{bmatrix} SS(X_1) & \mathrm{SCP}(X_1, X_2) & \ldots & \mathrm{SCP}(X_1, X_p) \\ \mathrm{SCP}(X_2, X_1) & SS(X_2) & \ldots & \mathrm{SCP}(X_2, X_p) \\ \vdots & \vdots & \vdots & \vdots \\ \mathrm{SCP}(X_p, X_1) & \mathrm{SCP}(X_p, X_2) & \ldots & SS(X_p) \end{bmatrix}. \qquad (16.5)$$

The diagonal elements of each **SCP** matrix are the sums of squares for $X_1, X_2, \ldots, X_p$. Each SS term is found as follows: $SS(X_i) = \sum (X_i - M_{X_i})^2$. The off-diagonal elements of each **SCP** matrix are the sums of cross products for all possible pairs of X variables. For X_i and X_j, $\mathrm{SCP} = \sum (X_i - M_{X_i}) \times (X_j - M_{Xj})$. Recall that in a univariate ANOVA, there was an assumption of homogeneity of variance; that is, the variance of the Y scores was assumed to be approximately equal across groups. In DA, the analogous assumption is that the elements of the variance/covariance matrix are homogeneous across groups.

Note that a sample sum of cross products matrix, **SCP**, can be used to obtain a sample estimate of the variance/matrix **S** for each group of data. To obtain the elements of **S** from the elements of **SCP**, we divide each element of **SCP** by the appropriate n or df term to convert each SS into a variance and each SCP into a covariance, as shown below.

$$\mathbf{S} = \begin{bmatrix} \mathrm{Var}(X_1) & \mathrm{Cov}(X_1, X_2) & \ldots & \mathrm{Cov}(X_1, X_p) \\ \mathrm{Cov}(X_2, X_1) & \mathrm{Var}(X_2) & \ldots & \mathrm{Cov}(X_2, X_p) \\ \vdots & \vdots & \vdots & \vdots \\ \mathrm{Cov}(X_p, X_1) & \mathrm{Cov}(X_p, X_2) & \ldots & \mathrm{Var}(X_p) \end{bmatrix}.$$

A sample variance/covariance matrix **S** is an estimate of a corresponding population variance covariance matrix denoted by $\mathbf{\Sigma}$. Each diagonal element of the $\mathbf{\Sigma}$ matrix corresponds to a population variance for each of the predictors—for example, $\sigma_1^2, \sigma_2^2, \ldots, \sigma_k^2$;

and the off-diagonal elements of the **Σ** matrix correspond to the population covariance for all possible pairs of X variables.

An assumption for DA is that these $\boldsymbol{\Sigma}_i$ population variance/covariance matrices are homogeneous across groups. This assumption about homogeneity includes the assumption that the variances for $X_1, X_2, \ldots, X_p$ are each homogeneous across groups and, in addition, the assumption that covariances between all pairs of the Xs are also homogeneous across groups. The assumption of **homogeneity of the variance/covariance matrices** across groups can be written as follows:

$$H_0 : \boldsymbol{\Sigma}_1 = \boldsymbol{\Sigma}_2 = \cdots = \boldsymbol{\Sigma}_k. \tag{16.6}$$

In words, this null hypothesis corresponds to the assumption that corresponding elements of these **Σ** matrices are equal across groups; in other words, the variances of each X variable are assumed to be equal across groups, and in addition, the covariances among all possible pairs of X variables are also assumed to be equal across groups. Box's M test (available as an optional test from SPSS) tests whether this assumption is significantly violated.

The Box M test can be problematic for at least two reasons. First, it is very sensitive to nonnormality of the distribution of scores. Also, when the number of cases in a DA is quite large, Box's M may indicate statistically significant violations of the homogeneity of the variance/covariance assumption, even when the departures from the assumed pattern are not serious enough to raise problems with the DA. On the other hand, when the number of cases in the DA is very small, the Box M may not be statistically significant at conventional alpha levels (such as $\alpha = .05$), even when the violation of assumptions is serious enough to cause problems with the DA. As with other preliminary tests of assumptions, because of the degrees of freedom involved, Box M is more sensitive to violations of the assumption of homogeneity in situations where this violation may have less impact on the validity of the results (i.e., when the overall N is large); also, Box M is less sensitive to violations of the assumption of homogeneity in situations where these violations may cause more problems (i.e., when the overall N is small). To compensate for this, when the overall number of cases is very large, it may be preferable to evaluate the statistical significance of Box M using a smaller alpha level (such as $\alpha = .01$ or $\alpha = .001$). When the overall number of cases involved in the DA is quite small, it may be preferable to evaluate the significance of Box M using a higher alpha level (such as $\alpha = .10$).

When there are serious violations of the assumption of multivariate normality and the homogeneity of variances and covariances across groups, it may be possible to remedy these violations by using data management strategies discussed in Chapter 4. For example, sometimes, the application of log transformations corrects for nonnormality of the distribution shape and also reduces the impact of outliers. Data analysts might consider dropping individual cases with extreme scores (because these may have inflated variances or covariances in the groups that they belonged to), dropping individual variables (which have unequal variances or covariances across groups), or dropping a group (if that group has variances and/or covariances that are quite different from those in other groups). However, as discussed in Chapter 4, decisions about the removal of outliers should be based on careful evaluation. Dropping cases for which the analysis makes a poor prediction may result in pruning the data to fit the model. Other possible remedies for violations of assumptions (such as log transformations of scores on variables that have nonnormal distributions) may be preferable. Complete information about the sum of cross products matrices within each of the groups can be requested as an optional output from the DA procedure.

16.4 ♦ Issues in Planning a Study

A study that involves DA requires design decisions about the number and size of groups and also requires the selection of discriminating variables. The goal of DA may be the development of a statistical model that makes accurate predictions of group memberships or the development of a model that provides theoretically meaningful information about the nature of differences between groups, or both. In the first situation, the selection of discriminating variables may be driven more by practical considerations (e.g., what information is inexpensive to obtain) while, in the second situation, theoretical issues may be more important in the selection of variables.

As in other one-way analyses, it is helpful, although not essential, to have equal numbers of cases in the groups. If the numbers in the groups are unequal, the researcher must choose between two different methods of classification of scores into predicted groups. The default method of classification in SPSS DA assumes that the model should classify the same proportion into each group regardless of the actual numbers; thus, if there are $k = 4$ groups, approximately one fourth or 25% of the cases will be classified as members of each group. An alternative method available in SPSS (called "priors = size") uses the proportions of cases in the sample as the basis for the classification rule; thus, if there were 4 groups with $n = 40$, 25, 25, and 10, then approximately 40% of the cases would be classified into Group 1, 25% into Group 2, and so forth.

The number of cases within each group should exceed the number of predictor variables (Tabachnick & Fidell, 2007, p. 250). To obtain adequate statistical power, larger numbers of cases than this minimum sample size are desirable. If n is less than the number of variances and covariances that need to be calculated within each group, there are not enough degrees of freedom to calculate independent estimates of these terms, and the **SCP** matrix will be singular (i.e., it will have a determinant of 0). In practice, as long as a pooled covariance matrix is used, SPSS can still carry out a DA even if some groups have a singular **SCP** sum of cross products matrix or **S** variance/covariance matrix. When the within-group numbers are very small, it is difficult to assess violations of the assumption about the homogeneity of variance/covariance matrices across groups (i.e., the Box M test does not have good statistical power to detect violations of this assumption when the sample sizes within groups are small).

The second major design issue is the selection of discriminating variables. In applied research, it may be important to choose predictor variables that can be measured easily at low cost. In basic or theory-driven research, the choice of predictors may be based on theory. As in any multivariate analysis, the inclusion of "garbage" variables (i.e., variables that are unrelated to group membership and/or are not theoretically meaningful) can reduce statistical power and, also, make the contributions of other variables difficult or impossible to interpret.

16.5 ♦ Equations for Discriminant Analysis

A DA produces several types of information about pattern in the data, and much of this information can be understood by analogy to the information we obtain from ANOVA and MR. The omnibus test statistic in DA that summarizes how well groups can be differentiated (using all discriminating variables and all discriminant functions) is called Wilks's Λ. Wilks's Λ is interpreted as the proportion of variance in discriminant function scores that is *not* predictable from group membership. Thus, Wilks's Λ is analogous to $(1 - \eta^2)$ in ANOVA or $(1 - R^2)$ in MR.

DA (like ANOVA) involves a partition of the total variances and covariances (for a set of *X* discriminating variables considered jointly as a set) into between-group and within-group variances and covariances. In one-way ANOVA, the partition of variance was as follows: $SS_{\text{total}} = SS_{\text{between}} + SS_{\text{within}}$. In DA, where **SCP** corresponds to a matrix that contains sums of squares and sums of cross products for all the predictor *X* variables, the corresponding matrix algebra expression is

$$\mathbf{SCP}_{\text{total}} = \mathbf{SCP}_{\text{between}} + \mathbf{SCP}_{\text{within}}. \tag{16.7}$$

The elements of $\mathbf{SCP}_{\text{total}}$ consist of the total sums of squares for each of the *X*s (in the diagonal elements) and the total sums of cross products for all pairs of *X* variables (in the off-diagonal elements) for the entire dataset with all groups combined. Next, within each group (let the group number be indicated by the subscript *i*), we obtain $\mathbf{SCP}_i$. The elements of $\mathbf{SCP}_i$ within each group consist of the sums of squares of the *X*s just for scores within Group *i* and the sums of cross products for all pairs of *X*s just for scores within Group *i*. We obtain a reasonable estimate of the error term for our analysis, $\mathbf{SCP}_{\text{within,}}$ by pooling, combining, or averaging these $\mathbf{SCP}_i$ matrices across groups; this pooling or averaging only makes sense if the assumption of homogeneity of **Σ** across groups is satisfied. We can compute the overall within-group variance/covariance matrix $\mathbf{S}_{\text{within}}$ from the **SCP** matrices within groups, as follows:

$$\mathbf{SCP}_{\text{within}} = [\mathbf{SCP}_1 + \mathbf{SCP}_2 + \ldots + \mathbf{SCP}_k]/[n_1 + n_2 + \ldots + n_k - k], \tag{16.8}$$

where *k* is the number of groups. Note that this is analogous to the computation of SS_{within} in a one-way ANOVA, as a sum of $SS_1, SS_2, \ldots, SS_k$, or to the computation of s^2_{pooled} for a *t* test, where s^2_{pooled} is the weighted average of s^2_1 and s^2_2.

In univariate ANOVA, to judge the relative magnitude of explained versus unexplained variance, an *F* ratio was examined ($F = MS_{\text{between}}/MS_{\text{within}}$). However, in DA we consider a ratio of the determinant of $\mathbf{SCP}_{\text{within}}$ to the determinant of $\mathbf{SCP}_{\text{total}}$. It is necessary to summarize the information about variance contained in each of these **SCP** matrices; this is done by taking the determinant of each matrix. One formula for the computation of Wilks's Λ is

$$\Lambda = |\mathbf{SCP}_{\text{within}}|/|\mathbf{SCP}_{\text{total}}|. \tag{16.9}$$

Recall (from the appendix to Chapter 14) that the determinant of a matrix is a single value that summarizes the variance of the *X*s while taking into account the correlations or covariances among the *X*s (so that redundant information isn't double counted). If there is just one discriminating variable, note that Wilks's Λ becomes

$$\Lambda = SS_{\text{within}}/SS_{\text{total}}. \tag{16.10}$$

Wilks's Λ is equivalent to the proportion of unexplained, or within-groups, variance in a one-way ANOVA. In a one-way ANOVA, Λ is equivalent to $(1 - \eta^2)$, and $\eta^2 = SS_{\text{between}}/SS_{\text{total}}$.

Note that an *F* ratio has two *df* terms because its distribution varies as a function of the degrees of freedom for within groups and the degrees of freedom for between groups. A Wilks's Λ statistic has three different *df* terms, which are based on the three factors that are involved in a DA:

N, the total number of subjects or cases,

k, the number of groups, and

p, the number of discriminating variables.

In principle, one could look up critical values for Wilks's Λ, but it is unwieldy to use tables that depend on three different *df* terms. It is more convenient to convert Wilks's Λ into a more familiar test statistic. SPSS Discriminant converts the Λ to an approximate chi-square with $df = p \times (k - 1)$.

$$\chi^2 = -\{N - [(p + k)/2] - 1\}\ln\ \Lambda, \tag{16.11}$$

where *N* is the number of cases, *p* the number of variables, and *k* the number of groups.

Finally, we convert Wilks's Λ to a familiar effect size index, η^2:

$$\eta^2 = 1 - \Lambda. \tag{16.12}$$

To summarize, all the preceding equations provide information about the omnibus test for DA. If Wilks's Λ is reasonably small, and if the corresponding omnibus *F* is statistically significant, we can conclude that group membership can be predicted at levels significantly better than chance when all the discriminant functions in the analysis are considered as a set.

Another question that may be of interest is the following: Which of the discriminating variables provide useful information about group membership? To answer this question, we need to look at sizes and signs of the discriminant function coefficients. The discriminant function coefficients for Discriminant function D_1 were given in Equation 16.4. More generally, there can be several Discriminant functions, denoted by $D_1, D_2, \ldots, D_{k-1}$. For each discriminant function D_i, there are separate discriminant function coefficients, as shown in the following expression: $D_i = d_{i1}\, z_1 + d_{i2}\, z_2 + \ldots + d_{ip}\, z_p$. That is, each discriminant function D_i is constructed from a different weighted linear combination of the *z* scores $z_1, z_2, \ldots, z_p$. For each discriminant function that differs significantly across groups (i.e., it is significantly predictive of group membership), the data analyst may want to interpret or discuss the meaning of the discriminant function by describing which variables were given the largest weights in computation of the discriminant function. These *d* coefficients are analogous to beta coefficients in an MR; they provide the optimal weighted linear composite of the *z* scores on the *X* predictor variables; that is, the *d* coefficients are the set of weights associated with the *z* scores on the predictor variables for which the value of each new discriminant function, D_i, has the largest possible SS_{between}

and the smallest possible SS_{within}. Other factors being equal, an X variable that has a d coefficient that is large (in absolute value) is judged to provide unique information that helps predict group membership when other discriminating variables are statistically controlled.

The discriminant function coefficients for a DA are obtained using a type of matrix algebra computation that we have not encountered before: the eigenvalue/eigenvector problem. An eigenvector is a vector or list of coefficients; these correspond to the weights that are given to predictor variables (e.g., discriminant function coefficients, factor loadings) when we form different weighted linear combinations of variables from a set of Xs. The elements of the eigenvector must be rescaled to be used as discriminant function coefficients. Each eigenvector has one corresponding **eigenvalue**. In DA, the eigenvalue (λ) for a particular discriminant function corresponds to the value of the goodness-of-fit measure $SS_{between}/SS_{within}$. In other words, the eigenvalue that corresponds to a discriminant function gives us the ratio of $SS_{between}$ to SS_{within} for a one-way ANOVA on scores on that discriminant function.

When properly rescaled, an eigenvector gives the d coefficients for one standardized discriminant function; like beta coefficients in an MR, these are applied to z scores. The eigenvalue (λ) yields an effect size measure (the **canonical correlation, r_c**) that describes how strongly its corresponding discriminant function scores are associated with group membership. The term *canonical* generally indicates that the correlation is between weighted linear composites that include multiple predictor and/or multiple outcome variables. The canonical correlation (r_c) between scores on a discriminant function and group membership is analogous to η in a one-way ANOVA. The subscript i is used to identify which of the discriminant functions each canonical correlation, r_c, is associated with; for example, r_{c2} is the canonical correlation that describes the strength of association between scores on D_2 and group membership. The eigenvalue λ_i for each discriminant function can be used to calculate a canonical correlation r_{ci}. For example, this canonical correlation r_{c1} describes how strongly scores on D_1 correlate with group membership:

$$r_{c1} = \sqrt{\frac{\lambda_1}{1 + \lambda_1}}. \tag{16.13}$$

The matrix algebra for the eigenvalue problem generates solutions with the following useful characteristics. First, the solutions are rank ordered by the size of the eigenvalues; that is, the first solution corresponds to the discriminant function with the largest possible λ_1 (and r_{c1}), the second corresponds to the next largest possible λ_2 (and r_{c2}), and so on. In addition, scores on the discriminant functions are constrained to be uncorrelated with each other; that is, the scores on D_1 are uncorrelated with the scores on D_2, and with scores on all other discriminant functions.

Because the discriminant function scores are uncorrelated with each other, it is relatively easy to combine the explained variance across discriminant functions and, therefore, to summarize how well an entire set of discriminant functions is associated with or predictive of group memberships. One way to decide how many discriminant functions to retain for interpretation is to set an alpha level (such as $\alpha = .05$), look at the p values associated with tests of sets of discriminant functions, and drop discriminant functions from the interpretation if they do not make a statistically significant contribution to the prediction of group membership, as discussed earlier in the section about "dimension reduction." Thus, in some research situations, it may be possible to discriminate among groups adequately using only the first one or two discriminant functions.

The notation used for the standardized discriminant function coefficients for each discriminant function (the subscript i is used to denote discriminant functions 1, 2, ..., and so forth, and the second subscript on each d coefficient indicates which of the X predictor variables that coefficient is associated with) is as follows:

$$D_i = d_{i1} z_1 + d_{i2} z_2 + \ldots + d_{ip} z_p. \tag{16.14}$$

These d coefficients, like beta coefficients in an MR, tend to lie between −1 and +1 (although they may lie outside this range). To provide a more concrete example of the optimal weighted linear combination represented by Equation 16.14, in the empirical example at the end of the chapter, we will compute and save discriminant function scores that are calculated from the formula for D_1 and then conduct a one-way ANOVA to see how scores on this new variable D_1 vary across groups. Comparison of the output of this one-way ANOVA with the DA results help make it clear that the DA involves comparison of the means for "new" outcome variables—that is, scores on one or more discriminant functions across groups.

The d coefficients can be interpreted like beta coefficients; each beta coefficient tells us how much weight is given to the z score on each X discriminating variable when all other X variables are statistically controlled. SPSS can also report raw score discriminant function coefficients; these are not generally used to interpret the nature of group differences, but they may be useful if the goal of the study is to generate a prediction formula that can be used to predict group membership easily from raw scores for future cases. SPSS also reports **structure coefficients**; that is, for each discriminating variable, its correlation with each discriminant function is reported. Some researchers base their discussion of the importance of discriminating variables on the magnitudes of the standardized discriminant function coefficients, while others prefer to make their interpretation based on the pattern of structure coefficients.

16.6 ♦ Conceptual Basis: Factors That Affect the Magnitude of Wilks's Λ

If the groups in a DA have high between-groups variance (and relatively low within-groups variance) on one or more of the discriminating variables, it is possible that the overall Wilks's Λ (the proportion of variance in discriminant scores that is not associated with group membership) will be relatively small. However, there are other factors that may influence the size of Wilks's Λ, including the strength and signs of correlations among the discriminating variables and whether combinations of variables differentiate groups much more effectively than any single variable (for a more detailed and technical discussion of this issue, see Bray & Maxwell, 1985). Figure 16.1 shows a graphic example in which 3 groups are clearly separated in a territorial map (scores on D_1 and scores on D_2). This clear spatial separation corresponds to a situation in which Wilks's Λ will turn out to be relatively small. On the other hand, Figure 16.2 illustrates a hypothetical situation in which there is no clear spatial separation among the 3 groups. The overlap among the groups in Figure 16.2 corresponded to a research situation in which Wilks's Λ would be quite large.

16.7 ♦ Effect Size

One question that can be answered using DA is the following: How well can group membership be predicted overall, using all discriminating variables and all discriminant functions? To answer this question, we look at the overall Wilks's Λ to find the corresponding effect size, η^2:

$$\eta^2 = 1 - \Lambda. \tag{16.15}$$

For each discriminant function, we can compute an effect size index (a canonical correlation, r_c) from its eigenvalue, λ_i:

$$r_{ci} = \sqrt{\frac{\lambda_i}{1 + \lambda_i}}. \tag{16.16}$$

There is no effect size index for each individual discriminating variable that can be interpreted as a proportion of variance (i.e., nothing like the sr^2_{unique} effect size estimate for each predictor in a regression). It is possible to evaluate the relative importance of the discriminating variables by looking at their structure coefficients, their standardized discriminant function coefficients, and possibly also the univariate *F*s (from a one-way ANOVA on each discriminating variable). However, these comparisons are qualitative, and the cutoff values used to decide which *X* predictors make "strong" predictive contributions are arbitrary.

16.8 ♦ Statistical Power and Sample Size Recommendations

Because the size of Wilks's Λ (and its corresponding χ^2 or *F* test of significance) is potentially influenced by such a complex set of factors (including not only the strength of the univariate associations of individual predictors with group membership but also the signs and magnitudes of correlations among predictors), it is difficult to assess the sample size requirements for adequate statistical power. Stevens (2002) cited Monte Carlo studies that indicate that estimates of the standardized discriminant function coefficients and the structure coefficients are unstable unless the sample size is large. Accordingly, Stevens recommended that the total number of cases for a DA should be at least 20 times as large as p, the number of discriminating variables. Another issue related to sample size is the requirement that the number of cases within each group be larger than the number of predictor variables, p (Tabachnick & Fidell, 2007).

Another set of guidelines for minimum sample size in a k group situation with p outcome variables is provided in the next chapter (see Table 17.1). This power table for multivariate analysis of variance (MANOVA) and DA, adapted from Stevens (2002), provides the minimum suggested sample sizes per group so as to achieve a power of .70 or greater using $\alpha = .05$ and three to six discriminating variables. The recommended sample size increases as the expected effect size becomes smaller; for further details, see Chapter 17 in this book or Stevens (2002). A much more detailed treatment of this problem is provided by Lauter (1978).

16.9 ♦ Follow-Up Tests to Assess What Pattern of Scores Best Differentiates Groups

If the overall DA model is not significant (i.e., if the chi-square associated with the overall Wilks's Λ for the set of all discriminant functions is nonsignificant), then it does not make sense to do follow-up analyses to assess which pairs of groups differed significantly and/or which discriminating variables provided useful information. If the overall model is significant, however, these follow-up analyses may be appropriate.

If there are only 2 groups in the DA, no further tests are required to assess contrasts. However, if there are more than 2 groups, post hoc comparisons of groups may be useful to identify which pairs of groups differ significantly in their scores on one or more discriminant functions. This can be done more easily using the MANOVA procedure that is introduced in the following chapter. Analysts may also want to evaluate which among the X predictor variables are predictive of group membership. A qualitative evaluation of the discriminant function coefficients and structure coefficients can be used to answer this question; variables with relatively large discriminant coefficients (in absolute value) are more predictive of group membership in a multivariate context, where correlations with the other X predictor variables are taken into consideration. It is also possible to examine the one-way ANOVA in which the means on each X variable are compared across groups (not controlling for the intercorrelations among the X variables), but these ANOVA results need to be interpreted with great caution for two reasons. First, the significance or p values reported for these ANOVAs by SPSS are not adjusted to limit the risk of Type I error that arises when a large number of significance tests are performed. The easiest way to control for inflated risk of Type I error that is likely to arise in this situation, where many significance tests are performed, is to use the Bonferroni correction. That is, for a set of q significance tests, to obtain an experiment-wise EW_α of .05, set the per-comparison PC_α at $.05/q$. Second, these tests do not control for the intercorrelations among predictor variables; as discussed in Chapter 10, an individual X predictor variable can have a quite different relationship with an outcome variable (both in magnitude and in sign) when you do control for other predictor variables than when you do not control for those other variables.

To assess the predictive usefulness of individual variables, the analyst may look at three types of information, including the magnitude and sign of the standardized discriminant function coefficients, the magnitude and sign of the structure coefficients, and the significance of the univariate F for each variable, along with the unadjusted group means for variables with significant overall Fs. The conclusions about the importance of an individual X_i predictor variable (and even about the nature of the relation between X_i and group membership, e.g., whether membership in one particular group is associated with high or low scores on X_i) may differ across these three sources of information. It is fairly easy to draw conclusions and summarize results when these three types of information appear to be consistent. When they are apparently inconsistent, the researcher must realize that the data indicate a situation where X_i's role as a predictor of group membership is different when other *X* predictors are statistically controlled (in the computation of discriminant function coefficients) compared with the situation when other *X* predictors are ignored (as in the computation of the univariate ANOVAs). Review the discussion of statistical control in Chapter 10, if necessary, to understand why a variable's predictive usefulness may change when other variables are controlled and to get an idea of possible explanations (such as spuriousness, confounding, and suppression) that may need to be considered.

Note that if none of the individual *X*s is a statistically significant predictor of group membership, the optimal linear combinations of *X*s that are formed for use as discriminant functions may not differ significantly across groups either. However, it is sometimes possible to discriminate significantly between groups using a combination of scores on two or more predictors even when no single predictor shows strong differences across groups (as in the example that appeared in Figures 16.3 and 16.4). On the other hand, usually, when there is a nonsignificant overall result for the DA, it generally means that none of the individual predictors was significantly related to group membership. However, there can be situations in which one or two individual predictors are significantly related to group membership, and yet, the overall DA is not significant. This result may occur when other "garbage" variables that are added to the model decrease the error degrees of freedom and reduce statistical power.

In clinical or practical prediction situations, it may be important to report the specific equation(s) that is(are) used to construct discriminant function scores (from standardized *z* scores and/or from raw scores) so that readers can use the predictive equation to make predictions in future situations where group membership is unknown. (An additional type of discriminant function coefficients that can be obtained, the Fisher linear discriminant function, is described in Note 1. The Fisher coefficients are not generally used to interpret the nature of group differences, but they provide a simpler method of prediction of group membership for each individual score, particularly in research situations where there are more than two categories or groups.) For example, a personnel director may carry out research to assess whether job outcome (success or failure) can be predicted from numerous tests that are given to job applicants. Based on data collected from people who are hired and observed over time until they succeed or fail at the job, it may be possible to develop an equation that does a reasonably good job of predicting job outcomes from test scores. This equation may then be applied to the test scores of future applicants to predict whether they are more likely to succeed or fail at the job in future. See Note 1 for a further discussion of the different types of equations used to predict group membership for individual cases.

16.10 ♦ Results

Selected cases and variables from the Project Talent Data (scores shown in Table 16.1) were used for this example. High school seniors were given a battery of achievement and aptitude tests; they were also asked to indicate their choice of future career. Three groups were included: Group 1, business careers ($n = 98$); Group 2, medicine ($n = 27$); and Group 3, teaching/social work ($n = 39$). Scores on the following six tests were used as discriminating variables: English, reading, mechanical reasoning, abstract reasoning, math, and office skills. The valid number for this analysis was $N = 164$. For better statistical power, it would be preferable to have larger numbers in the groups than the numbers in the dataset used in this example.

The SPSS menu selections to access the DA procedure appear in Figure 16.5. From the top level menu in the SPSS Data View Worksheet, make the following selections: <Analyze> → <Classify> → <Discriminant>, as shown in Figure 16.5. The name of the categorical group membership outcome variable, cargroup, is placed in the window under the heading Grouping Variable. The Define Range button was used to open up an additional window (not shown here) where the range of score codes for the groups included in the analysis was specified; scores that ranged from 1 to 3 on cargroup were included in this analysis. The decision to enter all predictor variables in one step was indicated by clicking the radio button marked Enter independents together in the DA dialog window in Figure 16.6. The Statistics button was clicked so that additional optional statistics could be requested.

From the list of optional statistics, the following selections were made (see Figure 16.7): Means, Univariate ANOVAs, Box's *M*, and Within-groups correlation matrices. The

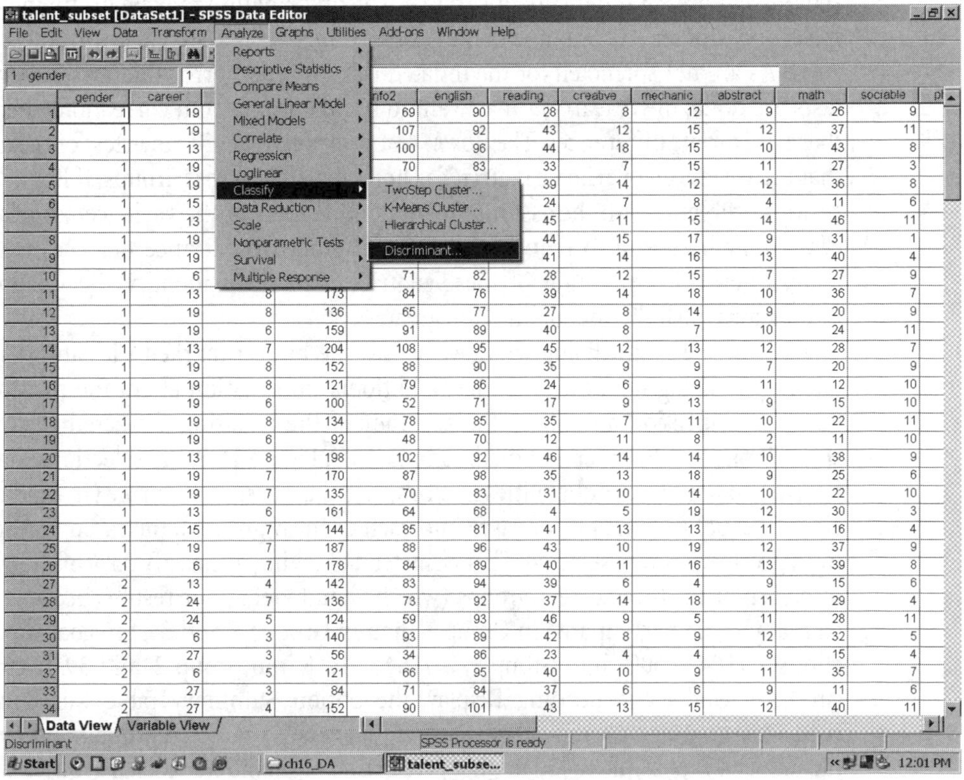

	gender	career			nfo2	english	reading	creative	mechanic	abstract	math	sociable
1	1	19			69	90	28	8	12	9	26	9
2	1	19			107	92	43	12	15	12	37	11
3	1	13			100	96	44	18	15	10	43	8
4	1	19			70	83	33	7	15	11	27	3
5	1	19					39	14	12	12	36	8
6	1	15					24	7	8	4	11	6
7	1	13					45	11	15	14	46	11
8	1	19					44	15	17	9	31	1
9	1	19					41	14	16	13	40	4
10	1	6			71	82	28	12	15	7	27	9
11	1	13	8	173	84	76	39	14	18	10	36	7
12	1	19	8	136	65	77	27	8	14	9	20	9
13	1	19	6	159	91	89	40	8	7	10	24	11
14	1	13	7	204	108	95	45	12	13	12	28	7
15	1	19	8	152	88	90	35	9	9	7	20	9
16	1	19	8	121	79	86	24	6	9	11	12	10
17	1	19	6	100	52	71	17	9	13	9	15	10
18	1	19	8	134	78	85	35	7	11	10	22	11
19	1	19	6	92	48	70	12	11	8	2	11	10
20	1	13	8	198	102	92	46	14	14	10	38	9
21	1	19	7	170	87	98	35	13	18	9	25	6
22	1	19	7	135	70	83	31	10	14	10	22	10
23	1	13	6	161	64	68	4	5	19	12	30	3
24	1	15	7	144	85	81	41	13	13	11	16	4
25	1	19	7	187	88	96	43	10	19	12	37	9
26	1	6	7	178	84	89	40	11	16	13	39	8
27	2	13	4	142	83	94	39	6	4	9	15	6
28	2	24	7	136	73	92	37	14	18	11	29	4
29	2	24	5	124	59	93	46	9	5	11	28	11
30	2	6	3	140	93	89	42	8	9	12	32	5
31	2	27	3	56	34	86	23	4	4	8	15	4
32	2	6	5	121	66	95	40	10	9	11	35	7
33	2	27	3	84	71	84	37	6	6	9	11	6
34	2	27	4	152	90	101	43	13	15	12	40	11

Figure 16.5 ♦ SPSS Menu Selections for Discriminant Analysis Procedure

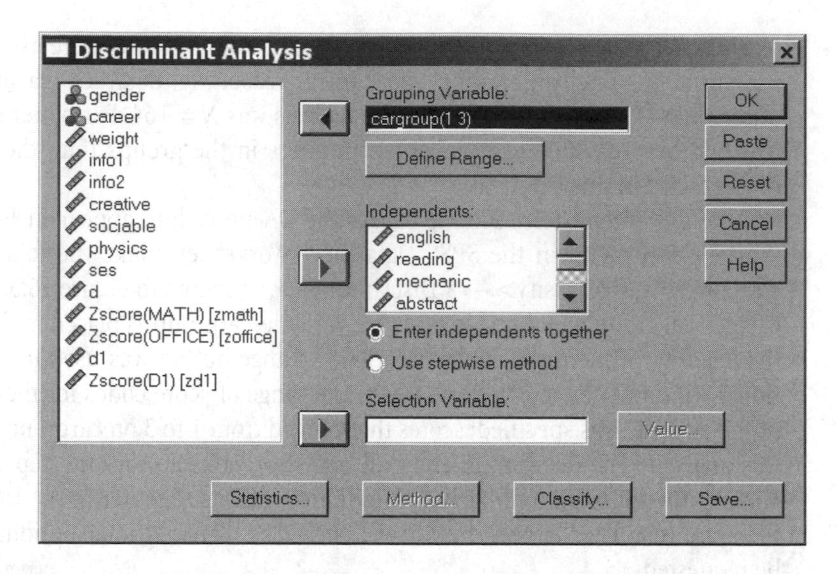

Figure 16.6 ♦ Main SPSS Dialog Window for Discriminant Analysis Procedure

univariate ANOVAs provide information about the differences in means across groups separately for each independent variable. This information may be useful in characterizing group differences, but note that the significance tests associated with these univariate ANOVAs are not corrected for the inflated risk of Type I error that arises when multiple tests are done; the researcher may want to use the Bonferroni correction when assessing these univariate differences. The Box M test provides a significance test of the assumption that the variance/covariance matrix $\boldsymbol{\Sigma}$ is homogeneous across groups. If Box's M indicates serious problems with the assumption of homogeneity of the variance/covariance matrices, it may be useful to request the separate-groups covariance matrices (these do not appear here). Examination of these matrices may help identify the variables and/or groups for which the inequalities are greatest.

Back in the main dialog window for DA, the button marked Classify was clicked to open up a dialog window to request additional information about the method used for group classification and about the accuracy of the resulting predicted group memberships. From the classification method options in Figure 16.8, the selections made were as follows: To use prior probabilities as a basis for the classification rule (instead of trying to classify equal proportions of cases into each of the 3 groups), the radio button Compute from group sizes was selected. That is, instead of trying to classify 33% of the participants into each of the three career groups (which would yield poor results, because there were so many more participants in Group 1 than in Groups 2 and 3), the goal was to classify about 98/164 = 60% into Group 1, 27/164 = 16% into Group 2, and 39/164 = 24% into Group 3. Under the heading Display, the option Summary table was checked. This requests a contingency table that shows actual versus predicted group membership; from this table, it is possible to judge what proportion of members of each group were correctly classified and, also, to see what types of classification errors (e.g., classifying future medicine aspirants as business career aspirants) were most frequent. Under the heading Use

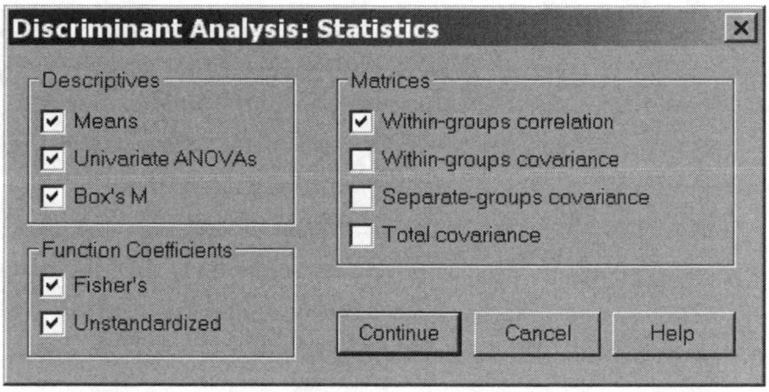

Figure 16.7 ♦ Statistics for Discriminant Analysis

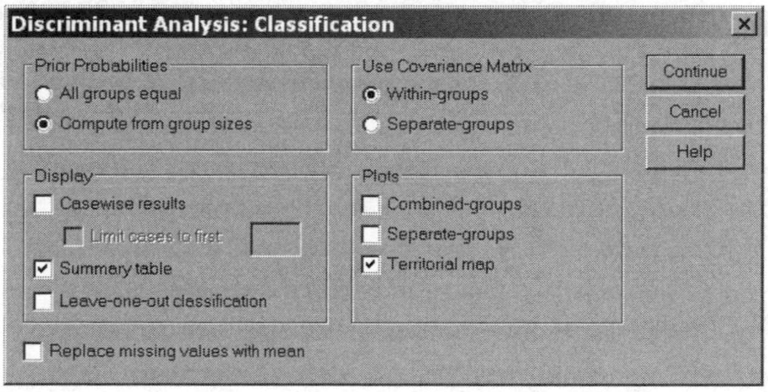

Figure 16.8 ♦ Classification Methods for Discriminant Analysis

Covariance Matrix, the option Within-groups was chosen; this means that the $\mathbf{SCP}_{\text{within}}$ matrix was pooled or averaged across groups. (If there are serious inequalities among the **SCP** matrices across groups, it may be preferable to select Separate-groups, i.e., to keep the **SCP** matrices separate instead of pooling them into a single estimate.) Under the heading Plots, the Territorial map was requested. This provides a graphic representation of how the members of the groups differ in their scores on Discriminant function 1 (D_1 scores on the X axis) and Discriminant function 2 (D_2 scores on the Y axis). A territorial map is useful because it helps us understand the nature of the differences among groups on each discriminant function—for example, what group(s) differed most clearly in their scores on D_1? Finally, the dialog window for the Save command was opened and a check box was used to request that the discriminant function scores should be saved as new variables in the SPSS worksheet (see Figure 16.9). The SPSS syntax that resulted from these menu selections is shown in Figure 16.10.

The output from the DA specified by the previous menu selections appears in Figures 16.11 through 16.23. The first two parts of the output report the simple univariate ANOVA results. That is, a univariate ANOVA was done to assess whether the means on each of the individual variables (English, reading, mechanical reasoning, abstract reasoning, math, and office skills) differed across the three levels of the career group choice outcome variable. The mean and

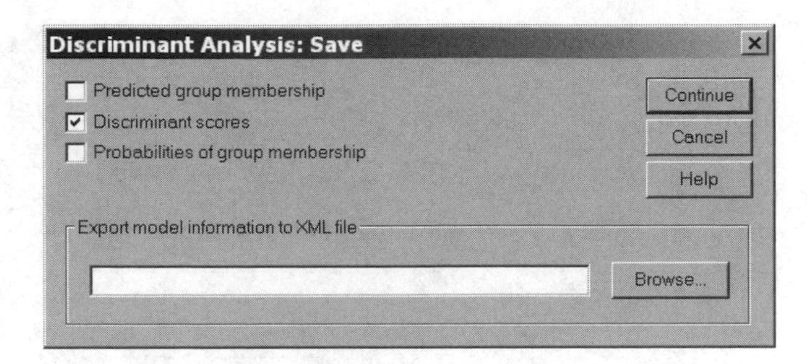

Figure 16.9 ♦ Command to Save the Computed Scores on Discriminant Functions 1 and 2 as New Variables in the SPSS Worksheet

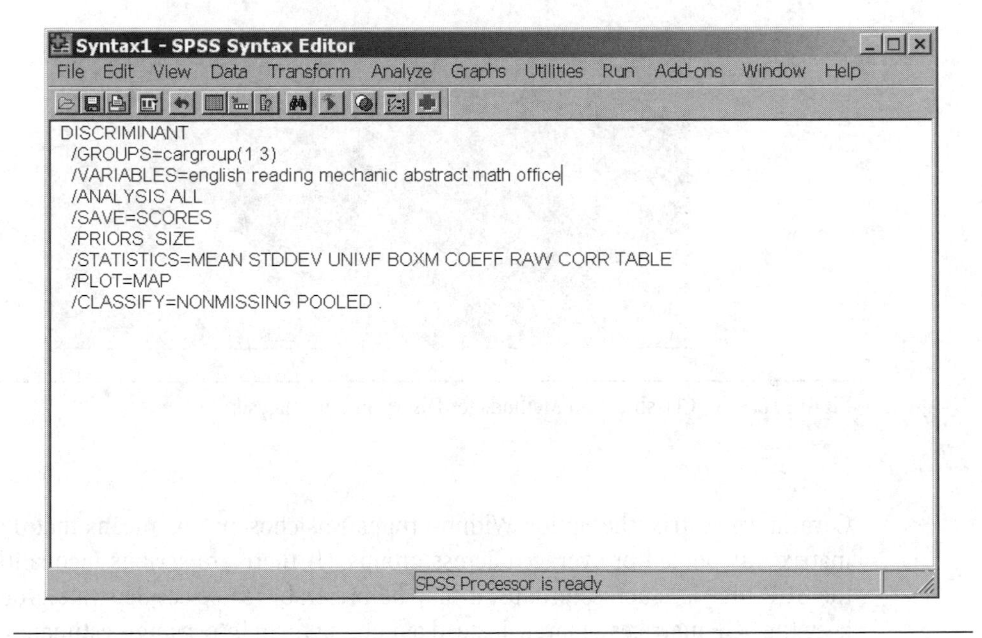

Figure 16.10 ♦ SPSS Syntax for Discriminant Analysis to Predict Cargroup (1 = Business, 2 = Medical, 3 = Teaching/Social Work) From Scores on English, Reading, Mechanical Reasoning, Abstract Reasoning, Math, and Office Skills, With All Predictor Variables Entered in One Step (Using Group Size as the Basis for Prior Probability of Group Membership)

standard deviation for each variable in each group appear in Figure 16.11. The *F* tests (which assess whether means differ significantly across the 3 groups, separately for each variable) appear in Figure 16.12. From the information in Figures 16.11 and 16.12, we can evaluate the group differences on each variable in isolation from the other variables that are included in the DA. For example, if we do not statistically control for the other five predictor variables, what is the nature of differences on English test scores across the 3 groups? When English scores are assessed in isolation, there is no significant difference in mean English scores across the 3 career groups: $F(2, 161) = .219, p = 803$.

(Text continues on page 685)

Group Statistics

cargroup		Mean	Std. Deviation	Valid N (listwise)	
				Unweighted	Weighted
business	english	88.44	8.688	98	98.000
	reading	32.44	8.489	98	98.000
	mechanic	9.12	3.681	98	98.000
	abstract	9.08	2.477	98	98.000
	math	20.05	7.159	98	98.000
	office	28.84	8.281	98	98.000
medicine	english	87.15	9.326	27	27.000
	reading	34.15	10.524	27	27.000
	mechanic	10.22	5.213	27	27.000
	abstract	9.70	2.614	27	27.000
	math	26.37	9.958	27	27.000
	office	18.37	9.199	27	27.000
teach_social	english	88.72	13.330	39	39.000
	reading	34.90	8.926	39	39.000
	mechanic	9.49	3.094	39	39.000
	abstract	9.69	3.229	39	39.000
	math	25.59	8.696	39	39.000
	office	20.62	8.583	39	39.000
Total	english	88.29	10.024	164	164.000
	reading	33.30	8.961	164	164.000
	mechanic	9.39	3.846	164	164.000
	abstract	9.33	2.695	164	164.000
	math	22.41	8.503	164	164.000
	office	25.16	9.601	164	164.000

Figure 16.11 ♦ Selected Output From a Discriminant Analysis (All Predictor Variables Entered in One Step); Univariate Means on Each Variable for Each Group

Tests of Equality of Group Means

	Wilks's Lambda	F	df1	df2	Sig.
english	.997	.219	2	161	.803
reading	.985	1.196	2	161	.305
mechanic	.989	.880	2	161	.417
abstract	.987	1.028	2	161	.360
math	.884	10.529	2	161	.000
office	.775	23.319	2	161	.000

Figure 16.12 ♦ Univariate ANOVA Results From the Discriminant Analysis

Pooled Within-Groups Matrices

		english	reading	mechanic	abstract	math	office
Correlation	english	1.000	.647	.212	.492	.537	-.006
	reading	.647	1.000	.380	.489	.586	-.140
	mechanic	.212	.380	1.000	.378	.551	-.259
	abstract	.492	.489	.378	1.000	.529	-.018
	math	.537	.586	.551	.529	1.000	-.217
	office	-.006	-.140	-.259	-.018	-.217	1.000

Figure 16.13 ♦ Pooled Within-Groups Correlation Matrix

Log Determinants

cargroup	Rank	Log Determinant
business	6	19.295
medicine	6	20.257
teach_social	6	20.501
Pooled within-groups	6	20.130

The ranks and natural logarithms of determinants printed are those of the group covariance matrices.

Test Results

Box's M		63.497
F	Approx.	1.398
	df1	42
	df2	20563.91
	Sig.	.045

Tests null hypothesis of equal population covariance matrices.

Figure 16.14 ♦ Box's Test of Equality of Variance/Covariance Matrices

Eigenvalues

Function	Eigenvalue	% of Variance	Cumulative %	Canonical Correlation
1	.448[a]	98.4	98.4	.556
2	.007[a]	1.6	100.0	.085

a. First 2 canonical discriminant functions were used in the analysis.

Wilks's Lambda

Test of Function(s)	Wilks's Lambda	Chi-square	df	Sig.
1 through 2	.685	59.879	12	.000
2	.993	1.156	5	.949

Figure 16.15 ♦ Summary of Canonical Discriminant Functions

Standardized Canonical Discriminant Function Coefficients

	Function	
	1	2
english	.420	.281
reading	.015	.533
mechanic	.443	-.937
abstract	-.116	-.027
math	-.794	.243
office	.748	.098

Figure 16.16 ♦ Standardized Canonical Discriminant Function Coefficients

Structure Matrix

	Function	
	1	2
office	.803*	.213
math	-.540*	.155
abstract	-.168*	.144
mechanic	-.138	-.577*
english	.035	.544*
reading	-.172	.474*

Pooled within-groups correlations between discriminating variables and standardized canonical discriminant functions. Variables ordered by absolute size of correlation within function.

*. Largest absolute correlation between each variable and any discriminant function

Figure 16.17 ♦ Structure Matrix: The Correlation of Each Predictor Variable With Each Discriminant Function (e.g., Scores on the Office Skills Test Correlate +.80 With Scores on Discriminant Function 1)

Canonical Discriminant Function Coefficients

	Function	
	1	2
english	.042	.028
reading	.002	.060
mechanic	.115	-.243
abstract	-.043	-.010
math	-.099	.030
office	.088	.012
(Constant)	-4.415	-3.035

Unstandardized coefficients

Figure 16.18 ♦ Raw Score Discriminant Function Coefficients

Classification Function Coefficients

	cargroup		
	business	medicine	teach_social
english	1.186	1.120	1.138
reading	-.312	-.323	-.306
mechanic	1.008	.869	.833
abstract	-.308	-.242	-.256
math	-.405	-.262	-.279
office	.394	.261	.286
(Constant)	-52.700	-47.289	-48.660

Fisher's linear discriminant functions

Figure 16.19 ♦ Fisher's Linear Discriminant Functions (One Function for Each Group)

Functions at Group Centroids

	Function	
cargroup	1	2
business	.541	-.008
medicine	-.954	-.147
teach_social	-.697	.123

Unstandardized canonical discriminant functions evaluated at group means

Figure 16.20 ♦ Means on Discriminant Functions 1 and 2 (Centroids) for Each Group

Prior Probabilities for Groups

		Cases Used in Analysis	
cargroup	Prior	Unweighted	Weighted
business	.598	98	98.000
medicine	.165	27	27.000
teach_social	.238	39	39.000
Total	1.000	164	164.000

Figure 16.21 ♦ Prior Probabilities for Groups (Based on Number of Cases in Each Group Because Priors = Size Was Selected)

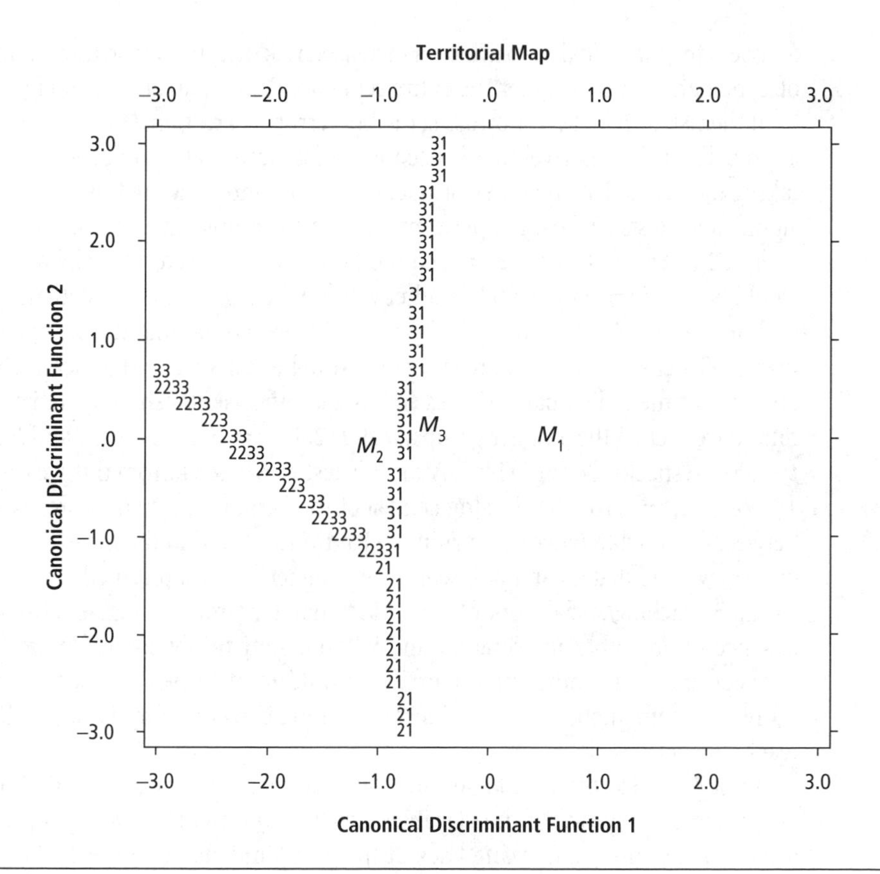

Figure 16.22 ♦ Territorial Map

NOTE: Symbols used: 1, Business; 2, Medicine; 3, Teach_social.

Classification Results[a]

			Predicted Group Membership			
		cargroup	business	medicine	teach_social	Total
Original	Count	business	91	2	5	98
		medicine	15	6	6	27
		teach_social	21	3	15	39
	%	business	92.9	2.0	5.1	100.0
		medicine	55.6	22.2	22.2	100.0
		teach_social	53.8	7.7	38.5	100.0

a. 68.3% of original grouped cases correctly classified.

Figure 16.23 ♦ Table of Original (Actual) Versus Predicted Group Membership

Reporting multiple significance tests (as in Figure 16.12) leads to an increased risk of Type I error, and the p values that appear on the SPSS printout are not adjusted in any way to correct for this problem. It is advisable to make an adjustment to the alpha levels that are used to assess the individual *F*s to avoid an increased risk of Type I error. If we want an overall risk of Type I error of $\alpha = .05$ for the entire set of six *F* tests that appear in Figure 16.12, we can limit the risk of Type I error by using the Bonferroni correction. For each individual *F* test, an appropriate PC alpha level that is used to assess statistical significance can be obtained by taking the overall EW Type I error rate, $EW_\alpha = .05$, and dividing this EW_α by the number of significance tests (q) that are performed. In this example, the number of significance tests included in Figure 16.12 is $q = 6$, so the Bonferroni-corrected PC alpha level for each test would be $EW_\alpha/q = .05/6 = .008$. In other words, when we assess the statistical significance of each univariate *F* test included in Figure 16.12, we require that the p value obtained is less than .008, before we judge a univariate *F* test to be statistically significant. By this criterion, only two of the individual variables (math and office skills) showed statistically significant differences across the 3 career groups, with $F(2, 161) = 10.53, p < .001$ and $F(2, 161) = 23.32, p < .001$, respectively. The Wilks's Λ for this test is $\Lambda = .884$; from this we can compute $\eta^2 = 1 - \Lambda$; so $\eta^2 = .116$. That is, almost 12% of the variance in math scores is associated with between-group differences. Returning to Figure 16.11 to find the means on math for these 3 groups, we find that mean math scores were higher for Group 2, medicine ($M_2 = 26.37$), and Group 3, teaching/social work ($M_3 = 25.59$), than for Group 1, business ($M_1 = 20.05$). In the absence of any other information and without statistically controlling for the correlation between math and other predictors, we would tend to predict that a student with low scores on math might choose business as a future career (Group 1) rather than medicine or teaching/social work.

To summarize the results so far, if we conduct a univariate ANOVA for each of the six variables, only two of the variables (math and office) showed significant differences across career groups. Looking back at the group means in Figure 16.11, we can see that Group 1 (business careers) scored lower than Groups 2 and 3 (medicine, teaching/social work) on math skills and higher than Groups 2 and 3 on office skills. If we used only our univariate test results, therefore, we could make the following preliminary guesses about group membership: Students with low scores on math and high scores on office may be more likely to be in Group 1, business. However, subsequent DA will provide us with more information about the nature of the differences among groups when intercorrelations among the p variables are taken into account; it will also provide us with summary information about our ability to predict group membership based on patterns of scores on the entire set of p predictor variables.

These univariate results are sometimes, although not always, included as part of the DA results. They provide a basis for comparison; the discriminant function coefficients tell us how much information about group membership is contributed by each predictor variable when we statistically control for correlations with all other predictors; the univariate results tell us how each variable is related to group membership when we do *not* control for other predictors. It can be useful to examine whether the nature of group differences changes (or remains essentially the same) when we control for correlations among predictors.

Figure 16.13 shows the correlations among all predictor variables; these correlations were calculated separately within each of the 3 groups, then pooled (or averaged) across groups. This information is analogous to the information obtained from the matrix of correlations among predictors in an MR. We do not want to see extremely high correlations among predictor variables because this indicates that the contributions of individual predictors can't be distinguished. On the other hand, if the predictors are almost entirely uncorrelated with each other, the multivariate analysis (DA) may not yield results that are very different from a set of univariate results. In rare situations where predictor variables are completely uncorrelated, it may be appropriate to just look at them individually using univariate methods. DA is most appropriate when there are weak to moderate correlations among predictors. In this case, none of the correlations among predictors exceeded .6 in absolute value; thus, there was no serious problem with multicollinearity among predictors.

Figure 16.14 shows the results of the Box *M* test. Recall that the null hypothesis that this statistic tests corresponds to the assumption of homogeneity of the sum of cross products matrices across groups:

$$H_0: \mathbf{\Sigma}_1 = \mathbf{\Sigma}_2 = \mathbf{\Sigma}_3.$$

We would prefer that this test be nonsignificant. If Box *M* is significant, it indicates a violation of an assumption that may make the results of DA misleading or uninterpretable. In this instance, Box $M = 63.497$, and the approximate *F* for Box $M = F(42, 20564) = 1.398$, $p = .045$. If the alpha level used to evaluate the significance of Box *M* is set at $\alpha = .01$ (because the *df* associated with the Box *M* test are so high), this value of Box *M* indicates no significant violation of the assumption of homogeneous variance/covariance matrices.

As in earlier situations where we tested for violations of assumptions, there is a dilemma here. The larger the number of cases, the less problematic the violations of assumptions tend to be. However, the larger the number of cases, the more statistical power we have for the detection of violations of assumptions. That is, in studies with relatively large numbers, we have a better chance of judging violations of assumptions to be statistically significant; and yet, these violations may create serious problems when *N* is large. When *N* is rather large, it is reasonable to use a relatively small alpha level to judge the significance of tests of assumptions; in this case, with $N = 164$, $\alpha = .01$ was used as the criterion for significance of Box *M*.

If the Box *M* test suggests a serious violation of the assumption of the equality of the $\mathbf{\Sigma}$s, the data analyst has several options. The application of log transformations to non-normally distributed variables (as recommended by Tabachnick & Fidell, 2007, chap. 4) may be an effective way of reducing differences in the variances of *X* across groups and the impact of multivariate outliers. An alternative statistical analysis that makes less restrictive assumptions about pattern in the data (such as binary logistic regression, see Chapter 21) may be preferred in situations where the Box *M* suggests that there are serious violations of assumptions. In this example, none of these potential remedies was applied; students may wish to experiment with the data to see whether any of these changes would correct the violation of the assumption of homogeneity of variance and covariance across groups.

The data included three groups based on the choice of future career (Group 1 = business, Group 2 = medicine, Group 3 = teaching/social work). There were six discriminating variables: scores on achievement and aptitude tests, including English, reading, mechanical reasoning, abstract reasoning, math, and office skills. The goal of the analysis was to assess whether future career choice was predictable from scores on some or all of these tests. The first DA was run with all six predictor variables entered in one step. The main results for the overall DA (with all the six predictor variables entered in one step) appear in Figure 16.15. In this example, with k = number of groups = 3 and p = number of discriminating variables = 6, there were two discriminant functions (the number of discriminant functions that can be calculated is usually determined by the minimum of $k - 1$ and p). The panel headed "Wilks's Lambda" contains significance tests for sets of discriminant functions. The first row, "1 through 2," tests whether group membership can be predicted using D_1 and D_2 combined—that is, it provides information about the statistical significance of the entire model, using all discriminant functions and predictor variables. For this omnibus test, $\Lambda = .685$ (and thus, $\eta^2 = 1 - \Lambda = .315$). That is, about 32% of the variance on the first two discriminant functions (considered jointly) is associated with between-groups differences. This was converted to a chi-square statistic; $\chi^2(12) = 59.88$, $p < .001$; thus, the overall model, including both Discriminant functions 1 and 2, significantly predicted group membership.

To answer the question about whether we need to include all the discriminant functions in our interpretation or just a subset of the discriminant functions, we look at the other tests of subsets of discriminant functions. In this example, there is only one possibility to consider; the second row of the Wilks's Lambda table tests the significance of Discriminant function 2 alone. For D_2 alone, $\Lambda = .993$; $\chi^2(5) = 1.16$, $p = .949$. Thus, D_2 alone did not contribute significantly to the prediction of group membership. We can limit our interpretation and discussion of group differences to the differences that involve D_1. Note that we do not have a direct significance test for D_1 alone in this set of DA results.

Additional information about the relative predictive usefulness of D_1 and D_2 is presented in the panel headed "Eigenvalues." For D_1, the eigenvalue $\lambda_1 = .448$ (recall that this corresponds to $SS_{between}/SS_{within}$ for a one-way ANOVA comparing groups on D_1 scores). This eigenvalue can be converted to a canonical correlation:

$$r_c = \sqrt{\frac{.448}{1 + .448}} = .556.$$

This r_c is analogous to an η correlation between group membership and scores on a quantitative variable in one-way ANOVA. If r_c is squared, we obtain information analogous to η^2—that is, the proportion of variance in scores on D_1 that is associated with between-group differences. It is clear that, in this case, D_1 is strongly related to group membership ($r_{c_1} = .556$) while D_2 is very weakly (and nonsignificantly) related to group membership ($r_{c_2} = .085$). The total variance that can be predicted by D_1 and D_2 combined is obtained by summing the eigenvalues ($.448 + .007 = .455$). The manner in which this predicted variance should be apportioned between D_1 and D_2 is assessed by dividing each eigenvalue by this sum of eigenvalues—that is, by the variance that can be predicted by D_1 and D_2 combined (.455); $.448/.455 = .984$ or 98.4% of that variance is due to D_1. It is clear, in this case, that D_1 significantly predicts group membership; D_2 provides very little additional information, and thus, D_2 should not be given much importance in the interpretation.

In the following written results, only the results for D_1 are included. To see what information was included in D_1, we look at the table in Figure 16.16. The values in this table correspond to the d coefficients in Equation 16.4. Column 1 gives the coefficients that provide the optimal weighted linear combination of z scores that are used to construct scores on D_1; Column 2 provides the same information for D_2. These *d* coefficients are analogous to beta coefficients in an MR; they tell us how much weight or importance was given to each *z* score in calculating the discriminant function when we control for or partial out correlations with other predictors. No tests of statistical significance are provided for these coefficients. Instead, the data analyst typically decides on an (arbitrary) cutoff value; a variable is viewed as useful if its standardized discriminant function coefficient exceeds this arbitrary cutoff (in absolute value). In this example, the arbitrary cutoff value used was .5. By this standard, just two of the six predictor variables were given a substantial amount of weight when computing D_1: math ($d_{1\text{math}} = -.784$) and office ($d_{1\text{office}} = +.748$). In some data analysis situations, it might be reasonable to choose a lower "cutoff" value when deciding which variables make an important contribution—for example, discriminant coefficients that are above .4 or even .3 in absolute value. In this situation, use of such low cutoff values would result in judging English and mechanical reasoning as "important" even though neither of these variables had a significant univariate *F*.

It is important to understand that, in DA as in MR, the strength of the association and even the direction of the association between scores on each predictor variable and the outcome variable can be quite different when the association is evaluated in isolation than when controlling for other predictor variables included in the analysis. In some situations, the same variables are "important" in both sets of results; that is, the magnitude and sign of discriminant function coefficients are consistent with the magnitude and direction of differences between group means in univariate ANOVAs.

In this research example, the combination of variables on which the 3 groups showed maximum between-group differences was D_1; high scores on D_1 were seen for students who had high scores on office (which had a positive coefficient) and low scores on math (which had a negative coefficient for D_1).

Figure 16.17 reports another type of information about the relationship between each individual predictor variable and each discriminant function coefficient; this is called the "structure matrix." The values in this table provide information about the correlation between each discriminating variable and each discriminant function. For example, for D_1 (Discriminant function 1), the two variables that had the largest correlations with this discriminant function were office and math; the correlation between scores on office and scores on D_1 was +.80, while the correlation between scores on math and scores on D_1 was −.54. These correlations do not have formal significance tests associated with them; they are generally evaluated relative to some arbitrary cutoff value. In this example, an arbitrary cutoff of .5 (in absolute value) was used. By this criterion, the two variables that were highly correlated with D_1 were office (with a correlation of +.80) and math (with a correlation of −.54). Note that the asterisks next to these structure coefficients do *not* indicate statistical significance. For each variable, the structure coefficient marked with an asterisk was just the largest correlation when correlations of variables were compared across discriminant functions.

Figure 16.18 reports the coefficients that are used to construct discriminant function scores from raw scores on the X_i predictor variables (see Note 1 for further explanation). These coefficients are analogous to the raw score slope coefficients in a raw score MR, as shown in Equation 16.1. While these coefficients are not usually interpreted as evidence of the strength of the predictive association of each X_i variable, they may be used to generate discriminant function scores that in turn may be used to predict group membership based on raw scores.

Figure 16.19 reports the Fisher classification coefficients. These are generally used in research situations in which the primary goal is prediction of group membership. Note that these differ from the discriminant functions discussed earlier; we obtain one Fisher classification function for each of the *k* groups. In situations where the number of groups is larger than two, the Fisher classification coefficients provide a more convenient way of predicting group membership (and generating estimates of probabilities of group membership based on scores on the X_i predictor variables). See Note 1 for further discussion. The Fisher coefficients are not generally used to interpret or describe the nature of group differences.

Figure 16.21 reports the "prior probability" of classification into each of the 3 groups. In this example, because the sizes of groups (numbers of cases in groups) were used as the basis for the classification procedure, the "prior probability" of classification into each group is equivalent to the proportion of cases in the sample that are in each group.

At this point, we have three types of information about the predictive usefulness of these six variables: the univariate *F*s, the standardized canonical discriminant function coefficients, and the structure coefficients. In this example, we reach the same conclusion about the association between scores on predictor variables and choice of career, no matter which of these three results we examine. Among the six variables, only the scores on math and office are related to group membership; scores on the other four variables are not related to group membership. Just as beta coefficients may differ in sign from zero-order Pearson correlations between a predictor and *Y*, and just as group means in an ANCOVA may change rank order when we adjust for a covariate, we can find in a DA that the nature of group differences on a particular variable looks quite different in the univariate ANOVA results compared with that in the DA results. That is, the relation between English and group membership might be quite different when we do not control for the other five predictors (in the univariate ANOVAs) than when we do control for the other five predictors (in the computation of the standardized discriminant function coefficients and structure coefficients). In situations like this, the discussion of results should include a discussion of how these two sets of results differ.

So far, we know that a high score on D_1 represents high scores on office and low scores on math. How are scores on D_1 related to group membership? This can be assessed by looking at the values of functions at group centroids (in Figure 16.20) and at the plot of the territorial map in Figure 16.22. A **centroid** is the mean of scores on each of the discriminant functions for members of 1 group. For the first discriminant function, the business group has a relatively high mean on D_1 (.541), while the other 2 groups have relatively low means (−.954 for medicine and −.697 for teaching/social work). In words, then, the business group had relatively high mean scores on D_1; members of this group scored high on office and low on math relative to the other 2 groups. Note that the means (or centroids) for D_2 were much closer together than the means on D_1 (centroids on D_2 were −.008, −.147, and +.123). This is consistent with earlier information (such as the small size of r_{c2}) that the groups were not very different on D_2 scores.

The territorial map (in Figure 16.22) graphically illustrates the decision or classification rule that is used to predict group membership for individuals based on their scores on the discriminant functions D_1 and D_2. Scores on D_1 are plotted on the horizontal axis; scores on D_2 are plotted on the vertical axis. The strings of 1s correspond to the boundary of the region for which we classify people as probable members of Group 1; the strings of 2s correspond to the boundary of the region for which we classify people as probable members of Group 2, and so forth. Roughly, anyone with a positive score on D_1 was classified as a probable member of Group 1, business. The other 2 groups were differentiated (but not well) by scores on D_2; higher scores on D_2 led us to classify people as members of Group 3, teaching/social work; lower scores on D_2 led us to classify people as members of Group 2, medicine. However, as we shall see from the table of predicted versus actual group membership in the last panel of the DA output, members of these 2 groups could not be reliably differentiated in this analysis.

Together with earlier information, this territorial map helps us interpret the meaning of D_1. High scores on D_1 corresponded to high scores on office skills and low scores on math achievement, and high scores on D_1 predicted membership in the business career group. The centroids of the groups appear as asterisks in the territorial map produced by SPSS; in Figure 16.22, these asterisks were replaced by the symbols M_1, M_2, and M_3 (to denote the centroids for Groups 1, 2, and 3, respectively). If the DA does a good job of discriminating among groups, these centroids should be far apart, and each centroid should fall well within the region for which people are classified into a particular group. In this example, the centroid for Group 1 is far from the centroids for Groups 2 and 3; this tells us that members of Group 1 could be distinguished from Groups 2 and 3 fairly well. However, the centroids of Groups 2 and 3 were relatively close together, and the centroid for Group 2 was not even within the region for Group 2 classification, which tells us that Group 2 members are not identified well at all by this DA. Note that in some DA solutions, it is possible that none of the cases is classified into a particular group, and when this happens, there is no region in the territorial map that corresponds to that group.

Finally, the classification results displayed in Figure 16.23 summarize the association between actual group ("Original" in the table in Figure 16.23) membership and predicted group membership. If the classification based on discriminant function scores was very accurate, most cases should fall into the diagonal cells of this table (i.e., the actual and predicted group membership should be identical). In fact, only 68% of the participants were correctly classified overall. Members of Group 1 were most often classified correctly (91 out of 98, or about 93%, of them were correctly classified). However, this analysis did not do a good job of identifying members of Groups 2 and 3; in fact, the majority of members of these groups were misclassified. Thus, even though the overall model was statistically significant (based on the chi-square for D_1 and D_2 combined), group membership was not predicted well at all for members of Groups 2 and 3. Only two of the six variables (scores on office and math) were useful in differentiating groups; high scores on office and low scores on math predicted a preference for business, but these scores provided no information about the choice between medicine and teaching. The following Results section illustrates how the information from the preceding DA could be written up for a journal article.

Results

A discriminant function analysis was done with all predictor variables entered in one step to assess how well future career choice could be predicted from scores on six aptitude and achievement tests. The 3 groups being compared were: Group 1, business ($N = 98$); Group 2, medicine ($N = 27$); and Group 3, teachers and social workers (denoted by teach_social in the SPSS printout) ($N = 39$). The six discriminating variables were scores on the following: English, reading, mechanical reasoning, abstract reasoning, math, and office skills. Preliminary screening indicated that scores on all six variables were approximately normally distributed.

A pooled within-groups correlation matrix was calculated to assess whether there was multicollinearity among these predictor variables; see Figure 16.13. No correlation exceeded .7 in absolute value, so multicollinearity was not a problem. Most test scores were modestly positively intercorrelated (with *r*s of the order of +.2 to +.6), except that scores on the office skills test were rather weakly correlated with scores on the other five variables. Using $\alpha = .01$, the Box *M* test for possible violation of the homogeneity of within-group variance/covariance matrices was judged to be nonsignificant.

Because there were 3 groups, two discriminant functions were created. Discriminant function 2 had a canonical correlation of .0853; thus, it was weakly related to group membership. The chi-square for Discriminant function 2 alone was not statistically significant: $\chi^2(5) = 1.156$, $p = .949$. Discriminant function 1 had a canonical correlation of .556; thus, it was moderately related to group membership. The test for the combined predictive value of Discriminant functions 1 and 2 combined was statistically significant: $\chi^2(12) = 59.879$, $p < .001$. Because the second discriminant function was not statistically significant, the coefficients for this function were not interpreted. Overall, the prediction of group membership was not very good; Wilks's Λ, which is analogous to $1 - \eta^2$ or the percentage of variance in the discriminant scores that is *not* explained by group membership, was .69. Thus, only 31% of the variance in discriminant scores was due to between-group differences.

Overall, 68% of the subjects were correctly classified into these 3 career groups by the DA (see Figure 16.23). This is actually not a very high rate of correct classification. Given that such a large proportion of the sample chose business careers, we could have achieved a 60% correct classification rate (98 out of 164) simply by classifying all the cases into the business group. When group sizes are unequal, the correct rate of classification that would be expected by chance depends on both the number of groups and the group sizes, as discussed in Tabachnick and Fidell (2007).

From Figure 16.23, it is evident that the discriminant function did a good job of classifying actual business career group subjects (91 out of 98 were correctly predicted to be in the business career group) but did a very poor job of predicting group membership for the other 2 groups. In fact, most persons in Groups 2 and 3 were classified incorrectly as members of other groups. The univariate *F*s and standardized discriminant function coefficients for the six discriminating variables are shown in Table 16.2.

Table 16.2 ♦ Summary of Information About Individual Predictor Variables for Discriminant Analysis: Differences Among Three Career Choice Groups (Business, Medicine, Teaching) on Six Achievement and Aptitude Tests

	Standardized Discriminant Function Coefficients[a] for			
Variable	*Function 1*	*Function 2*	*Univariate F*	*p Values[b] for Univariate F*
English	.42	.28	0.22	ns
Reading	.02	.53	1.20	ns
Mechanic	.44	−.94	0.88	ns
Abstract	−.12	−.03	1.03	ns
Math	−.79	.24	10.53	<.001
Office	.75	.10	23.32	<.001

a. Note that some authors present structure coefficients to show how each variable is correlated with each discriminant function (in addition to or instead of the standardized discriminant function coefficients).

b. To correct for the inflated risk of Type I error, Bonferroni-corrected alpha levels were used to assess the significance of each univariate *F*.

To achieve an overall risk of Type I error of .05 for the entire set of six *F* tests, each individual *F* ratio was judged significant only if its *p* value was less than .05/6, that is, for $p < .008$ (using Bonferroni corrected per-comparison alpha levels).

If an arbitrary cutoff of .50 is used to decide which of the standardized discriminant coefficients are large, then only two of the predictor variables have large coefficients for standardized Discriminant function 1: math and office. Because the second discriminant function alone was not statistically significant, coefficients on Discriminant function 2 were not interpreted.

Table 16.2 reports summary information about individual predictor variables, evaluated using both univariate methods (individual ANOVAs) and multivariate methods (DA).

The nature of the pattern suggested by the results in Table 16.2 (along with the pattern of univariate group means that appeared in Figure 16.11) was as follows. High scores on office skills and low scores on math corresponded to high scores on the first discriminant function. Office skills and math were also the only individual variables that had significant univariate *F*s. The nature of these mean differences among groups was as follows: Members of Group 1, business careers, had higher mean scores on the office skills test ($M_1 = 28.83$) than members of Group 2, medicine ($M_2 = 18.37$) or Group 3, teaching ($M_3 = 20.62$). The business career group had much lower math scores ($M_1 = 20.05$) than the medicine group ($M_2 = 26.37$) or the teaching group ($M_3 = 25.59$). Thus, high scores on office skills and low scores on math were associated with the choice of a business career.

(Continued)

(Continued)

Finally, the territorial map (Figure 16.22) indicated that the first discriminant function separates Group 1 (business) from Groups 2 and 3 (medicine, teaching). People with high scores on the first discriminant function (and, therefore, high scores on office and low scores on math) were predicted to be in the office group. The second discriminant function tended to separate the medicine and teaching groups, but these 2 groups were not well discriminated by the second function. This is evident because the group centroids for Groups 2 and 3 on Discriminant function 2 were quite close together and because the proportion of correct classifications of persons into Groups 2 and 3 was quite low.

A discriminant function that primarily represented information from the office and math scores predicted membership in the business career group at levels slightly better than chance. However, none of the variables was effective in discriminating between Group 2, medicine, and Group 3, teaching; the DA did not make accurate group membership predictions for members of these 2 groups. A reanalysis of this data could correct for this problem in a number of different ways; for example, it might be possible to identify some other variable that does discriminate between people who choose medicine versus teaching and social work as careers.

16.11 ♦ One-Way ANOVA on Scores on Discriminant Functions

One final analysis is included to help students understand what a DA does. This is not information that is typically reported in a journal article along with other DA results. One of the commands included in the preceding DA was the Save command; this saved the computed discriminant function scores for each person for both D_1 and D_2 as new variables in the SPSS worksheet. A second new variable was computed for each subject in this dataset; this was a simple unit-weighted composite of scores on the two variables that, according to the DA results, were the most useful variables for the prediction of group membership:

$$\text{Unit-weighted composite} = U = (+1 \times \text{Office}) + (-1 \times \text{Math}). \qquad (16.17)$$

The simple unit-weighted composite computed using Equation 16.17 is just the difference between each person's raw scores on office and math. Figure 16.24 shows how the scores on this new variable named "unitweightcomposite" or U were calculated using the SPSS Compute Variable procedure.

One-way ANOVA was performed to evaluate how the 3 career groups differed in mean scores on the following variables: D_1, the saved score on Discriminant function 1 from the preceding DA, and the unit-weighted composite or U, computed from Equation 16.17 above. The results of this one-way ANOVA appear in Figure 16.25.

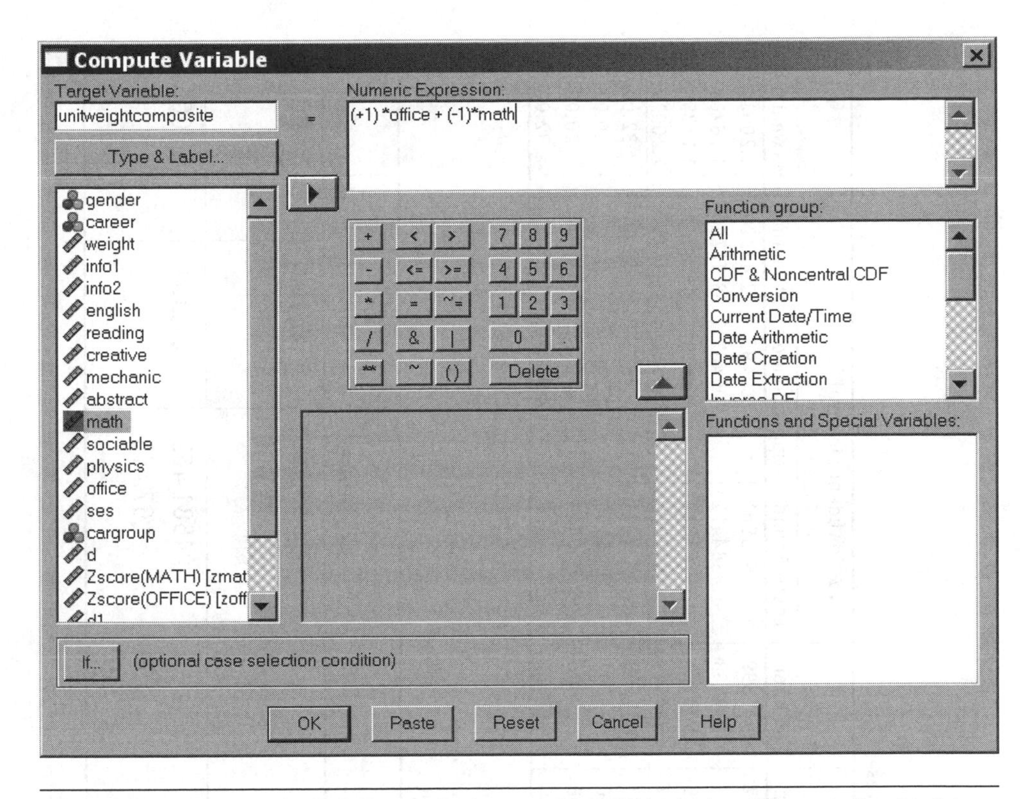

Figure 16.24 ♦ Creation of a Unit-Weighted Composite Variable: $(+1) \times$ Office $+ (-1) \times$ Math in the SPSS Compute Variable Dialog Window

For D_1, there was a statistically significant difference in means across groups; $F(2, 161) = 36.10$, $p < .001$. The corresponding value of eta squared (proportion of variance that is associated with between-group differences) was estimated as .309 or 31%. Refer back to Figure 16.15 to the table headed "Wilks's Lambda" and examine the value of Λ for the overall model; for the set of both discriminant functions D_1 and D_2, $\Lambda = .685$. (However, the second discriminant function did not make a significant contribution to discrimination between groups; therefore, most of the explained variance is associated with D_1 in this example.) Recall that $1 - \Lambda$ is equivalent to eta squared; in this preceding DA (see Figure 16.15), $1 - \Lambda = 1 - .685 = .315$. In the DA results, about 31.5% of the variance in the scores on the two discriminant functions considered jointly was not associated with group membership. In this particular example, the proportion of variance due to between-group differences on the D_1 variable in the one-way ANOVA (.3096) was slightly smaller than the value of $1 - \Lambda$ (.315) for the overall DA. The proportion of explained variance or eta squared in the one-way ANOVA for differences in D_1 across groups would have been closer to the value of $1 - \Lambda$ if the second discriminant function had an even weaker predictive power. The eta squared for the ANOVA on D_1 scores would have been much smaller than $1 - \Lambda$ if the second discriminant function had a much greater predictive power; this is the case because the ANOVA, unlike the DA, only took scores on D_1 (and not scores on D_2) into account.

Descriptives

		N	Mean	Std. Deviation	Std. Error	95% Confidence Interval for Mean		Minimum	Maximum
						Lower Bound	Upper Bound		
Discriminant Scores from Function 1 for Analysis 1	business	98	.5405178	.93837212	.0947899	.3523861	.7286495	-2.42326	2.20927
	medicine	27	-.9544909	1.223373	.2354383	-1.4384412	-.4705407	-3.37797	1.16265
	teach_social	39	-.6974227	.98240739	.1573111	-1.0158824	-.3789631	-2.70931	1.11760
	Total	164	.0000000	1.196107	.0934003	-.1844305	.1844305	-3.37797	2.20927
unitweightcomposite	business	98	8.7857	12.03367	1.21558	6.3731	11.1983	-31.00	32.00
	medicine	28	-8.3929	15.84260	2.99397	-14.5360	-2.2497	-34.00	20.00
	teach_social	40	-4.5250	12.88208	2.03684	-8.6449	-.4051	-38.00	20.00
	Total	166	2.6807	14.86466	1.15372	.4028	4.9587	-38.00	32.00

ANOVA

		Sum of Squares	df	Mean Square	F	Sig.
Discriminant Scores from Function 1 for Analysis 1	Between Groups	72.200	2	36.100	36.100	.000
	Within Groups	161.000	161	1.000		
	Total	233.200	163			
unitweightcomposite	Between Groups	9162.925	2	4581.462	27.359	.000
	Within Groups	27295.15	163	167.455		
	Total	36458.08	165			

Figure 16.25 ♦ One-Way ANOVA to Compare Scores Across 3 Career Groups

NOTE: Two different dependent variables are reported here. For the first analysis, the dependent variable was saved Discriminant scores from Function 1 (D_1). For the second analysis, the dependent variable was the unit-weighted composite computed as follows: U = office – math scores.

Now, examine the ANOVA results in Figure 16.25 for the dependent variable U—that is, the unweighted linear composite of scores (just the difference between office and math raw scores). For this ANOVA, $\eta^2 = .251$. About 25% of the variance in a simple unit-weighted linear composite of scores on just two of the variables, office and math, was associated with membership in the 3 career groups. This was less than the 31.5% of variance on D_1 and D_2 combined that was associated with group membership in the DA. This empirical example illustrates (but does not prove) that an optimally weighted linear composite of scores (derived from DA) provides a better prediction of group membership than a simple unit-weighted composite of the two "best" predictor variables.

In other words, the DA obtained the *optimal* weighted linear combination of scores on the X variables. For these data, the score on D_1 that had the largest possible differences between groups (and the minimum possible within-group variance) corresponded to $D_1 = +.42z_{\text{English}} + .01z_{\text{reading}} + .44z_{\text{mechanic}} - .12z_{\text{abstract}} - .79z_{\text{math}} + .75z_{\text{office}}$. (These standardized discriminant function coefficients were obtained from Figure 16.16.) These coefficients created the D_1 function that had the minimum possible Λ and the largest possible η^2 and F.

However, it is worth noting that we are able to show significant (although much smaller) differences between the groups by examining scores on the unit-weighted composite, which is obtained by assigning a weight of +1 to any variable with a large positive discriminant coefficient, a weight of −1 to any variable with a large negative discriminant function coefficient, and a weight of 0 to any variable with a small discriminant function coefficient. Although the between-group differences for this unit-weighted composite U (which in this example was $+1 \times \text{office} - 1 \times \text{math}$) were smaller than the between-group differences for the discriminant function D_1, in this case, there was still a significant difference between groups on this simpler unit-weighted composite.

This last analysis was presented for two reasons. First of all, it may help us understand what the DA actually does through a more concrete example using a simpler, more familiar analysis (one-way ANOVA). Second, there may be research situations in which it is important to present complex results in a simpler way, and in those situations, the DA may serve as a guide to the construction of simpler unit composite variables that may be adequate, in some situations, to describe group differences. Note that the discriminant function coefficients are optimized to predict group membership only for the sample(s) used to estimate these coefficients, and they may work less well when applied to new data.

16.12 ♦ Summary

In DA, scores on several quantitative variables are used to predict group membership; the researcher is usually interested in the amount of weight given to each predictor variable. When assumptions that are required for DA (such as the assumption of homogeneity of the variance/covariance matrix across groups) are violated, other analyses that involve the prediction of group membership from scores on several quantitative predictors (such as binary logit regression; see Chapter 21) may be preferred. Logit regression also provides information about relative risk or comparative probabilities of group membership that can be very useful in clinical prediction situations.

As in MR, it is possible to do a stepwise entry of predictor variables into a DA based on the predictive usefulness of variables. This is not recommended, particularly in situations

Table 16.3 ♦ Information Provided by Discriminant Analysis in Comparison With ANOVA and Multiple Regression

Type of Information	*ANOVA*	*Multiple Regression*	*Discriminant Analysis*
Overall effect size for entire model	η^2: Proportion of variance in *Y* outcome scores that is predictable from group membership	Multiple *R* and multiple R^2: Proportion of variance in *Y* outcome variable predictable from the *X* predictor variables	Wilks's Λ: Proportion of variance in discriminant function scores that is not associated with or predictive of group membership (equivalent to $1 - \eta^2$)
Overall significance test for entire model	$F = MS_{between}/MS_{within}$	$F = MS_{regression}/MS_{residual}$	Wilks's Λ is usually converted to an approximate χ^2 or *F*
Strength and nature of predictive contribution of each individual X_i Predictor variable	Not applicable	For each X_i predictor, examine the values of the *b* and β coefficients (and sr^2)	For each X_i predictor, examine the value of the raw score or standardized score discriminant function coefficients, or the structure coefficients
Statistical significance of each X_i individual predictor variable	Not applicable	For each X_i predictor, examine the value of the *t* ratio associated with its *b* raw score slope coefficient	No significance test for contribution of each X_i predictor provided by SPSS

where the goal of the analysis is to obtain a theoretically meaningful description of the nature of group differences.

It may be helpful to look back and see how the information that can be obtained from DA compares with the information that can be obtained from analyses discussed earlier in the textbook (ANOVA and MR). To highlight the similarities, Table 16.3 outlines the types of information that each of these analyses provide. In Chapter 17, we will see that MANOVA is another analysis that is even more closely related to DA. MANOVA is essentially DA in reverse; that is, in MANOVA, researchers usually speak of predicting a set of scores on a list of outcome variables for different groups (e.g., treatment groups in an experiment). In DA, researchers usually speak of predicting group membership from scores on a set of predictor variables. However, both DA and MANOVA fundamentally deal with the same data analysis problem: examining how scores on a categorical variable are related to or predictable from scores on several quantitative variables. DA, however, is generally limited to the consideration of just one categorical outcome variable, while MANOVA may include several categorical predictor variables or factors. When we discuss MANOVA in Chapter 17, we shall see that, sometimes, a DA provides a useful follow-up analysis that provides more detailed information about the nature of group differences found in a more complex MANOVA.

Appendix: Eigenvalue/Eigenvector Problem

Discriminant analysis requires the use of matrix algebra to solve the eigenvalue/eigenvector problem. In a sense, when we solve for eigenvectors and eigenvalues for a set of variables, we are "repackaging" the information about the variances and covariances of the variables. We do this by forming one (or more) weighted linear composites of the variables, such that these weighted linear composites are orthogonal to each other; in DA, each weighted linear composite is maximally predictive of group membership.

In DA, we begin with the following two matrices:

B = the between-groups sum of cross products matrix for the set of X predictors. Computations for this matrix were not described in this chapter, and can be found in Stevens (2002, chap. 5).

W = the pooled within-groups sum of cross products matrix for the set of X predictors. This matrix **W** is obtained by summing $\mathbf{SCP}_1 + \mathbf{SCP}_2 + \ldots + \mathbf{SCP}_k$, using the notation from an earlier section of this chapter.

The matrix algebra equivalent to division is multiplication by the inverse of a matrix, so $\mathbf{BW}^{-1}$ is the ratio of the between- versus within-groups sum of cross products matrices. Our goal is to "repackage" the information about the between- versus within-groups variances and covariances of X variables in the form of one or several discriminant functions. This is accomplished by finding the eigenvectors and eigenvalues for $\mathbf{BW}^{-1}$ and rescaling the eigenvectors so that they can be used as coefficients to construct the standardized discriminant function scores; the corresponding eigenvalue for each discriminant function, λ, provides information about the ratio $SS_{\text{between}}/SS_{\text{within}}$ for each discriminant function.

In DA, the matrix product $\mathbf{BW}^{-1}$ is the matrix for which we want to solve the eigenproblem. Each eigenvector will be a list of p values (one for each dependent variable). When properly scaled, the eigenvector elements will be interpretable as weights or coefficients for these variables—discriminant function coefficients, for instance. Each eigenvector has a corresponding constant associated with it called an eigenvalue. The eigenvalue, when properly scaled, will be an estimate of the amount of variance that is explained by the particular weighted linear combination of the dependent variables that you get by using the (rescaled) eigenvector elements to weight the dependent variable. When we extract solutions, we first extract the one solution with the largest possible eigenvalue; then, we extract the solution that has the second largest eigenvalue, which is orthogonal to the first solution, and so on, until all the possible solutions have been extracted.

Here is the matrix algebra for the eigenproblem in DA. Let $\mathbf{BW}^{-1}$ be the matrix for which you want to find one or more eigenvectors, each with a corresponding eigenvalue. We need to find one or more solutions to the equation

$$(\mathbf{BW}^{-1} - \lambda\mathbf{I})\mathbf{V} = 0.$$

Note that this equation implies that $\mathbf{BW}^{-1}\ \mathbf{V} = \lambda\mathbf{I} \times \mathbf{V}$, where λ corresponds to the one (or more) eigenvalues and **V** corresponds to the matrix of eigenvectors. This equation is the eigenproblem. Tabachnick and Fidell (2007, pp. 930–933) provide a worked numerical example for a 2×2 matrix; however, the matrix denoted **D** in their example is replaced by the matrix $\mathbf{BW}^{-1}$ as shown above in computation of a DA.

When we solve the eigenproblem, more than one solution may be possible; each solution consists of one eigenvector and its corresponding eigenvalue. In general, if there are k groups and p discriminating variables, the number of different solutions is the minimum of $(k - 1)$ and p. If there is perfect multicollinearity among the p predictor variables, then we must replace p, the number of predictors, by the rank of the matrix that represents the scores on the p predictor variables; the rank of the matrix tells us, after taking multicollinearity among predictors into account, how many distinguishable independent variables remain. Thus, for a design with 4 groups and six discriminating variables, there are potentially three solutions, each one consisting of an eigenvector (the elements of which are rescaled to obtain the d coefficients for a discriminant function) and its corresponding eigenvalue (λ), which can be used to find a canonical correlation that tells us how strongly scores on the discriminant function are associated with group membership.

For more extensive treatment of the eigenproblem and its multivariate applications, see Tatsuoka (1988, pp. 135–162).

Note

1. Two additional equations that are involved in discriminant function analysis are the following. These are used primarily in situations where the researcher's main interest is in the correct classification of cases (rather than a description of the nature of group differences or theory testing about possible "causal" variables).

The equation for Discriminant function 1 can be given in terms of z scores on the X predictor variables, as in Equation 16.4: $D_1 = d_{11} z_1 + d_{12} z_2 + \ldots + d_{1p} z_p$.

This equation is analogous to the standard score version of an MR prediction equation; for example, $z'_Y = \beta_1 z_{X_1} + \beta_2 z_{X_2} + \ldots + \beta_k z_{X_k}$.

We can also obtain a raw score version of the equation for each discriminant function:

$$D_1 = b_0 + b_1 X_1 + b_2 X_2 + \ldots + b_k X_k.$$

This is analogous to the raw score version of the prediction equation in MR (except that the coefficients in the DA are scaled differently than in the MR analysis):

$$Y' = b_0 + b_1 X_1 + b_2 X_2 + \ldots + b_k X_k.$$

In the SPSS printout, the coefficients for these raw score versions of the discriminant functions are found in the table headed "Canonical Discriminant Function Coefficients"; this is an optional statistic.

Note that the number of discriminant functions that can be obtained (whether they are stated in terms of z scores or raw X scores) is limited to the minimum of $(k - 1)$, that is, the number of groups on the dependent variable minus 1, and p, the number of X predictor variables that are not completely redundant with each other.

Fisher suggested another approach to group classification that is easier to apply in situations with multiple groups. In this approach, for each of the k groups, we compute a score for each participant on a Fisher classification coefficient that corresponds to each group (i.e., a function for Group 1, Group 2, . . . , Group k). The Fisher discriminant function, C_j, where j = Group number 1, 2, . . . , k, can be denoted as follows (Tabachnick & Fidell, 2007):

$$C_j = c_{j0} + c_{j1} X_1 + c_{j2} X_2 + \ldots + c_{jp} X_p.$$

A score for each case is computed for each group, and each participant is predicted to be a member of the group for which that participant has the highest C_j score. Values of C_j can be used to derive other useful information, such as the theoretical probability that the individual participant is a member of each group given his or her scores on the C_j functions. Adjustments that take group size into account, or that otherwise increase the likelihood of classification of cases into any individual group, can easily be made by modifying the size of the c_{j0} term (which is analogous to the b_0 concept or intercept in a regression equation). The values of coefficients for the C_j classification functions, often called Fisher's linear classification coefficients, appear in the SPSS table that has the heading "Classification Function Coefficients" in Figure 16.19. These coefficients are not generally interpreted or used to describe the nature of group differences, but they provide a more convenient method for classification of individual cases into groups when the number of groups, k, is greater than 2.

Comprehension Questions

1. Describe two research situations where DA can be used: one situation where the primary interest is in group classification and another situation where the primary interest is in the description of the nature of group differences.
2. What assumptions about the data should be satisfied for a DA? What types of data screening should be performed to assess whether assumptions are violated?
3. In general, if p is the number of predictor variables and k is the number of groups for the outcome variable in a DA, how many different discriminant functions can be obtained to differentiate among groups? (This question assumes that the X predictor variables have a determinant that is nonzero; that is, no individual X_i predictor variable can be perfectly predicted from scores on one or more other X predictor variables.)
4. Show how the canonical r_c for each discriminant function can be calculated from the eigenvalue (λ) for each function, and verbally interpret these canonical correlations. How is a squared canonical correlation related to an η^2?
5. From the territorial map, how can we describe the nature of group differences on one or more discriminant functions?
6. What information can a data analyst use to evaluate how much information about group membership is contributed by each individual variable? Be sure to distinguish between procedures that examine variables in isolation and procedures that take correlations among predictor variables into account.

CHAPTER 17

Multivariate Analysis of Variance

17.1 ♦ Research Situations and Research Questions

Multivariate analysis of variance (MANOVA) is a generalization of univariate analysis of variance (ANOVA). Recall that in a univariate one-way ANOVA, mean scores on just *one* quantitative Y were compared across groups; for example, in a one-way univariate ANOVA, the null hypothesis of interest was as follows:

$$H_0 : \mu_1 = \mu_2 = \cdots = \mu_k. \tag{17.1}$$

In Equation 17.1, each μ_i term corresponds to the population mean for the score on a single Y outcome variable in one of the k groups.

In a one-way MANOVA, mean scores on *multiple* quantitative outcome variables are compared for participants across 2 or more groups. When a MANOVA has only one factor or one categorical predictor variable, we will call it a one-way MANOVA with k levels. In other words, MANOVA is an extension of univariate ANOVA (ANOVA with just one outcome variable) to situations where there are multiple quantitative outcome variables; we will denote the number of quantitative outcome variables in a MANOVA as p. The set of p outcome variables in MANOVA can be represented as a list or vector of Y outcome variables:

$$H_0 = \begin{bmatrix} \mu_{11} \\ \mu_{12} \\ \vdots \\ \mu_{1p} \end{bmatrix} = \begin{bmatrix} \mu_{21} \\ \mu_{22} \\ \vdots \\ \mu_{2p} \end{bmatrix} = \cdots = \begin{bmatrix} \mu_{k1} \\ \mu_{k2} \\ \vdots \\ \mu_{kp} \end{bmatrix}. \tag{17.2}$$

In words, the null hypothesis in Equation 17.2 corresponds to the assumption that when the scores on all p of the Y outcome variables are considered jointly as a set, taking intercorrelations among the Y variables into account, the means for this set of p outcome variables do not differ across any of the populations that correspond to groups in the study. This is a multivariate extension of the univariate null hypothesis that appeared in Equation 17.1.

This multivariate null hypothesis can be written more compactly by using matrix algebra notation. We can use $\boldsymbol{\mu}_i$ to represent the vector of means for the Y outcome variables Y_1 through Y_p in each population ($i = 1, 2, \ldots, k$), which corresponds to each of the k sample groups in the study: $\boldsymbol{\mu}_i = [\mu_{i1}, \mu_{i2}, \mu_{i3}, \ldots, \mu_{ip}]$. When set in bold font, $\boldsymbol{\mu}_i$ corresponds to a vector or list of means on p outcome variables; when shown in regular font, μ_i corresponds to a population mean on a single Y outcome variable. The null hypothesis for a one-way MANOVA can be written out in full, using $\boldsymbol{\mu}_i$ to stand for the list of population means for each of the p outcome variables in each of the groups as follows:

$$H_0 : \boldsymbol{\mu}_1 = \boldsymbol{\mu}_2 = \cdots = \boldsymbol{\mu}_k. \tag{17.3}$$

In Equation 17.3, each $\boldsymbol{\mu}_i$ term represents the vector or set of means on p different Y outcome variables across k groups. In words, the null hypothesis for a one-way MANOVA is that the population means for the entire *set* of variables $Y_1, Y_2, \ldots, Y_p$ (considered jointly and taking their intercorrelations into account) do not differ across Groups $1, 2, \ldots, k$. If we obtain an omnibus F large enough to reject this null hypothesis, it usually implies at least one significant inequality between groups on at least one Y outcome variable. If the omnibus test is statistically significant, follow-up analyses may be performed to assess which contrasts between groups are significant and which of the individual Y variables (or combinations of Y variables) differ significantly between groups. However, data analysts need to be aware that even when the omnibus or multivariate test for MANOVA is statistically significant, there is an inflated risk of Type I error when numerous univariate follow up analyses are performed.

17.2 ♦ Introduction of the Initial Research Example: A One-Way MANOVA

MANOVA can be used to compare the means of multiple groups in nonexperimental research situations that evaluate differences in patterns of means on several Y outcome variables for naturally occurring groups. The empirical example presented in Chapter 16 as an illustration for discriminant analysis (DA) involved groups that corresponded to three different future career choices (complete data appear in Table 16.1). For the cargroup factor, these career choices were as follows: 1 = business career, 2 = medical career, and 3 = teaching or social work career. Each high school senior had scores on six aptitude and achievement tests. The goal of the DA in Chapter 16 was to see whether membership in one of these 3 future career choice groups could be predicted from each participant's set of scores on the six tests.

This same set of data can be examined using a MANOVA. The MANOVA examines whether the set of means on the six test scores differs significantly across any of the 3 career choice groups. Later in this chapter, a one-way MANOVA is reported for the data that were analyzed using DA in Chapter 16; this example illustrates that MANOVA and DA are essentially equivalent for one-way designs.

MANOVA can also be used to test the significance of differences of means in experiments where groups are formed by random assignment of participants to different treatment conditions. Consider the following hypothetical experiment as an example of a one-way MANOVA design. A researcher wants to assess how various types of stressors differ in their impact on physiological responses. Each participant is randomly assigned to one of the following conditions: Group 1 = No stress/baseline, Group 2 = Mental stress, Group 3 = Cold pressor, and Group 4 = Stressful social role play. The name of the factor in this one-way MANOVA is stress, and there are $k = 4$ levels of this factor. The following outcome measures are obtained: Y_1, self-reported anxiety; Y_2, heart rate; Y_3, systolic blood pressure; Y_4, diastolic blood pressure; and Y_5, electrodermal activity (palmar sweating). The research question is whether these 4 groups differ in the pattern of mean responses on these five outcome measures. For example, it is possible that the three stress inductions (Groups 2 through 4) all had higher scores on self-reported anxiety, Y_1, than the baseline condition (Group 1) and that Group 3, the pain induction by cold pressor, had higher systolic blood pressure and a higher heart rate than the other two stress inductions.

17.3 ♦ Why Include Multiple Outcome Measures?

Let us pause to consider two questions. First, what do researchers gain by including multiple outcome measures in a study instead of just one single outcome measure? And second, what are the advantages of using MANOVA to evaluate the nature of group differences on multiple outcome variables instead of doing a univariate ANOVA separately for each of the individual Y outcome variables in isolation?

There are several reasons why it may be advantageous to include multiple outcome variables in a study that makes comparisons across groups. For example, a stress intervention might cause increases in physiological responses such as blood pressure and heart rate, emotional responses such as self-reported anxiety, and observable behaviors such as grimacing and fidgeting. These measures might possibly be interpreted as multiple indicators of the same underlying construct (anxiety) using different measurement methods. In the stress intervention study, measuring anxiety using different types of methods (physiological, self-report, and observation of behavior) provides us with different kinds of information about a single outcome such as anxiety. If we use only a self-report measure of anxiety, it is possible that the outcome of the study will be artifactually influenced by some of the known weaknesses of self-report (such as social desirability response bias). When we include other types of measures that are not as vulnerable to social desirability response bias, we may be able show that conclusions about the impact of stress on anxiety are generalizable to anxiety assessments that are made using methods other than self-report.

It is also possible that an intervention has an impact on two or more conceptually distinct outcomes; for example, some types of stress interventions might arouse hostility in addition to or instead of anxiety. Whether the Y variables represent multiple measures of one single construct or measures of conceptually distinct outcomes, the inclusion of multiple outcome measures in a study potentially provides richer and more detailed information about the overall response pattern.

When we include more than one quantitative Y outcome variable in a study that involves comparisons of means across groups, one way that we might assess differences across groups would be to conduct a separate one-way ANOVA for each individual outcome measure. For example, in the hypothetical stress experiment described earlier, we could do a one-way ANOVA to see if mean self-reported anxiety differs across the four stress intervention conditions and a separate one-way ANOVA to assess whether mean heart rate differs across the four stress intervention conditions. However, running a series of p one-way ANOVAs (a univariate ANOVA for each of the Y outcome variables) raises a series of problems. The first issue involves the inflated risk of Type I error that arises when multiple significance tests are performed. For example, if a researcher measures 10 different outcome variables in an experiment and then performs 10 separate ANOVAs (one for each outcome measure), the probability of making at least one Type I error in the set of 10 tests is likely to be substantially higher than the conventional $\alpha = .05$ level that is often used as a criterion for significance, unless steps are taken to try to limit the risk of Type I error. (For example, Bonferroni-corrected per-comparison alpha levels can be used to assess statistical significance for the set of several univariate ANOVAs.) In contrast, if we report just one omnibus test for the entire set of 10 outcome variables and if the assumptions for MANOVA are reasonably well satisfied, then the risk of Type I error for the single omnibus test should not be inflated.

A second issue concerns the linear intercorrelations among the Y outcome variables. A series of separate univariate ANOVA tests do not take intercorrelations among the Y outcome variables into account, and therefore, the univariate tests do not make it clear whether each of the p outcome variables represents a conceptually distinct and independent outcome or whether the outcome variables are intercorrelated in a way that suggests that they may represent multiple measures of just one or two conceptually distinct outcomes. In the stress experiment mentioned earlier, for example, it seems likely that the measures would be positively intercorrelated and that they might all tap the same response (anxiety). When the outcome measures actually represent different ways of assessing the same underlying construct, reporting a separate univariate ANOVA for each variable may mislead us to think that the intervention had an impact on several unrelated responses, when, in fact, the response may be better described as an "anxiety syndrome" that includes elevated heart rate, increased self-report anxiety, and fidgeting behavior. Univariate tests for each individual Y outcome variable also ignore the possibility that statistically controlling for one of the Y variables may change the apparent association between scores on other Y variables and group membership (as discussed in Chapter 10).

A third reason why univariate ANOVAs may not be an adequate analysis is that there are situations in which groups may not differ significantly on any one individual outcome variable but their outcomes are distinguishable when the outcomes on two or more variables are considered jointly. For an example of a hypothetical research situation where this occurs, refer back to Figures 16.3 and 16.4; in this hypothetical example, skulls could be accurately classified using information about two different skull measurements, even though each individual measurement showed substantial overlap between the ape and human skull groups.

For all the preceding reasons, the results of a MANOVA may be more informative than the results of a series of univariate ANOVAs. In an ideal situation, running a single MANOVA yields an omnibus test of the overall null hypothesis in Equation 17.2 that does not have an inflated risk of Type I error (however, violations of assumptions for MANOVA can make the obtained p value a poor indication of the true level of risk of Type I error).

A MANOVA may be more powerful than a series of univariate ANOVAs; that is, sometimes a MANOVA can detect a significant difference in response pattern on several variables across groups even when none of the individual *Y* variables differ significantly across groups. However, MANOVA is not always more powerful than a series of univariate ANOVAs. MANOVA provides a summary of group differences that takes intercorrelations among the outcome variables into account, and this may provide a clearer understanding of the nature of group differences—that is, how many dimensions or weighted linear combinations of variables show significant differences across groups. For example, if the set of outcome variables in the stress study included measures of hostility as well as anxiety, the results of a MANOVA might suggest that one stress intervention elicits a pattern of response that is more hostile, while another stress intervention elicits a response pattern that is more anxious, and yet another stress intervention elicits both hostility and anxiety. It is sometimes useful to divide the list of *Y* outcome variables into different sets and then perform a separate MANOVA for each set of *Y* outcome variables to detect group differences (Stevens, 1992).

17.4 ♦ Equivalence of MANOVA and DA

A one-way MANOVA can be thought of as a DA in reverse. Each of these analyses assesses the association between one categorical or group membership variable and a set of several quantitative variables. DA is usually limited to the evaluation of one categorical variable or factor, while MANOVA analyses can be generalized to include multiple categorical variables or factors—for example, a factorial design. In MANOVA, we tend to speak of the categorical or group membership variable (or factor) as the predictor and the quantitative variables as outcomes; thus, in MANOVA, the outcome variables are typically denoted by $Y_1, Y_2, \ldots, Y_p$. In DA, we tend to speak of prediction of group membership from scores on a set of quantitative predictors; thus, in the discussion of DA in Chapter 16, the quantitative variables were denoted by $X_1, X_2, \ldots, X_p$. Apart from this difference in language and notation, however, DA and one-way MANOVA are essentially equivalent.

Let us reconsider the dataset introduced in Chapter 16 to illustrate DA. Six quantitative test scores (scores on English, reading, mechanical reasoning, abstract reasoning, math, and office skills) were used to predict future career choice; this categorical variable, cargroup, had three levels: 1 = Business career, 2 = Medical career, and 3 = Teaching or social work career. In Chapter 16, a DA was used to evaluate how well membership in these 3 career choice groups could be predicted from the set of scores on these six tests. We will reexamine this problem as a one-way MANOVA in this chapter; we will use MANOVA to ask whether there are any statistically significant differences across the 3 career groups for a set of one or more of these test scores. DA and MANOVA of these data yield identical results when we look at the omnibus tests that test the overall null hypothesis in Equation 17.2. However, the two analyses provide different kinds of supplemental information or follow-up analyses that can be used to understand the nature of group differences.

In MANOVA, the researcher tends to be primarily interested in *F* tests to assess which groups differ significantly; in DA, the researcher may be more interested in the assessment of the relative predictive usefulness of individual discriminating variables. However, the underlying computations for MANOVA and DA are identical, and the information that they provide is complementary (Bray & Maxwell, 1985). Thus, it is common to include some MANOVA results (such as omnibus *F* tests of group differences) even when a study is framed as a DA and to include DA results as part of a follow-up to a significant MANOVA.

The MANOVA procedures available in SPSS (through the SPSS general linear model, GLM, procedure) provide some tests that cannot be easily obtained from the DA procedure; for example, planned contrasts and/or post hoc tests can be requested as part of MANOVA to evaluate which specific pairs of groups differ significantly. On the other hand, DA provides some information that is not readily available in most MANOVA programs. For example, DA provides coefficients for discriminant functions that provide helpful information about the nature of the association between scores on each quantitative variable and group membership when intercorrelations among the entire set of quantitative variables are taken into account; DA also provides information about the accuracy of classification of individual cases into groups. It is fairly common for a researcher who finds a significant difference among groups using MANOVA to run DA on the same data to obtain more detailed information about the way in which individual quantitative variables differ across groups.

17.5 ♦ The General Linear Model

In SPSS, MANOVA is performed using GLM. In its most general form, SPSS GLM can handle any combination of multiple outcome measures, multiple factors, repeated measures, and multiple covariates. What is the GLM? In GLM, there can be one or multiple predictor variables, and these predictor variables can be either categorical or quantitative. There may also be one or several quantitative outcome variables and one or several quantitative covariates. In practice, the GLM procedure in SPSS permits us to include both multiple predictor and multiple outcome variables in an analysis, although there are some limitations to the ways in which we can combine categorical and quantitative variables. *Linear* indicates that for all pairs of quantitative variables, the nature of the relationship between variables is linear and, thus, can be handled by using correlation and linear regression methods. Many of the statistical techniques that are presented in an introductory statistics course are special cases of GLM. For example, an independent samples *t* test is GLM with one predictor variable that is dichotomous and one outcome variable that is quantitative. Pearson *r* is GLM with one quantitative predictor and one quantitative outcome variable. Multiple regression is GLM with one quantitative outcome variable and multiple predictor variables, which may include quantitative and dummy or dichotomous variables. All the GLM analyses encountered so far involve similar assumptions about data structure; for example, scores on the outcome variable must be independent, scores on quantitative variables should be at least approximately normally distributed, relations between pairs of quantitative variables should be linear, there should be no correlation between the error or residual from an analysis and the predicted score generated by the analysis, and so forth. Some specific forms of GLM involve additional assumptions; for example, ANCOVA assumes no treatment by covariate interaction, and DA assumes homogeneity of variance/covariance matrices across groups. The SPSS GLM procedure allows the user to set up an analysis that combines one or more quantitative dependent variables, one or more factors or categorical predictors, and possibly one or more quantitative covariates.

The questions that can be addressed in a MANOVA include some of the same questions described in Chapter 16 for DA. Using MANOVA and DA together, a researcher can assess which contrasts between groups are significant, which variables are the most useful in differentiating among groups, and the pattern of group means on each discriminant function. Issues involving accuracy of prediction of group membership, which were a major focus in the presentation of DA in Chapter 16, are usually not as important in MANOVA; the SPSS GLM procedure does not provide information about accuracy of classification into groups.

In addition, MANOVA can address other questions. Do group differences on mean vectors remain statistically significant when one or several covariates are statistically controlled? Do different treatment groups show different patterns or responses across time in repeated measures studies? Are there interaction effects such that the pattern of cell means on the discriminant functions is much different from what would be predicted based on the additive main effects of the factors? Do the discriminant functions that differentiate among levels on the B factor for the A_1 group have large coefficients for a different subset of the quantitative outcome variables from the subset for the discriminant functions that differentiate among levels of the B factor within the A_2 group?

Like DA and multiple regression, MANOVA takes intercorrelations among variables into account when assessing group differences. MANOVA is appropriate when the multiple outcome measures are moderately intercorrelated. If correlations among the outcome variables are extremely high (e.g., greater than .8 in absolute value), it is possible that the variables are all measures of the same construct; if this is the case, it may be simpler and more appropriate to combine the scores on the variables into a single summary index of that construct. In a typical research situation where correlations among the outcome measures are moderate in size, some outcome variables may appear to be less important when other outcome variables are statistically controlled; or the nature of the relationship between a quantitative variable and group membership may change when other variables are statistically controlled. The magnitude and signs of the correlations among outcome measures influence the statistical power of MANOVA and the interpretability of results in complex ways (see Bray & Maxwell, 1985, for further discussion).

17.6 ♦ Assumptions and Data Screening

The assumptions for MANOVA are essentially the same as those for DA (in Chapter 16) and for many other GLM procedures. We will refer to the multiple outcome measures as $Y_1, Y_2, \ldots, Y_p$.

1. Observations on the Y outcome variables should be collected in such a way that the scores of different participants on any one Y_i outcome variable are independent of each other (as discussed in Chapter 4). Systematic exceptions to this rule (such as the correlated observations that are obtained in repeated measures designs) require analytic methods that take these correlations among observations into account; an introduction to repeated measures ANOVA is presented in Chapter 20.

2. Each Y outcome variable should be quantitative and reasonably normally distributed. As discussed in earlier chapters, it is useful to examine univariate histograms to assess whether scores on each Y variable are approximately normally distributed. Univariate outliers should be evaluated; data transformations such as logs may remedy problems with univariate and even some multivariate outliers.

3. Associations between pairs of Y variables should be linear; this can be assessed by obtaining a matrix of scatter plots between all possible pairs of Y variables. More formally, the joint distribution of the entire set of Y variables should be multivariate normal within each group. In practice, it is difficult to assess multivariate normality, particularly when the overall number of scores in each group is small. It may be easier to evaluate univariate and bivariate normality, as discussed in Chapter 4. Some types of deviation from multivariate normality probably have a relatively small effect on Type I error (Stevens, 1992). However, MANOVA is not very robust to outliers. When there are extreme scores within groups, the data analyst should evaluate these carefully to decide whether the remedies for outliers discussed in Chapter 4, such as log transformations, would reduce the impact of extreme scores.

4. The variance/covariance matrices for the Y outcome variables (Σ) should be homogeneous across the populations that correspond to groups in the study. Recall from Chapter 16 that the variances and covariances that correspond to the elements of this population Σ matrix are as follows:

$$\Sigma = \begin{bmatrix} \text{Var}(Y_1) & \text{Cov}(Y_1, Y_2) & \dots & \text{Cov}(Y_1, Y_p) \\ \text{Cov}(Y_2, Y_1) & \text{Var}(Y_2) & \dots & \text{Cov}(Y_2, Y_p) \\ \vdots & \vdots & \vdots & \vdots \\ \text{Cov}(Y_p, Y_1) & \text{Cov}(Y_p, Y_2) & \dots & \text{Var}(Y_p) \end{bmatrix}. \tag{17.4}$$

The null hypothesis for the test of homogeneity of Σ across groups is as follows:

$$H_0 : \Sigma_1 = \Sigma_2 = \cdots = \Sigma_k, \text{ for Groups } 1, 2, \dots, k. \tag{17.5}$$

In words, this is the assumption that the variances for all the outcome variables (Y_1, Y_2, . . ., Y_p) are equal across populations and the covariances for all possible pairs of outcome variables, such as Y_1 with Y_2, Y_1 with Y_3, etc., are equal across populations. As in DA, the optional Box M test provides a test for whether there is a significant violation of this assumption; the Box M is a function of the sample sum of cross products (**SCP**) matrices calculated separately for each of the groups in the design.

What problems arise when the Box M test and/or the Levene test suggests serious violations of the assumption of homogeneity of variances and covariances across groups? Violations of this assumption may be problematic in two ways: They may alter the risk of committing a Type I error (so that the nominal alpha level does not provide accurate information about the actual risk of Type I error), and they may reduce statistical power. Whether the risk of Type I error increases or decreases relative to the nominal alpha level depends on the way in which the homogeneity of variance assumption is violated. When the group that has the largest variances has a small n relative to the other groups in the study, the test of significance has a higher risk of Type I error than is indicated by the nominal alpha level. On the other hand, when the group with the largest variance has a large *n* relative to the other groups in the study, it tends to result in a more conservative test (Stevens, 1992). Violations of the homogeneity of variance/covariance matrix pose a much more serious problem when the group *ns* are small and/or extremely unequal. Stevens (1992) states that if the Box *M* test is significant in a MANOVA where the *ns* in the groups are nearly equal, it may not have much effect on the risk of Type I error, but it may reduce statistical power.

What remedies may help reduce violations of the assumption of homogeneity of variances and covariances across groups? Transformations (such as logarithms) may be effective in reducing the differences in variances and covariances across groups and, thus, correcting for violations of the homogeneity of variance/covariance matrices. In addition, when the Box *M* test is significant, and particularly if there are unequal *ns* in the groups, the researcher may prefer to report **Pillai's trace** instead of Wilks's lambda (Λ) as the overall test statistic; Pillai's trace is more robust to violations of the homogeneity of variances and covariances.

Recall that in a factorial ANOVA (Chapter 13), unequal or imbalanced *ns* in the cells imply a nonorthogonal design; that is, some of the variance in scores that can be predicted from the *A* factor is also predictable from the *B* factor. The same problem arises in factorial MANOVA design: Unequal (or imbalanced) *ns* in the cells imply a nonorthogonal design; that is, there is an overlap in the variance accounted for by the *A* and *B* main effects and their interaction. As in univariate ANOVA, this overlap can be handled in two different ways. SPSS allows the user to choose among four methods of computation for sums of squares (*SS*) in the GLM procedure. Types II and IV methods of computation for sums of squares are rarely used and will not be discussed here. SS Type I computation of sums of squares partitions the sums of squares in a hierarchical fashion; that is, the effects or factors in the model are entered one at a time, and each effect is assessed statistically controlling for effects that entered in the earlier steps. For example, in an analysis with SS Type I, SS_A could be computed without controlling for *B* or the $A \times B$ interaction, SS_B could be computed controlling only for the *A* factor, and $SS_{A \times B}$ could be computed controlling for the main effects of both *A* and *B*. GLM SS Type III, which is similar to the variance partitioning in a standard or simultaneous multiple regression, computes an adjusted sum of squares for each effect in the model while controlling for all other predictors. Thus, for an $A \times B$ factorial design, SS_A would be calculated controlling for the main effect of *B* and the $A \times B$ interaction, SS_B would be calculated controlling for the main effect of *A* and the $A \times B$ interaction, and $SS_{A \times B}$ would be calculated controlling for the main effects of both *A* and *B*. The SS Type III method of variance partitioning usually yields a more conservative estimate of the proportion of variance uniquely predictable from each factor. If the cell *ns* are equal or balanced, the SS Types I and III computation methods produce identical results. The factorial MANOVAs presented as examples later in this chapter use SS Type III computation methods.

17.7 ♦ Issues in Planning a Study

The issues that were relevant in planning ANOVA designs and in planning regression or DA studies are also relevant for planning MANOVA.

The first design issue involves the number, composition, and size of groups and, also, the treatments that are administered to groups (in experimental research) and the basis for categorization into groups (in nonexperimental studies that compare naturally occurring groups). As in other ANOVA designs, it is desirable to have a reasonably small number of groups and to have a relatively large number of participants in each group. If a factorial design is employed, equal (or at least balanced) *ns* in the cells are desirable, to avoid confounds of the effects of factors and interactions. If treatments are compared, the researcher needs to think about the dosage levels (i.e., are the treatments different enough to produce detectable differences in outcomes?) and the number of levels of groups required (are 2 groups sufficient, or are multiple groups needed to map out a curvilinear dose-response relationship?) Chapters 6 and 13 review design issues in one-way and factorial ANOVA that are also applicable to MANOVA. MANOVA generally requires a larger within-cell *n* than the corresponding ANOVA, to have sufficient degrees of freedom to estimate the elements of the matrix **SCP** within each cell. As in other ANOVA designs, the cell *ns* in MANOVA should not be too small. It is desirable to have sufficient degrees of freedom in each cell to obtain independent estimates for all the elements of the **SCP** matrix within each group, although in practice, if this requirement is not met, SPSS GLM sets up an omnibus test based on pooled within-group **SCP** matrices across all groups. In practice, if there are at least 20 cases in each cell, it may be sufficient to ensure robustness of the univariate *F* tests (Tabachnick & Fidell, 2007, p. 250). Stevens (2002) recommended a minimum total *N* of at least 20*p*, where *p* is the number of outcome measures. Another suggested minimum requirement for sample size is that there should be more cases than dependent variables in every cell of the MANOVA design. Section 17.12 includes a table that can be used to estimate the per-group sample size that is needed to achieve reasonable power given the assumed values of the multivariate effect size; the number of cases, *N*; the number of outcome variables, *p*; and the alpha level used to assess statistical significance. Note also that grossly unequal and/or extremely small *ns* in the cells of a MANOVA will make the impact of violations of the homogeneity of the variance/covariance matrix across groups more serious, and this can result in an inflated risk of Type I error and/or a decrease in statistical power.

When quantitative outcome variables are selected, these should include variables that the researcher believes will be affected by the treatment (in an experiment) or variables that will show differences among naturally occurring groups (in nonexperimental studies). The researcher may have one construct in mind as the outcome of interest; in a study that involves manipulation of stress, for example, the outcome of interest might be anxiety. The researcher might include multiple measures of anxiety, perhaps using entirely different methods of measurement (e.g., a standardized paper-and-pencil test, observations of behaviors such as fidgeting or stammering, and physiological measures such as heart rate). Multiple operationalizations of a construct (also called **triangulation of measurement**) can greatly strengthen confidence in the results of a study. Different types of measurement (such as self-report, observations of behavior, and physiological monitoring) have different strengths and weaknesses. If only self-report measures are used, group differences might be due to some of the known weaknesses of self-report, such as social desirability bias, response to perceived experimenter expectancy or demand, or "**faking**." However, if the stress manipulation increases scores on three different types of measures (such as self-report, physiological and behavioral assessments), we can be more confident that we have a result that is not solely attributable to a response artifact associated with one type of measurement (such as the social desirability bias that often occurs in self-report). Thus, MANOVA can be used to analyze data with multiple measures of the same construct.

On the other hand, the multiple outcome measures included in a MANOVA may tap several different constructs. For example, a researcher might administer psychological tests to persons from several different cultures, including assessments of individualism/collectivism, self-esteem, internal/external locus of control, extraversion, openness to experience, conscientiousness, need for achievement, and so forth. It might turn out that the pattern of test scores that differentiates the United States from Japan is different from the pattern of scores that differentiates the United States from Brazil. MANOVA corrects for intercorrelations or overlaps among these test scores whether they are multiple measures of the same construct or measures of several different constructs.

17.8 ♦ Conceptual Basis of MANOVA and Some Formulas for MANOVA

Recall that in a univariate one-way ANOVA, SS_{total} for the Y outcome variable is partitioned into the following components:

$$SS_{total} = SS_{between} + SS_{within}. \quad (17.6)$$

In univariate ANOVA, the information about the variance of the single outcome variable Y can be summarized in a single SS term, SS_{total}. Recall that the variance of Y equals $SS_{total}/(N - 1)$. In MANOVA, because we have multiple outcome variables, we need information about the variance of each outcome variable $Y_1, Y_2, \ldots, Y_p$; and we also need information about the covariance of each pair of Y variables. Recall also that the SCP between Y_1 and Y_2 is computed as $\sum (Y_1 - M_{Y1})(Y_2 - M_{Y2})$ and that the covariance between Y_1 and Y_2 can be estimated by the ratio SCP/N. The total sum of cross products matrix (denoted by $\mathbf{SCP}_{\mathbf{total}}$) in a MANOVA has the following elements (all the scores in the entire design are included in the computation of each SS and SCP elements):

$$\mathbf{SCP}_{\text{total}} = \begin{bmatrix} SS(Y_1) & \text{SCP}(Y_1, Y_2) & \ldots & \text{SCP}(Y_1, Y_p) \\ \text{SCP}(Y_2, Y_1) & SS(Y_2) & \ldots & \text{SCP}(Y_2, Y_p) \\ M & M & M & M \\ \text{SCP}(Y_p, Y_1) & \text{SCP}(Y_p, Y_2) & \ldots & SS(Y_p) \end{bmatrix}. \quad (17.7)$$

If each element of the sample **SCP** matrix is divided by N or df, it yields a sample estimate **S** of the population variance/covariance matrix Σ (as shown in Equation 17.4).

In a one-way MANOVA, the total **SCP** matrix is partitioned into between-group and within-group components:

$$\mathbf{SCP}_{\text{total}} = \mathbf{SCP}_{\text{between}} + \mathbf{SCP}_{\text{within}}. \quad (17.8)$$

Note that the partition of **SCP** matrices in MANOVA (Equation 17.8) is analogous to the partition of sums of squares in univariate ANOVA (Equation 17.6), and in fact, Equation 17.8 would reduce to Equation 17.6 when *p*, the number of outcome variables, equals 1.

The same analogy holds for factorial ANOVA. In a balanced univariate factorial ANOVA, the SS_{total} is divided into components that represent the main effects and interactions:

$$SS_{\text{total}} = SS_A + SS_B + SS_{A\times B} + SS_{\text{within}}. \tag{17.9}$$

An analogous partition of the **SCP** matrix can be made in a factorial MANOVA:

$$\mathbf{SCP}_{\text{total}} = \mathbf{SCP}_A + \mathbf{SCP}_B + \mathbf{SCP}_{A\times B} + \mathbf{SCP}_{\text{within}}. \tag{17.10}$$

See Tabachnick and Fidell (2007, pp. 254–258) for a completely worked empirical example that demonstrates the computation of the elements of the total, between-group, and within-group **SCP** matrices.

The approach to significance testing (to assess the null hypothesis of no group differences) is rather different in MANOVA compared with univariate ANOVA. In univariate ANOVA, each *SS* term was converted to a mean square (*MS*) by dividing it by the appropriate degrees of freedom (*df*). In univariate ANOVA, the test statistic for each effect was $F_{\text{effect}} = MS_{\text{effect}}/MS_{\text{within}}$. *F* values large enough to exceed the critical values for *F* were interpreted as evidence for significant differences among group means; an eta-squared effect size was also reported to indicate what proportion of variance in scores on the individual *Y* outcome variable was associated with group membership in the sample data.

To summarize information about sums of squares and sums of cross products for a one-way MANOVA (Equation 17.8) with a set of *p* outcome variables, we need a single number that tells us how much variance for the set of *p* outcome variables is represented by each of the **SCP** matrix terms in Equation 17.8. The determinant of an **SCP** matrix provides the summary information that is needed. A nontechnical introduction to the idea of a determinant is therefore needed before we can continue to discuss significance testing in MANOVA.

The **|SCP|** notation refers to the determinant of the **SCP** matrix. The determinant of a matrix is a way of describing the variance for a set of variables considered jointly, controlling for or removing any duplicated or redundant information contained in the variables (due to linear correlations among the variables). For a 2 × 2 matrix with elements labeled *a*, *b*, *c*, and *d*, the determinant of the matrix is obtained by taking the products of the elements on the major diagonal of the matrix (from upper left to lower right) minus the products of the elements on the minor diagonal (from lower left to upper right), as in the following example:

$$\mathbf{A} = \begin{bmatrix} a & b \\ c & d \end{bmatrix}.$$

The determinant of **A**, $|\mathbf{A}| = (a \times d) - (c \times b)$.

One type of matrix for which a determinant can be computed is a correlation matrix, **R**. For two variables, Y_1 and Y_2, the correlation matrix **R** consists of the following elements:

$$\mathbf{R} = \begin{bmatrix} 1.00 & r_{12} \\ r_{12} & 1.00 \end{bmatrix},$$

and the determinant of $\mathbf{R}$, $|\mathbf{R}| = (1 \times 1) - (r_{12} \times r_{12}) = 1 - r^2_{12}$.

Note that $(1 - r^2_{12})$ describes the proportion of nonshared variance between Y_1 and Y_2; it allows us to summarize how much variance a weighted linear composite of Y_1 and Y_2 has after we correct for or remove the overlap or shared variance in this set of two variables that is represented by r^2_{12}. In a sense, the determinant is a single summary index of variance for a set of p dependent variables that corrects for multicollinearity or shared variance among the p variables. When a matrix is larger than 2×2, the computation of a determinant becomes more complex, but the summary information provided by the determinant remains essentially the same: A determinant tells us how much variance linear composites of the variables $Y_1, Y_2, \ldots, Y_p$ have when we correct for correlations or covariances among the Y outcome variables.

In MANOVA, our test statistic takes a different form from that in univariate ANOVA; in univariate ANOVA, F was the ratio $MS_{\text{effect}}/MS_{\text{error}}$. In MANOVA, the omnibus test statistic is a ratio of determinants of the matrix that represents "error" or within-group sums of cross products in the numerator, compared with the sum of the matrices that provide information about "effect plus error" in the denominator. In contrast to F, which is a signal-to-noise ratio that increases in magnitude as the magnitude of between-group differences becomes large relative to error, Wilks's Λ is an estimate of the proportion of variance in outcome variable scores that is *not* predictable from group membership. In a univariate one-way ANOVA, Wilks's Λ corresponds to $SS_{\text{within}}/SS_{\text{total}}$. For a univariate one-way ANOVA, Wilks's Λ is interpreted as the proportion of variance in scores for a single Y outcome variable that is *not* predictable from or associated with group membership. One computational formula for multivariate Wilks's Λ, which is essentially a ratio of the overall error or within-group variance to the total variance for the set of p intercorrelated outcome variables, is as follows:

$$\Lambda = \frac{|\mathbf{SCP}_{\text{error}}|}{|\mathbf{SCP}_{\text{effect}} + \mathbf{SCP}_{\text{error}}|}. \tag{17.11}$$

$|\mathbf{SCP}_{\text{error}}|$ is a measure of the generalized variance of the set of p dependent variables within groups; it is, therefore, an estimate of the within-group variance in scores for a set of one or more discriminant functions. $|\mathbf{SCP}_{\text{effect}} + \mathbf{SCP}_{\text{error}}|$ (which is equivalent to $|\mathbf{SCP}_{\text{total}}|$ in a one-way MANOVA) is a measure of the variance of the set of p dependent variables, including both within- and between-group sources of variation and correcting for intercorrelations among the variables. In a one-way MANOVA, Wilks's Λ estimates the ratio of unexplained variance to total variance; it is equivalent to $1 - \eta^2$. In a situation where the number of dependent variables, p, is equal to 1, Λ reduces to $SS_{\text{within}}/SS_{\text{total}}$ or $1 - \eta^2$. Usually, a researcher wants the value of Λ to be small (because a small Λ tends to correspond to relatively large between-group differences relative to the amount of variability of scores within groups). Lambda can be converted to an F (or a chi-square, as in the discussion of DA in Chapter 16) to assess statistical significance.

Thus, while a higher value of F in ANOVA typically indicates a *stronger* association between scores on the Y outcome variable and group membership (other factors being equal), a higher value of Wilks's Λ implies a *weaker* association between scores on the Y

outcome variables and group membership. We can calculate an estimate of partial eta squared—that is, the proportion of variance in scores on the entire set of outcome variables that is associated with group differences for one specific effect in our model, statistically controlling for all other effects in the model—as follows:

$$\eta^2_{\text{effect}} = 1 - \Lambda_{\text{effect}}. \quad (17.12)$$

SPSS GLM converts Wilks's Λ to an F ratio; this conversion is only approximate in many situations. Note that as the value of Λ decreases (other factors being equal), the value of the corresponding F ratio tends to increase. This conversion of Λ to F makes it possible to use critical values of the F distribution to assess the overall significance of the MANOVA—that is, to have an omnibus test of equality that involves all the $Y_1, Y_2, \ldots, Y_p$ variables and all k groups.

Tabachnick and Fidell (2007, pp. 259–260) provided formulas for this approximate conversion from Λ to F as follows. This information is provided here for the sake of completeness; SPSS GLM provides this conversion. First, it is necessary to calculate intermediate terms, denoted by s and y. In a one-way MANOVA with k groups, a total of N participants, and p outcome variables, $df_{\text{effect}} = k - 1$ and $df_{\text{error}} = N - k - 1$.

$$s = \sqrt{\frac{p^2 \times (df_{\text{effect}})^2 - 4}{p^2 + (df_{\text{effect}})^2 - 5}}, \quad (17.13)$$

$$y = \Lambda^{1/s}. \quad (17.14)$$

Next, the degrees of freedom for the F ratio, df_1 and df_2, are calculated as follows:

$$df_1 = p \times (df_{\text{effect}}), \quad (17.15)$$

$$df_2 = s \times \left[df_{\text{error}} - \frac{p - (df_{\text{effect}}) + 1}{2} \right] - \left[\frac{p \times (df_{\text{effect}}) - 2}{2} \right]. \quad (17.16)$$

Finally, the conversion from Λ to an approximate F with df_1, df_2 degrees of freedom is given by

$$= \text{Approximate } F(df_1, df_2) = \left(\frac{1 - y}{y} \right) \left(\frac{df_2}{df_1} \right). \quad (17.17)$$

If the overall F for an effect (such as the main effect for types of stress in the hypothetical research example described earlier) is statistically significant, it implies that there is at least one significant contrast between groups and that this difference can be detected when the entire set of outcome variables is examined. Follow-up analyses are required to assess which of the various contrasts among groups are significant and to evaluate what patterns of scores on outcome variables best describe the nature of the group differences.

17.9 ♦ Multivariate Test Statistics

SPSS GLM provides a set of four multivariate test statistics for the null hypotheses shown in Equations 17.2 and 17.3, including Wilks's lambda, **Hotelling's trace**, Pillai's trace, and **Roy's largest root**. In some research situations, one of the other three test statistics might be preferred to Wilks's Λ. In situations where a factor has only two levels and the effect has only 1 *df*, decisions about statistical significance based on these tests do not differ. In more complex situations where a larger number of groups are compared, and particularly in situations that involve violations of assumptions of multivariate normality, the decision whether to reject the null hypothesis may be different depending on which of the four multivariate test statistics the data analyst decides to report. When assumptions for MANOVA are seriously violated, researchers often choose to report Pillai's trace (which is more robust to violations of assumptions).

All the four multivariate test statistics are derived from the eigenvalues that are associated with the discriminant functions that best differentiate among groups. Refer back to Chapter 16 for a discussion of the way in which discriminant functions are created by forming optimal weighted linear composites of scores on the quantitative predictor variables and the information that is provided by the eigenvalue associated with each discriminant function. In MANOVA, as in DA, one or several discriminant functions are created by forming weighted linear composites of the quantitative variables. These weighted linear combinations of variables are constructed in a way that maximizes the values of $SS_{between}/SS_{within}$. That is, the first discriminant function is the weighted linear composite of $Y_1, Y_2, \ldots, Y_p$ that has the largest possible $SS_{between}/SS_{within}$ ratio. The second discriminant function is uncorrelated with the first discriminant function and has the maximum possible $SS_{between}/SS_{within}$. The number of discriminant functions that can be obtained for each effect in a factorial MANOVA is usually just the minimum of these two values: df_{effect} and p. The discriminant functions are obtained by computing eigenvalues and eigenvectors (see the appendix to Chapter 16). For each discriminant function D_i, we obtain a corresponding eigenvalue λ_i:

$$\lambda_i = SS_{between}/SS_{within}. \tag{17.18}$$

That is, the eigenvalue λ_i for each discriminant function tells us how strongly the scores on that discriminant function are related to group membership. When $SS_{between}/SS_{within}$ is large, it implies a stronger association of discriminant function scores with group membership. Thus, large eigenvalues are associated with discriminant functions that show relatively large between-group differences and that have relatively small within-group variances.

Because an eigenvalue contains information about the ratio of within- and between-group sums of squares, it is possible to derive a canonical correlation for each discriminant function from its eigenvalue—that is, a correlation describing the strength of the relation between group membership and scores on that particular weighted linear combination of variables.

$$r_{ci} = \sqrt{\frac{\lambda_i}{1 + \lambda_i}}. \tag{17.19}$$

To summarize how well a *set* of discriminant functions collectively explains variance, we need to combine the variance that is (or is not) explained by each of the individual discriminant functions. The summary can be created by multiplying terms that involve the eigenvalues, as in Wilks's Λ, or by adding terms that involve the eigenvalues, as in Pillai's trace or Hotelling's trace. The three multivariate test statistics Wilks's Λ, Pillai's trace, and Hotelling's trace represent three different ways of combining the information about between- versus within-group differences (contained in the eigenvalues) across multiple discriminant functions. Note that the value of Wilks's Λ obtained by multiplication in Equation 17.21 is equivalent to the Wilks's Λ obtained by taking a ratio of determinants of **SCP** matrices as shown earlier in Equation 17.11. Note that the symbol Π in Equation 17.20 indicates that the terms that follow should be multiplied:

$$\text{Wilk's } \Lambda = \Pi[1/(1 + \lambda_i)], \tag{17.20}$$

$$\text{Pillai's trace} = \Sigma \, [\lambda_i /(1 + \lambda_i)], \tag{17.21}$$

$$\text{Hotelling's trace } T = \Sigma \, \lambda_i. \tag{17.22}$$

The application of simple algebra helps to clarify just what information each of these multivariate tests includes (as described in Bray & Maxwell, 1985). First, consider Wilks's Λ. We know that $\lambda = SS_{\text{between}}/SS_{\text{within}}$. Thus, $1/(1 + \lambda) = 1/[1 + (SS_{\text{between}} \,/\, SS_{\text{within}})]$.

We can simplify this expression if we multiply both the numerator and the denominator by SS_{within} to obtain $SS_{\text{within}}/(SS_{\text{within}} + SS_{\text{between}})$; this ratio is equivalent to $(1 - \eta^2)$. Thus, the $1/(1 + \lambda)$ terms that are multiplicatively combined when we form Wilks's Λ each correspond to the proportion of error or within-group variance for the scores on one discriminant function. A larger value of Wilks's Λ therefore corresponds to a larger collective error (within-group variance) across the set of discriminant functions. A researcher usually hopes that the overall Wilks's Λ will be small.

Now, let's examine the terms included in the computation of Pillai's trace. Recall from Equation 17.19 that $\sqrt{\lambda/(1 + \lambda)} = r_{ci}$. Thus, Pillai's trace is just the sum of r^2_{ci} across all the discriminant functions. A larger value of Pillai's trace indicates a higher proportion of between-group or explained variance.

Hotelling's trace is the sum of the eigenvalues across discriminant functions; thus, a larger value of Hotelling's trace indicates a higher proportion of between-group or explained variance. Note that **Hotelling's T^2**, discussed in some multivariate statistics textbooks, is different from Hotelling's trace, although they are related and Hotelling's trace value can be converted to Hotelling's T^2. A researcher usually hopes that Wilks's Λ will be relatively small and that Pillai's trace and Hotelling's trace will be large, because these outcomes indicate that the proportion of between-group (or explained) variance is relatively high.

The fourth multivariate test statistic reported by SPSS GLM is called Roy's largest root (in some textbooks, such as Harris, 2001, it is called **Roy's greatest characteristic root,** or ***gcr***). Roy's largest root is unique among these summary statistics; unlike the first three statistics, which summarize explained variance across all the discriminant functions, Roy's largest root includes information only for the *first* discriminant function:

$$\text{Roy's largest root} = \lambda_1/(1 + \lambda_1). \tag{17.23}$$

Thus, Roy's largest root is equivalent to the squared canonical correlation for the first discriminant function alone. SPSS provides significance tests (either exact or approximate *F*s) for each of these multivariate test statistics. Among these four multivariate test statistics, Pillai's trace is believed to be the most robust to violations of assumptions; Wilks's Λ is the most widely used and therefore more likely to be familiar to readers. Roy's largest root is not very robust to violations of assumptions, and under some conditions it may also be less powerful than Wilks's Λ; however, Roy's largest root may be preferred in situations where the researcher wants a test of significance for only the first discriminant function.

To summarize, Wilks's Λ, Pillai's trace, and Hotelling's trace all test the same hypothesis: that a set of discriminant functions considered jointly provides significant discrimination across groups. When significant differences are found, it is typical to follow up this omnibus test with additional analyses to identify (a) which discriminating variables are the most useful, (b) which particular groups are most clearly differentiated, or (c) whether the weighted linear combinations that discriminate between some groups differ from the weighted linear combinations that best discriminate among other groups. All these issues may be considered in the context of more complex designs that may include repeated measures and covariates.

17.10 ♦ Factors That Influence the Magnitude of Wilks's Λ

As in DA, MANOVA is likely to yield significant differences between groups when there are one or several significant univariate differences on *Y* means, although the inclusion of several nonsignificant *Y* variables may lead to a nonsignificant overall MANOVA even when there are one or two *Y* variables that significantly differ between groups. MANOVA sometimes also reveals significant differences between groups that can be detected only by examining two or more of the outcome variables jointly. In MANOVA, as in multiple regression and many other multivariate analyses, adding "garbage" variables that do not provide predictive information may reduce statistical power. MANOVA, like DA, sometimes yields significant group differences for sets of variables even when no individual variable differs significantly across groups. The magnitude and signs of correlations among *Y* variables affect the power of MANOVA and the confidence of interpretations in experimental studies in complex ways (for further explanation, see Bray & Maxwell, 1985).

17.11 ♦ Effect Size for MANOVA

In a univariate ANOVA, Wilks's Λ corresponds to SS_{within}/SS_{total}; this, in turn, is equivalent to $1 - \eta^2$, the proportion of within-group (or error) variance in *Y* scores. In a univariate ANOVA, an estimate of η^2 may be obtained by taking

$$\eta^2 = 1 - \Lambda. \tag{17.24}$$

This equation may also be used to calculate an estimate of effect size for each effect in a MANOVA. Tabachnick and Fidell (2007, pp. 260–261) suggest a correction that yields a more conservative estimate of effect size in MANOVA, in situations where $s > 1$:

$$\text{Partial } \eta^2 = 1 - \Lambda^{1/s}, \tag{17.25}$$

where the value of *s* is given by Equation 17.13 above.

17.12 ♦ Statistical Power and Sample Size Decisions

Stevens (1992) provided a table of recommended minimum cell *ns* for MANOVA designs for a reasonably wide range of research situations; his table is reproduced here (Table 17.1). The recommended minimum *n* per group varies as a function of the following:

1. The (unknown) population multivariate effect size the researcher wants to be able to detect
2. The desired level of statistical power (Table 17.1 provides sample sizes that should yield statistical power of approximately .70)
3. The alpha level used for significance tests (often set at $\alpha = .05$)
4. The number of levels or groups (k) in the factor (usually on the order of 3–6)
5. The number of dependent variables, p (the table provided by Stevens provides values for p from 3 to 6)

Stevens (1992) provided algebraic indices of multivariate effect size, but it would be difficult to apply these formulas in many research situations. In practice, a researcher can choose one of the verbal effect size labels that Stevens suggested (these range from "small" to "very large") without resorting to computations that, in any case, would require information researchers generally do not have prior to undertaking data collection. Given specific values for effect size, α, power, k, and p, Table 17.1 (adapted from Stevens, 1992) can be used to look up reasonable minimum cell *ns* for many MANOVA research situations.

For example, suppose that a researcher sets $\alpha = .05$ for a MANOVA with $k = 3$ groups and $p = 6$ outcome variables and believes that the multivariate effect size is likely to be very large. Using Table 17.1, the number of participants per group would need to be about 12 to 16 in order to have statistical power of .70. If the effect size is only medium, then the number of participants required per group to achieve power of .70 would increase to 42 to 54.

Table 17.1 ♦ Sample Size Required Per Cell in a k-Group One-Way MANOVA, for Estimated Power of .70 Using $\alpha = .05$ With Three to Six Outcome Variables

	Number of Groups (k)			
Effect Size[a]	*3*	*4*	*5*	*6*
Very large	12–16	14–18	15–19	16–21
Large	25–32	28–36	31–40	33–44
Medium	42–54	48–62	54–70	58–76
Small	92–120	105–140	120–155	130–170

SOURCE: Adapted from Stevens (1992, p. 229).

a. Stevens (1992, p. 228) provided formal algebraic definitions for multivariate effect size based on work by Lauter (1978). In practice, researchers usually do not have enough information about the strength of effects to be able to substitute specific numerical values into these formulas to compute numerical estimates of effect size. Researchers who have enough information to guess the approximate magnitude of multivariate effect size (small, medium, or large) may use this table to make reasonable judgments about sample size.

Note that in MANOVA, as in many other tests, we have the same paradox. When *N* is large, we have greater statistical power for the Box *M* test and the Levene tests, which assess whether the assumptions about equality of variances and covariances across groups are violated. However, it is in studies where the group *n*s are small and unequal across groups that violations of these assumptions are more problematic. When the total *N* is small, it is advisable to use a relatively large alpha value for the Box *M* and Levene tests—for example, to decide that the homogeneity assumption is violated for $p < .10$ or $p < .20$; these larger alpha levels make it easier to detect violations of the assumptions in the small *N* situations where these violations are more problematic. On the other hand, when the total *N* is large, it may make sense to use a smaller alpha level for these tests of assumptions; when total *N* is large, we might require a Box *M* test with $p < .01$ or even $p < .001$ as evidence of significant violation of assumption of the homogeneity of Σ across groups.

17.13 ♦ SPSS Output for a One-Way MANOVA: Career Group Data From Chapter 16

The data for this example are from Table 16.1—that is, the same data that were used for the DA example in the previous chapter. Recall that in Chapter 16, a DA was performed for a small subset of data for high school seniors from the Project Talent survey. Students chose one of three different types of future careers. The career group factor had these three levels: Group 1, business; Group 2, medical; and Group 3, teaching/social work. Scores on six achievement and aptitude tests were used to predict membership in these future career groups. The DA reported in Chapter 16 corresponds to the following one-way MANOVA, which compares the set of means on these six test scores across the 3 career choice groups.

To run a MANOVA, the following menu selections are made: <General Linear Model> → <Multivariate> (see Figure 17.1). The choice of the <Multivariate> option makes it possible to include a list of several outcome variables. In Chapter 13, when SPSS GLM was used to run a univariate factorial ANOVA (i.e., a factorial ANOVA with only one *Y* outcome variable), the <Univariate> option was selected to limit the analysis to just a single outcome variable.

The main dialog window for GLM Multivariate appears in Figure 17.2. For this one-way MANOVA, the categorical variable cargroup was placed in the list of Fixed Factor(s), and the six test scores (English, reading, etc.) were placed in the list of Dependent Variables. To request optional statistics such as univariate means and univariate *F* tests and tests of the homogeneity of variance/covariance matrices across groups, we could click the Options button and select these statistics. This was omitted in the present example because this information is already available in the DA for this same set of data in Chapter 16. To request post hoc tests to compare all possible pairs of group means (in this instance, business vs. medical, medical vs. teaching/social work, and business vs. teaching/social work career groups), the Post Hoc button was used to open up the Post Hoc dialog window, which appears in Figure 17.3. The name of the factor for which post hoc tests are requested (cargroup) was placed in the right-hand window under the heading "Post Hoc Tests for," and a check box was used to select "Tukey—that is, the Tukey honestly significant difference test. Note that these post hoc tests are univariate—that is,

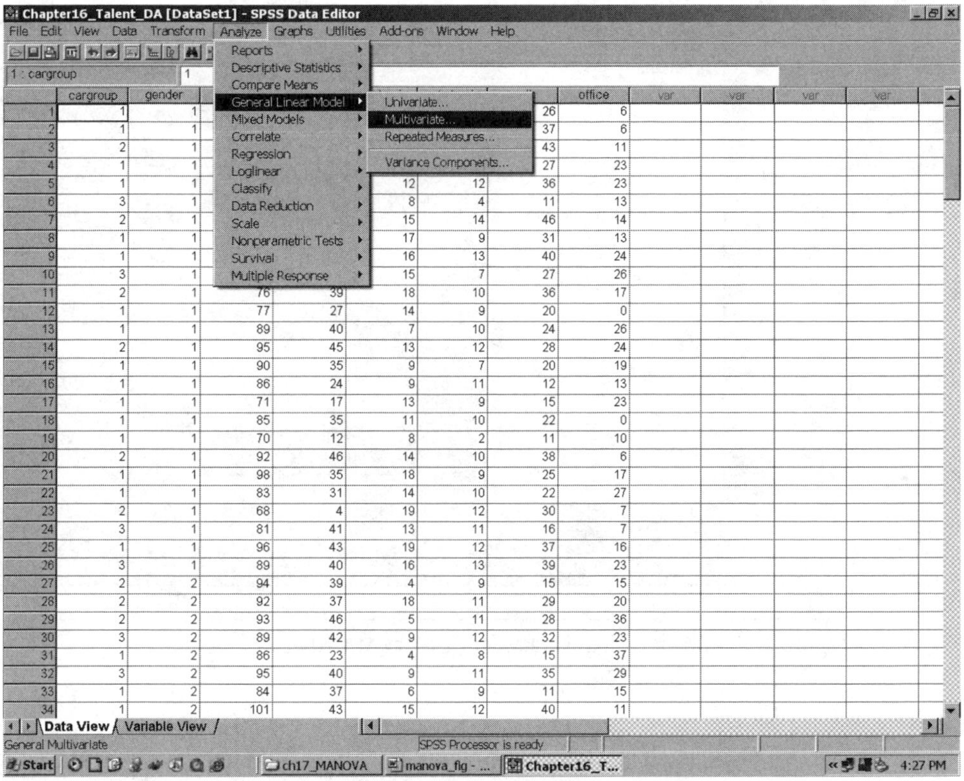

Figure 17.1 ♦ Menu Selections for a MANOVA in SPSS GLM

they assess which groups differ separately for each of the six outcome variables. The SPSS syntax obtained through these menu selections is given in Figure 17.4.

The output for the one-way MANOVA that compares six test scores (English, reading, and so forth) across 3 career groups (that correspond to levels of the factor named cargroup) is given in the next three figures. Figure 17.5 summarizes the multivariate tests for the null hypothesis that appears in Equation 17.2—that is, the null hypothesis that the entire list of means on all six outcome variables is equal across the populations that correspond to the 3 groups in this study. These results are consistent with the DA results for the same data presented in Chapter 16. In Chapter 16, the Box *M* test was requested; using $\alpha = .10$ as the criterion for statistical significance, the Box *M* indicated a statistically nonsignificant violation of the assumption of equality of variance/covariance matrices across these three populations: $p = .045$ (from Figure 16.14). If we set $\alpha = .05$, however, this Box *M* would be judged statistically significant. The violation of this assumption may decrease statistical power, and it may also make the "exact" *p* value on the SPSS printouts an inaccurate estimate of the true risk of committing a Type I error. Because Pillai's trace is somewhat more robust to violations of this assumption than the other multivariate test statistics, this is the test that will be reported. Figure 17.5 includes two sets of multivariate tests. The first part of the figure (effect = intercept) tests whether the vector of means on all *p* outcome variables is equal to 0 for all variables. This test is rarely of interest and it is

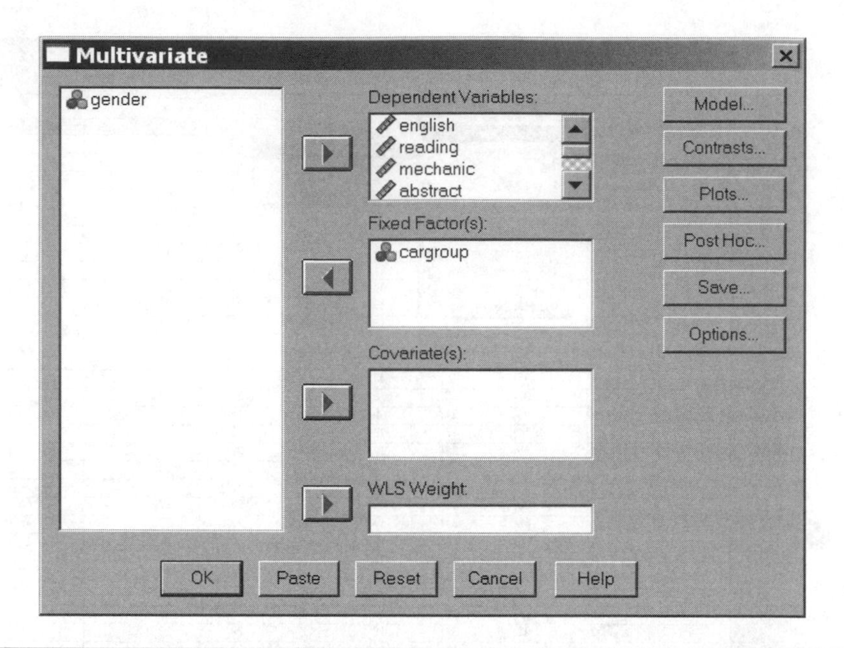

Figure 17.2 ♦ Main Dialog Window for One-Way MANOVA

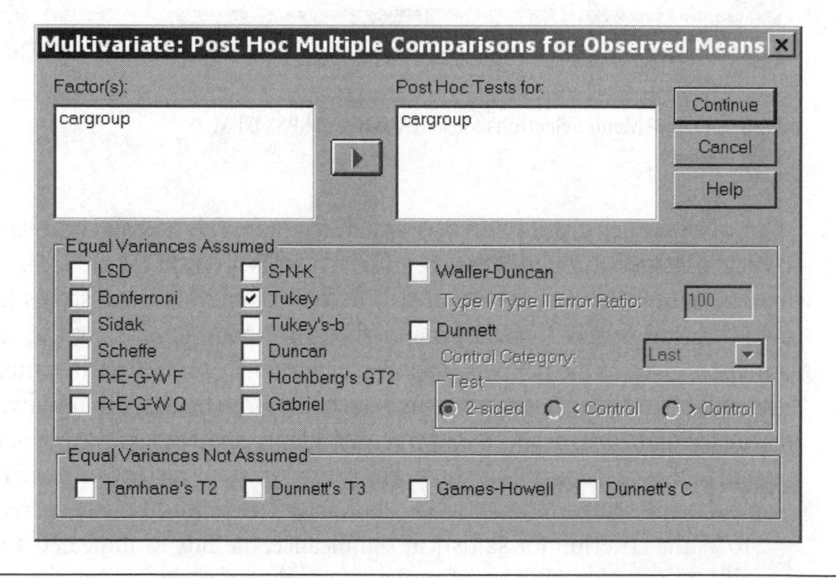

Figure 17.3 ♦ Tukey Honestly Significant Difference Post Hoc Tests Requested for One-Way MANOVA

not interpreted here. The second part of the figure (effect = cargroup) reports the four multivariate tests of the null hypothesis in Equation 17.2. The main effect for cargroup was statistically significant: Pillai's trace = .317, $F(12, 314) = 4.93$, partial $\eta^2 = .16$. This result suggests that it is likely that at least one pair of groups differs significantly on one outcome variable or on some combination of outcome variables; an effect size of $\eta^2 = .16$ could be

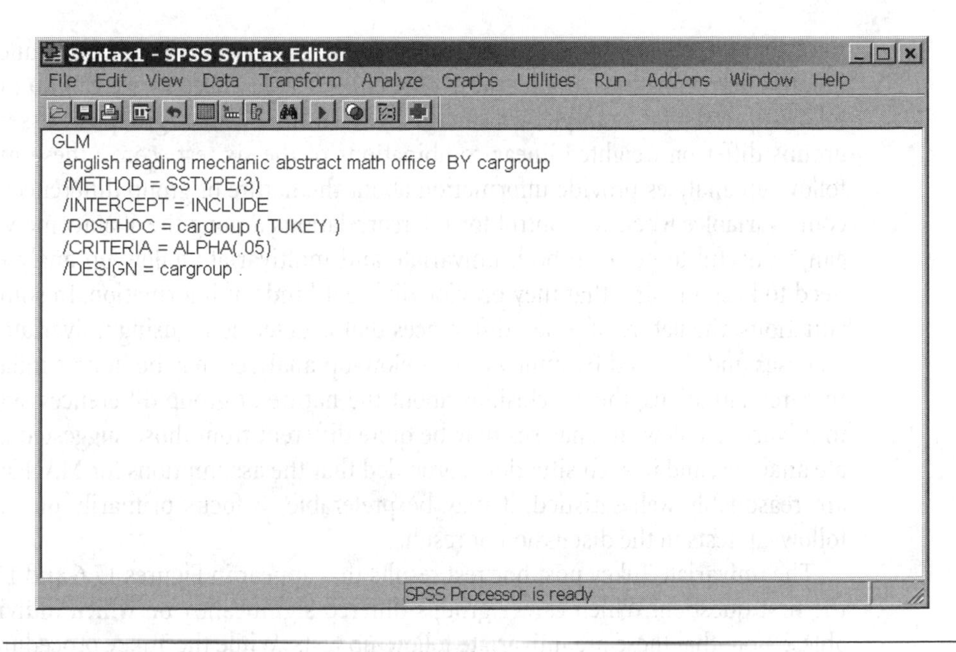

Figure 17.4 ♦ SPSS Syntax for One-Way MANOVA

Multivariate Tests[c]

Effect		Value	F	Hypothesis df	Error df	Sig.	Partial Eta Squared
Intercept	Pillai's Trace	.987	1917.575 [a]	6.000	156.000	.000	.987
	Wilks's Lambda	.013	1917.575 [a]	6.000	156.000	.000	.987
	Hotelling's Trace	73.753	1917.575 [a]	6.000	156.000	.000	.987
	Roy's Largest Root	73.753	1917.575 [a]	6.000	156.000	.000	.987
cargroup	Pillai's Trace	.317	4.926	12.000	314.000	.000	.158
	Wilks's Lambda	.685	5.406 [a]	12.000	312.000	.000	.172
	Hotelling's Trace	.456	5.887	12.000	310.000	.000	.186
	Roy's Largest Root	.448	11.734 [b]	6.000	157.000	.000	.310

a. Exact statistic

b. The statistic is an upper bound on F that yields a lower bound on the significance level.

c. Design: Intercept+cargroup

Figure 17.5 ♦ Output From One-Way MANOVA: Multivariate Tests for Main Effect on Cargroup Factor

considered large using the effect size labels suggested in Table 5.2 in this book. This omnibus test suggests that at least 1 group differs significantly from the other 2 groups (or perhaps all 3 groups differ significantly), and that this difference may involve one or several of the p outcome variables. Follow-up analyses are required to understand the nature of the cargroup main effect: which groups differed significantly and on which variable(s).

Follow-up questions about the nature of main effects can be examined in two different ways. First, we can examine univariate follow-up tests—for example, Tukey post hoc tests that provide information about whether scores on each individual Y outcome variable (such as scores on the math test) differ between each pair of career groups (such as the business and medicine career groups). As discussed previously, univariate follow-up tests raise many problems because they involve an inflated risk of Type I error and they also ignore intercorrelations among the Y outcome variables. Second, we can do multivariate follow-up tests; for example, we can do a DA on the 3 career groups to assess how the groups differ on weighted linear combinations of the six test scores; these multivariate follow-up analyses provide information about the nature of group differences on Y outcome variables when we control for intercorrelations among the Y outcome variables. It can be useful to perform both univariate and multivariate follow-up analyses; we just need to keep in mind that they provide different kinds of information. In some research situations, the nature of group differences that is detected by using univariate follow-up analyses and detected by multivariate follow-up analyses may be quite similar. In other research situations, the conclusions about the nature of group differences suggested by multivariate follow-up analyses may be quite different from those suggested by univariate analyses, and in such situations, provided that the assumptions for MANOVA and DA are reasonably well satisfied, it may be preferable to focus primarily on multivariate follow-up tests in the discussion of results.

The univariate Tukey post hoc test results that appear in Figures 17.6 and 17.7 answer the first question: Which career groups differed significantly on which individual variables? Note that these are univariate follow-up tests. While the Tukey procedure provides some protection from the inflated risk of Type I error that arises when we make multiple tests between pairs of groups, this procedure does not control for the inflated risk of Type I error that arises when we run six separate tests, one for each of the outcome variables (such as English, math, and so forth). There were no significant differences among the 3 career groups in scores on English, reading, mechanical reasoning, or abstract reasoning. However, group differences were found for scores on math and office skills. The homogeneous subsets reported in Figure 17.7 indicate that the business group scored significantly lower on math ($M = 20.05$) relative to the other 2 career groups ($M = 25.59$ for teaching, $M = 26.37$ for medicine); the last 2 groups did not differ significantly from each other. Also, the business group scored significantly higher on office skills ($M = 28.84$) compared with the other 2 groups (medicine, $M = 18.37$ and teaching/social work, $M = 20.62$); the last 2 groups did not differ significantly. The limitation of this follow-up analysis is that it does not correct for intercorrelations among the six test scores.

A multivariate follow-up analysis that takes intercorrelations among the six test scores into account is DA. A DA for this set of data was reported in Chapter 16 in Figures 16.11 through 16.23. For a more complete discussion of the DA results, refer back to Chapter 16. The highlights of the results are as follows: Two discriminant functions were formed, but the second discriminant function did not contribute significantly to discrimination between groups, and therefore the coefficients for the second discriminant function were not interpreted. The standardized discriminant function coefficients (from Table 16.2) that form the first discriminant function (D_1) based on the z scores on the six tests were as follows:

$$D_1 = +.42z_{\text{English}} + .02z_{\text{reading}} + .44z_{\text{mechanic}} - .12z_{\text{abstract}} - .79z_{\text{math}} + .75z_{\text{office}}.$$

If we use an arbitrary cutoff value of .50 as the criterion for a "large" discriminant function coefficient, only two variables had discriminant function coefficients that exceeded .5 in absolute value: z_{math}, with a coefficient of −.79, and z_{office}, with a coefficient of +.75. High scores on D_1 were thus associated with low z scores on math and high z scores on office skills. The business group had a much higher mean score on D_1 than the other 2

Multiple Comparisons

Tukey HSD

Dependent Variable	(I) cargroup	(J) cargroup	Mean Difference (I-J)	Std. Error	Sig.	95% Confidence Interval	
						Lower Bound	Upper Bound
english	business	medicine	1.29	2.189	.826	–3.89	6.47
		teach_social	–.28	1.907	.988	–4.79	4.23
	medicine	business	–1.29	2.189	.826	–6.47	3.89
		teach_social	–1.57	2.522	.808	–7.54	4.40
	teach_social	business	.28	1.907	.988	–4.23	4.79
		medicine	1.57	2.522	.808	–4.40	7.54
reading	business	medicine	–1.71	1.945	.655	–6.31	2.89
		teach_social	–2.46	1.695	.317	–6.47	1.55
	medicine	business	1.71	1.945	.655	–2.89	6.31
		teach_social	–.75	2.241	.940	–6.05	4.55
	teach_social	business	2.46	1.695	.317	–1.55	6.47
		medicine	.75	2.241	.940	–4.55	6.05
abstract	business	medicine	–.62	.586	.539	–2.01	.76
		teach_social	–.61	.510	.457	–1.82	.60
	medicine	business	.62	.586	.539	–.76	2.01
		teach_social	.01	.675	1.000	–1.58	1.61
	teach_social	business	.61	.510	.457	–.60	1.82
		medicine	–.01	.675	1.000	–1.61	1.58
math	business	medicine	–6.32*	1.749	.001	–10.46	–2.18
		teach_social	–5.54*	1.523	.001	–9.14	–1.94
	medicine	business	6.32*	1.749	.001	2.18	10.46
		teach_social	.78	2.014	.921	–3.98	5.55
	teach__social	business	5.54*	1.523	.001	1.94	9.14
		medicine	–.78	2.014	.921	–5.55	3.98
office	business	medicine	10.47*	1.849	.000	6.09	14.84
		teach_social	8.22"	1.611	.000	4.41	12.03
	medicine	business	–10.47*	1.849	.000	–14.84	–6.09
		teach_social	–2.25	2.130	.544	–7.28	2.79
	teach_social	business	–8.22*	1.611	.000	–12.03	–4.41
		medicine	2.25	2.130	.544	–2.79	7.28

Figure 17.6 ♦ Tukey Post Hoc Comparisons Among Means on Each of Six Outcome Test Scores for All Pairs of Career Groups

career groups; the medicine and teaching/social work groups had means on D_1 that were very close together. In this example, the DA leads us essentially to the same conclusion as the univariate ANOVAs: Whether or not we control for the other four test scores, the nature of the group differences was that the office group scored lower on math and higher on office than the other 2 career groups; the other 2 career groups (medicine and teaching/social work) did not differ significantly on any of the outcome variables. Note that it is possible for the nature of the differences among groups to appear quite different when discriminant functions are examined (as a multivariate follow-up) than when univariate ANOVAs and Tukey post hoc tests are examined (as univariate follow-ups). However, in this example, the multivariate and univariate follow-ups lead to similar conclusions about which groups differed and which variables differed across groups.

17.14 ♦ A 2 × 3 Factorial MANOVA of the Career Group Data

The career group data in Table 16.1 included information on one additional categorical variable or factor—that is, gender (coded 1= Male and 2 = Female). Having this information

english

Tukey HSD[a,b]

cargroup	N	Subset
		1
medicine	27	87.15
business	98	88.44
teach_social	39	88.72
Sig.		.760

Means for groups in homogeneous subsets are displayed.
Based on Type III Sum of Squares
The error term is Mean Square(Error) = 101.462.

a. Uses Harmonic Mean Sample Size = 41.162.

b. Alpha = .05.

mechanic

Tukey HSD[a,b]

cargroup	N	Subset
		1
business	98	9.12
teach_social	39	9.49
medicine	27	10.22
Sig.		.399

Means for groups in homogeneous subsets are displayed.
Based on Type III Sum of Squares
The error term is Mean Square(Error) = 14.813.

a. Uses Harmonic Mean Sample Size = 41.162.

b. Alpha = .05.

math

Tukey HSD[a,b]

cargroup	N	Subset	
		1	2
business	98	20.05	
teach_social	39		25.59
medicine	27		26.37
Sig.		1.000	.899

Means for groups in homogeneous subsets are displayed.
Based on Type III Sum of Squares
The error term is Mean Square(Error) = 64.736.

a. Uses Harmonic Mean Sample Size = 41.162.

b. Alpha = .05.

office

Tukey HSD[a,b]

cargroup	N	Subset	
		1	2
medicine	27	18.37	
teach_social	39	20.62	
business	98		28.84
Sig.		.457	1.000

Means for groups in homogeneous subsets are displayed.
Based on Type III Sum of Squares
The error term is Mean Square(Error) = 72.366.

a. Uses Harmonic Mean Sample Size = 41.162.

b. Alpha = .05.

Figure 17.7 ♦ Selected Homogeneous Subsets: Outcomes From Tukey Tests (for Four of the Six Outcome Test Scores)

about gender makes it possible to set up a 2 × 3 factorial MANOVA for the career group data (for which a one-way MANOVA was reported in the previous section and DA was reported in Chapter 16). This 2 × 3 factorial involves gender as the first factor (with two levels, 1 = Male and 2 = Female) and cargroup as the second factor (with 1 = business, 2 = medicine, and 3 = teaching or social work). The same six quantitative outcome variables that were used in the previous one-way MANOVA and DA examples are used in this 2 × 3 factorial MANOVA.

Prior to performing the factorial MANOVA, SPSS Crosstabs was used to set up a table to report the number of cases in each of the six cells of this design. Figure 17.8 shows the table of cell *n*s for this 2 × 3 factorial MANOVA design. Note that the cell *n*s in this factorial MANOVA are unequal (and not balanced). Therefore, this is a nonorthogonal factorial design; there is a confound between student gender and career choice. A higher proportion of females than males chose teaching/social work as a future career. Furthermore, the *n*s in two cells were below 10. These cell *n*s are well below any of the recommended minimum group sizes for MANOVA (see Table 17.1). The very small number of cases in some cells in this example illustrates the serious problems that can arise when some cells have small *n*s (although the overall *N* of 164 persons who had nonmissing scores on all the variables involved in this 2 × 3 MANOVA might have seemed reasonably adequate). This 2 × 3 MANOVA example illustrates some of the problems that can arise when there are insufficient data to obtain a clearly interpretable outcome. For instructional purposes, we will go ahead and examine the results of a 2 × 3 factorial MANOVA for this set of data. However, keep in mind that the very small cell *n*s in Figure 17.8 suggest that the results of factorial MANOVA are likely to be inconclusive for this set of data.

To run this 2 × 3 factorial MANOVA analysis using the GLM procedure in SPSS, the following menu selections were made: <Analyze> → <General Linear Model> → <Multivariate>.

Figure 17.9 shows the GLM dialog box for this MANOVA. Gender and cargroup were entered as fixed factors. The six dependent variables were entered into the box for Dependent Variables. The Options button was clicked to request optional statistics; the Options window appears in Figure 17.10. Descriptive statistics including means were requested; in addition, estimates of effect size and tests for possible violations of the homogeneity of variance/covariance matrices were requested. The SPSS syntax that corresponds to the preceding set of menu selections appears in Figure 17.11.

gender * cargroup Crosstabulation

			cargroup			
			business	medicine	teach_social	Total
gender	male	Count	16	6	4	26
		% within gender	61.5%	23.1%	15.4%	100.0%
	female	Count	84	23	36	143
		% within gender	58.7%	16.1%	25.2%	100.0%
Total		Count	100	29	40	169
		% within gender	59.2%	17.2%	23.7%	100.0%

Figure 17.8 ♦ Cell *n*s for the Gender by Cargroup Factorial MANOVA

NOTE: For two of the male groups (medicine and teaching careers), the cell *n* is less than 10.

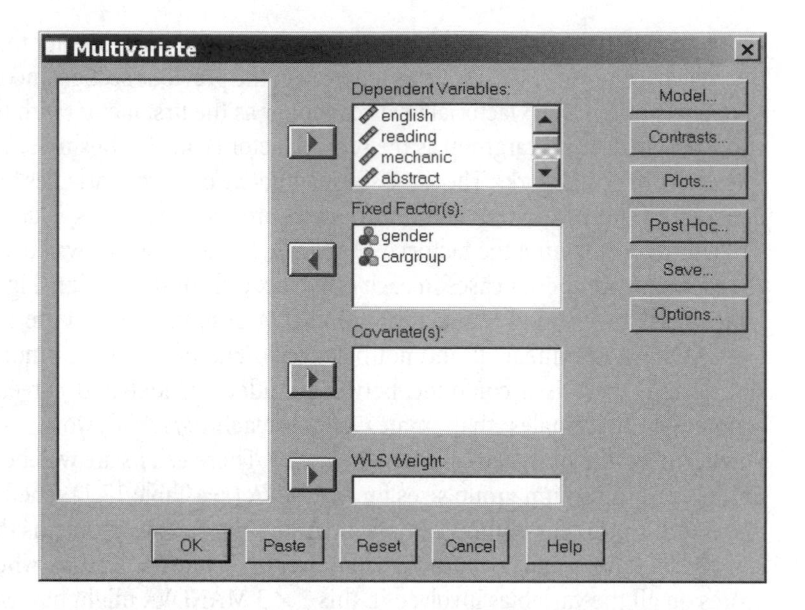

Figure 17.9 ♦ GLM Dialog Box for 2 × 3 Factorial MANOVA

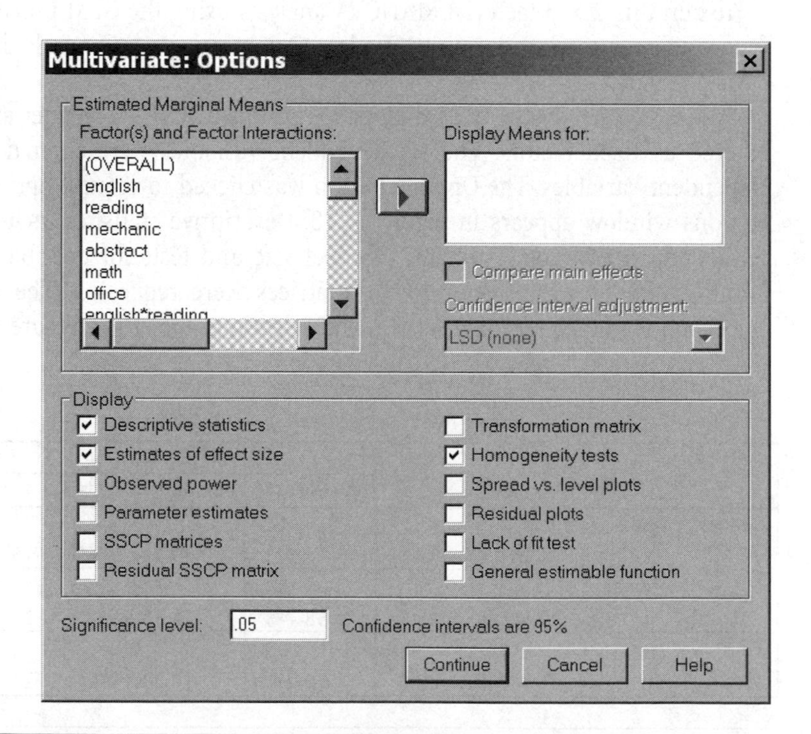

Figure 17.10 ♦ Optional Statistics Selected for MANOVA

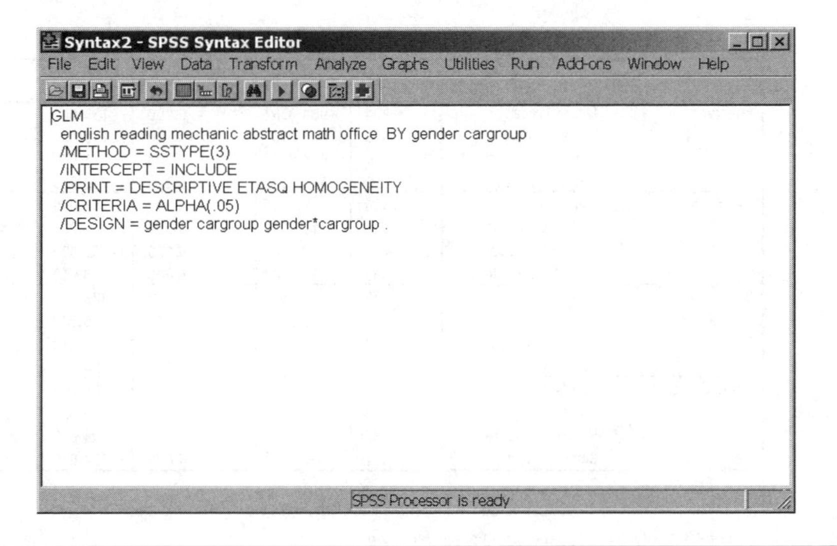

Figure 17.11 ♦ SPSS Syntax for GLM 2 × 3 Factorial MANOVA

Because of the small sample sizes in many cells, $\alpha = .20$ was used as the criterion for statistical significance; by this criterion, the Box *M* test (in Figure 17.12) did not indicate a statistically significant violation of the assumption of homogeneity of the variance/covariance matrices. In the following example, Wilks's Λ and its associated *F* values are reported for each effect in the 2 × 3 factorial model.

In a 2 × 3 factorial MANOVA (as in a 2 × 3 factorial ANOVA), we obtain statistical significance tests that address three questions. Is there a significant main effect for Factor *A*, in this case, gender? In other words, when the entire set of six test scores is considered jointly, do males and females differ significantly in their pattern of test scores? From Figure 17.13, for the main effect of gender, $\Lambda = .67$, $F(6, 153) = 12.53$, $p < .001$, partial $\eta^2 = .33$. Thus, there was a significant gender difference in the overall pattern of test scores on these six aptitude and achievement tests, and the effect size was large (33% of

Box's Test of Equality of Covariance Matrices[a]

Box's M	78.722
F	1.097
df1	63
df2	8330.280
Sig.	.279

Tests the null hypothesis that the observed covariance matrices of the dependent variables are equal across groups.

a. Design: Intercept+cargroup+gender+cargroup * gender

Figure 17.12 ♦ Selected Output From the 2 × 3 Factorial MANOVA: Box *M* Test of Equality of Covariance Matrices

Multivariate Tests[c]

Effect		Value	F	Hypothesis df	Error df	Sig.	Partial Eta Squared
Intercept	Pillai's Trace	.974	964.830 [a]	6.000	153.000	.000	.974
	Wilks's Lambda	.026	964.830 [a]	6.000	153.000	.000	.974
	Hotelling's Trace	37.836	964.830 [a]	6.000	153.000	.000	.974
	Roy's Largest Root	37.836	964.830 [a]	6.000	153.000	.000	.974
gender	Pillai's Trace	.330	12.534 [a]	6.000	153.000	.000	.330
	Wilks's Lambda	.670	12.534 [a]	6.000	153.000	.000	.330
	Hotelling's Trace	.492	12.534 [a]	6.000	153.000	.000	.330
	Roy's Largest Root	.492	12.534 [a]	6.000	153.000	.000	.330
cargroup	Pillai's Trace	.225	3.256	12.000	308.000	.000	.113
	Wilks's Lambda	.781	3.358[a]	12.000	306.000	.000	.116
	Hotelling's Trace	.273	3.459	12.000	304.000	.000	.120
	Roy's Largest Root	.241	6.198[b]	6.000	154.000	.000	.195
gender * cargroup	Pillai's Trace	.125	1.717	12.000	308.000	.062	.063
	Wilks's Lambda	.878	1.713[a]	12.000	306.000	.063	.063
	Hotelling's Trace	.135	1.709	12.000	304.000	.064	.063
	Roy's Largest Root	.092	2.361[b]	6.000	154.000	.033	.084

a. Exact statistic

b. The statistic is an upper bound on F that yields a lower bound on the significance level.

c. Design: Intercept+gender+cargroup+gender * cargroup

Figure 17.13 ♦ Omnibus Multivariate Test Results for the Set of Six Test Scores in the Gender by Cargroup MANOVA

the variance in scores on the single discriminant function that combined the six test scores was associated with gender).

Is there a significant main effect for Factor *B*, cargroup? From Figure 17.13, for the main effect of cargroup, we obtained $\Lambda = .78$, $F(12, 306) = 3.36$, $p < .001$, partial $\eta^2 = .12$. In other words, it appears that at least one of the 3 career groups is distinguishable from the other 2 groups based on some combination of scores on the six tests. Is there a statistically significant interaction between gender and cargroup? From Figure 17.13, the results for this interaction were as follows: $\Lambda = .88$, $F(12, 306) = 1.71$, $p = .063$, partial $\eta^2 = .063$. While a partial η^2 of .063 in the sample suggests a medium effect size for the data in this sample, this interaction was not statistically significant at the conventional $\alpha = .05$ level.

This 2 × 3 factorial MANOVA reveals a problem that is not resolvable using the available data. From Figure 17.8, we can see that the cell *n*s are quite small in several groups; therefore, we do not have good statistical power for detection of an interaction. Even if the gender by cargroup interaction had been statistically significant at the .05 level, we simply do not have large enough *n*s in the groups to make confident statements about the nature of an interaction. In this situation, we cannot reach a clear conclusion on whether there is an interaction, and if there is an interaction, we cannot be confident about understanding the nature of the interaction. There is enough indication of a possible interaction ($\eta^2 = .06$ in the sample) so that we cannot completely dismiss the possibility that a properly specified model needs to include a gender by cargroup interaction, and we should not proceed to interpret the main effects without taking possible interactions into account. On the other hand, we do not have sufficient information (cell *n*s are too small) to judge the interaction statistically significant or to provide a good description of the nature of the interaction.

An honest data analyst would decide that, because of the extremely small cell *n*s, the results of this 2 × 3 factorial MANOVA are inconclusive; these data are not publishable. There is not enough evidence to decide whether or not a gender by cargroup interaction is present and not enough information to describe the nature of the interaction between these factors. It would not be honest to backtrack and present one-way MANOVAS (using only cargroup, or only gender, as a factor) because this factorial MANOVA analysis has raised the possibility of an interaction between these factors; and if there is an interaction, a one-way MANOVA (using only gender or only cargroup) would not be a properly specified model. If the data analyst has the chance to recruit larger numbers of cases in the cells with small *ns*, analysis of a larger dataset might provide a clearer description of the data. The follow-up analyses that are presented in subsequent sections serve to illustrate what could be done to further understand the nature of main effects and interactions if a larger dataset were available, but these results would not be publishable with such small within-cell *ns*.

17.14.1 ♦ Potential Follow-Up Tests to Assess the Nature of Significant Main Effects

For instructional purposes only, a few additional follow-up analyses were performed to illustrate what could have been done if there had been enough data to decide whether to focus the interpretation on an interaction or on main effects. If the data had clearly indicated the absence of an interaction between gender and cargroup, then the following additional analyses could have been run to explore the nature of the significant main effects for gender and for cargroup. Multivariate assessment of the nature of group differences could be made by doing a DA separately for each factor. DA results describing the nature of differences across the 3 career groups represented by the cargroup factor were reported in Chapter 16, and a one-way MANOVA on the cargroup factor was discussed in the preceding section; information from those two analyses could be reported as follow-up information for the MANOVA in this chapter. In the DA, only two out of the six tests (math and office skills) had large weights in the discriminant function that differentiated the 3 career groups; the same two variables were also statistically significant when univariate post hoc Tukey tests were examined. The Tukey tests reported with the one-way MANOVA (see Figures 17.6 and 17.7) made the nature of the group differences clear: Members of the business career group scored significantly higher on office skills, and also significantly lower on math skills, than members of the medicine and teaching/social work career groups (and the last 2 groups did not differ from each other on any individual test score variables or on any discriminant functions formed from the test scores). Univariate *F*s were significant for math, $F(2, 158) = 6.83$, $p = .001$, and for office skills, $F(2, 158) = 7.47$, $p = .001$. Table 17.2 shows the univariate means for these variables across the 3 career groups. The group means shown in Table 17.2 were obtained from Figure 17.15, and the *F* tests and effect sizes were obtained from Figure 17.14.

Table 17.2 ♦ Significant Univariate Differences in Test Scores Across Levels of the Cargroup Factor: Means and Univariate *F* Ratios for Math and Office Scores

	Business	*Medicine*	*Teaching*	F^a	*p*	*Partial* η^2
Math	22.5	30.11	24.55	6.84	.001	.080
Office	24.14	16.51	19.13	7.47	.001	.086

a. Each *F* ratio in this table had (2, 158) *df*. The *p* values have not been adjusted for inflated risk of Type I error; that is, they correspond to the theoretical risk of Type I error for a single univariate *F* test. Only outcome variables that had a significant univariate main effect for cargroup at the $\alpha = .05$ level are included in this table.

Tests of Between-Subjects Effects

Source	Dependent Variable	Type III Sum of Squares	df	Mean Square	F	Sig.	Partial Eta Squared
Corrected Model	english	948.694[a]	5	189.739	1.943	.090	.058
	reading	283.835[b]	5	56.767	.700	.624	.022
	mechanic	573.758[c]	5	114.752	9.868	.000	.238
	abstract	45.917[d]	5	9.183	1.275	.277	.039
	math	2796.307[e]	5	559.261	9.830	.000	.237
	office	5707.152[f]	5	1141.430	19.353	.000	.380
Intercept	english	459975.928	1	459975.928	4709.674	.000	.968
	reading	72447.144	1	72447.144	893.926	.000	.850
	mechanic	8028.496	1	8028.496	690.429	.000	.814
	abstract	5800.541	1	5800.541	805.134	.000	.836
	math	41331.997	1	41331.997	726.468	.000	.821
	office	24808.601	1	24808.601	420.632	.000	.727
cargroup	english	313.822	2	156.911	1.607	.204	.020
	reading	87.673	2	43.837	.541	.583	.007
	mechanic	25.416	2	12.708	1.093	.338	.014
	abstract	13.202	2	6.601	.916	.402	.011
	math	777.823	2	388.912	6.836	.001	.080
	office	880.704	2	440.352	7.466	.001	.086
gender	english	540.513	1	540.513	5.534	.020	.034
	reading	15.104	1	15.104	.186	.667	.001
	mechanic	422.618	1	422.618	36.344	.000	.187
	abstract	5.519	1	5.519	.766	.383	.005
	math	543.873	1	543.873	9.559	.002	.057
	office	965.260	1	965.260	16.366	.000	.094
cargroup * gender	english	604.573	2	302.287	3.095	.048	.038
	reading	67.485	2	33.743	.416	.660	.005
	mechanic	23.902	2	11.951	1.028	.360	.013
	abstract	20.072	2	10.036	1.393	.251	.017
	math	524.025	2	262.013	4.605	.011	.055
	office	309.867	2	1 54 934	2.627	.075	.032
Error	english	15431.257	158	97.666			
	reading	12804.921	158	81.044			
	mechanic	1837.267	158	11.628			
	abstract	1138.302	158	7204			
	math	8989.321	158	56.894			
	office	9318.726	158	58.979			
Total	english	1294858.000	164				
	reading	195000.000	164				
	mechanic	16872.000	164				
	abstract	15458.000	164				
	math	94137.000	164				
	office	118830.000	164				
Corrected Total	english	16379.951	163				
	reading	13088.756	163				
	mechanic	2411.024	163				
	abstract	1184.220	163				
	math	11785.628	163				
	office	15025.878	163				

Figure 17.14 ♦ Tests of Univariate Differences for Each of the Six Test Scores in the 2 × 3 MANOVA (Gender by Cargroup)

The direction and magnitude of the differences between groups on these two variables (math and office) were essentially the same whether these variables were examined using DA or by doing two univariate ANOVAs, one on math scores across the 3 career groups and the other on office scores across the 3 career groups. In other words, this was not a situation in which controlling for other test scores changed our understanding of how the career groups differed with respect to scores on the math and office tests.

To assess the nature of main effect differences for gender, a DA was performed to see what weighted linear combination of test scores best differentiated males and females, and univariate differences on each test score are reported. Table 17.3 summarizes the male and female group means on the four tests that showed significant gender differences. The nature of the univariate differences was that men tended to score higher on tests of mechanical reasoning and math, whereas women scored higher on tests of English and office skills. In this sample, gender explained more variance in the test scores

1. Grand Mean

Dependent Variable	Mean	Std. Error	95% Confidence Interval	
			Lower Bound	Upper Bound
english	85.802	1.250	83.333	88.272
reading	34.052	1.139	31.803	36.301
mechanic	11.336	.431	10.484	12.188
abstract	9.635	.340	8.965	10.306
math	25.720	.954	23.835	27.605
office	19.927	.972	18.008	21.846

2. Cargroup

Dependent Variable	cargroup	Mean	Std.Error	95% Confidence Interval	
				Lower Bound	Upper Bound
english	business	87.839	1.426	85.022	90.657
	medicine	87.036	2.287	82.518	91.554
	teach_social	82.532	2.608	77.381	87.683
reading	business	32.762	1.299	30.195	35.328
	medicine	35.226	2.084	31.111	39.342
	teach_social	34.168	2.376	29.476	38.860
mechanic	business	10.798	.492	9.825	11.770
	medicine	12.167	.789	10.608	13.726
	teach_social	11.043	.900	9.265	12.820
abstract	business	9.345	.387	8.580	10.110
	medicine	10.286	.621	9.059	11.513
	teach_social	9.275	.708	7.876	10.674
math	business	22.500	1.089	20.350	24.650
	medicine	30.107	1.746	26.659	33.555
	teach_social	24.554	1.991	20.622	28.485
office	business	24.143	1.108	21.954	26.332
	medicine	16.512	1.778	13.001	20.023
	teach_social	19.125	2.027	15.122	23.128

3. gender

Dependent Variable	gender	Mean	Std.Error	95% Confidence Interval	
				Lower Bound	Upper Bound
english	male	82.861	2.301	78.316	87.407
	female	88.744	.978	86.813	90.675
reading	male	34.544	2.096	30.403	38.684
	female	33.560	.891	31.801	35.319
mechanic	male	13.937	.794	12.368	15.505
	female	8.735	.337	8.069	9.401
abstract	male	9.933	.625	8.698	11.167
	female	9.338	.266	8.814	9.863
math	male	28.671	1.757	25.201	32.140
	female	22.770	.746	21.296	24.244
office	male	15.996	1.788	12.464	19.528
	female	23.857	.760	22.356	25.358

Figure 17.15 ♦ Estimated Marginal Means for Univariate Outcomes *(Continued)*

4. cargroup × gender

Descriptive Statistics

	cargroup	gender	Mean	Std. Deviation	N
english	business	male	87.00	8.557	14
		female	88.68	8.737	84
		Total	88.44	8.688	98
	medicine	male	86.83	11.839	6
		female	87.24	8.831	21
		Total	87.15	9.326	27
	teach_social	male	74.75	18.839	4
		female	90.31	11.903	35
		Total	88.72	13.330	39
	Total	male	84.92	11.821	24
		female	88.87	9.613	140
		Total	88.29	10.024	164
reading	business	male	33.21	9.940	14
		female	32.31	8.284	84
		Total	32.44	8.489	98
	medicine	male	37.17	16.437	6
		female	33.29	8.539	21
		Total	34.15	10.524	27
	teach_social	male	33.25	8.539	4
		female	35.09	9.070	35
		Total	34.90	8.926	39
	Total	male	34.21	11.275	24
		female	33.15	8.542	140
		Total	33.30	8.961	164
mechanic	business	male	13.14	3.820	14
		female	8.45	3.220	84
		Total	9.12	3.681	98
	medicine	male	15.67	2.338	6
		female	8.67	4.747	21
		Total	10.22	5.213	27
	teach_social	male	13.00	3.559	4
		female	9.09	2.822	3
		Total	9.49	3.094	39
	Total	male	13.75	3.517	24
		female	8.64	3.384	140
		Total	9.39	3.846	164

Figure 17.15 ♦ (Continued)

Descriptive Statistics

	cargroup	gender	Mean	Std. Deviation	N
abstract	business	male	9.71	2.758	14
		female	8.98	2.430	84
		Total	9.08	2.477	98
	medicine	male	11.33	1.633	6
		female	9.24	2.682	21
		Total	9.70	2.614	27
	teach_social	male	8.75	4.031	4
		female	9.80	3.179	35
		Total	9.69	3.229	39
	Total	male	9.96	2.789	24
		female	9.22	2.674	140
		Total	9.33	2.695	164
math	business	male	25.93	9.458	14
		female	19.07	6.251	84
		Total	20.05	7.159	98
	medicine	male	36.83	7.055	6
		female	23.38	8.617	21
		Total	26.37	9.958	27
	teach_social	male	23.25	12.447	4
		female	25.86	8.374	35
		Total	25.59	8.696	39
	Total	male	28.21	10.413	24
		female	21.41	7.749	140
		Total	22.41	8.503	164
office	business	male	17.57	7.133	14
		female	30.71	6.870	84
		Total	28.84	8.281	98
	medicine	male	13.17	6.735	6
		female	19.86	9.393	21
		Total	18.37	9.199	27
	teach_social	male	17.25	8.808	4
		female	21.00	8.602	35
		Total	20.62	8.583	39
	Total	male	16.42	7.241	24
		female	26.66	9.163	140
		Total	25.16	9.601	164

Table 17.3 ◆ Univariate Gender Differences in Means for Four Achievements and Aptitude Tests

	Group Means					
	Males		*Females*	*F*[a]	*p*	*Partial* η^2
English	82.86	<	88.74	5.53	.020	.034
Office	16.00	<	23.86	16.37	<.001	.094
Mechanic	13.94	>	8.74	36.34	<.001	.187
Math	28.67	>	22.77	9.56	.002	.057

a. Each *F* ratio reported in this table has (1, 158) *df*. The *p* values have not been adjusted for inflated risk of Type I error; that is, they correspond to the theoretical risk of Type I error for a single univariate *F* test. Only the outcome variables that had a significant univariate main effect for gender at the $\alpha = .05$ level are included in this table.

than was explained by either career choice or the interaction between gender and career choice (partial $\eta^2 = .33$). Four of the six variables showed significant univariate *F*s for gender differences: for English, $F(1, 158) = 5.53, p = .02$; for mechanical reasoning, $F(1, 158) = 36.344, p < .001$; for math, $F(1, 158) = 9.56, p = .002$; and for office, $F(1, 158) = 16.37, p < .001$. There were no significant univariate gender differences on the reading or abstract reasoning tests.

A multivariate follow-up analysis to further evaluate the nature of gender differences in test scores is also presented (accompanied again by the warning that—for this dataset—we can't be sure whether we should be looking at an interaction of gender and career rather than at main effects for gender). DA using gender as the categorical variable provides a potential way to summarize the multivariate differences between males and females for the entire set of outcome measures, controlling for intercorrelations among the tests. The nature of the multivariate gender differences can be described by scores on one standardized discriminant function *D* with the following coefficients (see Figure 17.16):

$$D = +.50z_{\text{English}} + .20z_{\text{reading}} - .75z_{\text{mechanic}} - .02z_{\text{abstract}} - .31z_{\text{math}} + .44z_{\text{office}}.$$

If we require a discriminant function coefficient to be greater than or equal to .5 in absolute value to consider it "large," then only two variables had large discriminant function coefficients: there was a positive coefficient associated with z_{English} and a negative coefficient associated with z_{mechanic}. Thus, a higher score on *D* tends to be related to higher *z* scores on English and lower *z* scores on mechanical reasoning. When we look at the centroids in Figure 17.16 (i.e., the mean *D* scores for the male and female groups), we see that the female group scored higher on *D* (and therefore higher on z_{English} and lower on z_{mechanic}) than the male group. In this multivariate follow-up using DA, we might conclude that the gender differences are best described as higher scores for females on English and higher scores for males on mechanical reasoning; two variables for which statistically significant univariate *F*s were obtained (math and office) had standardized discriminant function coefficients that were less than +.5 in absolute value. A judgment call would be required. If we are willing to set our criterion for a large standardized discriminant function coefficient at a much lower level, such as +.3, we might conclude that—even when they are evaluated in a multivariate context—math and office scores continue to show differences across gender. If we adhere to the admittedly arbitrary standard chosen earlier (that a standardized discriminant function coefficent must equal or exceed .5 in absolute value to be considered large), then we would conclude that—when they are evaluated in a multivariate follow-up—math and office scores do not show differences between males and females.

17.14.2 ♦ Possible Follow-Up Tests to Assess the Nature of the Interaction

If a larger dataset had yielded clear evidence of a statistically significant gender by cargroup interaction in the previous 2 × 3 factorial MANOVA and if the *ns* in all the cells had been large enough (at least 20) to provide adequate information about group means for all six cells, then follow-up analyses could be performed to evaluate the nature of this interaction. For an interaction in a factorial MANOVA, there are more numerous options for the type of follow-up analysis. First of all, just as in a univariate factorial ANOVA (a factorial ANOVA with just one *Y* outcome variable), the researcher needs to decide whether to focus the discussion primarily on the interactions between factors or primarily on main effects. Sometimes both are included, but a statistically significant interaction that has a large effect size can be interpreted as evidence that the effects of different levels of Factor *B* differ substantially across groups based on levels of Factor *A*, and in that case,

Eigenvalues

Function	Eigenvalue	% of Variance	Cumulative %	Canonical Correlation
1	.546[a]	100.0	100.0	.594

a. First 1 canonical discriminant functions were used in the analysis.

Wilks's Lambda

Test of Function(s)	Wilks's Lambda	Chi-square	df	Sig
1	.647	69.227	6	.000

Standardized Canonical Discriminant Function Coefficients

	Function
	1
english	.502
reading	.196
mechanic	-.752
abstract	-.015
math	-.310
office	.444

Functions at Group Centroids

	Function
gender	1
male	-1.773
female	.304

Unstandardized canonical discriminant functions evaluated at group means

Figure 17.16 ♦ Selected Output From a Multivariate Follow-Up on the Significant Main Effect for Gender: Discriminant Analysis Comparing Scores on Six Achievement and Aptitude Tests Across Gender Groups (Gender Coded 1 = Male, 2 = Female)

a detailed discussion of main effects for the *A* and *B* factors may not be necessary or appropriate. If there is a significant $A \times B$ interaction in a MANOVA, the data analyst may choose to focus the discussion on that interaction. Two kinds of information can be used to understand the nature of the interaction in a MANOVA. At the univariate level, the researcher may report the means and *F* tests for the interaction for each individual *Y* outcome variable considered in isolation from the other *Y* outcome variables. In other words, the researcher may ask, If we examine the means on only Y_1 and Y_3, how do these differ across the cells of the MANOVA factorial design, and what is the nature of the interaction for each separate *Y* outcome variable? In the career group by gender factorial MANOVA data, for example, we might ask, Do males show a different pattern of scores on their math test scores across the 3 career groups from females?

A second approach to understanding the nature of an interaction in MANOVA is multivariate—that is, an examination of discriminant function scores across groups. We can do an "analysis of simple main effects" using DA. In an analysis of simple main effects, the researcher essentially conducts a one-way ANOVA or MANOVA across levels of the *B* factor, separately within each group on the *A* factor. In the empirical example presented here, we can do a MANOVA and a DA to compare scores on the entire set of *Y* outcome variables across the 3 career groups, separately for the male and female groups. To assess how differences among career groups in discriminant function scores may differ for males versus females, DA can be performed separately for the male and female groups. In this follow-up analysis, the question is whether the weighted linear combination of scores on the *Y* outcome variables that best differentiates career groups is different for males versus females, both in terms of which career groups can be discriminated and in terms of the signs and magnitudes of the discriminant function coefficients for the *Y* variables. It is possible, for example, that math scores better predict career choice for males and that abstract reasoning scores better predict career choice for females. It is possible that the 3 career groups do not differ significantly on *Y* test scores in the male sample but that at least one of the career groups differs significantly from the other 2 career groups in *Y* test scores in the female sample.

A univariate follow-up to explore the nature of the interaction involves examining the pattern of means across cells to describe gender differences in the predictability of career group from achievement test scores. At the univariate level, the individual *Y* variables that had statistically significant *F*s for the gender by cargroup interaction were English, $F(2, 158) = 3.095$, $p = .048$, and math, $F(2, 158) = 4.605$, $p = .011$. If we use Bonferroni-corrected per-comparison alpha levels, neither of these would be judged statistically significant. The cell means for these two variables (English and math) are given in Table 17.4. For example, there were significant univariate *F*s for the gender by career choice interaction for English, $F(2, 158) = 3.10$, $p = .048$, and for math, $F(2, 158) = 4.61$, $p = .011$. The nature of these univariate interactions was as follows. For males, those with low English scores were likely to choose teaching over business or medicine; for females, differences in English scores across career groups were small, but those with high English scores tended to choose teaching. For males, those with high scores on math tended to choose medicine as a career; for females, those with high scores on math were likely to choose teaching as a career. Thus, these data suggest that the pattern of differences in English and math scores across these 3 career groups might differ for males and females. However,

Table 17.4 ♦ Univariate Cell Means on English and Math Scores for the Gender by Career Interaction

	Business	*Medicine*	*Teaching*
English scores: $F^a(2, 158) = 3.095$, $p = .048$, partial $\eta^2 = .038$			
Male	87.00	86.83	74.75
Female	88.68	87.24	90.31
Math scores: $F^a(2, 158) = 4.605$, $p = .011$, partial $\eta^2 = .055$			
Male	25.93	36.83	23.25
Female	19.07	23.38	25.86

a. Each *F* ratio reported in this table has (2, 158) *df*. The *p* values have not been adjusted for inflated risk of Type I error; that is, they correspond to the theoretical risk of Type I error for a single univariate *F* test. Only the outcome variables that had a significant univariate interaction between the gender and cargroup factors at the $\alpha = .05$ level are included in this table.

given the very small *n*s in two of the cells and a nonsignificant *p* value for the multivariate interaction, this would not be a publishable outcome.

A multivariate follow-up to this finding of an interaction can be obtained by doing DA to assess differences among the 3 career groups separately for the male and female samples. This provides a way to describe the multivariate pattern of scores that best differentiates among the 3 career groups for the male and female participants. Selected output from two separate DAs across the three levels of the career group factor (one DA for male participants and a separate DA for female participants) appears in Figures 17.17 and 17.18.

These analyses provide information about simple main effects—that is, the differences in scores across the three levels of the cargroup factor separately for each gender group. In other words, does the pattern of test scores that predicts choice of career differ for females versus males? A DA to compare career group test scores for men indicated that for men, test scores were not significantly related to career group choice; the lack of significant differences among career groups for males could be due to low power (because of very small sample sizes in two of the cells). For females, the business group could be differentiated from the other 2 career groups. Females who chose business as a career tended to score higher on office and lower on math than females who chose the other two careers. It would be interesting to see whether a different pattern of career group differences would emerge if we could obtain data for a much larger number of male participants; the sample of male participants in this study was too small to provide an adequate assessment of career group differences for males.

17.14.3 ♦ Further Discussion of Problems With This 2 × 3 Factorial MANOVA

In the "real world," analysis of these 2 × 3 factorial data should be halted when the data analyst notices the small numbers of cases in several cells (in Figure 17.8) or when the data analyst obtains a *p* value of .063 for the interaction. Together, these outcomes make these data inconclusive; we cannot be sure whether or not an interaction is present. For instructional purposes, these data were used to demonstrate some possible ways to

Log Determinants[d]

CARGROUP	Rank	Log Determinant
business	6	17.017
medicine	.[a]	.[b]
teach_social	.[c]	.[b]
Pooled within-groups	6	20.139

The ranks and natural logarithms of determinants printed are those of the group covariance matrices.

a. Rank < 6

b. Too few cases to be non-singular

c. Rank < 4

d. GENDER = male

Test Results[a,b]

Tests null hypothesis of equal population covariance matrices.

a. No test can be performed with fewer than two nonsingular group covariance matrices.

b. GENDER = male

NOTE: The variance/covariance matrices for the medicine and teaching groups were "not of full rank" because the numbers of cases in these groups were too small relative to the number of variances and covariances among variables that needed to be estimated; there were not enough degrees of freedom to obtain independent estimates of these variances and covariances. In actual research, an analysis with such small *ns* would not be reported. This example was included here to illustrate the error messages that arise when within-group *ns* are too small.

Selected Discriminant Function Results Across Three Levels of Cargroup for Males

Eigenvalues[b]

Function	Eigenvalue	% of Variance	Cumulative %	Canonical Correlation
1	.554[a]	55.1	55.1	.597
2	.451[a]	44.9	100.0	.558

a. First 2 canonical discriminant functions were used in the analysis.

b. GENDER = male

Wilks's Lambda [a]

Test of Function(s)	Wilks's Lambda	Chi-square	df	Sig.
1 through 2	.443	15.050	12	.239
2	.689	6.889	5	.229

a. GENDER = male

Figure 17.17 ♦ Follow-Up to Assess the Nature of the Nonsignificant Gender by Cargroup Interaction: Discriminant Analysis of Scores on Six Tests Across 3 Career Groups (Only for Males)

Eigenvalues[b]

Function	Eigenvalue	% of Variance	Cumulative %	Canonical Correlation
1	.616[a]	97.3	97.3	.618
2	.017[a]	2.7	100.0	.130

a. First 2 canonical discriminant functions were used in the analysis.

b. GENDER = female

Wilks's Lambda[a]

Test of Function(s)	Wilks's Lambda	Chi-square	df	Sig.
1 through 2	.608	66.871	12	.000
2	.983	2.290	5	.808

a. GENDER = female

Standardized Canonical Discriminant Function Coefficients[a]

	Function	
	1	2
english	.347	.430
reading	.191	-.071
mechanic	.187	-.092
abstract	-.116	-.006
math	-.742	.723
office	.798	.440

a. gender = female

Functions at Group Centroids[a]

	Function	
CARGROUP	1	-5.18E-04
business	.634	2
medicine	-.959	-.264
teach_social	-.946	.160

Unstandardized canonical discriminant functions evaluated at group means

a. GENDER = female

Figure 17.18 ♦ Follow-Up Analysis to Assess the Nature of Nonsignificant Gender by Cargroup Interaction: Discriminant Analysis to Assess Differences in Six Test Scores Across Levels of Career Group (for Females Only)

examine the nature of main effects and interactions, but these results would not be publishable because of small cell sizes.

Note that the Discussion section following this Results section would need to address additional issues. Gender differences in contemporary data might be entirely different from any differences seen in those data collected prior to 1970. As Gergen (1973) pointed out in his seminal paper "Social Psychology as History," some findings may be specific to the time period when the data were collected; gender differences in achievement tests have tended to become smaller in recent years. Furthermore, causal inferences about either career group or gender differences would be inappropriate because neither of these were manipulated factors.

17.15 ♦ A Significant Interaction in a 3 × 6 MANOVA

A second empirical example is presented as an illustration of analysis of an interaction in MANOVA. Warner and Sugarman (1986) conducted an experiment to assess how person perceptions vary as a function of two factors. The first factor was type of information; the second factor was individual target persons. Six individual target persons were randomly selected from a set of 40 participants; for these target individuals, three different types of information about expressive behavior and physical appearance were obtained. A still photograph (head and shoulders) was taken to provide information about physical appearance and attractiveness. A standard selection of text from a textbook was given to all participants. A speech sample was obtained for each participant by tape-recording the participant reading the standard text selection out loud. A handwriting sample was obtained for each participant by having each participant copy the standard text sample in cursive handwriting. Thus, the basic experimental design was as follows: For each of the six target persons (with target person treated as a random factor), three different types of information were obtained: a facial photograph, a speech sample, and a handwriting sample. Thus, there were 18 different conditions in the study, such as a handwriting sample for Person 2 and a speech sample for Person 4.

A between-subjects experiment was performed using these 18 stimuli as targets for person perception. A total of 404 undergraduate students were recruited as participants; each participant made judgments about the personality of one target person based on one type of information about that target person (such as the handwriting sample of Person 3). Multiple-item measures that included items that assessed perceptions of friendliness, warmth, dominance, intelligence, and other individual differences were administered to each participant. For each participant rating each target person on one type of information (e.g., photograph, speech sample, or handwriting), scores on the following six dependent variables were obtained: ratings of social evaluation, intellectual evaluation, potency or dominance, activity, sociability, and emotionality.

Because the interaction between type of information and target was statistically significant, an analysis of simple main effects was performed to evaluate how the targets were differentiated within each type of information. The <Data> → <Split File> command in SPSS was used to divide the file into separate groups based on type of information (facial appearance, handwriting, speech); then, a DA was done separately for each type of information to assess the nature of differences in attributions of personality made to the six randomly selected target persons. As discussed in the following Results section, each of these three follow-up DAs told a different story about the nature of differences in personality attributions. When presented with information about physical appearance, people rated targets differently on social evaluation; when presented with information about handwriting, people differentiated targets on potency or dominance; and when given only a speech sample, people distinguished targets on perceived activity level.

Results

A 6 × 3 MANOVA was performed on the person perception data (Warner & Sugarman, 1986) using six personality scales (social evaluation, intellectual evaluation, potency, activity, sociability, and emotionality) as dependent variables. The factors were target person (a random factor with six levels) and type of information (a fixed factor with Level 1 = a slide of the facial photograph, Level 2 = a handwriting sample, and Level 3 = a speech sample for each target person). Each of the 18 cells in this design corresponded to one combination of target person and type of information—for example, the speech sample for Person 3. Although the 404 participants who rated these stimuli were each randomly assigned to one of the 18 cells in this 3 × 6 design (each participant rated only one type of information for one individual target person), the cell *ns* were slightly unequal. Type III sums of squares were used to correct for the minor confounding between factors that occurred because of the slightly unequal *ns* in the cells; that is, the *SS* term for each effect in this design was calculated controlling for all other effects.

Preliminary data screening did not indicate any serious violations of the assumption of multivariate normality or of the assumption of linearity of associations between quantitative outcome variables. The Box *M* test (using $\alpha = .10$ as the criterion for significance) did not indicate a significant violation of the assumption of homogeneity of variance/covariance matrices across conditions. Table 17.5 shows the pooled or averaged within-cell correlations among the six outcome variables. Intercorrelations between measures ranged from −.274 to +.541. None of the correlations among outcome variables was sufficiently large to raise concerns about multicollinearity.

Table 17.5 ♦ Pooled Within-Cell Correlation Matrix for Six Person Perception Outcome Scales in the 3 × 6 MANOVA Example

	Scale				
Scale	*1*	*2*	*3*	*4*	*5*
1. Social evaluation					
2. Intellectual evaluation	+.24				
3. Potency/dominance	+.14	+.26			
4. Activity	+.37	+.20	+.38		
5. Sociability/extraversion	+.50	+.01	+.22	+.54	
6. Emotionality	−.16	−.11	−.33	−.06	−.27

SOURCE: Warner and Sugarman (1986).

For the overall MANOVA, all three multivariate tests were statistically significant (using $\alpha = .05$ as the criterion). For the target person by type of information interaction, Wilks's $\Lambda = .634$, approximate $F(60, 1886) = 2.86$, $p < .001$. The corresponding η^2 effect size of .37 indicated a strong effect for this interaction. The main effect for target person was also statistically significant, with $\Lambda = .833$, approximate $F(30, 1438) = 2.25$, $p < .001$; this also corresponded to a fairly strong effect size ($\eta^2 = .17$). The main effect for type of information

(Continued)

(Continued)

was also statistically significant; because this factor was crossed with the random factor, target person, the interaction between target person and type of information would ordinarily be used as the error term for this test (see Chapter 13 for a brief description of fixed vs. random factors). However, for these data, a more conservative test of the main effect for type of information was obtained by using within-cell variances and covariances instead of interaction as the error term. The main effect for type of information was statistically significant, with $\Lambda = .463$, approximate $F(12, 718) = 28.12$, $p < .001$. This suggested that there was a difference in mean ratings of personality based on the type of information provided. For example, mean ratings of level of activity were higher for tape-recorded speech samples than for still photographs. This effect is not of interest in the present study, and therefore it is not discussed further.

Because the interaction was statistically significant and accounted for a relatively large proportion of variance, the follow-up analyses focused primarily on the nature of this interaction. An analysis of simple main effects was performed; that is, a one-way DA was performed to assess differences among the six target persons, separately for each level of the factor type of information. The goal of this analysis was to evaluate how the perceptions of the target persons differed depending on which type of information was provided. In other words, we wanted to evaluate whether people make different types of evaluations of individual differences among targets when they base their perceptions on facial appearance than when they base their perceptions on handwriting or speech samples.

Table 17.6 reports the standardized discriminant function coefficients that were obtained by performing a separate DA (across six target persons) for each type of information condition; a separate DA was performed on the ratings based on photographs of facial appearance (top panel), handwriting samples (middle panel), and tape-recorded speech sample (bottom panel). For each of these three DAs, five discriminant functions were obtained; coefficients for only the first discriminant function in each analysis are reported in Table 17.6. For all three DAs, the multivariate significance test for the entire set of five discriminant functions was statistically significant using $\alpha = .05$ as the criterion. Tests of sets of discriminant functions were examined to decide how many discriminant functions to retain for interpretation; because the set of discriminant functions from 2 through 5 was not statistically significant for two out of the three types of information, only the first discriminant function was retained for interpretation.

An arbitrary standard was used to evaluate which of the discriminant function coefficients were relatively large; an outcome variable was interpreted as an indication of meaningful differences across groups only if the standardized discriminant function coefficient had an absolute value >.50. *F* ratios were obtained for each of the individual outcome variables to see whether scores differed significantly on each type of evaluation in a univariate context as well as in a multivariate analysis that controls for intercorrelations among outcome measures.

Table 17.6 ♦ Analysis of Simple Main Effects to Describe the Nature of the Type of Information by Target Person Interaction: Discriminant Analysis of Differences in Ratings Received by Six Target Persons Separately for Each of the Three Types of Information

Scale	*Standardized Discriminant Function Coefficient for D_1*	*Univariate F*	*Univariate p*
Type of information Level 1: Facial appearance			
Social evaluation	**−.82**	**8.89**	**<.001**
Intellectual evaluation	+.35	3.03	.011
Potency/dominance	−.26	2.10	.065
Activity	−.18	**4.66**	**<.001**
Sociability/extraversion	−.11	**4.87**	**<.001**
Emotionality	−.03	0.45	.81
Type of information Level 2: Handwriting			
Social evaluation	+.31	1.19	.31
Intellectual evaluation	−.22	2.04	.07
Potency/dominance	**−.89**	**3.71**	**.003**
Activity	−.16	1.81	.11
Sociability/extraversion	+.20	1.50	.20
Emotionality	−.36	.27	.93
Type of information Level 3: Speech sample			
Social evaluation	+.46	**3.94**	**.002**
Intellectual evaluation	−.18	3.13	.009
Potency/dominance	−.43	.53	.78
Activity	**+1.07**	**7.92**	**<.001**
Sociability/extraversion	−.50	1.99	.08
Emotionality	+.02	0.75	.58

SOURCE: Warner and Sugarman (1986).

NOTES: Standardized discriminant function coefficients were reported only for the first discriminant function (D_1) for each discriminant analysis; a coefficient was judged large if it exceeded an arbitrary cutoff of .50 in absolute value. Within each type of information, the statistical significance for each univariate F test was assessed using Bonferroni-corrected alpha values; the experiment-wise (EW) alpha for each set of tests was set at $EW_a = .05$; there were $p = 6$ univariate F tests performed within each type of information, therefore an individual F was judged statistically significant only if its obtained p value was less than .05/6—that is, less than .008. Large discriminant function coefficients and significant univariate F values appear in bold font.

For the first type of information, facial appearance, only the outcome variable social evaluation (e.g., warmth, friendliness, attractiveness) had a large coefficient on the first discriminant function. In other words, when the participants made their person perception ratings based on facial appearance, the dimension along which the target persons were most clearly differentiated was

(Continued)

(Continued)

social evaluation. This is consistent with past findings that people tend to give globally positive overall evaluations to persons who are more facially attractive. The univariate *F* test for scores on social evaluation across the six target persons was also statistically significant. Thus, whether we statistically control for intercorrelations of social evaluation with other outcome variables or not, the variable that most clearly differentiated target persons when ratings were based on information about facial appearance only was the social evaluation scale. Table 17.6 shows significant univariate differences across target persons for several other outcome variables; using $\alpha = .05$ as the criterion, the ratings given to the six different target persons in the facial photograph condition differed significantly for intellectual evaluation, activity, and sociability (in addition to social evaluation) when univariate tests were examined. However, when intercorrelations among the six personality evaluation outcome variables were taken into account (by forming discriminant functions), the only variable that showed clear differences across targets was the rating of social evaluation. In other words, we may want to interpret the univariate differences in the ratings of intellect, activity, and sociability for the six photographs as having arisen only because these characteristics are correlated with social evaluation; when their correlations with social evaluation are taken into consideration and the six personality characteristics are evaluated in the context of a DA, the only variable that continued to be clearly differentiated across target persons was the social evaluation score.

For the second type of information, handwriting, only one variable (ratings of potency or dominance) had a large coefficient on the first discriminant function; it also had a significant univariate *F*, $p = .003$. In other words, when the raters based their evaluations only on samples of handwriting, the only personality characteristic or dimension along which the target persons were clearly distinguished was dominance or potency. Examination of the specific handwriting samples that received higher dominance ratings indicated that the outcome of this study was consistent with earlier studies suggesting that people tend to evaluate large handwriting as more "dominant" than smaller handwriting.

For the third type of information, speech samples, one variable had a large standardized discriminant function coefficient on the first discriminant function: ratings of activity. Sociability had a discriminant coefficient of −.50, and potency and social evaluation each had standardized discriminant function coefficients >.40 in absolute value. It was not clear whether differences on these three additional variables across targets should be interpreted. If we adhere strictly to the arbitrary standard stated earlier (that a discriminant function coefficient will be interpreted as large only if it exceeds .50 in absolute value), then we have a simpler outcome to discuss: Speech samples were more clearly differentiated in ratings of activity than on any other personality variables. Several outcome variables also had statistically significant univariate *F* values; only activity ratings had both a standardized discriminant function coefficient >.5 in absolute magnitude and a statistically significant univariate

F values; only activity ratings had both a standardized discriminant function coefficient >.5 in absolute magnitude and a statistically significant univariate *F* value ($p < .001$). Speech rate apparently influenced ratings of activity (e.g., people who read the standard selection more rapidly and in a louder tone of voice were rated more "active" than persons who read the text selection more slowly and less loudly). Something about the speech samples may also have influenced the ratings of sociability, social evaluation, and extraversion, but these results are far less clear. The decision was made to focus the interpretation of the ratings of speech samples on the one outcome variable that showed the strongest evidence of differences among targets in both the DA and the univariate *F* tests—that is, the ratings of activity.

The nature of the statistically significant target by type of information interaction can be summarized as follows. The judges in this study differentiated the six randomly selected target persons on different personality dimensions depending on which type of information was available for them to make their judgments. When only information about facial appearance was presented, targets were differentiated primarily on social evaluation. When only information about handwriting was made available, the targets were differentiated only on potency or dominance. When only information about speech was available, target persons were differentially rated on activity. These results are consistent with earlier research suggesting that different channels of communication or different types of expressive behavior are used to make attributions about different aspects or dimensions of personality. Judgments about social evaluation were based primarily on facial appearance. Judgments about potency were made based on handwriting. Differential evaluations of activity level were made only in the speech sample condition.

Because target persons were selected randomly from a group of 40 persons and not chosen in a manner that represented levels of some fixed factor (such as physical attractiveness), the values of univariate and multivariate means for specific targets are not reported. For example, it would not be informative to report that Person 1 was rated higher on social evaluation than Person 4 in the handwriting information condition.

17.16 ♦ Comparison of Univariate and Multivariate Follow-Up Analyses for MANOVA

The preceding examples have presented both univariate follow-up analyses (univariate *F*s and univariate Tukey post hoc tests) and multivariate follow-up analyses (discriminant functions that differ across groups). There are potential problems with univariate follow-up analyses. First, running multiple significance tests for univariate follow-ups may result in an inflated risk of Type I error (Bonferroni-corrected per-comparison alpha levels may be used to limit the risk of Type I error for large sets of significance tests). Second, these univariate analyses do not take intercorrelations among the individual *Y* outcome variables into account. This can be problematic in two ways. First, some of the *Y* variables may be highly correlated or confounded and may, thus, provide redundant information; univariate analysis may mislead us into thinking that we have detected significant differences on *p* separate *Y* outcome variables, when in fact, many of the outcome variables (or subsets of the *Y* outcome variables) are so highly correlated that they may actually be measures of the same underlying construct. Furthermore, as discussed in Chapter 10, we saw that the nature and strength of the association between a pair of variables can change substantially when we statistically control for one or more additional variables. In the context of MANOVA, this can lead to several different outcomes. Sometimes in MANOVA, the *Y* variables that show large univariate differences across groups when the variables are examined individually are the same variables that have the largest weights in forming discriminant functions that differ across groups. However, it is possible for a variable that shows large differences across groups in a univariate analysis to show smaller differences across groups when other *Y* outcome variables are statistically controlled. It is possible for suppression to occur; that is, a *Y* variable that was not significant in a univariate analysis may be given substantial weight in forming one of the discriminant functions, or the rank order of means on a Y_i outcome variable may be different across levels of an *A* factor when Y_i is examined in isolation in a univariate analysis than when Y_i is evaluated in the context of a multivariate analysis such as DA. Discussion of the nature of group differences is more complex when multivariate outcomes for some of the Y_i variables differ substantially from the univariate outcomes for these Y_i variables.

The dilemma that we face in trying to assess which individual quantitative outcome variables differ across groups in MANOVA is similar to the problem that we encounter in other analyses such as multiple regression. We can only obtain an accurate assessment of the nature of differences across groups on an individual *Y* outcome variable in the context of a correctly specified model—that is, an analysis that includes all the variables that need to be statistically controlled and that does not include variables that should not be controlled. We can never be certain that the model is correctly specified—that is, that the MANOVA includes just the right set of factors and outcome variables. However, we can sometimes obtain a better understanding of the nature of differences across groups on an individual Y_i outcome variable (such as math scores or ratings of dominance) when we control for other appropriate outcome variables. The results of any univariate follow-up analyses that are reported for a MANOVA need to be evaluated relative to the multivariate outcomes.

It is prudent to keep the number of outcome variables in a MANOVA relatively small for two reasons. First, there should be some theoretical rationale for inclusion of each outcome measure; each variable should represent some outcome that the researcher reasonably expects will be influenced by the treatment in the experiment. If some of the outcome variables are very highly correlated with each other, it may be preferable to consider whether those two or three variables are in fact all measures of the same construct; it may be better to average the raw scores or *z* scores on those highly correlated measures into a single, possibly more reliable outcome measure before doing the MANOVA. Chapter 19 discusses the issues involved in combining scores across multiple items or measures.

Whatever method of follow-up analysis you choose, keep this basic question in mind. For each contrast that is examined, you want to know whether the groups differ significantly (using an omnibus multivariate test) and what variable or combination of variables (if any) significantly differentiates the groups that are being compared. When we do multiple univariate significance tests in follow-up analyses to MANOVA, there is an inflated risk of Type I error associated with these tests. Finally, our understanding of the pattern of differences in values of an individual Y_i variable may be different when we examine a Y_i variable by itself in a univariate ANOVA than when we examine the Y_i variable in the context of a DA that takes correlations with other Y outcome variables into account. See Bray and Maxwell (1985) and Stevens (1992) for more detailed discussions of follow-up analyses that can clarify the nature of group differences in MANOVA.

17.17 ♦ Summary

Essentially, MANOVA is DA in reverse. Both MANOVA and DA assess how scores on multiple intercorrelated quantitative variables are associated with group membership. In MANOVA, the group membership variable is generally treated as the predictor, whereas in DA, the group membership variable is treated as the outcome. DA is generally limited to one-way designs, whereas MANOVA can involve multiple factors, repeated measures, and/or covariates. In addition, when one or more quantitative covariates are added to the analysis, MANOVA becomes **multivariate analysis of covariance, MANCOVA.**

MANOVA and DA provide complementary information, and they are often used together. In reporting MANOVA results, the focus tends to be on which groups differed significantly. In reporting DA, the focus tends to be on which discriminating variables were useful in differentiating groups and sometimes on the accuracy of classification of cases into groups. Thus, a DA is often reported as one of the follow-up analyses to describe the nature of significant group differences detected in MANOVA.

The research questions involved in a two-way factorial MANOVA include the following:

1. Which levels of the A factor differ significantly on the set of means for the outcome variables $Y_1, Y_2, \ldots, Y_p$?
2. Which levels of the B factor differ significantly on the set of means for $Y_1, Y_2, \ldots, Y_p$?
3. What do differences in the pattern of group means for the outcome variables $Y_1, Y_2, \ldots, Y_p$ tell us about the nature of $A \times B$ interaction?

In reporting a factorial ANOVA—for example, an $A \times B$ factorial—the following results are typically included. For each of these effects—main effects for A and B and the $A \times B$ interaction—researchers typically report the following:

1. An overall significance test (usually Wilks's Λ and its associated F, sometimes Pillai's trace or Roy's largest root)
2. An effect size—that is, a measure of the proportion of explained variance (1 – Wilks's Λ); this is comparable with $1 - \eta^2$ in an ANOVA
3. A description of the nature of the differences among groups; this can include the pattern of univariate means and/or DA results

Thus, when we look at the main effect for the *A* factor (cargroup), we are asking, What two discriminant functions can we form to discriminate among these 3 groups? (In general, if a factor has *k* levels or groups, then you can form $k - 1$ discriminant functions; however, the number of discriminant functions cannot exceed the number of dependent variables, so this can be another factor that limits the number of discriminant functions.) Can we form composites of the *z* scores on these six variables that are significantly different across these 3 career groups?

When we look at the main effect for the *B* factor (gender), we are asking, Can we form one weighted linear combination of these variables that is significantly different for men versus women? Is there a significant difference overall in the "test profile" for men versus women?

When we look at the $A \times B$ interaction effect, we are asking a somewhat more complex question: Is the pattern of differences in these test scores among the 3 career groups different for women than it is for men? If we have a significant $A \times B$ interaction, then several interpretations are possible. It could be that we can discriminate among men's career choices—but not among women's choices—using this set of six variables (or vice versa, perhaps we can discriminate career choice for women but not for men). Perhaps the pattern of test scores that predicts choice of a medical career for men gives a lot of weight to different variables compared with the pattern of test scores that predicts choice of a medical career for women. A significant interaction suggests that the nature of the differences in these six test scores across these 3 career groups is not the same for women as for men; there are many ways the patterns can be different.

MANOVA may provide substantial advantages over a series of univariate ANOVAs (i.e., a separate ANOVA for each *Y* outcome variable). Reporting a single *p* value for one omnibus test in MANOVA, instead of *p* values for multiple univariate significance tests for each *Y* variable, can limit the risk of Type I error. In addition, MANOVA may be more powerful than a series of univariate ANOVAs, particularly when groups differ in their patterns of response on several outcome variables.

+ − × ÷

Comprehension Questions

1. Design a factorial MANOVA study. Your design should clearly describe the nature of the study (experimental vs. nonexperimental), the factors (what they are and how many levels each factor has), and the outcome measures (whether they are multiple measures of the same construct or measures of different but correlated outcomes). What pattern of results would you predict, and what follow-up analyses do you think you would need?
2. In words, what null hypothesis is tested by the Box *M* value on the SPSS GLM printout for a MANOVA? Do you want this Box *M* test to be statistically significant? Give reasons for your answer. If Box *M* is statistically significant, what consequences might this have for the risk of Type I error and for statistical power? What might you do to remedy this problem?
3. For each of the following sentences, fill in the correct term (*increases*, *decreases*, or *stays the same*):

   As Wilks's Λ increases (and assuming all other factors remain constant), the multivariate *F* associated with Wilks's Λ ____________.

   As Pillai's trace increases, all other factors remaining constant, the associated value of *F* ____________.

   As Wilks's Λ increases, the associated value of partial η^2 ____________.

   As Wilks's Λ increases, the corresponding value of Pillai's trace ____________.
4. Suppose that your first discriminant function has an eigenvalue of $\lambda_1 = .5$.
   What is the value of r_c, the canonical correlation, for this function? What is the value of Roy's largest root for this function?
5. Suppose that you plan to do a one-way MANOVA with 5 groups, using $\alpha = .05$. There are four outcome variables, and you want to have power of .70. For each of the following effect sizes, what minimum number of cases per cell or group would be needed? (Use Table 17.1 to look up recommended sample sizes.)

   For a "very large" effect: _____

   For a "moderate" effect: _____

   For a "small" effect: _____
6. Find a journal article that reports a MANOVA, and write a brief (about three pages) critique, including the following:
   a. A brief description of the design: the nature of the study (experimental/nonexperimental), the factors and number of levels, a description of outcome variables
   b. Assessment: Was any information included about data screening and detection of any violations of assumptions?

c. Results: What results were reported? Do the interpretations seem reasonable?
d. Is there other information that the author could have reported that would have been useful?
e. Do you see any problems with this application of MANOVA?

7. Describe a hypothetical experimental research situation where MANOVA could be used.
8. Describe a hypothetical nonexperimental research situation in which MANOVA could be used.
9. In addition to the usual assumptions for parametric statistics (e.g., normally distributed quantitative variables, homogeneity of variance across groups), what additional assumptions should be tested when doing a MANOVA?
10. What information is provided by the elements of a within-group or within-cell **SCP** matrix?
11. How is the Wilks's Λ statistic obtained from the within-group and total **SCP** matrices?

    In a one-way ANOVA, what does Wilks's Λ correspond to? How is Wilks's Λ interpreted in a MANOVA?
12. Explain how MANOVA is similar to/different from ANOVA.
13. Explain how MANOVA is related to DA.
14. Using your own data or data provided by your instructor, conduct your own factorial MANOVA; use at least four outcome variables. If you obtain a statistically significant interaction, do appropriate follow-up analyses. If you do not obtain a statistically significant interaction, describe in detail what follow-up analyses you would need to do to describe the nature of an interaction in this type of design. Write up your results in the form of a journal article Results section.

CHAPTER 18

Principal Components and Factor Analysis

18.1 ♦ Research Situations

The term *factor analysis* actually refers to a group of related analytic methods. Programs such as SPSS offer numerous options for the set up of factor analysis using different computational procedures and decision rules. This chapter describes the basic concepts involved in only two of the analyses that are available through the SPSS factor procedure. The two analyses discussed here are called principal components (PC) analysis and principal axis factoring (PAF). PC is somewhat simpler and was developed earlier. PAF is one of the methods that is most widely reported in published journal articles. Section 18.11 explains the difference in computations for PC and PAF. Although the underlying model and the computational procedures differ for PC and PAF, these two analytic methods are frequently employed in similar research situations—that is, situations in which we want to evaluate whether the scores on a set of p individual measured X variables can be explained by a small number of **latent variables** called components (when the analysis is done using PC) or **factors** (when the analysis is done using PAF).

Up to this point, the analyses that were considered in this book have typically involved prediction of scores on one or more measured dependent variables from one or more measured independent variables. For example, in multiple regression, the score on a single quantitative Y dependent variable was predicted from a weighted linear combination of scores on several quantitative or dummy X_i predictor variables. For example, we might predict each individual faculty member's salary in dollars (Y) from number of years in the job (X_1), number of published journal articles (X_2), and a dummy coded variable that provides information about gender.

In PC and PAF, we will not typically identify some of the measured variables as X predictors and other measured variables as Y outcomes. Instead, we will examine a set of z scores on p measured X variables (p represents the number of variables) to see whether it would make sense to interpret the set of p measured variables as measures of some smaller number of underlying constructs or latent variables. Most of the path models presented here involve **standard scores** or **z scores** for a set of measured variables. For example, $z_{X_1}, z_{X_2}, z_{X_3}, \ldots, z_{X_p}$ represent z scores for a set of p different tests or measurements.

The information that we have to work with is the set of correlations among all these p measured variables. The inference that we typically want to make involves the question whether we can reasonably interpret some or all the measured variables as measures of the same latent variable or construct. For example, if $X_1, X_2, X_3, \ldots, X_p$ are scores on tests of vocabulary, reading comprehension, understanding analogies, geometry, and solving algebraic problems, we can examine the pattern of correlations among these variables to decide whether all these X variables can be interpreted as measures of a single underlying latent variable (such as "mental ability"), or whether we can obtain a more accurate understanding of the pattern of correlations among variables by interpreting some of the variables as a measure of one latent variable (such as "verbal ability") and other variables as measures of some different latent variable (such as "mathematical ability").

A latent variable can be defined as "an underlying characteristic that cannot be observed or measured directly; it is hypothesized to exist so as to explain [manifest] variables, such as behavior, that can be observed" (Vogt, 1999, pp. 154–155). A latent variable can also be called a component (in PC) or a factor (in factor analysis), and it is sometimes also described as a "dimension."

An example of a widely invoked latent variable is the concept of "intelligence." We cannot observe or measure intelligence directly; however, we can imagine a quality that we call "general intelligence" to make sense of patterns in scores on tests that are believed to measure specific mental abilities. A person who obtains high scores on numerous tests that are believed to be good indicators of intelligence is thought to be high on the dimension of intelligence; a person who obtains low scores on these tests is thought to occupy a low position on the dimension of intelligence. The concept of intelligence (or any latent variable, for that matter) is problematic in many ways. Researchers disagree about how many types or dimensions of intelligence there are and what they should be called; they disagree about the selection of indicator variables. Researchers disagree about which variety of factor analysis yields the best description of patterns in intelligence.

Once a latent variable such as intelligence has been given a name, it tends to become "reified"—that is, people think of it as a "real thing" and forget that it is a theoretical or imaginary construct. This reification can have social and political consequences. An individual person's score on an intelligence quotient (IQ) test may be used to make decisions about that person (e.g., whether to admit Kim to a program for intellectually gifted students). When we reduce information about a person to a single number (such as an IQ score of 145), we obviously choose to ignore many other qualities that the person has. If our measurement methods to obtain IQ scores are biased, the use of IQ scores to describe individual differences among persons and to make decisions about them can lead to systematic discrimination among different ethnic or cultural groups. See Gould (1996) for a history of IQ measurement that focuses on the problems that can arise when we overinterpret IQ scores—that is, when we take them to be more accurate, valid, and complete information about some theoretical construct (such as "intelligence") than they actually are.

In PC and PAF, the X_i scores are treated as multiple indicators or multiple measures of one or more underlying constructs or *latent variables* that are not directly observable. A latent variable is an "imaginary" variable, a variable that we create as part of a model that represents the structure of the relationships among the X_i measured variables; information about structure is contained in the **R** matrix of correlations among the X measured variables. When we use the PC method of **extraction**, each latent variable is called a component; when we use the PAF method of extraction, each latent variable is called a factor.

Many familiar applications of factor analysis involve X variables that represent responses to self-report measures of attitude or scores on mental ability tests. However, factor analysis can be used to understand a pattern in many different types of data. For instance, the X variables could be physiological measures; electrodermal activity, heart rate, salivation, and pupil dilation might be interpreted as indications of sympathetic nervous system arousal.

18.2 ♦ Path Model for Factor Analysis

In a factor analytic study, we examine a set of scores on p measured variables ($X_1, X_2, \ldots, X_p$); the goal of the analysis is not typically to predict scores on one measured variable from scores on other measured variables but, rather, to use the pattern of correlations among the X_i measured variables to make inferences about the structure of the data—that is, the number of latent variables that we need to account for the correlations among the variables. Part of a path model that corresponds to a factor analysis of z scores on $X_1, X_2, \ldots, X_p$ is shown in Figure 18.1.

Figure 18.1 corresponds to a theoretical model in which the outcome variables ($z_{X_1}, z_{X_2}, \ldots, z_{X_p}$) are standardized scores on a set of quantitative X measured variables, for example, z scores on p different tasks that assess various aspects of mental ability. The predictor variables are "latent variables" called factors (or sometimes components); factors are denoted by $F_1, F_2, \ldots, F_p$ in Figure 18.1. Each a_{ij} path in Figure 18.1 represents a correlation between a measured z_{X_i} variable and a latent F_j factor. In factor analysis, we

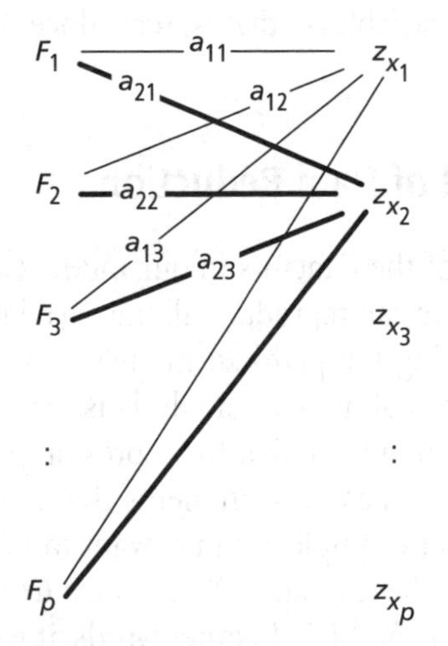
F_1
a_{11}
z_{x_1}
a_{21}
a_{12}
F_2
a_{22}
z_{x_2}
a_{13}
a_{23}
F_3
z_{x_3}
F_p
z_{x_p}
a_{ij} corresponds to the loading or correlation between variable x_{x_i} and Factor F_j
This diagram shows only the paths from z_{x1} and z_{x_2} to each of the Factors. The paths that connect $z_{x_3}, z_{x_4}, \ldots, z_{x_p}$ to the Factors are not shown.
To reproduce part of the correlation between X_1 and X_2 by tracing a path via F_1, we multiply $a_{11} \times a_{21}$.
To reproduce the part of the correlation between X_1 and X_2 that is explained by F_2, we multiple $a_{12} \times a_{22}$ and in general, to use the path via Factor j to reproduce part of the X_1, X_2 correlation, we multiply $a_{1j} \times a_{2j}$. The total correlation 1_{12} between X_1 and X_2 is the sum of all the aij products that correspond to all possible paths via Factors 1 to j.

Figure 18.1 ♦ Path Model for an Orthogonal Factor Analysis With p Measured Variables ($z_{X_1}, z_{X_2}, \ldots, z_{X_p}$) and p Factors ($F_1, F_2, \ldots, F_p$)

estimate correlations between all possible pairs of variables (z_{X_1} through z_{X_p}) and all pairs of possible latent variables or factors (F_1 through F_p). These correlations between measured variables and latent variables are also called "loadings."

Recall from Chapter 11 that the coefficients for paths in a "causal" model are estimated in such a way that they can be used to reconstruct the correlations among the measured X variables. For example, the beta coefficients for a standardized multiple regression with just two predictor variables (shown in Figure 11.11) were calculated from the correlations among the three variables (r_{1Y}, r_{2Y}, and r_{12}). By using the tracing rule to identify all possible paths from z_{X_1} to z_Y, and from z_{X_2} to z_Y, and by multiplying path coefficients along each path and summing contributions across different paths, we could perfectly reconstruct the observed correlations r_{1Y} and r_{2Y} from the regression path model including the beta path coefficients.

We can understand the factor analysis model that appears in Figure 18.1 as a representation of a set of multiple regressions. In the first multiple regression, the score on z_{X_1} is predicted from scores on all p of the factors ($F_1, F_2, \ldots, F_p$); in the next multiple regression, the score on z_{X_2} is predicted from scores on the same set of p factors; and so forth. The path model shown in Figure 18.1 differs from path models that appear in earlier chapters in one important way—that is, in path models in the earlier chapters, all the variables corresponded to things we could actually measure. In Figure 18.1, some of the variables (e.g., $F_1, F_2, \ldots, F_p$) represent theoretical or imaginary constructs or latent variables; the F latent variables are not directly measured or observed. We imagine a set of factors $F_1, F_2, \ldots, F_p$ to set up a theoretical model that will account for the underlying structure in the set of measured X variables—that is, reproduce the correlations among all pairs of measured X variables.

18.3 ♦ Factor Analysis as a Method of Data Reduction

We shall see that if we retain all p of the F factors in our model (where p represents the number of measured X variables), we can reproduce all the correlations among the measured X variables perfectly by tracing the paths in models similar to the models that appear in Figure 18.1. However, the goal in factor analysis is usually "data reduction" or "dimension reduction"; that is, we often hope that the scores on p measured X variables can be understood by thinking of them as measurements that are correlated with *fewer* than p latent variables or factors. For example, we may want to know "How well can we reconstruct the correlations among the measured X variables using only the paths that involve the first two latent variables F_1 and F_2?" In other words, if we drop latent variables F_3 through F_p from the model, can we use this simpler reduced model (with a smaller number of latent variables) to account for the correlation structure of the data? We usually hope that a model that includes a relatively small number of factors (on the order of 1–5 factors) will account for the pattern of correlations among the X variables. For example, if we can identify a set of two latent variables that do a reasonably good job of reconstructing the correlations among the p measured variables, we may be able to interpret the p variables as measures of just two latent variables, factors, or dimensions. For example, consider a set of mental ability tests: X_1, vocabulary; X_2, analogies; X_3, synonyms and antonyms; X_4, reading comprehension; X_5, algebra; X_6, trigonometry; and X_7, geometry. We would expect to see high positive correlations among variables X_1 through X_4 (because they all measure the same kind of ability, i.e., verbal skills) and high positive correlations among X_5 through X_7 (because they all assess related abilities, i.e., mathematical skills). On the other hand, correlations between the first set of measures (X_1 through X_4) and the second set of measures (X_5 through X_7) might be fairly low because these two sets of tests measure different abilities. If we perform a factor analysis on measurements on X_1 through X_7, we may find that we can reconstruct the correlations among these seven measured X variables through their correlations or relationships with just two latent constructs or factors. The first factor or latent variable, which we might label verbal ability, would have high correlations with variables X_1 through X_4 and low correlations with variables X_5 through X_7. The second latent variable or factor, which we might label mathematical skills, would have high correlations with variables X_5 through X_7 and low correlations with variables X_1 through X_4. If we can show that the correlations of these X predictors with just two factors (verbal and math abilities) are sufficient information to do a reasonably good job of reproducing the observed correlations among these X measured variables, this might lead us to conclude that the $p = 7$ test scores included in the set of measured X variables can be understood as information about just *two* latent variables that cannot be directly observed (verbal and math abilities).

Although we cannot directly measure or observe a latent variable, we can understand the nature of a latent variable by examining how it is correlated with measured X variables, and we can use the z scores for measured X variables to construct a score for each participant for each latent variable, factor, or dimension. The goal of factor analysis and related analytic methods such as PC is to assess the extent to which the various X variables in our dataset can be interpreted as measures of one or more underlying constructs or latent variables. In factor analysis, a large Pearson r between measures X_i and X_j is taken as an indication that X_i and X_j may both measure the same thing. For instance, if verbal Scholastic Aptitude Test (X_1) and grade in an English course (X_2) have a high positive correlation, we might assume that they are different measures of the same underlying ability (language skill).

A factor analytic study usually begins with the collection of measurements on many X variables that we think may all measure the "same thing" (e.g., a personality construct such as "extraversion") or, perhaps, a set of related constructs (such as different types of mental ability, e.g., verbal and mathematical). We have two basic questions in mind when we do a factor analysis on a set of X variables (p is the number of variables).

The first question is, How many different theoretical constructs (or components, factors, dimensions, or latent variables) do these p variables seem to measure? In the preceding hypothetical example, we saw that we could represent the correlation structure among a set of $p = 7$ measured variables by using a model that included just two factors or latent variables. The second question is, What (if anything) do these factors mean? Can we name the factors and interpret them as latent variables? In the preceding example, the two factors were named verbal ability and math ability; these labels were based on the type of tests each factor correlated with most highly.

We will answer these two questions (how many latent variables we need and what interpretation we can suggest for each latent variable) by using PC or PAF to convert the correlation matrix of correlations among our p measured variables into a matrix of loadings or correlations that tell us how strongly linearly related each measured X variable is with each component or factor. We can obtain scores for each participant on each latent variable or factor by computing specific weighted linear combinations of each person's scores on z_1 through z_x. In the examples presented in this chapter, the factors are constrained to be orthogonal to each other (i.e., scores on F_1 are almost perfectly uncorrelated with scores on F_2 for all possible pairs of factors).

Data reduction through factor analysis has potential advantages and disadvantages. First consider some potential advantages. Factor analysis can be helpful during the process of theory development and theory testing. Suppose that we analyze correlations among scores on 10 mental ability tests and decide that we can account for the pattern of correlations among scores by interpreting some of the variables as indicators of a latent verbal ability factor and other variables as indicators of a latent math ability factor. A factor analysis may help us to decide how many types of mental ability we need to measure (two types in this hypothetical example). Factor analysis can be an aid in theory development; however, it is only a good tool for making inferences about the "state of the world" if we are confident that the battery of tests included in the factor analysis comprehensively covers the entire domain of mental ability tests. Researchers need to understand that factor analysis and related methods such as PC tell us only how many factors (or components) are needed to account for the correlations among the variables that were included in the test battery in the study; this set of variables may not include all the mental abilities or traits that exist out in the real world.

One example of theory development that made extensive use of factor analysis was the development of the "Big Five" model of personality (Costa & McCrae, 1995, 1997). Factor analyses of large sets of personality measures suggested that much of the pattern in self-report personality data may be accounted for by a model that includes five latent variables, factors, or traits: openness to experience, conscientiousness, extraversion, agreeableness, and neuroticism.

In addition, factor analysis may have a practical application in some data analysis situations where a researcher wants to reduce a large number of predictors (in a multiple regression analysis, for example) to a smaller number of predictors. Suppose that the researcher wants to predict starting salary (Y) from the 10 mental ability test scores used in the previous examples. It might not be a good idea to use each of the 10 test scores as a separate predictor in one multiple regression for two reasons. First (as discussed in Chapter 11), when we include several predictor variables that are highly correlated with each other, to a great extent, they compete to predict or explain the same variance in the Y outcome variable. If we include all three of the verbal ability indicator variables (i.e., scores on vocabulary, reading comprehension, and understanding analogies tests, X_1, X_2, X_3) as predictors in a regression, we may find that none of these three predictors uniquely predicts a significant amount of variance in the Y outcome variable when its correlations with the other two verbal ability measures are statistically controlled. Instead, if we compute a single "verbal ability" **factor score** that summarizes the information contained in variables X_1, X_2, and X_3 and a single "math ability" factor score that summarizes the information contained in the math tests, we can then set up a regression to predict salary (Y) from just two variables: F_1, a factor score that summarizes information about each person's verbal ability; and F_2, a second factor score that summarizes information about each person's math ability. If the factors are orthogonal or uncorrelated, then the new predictor variables F_1 and F_2 will not compete with each other to explain the same variance; F_1 will have a correlation with F_2 that is near 0 if these factor scores are derived from an orthogonal factor analysis. There are potentially several reasons why it may be preferable to report a regression to predict Y from F_1 and F_2 instead of a regression to predict Y from a set of 10 predictor variables, X_1 through X_{10}. This two-predictor regression model may be easier to report and interpret; the predictor variables F_1 and F_2 will not compete with each other to explain the same variance; and this model has only two t tests for the significance of the individual predictors F_1 and F_2. In contrast, a regression with 10 predictors (X_1 through X_{10}) is more difficult to report and interpret; the analysis may involve high levels of multicollinearity among predictors; the degrees of freedom for the error term will be smaller, therefore, significance tests will have less power; and finally, conducting 10 t tests to assess the predictive contributions for each variable (X_1 through X_{10}) could lead to an inflated risk of Type I error. Thus, one practical use for factor analysis is in data reduction; in some research situations, a data analyst may be able to use a preliminary factor analysis to assess whether the information contained in the scores on a large set of individual measures (X_1 through X_p) can be summarized and replaced by a smaller set of factor scores, for example, scores on just two factors.

Factor analysis is often used during the process of development of multiple item scales to measure personality traits such as extraversion, political attitudes such as conservatism, or mental abilities such as intelligence; this is discussed more extensively in Chapter 19. Factor analysis can help a researcher to decide how many different latent variables are needed to understand the responses to test items and to decide which items are good indicators for each latent variable. Factor analysis is also useful for reexamination of existing measures. For example, the Type A personality was originally thought to be one single construct, but factor analysis of self-report items from Type A questionnaires suggested that time urgency, competitiveness/job involvement, and hostility were three separate components of Type A, and further research suggested that hostility was the one component of Type A that was most predictive of coronary heart disease. This resulted in a change in focus in this research domain, with greater interest in measures that are specific to hostility (Linden, 1987).

Factor analysis is also used for theory development and theory testing, although researchers need to be cautious when they try to interpret factors they see in their data as evidence of traits or abilities that exist in the real world. The history of mental ability or IQ testing is closely associated with the development of factor analysis as a statistical method; and the theories about the number of types of intelligence have changed over time as the test batteries became larger and more varied tests were added. There is often a kind of feedback process involved, such that researchers select the variables to include in a factor analysis based on a theory; they may then use the outcome of the factor analysis to help rethink the selection of variables in the next study.

However, there are also some potential disadvantages or problems with the use of factor analysis for data reduction. Factor analysis is sometimes used as a desperation tactic to look for a way to summarize the information in a messy dataset (where the selection of variables was not carefully planned). In research situations where long lists of questions were generated without much theoretical basis, factor analysis is sometimes used to try to make sense of the information contained in a poorly selected collection of variables and to reduce the number of variables that must be handled in later stages of analysis. In such situations, the outcome of the study is "data-driven." As in other data-driven approaches to analysis, such as statistical selection of predictor variables for multiple regression using forward methods of selection, we tend to obtain empirical results that represent peculiarities of the batch of data due to sampling error; these results may not make sense and may not replicate in different samples.

There are two reasons why some people are skeptical or critical of factor analysis. First, the use of factor analysis as a desperation tactic can lead to outcomes that are quite idiosyncratic to the selected variables and the sample. Second, researchers or readers sometimes mistakenly view the results of **exploratory factor analysis** as "proof" of the existence of latent variables; however, as noted earlier, the set of latent variables that are obtained in a factor analysis is highly dependent on the selection of variables that were measured. A researcher who includes only tests that measure verbal ability in the set of X measured variables will not be able to find evidence of mathematical ability or other potential abilities. The latent variables that emerge from a factor analysis are limited to a great extent by the set of measured variables that are input to the factor analysis procedure. Unless the set of X measured variables are selected in a manner that ensures that they cover the entire domain of interest (such as all known types of mental abilities), the outcome of the study tells us only how many types of mental ability were represented by the measurements included in the study and not how many different types of mental ability exist in the real world. Factor analysis and PC can be useful data analysis tools provided that their limitations are understood and acknowledged.

18.4 ♦ Introduction of an Empirical Example

The empirical example included here uses a subset of data collected by Robert Belli using a modified version of the Bem Sex Role Inventory (BSRI; Bem, 1974). The BSRI is a measure of sex role orientation. Participants were asked to rate how well each of 60 adjectives describe them using Likert-type scales (0 = Not at all to 4 = Extremely well). The items in the original BSRI were divided into masculine, feminine, and neutral filler items. Prior measures of masculinity, such as the Minnesota Multiphasic Personality Inventory, implicitly assumed that "masculine" and "feminine" were opposite endpoints of a single continuum—that is, to be "more masculine" a person had to be "less feminine." If this one-dimensional theory is correct, a factor analysis of items from a test such as the BSRI should yield one factor (which is positively correlated with masculine items and negatively correlated with feminine items). Bem (1974) argued that masculinity and femininity are two separate dimensions and that an individual may be high on one or high on both of these sex role orientations. Her theory suggested that a factor analysis of BSRI items should yield a two-factor solution; one factor should be correlated only with masculine items and the other factor should be correlated only with feminine items.

For the empirical example presented here, a subset of $p = 9$ items and $N = 100$ participants was taken from the original larger dataset collected by Belli. The nine items included self-ratings of the following characteristics: warm, loving, affectionate, nurturant, assertive, forceful, strong person, dominant, and aggressive. The complete dataset for the empirical examples presented in this chapter appears in Table 18.1. The correlations among the nine variables in this dataset are summarized in Figure 18.2.

Table 18.1 ♦ Modified Bem Sex Role Inventory Personality Ratings: Nine Items

V1	*V2*	*V3*	*V4*	*V5*	*V6*	*V7*	*V8*	*V9*
1	4	4	2	4	3	2	2	1
1	3	2	2	4	4	3	3	2
2	1	2	3	2	2	1	1	2
2	3	3	3	3	3	1	1	2
3	3	3	3	3	1	1	1	2
0	3	3	3	4	4	4	2	3
3	3	4	4	3	3	1	1	1
2	4	3	3	2	3	1	0	2
3	4	4	3	3	2	3	2	3
1	1	3	3	2	1	1	1	2
4	4	4	4	3	3	1	1	1
1	3	3	2	4	2	2	1	2
3	3	3	3	3	2	2	2	2
2	4	4	4	2	2	3	4	3
2	4	3	2	2	4	2	3	3
2	3	3	2	3	3	2	3	3
2	3	2	2	2	2	2	2	2
2	2	3	1	2	2	2	2	3
3	3	3	4	3	2	2	0	3
3	4	2	3	4	3	2	2	2
3	3	2	2	2	2	2	1	3
2	3	2	2	2	2	2	2	1
2	4	4	4	4	2	1	2	2
1	3	4	3	1	1	2	1	2
3	4	4	4	4	4	2	2	2
2	3	3	3	2	3	3	3	2
2	3	3	2	4	3	3	2	3
2	4	4	3	3	4	2	3	2
2	4	3	3	1	1	2	1	1
2	2	2	3	4	3	2	3	3
3	3	3	3	3	3	3	2	3
2	3	3	3	4	3	3	2	4
3	3	3	3	4	2	3	3	3
4	3	3	3	3	2	2	2	2
3	4	4	4	4	2	2	1	2
2	3	3	3	2	2	1	2	3
3	3	3	2	2	2	2	2	2
4	3	2	3	1	2	1	1	2
2	3	3	3	4	2	2	2	2
2	3	3	4	3	4	0	1	2
2	3	3	4	4	2	0	1	2
2	3	3	3	3	2	3	2	3
3	3	3	3	2	2	2	0	2
2	3	3	3	3	2	3	2	3
3	4	4	3	2	2	2	3	2
3	4	3	3	4	3	2	3	3
2	3	3	3	3	2	1	2	2
3	4	4	4	2	4	2	2	3
3	4	3	3	4	1	0	2	3
2	1	1	2	2	2	1	1	0
3	4	4	4	4	3	3	3	3
2	3	3	3	4	3	2	2	2

(Continued)

Table 18.1 ♦ (Continued)

V1	*V2*	*V3*	*V4*	*V5*	*V6*	*V7*	*V8*	*V9*
1	2	2	1	3	3	2	2	2
2	3	3	3	2	2	2	3	2
2	3	3	2	3	2	3	2	3
3	2	3	3	3	1	0	1	1
1	2	3	3	2	1	0	1	2
1	2	2	2	2	2	1	3	0
1	3	3	4	1	1	1	1	1
2	4	4	4	4	3	3	3	3
4	4	4	4	2	3	1	2	2
3	4	3	3	4	3	2	3	2
4	3	3	3	3	2	1	0	0
2	2	3	3	4	4	4	4	2
2	4	4	4	2	3	0	0	2
2	4	3	3	2	3	2	2	1
2	3	3	3	3	2	2	3	2
2	3	3	3	2	3	2	3	3
2	4	2	3	3	2	2	2	3
2	4	3	3	3	2	1	1	3
2	4	4	3	4	4	1	3	4
3	3	3	2	2	2	1	1	0
2	3	3	3	2	1	1	1	2
2	3	3	3	4	2	3	2	2
3	4	4	4	4	2	1	1	2
2	4	3	4	2	1	1	3	3
3	3	3	4	2	1	2	1	1
2	4	4	4	4	3	1	3	3
4	4	4	4	2	1	0	0	0
4	3	3	3	3	3	2	2	3
2	1	2	3	3	2	2	2	3
3	4	4	4	3	2	2	2	3
4	4	4	4	4	4	2	2	3
4	4	4	4	4	2	1	3	1
3	4	4	3	3	2	0	2	1
2	2	2	3	2	2	2	2	2
1	3	2	2	3	2	2	1	1
3	1	2	2	1	2	1	0	1
0	3	3	4	2	2	1	3	3
1	2	4	4	4	3	3	4	2
3	4	4	4	4	4	3	2	4
3	3	4	4	2	2	0	2	2
4	4	4	3	3	3	1	1	3
4	4	3	4	3	3	1	1	1
3	4	4	4	3	2	1	1	2
3	4	4	4	3	3	3	1	2
3	4	4	3	2	2	4	2	3
2	2	2	3	1	1	1	1	1
3	3	3	4	3	2	1	0	2
1	2	2	2	2	2	1	2	1

NOTE: V1, nurturant; V2, affectionate; V3, warm; V4, compassionate; V5, strong personality; V6, assertive; V7, forceful; V8, dominant; V9, aggressive.

Correlations

		nurturant	affectionate	warm	compassionate	strong personality	assertive	forceful	dominant	aggressive
nurturant	Pearson Correlation	1	.368 **	.303 **	.331 **	.062	.025	-.149	-.234 *	-.061
	Sig. (2-tailed)		.000	.002	.001	.543	.802	.139	.019	.544
	N	100	100	100	100	100	100	100	100	100
affectionate	Pearson Correlation	.368 **	1	.651 **	.432 **	.266 **	.266 **	.050	.106	.215 *
	Sig. (2-tailed)	.000		.000	.000	.007	.008	.622	.295	.032
	N	100	100	100	100	100	100	100	100	100
warm	Pearson Correlation	.303 **	.651 **	1	.603 **	.261 **	.209 *	.018	.106	.196
	Sig. (2-tailed)	.002	.000		.000	.009	.037	.856	.295	.050
	N	100	100	100	100	100	100	100	100	100
compassionate	Pearson Correlation	.331 **	.432 **	.603 **	1	.171	.053	-.175	-.059	.101
	Sig. (2-tailed)	.001	.000	.000		.088	.603	.081	.557	.315
	N	100	100	100	100	100	100	100	100	100
strongpersonality	Pearson Correlation	.062	.266 **	.261 **	.171	1	.453 **	.274 **	.312 **	.321 **
	Sig. (2-tailed)	.543	.007	.009	.088		.000	.006	.002	.001
	N	100	100	100	100	100	100	100	100	100
assertive	Pearson Correlation	.025	.266 **	.209 *	.053	.453 **	1	.350 **	.356 **	.322 **
	Sig. (2-tailed)	.802	.008	.037	.603	.000		.000	.000	.001
	N	100	100	100	100	100	100	100	100	100
forceful	Pearson Correlation	-.149	.050	.018	-.175	.274 **	.350 **	1	.467 **	.421 **
	Sig. (2-tailed)	.139	.622	.856	.081	.006	.000		.000	.000
	N	100	100	100	100	100	100	100	100	100
dominant	Pearson Correlation	-.234 *	.106	.106	-.059	.312 **	.356 **	.467 **	1	.381 **
	Sig. (2-tailed)	.019	.295	.295	.557	.002	.000	.000		.000
	N	100	100	100	100	100	100	100	100	100
aggressive	Pearson Correlation	-.061	.215 *	.196	.101	.321 **	.322 **	.421 **	.381 **	1
	Sig. (2-tailed)	.544	.032	.050	.315	.001	.001	.000	.000	
	N	100	100	100	100	100	100	100	100	100

**. Correlation is significant at the 0.01 level (2-tailed).

*. Correlation is significant at the 0.05 level (2-tailed).

Figure 18.2 ♦ Correlations Among Nine Items in the Modified Bem Sex Role Inventory Self-Rated Personality Data

To explain how PC and PAF work, we will examine a series of four analyses of the data in Table 18.1. Each of the following four analyses of the BSRI data introduces one or two important concepts. In Analysis 1, we will examine PC for a set of just three BSRI items. In this example, we shall see that when we retain three components we can reproduce all the correlations among the three X measured items perfectly. In Analysis 2, we will examine what happens when we retain only one component and use only one component to try to reproduce the correlations among the X variables. When the number of components is smaller than the number of variables (p), we cannot reproduce the correlations among the measured variables perfectly; we need to have a model that retains p components in order to reconstruct the correlations among p variables perfectly. In Analysis 3, a larger set of BSRI items is included ($p = 9$ items), and the PAF method of extraction is used. Finally, Analysis 4 shows how rotation of retained factors sometimes makes the outcome of the analysis more interpretable. In practice, data analysts are most likely to report an analysis similar to Analysis 4. The first three examples are presented for instructional purposes; they represent intermediate stages in the process of factor analysis. The last analysis, Analysis 4, can be used as a model for data analysis for research reports.

18.5 ♦ Screening for Violations of Assumptions

Principal components and factor analysis both begin with a correlation matrix **R**, which includes Pearson correlations between all possible pairs of the X variables included in the test battery (where p is the number of X variables that correspond to actual measurements):

$$\mathbf{R} = \begin{bmatrix} 1 & r_{12} & \dots & r_{1p} \\ r_{21} & 1 & \dots & r_{2p} \\ \vdots & \vdots & \vdots & \vdots \\ r_{p1} & r_{p2} & \dots & 1 \end{bmatrix}. \tag{18.1}$$

All the assumptions that are necessary for other uses of Pearson r should be satisfied. Scores on each X_i variable should be quantitative and reasonably normally distributed. Any association between a pair of X variables should be linear. Ideally, the joint distribution of this set of variables should be multivariate normal, but, in practice, it is difficult to assess multivariate normality. It is a good idea to examine histograms for each X_i variable (to make sure that the distribution of scores on each X variable is approximately normal, with no extreme outliers) and scatter plots for all X_i, X_j pairs of variables to make sure that all relations between pairs of X variables are linear with no bivariate outliers.

We also assume that **R** is not a **diagonal matrix**—that is, at least some of the r_{ij} correlations that are included in this matrix differ significantly from 0. If all the off-diagonal elements of **R** are 0, it does not make sense to try to represent the p variables using a smaller number of factors because none of the variables correlate highly enough with each other to indicate that they measure the same thing. A matrix is generally factorable when it includes a fairly large number of correlations that are at least moderate (>.3) in absolute magnitude.

It may be instructive to consider two extreme cases. If all the off-diagonal elements of **R** are 0, the set of p variables is already equivalent to p uncorrelated factors; no reduction in the number of factors is possible, because each X variable measures something completely unrelated to all the other variables. On the other hand, if all off-diagonal elements of **R** were 1.0, then the information contained in the X variables could be perfectly represented by just one factor, because all the variables provide perfectly equivalent information. Usually, researchers are interested in situations where there may be blocks, groups, or subsets of variables that are highly intercorrelated with the sets and have low correlations between sets; each set of highly intercorrelated variables will correspond to one factor.

Chapter 7 reviewed many factors that can make Pearson r misleading as a measure of association (e.g., nonlinearity of bivariate relationships, restricted range, attenuation of r due to unreliability). Because the input to factor analysis is a matrix of correlations, any problems that make Pearson r misleading as a description of the strength of the relationship between pairs of X variables will also lead to problems in factor analysis.

18.6 ♦ Issues in Planning a Factor Analytic Study

It should be apparent that the number of factors (and the nature of the factors) depends entirely on the set of X measurements that are selected for inclusion in the study. If a researcher sets out to study mental ability and sets up a battery of 10 mental ability tests, all of which are strongly correlated with verbal ability (and none of which are correlated with math ability), factor analysis is likely to yield a single factor that is interpretable as a verbal ability factor. On the other hand, if the researcher includes five verbal ability tests and five quantitative reasoning tests, factor analysis is likely to yield two factors: one that represents verbal ability and one that represents math ability. If other mental abilities exist "out in the world" that are not represented by the selection of tests included in the study, these other abilities are not likely to show up in the factor analysis. With factor analysis, it is quite clear that the nature of the outcome depends entirely on what variables you put in. (That limitation applies to all statistical analyses, of course.)

Thus, a key issue in planning a factor analysis study is the selection of variables or measures. In areas with strong theories, the theories provide a map of the domain that the items must cover. For example, Goldberg's International Personality Item Pool used Costa and McCrae's theoretical work on the Big Five personality traits to develop alternative measures of the Big Five traits. In a research area that lacks a clear theory, a researcher might begin by brainstorming items and sorting them into groups or categories that might correspond to factors, asking experts for their judgment about completeness of the content coverage, and asking research participants for open-ended responses that might provide content for additional items. For example, suppose a researcher is called on to develop a measure of patient satisfaction with medical care. Reading past research might suggest that satisfaction is not unidimensional, but rather, there may be three potentially separate components of satisfaction: satisfaction with practitioner competence, satisfaction with the interpersonal communication style or "bedside manner" of the practitioner, and satisfaction with cost and convenience of care. The researcher might first generate 10 or 20 items or questions to cover each of these three components of satisfaction, perhaps drawing on open-ended interviews with patients, research literature, or theoretical models to come up with specific wording. This pool of "candidate" questions would then be administered to a large number of patients, and factor analysis could be used to assess whether three separate components emerged clearly from the pattern of responses.

In practice, the development of test batteries or multiple-item tests using factor analysis is usually a multiple-pass process. Factor analysis results sometimes suggest that the number of factors (or the nature of the factors) may be different from what is anticipated. Some items may not correlate with other measures as expected; these may have to be dropped. Some factors that emerge may not have a large enough number of questions that tap that construct, and so additional items may need to be written to cover some components.

While there is no absolute requirement about sample size, most analysts agree that the number of subjects (N) should be large relative to the number of variables included in the factor analysis (p). In general, N should never be less than 100; it is desirable to have $N > 10p$. Correlations are unstable with small Ns, and this, in turn, makes the factor analysis results difficult to replicate. Furthermore, when you have p variables, you need to estimate $[p(p-1)]/2$ correlations among pairs of variables. You must have more degrees of freedom in your data than the number of correlations you are trying to estimate, or you can run into problems with the structure of this correlation matrix. Error messages such as "determinant = 0," "R is not positive definite," and so forth suggest that you do not have enough degrees of freedom (a large enough N) to estimate all the correlations needed independently. In general, for factor analysis studies, it is desirable to have the Ns as large as possible.

It is also important when designing the study to include X measures that have reasonably large variances. Restricted range on Xs tends to reduce the size of correlations, which will in turn lead to a less clear factor structure. Thus, for factor analytic studies, it is desirable to recruit participants who vary substantially on the measures; for instance, if your factor analysis deals with a set of mental ability tests, you need to have subjects who are heterogeneous on mental ability.

18.7 ♦ Computation of Loadings

A problem that has been raised implicitly in the preceding sections is the following: How can we obtain estimates of correlations between measured variables (X_1 through X_p) and latent variables or factors (F_1 through F_p)? We begin with a set of actual measured variables ($X_1, X_2, X_3, \ldots, X_p$)—for instance, a set of personality or ability test items or a set of physiological measures. We next calculate a matrix of correlations (**R**) among all possible pairs of items, as shown in Equation 18.1.

The matrix of loadings or correlations between each measured X variable and each factor is called a factor loading matrix; this matrix of correlations is usually denoted by **A**. Each element of this **A** matrix, a_{ij}, corresponds to the correlation of variable X_i with factor or component j, as shown in the path model in Figure 18.1:

$$A = \begin{array}{c} \\ X_1 \\ X_2 \\ \vdots \\ X_p \end{array} \begin{array}{c} \begin{array}{cccc} F_1 & F_2 & \cdots & F_p \end{array} \\ \begin{bmatrix} a_{11} & a_{12} & & a_{1p} \\ a_{21} & a_{22} & & a_{2p} \\ & & & \\ a_{p1} & a_{p2} & & a_{pp} \end{bmatrix} \end{array}. \tag{18.2}$$

The a_{ij} terms in Figure 18.1 and Equation 18.2 represent estimated correlations (or "loadings") between each measured X variable (X_i) and each latent variable or factor (F_j). How can we obtain estimates of these a_{ij} correlations between actual measured variables and latent or purely imaginary variables? We will address this problem in an intuitive manner, using logic very similar to the logic that was used in Chapter 11 to show how estimates of beta coefficients can be derived from correlations for a standardized regression that has just two predictor variables. We will apply the tracing rule (introduced in Chapter 11) to the path model in Figure 18.1. When we have traced all possible paths between each pair of measured z_X variables in Figure 18.1, the correlations or loadings between each z_X and each factor (F), that is, the values of all the a_{ij} terms, must be consistent with the correlations among the Xs. That is, we must be able to reconstruct all the r_{ij} correlations among the Xs (the **R** matrix) from the factor loadings (the a_{ij} terms that label the paths in Figure 18.1).

In the examples provided in this chapter (in which factors are orthogonal), a_{ij} is the correlation between variable i and factor j. For example, a_{25} is the correlation between X_2 and F_5. If we retain the entire set of correlations between all p factors and all p of the X variables as shown in the **A** matrix in Equation 18.2, we can use these correlations in the **A** matrix to reproduce the correlation between all possible pairs of measured X variables in the **R** matrix (such as X_1 and X_2) perfectly by tracing all possible paths from X_1 to X_2 via each of the p factors (see the path model in Figure 18.1).

For example, to reproduce the r_{12} correlation between X_1 and X_2, we multiply together the two path coefficients or loadings of X_1 and X_2 with each factor; we sum the paths that connect X_1 and X_2 across all p of the factors, and the resulting sum will exactly equal the overall correlation, r_{12} (see Figure 18.1 for the paths that correspond to each pair of a_{ij} terms):

$$r_{12} = (a_{11}\, a_{21}) + (a_{12}\, a_{22}) + (a_{13}\, a_{23}) + \ldots + (a_{1p}\, a_{2p}). \tag{18.3}$$

In practice, we try to reconstruct as much of the r_{12} correlation as we can with the loadings of the variables on the first factor F_1; we try to explain as much of the residual ($r_{12} - a_{11}a_{21}$) with the loadings of the variables on the second factor F_2, and so forth. We can perfectly reproduce the correlations among p variables with p factors. However, we hope that we can reproduce the observed correlations among X variables reasonably well using the correlations of variables with just a few of the factors. In other words, when we estimate the correlations of loadings of all the X variables with Factor 1, we want to obtain values for these loadings that make the product of the path coefficients that lead from X_1 to X_2 via F_1 ($a_{11}a_{21}$) as close to r_{12} as possible, make the product $a_{21}a_{31}$ as close to r_{23} as possible, and so on for all possible pairs of the Xs.

If we look at this set of relations, we will see that our path diagram implies proportional relations among the *a*s. This is only an approximation and not an exact solution for the computation of values of the a loadings. The actual computations are done using the algebra of eigenvectors and eigenvalues, and they include correlations with all p factors, not only the loadings on Factor 1, as in the following conceptual description. If we consider only the loadings on the first factor, we need to estimate loadings on Factor 1 that come as close as possible to reproducing the correlations between X_1, X_2 and X_1, X_3:

$$r_{12} = a_{11}a_{21}, \tag{18.4}$$

$$r_{13} = a_{11}a_{31}. \tag{18.5}$$

With the application of some simple algebra we can rearrange each of the two previous equations so that each equation gives a value of a_{11} in terms of a ratio of the other terms in Equations 18.4 and 18.5:

$$a_{11} = r_{12}/a_{21}, \tag{18.6}$$

$$a_{11} = r_{13}/a_{31}. \tag{18.7}$$

In the two previous equations, both r_{12}/a_{21} and r_{13}/a_{31} were equal to a_{11}. Thus, these ratios must also be equal to each other, which implies that the following proportional relationship should hold:

$$r_{12}/a_{21} = r_{13}/a_{31}. \tag{18.8}$$

We can rearrange the terms in Equation 18.8 so that we have (initially unknown) values of the a_{ij} loadings on one side of the equation and (known) correlations between observed X variables on the other side of the equation:

$$a_{31}/a_{21} = r_{13}/r_{12}. \tag{18.9}$$

For the *as* to be able to reconstruct the correlations among all pairs of variables, the values of the *a* coefficients must satisfy ratio relationships such as the one indicated in the previous equation. We can set up similar ratios for all pairs of variables in the model, and solve these equations to obtain values of the factor loadings (the *as*) in terms of the values of the computed correlations (the *rs*). In other words, any values of a_{31} and a_{21} that have the same ratio as r_{13}/r_{12} could be an acceptable approximate solution for loadings on the first factor. However, this approach to computation does not provide a unique solution for the factor loadings; in fact, there are an infinite number of pairs of values of a_{31} and a_{21} that would satisfy the ratio requirement given in Equation 18.9.

How do we obtain a unique solution for the values of the a_{ij} factor loadings? We do this by placing additional constraints or requirements on the values of the a_{ij} loadings. The value of each a_{ij} coefficient must be scaled so that it can be interpreted as a correlation and so that squared loadings can be interpreted as proportions of explained variance. We will see later in the chapter that the **sum of squared loadings (SSL)** for each X variable in a PC analysis, summed across all p components, must be equal to 1; this value of 1 is interpreted as the total variance of the standardized X variable and is called a **communality**.

Another constraint may be involved when loadings are calculated. The loadings that define the components or factors are initially estimated in a manner that makes the components or factors uncorrelated with each other (or orthogonal to each other). Finally, we will estimate the factor (or component) loadings subject to the constraint that the first factor will **reproduce the correlation matrix** as closely as possible, then we will take the residual matrix (the actual **R** matrix minus the **R** matrix predicted from loadings on the first factor) and estimate a second factor to reproduce that residual matrix as closely as possible, and then we will take the residuals from that step, fit a third factor, and so on. By the time we have extracted p factors, we have a $p \times p$ matrix (denoted by **A**), and the elements of this **A** matrix consist of the correlations of each of the p measured variables with each of the p factors. This **A** matrix of loadings can be used to exactly reproduce the observed **R** correlation matrix among the X variables.

When we place all the preceding constraints on the factor solution, we can obtain unique estimates for the factor loadings in the **A** matrix in Equation 18.2. However, the initial estimate of factor loadings is arbitrary; it is just one of an infinite set of solutions that would do equally well at reproducing the correlation matrix. An appendix to this chapter briefly describes how solving the eigenvalue/eigenvector problem for the **R** correlation matrix provides a means of obtaining estimated factor loadings.

18.8 ♦ Steps in the Computation of Principal Components or Factor Analysis

The term *factor analysis* refers to a family of related analytic techniques and not just a single form of specific analysis. When we do a PC or PAF analysis, there are a series of intermediate steps in the analysis. Some of these intermediate results are reported in SPSS printouts. At each point in the analysis the data analyst has choices about how to proceed. We are now ready to take on the problem of describing how a typical package program (such as SPSS) computes a PC or factor analysis and to identify the choices that are available to the data analyst to specify exactly how the analysis is to be performed.

18.8.1 ♦ Computation of the Correlation Matrix R

Most PC and PAF analyses begin with the computation of correlations among all possible pairs of the measured X variables; this **R** matrix (see Equation 18.1) summarizes these correlations. Subsequent analyses for PC and PAF are based on this **R** matrix. There are other types of factor analysis that are based on different kinds of information about data structure (see Harman, 1976, for further discussion).

It is useful to examine the correlation matrix before you perform a factor analysis and ask, "Is there any indication that there may be groups or sets of variables that are highly intercorrelated with each other, and not correlated with variables in other groups?" In the empirical example used in this chapter, nine items from a modified version of the BSRI were factor analyzed; the first result shown is the matrix of correlations among these nine items. These variables were intentionally listed so that the four items that all appeared to assess aspects of being caring (ratings on affectionate, compassionate, warm, and loving) appeared at the beginning of the list and five items that appeared to assess different types of strength (assertive, forceful, dominant, aggressive, and strong personality) were placed second on the list. Thus, in the correlation matrix that appears in Figure 18.2, it is possible to see that the first four items form a group with high intercorrelations, the last five items form a second group with high intercorrelations, and the intercorrelations between the first and second groups of items were quite low. This preliminary look at the correlation matrix allows us to guess that the factor analysis will probably end up with a two-factor solution: one factor to represent what is measured by each of these two sets of items. In general, of course, data analysts may not know a priori which items are intercorrelated, or even how many groups of items there might be, but it is useful to try to sort the items into groups before setting up a correlation matrix to see if a pattern can be detected at an early stage in the analysis.

18.8.2 ♦ Computation of the Initial Loading Matrix A

One important decision involves the choice of a method of extraction. SPSS provides numerous options. The discussion that follows considers only two of these methods of extraction: PC versus PAF. These two extraction methods differ both computationally and conceptually. The computational difference involves the values of the diagonal elements in the **R** correlation matrix. In PC, these diagonal elements have values of 1, as shown in Equation 18.1. These values of 1 correspond conceptually to the "total" variance of each measured variable (when each variable is expressed as a *z* score, it has a variance of 1). PC may be preferred by researchers who are primarily interested in reducing the information in a large set of variables down to scores on a smaller number of components (Tabachnick & Fidell, 2007). In PAF, the diagonal elements in **R** are replaced with estimates of the proportion of variance in each of the measured *X* variables that is predictable from or shared with other *X* variables; PAF is mathematically somewhat more complex than PC. PAF is more widely used because it makes it possible for researchers to ignore the unique or error variance associated with each measurement and to obtain factor loading estimates that are based on the variance that is shared among the measured variables.

When PC is specified as the method of extraction, the analysis is based on the **R** matrix as it appears in Equation 18.1, with values of 1 in the diagonal that represent the total variance of each z_X measured variable. When PAF is specified as the method of extraction, the 1s in the diagonal of the **R** matrix are replaced with "communality estimates" that provide information about the proportion of variance in each z_X measured variable that is shared with or predictable from other variables in the dataset. One possible way to obtain initial communality estimates is to run a regression to predict each X_i from all the other measured *X* variables. For example, the initial communality estimate for X_1 can be the R^2 for the prediction of X_1 from $X_2, X_3, \ldots, X_p$. These communality estimates can range from 0 to 1.00, but in general, they tend to be less than 1.00. When a PAF extraction is requested from SPSS, the 1s in the diagonal of the **R** correlation matrix are replaced with initial communality estimates, and the loadings (i.e., the correlations between factors and measured *X* variables) are estimated so that they reproduce this modified **R** matrix that has communality estimates that represent only shared variance in its diagonal. Unlike PC (where the loadings could be obtained in one step), the process of estimating factor loadings using PAF may involve multiple steps or **iterations**. An initial set of factor loadings (correlations between *p* measured variables and *p* factors) is estimated and this set of factor loadings is used to try to reproduce **R**, the matrix of correlations with initial communality estimates in the diagonal. However, the communalities implied by (or constructed from) this first set of factor loadings generally differ from the initial communality estimates that were obtained using the R^2 values. The initial communality estimates in the diagonal of **R** are replaced with the new communality estimates that are based on the factor loadings. Then the factor loadings are re-estimated from this new version of the correlation matrix **R**. This process is repeated ("iterated") until the communality estimates "converge"—that is, they do not change substantially from one iteration to the next. Occasionally, the values do not converge after a reasonable number of iterations (such as 25 iterations). In this case, SPSS reports an error message ("failure to converge").

In practice, the key difference between PC and PAF is the way they estimate communality for each variable. In PC, the initial set of loadings for p variables on p factors accounts for *all* the variance in each of the p variables. On the other hand, PAF attempts to reproduce only the variance in each measured variable that is shared with or predictable from other variables in the dataset.

Whether you use PC or PAF, the program initially calculates a set of estimated loadings that describe how all p of the X variables are correlated with all p of the components or factors. This corresponds to the complete set of estimated a_{ij} loadings in the **A** matrix that appears in Equation 18.2. If you have p variables, this complete loading matrix will have p rows (one for each X variable) and p columns (one for each component or factor). Most programs (such as SPSS) do not print out this intermediate result. In Analysis 1 reported below, the complete set of loadings for three variables on three components was requested to demonstrate that when we retain the loadings of p variables on p components, this complete set of loadings can perfectly reproduce the original correlation matrix.

18.8.3 ♦ Limiting the Number of Components or Factors

A key issue in both PC and PAF is that we usually hope that we can describe the correlation structure of the data using a number of components or factors that is smaller than p, the number of measured variables. The decision about the number of components or factors to retain can be made using an arbitrary criterion. For each component or factor, we have an eigenvalue that corresponds to the sum of the squared loadings for that component or factor. The eigenvalue provides information about how much of the variance in the set of p standardized measured variables can be reproduced by each component or factor. An eigenvalue of 1 corresponds to the variance of a single z score standardized variable. The default decision rule in SPSS is to retain only the components or factors that have eigenvalues greater than 1—that is, only latent variables that have a variance that is greater than the variance of a single standardized variable.

The eigenvalues can be graphed for each factor ($F_1, F_2, \ldots, F_p$). The factors are rank ordered by the magnitudes of eigenvalues—that is, F_1 has the largest eigenvalue, F_2 the next largest eigenvalue, and so forth. A graph of the eigenvalues (on the Y axis) across factor numbers (on the X axis) is called a **scree plot**. This term comes from geology; scree refers to the distribution of the rubble and debris at the foot of a hill. The shape of the curve in a plot of eigenvalues tends to decline rapidly for the first few factors and more slowly for the remaining factors, such that it resembles a graph of a side view of the scree at the foot of a hill. Sometimes data analysts visually examine the scree plot and look for a point of inflection—that is, they try to decide where the scree plot "flattens out," and they decide to retain only factors whose eigenvalues are large enough that they are distinguishable from the "flat" portion of the scree plot. An example of a scree plot appears in Figure 18.26.

Alternatively, it is possible to decide what number of components or factors to retain based on conceptual or theoretical issues. For example, if a researcher is working with a large set of personality trait items and has adopted the Big Five theoretical model of personality, the researcher would be likely to retain five factors or components (to correspond with the five factors in the theoretical model). In some cases, the decision about the number of components or factors to retain may take both empirical information (such as the magnitudes of eigenvalues) and conceptual background (the number of latent variables specified by a theory or by past research) into account. Usually, researchers hope that the number of retained factors will be relatively small. In practice, it is rather rare to see factor analyses reported that retain more than 5 or 10 factors; many reported analyses have as few as two or three retained factors.

Another consideration involves the number of items or measurements that have high correlations with (or high loadings on) each factor. In practice, data analysts typically want a minimum of three indicator variables for each factor, and more than three indicator variables is often considered preferable. Thus, a factor that has high correlations with only one or two measured variables might not be retained because there are not enough indicator variables in the dataset to provide adequate information about any latent variable that might correspond to that factor.

18.8.4 ♦ Optional: Rotation of Factors

If more than one component or factor is retained, researchers often find it useful to request a "rotated" solution. The goal of factor rotation is to make the pattern of correlations between variables and factors more interpretable. When correlations are close to 0, +1, or −1, these correlation values make it easy for the researcher to make binary decisions. For example, if the measured variable X_1 has a loading of .02 with F_1, the researcher would probably conclude that X_1 is not an indicator variable for F_1; whatever the F_1 factor or latent variable may represent conceptually, it is not related to the measured X_1 variable. On the other hand, if the measured variable X_2 has a loading of +.87 with F_1, the researcher would probably conclude that X_2 is a good indicator variable for the construct that is represented by F_1, and that we can make some inferences about what is measured by the latent variable F_1 by noting that F_1 is highly positively correlated with X_2. From the point of view of a data analyst, solutions that have loadings that are close to 0, +1, or −1 are usually the easiest to interpret. When many of the loadings have intermediate values (e.g., most loadings are on the order of +.30), the data analyst faces a more difficult interpretation problem. Usually, the data analyst would like to be able to make a binary decision: For each X variable, we want the loading to be large enough in absolute value to say that the X variable is related to Factor 1, or we want the loading to be close enough to 0 to say that the X variable is not related to Factor 1.

The initial set of component or factor loadings that are obtained are sometimes not easy to interpret. Often, many variables have moderate to large positive loadings on Factor 1; and many times the second factor has a mixture of both positive and negative loadings that are moderate in size. Numerous methods of "factor rotation" are available through SPSS; the most widely used is **varimax rotation**. Factor rotation is discussed more extensively in Section 18.13. The goal of factor rotation is to obtain a pattern of factor loadings that is easier to interpret.

If only one factor or component is retained, rotation is not performed. If two or more components or factors are retained, sometimes (but not always) the pattern of loadings becomes more interpretable after some type of rotation (such as varimax) is applied to the loadings.

18.8.5 ♦ Naming or Labeling Components or Factors

The last decision involves interpretation of the retained (and possibly also rotated) components or factors. How can we name a latent variable? We can make inferences about the nature of a latent variable by examining its pattern of correlations with measured X variables. If Factor 1 has high correlations with measures of vocabulary, reading comprehension, and comprehension of analogies, we would try to decide what (if anything) these three X variables have in common conceptually. If we decide that these three measurements all represent indicators of "verbal ability," then we might give Factor 1 a name such as verbal ability. Note, however, that different data analysts do not always see the same common denominator in the set of X variables. Also, it is possible for a set of variables to be highly correlated within a sample because of sampling error; when a set of X variables are all highly correlated with each other in a sample, they will also tend to be highly correlated with the same latent variable. It is possible for a latent variable to be uninterpretable. If a factor is highly correlated with three or four measured X variables that seem to have nothing in common with each other, it may be preferable to treat the solution as uninterpretable, rather than to concoct a wild post hoc explanation about why this oddly assorted set of variables are all correlated with the same factor in your FA.

18.9 ♦ Analysis 1: Principal Components Analysis of Three Items Retaining All Three Components

A series of four analyses will be presented for the data in Table 18.1. Each analysis is used to explain one or two important concepts. In practice, a data analyst is most likely to run an analysis similar to Analysis 4, the last example that is presented in this chapter. Each analysis takes the reasoning one or two steps further. The first analysis that is presented is the simplest one. In this analysis, a set of just three personality rating items are included; three components are extracted using the PC method and all three are retained. Examination of the results from the first analysis makes it possible to see how the component loadings can be used to obtain summary information about the proportion of variance that is explained for each measured variable (i.e., the communality for each variable); the proportion of variance that is explained by each component; and to demonstrate that, when we have p components that correspond to p measured variables, we can use the loadings on these components to reconstruct perfectly all the correlations among the measured variables (contained in the **R** correlation matrix in Equation 18.1).

In SPSS, the menu selections that are used to run a factor analysis are as follows. From the top-level menu on the Data View worksheet, select the menu options for <Analyze> → <Data Reduction> → <Factor>, as shown in Figure 18.3. The main dialog window for the Factor Analysis procedure is shown in Figure 18.4. The names of the three items that were selected for inclusion in Analysis 1 were moved into the right-hand side window under the heading "Variables." The button for Descriptive Statistics was used to open the Descriptive Statistics menu that appears in Figure 18.5. The following information was requested: **initial solution**, correlation coefficients, and the reproduced correlation coefficients. Next, the Extraction button in the main Factor Analysis dialog window was used to open up the Extraction window that appears in Figure 18.6. The pull-down menu at the top was used to select principal components as the method of extraction. Near the bottom of this window, the radio button next to Number of factors was clicked to make it possible for the data analyst to specify the number of components to retain; for this first example, we want to see the complete set of loadings (i.e., the correlations of all three measured variables with all three components), and therefore, the number of factors to be retained was specified to be "3." The SPSS Syntax that was obtained through these menu selections appears in Figure 18.7.

The output obtained using this syntax appears in Figures 18.8 through 18.12. Figure 18.8 shows the Pearson correlations among the set of three items that were included in Analysis 1—that is, the correlations among self-ratings on nurturant, affectionate, and compassionate; all these correlations were positive and the *r* values ranged from .33 to .43. These moderate positive correlations suggest that it is reasonable to interpret the ratings

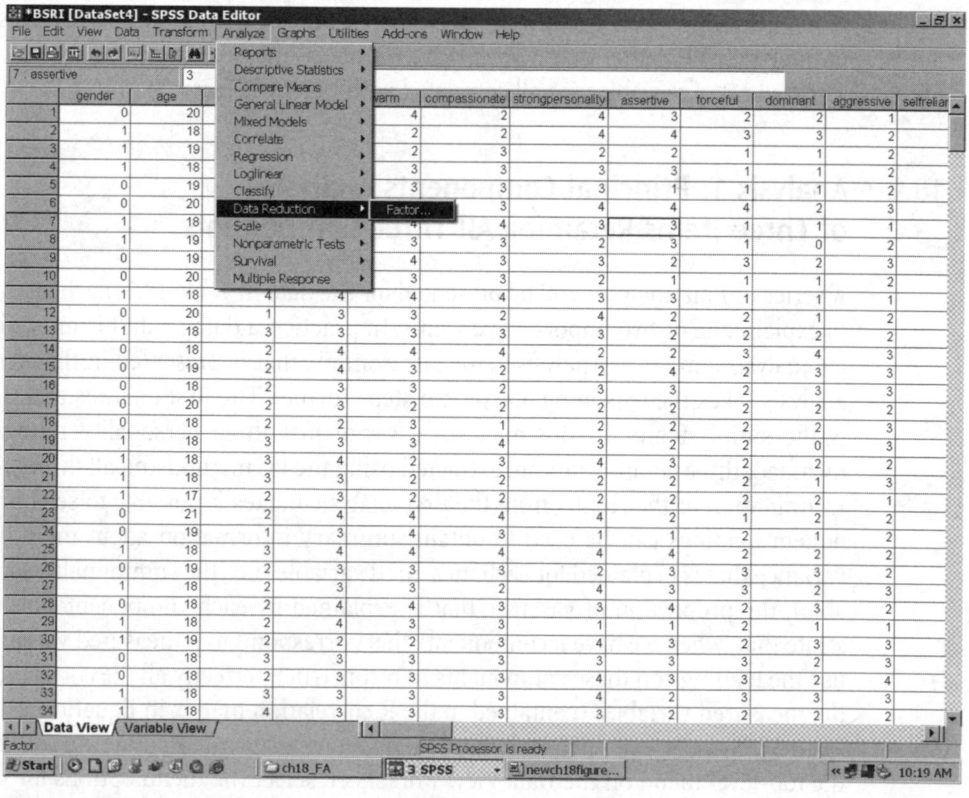

Figure 18.3 ♦ SPSS Menu Selections for Analysis 1: Principal Components Analysis of Three Items Without Rotation; All Three Components Retained

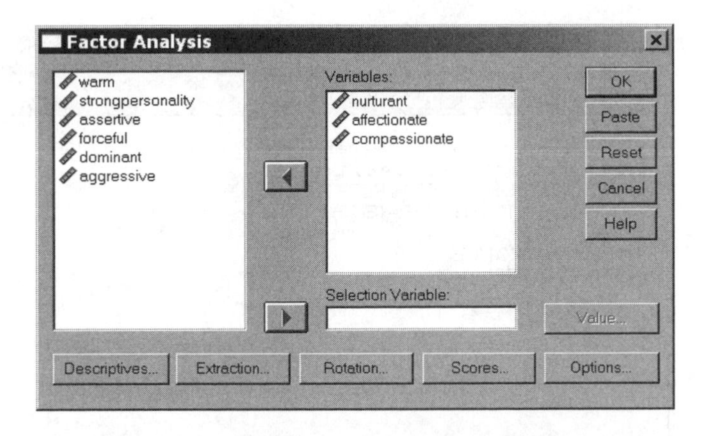

Figure 18.4 ♦ Analysis 1: Specification of Three Items to Include in Principal Components Analysis

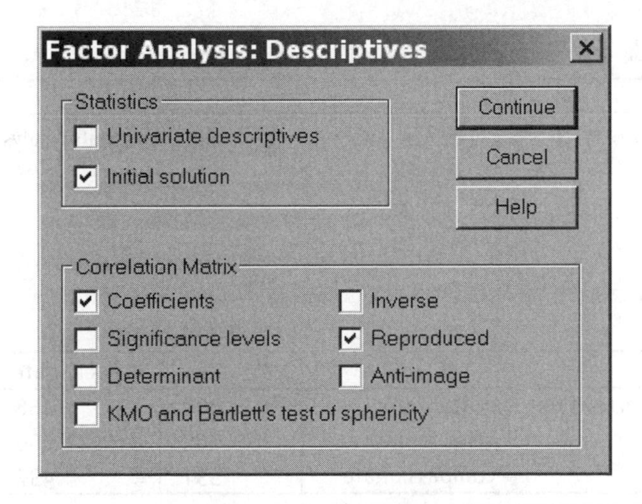

Figure 18.5 ♦ Analysis 1: Descriptive Statistics for Principal Components Analysis Including Correlations Reproduced From the Loading Matrix

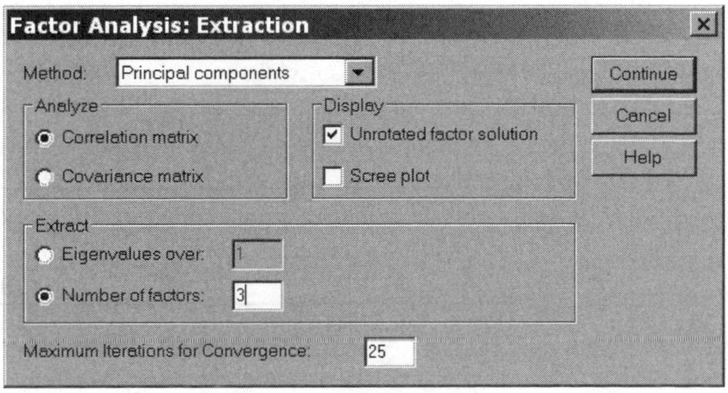

Factor Analysis: Extraction
Method: Principal components
Continue
Cancel
Help
Analyze
Correlation matrix
Covariance matrix
Display
Unrotated factor solution
Scree plot
Extract
Eigenvalues over: 1
Number of factors: 3
Maximum Iterations for Convergence: 25

Figure 18.6 ♦ Analysis 1: Instructions to Retain All Three Components in the Principal Components Analysis

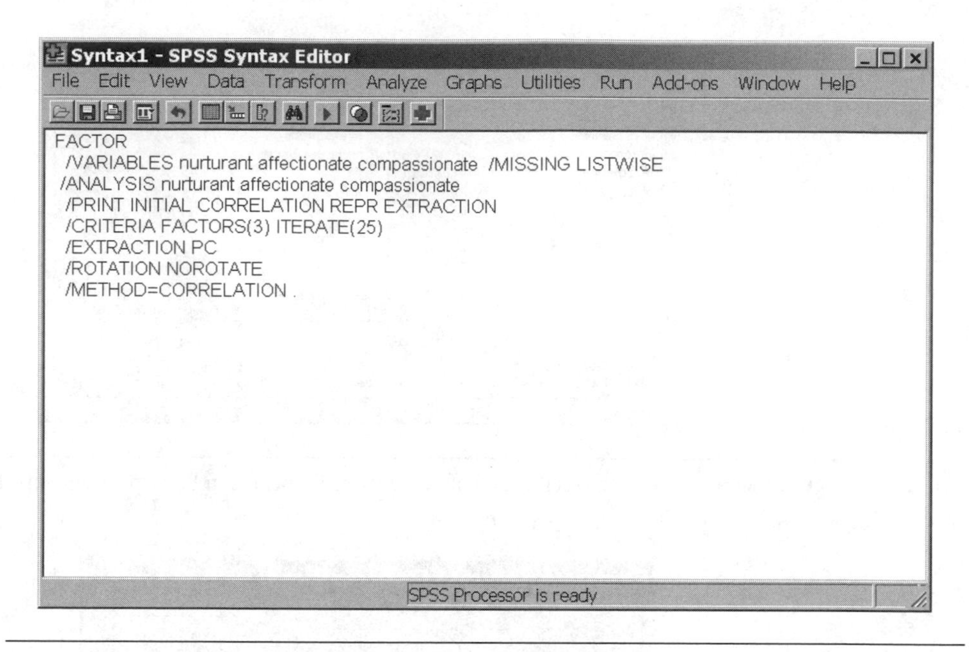

Figure 18.7 ♦ SPSS Syntax for Analysis 1: Principal Components Analysis of Three Items

Correlation Matrix

		nurturant	affectionate	compassionate
Correlation	nurturant	1.000	.368	.331
	affectionate	.368	1.000	.432
	compassionate	.331	.432	1.000

Figure 18.8 ♦ Output From Analysis 1: Correlations Among Three Items in Principal Components Analysis 1

Communalities

	Initial	Extraction
nurturant	1.000	1.000
affectionate	1.000	1.000
compassionate	1.000	1.000

Extraction Method: Principal Component Analysis.

Figure 18.9 ♦ Analysis 1: Communalities for Each of the Three Variables From the Principal Components Analysis With All Three Components Retained

Total Variance Explained

Component	Initial Eigenvalues			Extraction Sums of Squared Loadings		
	Total	% of Variance	Cumulative %	Total	% of Variance	Cumulative %
1	1.756	58.539	58.539	1.756	58.539	58.539
2	.681	22.684	81.223	.681	22.684	81.223
3	.563	18.777	100.000	.563	18.777	100.000

Extraction Method: Principal Component Analysis.

Figure 18.10 ♦ Analysis 1: Summary of Variance Explained or Reproduced by Each of the Three Components in Analysis 1

Componant Matrix[a]

	Component		
	1	2	3
nurturant	.726	.677	.123
affectionate	.795	-.205	-.571
compassionate	.773	-.425	.472

Extraction Method: Principal Component Analysis.
a. 3 components extracted.

Figure 18.11 ♦ Analysis 1: Loading of (or Correlation of) Each of the Three Measured Variables (Nurturant, Affectionate, and Compassionate) With Each of the Orthogonal Extracted Components (Components 1, 2, and 3)

Reproduced Correlations

		nurturant	affectionate	compassionate
Reproduced Correlation	nurturant	1.000[b]	.368	.331
	affectionate	.368	1.000[b]	.432
	compassionate	.331	.432	1.000[b]
Residual[a]	nurturant		-3.89E-016	5.0E-016
	affectionate	-4E-016		3.3E-016
	compassionate	5.0E-016	3.33E-016	

Extraction Method: Principal Component Analysis.

a. Residuals are computed between observed and reproduced correlations. There are 0 (.0%) nonredundant residuals with absolute values greater than 0.05.

b. Reproduced communalities

Figure 18.12 ♦ Analysis 1: Reproduced Correlations From Three-Component Principal Components Analysis Solution

NOTE: The residuals (differences between actual r_{ij} and predicted r'_{ij}) are all 0 to many decimal places.

on these three characteristics as measures of the "same thing," perhaps a self-perception of being a caring person.

18.9.1 ♦ Communality for Each Item Based on All Three Components

The communalities for each of the three variables in this PC analysis are summarized in Figure 18.9. Each communality is interpreted as "the proportion of variance in one of the measured variables that can be reproduced from the set of three uncorrelated components." The communality for each variable is obtained by squaring and summing the loadings (correlations) of that variable across the retained components. The component matrix appears in Figure 18.11. From Figure 18.11, we can see that the correlations of the variable nurturant with Components 1, 2, and 3, respectively, are .726, .677, and .123. Because each loading is a correlation, a squared loading corresponds to a proportion of explained or predicted variance. Because the three components are uncorrelated with each other, they do not compete to explain the same variance, and we can sum the squared correlations across components to summarize how much of the variance in self-rated "nurturance" is predictable from the set of Components 1, 2, and 3. In this example, the communality for nurturance is found by squaring its loadings on the three components and then summing these squared loadings: $.726^2 + .677^2 + .123^2 = .5271 + .4583 + .0151 = 1.00$. In other words, for self-ratings on nurturance, about 53% of the variance can be predicted based on the correlations with Component 1, about 46% of the variance can be predicted based on the correlation with Component 2, and about 1% of the variance can be predicted from the correlation with Component 3; when all three components are retained, together they account for 100% of the variance in scores on self-rated nurturance. When the PC method of extraction is used, the communalities for all variables will always equal 1 when all p of the components are retained. We shall see in the next analysis that the communalities associated with individual measured variables will decrease when we decide to retain only a few of the components.

18.9.2 ♦ Variance Reproduced by Each of the Three Components

Another way we can summarize information about the variance explained in Analysis 1 is to sum the loadings and square them for each of the components. To answer the question, How much of the total variance (for a set of three z score variables, each with a variance of 1.0) is represented by Component 1? we take the list of loadings for Component 1 from Figure 18.11, square them, and sum them as follows. For Component 1, the SSL $= .726^2 + .795^2 + .773^2 = .5271 + .6320 + .5975 \approx 1.756$. Note that this numerical value of 1.756 corresponds to the initial eigenvalue and also to the SSL for Component 1 that appears in Figure 18.10. The entire set of $p = 3$ z score variables has a variance of 3. The proportion of this total variance (because each of the three z scores has a variance of 1.0, the total variance for the set of three variables equals 3), represented by Component 1, is found by dividing the eigenvalue or SSL for Component 1 by the total variance. Therefore, in this example, we can see that 58.5% (1.756/3) of the variance in the data can be reproduced by Component 1. The sum of the SSL values for Components 1, 2, and 3 is 3; thus, for a set of three measured variables, a set of three components obtained by PC explains or reproduces 100% of the variance in the correlation matrix **R**.

18.9.3 ♦ Reproduction of Correlations From Loadings on All Three Components

The set of loadings in Figure 18.11 have to satisfy the requirement that they can reproduce the matrix of correlations, **R**, among the measured variables. Consider the path model in Figure 18.1; in this path model, the latent variables were labeled as "factors." A similar path model (see Figure 18.13) can be used to illustrate the pattern of relationships between components and *z* scores on measured variables.

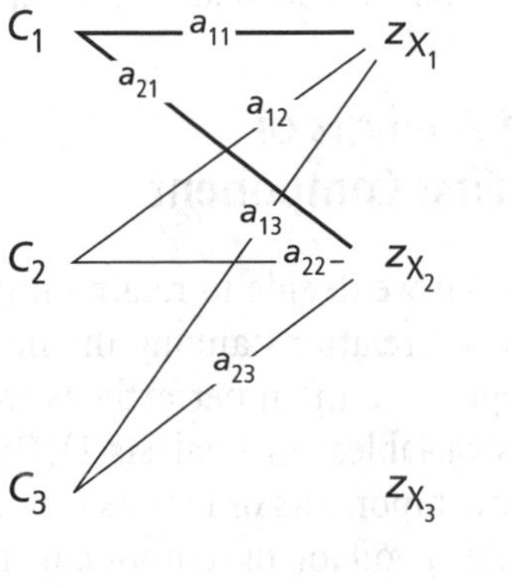

a_{ij} is the correlation between variable X_i and Component C_j

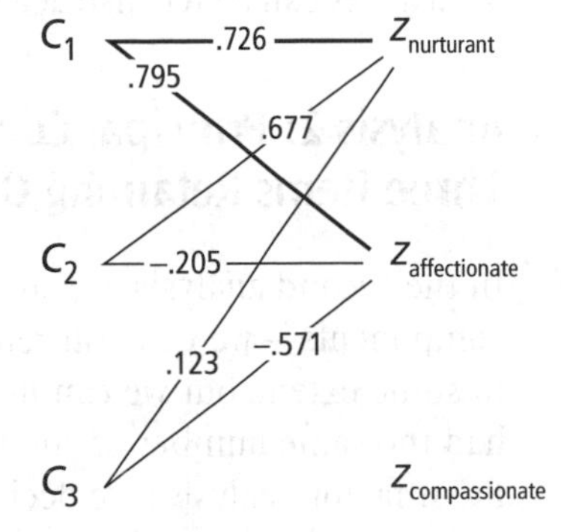

Reconstruction of correlation between nurturant and affectionate based on correlations with C_1, C_2 and C_3:

$r = .726 \times .795 + .677 \times (-.205) + .123 \times (-.571)$

$= .5772 \quad - \quad .1388 \quad - \quad .0702$

$= .368$

Figure 18.13 ♦ Path Model Showing Correlations Between the Three Variables and the Three Components Obtained in Analysis 1

The loadings in Figure 18.11 are the correlations or path coefficients for the PC model for three measured variables and the three components shown in Figure 18.13. To reproduce the correlation between one pair of variables—for example, the correlation between self-rated nurturance and self-rating on affection, we can apply the tracing rule that was explained in Chapter 11. There is one path from nurturance to affection via C_1, a second path via C_2, and a third path via C_3. We can reproduce the observed correlation between nurturance and affectionate ($r = .368$) by multiplying the correlations along each path and then summing the contributions for the three paths; for example, the part of the correlation that is reproduced by Component 1 ($= .726 \times .795$), the part of the correlation that is reproduced by Component 2 ($= .677 \times (-.205)$), and the part of the correlation that is reproduced by Component 3 ($= .123 \times (-.571)$). When these products are summed, the overall reproduced correlation is .368. When all three components are retained in the model, there is perfect agreement between the original correlation between each pair of variables (in Figure 18.8) and the reproduced correlation between each pair of variables (in Figure 18.12). Note that the residuals (difference between observed and reproduced correlations for each pair of variables) in Figure 18.12 are all 0 to several decimal places.

At this point, we have demonstrated that we can reproduce all the variances of the z scores on the three ratings, and the correlations among the three ratings, perfectly by using a three-component PC solution. However, we do not achieve any data reduction or simplification when the number of retained components (in this example, three components) is the same as the number of variables. The next step, in Analysis 2, will involve assessing how well we can reproduce the correlations when we drop two components from the PC model and retain only the first component.

Note that this is only a demonstration, not a formal proof, that the correlations among p variables can be reconstructed from a PC model that contains p components.

18.10 ◆ Analysis 2: Principal Component Analysis of Three Items Retaining Only the First Component

In the second analysis, we shall see that—if we decide to retain only one of the three components—we can still reproduce the correlations among the measured variables to some extent, but we can no longer reproduce them perfectly as we could when we had the same number of components as variables (in Analysis 1). One typical goal of PC or factor analysis is to decide how *few* components or factors we can retain and still have enough information to do a reasonably good job of reproducing the observed correlations among measured variables in the **R** correlation matrix. The second analysis reported here is identical to Analysis 1 except for one change in the procedure. Instead of instructing the program to retain all three components (as in Analysis 1), in Analysis 2, the decision rule that was used to decide how many components to retain was based on the magnitudes of the eigenvalues. In Figure 18.14, the radio button that corresponded to "Extract [retain Components with] Eigenvalues over 1" was selected. Apart from this, all the commands were identical to those used in Analysis 1. The SPSS syntax for Analysis 2 appears in Figure 18.15. The output for Analysis 2 appears in Figures 18.16 through 18.19.

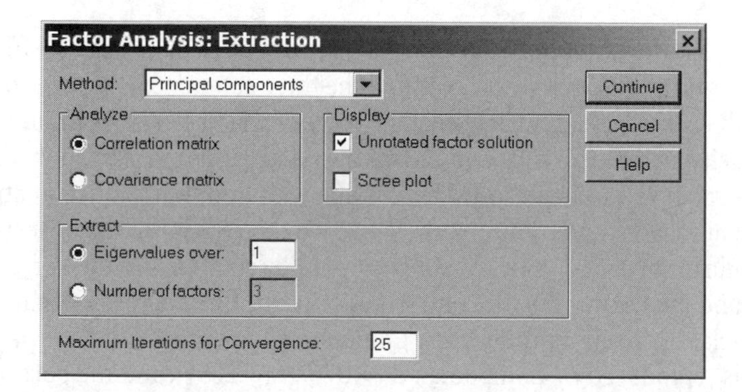

Figure 18.14 ◆ Analysis 2: Specifying Principal Components Method of Extraction and Retention of Only Components With Eigenvalues >1

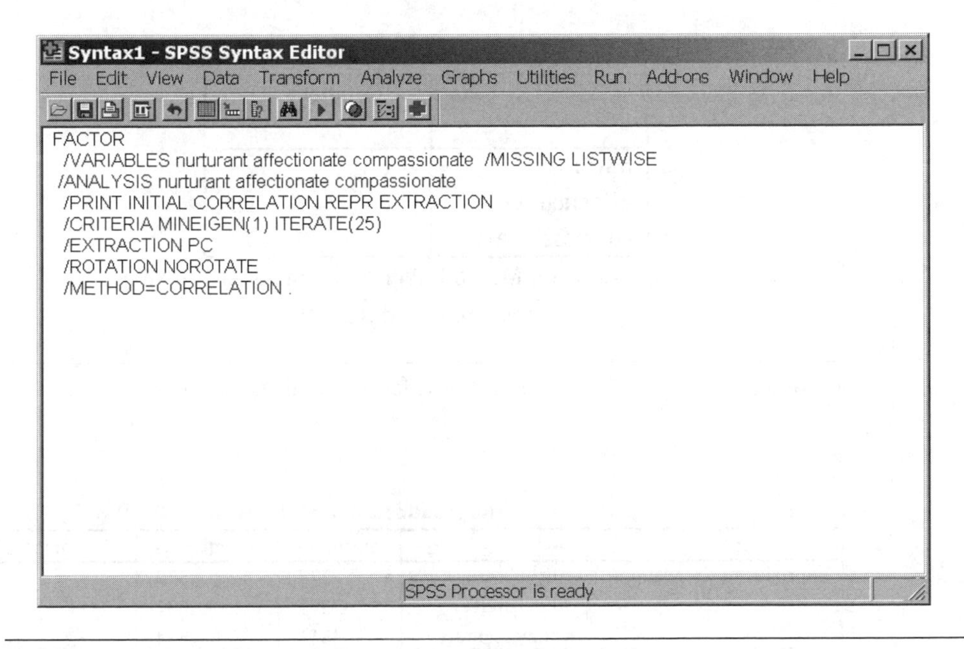

Figure 18.15 ♦ Analysis 2: SPSS Syntax

Communalities

	Initial	Extraction
nurturant	1.000	.527
affectionate	1.000	.632
compassionate	1.000	.597

Extraction Method: Principal Component Analysis.

Figure 18.16 ♦ Analysis 2: Communalities After Extraction When Only One Component Is Retained

Total Variance Explained

	Initial Eigenvalues			Extraction Sums of Squared Loadings		
Component	Total	% of Variance	Cumulative %	Total	% of Variance	Cumulative %
1	1.756	58.539	58.539	1.756	58.539	58.539
2	.681	22.684	81.223			
3	.563	18.777	100.000			

Extraction Method: Principal Component Analysis.

Figure 18.17 ♦ Analysis 2: Total Variance Explained by One Retained Component

Component Matrix[a]

	Component
	1
nurturant	.726
affectionate	.795
compassionate	.773

Extraction Method: Principal Component Analysis.

a. 1 components extracted.

Figure 18.18 ♦ Analysis 2: Component Matrix for One Retained Component

Reproduced Correlations

		nurturant	affectionate	compassionate
Reproduced Correlation	nurturant	.527[b]	.577	.561
	affectionate	.577	.632[b]	.614
	compassionate	.561	.614	.597[b]
Residual [a]	nurturant		-.209	-.229
	affectionate	-.209		-.182
	compassionate	-.229	-.182	

Extraction Method: Principal Component Analysis.

a. Residuals are computed between observed and reproduced correlations. There are 3 (100.0%) nonredundant residuals with absolute values greater than 0.05.

b. Reproduced communalities

Figure 18.19 ♦ Analysis 2: Correlations Reproduced From the One-Component Model

18.10.1 ♦ Communality for Each Item Based on One Component

The table of communalities in Figure 18.16 reports the communalities for the entire set of three components in the first column. Recall that p, the number of measured variables in this PC analysis, is equal to 3. As in Analysis 1, when we retain all three components, the communality (proportion of variance that can be predicted for each measured variable) equals 1. The second column in this table, under the heading "Extraction," tells us the values of the communalities after we make the decision to retain only Component 1. For example, the "after extraction" communality for nurturant in Figure 18.16, $\boldsymbol{h}^2 = .527$, was obtained by squaring the loading or correlation of the variable nurturant with the one retained component. The loadings for three variables on the one retained component appear in Figure 18.18. Because nurturant has a correlation of .726 with Component 1, its communality (the proportion of variance in nurturance ratings that can be reproduced from a one-component model) is $.726^2 = .527$. Almost 53% of the variance in self-ratings on nurturance can be explained when we represent the variables using a one-component model. We usually hope that communalities will be reasonably high for most of the measured variables included in the analysis although there is no agreed on standard for the minimum acceptable size of communality. In practice, a communality $<.10$ suggests that the variable is not very well predicted from the component or factor model; variables with such low communalities apparently do not measure the same constructs as the other variables included in the analysis, and it might make sense to drop variables with extremely low communalities and run the analysis without them to see whether a clearer pattern emerges when the "unrelated" variable is omitted.

18.10.2 ♦ Variance Reproduced by the First Component

How much of the variance in the original set of three variables (each with a standard score variance equal to 1) can be reproduced when we retain only one of the three components? To find out how much variance is explained by correlations of the variables with Component 1, all we need to do is square and sum the loadings for Component 1 only (in Figure 18.18). When we do this, we obtain the same numerical result as in Analysis 1; Component 1 has an SSL of 1.756 (see Figure 18.17); compared with the total variance for a set of $p = 3$ variables, this represents $1.756/3 = .585$ or almost 59% of the variance. There is no agreed on minimum value for the proportion or percentage of variance that should be reproduced by the retained component or components for a model to be judged adequate, but a higher proportion of variance tells us that the component does a relatively better job of reproducing variance. In practice, if the first component cannot explain much more than $1/p$ proportion of the variance (where p is the number of measured variables), then transforming the scores into components does not provide us with a useful way of summarizing the information in the data. In this example, with the number of variables $p = 3$, we want the first component to explain more than 33% of the variance.

18.10.3 ♦ Partial Reproduction of Correlations From Loadings on Only One Component

As in Analysis 1, we can try to reproduce the correlation between any pair of measured variables by tracing the paths that connect that pair of variables in the reduced model that has a limited number of components. In this example, we have limited the number of components to 1. Therefore, when we attempt to reproduce the correlation between one pair of variables (such as the correlation between nurturant and affectionate), the only information we can use is the correlation of each variable with Component 1. Refer back to the path model in Figure 18.13. If we use only Component 1, and trace the path from nurturant to affectionate via C_1, the predicted correlation between nurturant and affectionate is equal to the product of the correlations of these variables with Component 1—that is, the estimated correlation between nurturant and affection using a one-component model is $.726 \times .795 = .577$. The actual observed correlation between nurturant and affectionate (from Figure 18.8) was $r = .368$. The residual, or error of prediction, is the difference between the actual correlation and the predicted correlation based on the one-component model—in this case, the residual or prediction error for this correlation $= .368 - .577 = -.209$. The complete set of reproduced correlations based on this model, and differences between actual and reproduced correlations, appears in Figure 18.19. We would like these residuals or prediction errors to be reasonably small for most of the correlations in the **R** matrix. SPSS flags residuals that are greater than .05 in absolute value as "large," but this is an arbitrary criterion.

To summarize, for the PC model to be judged adequate after we drop one or more of the components from the model, we would want to see the following evidence:

1. Reasonably large communalities for all the measured variables (e.g., $h^2 > .10$ for all variables; however, this is an arbitrary standard).
2. A reasonably high proportion of the total variance p (where p corresponds to the number of z scores on measured variables) should be explained by the retained components.
3. The retained component or components should reproduce all the correlations between measured variables reasonably well; for example, we might require the residuals for each predicted correlation to be less than .05 in absolute value (however, this is an arbitrary criterion).
4. For all the preceding points to be true, we will also expect to see that many of the measured variables have reasonably large loadings (e.g., loadings greater than .30 in absolute magnitude) on at least one of the retained components.

Note that SPSS does not provide statistical significance tests for any of the estimated parameters (such as loadings), nor does it provide confidence intervals. Judgments about the adequacy of a one- or two-component model are not made based on statistical significance tests, but by making arbitrary judgments whether the model that is limited to just one or two components does an adequate job of reproducing the communalities (the variance in each individual measured x variable) and the correlations among variables (in the **R** correlation matrix).

18.11 ♦ Principal Components Versus Principal Axis Factoring

As noted earlier, the most widely used method in factor analysis is the PAF method. In practice, PC and PAF are based on slightly different versions of the **R** correlation matrix (which includes the entire set of correlations among measured X variables). PC analyzes and reproduces a version of the **R** matrix that has 1s in the diagonal. Each value of 1.00 corresponds to the total variance of one standardized measured variable, and the initial set of p components must have sums of squared correlations for each variable across all components that sum to 1.00. This is interpreted as evidence that a p-component PC model can reproduce all the variances of each standardized measured variable. In contrast, in PAF, we replace the 1s in the diagonal of the correlation matrix **R** with estimates of communality that represent the proportion of variance in each measured X variable that is predictable from or shared with other X variables in the dataset. Many programs use multiple regression to obtain an initial communality estimate for each variable; for example, an initial estimate of the communality of X_1 could be the R^2 for a regression that predicts X_1 from $X_2, X_3, \ldots, X_p$. However, after the first step in the analysis, communalities are defined as sums of squared factor loadings, and the estimation of communalities with a set of factor loadings that can do a reasonably good job of reproducing the entire **R** correlation matrix typically requires multiple iterations in PAF.

For some datasets, PC and PAF may yield similar results about the number and nature of components or factors. The conceptual approach involved in PAF treats each X variable as a measurement that, to some extent, may provide information about the same small set of factors or latent variables as other measured X variables, but at the same time, each X variable may also be influenced by unique sources of error. In PAF, the analysis of data structure focused on shared variance and not on sources of error that are unique to individual measurements. For many applications of factor analysis in the behavioral and social sciences, the conceptual approach involved in PAF (i.e., trying to understand the shared variance in a set of X measurements through a small set of latent variables called factors) may be more convenient than the mathematically simpler PC approach (which sets out to represent all of the variance in the X variables through a small set of components). Partly because of the conceptual basis (PAF models only the shared variance in a set of X measurements) and partly because it is more familiar to most readers, PAF is more commonly reported in social and behavioral science research reports than PC. The next two empirical examples illustrate application of PAF to nine items for the data in Table 18.1.

18.12 ♦ Analysis 3: PAF of Nine Items, Two Factors Retained, No Rotation

Analysis 3 differs from the first two analyses in several ways. First, Analysis 3 includes nine variables (rather than the set of three variables used in earlier analyses). Second, PAF is used as the method of extraction in Analysis 3. Finally, in Analysis 3, two factors were retained based on the sizes of their eigenvalues.

Figure 18.20 shows the initial Factor Analysis dialog window for Analysis 3, with nine self-rated characteristics included as variables (e.g., nurturant, affectionate, . . . , aggressive). None of the additional descriptive statistics (such as reproduced correlations) that were requested in Analyses 1 and 2 were also requested for Analysis 3. To perform the extraction as a PAF, the Extraction button was used to open the Factor Analysis: Extraction window that appears in Figure 18.21. From the pull-down menu near the top of this window, Principal axis factoring was selected as the method of extraction. The decision about the number of factors to retain was indicated by clicking the radio button for Eigenvalues over; the default minimum size generally used to decide which factors to retain is 1, and that was not changed. Under the Display heading, a box was checked to request a scree plot; a scree plot summarizes information about the magnitudes of the eigenvalues across all the factors, and sometimes the scree plot is examined when making decisions about the number of factors to retain. The Rotation button was used to open the Factor Analysis: Rotation window that appears in Figure 18.22. The default (indicated by a radio button) is no rotation ("None," under the heading for method), and that was not changed. The box for Loading plots under the heading Display was checked to request a plot of the factor loadings for all nine variables on the two retained (but not rotated) factors. The Options button opened up the window for Factor Analysis: Options in Figure 18.23; in this box, under the heading for Coefficient Display Format, a check was placed in the check box for the Sorted by size option. This does not change any computed results but it arranges the summary table of factor loadings so that variables that have large loadings on the same factor are grouped together, and this improves the readability of the output, particularly when the number of variables included in the analysis is large. The syntax for Analysis 3 that resulted from the menu selections just discussed appears in Figure 18.24. The results of this PAF analysis of nine variables appear in Figures 18.25 through 18.31.

The information that is reported in the summary tables for Analysis 3 includes loadings and sums of squared loadings. The complete 9×9 matrix that contains the correlations of all nine measured variables with all nine factors does not appear as part of the SPSS output; the tables that do appear in the printout summarize information that is

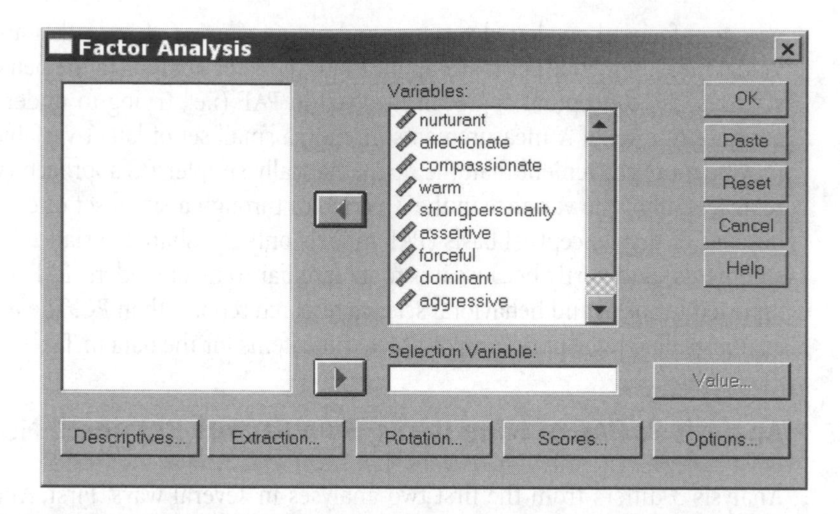

Figure 18.20 ♦ Analysis 3: Selection of All Nine Variables for Inclusion in Principal Axis Factoring Analysis

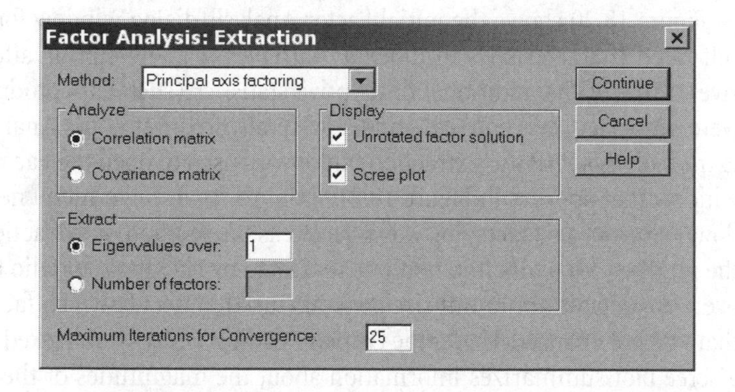

Figure 18.21 ♦ Analysis 3: Method of Extraction: Principal Axis Factoring Using Default Criterion (Retain Factors With Eigenvalues > 1); Request Scree Plot

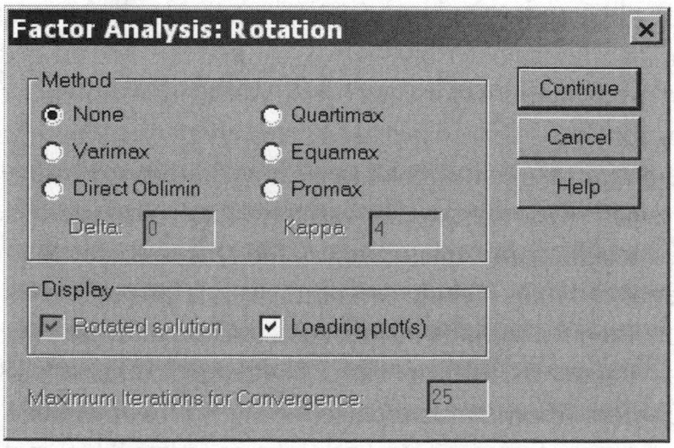

Factor Analysis: Rotation
Method
None
Quartimax
Varimax
Equamax
Direct Oblimin
Promax
Delta: 0
Kappa 4
Continue
Cancel
Help
Display
Rotated solution
Loading plot(s)
Maximum Iterations for Convergence: 25

Figure 18.22 ♦ Analysis 3: No Rotation Requested; Requested Plot of Unrotated Loadings on Factors 1 and 2

Factor Analysis: Options

Missing Values
- (•) Exclude cases listwise
- ( ) Exclude cases pairwise
- ( ) Replace with mean

Continue | Cancel | Help

Coefficient Display Format
- [✔] Sorted by size
- [ ] Suppress absolute values less than: .10

Figure 18.23 ♦ Analysis 3: Request for Factor Loadings Sorted by Size

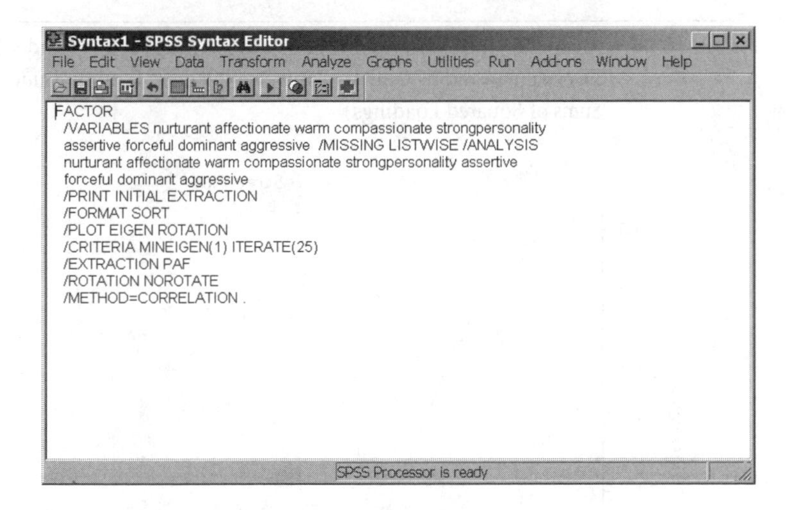

Syntax1 - SPSS Syntax Editor

File Edit View Data Transform Analyze Graphs Utilities Run Add-ons Window Help

```
FACTOR
  /VARIABLES nurturant affectionate warm compassionate strongpersonality
  assertive forceful dominant aggressive  /MISSING LISTWISE /ANALYSIS
  nurturant affectionate warm compassionate strongpersonality assertive
  forceful dominant aggressive
  /PRINT INITIAL EXTRACTION
  /FORMAT SORT
  /PLOT EIGEN ROTATION
  /CRITERIA MINEIGEN(1) ITERATE(25)
  /EXTRACTION PAF
  /ROTATION NOROTATE
  /METHOD=CORRELATION .
```

SPSS Processor is ready

Figure 18.24 ♦ Analysis 3: SPSS Syntax for PAF With Nine Items; Retain Factors With Eigenvalues >1; No Rotation

Communalities

	Initial	Extraction
nurturant	.242	.262
affectionate	.484	.570
compassionate	.427	.484
warm	.561	.694
strongpersonality	.285	.330
assertive	.309	.366
forceful	.349	.476
dominant	.339	.458
aggressive	.285	.355

Extraction Method: Principal Axis Factoring.

Figure 18.25 ♦ Analysis 3: Communalities for PAF on Nine Items

NOTE: Based on all nine factors, in column headed Initial; based on only two retained factors, in column headed Extraction.

Total Variance Explained

Factor	Initial Eigenvalues			Extraction Sums of Squared Loadings		
	Total	% of Variance	Cumulative %	Total	% of Variance	Cumulative %
1	2.863	31.807	31.807	2.346	26.070	26.070
2	2.194	24.380	56.187	1.649	18.327	44.397
3	.823	9.143	65.330			
4	.723	8.032	73.362			
5	.626	6.961	80.323			
6	.535	5.950	86.272			
7	.480	5.331	91.603			
8	.469	5.214	96.817			
9	.286	3.183	100.000			

Extraction Method: Principal Axis Factoring.

Figure 18.26 ♦ Analysis 3: Total Variance Explained by All Nine Factors (Under Heading Initial Eigenvalues) and by Only the First Two Retained Factors (Under Heading Extraction Sums of Squared Loadings)

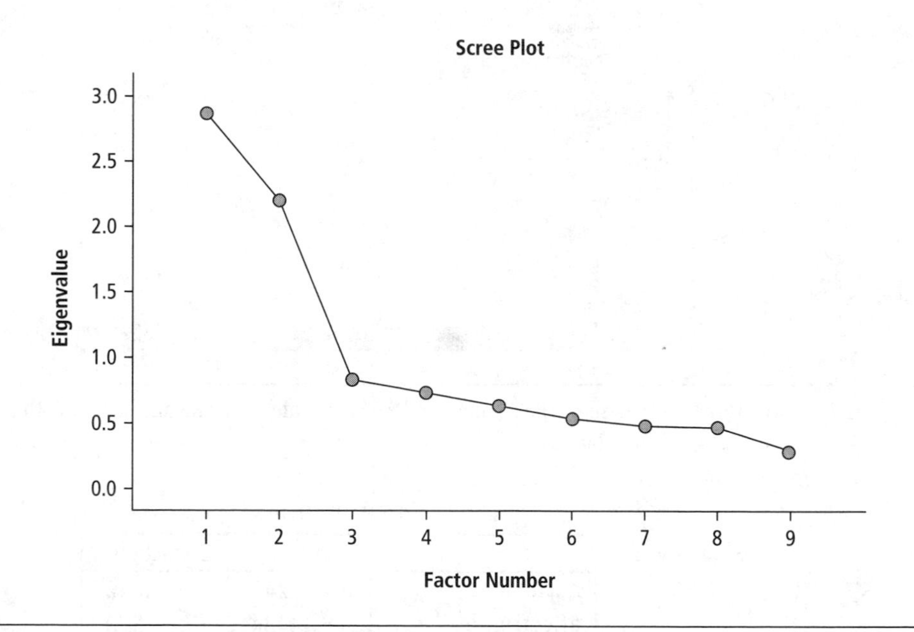

Figure 18.27 ♦ Analysis 3: Scree Plot of Eigenvalues for Factors 1 Through 9

contained in this larger 9 × 9 loading matrix. For example, each communality is the SSL for one variable across all retained factors. Each SSL for a retained factor corresponds to the SSL for all nine variables on that factor.

18.12.1 ♦ Communality for Each Item Based on Two Retained Factors

Figure 18.25 reports the communalities for each of the nine variables at two different stages in the analysis. The entries in the first column, headed Initial, tell us the proportion

Factor Matrix[a]

	Factor	
	1	2
warm	.725	-.411
affectionate	.682	-.323
strongpersonality	.534	.212
assertive	.518	.313
aggressive	.482	.351
forceful	.345	.597
dominant	.402	.544
compassionate	.481	-.503
nurturant	.240	-.452

Extraction Method: Principal Axis Factoring.

a. 2 factors extracted. 8 iterations required.

Figure 18.28 ♦ Analysis 3: Unrotated Factor Loadings on Two Retained Factors

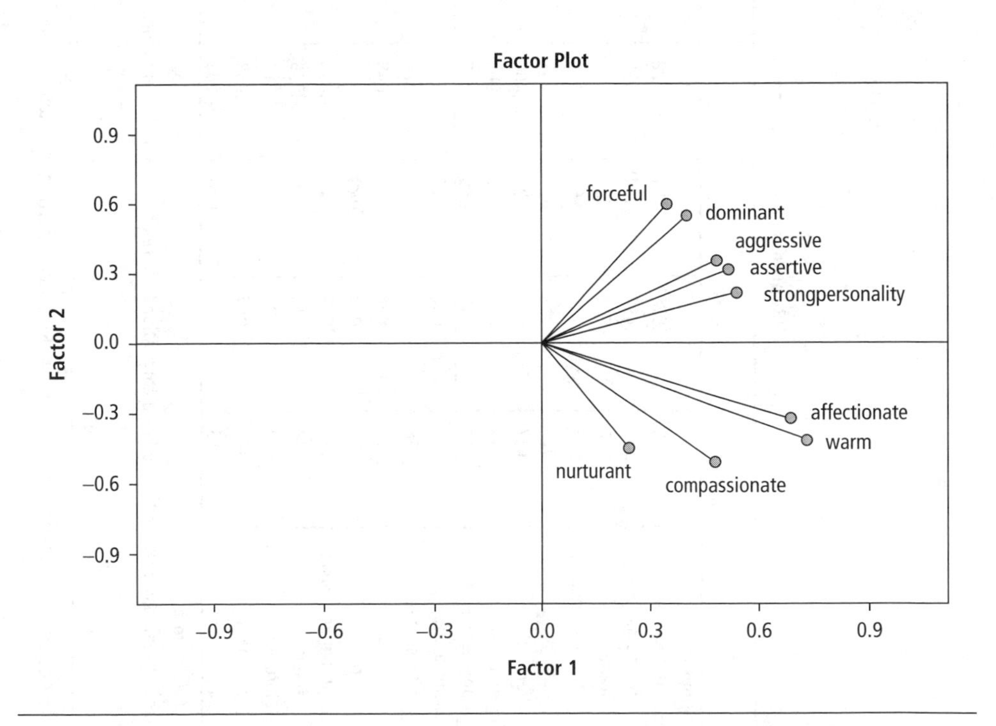

Figure 18.29 ♦ Analysis 3: Plot of the Factor Loadings for the Two Retained Unrotated Factors

NOTES: For example, "affectionate" had a loading of +.68 on Factor 1 and a loading of −.32 on Factor 2. The vectors (the lines that connect the dot labeled "affectionate" to the 0,0 origin in the graph) were added in a graphics program; they do not appear in the SPSS plot.

of variance in each measured variable that could be predicted from the other eight predictor variables in a preliminary multiple regression. The R^2 from this initial regression is used as the initial communality estimate. For example, a regression was performed

Reproduced Correlations

		nurturant	affectionate	compas sionate	warm	strong personality	assertive	forceful	dominant	aggressive
Reproduced Correlation	nurturant	.262[b]	.310	.343	.360	.032	-.017	-.187	-.149	-.043
	affectionate	.310	.570[b]	.491	.628	.296	.252	.042	.099	.215
	compassionate	.343	.491	.484[b]	.555	.150	.091	-.134	-.080	.055
	warm	.360	.628	.555	.694[b]	.300	.247	.005	.068	.205
	strongpersonality	.032	.296	.150	.300	.330[b]	.343	.311	.330	.332
	assertive	-.017	.252	.091	.247	.343	.366[b]	.366	.379	.359
	forceful	-.187	.042	-.134	.005	.311	.366	.476[b]	.464	.376
	dominant	-.149	.099	-.080	.068	.330	.379	.464	.458[b]	.385
	aggressive	-.043	.215	.055	.205	.332	.359	.376	.385	.355[b]
Residual [a]	nurturant		.058	-.011	-.057	.029	.043	.038	-.085	-.019
	affectionate	.058		-.058	.024	-.029	.014	.008	.007	.000
	compassionate	-.011	-.058		.048	.021	-.039	-.041	.021	.046
	warm	-.057	.024	.048		-.039	-.038	.014	.038	-.009
	strongpersonality	.029	-.029	.021	-.039		.110	-.037	-.019	-.011
	assertive	.043	.014	-.039	-.038	.110		-.016	-.022	-.038
	forceful	.038	.008	-.041	.014	-.037	-.016		.003	.045
	dominant	-.085	.007	.021	.038	-.019	-.022	.003		-.003
	aggressive	-.019	.000	.046	-.009	-.011	-.038	.045	-.003	

Extraction Method: Principal Axis Factoring.

a. Residuals are computed between observed and reproduced correlations. There are 5 (13.0%) nonredundant residuals with absolute values greater than 0.05.

b. Reproduced communalities

Figure 18.30 ♦ Analysis 3: Reproduced Correlations Based on Two Retained (Unrotated) Factors

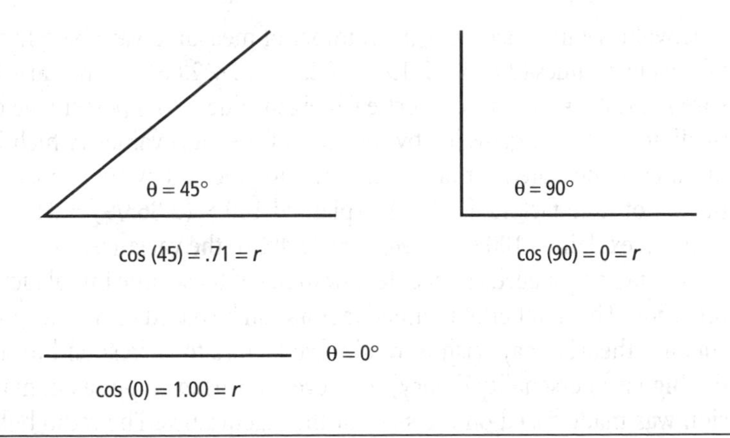

Figure 18.31 ♦ Geometric Representation of Correlation Between X_1 and X_2
$r = \cos(\theta)$, where θ is the angle between the vectors that represent the X_1 and X_2 vectors (given in degrees).

to predict scores on nurturant from the scores on the other eight measured variables ($R^2 = .242$ for this regression); therefore, the initial estimate of communality was .242, that is, 24.2% of the variance in nurturant could be predicted from the set of scores on the other eight variables. Note that in the PC analysis, the values of the communalities for the initial model were set to 1.00 for all variables. In the PAF analysis, however, these initial communalities are based on multiple regressions that tell us what proportion of the variance in each measured variable is predictable from, or shared with, other measured variables. The second column, headed Extraction, tells us what proportion of variance in each variable is explained after we extract nine factors and decide to retain only two factors. (Information about the number of factors retained appears in later parts of the SPSS output.) For example, in Figure 18.28, we can see that the variable affectionate had loadings or correlations with the two unrotated retained factors that were equal to .682 (for Factor 1) and −.323 (for Factor 2). The communality for affectionate—that is, the proportion of variance in scores on self-ratings of affectionate that can be predicted from the two retained factors—is obtained by summing these squared correlations: $(.682^2) + (-.323)^2 = .570$. This agrees with the communality value of .570 for the variable affectionate in the summary table in Figure 18.25. Thus, the communalities in the Extraction column of Figure 18.25 provide information about how well a PAF factor solution that retains only Factors 1 and 2 can reproduce the variance in the nine measured variables. The variable affectionate had a relatively high proportion of predicted variance (.570) while the variable nurturant had a much lower proportion of predicted variance (.262).

18.12.2 ♦ Variance Reproduced by Two Retained Factors

We also need summary information that tells us the extent to which the variances and correlations in the set of $p = 9$ variables could be reconstructed, Factor 1, Factor 2, . . ., Factor 9. This information appears in Figure 18.26. In the left-hand panel under the heading Initial Eigenvalues, for each of the nine factors, there is an eigenvalue (which is equivalent to an SSL). The factors are rank ordered by the sizes of their eigenvalues; thus, Factor 1 has the largest eigenvalue, Factor 2 the second largest eigenvalue, and so forth. The sum of the eigenvalues in the column Total under the banner heading Initial

Eigenvalues will always be p, the number of measured variables. In this example, the sum of the eigenvalues $(2.863 + 2.194 + .823 + \ldots + 2.86) = 9$, because there were nine measured variables. We can convert each eigenvalue into a percentage of explained variance by dividing each eigenvalue by the sum of the eigenvalues (which is equivalent to p, the number of measured variables) and multiplying this by 100. Thus, for example, in the initial set of nine factors, Factor 1 explained $100 \times (2.863/9) = 31.8\%$ of the variance and Factor 2 explained $100 \times (2.194/9) = 24.4\%$ of the variance.

A data analyst needs to decide how many of these nine initial factors to retain for interpretation. The number of retained factors can be based on a conceptual model (e.g., a personality theorist may wish to retain five factors to correspond to the five dimensions in the Big Five personality theory). However, in this example, as in many analyses, the decision was made based on the sizes of the eigenvalues. The radio button that was selected earlier requested retention of factors with eigenvalues >1.00. When we examine the list of eigenvalues for the initial set of nine factors in the left-hand column of Figure 18.26, we can see that only Factors 1 and 2 have eigenvalues >1; Factor 3 and all subsequent factors had eigenvalues <1. Using a cutoff value of 1.00 (which corresponds to the amount of variance in one z score or standardized variable) results in a decision to retain only Factors 1 and 2. The right-hand side of the table in Figure 18.26 shows the SSLs for only Factors 1 and 2. After the other seven factors are dropped, SPSS reestimates the factor loadings for the retained factors, and therefore the SSLs for Factors 1 and 2 changed when other factors are dropped from the model. After limiting the model to two factors, Factor 1 predicted or accounted for about 26% of the variance and Factor 2 accounted for about 18% of the variance; together, Factors 1 and 2 accounted for about 44% of the variance in the data (i.e., the variances in scores on measured variables and the pattern of correlations among measured variables).

The eigenvalues for the initial set of nine factors listed in the left-hand column of Figure 18.26 are also shown in a scree plot in Figure 18.27. Sometimes, data analysts use the scree plot as an aid in deciding how many factors to drop from the model. Essentially, we look for a point of inflection; the proportion of variance explained by the first few factors often tends to be large, and then, after two or three factors, the amount of variance accounted for by the remaining factors often levels off. Examination of the scree plot in Figure 18.27 would lead to the same decision as the decision that was made based on the sizes of the eigenvalues. Compared with Factors 1 and 2, Factors 3 through 9 all had quite small eigenvalues that corresponded to very small proportions of explained variance.

The unrotated factor loadings for the two retained factors appear in Figure 18.28 under the heading Factor Matrix. Each entry in this table corresponds to a factor loading or, in other words, a correlation between one of the measured variables (warm, affectionate, etc.) and one of the retained factors (Factor 1, Factor 2). The goal in factor analysis is often to identify sets or groups of variables, each of which have high correlations with only one factor. When this type of pattern is obtained, it is possible to interpret each factor and perhaps give it a name or label, based on the nature of the measured variables with which it has high correlations. The pattern of unrotated loadings that appears in Figure 18.28 is not easy to interpret. Most of the variables had rather large positive loadings on Factor 1; loadings on Factor 2 tended to be a mixture of positive and negative correlations. The plot in Figure 18.29 shows the factor loadings in a two-dimensional space. This pattern does not make it easy to differentiate Factor 1 and Factor 2 from each other and to identify them with different latent constructs. When more than one factor is retained, in fact, the unrotated loadings are often not very easy to interpret. The last empirical example, Analysis 4, will show how rotation of the factors can yield a more interpretable solution.

18.12.3 ♦ Partial Reproduction of Correlations From Loadings on Only Two Factors

A factor analysis, like a PC analysis, should be able to do a reasonably good job of reproducing the observed correlations between all pairs of measured variables. The matrix in Figure 18.30 summarizes the actual versus reproduced correlations for the two-factor model in Analysis 3. Each correlation (e.g., the reproduced correlation between nurturant and affectionate) is reproduced from the factor loadings by tracing the paths that connect these two variables via the latent variables Factor 1 (F_1) and Factor 2 (F_2). We multiply the path coefficients along each path and sum the contributions across the paths. The correlations of nurturant and affectionate with Factor 1, from the factor loading matrix in Figure 18.28, were .240 and .682, respectively; the product of these two correlations is (.24 × .682) = .1637. The correlations of nurturant and affectionate with Factor 2 were −.452 and −.323; the product of these two correlations was .1460. If we sum the predicted correlations based on F_1 (.1637) and F_2 (.1460), we obtain an overall predicted correlation between nurturant and affectionate that equals .1637 + .1460 = .31. This agrees with the reproduced correlation between nurturant and affectionate reported in Figure 18.30. A factor analysis, like a PC analysis, is supposed to do a reasonably good job of reproducing the correlations among observed variables (preferably using a relatively small number of factors).

The factor analysis reported as Analysis 3 has done a reasonably good job of reproducing the variance on each of the measured variables (with communalities that ranged from .262 to .570); it also did a reasonably good job of reproducing the correlations in the correlation matrix **R** (according to the footnote at the bottom of Figure 18.30, only 13% of the prediction errors or residuals for reproduced correlations were larger than .05 in absolute value). Analysis 3 suggests that we can do a reasonably good job of representing the information about the pattern in the **R** data matrix by using just two factors. However, the pattern of (unrotated) factor loadings for the two retained factors in Analysis 3 was not easy to interpret. Rotated factors are often, although not always, more interpretable. Before presenting a final example in which the factors are rotated, we will use a geometric representation of correlations between variables. This geometric representation will make it clear what it means to "rotate" factor axes.

18.13 ♦ Geometric Representation of Correlations Between Variables and Correlations Between Components or Factors

A correlation between two variables X_1 and X_2 can be represented geometrically by setting up two vectors (or arrows). Each vector represents one variable; the cosine of the angle between the vectors is equivalent to the correlation between the variables.

In other words, we can represent the correlation between any two variables by representing them as vectors separated by an angle θ (the angle is given in degrees). The angle between the two vectors that represent variables X_1 and X_2 is related to the correlation between variables X_1 and X_2 through the cosine function. Two variables that have a correlation of 1.00 can be represented by a pair of vectors separated by a 0° angle (as in the bottom example in Figure 18.31). Two variables that have a correlation of 0 (i.e., two variables that are uncorrelated or orthogonal to each other) are represented by a pair of vectors separated by a 90° angle, as in the upper right-hand drawing in Figure 18.31. Angles that are between 0° and 90° represent intermediate amounts of linear correlation; for example, if the vectors that represent variables X_1 and X_2 are at a 45° angle to each other, the correlation between variables X_1 and X_2 is equivalent to the cosine of 45°, or $r = .71$.

A factor loading is also a correlation, so we can represent the loadings (or correlations) between an individual X_1 variable and two factors, F_1 and F_2. So far, in all the examples we have considered, we have constrained the factors or components to be orthogonal to each other or uncorrelated with each other. Therefore, we can diagram Factor 1 and Factor 2 by using vectors that are separated by a 90° angle, as shown in Figure 18.32. We can represent the loadings or correlations of the X_1 variable with these two factors by plotting the X_1 variable as a point. For example, if X_1 has a correlation of .8 with F_1 and a correlation of .1 with F_2, we can plot a point that corresponds to +.8 along the horizontal (F_1) axis and +.1 along the vertical (F_2) axis. When we connect this point to the origin (the 0, 0) point in the graph, we obtain a vector that represents the "location" of the X_1 variable in a two-dimensional space relative to Factors 1 and 2. From this graph, we can see that X_1 is much more highly correlated with F_1 than it is with F_2—that is, the angle that separates the X_1 vector from the F_1 axis is much smaller than the angle that separates the X_1 vector from the F_2 axis.

Now let's think about the problem of showing the correlations among p variables geometrically. We can draw our first two vectors (representing X_1 and X_2) on a flat

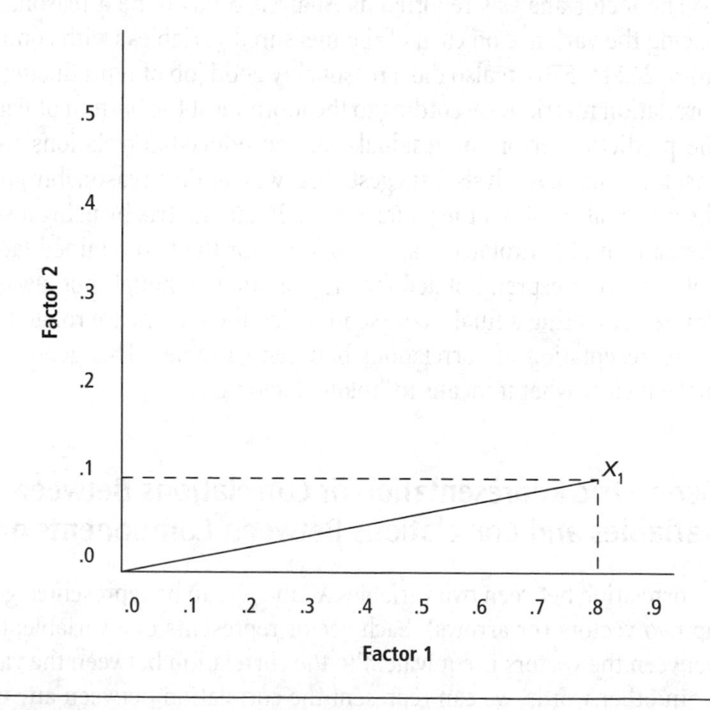

Figure 18.32 ♦ Graphic Representation of Correlation Between Variable X_1 and Factors F_1 and F_2

NOTE: Correlation of X_1 with $F_1 = +.80$, correlation of X_1 with $F_2 = +.1$.

plane. When we add a third variable X_3, we may or may not need to use a third dimension to set up angles between X_3 and X_1 and X_3 and X_2 that accurately represent the correlation of X_2 with the other variables. As we add more variables, we typically need a higher dimensional space if we want our "bundle" of vectors to be separated by angles that perfectly correspond to the correlations among all possible pairs of variables. In general, we can perfectly represent the correlations among p variables in a p-dimensional space. We can think of factor analysis as a reduction in the number of dimensions that we use to represent the correlations among variables. Imagine our set of p variables as a bundle of vectors in a p-dimensional space, like an imaginary set of p-dimensional umbrella spokes. We want to know: If we squeeze this bundle of vectors into a lower dimensional space (such as a two- or three-dimensional space), can we still fairly accurately represent the correlations among all pairs of variables by the angles among vectors? Or do we lose this information as we "flatten" the bundle of vectors into a lower dimensional space?

When we decide that a two-factor solution is adequate, our factor solution (a set of loadings, or of correlations between measured variables and imaginary factors) can be understood geometrically. We can represent each X_i variable in terms of its correlations with the axes that represent Factors 1 and 2. The vector that corresponds to X_1 will have a specific location in the two-dimensional space that is defined by the factor axes F_1 and F_2, as shown in Figure 18.32. If a two-dimensional representation is reasonably adequate, we will find that we can reconstruct the correlations among all pairs of measured variables (e.g., X_1 and X_2, X_1 and X_3, X_2 and X_3, etc.) reasonably well from the correlations of each of these variables with each of the two retained factors.

18.13.1 ♦ Factor Rotation

When we plotted the correlations or factor loadings for variables X_1 through X_9 (e.g., warm, nurturant, aggressive, etc.) with the two unrotated factors in Analysis 3, as shown in Figure 18.29, we could not see a clear pattern. We would like to be able to "name" each latent variable or factor by identifying a set of measured variables with which it has high correlations; and we would like to be able to differentiate two or more latent variables or factors by noticing that they correlated with different groups or sets of measured variables. Figure 18.33 illustrates the pattern that often arises in an unrotated factor solution (in the left-hand side graph). In an unrotated two-factor solution, we often find that all the measured X variables have at least moderate positive correlations with Factor 1, and that there is a mixture of smaller positive and negative correlations of variables with Factor 2, as seen in the hypothetical unrotated factor loadings in the left-hand side of Figure 18.33.

However, if our two-dimensional space is a "plane," and we want to be able to identify the location of vectors in that plane, our choice of reference axes to use is arbitrary. In the original unrotated solution, the factor loadings were calculated so that Factor 1 would explain as much of the variance in the measured X variables as possible; Factor 2 was constrained to be orthogonal to/uncorrelated with Factor 1, and to explain as much of the remaining variance as possible; and so forth. This strategy often yields a set of factor loadings that do a reasonably good job of reproducing all the correlations among measured X variables, but it often does not optimize the interpretability of the pattern.

We can think of factors as the axes that are used as points of reference to define locations of variables in our two-dimensional space. In the present example, F_1 and F_2 correspond to the unrotated factor axes that define a two-dimensional space. The fact that these factors are orthogonal or uncorrelated is represented by the 90° angle between the F_1 and F_2 factor axes. However, the location of the factor axes is arbitrary. We could define locations within this same plane relative to any two orthogonal F_1 and F_2 vectors in this plane.

When we "rotate" a factor solution, we simply move the factor axes to a different location in the imaginary plane and then recalculate the correlation of each X measured variable relative to these relocated or "rotated" axes. In the right-hand side of Figure 18.33, the rotated factor axes are labeled F_1' and F_2'. A factor is most easily interpretable if variables tend to either have very large loadings or loadings near 0—that is, you can say clearly that the factor either *is* or *is not* related to each variable. In this hypothetical example, variables have relatively high correlations with the new rotated F_1' axis (i.e., the vectors for these first three variables are separated from the F_1' axis by small angles). X_1, X_2, and X_3 have fairly low correlations with the F_2' axis. On the other hand, variables X_4, X_5, and X_6 are fairly highly correlated with F_2' and have very low correlations with F_1'. If we set out to interpret these rotated factors, then, we might say that the F_1' rotated factor is highly correlated with variables X_1, X_2, and X_3; we can try to name or label the latent variable that corresponds to this rotated factor F_1' by trying to summarize what common ability of trait the set of variables X_1, X_2, and X_3 might all measure.

There are many different approaches to factor rotation; the most popular rotation method is called varimax. According to the online help in SPSS 14, the varimax rotation method "minimizes the number of variables that have high loadings on each factor . . . this method simplifies the interpretation of factors." Varimax is one of several types of **"orthogonal" rotation.** In an orthogonal rotation, factor axes are relocated, but the factor axes retain 90° angles to each other after rotation. That is, when we use orthogonal rotation methods, the rotated factors remain orthogonal or uncorrelated to each other.

Many other rotation methods are available, and in some methods of rotation, we allow the rotated factor axes to become nonorthogonal—that is, we may decide we want to obtain factors that are correlated with each other to some degree. Varimax is the most popular and widely reported rotation method; see Harman (1976) for a discussion of other possible methods of rotation. When a nonorthogonal rotation method is used, the interpretation of factor loadings becomes more complex, and additional matrices are involved in the description of the complete factor solution; see Tabachnick and Fidell (2007) for a discussion of these issues.

A rotated solution is often (although not always) easier to interpret than the unrotated solution. Rotation simply means that the factor loadings or correlations of the X variables are recalculated relative to the new rotated factors. Varimax-rotated loadings will typically explain or reproduce the observed correlations as well as the unrotated loadings and so there is no mathematical basis to prefer one solution over the other. The choice between various possible sets of factor loadings (with or without various kinds of rotation) is based on theoretical considerations: Which solution gives you the most meaningful and interpretable pattern? One of the objections to factor analysis is the arbitrariness of the choice of one solution out of many solutions that are all equally good in the

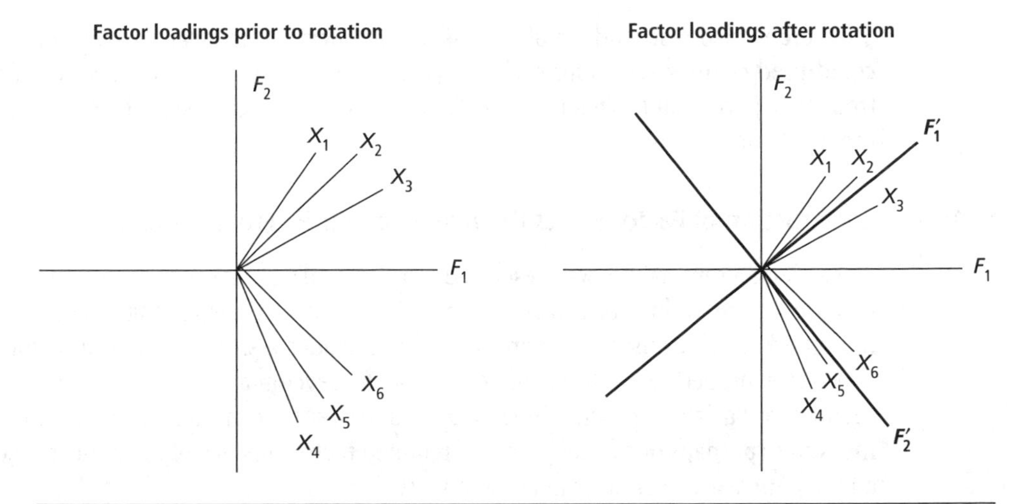

Figure 18.33 ♦ Hypothetical Graphs of Factor Loadings for Variables X_1 Through X_6 on Factors 1 and 2

mathematical sense (i.e., they all reproduce the observed correlations equally well). Analysis 4 (in a subsequent section) includes varimax **rotated factor loadings.**

Sometimes, even after rotation, one or more items still have relatively large loadings on two or more factors. An item that has a strong correlation with more than one factor can be called "factorially complex." For example, consider what might happen in a factor analysis of scores on the following mental ability tests: X_1 = Reading comprehension, X_2 = Vocabulary, X_3 = Algebra "story problems," X_4 = Linear algebra, and X_5 = Geometry. We might obtain a two-factor model such that X_1 and X_2 have high correlations with the first factor and X_4 and X_5 have high correlations with the second factor; we might label Factor 1 "verbal ability" and Factor 2 "math ability." However, the X_3 variable (scores on a test that involves algebra story problems) might have moderately high correlations or loadings for both Factors 1 and 2. The X_3 variable could be interpreted as factorially complex—that is, scores on this test are moderately predictable from both verbal ability and math ability; reasonably good reading comprehension is required to set up the problem in the form of equations, and reasonably good math skills are required to solve the equations. Thus, the X_3 test, story problems in algebra, might not be interpretable purely as a measure of verbal ability or purely as a measure of math ability; it might tap both these two different kinds of ability. If the goal of the data analyst is to identify items that each clearly measure only one of the two factors, the data analyst might drop this item from future analyses.

18.14 ♦ The Two Multiple Regressions

There is one more issue that needs to be considered before examining the final factor analysis example. It is possible to calculate and save a "factor score" on each factor for each participant. To understand how this is done, we need to think about factor analysis as two sets of multiple regressions. One set of regression models is used to construct factor scores for individual subjects from their z scores on the measured variables.

There is also a second set of regressions implicit in factor analysis; in principle, we could predict the z scores for each individual participant on each measured X variable from that participant's factor scores. This second set of regression does not have much practical application.

18.14.1 ♦ Construction of Factor Scores (Such as Score on F_1) From z Scores

Earlier, factor analysis was described as a method of "data reduction." If we can show that we can reproduce or explain the pattern of correlations among a large number of measured variables using a factor model that includes a small number of factors (e.g., a two-factor model), then it may be reasonable to develop a measure that corresponds to each latent variable or factor. In later data analyses, we can then use this smaller set of measures (perhaps one factor score per retained factor) instead of including the scores on all the p measured variables in our analyses.

A path diagram for the regression to construct a factor score on Factor 1 (F_1) from z scores on the measured X variables appears in Figure 18.34. If you request the "regression" method of computing saved factor scores, the **factor score coefficients** (denoted by beta in Figure 18.34) are used as weights. Factor score coefficients can be requested from SPSS as an optional output, and the factor scores for each participant can be generated

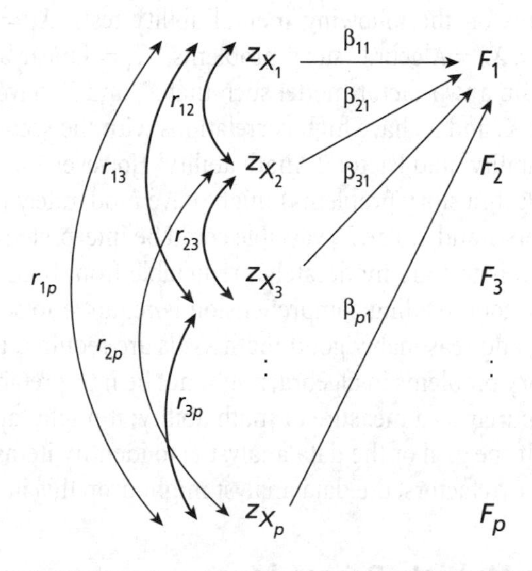

Figure 18.34 ♦ An Example of the First Set of Regressions: The Construction of a Factor Score on Factor 1 From Measured Items $z_{X_1}, z_{X_2}, \ldots, z_{X_p}$

NOTES:

1. Each β_{ij} coefficient in the diagram represents a factor score coefficient associated with a z_{Xj} predictor variable. To construct a score on F_1 for each participant in the study, we apply the β_{i1} factor score coefficients to that participant's scores on the standardized measured variables, z_{X_1} through z_{X_p}, as follows: Constructed factor score $F_1 = \beta_{11} z_{X_1} + \beta_{2_1} z_{X_2} + \beta_{3_1} z_{X_3} + \ldots + \beta_{p1} z_{X_p}$.
2. Typically, factor scores are constructed only for a small number of retained factors.

and saved as new variables in the SPSS data file. The factor score coefficients are similar to beta weights in a multiple regression. The z_X predictor variables are usually intercorrelated, and these intercorrelations must be taken into account when factor score coefficients are calculated. Factor scores on each of the rotated final factors can be saved for each subject and then used as variables in subsequent SPSS analyses. This can be quite useful in reducing the number of variables and/or obtaining orthogonal variables for use in later analyses.

Another possible follow-up to a factor analysis is to form a simple unit-weighted composite that corresponds to each factor. For example, if Factor 1 has high positive correlations with X_1, X_2, and X_3, we might combine the information in these three variables by summing their z scores (or possibly the raw scores). The construction and evaluation of unit-weighted composites is discussed in more detail in Chapter 19.

The last analysis that will be reported, Analysis 4, provides factor score coefficients that can be used to construct factor scores on the two retained factors. Figure 18.42 summarizes factor score coefficients from Analysis 4; these factor score coefficients correspond to the beta values in the path models in Figure 18.34. To construct a score for each person on Factor 1, the factor score coefficients in Figure 18.42 are applied to the z scores on the measured variables. A factor score for each participant on Factor 1 can be constructed as follows: $F_1 = .135 \times z_{\text{nurturant}} + .273 \times z_{\text{affectionate}} + .233 \times z_{\text{compassionate}} + .442 \times z_{\text{warm}} + .046 \times z_{\text{strong personality}} + .02 \times z_{\text{assertive}} - .109 \times z_{\text{forceful}} - .096 \times z_{\text{dominant}} - .004 \times z_{\text{aggressive}}$.

18.14.2 ♦ Prediction of Standard Scores on Variables (z_{X_i}) From Factors ($F_1, F_2, \ldots, F_p$)

We can view the factor loadings for an orthogonal solution as coefficients that we could use to predict an individual participant's scores on each individual measured variable from his or her scores on Factor 1, Factor 2, and any other retained factors. That is, $z_{X_1} = a_{11} F_1 + a_{12} F_2 + \ldots + a_{1p} F_p$. Examples of path models for regressions that could predict z scores on X_1 and X_2 measured variables from constructed factor scores F_1 and F_2 appear in Figure 18.35. Note that for these regressions, we can use the factor loadings (denoted by *as* in Figure 18.35) as regression coefficients, provided that the factors F_1 and F_2 are orthogonal to each other. See the appendix to this chapter for additional information about the matrix algebra involved in factor analysis; additional matrices are required when a nonorthogonal rotation is applied to factors.

18.15 ♦ Analysis 4: PAF With Varimax Rotation

The final analysis presented in this chapter is a replication of Analysis 3 (a factor analysis that uses the size of eigenvalues as the criterion for factor retention) with two additional specifications: varimax rotation is requested (to make the pattern of factor loadings more interpretable), and factor scores are saved for the two retained factors.

Figure 18.36 shows the radio button selection that was used to request varimax rotation. In addition, a Loading plot was requested. In the main Factor Analysis dialog window, the Save button was used to open the Factor Analysis: Factor Scores window that

Prediction of scores on z_{X_1} from scores on two retained orthogonal factors, F_1 and F_2

F_1 —a_{11}→ z_{X_1}

F_2 —a_{12}→ z_{X_1}

Prediction of score on z_{X_2} from scores on two retained orthogonal factors F_1, F_2

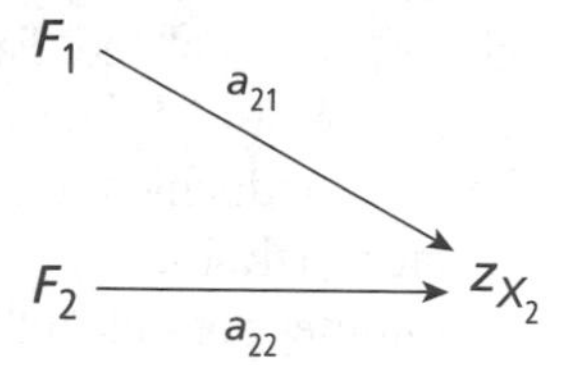

Figure 18.35 ♦ Two Examples From the Second Set of Multiple Regressions That Correspond to Factor Analysis

NOTE: The z score on each measured X variable can be perfectly predicted from the entire set of factor scores ($F_1, F_2, \ldots, F_p$) in the initial factor analysis (as shown in Figure 18.1). In the final stages of a factor analysis, we usually want to see how well we can predict the scores on the z_X items from scores on a small number of retained factors (in this example, two retained factors). This example shows uncorrelated or orthogonal factors. The communality for z_{X_1} is obtained by summing the squared correlations of z_{X_1} with F_1 and F_2. For example, the communality for z_{X_1} when we use only F_1 and F_2 as predictors = $a_{11}^2 + a_{12}^2$. To be judged adequate, one thing a factor analysis needs to do is reproduce or predict a reasonably high proportion of variance on the z_X variables. We can set up similar path models to show that scores on z_{X_3}, $z_{X_4}, \ldots, z_{X_p}$ can be predicted from F_1 and F_2.

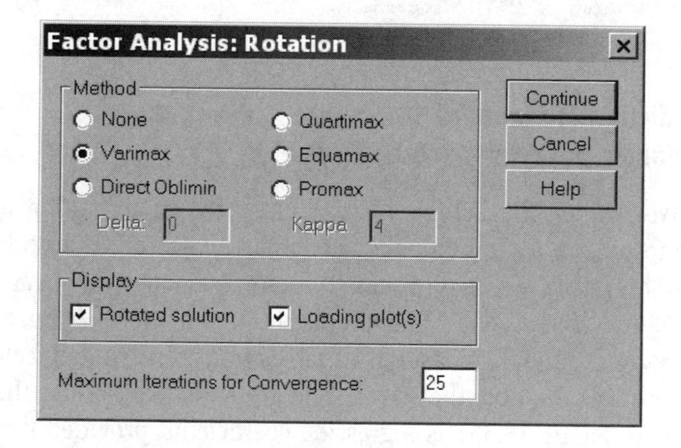

Figure 18.36 ♦ Analysis 4: Principal Axis Factoring With All Specifications the Same as in Analysis 3, With Varimax Rotation Requested

appears in Figure 18.37. Check box selections were made to save the factor scores as variables and to display the factor score coefficient matrix; a radio button selection was used to request the "regression" method of computation for the saved factor scores. The SPSS syntax for Analysis 4 appears in Figure 18.38. Selected output for Analysis 4 appears in Figures 18.39 through 18.43. Some parts of the results for Analysis 4 are identical to results from Analysis 3 (e.g., the communalities for individual items), and that information is not repeated here.

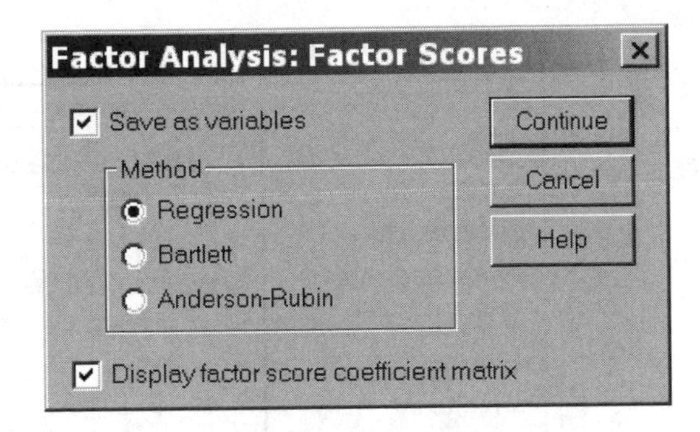

Figure 18.37 ♦ Request to Compute and Save Factor Scores From Analysis 4

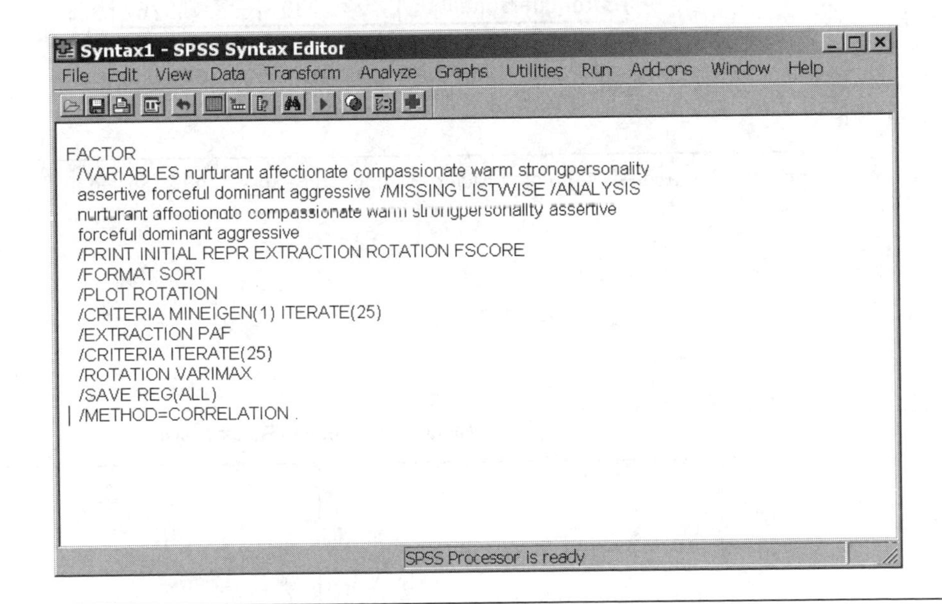

Figure 18.38 ♦ Analysis 4 SPSS Syntax: Nine Items; Method of Extraction, Principal Axis Factoring; Criterion for Retention of Factors Is Eigenvalue >1; Rotation = Varimax; Factor Scores Are Calculated and Saved

Total Variance Explained

	Initial Eigenvalues			Extraction Sums of Squared Loadings			Rotation Sums of Squared Loadings		
Factor	Total	% of Variance	Cumulative %	Total	% of Variance	Cumulative %	Total	% of Variance	Cumulative %
1	2.863	31.807	31.807	2.346	26.070	26.070	2.002	22.243	22.243
2	2.194	24.380	56.187	1.649	18.327	44.397	1.994	22.154	44.397
3	.823	9.143	65.330						
4	.723	8.032	73.362						
5	.626	6.961	80.323						
6	.535	5.950	86.272						
7	.480	5.331	91.603						
8	.469	5.214	96.817						
9	.286	3.183	100.000						

Extraction Method: Principal Axis Factoring.

Figure 18.39 ♦ Analysis 4 Principal Axis Factoring With Varimax Rotation: Summary of Total Variance Explained

Rotated Factor Matrix[a]

	Factor	
	1	2
warm	.804	.217
affectionate	.713	.250
compassionate	.696	-.020
nurturant	.489	-.153
dominant	-.096	.670
forceful	-.174	.667
aggressive	.096	.588
assertive	.148	.587
strongpersonality	.230	.526

Extraction Method: Principal Axis Factoring.
Rotation Method: Varimax with Kaiser Normalization.
a. Rotation converged in 3 iterations.

Figure 18.40 ♦ Analysis 4: Rotated Factor Loadings on Factors 1 and 2 (Sorted by Size)

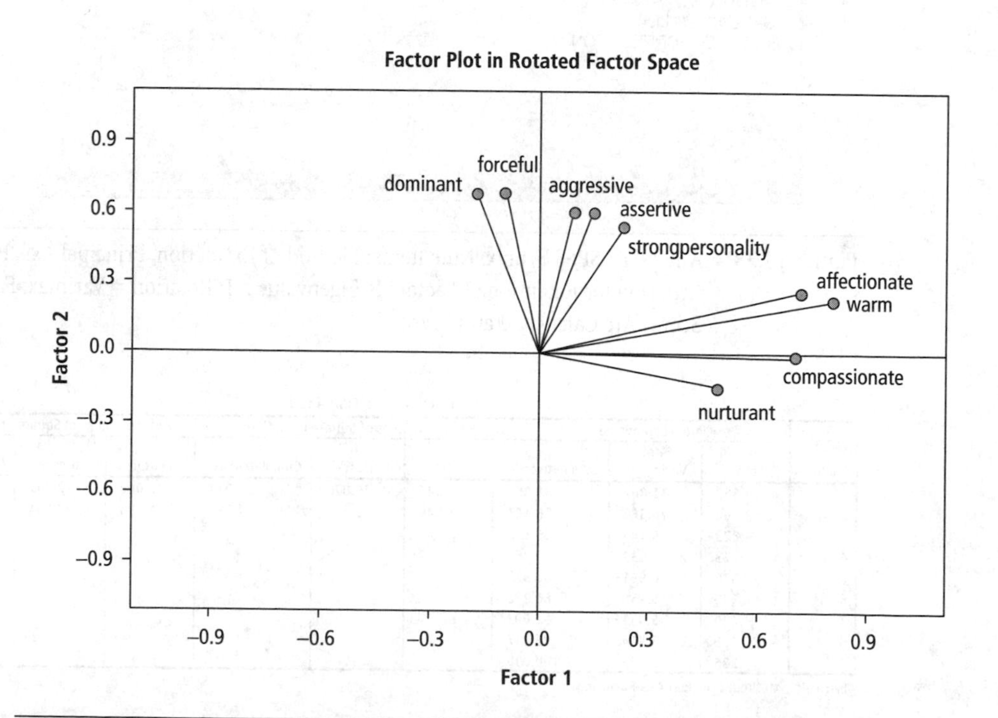

Figure 18.41 ♦ Analysis 4: Plots of Factor Loadings in Varimax Rotated Factor Space

NOTE: The plot in SPSS did not include the vectors, that is, the lines from the 0,0 origin that represent the correlation of each variable with the two factors; these vectors were added to the SPSS plot in a graphics program.

Factor Score Coefficient Matrix

	Factor	
	1	2
nurturant	.135	-.069
affectionate	.273	.067
compassionate	.233	-.055
warm	.442	.063
strongpersonality	.046	.174
assertive	.020	.207
forceful	-.109	.302
dominant	-.096	.290
aggressive	-.004	.203

Extraction Method: Principal Axis Factoring.
Rotation Method: Varimax with Kaiser Normalization.
Factor Scores Method: Regression.

Figure 18.42 ♦ Analysis 4: Factor Score Coefficients Used to Compute Saved Factor Scores

NOTES: These values correspond to the β_{ij} coefficients that appeared in Figure 18.35. β_{ij} corresponds to the factor score coefficient applied to z_{X_i} to construct a score on F_j. To construct a score for each person on Factor 1, we would do the following computation for each participant:

Constructed factor score on $F_1 = .135 \times z_{\text{nurturant}} + .273 \times z_{\text{affectionate}} + .233 \times z_{\text{compassionate}} + .442 \times z_{\text{warm}} + .046 \times z_{\text{strong personality}} + .02 \times z_{\text{assertive}} - .109 \times z_{\text{forceful}} - .096 \times z_{\text{dominant}} - .004 \times z_{\text{aggressive}}$.

In words, to obtain high scores on F_1, an individual would need to have relatively high scores on nurturant, affectionate, compassionate, and warm; the scores on the remaining five variables have such small coefficients that they do not make much contribution toward the value of F_1.

Rotated Factor Matrix[a]

	Factor	
	1	2
cold	-.804	-.217
affectionate	.713	.250
compassionate	.696	-.020
nurturant	.489	-.153
dominant	-.096	.670
forceful	-.174	.667
aggressive	.096	.588
assertive	.148	.587
strongpersonality	.230	.526

Extraction Method: Principal Axis Factoring.
Rotation Method: Varimax with Kaiser Normalization.
a. Rotation converged in 3 iterations.

Figure 18.43 ♦ Final Rotated Factor Loadings in Analysis 4 if "Warm" Is Replaced With "Cold"

NOTE: The variable cold was computed as follows: cold = 5 – warm. Thus, cold is just the reverse-scored version of warm. Note that when Analysis 4 was repeated using the variable cold instead of warm, the magnitudes of all factor loadings remained the same; however, now the variable cold has a negative sign on its factor loading. Thus, higher scores on rotated Factor 1 now correspond to high scores on affectionate, compassionate, and nurturant, and to a low score on cold.

18.15.1 ♦ Variance Reproduced by Each Factor at Three Stages in the Analysis

Figure 18.39 replicates the information in Figure 18.26 and adds a third panel to report the results from the varimax rotation. The total variance explained by factors is now reported at three stages in the analysis. In the left-hand panel of Figure 18.39, the Initial Eigenvalues provide information about the amount of variance explained by the entire set of nine factors extracted by PAF. Because only two of these factors had eigenvalues >1, only Factors 1 and 2 were retained. After the loadings for these two retained factors were reestimated, the SSLs that summarize the amount of information explained by just the two retained factors (prior to rotation) appear in the middle panel of Figure 18.39, under the heading Extraction Sum of Squared Loadings. Finally, the panel on the right-hand side of Figure 18.39 reports the amount of variance explained by each of the two retained factors after varimax rotation. Note that the sum of the variance explained by the two factors as a set is the same for the unrotated and the rotated solutions, but the way the variance is apportioned between the two factors is slightly different. After varimax rotation, rotated Factor 2 explained approximately as much variance as rotated Factor 1.

18.15.2 ♦ Rotated Factor Loadings

The varimax-rotated factor loadings appear in table form in Figure 18.40 and as a plot in Figure 18.41. It is now clear, when we examine the correlations or loadings in Figure 18.40, that the latent variable represented by Factor 1 is highly positively correlated with self-ratings on warm, affectionate, compassionate, and nurturant. The latent variable represented by Factor 2 is highly positively correlated with self-ratings on dominant, forceful, aggressive, assertive, and strong personality. (An arbitrary cutoff value of .40 was used to decide which factor loadings were "large" in absolute value.)

The plot in Figure 18.41 makes it clear how rotation has improved the interpretability of the factors. After rotation, Factor 1 corresponds to a bundle of intercorrelated variables that all seem to be related to caring (affectionate, warm, compassionate, nurturant). After rotation, Factor 2 corresponds to a bundle of intercorrelated vectors that represent variables that seem to be related to strength (dominant, forceful, aggressive, assertive, and strong personality). It is now easier to label or name the factors, because for each rotated factor, there is a clear group of variables that has high correlations with that factor.

The factor score coefficients that appear in Figure 18.42 can be used to compute and save a score on each of the retained factors. Because this type of information is often reported as a follow-up to a factor analysis, discussion of the factor scores appears in Section 18.18, after the results section for Analysis 4.

18.15.3 ♦ Example of a Reverse-Scored Item

Up to this point, we have considered only examples in which all the large factor loadings have been positive. However, researchers sometimes intentionally include **reverse-worded** or reverse-scored items. What would the rotated factor loadings look like if people had been asked to rate themselves on "coldness" instead of on warmth? Changing the wording of the item might alter people's responses. Here we will just consider what happens if we reverse the direction of the 0 to 4 rating scale scores on the variable warm. If we compute a new variable named cold by subtracting scores on warmth from the constant value 5, then scores on cold would be perfectly negatively correlated with scores on warmth. If we then rerun Analysis 4 including the new variable cold instead of warmth, we would obtain the rotated factor loadings that appear in Figure 18.43. Note that the rotated factor loading for cold is −.804; from Figure 18.40, the rotated factor loading for warmth was +.804. Reversing the direction of scoring did not change the magnitude of the loading, but it reversed the sign. We could interpret the rotated factor loadings in Figure 18.43 by saying that people who are high on the caring dimension that corresponds to Factor 1 tend to have high scores on nurturance, affectionate, and compassionate and low scores on coldness.

18.16 ♦ Questions to Address in the Interpretation of Factor Analysis

Whether the analysis is performed using PC or PAF, the data analyst typically hopes to answer the following questions about the structure of the data.

18.16.1 ♦ How Many Factors or Components or Latent Variables Are Needed to Account for (or Reconstruct) the Pattern of Correlations Among the Measured Variables?

To say this in another way, How many different constructs are measured by the set of X variables? In some cases, a single latent variable (such as general intelligence) may be sufficient to reconstruct correlations among scores on a battery of mental tests. In other research situations (such as the Big Five model of personality), we may need larger numbers of latent variables to account for data structure. The Big Five model of personality traits developed by Costa and McCrae (1995) suggests that people differ on five personality dimensions (openness, conscientiousness, extraversion, agreeableness, and neuroticism) and that a model that includes latent variables to represent each of these theoretical constructs can do a good job of reconstructing the correlations among responses to individual items in personality tests. The decision of how many factors or components to retain in the model can be made based on theoretical or conceptual issues. In exploratory analysis, in situations where the data analyst does not have a well-developed theory, empirical criteria (such as the number of factors with eigenvalues >1 or the scree plot) may be used to make decisions about the number of factors or components to retain. Sometimes decisions about number of factors to retain take both theoretical and empirical issues into account.

18.16.2 ♦ How "Important" Are the Factors or Components? How Much Variance Does Each Factor or Component Explain?

In the initial solution, each factor or component has an eigenvalue associated with it; the eigenvalue represents the amount of variance that can be reconstructed from the correlations of measured variables with that factor or component. In later stages of the analysis, we summarize information about the amount of variance accounted for by each component or factor by computing the SSL; that is, to find out how much variance is accounted for by Factor 1, we square the loadings of each of the p measured variables on Factor 1 and sum the squared loadings across all p variables. We can interpret an eigenvalue as a proportion or percentage of explained variance. If a set of p measured variables are all transformed into z scores or standard scores, and each standard score has a variance of 1, then the total variance for the set of p variables is represented by p. For example, if we do a factor analysis of nine measured variables, the total variance for that dataset is equal to 9. The SSL for the entire set of nine variables on nine factors is scaled so that it equals 9. We can interpret the proportion of variance explained by Factor 1 by dividing the eigenvalue or the SSL for Factor 1 by p, the number of measured variables. If we retain two or three factors in the model, we can report the proportion of percentage of variance that is accounted for by each factor and, also, the proportion of variance that is accounted for by the set of two or three factors. There is no standard criterion for how much variance a solution must explain to be considered adequate, but in general, you want the percentage of variance explained by your retained factors to be reasonably high, perhaps on the order of 40% to 70%. If the percentage of explained variance for a small set of retained factors is very low, this suggests that the effort to understand the set of measured variables as multiple indicators of a small number of latent variables has not been very successful and that the measured variables may be measuring a larger number of different constructs.

18.16.3 ♦ What, If Anything, Do the Retained Factors or Components Mean? Can We Label or Name Our Factors?

Usually, we answer this question by looking at the pattern of the rotated factor loadings (although occasionally, it is easier to make sense of unrotated loadings). A factor can be named by considering which measured variables it does (and does not) correlate with highly; and trying to identify a common denominator—that is, something all the variables with high loadings on that factor have in common. For instance, a factor with high loadings for tests of trigonometry, calculus, algebra, and geometry could be interpreted as a "math ability" factor. Note that a large negative loading can be just as informative as a large positive one, but we must take the sign into account when making interpretations. If respondents to the BSRI had rated themselves on the adjective cold, instead of warm, there would be a large negative loading for the variable cold on the first factor; this first factor had high positive correlations with variables such as affectionate and compassionate. We would say that high scores on Factor 1 were associated with high ratings on compassionate, affectionate, and loving and low ratings on coldness.

Before we try to interpret a component or factor as evidence that suggests the existence of a latent variable, each factor should have high correlations with at least three or four of the measured variables. If only two variables are correlated with a factor, the entire factor really represents just one correlation, and a single large correlation might arise just from sampling error. Generally, researchers want to have at least three to five measured variables that correlate with each retained factor; a larger number of items may be desirable in some cases. When factor analysis is used to help develop multiple-item measures for personality tests, the analyst may hope to identify 10 or 15 items that all correlate with a latent variable (e.g., "extraversion"). If this result is obtained, the data analyst is then in a position to construct a 10- or 15-item multiple-item scale to measure extraversion.

You should not try to force an interpretation if there is no clear common meaning among items that load on a factor. Sometimes results are simply due to sampling error and do not reflect any meaningful underlying pattern. Also avoid the error of reification; that is, do not assume that just because you see a pattern that looks like a factor in your data, there must be some "real thing" (an ability or personality trait) out in the world that corresponds to the factor you obtained in your data analysis.

18.16.4 ♦ How Adequately Do the Retained Components or Factors Reproduce the Structure in the Original Data—That Is, the Correlation Matrix?

To evaluate how well the factor model reproduces the original correlations between measured variables, we can examine the discrepancies between the reproduced **R** and the actual **R** correlation matrices. This information can be requested from SPSS. We want these discrepancies between observed and reconstructed correlations to be fairly small for most pairs of variables. You can also look at the final communality estimates for individual variables to see how well or poorly their variance is being reproduced. If the estimated communality for an individual variable (which is similar to an R^2 or a proportion of explained variance) is close to 1, then most of its variance is being reproduced; if variance is close to 0, little of the variance of that measured variable is reproduced by the retained components or factors. It may be helpful to drop variables with very low communality and rerun the factor analysis. This may result in a solution that is more interpretable and a solution that does a better job of reconstructing the observed correlations in the **R** matrix.

18.17 ♦ Results Section for Analysis 4: PAF With Varimax Rotation

The following Results section is based on Analysis 4. In this analysis, nine self-rated personality items were included, PAF was specified as the method of extraction, the rule used to decide how many factors to retain was to retain only factors with eigenvalues >1, and varimax rotation was requested.

Results

To assess the **dimensionality** of a set of nine items selected from a modified version of the BSRI, factor analysis was performed using PAF, the default criterion to retain only factors with eigenvalues greater than 1, and varimax rotation was requested. The items included consisted of self-reported ratings on the following adjectives: nurturant, affectionate, compassionate, warm, strong personality, assertive, forceful, dominant, and aggressive. The first four items were those identified as associated with feminine sex role stereotyped behavior and the last five items were associated with masculine sex role stereotyped behavior. Each item was rated on a 5-point scale that ranged from "0" (does not apply to me at all) to "4" (describes me extremely well).

The correlation matrix that appears in Figure 18.2 indicated that these nine items seemed to form two separate groups. There were moderately high positive correlations among the ratings on nurturant, affectionate, warm, and compassionate; these items were items that Bem identified as sex role stereotyped feminine. There were moderately high positive correlations among the items strong personality, assertive, forceful, dominant, and aggressive; these items were all identified as sex role stereotyped masculine.

(Continued)

(Continued)

The correlations between items in these two separate groups of items (e.g., the correlation between nurturant and assertive) tended to be small.

An exploratory factor analysis was performed to evaluate whether a two-factor model made sense for these data. The SPSS factor analysis procedure was used, the method of extraction was PAF, only factors with eigenvalues >1 were retained, and varimax rotation was requested. Two factors were retained and rotated. Results from this analysis, including rotated factor loadings, communalities, and SSL for the retained factors, are summarized in Table 18.2.

Table 18.2 ♦ Rotated Factor Loadings From Analysis 4: Principal Axis Factoring With Varimax Rotation

	Nine Self-Rated Personality Items		
	Factor 1: "Caring"	*Factor 2: "Strong"*	*Communality*
Warm	.80	.22	.69
Affectionate	.71	.25	.57
Compassionate	.70	−.02	.48
Nurturant	.49	−.15	.26
Dominant	−.10	.67	.46
Forceful	−.17	.67	.48
Aggressive	.10	.59	.46
Assertive	.15	.59	.37
Strong personality	.23	.43	.33
Sum of squared loadings	2.00	1.99	
% explained variance	22.24%	22.15%	

In the initial factor solution that consisted of nine factors, only two factors had eigenvalues greater than 1; therefore, only Factors 1 and 2 were retained and rotated. After varimax rotation, Factor 1 accounted for 22.24% of the variance and Factor 2 accounted for 22.15% of the variance; together, the first two factors accounted for 44.4% of the variance in this dataset (note that we can find these percentages of variance in the SPSS output in Figure 18.39; we can also compute the SSL for each column in Table 18.2 and then divide the SSL by p, the number of variables, to obtain the same information about proportion or percentage of explained variance for each factor. In Table 18.2, the SSL for Factor 1 can be reproduced as follows: $\text{SSL} = .80^2 + .71^2 + .70^2 + .49^2 + (-.10)^2 + (-.17)^2 + .10^2 + .15^2 + .23^2 = 2.00$.

Communalities for variables were generally reasonably high, ranging from a low of .26 for nurturant to a high of .69 for warm. (*Note:* These communalities were reported by SPSS in Figure 18.40 as part of Analysis 3. They can also be reproduced by squaring and summing the loadings in each row of Table 18.2; for example, warm had loadings of .80 and .22, therefore its communality is $.80^2 + .22^2 = .69$. The communality for each measured variable changes when we go from a nine-factor model to a two-factor model, but the communality for each measured variable does not change when the two-factor model is rotated.)

Rotated factor loadings (see Table 18.2) were examined to assess the nature of these two retained varimax-rotated factors. An arbitrary criterion was used to decide which factor loadings were large; a loading was interpreted as large if it exceeded .40 in absolute magnitude. The four items that had high loadings on the first factor (warm, affectionate, compassionate, and nurturant) were consistent with the feminine sex role stereotype. This first factor could be labeled "Femininity," but because this factor was based on such a limited number of BSRI femininity items, a more specific label was applied; for this analysis, Factor 1 was identified as "Caring." The five items that had high loadings on the second factor (dominant, forceful, aggressive, assertive, and strong personality) were all related to the masculine sex role stereotype, so this second factor could have been labeled "Masculinity." However, because only five items were included in Factor 2, a more specific label was chosen; Factor 2 was labeled "Strong." None of the items were factorially complex—that is, none of the items had large loadings on both factors.

18.18 ♦ Factor Scores Versus Unit-Weighted Composites

As a follow-up to this factor analysis, scores for caring and strength were calculated in two different ways. First, as part of Analysis 4, the regression-based factor scores for the two factors were saved as variables. For example, as noted previously, the factor score coefficients that were requested as part of the output in Analysis 4 (that appear in Figure 18.42) can be applied to the *z* scores on the measured variables to construct a factor score on each factor for each participant: $F_1 = .135 \times z_{\text{nurturant}} + .273 \times z_{\text{affectionate}} + .233 \times z_{\text{compassionate}} + .442 \times z_{\text{warm}} + .046 \times z_{\text{strong personality}} + .02 \times z_{\text{assertive}} - .109 \times z_{\text{forceful}} - .096 \times z_{\text{dominant}} - .004 \times z_{\text{aggressive}}$. We can see from this equation that people who have relatively high positive scores on nurturant, affectionate, compassionate, and warm will also tend to have high scores on the newly created score on Factor 1. We could say that a person who has a high score on Factor 1 is high on the caring dimension; we know that this person has high scores on at least some of the measured variables that have high correlations with the caring factor. (People's scores on strong personality, assertive, forceful, dominant, and aggressive are essentially unrelated to their factor scores on caring.)

A problem with factor score coefficients is that they are optimized to capture the most variance and create factor scores that are most nearly orthogonal only for the sample of participants whose data were used in the factor analysis. If we want to obtain scores for new participants on these nine self-ratings and form scores to summarize the locations of these new participants on the caring and strong dimensions, there are two potential disadvantages to creating these scores by using the factor score coefficients. First, it can be somewhat inconvenient to use factor score coefficients (reported to two or three decimal places) as weights when we combine scores across items. Second, when we apply these factor score coefficients to scores in new batches of data, the advantages that these factor score coefficients had in the data used in the factor analysis (that they generated factors that explained the largest possible amount of variance and that they generated scores on Factor 2 that had extremely small correlations with Factor 1) tend not to hold up when the factor scores are applied to new batches of data.

A different approach to forming scores that might represent caring and strong is much simpler. Instead of applying the factor score coefficients to *z* scores, we might form simple unit-weighted composites, using the factor analysis results to guide us in deciding whether to give each measured item a weight of 0, +1, or −1. For example, in Analysis 4, we decided that Factor 1 had "high" positive loadings with the variables warm, affectionate, compassionate, and nurturant and that it had loadings close to 0 with the remaining five measured variables. We could form a score that summarizes the information about the caring dimension contained in our nine variables by assigning weights of +1 to variables that have high correlations with Factor 1 and weights of 0 to variables that have low correlations with Factor 1. (If a variable has a large negative loading on Factor 1, we would assign it a weight of −1.) The unit-weighted score for caring based on raw scores on the measured items would be as follows:

$$\text{Caring} = (+1) \times \text{Warm} + (+1) \times \text{Affectionate} + (+1) \times \text{Compassionate} + (+1) \times \text{Nurturant} + 0 \times \text{Dominant} + 0 \times \text{Forceful} + 0 \times \text{Aggressive} + 0 \times \text{Strong personality}.$$

This can be reduced to

$$\text{Caring} = \text{Warm} + \text{Affectionate} + \text{Compassionate} + \text{Nurturant}.$$

Unit-weighted scores for new variables called caring and strong can be calculated in SPSS using the <Transform> → <Compute> menu selections to open up the computation window that appears in Figure 18.44. A unit-weighted composite called strong was computed by forming this sum: dominant + forceful + aggressive + assertive + strong personality. The SPSS data worksheet that appears in Figure 18.45 shows the new variables that were created as a follow-up to Factor Analysis 4. The columns headed "Fac1_1" and "Fac2_1" contain the values of the saved factor scores that were obtained by requesting saved factor scores in Figure 18.37; Fac1_1 corresponds to the following sum of z scores on the measured items: Fac1_1 = $.135 \times z_{\text{nurturant}} + .273 \times z_{\text{affectionate}} + .233 \times z_{\text{compassionate}} + .442 \times z_{\text{warm}} + .046 \times z_{\text{strong personality}} + .02 \times z_{\text{assertive}} - .109 \times z_{\text{forceful}} - .096 \times z_{\text{dominant}} - .004 \times z_{\text{aggressive}}$. The last two columns in the SPSS data view worksheet contain the unit-weighted scores that were computed for the group of variables with high loadings on Factor 1 (the caring factor) and the group of variables with high loadings on Factor 2 (the strong factor):

$$\text{Caring} = \text{Warm} + \text{Loving} + \text{Affectionate} + \text{Compassionate},$$
$$\text{Strong} = \text{Assertive} + \text{Aggressive} + \text{Dominant} + \text{Forceful} + \text{Strong personality}.$$

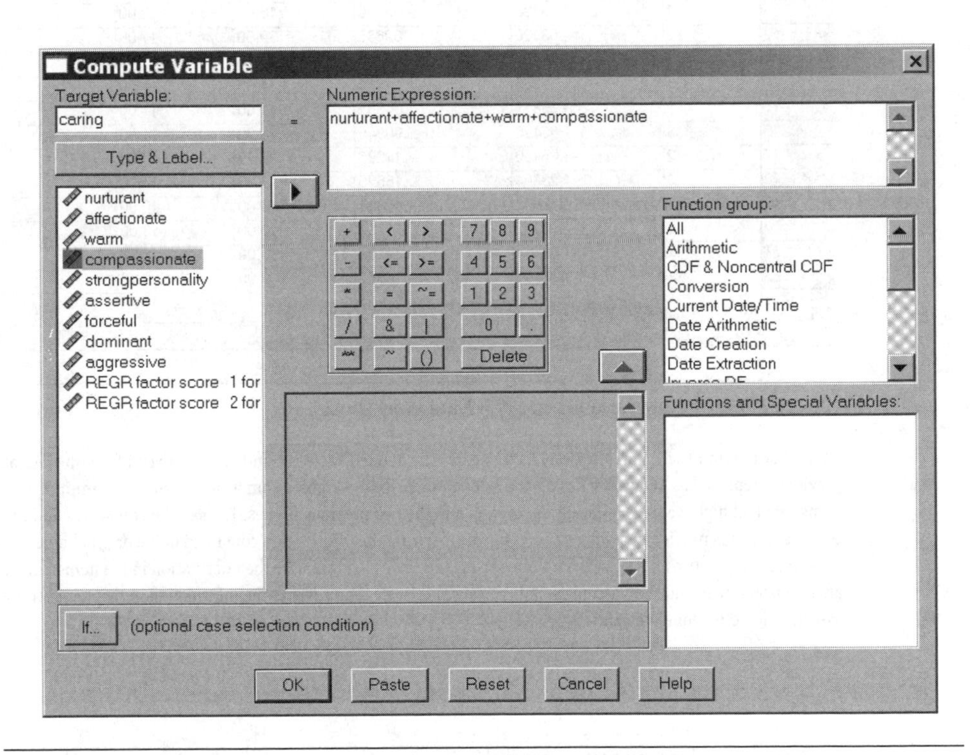

Figure 18.44 ♦ Computation of a Unit-Weighted Composite Score Consisting of Items With High Correlations With Factor 1

How much do the results differ for the saved factor scores (Fac1_1, Fac2_1) and the unit-weighted composites of raw scores (caring, strong)? We can assess this by obtaining Pearson correlations among all these variables; the Pearson correlations appear in Figure 18.46. In this empirical example, the saved factor score on Factor 1 (Fac1_1) correlated .966 with the unit-weighted composite score on caring; the saved factor score on Factor 2 (Fac2_1) correlated .988 with the unit-weighted score on strong. The correlations between factor scores and unit-weighted composite scores are not always as high as in this empirical example. In many research situations, scores that are obtained from

*shortbsri [DataSet1] - SPSS Data Editor

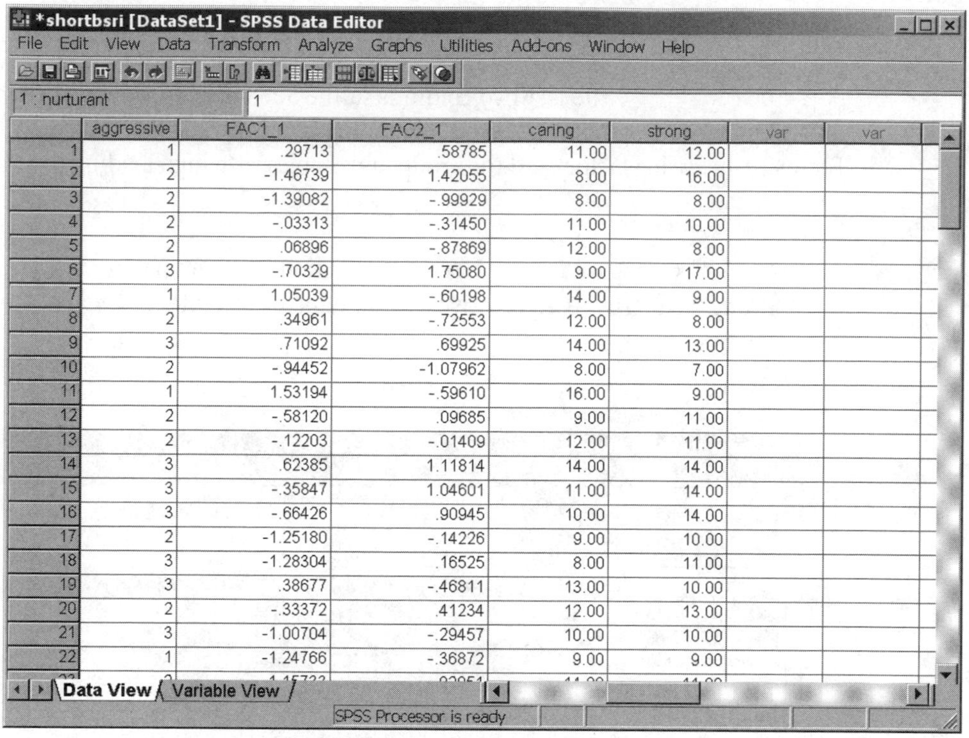

File Edit View Data Transform Analyze Graphs Utilities Add-ons Window Help

1 : nurturant 1

	aggressive	FAC1_1	FAC2_1	caring	strong	var	var
1	1	.29713	.58785	11.00	12.00		
2	2	-1.46739	1.42055	8.00	16.00		
3	2	-1.39082	-.99929	8.00	8.00		
4	2	-.03313	-.31450	11.00	10.00		
5	2	.06896	-.87869	12.00	8.00		
6	3	-.70329	1.75080	9.00	17.00		
7	1	1.05039	-.60198	14.00	9.00		
8	2	.34961	-.72553	12.00	8.00		
9	3	.71092	.69925	14.00	13.00		
10	2	-.94452	-1.07962	8.00	7.00		
11	1	1.53194	-.59610	16.00	9.00		
12	2	-.58120	.09685	9.00	11.00		
13	2	-.12203	-.01409	12.00	11.00		
14	3	.62385	1.11814	14.00	14.00		
15	3	-.35847	1.04601	11.00	14.00		
16	3	-.66426	.90945	10.00	14.00		
17	2	-1.25180	-.14226	9.00	10.00		
18	3	-1.28304	.16525	8.00	11.00		
19	3	.38677	-.46811	13.00	10.00		
20	2	-.33372	.41234	12.00	13.00		
21	3	-1.00704	-.29457	10.00	10.00		
22	1	-1.24766	-.36872	9.00	9.00		

Data View / Variable View

SPSS Processor is ready

Figure 18.45 ♦ New Variables in SPSS Data Worksheet

NOTE: Fac1_1 and Fac2_1 are the saved factor scores (for rotated Factors 1 and 2, respectively) from the PAF analysis previously reported as Analysis 4. Caring and strong are the unit-weighted composites formed by summing raw scores on the items that had high loadings on F_1 and F_2, as shown in the two previous figures. The saved factor scores and the unit-weighted composite scores provide two different ways of summarizing the information contained in the original nine personality items into scores on just two latent variables: Both Caring and Fac1_1 summarize the information in the items nurturant, affectionate, warm, and compassionate. Both Strong and Fac2_1 summarize the information in the items strong personality, assertive, forceful, dominant, and aggressive.

unit-weighted composites provide information very similar to scores that are based on weights derived from multivariate models—for example, saved factor scores (Fava & Velicer, 1992; Wainer, 1976).

Note that when all the items that are summed to form a weighted linear composite are measured in the same units, it may be acceptable to form a unit-weighted composite by summing the raw scores. In this example, all the adjectives were rated using the same "0" to "4" rating scale. However, if the variables that are summed have been measured in different units and have quite different variances, better results may be obtained by forming a unit-weighted composite of standard scores or z scores (rather than raw scores). For example, we could form a score on caring by summing unit-weighted z scores on warm, affectionate, compassionate, and nurturant:

$$\text{Caring} = z_{\text{warm}} + z_{\text{affectionate}} + z_{\text{compassionate}} + z_{\text{nurturant}}.$$

Correlations

		FAC1_1	FAC2_1	caring	strong
FAC1_1	Pearson Correlation	1	.046	.966(**)	.057
	Sig. (2 -tailed)		.652	.000	.575
	N	100	100	100	100
FAC2_1	Pearson Correlation	.046	1	.097	.988(**)
	Sig. (2 -tailed)	.652		.335	.000
	N	100	100	100	100
caring	Pearson Correlation	.966(**)	.097	1	.117
	Sig. (2 -tailed)	.000	.335		.244
	N	100	100	100	100
strong	Pearson Correlation	.057	.988(**)	.117	1
	Sig. (2 -tailed)	.575	.000	.244	
	N	100	100	100	100

**Correlation is significant at the 0.01 level (2-tailed).

Figure 18.46 ♦ Correlations Between Saved Factor Scores on Fac1 and Fac2 (From Analysis 4) and Unit-Weighted Composites

NOTES: Fac1_1 and caring are highly correlated; Fac2_1 and strong are highly correlated. Fac1_1 and Fac2_1 have a correlation that is near 0. The correlation between caring and strong is higher than the correlation between Fac1_1 and Fac2_1 but by a trivial amount.

18.19 ♦ Summary of Issues in Factor Analysis

Factor analysis is an enormously complex topic; this brief introduction does not include many important issues. For example, there are many different methods of factor extraction; only PC and PAF have been covered here. There are many different methods of factor rotation; only one type of orthogonal rotation, varimax rotation, was presented here. If you plan to use factor analysis extensively in your research, and particularly if you plan to use nonorthogonal rotation methods, you should consult more advanced source books for information (e.g., Harman, 1976).

Factor analysis is an exploratory analysis that is sometimes used as a form of data fishing. That is, when a data analyst has a messy set of too many variables, he or she may run a factor analysis to see if the variables can be reduced to a smaller set of variables. The analyst may also run a series of different factor analyses, varying the set of variables that are included and the choices of method of extraction and rotation until a "meaningful" result is obtained. When factor analysis is used in this way, the results may be due to Type I error; the researcher is likely to discover, and focus on, correlations that are large in this particular batch of data just due to sampling error. Results of factor analysis often replicate very poorly for new sets of data when the analysis has been done in this undisciplined manner. In general, it is preferable to plan thoughtfully, to select variables in a way that is theory driven, and to run a small number of planned analyses.

Results of factor analyses are sometimes "overinterpreted": The researcher may conclude that he or she has determined "the structure of intelligence," or how many types of mental ability there are, when in fact all we can ever do using factor analysis is assess how many factors we need to describe the correlations among the set of measures that we included in our research. The set of measures included in our research may or may not adequately represent the domain of all the measures that should be included in a comprehensive study (of mental ability, for instance). If we do a factor analysis of a test battery that includes only tests of verbal and quantitative ability, for instance, we cannot expect to obtain factors that represent musical ability or emotional intelligence. To a great extent, we get out of factor analysis exactly what we put in when we selected our variables. Conclusions should not be overstated: Factor analysis tells us about structure in our sample, our data, our selection of measures; and this is not necessarily a clear reflection of structure in "the world," particularly if our selection of participants and measures is in any way biased or incomplete.

We can do somewhat better if we map out the domain of interest thoroughly at the beginning of our research and think carefully about what measures should be included or excluded. Keep in mind that for any factor you want to describe, you should include a minimum of three to four measured variables, and ideally, a larger number of variables would be desirable. If we think we might be interested in a musical ability factor, for example, we need to include several measures that should reflect this ability. In research areas where theory is well developed, it may be possible to identify the factors you expect ahead of time and then systematically list the items or measures that you expect to load on each factor. When factor analysis is done this way, it is somewhat more "confirmatory." However, keep in mind that if the theory used to select measurements to include in the study is incomplete, it will lead us to omit constructs that might have emerged as additional factors.

Test developers often go through a process where they factor analyze huge lists of items (one study, e.g., collected all the items from self-report measures of love and did a factor analysis to see how many factors would emerge). Based on these initial results, the researchers may clarify their thinking about what factors are important. Items that do not load on any important factor may be dropped. New items may be added to provide additional indicator variables to enlarge factors that did not have enough items with high loadings in the initial exploratory analyses. It may be even more obvious in the case of factor analysis than for other analytic methods covered earlier in this book: The results you obtain are determined to a very great extent by the variables that you choose to include in (and exclude from) the analysis.

18.20 ♦ Optional: Brief Introduction to Concepts in Structural Equation Modeling

Up until about 1970, most factor analytic studies used the exploratory or data-driven approach to factor analysis described in this chapter. Work by Jöreskog (1969) on the general analysis of covariance structures led to the development of **confirmatory factor analysis (CFA)** and other types of structural equation (SEM) models. In the exploratory factor analyses described in this chapter, the number and nature of the retained factors is determined to a great extent by the sample data (e.g., often we retain only components or factors with eigenvalues greater than 1). In CFA, a researcher typically begins with a theoretically based model that specifies the number of latent variables and identifies which specific measured variables are believed to be indicators of each latent variable in the model. This model usually includes many additional constraints; for example, each time we assume a "0" correlation between one of the measured variables and one of the latent variables in the model, we reduce the number of free parameters in the model. In exploratory factor analysis, the initial p component or p factor solution always reproduced the correlation matrix **R** perfectly; in typical applications of exploratory factor analysis, there are more than enough free parameters (a $p \times p$ matrix of the correlations between p measured variables and p latent variables or factors) to perfectly reproduce the $p(p - 1)/2$ correlations among the measured variables.

CFA and SEM models, on the other hand, are usually set up to include so many constraints (and thus so few free parameters) that they cannot perfectly reproduce the correlations (or covariances) among the measured variables. Another difference between SEM models and the exploratory factor analysis described here is that most structural equation programs start with a matrix of variances and covariances among measured variables (rather than the correlation matrix **R**); and the adequacy of the model is judged by how well it reproduces the observed variance/covariance matrix (instead of a correlation matrix). In CFA, the data analyst may compare several different models or may evaluate one theory-based model to assess whether the limited number of relationships among latent and measured variables that are included in the specification of the model are sufficient to reproduce the observed variances and covariances reasonably well. For CFA and SEM models, statistical significance tests can be performed, both to assess whether the reproduced variance covariance matrix deviates significantly from the observed variances and covariances calculated for the measured variables and to assess whether the coefficients associated with specific paths in the model differ significantly from 0.

At later stages in research, when more is known about the latent variables of interest, researchers may prefer to test the fit of CFA or SEM models that are theory based instead of continuing to run exploratory analyses. While a complete description of SEM models is beyond the scope of this book, the following example provides a brief intuitive conceptual introduction to an SEM model.

An SEM model is often represented as a path model; the path model often represents two different kinds of assumptions. First, some parts of the SEM model correspond to a "measurement model." A measurement model, like a factor analysis, shows how a latent variable is correlated with scores on two or more measured variables that are believed to be "indicators" of that latent variable or construct. Consider a study done by Wallace, Bisconti, and Bergeman (2001). The study included nine measured variables: scores on measures of challenge, commitment, control, family support, friend support, perceived support, life satisfaction, positive affect, and negative affect. These nine measured variables were conceptualized as indicators of three latent variables; for example, the scores on measures of challenge, commitment, and control were theorized to be multiple indicators of a latent variable or construct called "hardiness"; family, friend, and perceived support were treated as indicators of a latent variable called "social support"; and life satisfaction and positive and negative affect were treated as indicators of a latent variable called "well-being."

The measurement model for the latent variable named hardiness is represented in Figure 18.47. On the left-hand side of this figure, the measurement model is shown graphically using the conventions of SEM programs such as AMOS. Each square corresponds to a measured variable. Each circle or oval corresponds to a latent variable. The arrows indicate "causal" connections, or in the case of the measurement model, assumptions that scores on specific indicator variables are "due to" some latent variable. The right-hand side of Figure 18.47 shows the same relationship among variables in the path model notation that has been used earlier in this chapter. The path coefficients in this model (.66, .78, .83) correspond to the correlations of each of the indicator variables with hardiness, or the "loadings" of each indicator variable on a hardiness factor. SEM models include additional notation to represent sources of error that were not included in the path models in this chapter. The measurement model that appears in Figure 18.47 specifies that the latent variable named hardiness should be associated with the measured variables challenge, commitment, and control. In the full SEM model that appears in Figure 18.49, you can see that we also assume that the other six measured variables are *not* direct indicators of hardiness. The absence of a direct path between, for example, the measured variable perceived support and the latent variable hardiness is as informative about the nature of the theory about relations among variables as the presence of a direct path between control and hardiness.

In Chapter 10, path models were introduced to show different possible kinds of causal and noncausal associations in three-variable research situations. One possible model to describe how X_1 and X_2 are related to a Y outcome variable was a "mediated" model—that is, a model in which the X_1 variable has part or all of its influence on the Y outcome variable through a path that involves a sequence of causal connections; first X_1 "causes" X_2 and

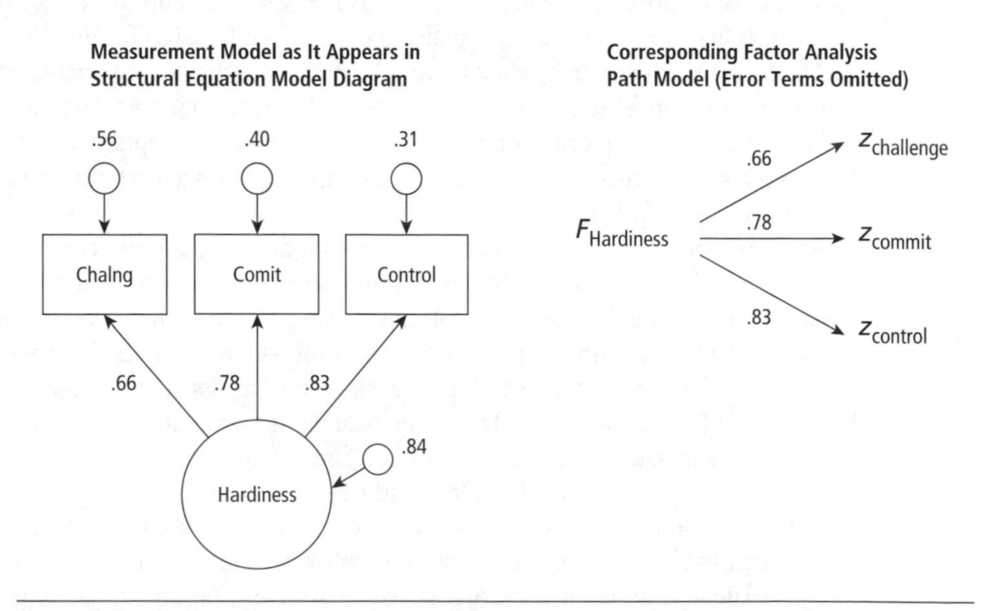

Figure 18.47 ◆ A Measurement Model for One Latent Variable ("Hardiness")

SOURCE: From Wallace, K. A., Bisconti, T. L., & Bergeman, C. S. (2001). The mediational effect of hardiness on social support and optimal outcomes in later life. *Basic and Applied Social Psychology, 23,* 267–279. Reprinted with permission from Lawrence Erlbaum Associates, Inc.

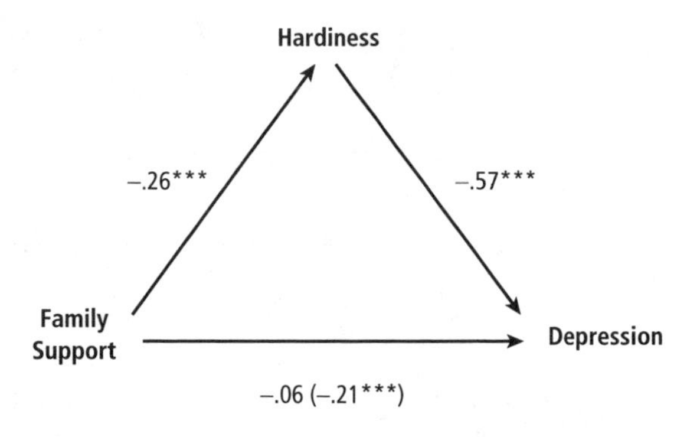

Figure 18.48 ♦ A Model That Represents Partly Mediated Causal Association

SOURCE: From Wallace, K. A., Bisconti, T. L., & Bergeman, C. S. (2001). The mediational effect of hardiness on social support and optimal outcomes in later life. *Basic and Applied Social Psychology, 23,* 267–279. Reprinted with permission from Lawrence Erlbaum Associates, Inc.

NOTE: The coefficient −.06 represents the strength of linear association between depression and family support when hardiness is controlled by including it as another predictor of depression; the coefficient −.21 represents the strength of the linear association between depression and family support when hardiness is not statistically controlled.

then X_2 "causes" Y. (The word "causes" is in quotes to remind us that we are talking about *hypothesized* causal connections. Analysis of nonexperimental data can yield results that are consistent or inconsistent with various "causal" models, but analysis of nonexperimental data cannot prove causality.)

Many SEM models involve similar types of paths to show theorized "causal" associations that involve latent variables. Figure 18.48 shows a path model that describes a possible pattern of associations among some of the measured variables in this study. The outcome or dependent variable is depression (elsewhere in their paper, this is called negative affect). Family support is a predictor; the path model in Figure 18.48 represents two possible ways in which family support might influence depression. First, family support may have direct effects on depression. The −.21 coefficient associated with the direct path that leads from family support to depression in Figure 18.48 represents the strength of association between z scores on family support and depression when the third variable, hardiness, is not statistically controlled; the −.06 coefficient for the path that leads directly from family support to depression represents the strength of association between these variables when the third variable, hardiness, is statistically controlled or partialled out.

However, the model in Figure 18.48 also includes an indirect or mediated path that leads from family support to depression via hardiness. The finding that the path coefficient for the direct path from family support to depression becomes nonsignificant when this indirect or mediated path is taken into account suggests that we might interpret the study as evidence that any "causal" impact of family support on depression might be largely mediated by the effects of family support on hardiness and the influence of hardiness on depression. We cannot prove any causal connections or mediated causal processes in this study because the study is nonexperimental. However, we assess whether the pattern of coefficients for the path model is consistent with what you would expect to see if mediation were present.

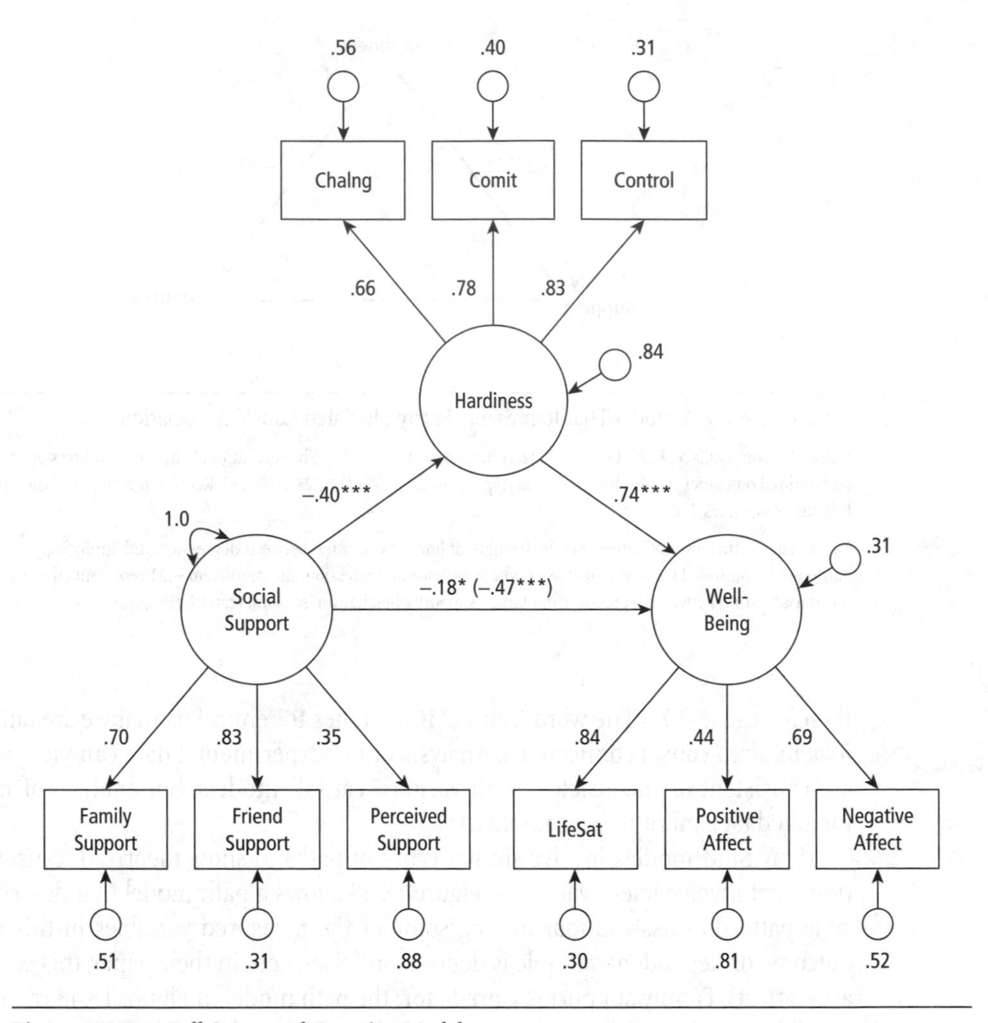

Figure 18.49 ♦ Full Structural Equation Model

SOURCE: From Wallace, K. A., Bisconti, T. L., & Bergeman, C. S. (2001). The mediational effect of hardiness on social support and optimal outcomes in later life. *Basic and Applied Social Psychology, 23,* 267–279. Reprinted with permission from Lawrence Erlbaum Associates, Inc.

NOTES: This includes a measurement model for each of the three latent variables (social support, hardiness, and well-being). Squares correspond to measured variables. Ovals and circles correspond to latent variables that are estimated in the SEM model so that they are consistent with the pattern of variances and covariances among the measured variables.

The model that appears in Figure 18.49 represents the full SEM model in Wallace et al. (2001). This model includes three measurement models: one for each of the latent variables (hardiness, social support, and well-being). It also includes a "causal" model that represents a theory that social support may have direct effects on well-being but that the effects of social support on well-being may be partly mediated by hardiness. An SEM model thus brings together or combines the idea of latent variables (for which our measured variables serve as multiple indicators), an idea that is central to factor analysis, and the idea of "causal" or path models that describe possible or theoretical causal and noncausal associations among variables (introduced in Chapter 10 and used extensively in the discussion of multiple regression).

The data in the Wallace et al. (2001) study could have been used to do a CFA (instead of testing "causal" models). See Figure 18.50 for examples of two different CFA models that could be evaluated using the set of nine measured variables in the Wallace et al. dataset. The top panel shows that a hypothesized factor structure with three latent variables is needed to model the structure; each of the three latent variables corresponds to three of the measured indicator variables (as indicated in the path model) and the three latent variables are intercorrelated or nonorthogonal. The model in the bottom panel is the same, except that in this more constrained model, the three factors or latent variables are assumed to be orthogonal or uncorrelated (the path coefficients between latent variables are constrained to be 0).

For a CFA or an SEM model to be judged acceptable, it must be possible to reconstruct the observed variances and covariances of the measured variables reasonably well using the (unstandardized) path coefficients that are estimated for the model. The algebra used to reconstruct variances and covariances from SEM model coefficients is analogous to the algebra that was used in Chapter 11 to reconstruct correlations from the beta coefficients in a path model that represented a regression analysis with correlated predictor variables.

Appendix: The Matrix Algebra of Factor Analysis

For a review of basic matrix algebra notation and simple matrix operations such as matrix multiplication, see Appendix 14.A. To perform a factor analysis, we begin with a correlation matrix **R** that contains the correlations between all pairs of measured *X* variables. For principal components, we leave 1s in the diagonal. For PAF, we place initial estimates of communality (usually the squared multiple correlations to predict each *X* from all the other *X*s) in the diagonal.

Next we solve the "eigenvalue/eigenvector" problem for this matrix **R**.

For the correlation matrix **R**, we want to solve for a list of eigenvalues ($\boldsymbol{\lambda}$) and a set of corresponding eigenvectors (**V**) such that $(\mathbf{R} - \boldsymbol{\lambda}\mathbf{I})\mathbf{V} = \mathbf{0}$.

This equation is formally called the "eigenproblem." This eigenvalue/eigenvector analysis in effect rearranges or "repackages" the information in **R**. When we create new weighted linear composites (factor scores) using the eigenvectors as weights, these new variates preserve the information in **R**. The variances of each of these new weighted linear composites are given by the eigenvalue that corresponds to each eigenvector. Our solution for a $p \times p$ correlation matrix **R** will consist of p eigenvectors, each one with a corresponding eigenvalue. The new weighted linear composites that can be created by summing the scores on the z_X standard scores on measured variables using weights that are proportional to the elements of each eigenvector are orthogonal to or uncorrelated with each other. Thus, a matrix of variances and covariances among them will have all 0s for the covariances.

$\boldsymbol{\lambda}\mathbf{I}$ stands for the matrix that includes the eigenvalues; its diagonal entries, the λs, correspond to the variances of these "new" variables (the factors) and the off-diagonal 0s stand for the 0 covariances among these new variates or factors.

When we find a set of eigenvectors and eigenvalues that are based on a correlation matrix **R** and that can reproduce **R** we can say that this eigenproblem solution "diagonalizes" the **R** matrix. We "repackage" the information in the **R** matrix as a set of eigenvalues

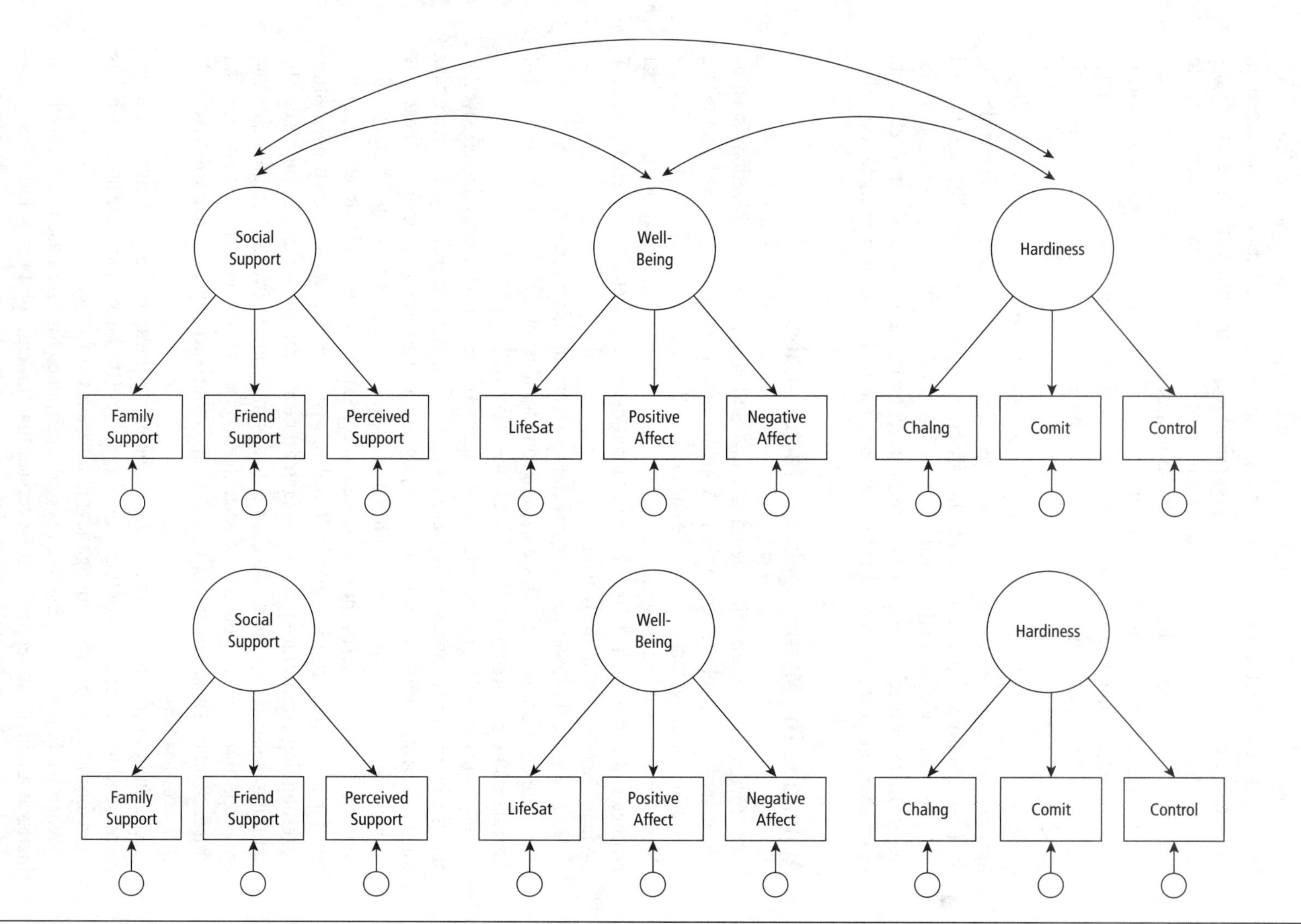

Figure 18.50 ♦ Two Possible Confirmatory Factor Analysis Model for the Wallace et al. (2001) Variables: Top, Three Intercorrelated or Nonorthogonal Factors; Bottom, Three Orthogonal Factors

SOURCE: From Wallace, K. A., Bisconti, T. L., & Bergeman, C. S. (2001). The mediational effect of hardiness on social support and optimal outcomes in later life. *Basic and Applied Social Psychology, 23,* 267–279. Reprinted with permission from Lawrence Erlbaum Associates, Inc.

and eigenvectors; the new variables represented by the eigenvector/eigenvalue pairs have a "diagonal" variance/covariance matrix—that is, in the **L** matrix, all off-diagonal entries are 0. We convert a set of (intercorrelated) X variables to a new set of variables (factors) that are uncorrelated with each other, but we do this in a way that preserves the information—that is, we can reconstruct the **R** matrix from its corresponding eigenvalues and eigenvectors. We can rewrite the eigenvalue/eigenvector problem as follows:

$$\mathbf{L} = \mathbf{V}'\mathbf{R}\mathbf{V},$$

where **L** is the eigenvalue matrix (denoted by $\boldsymbol{\lambda}\mathbf{I}$ earlier). The **L** matrix for a 4×4 case would be as follows:

$$\mathbf{L} = \begin{bmatrix} \lambda_1 & 0 & 0 & 0 \\ 0 & \lambda_2 & 0 & 0 \\ 0 & 0 & \lambda_3 & 0 \\ 0 & 0 & 0 & \lambda_4 \end{bmatrix}.$$

R is the correlation matrix. **V** is the matrix in which each column corresponds to one of the eigenvectors; and each eigenvector in the **V** matrix has a corresponding eigenvalue in the **L** matrix; for example, Column 1 of **V** and Λ_1 together constitute one of the solutions to the eigenvalue/eigenvector analysis of the **R** matrix.

As discussed earlier, the eigenvalue that corresponds to a factor is also equivalent to the SSL for that factor; the eigenvalue for each factor provides information about the amount of variance associated with each factor.

Diagonalization of **R** is accomplished by pre- and postmultiplying **R** by the matrix **V** and its transpose. A matrix is called "diagonal" when all its off-diagonal elements equal 0. Once we have "repackaged" the information about p intercorrelated variables by extracting p orthogonal factors, the correlations among these factors are 0, so we speak of this factor solution as diagonal.

We rescale the values of the eigenvectors such that they can be interpreted as correlations; this requires that $\mathbf{V}'\mathbf{V} = \mathbf{I}$. Thus, pre- and postmultiplying **R** by $\mathbf{V}'$ and **V** does not so much change it (multiplying by the identity matrix **I** is like multiplying by 1) as repackage it. Instead of p intercorrelated measured X variables, we now have p uncorrelated factors or components. We can rewrite the $\mathbf{L} = \mathbf{V}'\mathbf{R}\mathbf{V}$ equation given earlier in the following form:

$$\mathbf{R} = \mathbf{V}'\mathbf{L}\mathbf{V},$$

which is just a way of saying that we can reproduce **R**, the original observed correlation matrix, by multiplying together the eigenvectors and their corresponding eigenvalues.

The factor loadings for each factor are obtained by rescaling the values of each eigenvector such that they can be interpreted as correlations. This can be done by multiplying each eigenvector by the square root of its corresponding eigenvalue (the eigenvalue is equivalent to a variance). So we obtain a matrix of factor loadings, **A**, such that each column of **A** corresponds to the loadings of all p variables on one of the factors, from the following:

$$\mathbf{A} = \mathbf{V}\sqrt{\mathbf{L}}.$$

It follows that the product $\mathbf{AA'} = \mathbf{R}$—that is, we can reproduce the observed correlation matrix **R** from the factor loading matrix **A** and its transpose. $\mathbf{R} = \mathbf{AA'}$ is the "fundamental equation" for factor analysis. In words, $\mathbf{R} = \mathbf{AA'}$ tells us that the correlation matrix **R** can be perfectly reproduced from the loading matrix **A** (when we retain all p factors that were extracted from the set of p measured variables).

Returning to the path diagram shown near the beginning of this chapter, $\mathbf{AA'}$ implies that, for each pair of variables, when we trace all possible paths that connect all pairs of variables via their correlations with the factors, multiply these loadings together for each path, and sum the paths across all the p factors, the resulting values will exactly reproduce each of the bivariate correlations between pairs of measured X variables (provided that the model retains the same number of factors as variables).

Thus, to obtain the elements of the **A** matrix (which are the factor loadings), we do the following:

1. Calculate the sample correlation matrix **R** (and make whatever change in the diagonal entry of **R** that we wish to make, depending on our choice of PC, PAF, or other methods).
2. Obtain the p eigenvalue/eigenvector pairs that are the solutions for the eigenvalue problem for the **R** matrix.
3. Rescale the values in the eigenvectors by multiplying them by the square roots of the corresponding eigenvalues; this scales the values so that they can be interpreted as correlations or factor loadings.
4. In iterative solutions, such as PAF, we then reconstruct **R** from **A** by finding $\mathbf{R} = \mathbf{AA'}$. We compare the diagonal elements in our reconstructed **R** matrix (the communalities) with the communality estimates from the previous step. If they are the same (to some large number of decimal places), we say the solution has "converged" and we stop. If they differ, we put in our most recent communality estimates and redo the entire eigenvalue/eigenvector algebra to generate new loading estimates based on our most recent communalities. Occasionally, the solution does not converge; this leads to an error message.

Additional matrices may also be obtained as part of factor analysis. For instance, the calculation of factor scores (**F**) for each subject on each factor from each subject's z scores on the variables (**Z**) and the factor score coefficient matrix (**B**) can be written compactly as $\mathbf{F} = \mathbf{ZB}$. This equation says that we can construct a factor score for each participant on each factor by multiplying the z scores on each measured variable for each subject by the corresponding factor score coefficients, creating a weighted linear composite for each subject on each factor.

We can write an equation for the **B**s (the factor score coefficients, which are like beta coefficients in a multiple regression) in terms of the values of the factor loadings (correlations between the *X*s and the factors) and the intercorrelations among the *X*s, the **R** matrix:

$$\mathbf{B} = \mathbf{R}^{-1}\,\mathbf{A}.$$

This equation is analogous to our equation for the betas in multiple regression:

$$\mathbf{B}_i = \mathbf{R}_{ii}^{-1}\,\mathbf{R}_{iy}.$$

The multiple regression in the other direction, in which we predict (standardized) scores on the *X*s from factor scores, can be compactly written as $\mathbf{Z} = \mathbf{FA'}$. That is, we can multiply each individual subject's factor scores by the factor loadings of the variables and generate predicted scores on the individual variables. For instance, if we know that a subject has a high score on the math factor and low scores on the verbal ability factor, we would end up predicting high scores for that subject on tests that had high loadings on (or high correlations) with the math factor.

To convert the original factor loading matrix **A** to a rotated factor loading matrix, we multiply it by the "transformation matrix" $\mathbf{\Lambda}$:

$$\mathbf{A}_{\text{unrotated}}\mathbf{\Lambda} = \mathbf{A}_{\text{rotated}}.$$

The elements of this transformation matrix $\mathbf{\Lambda}$ are the cosines and sines of the angles through which the reference axes or factors are being rotated to give a more interpretable solution; usually, this is not of great intrinsic interest.

If you choose to perform an **oblique rotation** (i.e., one that results in rotated factors that are correlated to some degree), there are some additional matrices that you have to keep track of. You need a $\boldsymbol{\phi}$ matrix to provide information about the correlations among the oblique factors (for an orthogonal solution, these correlations among factors are constrained to be 0). When you have an oblique factor solution, you also have to distinguish between the factor loading or pattern matrix **A**, which tells you the regression-like weights that predict variables from factors, and the structure matrix **C**, which gives the correlations between the variables and the (correlated or oblique) factors. For an orthogonal solution, **A** = **C**. For an oblique solution, the **A** and **C** matrices differ.

For further discussion of the matrix algebra involved in factor analysis, see Tabachnick and Fidell (2007).

+−×÷

Comprehension Questions

1. For a set of three mental ability tests in the Project Talent Dataset that was used in Chapter 16, a principal components analysis was obtained and all three components were retained. Selected results from this analysis appear below:

Communalities

	Initial	Extraction
mechanic	1.000	1.000
abstract	1.000	1.000
math	1.000	1.000

Extraction Method: Principal Component Analysis.

Total Variance Explained

	Initial Eigenvalues			Extraction Sums of Squared Loadings		
Component	Total	% of Variance	Cumulative %	Total	% of Variance	Cumulative %
1	1.975	65.830	65.830	1.975	65.830	65.830
2	.622	20.740	86.570	.622	20.740	86.570
3	.403	13.430	100.000	.403	13.430	100.000

Extraction Method: Principal Component Analysis.

Component Matrix

	Component		
	1	2	3
mechanic	.790	-.534	.301
abstract	.776	.579	.251
math	.866	-.032	-.499

Extraction Method: Principal Component Analysis.
a. 3 components extracted.

a. Show that you can reconstruct the communality for the variable mechanic from the loadings of mechanic on all three components in the component matrix above.
b. Show that you can reconstruct the eigenvalue or SSL for the first component by using the loadings on Component 1 in the component matrix above.
c. If you had allowed SPSS to use the default method (retain only components with eigenvalues >1), how many components would have been retained?
d. For a one-component solution, what proportion of the variance in the data would be accounted for?
e. If you retained only one component, what proportion of the variance in the variable mechanic would be accounted for by just the first component?
f. Based on this analysis, do you think it is reasonable to infer that these three mental tests tap three distinctly different types of mental ability?

2. Using the data in Table 16.1, the Project Talent test score data:
   a. What kinds of preliminary data screening would you need to run to evaluate whether assumptions for factor analysis are reasonably well met? Run these analyses.

b. Using PAF as the method of extraction, do a factor analysis of this set of test scores: English, reading, mechanic, abstract, and math; the default criterion to retain factors with eigenvalues >1; varimax rotation; and the option to sort the factor loadings by size; also request that factor scores be computed using the regression method, and saved. Report and discuss your results. Do your results suggest more than one kind of mental ability? Do your results convince you that mental ability is "one-dimensional" and that there is no need to have theories about separate math and verbal dimensions of ability?

c. Compute unit-weighted scores to summarize scores on the variables that have high loadings on your first factor in the preceding analysis. Do this in two different ways: using raw scores on the measured variables and using z scores. (Recall that you can save the z scores for quantitative variables by checking a box to save standardized scores in the descriptive statistics procedure.) Run a Pearson correlation to assess how closely these three scores agree (the saved factor score from the factor analysis you ran in 2b, the sum of the raw scores, and the sum of the z scores). In future analyses, do you think it will make much difference which of these three different scores you use as a summary variable (the saved factor score, sum of raw scores, or sum of z scores)?

3. Describe a hypothetical study in which principal components or factor analysis would be useful either for theory development (e.g., trying to decide how many different dimensions are needed to describe personality traits) or for data reduction (e.g., trying to decide how many different latent variables might be needed to understand what is measured in a multiple-item survey of "patient satisfaction with medical care"). List the variables, describe the analyses that you would run, and describe what pattern of results you expect to see.

4. Using your own data or data provided by your instructor run a factor analysis using a set of variables that you expect to yield at least two factors, with at least four items that have high loadings on each factor. Write up a Results section for this analysis.

5. Locate a journal article that reports either a principal components or a principal axis factor analysis (avoid solutions that include nonorthogonal or oblique rotations or that use confirmatory factor analysis methods, because this chapter does not provide sufficient background for you to evaluate these). Evaluate the reported results and show that you can reproduce the communalities and SSLs from the table of factor loadings. Can you think of different verbal labels for the factors or components than the labels suggested by the authors? Identify one or more variables that—if they were omitted from the analysis—would greatly change the conclusions from the study.

CHAPTER 19

Reliability, Validity, and Multiple-Item Scales

19.1 ♦ Assessment of Measurement Quality

Whether the variables in a study are direct measurements of physical characteristics, such as height and weight, observer counts of behavior frequencies, or self-report responses to questions about attitudes and behaviors, researchers need to consider several measurement issues. The analyses described in previous chapters yield meaningful results only if the researcher has a well-specified model and if all the measurements for the variables are reliable and valid. This chapter outlines several important issues that should be taken into account when evaluating the quality of measurements. Self-report measures are used as examples throughout most of this chapter; however, the issues that are discussed (such as reliability and validity) are relevant for all types of measurements. When researchers use well-calibrated, widely used, well-established methods of measurement (whether these are physical instruments, such as scales, or self-report instruments, such as an intelligence test), they rely on an existing body of past knowledge about the quality of the measurement as evidence that the measurement method meets the quality standards that are outlined in this chapter. When researchers develop a new measurement method, or modify an existing method, they need to obtain evidence that the new (or modified) measurement method meets these quality standards. The development of a new measurement method (whether it is a physical device, an observer coding system, or a set of self-report questions) can be the focus of a study.

19.1.1 ♦ Reliability

A good measure should be reasonably reliable—that is, it should yield consistent results. Low reliability implies that the scores contain a great deal of measurement error. Other factors being equal, variables with low reliability tend to have low correlations with other variables. This chapter reviews basic methods for reliability assessment. To assess reliability, a researcher needs to obtain at least two sets of measurements. For example, to assess the reliability of temperature readings from an oral thermometer, the researcher needs to take two temperature readings for each individual in a sample. Reliability can then be assessed using relatively simple statistical methods; for example, the researcher can compute a Pearson *r* between the first and second temperature readings to assess the stability of temperature across the two times of measurement.

19.1.2 ♦ Validity

A measure is valid if it measures what it purports to measure (Kelley, 1927); in other words, a measure is valid if the scores provide information about the underlying construct or theoretical variable that it is intended to measure. Validity is generally more difficult to assess than reliability. For now, consider the following example: If a researcher wants to evaluate whether a new measure of depression really does assess depression, he or she would need to collect data to see if scores on the new depression measure are related to behaviors, group memberships, and test scores in a manner that would make sense if the test really measures depression. For example, scores on the new depression test should predict behaviors that are associated with depression, such as frequency of crying or suicide attempts; mean scores on the new depression test should be higher for a group of patients clinically diagnosed with depression than for a group of randomly selected college students; also, scores on the new depression test should be reasonably highly correlated with scores on existing, widely accepted measures of depression, such as the Beck Depression Inventory (BDI), the Center for Epidemiological Studies Depression (CESD) scale, the Hamilton Scale, and the Zung Scale. Conversely, if scores on the new depression test are not systematically related to depressed behaviors, to clinical diagnoses, or to scores on other tests that are generally believed to measure depression, the lack of relationship with these criteria should lead us to question whether the new test really measures depression.

19.1.3 ♦ Sensitivity

We would like scores to distinguish among people who have different characteristics. Some common problems that reduce measurement sensitivity include a limited number of response alternatives and ceiling or floor effects. For example, consider a question about depression that has just two response alternatives ("Are you depressed?" 1 = No, 2 = Yes) versus a rating scale for depression that provides a larger number of response alternatives ("How depressed are you?" on a scale from 0 = Not at all depressed to 10 = Extremely depressed). A question that provides a larger number of response alternatives potentially provides a measure of depression that is more sensitive to individual differences; the researcher may be able to distinguish among persons who vary in degree of depression, from mild to moderate to severe. A question that provides only two response alternatives (e.g., "Are you depressed?" no vs. yes) is less sensitive to individual differences in depression.

Some research has systematically examined differences in data obtained using different numbers of response alternatives; in some situations, a 5-point or 7-point rating scale may provide adequate sensitivity. Five-point degree of agreement rating scales (often called Likert scales) are widely used. A Likert scale item for an attitude scale consists of a "stem" statement that clearly represents either a positive or negative attitude about the object of the survey.

Here is an example of an attitude statement with Likert response alternatives for a survey about attitudes toward medical care. A high score on this item (a choice of strongly agree, or 5) would indicate a higher level of satisfaction with medical care.

My physician provides excellent health care.

1	2	3	4	5
Strongly disagree	Disagree	Neutral/ don't know	Agree	Strongly agree

Other rating scales use response alternatives to report the frequency of a feeling or behavior, instead of the degree of agreement, as in the CESD scale (see appendix to this chapter; Radloff, 1977). The test taker is instructed to report how often in the past week he or she behaved or felt a specific way, such as "I felt that I could not shake off the blues even with help from my family and friends." The response alternatives for this item are given in terms of frequency:

1 = Rarely or none of the time (less than 1 day)

2 = Some or a little of the time (1–2 days)

3 = Occasionally or for a moderate amount of the time (3–4 days)

4 = Most of the time or all the time (5–7 days)

The use of too few response alternatives on rating scales can lead to lack of sensitivity. Ceiling or floor effects are another possible source of lack of sensitivity. A ceiling effect occurs when scores pile up at the top end of a distribution; for instance, if scores on an examination have a possible range of 0 to 50 points, and more than 90% of the students in a class achieve scores between 45 and 50 points, the distribution of scores shows a ceiling effect. The examination was too easy, and thus, it does not make it possible for the teacher to distinguish differences in ability among the better students in the class. On the other hand, a floor effect occurs when most scores are near the minimum possible value; if the majority of students have scores between 0 and 10 points, then the examination was too difficult.

Note that the term *sensitivity* has a different meaning in the context of medical research. For example, consider the problem of making a diagnosis about the presence of breast cancer (0 = Breast cancer absent, 1 = Breast cancer present) based on visual inspection of a mammogram. Two different types of error are possible: the pathologist may report that breast cancer is present when the patient really does not have cancer (a false positive) or the pathologist may report that breast cancer is absent when the patient really does have cancer (a false negative). For a medical diagnostic test to be considered sensitive it should have low risks for both these possible types of errors.

19.1.4 ♦ Bias

A measurement is biased if the observed X scores are systematically too high or too low relative to the true value. For example, if a scale is calibrated incorrectly so that body weight readings are consistently 5 lb (2.27 kg) lighter than actual body weight, the scale yields biased weight measurements. When a researcher is primarily interested in correlations between variables, bias is not a major problem; correlations do not change when scores on one or both variables are biased. If measurements are used to make clinical diagnoses by comparing scores to cutoff values (e.g., diagnose high blood pressure if SBP > 130), then bias in measurement can lead to misdiagnosis.

19.2 ♦ Cost and Invasiveness of Measurements

19.2.1 ♦ Cost

Researchers generally prefer measures that are relatively low in cost (in terms of time requirements as well as money). For example, other factors being equal, a researcher would prefer a test with 20 questions to a test with 500 questions because of the amount of time required for the longer test. Including a much larger number of questions in a survey than really necessary increases the length of time required from participants. When a survey is very long, this can reduce return rate (i.e., a lower proportion of respondents tend to complete surveys when they are very time-consuming) and create problems with data quality (because respondents hurry through the questions or become bored and fatigued as they work on the survey). In addition, including many more questions in a survey than necessary can ultimately create problems in data analysis; if the data analyst conducts significance tests on large numbers of correlations, this will lead to an inflated risk of Type I error.

19.2.2 ♦ Invasiveness

A measurement procedure can be psychologically invasive (in the sense that it invades the individual's privacy by requesting private or potentially embarrassing information), or it can be physically invasive (e.g., when a blood sample is taken by venipuncture). In some research situations, the only way to obtain the information the researcher needs is through invasive procedures; however, professional ethical standards require researchers to minimize risk and discomfort to research participants. Sometimes, less direct or less invasive alternative methods are available. For example, a researcher might assess levels of the stress hormone cortisol by taking a sample of an infant's saliva instead of taking a blood sample. However, researchers need to be aware that less invasive measurements usually do not provide the same information that could be obtained from more invasive procedures; for example, analysis of infant saliva does not provide precise information about the levels of cortisol in the bloodstream. On the other hand, performing an invasive measure (such as taking a blood sample from a vein) may trigger changes in blood chemistry due to stress or anxiety.

19.2.3 ♦ Reactivity of Measurement

A measure is reactive when the act of measurement changes the behavior or mood or attribute that is being measured. For example, suppose a researcher asks a person to fill out an attitude survey on a public policy issue, but the person has not thought about the public policy issue. The act of filling out the survey may prompt the respondent to think about this issue for the first time, and the person may either form an opinion for the first time or report an opinion even if he or she really does not have an opinion. That is, the attitude that the researcher is measuring may be produced, in part, by the survey questions. This problem is not unique to psychology; Heisenberg's uncertainty principle pointed out that it was not possible to measure both the position and momentum of an electron simultaneously (because the researcher has to disturb, or interfere with, the system to make a measurement).

Reactivity has both methodological and ethical consequences. When a measure is highly reactive, the researcher may see behaviors or hear verbal responses that would not have occurred in the researcher's absence. At worst, when our methods are highly reactive, we may influence or even create the responses of participants. When people are aware that an observer is watching their behavior, they are likely to behave differently than they would have behaved if they believed themselves to be alone. One solution to this problem is not to tell participants that they are being observed, but of course, this raises ethical problems (invasion of privacy).

Answering questions can have unintended consequences for participants' everyday lives. When Rubin (1970) developed self-report measures for liking and loving, he administered his surveys to dating couples at the University of Michigan. One question he asked was, How likely is it that your dating relationship will lead to marriage? Many of the dating couples discussed this question after participating in the survey; some couples discovered that they were in agreement, while others learned that they did not agree. Rubin (1976) later reported that his participants told him that the discovery of disagreements in their opinions about the likelihood of future marriage led to the breakup of some dating relationships.

19.3 ♦ Empirical Examples of Reliability Assessment

19.3.1 ♦ Definition of Reliability

Reliability is defined as consistency of measurement results. To assess reliability, a researcher needs to make at least two measurements for each participant and calculate an appropriate statistic to assess the stability or consistency of the scores. The two measurements that are compared can be obtained in many different ways; some of these are given below:

1. A quantitative *X* variable, such as body temperature, may be measured for the same set of participants at two points in time (test-retest reliability). The retest may be done almost immediately (within a few minutes), or it may be done days or weeks later. Over longer periods of time, such as months or years, it becomes more likely that the characteristic that is being measured may actually change. One simple form of reliability is test-retest reliability. In this section, we will examine a Pearson correlation between body temperature measurements made at two points in time for the same set of 10 persons.
2. Scores on a categorical *X* variable (such as judgments about attachment style) may be reported by two different observers or raters. We can assess interobserver

reliability about classifications of persons or behaviors into categories by reporting the percentage agreement between a pair of judges. A reliability index presented later in this section (**Cohen's kappa**) provides an assessment of the level of agreement between observers that is corrected for levels of agreement that would be expected to occur by chance.

3. Scores on X can be obtained from questions in a multiple-item test, such as the CESD scale (Radloff, 1977). The 20-item CESD scale appears in the appendix to this chapter. In a later section of this chapter, factor analysis (FA) and the **Cronbach alpha** internal consistency reliability coefficient are discussed as two methods of assessing agreement about the level of depression measured by multiple test questions or items.

19.3.2 ♦ Test-Retest Reliability Assessment for a Quantitative Variable

In the first example, we will consider how to assess test-retest reliability for scores on a quantitative variable (temperature). The researcher obtains a sample of $N = 10$ participants and measures body temperature at two points in time (e.g., 8 a.m. and 9 a.m.). In general, Time 1 scores are denoted by X_1, and Time 2 scores are denoted by X_2. To assess the stability or consistency of the scores across the two times, a Pearson r can be computed. Figure 19.1 shows hypothetical data for this example. The variables Fahrenheit_1 and Fahrenheit_2 represent body temperatures in degrees Fahrenheit at Time 1 (8 a.m.) and Time 2 (9 a.m.) for the 10 persons included in this hypothetical study.

When Pearson r is used as a reliability coefficient, it often appears with a double subscript; for example, r_{XX} can be used to denote the test-retest reliability of X. An r_{XX} reliability coefficient can be obtained by calculating a Pearson r between measures of X at Time 1 (X_1) and measures on the same variable X for the same set of persons at Time 2 (X_2). For the data in this example, Pearson r was obtained by running the correlation procedure on the X_1 and X_2 variables (see Figure 19.2); the obtained r_{XX} value was $r = +.924$, which indicates very strong consistency or reliability. This high value of r tells us that there is a high degree of consistency between temperature readings at Times 1 and 2; for example, Fran had the highest temperature at both times.

There is no universally agreed on minimum standard for acceptable measurement reliability. Nunnally and Bernstein (1994, pp. 264–265) state that the requirement for reliability varies depending on how scores are used. They distinguished between two different situations: first, basic research, where the focus is on the size of correlations between X and Y variables or on the magnitude of differences in Y means across groups with different scores on X; second, clinical applications, where an individual person's score on an X measure is used to make treatment or placement decisions. In preliminary or exploratory basic research (where the focus is on the correlations between variables, rather than on the scores of individuals) modest measurement reliability (about .70) may be sufficient. If preliminary studies yield evidence that variables may be related, further work can be done to develop more reliable measures. Nunnally and Bernstein (1994) argue that, in basic research, increasing measurement reliabilities much beyond $r = .80$ may yield diminishing returns. However, when scores for an individual are used to make decisions that have important consequences (such as medical diagnosis or placement in a special

*tempexampl [DataSet1] - SPSS Data Editor

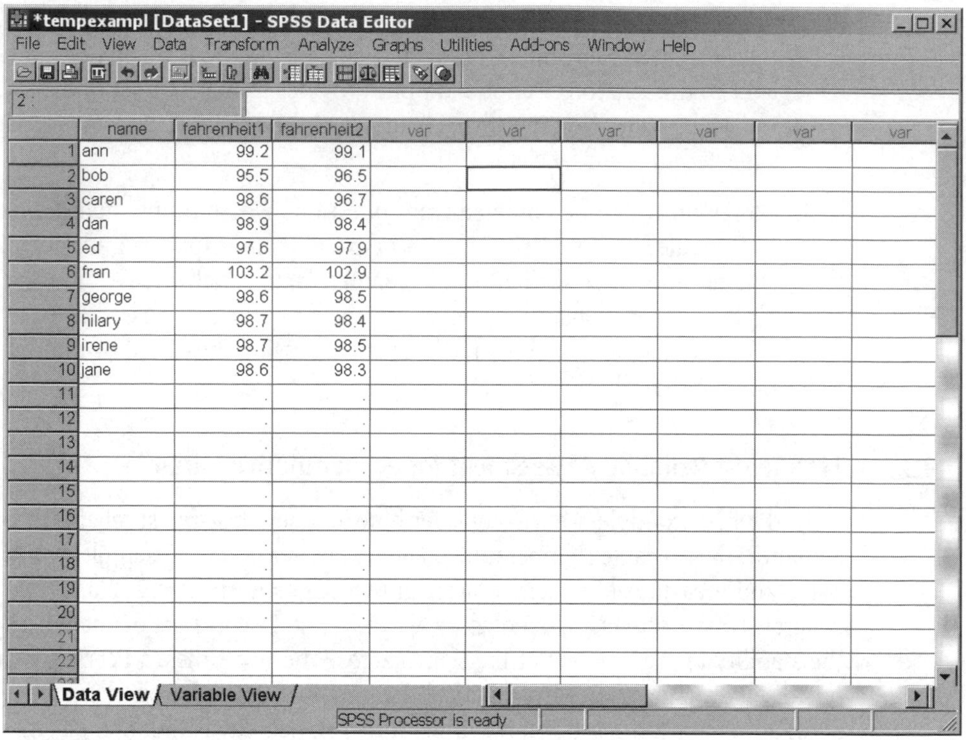

File Edit View Data Transform Analyze Graphs Utilities Add-ons Window Help

	name	fahrenheit1	fahrenheit2
1	ann	99.2	99.1
2	bob	95.5	96.5
3	caren	98.6	96.7
4	dan	98.9	98.4
5	ed	97.6	97.9
6	fran	103.2	102.9
7	george	98.6	98.5
8	hilary	98.7	98.4
9	irene	98.7	98.5
10	jane	98.6	98.3

Data View / Variable View

SPSS Processor is ready

Figure 19.1 ♦ Hypothetical Data for Test-Retest Reliability Study of Body Temperature

Correlations

		fahrenheit1	fahrenheit2
fahrenheit1	Pearson Correlation	1	.924**
	Sig. (2-tailed)		.000
	N	10	10
fahrenheit2	Pearson Correlation	.924**	1
	Sig. (2-tailed)	.000	
	N	10	10

**. Correlation is significant at the 0.01 level (2-tailed).

Figure 19.2 ♦ Pearson Correlation to Assess Test-Retest Reliability of Body Temperature in Degrees Fahrenheit

NOTE: Based on data in Figure 19.1.

class for children with low academic ability), a reliability of $r = .90$ is the bare minimum and $r = .95$ would be a desirable standard for reliability (because even very small measurement errors could result in mistakes that could be quite harmful to the individuals concerned).

Note that whether the correlation between X_1 and X_2 is low or high, there can be a difference in the means of X_1 and X_2. If we want to assess whether the mean level of scores (as well as the position of each participant's score within the list of scores) remains the same from Time 1 to Time 2, we need to examine a paired samples *t* test (see Chapter 20) to see whether the mean of X_2 differs from the mean of X_1, in addition to obtaining the Pearson *r* as an index of consistency of measurements across situations. In this example, there was a nonsignificant difference between mean temperatures at Time 1 ($M_1 = 98.76$) and Time 2 ($M_2 = 98.52$); $t(9) = 1.05, p = .321$.

In some research situations, the only kind of reliability or consistency that researchers are concerned about is the consistency represented by the value of Pearson *r*, but there are some real-life situations where a shift in the mean of measurements across times would be a matter of interest or concern. A difference in mean body temperature between Time 1 and Time 2 could represent a real change in the body temperature; the temperature might be higher at Time 2 if this second temperature reading is taken immediately after intense physical exercise. There can also be measurement artifacts that result in systematic changes in scores across times; for example, there is some evidence that when research participants fill out a physical symptom checklist every day for a week, they become more sensitized to symptoms and begin to report a larger number of symptoms over time. This kind of change in symptom reporting across time might be due to artifacts such as reactivity (frequent reporting about symptoms may make the participant focus more on physical feelings), social desirability, or demand (the participant may respond to the researcher's apparent interest in physical symptoms by helpfully reporting more symptoms), rather than due to an actual change in the level of physical symptoms.

19.3.3 ♦ Interobserver Reliability Assessment for Scores on a Categorical Variable

If a researcher wants to assess the reliability or consistency of scores on a categorical variable, he or she can initially set up a contingency table and examine the percentage of agreement. Suppose that attachment style (coded 1 = Anxious, 2 = Avoidant, and 3 = Secure) is rated independently by two different observers (Observer A and Observer B) for each of $N = 100$ children. Attachment style is a categorical variable; based on observations of every child's behavior, each observer classifies each child as anxious, avoidant, or secure. The hypothetical data in Table 19.1 show a possible pattern of results for a study in which each of these two observers makes an independent judgment about the attachment style for each of the 100 children in the study. Each row corresponds to the attachment score assigned by Observer A; each column corresponds to the attachment score assigned by Observer B. Thus, the value 14 in the cell in the upper left-hand corner of the table tells us that 14 children were classified as "anxiously attached" by both observers. The adjacent cell, with an observed frequency of 1, represents one child who was classified as "anxiously attached" by Observer A and "avoidantly attached" by Observer B.

A simple way to assess agreement or consistency in the assignment of children to attachment style categories at these two points in time is to add up the number of times the two judges were in agreement: we have 14 + 13 + 53 = 80; that is, 80 children were placed in the same attachment style category by both observers. This is divided by the total number of children (100) to yield 80/100 = .80 or 80% agreement in classification. Proportion or percentage of agreement is fairly widely used to assess consistency or reliability of scores on categorical variables.

Table 19.1 ♦ Data for Interobserver Reliability for Attachment Style Category

	Observer B			
Observer A	*1 = Anxious*	*2 = Avoidant*	*3 = Secure*	*Row Total (Proportion)*
1 = Anxious	14	1	8	23 (.23)
2 = Avoidant	2	13	4	19 (.19)
3 = Secure	1	4	53	58 (.58)
Column total (proportion)	17 (.17)	18 (.18)	65 (.65)	100

NOTES: Number of agreements between Observer A and Observer B: 14 + 13 + 53 = 80. Total number of judgments: 100 (sum of all cells in the table, all row totals, or all column totals). P_o = observed level of agreement = number of agreements/total number of judgments = 80/100 = .80. P_c = chance level of agreement for each category, summed across all three categories = (.17 × .23) + (.18 × .19) + (.65 × .58) = .0391 + .0342 + .377 = .4501.

$$\kappa = \frac{(P_o - P_c)}{(1 - P_c)} = \frac{(.80 - .4501)}{(1 - .4501)} = \frac{.3499}{.5499} = .636.$$

However, there is a problem with proportion or percentage of agreement as a reliability index; fairly high levels of agreement can arise purely by chance. Given that Observer A coded 58/100 (a proportion of .58) of the cases as secure and Observer B coded 65/100 (.65) of the cases as secure, we would expect them both to give the score "secure" to .58 × .65 = .377 of the cases in the sample just by chance. By similar reasoning, they should both code .17 × 23 = .0391 of the cases as anxious and .18 × .19 = .0342 of the cases as avoidant. When we sum these chance levels of agreement across the three categories, the overall level of agreement predicted just by chance in this case would be .377 + .0391 + .0342 = .4501.

Cohen's kappa (κ) takes the observed proportion of agreement (P_o, in this case, .80) and corrects for the chance level of agreement (P_c, in this case, .4501).

The formula for Cohen's κ is

$$\kappa = \frac{(P_o - P_c)}{(1 - P_c)} = \frac{(.80 - .4501)}{(1 - .4501)} = .636. \quad (19.1)$$

The Cohen's kappa index of interrater reliability in this case (corrected for chance levels of agreement) is .64. Landis and Koch (1977) suggested the following guidelines for evaluating the size of Cohen's κ: .21 to .40, fair reliability; .41 to .60, moderate reliability; .61 to .80, substantial reliability; and .81 to 1.00, almost perfect reliability. By these standards the obtained κ value of .64 in this example represents a substantial level of reliability. Cohen's kappa is available as one of the contingency table statistics in the SPSS Cross tabs procedure, and it can be used to assess the reliability for a pair of raters on categorical codes with any number of categories; it could also be used to assess test-retest reliability for a categorical measurement at two points in time.

19.4 ♦ Concepts From Classical Measurement Theory

The version of **measurement theory** presented here is a simplified version of classical measurement theory (presented in more detail by Carmines & Zeller, 1979). Contemporary measurement theory has moved on to more complex ideas that are beyond the scope of this book. The following is a brief introduction to some of the ideas in classical measurement theory. We can represent an individual observed score, X_i, as the sum of two theoretical components:

$$X_{ij} = T_i + e_{ij}, \tag{19.2}$$

where X_i represents the observed score on a test for person i at Time j, T_i is the theoretical "true" score of Person i, and e_{ij} is the error of measurement for Person i at Time j.

The first component, T, usually called the true score, denotes the part of the X score that is stable or consistent across measurements. In theory, this value is constant for each participant or test taker and should represent that person's true score on the measurement. We hope that the T component really is a valid measure of the construct of interest, but the stable components of observed scores can contain consistent sources of bias (i.e., systematic error). For example, if an intelligence quotient (IQ) test is culturally biased, then the high reliability of IQ scores might be due in part to the consistent effect of an individual's cultural background on that individual's test performance. It is generally more difficult to assess validity, that is, assess whether a scale really measures the construct it is supposed to measure, without any "contaminants," such as social desirability or cultural background, than to assess reliability (which is simply an assessment of score consistency or stability).

To see what the T and e components might look like for an individual person, suppose that Jim is Participant number 6 in a test-retest reliability study of the verbal SAT test; suppose Jim's score is $X_{61} = 570$ at Time 1 and $X_{62} = 590$ at Time 2. To estimate T_6, Jim's "true" score, we would find his mean score across measurements; in this case, $M = (570 + 590)/2 = 580$.

We can compute an error (e_{ij}) term for each of Jim's observed scores by subtracting T from each observed score; in this case, the error term for Jim at Time 1 is $570 - 580 = -10$ and the error at Time 2 is $590 - 580 = +10$. We can then use the model in Equation 19.2 to represent Jim's scores at Time 1 and Time 2 as follows:

$$X_{61} = T_6 + e_{61} = 580 - 10 = 570,$$
$$X_{62} = T_6 + e_{62} = 580 + 10 = 590.$$

Using this simplified form of classical measurement theory as a framework, we would then say that at Time 1, Jim's observed score is made up of his true score minus a 10-point error; at Time 2, Jim's observed score is made up of his true score plus a 10-point error. The measurement theory assumes that the mean of the measurement errors is 0 across all occasions of measurement in the study; and that the true score for each person corresponds to the part of the score that is stable or consistent for each person across measurements.

Errors may be even more likely to arise when we try to measure psychological characteristics (such as attitude or ability) than when we measure physical quantities (such as body temperature). Standardized test scores, such as the Scholastic Aptitude Test (SAT), are supposed to provide information about student academic aptitude. What factors could lead to measurement errors in Jim's SAT scores? There are numerous sources of measurement errors (the *e* component) in exam scores; for example, a student might score lower on an exam than would be expected given his or her ability and knowledge because of illness, anxiety, or lack of motivation. A student might score higher on an exam than his or her ability and general knowledge of the material would predict because of luck (the student just happened to study exactly the same things the teacher decided to put on the exam), cheating (the student copied answers from a better prepared student), randomly guessing some answers correctly, and so forth.

We can generalize this logic as follows. If we have measurements at two points in time, and X_{ij} is the observed score for Person *i* at Time *j*, we can compute the estimated true score for Person *i*, T_i, as $(X_1 + X_2)/2$. We can then find the estimated error terms (e_{i1} and e_{i2}) for Person *i* by subtracting this T_i term from each of the observed scores. When we partition each observed score (X_{ij}) into two parts (T_i and e_{ij}), we are using logic similar to ANOVA (Chapter 6). The *T* and *e* terms are uncorrelated with each other, and in theory, the error terms *e* have a mean of 0.

When we have divided each observed score into two components (*T* and *e*), we can summarize information about the *T* and *e* terms across persons by computing a variance for the values of *T* and a variance for the values of *e*. A reliability coefficient (e.g., a test-retest Pearson *r*) is the proportion of variance in the test scores that is due to *T* components that are consistent or stable across occasions of measurement. Reliability can be defined as the ratio of the variance of the "true score" components relative to the total variance of the observed *X* measurements as shown in Equation 19.3:

$$r_{XX} = \frac{s_T^2}{s_T^2 + s_e^2} = \frac{s_T^2}{s_X^2}, \qquad (19.3)$$

where s_X^2 is the variance of the observed *X* scores, s_T^2 is the variance of the *T* true scores, s_e^2 is the variance of the *e* or error terms, and $s_X^2 = s_T^2 + s_e^2$, that is, the total variance of the observed *X* scores is the sum of the variances of the *T* true score components and the *e* errors.

Equation 19.3 defines reliability as the proportion of variance in the observed *X* scores that is due to *T*, a component of scores that is stable across the two times of measurement (and different across persons). For a measure with a test-retest reliability of $r_{XX} = .80$, in theory, 80% of the variance in the observed *X* scores is due to *T*; the remaining 20% of the variance in the observed scores is due to error.

Note that it is r_{XX}, the test-retest correlation (rather than r^2, as in other situations where the Pearson correlation is used to predict scores on *Y* from *X*), that is interpreted as a proportion of variance in this case. To see why this is so, consider a theoretical path model that shows how the observed measurements (X_1 and X_2) are related to the stable characteristic *T*. The notation and tracing rules for path diagrams were introduced in Chapters 10 and 11.

Figure 19.3 corresponds to Equation 19.2; that is, the observed score on X_1 is predictable from T and e_1; the observed score on X_2 is predictable from T and e_2. The path coefficient that represents the strength of the relationship between T and X_1 is denoted by β; we assume that the strength of the relationship between T and X_1 is the same as the strength of the relationship between T and X_2, so the beta coefficients for the paths from $T \rightarrow X_1$ and $T \rightarrow X_2$ are equal.

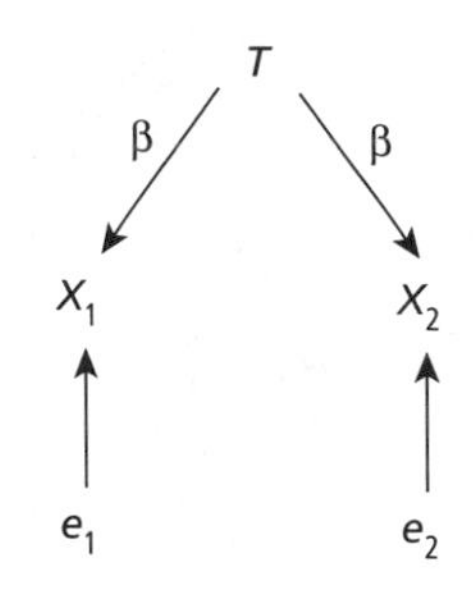

Figure 19.3 ♦ Classic Measurement Theory in Path Model Form

We would like to estimate beta, the strength of the relation between the observed scores X_1 and X_2 and the true score T. We can compute a test-retest r between X_1 and X_2 (we can denote this by r_{XX} or r_{12}). Based on the logic of path diagrams and the application of the tracing rule introduced in Chapter 11, we know that we can reproduce the observed value of r_{12} by tracing the path from X_1 to X_2 (by way of T) and multiplying the coefficients for each section of the path together. In this case, we see that (by the application of the tracing rule) $\beta \times \beta$ should equal r_{12}. This implies that the best estimate of β is $\sqrt{r_{12}}$. Thus, if we want to know what proportion of variance in X_1 (or in X_2) is associated with T, we square the path coefficient between X_1 and T; this yields r_{12} as our estimate of the proportion of shared or predicted variance. An examination of this path diagram helps explain why it is r_{12} (rather than r_{12}^2) that tells us what proportion of variance in X is predictable from T.

A related path diagram explains the phenomenon of attenuation of correlation between X and Y due to the unreliability of measurement. Suppose the real strength of relationship between the true scores on variables X and Y (i.e., the scores that would be obtained if they could be measured without error) is represented by ρ_{XY}. However, the reliabilities of X and Y (denoted by r_{XX} and r_{YY}, respectively) are less than 1.00; that is, the X and Y measurements used to compute the sample correlation do contain some measurement error; the smaller these reliabilities, the greater the measurement error. The observed correlation between the X and Y scores that contain measurement errors (denoted by r_{XY}) can be obtained from the data. The path diagram shown in Figure 19.4 tells us that the strength of the relationship between the observed X scores and the true score on X is denoted by $\sqrt{r_{XX}}$; the strength of the relationship between the observed Y scores and the true scores on Y is denoted by $\sqrt{r_{YY}}$. If we apply the tracing rule to this path model, we find that r_{XY} (the observed correlation between X and Y) should be reproducible by tracing the path from the observed X to the true X, from the true X to the true Y, and from the true Y to the observed Y (and multiplying these path coefficients). Thus, we would set up the following equation based on Figure 19.4:

$$r_{XY} = \rho_{XY}\sqrt{r_{XX}} \times \sqrt{r_{YY}} = \rho_{XY}\sqrt{r_{XX} \times r_{YY}}, \qquad (19.4)$$

where r_{XY} is the observed correlation between the observed scores on X and Y, ρ_{XY} is the correlation between the true scores on X and Y (i.e., the correlation between X and Y if they could be measured without error), and r_{XX} and r_{YY} are the reliabilities of the variables.

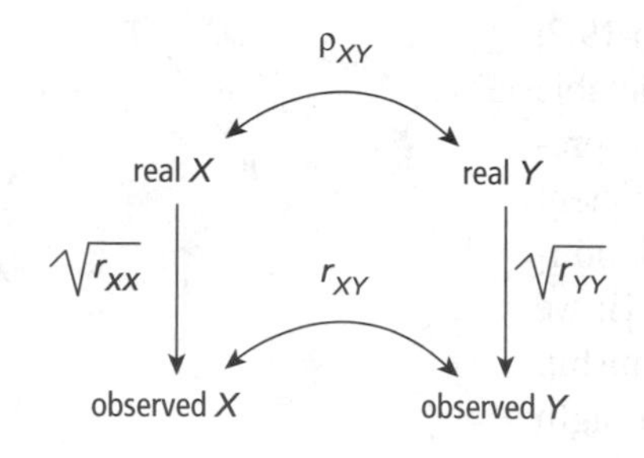

Figure 19.4 ♦ Path Model for Attenuation of Correlation between Observed X and Observed Y as a Function of Reliability of Measurement of X and Y

NOTE: ρ_{XY} is the "real" correlation between X and Y (i.e., the true strength of relationship between variables, if they were measured without error), r_{XY} the observed correlation between X and Y scores that contain measurement error, r_{XX} the estimated reliability of X, and r_{YY} the estimated reliability of Y.

In situations where there is measurement error, the reliability coefficients r_{XX} and r_{YY} will be less than 1. It follows that the observed correlation r_{XY} will be less than ρ_{XY} whenever there is measurement error, and in fact, the smaller the reliabilities of X and Y, the greater the reduction in the size of the observed r_{XY} correlation. This is called attenuation of correlation due to unreliability. In theory, the lower the reliabilities of X and Y, the smaller the observed r_{XY} correlation will tend to be.

It is theoretically possible to correct for attenuation, that is, to calculate an estimate ($\hat{\rho}_{XY}$) for the "true" correlation that would exist between X and Y, if they could be measured without error. An estimate of the correlation between X and Y that is adjusted for attenuation due to unreliability of measurement in X and Y is given by the following equation:

$$\hat{\rho}_{XY} = \frac{r_{XY}}{\sqrt{r_{XX} \times r_{YY}}}, \qquad (19.5)$$

where $\hat{\rho}_{XY}$ is the "attenuation-corrected" estimate of the correlation between X and Y, that is, an estimate of the strength of the relationship between X and Y if they could be measured with perfect reliability; r_{XY} is the observed correlation between the X and Y scores; r_{XX} is the estimated reliability coefficient for the X variable; and r_{YY} is the estimated reliability coefficient for the Y variable.

However, this attenuation-corrected estimate $\hat{\rho}_{XY}$ should be interpreted with great caution. If the reliability estimates r_{XX} and r_{YY} are inaccurate, this correction will yield misleading results. In particular, if the reliabilities of X and Y are underestimated, the attenuation-corrected estimate of the true strength of the association between X and Y will be inflated.

Poor reliability of measurement has two major negative consequences. First of all, reliability is a necessary (although not sufficient) condition for validity; thus, if measurements are not reliable, they cannot be valid. Second, when measurements of variables are unreliable, it tends to result in low correlations among variables. In traditional approaches to statistical analysis, using primarily the pre-1970 methods of analysis, we need to make an effort to obtain measurements for each variable in the analysis that are as reliable as possible; to the extent that we use unreliable measures, the outcomes from these traditional data analyses may be misleading. Another way in which we can handle the problems caused by unreliability of measurements is to use structural equation models (briefly introduced at the end of Chapter 18); in these models, the observed X variables are treated as multiple indicators of latent variables, and the analysis incorporates assumptions about errors of measurement for the observed X variables.

We can typically improve the reliability of measurement for an X variable in a sample in two different ways. One approach is to increase the differences in T across participants by working with samples that are more diverse on the characteristic that is being measured (i.e., we increase the magnitude of s_T^2). Another approach is to reduce the magnitude of errors of measurement by controlling extraneous variables that might influence measurements (i.e., we decrease the magnitude of s_e^2). Combining scores from multiple questions or measurements is another method that may result in more reliable scores; the use of scales that are formed by summing multiple measures is described in the next section.

19.5 ♦ Use of Multiple-Item Measures to Improve Measurement Reliability

For some variables, a single measurement or a single question is sufficient. For example, a single question is generally sufficient to obtain information about age or height. However, to assess personality traits, abilities, attitudes, or knowledge, it is usually necessary to include multiple questions. Using a score that corresponds to a sum (or an average) across multiple measurements provides a measure that is more reliable than a single measure. The inclusion of multiple measures also makes it possible to assess the reliability or consistency of responses. In many situations, we may also be more confident about the validity of a score that summarizes information across multiple measurements than regarding a score based on a single measurement; however, validity of measurement is more difficult to evaluate than reliability.

However, for variables that have complex definitions or that must be assessed indirectly, such as depression, the use of single-item measures can be problematic.

A one-item assessment of depression would have serious limitations. First of all, a measure with only two possible score values would provide very limited information about individual differences in the degree of depression; also, the scores could not be normally distributed (and normally distributed scores are desirable for many statistical analyses). Even a question with multiple response options, such as "Rate your degree of depression on a scale from 1 to 4" would produce a limited number of possible values and a nonnormal distribution. Second, when only one response is obtained, there is no way of assessing the reliability or consistency of responses. Third, a single question does not capture the complex nature of depression; depression includes not only feelings of sadness and hopelessness but also vegetative symptoms such as fatigue and sleep disturbance. A comprehensive measure of depression needs to include all relevant symptoms. In some surveys, the inclusion of multiple questions helps respondents better understand what kind of information the researcher is looking for. Responses to multiple, specific questions may be far more informative than responses to a single, global, nonspecific question. For instance, when people are asked a global, nonspecific question (such as Are you satisfied with your medical care?), they tend to report high levels of satisfaction; but when they are asked about specific potential problems (such as Did the physician use language you did not understand? Did you have enough time to explain the nature of your problems? Was the practitioner sympathetic to your concerns?), their responses typically become more informative.

To avoid the limitations of single-item measures, researchers often use multiple items or questions to obtain information about variables such as depression. For example, the CESD scale (Radloff, 1977) has 20 questions about depression (see the complete CESD scale in the appendix to this chapter). Scores for just the first five items from the CESD scale for a set of 97 participants are included in the dataset in Table 19.2; these data will be used to demonstrate the assessment of reliability of multiple-item scales. The last analysis of CESD scale reliability is based on the full 20-item scale; however, initial examples are limited to the assessment of Items 1 through 5 to keep the empirical examples brief and simple. Multiple-item measures often provide many advantages compared with single-item measures. They can potentially provide finer discriminations of the amount of depression, scores that are more nearly normally distributed, and scores that are generally more reliable than scores based on single-item assessments.

In many research situations, we can obtain more reliable measurements by measuring the variable of interest on multiple occasions, or using multiple methods or multiple items and, then, combining the information across multiple measures into a single summary score. Ideas derived from the measurement theory model presented in a previous section suggest that combining two or more scores should result in more reliable measures. As we sum a set of X_i measurements for a particular participant (which could correspond to a series of p different questions on a self-report survey, a series of behaviors, or a series of judgments by observers or raters), in theory, we obtain the following result:

$$
\begin{aligned}
X_1 &= T + e_1 \\
X_2 &= T + e_2 \\
X_3 &= T + e_3 \\
&\vdots \\
X_p &= T + e_p \\
\hline
\sum \mathbf{X} &= p \times T + \sum \mathbf{e}.
\end{aligned}
$$

The expected value of $\sum e$, particularly when errors are summed across many items, theoretically approaches 0. That is, we expect the errors to "cancel out" as we sum more and more items. When we use $\sum X$, the sum of scores on several questions, as a measure, it is likely that a large part of the score obtained from $\sum X$ will correspond to the true score for the participant (because we have $p \times T$ in the expression for $\sum X$). It is also likely that a small part of the score obtained by taking $\sum X$ will correspond to errors of measurement, because we expect that this sum approaches 0 when many errors are summed. Thus, the reliability of a score based on $\sum X$ (the percentage of variance in the value of $\sum X$ that is due to T) should be higher than the reliability of a score based on a single question.

When we use a sum (or a mean) of scores across multiple items, questions, or measures, there are several potential advantages:

1. Composite measures (scales) are generally more reliable than single scores, and the inclusion of multiple measures makes it possible to assess **internal consistency reliability.**

2. Scores on composite measures generally have greater variance than scores based on single items.

Table 19.2 ♦ Scores on Items 1 Through 5 of the CESD Scale for $N = 98$ Participants

Dep1	*Dep2*	*Dep3*	*Dep4*	*Dep5*	*RevDep4*
3	1	3	4	3	1
2	1	2	4	3	1
2	4	3	1	1	4
1	2	1	4	4	1
2	2	2	1	4	4
4	4	3	2	1	3
4	1	1	2	2	3
1	1	3	1	1	4
1	2	2	4	2	1
1	1	1	4	2	1
1	2	3	2	3	3
1	1	1	4	2	1
2	1	1	4	2	1
1	1	1	1	1	4
3	2	4	3	1	2
2	3	1	4	2	1
2	4	2	3	3	2
2	1	1	2	2	3
1	3	1	3	3	2
1	2	1	3	3	2
1	1	1	4	3	1
3	2	4	3	4	2
1	1	1	4	2	1
2	1	1	3	4	2
3	2	1	1	1	4
1	2	1	4	4	1
2	1	4	2	4	3
1	1	1	2	3	3
1	1	1	2	3	3
4	4	4	1	4	4
1	1	1	4	2	1
2	1	2	3	3	2
4	2	4	4	2	1
1	1	1	3	1	2
1	1	1	4	3	1
2	4	1	4	4	1
1	2	2	4	4	1
3	3	4	1	3	4
2	2	2	3	4	2
1	1	1	4	2	1
2	2	1	.	4	.
2	3	2	3	3	2
1	1	1	4	3	1
1	1	1	4	4	1
1	1	1	4	2	1
1	1	1	4	2	1
1	3	1	4	2	1
1	2	1	4	4	1
1	2	1	4	2	1
2	2	3	2	3	3

(Continued)

Table 19.2 ♦ (Continued)

Dep1	*Dep2*	*Dep3*	*Dep4*	*Dep5*	*RevDep4*
1	1	1	4	2	1
2	2	1	4	1	1
2	1	1	4	4	1
1	1	1	4	2	1
2	1	2	3	2	2
2	4	3	4	1	1
1	1	1	4	3	1
2	1	1	4	2	1
2	1	1	4	3	1
1	2	1	1	1	4
1	2	2	4	2	1
1	1	2	2	1	3
2	1	1	3	2	2
2	2	2	4	2	1
2	1	1	4	2	1
2	2	1	4	2	1
2	2	2	3	3	2
1	2	4	1	1	4
1	1	1	4	3	1
2	1	2	2	2	3
2	1	2	3	3	2
1	1	1	4	1	1
2	2	1	3	3	2
2	1	1	4	2	1
1	1	1	4	1	1
3	2	3	3	4	2
2	2	3	2	3	3
1	1	1	4	2	1
1	1	1	4	1	1
2	2	3	3	4	2
1	1	1	4	1	1
2	1	2	2	2	3
1	1	1	4	1	1
2	2	2	2	2	3
1	2	2	4	4	1
2	2	2	4	4	1
1	1	1	1	1	4
1	2	1	4	3	1
2	1	2	2	3	3
2	1	2	4	2	1
2	1	3	3	3	2
1	1	2	2	4	3
1	1	1	4	4	1
1	1	1	4	4	1
2	2	3	2	2	3
2	2	2	2	2	3
1	1	1	4	1	1
1	1	2	4	4	1

3. The distribution shape for measurements that are obtained using common methods, such as the 5-point Likert rating scales, are typically nonnormal. The distribution shape for scores formed by summing multiple measures that are positively correlated with each other tend to resemble a somewhat flattened normal distribution; the higher the correlation among items, the flatter the distribution for the sum of the scores (Nunnally & Bernstein, 1994). Many statistical tests work better with scores that are approximately normally distributed, and composite scores that are obtained by summing multiple-item measures sometimes approximate a normal distribution better than scores for single-item measures.

4. Scores on composite or multiple-item measures may be more sensitive to individual differences (because a larger number of different score values are possible than on single items that use 5-point rating scales, for example).

To summarize information across many items or questions, scores are often summed across items. We will define a summated scale as a sum (or mean) of scores across multiple occasions of measurement. For example, an attitude scale to measure patient satisfaction with medical treatment might include a set of 20 statements about various aspects of quality of medical care; patients may be asked to indicate the degree to which their care providers meet each of the quality standards included in the survey. It may make sense to compute a single score to summarize overall patient satisfaction with the quality of medical care by summing responses across all 20 items. However, it is also possible that, on closer examination, we might decide that the set of 20 items may provide information about more than one type of satisfaction. For example, some items might assess satisfaction with rapport, empathy, or bedside manner; some items might assess perceived technical competence of the practitioner; and other questions might assess issues of accessibility, for example, length of waiting time and difficulty in scheduling appointments.

Note that the use of multiple-item scales is not limited to self-report data. We can also combine other types of multiple measurements to form scales (such as physiological measures or ratings of behavior made by multiple judges or observers).

19.6 ♦ Three Methods for the Computation of Summated Scales

19.6.1 ♦ Implicit Assumption: All Items Measure the Same Construct and Are Scored in the Same Direction

When we add together scores on a list of measures or questions, we implicitly assume that all these scores measure the same underlying construct and that all the questions or items are scored in the same direction.

Consider the first assumption, the assumption that all items measure the same construct. What information would be obtained by a set of numbers that measured completely unrelated things? A sum of X_1 = height, X_2 = agreeableness, and X_3 = number of pairs of shoes owned by a person would be a meaningless number because the scores that are combined are not measures of the same underlying latent variable. In general, it does not make sense to summarize information across a set of *X* measured variables by summing them unless they are highly correlated with each other, and both the pattern of correlations and the nature of the items are consistent with the interpretation that all the individual *X* items are slightly different ways of measuring the same underlying latent variable (e.g., depression).

In Chapter 18, we saw that FA was one way of examining the pattern of correlations among a set of measured *X* variables to assess whether these variables can reasonably be interpreted as measures of one or several latent variables or factors. FA is one of the data analytic methods often used in the development of multiple-item measures; it can help us decide how many different latent variables or dimensions may be needed to account for the pattern of correlations among test items and aid us in characterizing these dimensions by examining the groups of items that have high correlations with each factor.

The items included in psychological tests such as the CESD scale are typically written so that they assess slightly different aspects of a complex variable such as depression (e.g., low self-esteem, fatigue, sadness). To evaluate empirically whether these items can reasonably be interpreted as measures of the same underlying latent variable or construct, we look for reasonably large correlations among the scores on the items. If the scores on a set of measurements or test items are highly correlated with each other, this evidence is consistent with the belief that the items may all be measures of the same underlying construct. However, high correlations among items can arise for other reasons and are not necessarily proof that the items measure the same underlying construct; for example, they may occur due to sampling error or may arise because the items have some kind of measurement artifact in common, such as a strong social desirability bias. Later in this chapter, we will see that the most widely reported method of evaluating reliability for summated scales, Cronbach's alpha, is based on the mean of the inter-item correlations.

19.6.2 ♦ Reverse-Worded Questions

Consider the second assumption: the assumption that all items are scored in the same direction. In the CESD scale in the appendix to this chapter, most of the items are worded in such a way that a higher score indicates a greater degree of depression. For example, for Question 3, "I felt that I could not shake off the blues even with help from my family or friends," the response that corresponds to 4 points ("I felt this way most of the time, 5–7 days per week") indicates a higher level of depression than the response that corresponds to 1 point ("I felt this way rarely or none of the time"). However, a few of the items (Numbers 4, 8, 12, and 16) are reverse worded. Question 4 asks how frequently the respondent "felt that I was just as good as other people." The response to this question that would indicate the highest level of depression corresponds to the lowest frequency of occurrence (1 = Rarely or none of the time). When reverse-worded items are included in a multiple-item measure, the scoring on these items needs to be recoded before we sum scores across items, such that a high score on each item corresponds to the same thing, that is, a higher level of depression.

When self-report methods are used, it is often desirable to include some reverse-worded questions. Self-report responses are prone to many types of bias, including **yea-saying** or **nay-saying bias** (some respondents tend to agree or disagree with all items), and social desirability bias (many people tend to report behaviors and attitudes that they believe are socially desirable); see Converse and Presser (1999) for further discussion. To avoid the yea-saying bias, some scales include reverse-worded items. For example, the CESD scale includes statements about feelings and behaviors, and respondents are asked to rate how frequently they experience each of these, using a scale from 1 (rarely or none of the time, less than 1 day a week) to 4 (most or all of the time, 5–7 days a week).

It is generally preferable to report final scores for a scale scored in a direction such that a higher score corresponds to "more" of the attitude or ability that the test is supposed to measure. For example, it is easier to talk about scores on a depression scale, and to interpret correlations of the depression scale with other variables, if a higher score corresponds to more severe depression. (If a depression scale was scored such that a high score corresponded to a low level of depression, then scores on the depression scale would correlate negatively with other measures of negative mood such as anxiety; this would be confusing for the data analyst and the reader.) Most of the items on the CESD scale are worded such that a high frequency of reported occurrence corresponds to a higher level of depression. For example, a high reported frequency of occurrence for the item "I had crying spells" corresponds to a higher level of depression. However, a few of the CESD scale items were reverse worded, for example, "I enjoyed life." For these reverse-worded items, a score of 1 or 2 indicating a *low* frequency of occurrence corresponds to a higher level of depression. Before combining scores across items that are worded in different directions (such that for some items, a high score corresponds to more depression, and for other items, a low score corresponds to more depression), it is necessary to recode the direction of scoring on reverse-worded items so that a higher score always corresponds to a higher level of depression. Items 4, 8, 12, and 16 in the appendix were reverse worded. Scores on these reverse-worded items must be recoded when we form a sum of the scores across all 20 items to serve as an overall measure of depression.

In the following example, dep4 is the name of the SPSS variable that corresponds to the reverse-worded depression item "I felt that I was just as good as other people" (item number 4 on the CESD scale). One simple method to reverse the scoring on this item (so that a higher score corresponds to more depression) is as follows: Create a new variable (revdep4) that corresponds to 5 – dep4. If you take a value that is one unit higher than the highest possible score on a measure (in this case, because the possible scores are 1, 2, 3, and 4, we use the value 5), and then subtract each person's score from that reference value, this reverses the direction of scoring. This can be done in SPSS by making the following menu selections: <Transform> → <Compute>.

In the dialog box for the Compute procedure (see Figure 19.5), the name of the new variable or Target Variable (revdep4) is placed in the left-hand side box. The equation to compute a score for this new variable as a function of the score on an existing variable is placed in the right-hand side box titled Numeric Expression (in this case, the numeric expression is 5 – dep4).

For a person with a score of 4 on dep4, the score on the reverse-coded version of the variable (revdep4) will be 5 – 4, or 1. (There was one participant with a system missing code on dep4; that person was assigned a system missing code on revdep4 also.) It is a good idea to paste the syntax for all recoding operations into an SPSS syntax window and save the syntax, so that you have a record of all the recoding operations that have been applied to the data. Figure 19.6 summarizes the SPSS syntax that reverses the direction of scoring on all the reverse-worded CESD scale items. It is also helpful to create a variable with a different name

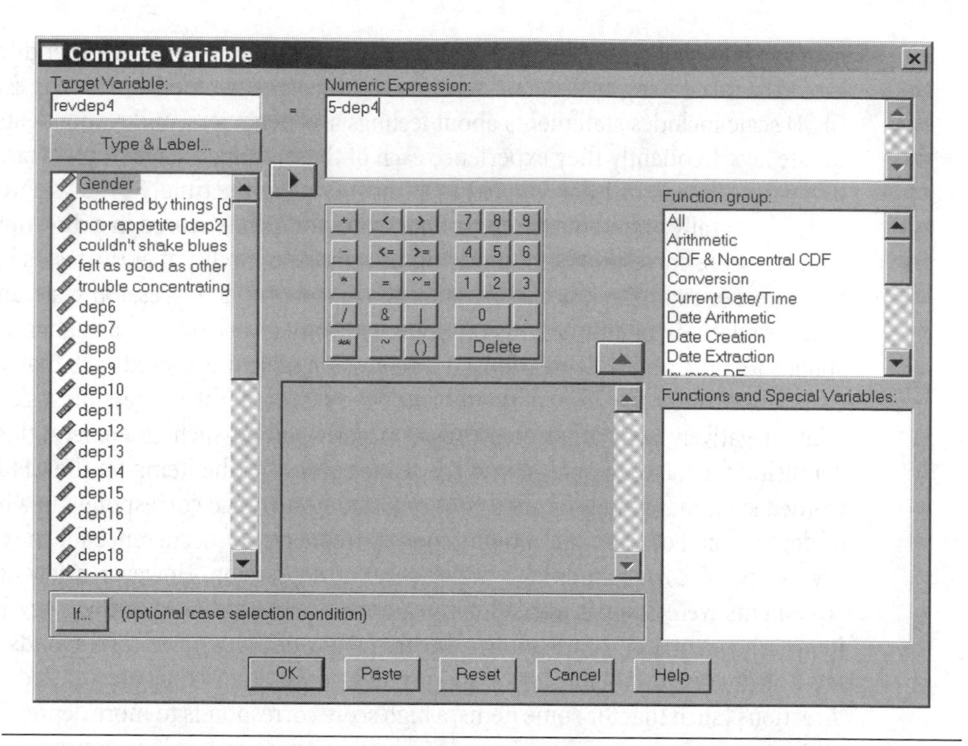

Figure 19.5 ♦ Computing a Reverse-Scored Variable for Dep4

for the reverse-coded score for each item (e.g., revdep4 is the reverse-coded score on Item dep4). If you change the direction of scoring by changing the original values and retain the original variable name (e.g., dep4 = 5 – dep4), it is easy to lose track of which items have been reverse scored.

Based on a preliminary examination of the data, the researcher evaluates whether the two assumptions required for simple summated scales are satisfied (i.e., scores on all the items are positively intercorrelated, and it makes sense to interpret all the items as measures of the same underlying construct or variable). If a FA yields a single factor that has high correlations with all the items or if a correlation matrix among all the items has correlations that are all positive and reasonably large in magnitude, this can be interpreted as evidence that the items might all be measures of the same underlying construct. If any of the items were reverse worded in the original survey, and the data analyst wants to form a simple summated scale, then the researcher must recode any reverse-scored items as necessary to make sure that all the items are scored in the same direction. In the present example, we want a higher number of points to correspond to higher levels of depression for all items.

There are three ways to form scales by summing items. The first two methods involve simple unit-weighted sums; the third method uses optimal weights that are derived from multivariate analyses such as FA or discriminant analysis (DA).

19.6.3 ♦ Method One: Simple Unit-Weighted Sum of Raw Scores

After recoding any reverse-worded items, you can create a total score for each scale by summing scores across items as shown in Figure 19.7. In this first example, a score for

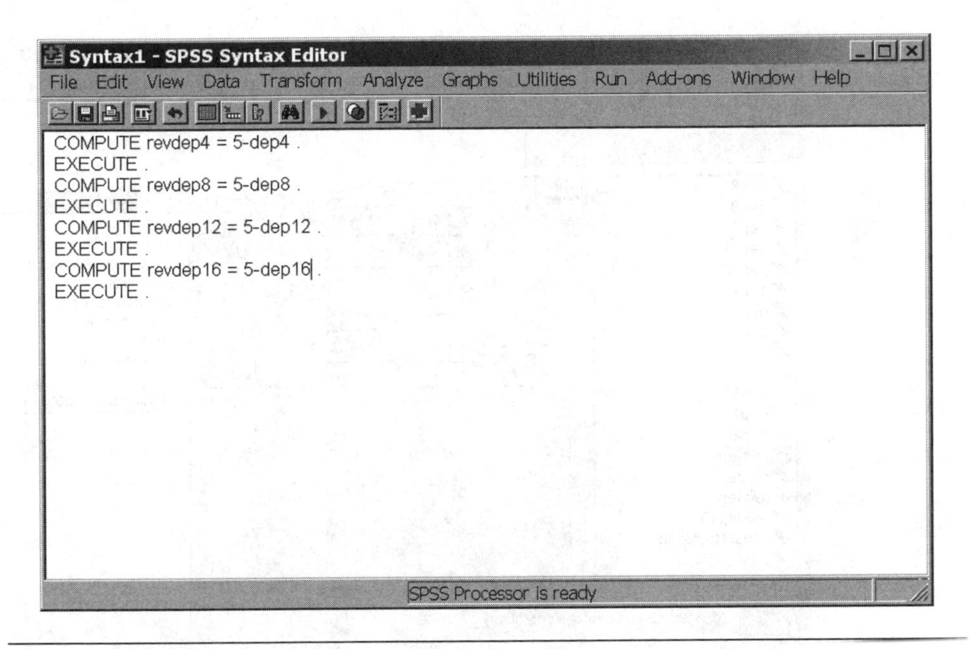

Figure 19.6 ♦ SPSS Syntax

NOTE: These commands compute new items (revdep4, revdep8, revdep12, and revdep16) that are reverse-scored versions of the original items dep4, dep8, dep12, and dep16.

selected items from the CESD scale was computed by summing the scores on Items 1 through 5 (with Item 4 reverse scored). The <Transform> and <Compute> menu selections open the SPSS Compute dialog window that appears in Figure 19.7. The name of the new variable (in this example, briefcesd) is placed in the left-hand side window under the Target Variable. The equation that specifies which scores are summed is placed in the Numeric Expression window. To form a score that is the sum of items named dep1 to dep5 (but using the reverse-scored version of Item 4 called revdep4), you can use the following numeric expression:

$$\text{briefcesd} = \text{dep1} + \text{dep2} + \text{dep3} + \text{revdep4} + \text{dep5}. \quad (19.6)$$

If an individual has a missing score on one or more individual items, this will result in a system missing code for the new scale total score. In this dataset, one participant had a system missing code on dep4 and revdep4; therefore, the number of scores is reduced from $N = 98$ in the entire SPSS data file to $N = 97$ for analyses that involve the variable briefcesd. If you want to obtain a score for people who have missing values on some items, you can use the "mean" function in SPSS; this returns the mean score, based on all nonmissing items. For example, if a person is missing a score on dep2, the numeric expression mean(dep1, dep2, dep3, revdep4, dep5) will return the mean for the scores on Items dep1, dep3, revdep4, and dep5. If you want to put the total score back into the units that you would have obtained by summing items, multiply the mean by the number of items in the scale (in this case, the number of items was 5).

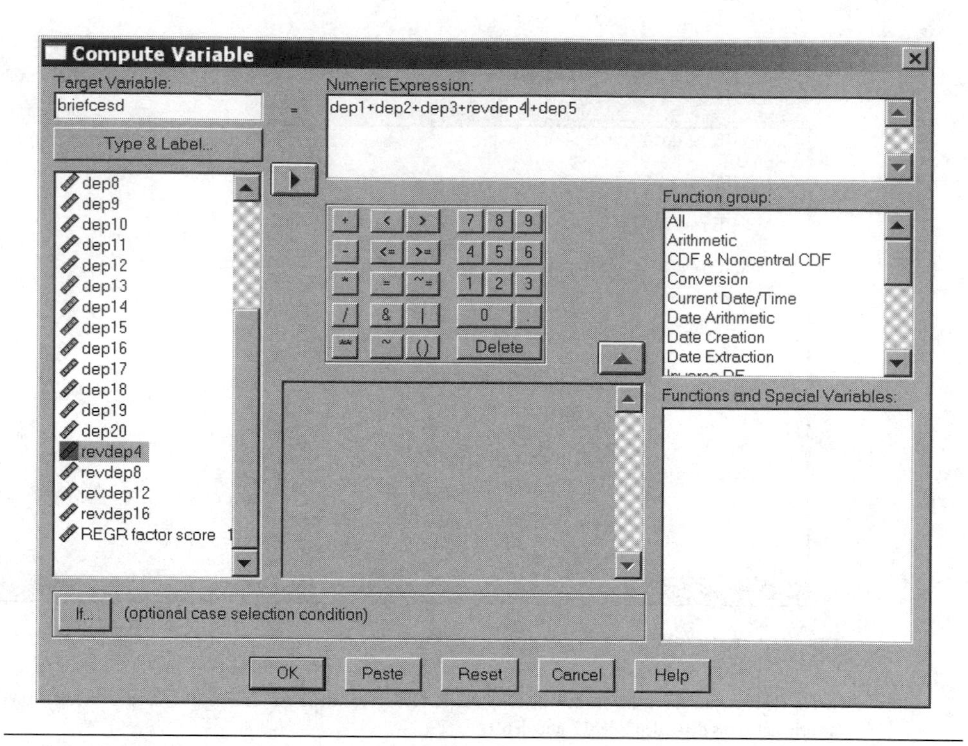

Figure 19.7 ♦ Computation of a Brief 5-Item Version of the Depression Scale

19.6.4 ♦ Method Two: Simple Unit-Weighted Sum of *z* Scores

Summing raw scores may be reasonable when the items are all scored using the same response alternatives or all measured in the same units. However, there are occasions when researchers want to combine information across variables that are measured in quite different units. Suppose a sociologist wants to create an overall index of socioeconomic status (SES) by combining information about the following measures: annual income in dollars, years of education, and occupational prestige rated on a scale from 0 to 100. If raw scores (in dollars, years, and points) were summed, the value of the total score would be dominated by the value of annual income. If we want to give these three factors (income, education, and occupational prestige) equal weight when we combine them, we can convert each variable to a *z* score or standard score and, then, form a unit-weighted composite of these *z* scores:

$$z_{\text{total}} = z_{X_1} + z_{X_2} + \cdots + z_{X_p}. \tag{19.7}$$

To create a composite of *z* scores on income, education, and occupational prestige so as to summarize information about SES, you could compute SES = $z_{\text{income}} + z_{\text{education}} + z_{\text{occupationprestige}}$.

19.6.5 ♦ Method Three: Optimally Weighted Linear Combinations

You may have noticed by now that whenever we encounter the problem of combining information from multiple predictor variables, as in multiple regression (MR), in DA, or from multiple outcome variables (as in MANOVA, multivariate analysis of variance), we have handled it in the same way: by forming a weighted linear composite of scores on the *X*s. This composite may be a sum of weighted raw scores or weighted *z* scores. For example, the goal in MR was to find the weighted linear composite $Y' = a + b_1 X_1 + b_2 X_2 + \ldots + b_k X_k$, such that the Y' values were as close as possible to the actual *Y* scores. The goal in DA was to find one (or several) discriminant function(s), $D_i = d_{i1} z_{X1} + d_{i2} z_{X2} + \ldots + d_{ip} z_{Xp}$, such that the new variable D_i had the largest possible SS_{between} and the smallest possible SS_{within}. The *b* coefficients in MR were often referred to as slopes because they describe slopes when the data are represented in the form of scatter plots; at this point, however, it is more helpful to think of coefficients such as *b* and *d* as *weights* that describe how much importance is given to scores on each predictor variable. Note also that the sign of the coefficient is important; some variables may be given positive weights and others negative weights when forming the optimally weighted linear composite.

In general, a multivariate analysis, such as regression, yields weights or coefficients that optimize the prediction of some outcome (such as scores for a quantitative *Y* outcome variable in a MR or group membership in a DA). We might apply this predictive equation to new cases, but in general, we cannot expect that predictions made using the regression or discriminant function coefficients will continue to be optimal when we apply these coefficients to data in new studies.

In factor analysis (Chapter 18), we also created new weighted linear composites of variables, called factor scores; these were created by summing *z* scores on all the variables using weights called factor score coefficients. The goal of FA was somewhat different from the goals of MR or DA. In MR and DA, the goal was to combine the scores on a set of *X* predictors in a manner that optimized the prediction of some outcome: a continuous *Y* score in MR or group membership in DA. In FA, the goal was to obtain a small set of factor scores, such that the scores on Factors 1, 2, and so forth are orthogonal to each other and in such a way that the *X* variables could be grouped into sets of variables, each of which had high correlations with one factor, and understood as measures of a small number of underlying factors or constructs. In other words, in FA, the correlations among *X*s were taken as a possible indication that the *X*s might measure the same construct, whereas in MR, correlations between *X*s and *Y* were usually interpreted to mean that *X* was useful as a predictor of *Y*, and we did not necessarily assume that the *X*s were measures of the same construct (although in some MR analyses, this might be the case).

All these multivariate methods (MR, DA, and FA) can calculate and save scores that are calculated using the optimal weighted linear combination of variables in that analysis; for example, factor scores can be saved as part of the results of a FA. It is possible to take the equation that was reported as the optimal solution (e.g., the standardized discriminant function coefficients in Chapter 16) and apply that equation to *z* scores in new data.

As a follow-up to a DA or an FA, a researcher might decide to create a composite or scale to summarize the information in several variables. One way to do this is to use the sample values of *d* (the coefficients for the standardized discriminant functions described in Chapter 16) or the factor score coefficients (described in Chapter 18) that provided the optimal prediction or fit in the sample that was used for the development of the model to form optimally weighted linear composites. However, these composites are optimized only for the sample for which they were calculated. Alternatively, the researcher might decide to group variables into three sets: those with large positive values for *d* or the factor score coefficients, those with coefficients close to 0, and those with large negative values for *d* or the factor score coefficients. The variables are assigned corresponding unit weights. For example, in DA, we might have an equation for a discriminant function as follows:

$$D_1 = .420 \times z_{\text{English}} + .015 \times z_{\text{reading}} + .443 \times z_{\text{mechanic}} - .116 \times z_{\text{abstract}} - .794 \times z_{\text{math}} + .748 \times z_{\text{office}}. \quad (19.8)$$

We could replace the optimally weighted linear composite given by Equation 19.8 with a simpler unit-weighted composite. In this example, the only variable with a large positive coefficient ($d > .5$ in absolute value) was office, and the only variable with a large negative coefficient ($d > .5$ in absolute value) was math; all other variables had d coefficients that were close to 0. Thus, we would assign a weight of +1 to z_{office}, a weight of −1 to z_{math}, and weights of 0 to all other variables to obtain this simple **unit-weighted linear composite** of the z scores (U_z). The letter U stands for "unit-weighted composite"; this is formed by summing scores with weights of +1, −1, or 0. The subscript for U indicates whether the sum is based on standard scores (as in Equation 19.9) or raw scores (as in Equation 19.10):

$$U_z = (-1) \times z_{\text{math}} + (+1) \times z_{\text{office}}. \quad (19.9)$$

Use of the z scores to create a composite ensures that the variances of the variables that are summed are scaled to be equal and, thus, that the variables are given equal weight in the new composite score. When the variables are measured in similar units and have approximately equal variances, it may be acceptable to combine the raw scores, as in the following example:

$$U_{\text{raw}} = (-1) \times \text{Math} + (+1) \times \text{Office} = \text{Office} - \text{Math}. \quad (19.10)$$

As we have seen in Chapter 16 (on DA), when we formed the simple composite variable U_{raw} = office − math, the differences among career groups on this unit-weighted composite were almost as large as the group differences on the optimally weighted linear composite given by Discriminant function 1 (D_1), specified by Equation 19.8. Similarly, in Chapter 18 (dealing with FA), the correlations between the factor scores (which were obtained by applying the factor score coefficients to z scores on all the Xs) and the unit-weighted composite scales (obtained by summing the raw scores for variables with high loadings on each factor) were in excess of .90. Fava and Velicer (1992) demonstrated that it often makes surprisingly little difference which of several possible methods are used to calculate factor scores, or whether we replace the optimally weighted linear combinations of variables that we obtain through multivariate analysis with simple unit-weighted combinations of raw scores or z scores; in their study, the correlations among linear composites obtained in these different ways averaged about .98.

19.6.6 ♦ Advantages and Disadvantages of Unit-Weighted Scores and Optimally Weighted Linear Composites

How do we decide whether to use unit-weighted composites (U_{raw} or U_z) or optimally weighted composites (such as discriminant function scores or saved factor scores) to combine the information across multiple items or variables? Each approach has potential advantages and disadvantages.

An advantage of using optimal weights (such as standardized discriminant function coefficients or factor score coefficients) when combining scores is that the performance of the weighted linear composite will be optimized for the sample of participants in the study. Discriminant function scores will show the largest possible differences across the groups in the study, while factor scores will (in theory) do the best possible job of summarizing the information in a set of p measured variables by creating scores on a limited number of factors. However, a disadvantage of using these optimal weights is that they are "optimal" only for the sample of participants whose data were used to estimate those coefficients. If we want to compute scores for new respondents, the discriminant function scores or factor score coefficients for these new participants will be less optimal in discriminating between groups or summarizing information about latent variables in the new group of participants than in the sample that was used to perform the original DA or FA. Another disadvantage of the use of optimal weights derived from multivariate analyses such as DA or FA is that this requires cumbersome computations. When a researcher creates a new multiple-item scale, it is often preferable to use simple methods of combining scores across items, so that the scale will be convenient for use by other researchers.

The advantage of using unit-weighted composites as the scoring method is that the unit-weighted scoring procedures are simpler to describe and apply in new situations. In many research situations, the scores that are obtained by applying optimal weights from a multivariate analysis are quite highly correlated with scores that are obtained using unit-weighted composites; when this is the case, it may make sense to prefer the unit-weighted composites because they provide essentially equivalent information and are easier to compute.

19.7 ♦ Assessment of Internal Homogeneity for Multiple-Item Measures

The internal consistency reliability of a multiple-item scale tells us the degree to which the items on the scale measure the same thing. If the items on a test all measure the same underlying construct or variable, and if all items are scored in the same direction, then the correlations among all the items should be positive. If we perform a FA on a set of five items that are all supposed to measure the same latent construct, we would expect the solution to consist of one factor that has large correlations with all five items on the test.

19.7.1 ♦ Factor Analysis of Five Items Selected From the CESD Scale

We could do a FA on all 20 items in the CESD scale. However, for this example, only Items 1 through 5 were included so as to keep the example brief and simple. Evaluation of results from a factor analysis provides a way to evaluate whether Items 1 through 5 on the CESD scale can all be interpreted as measures of the same underlying construct or latent variable. If all 5 items have large loadings or correlations with the same factor, it may make sense to interpret them all as measures of the same latent variable or underlying construct. A more detailed discussion of FA has appeared in Chapter 18. Only selected results are reported here.

The CESD scale items named dep1 through dep5 were entered in the FA (see the main Factor Analysis dialog window in Figure 19.8). The method of factor extraction requested was principal axis factoring. There was a user-determined decision to retain one factor (instead of relying on the default criterion to retain all factors with eigenvalues greater than 1). One factor was retained because, in theory, all these 5 items are supposed to be indicators of the same latent variable, depression. The SPSS Syntax for the FA appears in Figure 19.9 and the output from the FA appears in Figure 19.10.

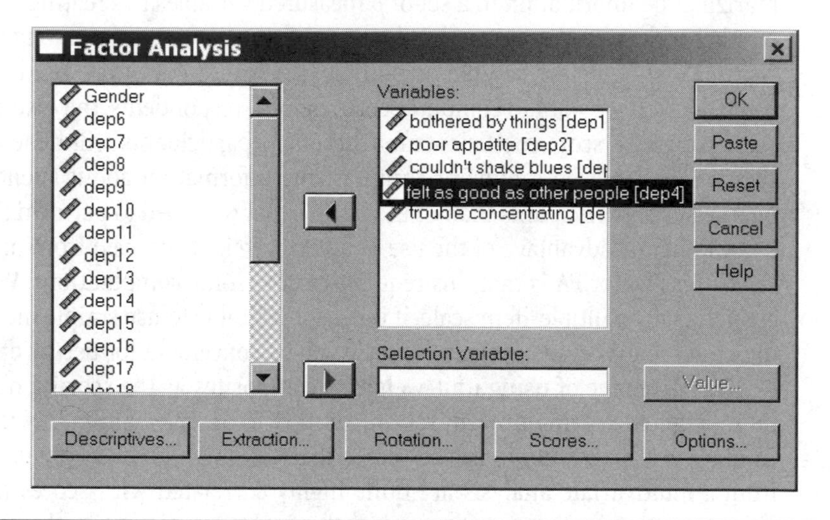

Figure 19.8 ♦ Factor Analysis of Items 1 Through 5 From the CESD Scales (See the Appendix to This Chapter for the Complete List of 20 Items)

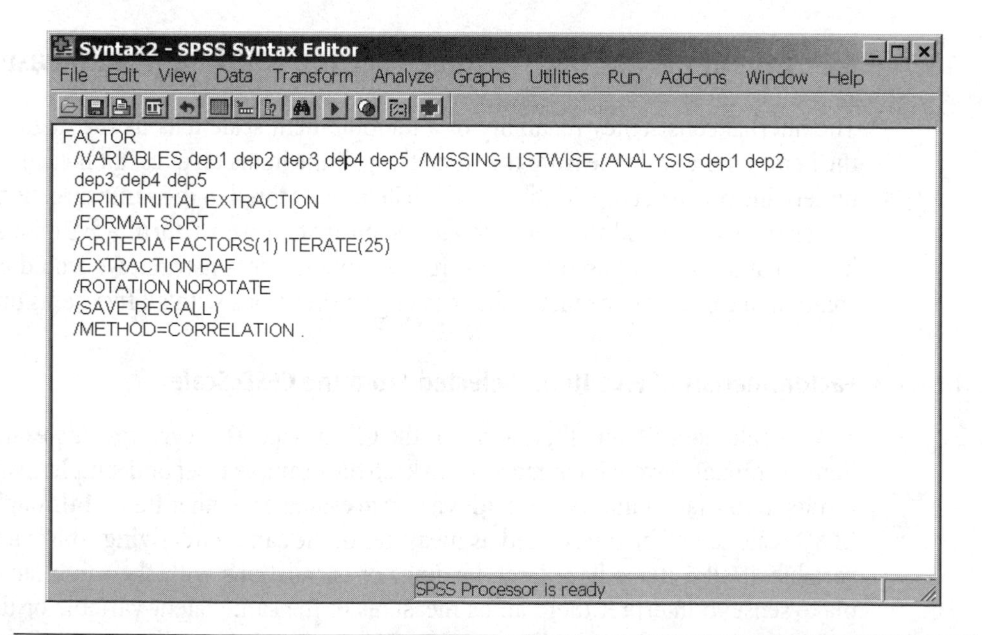

Figure 19.9 ♦ SPSS Syntax to Perform Principal Axis Factoring on Items Dep1 Through Dep5 From the CESD Scale and Retain One Factor

Communalities

	Initial	Extraction
bothered by things	.341	.456
poor appetite	.195	.247
couldn't shake blues	.434	.714
felt as good as other people	.237	.227
trouble concentrating	.055	.005

Extraction Method: Principal Axis Factoring.

Total Variance Explained

Factor	Initial Eigenvalues			Extraction Sums of Squared Loadings		
	Total	% of Variance	Cumulative %	Total	% of Variance	Cumulative %
1	2.170	43.400	43.400	1.649	32.982	32.982
2	1.089	21.775	65.175			
3	.737	14.743	79.918			
4	.605	12.109	92.026			
5	.399	7.974	100.000			

Extraction Method: Principal Axis Factoring.

Factor Matrix[a]

	Factor
	1
couldn't shake blues	.845
bothered by things	.676
poor appetite	.497
felt as good as other people	-.476
trouble concentrating	.070

Extraction Method: Principal Axis Factoring.
a. 1 factors extracted. 14 iterations required.

Figure 19.10 ♦ SPSS Output From Principal Axis FA of Items 1 Through 5 on the CESD Scale

The first factor accounted for approximately 43% of the variance (in the initial solution that included all five factors); after the model was reduced to one factor, the one retained factor accounted for approximately 33% of the variance. The unrotated factor loadings (which appear in Figure 19.10 under the heading "Factor Matrix") suggested the following interpretation. If we adopt an arbitrary cutoff of .40 in absolute value as the criterion for a "large" factor loading, then Items dep1 through dep4 all had large loadings on this single retained factor. For the items "couldn't shake blues," "bothered by things," and "poor appetite," the loadings or correlations were large and positive. The reverse-worded item dep4 ("I felt that I was just as good as other people") had a large negative correlation of −.48 on this single factor. The item "I had trouble concentrating" had a correlation with this retained factor that was close to 0. From this pattern of loadings or correlations we can make the following tentative inferences. As anticipated, the reverse-worded item dep4 ("felt as good as other people") correlated negatively with this factor. Prior to summing the scores on these five depression items, therefore, it is necessary to reverse the scoring on Item dep4. The item trouble concentrating has such a small correlation with the other 4 depression items in that it does not appear to measure the same latent variable as the other four. In later analyses, we will see additional evidence that this item trouble concentrating was not a "good" measure of depression; that is, it did not correlate highly with the other items in this set of questions about depression. To summarize, the very small factor loading for one of the items (trouble concentrating) suggests that that item may not be measuring the same construct as the other 4 items in this dataset; the negative sign associated with the loading or correlation associated with another item (dep4, felt as good as other people) indicates that this item is reverse scored relative to the other items (i.e., high scores on other items corresponded to higher levels of depression, but a high score on dep4 corresponded to a lower level of depression).

19.7.2 ♦ Cronbach Alpha Reliability Coefficient: Conceptual Basis

We can summarize information about positive intercorrelations between the items on a multiple-item test by calculating a Cronbach alpha reliability. The Cronbach alpha has become the most popular form of reliability assessment for multiple-item scales. As seen in an earlier section, as we sum a larger number of items for each participant, the expected value of $\sum e_i$ approaches 0, while the value of $p \times T$ increases. In theory, as the number of items (p) included in a scale increases, assuming other characteristics of the data remain the same, the reliability of the measure (the size of the $p \times T$ component compared with the size of the $\sum e$ component) also increases. The Cronbach alpha provides a reliability coefficient that tells us, in theory, how reliable our estimate of the "stable" entity that we are trying to measure is, when we combine scores from p test items (or behaviors or ratings by judges). The Cronbach alpha uses the mean of all the inter-item correlations (for all pairs of items or measures) to assess the stability or consistency of measurement.

The Cronbach alpha can be understood as a generalization of the **Spearman-Brown prophecy formula**; we calculate the mean inter-item correlation ($\bar{r}$) to assess the degree of agreement among individual test items, and then, we predict the reliability coefficient for a p-item test from the correlations among all these single-item measures. Another possible interpretation of the Cronbach alpha is that it is, essentially, the average of all possible split half reliabilities. Here is one formula for the Cronbach α from Carmines and Zeller (1979, p. 44):

$$\alpha = \frac{p\bar{r}}{[1 + \bar{r}(p-1)]}, \qquad (19.11)$$

where p is the number of items on the test and $\bar{r}$ the mean of the inter-item correlations.

The size of the Cronbach alpha depends on the following two factors:

As p (the number of items included in the composite scale) increases, and assuming that $\bar{r}$ stays the same, the value of the Cronbach alpha increases.

As $\bar{r}$ (the mean of the correlations among items or measures) increases, assuming that the number of items p remains the same, the Cronbach alpha increases.

It follows that we can increase the reliability of a scale by adding more items (but only if doing so does not decrease $\bar{r}$, the mean inter-item correlation) or by modifying items to increase $\bar{r}$ (either by dropping items with low item-total correlations or by writing new items that correlate highly with existing items). There is a trade-off: If the inter-item correlation is high, we may be able to construct a reasonably reliable scale with few items, and of course, a brief scale is less costly to use and less cumbersome to administer than a long scale. Note that all items must be scored in the same direction prior to summing. Items that are scored in the opposite direction relative to other items on the scale would have negative correlations with other items, and this would reduce the magnitude of the mean inter-item correlation.

Researchers usually hope to be able to construct a reasonably reliable scale that does not have an excessively large number of items. Many published measures of attitudes or personality traits include between 4 and 20 items for each trait. Ability or achievement tests (such as IQ) may require much larger numbers of measurements to produce reliable results.

Note that when the items are all dichotomous (such as true/false), the Cronbach alpha may still be used to assess the homogeneity of response across items. In this situation, it is sometimes called a **Kuder-Richardson 20 (KR-20)** reliability coefficient. However, the Cronbach alpha is not appropriate for use with items that have categorical responses with more than two categories.

19.7.3 ♦ Empirical Example: The Cronbach Alpha for Five Selected CESD Scale Items

Ninety-seven students filled out the 20-item CESD scale (items shown in the appendix to this chapter) as part of a survey. The names given to these 20 items in the SPSS data file that appears in Table 19.2 were dep1 to dep20. Questions 4, 8, 12, and 16 were reverse worded, and therefore, it was necessary to recode the scores on these items. The recoded values were placed in variables with the names revdep4, revdep8, revdep12, and revdep16. The SPSS reliability procedure was used to assess the internal consistency reliability of their responses. The value of the Cronbach alpha is an index of the internal consistency reliability of the depression score formed by summing the first 5 items. In this first example, only the first 5 items (dep1, dep2, dep3, revdev4, and dep5) were included. To run SPSS reliability, the following menu selections were made, starting from the top level menu for the SPSS data worksheet (see Figure 19.11): <Analyze> → <Scale> → <Reliability>.

The reliability procedure dialog box appears in Figure 19.12. The names of the 5 items on the CESD scale were moved into the variable list for this procedure. The Statistics button was clicked to request additional output; the Reliability Analysis: Statistics window appears in Figure 19.13. In this example, "Scale if item deleted" in the "Descriptives for" box and "Correlations" in the "Inter-Item" box were checked. The syntax for this procedure appears in Figure 19.14, and the output appears in Figure 19.15.

The Reliability Statistics panel in Figure 19.15 reports two versions of the Cronbach alpha statistic for the entire scale including all 5 items. For the sum dep1 + dep2 + dep3 + revdep4 + dep5, the Cronbach alpha estimates the proportion of the variance in this total that is due to $p \times T$, the part of the score that is stable or consistent for each participant across all 5 items. A score can be formed by summing raw scores (the sum of dep1,

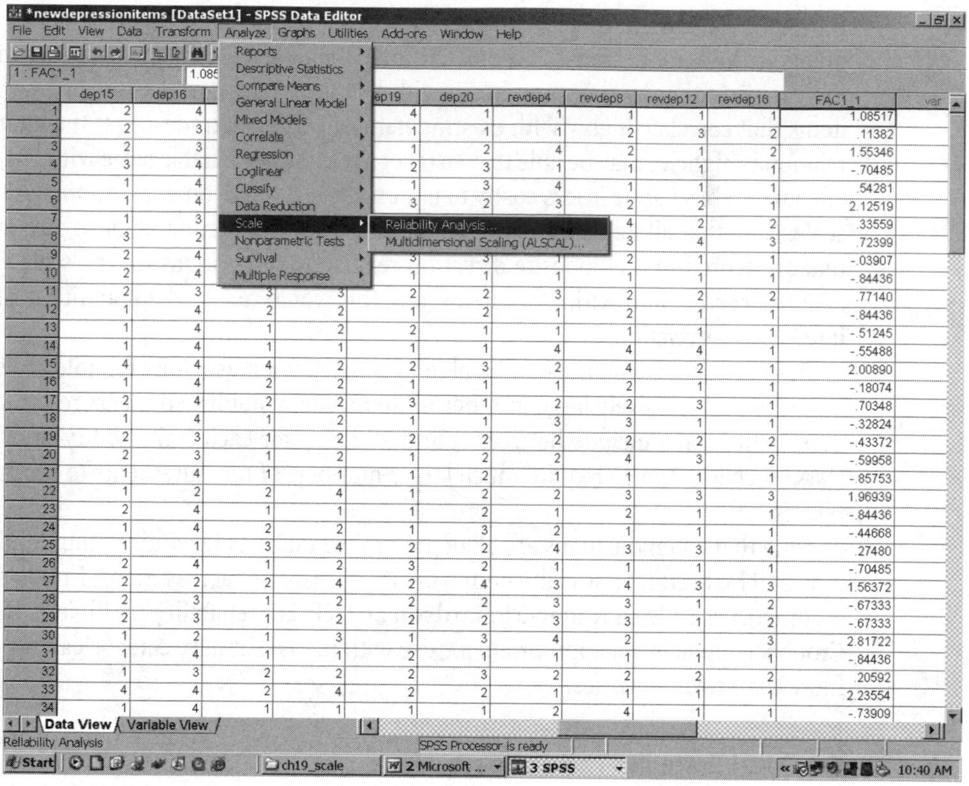

Figure 19.11 ♦ SPSS Menu Selections for the Reliability Procedure

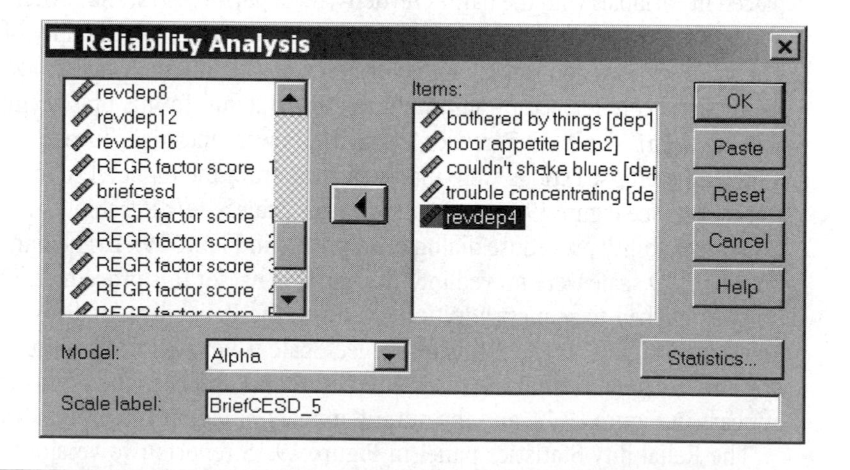

Figure 19.12 ♦ SPSS Reliability Analysis for 5 CESD Scale Items: Dep1, Dep2, Dep3, Dep5, and Revdep4

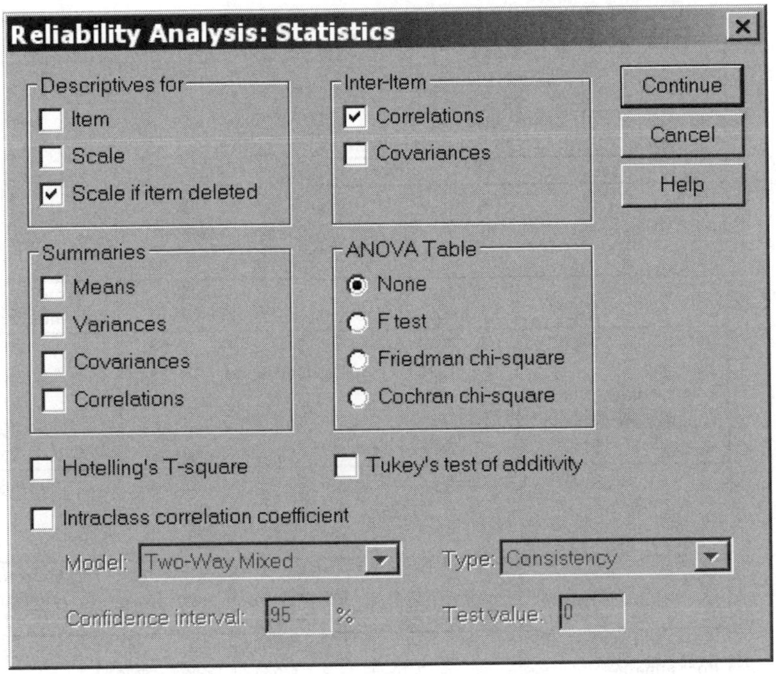

Reliability Analysis: Statistics

Descriptives for
- [ ] Item
- [ ] Scale
- [x] Scale if item deleted

Inter-Item
- [x] Correlations
- [ ] Covariances

Continue
Cancel
Help

Summaries
- [ ] Means
- [ ] Variances
- [ ] Covariances
- [ ] Correlations

ANOVA Table
- (•) None
- ( ) F test
- ( ) Friedman chi-square
- ( ) Cochran chi-square

- [ ] Hotelling's T-square
- [ ] Tukey's test of additivity
- [ ] Intraclass correlation coefficient

Model: Two-Way Mixed Type: Consistency

Confidence interval: 95 % Test value: 0

Figure 19.13 ♦ Statistics Selected for SPSS Reliability Analysis

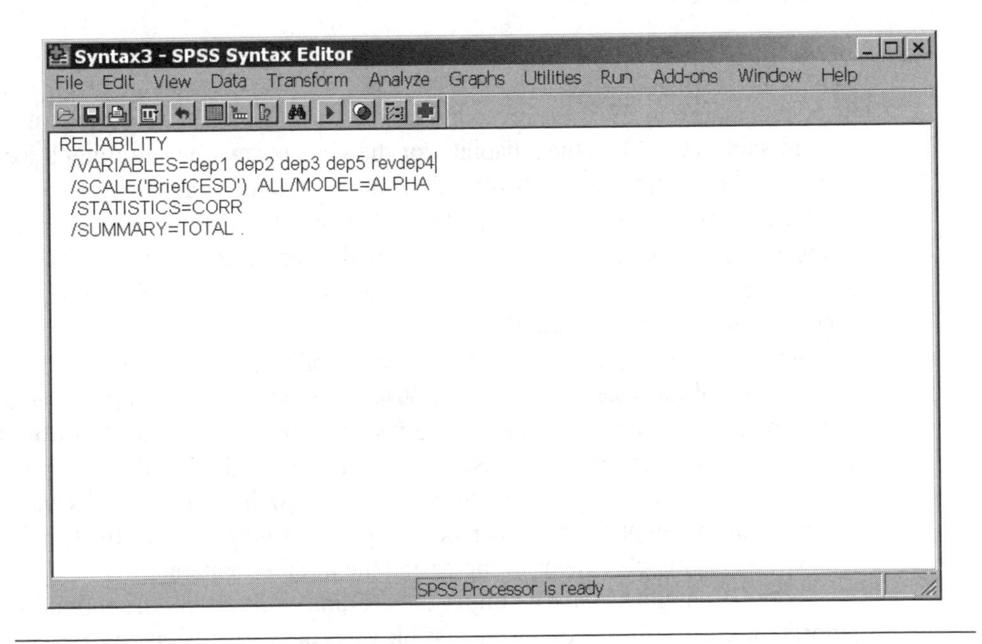

Figure 19.14 ♦ SPSS Syntax for Reliability Analysis

Reliability Statistics

Cronbach's Alpha	Cronbach's Alpha Based on Standardized Items	N of Items
.585	.614	5

Inter-Item Correlation Matrix

	bothered by things	poor appetite	couldn't shake blues	trouble concentrating	revdep4
bothered by things	1.000	.380	.555	.062	.302
poor appetite	.380	1.000	.394	.074	.213
couldn't shake blues	.555	.394	1.000	.115	.446
trouble concentrating	.062	.074	.115	1.000	-.129
revdep4	.302	.213	.446	-.129	1.000

Item-Total Statistics

	Scale Mean if Item Deleted	Scale Variance if Item Deleted	Corrected Item-Total Correlation	Squared Multiple Correlation	Cronbach's Alpha if Item Deleted
bothered by things	7.67	5.786	.511	.341	.455
poor appetite	7.70	5.941	.398	.195	.504
couldn't shake blues	7.61	4.845	.615	.434	.365
trouble concentrating	6.82	7.042	.032	.055	.703
revdep4	7.47	5.710	.294	.237	.562

Figure 19.15 ♦ SPSS Output From the First Reliability Procedure

NOTE: Scale: BriefCESD.

dep2, dep3, revdep4, and dep5), *z* scores, or standardized scores ($z_{dep1} + z_{dep2} + \ldots + z_{dep5}$). The first value, $\alpha = .59$, is the reliability for the scale formed by summing raw scores; the second value, $\alpha = .61$, is the reliability for the scale formed by summing *z* scores across items. In this example, these two versions of the Cronbach alpha (raw score and standardized score) are nearly identical. They generally differ from each other more when the items that are included in the sum are measured using different scales with different variances (as in the earlier example of an SES scale based on a sum of income, occupational prestige, and years of education).

Recall that the Cronbach alpha, like other reliability coefficients, can be interpreted as a proportion of variance. Approximately 60% of the variance in the total score for depression, which is obtained by summing the *z* scores on Items 1 through 5 from the CESD scale, is shared across these 5 items. A Cronbach α reliability coefficient of .61 would be considered unacceptably poor reliability in most research situations. Subsequent sections describe two different things researchers can do that may improve the Cronbach alpha reliability: deleting poor items or increasing the number of items.

A correlation matrix appears under the heading "Inter-Item Correlation Matrix." This reports the correlations between all possible pairs of items. If all items measure the same underlying construct, and if all items are scored in the same direction, then all the correlations in this matrix should be positive and reasonably large. Note that the same item that had a small loading on the depression factor in the preceding FA (trouble concentrating) also tended to have low or even negative correlations with the other 4 items. The Item-Total Statistics table shows how the statistics associated with the scale formed by summing all five items would change if each individual item were deleted from the scale. The Corrected Item-Total Correlation for each item is its correlation with the sum of the other 4 items in the scale; for example, for dep1, the correlation of dep1 with the "corrected total" (dep2 + dep3 + revdep4 + dep5) is shown. This total is called "corrected" because the score for dep1 is not included when we assess how dep1 is related to the total. If an individual item is a "good" measure, then it should be strongly related to the sum of all other items in the scale; conversely, a low item-total correlation is evidence that an individual item does not seem to measure the same construct as other items in the scale. The item that has the lowest item-total correlation with the other items is, once again, the question about trouble concentrating. This low item-total correlation is yet another piece of evidence that this item does not seem to measure the "same thing" as the other 4 items in this scale.

The last column in the Item-Total Statistics table reports Cronbach's Alpha if Item Deleted; that is, what is the Cronbach alpha for the scale if each individual item is deleted? For the item that corresponded to the question trouble concentrating, deletion of this item from the scale would increase the Cronbach α to .70. Sometimes the deletion of an item that has low correlations with other items on the scale results in an increase in α reliability. In this example, we can obtain slightly better reliability for the scale if we drop the item trouble concentrating, which tends to have small correlations with other items on this depression scale; the sum of the remaining 4 items has a Cronbach α of .70, which represents slightly better reliability.

19.7.4 ♦ Improving the Cronbach Alpha by Dropping a "Poor" Item

The SPSS reliability procedure was performed on the reduced set of 4 items: dep1, dep2, dep3, and revdep4. The output from this second reliability analysis (in Figure 19.16) shows that the reduced 4-item scale had Cronbach α reliabilities of .703 (for the sum of raw scores) and .712 (for the sum of *z* scores). A review of the column headed "Cronbach's Alpha if Item Deleted" in the new Item-Total Statistics table indicates that the reliability of the scale would become lower if any additional items were deleted from the scale. Thus, we have obtained slightly better reliability from the 4-item version of the scale (Figure 19.16) than for a 5-item version of the scale (Figure 19.15). The 4-item scale had better reliability because the mean inter-item correlation was higher after the item trouble concentrating was deleted.

19.7.5 ♦ Improving the Cronbach Alpha by Increasing the Number of Items

Other factors being equal, Cronbach alpha reliability tends to increase as p, the number of items in the scale, increases. For example, we obtain a higher Cronbach alpha when we use all 20 items in the full-length CESD scale than when we examine just the first 5 items. The output from the SPSS reliability procedure for the full 20-item CESD scale (with Items 4, 8, 12, and 16 reverse scored) appears in Figure 19.17. For the full scale formed by summing scores across all 20 items, the Cronbach α was .88.

Reliability Statistics

Cronbach's Alpha	Cronbach's Alpha Based on Standardized Items	N of Items
.703	.712	4

Inter-Item Correlation Matrix

	bothered by things	poor appetite	couldn't shake blues	revdep4
bothered by things	1.000	.380	.555	.302
poor appetite	.380	1.000	.394	.213
couldn't shake blues	.555	.394	1.000	.446
revdep4	.302	.213	.446	1.000

Item-Total Statistics

	Scale Mean if Item Deleted	Scale Variance if Item Deleted	Corrected Item-Total Correlation	Squared Multiple Correlation	Cronbach's Alpha if Item Deleted
bothered by things	5.18	4.625	.541	.341	.617
poor appetite	5.21	4.811	.407	.194	.686
couldn't shake blues	5.11	3.810	.633	.421	.542
revdep4	4.98	4.166	.410	.204	.702

Figure 19.16 ♦ Output for the Second Reliability Analysis: Scale Reduced to Four Items

NOTE: Item trouble concentrating has been dropped.

19.7.6 ♦ A Few Other Methods of Reliability Assessment for Multiple-Item Measures

19.7.6.1 ♦ *Split-Half Reliability*

A **split-half reliability** for a scale with p items is obtained by dividing the items into two sets (each with $p/2$ items). This can be done randomly or systematically; for example, the first set might consist of odd-numbered items and the second set might consist of even-numbered items. Separate scores are obtained for the sum of the Set 1 items (X_1) and the sum of the Set 2 items (X_2), and a Pearson r (r_{12}) is calculated between X_1 and X_2. However, this r_{12} correlation between X_1 and X_2 is the reliability for a test with only $p/2$ items; if we want to know the reliability for the full test that consists of twice as many items (all p items, in this example), we can "predict" the reliability of the longer test using the Spearman-Brown prophecy formula (Carmines & Zeller, 1979):

$$r_{XX} = \frac{2 \times r_{12}}{1 + r_{12}}, \qquad (19.12)$$

where r_{12} is the correlation between the scores based on split-half versions of the test (each with $p/2$ items), and r_{XX} is the reliability for a score based on all p items.

Case Processing Summary

		N	%
Cases	Valid	94	95.9
	Excluded [a]	4	4.1
	Total	98	100.0

a. Listwise deletion based on all variables in the procedure.

Reliability Statistics

Cronbach's Alpha	N of Items
.880	20

Figure 19.17 ♦ SPSS Output: Cronbach Alpha Reliability for the 20-Item CESD Scale

Depending on the way in which items are divided into sets, the value of the split-half reliability can vary. The Cronbach alpha can be interpreted as the mean of all possible different split-half reliabilities.

19.7.6.2 ♦ *Parallel Forms Reliability*

Sometimes it is desirable to have two versions of a test that include different questions but that yield comparable information; these are called parallel forms. Parallel forms of a test, such as the Eysenck Personality Inventory, are often designated Form A and Form B. Parallel forms are particularly useful in repeated measures studies where we would like to test some ability or attitude on two occasions, but we want to avoid repeating exactly the same questions. **Parallel forms reliability** is similar to split-half reliability, except that when parallel forms are developed, more attention is paid to matching items so that the two forms contain similar types of questions. For example, consider Eysenck's Extraversion scale. Both Form A and Form B include similar numbers of items that assess each aspect of extraversion—for instance, enjoyment of social gatherings, comfort in talking with strangers, sensation seeking, and so forth. A Pearson r between scores on Form A and Form B is a typical way of assessing reliability; in addition, however, a researcher wants scores on Form A and Form B to yield the same means, variances, and so forth, so these should also be assessed.

19.8 ♦ Correlations Among Scores Obtained Using Different Methods of Summing Items

In many empirical situations, it makes surprisingly little difference to the outcome whether the researcher uses weights derived from a multivariate analysis (such as MR, DA, MANOVA, or FA) or simply forms a unit-weighted composite of scores (Wainer, 1976). As the title of the classic paper by Wainer (1976) suggests, in many situations "It don't make no nevermind" whether researchers combine scores by just summing them, or by forming weighted composites. Simple unit-weighted composites and weighted composites that use coefficients from multivariate models sometimes yield nearly equivalent information, in practice.

The following example is a demonstration, not a formal proof. For the set of 5 items selected from the CESD scale, a summary scale was computed using each of the following methods:

Method 1: briefCESD = dep1 + dep2 + dep3 + revdep4 + dep5

Method 2: zbriefCESD = zdep1 + zdep2 + zdep3 + zrevdep4 + zdep5

Method 3: REGR factor score: Saved score on Factor 1 from the FA of 5 CESD scale items (see Figures 19.8 through 19.10).

Correlations among these three different versions of the 5-item depression scale were obtained; the results appear in Figure 19.18. For this set of data, all the intercorrelations among these scoring methods (sum of raw scores, sum of *z* scores, and saved factor scores) were greater than .90. This is consistent with the observation by Fava and Velicer (1992) and Wainer (1976) that, for some purposes, it may not make much difference whether we choose to use optimally weighted composites, such as factor scores, or simple unit-weighted composites obtained by summing raw scores or *z* scores. These different methods of scoring sometimes result in nearly equivalent summaries of the information in a set of items. Of course, a researcher should not assume that these three methods of scoring will always yield highly similar results, but it is easy to obtain correlations among scores derived from different scoring methods. When all three methods of scoring described here yield essentially equivalent information, then it may be sufficient to provide other users of the measurement with the simplest possible scoring instructions (i.e., simply sum the raw scores across items, making sure that scores on any reverse-worded items are appropriately recoded).

19.9 ♦ Validity Assessment

Validity of a measurement essentially refers to whether the measurement really measures what it purports to measure. In psychological and educational measurement, the degree to which scores on a measure correspond to the underlying construct that the measure is supposed to assess is called **construct validity**. (Some textbooks used to list construct validity as one of several types of measurement validity; in recent years, many authors use the term *construct validity* to subsume all the forms of validity assessment described below.)

For some types of measurement (such as direct measurements of simple physical characteristics), validity is reasonably self-evident. If a researcher uses a tape measure to obtain information about people's heights (whether the measurements are reported in centimeters, inches, feet, or other units), the researcher does not need to go to great lengths to persuade readers that this type of measurement is valid. However, there are many situations where the characteristic of interest is not directly observable, and researchers can only obtain indirect information about it. For example, we cannot directly observe intelligence (or depression); but we may infer that a person is intelligent (or depressed) if he or she

Correlations

		zbriefcesd	briefcesd	REGR factor score 1 for analysis 1
zbriefcesd	Pearson Correlation	1	.943**	.915**
	Sig. (2-tailed)		.000	.000
	N	97	97	97
briefcesd	Pearson Correlation	.943**	1	.947**
	Sig. (2-tailed)	.000		.000
	N	97	97	97
REGR factor score 1 for analysis 1	Pearson Correlation	.915**	.947**	1
	Sig. (2-tailed)	.000	.000	
	N	97	97	97

**. Correlation is significant at the 0.01 level (2-tailed).

Figure 19.18 ♦ Correlations Among Three Versions of the CESD Scale

gives certain types of responses to large numbers of questions that researchers agree are diagnostic of intelligence (or depression). A similar problem arises in medicine, for example, in the assessment of blood pressure. Arterial blood pressure could be measured directly by shunting the blood flow out of the person's artery through a pressure measurement system, but this procedure is invasive (and generally, less invasive measures are preferred). The commonly used method of blood pressure assessment uses an arm cuff; the cuff is inflated until the pressure in the cuff is high enough to occlude the blood flow; a human listener (or a microphone attached to a computerized system) listens for sounds in the brachial artery while the cuff is deflated. At the point when the sounds of blood flow are detectable (the Korotkoff sounds), the pressure on the arm cuff is read, and this number is used as the index of systolic blood pressure—that is, the blood pressure at the point in the cardiac cycle when the heart is pumping blood into the artery. The point of this example is that this common blood pressure measurement method is quite indirect; research had to be done to establish that measurements taken in this manner were highly correlated with measurements obtained more directly by shunting blood from a major artery into a pressure detection system. Similarly, it is possible to take satellite photographs and use the colors in these images to make inferences about the type of vegetation on the ground, but it is necessary to do validity studies to demonstrate that the type of vegetation that is identified using satellite images corresponds to the type of vegetation that is seen when direct observations are made at ground level.

As these examples illustrate, it is quite common in many fields (such as psychology, medicine, and natural resources) for researchers to use rather indirect assessment methods—either because the variable in question cannot be directly observed or because direct observation would be too invasive or too costly.

In cases such as these, whether the measurements are made through self-report questionnaires, by human observers, or by automated systems, validity cannot be assumed; we need to obtain evidence to show that measurements are valid.

For self-report questionnaire measurements, two types of evidence are used to assess validity. One type of evidence concerns the content of the questionnaire (**content** or **face validity**); the other type of evidence involves correlations of scores on the questionnaire with other variables (criterion-oriented validity).

19.9.1 ♦ Content and Face Validity

Both content and face validity are concerned with the content of the test or survey items. Content validity involves the question whether test items represent all theoretical dimensions or content areas. For example, if depression is theoretically defined to include low self-esteem, feelings of hopelessness, thoughts of suicide, lack of pleasure, and physical symptoms of fatigue, then a content-valid test of depression should include items that assess all these symptoms. Content validity may be assessed by mapping out the test contents in a systematic way and matching them to elements of a theory or by having expert judges decide whether the content coverage is complete.

A related issue is whether the instrument has *face validity*; that is, does it appear to measure what it says it measures? Face validity is sometimes desirable, when it is helpful for test takers to be able to see the relevance of the measurements to their concerns, as in some evaluation research studies where participants need to feel that their concerns are being taken into account.

If a test is an assessment of knowledge (e.g., knowledge about dietary guidelines for blood glucose management for diabetic patients), then content validity is crucial. Test questions should be systematically chosen so that they provide reasonably complete coverage of the information (e.g., What are the desirable goals for the proportions and amounts of carbohydrate, protein, and fat in each meal? When blood sugar is tested before and after meals, what ranges of values would be considered normal?).

When a psychological test is intended for use as a clinical diagnosis (of depression, for instance), clinical source books such as the *Diagnostic and Statistical Manual of Mental Disorders* (*DSM-IV*) might be used to guide item selection, to ensure that all relevant facets of depression are covered. More generally, a well-developed theory (about ability, personality, mood, or whatever else is being measured) can help a researcher map out the domain of behaviors, beliefs, or feelings that questions should cover to have a content-valid and comprehensive measure.

However, sometimes, it is important that test takers should *not* be able to guess the purpose of the assessment, particularly in situations where participants might be motivated to "fake good," "fake bad," lie, or give deceptive responses. There are two types of psychological tests that (intentionally) do not have high face validity: **projective tests** and empirically keyed objective tests. One well-known example of a projective test is the Rorschach test, in which people are asked to say what they see when they look at ink blots; a diagnosis of psychopathology is made if responses are bizarre. Another is the Thematic Apperception Test, in which people are asked to tell stories in response to ambiguous pictures; these stories are scored for themes such as need for achievement and need for affiliation. In projective tests, it is usually not obvious to participants what motives are being assessed, and because of this, test takers should not be able to engage in impression management or faking. Thus, projective tests intentionally have low face validity.

Some widely used psychological tests were constructed using **empirical keying** methods; that is, test items were chosen because the responses to those questions were empirically related to a psychiatric diagnosis (such as depression), even though the question did not appear to have anything to do with depression. For example, persons diagnosed with depression tend to respond "False" to the MMPI (Minnesota Multiphasic Personality Inventory) item "I sometimes tease animals"; this item was included in the MMPI depression scale because the response was (weakly) empirically related to a diagnosis of depression, although the item does not appear face valid as a question about depression (Wiggins, 1973).

Face validity can be problematic; people do not always agree about what underlying characteristic(s) a test question measures. Gergen, Hepburn, and Fisher (1986) demonstrated that when items taken from one psychological test (the Rotter Internal/External Locus of Control scale) were presented to people out of context and people were asked to say what trait they thought the questions assessed, they generated a wide variety of responses.

19.9.2 ♦ Criterion-Oriented Types of Validity

Content validity and face validity are assessed by looking inside a test to see what material it contains and what the questions appear to measure. Criterion-oriented validity is assessed by examining correlations of scores on the test with scores on other variables that should be related to it if the test really measures what it purports to measure. If the CESD scale really is a valid measure of depression, for example, scores on this scale should be correlated with scores on other existing measures of depression that are thought to be valid, and they should predict behaviors that are known or theorized to be associated with depression.

19.9.2.1 ♦ *Convergent Validity*

Convergent validity is assessed by checking to see if scores on a new test of some characteristic *X* correlate highly with scores on existing tests that are believed to be valid measures of that same characteristic. For example, do scores on a new brief IQ test correlate highly with scores on well-established IQ tests such as the WAIS or the Stanford-Binet? Are scores on the CESD scale closely related to scores on other depression measures such as the BDI? If a new measure of a construct has reasonably high correlations with existing measures that are generally viewed as valid, this is evidence of *convergent validity*.

19.9.2.2 ♦ *Discriminant Validity*

Equally important, scores on *X* should *not* correlate with things the test is not supposed to measure (**discriminant validity**). For instance, researchers sometimes try to demonstrate that scores on a new test are *not* contaminated by social desirability bias by showing that these scores are not significantly correlated with scores on the Crown-Marlowe Social Desirability scale or other measures of social desirability bias.

19.9.2.3 ♦ *Concurrent Validity*

As the name suggests, concurrent validity is evaluated by obtaining correlations between scores on the test with current behaviors or current group memberships. For example, if persons who are currently clinically diagnosed with depression have higher mean scores on the CESD scale than persons who are not currently diagnosed with depression, this would be one type of evidence for concurrent validity.

19.9.2.4 ♦ *Predictive Validity*

Another way of assessing validity is to ask whether scores on the test predict future behaviors or group membership. For example, are scores on the CESD scale higher for persons who later commit suicide than for people who do not commit suicide?

19.9.3 ♦ Construct Validity: Summary

Many types of evidence (including content, convergent, discriminant, concurrent, and **predictive validity**) may be required to establish that a measure has strong construct validity—that is, that it really measures what the test developer says it measures, and it predicts the behaviors and group memberships that it should be able to predict. Westen and Rosenthal (2003) suggested that researchers should compare a matrix of obtained validity coefficients or correlations with a target matrix of predicted correlations and compute a summary statistic to describe how well the observed pattern of correlations matches the predicted pattern. This provides a way of quantifying information about construct validity based on many different kinds of evidence.

Although the preceding examples have used psychological tests, validity questions certainly arise in other domains of measurement. For example, referring to the example discussed earlier, when the colors in satellite images are used to make inferences about the types and amounts of vegetation on the ground, are those inferences correct? Indirect assessments are sometimes used because they are less invasive (e.g., as discussed earlier, it is less invasive to use an inflatable arm cuff to measure blood pressure) and sometimes because they are less expensive (broad geographical regions can be surveyed more quickly by taking satellite photographs than by having observers on the ground). Whenever indirect methods of assessment are used, validity assessment is required.

Multiple-item assessments of some variables (such as depression) may be useful or even necessary to achieve validity as well as reliability. How can we best combine information from multiple measures? This brings us back to a theme that has arisen repeatedly throughout the book; that is, we can often summarize the information in a set of p variables or items by creating a weighted linear composite or, sometimes, just a unit weight sum of scores for the set of p variables.

19.10 ♦ Typical Scale Development Study

If an existing multiple-item measure is available for the variable of interest, such as depression, it is usually preferable to employ an existing measure for which we have good evidence about reliability and validity. However, occasionally, a researcher would like to

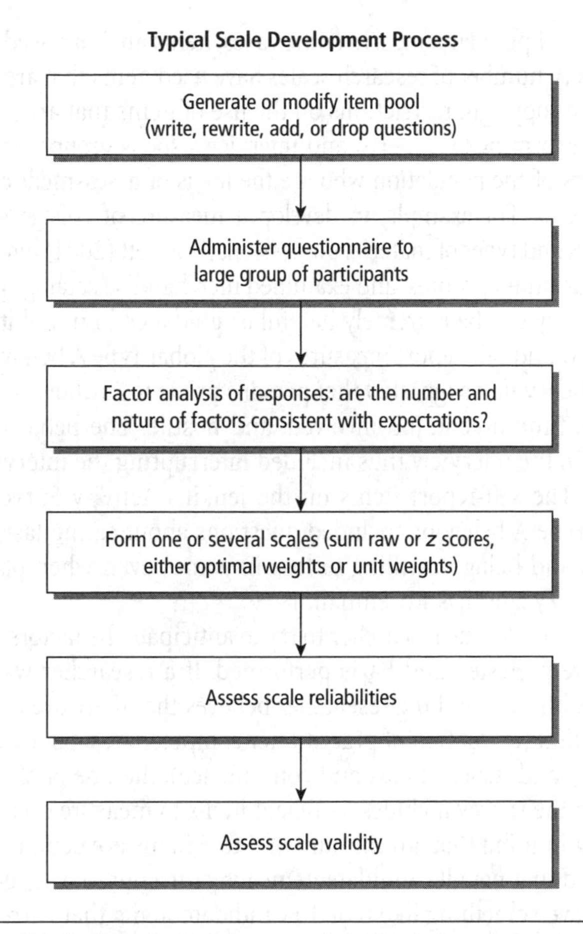
Typical Scale Development Process
Generate or modify item pool
(write, rewrite, add, or drop questions)
Administer questionnaire to
large group of participants
Factor analysis of responses: are the number and
nature of factors consistent with expectations?
Form one or several scales (sum raw or z scores,
either optimal weights or unit weights)
Assess scale reliabilities
Assess scale validity

Figure 19.19 ♦ Possible Steps in the Development of a Multiple-Item Scale

develop a measure for some construct that has not been measured before or develop a different way of measuring a construct for which the existing tests are flawed. An outline of a typical research process for scale development appears in Figure 19.19. In this section, the steps included in this diagram are discussed briefly. Although the examples provided involve self-report questionnaire data, comparable issues are involved in combining physiological measures or observational data.

19.10.1 ♦ Generating and Modifying the Pool of Items or Measures

When a researcher sets out to develop a measure for a new construct (for which there are no existing measures) or a different measure in a research domain where other measures have been developed, the first step is the generation of a pool of "candidate" items. There are many ways in which this can be done. For example, to develop a set of self-report items to measure "Machiavellianism" (a cynical, manipulative attitude toward people), Christie and Geis (1970) drew on the writings of Machiavelli for some items (and also on statements by P. T. Barnum, another notable cynic). To develop measures of love, Rubin (1970) drew on writings about love that ranged from the works of classic poets to the lyrics of popular songs. In some cases, items are borrowed from existing measures; for example, a number of research scales have used items that are part of the MMPI. However, there are copyright restrictions on the use of items that are part of **published tests**.

Brainstorming by experts, and interviews, focus groups, or open-ended questions with members of the population who are the focus of assessment can also provide useful ideas about items. For example, to develop a measure of college student life space, including numbers and types of material possessions, Brackett (2004) interviewed student informants, visited dormitory rooms, and examined merchandise catalogs popular in that age group.

A theory can be extremely helpful as guidance in initial item development. The early interview and self-report measures of the global Type A behavior pattern drew on a developing theory that suggested that persons prone to cardiovascular disease tend to be competitive, time urgent, job-involved, and hostile. The behaviors that were identified for coding in the interview thus included interrupting the interviewer and loud or explosive speech. The self-report items on the Jenkins Activity Survey, a self-report measure of global Type A behavior, included questions about eating fast, never having time to get a haircut, and being unwilling to lose in games even when playing checkers with a child (Jenkins, Zyzanski, & Rosenman, 1979).

It is useful for the researcher to try to anticipate the factors that will emerge when these items are pretested and FA is performed. If a researcher wants to measure satisfaction with health care, and the researcher believes that there are three separate components to satisfaction (evaluation of practitioner competence, satisfaction with rapport or "bedside manner," and issues of cost and convenience), then he or she should pause and evaluate whether the survey includes sufficient items to measure each of these three components. Keeping in mind that a minimum of 4 to 5 items are generally desired for each factor or scale and that not all candidate items may turn out to be good measures, it may be helpful to have something like 8 or 10 candidate items that correspond to each construct or factor that the researcher wants to measure.

19.10.2 ♦ Administer Survey to Participants

The survey containing all the candidate items should be pilot tested on a relatively small sample of participants; it may be desirable to interview or debrief participants to find out whether items seemed clear and plausible and whether response alternatives covered all the options people might want to report. A pilot test can also help the researcher judge how long it will take for participants to complete the survey. After making any changes judged necessary based on the initial pilot tests, the survey should be administered to a sample that is large enough to be used for FA (see Chapter 18 for sample size recommendations). Ideally, these participants should vary substantially on the characteristics that the scales are supposed to measure (because a restricted range of scores on T, the component of the X measures that taps stable individual differences among participants, will lead to lower inter-item correlations and lower scale reliabilities).

19.10.3 ♦ Factor Analyze Items to Assess the Number and Nature of Latent Variables or Constructs

Using the methods described in Chapter 18, FA can be performed on the scores. If the number of factors that are obtained and the nature of the factors (i.e., the groups of variables that have high loadings on each factor) are consistent with the researcher's expectations, then the researcher may want to go ahead and form one scale that corresponds to each factor. If the FA does not turn out as expected, for example, if the number of factors is different from what was anticipated or if the pattern of variables that load on each factor is not as expected, the researcher needs to make a decision. If the researcher wants to make the FA more consistent with a priori theoretical constructs, it may be necessary to go back to Step 1 to revise, add, and drop items. If the researcher sees patterns in the data that were not anticipated from theoretical evaluations (but the patterns make sense), he or she may want to use the empirical factor solution (instead of the original conceptual model) as a basis for grouping items into scales. Also, if a factor that was not anticipated emerges in the FA, but there are only a few items to represent that factor, the researcher may want to add or revise items to obtain a better set of questions for the new factor.

In practice, a researcher may have to go through these first three sets several times; that is, the researcher may run FA, modify items, gather additional data, and run a new FA several times until the results of the FA are clear, and the factors correspond to meaningful groups of items that can be summed to form scales.

Note that some scales are developed based on the predictive utility of items rather than on the factor structure; for these, DA (rather than FA) might be the data reduction method of choice. For example, items included in the Jenkins Activity Survey (Jenkins et al., 1979) were selected because they were useful predictors of a person having a future heart attack.

19.10.4 ♦ Development of Summated Scales

After FA (or DA), the researcher may want to form scales by combining scores on multiple measures or items. There are numerous options at this point.

1. One or several scales may be created (depending on whether the survey or test measures just one construct or several separate constructs).
2. Composition of scales (i.e., selection of items) may be dictated by conceptual grouping of items or by empirical groups of items that emerge from FA. In most scale development research, researchers hope that the items that are grouped to form scales can be justified both conceptually and empirically.
3. Scales may involve combining raw scores or standardized scores (z scores) on multiple items. Usually, if the variables use drastically different measurement units (as in the example above where an SES index was formed by combining income, years of education, and occupational prestige rating), z scores are used to ensure that each variable has equal importance.
4. Scales may be based on sums or means of scores across items.
5. Weights used to combine scores may be optimal weights (e.g., the factor score coefficients obtained through a FA) or unit weights ($+1, 0, -1$).

19.10.5 ♦ Assess Scale Reliability

At a minimum, the internal consistency of each scale is assessed, usually by obtaining a Cronbach alpha. Test-retest reliability should also be assessed if the construct is something that is expected to remain reasonably stable across time (such as a personality trait), but high test-retest reliability is not a requirement for measures of things that are expected to be unstable across time (such as moods).

19.10.6 ♦ Assess Scale Validity

If there are existing measures of the same theoretical construct, the researcher assesses convergent validity by checking to see whether scores on the new measure are reasonably highly correlated with scores on existing measures. If the researcher has defined the construct as something that should be independent of verbal ability or not influenced by social desirability, the researcher should assess discriminant validity by making sure that correlations with measures of verbal ability and social desirability are close to 0. To assess concurrent and predictive validity, scores on the scale can be used to predict current or future group membership and current or future behaviors, which it should be able to predict. For example, scores on Zick Rubin's Love Scale (Rubin, 1970) were evaluated to see if they predicted self-rated likelihood that the relationship would lead to marriage and whether scores predicted which dating couples would split up and which ones would stay together within the year or two following the initial survey.

19.10.7 ♦ Iterative Process

At any point in this process, if results are not satisfactory, the researcher may "cycle back" to an earlier point in the process; for example, if the factors that emerge from FA are not clear or if internal consistency reliability of scales is low, the researcher may want to generate new items and collect more data. In addition, particularly for scales that will be used in clinical diagnosis or selection decisions, normative data are required; that is, the mean, variance, and distribution shape of scores must be evaluated based on a large number of people (at least several thousand). This provides test users with a basis for evaluation. For example, for the BDI (Beck et al., 1961), the following interpretations for scores have been suggested based on normative data for thousands of test takers: scores from 5 to 9, normal mood variations; 10 to 18, mild to moderate depression; 19 to 29, moderate to severe depression; and 30 to 63, severe depression. Scores of 4 or below on the BDI may be interpreted as possible denial of depression or faking good; it is very unusual for people to have scores that are this low on the BDI.

19.10.8 ♦ Create Final Scale

When all the criteria for good quality measurement appear to be satisfied (i.e., the data analyst has obtained a reasonably brief list of items or measurements that appears to provide reliable and valid information about the construct of interest), a final version of the scale may be created. Often such scales are first published as tables or appendixes in journal articles. A complete report for a newly developed scale should include the instructions for the test respondents (e.g., what period of time should the test taker think about when reporting frequency of behaviors or feelings?); a complete list of items, statements, or questions; the specific response alternatives; indication whether any items need to be reverse coded; and scoring instructions. Usually, the scoring procedure consists of reversing the direction of scores for any reverse-worded items and then summing the raw scores across all items for each scale. If subsequent research provides additional evidence that the scale is reliable and valid, and if the scale measures something that has a reasonably wide application, at some point, the test author may copyright the test and perhaps have it distributed on a fee per use basis by a test publishing company. Of course, as years go by, the contents of some test items may become dated. Therefore, periodic revisions may be required to keep test item wording current.

19.11 ♦ Summary

To summarize, measurements need to be reliable. When measurements are unreliable, it leads to two problems. Low reliability may imply that the measure is not valid (if a measure does not detect *anything* consistently, it does not make much sense to ask *what* it is measuring). In addition, when researchers conduct statistical analyses, such as correlations, to assess how scores on an X variable are related to scores on other variables, the relationship of X to other variables becomes weaker as the reliability of X becomes smaller; the attenuation of correlation due to unreliability of measurement was discussed in Chapter 7. To put it more plainly, when a researcher has unreliable measures, relationships between variables usually appear to be weaker. It is also essential for measures to be valid: If a measure is not valid, then the study does not provide information about the theoretical constructs that are of real interest. It is also desirable for measures to be sensitive to individual differences, unbiased, relatively inexpensive, not very invasive, and not highly reactive.

Research methods textbooks point out that each type of measurement method (such as direct observation of behavior, self-report, physiological or physical measurements, and archival data) has strengths and weaknesses. For example, self-report is generally low cost, but such reports may be biased by social desirability (i.e., people report attitudes and behaviors that they believe are socially desirable, instead of honestly reporting their actual attitudes and behaviors). When it is possible to do so, a study can be made much stronger by including multiple types of measurements (this is called "triangulation" of measurement). For example, if a researcher wants to measure anxiety, it would be desirable to include direct observation of behavior (e.g., "um"s and "ah"s in speech and rapid blinking), self-report (answers to questions that ask about subjective anxiety), and physiological measures (such as heart rates and cortisol levels). If an experimental manipulation has similar effects on anxiety when it is assessed using behavioral, self-report, and physiological outcomes, the researcher can be more confident that the outcome of the study is not attributable to a methodological weakness associated with one form of measurement, such as self-report.

The development of a new measure can require a substantial amount of time and effort. It is relatively easy to demonstrate reliability for a new measurement, but the evaluation of validity is far more difficult and the validity of a measure can be a matter of controversy. When possible, researchers may prefer to use existing measures for which data on reliability and validity are already available.

For psychological testing, a useful online resource is the American Psychological Association FAQ on testing: www.apa.org/science/testing.html.

Another useful resource is a directory of published research tests on the Educational Testing Service (ETS) Test Link site www.ets.org/testcoll/index.html, which has information on about 20,000 published psychological tests.

Although most of the variables used as examples in this chapter were self-report measures, the issues discussed in this chapter (concerning reliability, validity, sensitivity, bias, cost effectiveness, invasiveness, and reactivity) are relevant for other types of data, including physical measurements, medical tests, and observations of behavior.

Appendix: The CESD Scale

INSTRUCTIONS: Using the scale below, please circle the number before each statement which best describes how often you felt or behaved this way DURING THE PAST WEEK.

1 = Rarely or none of the time (less than 1 day)

2 = Some or a little of the time (1–2 days)

3 = Occasionally or a moderate amount of time (3–4 days)

4 = Most of the time (5–7 days)

The total CESD depression score is the sum of the scores on the following twenty questions with Items 4, 8, 12, and 16 reverse scored.

1. I was bothered by things that usually don't bother me.
2. I did not feel like eating; my appetite was poor.
3. I felt that I could not shake off the blues even with help from my family or friends.
4. I felt that I was just as good as other people. (reverse worded)
5. I had trouble keeping my mind on what I was doing.
6. I felt depressed.
7. I felt that everything I did was an effort.
8. I felt hopeful about the future. (reverse worded)
9. I thought my life bad been a failure.
10. I felt fearful.
11. My sleep was restless.
12. I was happy. (reverse worded)

13. I talked less than usual.
14. I felt lonely.
15. People were unfriendly.
16. I enjoyed life. (reverse worded)
17. I had crying spells.
18. I felt sad.
19. I felt that people dislike me.
20. I could not get "going."

A total score on CESD is obtained by reversing the direction of scoring on the four reverse-worded items (4, 8, 12, and 16), so that a higher score on all items corresponds to a higher level of depression, and then summing the scores across all 20 items.

Appendix Source: Radloff, L. S. (1977). The CESD Scale: A self-report depression scale for research in the general population. *Applied Psychological Measurement, 1,* 385–401.

WWW Links: Resources on Psychological Measurement

American Psychological Association
www.apa.org/science/testing.html

ETS Test Collection
www.ets.org/testcoll/terms.html

Goldberg's International Personality Item Pool—royalty-free versions of scales that measure "Big Five" personality traits
http://ipip.ori.org/ipip/

Health and Psychosocial Instruments (HAPI)
Available through online access to databases at many universities.

Mental Measurements Yearbook Test Reviews online
http://buros.unl.edu/buros/jsp/search.jsp

PsychWeb information on psychological tests
www.psychweb.com/tests/psych_tests

Robinson et al. (1991). A full list of scales is available in their book *Measures of Personality and Social Psychological Attitudes*
www.rohan.sdsu.edu/~mstover/tests/robinson.html

Comprehension Questions

1. List and describe the most common methods of reliability assessment.
2. One way to interpret a reliability coefficient (often an r value) is as the proportion of variance in X scores that is stable across occasions of measurement. Why do we interpret r (rather than r^2) as the proportion of variance in X scores that is due to T, a component of the score that is stable across occasions of measurement? (See Figure 19.3.)
3. Explain the terms in this equation: $X = T + e$. You might use a specific example; for instance, suppose that X represents your score on a measure of life satisfaction and T represents your true level of life satisfaction.

The next set of questions focuses on the Cronbach alpha, an index of internal consistency reliability.

4. How is α related to p, the number of items, when the mean inter-item correlation is held constant?
5. How is α related to the mean inter-item correlation when p, the number of items in the scale, is held constant?
6. Why is it important to make certain that all the items in a scale are scored in the same direction before you enter your variables into the reliability program to compute a Cronbach alpha?
7. Why are reverse-scored questions often included in surveys and tests? For example, in a hostility scale, answering yes to some questions might indicate greater hostility, while on other questions, a "no" answer might indicate hostility.
8. What is the Kuder-Richardson 20 (KR-20) statistic?
9. How can you use the item-total statistics from the reliability program to decide whether reliability could be improved by dropping some items?
10. If a scale has 0 reliability, can it be valid? If a scale has high reliability, such as $\alpha = .9$, does that necessarily mean that it is valid? Explain.
11. What is the difference between α and standardized α? Why does it make sense to sum z scores rather than raw scores across items in some situations?
12. What is an "optimally weighted linear composite"? What is a "unit-weighted linear composite"? What are the relative advantages and disadvantages of these two ways of forming scores?
13. Three different methods were presented for the computation of scale scores: summing raw scores on the X variables, summing z scores on the X variables, and saving factor scores from a FA of a set of X variables. (Assume that all the X

variables are measures of a single latent construct.) How are the scores that are obtained using these three different methods of computation typically related?

14. What is meant by attenuation of correlation due to unreliability?
15. How can we correct for attenuation of correlation due to unreliability? Under what circumstances does correction for attenuation due to unreliability yield inaccurate results?
16. Discuss the kinds of information that you need to assess reliability of measurements.
17. Discuss the kinds of information that you need to assess the validity of a measure.
18. Why is it generally more difficult to establish the validity of a measure than its reliability?
19. In addition to reliability and validity, what other characteristics should good quality measurements have?

Data Analysis Exercise

Scale Construction and Internal Consistency Reliability Assessment

Select a set of at least 4 (or more) items that you think could form an internally consistent/unidimensional scale. Make sure that all items are scored in the same direction by creating reverse-scored versions of some items, if necessary. Create unit-weighted scores on your scale using the SPSS compute statement. Keep in mind that in some cases it may make more sense to sum *z* scores than raw scores. Run the SPSS Reliability program to assess the internal reliability or internal consistency of your scale. Also, run correlations between factor scores (created using the SAVE command) and a summated scale (created using the COMPUTE statement to add raw scores, *z* scores, or both).

Based on your findings, answer the following questions:

1. In terms of the Cronbach alpha, how reliable is your scale? Could the reliability of your scale be improved by dropping one or more items, and if so, which items?
2. How closely related are scores on the different versions of your scale—that is, factor scores saved from the FA versus scores created by summing raw scores and/or *z* scores?

CHAPTER 20

Analysis of Repeated Measures

20.1 ◆ Introduction

In previous chapters of this book, when groups of scores have been compared, we have assumed that the data were obtained from a **between-subjects (between-S)** or independent groups design. In a between-S or independent groups study, each group consists of a different set of participants, each participant is a member of only one group, and there is no systematic pairing or matching between participants. Under these conditions, assuming that data are collected in a manner that does not provide participants with opportunities to influence each other's behavior through processes such as competition or imitation or sharing of information, the observations should meet the basic assumption that is required for all independent samples analyses—that is, the assumption that each observation is independent of other observations. For further discussion of ways in which observations can become nonindependent, see Section 1.10.

This chapter considers the analysis of **within-subjects (within-S)** or repeated measures designs. In a within-S or repeated measures factor, each participant is tested or observed at multiple times or in several different treatment conditions. We also need to employ within-S or repeated measures data analytic methods in situations where participants are matched, paired, or **yoked** because these procedures also create correlated or nonindependent observations.

The analyses that have been presented so far, such as the independent samples *t* test and the between-S one-way analysis of variance (ANOVA), assume that the scores in different groups are independent of each other. Repeated measures scores generally are not independent across groups. Generally, when the same persons are tested under several different conditions, their scores are correlated across conditions. The existence of systematic correlations among observations violates a fundamental assumption of independent samples analyses such as the independent samples *t* test. However, we can include a factor that corresponds to the individual differences among participants in the model for data analysis. By incorporating individual participant differences as a variable in the analysis, we may be able to statistically control for the variability in scores that is due to individual participant differences. Modifying the data analysis method so that it explicitly takes participant differences into account can help us to achieve two goals. First, repeated measures analyses provide a way to deal with the violation of the assumption of independence of observations. Second, in many situations, a repeated measures analysis yields a smaller error term and, thus, a more powerful test for differences among group means than a comparable between-S design.

For each of the between-subjects or independent groups analysis that we have covered, there is a corresponding within-S or repeated measures analysis. Refer back to Table 1.4 to find names of the within-S or repeated measures analyses that correspond to several of the more widely used between-S or independent samples analyses. For example, consider research situations in which the independent variable X is a categorical variable that indicates group membership, and the outcome variable Y is a quantitative measure. When the design is between-S or independent groups, we can use the independent samples t test to compare means for just two groups or the one-way between-S ANOVA to compare means across any number of groups. When we have repeated measures or matched or paired samples, we will use the paired samples t test to evaluate whether means differ across scores collected under two different treatment conditions or between scores obtained at two different points in time. If there are more than two treatment conditions or times, we can perform a one-way repeated measures ANOVA to evaluate whether mean scores on a quantitative outcome variable differ across several conditions or times or trials.

20.2 ♦ Empirical Example: Experiment to Assess Effect of Stress on Heart Rate

A concrete example will illustrate how the organization of data differs for between-S and within-S designs. Suppose that a researcher wants to know how heart rate (HR) differs across the following four levels of a "stress" factor: baseline or no stress, a pain induction, a mental arithmetic task, and a stressful social role play. This study could be performed using a between-S design; in a between-S or independent samples version of this study, each participant would be randomly assigned to receive just one of the four levels of stress. Alternatively, this study could be performed using a within-S or repeated measures design, in which each participant's HR is measured under all four conditions that correspond to levels of the stress factor. In the examples that follow, there are 24 measurements of HR. In the first set of analyses, these 24 observations are organized as if they came from a between-S or independent samples study; we will apply an independent samples t test and a one-way between-S ANOVA to these scores. In later sections of this chapter, these 24 scores are reorganized as if they had been obtained using a within-S or repeated measures design; a paired samples t test and a one-way within-S or repeated measures ANOVA will be applied to these data. Comparison of the results from the between-S analyses in this section and the results from the within-S or repeated measures analyses later in the chapter illustrates an important potential advantage of within-S or repeated measures designs; often, the error term used to test for differences in mean outcomes across treatment groups is smaller for a within-S design and analysis than for a between-S analysis applied to the same set of scores.

Table 20.1 ♦ Data for a Hypothetical Between-S Study With Four Treatment Conditions or Types of Stress: 1 = No Stress/Baseline, 2 = Pain Induction, 3 = Mental Arithmetic Task, 4 = Role Play

Stress	*Name*	*Heart Rate*
1	Ann	70
1	Bob	78
1	Chris	61
1	Dale	70
1	Erin	89
1	Frank	79
	Mean HR for Group 1, Baseline/No Stress = 74.5	
2	Mary	86
2	Nan	82
2	Oriah	70
2	Penny	74
2	Quentin	105
2	Rose	84
	Mean HR for Group 2, Pain = 83.5	
3	George	80
3	Harriet	77
3	Irene	64
3	Jack	68
3	Kim	89
3	Leo	88
	Mean HR for Group 3, Mental Arithmetic = 77.7	
4	Tom	87
4	Ulrich	80
4	Vinny	72
4	Wendy	88
4	Xiao	101
4	Zara	87
	Mean HR for Group 4, Role Play = 85.8	

20.2.1 ♦ Analysis of Data From the Stress/HR Study as a Between-S or Independent Samples Design

First let's suppose that this study is conducted as a between-S or independent groups study. In the between-S version of the study, a total of $N = 24$ different participants are recruited; each participant is randomly assigned to one of the four treatment groups (no stress, pain, mental arithmetic, or stressful role play). Table 20.1 shows how the data would appear in a between-S version of this study. The first variable, stress, provides information about group membership (scores on the categorical variable named stress identify which of the four stress conditions each participant experienced). The second variable, HR, corresponds to each participant's score on the outcome variable.

Let's assume that the scores on HR meet the assumptions for use of the independent samples t test and one-way between-S ANOVA; that is, scores on HR are quantitative, are approximately normally distributed, and do not have substantially unequal variances across treatments. To compare mean HR between the baseline and pain induction conditions, we can do an independent samples t test (as described in Chapter 5); to assess whether mean HR differs across all four levels of the between-S stress factor, we can do a one-way between-S ANOVA (as described in Chapter 6). These results will provide a useful basis for comparison that will help us understand the potential advantages of the within-S or repeated measures analyses that are introduced in this chapter.

20.2.2 ♦ Independent Samples *t* Test for the Stress/HR Data

Output for an SPSS independent samples t test that compares just the means for just the first two groups (baseline and pain intervention) appears in Figure 20.1. The equal variances assumed version of the independent samples t test for these data was $t(10) = -1.42$, $p = .19$, $\eta^2 = .17$. Using the conventional $\alpha = .05$ criterion for statistical significance, the 9 beats per minute difference between mean HR for the baseline condition ($M = 74.5$) and the mean HR for the pain condition (83.5) was not statistically significant.

20.2.3 ♦ One-Way Between-S ANOVA for the Stress/HR Data

The entire dataset that appears in Figure 20.1 can be analyzed using one-way between-S ANOVA. The results for a one-way between-S ANOVA for these data appear in Figure 20.2. For a between-S ANOVA that examines differences in mean HR across four levels of a factor called stress (Level 1 = No stress/baseline, Level 2 = Pain induction, Level 3 = Mental arithmetic task, and Level 4 = Stressful social role play), $F(3, 20) = 1.485$, $p = .249$, $\eta^2 = .18$. From the independent samples t test and the one-way between-S ANOVA, we can conclude that if these HR observations were obtained in a between-S or **independent samples design**, mean HR does not differ significantly across different levels or types of stress.

20.3 ♦ Discussion of Sources of Within-Group Error in Between-S Versus Within-S Data

Chapter 6 described a simple model that is used to represent the scores in an independent samples design. This is the conceptual model that underlies both the independent samples t test and the between-S one-way ANOVA. The score for person j in treatment group i was denoted X_{ij}, where the first subscript identifies which treatment group the score comes from and the second subscript identifies the individual person; thus, X_{24} represents the score for the fourth person in the second treatment group. Because we often identify the first treatment factor in a between-S study as Factor A, the corresponding term that represents the effect for each level of the A treatment factor in the conceptual model is often denoted by α. The effect that corresponds to each level of the treatment factor is estimated by the difference between the sample mean for that group and the grand mean. If we denote the mean for Group 1 as M_1 and the grand mean as M_{grand}, then $\alpha_1 = M_1 - M_{grand}$. The model to predict an individual score in a between-S one-way ANOVA, from Equation 6.8 in Chapter 6, is as follows:

Group Statistics

	Stress	N	Mean	Std. Deviation	Std. Error Mean
HR	No Stress/ Baseline	6	74.50	9.649	3.939
	Pain	6	83.50	12.194	4.978

Independent Samples Test

		Levene's Test for Equality of Variances		t-test for Equality of Means						
									95% Confidence Interval of the Difference	
		F	Sig.	t	df	Sig. (2-tailed)	Mean Difference	Std. Error Difference	Lower	Upper
HR	Equal variances assumed	.028	.870	-1.418	10	.187	-9.000	6.348	-23.145	5.145
	Equal variances not assumed			-1.418	9.498	.188	-9.000	6.348	-23.247	5.247

Figure 20.1 ♦ Independent Samples *t* Test: Comparison of Mean HR Across Baseline Versus Pain Treatment Conditions for Independent Samples or Between-S Data in Table 20.1

Descriptives

HR

	N	Mean	Std. Deviation	Std. Error	95% Confidence Interval for Mean		Minimum	Maximum
					Lower Bound	Upper Bound		
No Stress/Baseline	6	74.50	9.649	3.939	64.37	84.63	61	89
Pain	6	83.50	12.194	4.978	70.70	96.30	70	105
Mental Arithmetic	6	77.67	10.211	4.169	66.95	88.38	64	89
Role Play	6	85.83	9.621	3.928	75.74	95.93	72	101
Total	24	80.38	10.798	2.204	75.82	84.93	61	105

ANOVA

HR

	Sum of Squares	df	Mean Square	F	Sig.
Between Groups	488.458	3	162.819	1.485	.249
Within Groups	2193.167	20	109.658		
Total	2681.625	23			

Figure 20.2 ◆ Output From One-Way Between-S ANOVA Applied to Data in Table 20.1

$$X_{ij} = \mu + \alpha_i + e_{ij}, \tag{20.1}$$

where

X_{ij} is the observed score for the jth subject in treatment group i,

μ is the grand mean (estimated by the sample grand mean M_{grand}),

α_i is the effect for group i (estimated by $M_i - M_{\text{grand}}$), and

e_{ij} is the residual or prediction error for participant j in group i (estimated by $X_{ij} - M_i$).

In Chapter 6, the residual was denoted ε_{ij} so that the notation used in Chapter 6 was consistent with the notation used in most ANOVA textbooks. In this chapter, the residual term in the ANOVA model is denoted e_{ij} so that it can be clearly distinguished from another common use of the symbol ε in discussions of repeated measures ANOVA; in Section 20.11, ε represents an index of the degree to which an important assumption about the pattern of variances and covariances among repeated measures is violated.

When we compute s_1^2 and s_2^2 for an independent samples t test, we summarize information about the sizes of the squared prediction errors (the sum of the squared e_{ij} terms) within each group. Recall from earlier discussions of the independent samples t test (in Chapter 5) and the one-way between-S ANOVA (in Chapter 6) that these within-group deviations from the group mean reflect the influence of many other variables (apart from the stress intervention in this study) that affect HR. For example, if the time of day, room temperature, and behavior of the experimenter vary across testing sessions, these extraneous variables may cause variability in the observed HR values within each treatment condition. However, one of the largest sources of variations in HR in this type of research may be stable individual differences among persons. Some persons tend to have high HRs across a variety of situations, while others may have lower HRs; stable individual differences in HR may be related to gender, anxiety, caffeine and drug use, aerobic fitness, age, and other differences among participants.

When we calculate an independent samples t test or a one-way between-S ANOVA, individual differences among persons in HR are included in the error component. The e_{ij} deviations from the group means that are used to compute s_1^2, s_2^2, and SE_{M1-M2} for the independent samples t test and MS_{error} for the between-S ANOVA thus include both the variance that is due to systematic or stable individual differences among participants and the variance due to variations in experimental procedure.

One of the reasons to set up a study as a within-S or repeated measures design is that it makes it possible, at least in theory, to identify and remove the variance in scores that is associated with stable individual differences among participants, which is part of the within-group error variance in a between-S design. In principle, when we analyze data as repeated measures, we can identify a component or part of each score that is predictable from individual differences among participants and compute a sum of squares (SS) that summarizes that source of variance. When we do that, the remaining SS_{error} term for a repeated measures analysis no longer includes the variance due to individual differences among participants. In principle, the SS_{error} term should usually be smaller for a within-S or repeated measures ANOVA than for a between-S or independent samples ANOVA or t test. The next section examines the conceptual model that underlies analysis of repeated measures (both the correlated samples t test and the repeated measures ANOVA) and compares this with the model for between-S designs to see how these data analyses differ in the way they handle the variance that is due to stable individual differences among participants.

20.4 ♦ The Conceptual Basis for the Paired Samples *t* Test and One-Way Repeated Measures ANOVA

Consider how the set of scores that appeared in Table 20.1 would be organized if the same set of $N = 24$ scores had been obtained from a study that was conducted as a within-S or repeated measures design. In a within-S or repeated measures design, the researcher measures each participant's HR under all four conditions (baseline, pain, arithmetic, and role play). The data for a hypothetical repeated measures study with $N = 6$ participants tested under $k = 4$ conditions appear in Table 20.2; the corresponding SPSS Data View worksheet appears in Figure 20.3. The total number of observations $= N \times k = 6 \times 4 = 24$.

Initial examination of the data in Table 20.2 reveals three things. First, if we find the mean for the HR scores under each of the four conditions (either by calculating a mean by hand or by running <Descriptive Statistics> in SPSS), we find that mean HR differs across the four conditions ($M = 74.5$ for baseline, $M = 83.5$ for pain, $M = 77.67$ for arithmetic, and $M = 85.83$ for role play). Second, if we compute a mean HR for each person, we find that the six people in this study had substantially different mean HRs, which ranged from a low mean HR of 66.75 for Chris to a high mean HR of 96.00 for Erin. Third, if we obtain Pearson correlations between the scores under baseline and arithmetic, baseline and pain, and baseline and role play, we find that these sets of scores have high positive correlations. For example, the correlation between the set of six baseline HR scores and the set of six arithmetic HR scores is $r = .86$. The high correlations among sets of scores in the columns of Table 20.2 provide evidence that the scores in this repeated measures study were not independent across treatments and that there is substantial consistency in the magnitude of HR for participants across conditions; that is, Erin's HR was always among the highest and Chris's score was always among the lowest scores across all four treatments.

We will examine three different ways to analyze these data. To compare mean HR across just two of the levels of stress (such as baseline vs. pain induction), we will do a paired samples t test; this is also sometimes called the correlated samples or direct difference t test. To assess whether mean HR differs across all four levels of the stress factor, we will do a univariate one-way repeated measures ANOVA. We will also analyze the scores using multivariate analysis of variance (MANOVA). One-way repeated measures ANOVA requires some assumptions about the structure of the repeated measures scores that are often violated in practice. An advantage of the MANOVA approach is that it requires less restrictive assumptions about data structure.

Our analysis of the within-S or repeated measures data in Table 20.2 will take advantage of the fact that each of the 24 individual HR scores is predictable from the treatment condition (baseline, pain, arithmetic, and role play) and also from the person (Ann, Bob, Chris, and so forth). When we did an independent samples t test or a one-way ANOVA on the HR scores in Table 20.1, our within-group error terms included variation in HR due

Table 20.2 ♦ Repeated Measures Data for HR Measured During Four Types of Stress

	Type of Stress Induction				
Name	*1 None (Baseline)*	*2 Pain*	*3 Arithmetic*	*4 Role Play*	*Mean for Each Participant*
Ann	70	86	80	87	80.75
Bob	78	82	77	80	79.25
Chris	61	70	64	72	66.75
Dale	70	74	68	88	75.00
Erin	89	105	89	101	96.00
Frank	79	84	88	87	84.50
Stress group means	74.5	83.5	77.7	85.8	$M_{grand} = 80.375$

NOTE: In this table, the scores from Table 20.1 are reorganized as they would appear if they had been obtained from a repeated measures or within-S study, with each participant tested under all four treatment conditions.

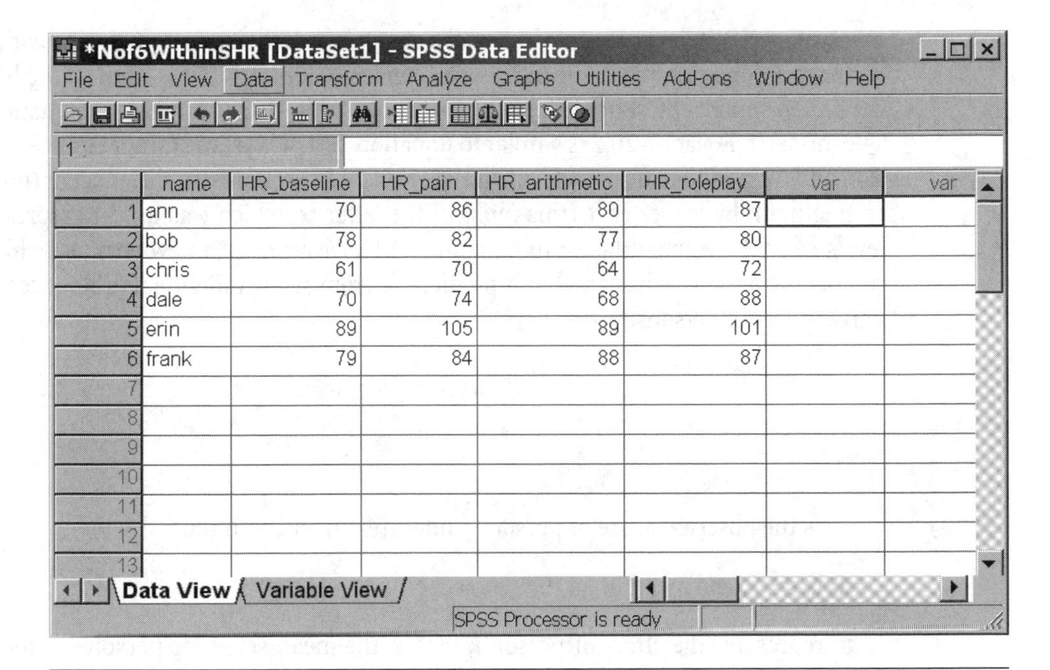

	name	HR_baseline	HR_pain	HR_arithmetic	HR_roleplay
1	ann	70	86	80	87
2	bob	78	82	77	80
3	chris	61	70	64	72
4	dale	70	74	68	88
5	erin	89	105	89	101
6	frank	79	84	88	87

Figure 20.3 ♦ SPSS Data Worksheet for One-Way Repeated Measures Analysis Comparing HR Across Four Levels of Stress: No Stress/Baseline, Pain, Mental Arithmetic Task, and Role Play

to variations in experimental procedures—including extraneous variables such as room temperature, time of day, and whether the experimenter behaves in an anxious and hurried manner—and the error term for the independent samples t test also included variation in HR due to individual differences among participants. When we do a paired samples t test or a repeated measures of the same set of 24 scores, the error term for this paired samples t test or the one-way repeated measures ANOVA does *not* include systematic differences in HR due to individual differences among participants; the error term for the paired samples t test, in theory, includes only variations due to other uncontrolled variables (e.g., room temperature, time of day, and experimenter behavior). Because the error term that we use to set up a paired samples t test does *not* include variation due to systematic individual differences among participants, the error term for a paired samples t test is usually smaller than the error term for an independent samples t test applied to the same set of scores.

When we analyze the HR data using a paired samples t test or direct difference t test, in theory, we can identify the part of each HR score that is associated with individual differences among persons, and thus, we can remove the variance due to individual differences among persons from the error term that we use to set up the t ratio. To see how this works, let's look at a formal model that represents how HR scores can be predicted from the components in a repeated measures design.

The model that we use to represent the structure of the data in a repeated measures design, which provides the conceptual basis for the paired samples t test and for repeated measures ANOVA, includes an additional term (π_i) that corresponds to the individual differences among persons in mean HR. The conceptual model that is used as the basis for the paired samples t test or one-way repeated measures ANOVA appears as Equation 20.2. The model in Equation 20.2 is similar to Equation 20.1, with some minor changes in notation and the addition of a new term. In Equation 20.2, we represent the effect of time, trial, or treatment by the letter τ (this replaces the letter α, which was used to represent the levels of an A treatment factor in Equation 20.1). We also add a new term, π_i, which represents the part of each score that is predictable from stable individual differences among participants or persons.

$$X_{ij} = \mu + \pi_i + \tau_j + e^*_{ij}, \tag{20.2}$$

where

X_{ij} is the observed score of person i under treatment condition j,

μ is the grand mean of all the scores,

π_i represents the effect of person i, that is, the mean score for person i – the grand mean,

τ_j represents the effect for time j or trial j or treatment j, and

e^*_{ij} is the error term that represents the effects of unique extraneous or error variables that influence person i's response at time j.

Note the difference between the error terms in the two models: the between-S or independent samples model in Equation 20.1 versus the within-S or repeated measures model in Equation 20.2. The e_{ij} term in the model for the independent samples design in Equation 20.1 includes variation due to stable individual differences among participants in HR (which we now represent as π) and, also, variation due to other variables such as nonstandard experimental procedures. The e^*_{ij} term for the repeated measures design in Equation 20.2 includes variation due to other variables such as experimental procedures; however, e^*_{ij} does *not* include the variance due to stable individual differences among participants. Thus, we would usually expect the values of e^*_{ij} to be smaller than the corresponding values of e_{ij}.

Theoretically, the e_{ij} error term in the independent samples t test or ANOVA model corresponds to the sum $e^*_{ij} + \pi_i$—that is, the sum of error due to all extraneous variables except individual differences (e^*_{ij}) and the variability among X_{ij} scores that is due to systematic individual differences (π_i). We usually expect the e^*_{ij} error terms to be smaller than the e_{ij} error terms because the e_{ij} error terms *do* include variations due to stable individual differences among participants, while the e^*_{ij} terms *do not* include variation due to individual differences among participants.

Table 20.3 shows the conceptual model that identifies the components of each score in a paired samples t test or a repeated measures ANOVA with just two levels. The first step in

Table 20.3 ♦ Formal Model for Components of Scores in a Repeated Measures or Paired Samples Design

	Treatment 1 Score	*Treatment 2 Score*	*Difference Score (Treatment 2 – Treatment 1 Score)*
Person 1	$\mu + \tau_1 + \pi_1 + e^*_{11}$	$\mu + \tau_2 + \pi_1 + e^*_{12}$	$(\mu + \tau_2 + \pi_1 + e^*_{12}) - (\mu + \tau_1 + \pi_1 + e^*_{11}) = (\tau_2 - \tau_1) + (e^*_{12} - e^*_{11})$
Person 2	$\mu + \tau_1 + \pi_2 + e^*_{21}$	$\mu + \tau_2 + \pi_2 + e^*_{22}$	$(\mu + \tau_2 + \pi_2 + e^*_{22}) - (\mu + \tau_1 + \pi_2 + e^*_{21}) = (\tau_2 - \tau_1) + (e^*_{22} - e^*_{21})$
Person 3	$\mu + \tau_1 + \pi_3 + e^*_{31}$	$\mu + \tau_2 + \pi_3 + e^*_{32}$	$(\mu + \tau_2 + \pi_3 + e^*_{32}) - (\mu + \tau_1 + \pi_3 + e^*_{31}) = (\tau_2 - \tau_1) + (e^*_{32} - e^*_{31})$
⋮			
Person *n*	$\mu + \tau_1 + \pi_n + e^*_{n1}$	$\mu + \tau_2 + \pi_n + e^*_{n2}$	$(\mu + \tau_2 + \pi_n + e^*_{n2}) - (\mu + \tau_1 + \pi_n + e^*_{n1}) = (\tau_2 - \tau_1) + (e^*_{n2} - e^*_{n1})$
Expected mean			$(\tau_2 - \tau_1) + 0$

NOTES: μ is the grand mean of scores on the outcome variable for all participants in the study. τ_1 and τ_2 are the "effects" of Treatments 1 and 2 or measures taken at Time 1 and Time 2. π_i is the systematic effect associated with person *i*. e^*_{ij} is the error term associated with the outcome for person *i* tested under treatment *j*, or at time *j*.

computation of a paired samples *t* test is obtaining a difference score, *d*, for each participant by subtracting each participant's Time 1 score from his or her Time 2 score. Note that when each difference score *d* is computed, the π_i term that corresponds to the individual difference in HR associated with each person, in theory, "cancels out." Consider the computation of a difference score *d* for Person 1, and examine the row that corresponds to the data for Person 1 in Table 20.3. For Person 1, the score under Treatment 1 is denoted ($\mu + \tau_1 + \pi_1 + e^*_{11}$), and the score under Treatment 2 is denoted ($\mu + \tau_2 + \pi_1 + e^*_{12}$). The difference score for Person 1 is obtained by subtracting the Time 1 score from the Time 2 score as follows: $(\mu + \tau_2 + \pi_1 + e^*_{12}) - (\mu + \tau_1 + \pi_1 + e^*_{11})$. The difference score corresponds to the following theoretical components: $(\tau_2 - \tau_1) + (e^*_{12} - e^*_{11})$. That is, the difference between Person 1's scores at Time 1 and Time 2 provides information about the difference between the effects of the Time 1 and Time 2 times or trials or treatments ($\tau_2 - \tau_1$); the difference score *d* also contains error variance due to variations in experimental procedure, represented by the term $e^*_{12} - e^*_{11}$. However, the difference terms that we obtain by calculating a *d* score for each person do not contain any variability associated with systematic individual differences among participants, represented by the terms $\pi_1, \pi_2, \ldots, \pi_n$. Bear in mind that this is purely theoretical and that real data may be more complex. For example, if the data show a person by treatment interaction, which is not represented in Equation 20.2, this simple interpretation of the difference score *d* will not be valid.

20.5 ♦ Computation of a Paired Samples *t* Test to Compare Mean HR Between Baseline and Pain Conditions

The paired samples *t* test can be used to compare means for groups of scores that are obtained by making repeated measurements on the same group of participants whose behaviors are assessed at two times or in two trials, before versus after an intervention, or under two different treatment conditions. This test is also appropriate when the scores in the two samples are correlated because the samples are matched or paired or yoked. The paired samples *t* test is also sometimes called the direct difference or correlated samples *t* test. The following computational formulas make it clear why it is sometimes called the direct difference *t* test; the first step in the computation of the paired samples *t* test involves finding a difference score (Time 2 score – Time 1 score) for each participant.

The paired samples or direct difference *t* test is done in the following way. This example uses the data for just the first two groups from Table 20.2—that is, the HR measurements for the 6 participants in the baseline/no-stress condition and the pain induction condition. First of all, we compute a difference score (*d*) for each person by taking the difference between the Treatment 1 and Treatment 2 scores for that person. Usually, the score for Treatment 1 is subtracted from the Treatment 2 score so that the difference can be interpreted as the change from Time 1 to Time 2. The numerical results obtained when difference scores are computed for each participant's data for the baseline and pain induction situations in Table 20.2 are presented in Table 20.4. Recall that in the independent samples *t* test reported earlier for the same set of scores, when we set up a *t* ratio, the research question was, Based on the magnitude of the differences between the sample mean

Table 20.4 ♦ Difference Scores for HR (Between Baseline and Pain Conditions) for the Data in Table 20.2

Name	*Baseline*	*Pain*	*d (Pain – Baseline)*
Ann	70	86	+16
Bob	78	82	+4
Chris	61	70	+9
Dale	70	74	+4
Erin	89	105	+16
Frank	79	84	+5
Mean of difference scores			+9

HRs M_{baseline} and M_{pain} and information about the sample sizes and variability of scores within groups, can we reject the null hypothesis that the corresponding population means μ_{baseline} and μ_{pain} are equal—in other words, the null hypothesis that the difference between the population means $\mu_{\text{baseline}} - \mu_{\text{pain}} = 0$? The null hypothesis for the paired samples *t* test is that the mean of these difference scores equals 0:

$$H_0: \mu_d = 0. \tag{20.3}$$

Thus, when we conduct a paired samples *t* test, we will set up a *t* ratio to examine the value of the sample mean of the difference scores (M_d) relative to sampling error; when we conduct a paired samples *t* test, we usually hope that the mean of the difference scores, M_d, will be large enough (relative to expected variations due to sampling error) for us to reject the null hypothesis in Equation 20.3.

To set up a paired samples *t* test, we need to find the mean of the differences scores (the mean of the *d* difference scores will be denoted M_d). We will also need to compute the variance (s_d^2) of these difference scores and use this information about variance to create an appropriate error term for the paired samples *t* test. The first step in computing a paired samples *t* test, once you have computed a difference score *d* for each participant as shown in Table 20.4, is to compute the mean of those difference scores:

$$M_d = (\Sigma d)/N. \tag{20.4}$$

Note that the value of *N* in Equation 20.4 corresponds to the number of participants (not the total number of observations). For the data in Table 20.4, the mean of the difference scores $M_d = 54/6 = +9.00$. In other words, the average amount of change in HR across participants when you subtract baseline HR from HR measured during the pain induction condition was a 9-point increase in HR for the pain condition relative to baseline. The next piece of information we need is the sample variance for the *d* difference scores, computed as follows:

$$s^2_d = \Sigma\ (d - M_d)^2\ /\ (N - 1). \tag{20.5}$$

For the data in Table 20.4, the variance of these difference scores is 32.80. Note that this value of s_d^2 is much lower than the within-group error variances of $s_1^2 = 93.10$ and $s_1^2 = 148.69$ that were obtained when the independent samples *t* test was applied to this same set of scores earlier. Provided that the repeated measures scores are positively correlated, s_d^2 is usually smaller than the pooled within-group error term for an independent samples *t* test based on the values of s_1^2 and s_2^2.

Next, we use the sample standard deviation to compute a standard error term (SE_{M_d}) for the difference scores as follows:

$$SE_{M_d} = \sqrt{\frac{s_d^2}{N}} = \sqrt{\frac{32.80}{6}} = \sqrt{5.467} = 2.34. \quad (20.6)$$

The degrees of freedom (*df*) for the paired samples *t* test are based on *N*, the number of difference scores:

$$df = N - 1 = 6 - 1 = 5. \quad (20.7)$$

Finally, we can set up the paired samples *t* ratio:

$$t = M_d / SE_{M_d} = \frac{9.00}{2.34} = 3.85 \quad \text{with 5 } df. \quad (20.8)$$

In this example, we would conclude that when we do a paired samples *t* test to evaluate whether mean HR differs significantly between the baseline and pain induction conditions, the 9-point increase in HR during the pain induction relative to the baseline or no-stress situation is statistically significant at the $\alpha = .05$ level. Recall that when the same batch of scores was analyzed using an independent samples *t* test in Section 20.2, the 9-point difference was not judged statistically significant. This is an example of a situation where the paired samples *t* test has better statistical power than the independent samples *t* test when these tests are applied to the same batch of data.

20.6 ♦ SPSS Example: Analysis of Stress/HR Data Using a Paired Samples *t* Test

To carry out the paired samples *t* test for the data in Table 20.2 using SPSS, the following menu selections (as shown in Figure 20.4) were made: <Analyze> → <Compare Means> → <Paired-Samples T Test>.

The paired samples *t* test dialog window appears in Figure 20.5. The names of the variables that represent the two repeated measures levels to be compared (in this case, HR_baseline and HR_pain) were double clicked to move the name for each of these variables into the list marked Current Selection. Then, the right arrow button was used to move the pair of selected variables into the window under the heading Paired

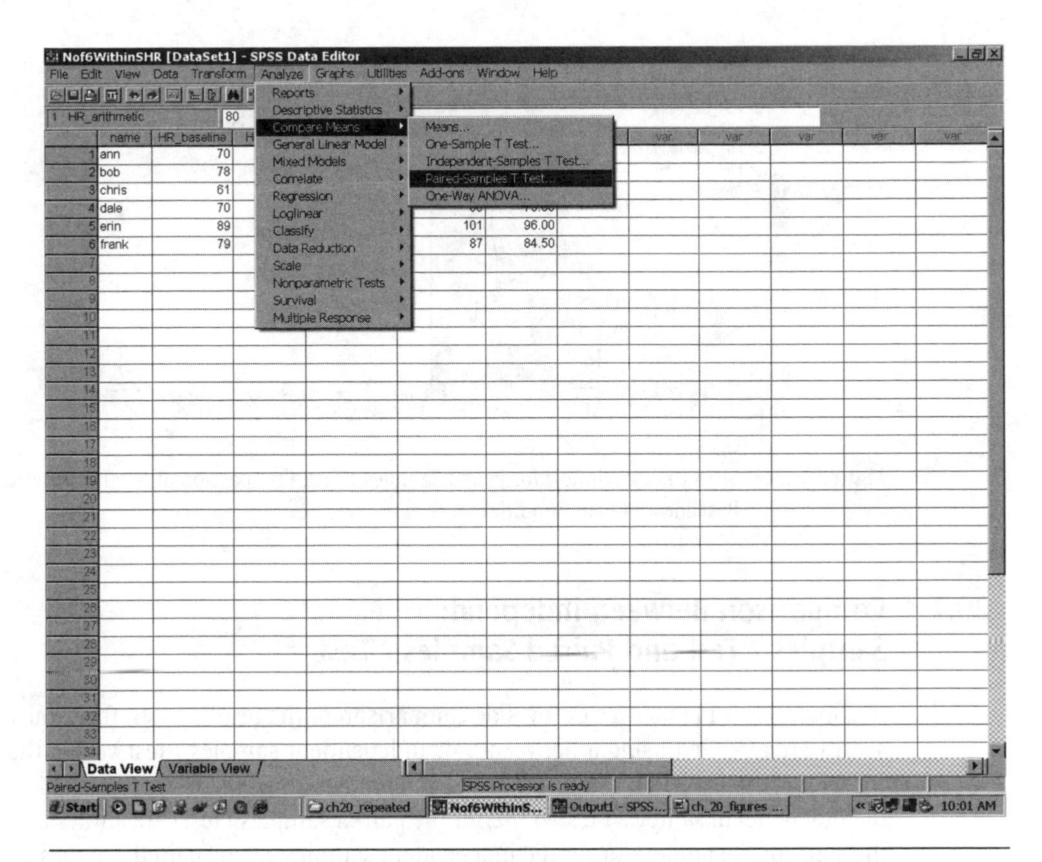

Figure 20.4 ♦ SPSS Menu Selections to Run Paired Samples *t* Test on Data in Table 20.2

Variables. Click OK to run the paired samples *t* test; the output from this procedure appears in Figure 20.6.

The output includes the means for each treatment group: M_1 = mean HR for the baseline condition = 74.50; M_2 = mean HR for the pain induction condition = 83.50. The correlation between the HR scores in the baseline condition and the HR scores in the pain condition was $r = +.888$. Because this *r* was large and positive, it is likely that a paired samples *t* test will have greater statistical power than an independent samples *t* test. The obtained paired samples *t* was $t(5) = -3.849$, $p = .012$. The mean for the difference scores, M_d, reported in the SPSS paired samples *t* test output was –9 because within SPSS, the difference score was obtained by subtracting the HR_pain scores from the HR_baseline scores. If we want to report a *t* test for difference scores computed in the more usual way (d = Time 2 score – Time 1 score), we would reverse the sign and report this as $t(5) = +3.849$, $p = .012$, two-tailed. Based on this analysis, we can conclude that the 9-point increase in mean HR for the pain induction condition relative to baseline (mean HR_pain = 83.50, mean HR_baseline = 74.50) was statistically significant at the $\alpha = .05$ level.

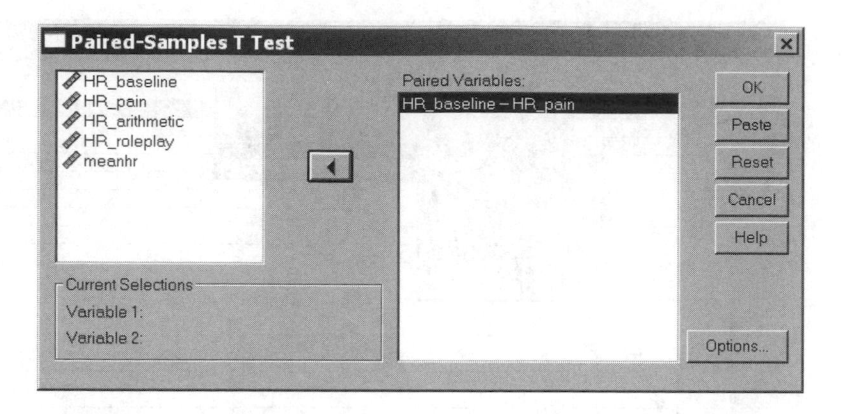

Figure 20.5 ♦ SPSS Dialog Window for Paired Samples *t* Test: Comparison of Mean HR Between Baseline and Pain Conditions

20.7 ♦ Comparison Between Independent Samples *t* Test and Paired Samples *t* Test

It is instructive to make a side-by-side comparison of the results when the same batch of scores are analyzed using a between-S or independent samples *t* test versus the paired samples *t* test. Table 20.5 summarizes these comparisons. The numerators ($M_1 - M_2$ for the independent samples *t* test vs. M_d for the paired samples *t* test) are always identical. If the signs of the numerator for the independent samples versus paired samples *t* tests are different, this is an artifact of the order in which the groups were placed when the differences between group means or the difference scores were calculated. For example, if we set up a *t* ratio using $M_{\text{pain}} - M_{\text{baseline}}$ as the numerator, the value of *t* will be positive because mean HR was 9 points higher in the pain situation than in the baseline condition. On the other hand, if we set up a *t* ratio with $M_{\text{baseline}} - M_{\text{pain}}$ as the numerator, the value of *t* will be negative. Similarly, if we compute each difference score as d = HR_ pain − HR_baseline, the mean of the *d* difference scores will be positive; on the other hand, if we compute each difference score as d = HR_baseline − HR_pain, the mean of the difference scores will be negative.

The divisor for the paired samples *t* test is usually smaller than the divisor for the independent samples *t* test because the sum of the squared error terms for the paired samples *t* test (i.e., the sum of the squared e^*_{ij} terms) excludes the variance due to stable individual differences among persons (the π term in the model in Equation 20.2), while the error variance for the independent samples *t* test (based on a sum of squared e_{ij} terms) includes the variance due to these individual differences. However, the *df* for the paired samples *t* test ($df = N - 1$, where N is the number of difference scores) is always smaller than the *df* for the independent samples *t* test ($df = N_1 + N_2 - 2$, where N_1 and N_2 are the numbers of scores in Treatment conditions 1 and 2). Because of this, the *df* term that is used to select the critical value of *t* required for statistical significance is smaller for the paired samples *t* test than for the independent samples *t* test, and the corresponding critical value of *t* is larger for the paired samples *t* test than for the independent samples *t* test. Therefore, a paired samples *t* test has better statistical power than an independent samples *t* test

Paired Samples Statistics

		Mean	N	Std. Deviation	Std. Error Mean
Pair 1	HR_baseline	74.50	6	9.649	3.939
	HR_pain	83.50	6	12.194	4.978

Paired Samples Correlations

		N	Correlation	Sig.
Pair 1	HR_baseline & HR_pain	6	.888	.018

Paired Samples Test

		Paired Differences					t	df	Sig. (2-tailed)
					95% Confidence Interval of the Difference				
		Mean	Std. Deviation	Std. Error Mean	Lower	Upper			
Pair 1	HR_baseline - HR_pain	-9.000	5.727	2.338	-15.010	-2.990	-3.849	5	.012

Figure 20.6 ♦ Output From SPSS Paired Samples t Test: Repeated Measures of Mean HR for $N = 6$ Participants, Baseline Versus Pain Condition, Data From Table 20.2

Table 20.5 ♦ Comparison of Independent Samples *t* Test Versus Paired Samples *t* Test (Applied to the Same Set of Scores)

	Independent Samples t Test	*Paired Samples t Test*	*Relative Sizes of Corresponding Terms*	*Comment*
Numerator	$M_1 - M_2$	M_d	$M_1 - M_2 = M_d$	These are always identical
Denominator	$SE_{M_1-M_2}$	SE_{Md}	$SE_{M_d} < SE_{M_1-M_2}$	Based on Equations 20.9 and 20.10, $SE_{M_d} < SE_{M1-M2}$ (as long as the correlation between the X_1 and X_2 scores is reasonably large and positive)
t	$t = (M_1 - M_2)/SE_{M_1-M_2}$	$t = M_d/SE_{Md}$	*t* ratio for a paired samples analysis is usually greater than *t* ratio obtained if the same data were analyzed using the independent samples *t* test	Because they have the same numerator and the paired samples *t* uses a smaller denominator, paired samples $t >$ independent samples *t*.
df	$N_1 + N_2 - 2$ (where N_1 and N_2 are the number of scores in Group 1 and Group 2, respectively)	$N - 1$ (where *N* is the number of difference scores)	*df* for the paired samples is less than *df* for the independent samples *t* test	The critical value of *t* required to reject H_0 is larger for the paired samples *t* test than for the independent samples *t* test
Power			Power for the paired samples *t* test is usually greater than power for the independent samples *t* test, provided that the reduction in the size of the denominator for the paired samples *t* test is great enough to compensate for the smaller *df*. In practice, power tends to be better for the paired samples *t* test when the correlation between scores at Time 1 and Time 2 is large and positive	

applied to the same batch of data only when the reduction in the sum of squared errors obtained using the paired samples *t* method leads to a relatively large decrease in the error term for the *t* ratio compared with the reduction in degrees of freedom used to evaluate the statistical significance of the *t* ratio. Under what circumstances does the paired samples *t* test provide better statistical power? The reduction in the size of the error variance that occurs when difference scores are calculated becomes greater as the correlation between the Time 1 and Time 2 scores increases. A precise statement of the amount of reduction of error variance that occurs when difference scores are calculated for correlated samples is given by Jaccard and Becker (2002). Recall from Chapter 5 that one version of the formula for $SE_{M_1-M_2}$ is

$$SE_{M_1-M_2} = \sqrt{\frac{s_1^2}{N_1} + \frac{s_2^2}{N_2}}, \quad (20.9)$$

where s_1^2 and s_2^2 are the sample variances of scores in Groups 1 (baseline) and 2 (pain induction) and N is the number of scores in each group. The SE of M_d, the mean of the difference scores, is given by

$$SE_{M_d} = \sqrt{\frac{s_1^2}{N} + \frac{s_2^2}{N} - \frac{2rs_1s_2}{N}}, \quad (20.10)$$

where r is the correlation between the scores in the paired samples. Assuming that r is positive, the stronger the correlation between the repeated measures (i.e., the higher the correlation between HR measures in the baseline condition and in the pain condition), the greater the reduction in error variance that we achieve by doing a paired samples *t* test using difference scores (rather than by doing an independent samples *t* test).

In this empirical example, $s_1 = 9.649$, $s_2 = 12.194$, $s_1^2 = 93.10$, $s_2^2 = 148.69$, $N = 6$, and $r = +.888$. We can compute SE_{M_d} by substituting these values into Equation 20.10, as follows:

$$SE_{M_d} = \sqrt{\frac{93.10}{6} + \frac{148.69}{6} - \frac{2 \times .888 \times 9.649 \times 12.194}{6}}$$
$$= \sqrt{15.5172 + 24.7823 - 34.8273} = \sqrt{5.4722} = 2.34.$$

Some parts of the results from the independent samples *t* test reported in Figure 20.1 are the same as the results for the paired samples *t* test in Figure 20.6, while other parts of the results differ. The mean HR for the baseline and pain conditions reported for the independent samples *t* test in Figure 20.1 are identical to the means for these two conditions reported in the paired samples *t* test (Figure 20.6). However, the *t* ratio obtained by doing an independent samples *t* test (from Figure 20.1, the independent samples *t* value was $t(20) = -1.142$, $p = .187$, two-tailed) was much smaller than the *t* ratio obtained by calculating a paired samples *t* test (from Figure 20.6, the paired samples *t* test for the HR data was $t(5) = -3.85$, $p = .012$, two-tailed). In this empirical example, the difference between mean HRs in the baseline versus pain conditions would be judged nonsignificant if the data were (inappropriately) analyzed as if they came from an independent groups design using an independent samples t test. On the other hand, when the scores were appropriately analyzed using a paired samples t test, the difference between mean HRs in the baseline versus pain situations was found to be statistically significant.

This empirical example illustrates one of the major potential advantages of repeated measures design: when each participant provides data in both of the treatment conditions that are compared, or when each participant serves as his or her own control, the variance due to stable individual differences among persons can be removed from the error term and the resulting repeated measures test often has better statistical power than an independent samples t test applied to the same set of scores.

20.8 ♦ SPSS Example: Analysis of Stress/HR Data Using a Univariate One-Way Repeated Measures ANOVA

When there are more than two repeated measures, the one-way repeated measures ANOVA is used to test this null hypothesis. For the data in Table 20.2, the null hypothesis would be

$$H_0: \mu_{\text{baseline}} = \mu_{\text{arithmetic}} = \mu_{\text{pain}} = \mu_{\text{roleplay}}. \tag{20.11}$$

In words, the null hypothesis is that the population mean HRs for persons who are tested under these four conditions (no stress/baseline, mental arithmetic, pain, and role play) are equal. In the more general case, with k = the number of repeated measures or treatments, the null hypothesis is written as

$$H_0: \mu_1 = \mu_2 = \mu_3 = \cdots = \mu_k. \tag{20.12}$$

The information needed to test this null hypothesis includes the set of sample means for HRs measured in these four treatment conditions, denoted by M_1 (mean HR during baseline), M_2 (pain), M_3 (arithmetic), and M_4 (role play). We also need information about error variance; in a repeated measures ANOVA, the error variance is what remains after we remove the variability associated with treatments and also the variability due to stable individual differences among persons from the overall SS_{total}. The partition of SS_{total} in a univariate one-way repeated measures ANOVA is given by the following equation:

$$SS_{\text{total}} = SS_{\text{treatment}} + SS_{\text{persons}} + SS_{\text{error}}. \tag{20.13}$$

In a one-way repeated measures ANOVA with N subjects and k treatments, there are a total of $N \times k$ (in our empirical example, $6 \times 4 = 24$) observations.

For SS_{total}, $df = (N \times k) - 1 = 23$.

$SS_{\text{treatment}}$ (or SS_{stress} or SS_{time}) represents the variability of mean scores on the outcome variable across different points in time, different trials, or different treatment conditions in the repeated measures study.

For $SS_{\text{treatment}}$, $df = k - 1$, where k is the number of within-S times or levels.

In the stress/HR study, $k = 4$, and therefore, $df_{\text{treatment}} = 3$.

For SS_{persons}, $df = N - 1$, where N is the number of persons in the study.
In the stress/HR study, $N = 6$; therefore, $df_{\text{persons}} = 5$.

For SS_{error}, $df = (N - 1)(k - 1)$.
In the stress/HR study, $df_{\text{error}} = 5 \times 3 = 15$.

Note how the partition of SS_{total} differs in a one-way repeated measures ANOVA compared with a one-way between-S ANOVA. In the independent samples or between-S ANOVA (as described in Chapter 6), the partition was

$$SS_{\text{total}} = SS_{\text{between}} + SS_{\text{within}}. \tag{20.14}$$

The SS_{within} in a between-S one-way ANOVA includes sources of error due to individual differences among persons (denoted SS_{persons} above), along with other sources of error. By comparison, in a one-way repeated measures ANOVA, the variance that is associated with individual differences among persons is not included in the error term; instead, a separate *SS* term was computed to capture the variance due to individual differences among participants in HR. In effect, the systematic individual differences among HR are no longer part of the SS_{error} term when the analysis is performed as repeated measures ANOVA. In other words, if we applied these two methods of computation of sums of squares to the same set of scores, SS_{error} for the repeated measures ANOVA (in Equation 20.13) will be smaller than SS_{within} for the independent groups ANOVA (in Equation 20.14) (just as SE_{M_d} is typically smaller than $SE_{M_1 - M_2}$ when we compare the paired samples and independent samples *t* tests).

As in earlier chapters, we begin the data analysis by finding a mean for each row and for each column of the data. In the general case, each row of the matrix has data for one person (denoted by $P_1, P_2, P_3, \ldots, P_N$, for *N* participants). Each column of the matrix has data for one treatment ($T_1, T_2, \ldots, T_k$, for *k* treatments). The mean for person *i* is denoted by M_{p_i}; the mean for each treatment *j* is denoted by M_{T_j}. The notation below can be used to represent repeated measures data for *N* persons across *k* times, trials, or treatments. Note that the data in Table 20.2 represent one specific case of this more general data matrix in which there are $N = 6$ persons and $k = 4$ treatments:

$$\begin{array}{c} \\ P_1 \\ P_2 \\ \vdots \\ P_N \\ \\ \end{array} \begin{array}{c} \begin{array}{ccccc} T_1 & T_2 & T_3 & \cdots & T_k \end{array} \\ \left[\begin{array}{ccccc} X_{11} & X_{12} & X_{13} & & X_{1k} \\ X_{21} & X_{22} & X_{23} & & X_{2k} \\ & & & & \\ X_{P1} & X_{P2} & X_{P3} & & X_{Pk} \end{array}\right] \\ \begin{array}{ccccc} M_{T_1} & M_{T_2} & M_{T_3} & & M_{T_k} \end{array} \end{array} \begin{array}{c} \\ M_{P1} \\ M_{P2} \\ \\ M_{P_N} \\ M_{\text{grand}} \end{array}.$$

This is the notation for a set of scores obtained by making repeated measures on Persons 1 through *N* across Treatments or Times or Trials 1 through *k*; this corresponds to the numerical data in Table 20.2. The score in each cell, X_{ij}, represents the HR measurement for person i at time j; for example, the first score in the upper left corner of Table 20.2 represents Ann's HR during the baseline condition. The mean for each row (M_{P_i}) represents the average HR for participant i (averaged across the four treatment conditions). The mean of each column (M_{T_j}) represents the average HR for each treatment (averaged across all six persons). In the lower right corner of the table, $M_{grand} = 80.375$ represents the grand mean HR (across all participants and times). To obtain a value for SS_{total}, as in any other ANOVA, we subtract the grand mean M_{grand} from each individual score, square the resulting deviations, and then sum the deviations (across all times and all treatments—in this case, across all 24 scores in the body of the table above):

$$SS_{total} = \Sigma\,(X_{ij} - M_{grand})^2. \tag{20.15}$$

In this example, we would obtain $SS_{total} = (70 - 80.375)^2 + (80 - 80.375)^2 + \cdots + (87 - 80.375)^2 = 2681.625$. To obtain a value for $SS_{treatment}$ or SS_{time}, we need to find the deviation of each column mean (for scores for treatment j) from the grand mean, square these deviations, and sum the squared deviations across all k times or treatments:

$$SS_{treatment} = N \times \Sigma\,(M_{T_j} - M_{grand})^2, \tag{20.16}$$

where

M_{T_j} is the mean score for treatment condition j,

M_{grand} is the grand mean of all the scores, and

N is the number of scores in each treatment condition (in this example, $N = 6$).

In this case, when we apply Equation 20.16 to the data in Table 20.2, we would have

$$SS_{treatment} = 6 \times [(74.5 - 80.375)^2 + (83.50 - 80.375)^2 + (77.67 - 80.375)^2 + (85.83 - 80.375)^2]$$

$$SS_{treatment} = 488.4583.$$

To obtain a value for $SS_{persons}$, we need to find the deviation of each row mean (i.e., the mean across all scores for person i) from the grand mean, square this deviation, and sum the squared deviations across all N persons:

$$SS_{persons} = k \times \Sigma\,(M_{P_i} - M_{grand})^2, \tag{20.17}$$

where

M_{P_i} is the mean of all the scores for person i,

M_{grand} is the grand mean, and

k is the number of treatment conditions or trials (in this example, $k = 4$).

In this case, when we apply Equation 20.17 to the data in Table 20.2, we obtain:

$$SS_{\text{persons}} = 4 \times [(80.75 - 80.375)^2 + (79.25 - 80.375)^2 + (66.75 - 80.375)^2 + (75 - 80.375)^2 + (96 - 80.375)^2 + (84.50 - 80.375)^2] = 1908.375.$$

Finally, the value for SS_{error} can be obtained by subtraction:

$$SS_{\text{error}} = SS_{\text{total}} - SS_{\text{treatment}} - SS_{\text{persons}}. \tag{20.18}$$

For the data in the preceding example, using the data from Table 20.2, we obtain the following:

$$SS_{\text{error}} = SS_{\text{total}} - SS_{\text{treatment}} - SS_{\text{persons}}$$

$$284.792 = 2681.6250 - 488.4583 - 1908.375.$$

Each *SS* is used to calculate a corresponding mean square (*MS*) term by dividing it by the appropriate *df* term. An *MS* term is not usually presented for the persons factor when summary information is reported for a repeated measures ANOVA.

$$MS_{\text{treatment}} = SS_{\text{treatment}}/(k - 1) = 488.4583/3 = 162.819 \tag{20.19}$$

$$MS_{\text{error}} = SS_{\text{error}}/(N - 1)(k - 1) = 284.7917/15 = 18.986. \tag{20.20}$$

Finally, an *F* ratio is set up; the *MS* for treatments is evaluated relative to *MS* for error:

$$F = MS_{\text{treatment}}/MS_{\text{error}}, \text{ with } k - 1 \text{ and } (N - 1)(k - 1)\ df. \tag{20.21}$$

For a one-way repeated measures ANOVA of the scores in Table 20.2, we obtain $F = 162.819/18.986 = 8.58$, with 3 and 15 *df*. The critical value of *F* for $\alpha = .05$ and (3, 15) *df* is 3.29. This result for this one-way repeated measures ANOVA can therefore be judged statistically significant at the $\alpha = .05$ level.

20.9 ♦ Using the SPSS GLM Procedure for Repeated Measures ANOVA

The SPSS procedure that provides the most flexible approach to handling repeated measures is the general linear model (GLM). A one-way repeated measures can also be obtained from the SPSS reliability procedure, but the reliability procedure does not provide for more complex designs that can be analyzed in SPSS GLM. To carry out a one-way repeated measures ANOVA on the scores shown in Table 20.2, the data were entered into an SPSS data file, as shown in Figure 20.3. The scores for each of the $N = 6$ participants appear on one line, and repeated measures of HR under each of the $k = 4$ different treatment conditions appear as variables. In this example, these variables are named HR_baseline, HR_pain, HR_arithmetic, and HR_roleplay.

The SPSS menu selections to run a univariate repeated measures ANOVA using GLM appear in Figure 20.7: <Analyze> → <General Linear Model> → <Repeated Measures>.

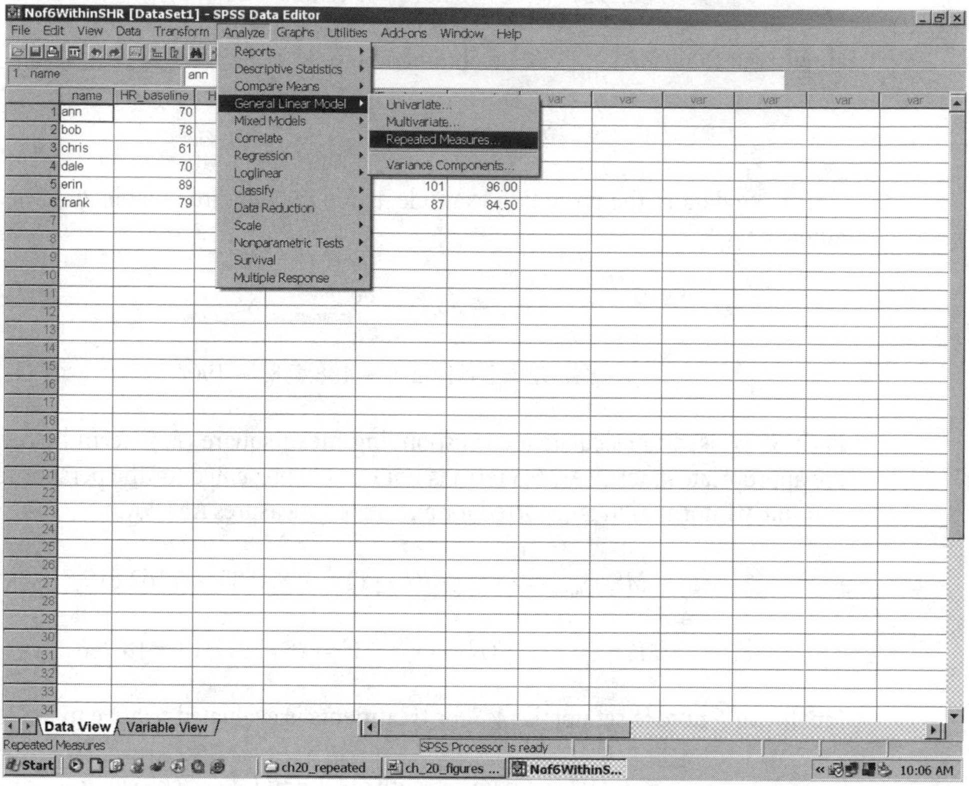

Figure 20.7 ♦ SPSS Menu Selections to Request Both Univariate Repeated Measures and MANOVA Analysis of Data in Figure 20.2 From SPSS GLM Repeated Measures

This series of menu selections opens up the Repeated Measures dialog window shown in Figure 20.8. Initially, the default name for the repeated measures factor was factor1. In this example, the default name for the within-S factor, factor1, was replaced by the name "Stress" by typing that new label in place of "factor1." The number of levels of the repeated measures factor (in this example, $k = 4$) was placed in the box for Number of Levels immediately below the name of the repeated measures factor. Clicking the button that says Add moves the new factor named Stress into the box or window that contains the list of repeated measures factors (this box is to the right and below the label Number of Levels). More than one repeated measures factor can be specified, but only one repeated measures factors was included in this example. Clicking on the button marked Define opens up a new dialog window, which appears in Figure 20.9. In this second dialog window, the user tells SPSS the name(s) of the variables that correspond to each level of the repeated measures factor by clicking on each blank name in the initial list under the heading Within Subjects Variables and using the right arrow to move each variable name into the list of levels. In this example, the first level of the repeated measures factor corresponds to the variable named HR_baseline, the second level corresponds to the variable named HR_pain, the third level corresponds to HR_arithmetic, and the fourth and last level corresponds to the variable named HR_roleplay. When all four of these variables have been specified as the levels of the within-S factor, the description of the basic repeated measures design is complete.

Additional optional information can be requested by opening up other windows. In this example, the Contrasts button was used to request planned comparisons among levels of the repeated measures factor (see Figure 20.10). In this example, "Simple" contrasts were selected from the pull-down menu of different types of contrasts. In addition, a radio button selection specified the first category (which corresponds to HR during baseline, in this example) as the reference group for all comparisons. In other words, planned contrasts were done to see whether the mean HRs for Level 2 (pain), Level 3 (arithmetic), and Level 4 (role play) each differed significantly from the mean HR on Level 1 of the repeated measures factor (baseline/no stress).

The Options button was used to open up the Options dialog window in Figure 20.11. Two check box selections were made to request descriptive statistics and estimates of effect size. Note that if you do not request descriptive statistics or place the names of factors in the window under the heading "Display Means for," means for the scores at each time or under each treatment condition are not provided as part of the GLM output. A plot of the means for HR across the four trials or types of stress was requested using the Profile Plots window (in Figure 20.12). The name of the within-S factor—in this example, Stress—was moved into

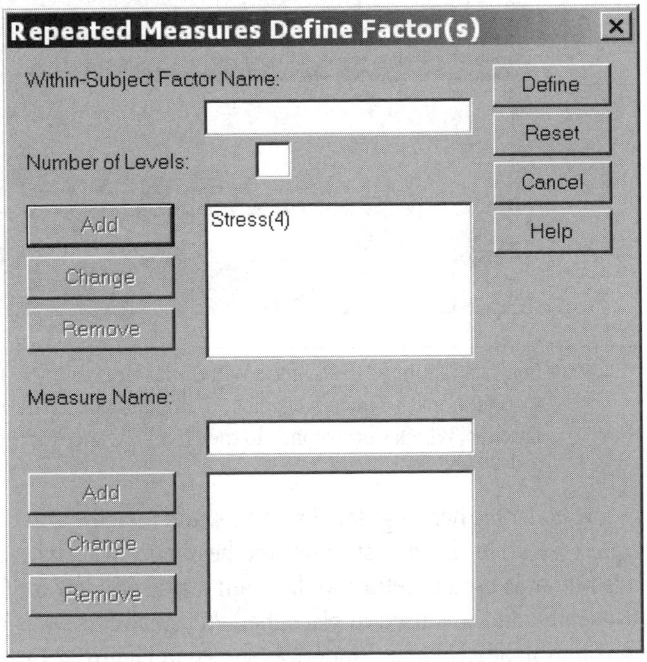

Repeated Measures Define Factor(s)

Within-Subject Factor Name:

Number of Levels:

Add | Change | Remove

Stress(4)

Measure Name:

Add | Change | Remove

Define | Reset | Cancel | Help

Figure 20.8 ♦ SPSS Repeated Measures: Define Factors

NOTE: Name for the within-subjects factor: Stress. Number of levels for within-subjects factor: 4.

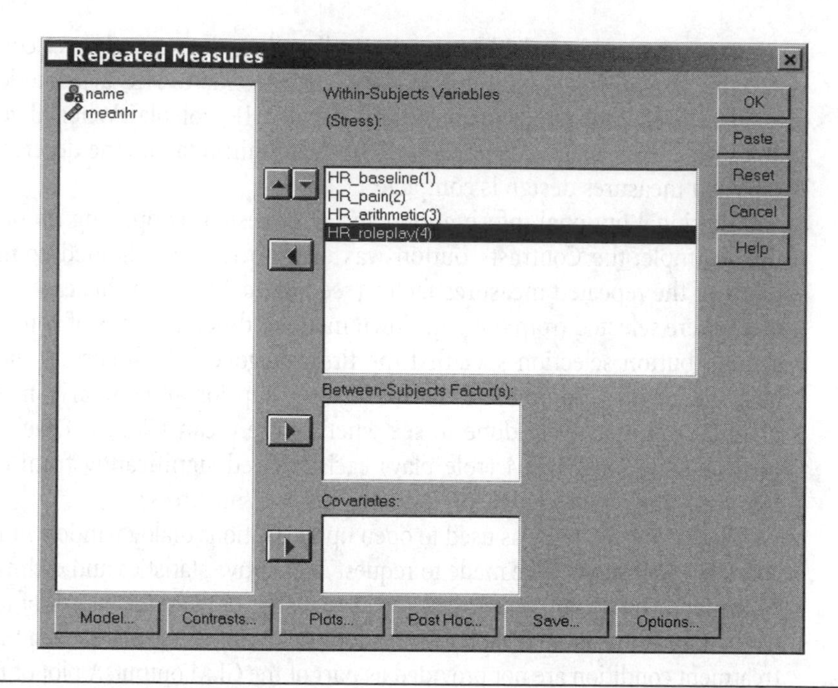

Figure 20.9 ♦ Identification of Variables That Correspond to Each of the Four Levels of the Within-Subjects Factor Stress

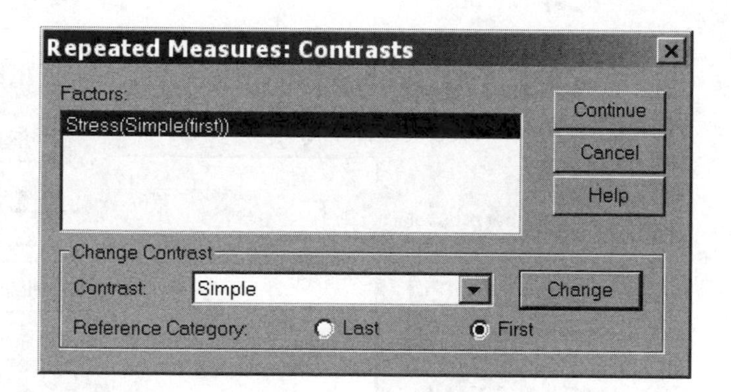

Figure 20.10 ♦ Contrasts Among Levels of the Within-Subjects Factor (Mean HR for the Stress Interventions at Times 2, 3, and 4, Each Contrasted With Mean HR for Time 1, Baseline, Which Corresponds to the First Reference Category)

the window under the heading Horizontal Axis, and then the Add button was clicked to place this request for a plot in the list under the heading Plots. For each of these windows, the Continue button is used to return to the main dialog window for GLM Repeated Measures. The Paste command was used to place the SPSS syntax that was generated by these menu selections into an SPSS Syntax window, as shown in Figure 20.13. When the data analyst has made menu selections similar to those in this example, GLM will do the following: run tests to assess whether the assumptions about the pattern in the set of variances and covariances for the repeated measures are violated; run a one-way repeated measures ANOVA that involves the computation of sums of squares and *F* ratios; and also, run a repeated measures analysis that is set up as a MANOVA (see section 20.12). If assumptions about the pattern of variances and covariances are seriously violated, data analysts sometimes report a one-way repeated measures with *df* terms that are adjusted to compensate for the violation of assumptions. Alternatively, when assumptions about the pattern of variances and covariances are seriously violated, they may prefer to report the MANOVA results because the MANOVA approach does not require such restrictive assumptions about data pattern. The differences between these two approaches are discussed in later sections of this chapter.

Basic descriptive statistics such as mean HR for each treatment condition appear in Figure 20.14. Keep in mind that the research question is whether mean HR is significantly higher during the three different types of stress intervention than during the baseline and, also, whether the three different types of stress differ significantly in their impact on HR. Preliminary examination of the means for the four different stress interventions in Figure 20.14 suggested that, as expected, mean HR was lowest during the baseline measurement period. Among the three types of stress, role play had the highest mean HR, pain had the second highest HR, and mental arithmetic had a mean HR that was only a little higher than HR during the no-stress/baseline condition.

The sequence of the output from the SPSS GLM procedure is as follows: The MANOVA version of the analysis appears in Figure 20.15; the test of assumptions about the pattern

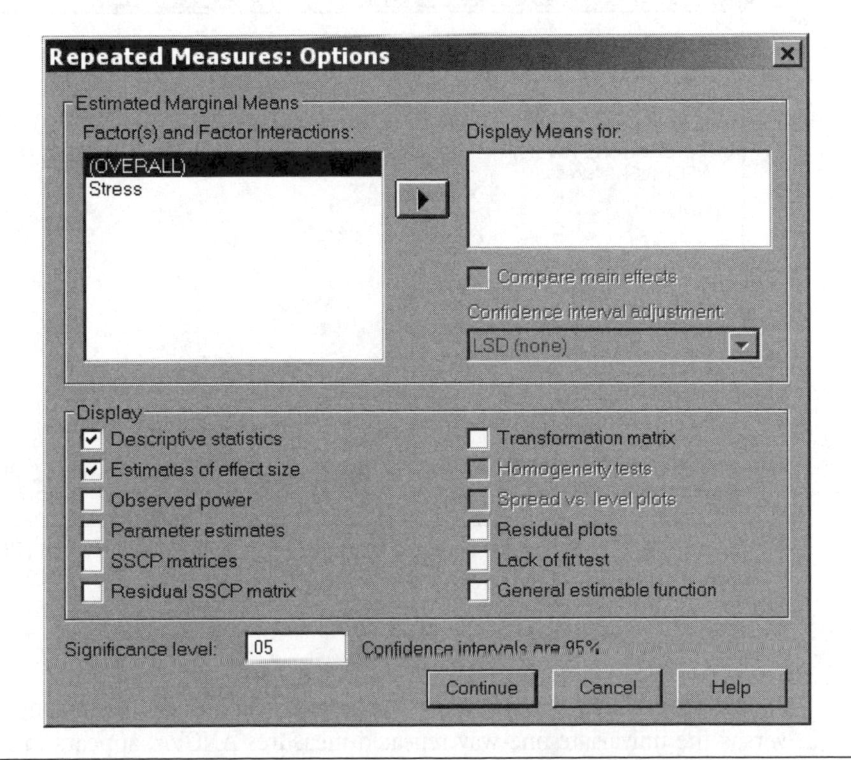

Figure 20.11 ♦ Optional Statistics Requested for SPSS GLM Repeated Measures

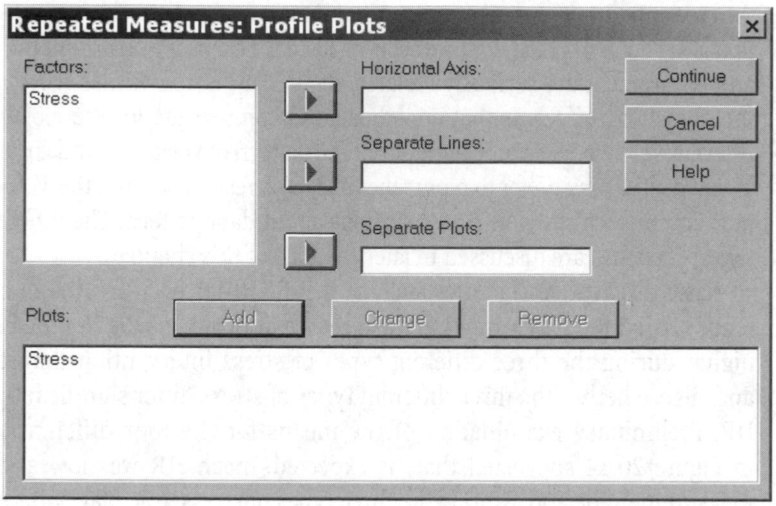

Repeated Measures: Profile Plots

Factors: Stress

Horizontal Axis:

Separate Lines:

Separate Plots:

Continue

Cancel

Help

Plots: Add Change Remove

Stress

Figure 20.12 ♦ Request of Plot of Mean HR Across All Four Levels of the Within-Subjects Stress Factor

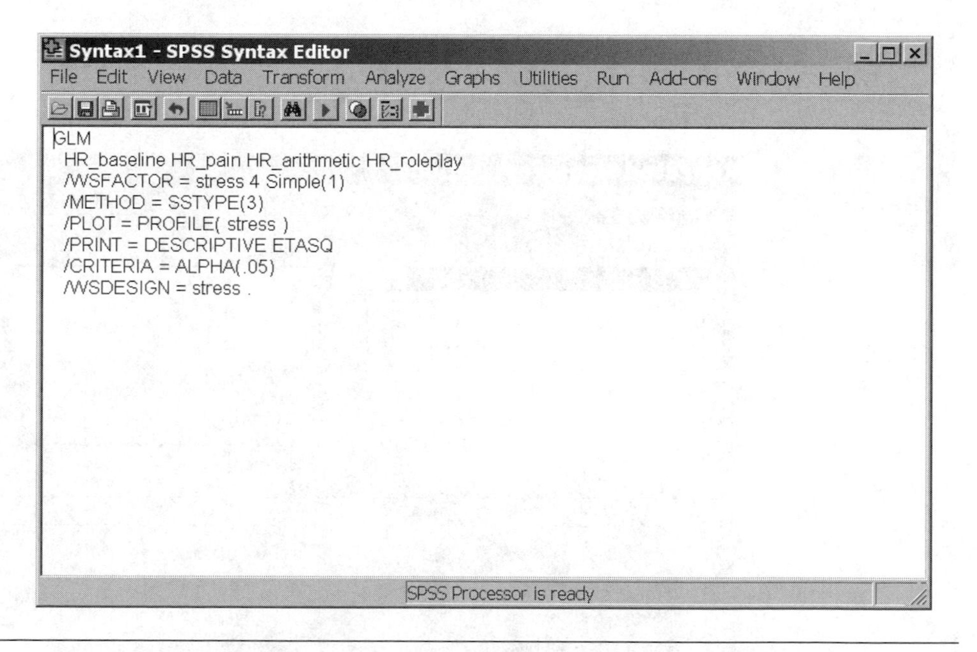

Figure 20.13 ♦ SPSS Syntax for Repeated Measures Analysis to Compare Mean HR Across Four Levels of the Stress Factor

of variances and covariances, which can be used to choose between reporting the MANOVA versus the univariate one-way repeated measures ANOVA, appears in Figure 20.16; the univariate repeated measures ANOVA results are presented in Figure 20.17; and the

Within-Subjects Factors

Measure: MEASURE_1

Stress	Dependent Variable
1	HR_baseline
2	HR_pain
3	HR_arithmetic
4	HR_roleplay

Descriptive Statistics

	Mean	Std. Deviation	N
HR_baseline	74.50	9.649	6
HR_pain	83.50	12.194	6
HR_arithmetic	77.67	10.211	6
HR_roleplay	85.83	9.621	6

Figure 20.14 ♦ SPSS Descriptive Statistics for GLM Repeated Measures ANOVA of Stress and HR Data in Table 20.2

Multivariate Tests[b]

Effect		Value	F	Hypothesis df	Error df	Sig.	Partial Eta Squared
Stress	Pillai's Trace	.840	5.269[a]	3.000	3.000	.103	.840
	Wilks's Lambda	.160	5.269[a]	3.000	3.000	.103	.840
	Hotelling's Trace	5.269	5.269[a]	3.000	3.000	.103	.840
	Roy's Largest Root	5.269	5.269[a]	3.000	3.000	.103	.840

a. Exact statistic

b. Design: Intercept
Within Subjects Design: Stress

Figure 20.15 ♦ MANOVA Approach to Analysis of Repeated Measures Data in Table 20.2

requested contrasts between mean HR in each of the three stress conditions and mean HR during baseline appear in Figure 20.18. Figure 20.19 presents a test for whether the grand mean for the measured outcome variable (in this case, HR) differed significantly from 0 for all the data across the entire study; this test is rarely of interest. Figure 20.20 presents the plot of mean HR across the four levels of stress.

For the univariate one-way repeated measures ANOVA reported in Figure 20.17, for each source of variance (treatment, error) there are four separate lines in the table; in all four lines, the *SS* terms are the same. The partition of sum of squares shown in this table

Mauchly's Test of Sphericity[b]

Measure: MEASURE_1

Within Subjects Effect	Mauchly's W	Approx. Chi-Square	df	Sig.	Epsilon[a] Greenhouse-Geisser	Huynh-Feldt	Lower-bound
Stress	.833	.681	5	.985	.897	1.000	.333

Tests the null hypothesis that the error covariance matrix of the orthonormalized transformed dependent variables is proportional to an identity matrix.

a. May be used to adjust the degrees of freedom for the averaged tests of significance. Corrected tests are displayed in the Tests of Within-Subjects Effects table.

b. Design: Intercept
Within Subjects Design: Stress

Figure 20.16 ♦ Test of Assumptions About Pattern in the Variance/Covariance Matrix for the Repeated Measures

Tests of Within-Subjects Effects

Measure: MEASURE_1

Source		Type III Sum of Squares	df	Mean Square	F	Sig.	Partial Eta Squared
Stress	Sphericity Assumed	488.458	3	162.819	8.576	.001	.632
	Greenhouse-Geisser	488.458	2.690	181.567	8.576	.002	.632
	Huynh-Feldt	488.458	3.000	162.819	8.576	.001	.632
	Lower-bound	488.458	1.000	488.458	8.576	.033	.632
Error(Stress)	Sphericity Assumed	284.792	15	18.986			
	Greenhouse-Geisser	284.792	13.451	21.172			
	Huynh-Feldt	284.792	15.000	18.986			
	Lower-bound	284.792	5.000	56.958			

Figure 20.17 ♦ Univariate ANOVA Approach to Repeated Measures ANOVA of Data in Table 20.2

Tests of Within-Subjects Contrasts

Measure: MEASURE_1

Source	Stress	Type III Sum of Squares	df	Mean Square	F	Sig.	Partial Eta Squared
Stress	Level 2 vs. Level 1	486.000	1	486.000	14.817	.012	.748
	Level 3 vs. Level 1	60.167	1	60.167	2.231	.195	.309
	Level 4 vs. Level 1	770.667	1	770.667	21.977	.005	.815
Error(Stress)	Level 2 vs. Level 1	164.000	5	32.800			
	Level 3 vs. Level 1	134.833	5	26.967			
	Level 4 vs. Level 1	175.333	5	35.067			

Figure 20.18 ♦ Contrasts to Compare Mean HR for Each Stress Intervention (Level 2 = Pain, Level 3 = Arithmetic, Level 4 = Role Play) With Baseline/No-Stress Condition (Level 1)

Tests of Between-Subjects Effects

Measure: MEASURE_1
Transformed Variable: Average

Source	Type III Sum of Squares	df	Mean Square	F	Sig.	Partial Eta Squared
Intercept	38760.844	1	38760.844	406.218	.000	.988
Error	477.094	5	95.419			

Figure 20.19 ♦ Test of Null Hypothesis That Grand Mean for HR = 0 Across the Entire Study

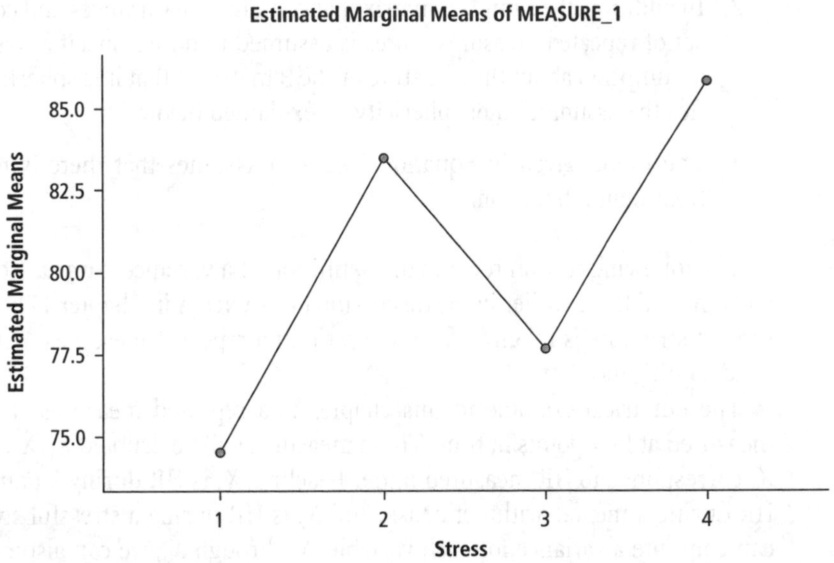

Estimated Marginal Means of MEASURE_1

85.0
82.5
80.0
77.5
75.0
Estimated Marginal Means
1 2 3 4
Stress

Figure 20.20 ♦ Plot of Mean HR Across Four Levels of Stress

corresponds to the values reported earlier for the by-hand computation; in Figure 20.17, SS_{stress} or $SS_{treatment}$ = 488.458 (the same value obtained using Equation 20.16 above), and SS_{error} = 284.792 (the same value obtained using Equation 20.18 above). The next section explains the four lines of information (**sphericity** assumed, Greenhouse-Geisser, Huynh-Feldt, and lower bound) that are listed for each source of variance in the design. The first line (sphericity assumed) corresponds to the outcome of the analysis if we assume that the assumptions about variances and covariances among repeated measures are satisfied and perform a univariate repeated measures ANOVA without making any adjustment to the *df* for the significance test for differences in mean HR across levels of the treatment factor stress. The following three lines report downwardly adjusted *df* values that may be used to evaluate the statistical significance of the *F* test for differences in HR across levels of stress when the assumptions for univariate repeated measures are violated.

20.10 ♦ Screening for Violations of Assumptions in Univariate Repeated Measures

For the *F* ratio in repeated measures ANOVA to provide an accurate indication of the true risk of Type I error, the assumptions for repeated measures ANOVA need to be reasonably well satisfied. For a repeated measures ANOVA, these assumptions include the following:

1. Scores on the outcome variables should be quantitative and approximately normally distributed. Scores on the repeated measures variables $X_1, X_2, \ldots, X_k$ should have a multivariate normal distribution.

2. Relationships among repeated measures should be linear.

3. In addition, the sample **S** matrix that contains the variances and covariances for the set of repeated measures scores is assumed to have a specific structure. The usual assumption about the structure of the **S** matrix is that it is spherical or that it satisfies the assumption of sphericity, as explained below.
4. The model given in Equation 20.2 also assumes that there is no participant by treatment interaction.

The following section reviews the definition of a variance/covariance matrix (we have encountered these earlier in the discussion of MANOVA in Chapter 17) and explains what type of structure is required for the *F* ratio in a repeated measures ANOVA to provide a valid significance test.

The empirical example in this chapter is a repeated measures study where HR is measured at four points in time. These measures will be denoted X_1, X_2, X_3, and X_4, where X_1 corresponds to HR measured under baseline, X_2 is HR during a pain induction, X_3 is HR during a mental arithmetic task, and X_4 is HR during a stressful social role play. We can compute a variance for each variable, X_1 through X_4; we can also compute a covariance between all possible pairs of variables, such as X_1 with X_2, X_1 with X_3, X_3 with X_4, and so forth. The variances and covariances for this set of four variables can be summarized as a sample variance covariance matrix **S**, as shown below:

$$\mathbf{S} = \begin{array}{c} \\ X_1 \\ X_2 \\ X_3 \\ X_4 \end{array} \begin{array}{c} \begin{array}{cccc} X_1 & X_2 & X_3 & X_4 \end{array} \\ \begin{bmatrix} s_1^2 & s_{12} & s_{13} & s_{14} \\ s_{21} & s_2^2 & s_{23} & s_{24} \\ s_{31} & s_{32} & s_3^2 & s_{34} \\ s_{41} & s_{42} & s_{43} & s_4^2 \end{bmatrix} \end{array}. \tag{20.22}$$

The elements in the main diagonal of the **S** matrix (which runs from the upper left to the lower right corner) are the sample variances for the four variables in this study—for example, s_1^2 is the sample variance of scores for X_1. The off-diagonal elements are the covariances for all possible pairs of variables—for example, s_{12} is the covariance between X_1 and X_2. The formula for a covariance between variables X_1 and X_2 is as follows:

$$s_{12} = \text{covariance of } (X_1, X_2) = \frac{\sum (X_1 - M_1)(X_2 - M_2)}{N}. \tag{20.23}$$

The corresponding population matrix, Σ, contains the population variances and covariances for the same set of variables:

$$\mathbf{\Sigma} = \begin{array}{c} \\ X_1 \\ X_2 \\ X_3 \\ X_4 \end{array} \begin{array}{c} \begin{array}{cccc} X_1 & X_2 & X_3 & X_4 \end{array} \\ \begin{bmatrix} \sigma_1^2 & \sigma_{12} & \sigma_{13} & \sigma_{14} \\ \sigma_{21} & \sigma_2^2 & \sigma_{23} & \sigma_{24} \\ \sigma_{31} & \sigma_{32} & \sigma_3^2 & \sigma_{34} \\ \sigma_{41} & \sigma_{42} & \sigma_{43} & \sigma_4^2 \end{bmatrix} \end{array}. \tag{20.24}$$

Two terms are widely used to describe the structure of a variance/covariance matrix—that is, the nature of the relationship among the elements of this matrix. **Compound symmetry** is a very simple structure; for **S** to have compound symmetry, all the variances on the diagonal must be equal:

$$H_0 : \sigma_1^2 = \sigma_2^2 = \cdots = \sigma_k^2. \tag{20.25}$$

In addition, for a variance/covariance matrix to conform to compound symmetry, all the off-diagonal elements that represent covariances must also be equal; that is

$$\sigma_{12} = \sigma_{13} = \sigma_{14} = \sigma_{23} = \sigma_{24} = \sigma_{34}, \tag{20.26}$$

where each σ_{ij} term represents the covariance between X_i and X_j; for example, σ_{23} is the covariance between X_2 and X_3. For compound symmetry to hold, it is not necessary for the variances to equal the covariances.

Note that compound symmetry is essentially a generalization of an assumption that was required for one-way between-S ANOVA; that is, Equation 20.25 is just the familiar assumption of homogeneity of variances across treatment populations presented when you first learned about between-S one-way ANOVA. However, for repeated measures ANOVA, we also need to make assumptions about the relationships among covariances (as shown in Equation 20.26). The assumption that $\boldsymbol{\Sigma}$ has compound symmetry thus combines an assumption of homogeneity of the variances for the repeated measures X_1, X_2, X_3, and X_4 and an assumption of equality of covariances for all possible pairs of the X variables. Unfortunately, real repeated measures data rarely show compound symmetry.

Fortunately, there is another, somewhat less restrictive pattern of relationships among covariances called sphericity, which is sufficient for the F ratio in repeated measures ANOVA to provide valid results. To assess sphericity, we need to look at differences or contrasts between the repeated measures—for example, $X_1 - X_2$, $X_1 - X_3$, and $X_1 - X_4$. If these differences have equal variances, then the assumption of sphericity is satisfied. The matrix shown below satisfies the sphericity assumption if the variances in the diagonal of the matrix are all equal to each other—that is, $\sigma_{1-2}^2 = \sigma_{1-3}^2 = \sigma_{1-4}^2$:

$$\begin{array}{c} \\ X_1 - X_2 \\ X_1 - X_3 \\ X_1 - X_4 \end{array} \begin{array}{c} \begin{array}{ccc} X_1 - X_2 & X_1 - X_3 & X_1 - X_4 \end{array} \\ \begin{bmatrix} \sigma_{1-2}^2 & & \\ & \sigma_{1-3}^2 & \\ & & \sigma_{1-4}^2 \end{bmatrix} \end{array}. \tag{20.27}$$

Note that if a matrix has compound symmetry, it must also satisfy the sphericity assumption, but a matrix that is spherical does not necessarily show compound symmetry. SPSS provides the Mauchly test to assess possible violations of the sphericity assumption (see the output in Figure 20.16). For the data used as an example in this chapter, there was no evidence of a significant violation of the sphericity assumption, with Mauchly's $W = .833$, $\chi^2\,(5) = .681$, $p = .985$. When the number of cases is small, this test is not robust to violations of the multivariate normality assumption, and the test may have too little power to detect departures from sphericity that could cause problems with the F test. On the other hand, when the number of cases is large, this test has too much power, and it may suggest a significant violation of sphericity even when the departure from sphericity is so small that it has very little effect on the F ratio. For these reasons, the Mauchly test may not be very helpful in evaluating whether there is a serious violation of the sphericity assumption. The most conservative course of action would be to assume that this assumption is violated unless there is strong evidence that it is not violated.

When the sphericity assumption is violated, the actual risk of Type I error associated with an F ratio for a repeated measures factor is likely to be higher than the nominal α level. There are two ways to remedy this problem. The first method involves using (downwardly) adjusted *df* values; this is described in Section 20.11. The second method involves using MANOVA rather than univariate repeated measures ANOVA; this method is described briefly in Section 20.12.

20.11 ♦ The Greenhouse-Geisser ε and Huynh-Feldt ε Correction Factors

When the sphericity assumption is met, the degrees of freedom that are used to determine the critical value of the F test for a one-way repeated measures ANOVA (with N participants and k levels on the repeated measures factor) are $(k - 1)$ and $(N - 1)(k - 1)$.

The **epsilon (ε)** statistic is a measure of the degree to which the sample variance covariance matrix **S** departs from sphericity. The highest possible value of $\varepsilon = 1.00$; when $\varepsilon = 1.00$, there is no departure from sphericity. In other words, the pattern of variances and covariances in the sample variance/covariance matrix **S** is consistent with the sphericity assumption. The lowest possible value of ε (for a repeated measures factor with k levels) is $1/(k - 1)$. In the example used throughout this chapter, with $k = 4$ levels, the lower-bound ε value is $1/(4 - 1) = .333$. Thus, we can evaluate the degree of violation of the sphericity assumption for a particular set of data by looking at ε. The closer the obtained value of ε is to the lower-bound value of ε, the worse the violation of the sphericity assumption; the closer ε is to 1.00, the less the sample variance/covariance matrix departs from sphericity.

SPSS provides two different versions of this ε index: the **Greenhouse-Geisser ε** and the **Huynh-Feldt ε.** For the empirical data used as an example throughout this chapter, the Greenhouse-Geisser ε was .897, and the Huynh-Feldt ε was 1.00 (as reported in Figure 20.16 along with the Mauchly test). The Greenhouse-Geisser version of ε is more conservative; the Huynh-Feldt ε is less conservative. The relatively large values of ε, along with the nonsignificant Mauchly test, suggest that any departure from sphericity for these data was relatively small.

These two ε values are used to make downward adjustments to the *df* used to determine the critical value of F for significance tests. The actual F ratio does not change, only the *df* values that are used to decide which F distribution should be used to evaluate significance. The *df*s for both the numerator and the denominator of F are multiplied by ε to obtain adjusted *df* values (this often results in noninteger values of *df*, of course). SPSS reports four different tests for repeated measures factors, as shown in Figure 20.17. For each source of variance (stress and error), there are four lines in the table. The first line, labeled Sphericity Assumed, gives the unadjusted *df*s for the F ratio. These unadjusted *df*s correspond to $(k - 1)$ *df* for the numerator and $(N - 1)(k - 1)$ *df* for the denominator, or in this case, 3 and 15 *df*. If the sphericity assumption is not violated, the obtained F test can be evaluated using critical values from an F distribution with 3 and 15 *df*. The second line, labeled Greenhouse-Geisser, reports *df* values that are obtained when the unadjusted *df*s are multiplied by the Greenhouse-Geisser ε (for both the numerator and the denominator). The Greenhouse-Geisser ε value was .897 for this set of data, so the adjusted *df*s for the Greenhouse-Geisser test were $3 \times .897 = 2.691$ and $15 \times .897 = 13.451$. The *SS*, *MS*, and F ratio reported as the Greenhouse-Geisser results are the same as in the Sphericity Assumed line of the table. The only thing that differs is the theoretical F distribution that is used to evaluate the significance of the obtained F. For the Greenhouse-Geisser test, the p value represents the tail area cut off by an obtained F of 8.576 relative to an F distribution with 2.690 and 13.451 *df*. In this example, because the Huynh-Feldt ε was 1.00, the Huynh-Feldt *df* values are the same as those in the first line of the table (the Sphericity Assumed results).

The most conservative adjustment to *df* would be obtained by taking the lower-limit value of ε (in this example, lower-limit $\varepsilon = .333$), and using that as a multiplier for the *df*, these adjusted *df*s would be $.333 \times 3 = 1$ for the numerator of F and $.333 \times 15 = 5$ for the denominator of F. In this case, using the Sphericity Assumed *df*, the overall F for the repeated measures ANOVA would be judged statistically significant: $F(3, 15) = 8.58$, $p = .001$. If the more conservative Greenhouse-Geisser adjusted *df*s are used, the result would be reported as $F(2.69, 13.451) = 8.58$, $p = .002$. If the most conservative possible adjustment to *df* is made using the lower-limit ε value, we would report $F(1, 5) = 8.58$, $p = .033$; this would still be judged statistically significant using the conventional $\alpha = .05$ criterion for significance.

For some datasets (as in this example), the decision about statistical significance is the same no matter which ε adjustment factor is used; in this example, no matter what adjustments are made to the *df*, the F ratio would be judged significant (using $\alpha = .05$ as the criterion). In other studies, no matter which set of *df*s are used to evaluate the F ratio, the outcome of the study would be judged nonsignificant. In these situations, the researcher may choose to report the sphericity-assumed test or one of the tests that uses adjusted *df*.

However, there are situations where the p value is less than .05 using the Sphericity Assumed *df* and the p value is greater than .05 for at least some of the tests that use adjusted *df*. These situations require careful consideration. The most conservative action would be to judge the results nonsignificant; another possible alternative is to use MANOVA instead of univariate repeated measures (as discussed in Section 20.12).

Field (1998) summarizes arguments that suggest that the Greenhouse-Geisser ε underestimates sphericity and that the Huynh-Feldt ε overestimates sphericity (in other words, the correction to the *df* using the Greenhouse-Geisser ε may be too conservative, while the correction to the *df* using the Huynh-Feldt ε may not be conservative enough, in many situations). There are diverse opinions about how to deal with this problem. Stevens (1992) suggested using an average of the Greenhouse-Geisser and Huynh-Feldt ε values to correct the *df*.

It is difficult to evaluate whether there might be a participant by treatment interaction. Within the SPSS reliability procedure, the Tukey test of nonadditivity is an available option that can be used to evaluate possible participant by treatment interaction; however, the SPSS reliability procedure can handle only one-way repeated measures designs, and the Tukey test does not detect all possible forms of nonadditivity or interaction.

20.12 ♦ MANOVA Approach to Analysis of Repeated Measures Data

Another way to analyze repeated measures data that violate the sphericity assumption is to do the analysis as a MANOVA instead of as a univariate repeated measures ANOVA. A univariate repeated measures ANOVA involves computation of sums of squares, mean squares, and *F* ratios, as described in Section 20.7. If the assumptions about the structure of the variance/covariance matrix are violated (as described in Section 20.10), then the significance tests provided by a univariate repeated measures ANOVA may have an inflated risk of Type I error.

A MANOVA approach handles the repeated measures data in a different manner. The scores on the variables (in this case, X_1 = HR during baseline, X_2 = HR during pain, X_3 = HR during mental arithmetic, and X_4 = HR during role play) are transformed into a set of $k - 1$ contrasts. SPSS provides a menu of several types of contrasts that may be set up (including polynomial, difference, simple, and Helmert); the final outcome for the overall MANOVA is the same no matter which type of contrast coding is used. In the empirical example presented in this chapter, "Simple" contrasts were requested, using the first level of the repeated measures factor as the reference group to which all other groups are compared. The simple contrasts were as follows:

Contrast 1: H_0: $\mu_1 - \mu_2 = 0$

In this example, this contrast assesses whether mean HR during the Time 2 task, pain, differs significantly from mean HR during the Time 1 task, baseline.

Contrast 2: H_0: $\mu_1 - \mu_3 = 0$

In words, this contrast tests whether mean HR at Time 3 (during the mental arithmetic task) differs from the mean HR averaged at Time 1 (baseline).

Contrast 3: H_0: $\mu_1 - \mu_4 = 0$

The third contrast asks if the mean HR during the fourth task (social role play) differs from the mean HR averaged during baseline.

(See Chapter 12 for further discussion of contrasts.) The MANOVA creates a vector of new outcome variables (C_1, C_2, C_3) that correspond to each of these contrasts:

$$\begin{bmatrix} C_1 = X_1 - X_2 \\ \\ C_2 = X_1 - X_3 \\ \\ C_3 = X_1 - X_4 \end{bmatrix}. \tag{20.28}$$

The null hypothesis for the MANOVA is that the corresponding population vector for these contrasts equals a vector with all elements equal to 0:

$$H_0: \begin{bmatrix} \mu_1 - \mu_2 \\ \mu_1 - \mu_3 \\ \mu_1 - \mu_4 \end{bmatrix} = \begin{bmatrix} 0 \\ 0 \\ 0 \end{bmatrix}. \quad (20.29)$$

If all three of these contrasts among the means are nonsignificant, then we can conclude there are no significant differences among the four means (μ_1, μ_2, μ_3, and μ_4). The contrasts that are included do not have to be orthogonal because MANOVA corrects for intercorrelations among dependent variables. The computations and test statistics involved in MANOVA were presented in Chapter 17. Instead of computing sums of squares and mean squares for a single outcome variable, MANOVA sets up sum of cross-products matrices or a variance/covariance matrix to summarize information about the variances and covariances of the dependent variables:

$$\mathbf{S} = \begin{matrix} & C_1 & C_2 & C_3 \\ C_1 & s_{c1}^2 & s_{12} & s_{13} \\ C_2 & s_{21} & s_{c2}^2 & s_{23} \\ C_3 & s_{31} & s_{32} & s_{c3}^2 \end{matrix}, \quad (20.30)$$

where s_{c1}^2 is the sample variance of C_1, s_{12} is the sample covariance of C_1 and C_2, and so forth. An assumption required for MANOVA is that these variance/covariance matrices, **S**, should be homogeneous across any groups that are compared in the design; that is, the corresponding variances, or s_{ci}^2 diagonal elements, in the **S** matrix should be equal across groups, and the corresponding covariances, or s_{ij} covariance elements, should be equal across groups. However, MANOVA does not require additional assumptions about equalities among the variance and covariance elements of each individual **S** matrix; a univariate repeated measures ANOVA does require these additional assumptions. In other words, the MANOVA approach to repeated measures does not require that the variances and covariances in the **S** matrix have a specific pattern (such as the pattern described earlier as sphericity). MANOVA requires only the less restrictive assumption that corresponding elements of the within-group **S** matrices be equal across groups. Therefore, in situations where the sphericity assumption is violated, MANOVA may be used instead of univariate repeated measures ANOVA.

The test statistics for MANOVA (e.g., Pillai's trace, Wilks's Lambda (Λ), Hotelling's trace, and Roy's largest root) were described in Chapter 17. For the HR/stress study data in Table 20.2, the MANOVA results from GLM were given in Figure 20.15. In this example, the overall MANOVA test was not significant: Wilks's $\Lambda = .160$, $F(3, 3) = 5.269$, $p = .103$.

20.13 ♦ Effect Size

In repeated measures ANOVA, as in other ANOVA designs, an η^2 effect size can be calculated. For a one-way within-S ANOVA, a partial η^2 (i.e., an effect size that has the variance due to individual differences among persons partialled out) is found by taking

$$\text{Partial } \eta^2 = SS_{\text{treatment}}/(SS_{\text{treatment}} + SS_{\text{error}}). \tag{20.31}$$

For the results reported in Figure 20.17, $SS_{\text{treatment}}$ or $SS_{\text{stress}} = 488.458$ and $SS_{\text{error}} = 284.782$, therefore partial $\eta^2 = 488.458/(488.458 + 284.782) = .632$. As in other situations, eta squared is interpreted as the proportion of variance in the scores that is predictable from the time or treatment factor (after individual person differences have been removed from the scores). Of the variance in HR that remains after individual differences in HR level are removed from the data, about 63% was associated with differences among the four treatment conditions.

When MANOVA is used to analyze repeated measures, the effect size measure is based on Wilks's Λ. Recall from Chapter 17 that Wilks's Λ is a proportion of variance that is not attributable to group differences; we can obtain an estimate of η^2 by taking $1 - \Lambda$. From Figure 20.15, then, the partial η^2 value of .84 is equal to $1 - \Lambda$ $(1 - .16)$. After the variance due to systematic individual differences among participants in HR is removed or partialled out, about 84% of the remaining variance in HR is predictable from type of stress.

20.14 ♦ Statistical Power

In many research situations, a repeated measures ANOVA has better statistical power than a between-S ANOVA. This advantage in power may occur when we use the repeated measures ANOVA to partial out or remove stable individual differences among persons (individual differences in HR, in the empirical example presented here) so that these are not included in the error term that is used to assess differences among treatment group means. For example, compare the values of the SS_{error} terms that were obtained by performing a between-S one-way ANOVA on the data in Table 20.1. The output from this one-way between-S ANOVA appeared in Figure 20.2. The partition of SS_{total} in this between-S analysis was as follows:

$$SS_{\text{total}} = SS_{\text{treatment}} + SS_{\text{within}} \tag{20.32}$$

$$2681.625 = 488.458 + 2193.167.$$

It follows from Equation 20.32 that in a between-S ANOVA,

$$SS_{\text{within}} = SS_{\text{total}} - SS_{\text{treatment}}$$

$$2193.167 = 2681.625 - 488.458.$$

The partition of SS_{total} that was obtained in the univariate repeated measures ANOVA, from the SPSS results in Figure 20.17 and the equations in Section 20.8, was as follows:

$$SS_{\text{total}} = SS_{\text{treatment}} + SS_{\text{persons}} + SS_{\text{error}} \tag{20.33}$$

$$2681.625 = 488.458 + 1908.375 + 284.792.$$

It follows from Equation 20.33 that in a within-S or repeated measures ANOVA,

$$SS_{\text{error}} = SS_{\text{total}} - SS_{\text{treatment}} - SS_{\text{persons}}$$

$$284.792 = 2681.625 - 488.458 - 1908.375.$$

Note that the sum of SS_{persons} and SS_{error} in Equation 20.33 ($SS_{\text{persons}} + SS_{\text{error}} = 1908.375 + 284.792 = 2193.167$) is equal to the SS_{within} error term in Equation 20.32; that is, $SS_{\text{within}} = 2193.167$. The repeated measures ANOVA in Equation 20.33 separated the variance due to stable differences among persons or participants from other sources of error variance when estimating SS_{error}; the between-S ANOVA in Equation 20.32 included variability due to individual differences among persons in the error term SS_{within}. Usually, the error term used to set up an F ratio to test for differences among treatment group means will be larger for a between-S ANOVA (as shown in Equation 20.32) than for a within-S ANOVA (as shown in Equation 20.33) if these different analyses are applied to the same set of scores.

In the repeated measures ANOVA, the SS_{within} term (from the between-S ANOVA) has been divided into variability due to stable individual differences in HR among persons (SS_{persons}) and variability due to other sources (SS_{error}). The F ratio for treatment group differences in the between-S ANOVA is

$$F = \frac{SS_{\text{treatment}}/(k-1)}{SS_{\text{within}}/(N-k-1)}, \quad \text{with } (k-1),\ (N-k-1)\ df. \qquad (20.34)$$

The F ratio for the repeated measures ANOVA is

$$F = \frac{SS_{\text{treatment}}/(k-1)}{SS_{\text{error}}/[(N-1)(k-1)]}, \quad \text{with } (k-1),\ (N-1)(k-1)\ df. \quad (20.35)$$

Which version of the F test is more likely to be large enough to yield a statistically significant outcome? That depends on two factors. The between-S F ratio in Equation 20.34 typically has a larger SS term in the divisor than the repeated measures F ratio, but the between-S F ratio also has a larger df term for the divisor. This larger df influences the size of MS_{error}, and it also determines which F distribution is used to obtain critical values for significance tests. When the decrease in error variance (i.e., the difference in size between SS_{error} and SS_{within}) is large enough to more than offset the reduction in degrees of freedom, the repeated measures ANOVA will have greater statistical power than the between-S ANOVA. Based on this reasoning, a repeated measures ANOVA will often, but not always, provide better statistical power than an independent samples ANOVA.

A comparison of statistical power can also be made between the two different methods of analysis for repeated measures data: the univariate repeated measures ANOVA (Section 20.8) and the MANOVA approach (Section 20.12). Under some circumstances, MANOVA may be more powerful than a univariate approach to repeated measures; Algina and Kesselman (1997) suggested the use of MANOVA under these conditions:

1. If the number of levels (k) is less than or equal to 4, use MANOVA if the number of participants is greater than $k + 15$.

2. If the number of levels (k) is between 5 and 8, use MANOVA if the number of participants is greater than $k + 30$.

When sample sizes are smaller than those suggested in these guidelines, MANOVA may be less powerful than univariate repeated measures. Power depends on many factors in this situation (such as the sizes of correlations among the repeated measures).

When the researcher has a reasonably good idea of the magnitude of the effect of the repeated measures treatment factor, statistical power tables can be used to look up the minimum number of participants required to achieve some desired level of statistical power (such as power of .80) for various values of k, the number of levels of the repeated measures factor.

For example, suppose that a researcher looks at effect size information from past research that has assessed the effect of similar stress manipulations on HR using similar types of participants; based on past research, the researcher guesses that the population effect size is $\eta^2 = .25$. The number of levels in the stress study that has been used as an example throughout this chapter is $k = 4$. Using Table 20.6, find the panel that has 3 *df* for the independent variable. Within this panel, look for the column that corresponds to the assumed population effect size ($\eta^2 = .25$) and the row that corresponds to the desired level of statistical power—in this example, power of .80. The cell entry in the table indicates that for a population effect size of $\eta^2 = .25$, we can obtain statistical power of .80 by using $N = 12$ participants. Remember that in many domains of research, effect sizes more typically correspond to η^2 values on the order of .10 to .20; in new areas of research, where there is not much information about effect size, it is more realistic to assume that population effect sizes are small. When in doubt, the more conservative course of action is to use a low value for assumed population effect size when looking up the sample sizes required to achieve reasonable statistical power.

20.15 ♦ Planned Contrasts

When a significant overall F is obtained for a repeated measures ANOVA, we can reject the null hypothesis that all the treatment means are equal. In the present empirical example, we can reject the null hypothesis that HR was equal across the four treatment conditions in the study (baseline, pain, arithmetic, and role play). Generally, researchers want to provide more specific information about the nature of differences. The SPSS GLM provides a set of contrasts among levels of the repeated measures factor by default. The default type of contrast provided by SPSS involves polynomial trend analysis; a pull-down menu includes other types of contrasts that may be requested. The choice of type of contrast makes no difference in the outcome of the overall F for the univariate repeated measures ANOVA or the omnibus multivariate tests such as Wilks's Λ for the MANOVA. However, if the contrasts that are requested correspond to meaningful comparisons between groups, the contrasts can be chosen to provide information about differences between specific levels of the repeated measures factor.

Table 20.6 ♦ Statistical Power for One-Way Repeated Measures ANOVA

Degrees of Freedom IV = 2, Alpha = .05

POPULATION ETA SQUARED

Power	*.01*	*.03*	*.05*	*.07*	*.10*	*.15*	*.20*	*.25*	*.30*	*.35*	*.40*	*.45*	*.50*	*.55*	*.60*	*.65*	*.70*	*.75*	*.80*
.10	32	11	7	5	4	3	2	2	2	2	—	—	—	—	—	—	—	—	—
.50	247	81	48	34	23	15	11	8	7	6	5	4	3	3	3	2	2	2	2
.70	332	125	74	52	36	23	16	13	10	8	7	6	5	4	4	3	3	2	2
.80	478	157	93	65	44	28	20	15	12	10	8	7	6	5	4	4	3	3	2
.90	627	206	121	85	58	37	26	20	16	13	10	9	7	6	5	4	4	3	3
.95	765	251	148	104	70	45	32	24	19	15	13	10	9	7	6	5	4	4	3
.99	1,060	347	204	143	97	62	44	33	26	21	17	14	12	10	8	7	6	5	4

Degrees of Freedom IV = 3, Alpha = .05

POPULATION ETA SQUARED

Power	*.01*	*.03*	*.05*	*.07*	*.10*	*.15*	*.20*	*.25*	*.30*	*.35*	*.40*	*.45*	*.50*	*.55*	*.60*	*.65*	*.70*	*.75*	*.80*
.10	27	9	6	4	3	2	2	2	2	—	—	—	—	—	—	—	—	—	—
.50	191	63	37	27	18	12	9	7	5	5	4	3	3	3	2	2	2	2	—
.70	291	96	57	40	27	18	13	10	8	6	5	5	4	3	3	3	2	2	2
.80	361	118	70	49	34	22	16	12	9	8	6	5	5	4	3	3	3	2	2
.90	469	154	91	64	44	28	20	15	12	10	8	7	6	5	4	4	3	3	2
.95	563	186	110	77	53	33	24	18	14	12	10	8	7	6	5	4	3	3	2
.99	777	254	150	105	72	45	32	25	19	16	13	11	9	7	6	5	4	4	3

(Continued)

Table 20.6 ♦ (Continued)

Degrees of Freedom IV = 4, Alpha = .05

POPULATION ETA SQUARED

Power	*.01*	*.03*	*.05*	*.07*	*.10*	*.15*	*.20*	*.25*	*.30*	*.35*	*.40*	*.45*	*.50*	*.55*	*.60*	*.65*	*.70*	*.75*	*.80*
.10	24	8	5	4	3	2	2	2	2	—	—	—	—	—	—	—	—	—	—
.50	160	53	31	22	15	10	7	6	5	4	3	3	3	2	2	2	2	2	—
.70	241	79	47	33	23	15	11	8	7	5	5	4	3	3	3	2	2	2	2
.80	297	98	58	41	28	18	13	10	8	7	5	5	4	3	3	3	2	2	2
.90	382	126	74	52	36	23	16	13	10	8	7	6	5	4	4	3	3	2	2
.95	461	151	89	63	43	27	20	15	12	10	8	7	6	5	4	3	3	3	2
.99	626	205	121	85	58	37	26	20	16	13	10	9	7	6	5	4	4	3	3

Degrees of Freedom IV = 5, Alpha = .05

POPULATION ETA SQUARED

Power	*.01*	*.03*	*.05*	*.07*	*.10*	*.15*	*.20*	*.25*	*.30*	*.35*	*.40*	*.45*	*.50*	*.55*	*.60*	*.65*	*.70*	*.75*	*.80*
.10	21	8	5	4	3	2	2	2	—	—	—	—	—	—	—	—	—	—	—
.50	139	46	28	20	14	9	7	5	4	4	3	3	2	2	2	2	2	—	—
70	208	69	41	29	20	13	9	7	6	5	4	4	3	3	2	2	2	2	2
.80	225	84	50	35	24	16	11	9	7	6	5	4	4	3	3	2	2	2	2
.90	327	108	64	45	31	20	14	11	9	7	6	5	4	4	3	3	2	2	2
.95	393	129	76	54	37	23	17	13	10	8	7	6	5	4	4	3	3	2	2
.99	530	174	103	72	49	31	22	17	13	11	9	8	6	5	5	4	3	3	2

Table 20.6 ♦ (Continued)

	Degrees of Freedom IV = 6, Alpha = .05																		
	POPULATION ETA SQUARED																		
Power	*.01*	*.03*	*.05*	*.07*	*.10*	*.15*	*.20*	*.25*	*.30*	*.35*	*.40*	*.45*	*.50*	*.55*	*.60*	*.65*	*.70*	*.75*	*.80*
.10	20	7	5	4	3	2	2	2	—	—	—	—	—	—	—	—	—	—	—
.50	125	41	25	18	12	8	6	5	4	3	3	3	2	2	2	2	2	—	—
.70	185	61	36	26	18	12	8	7	5	4	4	3	3	3	2	2	2	2	—
.80	225	74	44	31	21	14	10	8	6	5	4	4	3	3	3	2	2	2	2
.90	283	95	56	40	27	17	13	10	8	6	5	5	4	3	3	3	2	2	2
.95	345	113	67	47	32	21	15	11	9	7	6	5	4	4	3	3	2	2	2
.99	464	152	90	63	43	27	20	15	12	10	8	7	6	5	4	4	3	3	2

The types of contrasts available in SPSS GLM include the following:

Polynomial trend contrasts: For a within-S factor with $k = 4$ levels, this tests the following trends: linear, quadratic, and cubic.

Simple contrasts: The mean of each group is contrasted with the mean of one reference or "control" group. In the empirical example in this chapter, mean HR in each of the three stress conditions (pain, arithmetic, and role play) was contrasted with mean HR during baseline, which corresponded to the first level of the stress factor. This type of contrast makes sense when one group is a nontreatment control group.

Helmert contrasts: This compares Group 1 with the mean of all later groups, Group 2 with the mean of all later groups, and Group 3 with the mean of all later groups (Group 4).

In this empirical example, simple contrasts were requested using the first group as the reference group. The output for these contrasts appears in Figure 20.18. The mean HR during the arithmetic task was not significantly higher than the mean HR during baseline; however, mean HR during both the pain and role play tasks was significantly higher than mean HR during baseline.

Unfortunately, in some real-life studies, none of the contrasts available as menu options may correspond to research questions that are of interest. Custom-defined contrasts can be obtained by editing the GLM syntax.

20.16 ♦ Results

Usually, a researcher makes a decision to report just one of the two different analyses provided by SPSS: either the MANOVA or the univariate repeated measures ANOVA, which is based on computations of sums of squares and F ratios. When the univariate repeated measures ANOVA is reported, the researcher also typically makes a decision whether to use the uncorrected degrees of freedom or to report degrees of freedom that are corrected using either the Greenhouse-Geisser ε or the Huynh-Feldt ε. The choice about which of these analyses to report may depend on several issues. First of all, if there is evidence that the sphericity assumption for repeated measures ANOVA is violated, researchers typically try to correct for the violation of this assumption, either by reporting corrected degrees of freedom obtained by using one of the two versions of the ε correction factor or by reporting the MANOVA. MANOVA does not require the assumption of sphericity. However, when samples sizes are very small, the MANOVA approach may have quite weak statistical power.

Note that if the four levels of stress are presented in only one sequence, as in the preceding example (all participants were tested in this sequence: baseline, pain, arithmetic,

Results

To evaluate whether HR increases during stressful tasks, a one-way repeated measures ANOVA was performed. Six participants were tested under four treatment conditions: (1) baseline, (2) pain induction, (3) mental arithmetic task, and (4) a stressful social role play. The sample was too small to evaluate whether assumptions of multivariate normality were satisfied. The Mauchly test was performed to assess possible violation of the sphericity assumption; this was not significant: Mauchly's $W = .833$, $\chi^2(5) = .681$, $p = .985$; however, when the sample size is so small, this test may not have sufficient power to detect violations of sphericity. The Greenhouse-Geisser ε value of .897 suggested that the sample variance covariance matrix did not depart substantially from sphericity. Because the Greenhouse-Geisser ε value was close to 1.00, no correction was made to the degrees of freedom used to evaluate the significance of the *F* ratio.

The overall *F* for differences in mean HR across the four treatment conditions was statistically significant: $F(3, 15) = 8.58$, $p = .001$; the corresponding effect size was a partial η^2 of .632. In other words, after stable individual differences in HR are taken into account, about 63% of the variance in HR was related to treatment condition. Note that even if the ε correction factor is applied to the degrees of freedom for *F*, the obtained value of *F* for differences in mean HR among levels of stress remains statistically significant.

Planned contrasts were obtained to compare mean HR for each of the three stress conditions with the mean HR during baseline ($M_{baseline} = 74.5$). Mean HR during the arithmetic task ($M_{arithmetic} = 77.67$) was not significantly higher than baseline mean HR: $F(1, 5) = 2.231$, $p = .195$. Mean HR during the pain induction ($M_{pain} = 83.50$) was significantly higher than baseline mean HR: $F(1, 5) = 14.817$, $p = .012$. Mean HR during the stressful role play ($M_{roleplay} = 85.83$) was also significantly higher than baseline mean HR: $F(1, 5) = 21.977$, $p = .005$. A graph of these four treatment means appears in Figure 20.20. Thus, of the three stress interventions in this study, only two (the pain induction and the stressful social role play) had significantly higher mean HR than the no-stress/baseline condition.

role play), there is a confound between order of presentation of treatment and treatment effects. The following section discusses **order effects** and other problems that arise in repeated measures designs.

20.17 ♦ Design Problems in Repeated Measures Studies

The use of a repeated measures design can have some valuable advantages. When a researcher uses the same participants in all treatment conditions, potential confounds between type of treatment and participant characteristics (such as age, anxiety level, and drug use) are avoided. The same participants are tested in each treatment condition. In addition, the use of repeated measures design makes it possible, at least in theory, to identify and remove variance in scores that is due to stable individual differences in the outcome variable (such as HR), and this may result in a smaller error term (and a larger t or F ratio) than a between-S design. In addition, repeated measures can be more cost-effective; we obtain more data from each participant. This can be an important consideration when participants are difficult to recruit or require special training or background or preparation.

However, the use of repeated measures design also gives rise to some problems that were not an issue in between-S designs, including order effects, carryover and contrast effects, and sensitization. In the empirical example described so far, each participant's HR was measured four times:

Time 1: baseline

Time 2: pain induction

Time 3: mental arithmetic test

Time 4: stressful social role play

This simple design has a built-in confound between order of presentation (first, second, third, fourth) and type of treatment (none, pain, arithmetic, role play). If we observe the highest HR during the stressful role play, we cannot be certain whether this occurs because this is the most distressing of the four situations or because this treatment was the fourth in a series of (increasingly unpleasant) experiences.

Another situation that illustrates potential problems arising from order effects was provided by a magazine advertisement published several years ago. The advertisers (a company that sold rum) suggested the following taste test: First, taste a shot of whiskey (W); then, taste a shot of vodka (V); then, taste gin (G); and finally, taste rum (R). (Now, doesn't that taste great?) It is important to unconfound (or balance) order of presentation and type of treatment when designing repeated measures studies. This requires that the researcher present the treatments in different orders. The presentation treatments in different orders is called **counterbalancing** (what is being "balanced" is the type of treatment and the order of presentation). There are several ways that this can be done.

One method of counterbalancing, called complete counterbalancing, involves presenting the treatments in all possible orders. This is generally not practical, particularly for studies where the number of treatments (k) is large, because the number of possible orders in which k treatments can be presented is given by $k!$ or "k factorial," which is $k \times (k - 1) \times (k - 2) \times \ldots \times (1)$. Even if we have only $k = 4$ treatments, there are $4! = 4 \times 3 \times 2 \times 1 = 24$ different possible orders of presentation.

It is not necessary to run all possible orders in order to unconfound order with type of treatment. In fact, for k treatments, k treatment orders can be sufficient, provided that the treatment orders are worked out carefully. In setting up treatment orders, we want to achieve two goals. First, we want each treatment to occur once in each ordinal position. For example, in the liquor-tasting study, when we set up the order of beverages for different groups of participants, one group should taste rum first, one group should taste it second, one group should taste it third, and one group should taste it fourth (and last). This ensures that the taste ratings for rum are not entirely confounded with the fact that it is the last of four alcoholic beverages. In addition, we want to make sure that each treatment follows each other treatment just once; for example, one group should taste rum first, one group should taste rum immediately after tasting whiskey, one group should taste rum after vodka, and the last group should taste rum after gin. This controls for possible contrast effects; for example, rum might taste much more pleasant when it immediately follows a somewhat bitter-tasting beverage (gin) than when it follows a more neutral-tasting beverage (vodka).

Treatment orders that control for both ordinal position and sequence of treatments are called "Latin squares." It can be somewhat challenging to work out a Latin square for more than about three treatments; fortunately, textbooks on experimental methods often provide tables of Latin squares. For the four alcoholic beverages in this example, a Latin square design would involve the following four orders of presentation. Participants would be randomly assigned to four groups: Group 1 would receive the beverages in the first order, Group 2 in the second order, and so forth.

Order 1: WVGR

Order 2: VRWG

Order 3: RGVW

Order 4: GWRV

In these four orders, each treatment (such as W, whiskey) appears once in each ordinal position (first, second, third, fourth). Also, each treatment, such as whiskey, follows each other treatment in just one of the four orders. Using a Latin square, therefore, it is possible to control for order effects (i.e., to make sure that they are not confounded with type of treatment) using as few as k different orders of presentation for k treatments. However, there are additional potential problems.

Sometimes when treatments are presented in a series, if the time interval between treatments is too brief, the effects of one treatment do not have time to wear off before the next treatment is introduced; when this happens, we say there are "carryover effects." In the beverage-tasting example there could be two kinds of carryover: One, the taste of a beverage such as gin may still be stimulating the taste buds when the next "treatment" is introduced, and two, there could be cumulative effects on judgment from the alcohol consumption. Carryover effects can be avoided by allowing a sufficiently long time interval between treatments so that the effect of each treatment has time to "wear off" before the next treatment is introduced or to do something to try to neutralize treatment effects (at a wine tasting, for example, people may rinse their mouths with water or eat bread in between wines to clear the palate).

In any study where participants perform the same task or behavior repeatedly, there can be other changes in behavior as a function of time. If the task involves skill or knowledge, participants may improve their performance across trials with practice. If the task is arduous or dull, performance on the task may deteriorate across trials due to boredom or fatigue. In many studies where participants do a series of tasks and are exposed to a series of treatments, it becomes possible for them to see what features of the situation the researcher has set out to vary systematically; they may guess the purpose of the experiment, and they may try to do what they believe the researcher expects. Some self-report measures involve sensitization; for example, when people fill out a physical symptom checklist over and over, they begin to notice and report more physical symptoms over time. Finally, if the repeated measures study takes place over a relatively long period of time (such as months or years), maturation of participants may alter their responses; outside events may occur in between the interventions or treatments; and many participants may drop out of the study (they may move, lose interest, or even die). Campbell and Stanley's (2005) classic monograph on research design points out these and many other potential problems that may make the results of repeated measures studies difficult to interpret, and it remains essential reading for researchers who do repeated measures research under less than ideal conditions. For further discussion of analysis of repeated measures, see intermediate textbooks such as Keppel and Zedeck (1989) and advanced textbooks such as Kirk (1995) or Myers and Well (1995).

20.18 ♦ More Complex Designs

In previous chapters, we have seen that the design elements from various chapters can often be combined to create more complex designs. For example, when we begin with a one-way ANOVA that has just a single outcome variable measure, we can add another treatment factor (to set up a factorial ANOVA); we can add a covariate (to do an or analysis of covariance, or ANCOVA); or we can measure multiple outcome variables (MANOVA). The same variations are possible for designs that involve repeated measures. A factorial design may include one or more repeated measures factors along with any number of between-S factors. A repeated measures ANOVA may include a covariate. It is possible to do a repeated measures study in which multiple outcome measures are observed at each point in time. For example, consider the simple one-way repeated measures ANOVA design that was introduced at the beginning of the chapter: HR was measured at four points in time under different conditions (baseline, pain, arithmetic, and role play). A between-S factor (such as gender) could be added. A covariate (such as anxiety level) could be included in the analysis. In addition to measuring HR, we could also measure systolic blood pressure (SBP), diastolic blood pressure (DBP), and respiration rate in response to each treatment. Thus, we might do a study in which we assess how males and females differ in their pattern of responses across four levels of stress when multiple outcome measures are obtained under each treatment and controlling for one or more covariates.

A potential difficulty that arises in complex designs that include repeated measures is the choice of an appropriate MS_{error} term to use for the *F* test for each factor in the model. For simple designs, the default error terms reported by SPSS may be appropriate; however, for complex designs, researchers should consult advanced experimental analysis textbooks for guidance on setting up appropriate *F* ratios (e.g., Kirk, 1995; Myers and Well, 1995).

20.19 ♦ Alternative Analyses for Pretest and Posttest Scores

The analysis of repeated measures that represent a pretest and a posttest on the same variable for each participant before and after an intervention was discussed in the appendix to Chapter 15. In a simple pretest/posttest design, a researcher measures the outcome variance of interest before an intervention, then administers a treatment or intervention, and then measures the variance of interest again after the intervention. For example, a researcher might measure HR at Time 1 (HR_1), then administer a drug such as caffeine, then measure HR at Time 2 (HR_2). The means are compared between pretest HR_1 and posttest HR_2 to assess whether HR changed significantly during the study. Additional design factors may be present; for example, gender could be included as a between-S factor. The research question would then be whether males and females differ in their HR responses to caffeine. There are several ways in which pretest/posttest scores can be statistically analyzed and compared between different groups. Unfortunately, when groups differ in their scores at baseline or Time 1 and when different methods of data analysis are applied, it sometimes leads to substantially different conclusions. This is known as Lord's paradox (Lord, 1967; Wainer, 1991). Given a set of pretest scores (X_1) and a set of posttest scores (X_2), here are several different ways in which a researcher might carry out an analysis:

1. The researcher may compute a change score or gain score for each participant: $d = X_2 - X_1$. In a study that examines just one group of participants before and after an intervention, the mean change score for all participants can be tested to see if it differs significantly from 0; for example, in the HR/caffeine study, the grand mean of the change scores could be tested to see if it differed significantly from 0. If the study also includes a between-S factor such as gender, the mean HR change score for males can be compared with the mean HR change score for females to see if women and men responded differently to the intervention. Note that if a researcher does not explicitly compute the change score as a new variable, when a paired samples *t* test is applied to pretest/posttest data, in effect, the analysis implicitly examines change or gain scores.
2. The researcher may perform a repeated measures ANOVA with the pretest and posttest scores treated as two levels of a within-S factor that might be called time, treatment, or trials. If the design also includes a between-S factor such as gender, the presence of a statistically significant gender by time or treatment interaction indicates a gender difference in HR response to treatments.
3. The researcher can do an ANCOVA to compare mean scores on the posttest (such as HR_2) across gender groups, using the pretest score (such as HR_1) as a covariate. The application of ANCOVA to pretest/posttest designs was discussed in the appendix to Chapter 15.
4. The researcher might look at proportional change in HR across time—that is, look at the ratio of HR_2/HR_1.
5. The researcher might employ some form of growth curve analysis. If we have measurements at only two time points, this essentially involves a linear model for each participant's pretest score of X_1 as the intercept and the change score ($X_2 - X_1$) as the slope. If we have measures at multiple points in time, each participant's behavior trajectory over time can be modeled as a curvilinear function.

Lord (1967) pointed out that for some patterns of scores, the distances between group means (and sometimes even the rank ordering of group means) may appear to be different depending on which analysis is performed. In many cases, this potential problem does not actually arise; that is, in some empirical situations, if the researcher carries out all these analyses on the same dataset, the rank ordering of group means and the significant differences between groups remain the same. It is rather common for journal reviewers to ask a researcher who has reported difference scores whether the outcome would be the same using ANCOVA (and vice versa), so it is a good idea to look at the outcomes for several different analyses to see whether they suggest different interpretations. If the analyses all point to the same conclusions about the nature of group differences, it may be sufficient to report one analysis (such as a repeated measures) and then to indicate that alternative methods yielded essentially the same outcome.

However, researchers occasionally find themselves looking at data for which different analyses point toward substantially different conclusions about the nature of the impact of treatment. In these situations, researchers need to be able to provide a reasonable explanation why they prefer one analytic approach to other approaches. (If one approach yields a statistically significant outcome, and another analysis does not, it would not be intellectually honest to just report the statistically significant outcome without careful consideration of the situation.)

A potential problem with the use of change or gain scores is that the change score ($d = X_2 - X_1$) is often (although not always) artifactually negatively correlated with the pretest score X_1. To see why that sometimes happens, compare the simple gain score ($X_2 - X_1$) with the measure of change that is created when we use ANCOVA to assess posttest differences while statistically controlling for pretest scores. Let's assume that X_2 and X_1 are measures of the same variable (such as HR) before and after an intervention (such as a dose of caffeine) and that the correlation between X_1 and X_2 is less than 1.0. Recall that if we perform a bivariate regression to predict scores on X_2 from scores on X_1 (see Chapter 9), the bivariate regression has the following form:

$$X_2' = b_0 + b_1 X_1 \tag{20.36}$$

and

$$b_1 = r \times \frac{s_{X_2}}{s_{X_1}}. \tag{20.37}$$

If X_1 and X_2 are measured in similar units and have similar variances, the ratio of the standard deviation of X_2 to the standard deviation of X_1, s_{X2} / s_{X1}, is usually close to 1.00, and because $r < 1.00$, the value of b_1 that best predicts X_2 from X_1 will also be a value of b that is less than 1.00. In ANCOVA, when we want to obtain a change score X_2^* that is uncorrelated with the baseline score X_1, in effect, we use the residual from this regression:

$$X_2^* = X_2 - [b_0 + b_1 X_1]. \tag{20.38}$$

The residual, X_2^*, is uncorrelated with the predictor variable X_1 because the residual was obtained by subtracting $b_1 X_1$ from X_2. Assuming that $r < 1$ and that s_{X2} is approximately equal to s_{X1}, the b_1 coefficient used to calculate the adjusted X_2^* score in an ANCOVA is usually less than 1.00 in value.

By this reasoning, then, when we compute a simple gain score or change score by calculating $d = X_2 - 1 \times X_1$ (rather than $X_2 - b_1 \times X_1$), the change score d may represent an "overcorrection"; when we compute a difference score d, we subtract $1 \times X_1$ rather than $b_1 \times X_1$ to obtain d. For this reason, a simple change score d often (although not always) tends to correlate negatively with the pretest score X_1. In terms of the example using HR and stress, the d change scores might tend to be slightly smaller for persons whose baseline HR_1 value was higher.

Given that change scores may be slightly negatively correlated with pretest scores, does that make simple change scores an inappropriate index of change? Not necessarily. There may be logical reasons why we would expect people who already have higher scores on HR at baseline to show smaller increases in response to treatment than people who have lower scores at Time 1. Researchers need to think carefully about the way they want to conceptualize change; for example, should change be uncorrelated with baseline scores, or would one expect change scores to be correlated with baseline scores? There is no single correct answer to the question whether it is better to use change scores or ANCOVA (or one of the other methods listed above) to analyze pretest/posttest scores (Wainer, 1991). Each method is valid only under specific assumptions (and in practice, these assumptions may be difficult or impossible to test). Researchers whose primary focus will be the analysis of change need to familiarize themselves with the issues briefly identified in Wainer (1991) and the data analysis alternatives mapped out in more advanced treatments of the analysis of change over time, such as Gottman (1995).

In practice, whenever a study involves pretest/posttest comparisons, if there is no clear theoretical justification to prefer one type of analysis and one conceptual definition of change over the others, it may be helpful to evaluate the data using several different analyses:

1. Run repeated measures ANOVA.
2. Run ANCOVA using pretest scores as a covariate.
3. Obtain a Pearson r between the change score d ($X_2 - X_1$) and the pretest score X_1.

If the repeated measures ANOVA and ANCOVA point to the same pattern of group differences, and if there is no strong correlation between d and X_1, the researcher can reassure readers that the study is not one of the (relatively rare) situations in which the choice of method of analysis makes a substantial difference in the nature of conclusions about the outcome. On the other hand, if the results do differ substantially across these different methods of analysis, the researcher needs to think very carefully about the logic involved in ANCOVA and change scores, then make a reasonable case why one method of analysis is preferable to alternative methods. It is increasingly common for researchers to employ more complex methods of analysis of change over time, such as structural equation modeling or growth curve analysis, particularly when the study involves multiple predictor or multiple outcome measures, or multiple indicator variables for each latent variable, and in studies in which behavior is measured at multiple points in time. For further discussion of the analysis of change across time, see Gottman (1995). Time series analysis methods can be applied to repeated measures data, but these methods are generally used only when behaviors are observed or measured for a large number of points in time, at least 25 to 50 time series observations (Warner, 1998).

20.20 ◆ Summary

Repeated measures designs can offer substantial advantages to researchers. A repeated measures design can make it possible to obtain more data from each participant and to statistically control for stable individual differences among participants (and remove that variance from the error term) when testing treatment effects. However, repeated measures designs raise several problems. The study must be designed in a manner that avoids confounding order effects with treatment. Repeated measures ANOVA involves assumptions about the pattern of covariances among repeated measures scores, and if those assumptions are violated, it may be necessary to use corrected degrees of freedom to assess the significance of F ratios (or to use MANOVA rather than repeated measures ANOVA). Finally, the paired samples t test and repeated measures ANOVA may result in different conclusions about the nature of change than alternative methods (such as ANCOVA using pretest scores as a covariate); when the choice of statistical analysis makes a substantial difference, the researcher needs to consider carefully whether change scores or covariates provide a better way of assessing change across time.

Comprehension Questions

1. When you compute difference scores (for a paired samples t test), what happens to the parts of scores that are associated with stable individual differences among persons? (In the models in this chapter, these parts of scores were denoted $\pi_{i\cdot}$.) For example, in the study about stress and HR, after difference scores are computed, is there any variance among the difference scores that is due to individual differences in HR?
2. Why is the SS_{error} term in a repeated measures ANOVA typically smaller than the SS_{within} term for a between-S ANOVA?
3. How is the Greenhouse-Geisser ε interpreted? For example, consider a repeated measures design with $k = 6$ levels. What is the minimum possible value of ε for this situation? What approximate range of values for ε suggests that the sphericity assumption may be seriously violated? What value of ε would indicate that there was no violation of sphericity? How is the ε value used to adjust the df?
4. What is the compound symmetry assumption? What is sphericity? What information from the SPSS printout helps us to evaluate whether the sphericity assumption is violated?
5. Consider this study conducted by Kim Mooney (1990) at the University of New Hampshire.

   SBP was measured at six points in time using a Finapres noninvasive blood pressure monitor:

   Time 1: When participants arrived at the laboratory and before they signed an informed consent form

   Time 2: After participants read and signed an informed consent form that told them that they would be doing a stressful social role play (about being accused of shoplifting) that would be videotaped (they were told that they had 5 min to think about what they would say in this situation)

Time 3: After a 5-min period (during which they prepared what to say)

Time 4: During the stressful social role play

Time 5: Ten minutes after the role play was completed

Time 6: Twenty minutes after the role play was completed

At each point in time, for each of 65 participants, SBP was measured; the variables sys1, sys2, . . . , sys6 correspond to the measurements of blood pressure made at the six points in time just described. (The blood pressure monitor used in this study looked at peripheral arterial pressure in the middle finger, and so the blood pressure readings tended to be systematically higher than the readings usually obtained by more conventional blood pressure measurement methods.)

a. Before looking at the SPSS output for this study, answer these questions: What research questions can the researcher answer using these data? What form would the formal null hypothesis take?
b. After looking at the SPSS output for this study (see Figure 20.21), answer these questions. Also, write up the results as they would appear in a journal article. Is there any evidence that the sphericity assumption is violated? Overall, was there a statistically significant difference in SBP across the six points in time? At what time was SBP highest, and how great was the increase in SBP across time?

2. TIME

Measure: MEASURE_1

			95% Confidence Interval	
TIME	Mean	Std. Error	Lower Bound	Upper Bound
1	125.875	2.433	121.013	130.737
2	139.188	2.537	134.118	144.257
3	145.922	2.384	141.159	150.685
4	164.391	2.744	158.907	169.874
5	149.000	2.467	144.070	153.930
6	139.375	2.374	134.630	144.120

Mauchly's Test of Sphericity[b]

Measure: MEASURE_1

					Epsilon[a]		
Within Subjects Effect	Mauchly's W	Approx. Chi-Square	df	Sig.	Greenhouse-Geisser	Huynh-Feldt	Lower-bound
TIME	.253	84.022	14	.000	.666	.707	.200

Tests the null hypothesis that the error covariance matrix of the orthonormalized transformed dependent variables is proportional to an identity matrix.

a. May be used to adjust the degrees of freedom for the averaged tests of significance. Corrected tests are displayed in the Tests of Within-Subjects Effects table.

b. Design: Intercept
Within Subjects Design: TIME

Figure 20.21 ♦ SPSS Output for Comprehension Question 5 *(Continued)*

Multivariate Tests[b]

TIME		Value	F	Hypothesis df	Error df	Sig.	Partial Eta Squared
Effect	Pillai's Trace	.831	58.161[a]	5.000	59.000	.000	.831
	Wilks's Lambda	.169	58.161[a]	5.000	59.000	.000	.831
	Hotelling's Trace	4.929	58.161[a]	5.000	59.000	.000	.831
	Roy's Largest Root	4.929	58.161[a]	5.000	59.000	.000	.831

a. Exact statistic

b. Design: Intercept
Within Subjects Design: TIME

Tests of Within-Subjects Effects

Measure: MEASURE_1

Source		Type III Sum of Squares	df	Mean Square	F	Sig.	Partial Eta Squared
TIME	Sphericity Assumed	52321.740	5	10464.348	105.561	.000	.626
	Greenhouse-Geisser	52321.740	3.329	15715.768	105.561	.000	.626
	Huynh-Feldt	52321.740	3.537	14791.522	105.561	.000	.626
	Lower-bound	52321.740	1.000	52321.740	105.561	.000	.626
Error(TIME)	Sphericity Assumed	31226.260	315	99.131			
	Greenhouse-Geisser	31226.260	209.743	148.879			
	Huynh-Feldt	31226.260	222.849	140.123			
	Lower-bound	31226.260	63.000	495.655			

Tests of Within-Subjects Contrasts

Measure: MEASURE_1

Source	TIME	Type III Sum of Squares	df	Mean Square	F	Sig.	Partial Eta Squared
TIME	Level 2 vs. Level 1	11342.250	1	11342.250	82.649	.000	.567
	Level 3 vs. Level 1	25720.141	1	25720.141	118.250	.000	.652
	Level 4 vs. Level 1	94941.016	1	94941.016	277.657	.000	.815
	Level 5 vs. Level 1	34225.000	1	34225.000	148.242	.000	.702
	Level 6 vs. Level 1	11664.000	1	11664.000	59.568	.000	.486
Error(TIME)	Level 2 vs. Level 1	8645.750	63	137.234			
	Level 3 vs. Level 1	13702.859	63	217.506			
	Level 4 vs. Level 1	21541.984	63	341.936			
	Level 5 vs. Level 1	14545.000	63	230.873			
	Level 6 vs. Level 1	12336.000	63	195.810			

Multivariate Tests[b]

Effect		Value	F	Hypothesis df	Error df	Sig.	Partial Eta Squared
TIME	Pillai's Trace	.831	58.161[a]	5.000	59.000	.000	.831
	Wilks's Lambda	.169	58.161[a]	5.000	59.000	.000	.831
	Hotelling's Trace	4.929	58.161[a]	5.000	59.000	.000	.831
	Roy's Largest Root	4.929	58.161[a]	5.000	59.000	.000	.831

a. Exact statistic

b. Design: Intercept
Within Subjects Design: TIME

Figure 20.21 ♦ (Continued)

CHAPTER 21

Binary Logistic Regression

21.1 ♦ Research Situations

21.1.1 ♦ Types of Variables

Binary logistic regression can be used to analyze data in studies where the outcome variable is binary or dichotomous. In medical research, a typical binary outcome of interest is whether each patient survives or dies. The outcome variable, "death," can be coded 0 = Alive and 1 = Dead. The goal of the study is to predict membership in a target group (e.g., membership in the "dead" group) from scores on one or several predictor variables. Like multiple linear regression, a binary logistic regression model may include one or several predictor variables. The predictor variables may be quantitative variables, dummy-coded categorical variables, or both. For example, a researcher might want to predict risk of death from gender, age, drug dosage level, and severity of the patient's medical problem prior to treatment.

21.1.2 ♦ Research Questions

The basic research questions that arise in binary logistic regression studies are similar to the research questions that arise in multiple linear regression (see Chapter 14), except that the inclusion of a group membership outcome variable requires some modification to the approach to analysis. Researchers need to assess statistical significance and effect size for an overall model, including an entire set of predictor variables. They may test competing models that include different predictor variables. When an overall model is significant, the researcher examines the contribution of individual predictor variables to assess whether each predictor variable is significantly related to the outcome and the nature and strength of each predictor variable's association with the outcome.

Unlike the analyses presented in Chapters 5 through 20, logistic regression is not a special case of the general linear model; it uses different summary statistics to provide information about overall model fit and the nature of the relationship between predictors and group membership. In multiple linear regression (see Chapters 11, 12, and 14), the overall strength of prediction for a multiple linear regression model was indexed by multiple R, and the significance of multiple R was assessed by an F ratio; the nature of the predictive relationship between each predictor variable and the score on the outcome variable was indexed by a raw score or standardized slope coefficient (b or β), and the significance of each individual predictor was assessed by a t test. A logistic regression includes some familiar terms (such as raw score regression coefficients, denoted by B). In addition, it provides additional information about odds and odds ratios. Odds and odds ratios will be explained before examining an empirical example of binary logistic regression.

21.1.3 ♦ Assumptions Required for Linear Regression Versus Binary Logistic Regression

Analyses that are special cases of the general linear model (such as multiple linear regression and discriminant analysis [DA]) involve fairly restrictive assumptions. For a multiple linear regression in which scores on a quantitative Y variable are predicted from scores on quantitative X variables, the assumptions include the following: a multivariate normal distribution for the entire set of variables and, in particular, a univariate normal distribution for scores on Y; linear relations between scores on Y and scores on each X variable, and between each pair of X predictor variables; and uniform error variance across score values of the X variable (e.g., variance of Y is homoscedastic across score values of X). For a DA in which group membership is predicted from scores on several quantitative X variables, an additional assumption is required: homogeneity of elements of the variance/covariance matrix for the X variables across all groups. The use of a binary outcome variable in binary logistic regression clearly violates some of these assumptions. For example, scores on Y cannot be normally distributed if Y is dichotomous.

By contrast, logistic regression does not require such restrictive assumptions. The assumptions for logistic regression (from Wright, 1995) are as follows:

1. The outcome variable is dichotomous (usually the scores are coded "1" and "0").
2. Scores on the outcome variable must be statistically independent of each other.
3. The model must be correctly specified; that is, it should include all relevant predictors, and it should not include any irrelevant predictors. (Of course, this is essential for all statistical analyses, although textbooks do not always explicitly state this as a model assumption.)
4. The categories on the outcome variable are assumed to be exhaustive and mutually exclusive; that is, each person in the study is known to be a member of one group or the other but not both.

Note that binary logistic regression does not require normally distributed scores on the Y outcome variable, a linear relation between scores on Y and scores on quantitative X predictor variables, or homogeneous variance of Y across levels of X. Because it requires less restrictive assumptions, binary logistic regression is widely viewed as a more appropriate method of analysis than multiple linear regression or DA in many research situations where the outcome variable corresponds to membership in two groups. Even though fewer assumptions are involved in binary logistic regression, preliminary data screening is still useful and important. For example, it is a good idea to examine the expected frequency in each cell. Like the chi-square test for simple contingency tables, a binary logistic regression does not perform well when many cells have expected frequencies less than 5. In addition, extreme outliers on quantitative predictor variables should be identified. Recommendations for data screening are provided in Section 21.8.1.

21.2 ◆ Simple Empirical Example: Dog Ownership and Odds of Death

Friedmann et al. (1980) conducted a survey of 92 male patients who initially survived a first heart attack. Two variables from this study were examined in Chapter 8; Table 21.1 summarizes information about a binary predictor variable (Did the patient own a dog?) and the primary outcome of interest (Was the patient alive or dead at the end of the first year after the heart attack?). The data in Table 21.1 are contained in the SPSS file "dogdeath.sav," with the following variable names and value labels. The variable "dogowner" was coded 0 for persons who did not own dogs and 1 for persons who did own dogs. The variable "death" was coded 0 for persons who were alive at the end of 1 year and 1 for persons who had died by the end of the 1-year follow-up. The question whether dog ownership is statistically related to survival status was addressed in Chapter 8 using simple analyses, including computation of percentages and a chi-square test of association.

Table 21.1 ◆ Data From Friedmann et al. (1980) Survey: 92 Men Who Initially Survived a First Heart Attack

	Patient Does Not Own Dog (Y = 0)	*Patient Owns a Dog (Y = 1)*	*Total N*
Patient alive ($X = 0$)	28 (71.8%)	50 (94.3%)	78
Patient dead ($X = 1$)	11 (28.2%)	3 (5.7%)	14
Total N	39	53	92

NOTES: The X predictor variable was "dogowner," coded 0 for a patient who does not own dog and 1 for a patient who owns a dog. The Y outcome variable was "death," coded 0 for a patient who was alive at the end of 1 year and 1 for a patient who was dead at the end of 1 year.

In Chapter 8, the pattern of results in this table was described by comparing probabilities of death for dog owners and nonowners of dogs. Among the 39 people who did not own dogs, 11 died by the end of the year; thus, for nonowners of dogs, the probability of death in this sample was 11/39 = .282. Among the 53 people who owned dogs, 3 died; therefore, the probability of death for dog owners was 3/53 = .057. Based on a chi-square test of association ($\chi^2(1) = 8.85, p < .01$), this difference in proportions or probabilities was judged to be statistically significant. The researchers concluded that the risk of death was significantly lower for dog owners than for people who did not own dogs.

In this chapter, the same data will be used to illustrate the application of binary logistic regression. In binary logistic regression, the goal of the analysis is to predict the odds of death based on a person's scores on one or more predictor variables. Logistic regression with just one binary predictor variable is a special case that arises fairly often; therefore, it is useful to consider an example. In addition, it is easier to introduce concepts such as odds using the simplest possible data. A binary logistic regression analysis for the dog ownership/death data appears in Section 21.7. A second empirical example, which includes one quantitative and one categorical predictor variable, appears in Section 21.10.

21.3 ♦ Conceptual Basis for Binary Logistic Regression Analysis

Let's first think about a simple, but inadequate, way to set up an analysis of the dog ownership and death data using a simple multiple linear regression model. Consideration of the problems that arise in this simple model will help us understand why a different approach was developed for binary logistic regression. We could set up an equation to predict each individual's probability of membership in the target group, $\hat{p}$, (such as the dead patient group) from a weighted linear combination of that person's scores on one or several predictor variables as follows:

$$\hat{p}_i = B_0 + B_1 X_1 + B_2 X_2 + \cdots + B_k X_k, \quad (21.1)$$

where $\hat{p}_i$ is the estimated probability that person i is a member of the "target" outcome group that corresponds to a code of 1 (rather than the group that is coded 0). In this example, the target group of interest is the dead group and, therefore, $\hat{p}_i$ corresponds to the estimated probability that person i is dead given his or her scores on the predictor variables. B_0 is the intercept, and the B_i values are the regression coefficients that are applied to raw scores on the predictor variables. For example, we could predict how likely an individual is to be in the "dead" group based on the person's score on dog ownership (0 = Does not own a dog, 1 = Owns a dog). If we had information about additional variables, such as each individual's age, severity of coronary artery disease, and systolic blood pressure, we could add these variables to the model as predictors.

21.3.1 ♦ Why Ordinary Linear Regression Is Inadequate

The simple regression model shown in Equation 21.1 is inadequate for several reasons. One difficulty with this model is that probabilities, by definition, are limited to the range between 0 and 1. However, estimated values of $\hat{p}_i$ obtained by substituting a person's scores into Equation 21.1 would not necessarily be limited to the range 0 to 1. Equation 21.1 could produce estimated probabilities that are negative or greater than 1, and neither of these outcomes makes sense. We need to set up a model in a way that limits the probability estimates to a range between 0 and 1.

If one or more of the predictors are quantitative variables, additional potential problems arise. The relation between scores on a quantitative X predictor variable and the probability of membership in the "dead" group is likely to be nonlinear. For example, suppose that a researcher systematically administers various dosage levels of a toxic to animals to assess how risk of death is related to dosage level. Figure 21.1 shows results for hypothetical data. Level of drug dosage appears on the X axis, and probability of death appears on the Y axis.

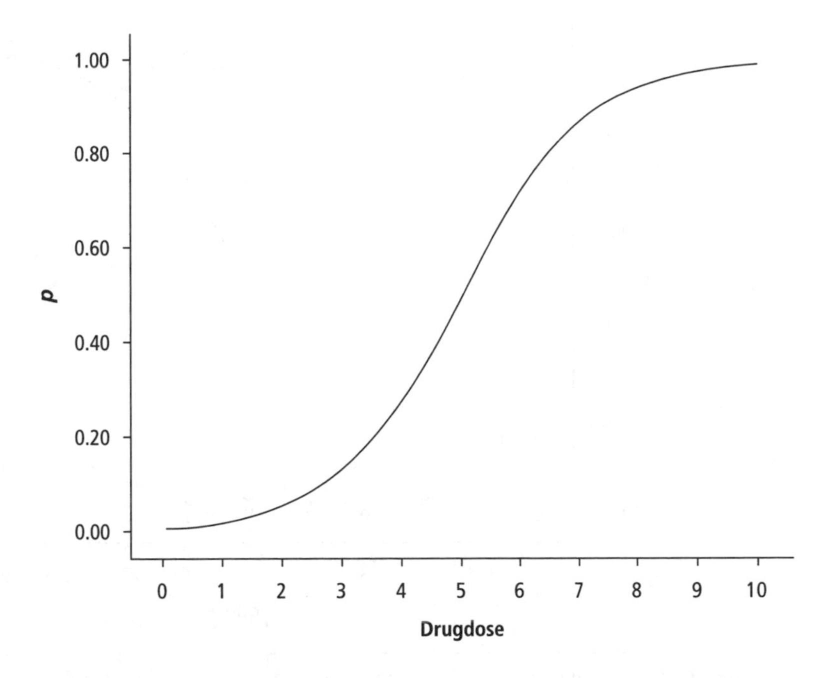

Figure 21.1 ♦ Sigmoidal Curve: Predicted Probability of Death (*Y* Axis) as a Function of Drug Dose (*X* Axis)

NOTES: *X* axis, drug dose in milligrams ranging from 0 to 10 mg; *Y* axis, predicted probability of death ranging from 0 to1.

The curve in Figure 21.1 is a **sigmoidal function** or S curve. If the underlying relation between the *X* predictor variable (such as drug dosage in milligrams) and the *Y* outcome variable (a binary-coded variable, 0 = Alive, 1 = Dead) has this form, then ordinary linear regression is not a good model. In this example, a linear model approximately fits the data for values of *X* between 3 and 7 mg, but a linear model does not capture the way the function flattens out for the lowest and highest scores on *X*. In this example, the relation between scores on the *X* variable and probability of membership in the target group is nonlinear; the nonlinearity of relationship violates one of the basic assumptions of linear regression.

A second problem is that when the underlying function has a sigmoidal form (and the *Y* scores can only take on two values, such as 0 and 1), the magnitudes of prediction errors are generally not uniform across scores on the *X* predictor variable. The actual outcome scores for *Y* can only take on two discrete values: $Y = 0$ or $Y = 1$. Figure 21.2 shows the actual *Y* scores for this hypothetical data superimposed on the curve that represents predicted probability of death at each drug dosage level. The actual *Y* values are much closer to the prediction line for drug dosage levels $X < 2$ and $X > 7$ than for intermediate drug dosage levels (*X* between 3 and 7). In other words, the probability of death is predicted much more accurately for animals that receive less than 2 mg or more than 8 mg of the drug. This pattern violates the assumption of homoscedasticity of error variance that is required for ordinary linear regression analysis.

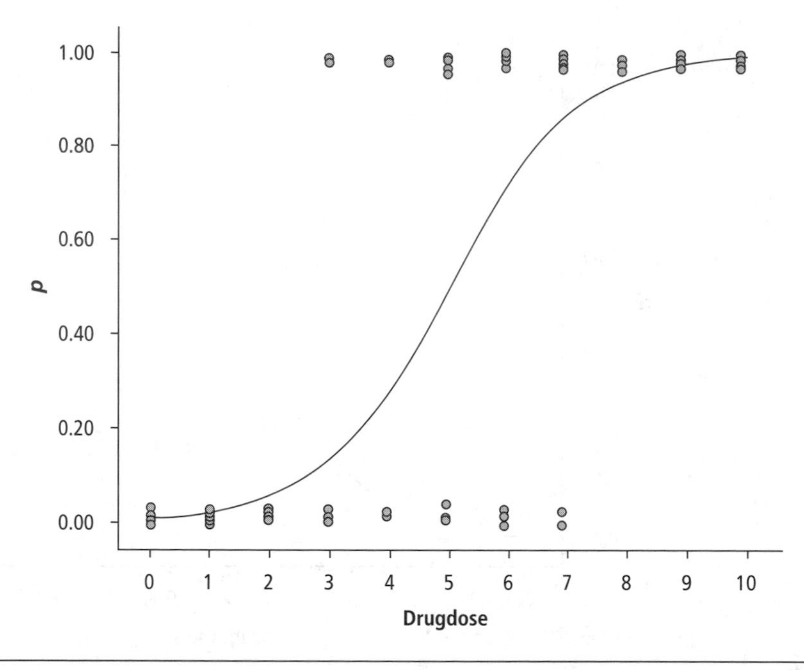

Figure 21.2 ♦ Actual Outcome Scores Superimposed on Predicted Probability Curve

NOTES: Dots represent actual Y outcomes ($Y = 0 =$ Alive, $Y = 1 =$ Dead) for each case. A small random error was added to each score to create a jittered plot; this makes it easier to see the number of cases at each drug dose level. This plot illustrates heteroscedasticity of error variance. The sigmoidal function predicts outcome (alive vs. dead) quite well for individuals who received very low doses of the drug (0–2 mg) and for individuals who received very high doses of the drug (8–10 mg). For individuals who received intermediate dosages (between 3 and 7 mg), the prediction errors are much larger. This violates one of the assumptions for ordinary multiple regression (the assumption of homoscedasticity of error variance across levels of X).

21.3.2 ♦ Modifying the Method of Analysis to Handle These Problems

How can we alter the model in Equation 21.1 in a way that avoids the potential problems just identified? We need to have a dependent variable in a form that avoids the problems just identified and a model that represents the potential nonlinearity of the relationship between scores on X and probability of group membership. At the same time, we want to be able to obtain information about probability of group membership from the scores on this new outcome variable. We need a transformed outcome variable that can give us predicted probabilities that are limited to a range between 0 and 1, that is approximately normally distributed, and that has a linear relationship to scores on quantitative X predictor variables. We can obtain an outcome variable for which the scores satisfy the requirements by using a **"logit"** as our outcome variable. In binary logistic regression, the outcome variable is called a logit; a logit is denoted by L_i in this chapter (in some textbooks, the logit is denoted by the symbol π). What is a logit (L_i)? How does this new form for the outcome variable solve the problems that have just been identified? How can we use the logit to make inferences about the probability of group membership as a function of scores on predictor variables?

A logit is a log of odds (or sometimes a log of a ratio of odds for two different groups or conditions). To understand the logit, we need to understand each of the terms involved in this expression. What are odds? How does the log transformation typically change the shape of a distribution of scores?

21.4 ♦ Definition and Interpretation of Odds

Odds are obtained by dividing the number of times an outcome of interest *does* happen by the number of times when it *does not* happen. For example, consider the data in Table 21.1. The main outcome of interest was whether a patient died (coded 0 = Patient survived, 1 = Patient died). In the entire sample of $N = 92$ people, 14 people died and 78 people survived. The odds of death for this entire group of 92 persons are found by dividing the number of dead people (14) by the number of survivors (78): 14/78 = .179. Conversely, we can calculate the odds of survival; to find this, we divide number of survivors by number of dead persons, 78/14 = 5.57. Note that 5.57 is just the reciprocal of .179. We can say that the odds of survival for patients in this study were greater than 5 to 1 or that the odds of death for patients in this study were less than 1 in 5.

What possible values can odds take on, and how are those values interpreted? The minimum value of odds is 0; this occurs when the frequency of the event in the numerator is 0. The maximum value for odds is $+\infty$; this occurs when the divisor is 0. To interpret odds, we ask whether the value of the odds is greater than 1 or less than 1. When odds are less than 1, the target event is less likely to happen than the alternative outcome. When odds are greater than 1, the target event is more likely to happen than the alternative outcome. When the odds exactly equal 1, the target event has an equal chance of happening versus not happening. To illustrate the interpretation of odds, consider a simple experiment in which a student tosses a coin 100 times and obtains heads 50 times and tails 50 times. In this case, the odds of getting heads is number of times when heads occur divided by the number of times when heads did not occur: 50/50 = 1.00. This is called "even odds"; an odds of 1 means that the two outcomes (heads or tails) are equally likely to occur. Consider another simple possibility. Suppose that a prize drawing is held and 10 tickets are sold. One ticket will win the prize, while the other nine tickets will not win. The odds of winning are 1/9 (or .1111); the odds of not winning are 9/1 (or 9). These examples show us how odds are interpreted. When odds are close to 1, the two outcomes (such as heads vs. tails for a coin toss) are equally likely to happen. When odds are less than 1, the outcome of interest is unlikely to happen, and the closer the odds get to 0, the less likely the outcome. When odds are greater than 1, the event of interest is more likely to occur than not; the higher the odds, the more likely the event.

Returning to the dog ownership and death data, the odds of death for the entire sample, given above as odds = .179, tells us that a randomly chosen person from this entire sample of 92 people is less likely to be dead than alive. Conversely, the value of the odds of surviving given above, odds = 5.57, tells us that a randomly selected individual from this group of $N = 92$ is much more likely to be alive than dead. A randomly selected person from this sample of 92 people is between five and six times as likely to be alive as to be dead.

The language of odds is familiar to gamblers. When the odds are greater than 1, the gambler is betting on an outcome that is likely to happen (and the larger the odds, the more likely the event). When the odds are less than 1, the gambler is betting on an outcome that is unlikely to happen (and the closer to zero the odds become, the less likely the event is).

An odds ratio is a comparison of the odds of some target event across two different groups or conditions. We can compute odds of death separately for each of the two groups in Table 21.1. For the nonowners of dogs, the odds of death are 11/28 = .393. For the owners of dogs, the odds of death are 3/50 = .060. To compare owners and nonowners of dogs, we set up a ratio of these two odds. Throughout this process, we have to be careful to keep track of which group we place in the numerator. Suppose that we use the group of nonowners of dogs as the reference group when we set up a ratio of odds. The ratio of odds that describes the relatively higher odds of death for nonowners of dogs compared with owners of dogs is found by dividing the odds of death for nonowners (.393) by the odds of death for owners (.060). Therefore, the odds ratio in this example is .393/.060 = 6.55. In words, the odds of death are about 6.5 times greater for nonowners of dogs than for dog owners. If we set up the odds ratio the other way, that is, if we divide the odds of death for dog owners by the odds of death for nonowners, .060/.393 = .153, we can say that the odds of death are less than one sixth as high for owners of dogs as for nonowners of dogs in this sample. Whichever group we choose to use in the denominator, the nature of the relationship remains the same, of course: Dog owners in this study were about six times less likely to die than people who did not own dogs.

An odds ratio has an advantage over a probability ($\hat{p}_i$). Unlike a probability (which cannot exceed 1), an odds ratio has no fixed upper limit. However, an odds ratio still has a fixed lower limit of 0; values of an odds ratio do not tend to be normally distributed, and values of the odds ratio do not tend to be linearly related to scores on quantitative predictor variables (Pampel, 2000). These are not desirable characteristics for an outcome variable, but these problems can be remedied by applying a simple data transformation: the **natural logarithm.** Logarithmic transformations were discussed in Chapter 4; they are reviewed briefly here. The base 10 logarithm of X is the power to which 10 must be taken to obtain the value of X. For example, suppose we have a score value X (human body weight in pounds)—for instance, $X = 100$ lb. The base 10 log of $X = 100$ is the power to which 10 must be raised to equal $X = 100$. In this case, the log is 2; $10^2 = 100$.

Natural logarithms use a different base. Instead of 10, natural logs are based on the mathematical constant "e"; the value of "e" to five decimal places is 2.71828. The natural log of X, usually denoted by $\ln(X)$, is the power to which "e" must be raised to obtain X. Examples of X and the corresponding values of $\ln(X)$ appear in Table 21.2.

The operation that is the inverse of natural log is called the **exponential function**; the exponential of X is usually denoted by either e^X or $\exp(X)$. By applying the exponential function, we can convert the $\ln(X)$ value back into the X score. For example, $e^0 = 1$, $e^{.5} = 1.64872$, $e^1 = 2.71828$, and $e^2 = 7.38905$. Some calculators have the $\ln(X)$ function (input

Table 21.2 ♦ Examples of Values of X, $a = \ln(X)$, and $\exp(a)$

X	$a = \ln(X)$	$\exp(a)$ or e^a
0	Undefined	
1	0	1
1.64872 ($\sqrt{e}$)	.5	1.64872 ($\sqrt{e}$)
2.71828 (e)	1	2.71828 (e)
7.38905 (e^2)	2	7.38905 (e^2)

X, output the natural logarithm of X) and an "exp" or e^X function (input X, output "e" raised to the Xth power). Online calculators are also available to provide these values. Functions that involve the mathematical constant "e" are often encountered in the biological and social sciences; a familiar function that involves the constant "e" is the formula for the normal or Gaussian distribution. Functions that involve "e" often have very useful properties. For example, although the normal curve is defined for values of X that range from $-\infty$ to $+\infty$, the total area under the curve is finite and is scaled to equal 1.00. This property is useful because slices of the (finite) area under the normal curve can then be interpreted as proportions or probabilities. The mathematical constant "e" is often included in curves that model growth or that involve sigmoidal or S-shaped curves (similar to Figure 21.1). To summarize, taking the natural log of an odds (or a ratio of odds) to create a new outcome variable called the logit, denoted by L_i, provides an outcome variable that has properties that are much more suitable for use as a dependent variable in a regression model. SPSS and other computer programs provide regression coefficients and the exponential of each regression coefficient as part of the output for binary logistic regression. A data analyst is not likely to need to compute natural logs or exponentials by hand when conducting binary logistic regression because these transformations are usually provided where they are needed as part of the output from computer programs.

21.5 ♦ A New Type of Dependent Variable: The Logit

Taking the natural log of an odds (or a ratio of odds) converts it to a new form (L_i, called the logit) that has much more desirable properties. It can be shown that L_i scores have no fixed upper or lower limit, L_i scores tend to be normally distributed, and in many research situations, L_i is linearly related to scores on quantitative predictor variables (see Pampel, 2000). Thus, scores on L_i satisfy the assumptions for a linear model, and we can set up a linear equation to predict L_i scores from scores for one or several predictor variables. One disadvantage of using L_i as the dependent variable is that values of L_i do not have a direct intuitive interpretation. However, predicted values of L_i from a binary logistic regression can be used to obtain information about odds and to predict the probability of membership in the target group ($\hat{p}_i$) for persons with specific values of scores on predictor variables. These probability estimates are more readily interpretable.

At this point, we can modify Equation 21.1. Based on the reasoning just described, we will use L_i (the logit) rather than $\hat{p}_i$ (a predicted probability) as the dependent variable:

$$L_i = B_0 + B_1 X_1 + \cdots + B_k X_k. \qquad (21.2)$$

Equation 21.2 shows that the scores for the logit (L_i) are predicted as a linear function of scores on one or several predictor variables, $X_1, \ldots, X_k$. As in a multiple linear regression, each predictor can be either a quantitative variable or a binary categorical variable. For example, in a medical study where the outcome is patient death versus patient survival, and the predictors include patient gender, age, dosage of drug, and severity of initial symptoms, we would assess the overall fit of the model represented by Equation 21.2; we can use the coefficients associated with the variables in Equation 21.2 to obtain information about the nature and strength of the association of each predictor variable with patient outcome.

Logistic regression is often used to compare two or more different models. The first model examined here, called a **"null"** or **"constant-only" model**, generates one predicted odds value that is the same for all the people in the entire study; it does not take scores on any X predictor variables into account. A comparison model or "full model" includes scores on all the X predictor variables and predicts different odds and probabilities for people who have different scores on the predictor variables. A researcher may evaluate several models that include different sets of predictors (King, 2003). The goal is generally to identify a small set of predictor variables that makes sense in terms of our theoretical understanding and that does a good job of predicting group membership for most individuals in the sample.

21.6 ♦ Terms Involved in Binary Logistic Regression Analysis

In the dog ownership/death study (Table 21.1), the null or constant-only model for this research problem can be written as follows:

$$L_i = B_0. \tag{21.3}$$

That is, the null model predicts a logit score (L_i) that is the same for all members of the sample (and that does not differ for owners versus nonowners of dogs). The full model for this first example in this chapter, where we want to predict odds of death from an X predictor variable that represents dog ownership, takes the following form:

$$L_i = B_0 + B_1X. \tag{21.4}$$

Note that the numerical value of the B_0 coefficient in Equation 21.4 will generally not be equal to the numerical value of B_0 in Equation 21.3. Equation 21.4 predicts different values of log odds of death for people whose X score is 0 (people who don't own dogs) and people whose X score is 1 (people who own dogs). The two different predicted values of L_i are obtained by substituting in the two possible values of X ($X = 0$ and $X = 1$) into Equation 21.4 and then simplifying the expression. For people with scores of $X = 0$, $L_i = B_0$. For people with scores of $X = 1$, $L_i = B_0 + B_1$. Of course, in research situations where the X predictor variable has more than two possible values, the model generates a different L_i value for each possible score value on the X variable. In a later section, SPSS will be used to obtain estimates of the coefficients in Equations 21.3 and 21.4. In this simple example (with only one binary predictor variable), it is instructive to do some by-hand computations to see how the logit L_i is related to the odds of death and to the probability of death in each group. The logit (L_i) is related to the predicted probability ($\hat{p}_i$) of membership in the target group (in this empirical example, the probability of death) through the following equation:

$$L_i = \ln\left[\frac{\hat{p}_i}{1 - \hat{p}_i}\right]. \tag{21.5}$$

Conversely, given a value of L_i, we can find the corresponding value of $\hat{p}_i$:

$$\hat{p}_i = \frac{e^{L_i}}{1 + e^{L_i}}. \quad (21.6)$$

For a specific numerical example of a value of L_i, consider the null model in Equation 21.3. The null model predicts the same odds of death for all cases; this model does not include information about dog ownership when making a prediction of odds. For the entire set of 92 people in this study, 14 were dead and 78 were alive. The odds of death for the entire sample is number of dead persons (14) divided by number of persons who survived (78) = 14/78 = .179. L_i, the log odds of death for the null model, can be obtained in either of two ways. It can be obtained by taking the natural log of the odds (odds = .179): $L_i = \ln(\text{odds of death}) = \ln(.179) = -1.718$.

In this example, the probability of death (p) for the overall sample of $N = 92$ is found by dividing the number of dead persons (14) by the total number of persons: 14/92 = .152. We can use Equation 21.5 to find the value of L_i from this probability:

$$L_i = \ln\left[\frac{\hat{p}_i}{1 - \hat{p}_i}\right] = \ln\left[\frac{.152}{1 - .152}\right] = \ln\left[\frac{.152}{.848}\right] = \ln(.179) = -1.718.$$

Look again at the data in Table 21.1. The L_i value of −1.718 for the null model is consistent with the odds of dying (odds = .179) observed in the overall sample of $N = 92$. When we use the null model, we can say that (if we ignore information about dog ownership) the odds of death for any individual in this sample is .179—that is, less than 1 in 5. Alternatively, we can report that the probability of death for the entire sample is $p = .152$.

When we add a predictor variable (dog ownership), we want to see if we can improve the prediction of odds of death by using different predicted odds for the groups owners of dogs and nonowners of dogs. When we examine the SPSS output for this binary logistic regression, we will demonstrate that the coefficients for this model are consistent with the observed odds and probabilities of death for each group (the group of people who own dogs versus the group of people who do not own dogs).

21.6.1 ♦ Estimation of Coefficients for a Binary Logistic Regression Model

Notice that L_i is not a directly measured variable; scores on L_i are constructed in such a way that they maximize the goodness of fit—that is, the best possible match between predicted and actual group memberships. The problem is to find the values of B_0, B_1, B_2, . . . , B_k that generate values of L_i, which in turn generate probabilities of group membership that correspond as closely as possible to actual group membership. (For example, a person whose predicted probability of being dead is .87 should be a member of the dead group; a person whose predicted probability of being dead is .12 should be a member of the surviving group.) There is no simple ordinary least squares (OLS) analytic solution for this problem. Therefore, estimation procedures such as maximum likelihood estimation (MLE) are used. MLE involves brute force empirical methods. One simple MLE approach involves grid search. For example, if a person did not know that the formula for the sample mean M was $\sum X/M$ but the person did know that the best estimate of M was the value for which the sum of squared deviations had the minimum possible value, the person could obtain an MLE estimate of M by systematically computing the sum of squared deviations for all possible values of M within some reasonable range and choosing the value of M that resulted in the smallest possible sum of squared deviations. In situations for which we have simple analytic procedures (such as ordinary linear regression), we can compute the coefficients that provide the optimal model fit by simply carrying out some matrix algebra. For models that do not have simple analytic solutions, such as logistic regression, computer programs search for the combination of coefficient values that produce the best overall model fit.

21.6.2 ♦ Assessment of Overall Goodness of Fit for a Binary Logistic Regression Model

The overall "fit" measure that is most often used to assess binary logistic regression models is the **log likelihood (LL)** function. LL is analogous to the sum of squared residuals in multiple linear regression; that is, the larger the (absolute) value of LL, the poorer the agreement between the probabilities of group membership that are generated by the logistic regression model and the actual group membership. Although LL is called a "goodness-of-fit" measure, this description is potentially confusing. Like many other goodness-of-fit measures, a larger absolute value of LL actually indicates worse model fit; it would be more intuitively natural to think of LL as a "badness-of-fit" measure. Each case in the sample has an actual group membership on the outcome variable; in the dog ownership/survival status example, the outcome variable Y had values of 0 for patients who survived and 1 for patients who died. SPSS can be used to find numerical estimates for the coefficients in Equation 21.3, and Equation 21.4 can then be used to obtain an estimate of the probability of being a member of the "dead" group for each individual patient. The LL function essentially compares the actual Y scores (0, 1) with the logs of the estimated probabilities for individual patients:

$$\text{LL} = \Sigma[(Y_i * \ln \hat{p}_i) + (1 - Y_i) * \ln(1 - \hat{p}_i)], \tag{21.7}$$

where Y_i is the actual group membership score for person i (coded $Y_i = 0$ for those who survive and $Y_i = 1$ for those who died) and $\hat{p}_i$ is the predicted probability of membership in the target group—in this example, the "dead" group—for each person. A different predicted probability is obtained for each score value on one or more X predictor variables.

In this example, there is only one predictor variable (dog ownership). For each person, the value of $\hat{p}_i$ was obtained by using Equation 21.4 to find the LL L_i from that individual's score on the predictor variable; then, Equation 21.5 can be used to convert the obtained L_i value into a $\hat{p}_i$ value. Because there is only one predictor variable and it has only two possible values, the equation will generate just two probabilities: the probability of death for owners of dogs and the probability of death for nonowners of dogs. When many predictor variables are included in the analysis, of course, different L_i and $\hat{p}_i$ probability estimates are obtained for all possible combinations of scores on the predictor variables.

Taking the natural log of the $\hat{p}_i$ values, $\ln(\hat{p}_i)$, always results in a negative value (because the value of $\hat{p}_i$ must be less than 1 and the natural log of scores below 1 are all negative numbers). Thus, LL is a sum of negative values, and it is always negative.

When LL values are compared, an LL that is larger in absolute value indicates poorer model fit. That is, LL tends to be closer to zero when actual group memberships are close to the predicted probabilities for most cases; as predicted probabilities get further away from actual group memberships, the negative values of LL become larger in absolute value.

Multiplying LL by −2 converts it to a chi-square distributed variable. SPSS can report **−2LL** for the null or constant-only model (if the "iteration requested" option is selected) as well as for the full model that includes all predictor variables. The null or constant-only model corresponds to Equation 21.3. This model predicts that all the cases are members of the outcome group that has a larger *N* (regardless of scores on other variables). The general case of a full model that includes several predictor variables corresponds to Equation 21.2; a model with just one predictor variable, as in the study of dog ownership and death, corresponds to Equation 21.4.

Taking the difference between −2LL for the full model and −2LL for the null model yields a chi-square statistic with *k* degrees of freedom (*df*), where *k* is the number of predictors included in the full model:

$$\chi^2 = -2(\text{LL}_{\text{null model}} - \text{LL}_{\text{full model}}) \text{ or } (-2\text{LL}_{\text{null}} - (-2\text{LL}_{\text{full}})). \tag{21.8}$$

Researchers usually hope that this chi-square statistic will be large enough to be judged statistically significant as evidence that the full model produces significantly less prediction error than the null model. In words, the null hypothesis that is tested by this chi-square statistic is that the probabilities of group membership derived from the full binary logistic regression model are not significantly closer to actual group membership than probabilities of group membership derived from the null model. When a large chi-square is obtained by taking the difference between LL for the full and null models, and the obtained chi-square exceeds conventional critical values from the table of the chi-square distribution, the researcher can conclude that the full model provides significantly better prediction of group membership than the null model. Obtaining a large chi-square for the improvement in binary logistic regression model fit is analogous to obtaining a large *F* ratio for the test of the significance of the increase in multiple *R* in a multiple linear regression model as predictor variables are added to the model; this was discussed in Chapter 14.

21.6.3 ♦ Alternative Assessments of Overall Goodness of Fit

Other statistics have been proposed to describe the goodness of fit of logistic regression models. Only widely used statistics that are reported by SPSS are discussed here; for a more thorough discussion of measures of model fit, see Peng, Lee, and Ingersoll (2002). In OLS multiple linear regression, researchers generally report a multiple *R* (the correlation between the *Y′* predicted scores generated by the regression equation and the actual *Y* scores) to indicate how well the overall regression model predicts scores on the dependent variable; in OLS multiple linear regression, the overall R^2 is interpreted as the percentage of variance in scores on the dependent variable that can be predicted by the model. Binary logistic regression does not actually generate predicted *Y′* scores; instead, it generates predicted probability values ($\hat{p}_i$) for each case, and so the statistic that assesses overall accuracy of prediction must be based on the $\hat{p}_i$ estimates. Binary logistic regression does not yield a true multiple *R* value, but SPSS provides **pseudo *R*** values that are (somewhat) comparable to a multiple *R*. **Cox and Snell's R^2** was developed to provide something similar to the "percentage of explained variance" information available from a true multiple R^2, but its maximum value is often less than 1.0. Similar to a ϕ coefficient for a 2×2 table, the Cox and Snell R^2 has a restricted range when the marginal distributions of the predictor and outcome variables differ. **Nagelkerke's R^2** is more widely reported; it is a modified version of the Cox and Snell R^2. Nagelkerke's R^2 is obtained by dividing the obtained Cox and Snell R^2 by the maximum possible value for the Cox and Snell R^2 given the marginal distributions of scores on the predictor and outcome variables; thus, Nagelkerke's R^2 can take on a maximum value of 1.0.

21.6.4 ♦ Information About Predictive Usefulness of Individual Predictor Variables

If the overall model is significant, information about the contribution of individual predictor variables can be examined to assess which of them make a statistically significant contribution to the prediction of group membership and the nature of their relationship to group membership (e.g., As age increases, does probability of death increase or decrease?). As in a multiple linear regression, it is possible to estimate coefficients for a raw score predictive equation. However, because binary logistic regression has such a different type of outcome variable (a logit), the interpretation of the *B* slope coefficients for a logistic regression is quite different from the interpretation of *b* raw score slope coefficients in ordinary linear regression. The *B* coefficient in a logistic regression tells us by how many units the log odds ratio increases for a one-unit increase in score on the *X* predictor variable.

For statistical significance tests of individual predictors, the null hypothesis can be stated as follows:

$$H_0: B_i = 0. \tag{21.9}$$

That is, the null hypothesis for each of the X_i predictor variables $(X_1, X_2, \ldots, X_k)$ is that the variable is unrelated to the log odds ratio outcome variable L_i. In ordinary multiple linear regression (Chapter 14), a similar type of null hypothesis was tested by setting up a *t* ratio:

$$t = \frac{b_i}{SE_{b_i}}. \tag{21.10}$$

For binary logistic regression, the test that SPSS provides for the null hypothesis in Equation 21.9 is the **Wald chi-square statistic**:

$$W = \left[\frac{B_i}{SE_{B_i}}\right]^2. \tag{21.11}$$

To test the significance of each raw score slope coefficient, we divide the slope coefficient (B_i) by its standard error (SE_{B_i}) and square the ratio; the resulting value has a chi-square distribution with 1 *df*, and SPSS reports an associated *p* value. The *B* coefficient is not easy to interpret directly, because it tells us how much L_i (the logit or log odds) is predicted to change for each one-unit change in the raw score on the predictor variable. It would be more useful to be able to talk about a change in the odds than a change in the log of the odds. The operation that must be applied to L_i to convert it from a log odds to an odds is the exponential function. That is, e^{L_i} provides information about odds. To evaluate how a one-unit change in X_i is related to change in the odds ratio, we need to look at e^{B_i}, also called $\exp(B_i)$. The value of e^{B_i} can be interpreted more directly. For a one-unit increase in the raw score on X_i, the predicted odds changes by e^{B_i} units. If the value of e^{B_i} is less than 1, the odds of membership in the target group go down as scores on X_i increase; if the value of e^{B_i} equals 1, the odds of membership in the target group do not change as X_i increases; and if the value of e^{B_i} is greater than 1, the odds of membership in the target group increase as X_i increases. The distance of e^{B_i} from 1 indicates the size of the effect (Pampel, 2000). The percentage of change (%Δ) in the odds ratio that is associated with a one-unit increase in the raw score on X_i can be obtained as follows:

$$\%\Delta = (e^{B_i} - 1) \times 100. \tag{21.12}$$

For example, if $e^{B_i} = 1.35$, then $\%\Delta = (e^{B_i} \quad 1) \times 100 = (1.35 - 1) \times 100 = 35\%$; that is, the odds of death increase by 35% for a one-unit increase in the score on X_i. In addition to tests of statistical significance, researchers should also evaluate whether the effect size or strength of predictive relationship for each predictor variable in the model is sufficiently large to be considered practically and/or clinically significant (Kirk, 1995; Thompson, 1999).

21.6.5 ♦ Evaluating Accuracy of Group Classification

One additional way to assess the adequacy of the model is to assess the accuracy of classification into groups that is achieved when the model is applied. This involves setting up a contingency table to see how well actual group membership corresponds to predicted group membership. For each case in the sample, once we have coefficients for the equations above, we can compute $\hat{p}_i$, the predicted probability of membership in the "target" group (such as the "dead patient" group) for each case. We can then classify each case into target group if $\hat{p}_i > .50$ and into the other group if $\hat{p}_i < .50$. Finally, we can set up a contingency table to summarize the correspondence between actual and predicted group memberships. For a binary logistic regression, this is a 2 × 2 table. The percentage of correctly classified cases can be obtained for each group and for the overall sample. However, in the dog owner/death dataset, both owners and nonowners of dogs are more likely to be alive than to be dead at the end of 1 year. Thus, when we use the model to predict group membership, the predicted group membership for all $N = 92$ members of the sample is the "alive" group. This is an example of a dataset for which accuracy of prediction of group membership is poor even though there is a strong association between odds of death and dog ownership. In this sample, people who do not own dogs have odds of death six times as high as people who do not own dogs, which seems to be a substantial difference; however, even the group of people who do not own dogs are much more likely to survive than to die. The second empirical example (presented in Section 21.10) is an example of a research situation where group membership can be predicted more accurately.

In addition, residuals can be examined for individual cases (this information was not requested in the following example of binary logistic regression analysis). Examination of individual cases for which the model makes incorrect predictions about group membership can sometimes be helpful in detecting multivariate outliers, or perhaps in identifying additional predictor variables that are needed (Tabachnick & Fidell, 2007).

21.7 ♦ Analysis of Data for First Empirical Example: Dog Ownership/Death Study

21.7.1 ♦ SPSS Menu Selections and Dialog Windows

To run a binary logistic regression for the data shown in Table 21.1, the following menu selections are made: <Analyze> → <Regression> → <Binary Logistic> (as shown in Figure 21.3).

The main dialog window for the binary logistic regression procedure appears in Figure 21.4. The name of the dependent variable (in this example, "death") is placed in the window for the dependent variable. The name for each predictor variable (in this example, "dogowner") is placed in the window for Covariates. If a predictor is a categorical variable, it is necessary to define it as categorical and to specify which group will be treated as the reference group when odds ratios are set up.

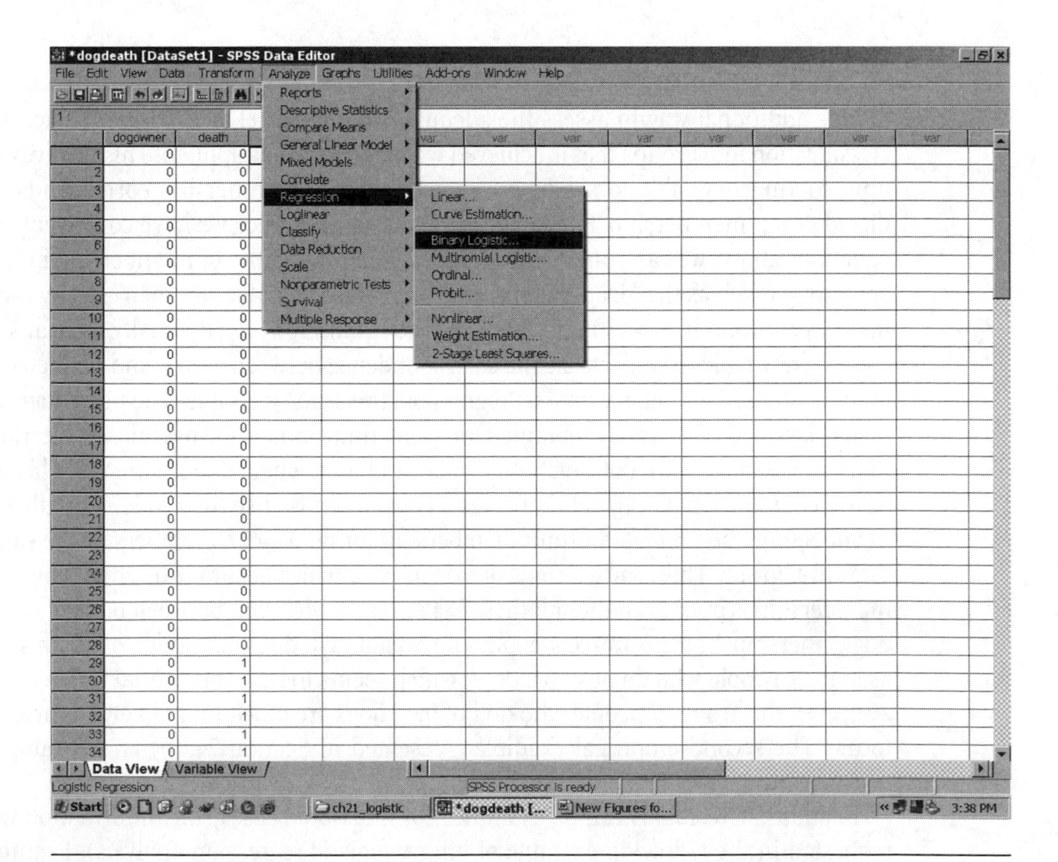

Figure 21.3 ♦ SPSS Menu Selections to Run Binary Logistic Regression Procedure

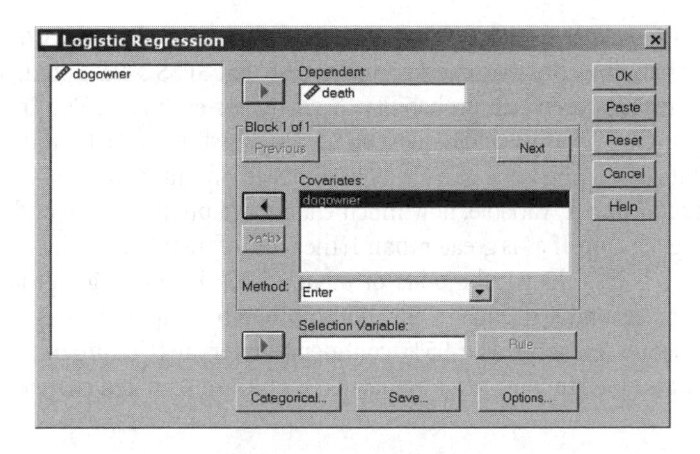

Figure 21.4 ♦ Main Dialog Window for SPSS Binary Logistic Regression

The Categorical, Save, and Options buttons were clicked to open additional SPSS dialog windows. In the Categorical dialog window that appears in Figure 21.5, the independent variable X, the variable "dogowner," was identified as a categorical predictor variable. The radio button Reference Category "Last" was selected and the Change button was clicked. This instructs SPSS to use the group that has the largest numerical code (in this example, $X = 1$, the person is a dog owner) as the reference group when setting up a ratio of odds. Note that the interpretation of outcomes must be consistent with the way scores are coded on the binary outcome variable (e.g., 1 = Dead, 0 = Alive) and the choice of which group to use as the reference group when setting up a ratio of odds to compare groups on a categorical variable. In this example, the analysis was set up so that the coefficients in the binary logistic regression can be interpreted as information about the comparative odds of death for people who do *not* own dogs compared with people who *do* own dogs. Different computer programs may have different default rules for the correspondence of the 0 and 1 score codes to the reference group and the comparison group, respectively. When a data analyst is uncertain what comparison the program provides when a particular menu selection is used to identify the reference group, applying the commands to a dataset for which the direction of differences in odds between groups is known (such as this dog ownership/death dataset) can help clear up possible confusion.

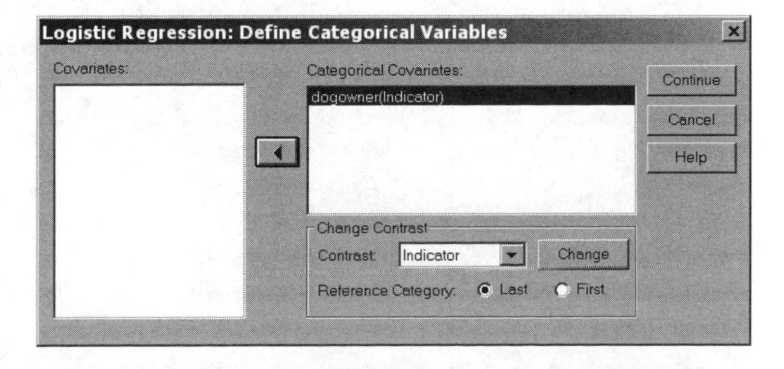

Figure 21.5 ♦ Defining Categorical Predictor Variables for Binary Logistic Regression

In the Save dialog window that appears in Figure 21.6, the Predicted Values: Probabilities box was checked to request that SPSS save a predicted probability (in this example, a predicted probability of death) for each case. The Options dialog window in Figure 21.7 was used to request an Iteration history and a Confidence interval for the estimate of e^{B_i}. The value of e^{B_i} tells us, for each one-unit change in the raw score on the corresponding X_i variable, how much change we predict in the odds of membership in the target group. If e^{B_i} is greater than 1, the odds of membership increase as X scores increase; if e^{B_i} is less than 1, the odds of membership in the target group decrease as X scores increase; and if e^{B_i} equals 1, the odds of membership in the target group do not change as X scores increase. The 95% confidence interval (CI) provides information about the amount of sampling error associated with this estimated change in odds.

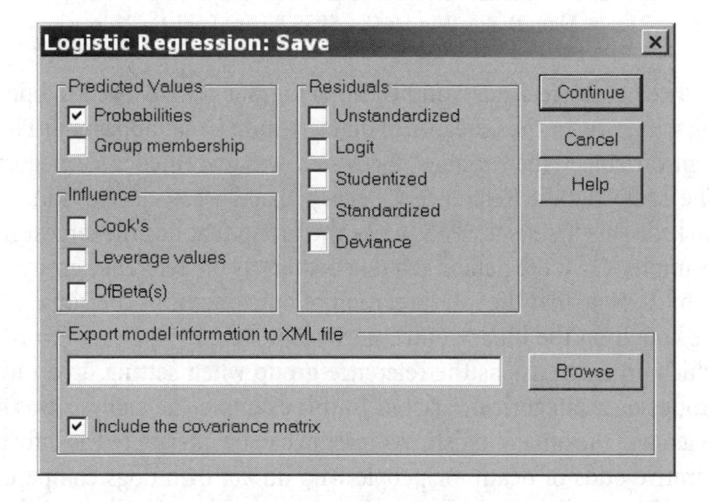

Figure 21.6 ♦ Save Predicted Probability (of Death) for Each Case

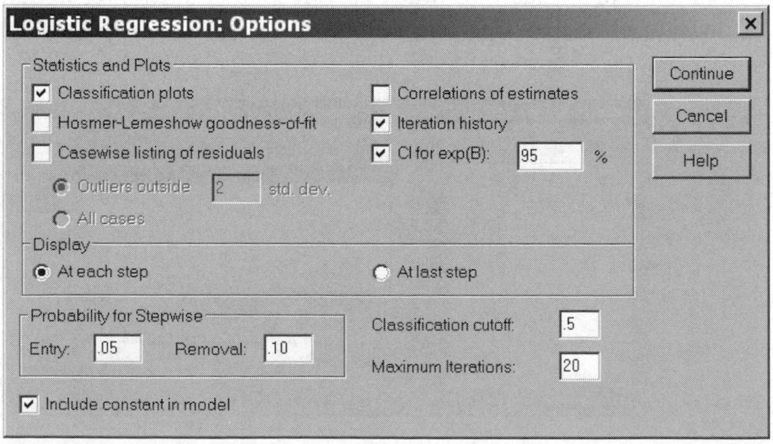

Figure 21.7 ♦ Options for Binary Logistic Regression

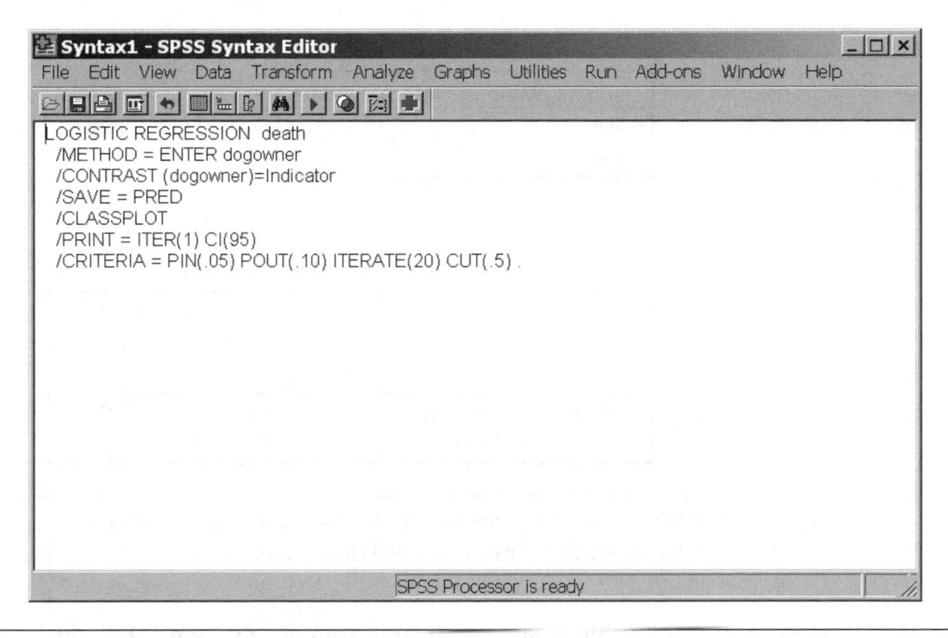

Figure 21.8 ♦ SPSS Syntax for Binary Logistic Regression to Predict Death From Dog Ownership

Finally, the Paste button was used to display the SPSS syntax that was generated by these menu selections in the syntax window that appears in Figure 21.8. It is helpful to save the syntax as documentation; this provides a record of the commands that were used.

21.7.2 ♦ SPSS Output

Selected SPSS output generated by these commands appears in Figures 21.9 through 21.17. Figure 21.9 provides information about the coding for the dependent variable, death; for this variable, a score of 0 corresponded to "alive" and a score of 1 corresponded to "dead." In other words, the target outcome of interest was death, and the results of the model can be interpreted in terms of odds or probabilities of death. In this example, the odds were calculated using the group of nonowners of dogs as the reference group. The coefficients from the binary logistic regression, therefore, are interpreted as information about the comparative odds of death for the dog-owner group relative to the group of people who did not own dogs.

21.7.2.1 ♦ *Null Model*

The Block 0 results that appear in Figures 21.10 and 21.11 are results for the null or constant-only model, which correspond to Equation 21.3. The null model is generally used only as a baseline to evaluate how much predicted odds change when one or more predictor variables are added to the model. The null hypothesis for the null model is that the odds of being alive versus being dead for the entire sample of $N = 92$ men, ignoring dog ownership status, equal 1. Thus, if 50% of the sample is in the alive group and 50% in the dead group, the value of B_0 would be 0, and $\log(B_0)$ would equal 1. $\text{Log}(B_0)$ is interpreted as the odds of death when group membership on categorical predictor variables

Dependent Variable Encoding

Original Value	Internal Value
alive	0
dead	1

Categorical Variables Codings

		Frequency	Parameter coding (1)
dogowner	no dog	39	1.000
	own dog	53	.000

Figure 21.9 ♦ SPSS Output for Binary Logistic Regression to Predict Death From Dog Ownership: Internal Coding for Predictor and Outcome Variables

such as dog ownership is ignored. In this example, the value of B_0 differs significantly from 0 ($B_0 = -1.718$, Wald $\chi^2(1) = 35.019$, $p < .001$). This tells us that the odds of death for the overall sample differed significantly from 1.00 or, in other words, that the probability of death for the overall sample differed significantly from .50. Exp(B_0) = .179 tells us that the overall odds of death (for the entire sample of $N = 92$) was .179. In other words, approximately one sixth of the 92 participants in this entire sample died. This null model (that does not include any predictor variables) is used as a baseline to evaluate how much closer the predicted probabilities of group membership become to actual group membership when one or more predictor variables are added to the model.

Block 0: Beginning Block

Iteration History[a,b,c]

Iteration		-2 Log likelihood	Coefficients Constant
Step 0	1	79.831	-1.391
	2	78.481	-1.686
	3	78.469	-1.717
	4	78.469	-1.718

a. Constant is included in the model.

b. Initial -2 Log Likelihood: 78.469

c. Estimation terminated at iteration number 4 because parameter estimates changed by less than .001.

Figure 21.10 ♦ SPSS Output for Binary Logistic Regression to Predict Death From Dog Ownership: Null Model

NOTE: No predictors included.

Variables in the Equation

		B	S.E.	Wald	df	Sig.	Exp(B)
Step 0	Constant	-1.718	.290	35.019	1	.000	.179

Figure 21.11 ♦ Coefficients for Binary Logistic Regression Null Model

21.7.2.2 ♦ *Full Model*

The Block 1 model in this example corresponds to a model that uses one categorical predictor variable (does each participant own a dog, yes or no) to predict odds of death. The fit of this one-predictor model is assessed by evaluating whether the goodness of fit for this one-predictor model is significantly better than the fit for the null model. Selected SPSS output for the Block 1 model appears in Figures 21.12 through 21.15. The first question to consider in the evaluation of the results is, Was the overall model statistically significant? When dog ownership was added to the model, did the LL goodness- (or "badness-") of-fit measure that assessed differences between predicted and actual group memberships decrease significantly? The chi-square test for the improvement in fit for the full model compared with the null model reported in Figure 21.13 was obtained by subtracting the LL value for the null model from the LL value for the full model and multiplying this difference by –2 (see Equation 21.8). The *df* for this improvement in model fit chi-square is *k*, where *k* is the number of predictor variables in the full model; in this example, $df = 1$. The omnibus test in Figure 21.13, $\chi^2(1) = 9.011$, $p = .003$, indicated that the full predictive model had a significantly smaller LL "badness-of-fit" measure than the null model. If the conventional $\alpha = .05$ is used as the criterion for statistical significance, this model would be judged statistically significant. In other words, the full predictive model with just one predictor variable, dog ownership, predicts odds of death significantly better than a null model that does not include any predictors. Other goodness-of-fit statistics such as the Nagelkerke R^2 value of .163 appear in Figure 21.14. Figure 21.15 shows the table of actual versus predicted group memberships; in this study, members of both groups (owners and nonowners of dogs) were more likely to survive than to die, and therefore, the best prediction about group membership for both owners and nonowners of dogs was that the person would be alive. When we are trying to predict an outcome, such as death, that has a relatively low rate of occurrence across both groups in a study, an outcome similar to this one is common. Our analysis provides useful information about differences in odds of death for the two groups (people who do not own dogs have a higher odds of death), but it does not make good predictions about death outcomes for individual cases.

The data analyst can go on to ask whether each individual predictor variable makes a statistically significant contribution to the prediction of group membership and how much the odds of membership in the target group (dead) differ as scores on predictor variables change. In this empirical example, the question is whether the odds of death are higher (or lower) for nonowners of dogs than for dog owners. The coefficients for the estimated predictive model appear in Figure 21.16. The line that begins with the label Constant provides the estimated value for the B_0 coefficient; the B_0 coefficient is used to obtain information about odds of death for the reference group (in this example, the odds

Iteration History[a,b,c,d]

Iteration		-2 Log likelihood	Coefficients	
			Constant	dogowner(1)
Step 1	1	73.691	-1.774	.902
	2	69.790	-2.487	1.553
	3	69.463	-2.772	1.837
	4	69.458	-2.813	1.878
	5	69.458	-2.813	1.879

a. Method: Enter

b. Constant is included in the model.

c. Initial -2 Log Likelihood: 78.469

d. Estimation terminated at iteration number 5 because parameter estimates changed by less than .001.

Figure 21.12 ♦ SPSS Output for Binary Logistic Regression to Predict Death From Dog Ownership Model That Includes "Dog Owner" Variable as Predictor

Omnibus Tests of Model Coefficients

		Chi-square	df	Sig.
Step 1	Step	9.011	1	.003
	Block	9.011	1	.003
	Model	9.011	1	.003

Figure 21.13 ♦ Chi-Square Test for Improvement in Model Fit Relative to Null Model

Model Summary

Step	-2 Log likelihood	Cox & Snell R Square	Nagelkerke R Square
1	69.458[a]	.093	.163

a. Estimation terminated at iteration number 5 because parameter estimates changed by less than .001.

Figure 21.14 ♦ Model Summary: Additional Goodness-of-Fit Measures

of death for dog owners). The line that begins with the name of the predictor variable, dogowner(1), provides the B_1 coefficient associated with the categorical predictor variable. This B_1 coefficient is used to obtain information about the difference in odds of death for the other group (nonowners of dogs) compared with the reference group (dog owners). The numerical values for the predictive model in Equation 21.5, $L_i = B_0 + B_1X$,

Classification Table[a,b]

			Predicted		
			death		Percentage Correct
	Observed		alive	dead	
Step 0	death	alive	78	0	100.0
		dead	14	0	.0
	Overall Percentage				84.8

a. Constant is included in the model.

b. The cut value is .500

Figure 21.15 ♦ Classification Table for Dog Ownership/Death Study

were as follows. From Figure 21.16, we find that the coefficients for this model are $L_i = -2.813 + 1.879 \times X$. For each coefficient, a Wald chi-square was calculated using Equation 21.11. Both coefficients were statistically significant; for B_0, Wald $\chi^2 = 7.357, p = .007$; for B_1, Wald $\chi^2 = 22.402, p < .001$.

Variables in the Equation

		B	S.E.	Wald	df	Sig.	Exp(B)	95.0% C.I.for EXP(B) Lower	Upper
Step 1[a]	dogowner(1)	1.879	.693	7.357	1	.007	6.548	1.684	25.456
	Constant	-2.813	.594	22.402	1	.000	.060		

a. Variable(s) entered on step 1: dogowner.

Figure 21.16 ♦ SPSS Output for Binary Logistic Regression to Predict Death From Dog Ownership Predictive Equation

It is easier to interpret **exp(*B*)** (also denoted by e^{B_i}) than to interpret *B* because the exp(*B*) value is directly interpretable as a change in odds (while *B* itself represents a change in log odds). Because B_0 was significantly different from 0, we know that the odds of death in the "dog owner" group were significantly different from "even odds"—that is, the risk of death was different from 50%. Because $\exp(B_0) = .060$, we know that the odds of death for the dog owner group were .060. Because this odds is much less than 1.00, it tells us that dog owners were less likely to be dead than to be alive.

The B_1 coefficient was also statistically significant, which tells us that the odds of death were significantly different for the group of nonowners of dogs than for the dog owner group. The $\exp(B_1) = 6.548$ value tells us that the odds of death were about 6.5 times higher for nonowners than for owners of dogs. In other words, people who did not own dogs had much higher odds of death (more than six times as high) compared with people who did own dogs. A 95% CI for the value of $\exp(B_1)$ was also provided; this had a lower limit of 1.684 and an upper limit of 25.456. The fact that this CI was so wide suggests that because of the rather small sample size, $N = 92$, the estimate of change in odds is not very precise. This CI suggests that the actual difference in odds of death for nonowners of dogs compared with dog owners could be as low as 1.7 or as high as 25.

One issue to keep in mind in interpretation of results is the nature of the original research design. This study was not an experiment; dog ownership was assessed in a survey, and dog ownership was not experimentally manipulated. Therefore, we cannot interpret the results as evidence that dog ownership causally influences death. We would make causal interpretations based on a logistic regression analysis only if the data came from a well-controlled experiment. A second issue involves evaluation whether the difference in odds is large enough to have any practical or clinical significance (Thompson, 1999). In this sample, the odds of death were more than six times as high for people who did not own dogs compared with people who did own dogs. That sounds like a substantial difference in odds. However, information about odds ratios should be accompanied by information about probabilities. In this case, it is helpful to include the sample probabilities of death for the two groups. From Table 21.1, approximately 6% of dog owners died in the first year after a heart attack, while about 28% of nonowners of dogs died within the first year after a heart attack. This additional information makes it clear that for both groups survival was a higher probability outcome than death. In addition, the probability of death was substantially higher for people who did not own dogs than for people who did own dogs.

Recommendations for reporting logistic regression outlined by Peng et al. (2002) state that reports of logistic regression should include predicted probabilities for selected values of each X predictor variable, in addition to information about the statistical significance and overall fit of the model, and the coefficients associated with individual predictor variables. In this example, the SPSS Save procedure created a new variable (Pre_1); this is the predicted probability of death for each case. These predicted probabilities should be examined and reported. Figure 21.17 shows the saved scores for this new variable Pre_1. For each member of the dog-owner group, the value of Pre_1 corresponds to a predicted probability of death = .057; for each member of the non-dog-owning group, Pre_1 corresponded to a predicted probability of death = .282. The predicted values of death for each score on the categorical predictor variable (dog ownership) are summarized in the table in Figure 21.18. These correspond to the proportions of cases in the original data in Table 21.1.

For this dog ownership/death data, it is relatively easy to show how predicted probabilities are obtained from the binary logistic regression results. First, we substitute the possible score values for X ($X = 0$ and $X = 1$) into the logistic regression equation:

$$L_i = B_0 + B_1 X = -2.813 + 1.879X.$$

When we do this, we find that the predicted value of the logit L_0 for the $X = 0$ reference group (dog owners) is −2.813; the predicted value of the logit L_1 for the $X = 1$ group (nonowners) is −2.813 + 1.879 = −.934. We can then use Equation 21.6 to find the estimated probability of death for members of each group based on the value of L_i for each group. For $X = 0$, participants who are dog owners ($L_0 = -2.813$):

$$\hat{p}_0 = \frac{e^{L_0}}{1 + e^{L_0}} = \frac{e^{-2.813}}{1 + e^{-2.813}} = .0556.$$

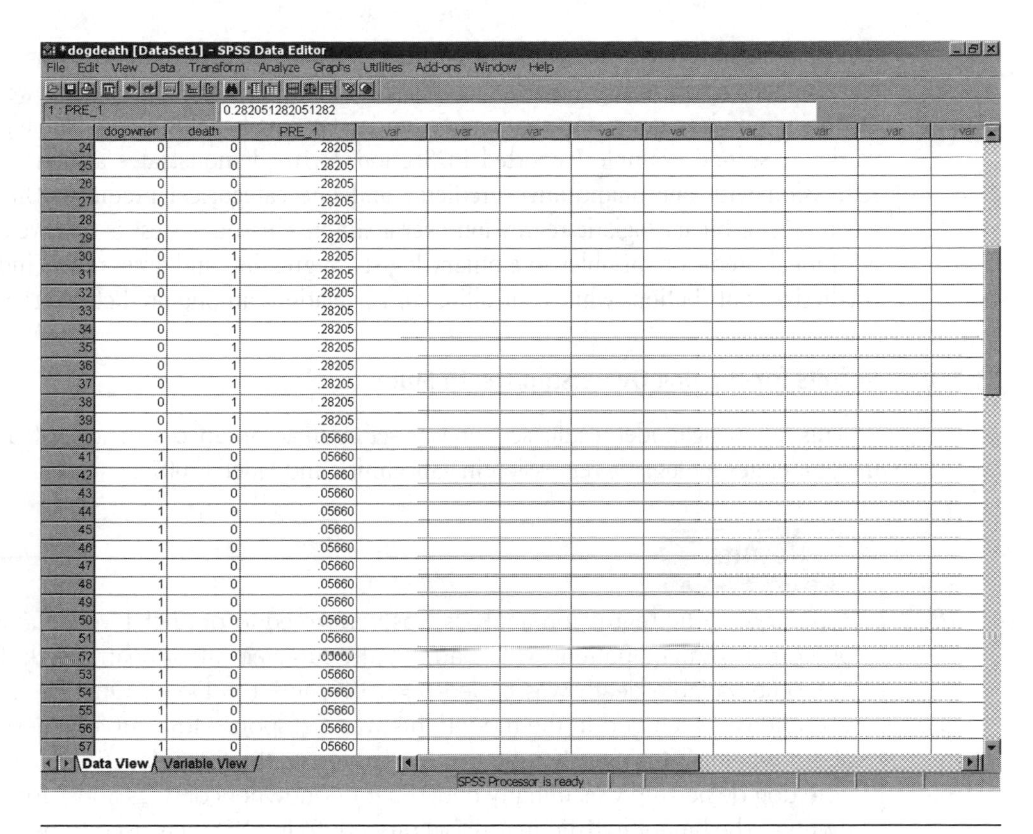
*dogdeath [DataSet1] - SPSS Data Editor

File Edit View Data Transform Analyze Graphs Utilities Add-ons Window Help

1 : PRE_1 0.282051282051282

	dogowner	death	PRE_1	var	var	var	var	var	var	var	var	var
24	0	0	.28205									
25	0	0	.28205									
26	0	0	.28205									
27	0	0	.28205									
28	0	0	.28205									
29	0	1	.28205									
30	0	1	.28205									
31	0	1	.28205									
32	0	1	.28205									
33	0	1	.28205									
34	0	1	.28205									
35	0	1	.28205									
36	0	1	.28205									
37	0	1	.28205									
38	0	1	.28205									
39	0	1	.28205									
40	1	0	.05660									
41	1	0	.05660									
42	1	0	.05660									
43	1	0	.05660									
44	1	0	.05660									
45	1	0	.05660									
46	1	0	.05660									
47	1	0	.05660									
48	1	0	.05660									
49	1	0	.05660									
50	1	0	.05660									
51	1	0	.05660									
52	1	0	.05660									
53	1	0	.05660									
54	1	0	.05660									
55	1	0	.05660									
56	1	0	.05660									
57	1	0	.05660									

Data View / Variable View

SPSS Processor is ready

Figure 21.17 ♦ SPSS Data View Worksheet With Saved Estimated Probabilities of Death

no dog	Predicted probability	.28205
own dog	Predicted probability	.05660

Figure 21.18 ♦ Predicted Probability of Death for Two Groups in the Friedmann et al. (1980) Study: Nonowners Versus Owners of Dogs

For $X = 1$, participants who are not dog owners ($L_1 = -.934$):

$$\hat{p}_1 = \frac{e^{L_1}}{1 + e^{L_1}} = \frac{e^{-.934}}{1 + e^{-.934}} = .2821.$$

Note that the predicted probabilities generated from the logistic regression model are the same as the probabilities that are obtained directly from the frequencies in Table 21.1: The proportion of dog owners who died is equal to the number of dead dog owners divided by the total number of dog owners = 3/53 = .0556. The proportion of nonowners of dogs who died is equal to the number of dead nonowners divided by the total number of nonowners = 11/39 = .2821. The preceding example demonstrates how binary logistic regression can be used to analyze simple data (the dog ownership/death data) that were analyzed earlier in the textbook using simpler methods. A chi-square test of association for these data was reported in Chapter 8. When the data are this simple, we may not gain a great deal by performing the more complicated binary logistic regression. However, this example provides a good starting point for understanding how binary logistic regression works. A second example (reported in Section 21.10) demonstrates a binary logistic regression with one quantitative predictor and one categorical predictor. One major advantage of binary logistic regression over a simple chi-square test is that we can add additional predictor variables to a binary logistic regression and assess their individual predictive contributions while controlling for correlations among predictor variables.

21.7.3 ♦ Results for the Dog Ownership/Death Study

This following model results section was set up in accordance with suggested reporting guidelines for logistic regression in Tabachnick and Fidell (2007).

Results

A binary logistic regression analysis was performed to predict 1-year survival outcomes for male patients who had initially survived a first heart attack. The outcome variable death was coded 0 = Alive and 1 = Dead. One predictor variable was included in the model; this was a response to a survey question about whether or not each patient owned a dog. In the SPSS data file, the variable dog ownership was initially coded 0 for nonowners of dogs and 1 for dog owners. The binary logistic regression procedure in SPSS was used to perform the analysis. Data from 92 cases were included in this analysis.

A test of the full model (with dog ownership as the predictor variable) compared with a constant-only or null model was statistically significant, $\chi^2(1) = 9.011$, $p = .003$. The strength of the association between dog ownership and death was relatively weak with Cox and Snell's $R^2 = .093$ and Nagelkerke's $R^2 = .163$.

Table 21.3 summarizes the raw score binary logistic regression coefficients, Wald statistics, and the estimated change in odds of death for nonowners of dogs compared with dog owners, along with a 95% CI. Table 21.4 reports the odds and the predicted probability of death for each of the two groups (owners and nonowners of dogs). The odds of death for dog owners was .179 (e.g., number of dead owners divided by number of surviving owners = 3/50 = .179). The odds of death for nonowners of dogs was .393 (number of dead divided by number of surviving persons in the group of nonowners = 11/28 = .393). Thus, the odds of death by the end of the first year were approximately 6.5 times higher for people who did not own dogs than for people who did own dogs (.393/.06 = 6.5). In other words, dog ownership was associated with a lower risk of death. However, both owners and nonowners of dogs were more likely to be alive than to be dead at the end of the 1-year follow-up. For dog owners, the probability of death during the 1-year follow-up period was .057; for nonowners of dogs, the probability of death was .282. Note that the probabilities derived from this binary logistic regression analysis correspond to the row percentages (dead vs. alive) in the original contingency table in Table 21.1.

Table 21.3 ♦ Binary Logistic Regression Analysis: Prediction of Death 1 Year After Heart Attack From Dog Ownership

Predictor Variable	*B*	*Wald Chi-Square Test*	*P*	*exp(B)*	*95% Confidence Interval for exp(B)*	
					Lower	*Upper*
Dog owner	1.879	7.36	.007	6.55	1.68	25.46
Constant	−2.813	22.402	<.001	.06		

Table 21.4 ♦ Predicted Odds and Probability of Death for Nonowners Versus Owners of Dogs

	Odds of Death	*Probability of Death*
Nonowners	.393	.282
Owners	.060	.057
Total sample	.179	.152

21.8 ♦ Issues in Planning and Conducting a Study

21.8.1 ♦ Preliminary Data Screening

Binary logistic regression involves at least one categorical variable (the *Y* outcome variable). One of the most important issues that should be addressed in preliminary data screening is the distribution of scores on the binary outcome variable, *Y*. When the outcome variable *Y* is binary it has only two possible values. If the proportion of people in the two groups deviates greatly from a 50/50 split, and if the total *N* of cases in the study is very small, the number of cases in the smaller outcome group may simply be too small to obtain meaningful results. For example, suppose that the outcome of interest is the occurrence of a relatively rare disease. The researcher has a total *N* of 100; 97 of the people in this sample do not have the disease and only 3 people in the sample have the disease. No matter what kind of statistical analysis is performed, the outcome of the analysis will be largely determined by the characteristics of the 3 people in the "disease" group. If a researcher is interested in a disease that is so rare that it occurs only three times per 100 people, the researcher needs to obtain a much larger total *N*, or to sample a much larger number of cases in the disease group, to obtain enough data for a binary logistic regression analysis.

Binary logistic regression sometimes includes one or more categorical predictor variables. For any pair of categorical variables, it is useful to set up a table to show the cell frequencies. Like the chi-square test of association for contingency tables, binary logistic regression may not produce valid results when there are one or several cells that have expected cell frequencies <5. If a preliminary look at tables reveals that more than 20% of the cells have expected values <5, this situation should be remedied. Groups for which one or more expected frequencies are <5 can either be combined with other groups or dropped from the analysis, whichever decision seems more reasonable.

Binary logistic regression may also involve one or several quantitative predictor variables. If there are extreme outliers on any of these variables, the data analyst should evaluate whether these extreme scores are incorrect or whether the scores indicate individuals who differ so much from the majority of cases in the sample that they might be thought of as members of a different population. As discussed in Chapter 4, researchers should make reasonable judgments about the handling of outliers. In some situations, the analysis will produce more believable results if extreme scores are removed from the sample or made less extreme by changing the scores. For example, suppose that the predictor variable X is systolic blood pressure and that the majority of participants have X scores between 100 and 130 (i.e., within the "normal" range). If one participant has a systolic blood pressure of 230, that score could be viewed as an outlier. This large value might be due to measurement error or it could represent a valid case of unusually high blood pressure. If the researcher defines the population of interest in the study as people with normal scores on systolic blood pressure, the blood pressure score of 230 could be dropped from the dataset (because that individual is not a member of the population of interest, i.e., persons with normal blood pressure). Another possible way to handle outlier scores is to recode them to the value of the next highest score; if the second highest blood pressure in the sample is 130, the score of $X = 230$ could be recoded to $X = 130$.

21.8.2 ♦ Design Decisions

Binary logistic regression uses MLE rather than OLS to estimate model coefficients. If there are few cases for each observed combination of scores on predictor variables, the reliability of estimates tends to be low. Peduzzi, Concato, Kemper, Holford, and Feinstein (1996) have suggested a minimum N that is at least 10 times k, where k is the number of independent variables in the model. In addition, as described in the previous section, when contingency tables are set up for each pair of categorical variables in the model (including the binary outcome variable), there should be few cells with expected frequencies below 5. If many cells have expected frequencies <5, the researcher may need to drop variables, combine scores for some categories, or omit some categories.

Statistical power in multivariate analyses depends on many factors, including the strength of the association between each predictor variable and the outcome, the degree to which assumptions are violated, the size and sign of correlations among predictor variables, and the sample size (not only the total N but also the sizes of Ns within cells in the design). Therefore, it is difficult to provide recommendations about the sample size required to have adequate statistical power in binary logistic regression. The sample sizes suggested above (at least 10 times as many cases as predictor variables and few cells with expected cell frequencies <5) are minimal requirements, and larger Ns may be required to have acceptable statistical power.

Another design decision that has an effect on statistical power involves the choice of dosage levels for categorical or interval-level predictor variables in the model. The second empirical example in this chapter examines how increasing dosages of a drug are related to death. In this hypothetical dataset, the drug dosage levels range from 0 to 10 mg in increments of 1 mg. For some drugs, a 1-mg increase might be large enough to have a noticeable impact. However, for other drugs, increases of 10 mg or even 1 g might be required to see noticeable changes in outcomes across different dosage levels. The same issue was discussed in Chapters 5 and 6 in connection with the choice of treatment dosage levels for studies that involve comparing group means using *t* tests and ANOVA.

Other factors being equal, a researcher is more likely to obtain a significant increase in odds as *N*, the overall sample size, increases; when all the groups formed by combinations of categorical variables have expected cell frequencies >5; and when the dosage difference that corresponds to each one-point increase in a predictor variable is large enough to produce a noticeable effect on the outcome.

21.8.3 ♦ Coding Scores on Binary Variables

It is conventional to code group membership on binary outcome variables using values of 0 and 1 to represent group membership. Using values of 0 versus 1 makes the coefficients of the logistic regression model easier to interpret. It is helpful to think about the nature of the research question and the clearest way to phrase the outcome when assigning these codes to groups. Researchers in social science and medicine often want to report how much a risk factor (such as smoking) increases the odds of some negative outcome (such as death) or how much a treatment or intervention (such as a low-dose aspirin) decreases the odds of some negative outcome (such as a heart attack). The "target" outcome, the one for which we want to report odds of occurrence, is very often a negative outcome. Interpretation of the results from a logistic regression depends on how scores are coded on the binary outcome variable and on which score value the computer program treats as the "target" outcome. In the preceding example, the outcome variable death was coded 0 = Alive and 1 = Dead; SPSS reports odds for membership in the target group that has a score value of "1" on the dependent variable, so in this empirical example, the odds that were reported were odds of death. Results are usually more interpretable if a code of 1 is assigned to the group with the more negative outcome (death, heart attack, dropping out of school, etc.), and a code of 0 is assigned to the group with a more positive outcome (survival, no heart attack, graduation from school, etc.).

The coding of categorical predictor variables in binary logistic regression is also potentially confusing, and the handling of these codes may differ across programs. When a predictor variable is identified as categorical, users of the SPSS binary logistic regression procedure use a radio button selection to identify whether the "first" or "last" group will be used as the reference group when odds ratios are set up to compare risks across groups. In SPSS, the "first" group is the group with a lower score on the binary predictor variable; the "last" group is the group with a higher score on the binary predictor. In the dog owner/death example presented earlier, we wanted to be able to evaluate how much higher the odds of death were for nonowners of dogs compared with dog owners. That is, we wanted to use the dog owner group as the "reference group" when we set up odds. In the original SPSS data file "dogdeath.sav," the variable dogowner was coded 0 = Does not own dog and 1 = Owns a dog. If we want to use the dog-owner group as the reference group, we need to click the radio button that corresponds to Reference Group: Last. After this selection was made, SPSS assigned new internal code values to the categorical predictor variable; in this example, the internal codes used by SPSS that appear in Figure 21.9 were 1 for "no dog" and 0 for "owns dog." Given these codes, the value of $\exp(B_0)$ is interpreted as information about odds of death for the reference group (dog owners) and the value of $\exp(B_1)$ is interpreted as the increase (or decrease) in odds of death for the comparison group of nonowners of dogs.

In many applications of binary logistic regression, a risk factor is assessed to see how much it increases the odds of a negative outcome (such as death). When a binary predictor variable represents the presence versus absence of a risk factor, it often makes more sense to set up the internal coding for this variable in SPSS so that the group *without* the risk factor is used as the *reference* group (and has an internal SPSS code of 0) and the group that *has* the risk factor is compared with the reference group. (It is less confusing to code scores initially as Yes/No, 0 vs. 1.) For the variable dog owner, a value of 0 was given to persons who did not own a dog and a value of 1 was given to persons who did own a dog. To make sure that SPSS used the dog-owner group as the reference group when setting up ratios of odds, a radio button selection was made to identify the last group (the group with the higher score) as the reference group.

If binary logistic regression were used to assess the association between being a smoker versus being a nonsmoker and having lung cancer versus not having lung cancer, data analysts who use SPSS would probably want to set up the scores on the binary categorical variables so that the undesirable medical outcome (lung cancer) corresponds to a score of "1" on the outcome variable and to use nonsmokers (i.e., people who do not have the risk factor of interest) as the reference group. It is important to look carefully at the codes that SPSS uses internally (as shown in Figure 21.9) to make certain that they correspond to the comparisons that you want to make. The group with an internal SPSS code of 0 is used as the reference group; the other group is compared with this reference group. In other common applications of binary logistic regression, the presence of a treatment or intervention is evaluated to see how much it decreases the risk of a negative outcome (such as death). Consider the well-known experiment that assessed the effects of low-dose aspirin on risk of heart attack (Steering Committee of the Physicians' Health Study Research Group, 1989). Results from this study appear in Table 21.5.

The predictor variable was the treatment group: patients were either given a placebo (Treatment = 0) or low-dose aspirin (Treatment = 1). The outcome variable was occurrence of a heart attack (No heart attack = 0, Heart attack = 1). If the researcher wants to report results in the form of a statement such as "receiving aspirin reduces the risk of having a heart attack," then it is more convenient to assign a code of 1 to persons who have heart attacks (i.e., to make having a heart attack the target outcome). It is also more convenient in this case to specify the first group on the predictor variable (i.e., the people who receive placebo) as the reference group. These code assignments ensure that the risk of death is the target outcome for which odds and probabilities are calculated and that the odds ratio will indicate how much lower the risk of heart attack is for people who take aspirin compared with the reference group of people who did not take aspirin.

Users of programs other than SPSS need to be aware that the handling of binary-coded predictor and outcome variables differs across programs. Comparing the results from the binary logistic regression (such as predicted probabilities) to simpler statistics based on the original 2×2 contingency table can be a helpful way to verify that the odds ratios in the SPSS output correspond to the comparisons that the data analyst had in mind. The data analyst must keep track of the codes on binary variables and understand how the computer program uses these codes to identify the target response and the reference group when setting up odds to interpret odds derived from the analysis correctly.

Table 21.5 ♦ Data From an Experiment to Assess Effect of Low-Dose Aspirin on Risk of Heart Attack in Men

	Placebo (X = 0)	*Low-Dose Aspirin (X = 1)*
No heart attack ($Y = 0$)	10,845	10,933
Heart attack ($Y = 1$)	189	104

SOURCE: Steering Committee of the Physicians' Health Study Research Group (1989).

21.9 ♦ More Complex Models

The empirical example reported above represented the simplest possible type of binary logistic regression with just one binary predictor variable. There are several ways in which binary logistic regression analysis can be made more complex. These can generally be understood by analogy to complex models in multiple linear regression:

1. Binary logistic regression may include a categorical predictor variable that has more than two possible categories. For example, suppose we wanted to report risk of a death across four different disease diagnosis categories, such as Stage 1 through Stage 4 lung cancer. In the initial SPSS data file, stages of lung cancer could be coded as 1, 2, 3, and 4. However, the researcher might not want to treat this as a quantitative variable; instead, the researcher may want to compare the groups. This can be done in SPSS by identifying lung cancer stage as a categorical variable and then indicating which of the four categories should be used as the reference group. If the first group (Stage 1) is used as the reference category, then the reported odds ratios will tell us how much higher the odds of death are for Stages 2, 3, and 4 compared with Stage 1.

2. Binary logistic regression may use quantitative predictor variables such as age, body mass index, and drug dosage. If age is used to predict risk of death from heart disease, and if age is entered in years, binary logistic regression will tell us how much the odds of death increase for each 1 year increase in age. (The association between age and odds of death would appear greater, of course, if we used scores that corresponded to age groups, e.g., $1 = 0$ to $10, 2 = 20$ to $29, \ldots, 9 = 90$ to 99 years old.)

3. Binary logistic regression may use more than one predictor variable (and these variables may be categorical, quantitative, or a mixture of both). In this type of model, as in a multiple linear regression, we estimate the increase in odds ratio associated with a one-unit increase in score on each individual predictor variable, statistically controlling for all the other predictor variables in the model. Note that if all the predictor variables are categorical, and the outcome variable is also categorical, log linear analysis might be used rather than logistic regression.

4. Interaction terms may be added to a binary logistic regression model by forming product terms between pairs of predictor variables (e.g., see Chapter 12). The potential problems that arise when interaction terms are included in other analyses are similar to the issues that arise when including interaction terms in multiple linear regression (Jaccard, Turrisi, & Wan, 1990).

5. All predictors may be entered in one step as in a direct or simultaneous or standard multiple linear regression (as discussed in Chapter 14). Alternatively, predictor variables may be entered one at a time or in groups, in a series of steps, in a sequence that is determined by the data analyst; this is analogous to hierarchical multiple linear regression (as discussed in Chapter 14). Or statistical decision rules may be used to decide on the order of entry of predictor variables and to decide which variables will be excluded from the final model; this is analogous to statistical regression (as discussed in Chapter 14). As described in Chapter 14, statistical methods of predictor variable selection such as forward, backward, or stepwise regression can substantially increase the risk of Type I error so that the real risk of Type I error is much higher than the nominal *p* values that appear on the SPSS output. SPSS does not adjust the *p* values to correct for the possible inflated risk of Type I error that may arise when a program selects a relatively small number of predictor variables out of a relatively large set of candidate predictor variables. If statistical methods are used to enter predictor variables into a binary logistic regression model, then the *p* values on the SPSS printout may substantially underestimate the true risk of Type I error. King (2003) describes alternative methods for selecting best subsets of predictors in logistic regression.
6. There is a generalization of the binary logistic regression procedure to situations where the outcome variable has more than two possible outcomes; this is sometimes called polytomous logistic regression (see Hosmer & Lemeshow, 2000).

The list above is not exhaustive, but it includes the most widely used variations of binary logistic regression. For further discussion and detailed presentation of empirical examples with larger numbers of predictor variables, see Hosmer and Lemeshow (2000) or Tabachnick and Fidell (2007).

21.10 ♦ Binary Logistic Regression for Second Empirical Analysis: Drug Dose and Gender as Predictors of Odds of Death

The second empirical example includes one quantitative variable and one binary predictor variable. While this is still a very simple case, it illustrates a few additional basic concepts. Consider a hypothetical experiment to assess the increase in risk of death as a function of increased drug dosage of a toxic drug. The researcher assigns 10 rats (5 male, 5 female) to each of the following dosage levels of the drug: 0, 1, 2, . . . , 10 mg. The total number of animals was $N = 110$ (10 animals at each of the 11 dosage levels). The outcome variable that is recorded is whether the rat survives (0) or dies (1) after receiving the drug. The predictor variable gender is categorical and it is coded 0 = Male and 1 = Female. The binary logistic regression model has the following form:

$$L_i = B_0 + B_1 \times \text{Drug dose} + B_2 \times \text{Gender},$$

where L_i is the logit, as described earlier in this chapter.

Like the previous example, this equation includes a categorical predictor. Gender was identified as a categorical predictor variable. The last gender group (i.e., the gender group that had a code of 1, female) was used as the reference category. For each animal in the study, the model will be used to calculate an estimated value of L_i (based on the animal's gender and the drug dosage received) and then to use the value of L_i to estimate probability of death for each combination of gender and drug dosage.

Prior to conducting a binary logistic regression, contingency tables were set up to assess whether any of the cells in this design had expected frequencies less than 5 and also to see whether the two predictor variables (gender and drug dosage) were independent of each other. The results appear in Figure 21.19. Visual examination of the frequencies in the drug dose by death table indicates that as drug dosage increased from 0 to 10 mg the proportion of rats that died also increased. The cross-tabulation of gender and drug dosage indicates that the design was balanced—that is, the same proportions of male and female rats were tested at each drug dosage level and, therefore, the predictor variables in this experiment were not confounded or correlated with each other. The pattern of cell frequencies in the gender by death table indicates a slightly higher proportion of deaths among male rats than female rats. The logistic regression model will provide information about the increase in odds of death for each 1-mg increase in drug dosage. It can also be used to generate a predicted probability of death for each animal as a function of two variables: drug dosage and gender.

An additional overall goodness-of-fit measure that is frequently reported when the model includes quantitative predictor variables is the Hosmer and Lemeshow (2000) goodness-of-fit test. This test is done by rank ordering cases by their scores on predicted probability of membership in the target group ($\hat{p}_i$), dividing them into deciles (e.g., highest 10%, next 10%, . . . , lowest 10%), and doing a chi-square to assess whether the

drugdose * death Crosstabulation

Count

		death		
		survived	died	Total
drugdose	0	10	0	10
	1	10	0	10
	2	10	0	10
	3	8	2	10
	4	6	4	10
	5	5	5	10
	6	4	6	10
	7	2	8	10
	8	0	10	10
	9	0	10	10
	10	0	10	10
Total		55	55	110

drugdose * gender Crosstabulation

Count

		gender		
		male	female	Total
drugdose	0	5	5	10
	1	5	5	10
	2	5	5	10
	3	5	5	10
	4	5	5	10
	5	5	5	10
	6	5	5	10
	7	5	5	10
	8	5	5	10
	9	5	5	10
	10	5	5	10
Total		55	55	110

gender * death Crosstabulation

Count

		death		
		survived	died	Total
gender	male	26	29	55
	female	29	26	55
Total		55	55	110

Figure 21.19 ♦ Preliminary Cross-Tabulations of Variables in Drug Dose/Death Study

observed frequencies of membership in the target group are closely related to the predicted frequencies based on the model within each decile. A large value of chi-square for the Hosmer and Lemeshow goodness-of-fit test (and a correspondingly small p value, i.e., $p < .05$) would indicate that the predicted group memberships generated by the model deviate significantly from the actual group memberships; in other words, a large chi-square for the Hosmer and Lemeshow goodness-of-fit test indicates poor model fit. Usually, the data analyst hopes that this chi-square will be small and that its corresponding p value will be large (i.e., $p > .05$).

The commands that are required to set up the logistic regression are similar to those used in the previous example except that two variables were included in the list of covariate or predictor variables (drug dose and gender). When more than one predictor variable is included in the model, the data analyst can decide to enter all the predictor variables in one step by leaving the method of entry at the default choice "Enter"; this is similar to standard multiple linear regression, as described in Chapter 14. In this analysis, both the predictor variables were entered in the same step.

The SPSS commands and syntax for this binary logistic regression appear in Figures 21.20 through 21.24. The commands were similar to the ones for the previous simpler example, with the following exceptions: Two predictor variables, drug dose and gender, were identified by placing them in the list of covariates. Only one of these variables, gender, was further defined as categorical. To open the Options dialog window, click the button marked Options in the main dialog window for Logistic Regression.

Selected SPSS output from this binary logit regression with two predictors appears in Figures 21.25 through 21.34. Figure 21.25 shows that the internal SPSS coding for the variable gender is Male = 1 and Female = 0. This means that the female group will be used as the reference group; the odds ratio associated with gender will tell us if males have higher or lower odds of death relative to females. Figure 21.29 provides the omnibus

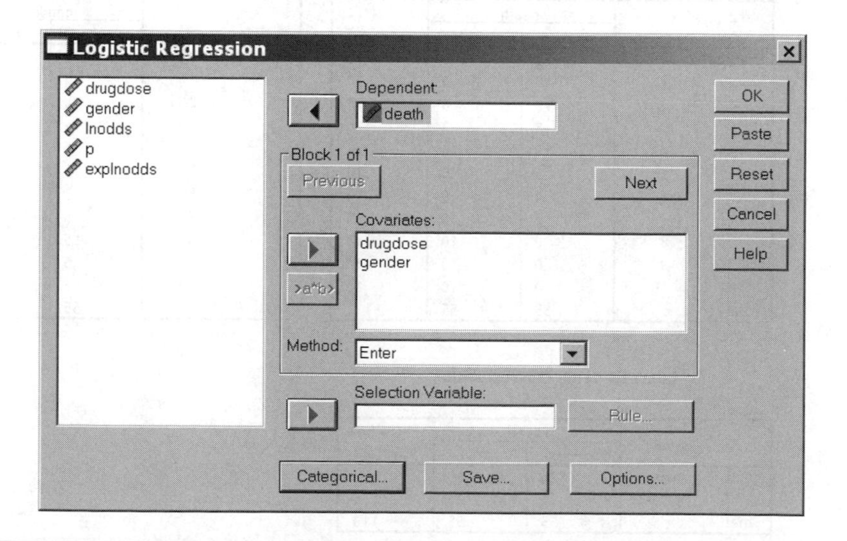

Figure 21.20 ♦ Variables for Second Binary Logistic Regression to Predict Death From Drug Dose and Gender

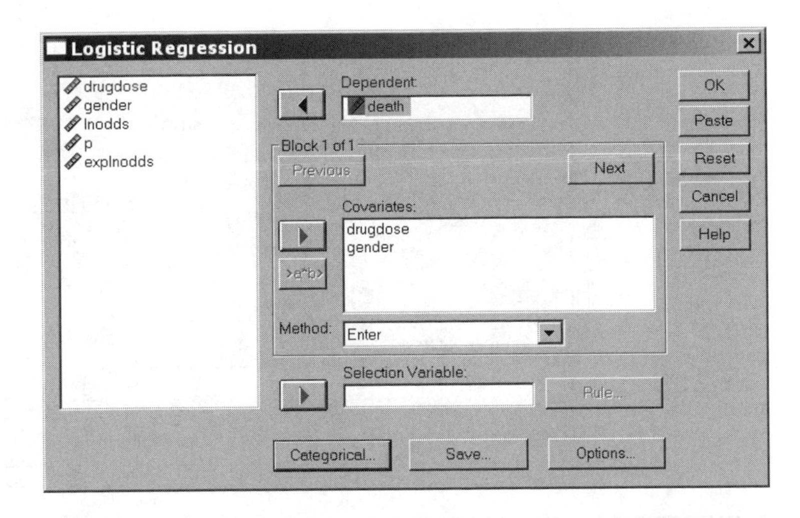

Figure 21.21 ♦ Defining Gender as a Categorical Predictor Variable in the Second Binary Logistic Regression Example

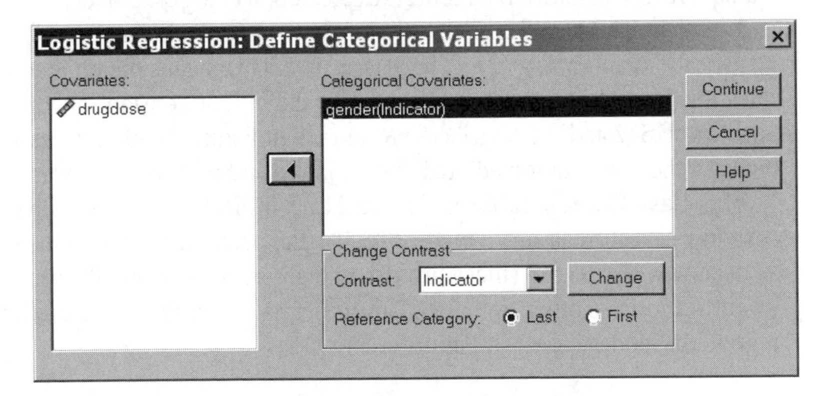

Figure 21.22 ♦ Options for Second Binary Logistic Regression Example

Figure 21.23 ♦ Command to Save Predicted Probabilities for Second Binary Logistic Regression Example

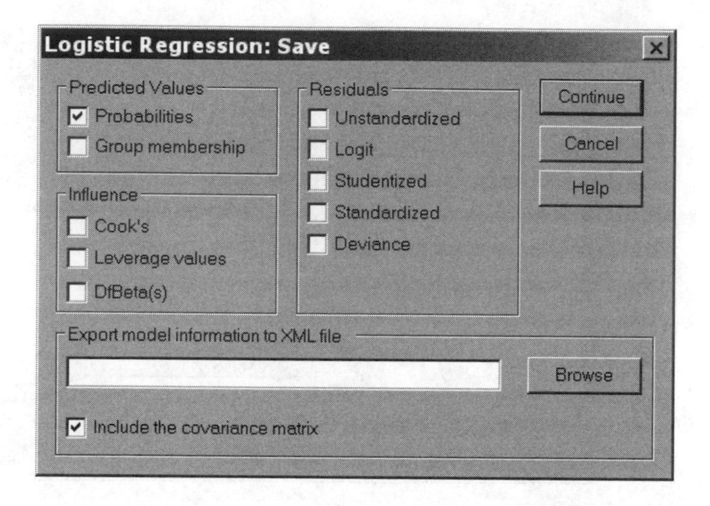

Figure 21.24 ♦ SPSS Syntax for Second Binary Logistic Regression Example

chi-square test for the overall model (i.e., prediction of log odds of death from both drug dose and gender). The overall model, including both drug dosage and gender as predictors, was statistically significant, $\chi^2(2) = 86.998, p < .001$. This means the LL "badness of fit" was significantly lower for this full model than for the null or constant-only model. The Cox and Snell R^2 of .547 and the Nagelkerke R^2 of .729 in Figure 21.30 indicate that the association between these two combined predictor variables and odds of death was strong.

The **classification table** in Figure 21.33 indicates that the percentage of animals correctly classified as surviving was 85.5%; the percentage of animals correctly classified as dead was also 85.5% (these two percentages of correct classification are not necessarily always equal in other samples). The numerical estimates of the coefficients for the logit regression model appear in Figure 21.34

$$\ln(\text{Odds}) = L_i = -5.220 + .984 \times \text{Drug dose} + .596 \times \text{Gender}.$$

Dependent Variable Encoding

Original Value	Internal Value
survived	0
died	1

Categorical Variables Codings

		Frequency	Parameter coding (1)
gender	male	55	1.000
	female	55	.000

Figure 21.25 ♦ Coding of Categorical Variables for Second Binary Logistic Regression Example: Prediction of Death From Drug Dose and Gender

Iteration History[a,b,c]

Iteration	-2 Log likelihood	Coefficients
		Constant
Step 0 1	152.492	.000

a. Constant is included in the model.

b. Initial -2 Log Likelihood: 152.492

c. Estimation terminated at iteration number 1 because parameter estimates changed by less than .001.

Figure 21.26 ♦ Null Model for Second Binary Logistic Regression Example

Variables in the Equation

		B	S.E.	Wald	df	Sig.	Exp(B)
Step 0	Constant	.000	.191	.000	1	1.000	1.000

Figure 21.27 ♦ Regression Coefficients for the Null Model for the Second Binary Logistic Regression Example

Iteration History[a,b,c,d]

Iteration		-2 Log likelihood	Coefficients		
			Constant	drugdose	gender(1)
Step 1	1	77.997	-2.545	.487	.218
	2	67.479	-3.984	.756	.411
	3	65.606	-4.903	.926	.548
	4	65.495	-5.197	.980	.592
	5	65.494	-5.220	.984	.596
	6	65.494	-5.220	.984	.596

a. Method: Enter

b. Constant is included in the model.

c. Initial -2 Log Likelihood: 152.492

d. Estimation terminated at iteration number 6 because parameter estimates changed by less than .001.

Figure 21.28 ♦ Output for Full Binary Logistic Model to Predict Death From Drug Dose and Gender

Omnibus Tests of Model Coefficients

		Chi-square	df	Sig.
Step 1	Step	86.998	2	.000
	Block	86.998	2	.000
	Model	86.998	2	.000

Figure 21.29 ♦ Chi-Square Test for Improvement in Model Fit: Full Model Compared With Null Model

Model Summary

Step	-2 Log likelihood	Cox & Snell R Square	Nagelkerke R Square
1	65.494[a]	.547	.729

a. Estimation terminated at iteration number 6 because parameter estimates changed by less than .001.

Figure 21.30 ♦ Additional Goodness-of-Fit Measures for Second Binary Logistic Regression Example

Hosmer and Lemeshow Test

Step	Chi-square	df	Sig.
1	4.174	8	.841

Figure 21.31 ♦ Hosmer and Lemeshow Test From Second Binary Logistic Regression Example

Contingency Table for Hosmer and Lemeshow Test

		death = survived		death = died		
		Observed	Expected	Observed	Expected	Total
Step 1	1	10	9.925	0	.075	10
	2	10	9.801	0	.199	10
	3	10	9.485	0	.515	10
	4	8	8.739	2	1.261	10
	5	6	7.240	4	2.760	10
	6	5	5.000	5	5.000	10
	7	4	2.760	6	7.240	10
	8	2	1.261	8	8.739	10
	9	0	.515	10	9.485	10
	10	0	.275	20	19.725	20

Figure 21.32 ♦ Contingency Table for Hosmer and Lemeshow Test for Second Binary Logistic Regression Example

Classification Table[a]

Observed			Predicted: death survived	Predicted: death died	Percentage Correct
Step 1	death	survived	47	8	85.5
		died	8	47	85.5
	Overall Percentage				85.5

a. The cut value is .500

Figure 21.33 ♦ Classification Table for Second Binary Logistic Regression Model

Variables in the Equation

		B	S.E.	Wald	df	Sig.	Exp(B)	95.0% C.I.for EXP(B) Lower	95.0% C.I.for EXP(B) Upper
Step 1[a]	drugdose	.984	.181	29.612	1	.000	2.676	1.877	3.815
	gender(1)	.596	.639	.870	1	.351	1.815	.519	6.347
	Constant	-5.220	1.055	24.464	1	.000	.005		

a. Variable(s) entered on step 1: drugdose, gender.

Figure 21.34 ♦ Final Binary Logistic Model to Predict Death From Drug Dose and Gender

The Wald ratio for the coefficient associated with gender was not statistically significant, $\chi^2(df = 1) = .870, p = .351$. Thus, there was no significant difference in odds of death for male versus female animals. Exp(B) for gender was 1.815; this indicates that the odds of death for the male group tended to be higher than the odds of death for the female group, but the nonsignificant Wald test tells us that this difference was too small to be judged statistically significant. The Wald chi-square for the coefficient associated with drug dose was statistically significant, $B = +.984, \chi^2(1) = 29.612, p < .001$. Exp($B$) for drug dose was 2.676. This indicates that for each 1-mg increase in drug dose the predicted odds of death almost tripled. The 95% CI for exp(B) ranged from 1.877 to 3.815. We can conclude that increasing the drug dosage significantly increased the odds of death.

It is useful to include predicted probabilities (either in tabular form or as a graph) when the results of logistic regression are reported, because predicted probabilities are much easier to understand than odds. The predicted probabilities of death for each drug dosage level appeared as a graph in Figure 21.1. This graph makes the nature of the relationship represented by the model very clear. At a drug dosage of 0 mg, the predicted probability of death is approximately 0. As drug dosage increases from 0 to 1 to 2 mg, the predicted probability of death increases very slowly. For drug dosages between 3 and 7 mg, the predicted probability of death increases rapidly as a function of drug dose—for example, 50% of the animals who received 5 mg of the drug died—and the curve passes through this point. As the drug dosage continues to increase from 8 to 10 mg, the predicted probability of death increases more slowly (and levels off at .99, within rounding

Drug Dose:	Gender: male	female
0	.00972	.00538
1	.02558	.01426
2	.06562	.03726
3	.15816	.09381
4	.33448	.21687
5	.57346	.42556
6	.78245	.66463
7	.90585	.84131
8	.96260	.93413
9	.98568	.97432
10	.99460	.99024

Figure 21.35 ♦ Summary Table: Probability of Death Tabled By Drug Dose and Gender

error of 1.00). The predicted probabilities of death for each participant (based on drug dose and gender) can also be summarized in table form as shown in Figure 21.35.

21.11 ♦ Comparison of Discriminant Analysis to Binary Logistic Regression

Chapter 16 described the use of DA to predict group membership. DA is one of the forms of the general linear model. DA requires some rather restrictive assumptions about data structure, and the use of DA is problematic when the sizes of the two groups are drastically unequal (e.g., when there is a 95% vs. 5% split of cases into Group 0 and Group 1) and in situations where the underlying relationship between scores on the predictor and scores on the outcome is nonlinear. Many research situations (such as diagnosis of uncommon diseases) do involve small numbers in one of the groups. Logistic regression and DA can both be used to predict membership in one of two groups; however, logistic regression offers some potential advantages compared with DA. Logistic regression does not require such stringent assumptions about data structure, and it provides a better model for situations where the probabilities of group membership become small (e.g., 10% and less). Logistic regression provides information about odds and differences in odds. In some ways, having the information in this form is an advantage; researchers can readily provide effect size information for individual predictors by reporting how much higher the odds of disease becomes for each one-unit increase in scores on each predictor variable. However, there are two potential disadvantages of logistic regression. The values of coefficients in a logistic regression cannot be obtained using OLS estimation methods; instead, MLE methods are required. Maximum likelihood estimates tend to be unstable unless sample sizes are fairly large. In addition, because the concept of odds is relatively unfamiliar to many readers, reports given in terms of odds may be more difficult for some readers to understand.

In situations where the sizes of the groups are substantially unequal (e.g., a 95% vs. 5% split between groups 0 and 1), the overall sample size is large, and the multivariate normality assumptions and homogeneity of variance/covariance matrix assumptions that are required for DA are violated, logistic regression provides a better method for prediction of group membership. In practice, many researchers have developed a strong preference for the use of logistic regression rather than DA in clinical prediction studies.

21.12 ♦ Summary

As noted in an earlier section, incomplete reports about the results of logistic regression analyses can lead to misleading impressions about the strength of relationships between variables. Researchers need to include probabilities along with odds when they report results. This information can help lay readers (and professional audiences) better understand the true magnitude of increase in risk in studies that examine risk factors as predictors of negative outcomes and the true strength of the benefits in studies that examine how treatments or interventions reduce the likelihood of negative outcomes.

There are two reasons why reporting only information about changes in odds can be misleading. First, most people do not have a good intuitive grasp of the concepts of odds ratios. Second, information about changes in odds can be difficult to interpret without the knowledge of baseline levels of risk—for example, the probability of the negative event when treatment is withheld. For example, suppose that a new cholesterol-reducing drug cuts a person's risk of heart attack in half. The practical or clinical significance of the drug for individual patients is arguably different if the drug reduces a preexisting probability of heart attack of 50% to 25% than if the drug reduces a baseline risk or probability of heart attack of about 1% to .5%. Converting log odds to probability estimates for selected cases with representative scores on the predictor variables provides extremely useful information.

An illustrative example was reported in a comment published in the *New England Journal of Medicine.* Schwartz, Woloshin, and Welch (1999) discussed mass media reports about a study (Schulman et al., 1999) that examined how often physicians recommend cardiac catheterization for hypothetical patients with chest pain; predictor variables included patient ethnicity and gender. The odds ratio for referral for catheterization for black compared with white patients was reported in the journal as .60. The outcome of the Schulman et al. study was reported in mass media as evidence that black patients were "40% less likely" to be referred for cardiac testing than white patients. Schwartz et al. (1999) argued that this mass media statement, in isolation, was misleading to many readers; they argued that simpler statistics (84.7% of hypothetical black patients vs. 90.6% of hypothetical white patients were referred for catheterization) would provide readers with a better sense of the real magnitude of the difference in referral rates. The difficulty involved in translating odds ratios into everyday language in a way that is easy for people to understand is further discussed by Schwartz (2003).

Based on recommendations by Tabachnick and Fidell (2007) and Peng et al. (2002), the following information should be included in reports of results for binary logistic regression analysis:

1. Preliminary data screening and a clear statement about the variables and options involved in the analysis.

2. Overall model fit. This should include information about statistical significance, usually in the form of a chi-square test. It should also include some discussion of effect size. For the overall model, statistics such as Nagelkerke's R^2 provide information about effect size.

3. Information about individual predictors should be presented. This includes the model coefficients, which can be used to derive odds and estimated probabilities; statistical significance tests; a discussion of the nature or direction of the association (e.g., as drug dose increases, do the odds of death increase or decrease?). CIs should be reported. Effect size information should be provided, and researchers need to evaluate whether the association between the predictor and outcome variables is strong enough to be of any practical or clinical importance. In the dog ownership/death sample, for example, odds of death were about six times higher for nonowners of dogs than for dog owners.

4. Other goodness-of-fit information may be useful if there is at least one quantitative predictor variable—for example, the Hosmer and Lemeshow goodness-of-fit test.

5. Predicted probabilities as a function of reasonable and representative score values can be summarized either as a figure (as in Figure 21.1) or in table form (as in Figure 21.35). For example, a report of the dog ownership/death study should make it clear that death was not a highly likely outcome even for the group that had higher odds of death (nonowners of dogs).

6. Additional information can also be provided; for example, an analysis of residuals can be useful in identifying any multivariate outliers or in detecting violations of assumptions.

This chapter presented two simple examples of binary logistic regression. The first example used one binary predictor (dog ownership) to predict odds of death. The second example used one binary predictor (gender) and one quantitative variable (dosage of a toxic drug) to predict odds of death. Either of these analyses could be extended in several ways. For example, we might want to add a term to test for a possible drug dose by gender to the second analysis. We might want to use multiple predictor variables to assess their relative contributions; for example, among a set of risk factors that includes age, gender, smoking, high-density lipoprotein cholesterol, and hostility, how strongly are scores on each of these predictors related to odds of death from heart attack when other correlated predictors are statistically controlled? Other forms of logistic regression provide ways to analyze situations where the categorical outcome variable has more than two possible outcomes; this is sometimes called polytomous logistic regression (Hosmer & Lemeshow, 2000).

+ − × ÷

Comprehension Questions

1. Examine the data in Table 21.5, which report the outcomes for the Physician's Health Study that assessed the effect of a low-dose aspirin regimen on risk of heart attack.
   a. For the group of patients that had the low-dose aspirin regimen, calculate the odds of death.
   b. For the group of patients that received placebo, calculate the odds of death.
   c. Compute the odds ratio that tells us how much more likely a heart attack was for patients in the placebo group (compared with the aspirin group).
   d. Compute the odds ratio that tells us how much less likely a heart attack was for patients in the aspirin group (compared with the placebo group).
   e. In simple language, how did the aspirin regimen change the odds of heart attack?
   f. Now look at the data in a different way. What percentage of the aspirin regimen group had a heart attack? What percentage of the placebo group had a heart attack? Does looking at these two percentages give you a different impression about the impact of aspirin?
   g. Describe two different statistical analyses that you could apply to these data to assess whether heart attack is significantly related to drug regimen. (Hint: for a 2×2 table, simpler analyses were described in earlier chapters.)

2. The data summarized in Table 21.6 (from a study by Goodwin, Schylsinger, Hermansen, Guze, & Winokur, 1973) show how "becoming a heavy smoker" is related to "having a biological parent who was diagnosed alcoholic." Either enter these data into SPSS by hand or use the dataset goodwin.sav (available on the Web site for the textbook). Run a binary logistic regression to predict smoking status from parental alcohol diagnosis and write up your results, including tables that summarize all relevant information. In simple language, do people whose biological parent is diagnosed as alcoholic have significantly higher odds of being heavy smokers?

Table 21.6 ♦ Diagnosis of Parental Alcoholism as Predictor of Heavy Smoking

	Parent Not Diagnosed as Alcoholic, $X = 0$	*Parent Diagnosed as Alcoholic, $Y = 1$*	
Adult child not a heavy smoker, $Y = 0$	41	6	47
Adult child is a heavy smoker, $Y = 1$	37	49	86
	78	55	133

SOURCE: Goodwin, Schylsinger, Hermansen, Guze, and Winokur (1973).

3. Table 21.7 shows how values of p_i are related to values of odds and values of the logit L_i. To do this exercise, you will need a calculator that has the natural log function; you can find a calculator that has this function online (e.g., at http://eri.gg.uwyo.edu/toolbar/calculator/natlog.htm).

Table 21.7 ♦ Selected Corresponding Values of Probabilities (p), Odds, and the Logit (L_i)

p_i	$(1 - p_i)$	$Odds = p_i/(1 - p_i)$	$Logit = L_i = ln[p_i/(1 - p_i)]$
.1		.111	−2.20
.2	.8		−1.39
.3	.7	.429	−.847
.4	.6	.667	
.5	.5		0
.6	.4	1.50	
.7	.3	2.33	.847
.8	.2	4.00	1.39
.9	.1	9.00	2.20

SOURCE: Adapted from Pampel (2000, p. 14).

NOTE: For one of the comprehension questions, the task is to fill in the blank cells of this table.

a. Fill in the values in the empty cells in Table 21.7.
b. Consider the set of scores on p_i. Do these scores satisfy the assumptions that are generally required for scores on a Y outcome variable in a multiple linear regression? In what way do these scores violate these assumptions?
c. An odds of 1.00 corresponds to values of p = ________ and $(1 - p)$ = ________.
d. An odds of 1.00 corresponds to a logit L_i value of ________.
e. Look at the column of scores for L_i, the logit. Do these scores satisfy the assumptions that are generally required for scores on a Y outcome variable in a multiple linear regression? Can you see any way that these scores violate these assumptions?
f. Based on this table, what advantages are there to using L_i, rather than p_i, as the dependent variable in a logistic regression? Hint: You might want to graph the values in the last column of Table 21.7 to see what distribution shape they suggest.

Appendix A

Proportions of Area Under Standard Normal Curve

A z	B	C	A z	B	C	A z	B	C
0.00	.0000	.5000	0.30	.1179	.3821	0.60	.2257	.2743
0.01	.0040	.4960	0.31	.1271	.3783	0.61	.2291	.2709
0.02	.0080	.4920	0.32	.1255	.3745	0.62	.2324	.2676
0.03	.0120	.4880	0.33	.1293	.3707	0.63	.2357	.2643
0.04	.0160	.4840	0.34	.1331	.3669	0.64	.2389	.2611
0.05	.0199	.4801	0.35	.1368	.3632	0.65	.2422	.2578
0.06	.0239	.4761	0.36	.1406	.3594	0.66	.2454	.2546
0.07	.0279	.4721	0.37	.1443	.3557	0.67	.2486	.2514
0.08	.0319	.4681	0.38	.1480	.3520	0.68	.2517	.2483
0.09	.0359	.4641	0.39	.1517	.3483	0.69	.2549	.2451
0.10	.0398	.4602	0.40	.1554	.3446	0.70	.2580	.2420
0.11	.0438	.4562	0.41	.1591	.3409	0.71	.2611	.2389
0.12	.0478	.4522	0.42	.1628	.3372	0.72	.2642	.2358
0.13	.0517	.4483	0.43	.1664	.3336	0.73	.2673	.2327
0.14	.0557	.4443	0.44	.1700	.3300	0.74	.2704	.2296
0.15	.0596	.4404	0.45	.1736	.3264	0.75	.2734	.2266
0.16	.0636	.4364	0.46	.1772	.3228	0.76	.2764	.2236
0.17	.0675	.4325	0.47	.1808	.3192	0.77	.2794	.2206
0.18	.0714	.4286	0.48	.1844	.3156	0.78	.2823	.2177
0.19	.0753	.4247	0.49	.1879	.3121	0.79	.2852	.2148
0.20	.0793	.4207	0.50	.1915	.3085	0.80	.2881	.2119
0.21	.0832	.4168	0.51	.1950	.3050	0.81	.2910	.2090
0.22	.0871	.4129	0.52	.1985	.3015	0.82	.2939	.2061
0.23	.0910	.4090	0.53	.2019	.2981	0.83	.2967	.2033
0.24	.0948	.4052	0.54	.2054	.2946	0.84	.2995	.2005
0.25	.0987	.4013	0.55	.2088	.2912	0.85	.3023	.1977
0.26	.1026	.3974	0.56	.2123	.2877	0.86	.3051	.1949
0.27	.1064	.3936	0.57	.2157	.2843	0.87	.3078	.1922
0.28	.1103	.3897	0.58	.2190	.2810	0.88	.3106	.1894
0.29	.1141	.3859	0.59	.2224	.2776	0.89	.3133	.1867
A −z	B	C	A −z	B	C	A −z	B	C

A	B	C	A	B	C	A	B	C
z			z			z		
0.90	.3159	.1841	1.21	.3869	.1131	1.52	.4357	.0643
0.91	.3186	.1814	1.22	.3888	.1112	1.53	.4370	.0630
0.92	.3212	.1788	1.23	.3907	.1093	1.54	.4382	.0618
0.93	.3238	.1772	1.24	.3925	.1075	1.55	.4394	.0606
0.94	.3264	.1736	1.25	.3944	.1056	1.56	.4406	.0594
0.95	.3289	.1711	1.26	.3962	.1038	1.57	.4418	.0582
0.96	.3315	.1685	1.27	.3980	.1020	1.58	.4429	.0571
0.97	.3340	.1660	1.28	.3997	.1003	1.59	.4441	.0559
0.98	.3365	.1635	1.29	.4015	.0985	1.60	.4452	.0548
0.99	.3389	.1611	1.30	.4032	.0968	1.61	.4463	.0537
1.00	.3413	.1587	1.31	.4049	.0951	1.62	.4474	.0526
1.01	.3438	.1562	1.32	.4066	.0934	1.63	.4484	.0516
1.02	.3461	.1539	1.33	.4082	.0918	1.64	.4495	.0505
1.03	.3485	.1515	1.34	.4099	.0901	1.65	.4505	.0495
1.04	.3508	.1492	1.35	.4115	.0885	1.66	.4515	.0485
1.05	.3531	.1469	1.36	.4131	.0869	1.67	.4525	.0475
1.06	.3554	.1446	1.37	.4147	.0853	1.68	.4535	.0465
1.07	.3577	.1423	1.38	.4162	.0838	1.69	.4545	.0455
1.08	.3599	.1401	1.39	.4177	.0823	1.70	.4554	.0446
1.09	.3621	.1379	1.40	.4192	.0808	1.71	.4564	.0436
1.10	.3643	.1357	1.41	.4207	.0793	1.72	.4573	.0427
1.11	.3665	.1335	1.42	.4222	.0778	1.73	.4582	.0418
1.12	.3686	.1314	1.43	.4236	.0764	1.74	.4591	.0409
1.13	.3708	.1292	1.44	.4251	.0749	1.75	.4599	.0401
1.14	.3729	.1271	1.45	.4265	.0735	1.76	.4608	.0392
1.15	.3749	.1251	1.46	.4279	.0721	1.77	.4616	.0384
1.16	.3770	.1230	1.47	.4292	.0708	1.78	.4625	.0375
1.17	.3790	.1210	1.48	.4306	.0694	1.79	.4633	.0367
1.18	.3810	.1190	1.49	.4319	.0681	1.80	.4641	.0359
1.19	.3830	.1170	1.50	.4332	.0668	1.81	.4649	.0351
1.20	.3849	.1151	1.51	.4345	.0655	1.82	.4656	.0344
A	B	C	A	B	C	A	B	C
−z			−z			−z		

(Continued)

(Continued)

A z	B	C	A z	B	C	A z	B	C
1.83	.4664	.0336	2.13	.4834	.0166	2.43	.4925	.0075
1.84	.4671	.0329	2.14	.4838	.0162	2.44	.4927	.0073
1.85	.4678	.0322	2.15	.4842	.0158	2.45	.4929	.0071
1.86	.4686	.0314	2.16	.4846	.0154	2.46	.4931	.0069
1.87	.4693	.0307	2.17	.4850	.0150	2.47	.4932	.0068
1.88	.4699	.0301	2.18	.4854	.0146	2.48	.4934	.0066
1.89	.4706	.0294	2.19	.4857	.0143	2.49	.4936	.0064
1.90	.4713	.0287	2.20	.4861	.0139	2.50	.4938	.0062
1.91	.4719	.0281	2.21	.4864	.0136	2.51	.4940	.0060
1.92	.4726	.0274	2.22	.4868	.0132	2.52	.4941	.0059
1.93	.4732	.0268	2.23	.4871	.0129	2.53	.4943	.0057
1.94	.4738	.0262	2.24	.4875	.0125	2.54	.4945	.0055
1.95	.4744	.0256	2.25	.4878	.0122	2.55	.4946	.0054
1.96	.4750	.0250	2.26	.4881	.0119	2.56	.4948	.0052
1.97	.4756	.0244	2.27	.4884	.0116	2.57	.4949	.0051
1.98	.4761	.0239	2.28	.4887	.0113	2.58	.4951	.0049
1.99	.4767	.0233	2.29	.4890	.0110	2.59	.4952	.0048
2.00	.4772	.0228	2.30	.4893	.0107	2.60	.4953	.0047
2.01	.4778	.0222	2.31	.4896	.0104	2.61	.4955	.0045
2.02	.4783	.0217	2.32	.4898	.0102	2.62	.4956	.0044
2.03	.4788	.0212	2.33	.4901	.0099	2.63	.4957	.0043
2.04	.4793	.0207	2.34	.4904	.0096	2.64	.4959	.0041
2.05	.4798	.0202	2.35	.4906	.0094	2.65	.4960	.0040
2.06	.4803	.0197	2.36	.4909	.0091	2.66	.4961	.0039
2.07	.4808	.0192	2.37	.4911	.0089	2.67	.4962	.0038
2.08	.4812	.0188	2.38	.4913	.0087	2.68	.4963	.0037
2.09	.4817	.0183	2.39	.4916	.0084	2.69	.4964	.0036
2.10	.4821	.0179	2.40	.4918	.0082	2.70	.4965	.0035
2.11	.4826	.0174	2.41	.4920	.0080	2.71	.4966	.0034
2.12	.4830	.0170	2.42	.4922	.0078	2.72	.4967	.0033
A −z	B	C	A −z	B	C	A −z	B	C

A z	B	C	A z	B	C	A z	B	C
2.73	.4968	.0032	2.94	.4984	.0016	3.15	.4992	.0008
2.74	.4969	.0031	2.95	.4984	.0016	3.16	.4992	.0008
2.75	.4970	.0030	2.96	.4985	.0015	3.17	.4992	.0008
2.76	.4971	.0029	2.97	.4985	.0015	3.18	.4993	.0007
2.77	.4972	.0028	2.98	.4986	.0014	3.19	.4993	.0007
2.78	.4973	.0027	2.99	.4986	.0014	3.20	.4993	.0007
2.79	.4974	.0026	3.00	.4987	.0013	3.21	.4993	.0007
2.80	.4974	.0026	3.01	.4987	.0013	3.22	.4994	.0006
2.81	.4975	.0025	3.02	.4987	.0013	3.23	.4994	.0006
2.82	.4976	.0024	3.03	.4988	.0012	3.24	.4994	.0006
2.83	.4977	.0023	3.04	.4988	.0012	3.25	.4994	.0006
2.84	.4977	.0023	3.05	.4989	.0011	3.30	.4995	.0005
2.85	.4978	.0022	3.06	.4989	.0011	3.35	.4996	.0004
2.86	.4979	.0021	3.07	.4989	.0011	3.40	.4997	.0003
2.87	.4979	.0021	3.08	.4990	.0010	3.45	.4997	.0003
2.88	.4980	.0020	3.09	.4990	.0010	3.50	.4998	.0002
2.89	.4981	.0019	3.10	.4990	.0010	3.60	.4998	.0002
2.90	.4981	.0019	3.11	.4991	.0009	3.70	.4999	.0001
2.91	.4982	.0018	3.12	.4991	.0009	3.80	.4999	.0001
2.92	.4982	.0018	3.13	.4991	.0009	3.90	.4999	.0000
2.93	.4983	.0017	3.14	.4992	.0008	4.00	.4999	.0000
A −z	B	C	A −z	B	C	A −z	B	C

SOURCE: Abridged from Fisher and Yates (1974).

Appendix B

Critical Values for t Distribution

	Confidence Intervals (%)					
	80	*90*	*95*	*98*	*99*	*99.9*
	Level of Significance for One-Tailed Test					
	.10	*.05*	*.025*	*.01*	*.005*	*.0005*
	Level of Significance for Two-Tailed Test					
df	*.20*	*.10*	*.05*	*.02*	*.01*	*.001*
1	3.078	6.314	12.706	31.821	63.657	636.619
2	1.886	2.920	4.303	6.965	9.925	31.598
3	1.638	2.353	3.182	4.541	5.841	12.941
4	1.533	2.132	2.776	3.747	4.604	8.610
5	1.476	2.015	2.571	3.365	4.032	6.859
6	1.440	1.943	2.447	3.143	3.707	5.959
7	1.415	1.895	2.365	2.998	3.499	5.405
8	1.397	1.860	2.306	2.896	3.355	5.041
9	1.383	1.833	2.262	2.821	3.250	4.781
10	1.372	1.812	2.228	2.764	3.169	4.587
11	1.363	1.796	2.201	2.718	3.106	4.437
12	1.356	1.782	2.179	2.681	3.055	4.318
13	1.350	1.771	2.160	2.650	3.012	4.221
14	1.345	1.761	2.145	2.624	2.977	4.140
15	1.341	1.753	2.131	2.602	2.947	4.073
16	1.337	1.746	2.120	2.583	2.921	4.015
17	1.333	1.740	2.110	2.567	2.898	3.965
18	1.330	1.734	2.101	2.552	2.878	3.922
19	1.328	1.729	2.093	2.539	2.861	3.883
20	1.325	1.725	2.086	2.528	2.845	3.850
21	1.323	1.721	2.080	2.518	2.831	3.819
22	1.321	1.717	2.074	2.508	2.819	3.792
23	1.319	1.714	2.069	2.500	2.807	3.767
24	1.318	1.711	2.064	2.492	2.797	3.745
25	1.316	1.708	2.060	2.485	2.787	3.725
26	1.315	1.706	2.056	2.479	2.779	3.707
27	1.314	1.703	2.052	2.473	2.771	3.690
28	1.313	1.701	2.048	2.467	2.763	3.674
29	1.311	1.699	2.045	2.462	2.756	3.659
30	1.310	1.697	2.042	2.457	2.750	3.646
40	1.303	1.684	2.021	2.423	2.704	3.551
60	1.296	1.671	2.000	2.390	2.660	3.460
120	1.289	1.658	1.980	2.358	2.617	3.373
∞	1.282	1.645	1.960	2.326	2.576	3.291

SOURCE: Abridged from Fisher and Yates (1974, Table V).

Appendix C

Critical Values of F

Critical Values of F for $\alpha = .05$

	$df_{numerator}$									
$df_{denominator}$	1	2	3	4	5	6	8	12	24	∞
1	161.40	199.50	215.70	224.60	230.20	234.00	238.90	243.90	249.00	254.30
2	18.51	19.00	19.16	19.25	19.30	19.33	19.37	19.41	19.45	19.50
3	10.13	9.55	9.28	9.12	9.01	8.94	8.84	8.74	8.64	8.53
4	7.71	6.94	6.59	6.39	6.26	6.16	6.04	5.91	5.77	5.63
5	6.61	5.79	5.41	5.19	5.05	4.95	4.82	4.68	4.53	4.36
6	5.99	5.14	4.76	4.53	4.39	4.28	4.15	4.00	3.84	3.67
7	5.59	4.74	4.35	4.12	3.97	3.87	3.73	3.57	3.41	3.23
8	5.32	4.46	4.07	3.84	3.69	3.58	3.44	3.28	3.12	2.93
9	5.12	4.26	3.86	3.63	3.48	3.37	3.23	3.07	2.90	2.71
10	4.96	4.10	3.71	3.48	3.33	3.22	3.07	2.91	2.74	2.54
11	4.84	3.98	3.59	3.36	3.20	3.09	2.95	2.79	2.61	2.40
12	4.75	3.88	3.49	3.26	3.11	3.00	2.85	2.69	2.50	2.30
13	4.67	3.80	3.41	3.18	3.02	2.92	2.77	2.60	2.42	2.21
14	4.60	3.74	3.34	3.11	2.96	2.85	2.70	2.53	2.35	2.13
15	4.54	3.68	3.29	3.06	2.90	2.79	2.64	2.48	2.29	2.07
16	4.49	3.63	3.24	3.01	2.85	2.74	2.59	2.42	2.24	2.01
17	4.45	3.59	3.20	2.96	2.81	2.70	2.55	2.38	2.19	1.96
18	4.41	3.55	3.16	2.93	2.77	2.66	2.51	2.34	2.15	1.92
19	4.38	3.52	3.13	2.90	2.74	2.63	2.48	2.31	2.11	1.88
20	4.35	3.49	3.10	2.87	2.71	2.60	2.45	2.28	2.08	1.84
21	4.32	3.47	3.07	2.84	2.68	2.57	2.42	2.25	2.05	1.81
22	4.30	3.44	3.05	2.82	2.66	2.55	2.40	2.23	2.03	1.78
23	4.28	3.42	3.03	2.80	2.64	2.53	2.38	2.20	2.00	1.76
24	4.26	3.40	3.01	2.78	2.62	2.51	2.36	2.18	1.98	1.73
25	4.24	3.38	2.99	2.76	2.60	2.49	2.34	2.16	1.96	1.71
26	4.22	3.37	2.98	2.74	2.59	2.47	2.32	2.15	1.95	1.69
27	4.21	3.35	2.96	2.73	2.57	2.46	2.30	2.13	1.93	1.67
28	4.20	3.34	2.95	2.71	2.56	2.44	2.29	2.12	1.91	1.65
29	4.18	3.33	2.93	2.70	2.54	2.43	2.28	2.10	1.90	1.64
30	4.17	3.32	2.92	2.69	2.53	2.42	2.27	2.09	1.89	1.62
40	4.08	3.23	2.84	2.61	2.45	2.34	2.18	2.00	1.79	1.51
60	4.00	3.15	2.76	2.52	2.37	2.25	2.10	1.92	1.70	1.39
120	3.92	3.07	2.68	2.45	2.29	2.17	2.02	1.83	1.61	1.25
∞	3.84	2.99	2.60	2.37	2.21	2.09	1.94	1.75	1.52	1.00

SOURCE: Abridged from Fisher and Yates (1974, Table V).

NOTE: In ANOVA, $df_{numerator} = df_{between}$; $df_{denominator} = df_{within}$. In regression, $df_{numerator} = df_{regression}$; $df_{denominator} = df_{residual}$.

Critical Values of F for $\alpha = .01$

$df_{denominator}$ \ $df_{numerator}$	1	2	3	4	5	6	8	12	24	∞
1	4052	4999	5403	5625	5764	5859	5981	6106	6234	366
2	98.49	99.01	99.17	99.2	99.3	99.3	99.3	99.4	99.4	9.50
3	34.12	30.81	29.46	28.71	28.24	27.91	27.49	27.05	26.60	26.12
4	21.20	18.00	16.69	15.98	15.52	15.21	14.80	14.37	13.93	13.46
5	16.26	13.27	12.06	11.39	10.97	10.67	10.27	9.89	9.47	9.02
6	13.74	10.92	9.78	9.15	8.75	8.47	8.10	7.72	7.31	6.88
7	12.25	9.55	8.45	7.85	7.46	7.19	6.84	6.47	6.07	5.65
8	11.26	8.65	7.59	7.0	6.63	6.37	6.03	5.67	5.28	4.86
9	10.56	8.02	6.99	6.42	6.06	5.80	5.47	5.11	4.73	4.31
10	10.04	7.56	6.55	5.99	5.64	5.39	5.06	4.71	4.33	3.91
11	9.65	7.20	6.22	5.67	5.32	5.07	4.74	4.40	4.02	3.60
12	9.33	6.93	5.95	5.41	5.06	4.82	4.50	4.16	3.78	3.36
13	9.07	6.70	5.74	5.20	4.86	4.62	4.30	3.96	3.59	3.16
14	8.86	6.51	5.56	5.03	4.69	4.46	4.14	3.80	3.43	3.00
15	8.68	6.36	5.42	4.89	4.56	4.32	4.00	3.67	3.29	2.87
16	8.53	6.23	5.29	4.77	4.44	4.20	3.89	3.55	3.18	2.75
17	8.40	6.11	5.18	4.67	4.34	4.10	3.79	3.45	3.08	2.65
18	8.28	6.01	5.09	4.58	4.25	4.01	3.71	3.37	3.00	2.57
19	8.18	5.93	5.01	4.50	4.17	3.94	3.63	3.30	2.92	2.49
20	8.10	5.85	4.94	4.43	4.10	3.87	3.56	3.23	2.86	2.42
21	8.02	5.78	4.87	4.37	4.04	3.81	3.51	3.17	2.80	2.36
22	7.94	5.72	4.82	4.31	3.99	3.76	3.45	3.12	2.75	2.31
23	7.88	5.66	4.76	4.26	3.94	3.71	3.41	3.07	2.70	2.26
24	7.82	5.61	4.72	4.22	3.90	3.67	3.36	3.03	2.66	2.21
25	7.77	5.57	4.68	4.18	3.86	3.63	3.32	2.99	2.62	2.17
26	7.72	5.53	4.64	4.14	3.82	3.59	3.29	2.96	2.58	2.13
27	7.68	5.49	4.60	4.11	3.78	3.56	3.26	2.93	2.55	2.10
28	7.64	5.45	4.57	4.07	3.75	3.53	3.23	2.90	2.52	2.06
29	7.60	5.42	4.54	4.04	3.73	3.50	3.20	2.87	2.49	2.03
30	7.56	5.39	4.51	4.02	3.70	3.47	3.17	2.84	2.47	2.01
40	7.31	5.18	4.31	3.83	3.51	3.29	2.99	2.66	2.29	1.80
60	7.08	4.98	4.13	3.65	3.34	3.12	2.82	2.50	2.12	1.60
120	6.85	4.79	3.95	3.48	3.17	2.96	2.66	2.34	1.95	1.38
∞	6.64	4.60	3.78	3.32	3.02	2.80	2.51	2.18	1.79	1.00

SOURCE: Abridged from Fisher and Yates (1974, Table V).

NOTE: In ANOVA, $df_{numerator} = df_{between}$; $df_{denominator} = df_{within}$. In regression, $df_{numerator} = df_{regression}$; $df_{denominator} = df_{residual}$.

Critical Values of F for $\alpha = .001$

$df_{denominator}$ \ $df_{numerator}$	1	2	3	4	5	6	8	12	24	∞
1	405284	500000	540379	562500	576405	585937	598144	610667	623497	636619
2	998.5	999.0	999.2	999.2	999.3	999.3	999.4	999.4	999.5	999.5
3	167.5	148.5	141.1	137.1	134.6	132.8	130.6	128.3	125.9	123.5
4	74.14	61.25	56.18	53.44	51.71	50.53	49.00	47.41	45.77	44.05
5	47.04	36.61	33.20	31.09	29.75	28.84	27.64	26.42	25.14	23.78
6	35.51	27.00	23.70	21.90	20.81	20.03	19.03	17.99	16.89	15.75
7	29.22	21.69	18.77	17.19	16.21	15.52	14.63	13.71	12.73	11.69
8	25.42	18.49	15.83	14.39	13.49	12.86	12.04	11.19	10.30	9.34
9	22.86	16.39	13.90	12.56	11.71	11.13	10.37	9.57	8.72	7.81
10	21.04	14.91	12.55	11.28	10.48	9.92	9.20	8.45	7.64	6.76
11	19.69	13.81	11.56	10.35	9.58	9.05	8.35	7.63	6.85	6.00
12	18.64	12.97	10.80	9.63	8.89	8.38	7.71	7.00	6.25	5.42
13	17.81	12.31	10.21	9.07	8.35	7.86	7.21	6.52	5.78	4.97
14	17.14	11.78	9.73	8.62	7.92	7.43	6.80	6.13	5.41	4.60
15	16.59	11.34	9.34	8.25	7.57	7.09	6.47	5.81	5.10	4.31
16	16.12	10.97	9.00	7.94	7.27	6.81	6.19	5.55	4.85	4.06
17	15.72	10.66	8.73	7.68	7.02	6.56	5.96	5.32	4.63	3.85
18	15.38	10.39	8.49	7.46	6.81	6.35	5.76	5.13	4.45	3.67
19	15.08	10.16	8.28	7.26	6.61	6.18	5.59	4.97	4.29	3.52
20	14.82	9.95	8.10	7.10	6.46	6.02	5.44	4.82	4.15	3.38
21	14.59	9.77	7.94	6.95	6.32	5.88	5.31	4.70	4.03	3.26
22	14.38	9.61	7.80	6.81	6.19	5.76	5.19	4.58	3.92	3.15
23	14.19	9.47	7.67	6.69	6.08	5.65	5.09	4.48	3.82	3.05
24	14.03	9.34	7.55	6.59	5.98	5.55	4.99	4.39	3.74	2.97
25	13.88	9.22	7.45	6.49	5.88	5.46	4.91	4.31	3.66	2.89
26	13.74	9.12	7.36	6.41	5.80	5.38	4.83	4.24	3.59	2.82
27	13.61	9.02	7.27	6.33	5.73	5.31	4.76	4.17	3.52	2.75
28	13.50	8.93	7.19	6.25	5.66	5.24	4.69	4.11	3.46	2.70
29	13.39	8.85	7.12	6.19	5.59	5.18	4.64	4.05	3.41	2.64
30	13.29	8.77	7.05	6.12	5.53	5.12	4.58	4.00	3.36	2.59
40	12.61	8.25	6.60	5.70	5.13	4.73	4.21	3.64	3.01	2.23
60	11.97	7.76	6.17	5.31	4.76	4.37	3.87	3.31	2.69	1.90
120	11.38	7.31	5.79	4.95	4.42	4.04	3.55	3.02	2.40	1.56
∞	10.83	6.91	5.42	4.62	4.10	3.74	3.27	2.74	2.13	1.00

SOURCE: Abridged from Fisher and Yates (1974, Table V).

NOTE: In ANOVA, $df_{numerator} = df_{between}$; $df_{denominator} = df_{within}$. In regression, $df_{numerator} = df_{regression}$; $df_{denominator} = df_{residual}$.

Appendix D

Critical Values of Chi-Square

df	.10	.05	.01	.001
1	2.71	3.84	6.64	10.83
2	4.60	5.99	9.21	13.82
3	6.25	7.81	11.34	16.27
4	7.78	9.49	13.28	18.47
5	9.24	11.07	15.09	20.52
6	10.64	12.59	16.81	22.46
7	12.02	14.07	18.48	24.32
8	13.36	15.51	20.09	26.12
9	14.68	16.92	21.67	27.88
10	15.99	18.31	23.21	29.59
11	17.28	19.68	24.72	31.26
12	18.55	21.03	26.22	32.91
13	19.81	22.36	27.69	34.53
14	21.06	23.68	29.14	36.12
15	22.31	25.00	30.58	37.70
16	23.54	26.30	32.00	39.25
17	24.77	27.59	33.41	40.79
18	25.99	28.87	34.80	42.31
19	27.20	30.14	36.19	43.82
20	28.41	31.41	37.57	45.32
21	29.62	32.67	38.93	46.80
22	30.81	33.92	40.29	48.27
23	32.01	35.17	41.64	49.73
24	33.20	36.42	42.98	51.18
25	34.38	37.65	44.31	52.62
26	35.56	38.88	45.64	54.05
27	36.74	40.11	46.96	55.48
28	37.92	41.34	48.28	56.89
29	39.09	42.56	49.59	58.30
30	40.26	43.77	50.89	59.70
40	51.80	55.76	63.69	73.40
50	63.17	67.50	76.15	86.66
60	74.40	79.08	88.38	99.61
70	85.53	90.53	100.42	112.32

SOURCE: Abridged from Fisher and Yates (1974).

Appendix E

Critical Values of the Correlation Coefficient

	Level of Significance for One-Tailed Test			
	.05	*.025*	*.01*	*.005*
	Level of Significance for Two-Tailed Test			
df	*.10*	*.05*	*.02*	*.01*
1	.988	.997	.9995	.9999
2	.900	.950	.980	.990
3	.805	.878	.934	.959
4	.729	.811	.882	.917
5	.669	.754	.833	.874
6	.622	.707	.789	.834
7	.582	.666	.750	.798
8	.549	.632	.716	.765
9	.521	.602	.685	.735
10	.497	.576	.658	.708
11	.476	.553	.634	.684
12	.458	.532	.612	.661
13	.441	.514	.592	.641
14	.426	.497	.574	.623
15	.412	.482	.558	.606
16	.400	.468	.542	.590
17	.389	.456	.528	.575
18	.378	.444	.516	.561
19	.369	.433	.503	.549
20	.360	.423	.492	.537
21	.352	.413	.482	.526
22	.344	.404	.472	.515
23	.337	.396	.462	.505
24	.330	.388	.453	.496
25	.323	.381	.445	.487
26	.317	.374	.437	.479
27	.311	.367	.430	.471
28	.306	.361	.423	.463
29	.301	.355	.416	.456
30	.296	.349	.409	.449
35	.275	.325	.381	.418
40	.257	.304	.358	.393
45	.243	.288	.338	.372
50	.231	.273	.322	.354
60	.211	.250	.295	.325
70	.195	.232	.274	.303
80	.183	.217	.256	.283
90	.173	.205	.242	.267
100	.164	.195	.230	.254

SOURCE: Abridged from Fisher and Yates (1974, Table VII).

Appendix F

Critical Values of the Studentized Range Statistic

		Number of Groups								
df_{within}	α	*2*	*3*	*4*	*5*	*6*	*7*	*8*	*9*	*10*
5	.05	3.64	4.60	5.22	5.67	6.03	6.33	6.58	6.80	6.99
	.01	5.70	6.98	7.80	8.42	8.91	9.32	9.67	9.97	10.24
6	.05	3.46	4.34	4.90	5.30	5.63	5.90	6.12	6.32	6.49
	.01	5.24	6.33	7.03	7.56	7.97	8.32	8.61	8.87	9.10
7	.05	3.34	4.16	4.68	5.06	5.36	5.61	5.82	6.00	6.16
	.01	4.95	5.92	6.54	7.01	7.37	7.68	7.94	8.17	8.37
8	.05	3.26	4.04	4.53	4.89	5.17	5.40	5.60	5.77	5.92
	.01	4.75	5.64	6.20	6.62	6.96	7.24	7.47	7.68	7.86
9	.05	3.20	3.95	4.41	4.76	5.02	5.24	5.43	5.59	5.74
	.01	4.60	5.43	5.96	6.35	6.66	6.91	7.13	7.33	7.49
10	.05	3.15	3.88	4.33	4.65	4.91	5.12	5.30	5.46	5.60
	.01	4.48	5.27	5.77	6.14	6.43	6.67	6.87	7.05	7.21
11	.05	3.11	3.82	4.26	4.57	4.82	5.03	5.20	5.35	5.49
	.01	4.39	5.15	5.62	5.97	6.25	6.48	6.67	6.84	6.99
12	.05	3.08	3.77	4.20	4.51	4.75	4.95	5.12	5.27	5.39
	.01	4.32	5.05	5.50	5.84	6.10	6.32	6.51	6.67	6.81
13	.05	3.06	3.73	4.15	4.45	4.69	4.88	5.05	5.19	5.32
	.01	4.26	4.96	5.40	5.73	5.98	6.19	6.37	6.53	6.67
14	.05	3.03	3.70	4.11	4.41	4.64	4.83	4.99	5.13	5.25
	.01	4.21	4.89	5.32	5.63	5.88	6.08	6.26	6.41	6.54
15	.05	3.01	3.67	4.08	4.37	4.59	4.78	4.94	5.08	5.20
	.01	4.17	4.84	5.25	5.56	5.80	5.99	6.16	6.31	6.44
16	.05	3.00	3.65	4.05	4.33	4.56	4.74	4.90	5.03	5.15
	.01	4.13	4.79	5.19	5.49	5.72	5.92	6.08	6.22	6.35
17	.05	2.98	3.63	4.02	4.30	4.52	4.70	4.86	4.99	5.11
	.01	4.10	4.74	5.14	5.43	5.66	5.85	6.01	6.15	6.27
18	.05	2.97	3.61	4.00	4.28	4.49	4.67	4.82	4.96	5.07
	.01	4.07	4.70	5.09	5.38	5.60	5.79	5.94	6.08	6.20
19	.05	2.96	3.59	3.98	4.25	4.47	4.65	4.79	4.92	5.04
	.01	4.05	4.67	5.05	5.33	5.55	5.73	5.89	6.02	6.14
20	.05	2.95	3.58	3.96	4.23	4.45	4.62	4.77	4.90	5.01
	.01	4.02	4.64	5.02	5.29	5.51	5.69	5.84	5.97	6.09
24	.05	2.92	3.53	3.90	4.17	4.37	4.54	4.68	4.81	4.92
	.01	3.96	4.55	4.91	5.17	5.37	5.54	5.69	5.81	5.92
30	.05	2.89	3.49	3.85	4.10	4.30	4.46	4.60	4.72	4.82
	.01	3.89	4.45	4.80	5.05	5.24	5.40	5.54	5.65	5.76
40	.05	2.86	3.44	3.79	4.04	4.23	4.39	4.52	4.63	4.73
	.01	3.82	4.37	4.70	4.93	5.11	5.26	5.39	5.50	5.60
60	.05	2.83	3.40	3.74	3.98	4.16	4.31	4.44	4.55	4.65
	.01	3.76	4.28	4.59	4.82	4.99	5.13	5.25	5.36	5.45
120	.05	2.80	3.36	3.68	3.92	4.10	4.24	4.36	4.47	4.56
	.01	3.70	4.20	4.50	4.71	4.87	5.01	5.12	5.21	5.30
∞	.05	2.77	3.31	3.63	3.86	4.03	4.17	4.29	4.39	4.47
	.01	3.64	4.12	4.40	4.60	4.76	4.88	4.99	5.08	5.16

SOURCE: Abridged from Pearson and Hartley (1970, Table 29). Retrieved from http://fsweb.berry.edu/academic/education/vbissonnette/tables/posthoc.pdf.

Appendix G

Transformation of *r* (Pearson Correlation) to Fisher *Z*

r	Z′	r	Z′	r	Z′	r	Z′
.00	.000	.30	.310	.60	.693	.850	1.256
.01	.010	.31	.321	.61	.709	.855	1.274
.02	.020	.32	.332	.62	.725	.860	1.293
.03	.030	.33	.343	.63	.741	.865	1.313
.04	.040	.34	.354	.64	.758	.870	1.333
.05	.050	.35	.365	.65	.775	.875	1.354
.06	.060	.36	.377	.66	.793	.880	1.376
.07	.070	.37	.388	.67	.811	.885	1.398
.08	.080	.38	.400	.68	.829	.890	1.422
.09	.090	.39	.412	.69	.848	.895	1.447
.10	.100	.40	.424	.70	.867	.900	1.472
.11	.110	.41	.436	.71	.887	.905	1.499
.12	.121	.42	.448	.72	.908	.910	1.528
.13	.131	.43	.460	.73	.929	.915	1.557
.14	.141	.44	.472	.74	.950	.920	1.589
.15	.151	.45	.485	.75	.973	.925	1.623
.16	.161	.46	.497	.76	.996	.930	1.658
.17	.172	.47	.510	.77	1.020	.935	1.697
.18	.182	.48	.523	.78	1.045	.940	1.738
.19	.192	.49	.536	.79	1.071	.945	1.783
.20	.203	.50	.549	.800	1.099	.950	1.832
.21	.213	.51	.563	.805	1.113	.955	1.886
.22	.224	.52	.576	.810	1.127	.960	1.946
.23	.234	.53	.590	.815	1.142	.965	2.014
.24	.245	.54	.604	.820	1.157	.970	2.092
.25	.255	.55	.618	.825	1.172	.975	2.185
.26	.266	.56	.633	.830	1.188	.980	2.298
.27	.277	.57	.648	.835	1.204	.985	2.443
.28	.288	.58	.662	.840	1.221	.990	2.647
.29	.299	.59	.678	.845	1.238	.995	2.994

SOURCE: From Cohen, J., & Cohen, P. (1983). *Applied Multiple Regression/Correlation Analysis for the Behavioral Sciences, 2nd Edition,*

NOTE: Some books denote Fisher *Z* using lowercase *z*. Uppercase *Z* is used to represent Fisher *Z* scores in this book to avoid confusion with the more familiar use of lowercase *z* to represent standardized scores or significance tests that use the standard normal distribution for critical values.

Glossary

Accidental sample (or convenience sample). This is a sample that is not obtained through random selection but through ease of access. Accidental or convenience samples are often not representative of any "real" population. *See also* Convenience sample

Adjusted degrees of freedom. When there is evidence for violation of the sphericity assumption, the *df* used to look up the critical values of *F* to assess statistical significance may be downwardly adjusted to correct for the violation of this assumption. In repeated measures ANOVA, this adjustment involves multiplying *df* by an adjustment factor, ε.

Adjusted means. In an ANCOVA, adjusted *Y* means are created by subtracting off the part of the scores that are related to or predictable from the covariates; this adjustment may correct for any confound of the covariates with treatment, and/or this adjustment may suppress some of the error variance or noise and increase the statistical power of tests of group differences.

Adjusted or partial eta squared (η^2). This is an eta squared that has variance associated with a covariate (such as X_c) removed from the divisor:

$$\text{partial } \eta^2 = \frac{SS_{\text{effect}}}{(SS_{\text{effect}} + SS_{\text{residual}})}.$$

Adjusted R^2. This is an adjusted estimate of the proportion of variance that is explained by a regression equation; it is adjusted downward to make it a more conservative estimate of the population R^2. The downward adjustment is greater in cases where *N* is relatively small and *k*, the number of predictor variables, is relatively large.

Aggregated data. These are data points that represent group means instead of scores for individual cases.

Alpha (α). This is the theoretical risk that a Type I error will be committed using significance test procedures; that is, the probability that H_0 will be rejected when it is correct.

Alternative hypothesis (or research hypothesis). This is a hypothesis that represents an outcome different from the one specified in the null hypothesis, often denoted H_1.

Analysis of variance (ANOVA). This is a statistical analysis that tests whether there are statistically significant differences between group means on scores on a quantitative outcome variable across two or more groups. The test statistic, an *F* ratio, compares the magnitude of differences among group means (as indexed by $MS_{between}$) with the amount of variability of scores within groups that arises due to the influence of error variables (indexed by MS_{within}).

ANCOVA. This is an analysis in which group differences on means for one continuous *Y* outcome variable are assessed, statistically controlling for or removing any part of *Y* that is predictable from one or more covariates.

ANOVA. *See* Analysis of variance

A priori comparison. *See* Contrast

Asymmetrical index of association. An index of association is asymmetrical if the value of the index differs depending on which variable is designated as the predictor. Most of the indexes of association described in Chapters 7 and 8 are symmetrical; for example, the value of the Pearson *r* for the prediction of *Y* from *X* is the same as the Pearson *r* for the prediction of *X* from *Y*. There are a few contingency table statistics that are asymmetrical; that is, the strength of association may be different when predictions are made from *X* to *Y* than when predictions are made from *Y* to *X*. Somers's *d* and lambda are examples of asymmetrical indexes of association available through the SPSS Crosstabs procedure.

Attenuation (of correlation) due to unreliability. This refers to the tendency for correlations (or other statistical indexes of strength of relationship between variables) to become smaller when one or both variables have poor measurement reliability. That is, when *X* or *Y* is unreliably measured, the sample estimate of r_{XY} will underestimate the true strength of relationship ρ_{XY}; the lower the reliabilities, the greater the reduction in the size of the observed sample r_{XY} relative to the true strength of the relationship.

***B*.** *B* denotes the coefficient for a predictor variable (*X*) in a logistic regression model. For a one-unit increase in *X*, there is a *B* unit change in the log odds ratio. The value of *B* is not easy to interpret directly; instead, *B* is used to calculate other values that are more directly interpretable, such as the odds ratio and the probability of membership in each group.

***b*.** This symbol stands for the slope or raw score regression coefficient and is used to generate a predicted raw score on *Y* from a raw score on *X* in multiple regression. It corresponds to the predicted number of units of change in *Y* for each one-unit increase in *X* (e.g., the number of dollars of increase in salary for each 1-year increase in job experience). In some textbooks this slope is denoted by other characters; for example, sometimes a linear regression to predict *Y* from *X* is written as $Y' = mX + b$ or $Y' = a + bX$. In this textbook, the notation used for raw score regression equations is $Y' = b_0 + b_1X_1 + b_2X_2$, because this notation is more easily generalized to situations where there is more than one *X* predictor variable.

Backward method of entry. This is a form of statistical regression that begins with all candidate predictor variables included in the equation; in each step, the predictor variable for which the removal leads to the smallest decrease in the R^2 for the model is dropped. The process of variable deletion ends when dropping a variable would result in a significant decrease in the total R^2.

Bar chart. This is a type of graph that provides information about the number or proportion of cases in each group on a categorical variable. Typically, group membership labels appear on the X axis. For each group, the height of the vertical bar corresponds to the frequency (or proportion) of scores in that group. Conventionally, the bars in a bar graph do not touch each other. The reference scale used to evaluate the information provided by the height of each bar is indicated on the Y axis.

Beta (β). Beta denotes the probability that a Type II error will be committed—that is, that H_0 will not be rejected when it is incorrect.

Beta coefficient (β). This is the standardized regression coefficient, used to predict standardized or z scores on Y from z scores on X. For bivariate regression, $\beta = r$.

Between-S design. This is a design in which the groups that are compared are composed of different participants; that is, each participant is assigned to one and only one group, and the behavior of each subject is independent of the behavior of subjects in other groups.

Bias. This is a systematic error in scores, such that the latter are consistently either too low or too high relative to the actual value. For example, a scale that gives weight measurements that are consistently 5 lb too heavy is biased.

Biased sample. This is a nonrepresentative sample; that is, it over- or underrepresents some types of cases included in the population, often because the method of selection of members in the sample favors certain types of cases over others. For example, a sample that consists of an introductory psychology class at an elite university is a biased sample relative to the entire adult population of the United States. Compared with the overall U.S. population, the sample overrepresents people who are 18 to 20 years old and underrepresents people of other ages. The mean, variance, and other characteristics of such samples are usually not similar to those of the population.

Binary logistic regression. This is a model that uses scores on one or several predictor variables (which may be either dichotomous dummy variables or quantitative predictors) to estimate the odds of a specific outcome (such as having a heart attack) for a dependent variable that has just two possible outcomes, generally coded as 0 and 1 (e.g., participant has a heart attack/does not have a heart attack).

Biserial correlation (r_b). A correlation coefficient used to assess the relationship between an artificial dichotomy and a quantitative variable (it is not a form of the Pearson r). The computational formula attempts to adjust for the information lost by artificial dichotomization.

Bivariate normality. This is a joint distribution between two quantitative variables, X and Y, such that Y is normally distributed at each level of X and X is normally distributed for each specific value of Y.

Bivariate outlier. A bivariate outlier is a combination of X and Y scores that is not necessarily extreme on either X or Y but that represents an unusual combination of scores. In a scatter plot of scores for a pair of quantitative X and Y variables, a bivariate outlier is a data point that falls outside the area of the plot that contains most of the data points (see Figures 7.10 and 7.11 for examples).

Bivariate statistics. These are statistics that describe the relation between two variables. Often, one of the variables is identified as a predictor or independent variable and the other variable is identified as an outcome or dependent variable. Examples include the Pearson correlation, the independent samples *t* test, one-way ANOVA, and the chi-square test of association.

Blocking variable. This is a factor that is included in an ANOVA to represent some naturally occurring participant characteristic (such as gender or grade level in school). When a one-way ANOVA with a single *A* factor is modified to create a factorial by adding a blocking factor, the blocking factor *B* may suppress error variance (and, therefore, increase the statistical power for detection of an *A* main effect). Inclusion of a blocking factor also makes it possible to detect a potential $A \times B$ interaction.

Bonferroni procedure. This is a method used to limit inflated risk of Type I error when a large number of significance tests is reported. If an overall error rate of $EW_\alpha = .05$ is set for the entire set of *k* tests, the researcher uses $.05/k$ as the per-comparison alpha (PC_α). That is, each individual correlation (in the set of *k* correlations) is judged significant only if its obtained *p* value is less than the PC_α value of $.05/k$, where *k* is the number of significance tests.

Box and whiskers plot (or box plot). This is a graph that shows the distribution of scores on a quantitative variable within one or more groups. The center of the distribution of scores in each group is indicated by a median, and the "box" in the plot corresponds to the middle 50% of the distribution of scores (i.e., the range between the 25th and 75th percentiles).

b_0. This symbol stands for the intercept or constant included in a regression equation; b_0 is the predicted score on *Y* when $X = 0$ and is the point on the *Y* axis where the regression line intersects it. Some textbooks denote the intercept by *a* rather than b_0.

Canonical correlation (r_c). This refers to the correlation between scores on one discriminant function and group membership. This is analogous to an η coefficient in an ANOVA. The term *canonical* implies that there are multiple variables on one (or both) sides of the model. The canonical correlation, r_c, for each discriminant function is related to the eigenvalue for each function.

Carryover effects. In a repeated measures design, a carryover effect occurs if the effect of one treatment has not worn off before the next treatment in the series is introduced. For many types of treatments, carryover effects can be avoided by making the time intervals between treatments sufficiently long.

Categorical variable. This is a variable that identifies group membership; the numerical scores for the categorical variable serve merely as labels and convey no information about rank order or quantity.

Causal model. A causal model is often represented by a path diagram; unidirectional arrows represent hypotheses that variables are causally related, and bidirectional arrows represent hypotheses that variables are noncausally associated. These models are sometimes used to help data analysts decide what partial correlation and regression analyses to perform on data obtained from nonexperimental research. The term *causal model* is potentially misleading; we cannot make causal inferences about variables based on nonexperimental data. A causal model is purely theoretical; it represents *hypotheses* about the ways in which two or more variables may be related to each other, and these hypothesized relations may include both causal and noncausal associations. An example of a simple hypothetical causal model is as follows:

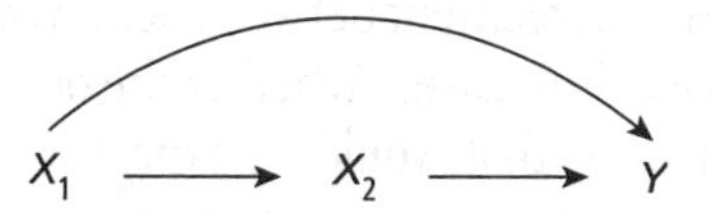

In words, this model corresponds to the following theory about the relations among these variables: X_1 directly causes or influences X_2; X_2 directly causes or influences Y; part of the influence of X_1 on Y is direct, and part of it is indirect influence mediated by the X_2 variable. When we calculate partial correlations and regression coefficients to describe the relations among X_1, X_2, and Y, the outcome of the analysis may be either consistent or inconsistent with a specific causal model. However, the empirical outcome cannot be interpreted as proof that one particular hypothetical causal model is the correct one.

Causal path. In a causal model, a unidirectional arrow that points from X toward Y denotes the theoretical existence of a causal connection in which X causes or influences Y ($X \rightarrow Y$).

Central limit theorem (CLT). The CLT describes what happens when numerous random samples of size N are drawn from a population distribution that has a population mean μ and population standard deviation σ, and the sample mean M is calculated for each of these many random samples. The shape of the distribution of values of M across many samples (called the sampling distribution of M) becomes increasingly close to the normal distribution as N increases, even if the distribution of the population of individual X scores from which the samples are drawn does not have a normal shape. The sampling distribution of values of M tends to be approximately normal in shape if the N in each sample is reasonably large ($N > 30$). This distribution of values of M is predicted to have a mean of μ and a standard deviation or standard error equal to $\sigma/\sqrt{N}$.

Centroid. A multivariate mean; in discriminant analysis, the centroid of a group corresponds to the mean values of the scores on D_1, D_2, and any other discriminant functions. In a graph, the centroid is the midpoint of the cluster of data points in the space defined by axes that represent the discriminant functions.

Change score (or difference score or gain score). In a simple pretest-posttest design in which the same behavior or attitude is measured twice, a change score (sometimes denoted by d for difference) is computed by subtracting the pretest score from the posttest score: $d = Y_{post} - Y_{pre}$. The null hypothesis of interest is either H_0: $d = 0$ (for studies of a single group) or H_0: $d_1 = d_2 = \cdots = d_k$ (for comparisons of treatment impact across k groups).

Chi-square test for binary logistic regression. For the overall logistic regression model, the chi-square test tests the null hypothesis that the overall model including all predictor variables is not predictive of group membership. Typically, the value of chi-square is obtained by taking the difference between the −2LL value for the model that includes all predictor variables of interest and the −2LL value for a model that includes only a constant.

Chi-square test of association (χ^2). This is a test statistic that compares observed and expected values (often, cell frequencies) to see if they are discrepant; a large chi-square indicates a poor level of agreement between observed and expected frequencies. The most common use of this test is to assess whether the observed cell frequencies are equal to the pattern of cell frequencies that would be expected if group membership on the row and column variables in a contingency table were unrelated. For example, we could assess whether group membership on one categorical variable (such as whether a person is a smoker) is predictable from group membership on another categorical variable (such as gender). A large chi-square indicates that there is a relationship between group membership on the two variables.

Classification errors. When the predicted group membership for a case (based on the values of discriminant function scores or a binary logistic regression) is not the same as the actual group membership for that case, this is referred to as a classification error.

Classification table. To assess how well a logistic regression or a discriminant analysis predicts group membership, a contingency table can be set up to show how well or how poorly the predicted group memberships correspond to the actual group memberships.

Clinical significance. *See* Practical significance

CLT. *See* Central limit theorem

Cohen's *d*. This is a popular effect size measure associated with the t test; it is given by the distance between group means ($M_1 - M_2$) divided by the pooled within-group standard deviation (s_p). It is interpreted as the number of standard deviations between the group means; thus, if $d = .5$, the group means are half a standard deviation apart. Table 5.2 provides suggested verbal labels for the strength of relationship implied by different values of d.

Cohen's kappa (κ). This is a statistic that assesses the degree of agreement in the assignment of categories made by two judges or observers (correcting for chance levels of agreement).

Communality (sometimes denoted by h^2). A communality is an estimate of the proportion of variance in each of the original p variables that is reproduced by a set of retained components or factors. Communality is obtained for each variable by summing and squaring the correlations (or loadings) for that variable across all the retained factors.

Compound symmetry. This term describes the relationships among elements in a variance/covariance matrix. A variance/covariance matrix has compound symmetry if all the variances on the main diagonal are equal to each other and all the covariances in the off-diagonal cells are equal to each other. (Note that the covariances do not have to be equal to the variances.) When scores in a repeated measures or within-S design show compound symmetry, an F ratio from a repeated measures ANOVA should provide a valid significance test. A less restrictive assumption about the pattern of variances and covariances among repeated measures, called Pattern H, is sufficient for the F test to be valid.

Concordant pairs. This refers to scores for participants who have either low scores on both X and Y or high scores on both X and Y.

Conditional probability. A conditional probability is just a probability of an outcome on one variable (e.g., survival vs. death) for some subgroup based on scores on a predictor variable (e.g., dog owners). The "conditional probability of survival given the condition that the person is a dog owner" is simply the proportion of survivors among the dog owner group; this corresponds to the row percentages in Table 8.4. The conditional probability of survival given that the person belongs to the dog owner group was 50/53 = .94; the conditional probability of survival given that the person is not a member of the dog owner group was 28/39 = .72.

Confidence interval (CI). This is a range of values above and below a sample statistic that is used as an interval estimate of a corresponding population parameter. Given a specific confidence level (such as 95%), then, over the long run, when the process is repeated across hundreds of samples drawn randomly from the same population, 95% of the confidence intervals that are set up following the specified procedures should contain the value that corresponds to the true population parameter (and 5% of the CIs will not contain the true population parameter).

Confidence interval for the sample mean. This refers to an interval around the sample mean, M, that is (theoretically) likely to include μ, the true population mean. When the confidence level is set at a specific numerical level, such as 95%, it theoretically describes the performance of a large number of CIs. When the confidence level is 95%, in theory, approximately 95 out of 100 CIs that are set up using the specified procedures will contain μ, and approximately 5 out of 100 CIs will not contain μ.

Confidence level. This is the theoretical long-term probability (arbitrarily selected by the data analyst) that a confidence interval set up using sample data will contain the true population parameter, across many samples drawn randomly from the same population. The most commonly used levels of confidence are 90%, 95%, and 99%. As the confidence level increases, and if other terms (such as the sample size and standard deviation) remain fixed, the width of the confidence interval increases.

Confirmatory factor analysis (CFA). An analysis in which the data analyst develops a constrained model that specifies the number of latent variables needed to explain the variances and covariances in the data and also specifies which measured variables are indicators of each factor. Because this model includes many constraints, it generally cannot reproduce the variance/covariance matrix perfectly. Competing models can be compared by asking how much they differ in their ability to reconstruct the original data structure. Structural equation modeling programs, such as AMOS, EQS, or LISREL, can be used to estimate parameters for CFA and to assess how well CFA models reproduce the variances and covariances among measured variables.

Confound (or confounded variable). In an experiment, if another variable systematically co-occurs with a specific type of treatment, that variable is confounded with treatment, and therefore it is impossible to determine whether the outcomes for that group are due to the type of treatment or the confound. If participants who receive 150 mg of caffeine are assessed just prior to an examination and the participants who receive no caffeine are assessed just before departure for vacation, the situational factor (exam vs. vacation) is confounded with the treatment variable (drug dosage level). If the first group scores higher on anxiety, the researcher cannot tell whether this is due to the influence of caffeine, the exam, or both. In well-designed experiments, researchers try to make certain that no other variables are confounded with the treatment variable. In nonexperimental research situations, confounds often cannot be avoided.

Confounded variable. *See* Confound

Conservative test. One statistical test procedure is more conservative compared with another when it involves decision rules that make it more difficult to reject the null hypothesis (and that therefore reduce the risk of Type I error). For example, a z test that uses $\alpha = .01$, two-tailed, is more conservative than a z test that uses $\alpha = .05$, two-tailed.

Constant-only model. *See* Null model

Construct validity. Construct validity is the degree to which an X variable really measures the construct that it is supposed to measure. In practice, construct validity can be assessed by examining whether scores on X have the sizes and signs of correlations with other variables (which may include experimental manipulations) that they should have, according to theories about the nature of the variables and their interrelationships.

Content validity. The degree to which the content of questions in a self-report measure covers the entire domain of material that should be included (based on theory or assessments by experts).

Contingency. A Y variable is contingent on an X variable if scores on the Y variable are predictable to some extent from scores on the X variable. A more precise definition of contingency can be given in terms of the conditional and unconditional probabilities. If the conditional probabilities of outcome Y based on group membership X are different, then Y is contingent on X. For example, because the conditional probability of survival was higher for dog owners (.94) than for nonowners of dogs (.72) in Table 8.4, this is evidence that survival is contingent on dog ownership. The chi-square test of association provides a way of evaluating whether the difference between these conditional probabilities is sufficiently large to be judged statistically significant.

Contingency table. A table in which each row corresponds to group membership on one categorical variable, each column corresponds to group membership on a second categorical variable, and each cell contains the number of cases that occurred for each group. For example, a cross-tabulation of gender by political party would include the number of male Republicans, female Independents, and so forth. A contingency table for gender and smoking could take the following form, where the number in each cell represents a frequency count (e.g., 17 is the number of female Republicans):

	Republican	*Democrat*
Female	17	44
Male	51	9

Continuity-adjusted chi-square. *See* Yates's chi-square

Continuity correction. When sample *Ns* are small, the actual sampling distribution of the chi-square statistic is not a smooth and continuous curve (as defined by the mathematical function for χ^2). Instead, the possible values of χ^2 that can actually occur in sample data correspond to a set of discrete outcomes, as shown at right in this illustration of a discrete chi-square distribution with 1 degree of freedom. When we use critical values of chi-square from tables that are based on the theoretical continuous distribution, the shape of that continuous distribution used to develop the tables of critical values does not exactly correspond to the empirical sampling distribution, which is discrete or discontinuous. To correct for possible problems due to this mismatch between the continuous sampling distribution used to set up tables and the discrete sampling distributions that can arise in small samples, Yates proposed a continuity correction; this involves subtracting .5 from each $(O - E)$ difference prior to squaring it. This continuity-corrected version of the chi-square, often called Yates's chi-square, is somewhat more conservative than the uncorrected (or Pearson) chi-square.

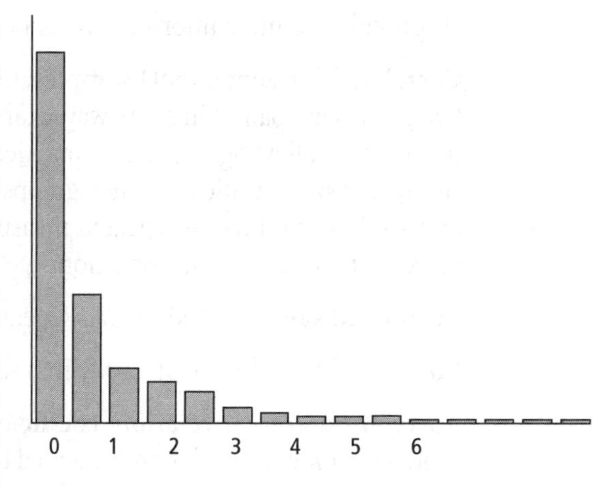

Contrast (also called planned contrast or a priori comparison). A contrast is a comparison between two group means (or among several group means). Planned contrasts are decided on prior to looking at the data (based on theory), and if only a limited number of theoretically predicted differences is tested, most researchers do not worry about inflated risk of Type I error. In the SPSS one-way ANOVA procedure, a custom contrast is requested by typing in a list of contrast coefficients (usually small integers that sum to 0, e.g., −1, 0, +1). There must be one coefficient for each group in the one-way ANOVA.

Convenience sample. A sample that includes readily available cases, rather than cases randomly selected from a specific population. *See also* Accidental sample

Convergence. When principal axis factoring is performed, the program must estimate the matrix of factor loadings, use it to calculate the communalities, and then reestimate the factor loading matrix using the new communality values; this is done repeatedly until, from one iteration to the next, the communality values remain about the same; that is, the estimates converge to a value that does not change on subsequent reestimation steps.

Convergent validity. The degree to which a new measure, X', correlates with an existing measure, X, that is supposed to measure the same construct.

Correctly specified model. A model is correctly specified if it includes all the variables that should be taken into account (e.g., all variables that are confounded with or causally related to X_1 or Y, or all variables that interact with X_1 or Y) and, also, if it does not include any variable that should not be included. A well-developed theory can be helpful in deciding what variables need to be included in a model; however, we can never be certain that the models we use to try to explain relations among variables in nonexperimental studies are correctly specified. Thus, we can never be certain that we have taken all potential rival explanatory variables into account; also, we can never be certain that our statistical analysis provides accurate information about the nature of the relation between X_1 and Y.

Correlated (or dependent) samples. Correlated samples arise when the participants in two groups are paired in some way. Examples of designs that produce correlated samples include the following: studies of married or dating couples, where Group 1 = Husband, Group 2 = Wife; studies in which groups are matched on some participant characteristic; and studies that involve repeated measures obtained from the same participants under more than one treatment condition.

Correlated samples design. *See* Repeated measures ANOVA

Correlated samples *t* test. *See* Paired samples *t* test

Correlation (or correlation coefficient). Usually, this term refers to the Pearson product-moment correlation presented in Chapter 7. However, there are many other types of correlations, such as the phi (ϕ) coefficient, tetrachoric, point biserial r, biserial r, Spearman r, and η, the correlation ratio (in Chapter 8). The word *correlation* is sometimes used generically to refer to any statistical association between a pair of variables.

Correlation coefficient. *See* Correlation

Counterbalancing. In repeated measures experiments, counterbalancing involves presenting the treatments in different orders to avoid a confound between order of presentation of treatment and type of treatment.

Covariance. An unstandardized version of correlation; it provides information about the tendency for scores on a pair of variables to co-occur, but unlike the Pearson r, it is not standardized.

Covariate. In ANCOVA or a multiple regression, a covariate is a variable that is included so that its association with the Y outcome and its confound with other predictors can be statistically controlled when assessing the predictive importance of other variables that are of greater interest. Covariates are usually continuous variables (but may be dummy variables).

Cox and Snell R^2. This is one of the pseudo R^2 values provided by SPSS as an overall index of the strength of prediction for a binary logistic regression model.

Cramer's *V*. This is an index of association that is often reported as effect size information for a chi-square test of association; it can be used for tables with any number of rows and columns and can be computed from the chi-square value. For a 2 × 2 table, Cramer's V is equivalent to the absolute value of the ϕ coefficient. Cramer's V can range from 0 to +1; values close to 0 indicate no relationship between the categorical variables that identify the rows and columns in the table.

Critical value. This is a value from a sampling distribution (such as a normal or t distribution) that corresponds to some arbitrarily specified area(s) under the curve or the proportion of outcomes included in the distribution. When researchers set up confidence intervals, they generally want to know what critical values of z or t correspond to the middle 90%, 95%, or 99% of the z or t distribution. Critical values are also used to make decisions in null hypothesis significance tests.

Critical value of *t*. This is a value of t (i.e., a distance from the mean in standard score units) that corresponds to a specific proportion or percentage of the area. It can be obtained from the table in Appendix B. This critical or cutoff value of t is used as the criterion in making decisions about statistical significance. When a sample value of t (based on information in the sample including M, s, and N) is larger than the critical value of t (obtained by looking up an appropriate critical value of t based on the α level and depending on the use of a one- or two-tailed test and the degrees of freedom), the researcher can decide to reject the null hypothesis. In theory, if all the assumptions for significance testing are satisfied, the probability of committing a Type I error using this decision rule should be approximately equal to the alpha level used to look up the critical value of t.

Critical value of *z*. This is a value of z (i.e., a distance from the mean in standard score units) that corresponds to a specific proportion or percentage of the area. For example, z score values of ±1.96 correspond to the boundaries between the middle 95% of the area under the normal distribution and the bottom and top 2.5% of the area. That is, 2.5% of the area in a normal distribution lies below $z = -1.96$, 95% of the area lies between $z = -1.96$ and $z = +1.96$, and 2.5% of the area lies above $z = +1.96$. The Java applet "NormalProb" found at www.bolderstats.com/jmsl/doc/ can be used to see how changes in the value of z are related to changes in the corresponding areas under a normal curve. The critical value of z can be obtained from the table of the standard normal distribution in Appendix A. This critical or cutoff value of z is used as the criterion in making decisions about statistical significance. When a sample value of z (based on information in the sample including M and N) is larger than the critical value of z (obtained by looking up an appropriate critical value of z based on the α level and depending on the use of a one- or two-tailed test), the researcher can decide to reject the null hypothesis. In theory, if all the assumptions for significance testing are satisfied, the probability of committing a Type I error using this decision rule should be approximately equal to the alpha level used to look up the critical value of z.

Cronbach alpha (α). An index of internal consistency reliability that assesses the degree to which responses are consistent across a set of multiple measures of the same construct—usually self-report items.

Cross-tabulation. *See* Contingency table

Custom model. In SPSS GLM, if the user wants to run an analysis that includes terms that are not included in the default model, these terms can be requested by using the Custom Model procedures. For example, the default model for ANCOVA does not include Treatment × Covariates interactions; these may be added to the model so that the assumption of "no treatment by covariate interaction" can be assessed empirically.

Cutoff value. In binary logistic regression analysis or in discriminant analysis, this refers to the cutoff value of the estimated probability that is used to classify an individual case into a group. When the outcome variable involves only two groups, usually, the cutoff value is $p = .5$; that is, a case is assigned to Group 1 if it has a probability greater than .5 of being a member of Group 1.

Data-driven order of entry. Some methods of regression (which are referred to here as "statistical" methods) involve allowing the computer program to select variables according to the order of their predictive usefulness; at each step in a forward statistical regression, for example, the variable that enters the regression equation is the one for which the increase in R^2 is the largest. An advantage of such procedures is that they can yield a regression model with the maximum possible R^2 based on the minimum possible number of predictors. However, a disadvantage of data-driven procedures is that they often lead to an inflated risk of Type I error. Variables that have large correlations with the Y outcome variable in the sample due to sampling error will be selected as predictors, whether or not the variables make any theoretical sense.

Degrees of freedom (*df*). This is the number of independent pieces of information that a statistic is based on. For example, the sum of squares, *SS* (and the sample statistics that are computed using *SS*, including the sample standard deviation s and the sample variance s^2), has $N - 1$ degrees of freedom, where N is the number of cases in the sample.

Dependent samples. *See* Correlated samples

Dependent (or outcome) variable. This is a variable that is measured as the outcome of an experiment or predicted in a statistical analysis.

Descriptive statistics. Statistics that are reported merely as information about the sample of observations included in the study and that are not used to make inferences about some larger population.

Determinant of a matrix. The determinant of a sum of cross products, or correlation matrix, is a single-number summary of the variance for a set of variables when intercorrelations among variables are taken into account. The computation of a determinant $|\mathbf{A}|$ for a 2×2 matrix $\mathbf{A}$ is quite simple:

$$\text{For } \mathbf{A} = \begin{bmatrix} a & b \\ c & d \end{bmatrix}, \text{ the determinant } |\mathbf{A}| = (a \times d) - (c \times b).$$

Deviance chi-square. *See* −2LL

***df*.** *See* Degrees of freedom

Diagonal matrix. A matrix in which all the off-diagonal elements are 0, as in the following example:

$$\begin{bmatrix} 12 & 0 & 0 & 0 \\ 0 & 7 & 0 & 0 \\ 0 & 0 & 18 & 0 \\ 0 & 0 & 0 & 14 \end{bmatrix}.$$

Dichotomous variable (or dichotomy). A categorical variable that can take only two values. Usually, these are coded 0, 1 or 1, 2. For example, gender may be coded as a dichotomous variable with 1 = Male, 2 = Female.

Dichotomy. *See* Dichotomous variable

Dichotomy—true versus artificial. A true dichotomy is a naturally occurring group membership (e.g., male vs. female). An artificial dichotomy arises when an arbitrary cutoff value is applied to an underlying continuous variable (such as a pass/fail cutoff for exam scores).

Difference score (*d*). *See* Change score

Dimensionality. The dimensionality of a set of X variables involves the following question: How many different constructs do these X variables seem to measure? The most common way of assessing dimensionality is to do factor analysis or principal components analysis and to decide how many factors or components are needed to reproduce the correlation matrix R adequately. If these factors or components are meaningful, they may be interpreted as dimensions, latent variables, or measures of different constructs (such as different types of mental ability or different personality traits).

Dimension reduction analysis. When a discriminant analysis yields several discriminant functions, the data analyst may think of each discriminant analysis as a "dimension" along which groups differ. The discriminant functions are rank ordered in terms of their strength of association with group membership, and sometimes, only the first one or two discriminant functions provide useful information about the prediction of group membership. When a researcher looks at significance tests for sets of discriminant functions and makes the judgment to drop some of the weaker discriminant functions from the interpretation and retain only the first few discriminant functions, it is called a dimension reduction analysis. The descriptions of differences among groups are then described only in terms of their differences on the dimensions that correspond to the retained discriminant functions.

Direct difference *t* test. *See* Paired samples *t* test

Direct entry. When all predictor variables are entered at one step in multiple regression or in a logistic regression, it is called a direct entry; this is equivalent to standard or simultaneous multiple linear regression as described in Chapter 14.

Directional test. *See* One-tailed test

Discordant pairs. Cases that have a high score on X paired with a low score on Y or a low score on X paired with a high score on Y are discordant (defined in Chapter 7).

Discriminant function. This is a weighted linear combination of scores on discriminating variables; the weights are calculated so that the discriminant function has the maximum possible between-groups variance and the minimum possible within-groups variance. For convenience, discriminant function scores are generally scaled so that, like z scores, they have variance equal to 1.0.

Discriminant function coefficients. These are the weights given to individual predictor variables when we compute scores on discriminant functions. Like regression coefficients, these discriminant function coefficients can be obtained for either a raw score or standard (z) score version of the model; the coefficients that are applied to standardized scores are generally more interpretable.

Discriminant validity. When our theories tell us that a measure X should be unrelated to other variables such as Y, a correlation of near 0 is taken as evidence of discriminant validity—that is, evidence that X does not measure things it should not be measuring.

Disproportionately influential score. A score is disproportionately influential if the value of a summary statistic, such as a group mean or a Pearson correlation, is strikingly different when the analysis is done with the score included from the value when the analysis is done with the score omitted.

Dummy-coded dummy variable. Dummy coding involves treating each dummy variable as a yes/no question about group membership; a "yes" response is coded as "1" and a "no" response is coded as "0." In dummy coding of dummy variables, members of the last group receive codes of "0" on all the dummy variables.

Dummy variable. A dichotomous predictor variable in multiple regression with a score, usually a 0, +1, or −1, that represents membership in one of two groups. A dummy variable corresponds to a yes/no question about group membership; for example, "Is the participant male?" 0 = No, 1 = Yes.

e. A mathematical constant (approximately equal to 2.718281828), sometimes called Euler's number or Napier's constant. It is often used in models of growth or decay processes, and it appears in many statistical functions (such as the equation that defines the normal function).

Effect. In ANOVA this term has a specific technical meaning. The "effect" of a particular treatment, the treatment given to Group i, is estimated by the mean score on the outcome variable for that group (M_i) minus the grand mean (M_Y). The effect for Group i is usually denoted by α_i; this should not be confused with other uses of alpha (alpha is used elsewhere to refer to the risk of Type I error in significance test procedures and to the Cronbach alpha reliability coefficient). The "effect" α_i represents the "impact" of the treatment or participant characteristics for Group i. When $\alpha_1 = +9.5$, for example, this tells us that the mean for Group 1 was 9.5 points higher than the grand mean. In the context of a well-controlled experiment, this may be interpreted as the magnitude of treatment effect. In studies that involve comparisons of naturally occurring groups, the *effect* term is not interpreted causally but, rather, as descriptive information about the difference between each group mean and the grand mean.

Effect-coded dummy variable. Effect coding involves treating each dummy variable as a yes/no question about group membership; a "yes" response is coded as "1" and a "no" response is coded as "0." However, the last group (the group that does not correspond to a yes answer on any of the effect-coded dummy variables) receives codes of −1 on all the dummy variables.

Eigenvalue (λ). In the context of discriminant analysis, an eigenvalue is a value associated with each discriminant function; for each discriminant function, D_i, its corresponding eigenvalue, λ_i, equals $SS_{between}/SS_{within}$. When discriminant functions are being calculated, the goal is to find functions that have the maximum possible values of λ—that is, discriminant functions that have the maximum possible between-group variance and the minimum possible within-group variance. In the context of factor analysis or principal components analysis, an eigenvalue is a constant value that is associated with one of the factors in a factor analysis (or one of the components in a principal components analysis). The eigenvalue is equal to the sum of the squared loadings on a factor (or on a component). When the eigenvalue for a factor or component is divided by k, where k is the number of variables included in the factor analysis, it yields the proportion of the total variance in the data that is accounted for or reproducible from the associated factor or component.

Empirical keying. This is a method of scale construction in which items are selected for inclusion in the scale because they have high correlations with the criterion of interest (e.g., on the MMPI, items for the depression scale were selected because they predicted a clinical diagnosis of depression). Items selected in this way may not have face validity; that is, it may not be apparent to the test taker that the item has anything to do with depression.

Epsilon (ε). This is a measure of the departure of the sample variance/covariance matrix **S** from a pattern that would satisfy the sphericity assumption. When $\varepsilon = 1$ there is no departure; the smallest possible value of ε is $1/(k - 1)$, where k is the number of levels of the repeated measures factor, and smaller values of ε indicate more serious violations of the sphericity assumption. Different versions of ε have been proposed (Greenhouse-Geisser, Huynh-Feldt). Multiplying *df* by ε provides an adjusted *df* that is used to determine the critical value of F for significance tests.

Error variance suppression. When a covariate is added to a model (such as an ANCOVA or a regression), $SS_{residual}$ or SS_{within} may go down substantially. If the decrease in $SS_{residual}$ is large compared with the change in degrees of freedom, the F ratios for tests of variables that are of greater interest may decrease. When a covariate reduces $SS_{residual}$ substantially, we say that it is a noise suppressor or an error variance suppressor. *See also* Noise suppression

Eta squared (η^2). This is an effect size statistic that can be used with the t test or ANOVA; it is interpreted as the proportion of variance in the scores on the outcome variable that is predictable from group membership.

EW_α. *See* Experiment-wise alpha

Exact p value. For a given statistical outcome (such as a value of the independent samples t test), the exact p value is the area in one or both of the tails of the t distribution that is more than t units away from the center of the distribution.

exp(B). The exponential of B, which can also be represented as e^B.

Expected cell frequency. In a contingency table, the expected frequency for each cell (i.e., the number of cases that you would expect to see in each cell if membership in the row category is not related to membership in the column category) is found by multiplying the row total by the column total and dividing the result by the table total *N*. The chi-square test of association should be used only if all cells have expected frequencies greater than 5. When one or more cells in a 2 × 2 table have expected frequencies less than 5, a χ^2 test is not appropriate. Alternative analyses such as the Fisher exact test are more appropriate when some expected cell frequencies are very small.

Experimental control. Researchers may use a variety of strategies to "control" variables other than the manipulated independent variable, which could influence their results. For example, consider the possible influence of gender as a third variable in the study of the effects of caffeine on anxiety. Females and males may differ in baseline anxiety, and they may differ in the way they respond to caffeine. To get rid of or control for gender effects, a researcher could hold gender constant—that is, conduct the study only on males (or only on females). Limiting the participants to just one gender has a major drawback; that is, it limits the generalizability of the findings (the results of a study run only with male participants may not be generalizable to females). The researcher might also include gender as a variable in the study; for example, the researcher might assign 10 women and 10 men to each group in the study (the caffeine group and the no caffeine group). See Cozby (2004) or other basic research methods textbooks for a further discussion of the many forms of experimental control.

Experimental design. Experiments typically involve the manipulation of one or more independent variables by the researcher and experimental control over other variables that might influence the outcomes or responses of participants. Often experiments involve comparisons of mean scores on one or more outcome variables across groups that have received different types or amounts of treatments.

Experiment-wise alpha (EW_α). In the Bonferroni procedure, this is the risk of having at least one Type I error in the entire set of significance tests. Sometimes this is set at EW_α = .05, but it may also be set to higher levels such as EW_α = .10 if the researcher is willing to tolerate higher levels of risk of Type I error.

Exploratory factor analysis. If a researcher extracts factors from a dataset without placing any constraints on the number of factors or the pattern of correlations between factors and measured variables, and decisions, such as the number of factors to be retained, are made on the basis of arbitrary criteria, such as the magnitude of eigenvalues, the factor analysis is exploratory. The SPSS Factor procedure provides exploratory forms of factor analysis.

Exponential function. The exponential function applied to a specific numerical value such as *a* is simply e^a (i.e., the exponential of *a* is given by the mathematical constant e raised to the *a*th power). It is sometimes written as Exp(*a*).

External validity. The degree to which research results are generalizable to participants, settings, and materials beyond those actually included in the study is called external validity.

Extraction. This is a computation that involves estimating a set of factor or component loadings; loadings are correlations between actual measured variables and latent factors or components.

Face validity. This refers to the degree to which it is obvious what attitudes or abilities a test measures from the content of the questions posed. High face validity may be desirable in some contexts (when respondents need to feel that the survey is relevant to their concerns), but low face validity may be preferable in other situations (when it is desirable to keep respondents from knowing what is being measured in order to avoid faking, social desirability bias, or deception).

Factor (in ANOVA). In the context of ANOVA, a categorical predictor variable is usually called a factor. In an experiment, the levels of a factor typically correspond to different types of treatment, or different dosage levels of the same treatment, administered to participants by the researcher. In nonexperimental studies, the levels of a factor can correspond to different naturally occurring groups, such as political party or religious affiliation.

Factor (in factor analysis). In the context of factor analysis, a factor refers to a latent or imaginary variable that can be used to reconstruct the observed correlations among measured *X* variables. By looking at the subset of *X* variables that have the highest correlation with a factor, the researcher decides what underlying dimension or construct might correspond to that factor. For example, if a set of mental ability test scores are factor analyzed, and the first factor has high positive loadings (positive correlations) with the scores on the variables vocabulary, reading comprehension, analogies, and anagrams, the researcher might decide to label this factor "verbal ability."

Factorial design. A design in which there is more than one factor or categorical predictor variable; see Chapter 13 for a discussion of factorial ANOVA.

Factor loading. *See* Loading

Factor score. This is a score for each individual participant that is calculated by applying the factor score coefficients for each factor to *z* scores on all items.

Factor score coefficients. Coefficients that are used to construct scores on each factor for each participant from *z* scores on each individual *X* variable are called factor score coefficients. These are analogous to β coefficients in a multiple regression.

Faking. Sometimes test takers make an effort to try to obtain either a high or low score on an instrument by giving answers that they believe will lead to high (or low) scores; informally, this can be called faking good (or faking bad).

Family of distributions. The set of distributions obtained by substituting different parameters into one general equation—for example, the set of *t* distributions that is obtained for various degrees of freedom or the set of normal distributions for varying values of μ and σ.

First-order partial correlation. In a first-order partial correlation, just one other variable is statistically controlled when the relation between X_1 and Y is assessed.

Fisher exact test. This test is sometimes applied to 2×2 tables (instead of a chi-square test of association) when expected cell frequencies are below 5. The Fisher exact test yields a *p* value that tells you how "unlikely" it is for the observed arrangement of cases in cells to occur just by chance, when you consider all possible arrangements of cases in cells that could occur, given the marginal frequencies in the table. The Fisher exact test is itself a "*p*" value or probability. If you use the conventional $\alpha = .05$ criterion for statistical significance, therefore, you conclude that the variables are significantly related when the

Fisher exact test result is less than .05. (The Fisher exact test can be calculated for tables of dimensions larger than 2 × 2, but it is computationally unwieldy; SPSS provides it only for 2 × 2 tables.)

Fisher *r* to *Z* transformation. This is a transformation that is applied to *r* values in order to make their distribution more normal in shape; this is necessary when setting up some significance tests about *r* values and when setting up confidence intervals. Note that lowercase *z* is used in this book to stand for a normally distributed variable, while uppercase *Z* is always used in this book to refer to the Fisher *r* to *Z* transformation. (Some textbooks and tables denote the Fisher *r* to *Z* transformed score as *z* or *z*′.)

Fixed factor. A factor in an ANOVA is "fixed" if the levels of the factor that are included in the study include all the possible levels for that factor or if the levels of the factor included in the study are systematically selected to cover the entire range of "dosage levels" that is of interest to the researcher. For example, if we code gender as 1 = Male, 2 = Female and use these two levels of gender in a factorial study, gender would be treated as a fixed factor. If we select equally spaced dosage levels of caffeine that cover the entire range of interest (e.g., 0, 100, 200, 300, and 400 mg), then caffeine would be treated as a fixed factor.

Forward method of entry. In this method of statistical regression, the analysis begins with none of the candidate predictor variables included; at each step, the predictor that contributes the largest R^2 increment is entered. The addition of predictor variables stops when we arrive at the situation where none of the remaining predictors would cause a significant increase in R^2 (the criterion for significance to enter may be either $p < .05$ or a user-specified *F*-to-enter).

***F* ratio.** In ANOVA and other analyses, an *F* ratio is obtained by taking a mean square that represents variability that can be predicted from the independent variable (in the case of ANOVA, this provides information about differences among group means) and dividing it by another mean square that provides information about the variability that is due to other variables or "error." If *F* is much greater than 1 and if it exceeds the tabulated critical values for *F*, the researcher concludes that the independent variable had a significant predictive relationship with the outcome variable. (It was named the *F* ratio in honor of Sir Ronald Fisher, one of the major contributors to the development of modern statistics.)

Frequency distribution. A list of all possible scores on a variable, along with the numbers of persons who received each possible score, is called a frequency distribution. For example, a frequency table for the variable "Type of tobacco used" could be as follows:

Type of Tobacco	*Number (Frequency) of Persons*
None	43
Cigarette	41
Pipe	6
Chewing tobacco	11

F-to-enter. In the forward method of statistical regression, this is the minimum value of F that a variable has to have for its R^2 increment before it is entered into the equations.

Fully redundant predictor variable. Consider an example in which data are obtained from 100 countries: X_1 is per-person income, X_2 is ease of access to mass media such as television, and Y is the life expectancy. When you control for X_1 (per-person income), X_2 (access to television) is no longer significantly predictive of life expectancy. When you control for access to television, however, income is still significantly predictive of life expectancy. In this situation, we might reason that wealthy countries (those with high per capita income) also tend to have better availability of television and mass media. We might reasonably hypothesize that only income influences life expectancy. Access to television may not have any direct causal influence on life expectancy; it may be correlated with life expectancy only because access to television is associated with or caused by a variable (income) that does influence life expectancy.

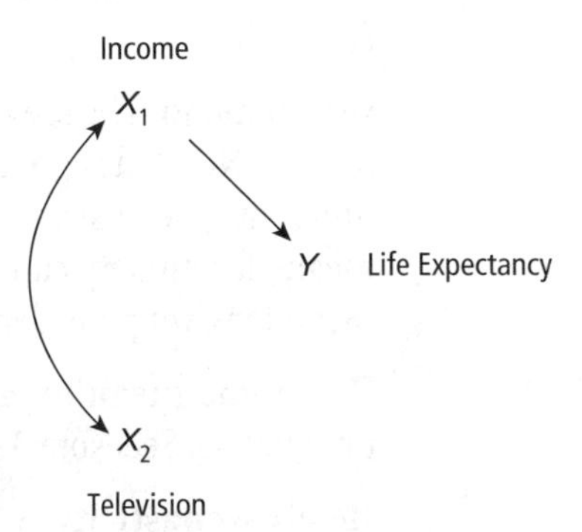

In this case, when you control for X_1 (income), the partial correlation of X_2 (access to television) and Y (life expectancy) drops to 0. We could reasonably conclude that the X_2, Y correlation was spurious; X_2 and Y are correlated only because X_2 is associated with X_1, and X_1 influences Y. We could also say that the predictive information (about life expectancy) that is contained in the X_2 (access to television) variable is "completely accounted for by," "completely explained away by," or "fully redundant with" the X_1 variable (income).

Gain score. *See* Change score

Gamma. This is a symmetric measure of association between two ordinal variables and ranges between −1 and 1. Values close to an absolute value of 1 indicate a strong relationship between the two variables. Values close to 0 indicate little or no relationship. For two-way tables, zero-order gammas are displayed. For three-way to n-way tables, conditional gammas are displayed.

gcr. *See* Roy's largest root

General linear model (GLM). The most general case of this model is one in which one or several predictor variables (which may be categorical or continuous) are used to predict outcomes on one or several outcome variables (which may be categorical or continuous). Most of the analyses taught in introductory statistics courses are special cases of this more general model; for example, in one-way ANOVA, scores on one continuous outcome variable are predicted from one categorical variable. All these analyses involve the computation of similar terms (e.g., sums of squares).

GLM. *See* General linear model

Goodness of fit. Indexes such as LL (the log likelihood) and χ^2 that provide information about agreement between model predictions and actual outcomes are often called "goodness-of-fit" indexes. However, this terminology is confusing, because for many "goodness-of-fit" indexes, larger values indicate a poorer model fit. (It makes more intuitive sense, in some cases, to think of statistics such as LL and χ^2 as "badness-of-fit" measures.)

Grand mean. The mean for all the scores in an entire study, denoted by M_Y or M_{grand}.

Greenhouse-Geisser ε adjustment. The more conservative of the two widely used methods of adjusting *df* in repeated measures ANOVA, to compensate for violations of sphericity.

Group mean. The mean of all the scores within one level of the factor or within one group in the ANOVA design; for example, imagine a study in which each group is exposed to a different type of stress, and the outcome measure is self-reported anxiety. Separate group means for anxiety can be calculated for the group that received each type of stress. The mean for Group *i* is denoted by M_i; for example, the mean for Group 1 is denoted by M_1.

Harmonic mean of *n*. A method of computing an average *n* across groups that have unequal *n*s. See Note 3 in Chapter 6 for the formula.

Heteroscedasticity. A *Y* (dependent) variable is heteroscedastic if its variance differs as a function of the value of *X*. This is a violation of an assumption (homoscedasticity or uniform variance of *Y* across levels of *X*) for the use of the Pearson *r*. In ANOVA, heteroscedasticity refers to differences in variances across populations or groups.

Hinges. The hinges are the 25th and 75th percentile points for a distribution of scores on quantitative variables shown in a box plot or box and whiskers plot.

Histogram. This is a type of graph that provides information about the number or proportion of people with a particular score, for each possible score or interval of score values on a quantitative variable (*X*). Typically, *X* score values are indicated by tick marks on the *X* axis. For each *X* score, the height of the vertical bar corresponds to the frequency (or proportion) of people in the sample who have that score value for *X* (or a score for *X* that falls within the indicated score interval). The reference scale used to evaluate the information provided by the height of each bar is indicated on the *Y* axis, which is usually labeled in terms of either frequencies or proportions of cases. Conventionally, the bars in a histogram touch each other.

Homogeneity of regression assumption. In ANCOVA, we assume no treatment by covariate interactions; this is also called the homogeneity of regression assumption. (The raw scores slope *b* to predict *Y* from X_c is assumed to be equal or homogeneous across all groups in the ANCOVA.)

Homogeneity of variance assumption. This refers to the assumption that variances of the populations being compared (using the *t* test or ANOVA) are equal. For a *t* test or ANOVA, possible violations of this assumption can be detected using the Levene test or other test statistics that compare the sample variances across groups. In regression or correlation, homogeneity of variance refers to an assumption of uniform variance of *Y* scores across levels of *X*.

Homogeneity of variance/covariance matrices. Just as univariate ANOVA assumes equality of the population variances of the dependent variable across groups, discriminant analysis (DA) assumes homogeneity of Σ, the population variance/covariance matrix, across groups.

Homoscedasticity. This refers to equal variances of *Y* scores at each level of the *X* variable. This is an assumption for the use of the Pearson *r*. *See also* Homogeneity of variance assumption

Hotelling's T^2. Just as MANOVA is a multivariate generalization of ANOVA (i.e., MANOVA compares vectors of means across multiple groups), Hotelling's T^2 is a multivariate generalization of the *t* test (Hotelling's T^2 compares vectors of means on *p* variables across two groups). Note, however, that Hotelling's T^2 is not the same statistic as Hotelling's trace, which is part of the SPSS GLM output.

Hotelling's trace. This is one of the multivariate test statistics provided by SPSS GLM to test the null hypothesis shown in Equation 17.3: H_0: $\mu_1 = \mu_2 = \cdots = \mu_k$. Hotelling's trace is the sum of the eigenvalues across all the discriminant functions; a larger value of Hotelling's trace indicates larger between-group differences on mean vectors.

H-spread. The H-spread is (usually) defined as 1.5 times the distance between the hinges—that is, 1.5 times the distance between the 25th and 75th percentiles for a distribution of quantitative scores in a box plot.

***h^2*.** *See* Communality

Huynh-Feldt ε adjustment. This is the less conservative of the two widely used methods of adjusting *df* in a repeated measures ANOVA to compensate for violations of sphericity.

Hypothetical population. The (often imprecisely defined) population to which generalizations are made when research is based on an accidental or convenience sample is called the hypothetical population. For instance, a researcher who studies the effect of caffeine on anxiety in a convenience sample of college students may want to generalize the results to all healthy young adults; this broader population is purely hypothetical.

H_0. *See* Null hypothesis

Impossible values. Such values refer to responses that do not correspond to any of the response alternatives provided on self-report or that are too extreme to be believed (such as a height of 120 in.).

Imputation (of missing data). *See* Missing values imputation

Inconsistent responses. These responses are self-contradictory or logically inconsistent responses by an individual participant across questions—for example, when a person reports no smoking on one question and substantial smoking on a differently worded question.

Incremental sr^2 (sr^2_{inc}). In a sequential or statistical regression, this is the additional proportion of variance explained by each predictor variable at the step when it first enters the analysis. When only one variable enters at a step, sr^2_{inc} is equivalent to the R^2 increment, or R^2_{inc}.

Independence of observations. Observations in a sample are not independent of each other when the participants in a study influence each other's behavior, through processes such as imitation, competition, cooperation, discussion, and so forth. Nonindependence of observations often means that the sample variances seriously underestimate the population variance. This in turn means that the error terms for *t* tests are too small and that the true risk of Type I error is higher than the nominal *p* values that are obtained. When scores are dependent or related to each other in some way, the nature of this interdependence needs to be evaluated (Kenny & Judd, 1996), and the analysis must take this interdependence into account (Kenny et al., 2002).

Independent samples design. In such a design, responses of participants are not related or correlated between groups. Usually, between-S designs (in which each participant is assigned to only one treatment group) generate independent samples. If groups are formed by using matched or paired participants, the samples are not independent. The independent samples *t* test (Chapter 5) and the one-way between-S ANOVA (Chapter 6) require independent samples. If samples are not independent, repeated measures analysis is required (see Chapter 20). *See also* Between-S design

Independent (or predictor) variable. These are the variable or variables that are used to predict scores on one or more outcome variables. In an experiment, the independent variables are the factors that are manipulated by the researcher. In nonexperimental research, the designation of some variables as independent may be based on implicit causal theories. In some research situations, there is no clear distinction between predictor and outcome variables; in these situations, the designation of one variable as a predictor may be entirely arbitrary.

Inferential statistics. Inferential statistics involve using a descriptive statistic for a sample to make inferences or estimates about the value of a corresponding population parameter; for example, the use of a sample mean *M* to test hypotheses about the value of the corresponding population mean μ or to construct a confidence interval estimate for μ.

Inflated risk of Type I error. When multiple significance tests are performed, the probability that there will be at least one instance of a Type I error in the set of decisions goes up as a function of the number of tests that are performed; this is called inflated risk of Type I error, and it means that the likelihood of having at least one Type I error in the entire set of analyses is greater than the nominal $\alpha = .05$ Type I error rate associated with any of the individual tests.

Initial solution. When a set of *p* variables is factor analyzed, the initial solution consists of *p* factors. This full solution (same number of factors as variables) is called the initial solution. In later stages, the number of factors is usually reduced and the solution may be rotated to improve interpretability. The factor loadings for the initial solution are not usually reported.

Institutional review board (IRB). This committee reviews and evaluates all proposed research that involves human participants in the United States. Researchers must obtain IRB approval before collecting data from human participants. The corresponding committee that reviews and evaluates research that involves nonhuman animal subjects is the Institutional Animal Care and Use Committee (IACUC).

Interaction. An interaction between predictors can be incorporated into a regression equation by using the product of the two variables that interact as a predictor. When one or both of the variables are dichotomous, the interpretation of the nature of the interaction (in terms of different slopes predicted for different groups) is straightforward. When both variables involved in an interaction are quantitative, product terms for interactions are more difficult to interpret. Note that another widely used term for interaction is "moderation" (see Chapter 12). We can say that X_1 and X_2 interact to predict Y or that X_2 moderates the predictive relationship between X_1 and Y.

Interaction effect. This is a pattern of cell means in a factorial ANOVA that is different from what would be predicted by summing the grand mean, the row effect, and the column effect. When there is a significant interaction, the lines that connect cell means in a graph are not parallel; in other words, for members of the A_1 group, changes in scores on the dependent variable across levels of the B factor are not the same as the changes in the A_2 group. Interaction effects correspond to a pattern of cell means that cannot be reproduced just by summing the main effects of the row and column factors. Interaction is equivalent to "moderation," as discussed in Chapter 12.

Intercept. *See* b_0

Internal consistency reliability. This refers to consistency or agreement across a number of measures of the same construct, usually multiple items on a self-report test. The Cronbach alpha is the most popular measure of internal consistency reliability when each measure is quantitative. When scores are dichotomous, the internal consistency reliability coefficient is sometimes called the Kuder-Richardson 20, or KR-20, but this is just a special case of Cronbach alpha.

Internal homogeneity. *See* Internal consistency reliability

Internal validity. The degree to which results from a study can be used as evidence of a causal connection between variables is called internal validity. Typically, well-controlled experiments can provide stronger support for causal inference than nonexperimental studies.

Interrater reliability. This is an assessment of consistency or agreement for two or more raters, coders, or observers. If the ratings involve categorical variables, percentage agreement and Cohen's kappa (κ) may be used to quantify agreement; if ratings involve dichotomous (yes/no) judgments or quantitative ratings, then the Cronbach alpha or KR-20 may be used to assess reliability.

Interval level of measurement. Scores have interval level of measurement properties when differences between any pair of score values correspond to the same magnitude of difference on the underlying variable or property that the scores measure. A classic example of interval level measurement is temperature measured using either the Fahrenheit or Centigrade scale. The 10-point difference between 20 and 30 degrees is considered equivalent to the 10-point difference between 50 and 60 degrees. By contrast, consider a common type of score that probably does *not* meet the assumption that equal differences between scores correspond to equal intervals in the underlying construct that the scores are supposed to measure. When degree of agreement with an attitude statement is rated on a 5-point scale (1 = strongly disagree, 2 = disagree, 3 = neutral, 4 = agree, and 5 = strongly agree), the difference between scores of 2 and 3 probably does not correspond to exactly the same amount of change in agreement as the difference between scores of 4 and 5. Thus, 5-point rating scales probably do not satisfy the requirement that equal differences between scores correspond to precisely equal differences in the underlying construct that is measured, such as degree of agreement. Interval level scores are not required to have a true zero point (e.g., 0 degrees in temperature on the Fahrenheit or

Centigrade scale does not correspond to a total absence of heat, but these temperature scales are generally considered good examples of interval level of measurement).

IRB. *See* Institutional review board

Iteration. In this process, model parameters are estimated multiple times until some goodness-of-fit criterion is reached. In principal axis factoring, an iterative process occurs; communality for each variable is initially estimated by the calculation of an R^2 (to predict each X_i from all the other X variables). Then, factor loadings are obtained, and the communality for each X_i is reestimated by squaring and summing the loadings of X_i across all factors. The loadings are then reestimated using these new communality estimates as the diagonal elements of **R**, the correlation matrix that has to be reproduced by the factor loadings. This iterative process ends when the estimates of communality do not change (by more than .001) at subsequent steps; that is, the estimates converge. By default, SPSS allows 25 iterations.

Kendall's tau-b (τ-b). This is a nonparametric correlation that is suitable for use with ordinal or rank scores; it takes tied scores into account. As with the Pearson *r*, the sign of tau indicates the direction of the association between *X* and *Y* (when tau is positive, high ranks on *X* tend to be associated with high ranks on *Y*; when tau is negative, high ranks on *X* tend to be associated with low ranks on *Y*). The absolute magnitude of tau provides information about the strength of association (when τ is close to 0, the association is weak; when tau is close to +1 or −1, the association is strong). However, in other respects, tau cannot be interpreted in the same manner as the Pearson *r*; for example, tau squared cannot be interpreted as a proportion of explained variance. Possible values range from −1 to +1, but a value of −1 or +1 can be obtained only for tables where the number of rows is equal to the number of columns.

Kendall's tau-c (τ-c). This is a nonparametric measure of association for ordinal variables that, unlike tau-b, ignores ties. The sign of the coefficient indicates the direction of the relationship, and its absolute value indicates the strength, with larger absolute values indicating stronger relationships. Possible values range from −1 to +1, but a value of −1 or +1 can be obtained only from square tables.

KR-20. *See* Kuder-Richardson 20

Kuder-Richardson 20 (KR-20). This is the name given to the Cronbach alpha when all items are dichotomous. *See also* Internal consistency reliability

Kurtosis. The degree to which a distribution deviates from the "peakedness" of an ideal normal distribution is called the kurtosis; an empirical distribution that has a flatter peak than a normal distribution is described as platykurtic, while a distribution that has a narrower, taller peak is described as leptokurtic.

Lambda (λ). This is a measure of association that reflects the proportional reduction in error when values of the independent variable are used to predict values of the dependent variable. A value of 1 means that the independent variable perfectly predicts the dependent variable; a value of 0 means that the independent variable is of no help in predicting the dependent variable.

Latent variable. An "imaginary" variable that we use to understand why our observed *X*s are correlated. In factor analysis, we assume that our measured *X* scores are correlated because they measure the same underlying construct or latent variable. Each factor potentially corresponds to a different latent variable. However, factor solutions do not always have meaningful interpretations in terms of latent variables.

Level of confidence. *See* Confidence level

Level of significance. *See* Significance level

Levels of a factor. Each group in an ANOVA corresponds to a level of the factor. Depending on the nature of the study, levels of a factor may correspond to different amounts of treatment (for instance, if a researcher manipulates the dosage of caffeine, the levels of the caffeine factor could be 0, 100, 200, and 300 mg of caffeine). In other cases, levels of a factor may correspond to qualitatively different types of treatment (for instance, a study might compare Rogerian, cognitive/behavioral, and Freudian therapy as the three levels of a factor called "Type of Psychotherapy"). In some studies where naturally occurring groups are compared, the levels of a factor correspond to naturally occurring group memberships (e.g., gender, political party).

Levels of measurement. These are the formal levels of measurement (i.e., nominal, ordinal, interval, and ratio) described by Stevens (1946, 1951). See Note 3 at the end of Chapter 1 for details.

Levene test. This is a test of the assumption of equal variances that is reported by SPSS ANOVA programs; it is in the form of an *F* ratio. If the Levene test value is statistically significant, there is evidence that the equality of variance assumption may be violated.

Leverage. This is an index that indicates whether a case is disproportionately influential; when a case with a large leverage value is dropped, one or more regression slopes change substantially.

L_i. *See* Logit

Likert scale. Rensis Likert, a sociologist, devised this rating scale format where respondents report the degree of agreement with a statement about an attitude or belief using a multiple-point rating scale (usually 5 points, with labels that range from 1 = *Strongly disagree* to 5 = *Strongly agree*).

Linear transformation. Any transformation that involves only addition, subtraction, multiplication, and division by constants is called linear. The most frequently used linear transformation is the transformation of a raw *X* score to a standard *z* score; $z = (X - M)/SD$. Note that a linear transformation does not change the shape of the distribution of *X* or the correlation of *X* with other variables.

Listwise deletion. This is a method of handling missing data in SPSS (and many other programs); if a participant has missing data for any of the variables included in an analysis, that participant's data are excluded from all the computations for that analysis. *See also* Pairwise deletion

LL. *See* Log likelihood

ln. *See* Natural log

Loading. The correlation between a measured X_i variable and each factor or latent variable is called a loading or factor loading. The loading of variable *i* on factor *j* is denoted by a_{ij}.

In an orthogonal factor analysis, because the factors are uncorrelated with each other, these loadings may be used to predict z scores on each X variable from factor scores.

Logarithmic function. A logarithmic function is specified by using a base; the base of the so-called natural logarithm is the mathematical constant e; the base of the base 10 logarithm is 10. The logarithm of a particular numerical value a corresponds to the power to which the base must be raised to obtain a. For example, if $a = 100$, the base 10 log of $a = 2$, because $10^2 = 100$. Prior to the development of high-speed computers, logarithms were often used to simplify computations. Taking the logarithm of a distribution of numerical values $(X_1, X_2, X_3, \ldots, X_n)$ tends to have the following useful outcomes: If the raw X scores are strongly positively skewed, that is, there are a few extreme outliers on the upper end of the distribution, the log of X may tend to be more nearly normally distributed. Thus, log transformations are often used to minimize the impact of outliers and/or to make the distributions of scores more nearly normal in shape (see Chapter 4). There are also situations where the raw X and Y scores are nonlinearly related but the log of X has a nearly perfectly linear relation to the log of Y. Thus, there are some situations where applying a log transformation to X or Y, or both, can make the relation more nearly linear.

Logistic regression. A regression analysis for which the outcome variable is categorical; the goal of the analysis is prediction of group membership. Because categorical variables do not conform to ordinary linear regression assumptions, different computational procedures are required.

Logit (log odds ratio, L_i). A logit is a natural logarithm of an odds (in this book, a logit is denoted by L_i; in some texts, the logit is denoted by π). The dependent variable in a binary logistic regression takes the form of a logit.

Log likelihood (LL). In binary logistic regression, the LL is an index of goodness of fit. For each case, the log of the predicted probability of group membership (such as the log of $p = .89$) is multiplied by the actual group membership code (e.g., 0 or 1). The sum of these products across all cases is a negative value (because probabilities must be less than 1 and the natural logs of values less than 1 are negative numbers). The larger the absolute value of the LL, the worse the agreement between the probabilities of group membership generated by the logistic regression model and the actual group memberships. When coefficients for the logistic regression are calculated using maximum likelihood estimation, the aim is to find the values of coefficients that yield an LL value that is small in absolute value (and that, therefore, correspond to estimated probabilities that agree relatively well with actual group membership). SPSS reports −2LL for each model and for the change in fit between models.

Log odds ratio (L_i). *See* Logit

Log transformation. The most common types of log transformation are base 10 and natural logarithms. If $X = 10^p$, the log 10 transformation of X is p. For instance, the base 10 log of 10^3 is 3. If $X = e^c$, then the natural log of X is c, the power to which the mathematical constant e must be raised to get the raw score value of X.

Lord's paradox. For some batches of pretest and posttest data, the nature of the conclusions about group differences are entirely different for an ANOVA on gain scores compared with an ANCOVA (Lord, 1967; Wainer, 1991). This outcome is uncommon, but the possibility that different analyses will lead to different conclusions should be carefully considered in research that involves pretest and posttest data. See the appendix to Chapter 15 for further discussion.

M. *See* Mean

Mahalanobis *d*. This is a value that indicates the degree to which a score is a multivariate outlier.

Main effect. Main effects in factorial ANOVA correspond to differences among row means or among column means.

MANCOVA. This is a version of MANOVA that adds one (or more) covariates to the model. It is a multivariate generalization of ANCOVA.

MANOVA. *See* Multivariate analysis of variance

Marginal dependent index of association. Many indexes of association (such as the phi coefficient) can take on the maximum value of +1 only when the marginal distributions for the row and column variables are the same—for example, when there is a 60%/40% split on *X* and a 60%/40% split on *Y*. (See Section 8.4.2 for a more detailed discussion.)

Marginal frequencies. In a contingency table, these are the total number of cases in each row or each column, obtained by summing cell frequencies within each row or column.

Matched samples. Samples that are set up by matching participants on their scores on some important participant characteristic; for example, to create two samples that are matched on IQ, the researcher would obtain IQ scores for all participants, rank order participants by scores, divide this list into pairs (the top two scores, the next two scores, and so forth), and then randomly assign one participant from each pair to Sample 1 and the other to Sample 2. The resulting samples would be matched on IQ in that each member of Sample 1 would correspond to a member of Sample 2 with a nearly equal score. Data from this type of design should be analyzed using a paired samples *t* test (for two samples) or a one-way repeated measures ANOVA (for more than two samples) because the scores on the outcome variable are likely to be correlated across samples.

McNemar test. A test that is applied to 2 × 2 tables that include repeated measures—for example, a dichotomous response outcome assessed at two points in time (such as before and after an intervention).

Mean (*M*). A measure of central tendency that is obtained by summing the scores in a sample and dividing by the number of scores.

Mean square (*MS*). A mean square is calculated by dividing a sum of squares by its degrees of freedom. *MS* is similar to a variance.

Measurement theory. The set of mathematical assumptions that underlie the development of reliability indexes is called measurement theory. In Chapter 19, only one aspect of basic measurement theory is discussed. The equation $X_i = T + e_i$ represents the assumption that each observed *X* score consists of a *T* component that is stable across occasions of measurement and an e_i component that represents random errors that are unique to each occasion of measurement.

Median. A measure of central tendency that is obtained by ranking the scores in a sample from lowest to highest and identifying the score that has 50% of the scores below it and 50% of the scores above it.

Mediated relationship (or mediation). X_2 is said to mediate the relationship between X_1 and *Y* only if there is a two-step causal sequence. For example, if X_1 first causes X_2 and then

X_2 causes Y, we would say that the relationship between X_1 and Y is mediated by X_2. A relationship can be either partially mediated or fully mediated by an X_2 variable. Many conditions must be satisfied before we can interpret the outcome of a correlational study as evidence that is consistent with a mediated model. If we hypothesize that the relationship between X_1 and Y is completely mediated by X_2, for example, we need to meet the following conditions:

1. It must make theoretical sense that X_1 could causally influence X_2 and that X_2 could causally influence Y.
2. There should be appropriate temporal precedence; that is, X_1 should occur or be measured prior to X_2, and X_2 should occur or be measured prior to Y.
3. The correlations for r_{1Y}, r_{12}, and r_{Y2} should be statistically significant and large enough to be of some practical or theoretical importance (and the signs should be in the theoretically expected direction).
4. When you control for X_2, the partial correlation $r_{Y1.2}$ should be close to 0 and/or the slope coefficient for X_1 in a regression in which Y is predicted from X_1 while controlling for X_2 should be close to 0.

Data from a nonexperimental study cannot prove hypotheses about mediated causation. Better evidence in support of causal hypotheses can be obtained from experimental studies in which both X_1 and X_2 are manipulated, or X_1 is manipulated while controlling for X_2.

Mediating variable. *See* Mediator

Mediation. *See* Mediated relationship

Mediator (or mediating variable). This is an intervening variable in a causal sequence. If X_1 causes X_2 and then X_2 causes Y, X_2 is the mediator or mediating variable. The following is an example of a theory that involves mediation: X_1 (anxiety) may cause X_2 (release of stress hormones), which in turn leads to Y (reduced immune system competence). Anxiety may reduce immune system strength through its effects on stress hormones such as cortisol, which in turn have an influence on the functioning of the immune system.

Minimum expected cell frequency. The chi-square test of association is considered inappropriate in situations where the total number of cases in a contingency table is less than 20 and/or when one or more cells in a 2 × 2 table have expected cell frequencies less than 5. For larger tables (e.g., of dimensions 2 × 3), recommendations about the minimum expected cell frequencies differ.

Missing value. A number (or blank) in a cell in an SPSS data sheet that represents a missing response is called a system missing value; such values are excluded from computations.

Missing values imputation. This refers to a systematic method of replacing missing scores with reasonable estimates. Estimated scores for missing values are obtained by making use of relationships among variables and a person's scores on nonmissing variables. This can be quite simple (e.g., replace all missing values of blood pressure with the mean blood pressure across all subjects) or complex (e.g., calculate a different estimated blood pressure for each subject based on that subject's scores on other variables that are predictively related to blood pressure).

Mixed model factorial ANOVA. A factorial design that includes at least one between-S factor and at least one within-S or repeated measures factor.

Mode. A measure of central tendency that is obtained by finding the score in a sample that has the highest frequency of occurrence.

Model. The term *model* can have many different meanings in different contexts. In this textbook, the term *model* is used to refer to two different things. Diagrams similar to those in Figure 11.13 are examples of causal models; in a causal model, directional and bidirectional arrows are used to represent hypothesized causal and noncausal associations between variables. Regression models are predictive equations in which one or more X variables are used to predict scores on a quantitative Y variable; for example, $Y' = b_0 + b_1X_1 + b_2X_2$ is a regression model to predict Y from scores on X_1 and X_2. The hypotheses about the nature of relationships among variables that are represented by theoretical "causal models" are sometimes used to decide which variables to include as predictors and which to include as dependent variables in one or more multiple regression models. The empirical results for a regression model (such as the slope coefficients) can be evaluated as evidence that is consistent or inconsistent with various causal models; however, when regression is applied to nonexperimental data, regression results that are consistent with a causal model do not prove that the causal model is correct.

Moderated relationship (or moderation). A relationship between X_1 and Y is said to be moderated by X_2 if the slope to predict Y from X_1 differs significantly across groups that are formed by looking at scores on the X_2 control variable. Moderation is the same as interaction. If the slope to predict Y from X_1 differs significantly across levels of X_2, we can say that X_2 moderates the X_1, Y relationship or that X_1 and X_2 show an interaction as predictors of Y. In Chapter 10, a simple method is provided to look for evidence of possible moderation; the X_1, Y relationship is examined separately for each group of people with different scores on the X_2 control variable. Chapter 12 provides methods to assess whether any difference in the slope across groups is large enough to be judged statistically significant. If moderation is present, then it is quite misleading to ignore the X_2 variable when describing how X_1 and Y are related.

Moderation. *See* Moderated relationship

Monotonic relationship. Scores on X and Y are monotonically related if an increase in X is always associated with an increase in Y.

Monte Carlo simulation. A method of evaluating how violations of assumptions about the distributions of scores in populations affect the risk of Type I error for significance tests such as the independent samples t test.

MS. *See* Mean square

Multicollinearity. Multicollinearity refers to the degree of intercorrelation among predictor variables; perfect multicollinearity exists when one variable is completely predictable from one or more other variables.

Multiple measures in repeated measures designs. This is a repeated measures design in which more than one outcome variable is measured at each point in time or under each treatment condition; for example, in the stress study used as an example in Chapter 20, measures of systolic and diastolic blood pressure could be obtained, in addition to the measure of heart rate.

Multiple point rating scale. Rating scale items may have any number of response alternatives, and many different types of labels can be used for response alternatives (such as frequencies of behaviors or intensities of feelings). A rating scale is called a Likert scale if it involves a 5-point degree of agreement rating.

Multiple *R*. This is a correlation between actual Y scores and Y' scores that are predicted from a regression equation that uses one or several X predictor variables.

Multiple regression. In a multiple regression, more than one predictor variable is included in an equation to predict scores on a quantitative Y variable. For example, we could predict scores on life expectancy for individuals (Y) from variables such as cholesterol level (X_1), gender (X_2), income (X_3), physical fitness (X_4), and so forth: $Y' = b_0 + b_1X_1 + b_2X_2 + b_3X_3 + b_4X_4$. Chapter 11 describes regression models that include two predictor variables. Chapter 14 discusses how multiple regression works when there are more than two predictor variables. In a standard multiple regression model, the association of each individual predictor variable (such as X_1) with Y is assessed while statistically controlling for or partialling out all the other predictor variables (e.g., X_2, X_3, and X_4).

Multivariate analysis of variance (MANOVA). This is a multivariate generalization of ANOVA. Like ANOVA, it involves comparisons of means across groups; however, univariate ANOVA examines means on just one outcome variable, Y, while MANOVA compares a vector or list of means on p outcome variables across groups ($Y_1, Y_2, \ldots, Y_p$).

Multivariate outlier. A score that represents an unusual combination of values on $Y, X_1, X_2, \ldots, X_p$ (although it need not be extreme on any one variable). If one could examine a scatter plot of the scores in a $(p + 1)$-dimensional space, most points would form a cluster in this space, and a multivariate outlier would be an isolated point outside this cluster. In practice, it is easier to identify multivariate outliers through identification of outliers in the predicted scores and/or residuals from a regression or by examination of individual case statistics, such as Mahalanobis *d*.

***N*.** The total number of observations in a sample is denoted by N.

Nagelkerke R^2. In binary logistic regression, this is one of the pseudo R^2 values provided by SPSS as an overall index of the strength of prediction for the entire model.

Natural log (or natural logarithm; often abbreviated as ln). This is the base e logarithm; that is, the natural log of the value a is the power to which the mathematical constant e must be raised to obtain the number a.

Natural logarithm. *See* Natural log

Nay-saying bias. This refers to the tendency to respond "no" or to indicate high levels of disagreement with many questions on a questionnaire, regardless of question content.

−2LL (or −2 × log likelihood; also called the deviance chi-square). −2LL is analogous to the sum of squared residuals in a multiple linear regression; the larger the absolute value of −2LL, the worse the agreement between predicted probabilities and actual group membership. LL values can be computed for different models. Multiplying an LL value by −2 converts it to a chi-square distributed variable. Critical values of the chi-square distribution can be used to assess the statistical significance of one model or the significance of change in goodness of fit when variables are added to a model. *See also* Log likelihood

Nested models. Model A (e.g., $Y_i = a + b_1 \times X_1$) is nested in Model B (e.g., $Y_i = a + b_1 \times X_1 + b_2 \times X_2$) if all the variables in Model A are also included in Model B, but Model B has one or more additional variables that are not included in Model A. Researchers often want to test whether the fit of Model B is significantly better than the fit of Model A. In the simplest application of binary logistic regression, two nested models are compared. In some situations, Model A is a null model with no predictor variables, while Model B is a full model with all predictor variables included. For each model, an index of fit is obtained (−2LL). The difference between $-2LL_B$ and $-2LL_A$ has a χ^2 distribution with k degrees of freedom, where k is the number of predictors in Model B. If this χ^2 is large enough to be judged statistically significant, the data analyst can conclude that the overall model, including the entire set of independent variables, predicted group membership significantly better than a null model.

NHST. *See* Null hypothesis significance testing

Noise suppression. When one or more additional predictor variables are added to an analysis, the value of SS_{within} or SS_{error} or $SS_{residual}$ may go down. Depending on how large the reduction in SS_{error} is, relative to reductions in the degrees of freedom for error, the overall effect may be such that there is an increase in F ratios for the tests of some effects. When this happens, we can say that the added variable or variables suppressed some of the error variance or noise. *See also* Error variance suppression

Nominal level of measurement. In a nominal level of measurement, numbers serve only as names or labels for group membership and do not convey any information about rank order or quantity; this is also called a categorical variable.

Noncausal path. In a causal model, a bidirectional arrow is used to represent a situation where two variables are noncausally associated with each other; they are correlated, confounded, or redundant. In this model, the double-headed arrow corresponds to the hypothesis that scores on X_1 and Y are noncausally associated.

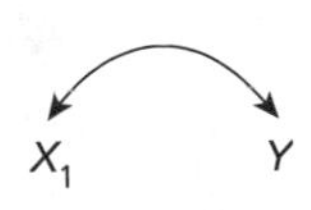

Nondirectional test. This is a significance test that uses an alternative hypothesis that does not specify a directional difference. For H_0: $\mu = 100$, the nondirectional alternative hypothesis is H_1: $\mu \neq 100$. For a nondirectional test, the reject regions include both the upper and lower tails of the z or t distribution.

Nonequivalent comparison group. *See* Nonequivalent control group

Nonequivalent control group. When individual participants cannot be randomly assigned to treatment and/or control groups, we often find that these groups are nonequivalent; that is, they are unequal on their scores on many participant characteristics prior to the administration of treatment. Even when a random assignment of participants to groups occurs, sometimes, nonequivalence among groups occurs due to "unlucky randomization." If it is not possible to use experimental controls (such as matching) to ensure equivalence, ANCOVA is often used to try to correct for or remove this type of nonequivalence. However, the statistical control for one or more covariates in ANCOVA is not guaranteed to correct for all sources of nonequivalence; also, if assumptions of ANCOVA are violated, the adjustments it makes for covariates may be incorrect.

Nonexperimental research design. In nonexperimental research, the investigator does not manipulate an independent variable and does not have experimental control over other variables that might influence the outcome of the study.

Nonlinear transformation. This transformation uses nonlinear operations (such as the replacement of raw *X* scores with X^2 or X^3, square root of *X*, base 10 log of *X*). In principle, some transformations (such as log of *X*) may make the distribution shape somewhat closer to normal or the relation between a pair of variables closer to linear.

Nonorthogonal factorial design. This is a factorial design in which the cell numbers are unequal, such that there are different proportions of members in each *B* group across levels of *A*. In a nonorthogonal design, the effects of factors are confounded to some degree. This confounding should be handled by making adjustments to the sums of squares; these adjustments remove or partial out effects of other factors when making an assessment of the unique contribution of each factor, that is, when computing *SS* terms for each factor.

Nonparametric statistics. Statistics that do not require the assumptions that are required for parametric statistics are nonparametric; typically, parametric statistics do not require interval-ratio level of measurement, normal distributions of scores, or equal variances of scores in different groups. Because scores are converted to ranks for many nonparametric procedures, outliers have little impact on results. The most familiar examples of nonparametric statistics are the Spearman *r*, the chi-square test, the Wilcoxon test, the sign test, and the Friedman one-way ANOVA by ranks.

Nonreactive measures. These are measurement methods that do not change the behavior that is being observed—for example, the use of physical trace data or observations of people without their knowledge.

Normal distribution (or normally distributed variable). An ideal or theoretical distribution defined by a specific equation. This distribution is a bell-shaped symmetrical curve. There is a fixed relationship between distance from the mean (in standard deviation units) and area under the curve, as shown in Figure 1.4. Some observed measurements and sample statistics have an approximately normal distribution.

Normal distribution assumption. Most of the parametric analyses covered in introductory statistics books, and in this book, assume that scores on quantitative variables are approximately normally distributed.

Normally distributed variable. *See* Normal distribution

Nuisance or error variable. This is a variable that is not of primary interest to the researcher. For example, in a study of the effects of caffeine on anxiety, other variables (such as the timing of examinations, smoking, and use of other drugs) may also influence anxiety. A nuisance variable may be confounded with the independent variable and may influence scores on the outcome variable. Nuisance variables can be handled by experimental control (e.g., a nuisance variable can be held constant; if a researcher wants to get rid of any possible influence of cigarette smoking on anxiety, the researcher may decide to include only nonsmokers as participants). Another way of handling nuisance variables is to measure them and include them in the statistical analysis.

Null hypothesis (H_0). This is an algebraic statement that some population parameter has a specific value. For example, H_0 for the one-sample *z* test is usually of the form H_0: $\mu = c$ where *c* is a specific numerical value. In other words, H_0 is the assumption that the population mean on a variable corresponds to a specific numerical value *c*.

Null hypothesis significance testing (NHST). NHST involves the selection of an alpha level to limit the risk of Type I error, statements of null and alternative hypotheses, and the evaluation of obtained research results (such as a t ratio) to decide whether or not to reject the null hypothesis. In theory, if all the assumptions for NHST are satisfied, the risk of committing a Type I error is (theoretically) equal to alpha. In practice, some assumptions of NHST are often violated, so obtained p values may not accurately estimate the true risk of Type I error in many studies.

Null model. In binary logistic regression, the null model represents prediction of group membership that does not use information about any predictor variables. As one or more predictor variables are added to the logistic regression, significance tests are used to evaluate whether these predictors significantly improve prediction of group membership compared with the null model.

Null result. A study that yields a test statistic (such as t) that falls within the "do not reject H_0" region or a study that yields an exact p value that is greater than the preselected alpha value. A null result should not be interpreted as proof that the null hypothesis is correct: the researcher should not accept H_0 but rather "fail to reject H_0."

Numerical variable. These are variables that have quantitative or numerical scores.

Oblique rotation. A rotation in which we allow the factors to become correlated to some degree is called an oblique rotation. This is useful in situations where the constructs that we think we are measuring are correlated to some degree, and oblique rotation may enhance interpretability.

Odds. Odds involve comparison of the number of cases for two possible outcomes. For example, in a study of cancer patients, if 40 patients die and 10 patients survive, the odds of death in that sample are found by dividing the number of deaths by the number of survivals, 40/10, to obtain an odds of 4 to 1 (an individual patient selected at random is four times as likely to die, as to survive). As another example, in the data in Table 21.1 there were two possible outcomes: an individual could be either alive or dead at the end of the first year after the heart attack. For the dog-owning group, the odds of survival were found by dividing the number of surviving owners of dogs (50) by the number of dead owners of dogs (3) = 50/3 = 16.67. In this sample, a randomly selected dog owner was about 17 times as likely to survive as to die. In contrast, for the group of people who did not own dogs, where 28 survived and 11 died, the odds of survival were 28/11 or 2.545. Logistic regression involves the estimation of the odds for outcomes on a categorical variable.

Odds ratio. An odds ratio is a ratio of the odds for members of two different groups, often used to summarize information about outcomes when the outcome variable is a true dichotomy. The odds themselves are also a ratio. In the study of survival status among owners and nonowners of dogs, we could set up an odds ratio to describe how much more likely survival is for a dog owner than for a nonowner by taking the ratio of the odds of survival for dog owners (16.67) to the odds of survival for a nonowner (2.545). This ratio, 16.67/2.54 = 6.56, tells us that, in this sample, the odds of survival were more than six times as high for dog owners than for nonowners. Chapter 21 in this book discusses the use of binary logistic regression to estimate odds ratios for the dog owner/survival data.

OLS. *See* Ordinary least squares

Omnibus test. This is a test of the significance of an overall model (such as a multiple regression) that includes all predictor variables. For example, the F ratio that tests the null hypothesis that multiple R equals 0 is the omnibus test for a multiple regression. An F test that tests whether all population means that correspond to the groups in a study are equal to each other is the omnibus test in a one-way ANOVA.

One-tailed test. This is a significance test that uses an alternative hypothesis that specifies an alternative value of μ, which is either less than or greater than the value of μ stated in the null hypothesis. For example, if H_0: $\mu = 100$, then the two possible directional alternative hypotheses are H_1: $\mu < 100$ and H_1: $\mu > 100$. The reject region consists of just one tail of the normal or *t* distribution. The lower tail is used as the reject region for H_1: $\mu < 100$, and the upper tail is used as the reject region for H_1: $\mu > 100$.One-tailed tests are used when there is a directional alternative hypothesis.

Optimal weighted linear composite. In analyses such as regression and discriminant analysis, raw scores or *z* scores on variables are summed using a set of optimal weights (regression coefficients, in the case of multiple regression, and discriminant function coefficients, in the case of discriminant analysis). In regression, the regression coefficients are calculated so that the multiple correlation *R* between the weighted linear composite of scores on *X* predictor variables has the maximum possible value and the sum of squared prediction errors has the smallest possible value. In a multiple regression of the form $Y' = b_0 + b_1X_1 + b_2X_2 + b_3X_3$, the weights b_1, b_2, and b_3 are calculated so that the multiple correlation *R* between actual *Y* scores and predicted *Y′* scores has the maximum possible value. In discriminant function analysis, the coefficients for the computation of discriminant function scores from raw scores on the *X* predictor variables are estimated such that the resulting discriminant function scores have the maximum possible sum of squares between groups and the minimum possible sum of squares within groups.

Order effects. In repeated measures studies, if each participant receives two or more different treatments, it is possible that participant responses depend on the order in which treatments are administered as well as the type of treatment. Factors such as practice, boredom, fatigue, and sensitization to what is being measured may lead to differences in the outcome measures taken at Times 1, 2, and so forth. If all participants experience the treatments in the same order, there is a confound between order effect and treatment effect. To prevent this confound, in most repeated measures studies, researchers vary the order in which treatments are administered.

Ordinal level of measurement. In Stevens's (1946, 1951) description of levels of measurement, ordinal measures are those that contain information only about rank.

Ordinary least squares (OLS). A statistic is the best OLS estimate if it minimizes the sum of squared prediction errors; for example, *M* is the best OLS estimate of the sample mean because it minimizes the sum of squared prediction errors, $\sum (X - M)^2$, that arises if we use *M* to predict the value of any one score in the sample chosen at random.

Orthogonal. The term *orthogonal* means that the contrasts are independent or uncorrelated. If you correlate a pair of variables (O_1 and O_2) that are coded as orthogonal contrasts, the correlation between O_1 and O_2 should be 0.

Orthogonal factorial design. This is a factorial design in which the numbers of cases in the cells are equal (or proportional in such a way that the percentage of members in each *B* group is equal across levels of *A*). In an orthogonal design, the effects of factors are not confounded.

Orthogonally coded dummy variable. This refers to dummy variables that are coded so that they represent independent contrasts. Tables of orthogonal polynomials can be used as a source of coefficients that have already been worked out to be orthogonal. Orthogonality of contrasts can also be verified by either finding a 0 correlation between the dummy variables that represent the contrasts or by cross multiplying the corresponding contrast coefficients.

Orthogonal rotation. A factor rotation in which the factors are constrained to remain uncorrelated or orthogonal to each other is called an orthogonal rotation.

Outcome variable. *See* Dependent variable

Outer fence. The outer fence is identified in a box plot by the horizontal lines at the ends of the "whiskers."

Outlier. A score that is extreme or unusual relative to the sample distribution is called an outlier. There are many standards that may be used to decide which scores are outliers; for example, a researcher might judge any case with a *z* score greater than 3.3 to be an outlier. Alternatively, scores that lie outside the outer fences of a box plot might also be designated as outliers.

***p* (or *p* value).** *p* represents the theoretical probability of obtaining a research result (such as a *t* value) equal to or greater than the one obtained in the study, when H_0 is correct. It thus represents the "surprise value" (Hays, 1973) of the following result: If H_0 is correct, how surprising or unlikely is the outcome of the study? When a small *p* value is obtained (*p* less than a preselected α value), then the researcher may decide to reject the null hypothesis. Another definition for *p* would be the answer to the following question: How likely is it that, when H_0 is correct, a study would yield a result (such as a *t* value) equal to or greater than the observed *t* value just due to sampling error? A *p* value provides an accurate estimate of the risk of committing a Type I error only when all the assumptions required for NHST are satisfied.

Paired sample. A pair of scores that come from the same person tested at two points in time or the same person tested under two different treatment situations is a paired sample; the scores that are obtained using matched samples or naturally occurring pairs (e.g., husbands and wives) are also paired samples.

Paired samples *t* test (also called the correlated samples *t* test or direct difference *t* test). This is a form of the *t* test that is appropriate when scores come from a repeated measures study, pretest/posttest design, matched samples, or other designs where scores are paired in some manner (e.g., if a researcher had pairs of husbands and wives and wanted to see if they differed in scores on marital satisfaction, he or she would need to use the paired samples *t* test; see Chapter 20 in this textbook).

Pairwise deletion. A method of handling missing data in SPSS, such that the program uses all the available information for each computation; for example, when computing many correlations as a preliminary step in regression, pairwise deletion would calculate each correlation using all the participants who had nonmissing values on that pair of variables. *See also* Listwise deletion

Parallel forms reliability. When a test developer creates two versions of a test (which contain different questions but are constructed to include items that are matched in content), parallel forms reliability is assessed by giving the same group of people both Form A and Form B and correlating scores on these two forms.

Parametric statistics. These are statistics that require assumptions about levels of measurement and distribution shapes; the most familiar parametric statistics include *t*, *F*, and the Pearson *r*. Ideally, when parametric statistics are employed, the scores on quantitative variables should have interval-ratio level of measurement, and they should be normally distributed.

"Part" correlation. This is another name for the semipartial correlation. The value that SPSS denotes as part correlation on a regression printout corresponds to the semipartial correlation.

Partial correlation. The first-order partial correlation between X_1 and Y controlling for X_2 represents the strength of the association between X_1 and Y when the parts of the X_1 and Y scores that are predictable from, or related to, X_2 are completely removed.

Partial eta squared (η^2). *See* Adjusted or partial eta squared (η^2)

Partially confounded or partly redundant predictor variables. Predictor variables that are correlated with or noncausally associated with each other are "confounded" or redundant. Consider an example in which data are obtained from 100 countries. X_1 is per-person income, X_2 is access to medical care, and Y is life expectancy. X_1 and X_2 are partially confounded predictors of Y. X_1 and X_2 are correlated with each other, but both X_1 and X_2 are significantly predictive of Y even when the other predictor is statistically controlled (e.g., by doing a partial correlation). In this situation, we might reason that wealthy countries (those with high per capita income) also tend to have better availability of medical care. However, each of these variables (income and medical care) may be significantly associated with Y even when the confound with the other predictor is taken into account. As noted by Cohen et al. (2003), models that involve partially confounded predictors, similar to this example, are very common.

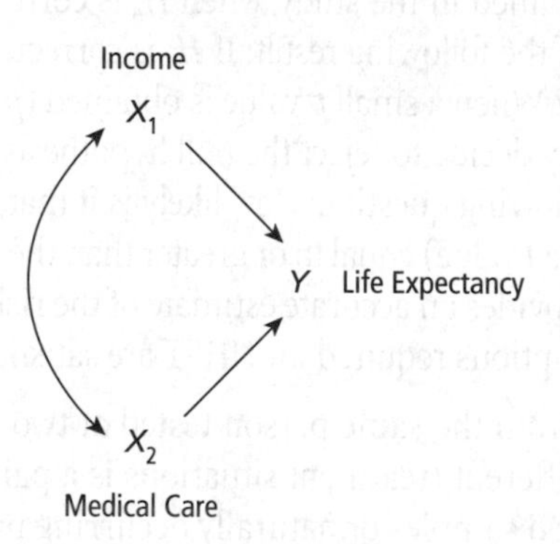

Partition of variance. The variability of scores (as indexed by their sum of squares) can be partitioned or separated into two parts: the variance explained by between-group differences (or treatment) and the variance not predictable from group membership (due to extraneous variables). The ratio of between-group (or explained) variation to total variation, eta squared, is called the proportion of explained variance. Researchers usually hope to explain or predict a reasonably large proportion of the variance. The partition of variance was demonstrated in Chapter 6 in terms of one-way ANOVA, but partition of variance is also provided by bivariate regression (Chapter 9) and multivariate analyses.

PC_α. *See* Per-comparison alpha

Pearson chi-square. When not otherwise specified, the version of chi-square that is reported is generally Pearson's chi-square (without the Yates correction for continuity).

Pearson product-moment correlation. *See* Pearson *r*

Pearson *r* (also called Pearson product-moment correlation). This is a parametric correlation statistic that provides information about the strength of relationship between two quantitative variables; it should be used only when the variables are normally distributed, linearly related, and at least approximately interval-ratio level of measurement. When not otherwise specified, the term *correlation* usually refers to the Pearson product-moment correlation that was discussed in Chapter 7. Chapter 8 discusses alternative forms of correlation coefficients that are appropriate for various types of data; some of these alternative forms (such as the phi coefficient) are equivalent to the Pearson *r*, while others (such as Kendall's τ) are not equivalent.

Per-comparison alpha (PC_α). When the Bonferroni procedure is used, this is the more conservative alpha level used to test individual correlations so that the overall error rate for the entire set of correlations can be limited. $PC_\alpha = EW_\alpha/k$, where EW_α is the experiment-wise alpha and *k* is the number of significance tests.

Perfect association. *X* and *Y* are "perfectly associated" if the following statements are true: When *X* occurs, *Y* always occurs; when *X* does not occur, *Y* never occurs. A hypothetical example of a perfect association between *X* (smoking) and *Y* (lung cancer) is as follows. If all smokers develop lung cancer and if all people who develop lung cancer are smokers, then smoking and lung cancer would be perfectly associated. (In fact, these variables are not perfectly associated. Some smokers do not develop lung cancer; some people who develop lung cancer never smoked.)

Phi and Cramer's *V*. Phi is a chi-square-based measure of association that involves dividing the chi-square statistic by the sample size and taking the square root of the result. Cramer's *V* is a measure of association based on chi-square.

Phi coefficient (ϕ). This is a correlation that indexes the strength of association between scores on two true dichotomous variables; it is equivalent to a Pearson *r*.

Pillai's trace. This is one of the multivariate test statistics used to test the null hypothesis shown in Equation 17.3: H_0: $\mu_1 = \mu_2 = \ldots = \mu_k$. It is the sum of the squared canonical correlations (each canonical correlation is a correlation between scores on one discriminant function and group membership, as described in Chapter 16) across all the discriminant functions. A larger value of Pillai's trace implies that the vectors of means show larger differences across groups. Among the multivariate test statistics, Pillai's trace is thought to be generally more robust to violations of assumptions (although it is not the most robust test under all conditions).

Planned contrast. *See* Contrast

Point biserial correlation (r_{pb}). This is a correlation that is used to show how a true dichotomous variable is related to a quantitative variable; it is equivalent to a Pearson *r*.

Pooled variances *t* test. When there is no evidence that the homogeneity of variance assumption is violated, this is the version of the *t* test that is preferred. (This is the version of *t* that is usually taught in introductory statistics courses.)

Population. This refers to the larger group of people or cases that the researcher wants to generalize about. In most behavioral and social science research, this population is not clearly defined.

Population value of standard error of *M* (σ_M). This is, in effect, the "standard deviation" of the sampling distribution of *M*. When different values of *M* are obtained for all possible samples of size N from a population with a known population standard deviation of σ, $\sigma_M = \sigma/\sqrt{N}$. The central limit theorem says that the values of M are normally distributed around μ, and 95% of the sample M values should lie within $\pm 1.96\sigma_M$ units of μ.

Population variance (σ^2). This is the variance of scores in a population that provides information about the dispersion or distance of scores from the population mean; in most research situations, σ^2 is not known and s^2 is used to estimate it.

Post hoc test. *See* Protected test

Power. Power is the probability of obtaining a test statistic, such as a t ratio, that is large enough to reject the null hypothesis when the null hypothesis is actually false. Power varies as a function of the α level used for significance tests (as alpha decreases, and if other factors remain the same, power decreases) and effect sizes such as eta squared (as eta squared increases, and if other factors remain the same, then power increases); it also varies as a function of n (as n increases, if other factors remain the same, then power increases). A researcher can make an educated guess about the effect size that he or she is trying to detect and then look up statistical power for that assumed effect size as a function of the sample size; this makes it possible to judge whether the planned sample size is adequate to provide good statistical power. A common standard is to choose a minimum sample size that would give a power of 80% for the assumed effect size. It is desirable to have a fairly high level of power; if power = .80, this means that the researcher has (in theory) an 80% chance of rejecting the null hypothesis (if all the assumptions used to obtain the power estimate are correct).

Practical significance. A subjective judgment as to whether the value of the $M_1 - M_2$ observed difference between group means in a study represents a difference that is large enough to be of any practical or clinical significance; in practice, to make this judgment, a person needs to know something about the possible range of values on the outcome variable and perhaps also the sizes of differences in means that have been obtained in other studies.

Predictive validity. The ability of a test to predict a future behavior or future group membership that should occur if the test is a valid measure of what it purports to measure; for example, a high score on a test of depression should predict future behaviors or future outcomes such as seeking counseling for depression or being diagnosed with depression.

Predictor variable. *See* Independent variable

Pretest/posttest design. In this design, the same response is measured for the same participants both before and after an intervention. This may be in the context of a true experiment (in which participants are randomly assigned to different treatments) or a quasi-experiment (one that lacks comparison groups or has nonequivalent comparison groups). If the study has just one group, the data may be analyzed using the direct difference t test, repeated measures ANOVA, or ANCOVA. If the study has multiple groups, then an ANOVA of gain scores, a repeated measures ANOVA, or an ANCOVA may be performed. If the conclusions about the nature of the effects of treatments differ when different analyses are run, then the researcher needs to think carefully about possible violations of assumptions and the choice of statistical method.

Projective tests. These are tests that involve the presentation of ambiguous stimuli (such as Rorschach inkblots or Thematic Apperception Test drawings). Participants are asked to interpret or tell stories about the ambiguous stimulus. Because it is not obvious to the test taker what the test is about, the responses to the stimuli should not be influenced by social desirability or self-presentation style. For example, Thematic Apperception Test stories are scored to assess how much achievement imagery a respondent's story includes.

Protected test. A protected test is a test that reduces the risk of Type I error by using more conservative procedures. Some examples include the Bonferroni procedure and the Tukey HSD test.

Public domain tests. *See* Unpublished or public domain test

Published tests. This refers to psychological tests that have been copyrighted and published by a test publishing company (such as EDITS or PAR). It is a copyright violation to use these tests unless the user pays royalty fees to the publisher; some published tests are available only to persons who have specific professional credentials such as a PhD in clinical psychology.

***p* value.** *See p*

Quantitative variable. This is a variable that contains information about the quantity or amount of some underlying characteristic—for example, age in years or salary in dollars. This includes the levels of measurement that Stevens (1946, 1951) called interval and ratio.

Quasi-experimental design. This is a research design that involves pretest/posttest comparisons, or comparisons of treatment groups, but lacks one or more of the features of a true, well-controlled experiment. For example, in a true experiment, participants are randomly or systematically assigned to treatment groups in a way that makes the groups as nearly equivalent as possible with respect to participant characteristics; in a quasi-experiment, random assignment of participants to treatments is often not possible, and therefore, the groups are often not equivalent with respect to participant characteristics prior to treatment. In a true experiment, the intervention is under the control of the researcher; a quasi-experiment often assesses the impact of an intervention that is not under the direct control of the researchers. An example of a quasi-experiment is a study that compares mean scores on an outcome variable for two groups that have received different treatments, in a situation where the researcher did not have control over the assignment of participants to groups and/or does not have control over other variables that might influence the outcome of the study. ANCOVA is often used to analyze data from quasi-experimental designs that include pretest/posttest comparisons and/or nonequivalent control groups.

***R*.** *See* Multiple *R*

R. A matrix of correlations among a set of variables such as $X_1, X_2, \ldots, X_p$. (*Note:* **R** should be set boldface when used to represent a matrix.)

Random factor. A factor in an ANOVA is random if the levels included in the study represent an extremely small proportion of all the possible levels for that factor. For example, if a researcher randomly selected 10 photographs as stimuli in a perception task, the factor that corresponded to the 10 individual photographs would be a random factor. In a factorial ANOVA, the *F* ratio to test the significance for a factor that is crossed with a random factor is based on an error term that involves an interaction between factors.

Random sample. A random sample is a subset of cases from a population, which is selected in a manner that gives each member of the population an equal chance of being included in the sample.

Ratio level of measurement. Scores that have interval level of measurement properties and also have a true zero point are described as having ratio level of measurement properties. Height is an example of a measure that meets the requirements for ratio level of measurement. When measurements satisfy the requirements for ratio level of measurement, ratios of scores are meaningful. For example, a person who is 6 feet tall is twice as tall as a person who is 3 feet tall (and this makes sense only because a height of 0 represents a true zero point). On the other hand, a substance that has a temperature of 40 degrees Fahrenheit is not "twice as hot" as a substance that has a temperature of 20 degrees Fahrenheit, because 0 degrees Fahrenheit does not correspond to a complete absence of heat. Because most parametric statistics involve the use of arithmetic operations that include multiplication and division, as well as addition and subtraction, a conservative approach to application of statistics may limit the application of parametric statistics (such as mean and variance) to scores that have true interval and ratio level of measurement properties. In practice, however, many statisticians believe that statistics such as mean and variance yield useful results even when they are calculated for scores that fall short of meeting the assumptions required for the interval and ratio levels of measurement outlined by Stevens (1946, 1951), and parametric statistics are often applied to scores (such as 5-point ratings of degree of agreement) that do not satisfy the assumptions for interval/ratio level of measurement. This point is discussed in more detail in Chapter 1.

Raw scores. Scores reported in their original units of measurement, such as dollars, feet, or years (as opposed to standardized or *z* scores, given in unit-free terms).

r_b. *See* Biserial correlation

r_c. *See* Canonical correlation

Reactivity of measurement. A measurement is reactive if the very act of measuring a behavior or attitude changes it. For example, asking a leading or biased question may influence the response or create an attitude that was not present before the question was asked. Self-report measures, interviews, and observation of participants who know that they are being observed tend to be highly reactive.

Regression plane. When two *X* variables are used to predict scores on *Y*, a three-dimensional space is needed to plot the scores on these three variables. The equation to predict *Y* from X_1 and X_2 corresponds to a plane that intersects this three-dimensional space, called the regression plane.

Regression toward the mean. This refers to the tendency for predicted scores on an outcome variable to be closer to their mean (in *z* score units) than the scores on the predictor variable were to their mean.

Repeated measures. This is a design in which each participant is tested at every point in time, or under every treatment condition; because the same participants contribute scores for all the treatments, participant characteristics are held constant across treatments (which avoids confounds), and variance due to participant characteristics can, in theory, be removed from the error term used to assess the significance of differences of treatment group means.

Repeated measures ANOVA (or within-S design). A study in which each participant has the same response measured at least twice and, possibly, several times. A between-S ANOVA assumes that scores are uncorrelated; repeated measures are almost always correlated, and therefore, different analytic methods are required for repeated measures data (see Chapter 20).

Representative sample. A sample is representative of a population if the characteristics of the sample are similar to those of the population in all important respects; in theory, random sampling of cases from a population should result in a representative sample.

Reproduced correlation matrix. Squaring the factor loading matrix **A** (i.e., finding the matrix product **A′A**) reproduces the correlation matrix **R**. If we retain p factors for a set of p variables, we can reproduce **R** exactly. What we usually want to know, in factor analysis, is whether a reduced set of factors reproduces **R** reasonably well.

Research hypothesis. *See* Alternative hypothesis

Residual. A residual is a difference between an actual individual score and the score that you would predict for that individual from your analysis. It is denoted by e_{ij} and calculated as $e_{ij} = Y_{ij} - M_i$, where e_{ij} is the residual for Person j in Group i, Y_{ij} the score for Person j in Group i, and M_i the mean of scores for all participants in Group i. The residual theoretically reflects that part of an individual's behavior that is influenced by variables other than the group difference variable. For instance, in a comparison of HR scores for male and female groups, the residual for each participant represents other factors apart from gender that influence that individual's heart rate (drug use, smoking, fitness, cardiovascular disease, anxiety, and so forth). Researchers usually want residuals to be reasonably small because they represent the part of people's behavior that is not accounted for by the analysis or predicted by the independent variable.

Reverse-worded questions. These are questions that are worded in such a way that a strong level of *disagreement* with the statement indicates more of the trait or attitude that the test is supposed to measure. For example, a reverse-worded item on a depression scale could be "I am happy"; strong disagreement with this statement would indicate a higher level of depression. Reverse-worded questions are included to avoid inflation of scores because of yea-saying bias.

Rho (ρ). The symbol used to represent the true (usually unknown) value of the correlation between variables in the population (ρ is to r as μ is to M).

Rival explanations. Rigorously controlled experiments are set up so that, ideally, the manipulated X_1 variable is the only reasonable or likely explanation for any observed changes in the Y outcome variable; experimental controls, such as random assignment of participants to treatment groups and standardization of procedures, are among the many ways in which experimenters try to eliminate or systematize the influence of rival explanatory variables, so that their effects are not confounded with the effects of the X_1 manipulated causal variable that is of interest. In practice, experimenters are not always completely successful in controlling for rival explanatory variables, but in principle, a well-controlled experiment makes it possible to rule out many rival explanatory variables. In nonexperimental research, it is typically the case that for any predictor variable of interest (X_1) there are many other potential predictors of Y that are correlated with or confounded with X_1. The existence of numerous rival explanations that cannot be completely ruled out is the primary reason why we say "correlation does not indicate causation." This caveat can be worded more precisely: "Correlational (or nonexperimental) research does not provide a basis for making confident causal inferences, because there is no way to completely rule out all possible rival explanatory variables in nonexperimental studies." A nonexperimental researcher can identify a few rival, important explanatory variables and attempt to control for their influence by doing partial correlations; but at best, this can be done for only a few rival variables, whereas in real-world situations, there can potentially be hundreds of rival explanatory variables that would need to be ruled out before we could conclude that X_1 influences Y.

Robustness. One way that robustness is assessed is by examining the degree to which Type I error rates are affected by violations of assumptions, such as nonnormal distribution shapes. A statistic is said to be robust if it yields an actual Type I error rate close to the nominal α level, even when assumptions such as normality have been violated. We can also say that a statistic (such as the sample median) is robust if it is not very sensitive to the presence of outliers. For example, adding one extreme outlier to a small sample of scores can change the value of the sample mean, M, very substantially, but one extreme outlier typically has a much smaller impact on the size of the sample median. The mean is thus less robust than the median with respect to the impact of outliers.

Rotated factor loadings. These are factor loadings (or correlations between X variables and factors) that have been reestimated relative to rotated or relocated factor axes. The goal of factor rotation is to improve the interpretability of results.

Roy's greatest characteristic root (gcr). *See* Roy's largest root

Roy's largest root. This is a statistic that tests the significance of just the first discriminant function; it is equivalent to the squared canonical correlation of scores on the first discriminant function with group membership. It is generally thought to be less robust to violations of assumptions than the other three multivariate test statistics reported by SPSS GLM (Hotelling's trace, Pillai's trace, and Wilks's Λ).

r_{pb}. *See* Point biserial correlation

r_s. *See* Spearman r

R^2. This is the squared multiple R and can be interpreted as the proportion of variance in Y that can be predicted from X.

R^2_{inc} **(increment in** R^2**).** This is defined as the increase in R^2 from one step in a sequential or statistical regression to the next step; it represents the contribution of the one or more variables added at that step.

r_{tet}. *See* Tetrachoric correlation

S. The sample variance/covariance matrix for a set of repeated measures $X_1, X_2, \ldots, X_k$. The diagonal elements of **S** are the sample variances (of $X_1, X_2, \ldots, X_k$), while the off-diagonal elements of **S** are the sample covariances for all possible pairs of Xs (e.g., the covariance of X_1 with X_2, X_2 with X_3, etc.).

Sample. In formal or ideal descriptions of research methods, a sample refers to a subset of cases drawn from the population of interest (often using combinations of sampling methods, such as random sampling and stratification). In actual practice, samples often consist of readily available cases that were not drawn from a well-defined broader population.

Sample estimate of standard error of the sampling distribution of M **across many samples (**SE_M**).** A standard error describes the variability in the values of a sample statistic (such as M) across hundreds or thousands of different samples from a specific population. When σ is known, we can find the population standard error σ_M. When σ is not known, we can estimate σ_M by computing $SE_M = s/\sqrt{N}$. SE_M is to M as s is to X; that is, the standard error is the standard deviation of the distribution of M values.

Sample mean. *See* Mean

Sample standard deviation (s). This is the square root of the sample variance; it can be approximately understood as the typical distance of a randomly selected score from the mean of the distribution.

Sampling distribution. This is a distribution of outcomes that is obtained when thousands of random samples are taken from a population and a statistic (such as M) is calculated for each sample.

Sampling error. When hundreds or thousands of random samples are drawn from the same population, and a sample statistic such as M is calculated for each sample, the value of M varies across samples. This variation across samples is called sampling error. It occurs because, just by chance, some samples contain a few unusually high or low scores.

Scale. A sum (or average) of scores on several X variables is called a scale; often a scale is developed by summing or averaging items that have large loadings on the same factor in a factor analysis.

Scatter plot. A graph that shows the relation between two quantitative variables, X and Y, is called a scatter plot. Typically, the value of X is indicated by tick marks along the X or horizontal axis; the value of Y is indicated by tick marks along the Y or vertical axis. Each participant's score is represented by a dot or point that corresponds to that person's score values on X and Y.

SCP. *See* Sum of cross products matrix

Scree plot. A plot of the eigenvalues (on the Y axis) by factor number $1, 2, \ldots, p$ (on the X axis). "Scree" refers to the rubble at the foot of a hill, which tends to level off abruptly at some point. Sometimes, visual examination of the scree plot is the basis for the decision regarding the number of factors to retain. Factors that have low eigenvalues (and thus, correspond to the point where the scree plot levels off) may be dropped.

SE_b. *See* Standard error of the raw score slope coefficient (SE_b) in a multiple regression equation

SE_B. The standard error of a B coefficient associated with one of the predictors in a binary logistic regression. The value of SE_B may be used to test the significance of B and/or to set up a confidence interval for B.

Second-order partial correlation. A second-order partial correlation between X_1 and Y involves statistically controlling for two variables (such as X_2 and X_3) while assessing the relationship between X_1 and Y.

SE_{est}. *See* Standard error of the estimate

SE_M. *See* Sample estimate of standard error of the sampling distribution of M across many samples

Semipartial correlation (sr). This is a correlation between scores on an outcome variable Y and scores for a predictor variable X_1 after the portion of each X_1 score that is predictable from X_2, a control variable, has been removed from each X_1 score (but X_1 is not removed from or partialled out of Y).

Separate variances *t* test. This is a modified version of the independent samples *t* test that is preferred in situations where there is evidence that the homogeneity of variance assumption is badly violated; it has adjusted estimated degrees of freedom that usually provide a more conservative test. This statistic is automatically provided by SPSS and many other programs, but it is usually not covered in introductory statistics textbooks.

Sequential regression. In this form of regression, variables are entered one at a time (or in groups or blocks) in an order determined by the researcher. Sometimes this is called hierarchical regression.

Shrunken R^2. *See* Adjusted R^2

"Sig." On SPSS printouts, the exact *p* value that corresponds to a test statistic such as a *t* ratio is often reported as "sig," which is an abbreviation for significance level.

Sigma (Σ). This symbol represents a population variance/covariance matrix for a set of *X* variables. The diagonal elements correspond to the population variances ($\sigma_1^2, \sigma_2^2, \ldots, \sigma_\kappa^2$). The off-diagonal elements correspond to the population covariances for all possible pairs of the *X* variables. A corresponding sample variance/covariance matrix **S** is obtained by dividing each element of the sample **SCP** matrix by *N*, the number of cases.

Sigmoidal function. This is a function that has an S-shaped curve, similar to the example shown in Figure 21.1. This type of function usually occurs when the variable plotted on the *X* axis is a probability, because probabilities must fall between 0 and 1. For example, consider an experiment to assess drug toxicity. The manipulated independent variable is level of drug dosage, and the outcome variable is whether the animal dies or survives. At very low doses, the probability of death would be near 0; at extremely high doses, the probability of death would be near 1; at intermediate dosage levels, the probability would increase as a function of dosage level.

Significance level. This term corresponds to alpha (when the results of a study are used to make a binary decision to reject or not reject H_0) or to *p* (when an exact *p* value is reported instead of, or in addition to, a binary decision).

Simultaneous multiple regression. *See* Standard multiple regression

Simultaneous regression. This describes a method of entry of predictors for multiple regression; in simultaneous regression, all predictors are entered at one step. This term is synonymous with standard or direct methods of entry in regression. In a simultaneous regression, the predictive usefulness of each X_i predictor variable is assessed while statistically controlling for all the other predictor variables.

Skewness. This is the degree to which a distribution is asymmetric (and, therefore, departs from the ideal normal distribution shape).

Slope coefficient (raw score version). *See b*

Social desirability bias. This refers to the tendency for test takers to report the responses that they believe are socially approved, rather than responses that reflect their actual beliefs or behaviors. This is a common source of bias in self-report measures.

Somers's *d*. This is a measure of association between two ordinal variables that ranges from −1 to +1. Values close to an absolute value of 1 indicate a strong relationship between the two variables, and values close to 0 indicate little or no relationship between the variables. Somers's *d* is an asymmetric extension of gamma that differs only in the inclusion of the number of pairs not tied on the independent variable.

Spearman-Brown prophecy formula. When a correlation is obtained to index split-half reliability, that correlation actually indicates the reliability or consistency of a scale with *p*/2 items. If the researcher plans to use a score based on all *p* items, this score should be more reliable than a score that is based on *p*/2 items. The Spearman-Brown prophecy formula "predicts" the reliability of a scale with *p* items from the split-half reliability.

Spearman *r* (r_s). This is a nonparametric correlation that can be used to assess the strength of relationship between two ordinal variables. It is applied when scores come in the form of ranks, or are converted into ranks, to get rid of problems such as outliers and extremely nonnormal distribution shapes.

Sphericity. An assumption (about the pattern of variances and covariances among scores on repeated measures) that must be satisfied for the *F* ratio in a univariate repeated measures ANOVA to correspond to an accurate Type I error risk.

Split-half reliability. This is a type of internal consistency reliability assessment that is used with multiple-item scales. The set of *p* items in the scale is divided (either randomly or systematically) into two sets of *p*/2 items, a score is computed for each set, and a correlation is calculated between the scores on the two sets to index split-half reliability. This correlation can then be used to "predict" the reliability of the full *p* item scale by applying the Spearman-Brown prophecy formula.

SPSS®. This acronym refers to the Statistical Package for the Social Sciences; version 14 of this statistical program is used in this textbook. SPSS is fairly similar in menu structure and output to other widely used program packages, such as SAS, STATA, and SYSTAT.

SPSS GLM procedure. GLM refers to the general linear model. This is a very general model that may include any or all of the following: multiple predictors (which may include multiple group memberships or factors, and/or multiple quantitative predictor variables, which are entered as covariates) and/or multiple outcome measures (which are generally quantitative variables). Many of the analyses described in this textbook are special cases of GLM; for example, an ANCOVA as presented in Chapter 15 is a GLM with one or more categorical predictor variables, one or more quantitative covariates, and one quantitative *Y* outcome variable.

Spurious correlation. A correlation between X_1 and *Y* is said to be spurious if the correlation between X_1 and *Y* drops to 0 when you control for an appropriate X_2 variable, and it does not make sense to propose a causal model in which X_2 mediates a causal connection between X_1 and *Y*.

***sr*.** *See* Semipartial correlation

***sr*^{2}.** A common notation for a squared semipartial (or squared part) correlation.

sr^2_{inc}. *See* Incremental sr^2

sr^2_{unique}. In a standard multiple regression, this is the proportion of variance uniquely predictable from each independent variable (controlling for all other predictors); this is also the squared part (or semipartial) correlation for each predictor.

SS. *See* Sum of squares

SSL. *See* Sum of squared loadings

SS Type I. *See* Type I sum of squares

SS Type II. *See* Type II sum of squares

Standard error of the estimate (SE_{est}). In regression, this corresponds to the standard deviation of the distribution of actual Y scores relative to the predicted Y' at each individual X score value (as shown in Figure 9.7). SE_{est} provides information about the typical magnitude of the prediction error (difference between actual and predicted Y) in regression. Smaller values of SE_{est} are associated with smaller prediction errors and, thus, with more accurate predictions of Y.

Standard error of the raw score slope coefficient (SE_b) in a multiple regression equation. Each b coefficient in a multiple regression has a corresponding standard error estimate; this standard error estimates how much the value of b should vary across different samples drawn from the same population. The SE_b term for each b coefficient can be used to set up a confidence interval estimate for b or to set up a t ratio to test the following null hypothesis: $b = 0$; $t = (b - 0)/SE_b$.

Standardized scores. These are scores expressed in z score units—that is, as unit-free distances from the mean. For example, the standard score version of X, z_X, is obtained from $z_X = (M - M_X)/s_X$.

Standard multiple regression. In standard multiple regression, no matter how many X_i predictors are included in the equation, each X_i predictor variable's unique contribution to explained variance is assessed controlling for or partialling out *all* other X predictor variables. In Figure 11.2, for example, the X_1 predictor variable is given credit for predicting only the variance that corresponds to Area *a*; the X_2 predictor variable is given credit for predicting only the variance that corresponds to Area *b*; and Area *c*, which could be predicted by either X_1 or X_2, is not attributed to either predictor variable.

Standard regression. A method of regression in which all predictor variables are entered into the equation at one step, and the proportion of variance uniquely explained by each predictor is assessed controlling for all other predictors. It is also called simultaneous (or sometimes direct) regression.

Standard score (z score). This is the distance of an individual score from the mean of a distribution expressed in unit-free terms (i.e., in terms of the number of standard deviations from the mean). If μ and σ are known, the z score is given by $z = (X - \mu)/\sigma$. When μ and σ are not known, a distance from the mean can be computed using the corresponding sample statistics, M and s. If the distribution of scores has a normal shape, a table of the standard normal distribution can be used to assess how distance from the mean (given in z score units) corresponds to proportions of area or proportions of cases in the sample that correspond to distances of z units above or below the mean.

Statistically significant association. We evaluate whether X and Y are significantly associated by obtaining an appropriate statistical index of strength of association (such as the Pearson r) and by evaluating the statistical significance of the statistic to rule out chance as the most likely explanation. For example, if the proportion of smokers who develop lung cancer is significantly higher than the proportion of non-smokers who develop lung cancer, we may conclude that smoking and lung cancer are associated to some degree or that the risk of lung cancer is higher for smokers than for nonsmokers.

Statistically significant outcome. A result is statistically significant if a test statistic (such as t) falls into the reject region that is set up using the preselected alpha level and the appropriate distribution. An equivalent way to say this is as follows: The exact p value associated with the obtained t value is less than the preselected alpha level. That is, we reject H_0 and conclude that the outcome is statistically significant when t is larger than a cutoff value obtained from the t distribution table and/or when p is smaller than the preselected alpha level.

Statistical power. *See* Power

Statistical regression. This is a method of regression in which the decisions to add or drop predictors from a multiple regression are made on the basis of statistical criteria (such as the increment in R^2 when a predictor is entered). This is sometimes called "stepwise," but that term should be reserved for a specific method of entry of predictor variables. *See also* Stepwise method of entry

Stepwise method of entry. This method of statistical (data-driven) regression combines forward and backward methods. At each step, the predictor variable in the pool of candidate variables that provides the largest increase in R^2 is added to the model; also, at each step, variables that have entered are reevaluated to assess whether their unique contribution has become nonsignificant after other variables are added to the model (and dropped, if they are no longer significant).

Stratified sampling. These methods of sampling involve dividing a population into groups (or strata) on the basis of characteristics such as gender, age, ethnicity, and so forth; random samples are drawn within each stratum. This can be used to construct a sample that has either the same proportional representation of these groups as the population or some minimum number of cases taken from each stratum in the population. This is often used in large-scale surveys. When this method is used in laboratory research (e.g., to ensure equal numbers of male and female participants), it is generally called "blocking" (on gender, for instance).

Structure coefficient. This is the correlation between a discriminant function and an individual discriminating variable (this is analogous to a factor loading); it tells us how closely related a discriminant function is to each individual predictor variable. These structure coefficients can sometimes be interpreted in a manner similar to factor loading to name the dimension represented by each discriminant function. However, discriminant functions are not always easily interpretable.

Sum of cross products (SCP). For variables X and Y with means of M_x and M_y, the sample cross product is

$$SCP = \Sigma (X - M_x)(Y - M_y).$$

Sum of cross products matrix (SCP). A sum of cross products matrix (**SCP**) for a multivariate analysis such as DA or MANOVA includes the sum of squares (SS) for each of the quantitative variables and the sum of cross products (SCP) for each pair of quantitative variables. In DA the quantitative variables are typically denoted X; in MANOVA they are typically denoted Y. SS terms appear in the diagonal of the **SCP** matrix; SCP terms are the off-diagonal elements. In MANOVA, evaluation of the **SCPs** computed separately within each group, compared to the **SCP** for the total set of data, provides the information needed for multivariate tests to evaluate whether lists of means on the Y variables differ significantly across groups, taking into account the covariances or correlations among the Y variables. Note that dividing each element of the **SCP** matrix by *df* yields the sample variance/covariance matrix **S**.

Sum of squared loadings (SSL). For each factor, the SSL is obtained by squaring and summing the loadings of all variables with that factor. In the initial solution, the SSL for a factor equals the eigenvalue for that factor. The proportion of the total variance in the p variables that is accounted for by one factor is estimated by dividing the SSL for that factor by p.

Sum of squares (*SS*). This is a sum of squared deviations of scores relative to a mean; this term provides information about variability. For a variable X with a sample mean of M_X, $SS = \Sigma(X - M_X)^2$. If $SS = 0$, the set of X terms being assessed are all equal; as variability among X scores increases, *SS* becomes larger. *SS* terms are the building blocks not only for ANOVA but also for other analyses such as multiple regression.

Suppressor variable. An X_2 variable is called a suppressor variable relative to X_1 and Y if the partial correlation between X_1 and Y controlling for X_2 ($r_{1Y.2}$) is larger in absolute magnitude than r_{1Y} or if $r_{1Y.2}$ is significantly different from 0 and opposite in sign to r_{1Y}. In such situations, we might say that the effect of the X_2 variable is to suppress or conceal the true nature of the relationship between X_1 and Y; it is only when we statistically control for X_2 that we can see the true strength or true sign of the X_1, Y association.

Symmetrical index of association. This is an index of the strength of association between two variables that has the same value no matter which variable is designated as the predictor (in contrast, for asymmetrical indexes of association, the strength of the prediction from X to Y may differ from the strength of the predictive association between Y and X). Most of the correlations discussed in Chapters 7 and 8 are symmetrical indexes of strength of association.

System missing value. This is the label that SPSS gives to any scores that are entered as blanks in the SPSS data worksheet.

***t* distribution.** This is a family of distributions that resemble the normal distribution (e.g., the *t* distribution is symmetrical and roughly bell shaped). However, relative to the normal distribution, *t* distributions are flatter with thicker tails. There is a family of *t* distributions; that is, the *t* distribution has a different shape for each value of the corresponding degrees of freedom. Thus, the distance from the mean that corresponds to the middle 95% of the *t* distribution changes as a function of *df*; the smaller the *df*, the larger the distance from the mean that is required to include the middle 95% of the distribution. For example, when $df = 3$, the middle 95% of the area corresponds to values of t that range from −3.18 to +3.18; when $df = 6$, the middle 95% of the area corresponds to values of t that range from −2.45 to +2.45; when $df = \infty$ (in practice, when $N > 100$), the middle 95% of the area lies between $t = -1.96$ and $t = +1.96$ (and that is the same distance that corresponds to the middle 95% of the area for a standard normal distribution). These distributions are used to describe sampling errors for sample statistics (such as M) in situations where σ is not known, and a sample standard deviation s is used to compute the estimated standard error. The sample standard deviation s has associated degrees of freedom ($df = N - 1$, where N is the number of scores in the sample). The sampling error increases as N and df decrease; as the sampling error increases, the t distribution becomes flatter and the tails become thicker. If N is large ($N > 100$), the shape of the t distribution is essentially identical to that of a normal distribution. When setting up confidence intervals, researchers need to know how much sampling error is associated with their sample mean; in practice, when σ is known or when N is greater than 100, researchers use the standard normal distribution to look up the distances from the mean that correspond to the middle 95% or 90% of the sampling distribution. When σ is unknown, and N is less than or equal to 100, researchers use a t distribution with $N - 1$ degrees of freedom to look up distances from the mean that correspond to the middle 95% or 90% of the sampling distribution for M.

Temporal precedence. Temporal precedence refers to the order of events in time. If X_1 happens before Y, or if X_1 is measured or observed before Y, we say that X_1 has temporal precedence relative to Y. For X_1 to be considered as a possible cause of Y, X_1 must occur earlier in time than Y. (However, temporal precedence alone is not a sufficient basis to conclude that X_1 causes Y.)

Territorial map. This is a graphical representation for the scores of individual cases on discriminant functions (SPSS plots only scores on the first two discriminant functions, D_1 and D_2). This map also shows the boundaries that correspond to the classification rules that use scores on D_1 and D_2 to predict group membership for each case.

Tetrachoric correlation (r_{tet}). This is a correlation between two artificially dichotomous variables. It is rarely used.

Tolerance. For each X_i predictor variable, this is the proportion of variance in X_i that is not predictable from other predictor variables already in the equation (i.e., tolerance = $1 - R^2$ for a regression to predict X_i from all the other Xs already included in the regression equation). A tolerance of 0 would indicate perfect multicollinearity, and a variable with zero tolerance cannot add any new predictive information to a regression analysis. Generally, predictor variables with higher values of tolerance may possibly contribute more useful predictive information. The maximum possible tolerance is 1.00, and this would occur when X_i is completely uncorrelated with other predictors.

***t* ratio.** The most general form of a t ratio, which is used to assess statistical significance, is as follows:

$$t = \frac{\text{sample statistic} - \text{hypothesized population parameter}}{SE_{\text{sample statistic}}}.$$

A large value of t is usually interpreted as evidence that the value of the sample statistic is one that would be unlikely to be observed if the null hypothesis were true.

Treatment by covariate interaction. An assumption required for ANCOVA is that there must not be a treatment (A) by covariate (X_c) interaction; in other words, the slope that predicts Y from X_c should be the same within each group or level of A.

Triangulation of measurement. This refers to the use of multiple and different types of measurement to tap the same construct—for example, direct observation of behavior, self-report, and physiological measures. We do this because different methods of measurement have different advantages and disadvantages, and we can be more confident about results that replicate across different types of measures.

***t* test.** This is a statistic that can be used to test many different hypotheses about scores on quantitative variables—for example, whether the means on a quantitative variable Y differ between two groups.

Two-tailed test. This is a significance testing procedure that uses both tails of a distribution (such as the t distribution) as the "reject regions." Two-tailed tests are used when the alternative hypothesis is nondirectional.

Type I error. This is a decision to reject H_0 when H_0 is correct.

Type I sum of squares. This method of variance partitioning in SPSS GLM is essentially equivalent to the method of variance partitioning in sequential or hierarchical multiple regression discussed in Chapter 14. Each predictor is assessed controlling only for other predictors that are entered at the same step or in earlier steps of the analysis.

Type II error. This is a decision not to reject H_0 when H_0 is incorrect.

Type III sum of squares. This is a method of variance partitioning in SPSS GLM that is essentially equivalent to the method of variance partitioning in standard or simultaneous multiple regression (as discussed in Chapter 14). Each predictor is assessed controlling for all other predictors in the analysis. This corrects for any confounds between treatments (as in nonorthogonal or unbalanced factorial designs), for any confounds between treatments and covariates, and for any correlations among covariates. Usually, Type III SS provides the most conservative assessment of the predictive usefulness of each factor and the covariates. This is the default method for computation of sums of squares in SPSS GLM.

Types of variables. It is sufficient (for some purposes) to distinguish between just two types of variables: categorical and quantitative. Some authors distinguish between four levels of measurement (nominal, ordinal, interval, and ratio). Nominal level of measurement is equivalent to categorical variable type. Quantitative variables may correspond to ordinal, interval, or ratio level of measurement.

Unbiased estimator. This is a sample statistic that yields a value that, on average, across many samples, is not systematically higher or lower than the corresponding population parameter. For example, the sample mean, M, is an unbiased estimator of μ, the true population mean; the average value of M across hundreds of samples drawn randomly from the same population equals μ. To obtain a value of s that is an unbiased estimator of σ, it is necessary to use $(N-1)$ rather than N in the formula for the sample standard deviation s. When it is computed using $s = \sqrt{SS/(N-1)}$, the average value of s is equal to σ across hundreds of random samples from a specific population.

Unconditional probability. The unconditional probability is the overall probability of some outcome (such as survival vs. death) for the entire sample, ignoring membership on any other categorical variables. For the dog owner survival data in Chapter 8, the unconditional probability of survival is the total number of survivors in the entire sample (78) divided by the total N in the entire study (92), which is equal to .85.

Unit-weighted linear composite. A unit-weighted linear composite is a sum of raw X scores or standardized z_x scores across several variables.

Unpublished or public domain test. This term refers to a test or scale that is not distributed by a test publishing company. Tests that appear in journal articles, books, or on Web sites are usually considered "unpublished" or noncommercial in the sense that they have not been published by a test publishing company. However, tests that originally appear in journal articles are sometimes subsequently acquired by test publishers. Thus, it is often unclear whether test material is public domain or whether you must pay a test publisher to use the test. When in doubt, it is best to contact the test author and/or test publisher for specific information. Even for material in the public domain, users should obtain the permission of the test author as a courtesy and must credit the test author by proper citation of the source. See the FAQ on the American Psychological Association Web site for further guidelines about the selection and use of existing psychological tests, both published and unpublished: www.apa.org/science/faq-findtests.html.

Unweighted mean. This is a mean that combines information across several groups or cells (such as the mean for all the scores in one row of a factorial ANOVA) and is calculated by averaging cell means together without taking the numbers of scores within each cell or group into account.

User-determined order of entry. Some methods of entering variables into a series of regression analyses allow the data analyst to specify the order in which the predictor variables enter the analysis, based on theoretical or temporal priority of variables (also called sequential or hierarchical methods of entry).

Variance. Differences among scores for the participants in a study (this may be scores in a sample or in a population) are called variance. Sample variance is denoted by s^2; population variance is denoted by σ^2.

Variance/covariance matrix. This is a matrix that summarizes all the variances and all the possible covariances for a list of variables. The population matrix is denoted by $\mathbf{\Sigma}$; the corresponding sample variance/covariance matrix is denoted by **S**.

Varimax rotation. This is the most popular method of factor rotation. The goal is to maximize the variance of the absolute values of loadings of variables within each factor—that is, to relocate the factors in such a way that, for each variable, the loading on the relocated factors is as close to 0, or as close to +1 or −1, as possible. This makes it easy to determine which variables are related to a factor (those with large loadings) and which variables are not related to each factor (those with loadings close to 0), which improves the interpretability of the solution.

Wald χ^2 statistic. For each *B* coefficient in a logistic regression model, the corresponding Wald function tests the null hypothesis that $B = 0$. The Wald χ^2 is given as $[B/SE_B]^2$; the sampling distribution for the Wald function is a chi-square distribution with one degree of freedom. Note that in an ordinary linear regression, the null hypothesis that $b = 0$ is tested by setting up a *t* ratio where $t = b/SE_b$; the Wald chi-square used in binary logistic regression is comparable to the square of this *t* ratio.

Weighted mean. This is a mean that combines information across several groups or cells (such as the mean for all the scores in one row of a factorial ANOVA) and is calculated by weighting each cell mean by its corresponding number, *n*, of cases. For example, if Group 1 has n_1 cases and a mean of M_1 and Group 2 has n_2 cases and a mean of M_2, the weighted mean $M_{weighted}$ is calculated as follows: $M_{weighted} = [(n_1M_1) + (n_2M_2)]/(n_1 + n_2)$.

Whiskers. In a box and whiskers plot, these are the vertical lines that extend beyond the hinges out to the adjacent values. Any scores that lie beyond the whiskers are labeled as outliers.

Wilcoxon rank sum test. This is a nonparametric alternative to the independent samples *t* test. It is used to compare mean ranks for independent samples in situations where the assumptions required for the use of the parametric test (independent samples *t* test) are violated.

Wilks's lambda (Λ). This is an overall goodness-of-fit measure used in discriminant analysis. It is the most widely used multivariate test statistic for the null hypothesis shown in Equation 17.3: H_0: $\mu_1 = \mu_2 = \ldots = \mu_k$. Unlike Hotelling's trace and Pillai's trace, larger values of Wilks's Λ indicate smaller differences across groups. Wilks's Λ may be calculated in two different ways that yield the same result: as a ratio of determinants of sum of cross product matrices (see Equation 17.11) or as the product of $(1/1 + \lambda_i)$ across all the discriminant functions (Equation 17.20). It can be converted to an estimate of effect size, $\eta^2 = 1 - \Lambda$. In a univariate ANOVA, Wilks's Λ is equivalent to $(1 - \eta^2)$, that is, the proportion of unexplained or within-group variance on outcome variable scores. Like Hotelling's trace and Pillai's trace, it summarizes information about the magnitude of between-group differences on scores across all discriminant functions. Unlike these two other statistics, however, larger values of Wilks's Λ indicate smaller between-group differences on vectors of means or on discriminant scores. In a discriminant analysis, the goal is to *minimize* the size of Wilks's Λ.

Within-S design. See Repeated measures ANOVA.

Yates's chi-square (often called the continuity-adjusted chi-square). In the computation of Yates's chi-square, a small correction factor of .5 is subtracted from each $(O - E)$ term before squaring it. This results in a more conservative test. This correction is applied only for 2×2 tables. Some authors recommend against use of this correction.

Yea-saying bias. This refers to a tendency to say yes to most questions on a test, or to indicate high levels of agreement with many statements on a questionnaire, regardless of question content. Test developers often try to avoid this type of bias by including at least some reverse-worded items in the test, so that a person does not recevice a high score on the test simply by giving "yes" or "often" answers to all the questions on the test.

Yoked samples. Participants in a study are yoked if they receive identical outcomes. For example, studies of stress sometimes examine yoked pairs of monkeys. One monkey in each pair has a lever that can be used to turn off impending electric shocks; the other monkey in the pair also has a lever, but the second monkey's lever does not turn off impending electric shocks. Both monkeys receive exactly the same number and intensity of electrical shocks; the difference between them is in whether or not the animal had control over the shocks recevied.

Zero-order correlation. A simple Pearson correlation between X_1 and Y (which does not involve statistical control for any other variables) is called a zero-order correlation. The answer to the question "How many variables were statistically controlled while assesing the relationship between X_1 and Y'?" is 0.

***ZPRED.** This is the standardized or *z* score version of the predicted value of Y (Y') from a multiple regression. This is one of the new variables that can be computed and saved into the SPSS worksheet in SPSS multiple regression.

***ZRESID.** This is the standardized or *z* score version of the residuals from a multiple regression ($Y - Y'$). If any of these lie outside the range that includes the middle 99% of the standard normal distribution, these cases should be examined as possible multivariate outliers.

***z* score.** *See* Standard score

References

Algina, J., & Kesselman, H. J. (1997). Detecting repeated measures effects with univariate and multivariate statistics. *Psychological Methods, 2,* 208–218.

American Psychological Association. (2001). *Publication manual of the American Psychological Association* (5th ed.). Washington, DC: Author.

Anderson, C. A., & Bushman, B. J. (2001). Effects of v'iolent video games on aggressive behavior, aggressive cognition, aggressive affect, physiological arousal, and prosocial behavior: A meta-analytic review of the scientific literature. *Psychological Science, 12,* 353–359.

Aron, A., & Aron, E. (2002). *Statistics for the behavioral and social sciences* (3rd ed.). New York: Prentice Hall.

Aronson, E., & Mills, J. (1959). The effect of severity of initiation on liking for a group. *Journal of Abnormal and Social Psychology, 59,* 177–181.

Aspland, H., & Gardner, F. (2003). Observational measures of parent-child interaction: An introductory review. *Child and Adolescent Mental Health, 8,* 136–143.

Bakeman, R. (2000). Behavioral observation and coding. In H. T. Reis & C. M. Judd (Eds.), *Handbook of research methods in social and personality psychology* (pp. 138–159). New York: Cambridge University Press.

Baron, R. M., & Kenny, D. A. (1986). The moderator-mediator variable distinction in social psychological research: Conceptual, strategic and statistical considerations. *Journal of Personality and Social Psychology, 51,* 1173–1182.

Baum, A., Gatchel, R. J., & Schaeffer, M. A. (1983). Emotional, behavioral, and physiological effects of chronic stress at Three Mile Island. *Journal of Consulting and Clinical Psychology, 51,* 565–572.

Beck, A. T., Steer, R. A., & Brown, G. K. (1996). *Manual for the Beck Depression Inventory–II.* San Antonio, TX: Psychological Corporation.

Beck, A. T., Ward, C. H., Mendelson, M., Mock, J., & Erbaugh, J. (1961). An inventory for measuring depression. *Archives of General Psychiatry, 4,* 561–571.

Bem, S. L. (1974). The measurement of psychological androgyny. *Journal of Consulting and Clinical Psychology, 42,* 155–162.

Berry, W. D. (1993). *Understanding regression assumptions* (Quantitative Applications in the Social Sciences, No. 92). Thousand Oaks, CA: Sage.

Bohrnstedt, G. W., & Carter, T. M. (1971). Robustness in regression analysis. *Sociological Methodology, 12,* 118–146.

Brackett, M. A. (2004). Conceptualizing and measuring the life space and its relation to openness to experience (Dissertation at University of New Hampshire, Durham, NH). *Dissertation Abstracts International, 64*(7), 3569B.

Brackett, M. A., Mayer, J. D., & Warner, R. M. (2004). Emotional intelligence and its relation to everyday behaviour. *Personality and Individual Differences, 36,* 1387–1402.

Bray, J. H., & Maxwell, S. E. (1985). *Multivariate analysis of variance* (Quantitative Applications in the Social Sciences, No. 54). Beverly Hills, CA: Sage.

Burish, T. G. (1981). EMG biofeedback in the treatment of stress-related disorders. In C. Prokop & L. Bradley (Eds.), *Medical psychology* (pp. 395–421). New York: Academic Press.

Cacioppo, J. T., Tassinary, L. G., & Berntson, G. G. (2000). *Handbook of psychophysiology* (2nd ed.). Cambridge, UK: Cambridge University Press.

Campbell, D. T., & Stanley, J. S. (2005). *Experimental and quasi-experimental designs for research.* Boston: Houghton Mifflin.

Carmines, E. G., & Zeller, R. A. (1979). *Reliability and validity assessment* (Quantitative Applications in the Social Sciences, No. 17). Beverly Hills, CA: Sage.

Christie, R., & Geis, F. (1970). *Studies in Machiavellianism.* New York: Academic Press.

Cohen, J. (1977). *Statistical power analysis for the behavioral sciences.* Hillsdale, NJ: Lawrence Erlbaum.

Cohen, J. (1988). *Statistical power analysis for the behavioral sciences* (2nd ed.). Hillsdale, NJ: Lawrence Erlbaum.

Cohen, J. (1990). Things I have learned (so far). *American Psychologist, 45,* 1304–1312.

Cohen, J. (1994). The earth is round ($p < .05$). *American Psychologist, 49,* 997–1003.

Cohen, J., & Cohen, P. (1983). *Applied multiple regression/correlation analysis for the behavioral sciences* (2nd ed.). Hillsdale, NJ: Lawrence Erlbaum.

Cohen, J., Cohen, P., West, S. G., & Aiken, L. S. (2003). *Applied multiple regression/correlation analysis for the behavioral sciences* (3rd ed.). Mahwah, NJ: Lawrence Erlbaum.

Converse, J. M. M., & Presser, S. (1999). *Survey questions: Handcrafting the standardized questionnaire* (Quantitative Applications in the Social Sciences, No. 20). Thousand Oaks, CA: Sage.

Cook, T. D., & Campbell, D. T. (1979). *Quasi-experimentation: Design and analysis issues for field settings.* Boston: Houghton Mifflin.

Costa, P. T., & McCrae, R. R. (1995). Domains and facets: Hierarchical personality assessment using the Revised NEO Personality Inventory. *Journal of Personality Assessment, 64,* 21–50.

Costa, P. T., & McCrae, R. R. (1997). Stability and change in personality assessment: The Revised NEO Personality Inventory in the year 2000. *Journal of Personality Assessment, 68,* 86–94.

Cozby, P. C. (2004). *Methods in behavioral research* (8th ed.). Boston: McGraw-Hill.

Davis, J. (1971). *Elementary survey analysis.* Englewood Cliffs, NJ: Prentice Hall.

deGroot, M. H., & Schervish, M. J. (2001). *Probability and statistics* (3rd ed.). New York: Addison-Wesley.

Doob, A. N., & Gross, A. E. (1968). Status of frustrator as an inhibitor of horn-honking responses. *Journal of Social Psychology, 76,* 213–218.

Einspruch, E. L. (2005). *An introductory guide to SPSS for Windows* (2nd ed.). Thousand Oaks, CA: Sage.

Everitt, B. S. (1977). *The analysis of contingency tables.* New York: Wiley.

Fava, J. L., & Velicer, W. F. (1992). An empirical comparison of factor, image, component, and scale scores. *Multivariate Behavioral Research, 27,* 301–322.

Field, A. P. (1998). A bluffer's guide to . . . sphericity. *Newsletter of the Mathematical, Statistical and Computing Section of the British Psychological Society, 6,* 13–22.

Fisher, R. A., & Yates, F. (1974). *Statistical tables for biological, agricultural and medical research* (6th ed.). Reading, MA: Addison-Wesley.

Freedman, J. L. (1975). *Crowding and behavior.* New York: Viking.

Friedmann, E., Katcher, A. H., Lynch, J. J., & Thomas, S. A. (1980). Animal companions and one year survival of patients after discharge from a coronary care unit. *Public Health Reports, 95,* 307–312.

Gaito, J. (1980). Measurement scales and statistics: Resurgence of an old misconception. *Psychological Bulletin, 87,* 564–567.

Gergen, K. J. (1973). Social psychology as history. *Journal of Personality and Social Psychology, 26,* 309–320.

Gergen, K. J., Hepburn, A., & Fisher, D. C. (1986). Hermeneutics of personality description. *Journal of Personality and Social Psychology, 50,* 1261–1270.

Goldberg, L. R. (1999). A broad-bandwidth, public domain, personality inventory measuring the lower-level facets of several five-factor models. In I. Mervielde, I. J. Deary, F. De Fruyt, & F. Ostendorf (Eds.), *Personality psychology in Europe* (Vol. 7, pp. 7–28). Tilburg, The Netherlands: Tilburg University Press.

Goodwin, D. W., Schylsinger, F., Hermansen, L., Guze, S. B., & Winokur, G. (1973). Alcohol problems in adoptees raised apart from alcoholic biological parents. *Archives of General Psychiatry, 28,* 238–243.

Gottman, J. M. (1995). *The analysis of change.* Mahwah, NJ: Lawrence Erlbaum.

Gottman, J. M., & Notarius, C. I. (2002). Marital research in the 20th century and a research agenda for the 21st century. *Family Process, 41,* 159–197.

Gould, S. J. (1996). *The mismeasure of man.* New York: Norton.

Green, S. B. (1991). How many subjects does it take to do a regression analysis? *Multivariate Behavioral Research, 26,* 449–510.

Greenwald, A. G., Gonzalez, R., Harris, R. J., & Guthrie, D. (1996). Effect sizes and *p* values: What should be reported and what should be replicated? *Psychophysiology, 33,* 175–183.

Guber, D. L. (1999). Getting what you pay for: The debate over equity in public school expenditures. *Journal of Statistics Education, 7*(2). Retrieved November 24, 2006, from www.amstat.org/publications/jse/secure/v7n2/datasets.guber.cfm

Hansell, S., Sparacino, J., & Ronchi, D. (1982). Physical attractiveness and blood pressure: Sex and age differences. *Personality and Social Psychology Bulletin, 8,* 113–121.

Harker, L., & Keltner, D. (2001). Expressions of positive emotion in women's college yearbook pictures and their relationship to personality and life outcomes across adulthood. *Journal of Personality and Social Psychology, 80,* 112–124.

Harman, H. H. (1976). *Modern factor analysis* (2nd ed.). Chicago: University of Chicago Press.

Harris, R. J. (2001). *A primer of multivariate statistics* (3rd ed.). Mahwah, NJ: Lawrence Erlbaum.

Hayes, D. P., & Meltzer, L. (1972). Interpersonal judgments based on talkativeness: I. Fact or artifact? *Sociometry, 35,* 538–561.

Hays, W. L. (1973). *Statistics for the social sciences* (2nd ed.). New York: Holt, Rinehart & Winston.

Hays, W. (1994). *Statistics* (5th ed.). Fort Worth, TX: Harcourt Brace.

Hinkle, D. E., Wiersma, W., & Jurs, S. G. (1994). *Applied statistics for the behavioral sciences.* New York: Houghton Mifflin.

Hodges, J. L., Jr., Krech, D., & Crutchfield, R. S. (1975). *Statlab: An empirical introduction to statistics.* New York: McGraw-Hill.

Horton, R. L. (1978). *The general linear model: Data analysis in the social and behavioral sciences.* New York: McGraw-Hill.

Hosmer, D. W., & Lemeshow, S. (2000). *Applied logistic regression* (2nd ed.). New York: Wiley.

Howell, D. C. (1992). *Statistical methods for psychology* (3rd ed.). Boston: PWS-Kent.

Hubbard, R., & Armstrong, J. S. (1992). Are null results becoming an endangered species in marketing? *Marketing Letters, 3,* 127–136.

Huberty, C. J. (1994). *Applied discriminant analysis.* New York: Wiley.

Huberty, C. J., & Morris, D. J. (1989). Multivariate analysis versus multiple univariate analyses. *Psychological Bulletin, 105,* 302–308.

International personality item pool: A scientific collaboratory for the development of advanced measures of personality traits and other individual differences. (2001). Retrieved December 5, 2006, from http://ipip.ori.org

Jaccard, J., & Becker, M. A. (1997). *Statistics for the behavioral sciences* (3rd ed.). Pacific Grove, CA: Brooks/Cole.

Jaccard, J., & Becker, M. A. (2002). *Statistics for the behavioral sciences* (4th ed.). Pacific Grove, CA: Brooks/Cole.

Jaccard, J., Turrisi, R., & Wan, C. K. (1990). *Interaction effects in multiple regression* (Quantitative Applications in the Social Sciences, Vol.118). Thousand Oaks, CA: Sage.

Jaccard, J., Wan, C. W., & Turrisi, R. (1990). The detection and interpretation of interaction effects between continuous variables in multiple regression. *Multivariate Behavioral Research, 25,* 467–478.

Jenkins, C. D., Zyzanski, S. J., & Rosenman, R. H. (1979). *Jenkins Activity Survey manual.* New York: Psychological Corp.

Joreskog, K. G. (1969). A general approach to confirmatory maximum likelihood factor analysis. *Psychometrika, 34,* 183–202.

Kalton, G. (1983). *Introduction to survey sampling* (Quantitative Applications in the Social Sciences, No. 35). Beverly Hills, CA: Sage.

Kelley, T. L. (1927). *Interpretation of educational measurements.* Yonkers-on-Hudson, NY: World Book.

Kenny, D. A. (1979). *Correlation and causality.* New York: Wiley.

Kenny, D. A., & Judd, C. M. (1996). A general method for the estimation of interdependence. *Psychological Bulletin, 119,* 138–148.

Kenny, D. A., Mannetti, L., Pierro, A., Livi, S., & Kashy, D. A. (2002). The statistical analysis of data from small groups. *Journal of Personality and Social Psychology, 83,* 126–137.

Keppel, G. (1991). *Design and analysis: A researcher's handbook* (3rd ed.). New York: Prentice Hall.

Keppel, G., & Zedeck, S. (1989). *Data analysis for research designs: Analysis of variance and multiple regression/correlation approaches.* New York: W. H. Freeman.

Keys, A. B. (1980). *Seven countries: A multivariate analysis of death and coronary heart disease.* Cambridge, MA: Harvard University Press.

King, J. E. (2003). Running a best-subsets logistic regression: An alternative to stepwise methods. *Educational and Psychological Measurement, 63,* 393–403.

Kirk, R. E. (1995). *Experimental design: Procedures for the behavioral sciences* (3rd ed.). New York: Wadsworth.

Kirk, R. (1996). Practical significance: A concept whose time has come. *Educational and Psychological Measurement, 56,* 746–759.

Kline, R. B. (2004). *Beyond significance testing: Reforming data analysis in behavioral research.* Washington, DC: American Psychological Association.

Kling, K. C., Hyde, J. S., Showers, C. J., & Buswell, B. N. (1999). Gender differences in self-esteem: A meta-analysis. *Psychological Bulletin, 125,* 470–500.

Krueger, J. (2001). Null hypothesis significance testing: On the survival of a flawed method. *American Psychologist, 56,* 16–26.

Landis, J. R., & Koch, G. G. (1977). The measurement of observer agreement for categorical data. *Biometrics, 33,* 159–174.

Lane, D. M. (2001). *Hyperstat* (2nd ed.). Cincinnati, OH: Atomic Dog Publishing.

Lauter, J. (1978). Sample size requirements for the t^2 test of MANOVA (tables for one-way classification). *Biometrical Journal, 20,* 389–406.

Levine, J. D., Gordon, N. C., & Fields, H. L. (1978). The mechanism of placebo analgesia. *Lancet, 2*(8091), 654–657.

Liebetrau, A. M. (1983). *Measures of association* (Quantitative Applications in the Social Sciences, No. 32). Beverly Hills, CA: Sage.

Lindeman, R. H., Merenda, P. F., & Gold, R. Z. (1980). *Introduction to bivariate and multivariate analysis.* Glenview, IL: Scott, Foresman.

Linden, W. (1987). On the impending death of the Type A construct: Or is there a phoenix rising from the ashes? *Canadian Journal of Behavioural Science, 19,* 177–190.

Lohnes, P. R. (1966). *Measuring adolescent personality.* Pittsburgh, PA: University of Pittsburgh Press, Project TALENT.

Lord, F. M. (1967). A paradox in the interpretation of group comparisons. *Psychological Bulletin, 72,* 304–305.

Lyon, D., & Greenberg, J. (1991). Evidence of codependency in women with an alcoholic parent: Helping out Mr. Wrong. *Journal of Personality and Social Psychology, 61,* 435–439.

MacKinnon, D. P., Lockwood, C. M., Hoffman, J. M., West, S. G., & Sheets, V. (2002). A comparison of methods to test mediation and other intervening variable effects. *Psychological Methods, 7,* 83–104.

Mackowiak, P. A., Wasserman, S. S., & Levine, M. M. (1992). A critical appraisal of 98.6 degrees F, the upper limit of the normal body temperature, and other legacies of Carl Reinhold August Wunderlich. *Journal of the American Medical Association, 268,* 1578–1580.

Maxwell, S. E. (2004). The persistence of underpowered studies in psychological research: Causes, consequences, and remedies. *Psychological Methods, 9,* 147–163.

Meehl, P. E. (1978). Theoretical risks and tabular asterisks: Sir Karl, Sir Ronald, and the slow progress of soft psychology. *Journal of Consulting and Clinical Psychology, 46,* 806–834.

Menard, S. (2001). *Applied logistic regression analysis* (Quantitative Applications in the Social Sciences, No. 07-106). Thousand Oaks, CA: Sage.

Mooney, K. M. (1990). Assertiveness, family history of hypertension, and other psychological and biophysical variables as predictors of cardiovascular reactivity to social stress. *Dissertation Abstracts International, 51*(3-B), 1548–1549.

Myers, J. L., & Well, A. D. (1991). *Research design and statistical analysis.* New York: Harper Collins.

Myers, J. L., & Well, A. D. (1995). *Research design and statistical analysis.* Mahwah, NJ: Lawrence Erlbaum.

Norusis, M. (2005a). *SPSS 14.0 advanced statistical procedures companion.* New York: Prentice Hall.

Norusis, M. (2005b). *SPSS 14.0 statistical procedures companion.* New York: Prentice Hall.

Nunnally, J. C., & Bernstein, I. (1994). *Psychometric theory* (3rd ed.). New York: McGraw-Hill.

Olkin, I., & Finn, J. D. (1995). Correlations redux. *Psychological Bulletin, 118,* 155–164.

Pampel, F. C. (2000). *Logistic regression: A primer* (Quantitative Applications in the Social Sciences, No. 07-132). Thousand Oaks, CA: Sage.

Pavkov, T. P. (2005). *Ready, set, go! A student guide to SPSS 12.0 for Windows.* New York: McGraw-Hill.

Pearson, E. S., & Hartley, H. O. (Eds.). (1970). *Biometrika tables for statisticians* (3rd ed., Vol. 1). Cambridge, UK: Cambridge University Press.

Peduzzi, P., Concato, J., Kemper, E., Holford, T. R., & Feinstein, A. (1996). A simulation of the number of events per variable in logistic regression analysis. *Journal of Clinical Epidemiology, 99,* 1373–1379.

Peng, C. J., Lee, K. L., & Ingersoll, G. M. (2002). An introduction to logistic regression analysis and reporting. *Journal of Educational Research, 96*(1), 3–14.

Powers, D. E., & Rock, D. A. (1993). Coaching for the SAT: A summary of the summaries and an update. *Educational Measurement Issues and Practice, 12,* 24–30.

Radloff, L. S. (1977). The CES-D scale: A self-report depression scale for research in the general population. *Applied Psychological Measurement, 1,* 385–401.

Record, R. G., McKeown, T., & Edwards, J. H. (1970). An investigation of the difference in measured intelligence between twins and single births. *Annals of Human Genetics, 34,* 11–20.

Reis, H. T., & Gable, S. L. (2000). Event-sampling and other methods for studying everyday experience. In H. T. Reis & C. M. Judd (Eds.), *Handbook of research methods in social and personality psychology* (pp. 190–222). New York: Cambridge University Press.

Robinson, J. P., Shaver, P. R., & Wrightsman, L. S. (Eds.). (1991). *Measures of personality and social psychological attitudes.* San Diego: Academic Press.

Rosenthal, R. (1966). *Experimenter effects in behavioral research.* New York: Appleton-Century-Crofts.

Rosenthal, R., & Rosnow, R. L. (1980). Summarizing 345 studies of interpersonal expectancy effects. In R. Rosenthal (Ed.), *New directions for methodology of social and behavioral science: Quantitative assessment of research domains* (Vol. 5, pp. 79–95). San Francisco: Jossey-Bass.

Rosenthal, R., & Rosnow, R. L. (1991). *Essentials of behavioral research: Methods and data analysis* (2nd ed.). New York: McGraw-Hill.

Rozeboom, W. W. (1960). The fallacy of the null-hypothesis significance test. *Psychological Bulletin, 57,* 416–428.

Rubin, Z. (1970). Measurement of romantic love. *Journal of Personality and Social Psychology, 16,* 265–273.

Rubin, Z. (1976). On studying love: Notes on the researcher-subject relationship. In M. P. Golden (Ed.), *The research experience* (pp. 508–513). Itasca, IL: Peacock.

Schafer, J. L. (1997). *Analysis of incomplete multivariate data* (Chapman & Hall Series Monographs on Statistics and Applied Probability, Book 72). London: Chapman & Hall.

Schafer, J. L. (1999). Multiple imputation: A primer. *Statistical Methods in Medical Research, 8,* 3–15.

Schafer, J. L., & Olsen, M. K. (1998). Multiple imputation for multivariate missing-data problems: A data analyst's perspective. *Multivariate Behavioral Research, 33,* 545–571.

Schulman, K. A., Berlin, J. A., Harless, W., Kerner, J. F., Sistrunk, S., Gersh, B. J., et al. (1999). The effect of race and sex on physicians' recommendations for cardiac catheterization. *New England Journal of Medicine, 340,* 618–626.

Schwartz, A. J. (2003). A note on logistic regression and odds ratios. *Journal of American College Health, 51,* 169–170.

Schwartz, L. M., Woloshin, S., & Welch, H. G. (1999). Misunderstandings about the effects of race and sex on physicians' referrals for cardiac catheterization. *New England Journal of Medicine, 341,* 279–283.

Sears, D. O. (1986). College sophomores in the laboratory: Influences of a narrow data base on psychology's view of human nature. *Journal of Personality and Social Psychology, 51,* 515–530.

Shadish, W. R., Cook, T. D., & Campbell, D. T. (2001). *Experimental and quasi-experimental designs for generalized causal inference.* Boston: Houghton Mifflin.

Shaughnessy, J. J., Zechmeister, E. B., & Zechmeister, J. S. (2003). *Research methods in psychology* (6th ed.). New York: McGraw-Hill.

Shoemaker, A. L. (1996). What's normal? Temperature, gender, and heart rate. *Journal of Statistics Education, 4.* Retrieved June 27, 2006, from www.amstat.org/publications/jse/v4n2/datasets .shoemaker.html

Siegel, S., & Castellan, N. J. (1988). *Nonparametric statistics for the behavioral sciences* (2nd ed.). New York: McGraw-Hill.

Sigall, H., & Ostrove, N. (1975). Beautiful but dangerous: Effects of offender attractiveness and nature of the crime on juridic judgment. *Journal of Personality and Social Psychology, 31,* 410–414.

Sobel, M. E. (1982). Asymptotic confidence intervals for indirect effects in structural equation models. In S. Leinhardt (Ed.), *Sociological methodology* (pp. 290–312). Washington, DC: American Sociological Association.

Spring, B., Chiodo, J., & Bowen, D. J. (1987). Carbohydrates, tryptophan, and behavior: A methodological review. *Psychological Bulletin, 102,* 234–256.

Steering Committee of the Physicians' Health Study Research Group. (1989). Final report on the aspirin component of the ongoing Physician's Health Study. *New England Journal of Medicine, 321,* 129–135.

Sternberg, R. J. (1997). Construct validation of a triangular love scale. *European Journal of Social Psychology, 27,* 313–335.

Stevens, J. (1992). *Applied multivariate statistics for the social sciences* (2nd ed.). Hillsdale, NJ: Lawrence Erlbaum.

Stevens, J. P. (2002). *Applied multivariate statistics for the social sciences* (4th ed.). Mahwah, NJ: Lawrence Erlbaum.

Stevens, S. (1946). On the theory of scales of measurement. *Science, 103,* 677–680.

Stevens, S. (1951). Mathematics, measurement, and psychophysics. In S. Stevens (Ed.), *Handbook of experimental psychology* (pp. 1–49). New York: Wiley.

Stone, A. A., Turkkan, J. S., Kurtzman, K. S., Bachrach, C., & Jobe, J. B. (Eds.). (1999). *Science of self report: Implications for research and practice.* Mahwah, NJ: Lawrence Erlbaum.

Tabachnick, B. G., & Fidell, L. S. (2007). *Using multivariate statistics* (5th ed.). Boston: Pearson.

Tatsuoka, M. M. (1988). *Multivariate analysis: Techniques for educational and psychological research* (2nd ed.). New York: Macmillan.

Thompson, B. (1999). Statistical significance tests, effect size reporting, and the vain pursuit of pseudo-objectivity. *Theory and Psychology, 9,* 191–196.

Trochim, W. M. K. (2001). *The research methods knowledge base* (2nd ed.). Cincinnati, OH: Atomic Dog.

Tufte, E. R. (1983). *The visual display of quantitative information.* Cheshire, CT: Graphics Press.

Vacha-Haase, T. (2001). Statistical significance should not be considered one of life's guarantees: Effect sizes are needed. *Educational & Psychological Measurement, 61,* 219–224.

Vogt, W. P. (1999). *Dictionary of statistics and methodology: A nontechnical guide for the social sciences.* Thousand Oaks, CA: Sage.

Wainer, H. (1976). Estimating coefficients in multivariate models: It don't make no nevermind. *Psychological Bulletin, 83,* 213–217.

Wainer, H. (1991). Adjusting for differential base rates: Lord's paradox again. *Psychological Bulletin, 109,* 147–151.

Wallace, K. A., Bisconti, T. L., & Bergeman, C. S. (2001). The mediational effect of hardiness on social support and optimal outcomes in later life. *Basic and Applied Social Psychology, 23,* 267–279.

Warner, B., & Rutledge, J. (1999). Checking the Chips Ahoy! guarantee. *Chance, 12,* 10–14.

Warner, R. M. (1998). *Spectral analysis of time-series data.* New York: Guilford.

Warner, R. M., & Sugarman, D. B. (1986). Attributions of personality based on physical appearance, speech, and handwriting. *Journal of Personality and Social Psychology, 50,* 792–799.

Westen, D., & Rosenthal, R. (2003). Quantifiying construct validity: Two simple measures. *Journal of Personality and Social Psychology, 84,* 608–618.

Wiggins, J. S. (1973). *Personality and prediction: Principles of personality assessment.* New York: Random House.

Wilkinson, L., & Dallal, G. E. (1981). Tests of significance in forward selection regression with an F-to-enter stopping rule. *Technometrics, 23,* 377–380.

Wilkinson, L., & Task Force on Statistical Inference, APA Board of Scientific Affairs. (1999). Statistical methods in psychology journals: Guidelines and explanations. *American Psychologist, 54,* 594–604.

Winer, B. J., Brown, D. R., & Michels, K. M. (1991). *Statistical principles in experimental design* (3rd ed.). New York: McGraw-Hill.

Wright, D. B. (2003). Making friends with your data: Improving how statistics are conducted and reported. *British Journal of Educational Psychology, 73,* 123–136.

Wright, R. E. (1995). Logistic regression. In L. G. Grimm & P. R. Yarnold (Eds.), *Reading and understanding multivariate statistics* (pp. 217–244). Washington, DC: American Psychological Association.

Zumbo, B. D., & Zimmerman, D. W. (1993). Is the selection of statistical methods governed by level of measurement? *Canadian Psychology, 34,* 390–400.

Index

About the Author

Rebecca M. Warner is Professor in the Department of Psychology at the University of New Hampshire. She has taught statistics for more than 25 years; her courses have included Introductory and Intermediate Statistics as well as seminars in Multivariate Statistics, Structural Equation Modeling, and Time Series Analysis. She received a UNH Liberal Arts Excellence in Teaching Award in 1992; is a fellow in the Association for Psychological Science; and is a member of the American Psychological Association, the International Association for Relationships Research, the Society of Experimental Social Psychology, and the Society for Personality and Social Psychology. She has consulted on statistics and data management for the Carsey Center at the University of New Hampshire and the World Health Organization in Geneva and served as a visiting faculty member at Shandong Medical University in China. Her previous book, *The Spectral Analysis of Time-Series Data*, was published in 1998. She has published articles on statistics, health psychology, and social psychology in numerous journals, including the *Journal of Personality and Social Psychology*; she has served as a reviewer for many journals, including *Psychological Bulletin, Psychological Methods, Personal Relationships,* and *Psychometrika.* Dr. Warner has taught in the Semester at Sea program and worked with Project Orbis to set up a data management system for an airplane-based surgical education program. She has written a novel (*A.D. 62: Pompeii*) under the pen name Rebecca East. She received a BA from Carnegie Mellon University in social relations in 1973 and a PhD in social psychology from Harvard in 1978.